DIRECTORY OF ON-GOING RESEARCH IN CANCER EPIDEMIOLOGY

1994

This Directory is produced jointly by the International Agency for Research on Cancer (IARC), Lyon, France, and the German Cancer Research Centre (DKFZ), Heidelberg, Germany. The 'Europe against cancer' programme of the Commission of the European Communities partially supports the Directory.

INTERNATIONAL AGENCY FOR RESEARCH ON CANCER

The International Agency for Research on Cancer (IARC) was established in 1965 by the World Health Assembly, as an independently financed organization of the World Health Organization. The headquarters of the Agency are at Lyon, France.

The Agency conducts a programme of research concentrating particularly on the epidemiology of cancer and the study of potential carcinogens in the human environment. Its field studies are supplemented by biological and chemical research carried out in the Agency's laboratories in Lyon and, through collaborative research agreements, in national research institutions in many countries. The Agency also conducts a programme for the education and training of personnel for cancer research.

The publications of the Agency are intended to contribute to the dissemination of authoritative information on different aspects of cancer research. A complete list is printed at the back of this book.

DIRECTORY OF ON-GOING RESEARCH IN CANCER EPIDEMIOLOGY

1994

IARC
DKFZ

R. Sankaranarayanan
J. Wahrendorf
E. Demaret

In collaboration with

H.J. Baur

Published by the International Agency for Research on Cancer,

150 cours Albert Thomas, 69372 Lyon Cedex 08, France

Distributed by Oxford University Press, Walton Street, Oxford OX2 6DP, UK

Distributed in the USA by Oxford University Press, New York

ISBN 92 832 2130 3

ISSN 0300 5085

Printed in the United Kingdom

TABLE OF CONTENTS

PREFACE vii

INTRODUCTION ix

Selection of material ix

Content and Quality of Abstracts ix

Keeping the Directory Current x

References x

Addresses x

Directories 1976–1994 x

Indexing xi

Cancer Registries xii

Biological Materials Banks xii

Acknowledgements xii

HOW TO USE THE DIRECTORY xiii

LIST OF ABBREVIATIONS xv

USER SURVEY xvii

PROJECTS

Algeria 3
Australia 4
Austria 22
Bangladesh 24
Belgium 25
Brazil 29
Canada 31
Chile 54
China 55
Costa Rica 79
Croatia 80
Cuba 81
Czech Republic 82
Denmark 84
Egypt 97
Estonia 98
Finland 99
France 107
Gambia 142
Germany 143
Greece 157
Hong Kong 160
Hungary 161
Iceland 165
India 168
Indonesia 185
Ireland 187
Israel 188
Italy 197
Japan 226
Kyrghyzstan 247
Malaysia 248
Mexico 250
Netherlands 252
New Caledonia 266
New Zealand 267

Nigeria 272
Norway 274
Pakistan 279
Poland 281
Romania 286
Russian Federation 287
Slovak Republic 298
Slovenia 300
South Africa 302
Spain 306
Sri Lanka 309
Sweden 310
Switzerland 334
Thailand 337
Turkey 339
United Kingdom 340
United States of America 388
Uruguay 507
Yugoslavia 510

INDEXES

Investigators 515
Terms 549
Sites 619
Types of Study 679
Chemicals 691
Occupations 719
Countries 737
Cancer Registries 759

LIST OF CANCER REGISTRIES 765

PREFACE

The 1994 Directory of On–Going Research in Cancer Epidemiology introduces the new biennial publication cycle. It contains 1246 projects carried out by some 900 scientists all over the world, involving 80 countries. Although the number of projects reported has remained fairly constant in recent years, an expansion of epidemiological effort is seen in several countries, notably in China, India and Eastern Europe.

The main feature of the 1994 Directory is the increasing interest in genetic epidemiology, intervention, and biomarkers. The number of studies on the role of viruses in the aetiology of cancer has also increased, as well as those on the role of Helicobacter pylori in stomach cancer. Radiation is an area of considerable interest and the health effects of radiation, in particular background and diagnostic radiation and electromagnetic fields, are increasingly being studied.

Diet, tobacco, and occupational exposures continue to be major research topics and the number of such studies have quadrupled since the first Directory in 1975. Little variation has been seen in the cancer sites under study throughout the years and the lung, the gastrointestinal tract and the female breast are still the most frequently studied sites.

Collecting and storing biological materials for use in future studies are becoming more common and such collections can be easily identified in the Directory. The involvement of population–based cancer registries in epidemiologcial research has also increased.

A survey was carried out in 1992 among all the contributors to the 1991 and 1992 Directories, to find out how the Directory is being used and how it can be improved. The detailed results are presented in the Introduction. Following these results it was decided to abandon PROSE, the electronic tool which enabled searching the Directory on a personal computer. Clearly Directory users prefer to consult the book. Other modifications will be implemented for the 1996 edition. It is of particular interest to note that, following use of the Directory, 57% of the respondents had developed or changed a research idea, and 6% had abandoned an idea. The Editors were pleased to note that the biennial cycle is well accepted and that 97% of the respondents find that the value of the Directory to them justifies completing the abstract form. We are therefore confident that epidemiologists will continue to report their work, hence contributing to making the scientific community aware of what is going on in the field and supporting the Editors in their continuous effort to increase the Directory coverage and standards.

Dr L. Tomatis
Director
International Agency for
Research on Cancer (IARC)

Prof. H. zur Hausen
Scientific Director
German Cancer Research
Centre (DKFZ)

INTRODUCTION

The Directory of On–Going Research in Cancer Epidemiology has three aims:

(1) to inform cancer epidemiologists and other interested scientists about current work in the field of cancer epidemiology;

(2) to facilitate direct contacts between research workers;

(3) to enable unnecessary duplication of work to be avoided.

The Directory is produced jointly by the International Agency for Research on Cancer (IARC), Lyon, France and the German Cancer Research Centre (Deutsches Krebsforschungszentrum – DKFZ), Heidelberg, Germany.

Selection of Material

The main criterion for inclusion of a project in the Directory is that it deals with cancer epidemiology. Originally, "epidemiology" was interpreted fairly broadly, but editorial policy has been more restrictive in recent years, partly due to increasing pressure on Directory space. The large number of projects received which are judged to fall outside the scope of the Directory has also prompted us to attempt a more precise working definition of cancer epidemiology.

For the purposes of inclusion in the Directory, "cancer epidemiology" is taken to refer to studies of cancer which are designed to estimate cancer morbidity or mortality in human populations, or to investigate potential risk factors in cancer aetiology, or to evaluate methods of cancer prevention. Studies in the field of genetic and molecular epidemiology are also accepted.

Purely descriptive studies from developed countries, e.g. of the incidence of cancer in the general population, are not usually included, while such studies are still accepted from developing countries, where analytical studies may be difficult to carry out. Two other types of project are also accepted, namely those which involve:

(1) development of methods to assess human exposures and to test the application of such methods in epidemiological studies. Methods of assessing exposure to potential carcinogens in large numbers of individuals, e.g. by the use of DNA adducts, are undergoing rapid development, and are increasingly important in monitoring of actual rather than indirectly estimated exposure;

(2) biological materials banks (BMB): biological material is increasingly being used, or stored for later use, in epidemiological studies, to look for tumour markers or biochemical markers of premalignant lesions. More detailed descriptions of biological materials banks are accepted as project abstracts, provided the project also comprises a current epidemiological study.

Studies which primarily deal with diagnosis, treatment, prognosis, survival, health care programmes and, of course, animal studies will not normally be included. Any of these topics may well form part of a study included in the Directory, but the primary purpose of the study should fall within the definition of cancer epidemiology given above.

Content and Quality of Abstracts

The amount of information given for each project abstract is variable, but all abstracts should normally cover the following key points:

- aim of the study or hypothesis being tested;
- study type or design;

– size of study population;

– nature of any control group; and

– outline of methods of data collection and analysis.

Investigators should limit background information (e.g. the rationale for the study) to a bare minimum. Directory users are mainly interested in what is being done; they will normally know why it is being done.

In a proportion of all new projects reported, essential information is missing. In such cases the principal investigator is contacted. If no reply is received the project is excluded.

We must stress that the Directory editors do not act as referees of the quality of research. We simply try to include clear, concise accounts of relevant research, in order to make these available to the scientific community, since one of the aims of the Directory is to reflect, as far as possible, what is being done in the field. Many excellent research projects are not included only because they fall outside the scope of the Directory, while inclusion in the Directory implies only an adequate description of relevant research, not an assessment of its quality.

Keeping the Directory Current

To keep the Directory free from studies which have already been completed and published, all contributors to the current Directory are contacted every year and asked to update their contribution.

Investigators of new projects with abstracts which seemed to report completed work, were contacted to ensure that the study had not been published. Any projects for which apparently final results have been published were removed.

We would like to stress that investigators can report studies at any time. Abstract forms can be obtained from Mrs E. Demaret, International Agency for Research on Cancer, 150 cours Albert Thomas, 69372 Lyon Cedex 08, France.

References

Up to three references to interim publications from a study are accepted in the project abstracts. It is not necessary to reference background information, which should in any case be very brief. Journal articles and book chapters are accepted, but not meeting abstracts, letters or editorials, or articles submitted or in press. The reference will give the journal title, volume number, page number and year of publication in that order. For the sake of brevity, the title of the article will be excluded, and the name of the first author will be given only if other than the principal investigator cited in the Directory abstract.

Addresses

Addresses of potential contributors are extracted from various sources — literature data bases, annual reports from research institutes, lists of participants at meetings, membership lists of scientific associations, etc. Some 2500 research workers are contacted every year.

An important part of keeping the Directory current is constant vigilance over the accuracy and usefulness of addresses. Our address list is constantly updated, in order to avoid, for example, sending an invitation to an erroneous address, or recontacting investigators who have informed us that they do not work in cancer epidemiology, or have retired.

Directories 1976–1994

The table below shows, for each issue of the Directory, the number of projects included, the number and percentage of completely new studies and the number of countries in which data were being collected.

Year	No. of projects	New projects (%)	No. of countries
1976	622	622 (100.0)	65
1977	908	467 (51.4)	69
1978	1025	341 (33.3)	70
1979	1092	295 (27.0)	74
1980	1261	353 (28.0)	78
1981	1313	299 (22.8)	80
1982	1247	275 (22.1)	74
1983	1302	256 (19.7)	80
1984	1213	200 (16.5)	80
1985	1229	261 (21.2)	88
1986	1352	334 (24.7)	89
1987	1320	240 (18.2)	84
1988	1237	173 (14.0)	84
1989/90	1300	278 (21.3)	86
1991	1147	208 (18.1)	86
1992	1197	284 (23.7)	70
1994	1246	275 (22.1)	80

Indexing

Each project has a unique serial number, which is used for all references to that project in the indexes. This number appears at the top left of each entry.

Projects are listed by serial number in the main body of the Directory. They are in alphabetic order of the country from which they are reported, usually the same as the country in which they are being carried out. Within a country, the listing is alphabetic by town, and within town, alphabetic by principal investigator.

In order to find projects which fit any particular description, eight separate indexes can be used:

Name of investigator

Term (key–word)

Cancer site

Study type

Specific chemical exposure(s)

Specific occupation(s)

Country in which data are being collected

Cancer registry providing data for a study.

Every project is indexed in at least five ways: by name of each investigator; by key–word(s); by cancer site; by study type; and by country. Where relevant, studies are also indexed to chemical exposure or occupation, and to collaborating cancer registries.

All index headings currently in use are listed at the front of each index, including cross–references to more specific index headings and to other indexes. For example, the list of headings at the front of the TERMS (key–words) index shows that studies indexed to the generic term "Reproductive Factors" may also be sought under any of the more specific headings "Abortion", "Birthweight", "Menarche", etc., and that studies indexed to the general term "Metals" may also be searched for in the CHEMICALS index under any of the specific metals "Aluminium", "Beryllium", etc. The lists of index headings should simplify the selection of efficient search terms. The main features of the indexes, together with major changes in 1992, are outlined at the front of each index.

Cancer Registries

The index of Cancer Registries was created in 1989 and identifies each project in the Directory in which a given cancer registry is involved. "Involvement" implies either that the project is carried out within the cancer registry or that the registry releases individual data for a project carried out by an external investigator. Each registry has been given a short name, by which it appears in the index. The short name of the registry also appears beneath the abstract in the body of the Directory, under the heading REGI.

The address list of population–based cancer registries, which follows the index, enables registries to be readily contacted. This list contains general and specialized registries. The registries appear alphabetically by short name within country. Registries which are members of the International Association of Cancer Registries are marked with an asterisk.

Biological Materials Banks

For several years a section of biological materials collections has been included in the Directory. Unfortunately, it has not been possible to update this section for the last two issues and the editors considered it to be of little value to repeat it a third time without modification. It is planned to revise and expand this section for the 1996 edition.

It should be noted that studies collecting biological materials for later use, or using existing collections, are coded to the TERM "BMB", as well as to the material collected (e.g. Blood, Urine, Semen) and can thus be identified through the TERM Index.

Acknowledgements

Production of this Directory would not have been possible without the contribution of all research workers and we would like to thank them for reporting their studies and thus ensuring that other researchers are aware of their work. We apologize to scientists whose reports arrived too late for inclusion in this edition (they will be considered for inclusion in the next issue), and to those scientists whom we have yet to contact.

Lyon and Heidelberg, January 1994

R. Sankaranarayanan
J. Wahrendorf
E. Démaret

HOW TO USE THE DIRECTORY

Layout of Information about a Study

The layout of a typical project abstract is shown in Fig. 1. The numbers circled in the figure correspond to the numbers in the description below.

1. Serial number (an asterisk denotes a project included for the first time)
2. Principal investigator or study co−ordinator
3. Address of principal investigator or study co−ordinator. Telephone, telex and telefax numbers are given when available.
4. Project collaborator(s)
5. Title of project
6. Abstract of project
7. TYPE: The type of study, e.g. Cohort, Cross−Sectional, etc. Some projects involve more than one type of study.
8. TERM: The key−words summarizing the main features of the study.
9. SITE: The cancer sites or types under study. For many studies, particularly prospective ones, all sites of cancer are of interest, and hence "All Sites" is a frequent entry in this section.
10. CHEM: Specific chemical(s) under study.
11. OCCU: Specific occupation(s) under study.
12. LOCA: The country (location) where the data are being collected. Only shown if different from country of investigator reporting the project, or if more than one country.
13. REGI: Cancer registry involved in the study.
14. TIME: The period over which the investigators consider their study will take place. We have tried to avoid open−ended studies, but this has not always been possible, particularly for long−term prospective studies or registries. When no time period was given the investigator was contacted, but we did not always receive a reply.
15. Unique identification number, which remains the same throughout the "life" of the project. For internal use.

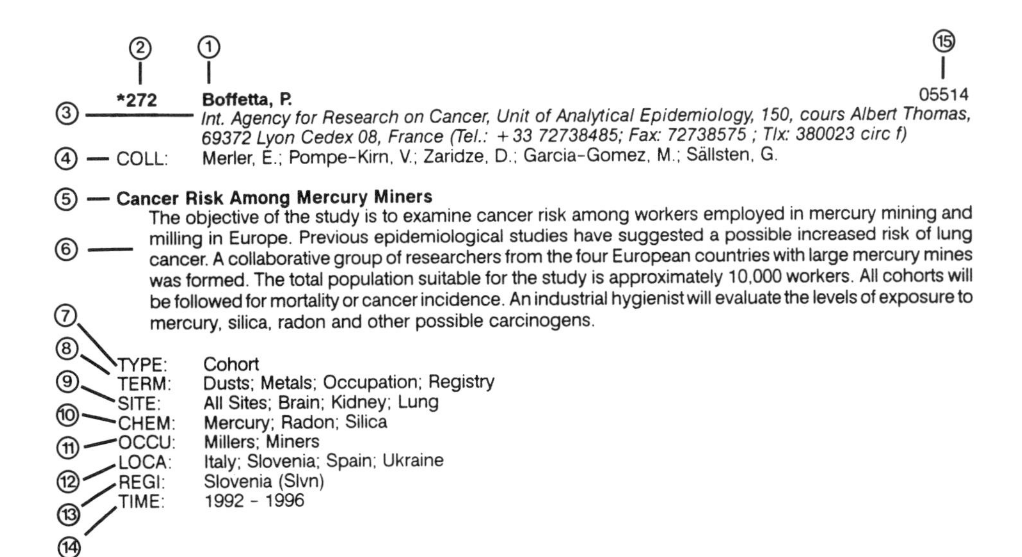

Fig. 1 Layout of a sample project

LIST OF ABBREVIATIONS

This list includes the most common acronyms and abbreviations used in the Directory. The meaning of other abbreviations is given in full in the abstract in which they appear.

AFP	Alphafoetoprotein
AHH	Aryl Hydrocarbon Hydroxylase
AIDS	Acquired Immunodeficiency Syndrome
ALL	Acute Lymphocytic Leukaemia
AML	Acute Myeloid Leukaemia
ANLL	Acute Non–Lymphoblastic Leukaemia
ATLL	Acute T–Cell Leukaemia–Lymphoma
BCC	Basal Cell Carcinoma
BCG	Bacille Calmette–Guerin
BCME	Bis(chloromethyl)Ether
BL	Burkitt's Lymphoma
BMB	Biological Materials Banks
CEA	Carcinoembryonic Antigen
CIN	Cervical Intraepithelial Neoplasia
CIS	Carcinoma In–Situ
CLL	Chronic Lymphocytic Leukaemia
CML	Chronic Myeloid Leukaemia
CMME	Chloromethyl Methyl Ether
CMV	Cytomegalovirus
CNRS	Centre National de la Recherche Scientifique (France)
CNS	Central Nervous System
2,4–D	2,4–Dichlorophenoxyacetic Acid
DBCP	Dibromochloropropane
DCLG	Dutch Childhood Leukaemia Group
DES	Diethylstilboestrol
DKFZ	Deutsches Krebsforschungzentrum (Germany)
DMPA	Depot Medroxyprogesterone Acetate
DNA	Deoxyribonucleic Acid
EBV	Epstein–Barr Virus
ELF	Extremely Low Frequency (0–300 Hz)
ELISA	Enzyme–Linked Immuno–Sorbent Assay
EMF	Electromagnetic Fields
ENT	Ear, Nose and Throat
EORTC	European Organisation for Research and Treatment of Cancer
EPA	Environmental Protection Agency (USA)
FAB	French–American–British Classification
FAP	Familial Adenomatous Polyposis
FDA	Food & Drug Administration (USA)
FSH	Follicle Stimulating Hormone
GC	Gas Chromatography
G6PD	Glucose–6–Phosphate Dehydrogenase
HBsAg	Hepatitis B Surface Antigen
HBV	Hepatitis B Virus
HCG	Human Chorionic Gonadotrophin
HCV	Hepatitis C Virus
HD	Hodgkin's Disease
HDV	Hepatitis D Virus
HIV	Human Immunodeficiency Virus
HLA	Histocompatibility Locus Antigen

HPLC	High Pressure Liquid Chromatography
HPV	Human Papilloma Virus
HSV	Herpes Simplex Virus
HTLV	Human T–Cell Leukaemia Virus
IARC	International Agency for Research on Cancer
ICD	International Classification of Diseases
ICD–O	International Classification of Diseases – Oncology
ILO	International Labour Office
INSEE	Institut National de la Statistique des Etudes Economiques (France)
INSERM	Institut National de la Santé et de la Recherche Médicale (France)
ISIC	International Standard Industrial Classification
KS	Kaposi's Sarcoma
LH	Luteinizing Hormone
MDA	4,4'–Methylene Dianiline
MOCA	4,4–Methylene–bis(2–Chloroaniline)
MRC	Medical Research Council (UK)
MS	Mass Spectrometry
NCI	National Cancer Institute (USA)
NCIC	National Cancer Institute of Canada
NHL	Non–Hodgkin's Lymphoma
NIEHS	National Institute of Environmental Health Sciences (USA)
NIH	National Institutes of Health (USA)
NIOSH	National Institute for Occupational Safety and Health (USA)
NPC	Nasopharyngeal Cancer
OPCS	Office of Population Censuses and Surveys (England & Wales)
OSHA	Occupational Safety and Health Administration (USA)
PAH	Polycyclic Aromatic Hydrocarbons
PBB	Polybrominated Biphenyls
PCB	Polychlorinated Biphenyls
PCR	Polymerase Chain Reaction
PHC	Primary Hepatocellular (Liver) Carcinoma
PMR	Proportional Mortality (or Morbidity) Ratio
PVC	Polyvinyl Chloride
RFLP	Restriction Fragment Length Polymorphism
RIA	Radioimmunoassay
RNA	Ribonucleic Acid
SCE	Sister Chromatid Exchange
SEARCH	Surveillance of Environmental Aspects Related to Cancer in Humans (IARC programme)
SEER	Surveillance, Epidemiology and End Results program (USA)
SHBG	Sex Hormone Binding Globulins
SIR	Standardized Incidence Ratio
SMR	Standardized Mortality Ratio
2,4,5–T	2,4,5–Trichlorophenoxyacetic Acid
TLV	Threshold Limit Value
TNM	Tumour–Nodes–Metastasis classification (UICC)
TSH	Thyroid Stimulating Hormone
UICC	Union Internationale Contre le Cancer (Switzerland)
WHO	World Health Organization

DIRECTORY OF ON–GOING RESEARCH IN CANCER EPIDEMIOLOGY

User survey

A survey was carried out in 1992 with the aim to better appreciate the use being made of the Directory and, if necessary, to improve it. Questionnaires were sent to 1005 principal investigators in the 1991 and 1992 Directories. 486 replies were received, a response rate of 48%, of which 470 were included in the analysis. The questionnaire is reproduced below (Figs 1 and 2) and gives the numbers and percentages for each question. Please note that for some questions no reply was given and the percentage does not add up to 100.

The majority of the respondents (45%) used the Directory 1–4 times in the past year, 26% used it 5–9 times, and 20% used it more than 10 times (Fig. 3). Of the 39 investigators who said they had never used the Directory in the past year, 10 stated that they had not received the book. User frequency by country is shown in Fig. 4.

Most investigators (91%) used the book to find out what is being done on a particular subject. 61% also used it to see what a particular individual was doing and 49% to obtain addresses or telephone numbers of investigators. Other purposes for which the Directory has been used are listed in Fig. 5.

Following use of the Directory 175 investigators (37%) developed a research idea, 96 (20%) changed an idea and 29 (6%) abandoned it.

From this survey it is clear that PROSE has not been the success we hoped when implementing it some years ago and 71% of the respondents never used it. Various reasons for this were given and are listed below (Fig. 7). It is interesting to note that, while 50% of the respondents from China had used PROSE, only 18% in, for example, UK or USA had.

Investigators were asked to rate the usefulness of each index. The Investigator, Term and Site indexes were all considered by a majority of respondents to be of great importance, while those of Type of Study, Chemical, Occupation, Country and Cancer Registry were considered of moderate importance. The section of Biological Materials Banks seems to be little used.

Although 75% of the respondents find the abstracts of great importance, 55% are in favour of replacing the current free–text abstract by a more structured one prepared from responses to a short questionnaire. A revision of the abstract form is being envisaged for the subsequent issue of the Directory.

The switch to biennial cycle seems to be well accepted, 84% of the respondents considering that this will have no or little effect on the usefulness of the Directory.

The Editors were pleased to see that 94% of the respondents feel that the usefulness of the Directory to them justifies completing the form, and we look forward to continued collaboration. To the 19 investigators who feel that the value of the Directory does not justify filling in the form, we express our thanks for nevertheless contributing, some for many years, thus helping us to improve Directory coverage of research being done in this field and keeping the scientific community informed about their work.

RESULTS

DIRECTORY OF ON-GOING RESEARCH IN CANCER EPIDEMIOLOGY

User survey

---

1. About how many times have you used the Directory in the past year?

	No.	%
10 or more	93	20
5-9	120	26
1-4	213	45
None	39	8

2. Have you ever used the Directory for any of the following purposes?

	Yes		No	
	No.	%	No.	%
- to find out what research was being done on a particular subject?	427	91	42	9
- to find out what a particular individual was doing?	288	61	180	38
- to obtain an address or telephone number?	229	49	237	50
- other purposes (please specify)________________ ________________	64	14	379	81

3. Has reference to the Directory ever caused you to develop, change, or abandon a research idea?

	Yes		No	
	No.	%	No.	%
- Develop	175	37	288	61
- Change	96	20	364	77
- Abandon	29	6	430	91

4. Have you ever used the PROSE diskette to search the Directory on your PC?

Yes		No	
No.	%	No.	%
129	27	333	71

If no, why not? ________________

If yes, do you have any suggestions as to how it could be made more useful or easier to use? ________________

Figure 1. Questionnaire

5. Please rate the importance to you of the following components of the Directory:

	Great importance	Moderate importance	Little importance	No importance
	%	%	%	%
Project abstracts	75	17	3	2
Index of investigators	51	33	10	3
Index of key-terms	49	29	13	4
Index of cancer sites	57	29	7	3
Index of types of study	29	37	23	7
Index of chemicals	26	32	26	9
Index of occupations	25	35	26	9
Index of countries	23	33	28	10
Index of cancer registries	23	35	26	11
Cancer registry address list	25	32	27	11
Biological materials banks	15	26	33	19

6. What effect do you think **biennial publication** will have on the usefulness of the Directory to you?

	No.	**%**
No effect	191	41
Small reduction in usefulness............	201	43
Moderate reduction in usefulness.........	60	13
Large reduction in usefulness............	15	3

7. Considering both the use you make of Directory abstracts and the work you put into preparing your own abstracts, what would be your opinion of replacing the current free-text abstract by a structured abstract, prepared from responses to a short questionnaire?

	No.	**%**
Favourable..............................	258	55
No opinion..............................	123	26
Unfavourable............................	81	17

8. Does the usefulness of the Directory to you justify your time to complete a short questionnaire or update form every two years?

Yes		**No**	
No.	**%**	**No.**	**%**
442	94	19	4

Figure 2. Questionnaire (continued)

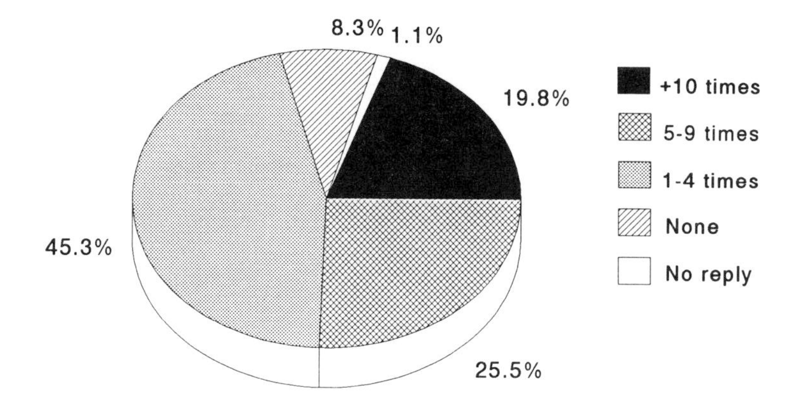

Figure 3. User frequency

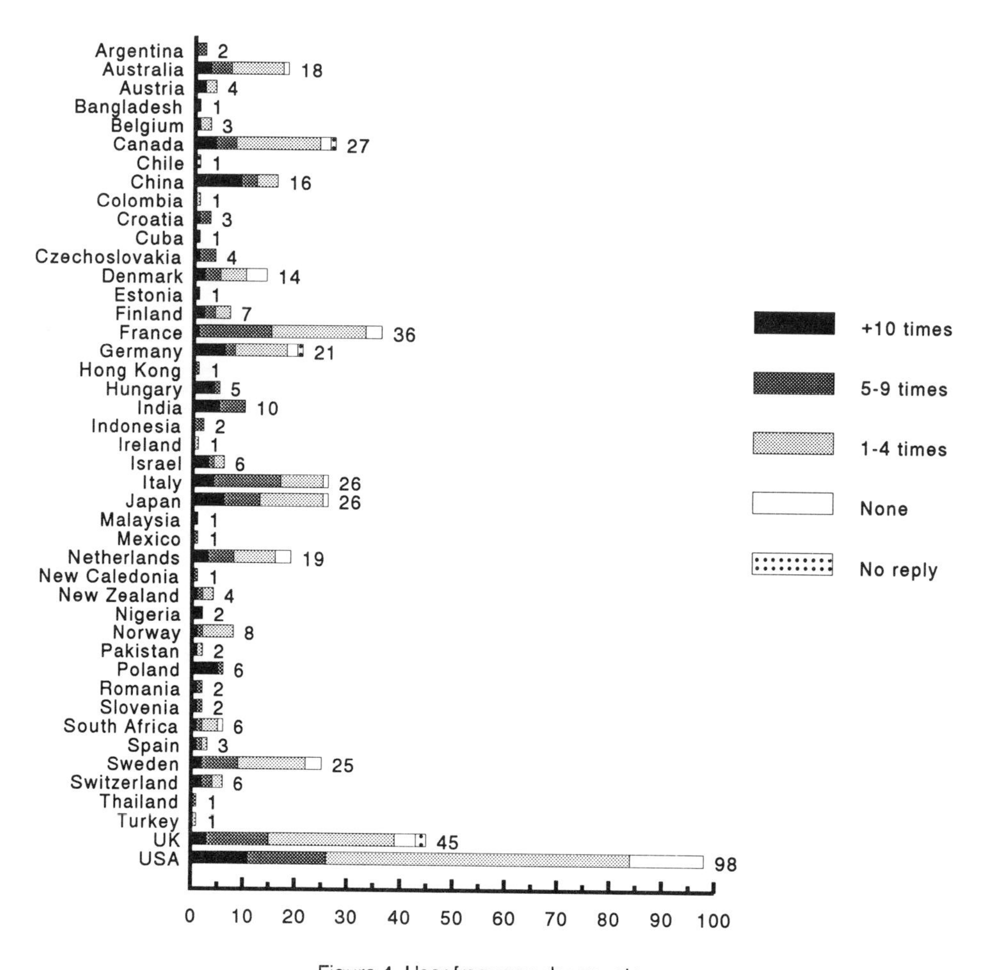

Figure 4. User frequency by country

USER SURVEY COMMENTS

The Directory has been used for the following purposes:

- to find out what is going on in the world/specific country
- to derive statistics of the type of research going on in a field
- to get a profile of the work carried out in an institute or department, e.g. when planning a visit
- to analyse trends in epidemiological research (using previous issues)
- to assess progress
- to see the exact indicators and parameters researchers are using
- to see who is doing research in a particular field
- to identify areas or topics where no research is being done
- to find out the type of populations research is done on
- to find out about study methods
- to determine number of projects on a certain topic
- to ask for reprints of published studies
- to locate cancer registries
- to communicate with others
- state of the arts concerning particular associations
- intorudction of cancer epidemiology for education

Figure 5. Purposes for which the Directory has been used

LIST OF PROJECTS

Sétif

1 **Hamdi-Chérif, M.** 04702

Centre Hospitalier de Sétif, Serv. de Médecine Prév. & d'Epidémiologie, Sétif 19000, Algeria (Tel.: +213 5 900810; Fax: 903993; Tlx: 86958)

COLL: Coleman, M.P.; Sekfali, N.; Benlatreche, K.; Allouache, A.; Rahal, D.

Retrospective Study of Cancer Morbidity in Sétif, Algeria

Cancer is a major public health problem in Algeria. This project has two major objectives: to provide an accurate description of cancer morbidity in the wilaya (province) of Setif, and to serve as a feasibility study for a population-based cancer registry in Setif. A team of 12 public health interns has been formed to collect a simple data set on each tumour diagnosed in the 3-year period 1 January 1986 to 31 December 1988 in persons normally resident in the wilaya. Data sources include hospital records, pathology laboratories, private physicians and town halls (death certificates), and referral hospitals outside the wilaya. Only 15 data items will be recorded, including sex, birthdate and commune of residence of the person; and date of diagnosis, site, morphology and basis of diagnosis of the tumour. These data will enable incidence rates to be calculated by sex, age and cancer site and by geographical region within the county. The study is expected to identify over 6,000 cancers, and to establish the contacts and data collection mechanisms required for setting up the cancer registry, which will record prospectively, from 1 January 1990, all cancers diagnosed in the wilaya of Setif.

TYPE: Incidence
TERM: Geographic Factors; Registry
SITE: All Sites
TIME: 1989 - 1994

***2** **Hamdi-Chérif, M.** 05404

Centre Hospitalier de Sétif, Serv. de Médecine Prév. & d'Epidémiologie, Sétif 19000, Algeria (Tel.: +213 5 900810; Fax: 903993; Tlx: 86958)

COLL: Nouasria-Sekfali, N.; Touabti, A.; Aït Hamouda, R.; Laouamri, S.; Hubert, A.; de Thé, G.; Jeannel, D.

Nasopharyngeal Carcinoma: Diet, Lifestyle and Early Detection

Nasopharyngeal carcinoma (NPC) is the most frequent ear, nose and throat cancer in the countries of the Maghreb, with an age standardised incidence rate of 4.2/100.000 for males and 2.4/100.000 for females in Algeria. This project has two major objectives: (1) identification of aetiological factors associated with NPC, including EBV infection. A case-control study is being conducted to assess the role of diet, dietary patterns and lifestyle. Cases are identified in the Cancer Registry of Setif. Neighbourhood controls are selected from the cases' place of residence. Information on socio-economic factors, housing, and past and present diet is collected; (2) screening of NPC, using techniques for detection of EBV IgA/EA and IgA/VCA. NPC can be detected in an early stage and treated. A follow-up study on EBV IgA/VCA antibody positive persons is carried out. Clinical and histological examinations are carried out once a year for four years.

TYPE: Case-Control; Incidence
TERM: Antibodies; Diet; EBV; Lifestyle; Registry; Screening; Socio-Economic Factors
SITE: Nasopharynx
REGI: Sétif (Alg)
TIME: 1993 - 1997

AUSTRALIA

Adelaide

3 **Woodward, A.** 01999
Univ. of Adelaide, Dept. of Community Medicine, Box 498 G.P.O., Adelaide SA 5001, Australia (Tel.: +61 8 2284637; Fax: 2234075)
COLL: McMichael, A.J.; Roder, D.M.; Crouch, P.; Mylvaganam, A.

Lung Cancer and Other Cancers among Uranium Miners, as Related to Radiation Exposure

2,580 individuals were employed in the South Australian Radium Hill uranium mine during its operational period from 1951 to 1961. Death records and cancer registry records are being examined progressively to identify lung cancers and other cancer cases among these individuals, by occupational category at Radium Hill. Electoral rolls, driving licences and health insurance records have been employed to identify the miners still alive and resident in Australia. Accessible miners, or next-of-kin have been interviewed to determine their smoking and occupational histories. Individual workers' exposures to radiation are being estimated from historical records of radon gas levels in the mine, and detailed job histories. Smoking histories obtained from miners who are still alive show a slightly greater smoking prevalence among underground workers but not an association of smoking with estimated exposure to radon daughters at Radium Hill. By the end of 1988, about 25% of the Radium Hill workforce were found to have died. Preliminary analyses show that workers receiving 40 Working Level Months of radiation or more have experienced approximately four times the rate of lung cancer among surface workers at the mine. A further round of death record searches will occur in 1993-1994.

TYPE: Case-Control; Cohort
TERM: Metals; Mining; Occupation; Radiation, Ionizing; Registry; Tobacco (Smoking)
SITE: Lung
CHEM: Uranium
OCCU: Miners, Uranium
REGI: S. Australia (Aus)
TIME: 1981 - 1994

***4** **Woodward, A.** 05420
Univ. of Adelaide, Dept. of Community Medicine, Box 498 G.P.O., Adelaide SA 5001, Australia (Tel.: +61 8 2284637; Fax: 2234075)
COLL: Shanahan, E.M.; Morley, A.A.; McMichael, A.J.

Somatic Mutations and Exposure to Radon Progeny in a Cohort of Former Uranium Miners

The aim of the study is to compare the frequency of selected mutations in peripheral blood cells collected from a cohort of former uranium miners with exposures to radon progeny experience in the past. Participants include 253 men who worked in the Radium Hill uranium mine between 1953 and 1962. Somatic mutations are being studied by the glycophorin-A, hypoxanthine phosphoribosyl transferase and HLA-A assays. Information on other exposures which may influence the frequency of mutations is being collected by interview. Radon progeny exposures have been estimated from historical records of radon gas levels, combined with production records and individual job histories.

TYPE: Cohort; Molecular Epidemiology
TERM: BMB; Biomarkers; Mutation, Somatic; Occupation; Radiation, Ionizing
SITE: Inapplicable
CHEM: Radon
OCCU: Miners, Uranium
TIME: 1991 - 1994

Brisbane

5 **Battistutta, D.** 04340
Queensland Inst. of Medical Research, Dept. of Cancer Epidemology, 300 Herston Rd , Herston, Brisbane Qld 4029, Australia (Tel.: +61 7 2536222; Tlx: AA145420)
COLL: Knight, N.; MacLennan, R.

Population-Based Registry of Colorectal Familial Adenomatous Polyposis

A register has been established of all affected and at-risk members of familial adenomatous polyposis (FAP) families in Queensland. It has identified 36 FAP families, of which 65 individuals are affected and 122 at risk for FAP. In collaboration with the responsible clinician, a screening protocol for each at-risk individual is developed. The register provides reminders to clinicians and at-risk individuals of forthcoming screening tests, traces at-risk individuals lost to follow-up, and acts as a source of information on FAP for family members. Similar registers in other countries (UK, Scandinavia) have been shown to be very effective in preventing large bowel cancer in persons identified to be at risk for FAP. In collaboration with similar registers being set up in other Australian states, it is envisaged that research into the genetic and environmental factors that play a role in the development of FAP, and ultimately large bowel cancer, will be possible.

TYPE: Incidence; Registry
TERM: Environmental Factors; Familial Adenomatous Polyposis; Familial Factors; High-Risk Groups; Premalignant Lesion
SITE: Colon; Rectum
REGI: Brisbane II (Aus)
TIME: 1988 - 1994

6 **Khoo, S.K.** 02575
Univ. of Queensland, Royal Brisbane Hosp., Dept. of Obstetrics & Gynecology, Clinical Sciences Bldg, Brisbane Qld 4029, Australia (Tel.: +61 7 2535211)

Registry of Gestational Trophoblastic Disease in Queensland

The rarity of the disease in Australia necessitates centralization of facilities and collection of information. The Registry continues to function with the support of the Queensland Cancer Fund. Since June 1976, 431 patients have been registered. The objectives of the Registry are to: (1) study the incidence of hydatidiform mole and other neoplastic disorders of the trophoblast in Queensland; (2) provide a comprehensive follow-up programme, using HCG-beta measurements in blood; (3) centralize facilities for consultation and treatment of complications of the diseases; (4) determine long-term effects of cytotoxic chemotherapy in successfully treated patients in terms of fertility and pregnancy outcome; (5) evaluate various factors as a guide to prognosis. Patients with gestational trophoblastic disease are registered and the clinical data recorded by the Registry. Follow-up studies are done with the cooperation of the doctors in Queensland. Papers appeared in Aust. NZ J Obstet. Gynaecol. 20:35-42, 1980; 141-150, 1982 and 26:129-135, 1986.

TYPE: Incidence; Registry
TERM: Chemotherapy; Fertility; Pregnancy; Premalignant Lesion; Prognosis
SITE: Choriocarcinoma; Hydatidiform Mole
TIME: 1976 - 1994

7 **MacLennan, R.** 03493
Queensland Inst. Medical Research, Cancer Epidemiology Unit, Bramston Terrace, Herston, Brisbane Qld 4006, Australia (Tel.: +61 7 2536247; Fax: 2525499)
COLL: McLeod, G.R.C.; Little, J.

Fluorescent Light and the Risk of Malignant Melanoma in Brisbane

A case-control study of melanoma of the skin is being conducted in Brisbane, Australia, to estimate the possible risk of prior exposure to fluorescent light. Due to the high incidence rate it has been feasible to recruit over 400 new cases for a study powerful enough to detect risks smaller than those reported in previous studies. All incident cases in the Greater Brisbane Metropolitan area have been included in the sample approached with permission of their doctors. Controls are a sample of the general population matched by sex and age. Data collection includes a detailed history of exposure to fluorescent light. The history will be validated by information on lighting at former work places. The study includes all significant risk factors found in recent studies in Australia and North America, to enable control of potential confounding factors which might falsely produce apparently increased risk for exposure to fluorescent light.

TYPE: Case-Control
TERM: Fluorescent Light; Radiation, Ultraviolet
SITE: Melanoma
TIME: 1985 - 1994

8 **MacLennan, R.** 03494

Queensland Inst. Medical Research, Cancer Epidemiology Unit, Bramston Terrace, Herston, Brisbane Qld 4006, Australia (Tel.: +61 7 2536247; Fax: 2525499)

COLL: Battistutta, D.; Grattan, H.; Ward, M.; Cowen, A.; Bain, C.J.; Goulston, K.; Bokey, L.; Chapuis, P.; Killingback, M.; Lambert, J.; Korman, M.; Eaves, E.R.; McIntyre, O.R.; McLeish, J.; Macrae, F.A.; Penfold, J.C.B.; St John, D.J.B.; Wahlqvist, M.L.

Multicentre Collaborative Clinical Trial for Prevention of Large Bowel Adenomas

This multicentre clinical trial aims to assess the effects of dietary modification on the incidence and growth of colorectal adenomas, the precursors of one of the most common cancers. Subjects with histologically confirmed adenomatous polyps will be selected. In this high risk and well motivated group of people, the effects of a low fat diet, increased dietary fibre, and beta carotene on the incidence and size of adenomatous polyps after two years follow-up will be measured. This has significant implications for the frequency of colonoscopy and related health service costs. The design will be 2x2x2 factorial with at least 420 subjects. 400 subjects had been recruited by the end of 1987. The trial should be concluded by 1989 and analysis by 1990.

TYPE: Incidence
TERM: Diet; Fat; Fibre; Premalignant Lesion; Prevention; Vitamins
SITE: Rectum
CHEM: Beta Carotene
TIME: 1987 - 1994

9 **Martin, N.G.** 04193

Queensland Inst. of Medical Research, Cancer Epidemiology Unit, Bramston Terrace, Herston, Brisbane Qld 4006, Australia (Tel.: +61 7 2536278; Fax: 2525499)

COLL: Green, A.C.; MacLennan, R.; Aitken, J.F.; McLeod, G.R.C.; Little, J.H.; Ring, I.

Genotype and Environment Interactions in the Aetiology of Malignant Melanoma

This collaborative project with population-based familial registries of melanoma in Queensland and New South Wales will contact all patients with melanoma reported to population-based cancer registries to establish if there is a family history of melanoma and whether the patient is a twin. If either is present, patients will be further investigated together with first degree relatives. The combined populations of Queensland and New South Wales yield some 2,000 new cases of melanoma per year, of which some 300 will be familial, and 50 will be a twin. Cases back to 1982 will be included in the initial survey.

TYPE: Case Series
TERM: Familial Factors; Genetic Factors; Genetic Markers; Registry; Twins
SITE: Melanoma
REGI: NSW (Aus); Queensland (Aus)
TIME: 1987 - 1995

10 **Muller, J.M.** 05152

Queensland Dept. of Health, Epidemiology and Prevention Unit, 147-163 Charlotte St., G.P.O. Box 48, Brisbane Qld 4000, Australia (Tel.: +61 7 2340907; Fax: 2210951)

COLL: Ring, I.; Balanda, K.P.; Cannon, L.; Clutton, S.

Cervical Cancer Screening in Queensland

A range of studies has been initiated to evaluate several aspects of cervical cancer screening, using Pap smears, as a means of reducing morbidity and mortality from cancer of the cervix among Queensland women. The primary aim of these studies is to develop and evaluate strategies to increase screening coverage among the most at-risk women in the community, a secondary aim being to increase overall levels of coverage among all eligible women aged 20-69 years. The various components of this research include: cross-sectional postal surveys of 2,500 randomly selected Queensland women, covering knowledge, attitudes, psychosocial issues and behaviour; a cohort study of four groups of at-risk women, being women who have not had a Pap smear for two years or more, exposed to different strategies to promote Pap smear screening; evaluation of a mass media campaign using television, through analysis of pathology laboratory data and health services data; and the evaluation of specific programmes piloted for women living in rural areas and Aboriginal and Islander women which focus on service provision issues.

TYPE: Cohort; Methodology
TERM: Cytology; Screening
SITE: Uterus (Cervix)
TIME: 1989 - 1994

11 **Muller, J.M.** 05153
Queensland Dept. of Health, Epidemiology and Prevention Unit, 147-163 Charlotte St., G.P.O. Box 48, Brisbane Qld 4000, Australia (Tel.: +61 7 2340907; Fax: 2210951)
COLL: Ring, I.; Balanda, K.P.; Cannon, L.; Clutton, S.

Breast Cancer Screening in Queensland

This project will evaluate various aspects of breast cancer screening, using mammography, as a mean of reducing morbidity and mortality from breast cancer among Queensland women. The primary aim is to assist in the development and implementation of the breast cancer prevention programme. The project includes monitoring and evaluation of breast screening services using routine data collected at the clinics; an assessment of women's satisfaction with the services and experience of mammography from point of service surveys; an examination of factors that influence and motivate attendance for screening, through point-of-service surveys; and the monitoring and evaluation of non-individualised strategies used to recruit eligible women from the target population, through population-based community surveys which will include an examination of knowledge, attitudes, psychosocial issues and behaviour.

TYPE: Methodology
TERM: Mammography; Screening
SITE: Breast (F)
TIME: 1989 - 1994

12 **Ring, I.** 04144
Queensland Dept. of Health, Epidemiology and Health Information Br., GPO Box 48, Brisbane Qld 4001, Australia (Tel.: +61 7 2344187; Fax: 2210951 ; Tlx: AA 42531)
COLL: Williams, M.; MacLennan, R.; Chick, J.

Estimated Incidence of Skin Cancer in Queensland

The aim of this project is to estimate the incidence of diagnosed skin cancer in Queensland both for selected areas of the state and for the state as a whole. Queensland has probably the highest incidence of skin cancer in the world and this project will provide basic information not presently available on the level and pattern of this important disease. This knowledge is a basic prerequisite of programmes designed to have an impact on skin cancer. The Queensland Cancer Registry now collects histologically verified squamous cell carcinomas and basal cell carcinomas and a survey is to be conducted to obtain an estimate of clinically diagnosed non-melanoma skin cancers, to combine with the Registry data to provide an overall estimate of diagnosed skin cancer. Fourty doctors and a hospital outpatient department were selected randomly in each of four parts of the state, so that the four areas combine to form a state estimate.

TYPE: Incidence
TERM: Histology; Registry
SITE: Skin
REGI: Queensland (Aus)
TIME: 1984 - 1994

Darwin

13 **Mathews, J.D.** 04067
Menzies School of Health Research, P.O. Box 41096, Casuarina, Darwin NT 5792, Australia
COLL: Parsons, W.J.; Giles, G.G.

Mortality and Cancer Incidence in a Large Cohort of Men Employed as Herbicide Sprayers in Victoria

From the 1950s, phenoxyherbicides have been used in increasing quantities for weed control in Victoria, Australia. Spraying on government land and on some private land was carried out by men employed by the Victorian Department of Crown Lands. A mortality follow-up of over 2,000 men to 1983 showed that there was no significant increase in cancer overall, nor in soft-tissue sarcoma or lymphoma mortality. Data are being linked to the Victoria Cancer Registry to look for possible effects on cancer incidence.

TYPE: Cohort
TERM: Chemical Exposure; Herbicides; Occupation; Pesticides; Registry
SITE: Lymphoma; Sarcoma
CHEM: Phenoxy Acids
OCCU: Herbicide Manufacturers; Herbicide Sprayers
REGI: Victoria (Aus)
TIME: 1982 - 1994

14 **Mathews, J.D.** 02632
Menzies School of Health Research, P.O. Box 41096, Casuarina, Darwin NT 5792, Australia
COLL: Clifford, C.; Hopper, J.L.; Giles, G.G.

Australian Twin Registry

Previous studies have shown that cancer concordance rates in monozygous (MZ) twins are quite low, and this evidence has sometimes been cited to support the view that genetic factors are unimportant in the origin of cancer. The alternative view is that genetic factors are necessary, but not sufficient, as causes of cancer; the low concordance rates in MZ twins would then be attributed to the effect of environmental differences and/or to the effects of chance (stochastic) events which may influence the age of onset of cancer. The Australian National Health and Medical Research Council Twin Registry comprises 15,000 pairs of twins (volunteers); the incidence of cancer will be followed by linking Twin Registry information to Cancer Registry data and by follow-up of twins. On a collaborative basis, it is hoped to obtain additional information from other twin registries on the age of onset of cancer in twins. Data on the age of onset of cancer in twins will be used to develop and test statistical models for the origin of cancer which allow for the effect of genetic and environmental factors as well as for the effects of random (stochastic) processes.

TYPE: Incidence
TERM: Environmental Factors; Genetic Factors; Mathematical Models; Record Linkage; Registry; Twins
SITE: All Sites
TIME: 1981 - 1994

Herston

***15** **Smith, P.J.** 05380
Univ. of Queensland Medical School, Pathology Dept., Herston Rd, Herston Qld 4006, Australia (Tel.: +61 7 3655340; Fax: 3655511)

Genetic Basis of Wilms' Tumour

The aim of this study is to define loci important in the development of Wilms' tumour and to study mutation and genomic imprinting at those loci. It is hoped to determine the relative importance of genetically determined versus environmentally acquired lesions in the development of this form of paediatric neoplasm. The project has accumulated a materials bank of DNA from over 90 Wilms' Tumour. Tumour tissue, autologous kidney, RNA and parental DNA is available for many of the specimens.

TYPE: Genetic Epidemiology
TERM: BMB; Chromosome Effects; DNA; Tissue
SITE: Wilms' Tumour
TIME: 1986 - 1994

Hobart

16 **Dwyer, T.** 04992
Univ. of Tasmania, Menzies Center for Population Health Research, Tasmanian Cancer Registry, 43 Collins St., Hobart Tas. 7000, Australia (Tel.: +61 2 354880; Fax: 354816)
COLL: Berwick, M.; Roy, C.R.; Gies, H.P.

Human Exposure to Ultraviolet Radiation at Varying Latitudes

Currently available data linking environmental exposure to ultraviolet radiation and the incidence of malignant melanoma in humans lacks objective validation. This study aims to measure ultraviolet dose in a cohort of people and to validate an exposure questionnaire. Individual environmental exposure to ultraviolet radiation during specific activities will be measured by polysulphone film badges located on seven anatomical sites (hand, shoulder, thigh, calf, cheek, back and chest). An extra badge is placed on the surface where the individual undertook the activity. The activities will cover leisure activities (e.g. boating, swimming, tennis) and work-related exposures, e.g. outdoor work, office work). Current and typical sun exposure patterns, duration of activity, dress and use of sunscreen (blockouts) will be measured by questionnaire and related back to the actual exposure recorded by the badge and by environmental monitoring of ultraviolet radiation. Measurements will take place during different seasons and at different latitudes in Australia. The age range of the subjects will be 5-69 years and the sample size will be 1,800 of which 1,200 will be in Hobart.

TYPE: Methodology
TERM: Radiation, Ultraviolet
SITE: Melanoma
LOCA: Australia; United States of America
TIME: 1992 - 1994

17 **Lowenthal, R.M.** 04244
Royal Hobart Hosp., Univ. of Tasmania, Liverpool St., Hobart Tas. 7001, Australia (Tel.: +61 2 388157; Fax: 354894)
COLL: Marsden, K.; Bicevskis, M.; Jupe, D.M.L.; Beamish, M.

Environmental and Occupational Exposures of Patients with Myelodysplastic Syndromes

The aim of the study is to identify occupational and other environmental factors which may be associated with an increased risk of developing a myelodysplastic syndrome (MDS). Since 1988, all patients diagnosed with MDS in southern Tasmania have been identified and categorised by a series of laboratory tests. Since 1 January 1990, all surviving and new patients in southern Tasmania have been interviewed, using a detailed questionnaire about their life time exposure to occupational and other possible aetiological agents. This will enable a Siemiatycki exposure index to be calculated. To date, 105 patients and two age- and sex-matched controls per patient have been interviewed. Patients visiting hospital outpatient clinics serve as controls for hospital MDS patients, and patients of private medical practitioners as controls for non-hospital patients. In January 1993, the study was extended to include the whole of the state of Tasmania.

TYPE: Case-Control
TERM: Chemical Exposure; Environmental Factors; Occupation; Premalignant Lesion
SITE: Myelodysplastic Syndrome
REGI: Tasmania (Aus)
TIME: 1988 - 1994

***18** **Lowenthal, R.M.** 05460
Royal Hobart Hosp., Univ. of Tasmania, Liverpool St., Hobart Tas. 7001, Australia (Tel.: +61 2 388157; Fax: 354894)
COLL: Tuck, D.M.; Blizzard, C.L.

Exposure to Hydro-Electric Transmission Power Lines in Patients with Myeloproliferative and Lymphoproliferative Diseases

The question of the possible relationship between residence in close proximity to high voltage electricity transmission lines and the development of cancer is of considerable recent interest. During the years 1972-1980, 857 patients were diagnosed with myeloproliferative and lymphoproliferative diseases in Tasmania, an island state with a population of 450,000. The diagnoses included acute and chronic leukaemia, lymphoma and myeloma. 786 were adults (436 males/350 females) and 71 were children 16 years or younger (45 males/26 females). At diagnosis all patients or significant others were interviewed, enabling a life history to be documented. Included in the history were residential addresses from birth to diagnosis. Each of the patients had an age- and sex-matched control, who provided information on identical questionnaires. In the current study lifetime addresses and their proximity to Hydro-Electric Commission transmission lines equal to or more than 88,000 volts are being looked at. In all, over 9,000 addresses are being checked All addresses within 300 meters either side of the lines are considered to be of importance. The incidence and duration of living within 300 meters of a transmission line will be statistically analysed. An estimate of historical field exposure in each case will be calculated.

TYPE: Case-Control
TERM: Electromagnetic Fields
SITE: Leukaemia; Lymphoma; Multiple Myeloma
TIME: 1992 - 1994

AUSTRALIA

Melbourne

19 **Christie, D.G.** 02892
Univ. of Melbourne Dept. of Community Medicine, 159 Barry St. Carlton, Melbourne Vict. 3053, Australia (Tel.: +61 3 3447449; Fax: 3476136 ; Tlx: AA35185)
COLL: Robinson, K.; Gordon, I.; Potter, A.

Health Watch: The Australian Petroleum Industry Health Surveillance Programme

This prospective study within the Australian oil industry is designed to investigate possible associations between cancer mortality/morbidity and occupationally determined exposure to hydrocarbon chemicals. Data are collected on all deaths; cancer deaths are confirmed by reference to biopsy or autopsy reports. Subjects have been classified by industrial hygienists into four ranked exposure groups. Study members have contributed 50,000 person-years at risk to the mortality analyses with 170 deaths. The all-cause SMR for males is 0.7; for all malignant neopolasms in males the SMR is 0.9. 40,000 person-years have accumulated for cancer incidence with 116 incident cancers; the all-cancer SIR in males is 0.9. Twenty incident cancers of the lymphatic and haematopoietic tissue have been reported and the SIR for this group of cancer is 2.0 (95%C.I. 1.2-3.2). A case-control study of lympho-haematopoietic cancer is underway. A paper has been published in Med. J. Aust. 147:222-225, 1987.

TYPE: Case-Control; Cohort
TERM: Chemical Exposure; Occupation; Petroleum Products
SITE: All Sites; Haemopoietic; Lymphoma
CHEM: Hydrocarbons
OCCU: Petroleum Workers
TIME: 1981 - 1994

20 **Giles, G.G.** 03862
Anti-Cancer Council of Victoria, Cancer Epidemiology Centre, 1 Rathdowne St., Carlton South, Melbourne Vict. 3053, Australia (Tel.: +61 3 2791111; Fax: 2791270)
COLL: Marks, R.; Staples, M.

National Skin Cancer Survey

The aims are to estimate the incidence of skin cancers, principally basal cell carcinoma and squamous cell carcinoma, in Australia. This survey was first conducted in 1985 and is repeated every five years to assess trends. In the 1990 re-survey 63,450 persons in a random population-based sample across Australia were interviewed concerning their history of skin cancer treatment in the last 12 months, and their tanning ability. Medical details with regard to the site and histology of the lesions treated were confirmed with the treating medical practitioners. Demographic data, including lifestyle factors, smoking, occupation, and socio-economic status, were also collected. The 1985 survey was published in Br. Med. J. 296:13-17, 1988. The 1990 survey was published in Int. J. Cancer 53:1-6, 1993.

TYPE: Correlation; Incidence
TERM: Geographic Factors; Lifestyle; Occupation; Premalignant Lesion; Prevalence; Radiation, Ultraviolet; Registry; Socio-Economic Factors; Tobacco (Smoking)
SITE: Melanoma; Skin
REGI: Victoria (Aus)
TIME: 1985 - 1996

21 **Giles, G.G.** 04745
Anti-Cancer Council of Victoria, Cancer Epidemiology Centre, 1 Rathdowne St., Carlton South, Melbourne Vict. 3053, Australia (Tel.: +61 3 2791111; Fax: 2791270)
COLL: Hopper, J.L.; Ireland, P.; Ktenas, D.; Larkins, R.; O'Dea, K.; Potter, J.; Powles, J.W.; Proietto, J.; Wahlqvist, M.L.; Williams, J.

Melbourne Collaborative Cohort Study

The aim of the programme is to investigate the relationship between dietary intakes and specific disease outcomes, particularly cancer of the colon and rectum, breast and prostate. Emphasis is placed on the role of lipids, fibre and free radical scavengers and their food sources. The study population is a mix of Italian-born, Greek-born and Australian-born residents of the Melbourne statistical division aged between 40 and 69 years at recruitment. The cohort will eventually comprise 50,000 persons comprising approximately equal numbers of males and females and a higher number of migrants than of

Australian-born. Optically scannable questionnaires are administered at the survey center. The questionnaires include questions on demographic, medical and dietary and other lifestyle factors. A food frequency questionnaire is also completed. The physical examination includes anthropometry, body impedance and blood pressure measurements. Blood samples are taken and total cholesterol and glucose are assayed immediately. Two 1ml aliquots of plasma are stored in liquid nitrogen. Dried blood clots are collected on Gutherie paper for eventual analysis of genetic markers. Each participant is followed up on the third anniversary of recruitment to collect details of health events and dietary status. Follow-up is planned for 20 years. Record linkage to the Victorian Cancer Registry and to all death certificates is performed annually. A paper describing the study was published in Proc. Nutr. Soc. Aust. 15:61-68, 1990.

TYPE: Cohort
TERM: BMB; Blood; Diet; Lifestyle; Lipids; Migrants; Plasma
SITE: Breast (F); Colon; Prostate; Rectum
REGI: Victoria (Aus)
TIME: 1990 - 2010

22 **Giles, G.G.** 05054
Anti-Cancer Council of Victoria, Cancer Epidemiology Centre, 1 Rathdowne St., Carlton South, Melbourne Vict. 3053, Australia (Tel.: +61 3 2791111; Fax: 2791270)
COLL: Kaye, A.; Gonzales, M.

Australian Brain Tumour Register

The register includes all intracranial tumours and all tumours of the spinal cord. Its broad aim is to describe the incidence of these tumours in more detail than is usually possible in cancer registries, with a view to increasing their descriptive epidemiology and to monitor changes over time. The register records about 1,600 tumours annually. Notification is usually via neurosurgeons, neurosurgical units and neuropathologists. Follow-up details are obtained by writing to surgeons. Detailed socio-demographic data are collected in addition to review diagnoses and clinical information and treatment details. It is anticipated that the register will serve as a resource for research into the epidemiology and pathology of these tumours. To this end, a database and repository of slides is kept centrally.

TYPE: Incidence; Registry
TERM: Demographic Factors; Registry; Survival; Trends
SITE: Brain; Spinal Cord
REGI: Victoria (Aus)
TIME: 1986 - 1999

23 **Giles, G.G.** 05055
Anti-Cancer Council of Victoria, Cancer Epidemiology Centre, 1 Rathdowne St., Carlton South, Melbourne Vict. 3053, Australia (Tel.: +61 3 2791111; Fax: 2791270)
COLL: Russell, I.; Reed, R.

Register of In Situ and Small Size Breast Cancers

The register was established prior to the widespread use of mammographic screening in Victoria, primarily to elucidate the classification, treatment and natural history of in situ carcinoma and invasive cancers less than 10mm in diameter. These data will be useful in evaluating the impact of mammographic screening. Eligible cases (currently about 150 annually) are identified via pathology notifications to the Victoria Cancer Registry and demographic, diagnostic, clinical and pathological data are then collected from the treating surgeons and pathologists. Annual follow-up of recurrences and new primary tumours is conducted via the treating surgeon.

TYPE: Incidence
TERM: In Situ Carcinoma; Mammography; Registry; Survival
SITE: Breast (F)
REGI: Victoria (Aus)
TIME: 1988 - 1999

24 **Giles, G.G.** 05056
Anti-Cancer Council of Victoria, Cancer Epidemiology Centre, 1 Rathdowne St., Carlton South, Melbourne Vict. 3053, Australia (Tel.: +61 3 2791111; Fax: 2791270)
COLL: St John, D.J.B.; Macrae, F.A.; Carden, A.; Bankier, A.; Watts, C.

Familial Adenomatous Polyposis Register for Victoria

The principal aim of the registry is to prevent death from colorectal cancer in a small but high-risk group of individuals who have a family history of familial adenomatous polyposis (FAP). It is also expected to contribute to research into the natural history of FAP and the environmental and genetic causes of cancer. The registry and data collection protocols have been developed in close consultation with the Leeds Castle group and other State registers. Cases of FAP are usually recruited via their doctors. The registrar then arranges for family members affected and at risk to be entered on the register; clinicians and persons are then sent reminders of forthcoming screening tests. The register also serves as a source of information and education about FAP for doctors and families. Currently 55 pedigrees, including 123 affected and 188 persons at risk of FAP are registered. A further 95 people aged less than 15 years are also registered.

TYPE: Incidence; Registry
TERM: Environmental Factors; Familial Adenomatous Polyposis; Familial Factors; High-Risk Groups; Registry
SITE: Colon; Rectum
REGI: Victoria (Aus)
TIME: 1987 - 1999

25 **Hopper, J.L.** 05111
Univ. of Melbourne, Epidemiology Unit, 151 Barry St., Carlton Melbourne Vict. 3053, Australia (Tel.: +61 3 3446990; Fax: 3447014)
COLL: Flander, L.; Giles, G.G.; Carlin, J.; Green, M.; Collins, J.; Russell, I.

Breast Cancer in Families

The study aims to measure the risk of breast cancer in genetically related relatives of breast cancer patients, and specifically to measure the variation in risk by age at onset and laterality. It will also determine the extent to which risk is a consequence of other factors including current and past weight, reproductive, contraceptive and menstrual history, and use of alcohol. Patients with a recent diagnosis of breast cancer are selected at random, stratified by early (age < 40) and late onset, from new registrations with the Victoria Cancer Registry. With the consent of both the treating surgeon and the patient, a questionnaire on major risk factors is administered at face-to-face interview or by telephone. Each patient is asked to identify and encourage the cooperation of living relatives. If permission is given by the patient, these relatives will be asked to complete the same questionnaire. A similar process will be undertaken to study the relatives of the spouse or partner of the patient, to obtain "control" pedigrees. Stored DNA will allow testing of hypotheses relating susceptibility to putative genetic markers.

TYPE: Case Series; Genetic Epidemiology
TERM: Alcohol; BMB; Contraception; DNA; Genetic Factors; Genetic Markers; Menstruation; Physical Factors; Registry; Reproductive Factors
SITE: Breast (F)
REGI: Victoria (Aus)
TIME: 1990 - 1995

***26** **Hopper, J.L.** 05436
Univ. of Melbourne, Epidemiology Unit, 151 Barry St., Carlton Melbourne Vict. 3053, Australia (Tel.: +61 3 3446990; Fax: 3447014)
COLL: McCredie, M.R.E.; Giles, G.G.

Oral Contraceptive Use and Family History in Early Onset Breast Cancer

The aims of this study are to determine the extent to which the risk of early onset breast cancer is associated with oral contraceptive use and whether the risk is limited or increased in women with a family history of breast cancer. The design is a case-control study of families. Cases will be between 18 and 39 years of age at diagnosis, residing in the Melbourne or Sydney Statistical Divisions, and reported to the cancer registries. Controls will be obtained from the electoral roll by age stratified random sampling, frequency matched to 5-year age-groups of the cases. It is expected to recruit 300 cases in each city over a 3-year period and an equal number of controls. Cases and controls will serve as probands for sampling families. The mother and sisters, grandmothers and daughters aged over 18 will be included. Any relative reporting breast cancer will have her female relatives sampled in the same way. A structured risk factor questionnaire (including details of OC use) will be administered to all cases and controls. A pedigree questionnaire will be used to collect demographic and cancer details on all family members including males. Dried blood spots will be collected on all family members including males for future

DNA analysis using PCR technology. Reported cases of cancer will be validated in the cancer registries and the Registrars of Births, Deaths & Marriages.

TYPE: Case-Control
TERM: BMB; Blood; Environmental Factors; Familial Factors; Family History; Genetic Factors; Genetic Markers; Oral Contraceptives; Pedigree; Registry
SITE: Breast (F)
REGI: Victoria (Aus)
TIME: 1993 - 1996

***27 Hopper, J.L.** 05437
Univ. of Melbourne, Epidemiology Unit, 151 Barry St., Carlton Melbourne Vict. 3053, Australia (Tel.: +61 3 3446990; Fax: 3447014)
COLL: Giles, G.G.; St John, D.J.B.; Collins, J.; Green, M.; Macrae, F.A.

Cancer in Families

The aims of this study are to quantify; (1) the extent of familial clustering of breast and colorectal cancers in the Victorian population; (2) the proportion of cancer incidence associated with familial risk factors; (3) the increased cancer risks to relatives; (4) the proportion of risk explained by non-genetic familial exposures; (5) the proportion of risk explained by genetic markers; (6) the extent to which family history affects prognosis. The hypothesis is that risk associated with exposure to exogenous factors is increased in persons with a family history and that prognosis in persons with a family history is different to those without one. The design is a mixed retrospective/prospective population-based cohort of case and control families, recruited by random age stratified (18-70) sampling of cancer probands registered in the Victorian Cancer Registry. Controls will be families recruited through probands' spouses or partners. Families will consist of all first and second degree relatives of the index person (proband or partner/spouse) and other relatives selected sequentially according to the disease status of individuals in the first family. A total of 700 breast and 700 colorectal probands will be interviewed over 5 years. Blood will be taken from all probands and from selected families. Dried blood blots will be collected from as many family members as possible. Family members will be asked to complete a questionnaire covering relevant lifestyle exposures and events.

TYPE: Case-Control; Cohort
TERM: BMB; Blood; Environmental Factors; Familial Factors; Family History; Genetic Factors; Genetic Markers; Lifestyle; Pedigree; Prognosis; Registry
SITE: Breast (F); Colon; Rectum
REGI: Victoria (Aus)
TIME: 1993 - 1998

28 St John, D.J.B. 02984
Royal Melbourne Hosp., Dept. of Gastroenterology, c/o Post Office, Melbourne Vict. 3050, Australia (Tel.: +61 3 3427470; Fax: 3427802)
COLL: Young, G.P.; Macrae, F.A.; Alexeyeff, M.; Deacon, M.C.; Evans, G.; Fenwick, M.; Gange, D.; Hankinson, D.

Familial Colorectal Cancer and Early Diagnosis of Colorectal Cancer

A programme of selective screening for colorectal cancer was established in the Department of Gastroenterology at The Royal Melbourne Hospital in 1979. The register includes over 1,500 individuals who are members of families with hereditary non-polyposis colorectal cancer or who have close relatives with common colorectal cancer. The investigators are also closely involved with the state-wide ESSO Familial Polyposis Register, established within the Victorian Cancer Registry in April 1988. Activities include evaluation of screening strategies and assessment of faecal occult blood tests for use in screening programmes. Recent activities include a case-control study to assess cancer risk in relatives of patients with common colorectal cancer. A further study of cancer in families was started in 1993, using population-based cohorts of case families and control families. Recent papers were published in Ann. Intern. Med. 117:376-382, 1992 and in J. Gastroenterol. Hepatol. 5:94-203, 1990 (Young et al.).

TYPE: Case-Control; Intervention; Registry
TERM: Familial Adenomatous Polyposis; Familial Factors; Genetic Factors; High-Risk Groups; Registry; Screening
SITE: Colon; Rectum
REGI: Brisbane II (Aus)
TIME: 1979 - 1994

AUSTRALIA

Newcastle

***29** **Christie, D.G.** 05308
Univ. of Newcastle, Discipline of Environmental & Occupational Health, 86 Platt St., Newcastle NSW 2298, Australia (Tel.: +61 49 211234; Fax: 601197)
COLL: Brown, A.M.; Taylor, R.

New South Wales Colliary Cancer Surveillance Project

The aims of this historical cohort study is to describe the patterns of cancer incidence and mortality (1972-1993) in coalminers in New South Wales (NSW). The cohort consists of all men employed in the NSW coal mining industry between 1 January 1972 and 31 December 1992 (n=78,000). The Joint Coal Board database will be matched with the NSW Cancer Registry, national pooled registry data and the National Death Index.

TYPE: Cohort
TERM: Mining; Occupation; Registry
SITE: All Sites
OCCU: Miners, Coal
REGI: NSW (Aus)
TIME: 1992 - 2000

Perth

30 **de Klerk, N.H.** 04825
Univ. of Western Australia, Dept. of Social and Prev. Med., Queen Elizabeth II Med. Cent., Verdun St., Nedlands, Perth WA 6009, Australia (Tel.: +61 9 3893456)
COLL: Musk, A.W.; Hobbs, M.S.T.; Eccles, J.

Occupational Exposures in Subjects with Mesothelioma but no Known Asbestos Exposure

The aim of the study is to determine if subjects with malignant mesothelioma but no history of asbestos exposure have other exposures which may be responsible for their disease. All subjects reported to the Mesothelioma Registry of Western Australia and in whom there is no evidence of asbestos exposure will be included as cases. An interviewer-administered questionnaire on occupational and environmental exposures will be conducted. Comparison subjects will include subjects with other (nonasbestos-related) cancers who will receive the same questionnaires.

TYPE: Case-Control
TERM: Environmental Factors; Occupation; Registry
SITE: Mesothelioma
REGI: Sydney (Aus); W. Australia (Aus)
TIME: 1990 - 1995

31 **English, D.R.** 03033
Univ. of Western Australia, Dept. of Public Health, Nedlands, Perth WA 6009, Australia (Tel.: +61 9 3893134; Fax: 3893648)
COLL: Armstrong, B.K.

Aetiology of Benign Pigmented Naevi in Children and Adolescents

The aim of this study is to identify factors related to the development and disappearance of benign pigmented naevi in young adults. 1,000 white children, aged 8-9 years, were interviewed in 1985 and had their backs photographed. They will be re-examined in 1994. The data analysis will focus upon the appearance of new naevi in relationship to patterns of exposure to the sun. The cohort will be enlarged in 1994 to enable further longitudinal studies of the disappearance of naevi in adult life.

TYPE: Cohort
TERM: Lifestyle; Naevi; Pigmentation; Premalignant Lesion; Radiation, Ultraviolet
SITE: Benign Tumours; Melanoma
TIME: 1984 - 1996

32 **English, D.R.** 04837
Univ. of Western Australia, Dept. of Public Health, Nedlands, Perth WA 6009, Australia (Tel.: +61 9 3893134; Fax: 3893648)
COLL: Armstrong, B.K.; Kricker, A.; Randell, P.; Heenan, P.

Incidence of Non-Melanocytic Skin Cancer

The aim of this study is to determine the incidence of non-melanocytic skin cancer in the city of Geraldton, Western Australia. In 1987, 4,176 persons resident in the city of Geraldton participated in a prevalence survey of skin cancer. Approximately 3,000 of the original participants took part in a second prevalence survey which was conducted in 1992. On two occasions in the intervening period subjects were sent questionnaires seeking details of skin cancers which had been treated. The incidence of skin cancer in the five-year period will be calculated and related to measures of sun exposure and to host characteristics such as sensitivity to sunlight.

TYPE: Cohort
TERM: Radiation, Ultraviolet
SITE: Skin
TIME: 1987 - 1994

33 **Jacobs, I.** 04326
Curtin Univ., School of Public Health, Dept. of Epidemiology & Biostatistics, Kent St., Bentley, Perth WA 6102, Australia (Tel.: +61 9 3512816; Fax: 3512958)
COLL: Phillips, M.; Spickett, J.T.

Health Effects of Agricultural Chemicals

This is a prospective study of 8,178 cereal farmers and their families in Western Australia, to determine if there are any long term adverse effects associated with exposure to herbicides or other agricultural pesticides. Initial health status was assessed by postal questionnaire. Follow-up of health assessments will be undertaken by postal questionnaire, together with tracing of national registry information on cancers, birth anomalies and deaths. Approximately 100 incident cases of cancer are expected in the first five years of follow-up. Cancers of all sites are of interest, with special emphasis on those of the soft tissue and leukaemias. Base-line data on pesticide usage and farm practices were collected in 1984, prior to administration of the health questionnaires, by postal questionnaire and validated against governmental agricultural statistics. The use of pesticides and related farm practices since collection of the base-line data is monitored by annual questionnaires.

TYPE: Cohort
TERM: Chemical Exposure; Herbicides; Occupation; Pesticides; Registry
SITE: All Sites; Leukaemia; Soft Tissue
OCCU: Farmers
TIME: 1984 - 1994

34 **Musk, A.W.** 00919
Sir Charles Gairdner Hosp., Dept. of Respiratory Medicine, Verdun St., Nedlands, Perth WA 6009, Australia (Tel.: +61 9 3983251)
COLL: Hobbs, M.S.T.; de Klerk, N.H.; Hansen, J.; Eccles, J.

Occupational and Environmental Exposure to Crocidolite

This study aims to describe the mortality of persons formerly in the workforce of the crocidolite mine and mill at Wittenoom, Western Australia and relate their mortality to measures of crocidolite exposure. The mortality from asbestos-related diseases of persons in the previous resident population of the township of Wittenoom, who had no occupational exposure to asbestos will also be described. Interactions between smoking and exposure to crocidolite in causing disease will be investigated. It is intended to determine whether pleural thickening or pleural plaque formation are antecedents of malignant pleural mesothelioma independently of level of exposure to crocidolite, whether asbestosis is a necessary precursor of crocidolite-caused lung cancer, and to describe other aspects of the causation, natural history and outcome of asbestos-related disease in those exposed to crocidolite at Wittenoom.

TYPE: Cohort
TERM: Dusts; Environmental Factors; Mining; Occupation; Tobacco (Smoking)
SITE: Colon; Lung; Mesothelioma; Peritoneum; Stomach
CHEM: Asbestos, Crocidolite
OCCU: Miners, Asbestos
TIME: 1975 - 1995

35 **Musk, A.W.** 04467
Sir Charles Gairdner Hosp., Dept. of Respiratory Medicine, Verdun St., Nedlands, Perth WA 6009, Australia (Tel.: +61 9 3983251)
COLL: Hobbs, M.S.T.; de Klerk, N.H.

Prevention of Malignant Disease among Workers Exposed to Crocidolite at Wittenoom Gorge

The study aims to determine the effect of dietary supplements of beta carotene or retinol on the incidence of malignancy in subjects who were exposed to crocidolite at Wittenoom Gorge in Western Australia between 1943 and 1966. Over 1,000 subjects have been enrolled since 1990. Following initial assessment with chest radiography, dietary and smoking histories, and measurement of serum beta carotene and retinol, subjects have been randomly assigned to receive supplements of beta carotene or retinol. They will be reviewed annually for 5 years, with periodic measurement of plasma beta carotene and retinol levels. The incidence of malignancies of all sites will be compared between the two groups and with the expected incidence of cancer derived from previous calculations.

TYPE: Intervention
TERM: Diet; Dusts; Occupation; Prevention; Tobacco (Smoking); Vitamins
SITE: Lung; Mesothelioma
CHEM: Asbestos, Crocidolite; Beta Carotene; Retinoids
OCCU: Asbestos Workers; Miners, Asbestos; Transport Workers
TIME: 1988 - 1995

36 **Musk, A.W.** 05120
Sir Charles Gairdner Hosp., Dept. of Respiratory Medicine, Verdun St., Nedlands, Perth WA 6009, Australia (Tel.: +61 9 3983251)
COLL: Hobbs, M.S.T.; de Klerk, N.H.; Eccles, J.

Mortality in Gold Miners

A group of 2,000 Kalgoorlie gold miners were included in a cross-sectional morbidity study of respiratory symptoms and lung function in 1966. Details of their occupational histories and smoking histories were obtained. It is proposed to ascertain the vital status and cause of death of all members of the cohort and, in addition, the incidence of lung cancer, through the Registrar General's Department of Births, Deaths and Marriages to obtain death certificates and cancer registries throughout Australia to ascertain the incidence of lung cancer. This study will test the hypothesis that exposure to silica is associated with an increased rate of lung cancer.

TYPE: Cohort
TERM: Dusts; Mining; Occupation; Registry
SITE: Lung
CHEM: Silica
OCCU: Miners, Gold
REGI: W. Australia (Aus)
TIME: 1991 - 1994

Prahran

***37** **Fairley, C.K.** 05452
Monash Univ., Alfred Hosp., Dept. of Social and Preventive Medicine, Commercial Rd, Prahran Vict. 3181, Australia (Tel.: +61 3 2762645; Fax: 5298580)
COLL: McNeil, J.J.; Garland, S.; Baghurst, P.A.

Efficacy of Oral Beta Carotene for the Treatment of Cervical HPV Infection

This is a randomised clinical trial to determine the efficacy of oral beta carotene for the treatment of cytological HPV infection. 30 mg of oral beta carotene or Lecithin placebo are administered to 160 women (80 women in each group). The end points of the study are cytological progression to CIN, regression to normal and quantitative measurements of HPV copy numbers.

TYPE: Intervention
TERM: BMB; HPV
SITE: Uterus (Cervix)
CHEM: Beta Carotene
TIME: 1992 - 1994

38 **Hurley, S.** 05069
Monash Univ., Alfred Hosp., Dept. of Social and Preventive Medicine, Commercial Rd, Prahran Vict. 3181, Australia (Tel.: +61 3 2762651; Fax: 5298580)
COLL: McNeil, J.J.; Donnan, G.A.; Giles, G.G.

Adult Cerebral Glioma and Occupational Exposures

A case-control study of malignant brain tumours is being conducted, testing the following hypotheses about occupational exposure: (1) certain workplace chemicals (formaldehyde, organic solvents, PAHs, phenols, vinyl chloride, acrylonitrile) are associated with malignant brain tumours; (2) persons who have worked in occupations with presumed high exposure to electromagnetic fields, particularly in the extremely low frequency range, are at elevated risk of malignant brain tumours. Approximately 400 cases and 400 controls will be interviewed, using a structured interview, administered by a nurse interviewer. The usefulness of expert panels in performing retrospective assessments of occupational exposures will be assessed. This study includes tests of (a) the feasibility of using such a panel; and (b) validity and reliability.

TYPE: Case-Control
TERM: Electromagnetic Fields; Occupation; Pesticides; Plastics; Solvents
SITE: Brain
CHEM: Acrylonitrile; Formaldehyde; PAH; Phenol; Vinyl Chloride
TIME: 1989 - 1994

Sydney

39 **Coates, M.S.** 04342
New South Wales Central Cancer Registry, P.O.Box 380, North Ryde Sydney NSW 2113, Australia (Tel.: +61 2 8875634; Fax: 8887210 ; Tlx: 71675)
COLL: MacLennan, R.

New South Wales Familial Melanoma Registry

The aim is to investigate genetic and environmental factors and their interaction in the aetiology of melanoma by contacting all cases of melanoma incident in New South Wales in the study period (1,000). Postal questionnaires will be used to ask about melanoma in first degree and other relatives, and spouses; twin status; morbidity and risk factors, e.g. sun exposure for the index case and his or her spouse and first degree relatives. The validity of postal questionnaires will be assessed by interview.

TYPE: Cross-Sectional
TERM: Environmental Factors; Familial Factors; Radiation, Ultraviolet; Registry; Twins
SITE: Melanoma; Skin
REGI: NSW (Aus)
TIME: 1987 - 1994

40 **Corbett, S.J.** 04588
New South Wales Dept. of Health, Epidemiology Branch, Locked Mail Bag 961 , Sydney NSW 2059, Australia (Tel.: +61 2 3919207; Fax: 3919232)
COLL: O'Neill, B.J.

Lymphoid Malignancy and Occupation: A Population-Based Case-Referent Study in a Coal and Steel Producing Region

An apparent occupational outbreak of non Hodgkin's lymphoma among the employees of one of the underground coal mines on the southern New South Wales Coalfield prompted this investigation. The primary aim of the study is to explore any relationship between a history of work in an underground coal mine and the risk of developing one of these malignancies. The Illawarra region of Southern New South

Wales is a coal-mining and steel-producing area with a population of approximately 250,000. All male cases of Hodgkin's Disease, non-Hodgkin's lymphoma and chronic lymphocytic leukaemia (ICD9 codes 200, 202 and 204.1) occurring in the region between 1977 and 1986 were identified from hospital and general practitioner records and from the New South Wales Central Cancer Registry. 155 male cases have been identified. For each case two population controls from the region were chosen from a stratified sample of the national medical insurance database, which contains the names, addresses and dates of birth of 96% of the Australian population. Personal interviews were conducted with each of the subjects enrolled in the study, using a structured questionnaire to obtain a consecutive occupational history and details of specific occupational exposures, and a medical, family and residential history. Proxy interviews with the nearest family members were conducted for deceased cases and their matched controls. Validation of occupational histories with respect to coal mining will be made by reference to the NSW Miners Superannuation Fund.

TYPE: Case-Control
TERM: Coal; Mining; Occupation; Registry
SITE: Hodgkin's Disease; Leukaemia (CLL); Non-Hodgkin's Lymphoma
OCCU: Miners, Coal
REGI: NSW (Aus)
TIME: 1986 - 1994

41 **Kaldor, J.** 05200
National Centre in HIV Epidemiology and Clinical Research Univ. of New South Wales, 376 Victoria St., Sydney NSW 2010, Australia (Tel.: +61 2 3324648; Fax: 3321837)
COLL: Cooper, D.

Aetiology of HIV-Related Malignancies

Kaposi's sarcoma (KS), specific forms of lymphoma and perhaps other cancers occur at an increased risk in people with HIV infection. KS is particularly frequent among people whose HIV infection was acquired through homosexual contact. A study of cancer in people with HIV infection is being undertaken in two parts. First, people newly diagnosed with AIDS are being interviewed with regard to history of specific sexual activities and sexually transmissible disease, and other factors. Comparison will then be made between those who develop KS and those who do not. About 75 cases of KS are expected in the study. A similar comparison will be made between those who develop lymphoma and those who do not. In the second part of the study, people who have been treated with the anti-HIV drug zidovudine are being followed up for the occurrence of malignancy, in particular leukaemia.

TYPE: Case-Control; Cohort
TERM: AIDS; BMB; Drugs; HIV; Sexual Activity; Sexually Transmitted Diseases; Treatment
SITE: Kaposi's Sarcoma; Lymphoma
TIME: 1991 - 1994

42 **Leigh, J.** 02454
National Inst. of Occupational Health and Safety, GPO Box 58, Camperdown, Sydney NSW 2001, Australia (Tel.: +61 2 5659555; Fax: 5659300 ; Tlx: 177243)
COLL: Driscoll, T.R.; Corvalan, C.F.; Copland, P.I.

Australian Mesothelioma Surveillance Program. Australian Mesothelioma Register

The Mesothelioma Surveillance Program began on 9 July 1980 with the aim of standardizing the diagnosis, classification and further epidemiological knowledge of the disease. Particularly it aims to identify potential causal agents other than asbestos, to determine the relative importance of different fibre types in causation and to establish dose-response relationships. The Program sought notification of all cases of mesothelioma in Australia from 1 January 1980 to 31 December 1985 (903 case). Since that time the Register has collected less detailed information on all incident cases of mesothelioma in Australia. Recent papers arising from these data appeared in Am. J. Ind. Med. 20:643-655, 1991, Cancer 67:1912-1920 and 68:135-141, 1991 and in J. Occup. Health Safety-Austr. NZ 7:365-371, 1991.

TYPE: Incidence; Registry
TERM: Chemical Exposure; Dose-Response; Dusts; Environmental Factors; Mining; Occupation
SITE: All Sites; Mesothelioma; Peritoneum; Pleura
CHEM: Asbestos; Mineral Fibres
REGI: Sydney (Aus)
TIME: 1981 - 2000

***43** **Leigh, J.** 05378
National Inst. of Occupational Health and Safety, GPO Box 58, Camperdown, Sydney NSW 2001, Australia (Tel.: +61 2 5659555; Fax: 5659300 ; Tlx: 177243)
COLL: Nurminen, M.; Corvalan, C.F.; Baker, G.

Prediction of Lung Cancer in the Australian Labour Force Exposed to Silica

Empirical models for risk based on recently published epidemiologic data and simple prediction formulas are being used to predict the occurrence of lung cancer in the Australian labour force currently exposed to crystalline silica dust. A recent paper arising from this study appeared in Scand. J. Work. Environ. Health, 18:393-399, 1992.

TYPE: Methodology
TERM: Dusts; Mathematical Models; Occupation
SITE: Lung
CHEM: Silica
TIME: 1992 - 1994

44 **McCredie, M.R.E.** 04330
NSW Cancer Council, Cancer Epidemiology Research Unit, 153 Dowling St. , Woolloomooloo Sydney NSW 2011, Australia (Tel.: +61 2 3341909; Fax: 3572676)
COLL: Ford, J.M.

Case-Control Study of Childhood Brain Tumours in New South Wales

A case-control study of childhood brain tumours is being conducted in New South Wales (NSW). Potential cases comprise all children aged 0-19 years diagnosed in 1985-1989, reported to the NSW Central Cancer Registry and living in the Sydney Metropolitan area, Newcastle, Wollongong or the Australian Capital Territory. Two controls for each case, matched for age and sex are being obtained from the electoral rolls. Mothers and fathers of 100 cases and 200 controls are to be interviewed with the standard SEARCH questionnaires. Questions relate to parental and childhood exposures to N-nitroso compounds, ionizing radiation, genetic predisposition, head trauma, barbiturate consumption and parental occupational exposures. Plans for analyses of pooled data from alal centres are currently being finalised.

TYPE: Case-Control
TERM: Childhood; Drugs; Intra-Uterine Exposure; Radiation, Ionizing; Registry; Trauma
SITE: Brain
CHEM: Barbiturates; N-Nitroso Compounds
REGI: NSW (Aus)
TIME: 1988 - 1994

45 **McCredie, M.R.E.** 04554
NSW Cancer Council, Cancer Epidemiology Research Unit, 153 Dowling St. , Woolloomooloo Sydney NSW 2011, Australia (Tel.: +61 2 3341909; Fax: 3572676)
COLL: McLaughlin, J.K.; Adami, H.O.; Wahrendorf, J.

Case-Control Study of Renal Adenocarcinoma

The main aims of this study are to confirm an increased risk of renal adenocarcinoma associated with use of diuretics, to re-evaluate the effect of analgesics (particularly paracetamol); and to investigate the relationship with occupation in the petroleum industry or serving in the Armed Forces. Cases will comprise all persons aged 20-79 years diagnosed in 1989-1990 with renal adenocarcinoma, reported to NSW Central Cancer Registry. Controls from the electoral rolls will be frequency-matched to the cases. 600 cases and 600 controls will be interviewed by telephone over the two-year period using a standard questionnaire. Questions relate to use of analgesics, diuretics, oestrogens and diet pills; tobacco use and passive smoking; consumption of alcohol and coffee; reproductive history; medical history relating to diseases of the kidney, heart and thyroid; diabetes and other cancers; occupation (current and usual). Standard methods for analysis of case-control studies will be used. Pooled data from all centres are currently being analysed. Papers were published in Eur. J. Cancer 28A:2050-2054, 1992 and Cancer Causes and Control 3:323-331, 1992.

TYPE: Case-Control
TERM: Alcohol; Analgesics; Coffee; Diabetes; Drugs; Hormones; Obesity; Occupation; Passive Smoking; Registry; Reproductive Factors; Tobacco (Smoking)
SITE: Kidney
CHEM: Oestrogens; Paracetamol
OCCU: Petroleum Workers
LOCA: Australia; Denmark; Germany; Sweden; United States of America
REGI: NSW (Aus)
TIME: 1989 - 1994

***46 McCredie, M.R.E.** 05342
NSW Cancer Council, Cancer Epidemiology Research Unit, 153 Dowling St. , Woolloomooloo Sydney NSW 2011, Australia (Tel.: +61 2 3341909; Fax: 3572676)
COLL: Hopper, J.L.; Boyle, P.; Birch, J.M.

Oral Contraceptive Use and Family History in Early Onset Breast Cancer

The main aim of this population-based case-control family study is to determine the relationship between oral contraceptive (OC) use and family history of breast cancer in early onset breast cancer. Cases will be all women aged <40 years, diagnosed with breast cancer in 1993-1995, reported to the NSW Central Cancer Registry and living in Sydney. Controls from the electoral rolls will be frequency-matched to the cases. 300 cases and 300 controls will be interviewed face-to-face over a two and a half year period using a standard questionnaire. Questions relate to use of OC, other hormones, drugs for infertility, consumption of alcohol and tobacco, obesity, physical exercise, reproductive history, breast disease and family history of cancer. Similar questions will be asked of close relatives following the Cannings-Thompson ascertainment scheme. Blood (for DNA) will be obtained from all cases and controls and some relatives. Standard methods for analysis of case-control studies will be used as well as a new method which allows information from relatives to be taken into account in assessment of risk. Pooled analyses of data from Australia (600 cases, 600 controls) and from all three centres is envisaged.

TYPE: Case-Control
TERM: Alcohol; BMB; Blood; Drugs; Familial Factors; Obesity; Oral Contraceptives; Physical Activity; Registry; Reproductive Factors; Tobacco (Smoking)
SITE: Breast (F)
CHEM: Oestrogens
LOCA: Australia; United Kingdom
REGI: NSW (Aus)
TIME: 1993 - 1995

47 Shaw, H.M. 04713
Univ. of Sydney, Royal Prince Alfred Hosp., Melanoma Unit, Missenden Rd, Camperdown, Sydney NSW 2050, Australia (Tel.: +61 2 5167156; Fax: 5506316)
COLL: Kefford, R.F.; McCarthy, W.H.

Molecular Genetics of Melanoma

The hypotheses being tested are that (1) hereditary melanoma and the dysplastic naevus syndrome are pleiotropic effects of an autosomally inherited gene, or gene defect, located on the short arm of chromosome 1 (1p), and (2) that sporadic (non-hereditary) melanoma is associated with somatic defects on the same gene. This study aims to test the above hypotheses by (1) documentation of the incidence of melanoma and dysplastic naevus syndrome in 300 kindreds with a family history of melanoma and (2) genetic linkage analysis of a carefully documented group of Australian kindreds affected by hereditary melanoma/dysplastic naevus syndrome using RFLP analysis of the inheritance within those kindreds of alleles detected by a series of probes mapped to the short arm of chromosome 1, and (3) analysis of loss of constitutional heterozygocity in sporadic and familial melanoma tumour samples using this analysis on chromosome 1p.

TYPE: Genetic Epidemiology
TERM: Genetic Markers; Heredity; RFLP
SITE: Melanoma
TIME: 1985 - 1994

48 **Stacey, N.** 03416
National Inst. of Occupational Health and Safety, G.P.O.Box 58, Sydney NSW 2001, Australia (Tel.: +61 2 5659555; Fax: 5659300 ; Tlx: 177243)
COLL: Leigh, J.; Koemeyer, H.; Bonin, A.; Qu, S.

Carcinogenic Effects of Diesel Exhaust in Coal Workers

DNA adducts of diesel emission products will be determined in peripheral blood lymphocytes in underground coal miners exposed to various levels of diesel exhausts. Surface workers will be used as controls. 32p-post labelling techniques will be used. Power calculations suggest that measurements in four groups of 30 at nil, low, medium and high exposure will enable detection of significant exposure related outcomes at alpha = 0.05, beta = 0.70.

TYPE: Correlation
TERM: Chemical Exposure; Coal; DNA Adducts; Environmental Factors; Mining; Occupation
SITE: Inapplicable
CHEM: Diesel Exhaust; PAH
OCCU: Miners, Coal
TIME: 1993 - 1994

AUSTRIA

Graz

49 **Marth, E.** 05119
Hygiene-Inst. der Karl-Franzens Univ., Universitätsplatz 4, 8010 Graz, Austria (Tel.: +43 316 3804360; Fax: 382218)
COLL: Pfeiffer, K.P.; Möse, J.R.; Köck, M.; Pichler-Semmelrock, F.

Environmental, Dietary and Social Factors in Regional Cancer Mortality in Styria

This is an analysis of cancer mortality between 1978 and 1987 in small regions of Styria, Austria, with different respiratory cancer mortality rates in regions with substantial environmental pollution (air and/or water pollution), compared with small regions without substantial environmental pollution. In some regions the respiratory cancer mortality rate was twice as high as in regions without substantial environmental pollution (Zbl. Bakt. Hyg. 1991). Starting from these results and considering additional socio-economic data in the next few years, studies about the multifactorial genesis of cancer will be conducted, which also consider the social situation and food habits together with environmental conditions. With multivariate statistical methods the importance of single factors and the interaction between these factors will be studied.

TYPE: Correlation; Mortality
TERM: Air Pollution; Diet; Environmental Factors; Socio-Economic Factors; Water
SITE: All Sites
TIME: 1990 - 1994

Vienna

***50** **Haidinger, G.** 05326
Univ. of Vienna, Inst. of Social Medicine, Alser Str. 21, 1080 Vienna, Austria (Tel.: +43 222 4085681; Fax: 4088833)
COLL: Vutuc, C.; Kunze, M.J.

Lung Cancer in Austria - Differences in Exposure to Carcinogens between 1980 and 1990

The objectives of the present study are to identify differences in exposure to carcinogens between patients with lung cancer in the year 1980 and now. For that purpose, a case-control study is presently carried out in Austria. Interviews of patients with lung cancer started in January 1990. Data on approximately 250 histologically confirmed incident cases, including detailed information about occupational exposure smoking habits, alcohol intake and personal dietary habits, have been collected. Controls were randomly selected from the general population and will be matched with cases for age and sex. Both cases and controls were interviewed by trained interviewers with a structured questionnaire. Data on exposure to carcinogens (smoking and occupation), obtained from this study, will be compared with results of a study carried out in 1980. One objective will be to assess whether any changes have occurred with the launch of Less Harmful Cigarettes in 1970.

TYPE: Case-Control
TERM: Alcohol; Diet; Occupation; Tobacco (Smoking)
SITE: Lung
TIME: 1990 - 1994

51 **Karner-Hanusch, J.** 04721
Univ. of Vienna, Dept. of Surgery, Alserstr. 4, 1097 Vienna, Austria (Tel.: +43 222 40400/2243)
COLL: Roth, E.; Hufnagl, A.

Genetic Alterations in Patients with Colorectal Tumours

The aim is to monitor genetic alterations in colorectal tumours in order to examine allelic losses in (1) patients with spontaneous tumours; (2) patients with a positive family history of colorectal tumours; (3) patients with adenomas. A second objective is to identify genetic alterations in normal colonic mucosae in the three groups in order to reveal chromosomal defects even in normal mucosae. Peripheral blood leucocytes are used as a source of normal DNA. This should make it possible to detect patients at risk of developing colorectal cancer. Tumour tissue, normal mucosa and blood from every patient operated on for colorectal tumours are analysed for RFLP. 20 samples from which DNA has been successfully collected have been taken to date.

TYPE: Genetic Epidemiology
TERM: Chromosome Effects; DNA; Familial Factors; RFLP
SITE: Colon; Rectum
TIME: 1990 - 1995

52 **Neuberger, M.** 04755
Univ. of Vienna, Inst. of Environmental Hygiene, Dept. of Preventive Medicine, Kinderspitalgasse 15, 1090 Vienna, Austria (Tel.: +43 222 431595)
COLL: Haider, M.; Kundi, M.

Individual Asbestos Exposure, Smoking and Mortality

In the oldest asbestos cement factory of the world, a historical prospective cohort study started in 1973 including all persons employed in production in 1950-1981 for at least three years. From 2,816 persons eligible for the study, factory records, measurements of dust and fibres and interview-based smoking histories were used to estimate exposures over time. Underlying causes of death are compared with regional and national rates. In addition, best available diagnoses from autopsy records, hospital records and other sources are collected. SMR and life-table methods are used to calculate lung cancer risks compared to the general population and between different exposure groups within the cohort. Nested case-control studies are used for analysing details of exposure such as fibre type. A paper was published in Br. J. Indust. Med. 47:615-620, 1990.

TYPE: Case-Control; Cohort
TERM: Clinical Records; Dusts; Occupation; Tobacco (Smoking)
SITE: All Sites; Lung; Mesothelioma
CHEM: Asbestos, Chrysotile; Asbestos, Crocidolite; Mineral Fibres
OCCU: Asbestos Workers
TIME: 1973 - 1994

Wiener Neustadt

***53** **Niessner, H.** 05409
General Hosp. Wiener Neustadt, Dept. of Internal Medicine, Corvinusring 3-5 , 2700 Wiener Neustadt, Austria (Tel.: +43 2622 23521; Fax: 2777)
COLL: Schulte-Hermann, R.; Vutuc, C.; Waldhör, T.; Rist, M.; König, K.

Cancer of the Colon and Rectum in the South-East of Lower Austria

The aim is to demonstrate a possible correlation between several factors (e.g. drinking water contamination with halogenated hydrocarbons, nutrition, heredity, laxatives, alcohol, tobacco) and the genesis of colorectal cancer. A descriptive study has been completed and a case-control study is now being carried out. Cases and controls will be interviewed with standardized questionnaires, containing questions about the factors mentioned above. The study will include approximately 400 cases with histologically verified colorectal cancer (from the hospitals of Wiener Neustadt, Modling and Baden) and at least the same number of controls from the same hospitals.

TYPE: Case-Control
TERM: Alcohol; Heredity; Nutrition; Tobacco (Smoking); Water
SITE: Colon; Rectum
CHEM: Hydrocarbons, Halogenated
TIME: 1992 - 1995

BANGLADESH

Bogra

***54** **Rahim, M.A.** 05410

Cancer Epidemiology Research Program, Nishindhara Housing Estate, Main Rd, Plot 16, Bogra 5800, Bangladesh (Tel.: +880 51 5981)

COLL: Ahmed, Q.A.; Hossain, M.A.; Miah, M.N.I.; Elahi, M.F.; Rahim, M.M.; Mohammad, G.

Cancer Epidemiology Research Program (CERP)

CERP is a non-profit, non-governmental organisation of the Bangladesh Association of Cancer Research. Its aim is to study the frequency of cancer of all anatomical sites. A tumour registry has been in operation since 1970. Sources of registration are the radiotherapy centers, where all oncological activities are carried out. Other hospitals may participate on a voluntary basis, as cancer registration is not compulsory in Bangladesh. Completed forms are collected by the central office of CERP. The data are coded for site and histology and validated. Results are published regularly in Bangladesh Cancer Reports. Other activities of CERP include studies of (1) the increase in male breast cancer compared with female breast cancer; (2) the increase in cancer of the oropharynx in females compared with males; (3) the geographical pathology of cancer in Bangladesh, and (4) childhood cancer clustering.

TYPE: Incidence; Registry
TERM: Cluster; Geographic Factors
SITE: All Sites; Breast (M); Childhood Neoplasms; Oropharynx
TIME: 1970 - 1994

Antwerp

55 **Engels, H.** 04619
Univ. of Antwerp, Dept. of Epidemiology and Community Medicine, Universiteitsplein 1 , 2610 Antwerp, Belgium (Tel.: +32 3 8202523; Fax: 8202640; Tlx: 33646)
COLL: Eylenbosch, W.J.; Weyler, J.; Van Marck, E.; Ramael, M.; Temmerman; Nyong'o, A.

Genital HPV Infection and Cervical Cancer

Prevalence of genital HPV-infection is being estimated among 1,000 women participating in a cervical screening programme, and among 500 women at risk for sexually transmitted diseases (STDs). Different detection methods are used: physical examination, colposcopy, cytology and HPV-DNA detection techniques (PCR, In Situ Hybridisation). Prevalence will be compared by method of detection. A case-control study will also be conducted in Nairobi, Kenya. The hypothesis that HPV infection is a more important risk factor for precancerous lesions than other STDs (HSV, Chlamydia T., Gonorrhoea, Syphilis) will be tested. 100 cases (women with abnormal PAP smears) and 200 controls will be recruited from women participating in a cervical cancer screening programme in city council clinics and STD clinics in Nairobi. A feasibility study is now being carried out.

TYPE: Case-Control; Cross-Sectional
TERM: Cytology; HPV; HSV; PCR; Prevalence; Screening; Sexually Transmitted Diseases
SITE: Uterus (Cervix)
LOCA: Belgium; Kenya
REGI: Belgium (Bel)
TIME: 1988 - 1994

56 **Engels, H.** 05114
Univ. of Antwerp, Dept. of Epidemiology and Community Medicine, Universiteitsplein 1 , 2610 Antwerp, Belgium (Tel.: +32 3 8202523; Fax: 8202640; Tlx: 33646)
COLL: Holmstock, L.; Van Mieghem, E.; Laleman, G.R.

Cancer Risk among Workers Employed at a Nuclear Research Centre

The primary objective of this retrospective cohort study with a nested case-control study is to examine whether an excess in cancer morbidity and mortality can be observed among 4,000 workers employed between 1954 and 1990 at the Nuclear Research Centre (S.C.K./C.E.N.) Belgonucleaire (nuclear fuel) and Belgoprocess (nuclear waste) in the Mol-Dessel region. The second objective is to examine whether an excess in leukaemia or lymphoma morbidity and mortality can be observed among the children of these workers. The personnel register and the medical records will be used to define the study cohort and to collect data on exposure to low doses of ionising radiation. Data on morbidity, mortality and possible confounding factors will be collected through National Registries, questionnaires, home visits and contacting physicians. The experience of the cohort will be compared with figures from the National Institute of Statistics, the National Cancer Registry and eventually morbidity and mortality registers of neighbouring countries. The study is part of the IARC 'International Collaborative Study of Cancer Risk among Nuclear Industry Workers'.

TYPE: Cohort
TERM: Childhood; Occupation; Radiation, Ionizing; Registry
SITE: All Sites; Leukaemia; Lymphoma
OCCU: Power Plant Workers
REGI: Belgium (Bel)
TIME: 1991 - 1997

Brussels

57 **De Quint, P.** 05105
Free Univ. of Brussels, Faculty of Medicine and Pharmacy, Sect. of Public Health, Laarbeeklaan 103, 1090 Brussels, Belgium (Tel.: +32 2 4774213; Fax: 4774311)
COLL: Depoorter, A.M.

Parental Occupational Exposures and the Development of Cancer in the Offspring

The aim of the study is to investigate the association between parental occupational exposures and the incidence of leukaemia, lymphoma, brain tumours and neuroblastoma in their offspring. The cases and the controls are recruited in two large hospitals for children in Brussels. The case group includes children less than 15 years of age with newly diagnosed tumours. The controls are surgery patients, pair-matched to the cases on sex and age at diagnosis. A face-to-face interview of the parents is used to assess information on the personal and familial antecedents of the index child, the family habits, their living environment and the occupations and occupational exposures of both parents. Information on the occupational exposures is also obtained from occupational medical officers, specialists in industrial techniques and industrial hygienists. At least 50 cases and 50 controls will be included in the study. A conditional logistic regression model for matched sets will be used to analyse the data.

TYPE: Case-Control
TERM: Childhood; Parental Occupation; Registry
SITE: Brain; Leukaemia; Lymphoma; Neuroblastoma
REGI: Belgium (Bel)
TIME: 1988 - 1994

58 **Lechat, M.F.** 03936
Catholic Univ. of Louvain, Dept. of Epidemiology, School of Public Health, Clos Chapelle-aux-Champs 30, UCL 30.34, 1200 Brussels, Belgium (Tel.: +32 2 215878)
COLL: De Wals, P.; Weatherall, J.W.; Beckers, R.; Borlee, I.; Goujard, J.; Stoll, C.; Karkut, G.; Lillis, D.F.; Radic, A.; Calabro, A.; Calzolari, E.; Galanti, C.; Hansen-Koenig, D.; Ten Kate, L.P.; Stone, D.; Harris, F.; Nevin, N.; Pexieder, T.; Svel, I.; Laurence, K.; Tenconi, R.; Ayme, S.; Cuschieri, A.; Mastroiacovo, P.

EUROCAT - Registration of Congenital Anomalies

The long term objective of the EUROCAT project is to test the feasibility of carrying out epidemiological surveillance in the countries of the EEC taking congenital anomalies as an example. The specific objectives of the study are: (1) to establish in each country of the EEC one or several regional registers of congenital anomalies providing reliable epidemiological information on the occurrence of the registered conditions in the progeny of the population living in a defined geographical area; (2) to harmonize the methods of diagnosis and of data collection in order to ensure the comparability of statistics between participating centres; (3) to monitor the occurrence rate of congenital anomalies in the progeny of different population groups in order to identify any expected frequency; (4) to investigate the risk associated with possible teratogenic factors; (5) to create in each country an area where reporting is reliable, so that base line rates are available for calibrating any information system established at national level for the detection of adverse environmental influences; (6) to evaluate the effectiveness and efficiency of screening programmes, of preventive measures, and of treatment methods; (7) to provide a well documented set of cases recorded in a defined population for clinical research. (EUROCAT Report No. 2, Surveillance of Congenital Anomalies. Years 1980-1984 De Wals, P. and Lechat, M.F. (eds), Dept. of Epidemiology, Chatholic University of Louvain, Brussels, 1987.)

TYPE: Registry
TERM: Chemical Exposure; Congenital Abnormalities; Mutation, Germinal; Mutation, Somatic; Prevention; Registry; Screening
SITE: Inapplicable
LOCA: Belgium; Denmark; France; Germany; Ireland; Italy; Luxembourg; Netherlands; Switzerland; United Kingdom; Yugoslavia
TIME: 1979 - 1994

59 **Van den Oever, R.** 03141
National Confederation of Christian Sickness Funds, Medical Direction, Wetstraat 121, 1040 Brussels, Belgium
COLL: Van den Berghe, H.; Lahaye, D.

Detection of Occupational Cancer Based on the Health Insurance Cancer Registry

Of the total of 10 million Belgians, 46% are affiliated with the Confederation of Christian Sickness Funds (CCSF) where a cancer morbidity register has been kept since 1963. Studies of the complete occupational history of each new case of cancer of the nasal cavity and sinuses and of mesothelioma reported each year to the CCSF registry have been carried out since 1978 (in 1982 a total of 155 nasal cancers and 104 mesotheliomas). The relationship between these malignant neoplasms and exposure to occupational and extra-occupational risks has been shown in numerous published reports. Since

1980 a total of 532 new cases of cancer of the larynx have been included in the study to detect occupational influences (e.g., asbestos) as opposed to alcohol and tobacco use. Occupational exposures to arsenic, asbestos, asphalt, chromates, mineral oil, nickel and wood dust are specifically investigated. Detailed clinical information on histology, site, stage and treatment of the cancer is obtained from the hospital and the attending physicians. If occupational exposure to a carcinogen is shown, an application is made to the Belgian Fund of Occupational Diseases for compensation. The results of these continuing epidemiological investigations are published for each type of cancer separately at regular intervals.

TYPE: Incidence
TERM: Alcohol; Dusts; Histology; Metals; Occupation; Registry; Stage; Tobacco (Smoking); Treatment; Wood
SITE: Larynx; Mesothelioma; Nasopharynx
CHEM: Arsenic; Asbestos; Chromium; Mineral Oil; Nickel; PAH; Tars
TIME: 1978 - 1994

Ghent

60 **Bleyen, L.J.** 05044
State Univ. of Ghent, Dept. of Hygiene and Social Medicine, Block A, De Pintelaan 185, 9000 Ghent, Belgium (Tel.: +32 91 403636; Fax: 404994)
COLL: de Backer, G.; Roels, H.; Vandevelde, E.; De Maeseneer, J.

Feasibility and Efficiency of Breast Cancer Screening

The objectives of this prospective cohort study in Ghent are: (a) to assess whether screening for breast cancer is as feasible and efficient using the existing infrastructure (private and state), in order to obtain a given reduction in mortality, as with the set-up of a special unit; (b) a study of risk factors in breast cancer aetiology (weight, height, Quetelet-index and body-surface, age at menarche, children and age at each birth, breastfeeding, personal and family history, mammographical breast pattern, etc.); (c) the development of measures and strategies, in order to increase the attendance rate and to maintain compliance with the programme; and (d) the contribution of breast examination and breast self-examination towards mammography, corrected for age, and the impact on interval cancer mortality. The target population consists of 26,000 women aged 25-39, and 42,000 women aged 40-69. The control group will either be a historical control, or the female population aged 25-69 in another city with similar characteristics. The data collection will be through specially designed registration forms, surveys, and through the National Mortality Statistics and the National Cancer Registry.

TYPE: Cohort; Methodology
TERM: Lactation; Menarche; Menopause; Parity; Physical Factors; Registry; Screening
SITE: Breast (F)
REGI: Belgium (Bel)
TIME: 1991 - 1998

Leuven

61 **Kesteloot, H.** 05197
University Hospital St. Rafaël, School of Public Health, Dept. of Epidemiology, Capucijnenvoer 33, 3000 Leuven, Belgium (Tel.: +32 16 216894; Fax: 215500 ; Tlx: 24484 AZRAFL)
COLL: Lesaffre, E.; Joossens, J.V.

Relationship between Nutrients and Cancer

This is a study of the relationship between macro-nutrients and organ-specific and total cancer mortality. Statistical methods are being used to study the relations involved, both univariate and multivariate. The studies are both cross-sectional and longitudinal and use data obtained from a total population of more than 1 billion people. Use is made of the Leuven Mortality Monitoring System and of data provided by FAO and other sources. Each country is compared with other countries (which can be considered as control or reference populations). Changes in nutrient intake are also related to changes in organ-specific cancer mortality. Special attention is given to the role of fat (saturated, mono-unsaturated, poly-unsaturated) for all cancers and to the role of salt in stomach cancer in particular. Papers have been published in Cancer Causes and Control 2:79-83, 1991 and Prev. Med. 20:226-236, 1991.

TYPE: Correlation; Mortality
TERM: Diet; Fat; Nutrition
SITE: All Sites
TIME: 1987 - 1994

62 **Van den Berghe, H.** 03590
Univ. of Louvain, Center for Human Genetics, Gasthuisberg Campus 0 en N, Herestr. 49, 3000 Leuven, Belgium (Tel.: +32 16 345878; Fax: 215992)
COLL: Cassiman, J.J.; Fryns, J.P.; Vlietinck, R.; Oosterlinck, A.; David, G.; Van Leeuwen, F.E.; Marynen, P.; Mecucci, C.; Vercauteren, P.; van de Ven, W.; Dalcin, P.

Cancer Genetics and Cytogenetics

This project involves: (1) A cytogenetic investigation of patients with leukaemia, preleukaemia, lymphoma and allied disorders, as well as solid tumours. Specific or non-randomly occurring chromosome aberrations in the malignant cells are being identified and the relation between the chromosome pattern observed and the professional activities and/or exposure of the individuals with the disease studied. 1,500 patients are being studied per year. A control group has not been set up so far; (2) Assessment of the relation between the observed chromosome aberrations and oncogene expression; (3) Monitoring with cytogenetic techniques of workers exposed to ionizing radiation (chromosome breakage and SCE); (4) A computer program to score SCE and statistical treatment of the data; (5) A semi-automated cytogenetic analysis allowing for large-scale investigations. Recent papers appeared in Cancer Genet. Cytogenet. 25:233-245, 1987, 26:5-13 and 51-58, 1987 and in "Leukaemia", (J.A. Whittaker and I.W. Delamore, (eds)), Blackwell Scientific Publications, London, Boston, Melbourne, 1987, pp. 137-151.

TYPE: Molecular Epidemiology
TERM: Chromosome Effects; Monitoring; Mutation, Somatic; Occupation; Radiation, Ionizing; SCE
SITE: All Sites; Leukaemia; Lymphoma
TIME: 1970 - 1994

Brasilia

63 **Sichieri, R.** 05207
Assoc. das Pioneiras Sociais, Hosp. do Aparelho Locomotor, HAL/SARAH, CENEI, SMHS-Quadra 5010-Conjunto "A", 70.330-150 Brasilia DF, Brazil (Tel.: +55 61 2253763; Fax: 2265280)
COLL: Everhart, J.

Anthropometric Measures and Diet as Risk Factors for Chronic Diseases: an Ecological Study

This study comprises an ecological analysis of anthropometric and dietary data from a national survey conducted in 1974 among 55,000 households in the 25 state capitals of Brazil, as predictors for cancer mortality of the following sites: stomach, oesophagus, lung, breast and large intestine. Dietary intake was measured for a period of seven days. Units of analysis will be the 25 capitals. For each state, age-adjusted mortality rates and proportional mortality are being calculated for the age-group 30 years and older. Predictive variables include height, body mass index, and the dietary factors fat composition (animal and vegetable), amount of cereals, type of staple food, consumption of sugar, alcohol, vitamin A, vitamin C, calcium, meat, fish. Measures of association between the predictors of diet or body composition and the outcome variables will be based on Spearman correlation coefficients.

TYPE: Correlation
TERM: Alcohol; Diet; Fat; Physical Factors; Vitamins
SITE: Breast (F); Colon; Lung; Oesophagus; Stomach
CHEM: Calcium
TIME: 1991 - 1994

Rio de Janeiro

64 **Pinto, C.B.** 05080
Oswaldo Cruz Foundation, Health Information Centre, Av. Brasil 4365, Manguinhos, 21040 Rio de Janeiro, Brazil (Tel.: +55 21 2901696; Fax: 5909741 ; Tlx: 23239)
COLL: Szwarcwald, C.L.; Castilho, E.A.

Cancer Mortality in Rio de Janeiro

The main purposes of this study are (1) to describe cancer mortality for sites of the digestive tract and (2) to analyse regional differentials. Average annual age- and sex-specific mortality rates per 100,000 were calculated for the period 1979-1981 for each cancer site and for five-year age groups up to 79 years and for 80 years and above, for each of the 64 municipalities of the state. Age-standardised rates, cumulative rates and years of life lost were calculated. Standardised mortality ratios (SMR) were also calculated to compare different geographical areas. Factor and cluster analysis will be used. As soon as population data for 1990 are available from the census, temporal trends will be evaluated.

TYPE: Mortality
TERM: Cluster; Geographic Factors; Time Factors; Trends
SITE: Gastrointestinal
TIME: 1989 - 1994

65 **Pinto, C.B.** 05081
Oswaldo Cruz Foundation, Health Information Centre, Av. Brasil 4365, Manguinhos, 21040 Rio de Janeiro, Brazil (Tel.: +55 21 2901696; Fax: 5909741 ; Tlx: 23239)
COLL: Szwarcwald, C.L.; Castilho, E.A.

Cancer Mortality in Brazilian Metropolitan Areas

Brazil is the fifth largest country in the world, with a wide variety of socio-economic conditions, and national morbidity and mortality data conceal a considerable range of regional variations for many diseases, including cancer. There are nine so-called "metropolitan regions" consisting of the capitals of nine states and their neighbouring municipalities. These regions are supposed to have more reliable mortality data and to have socio-demographic characteristics representative of their regions. The objective of this study is to describe the most important types of cancer for each region. Sex-specific age-standardised (world) mortality rates (ASR) per 100,000 will be calculated for the period 1977-1987, in order to analyse temporal trends. Some other variables from death certification, such as place of birth, may also be analysed, to assess any effect of migration.

TYPE: Mortality
TERM: Trends
SITE: Breast (F); Colon; Liver; Lung; Oesophagus; Prostate; Rectum; Stomach; Uterus (Cervix)
TIME: 1989 - 1994

Sao Paulo

66 **Brasilino de Carvalho, M.** 04103
Heliopolis Hosp., Dept. of Head & Neck Surgery, Rua Manoel da Nobnega, 518, apto 72, 04001-001 Sao Paulo, Brazil (Tel.: +55 11 8844595)
COLL: Sobrinho, J.A.; Franco, E.L.F.; Rapoport, A.; Fava, A.S.; Gois Filho, J.F.; Chagas, J.F.S.; Kowalski, L.P.; Kanda, J.L.

Relationship between Cancer of the Upper Aero-Digestive Tract and Consumption of Alcohol and Tobacco

Although the relation between the role of alcohol and tobacco consumption in the development of cancer of the upper aero-digestive tract is today well established, the carcinogenic mechanisms of these agents remain obscure. The objective of this study is to follow up, for a period of ten years, asymptomatic subjects belonging to a high risk group due to their alcohol and tobacco habits, and to identify the various factors directly or indirectly related to the development of tumours. 300 chronic alcoholics, smokers or non-smokers with no symptoms of disease of the upper aero-digestive tract, will be examined every six months at the Department of the Head and Neck Surgery of the Hospital Heliopolis in Sao Paulo, Brazil. At the first examination a specific questionnaire will be completed and at each subsequent visit new data will be reported. A physical examination will also be carried out. If a neoplastic lesion is found the patient will be treated. If the examination is normal, a cytological smear of the mouth and oropharynx will be taken. At the end of the ten-year period an evaluation of the influence of alcohol and tobacco on the development of neoplasms at the various sites will be made, as an assessment of the influence of initiation and duration of consumption.

TYPE: Cohort; Incidence
TERM: Alcohol; Cytology; High-Risk Groups; Tobacco (Smoking)
SITE: Oral Cavity; Oropharynx
TIME: 1986 - 1996

Edmonton

67 **Fincham, S.F.** 03916
Alberta Cancer Board, Div. of Epidemiology and Preventive Oncology, 9707-110th St., Edmonton Alberta, Canada T5K 2L9 (Tel.: +1 403 4829375; Fax: 4887809)

Occupation as a Factor in Alberta Cancer Patients

A long-term study programme has been established to monitor cancer incidence by occupation, using the Alberta Cancer Registry as a source of cases. A detailed occupational and exposure history, demographic data and histories of smoking and alcohol consumption are routinely collected, using a self-reporting questionnaire. Subjects are all new cancer registrants, aged 25-74, in Alberta. Data are analysed annually, using descriptive and analytical statistics for effect estimation and control of confounders, for all sites.

TYPE: Incidence
TERM: Alcohol; Monitoring; Occupation; Registry; Tobacco (Smoking)
SITE: All Sites
REGI: Alberta (Can)
TIME: 1983 - 1994

68 **Fincham, S.M.** 02142
Alberta Cancer Board, Dept. of Epidemiology and Preventive Oncology, 9707-110th St., Edmonton Alberta, Canada T5K 2L9 (Tel.: +1 403 4829375)
COLL: Hanson, J.

Occupational Diseases in Alberta Workers Exposed to Chemicals

A data base was established on a cohort of 5,777 Alberta workers occupationally exposed to chemicals and 1,645 workers not exposed. The data file wili be matched periodically with the Alberta Cancer Registry using record-linkage techniques. Registry linkage in 1988 permitted cancer incidence risk estimation after a duration of 10 years and linkage with provincial vital statistics files for other causes of death is planned.

TYPE: Cohort
TERM: Chemical Exposure; Occupation; Record Linkage; Registry
SITE: All Sites
REGI: Alberta (Can)
TIME: 1978 - 1994

Hamilton

69 **Walter, S.D.** 04513
McMaster Univ., Health Science Centre, Room 3H4, 1200 Main St. W., Hamilton Ontario, Canada L8N 3Z5 (Tel.: +1 416 5259140; Fax: 5770017 ; Tlx: 0618347)
COLL: Taylor, S.M.; Davies, J.A.; Marrett, L.D.; Drake, J.J.; Hayes, M.V.

Geographical Variation in Ontario Cancer Rates

Site-specific and total cancer incidence and mortality data for Ontario in the period 1964-86 will be examined for spatial and temporal variation. The accuracy and completeness of address registration in the population-based provincial registry will be evaluated using capture-recapture and related methodology. Various methods of address imputation and correction will be considered, together with an assessment of their effects on spatial variation in cancer rates across the province. Several statistical indices will be computed to measure the degree of spatial auto-correlation in the data, and the persistence of spatial auto-correlation over time. Finally, selected ecologic comparisons will be made between cancer rates, demographic profiles, and environmental variables derived from the census and other data sources; the comparisons will use weighted maximum likelihood regression, taking into account systematic and statistical variation between small areas of the region.

TYPE: Correlation; Incidence; Methodology; Mortality
TERM: Geographic Factors; Registry
SITE: All Sites
REGI: Ontario (Can)
TIME: 1988 - 1994

Laval-des-Rapides

70 **Franco, E.L.F.** 04232
Univ. de Québec, Inst. Armand-Frappier, Service d'Epidémiology, 531 Blvd des Prairies, Laval-des-Rapides Québec, Canada H7V 1B7 (Tel.: +1 514 6875010; Fax: 6865501)
COLL: Kowalski, L.P.; Curado, M.P.; Oliveira, B.V.; Fava, A.S.; Kanda, J.L.

Risk Factors for Upper Respiratory and Digestive System Cancers in Brazil

A case-control study has been used to analyse risk factors for upper aero-digestive system cancers in three metropolitan areas in Southeastern Brazil (Sao Paulo, Curitiba, Goiania). Between February 1986 and December 1988, complete interviews have been obtained from 784 cases and 1,568 age- and sex-matched hospital controls. The distribution of cases according to anatomical site was follows: 67 lip, 306 oral cavity, 217 pharynx and 194 larynx cancers. Preliminary analyses for the effect to tobacco and alcohol and definitive results for cancers of the oral cavity have been published (Rev. Bras. Cir. Cab. Pesc. 11:23-33, 1987; Int. J. Cancer 43:992-1000, 1989). The substantive analyses for the other sites are currently being performed. A cohort of patients admitted to one of the participating hospitals in Sao Paulo has been actively followed up to monitor for the occurrence of second tumours and their determinants. The analysis of the effects of non-alcoholic beverages (coffee, tea and mate) and of the use of wood stoves for cooking and heating is now being completed.

TYPE: Case-Control
TERM: Alcohol; Diet; Multiple Primary; Tobacco (Smoking)
SITE: Larynx; Lip; Oral Cavity; Pharynx
TIME: 1987 - 1994

71 **Franco, E.L.F.** 04233
Univ. de Québec, Inst. Armand-Frappier, Service d'Epidémiology, 531 Blvd des Prairies, Laval-des-Rapides Québec, Canada H7V 1B7 (Tel.: +1 514 6875010; Fax: 6865501)
COLL: Camargo, B.; Lopes, L.F.; Barreto, J.H.S.; Johnsson, R.R.; Mauad, M.A.; Saba, L.M.B.; Sharpe, C.

Risk Factors for Wilms' Tumour in Brazil

The aim was to investigate risk factors for Wilms' tumour in different areas in Brazil as a parallel project to the Brazilian Cooperative Group for the Treatment of Wilms' Tumour (GCBTTW). Hypotheses to be investigated are: parental occupation with particular emphasis on chemical and crude oil refinery work, maternal exposure to medications from the most common pharmacological groups during pregnancy, food and beverage consumption for selected items (carotene-containing vegetables and fruits, tea and coffee, and alcoholic beverages), smoking history, history of common infectious diseases, herbicide exposure, and general demographic and socioeconomic information. The study uses a case-control design which identifies prospectively all incident cases of histologically confirmed Wilms' tumour in three large metropolitan areas in Brazil during a period of two years. Two hospital controls were identified for each case matched for age (+/- one year), sex, hospital catchment area, and trimester of admission. Controls were children without neoplastic diseases attending both outpatient and inpatient clinics in hospitals in the three areas. Between April 1987 and January 1989 complete interviews were obtained with the parents of 112 cases and 224 controls. Ninety-seven of the Wilms' tumour patients (86.6%) were identified through the GCBTTW trial. In-depth histopathologic typing and staging have been completed and the substantive analysis of risk factor information is being currently performed. The analysis of epidemiological correlates of genetic characteristics of Wilms' tumour patients was published in Int. J. Cancer 48:641-646, 1991. The effect of environmental exposures are now being analysed.

TYPE: Case-Control
TERM: Age; Alcohol; Chemical Exposure; Coffee; Diet; Drugs; Fruit; Herbicides; Infection; Occupation; Socio-Economic Factors; Tea; Tobacco (Smoking); Vegetables
SITE: Wilms' Tumour
OCCU: Chemical Industry Workers; Dyestuff Workers; Leather Workers; Petroleum Workers
LOCA: Brazil
TIME: 1987 - 1994

72 **Siemiatycki, J.** 04785
Inst. Armand-Frappier, Univ. du Québec, 531, Blvd des Prairies , Laval-des-Rapides Québec, Canada H7N 4Z3 (Tel.: +1 514 6875010; Fax: 6865501)
COLL: Camus, M.

Health Risks due to Chrysotile Asbestos in the Non-Occupational Environment

This ecological study of the effects of non-occupational exposures to chrysotile asbestos, will (1) determine whether women having lived near chrysotile asbestos mines have had excess mortality from lung diseases and asbestos-related cancers; and (2) evaluate the EPA linear risk model of lung cancer, estimating the mean cumulative exposure in women exposed and comparing the risk projected by the model with that observed at the estimated dose. The exposed population consists of 10,000 women > =30 years of age, in the two asbestos areas in Quebec Province, from 1950 to 1989 (400,000 person-years). 600,000 women from other comparable regions of Quebec will serve as control population. The study will have enough power (alpha=5%, beta=80%) to detect SMRs of 108 for all cancers, 114 for gastric and peritoneal cancers, 127 for lung cancer and 117 for non neoplastic respiratory diseases. Data will be obtained from death certificates for 1950 to 1989, from population census figures by municipality and age for 1951 to 1986, and from a survey of residential history of 1,000 women aged +50 years. Past exposures will be estimated applying regression techniques, using information on fibre concentrations in 1984, dust measures (1972-1990), volume and granulometry of fibres retained in chimney filters in asbestos mills, computer simulation of aerosol dispersion, and eye-witnesses. Using these data, five specialists will infer a probabilistic distribution of probable past atmospheric concentrations.

TYPE: Incidence; Mortality
TERM: Dusts; Environmental Factors; Female
SITE: Gastrointestinal; Lung; Mesothelioma
CHEM: Asbestos, Chrysotile
TIME: 1989 - 1994

73 **Siemiatycki, J.** 05222
Inst. Armand-Frappier, Univ. du Québec, 531, Blvd des Prairies , Laval-des-Rapides Québec, Canada H7N 4Z3 (Tel.: +1 514 6875010; Fax: 6865501)
COLL: Gérin, M.; Richardson, L.; Nadon, L.; Lakhani, R.

Multivariate Regression Analyses of Occupational Risk Factors for Cancer

Among recognized human carcinogenic agents and circumstances, about half are substances which were first found to be carcinogenic in the occupational milieu. Although it it important to discover occupational carcinogens for the sake of preventing occupational cancer, the potential benefit of such discoveries goes beyond the factory walls since most occupational exposures find their way into the general environment, sometimes at higher concentrations than in the workplace. In 1979, a large case-control study was undertaken in Montreal, designed to provide evidence on the associations between hundreds of relatively common occupational exposures and many sites of cancer. Between 1979 and 1986, interviews were carried out with 3,730 cancer cases and 533 population controls. Each subject's job history was scrutinized by a team of specially trained chemists and hygienists, who inferred a list of chemical exposures. The first several years of the study were devoted to development and evaluation of methods, and to data collection. Most recently a sweeping analysis of the possible interactions between all types of cancer in the data set and 294 occupational exposures has been completed and published in a book by CRC Press. Further analyses will be carried out by logistic regression methods and will entail a tailored site-by-site approach rather than the totally mechanised analysis that was possible at the first stage. The objective of this stage of the study is to exploit the existing database from the case-control study to detect and characterize occupational risk factors for the main types of cancer in the database.

TYPE: Case-Control
TERM: Chemical Exposure; Dusts; Metals; Occupation; Petroleum Products; Plastics; Solvents
SITE: Bladder; Colon; Kidney; Lung; Lymphoma; Melanoma; Oesophagus; Pancreas; Prostate; Rectum; Stomach
CHEM: Asbestos; Chromium; Formaldehyde; Mineral Oil; Nickel; PAH; Silica; Sulphuric Acid
TIME: 1979 - 1994

Montréal

74 **Ghadirian, P.** 04346
Hôtel-Dieu de Montréal, Unité de Recherche en Epidémiologie, 3840, rue St-Urbain, Montréal Québec, Canada H2W 1T8 (Tel.: +1 514 8432742; Fax: 8496880 ; Tlx: 05562117 hdm)
COLL: Lacroix, A.; Bhat, B.; Gavino, V.

Diet and Cancers of the Breast, Colon and Prostate: Biological Markers

Breast cancer is the first common cancer among women in Canada; colorectal cancer is the second most common cancer in females and the third in males, after prostate cancer. Therefore these three nutritional-fat-related cancers make up a very large proportion of total cancer incidence and mortality rates. Although several causative factors have been suggested, the aetiology of these diseases, in particular the role of diet and nutrition are poorly understood. A total of 1,044 cases (400 colon, 230 prostate and 414 breast) and 1,325 population-based controls (666 for colon, 230 for prostate and 429 for breast cancer) have been interviewed. Of these, 49.5% of cases and 2.4% of controls had biological specimens taken, such as blood, adipose tissue, cheek cells and toenails will be collected from both case and control groups for determination of: (1) fatty acid patterns in fat biopsies, red blood cells and cheek cells; (2) carotenoids in plasma, fat biopsies and cheek cells; (3) tocopherols in plasma and fat biopsies; (4) selenium in toenail clippings and glutathione peroxidase activity in red blood cells. The results of biological specimens will be compared with the information in the food frequency questionnaire. Biological specimens was collected prior to surgery or initiation of therapy.

TYPE: Case-Control
TERM: BMB; Biomarkers; Blood; Diet; Metals; Tissue; Toenails
SITE: Breast (F); Colon; Prostate
CHEM: Selenium; Tocopherol
TIME: 1987 - 1994

75 **Infante-Rivard, C.** 04951
McGill Univ., School of Occupational Health, 1130 Pine Ave. W., Montréal Québec, Canada H3A 1A3 (Tel.: +1 514 3984231; Fax: 3987435)
COLL: Siemiatycki, J.

Parental Occupation and Incidence of Acute Lymphoblastic Leukaemia in Children

It has been postulated that parental exposure to solvents during the course of their work, and in particular during the months before conception for both parents, and during pregnancy for the mother, is a risk factor for the development of acute lymphoblastic leukaemia (ALL) in their offspring. To study this hypothesis, a case-control study will be carried out. Approximately 500 cases of ALL, up to nine years of age, diagnosed between 1980 and 1993 will be included. Cases from six regions of the province of Quebec, accounting for 85% to 90% of the total Quebec population, will be ascertained. Two controls per case, matched for age, will be chosen: one from all other cancers and serious haematological disorders diagnosed at the same hospital as the case, and one population control from the same geographical area as the case at time of diagnosis. Data will be collected using a telephone interview. Exposures will be coded by chemists according to the method developed by Siemiatycki. The children's exposures to pesticides and passive smoking are also of interest.

TYPE: Case-Control
TERM: Chemical Exposure; Intra-Uterine Exposure; Parental Occupation; Passive Smoking; Pesticides; Solvents
SITE: Leukaemia (ALL)
TIME: 1990 - 1994

76 **McDonald, J.C.** 05142
McGill Univ., School of Occupational Health, 1130 Pine Ave. W., Montréal Québec, Canada H3A 1A3 (Tel.: +1 514 3984238; Fax: 3988981)
COLL: Liddell, F.D.K.; McDonald, A.D.

Cancer Mortality in Quebec Chrysotile Miners and Millers

This study will extend follow-up to the end of 1989 of a cohort of over 11,000 chrysotile miners and millers, previously followed through 1975. The specific objectives are (1) to estimate exposure-response relationships between fibre and dust measurments of workers and mortality from malignant diseases at several sites including lung, pleura and peritoneum, gastrointestinal tract, kidney, and larynx, accounting for duration and intensity of exposure particularly as regards industrial process, time-related variables, and low exposure levels; (2) to examine the interaction between smoking and asbestos exposure in lung cancer; (3) to explore the relationship between laryngeal cancer mortality, alcohol consumption, smoking, and asbestos. These questions will be answered by a series of case-control studies nested within the cohort, where cases will comprise those deceased from the cause of interest and controls for each case will be selected from those who survived the case.

TYPE: Case-Control; Cohort
TERM: Alcohol; Dose-Response; Dusts; Occupation; Time Factors; Tobacco (Smoking)
SITE: Gastrointestinal; Kidney; Larynx; Lung; Peritoneum; Pleura
CHEM: Asbestos, Chrysotile; Mineral Fibres
OCCU: Miners, Asbestos
TIME: 1989 - 1994

77 **Messing, K.** 04403
Univ. du Québec à Montréal, Groupe de Recherche-Action en Biologie du Travail, C.P. 8888, Succ. "A", Montréal Québec, Canada (Tel.: +1 514 2823334)
COLL: Bradley, W.E.C.; Dubeau, H.; Arsenault, A.; Dupras, G.

Frequencies and Types of Lymphocyte Mutants Induced in Humans Exposed to Ionizing Radiation

In Project 1, 60 workers in a nuclear power plant exposed to 100 mrem-5 rem annual doses of radiation, are being compared with 20 office workers at the same plant and 20 laboratory controls. HPRT-mutant T-lymphocytes are cloned and frequency determined by the Albertini method. DNA of mutant lymphocytes is examined by Southern transfer analysis to determine whether damage is typical of exposure to radiation. In Project 2, 25 nuclear medicine patients exposed to 5 mCi of Th. 201 are studied before and after exposure using the same techniques as above. A paper was published in Mutat. Res. 262:1-6, 1991.

TYPE: Cross-Sectional; Molecular Epidemiology
TERM: DNA; HPRT-Mutants; Lymphocytes; Metals; Occupation; Radiation, Ionizing
SITE: Inapplicable
CHEM: Technetium
OCCU: Radiation Workers
TIME: 1987 - 1994

***78** **Narod, S.A.** 05494
McGill Univ., 1650 Cedar Ave., Room L10-116, Montréal Québec, Canada H3G 1A4 (Tel.: +1 514 9376011; Fax: 9348273)
COLL: Easton, D.; Lynch, H.T.; Watson, P.; Goldgar, D.; Lenoir, G.

Identification of Risk Factors Among a Cohort of Women who Carry Mutations in the Gene for Hereditary Breast-Ovarian Cancer

A gene for hereditary breast-ovarian cancer, BRCA1, has been mapped to chromosome 17q21. It appears that most of the family clusters of breast and ovarian cancer, and about one-half of the clusters of breast cancer alone are due to mutations of this gene. It is estimated that carriers of mutations in BRCA1 have about an 80% lifetime risk of developing breast or ovarian cancer and that 40% of carriers will develop cancer between the ages of 40 and 50. DNA marker technology permits, in selected families, the identification of women who are BRCA1 gene carriers from birth. The goal is to identify a cohort of 400 women, world-wide, who carry mutations in the BRCA1 gene, who are above the age of 20 and who are currently healthy. This cohort will be studied prospectively for cancer risk and for factors which modify the risk, including reproductive history and the use of oral contraceptives and hormone replacement therapy. The relative value of screening by mammography or ultrasound and prophylactic surgery in preventing death from cancer in this high-risk cohort will also be evaluated.

TYPE: Cohort; Genetic Epidemiology
TERM: Chromosome Effects; Cluster; Familial Factors; Genetic Markers; Heredity; High-Risk Groups; Prevention; Screening
SITE: Breast (F); Ovary
LOCA: Canada; France; United Kingdom; United States of America
TIME: 1993 - 2005

***79** **Parent, M.E.** 05496

Hôtel Dieu de Montréal, Unité de Recherche en Epidémiologie, 3850, rue St-Urbain, Montréal Québec, Canada H2W 1T8 (Tel.: +1 514 8432742; Fax: 8496880)

COLL: Ghadirian, P.; Lacroix, A.

Weight History and Familial Breast Cancer

Both obesity and breast cancer demonstrate a high degree of familial clustering. The sharing by family members of a wide variety of factors which could influence the development and maintenance of obesity among relatives could explain, at least in part, the familial aggregation of breast cancer. To test this hypothesis, a case-control study to assess the weight history of predisposed families (i.e. having at least two first-degree relatives affected) and of cancer free families was carried out. Weight history data were obtained via a self-administered questionnaire, including a system of body silhouettes. It has been applied to establish the lifetime body size of all male and female first-degree relatives of 60 recently diagnosed primary breast cancer cases, who had a positive first-degree family history of breast cancer and of 120 healthy population-based sex- and age-matched controls with negative family history. Respondents provided weight history data about themselves and their relatives, totalling 1,830 individuals. Weight changes during childhood, adolescence and adulthood, as well as body fat distribution, were considered. Analyses will be carried out to compare the weight history of families with and without breast cancer predispositon, and to compare affected and unaffected relatives.

TYPE: Case-Control
TERM: Familial Factors; Family History; High-Risk Groups; Obesity
SITE: Breast (F)
TIME: 1991 - 1994

80 **Thériault, G.P.** 04502

McGill Univ., School of Occupational Health, 1130 Pine Ave. W., Montréal Québec, Canada H3A 1A3 (Tel.: +1 514 3984227; Fax: 3987435)

COLL: Goldberg, M.; Miller, A.B.

Long-Term Effects of to 50/60 Hz Electric and Magnetic Fields

A case/control study will be carried out to determine whether there exists an aetiological relationship between exposure to 50 and 60 hertz electric and magnetic fields and cancer among exposed workers at Electricite de France (EDF), Hydro-Quebec (HQ), and Ontario Hydro (OH). The cohort will consist of 150,000 employees from the three utilities. Cases will comprise all cancers identified in the relevant period (1978-1988 for EDF, 1970-1988 for HQ and OH). They will be ascertained through the existing records at the company (and from tumour registry data or death certificates for HQ and OH). One to four controls will be selected from the files of the corresponding elective utility, according to cancer site (or electric utility). Analysis will be performed using methods appropriate to the matched design of the study. The validity of combining the three data sets will be tested by standard epidemiological and biostatistical procedures.

TYPE: Case-Control; Cohort
TERM: Electromagnetic Fields; Occupation
SITE: All Sites
LOCA: Canada; France
TIME: 1989 - 1994

Ottawa

81 **Ashmore, J.P.** 04520

Bureau of Radiation and Med. Devices, National Dose Registry, Occupational Radiation Hazards Div., 775 Brookfield Rd, Ottawa Ontario, Canada K1A 1C1 (Tel.: +1 613 9546660; Fax: 9578698 ; Tlx: 0533679)

COLL: Grogan, D.; Krewski, D.; Wigle, D.T.; Mao, Y.S.; Semenciw, R.; Fair, M.E.; Werner, M.

Mortality Study of Radiation Workers

The study involves the computer linkage of the Department of National Health and Welfare's National Dose Registry for Radiation Workers with the Mortality Data Base at Statistics Canada. The National Dose Registry is a centralised occupational radiation dose record system which includes annual dose summaries of all monitored radiation workers in Canada from 1951 to the present. Dose records originate from the Department's National Dosimetry Services and from a number of Nuclear Power facilities which perform their own dosimetry. The Registry covers 80 different occupations at over 17,500 different organisations. The Mortality Data Base at Statistics Canada includes a record of all deaths in Canada from 1951 to the present. The cohort consists of an estimated 325,000 individuals representing a collective dose of 1,750 person-sieverts and 1,500,000 person-years of radiation monitoring. Computer linkages are complete and analysis is in progress.

TYPE: Cohort
TERM: Occupation; Radiation, Ionizing; Registry
SITE: All Sites
OCCU: Radiation Workers
REGI: Canada (Can)
TIME: 1984 - 1994

82 **Gaudette, L.A.** 04522
Statistics Canada, Tunney's Pasture, Ottawa Ontario, Canada (Tel.: +1 613 9571765)
COLL: Miller, A.B.; Freitag, S.; Ball, D.; Dufour, R.

Cancer Among the Inuit of Canada

The first phase of the study is to produce a valid account of the descriptive epidemiology of cancer in the Inuit of Canada for comparison with other circumpolar Inuit populations. The registry so created will then be available for analytical epidemiological studies and (potentially) intervention studies to reduce the prevalence of current cancer public health problems in the Inuit and to prevent lifestyle changes which could lead to further increases. Data on about 500 Inuit cases are being identified for the time period 1969 to present from cancer registries in the Northwest Territories, Quebec and Newfoundland for inclusion in an international monograph. The distribution by site of observed cases will be compared to that expected based on Canadian rates. The registry will be maintained by the Vital Statistics and Disease Registries Section, Health Division, Statistics Canada.

TYPE: Incidence
TERM: Eskimos; Ethnic Group; Registry
SITE: All Sites
REGI: N.W. Territory (Can); Newfoundland (Can); Quebec (Can)
TIME: 1987 - 1994

83 **Mao, Y.S.** 02296
Health and Welfare Canada, Lab. Centre for Disease Control, Bureau of Chronic Disease Epidemiology, Tunney's Pasture, Ottawa Ontario, Canada K1A 0L2 (Tel.: +1 613 9571765; Fax: 9527009)
COLL: Semenciw, R.; Wigle, D.T.

Canadian Farm Operators Mortality Study

The objective of this record-linkage study is to examine the mortality patterns in a cohort of about 360,000 men identified as farm operators in 1971, relative to the referent population(s). Record linkage of census records to the Canadian Mortality Data Base is underway at Statistics Canada, using the Generalized Iterative Record Linkage System. rates of death from particular causes which exceed the corresponding rates of death in the referent population(s) will be examined carefully for possible associations with certain variables, including age of the farm operators, the type and size of farming operations, use of agricultural chemicals (herbicides and insecticides or fertilizers) the amount of fuel used for farm machinery, geographical location, and some non-farming aspects such as off-farm wor or off-farm residence. Poisson regression analysis will be conducted using both internal and external control groups. While the zero exposure group within the cohort will be used as an internal control, the general population will serve as an external control group.

TYPE: Case-Control; Cohort
TERM: Fertilizers; Geographic Factors; Herbicides; Insecticides; Record Linkage
SITE: All Sites; Sarcoma
TIME: 1985 - 1994

84 **Mao, Y.S.** 05236
Health and Welfare Canada, Lab. Centre for Disease Control, Bureau of Chronic Disease Epidemiology, Tunney's Pasture, Ottawa Ontario, Canada K1A 0L2 (Tel.: +1 613 9571765; Fax: 9527009)
COLL: Semenciw, R.; Mills, C.; Eliasziw, M.; Marrett, L.D.

Great Lakes Cancer Risk Assessment Study

The first part of this study is an ecological study of water quality and cancer among Great Lakes communities. Descriptive statistics regarding water quality, socio-demographic information and cancer mortality for roughly 70 communities which border on or obtain drinking water from the Great Lakes will be examined. Water quality information is available on 35 of these communities. Secondly, a pilot case-control study is under way which will examine the following cancers: colorectal, stomach, bladder, kidney, brain, leukaemia, prostate, and lymphoma in the light of contaminants in drinking water.

TYPE: Case-Control; Correlation
TERM: Chemical Exposure; Chlorination; Geographic Factors; Water
SITE: All Sites; Bladder; Brain; Colon; Kidney; Leukaemia; Lymphoma; Prostate; Rectum; Stomach
TIME: 1989 - 1994

85 **Mao, Y.S.** 05237
Health and Welfare Canada, Lab. Centre for Disease Control, Bureau of Chronic Disease Epidemiology, Tunney's Pasture, Ottawa Ontario, Canada K1A 0L2 (Tel.: +1 613 9571765; Fax: 9527009)
COLL: Desmeules, M.; Robson, D.; Mikkelsen, T.; Gaudette, L.A.; Hill, G.B.; Thomas, J.; Montpetit, V.; Lach, B.; Downey, R.

Brain Cancer Study

This is a four-part study of brain cancer in Canada. The first part will consist of a descriptive study of temporal trends in brain cancer. The second part will examine the role of changes in diagnostic practices as a possible explanation for upward trends in brain cancer mortality. This will involve a chart review of roughly 350 cases of individuals with neurological disorders (about 200 brain cancer cases) in which diagnostic information from computerized tomopgraphy and magnetic resonance imaging has been removed. A case-control study of brain cancer in which prescription drug use is determined by way of record linkage to the Saskatchewan drug plan dataset is under way. A slide review of roughly 800 brain cancer cases from two time periods (1960's versus 1980's) has also been undertaken in order to investigate changes in the relative frequency of brain cancer histological types.

TYPE: Case-Control; Incidence
TERM: Clinical Records; Drugs; Histology; Trends
SITE: Brain
TIME: 1991 - 1994

86 **Morrison, H.** 05239
Health and Welfare Canada, Lab. Centre for Disease Control Bureau of Chronic Disease Epidemiology, Tunney's Pasture, LCDC Bldg, Ottawa Ontario, Canada K1A 0L2 (Tel.: +1 613 9411286; Fax: 9527009)
COLL: Semenciw, R.; Wigle, D.T.; Morison, D.; Bartlett, S.

Canadian Farm Operators Study

The objective of this record-linkage study is to examine mortality patterns in a cohort of about 360,000 men identified as farm operators in 1971, relative to referent populations. Record linkage of census records to the Canadian Mortality database has been completed by Statistics Canada using the Generalized Iterative Record Linkage System. Linkage to cancer incidence records is in progress. Selected cancers which have been reported as being increased among farm populations will be examined for possible associations with farm and socio-demographic variables, including age of the farm operators, the type and size of farming operations, use of agricultural chemicals (herbicides, insecticides or fertilizers), fuel/oil use, numbers of farm animals and location.

TYPE: Case-Control; Cohort
TERM: Fertilizers; Herbicides; Insecticides; Occupation; Pesticides; Petroleum Products; Record Linkage
SITE: All Sites; Brain; Leukaemia; Multiple Myeloma; Non-Hodgkin's Lymphoma; Prostate
CHEM: Phenoxy Acids
OCCU: Farmers
TIME: 1985 - 1994

87 **Morrison, H.** 05240

Health and Welfare Canada, Lab. Centre for Disease Control Bureau of Chronic Disease Epidemiology, Tunney's Pasture, LCDC Bldg, Ottawa Ontario, Canada K1A 0L2 (Tel.: +1 613 9411286; Fax: 9527009)
COLL: Semenciw, R.; Stocker, H.; Mao, Y.S.; Villeneuve, P.

Fluorspar Miners Mortality Update Study

A cohort of approximately 2,000 fluorspar miners occupationally exposed to radon has been identified. A previous analysis of this cohort was based on mortality follow-up to the end of 1984. An update is under way to extend the follow-up period to the end of 1990. A Poisson regression analysis will be performed.

TYPE: Cohort
TERM: Occupation; Radiation, Ionizing
SITE: All Sites; Lung
CHEM: Radon
OCCU: Miners, Fluorspar
TIME: 1991 - 1994

88 **Semenciw, R.** 03346

Health and Welfare Canada, Lab. Centre for Disease Control Bureau of Chronic Disease Epidemiology, Tunney's Pasture, Ottawa Ontario, Canada K1A 0L2 (Tel.: +1 613 9571768; Fax: 9527009)
COLL: Mao, Y.S.; Wigle, D.T.; Morrison, H.

Cancer Surveillance in the Laboratory Centre for Disease Control

The Cancer Section of the Laboratory Centre for Disease Control was established in 1977 to: (1) monitor the risk of cancer in the Canadian population to identify high-risk groups and to determine if cancer risks change significantly over time; (2) evaluate factors (lifestyle, environment, human biology) which contribute to the risk of cancer; (3) provide epidemiological and biostatistical expertise to other programmes in the Health Protection Branch and to other government agencies and institutions; (4) disseminate cancer surveillance information to health planners, administrators and research workers (5) establish national priorities for cancer research and control programmes; (6) assess progress in disease control in Canada and identify countries with apparently successful programmes to learn from their experience; (7) assess trends in incidence, including the potential role of changes in diagnostic technique; (8) develop improved methods for the national surveillance of cancer. Current work is concentrated on major cancer risk factors. The main data source is Statistics Canada, which operates mortality, cancer incidence and hospital morbidity data bases.

TYPE: Cross-Sectional
TERM: Diet; Environmental Factors; High-Risk Groups; Lifestyle; Occupation
SITE: All Sites
TIME: 1977 - 1994

89 **Semenciw, R.** 05242

Health and Welfare Canada, Lab. Centre for Disease Control Bureau of Chronic Disease Epidemiology, Tunney's Pasture, Ottawa Ontario, Canada K1A 0L2 (Tel.: +1 613 9571768; Fax: 9527009)
COLL: Morrison, H.; Fair, M.E.; Verdier, P.; Mao, Y.S.

Cancer Incidence Follow-Up of the Nutrition Canada Survey Cohort

A nutrition survey of about 12,000 Canadians was conducted during the period 1970-1972. Dietary history, medical history and anthropometric measurements are available for participants. The cancer incidence experience of this cohort is being determined by way of record linkage using the Generalized Iterative Record Linkage System. The objective of this study is to investigate the relationship between potential dietary risk factors and major cancers.

TYPE: Cohort
TERM: Alcohol; Diet; Nutrition; Obesity; Registry; Tobacco (Smoking)
SITE: All Sites; Breast (F); Colon
REGI: Canada (Can)
TIME: 1991 - 1994

90 **Silins, J.** 03063

Health Division, Vital Statistics & Disease Registries Section, 18-A R.H. Coats Bldg, Tunney's Pasture, Ottawa Ontario, Canada K1A 0T6 (Tel.: +1 613 9518553; Tlx: 0533585)

COLL: Fair, M.E.; Coppock, E.A.; Carpenter, M.; Jordan-Simpson, D.; Brancker, A.; Gaudette, L.A.

Vital Statistics and Disease Registries Section - Statistics Canada

Statistics Canada has developed the files and facilities for (and is currently collaborating with a number of agencies in carrying out) a variety of long term medical follow-up studies, including cancer incidence and mortality related to occupational and other defined population groups. In particular, three inter-related computer systems have been developed to permit optional use of a number of different records for the entire country for health-related research. These include: (1) a computerized generalized record linkage system which uses probabilistic matching techniques to link occupational or other records with mortality or cancer incidence records; (2) the Mortality Data Base which contains all Canadian death records, including coded cause of death, on magnetic tape from 1950 onwards; and (3) the National Cancer Incidence Reporting System which includes all incident cases reported by provincial cancer registries since 1969. Quality assessment programmes and trends in cancer incidence and mortality are also carried out. Mortality atlases have been prepared to illustrate the spatial variation of cancer mortality rates in Canada in order to facilitate the detection of high risk regions and any general patterns of disease distribution. A number of epidemiological studies, primarily concerned with excess cancer risk are currently being conducted in five broad areas including: (1) follow-up of occupational groups such as uranium and other miners, nickel workers, farmers, radiation workers and asbestos workers; (2) long-term follow-up of patients with specific medical diagnoses and/or receiving various treatments; (3) reproductive outcomes; (4) follow-up of persons categorized by various lifestyle or environmental exposures; and (5) death clearance of cancer registry files. Several of these projects are carried out on a cost-recovery basis for a number of different agencies (e.g., NCIC, Health and Welfare, universities, provincial health departments, private industries, etc.). Further work is underway to investigate the usefulness of a number of starting point files for follow-up of certain population groups, to evaluate the accuracy and improve the strategies of the record linkage procedures, and to further develop cancer and reproductive outcome files.

TYPE: Incidence; Mortality
TERM: Chemical Exposure; Dusts; Environmental Factors; Geographic Factors; Lifestyle; Mapping; Metals; Mining; Occupation; Pesticides; Petroleum Products; Plastics; Radiation, Ionizing; Record Linkage; Registry; Reproductive Factors
SITE: All Sites
CHEM: Asbestos; Formaldehyde; Glass Fibres; Gold; Nickel; Uranium; Vinyl Chloride
OCCU: Asbestos Workers; Chemical Industry Workers; Farmers; Firemen; Mineral Fibre Workers; Miners; Miners, Uranium; Morticians; Nickel Workers; Radiation Workers
LOCA: Canada; United States of America
REGI: Canada (Can)
TIME: 1974 - 1994

91 **Wigle, D.T.** 04319

Dept. of National Health & Welfare, Health Protection Branch, Lab. Center for Disease Control, Tunney's Pasture, Ottawa Ontario, Canada K1A 0L2 (Tel.: +1 613 9571852; Tlx: 0533679)

COLL: Mao, Y.S.

Mortality Follow-Up of Canada Health Survey Cohort

A health survey of 33,000 Canadians was conducted in 1978-1979. The survey involved a detailed interview consisting of medical history, physical examination and biological tests. The mortality experience of the survey cohort to the end of 1985 has been determined by linkage to the National Death Index of Statistics Canada. The objectives of this study are to investigate the associations between potential risk factors (alcohol, smoking, obesity, socio-economic status) and major causes of death, and to determine the number of deaths, attributable to smoking and passive smoking. Multivariate analysis will be used to control for confounding factors. Mortality follow-up to the end of 1988 was completed during 1990.

TYPE: Cohort
TERM: Alcohol; Passive Smoking; Record Linkage; Socio-Economic Factors; Tobacco (Smoking)
SITE: All Sites
TIME: 1987 - 1994

Québec

92 **Meyer, F.** 04881

Hôp. du Saint-Sacrement, Centre de Recherche, 1050, chemin Ste-Foy , Québec Québec, Canada G1S 4L8 (Tel.: +1 418 6565186; Fax: 6562627)

COLL: Bairati, I.; Fradet, Y.

Influence of Diet on the Occurrence and Progression of Prostate Cancer

This case-control study will assess the relationship between diet and prostate cancer. Cases (about 410 men) will be all newly diagnosed prostate cancer patients, aged 55 to 75, treated in seven Quebec City hospitals from September 1990 to December 1992. Controls (about 460 men) will be randomly selected among patients of similar ages treated in the same hospitals for a first transurethral resection for benign prostatic hyperplasia. Nutritionists, blinded to the disease status, will interview all subjects on their dietary intakes during the past year using a detailed food frequency questionnaire and food models. All pathological samples will be reviewed to assess histological grade of cancer and presence and severity of precancerous lesions. Diet, especially fat and carotene intakes, will be compared using polytomous logistic regression across six groups of subjects. These will include patients with cancer at various stages (latent, localised, regional and metastatic), those with precancerous lesions and men free of both cancer and precancerous lesions.

TYPE: Case-Control
TERM: Diet; Fat; Premalignant Lesion; Stage; Vitamins
SITE: Prostate
CHEM: Beta Carotene
TIME: 1990 - 1994

Saskatoon

93 **Dosman, J.** 04741

Univ. of Saskatchewan, Dept. of Medicine, Centre for Agricultural Medicine, Room 539, Ellis Hall, Saskatoon Saskatchewan, Canada S7N OXO (Tel.: +1 306 9668286; Tlx: 9668799)

COLL: McDuffie, H.H.; Hill, G.B.; Thériault, G.P.; McLaughlin, J.K.; Choi, N.W.; Robson, D.; Fincham, S.F.; Skinnider, L.; Pahwa, P.; White, D.; To, T.; West, R.; Spinelli, J.J.

National Study of Pesticide Exposure and Health

Recent studies suggest that use of agricultural pesticides is associated with an increased risk of developing soft tissue sarcoma. A group of investigators from Alberta, Saskatchewan, Manitoba, Quebec, Ontario, Newfoundland and British Columbia proposes to conduct a case-control study involving about 1,600 male patients, newly diagnosed as having one of the above types of cancer, together with an equal number of male controls selected from the general population. Patients and controls will be sent questionnaires to ascertain exposure to pesticides and the presence of other relevant factors. Subjects who indicate exposure to pesticides will be interviewed to determine the type and level of exposure. The risk of each of the above types of cancer in relation to pesticide exposure will be derived and the attributable proportion of such cancers among males will be estimated.

TYPE: Case-Control
TERM: Occupation; Pesticides
SITE: Hodgkin's Disease; Multiple Myeloma; Non-Hodgkin's Lymphoma; Sarcoma; Soft Tissue
OCCU: Farmers; Forest Workers; Horticulturists
TIME: 1989 - 1994

94 **McDuffie, H.H.** 03239

Univ. of Saskatchewan, Centre for Agricultural Medicine, Dept. of Medicine, Royal Univ. Hosp., Saskatoon Saskatchewan, Canada S7N OXO (Tel.: +1 306 9666154; Fax: 9668799)

COLL: Dosman, J.; Klaassen, D.J.

Collaborative Studies of Primary Lung Cancer

A series of case-control studies is being conducted to investigate the relationship between various risk factors (positive family history, occupational exposures, atopy, gender, radon levels in homes) and cigarette smoking habits and the aetiology of primary lung cancer. Data recorded by the cancer registry have been compiled on over 2,900 cases and postal questionnaires or personal interviews conducted with more than 1,500 cases and similar number of controls. Manuscripts concerning histology (Chest 98:1187-1193, 1990 and 99:404-407, 1991) and family history (J. Clin. Epidemiol. 44:69-76, 1991) have recently been published.

TYPE: Case-Control
TERM: Chemical Exposure; Dusts; Enzymes; Familial Factors; Fungicides; Genetic Factors; Herbicides; Insecticides; Occupation; Tobacco (Smoking)
SITE: Lung
CHEM: Radon
OCCU: Agricultural Workers; Farmers
REGI: Saskatchewan (Can)
TIME: 1982 - 1994

95 **McDuffie, H.H.** 04752
Univ. of Saskatchewan, Centre for Agricultural Medicine, Dept. of Medicine, Royal Univ. Hosp., Saskatoon Saskatchewan, Canada S7N OXO (Tel.: +1 306 9666154; Fax: 9668799)

Radon Exposure in Homes and Primary Lung Cancer in Women

Various mechanisms have been proposed to explain sex differences in the occurrence of primary lung cancer. Women patients have lower exposure to cigarette smoke and putative occupational carcinogens, while developing this disease at younger ages than men. There is limited evidence that decreases in pulmonary function related to cigarette smoking may be one biological mechanism which helps to explain these results. Exposure in the home to alpha radiation derived from radon and radon progeny is a viable alternative hypothesis. In Saskatchewan, exposure to this naturally occurring radiation in homes is possible over a wide geographical area. Women in the age group most at risk of developing lung cancer have typically spent more time in the home than men. Women have two to three times the levels of body contamination than men exposed to the same ambient living-area radon concentrations. Personal interviews are being conducted and alpha radiation levels being measured in the homes of female patients with primary lung cancer and age-matched control women.

TYPE: Case-Control
TERM: Female; Radiation, Ionizing
SITE: Lung
CHEM: Radon
REGI: Saskatchewan (Can)
TIME: 1989 - 1994

96 **McDuffie, H.H.** 04843
Univ. of Saskatchewan, Centre for Agricultural Medicine, Dept. of Medicine, Royal Univ. Hosp., Saskatoon Saskatchewan, Canada S7N OXO (Tel.: +1 306 9666154; Fax: 9668799)
COLL: White, M.D

Pesticide Exposure and Cancer in Females

Several cancer sites have been epidemiologically associated with rural residence, farming, specific farm practices and pesticide exposure in men. Women are often arbitrarily excluded from studies of putative carcinogens when occupational exposure is the primary focus. A population-based case-control study (45 soft tissue sarcoma, 144 non-Hodgkin's lymphoma, 16 gliomas, 255 community controls, 128 sisters of patients) is being conducted, using a postal questionnaire to ascertain medical history, reproductive factors, pesticide exposure and smoking patterns. Cancer Registry records are used to access cases and verify the diagnosis, controls are obtained through the Saskatchewan Hospital Services Plan. Age - adjusted odds ratios will be used to assess the hypothesis regarding increased risk of site specific tumours and pesticide exposure.

TYPE: Case-Control
TERM: Female; Occupation; Pesticides; Reproductive Factors; Tobacco (Smoking)
SITE: Brain; Multiple Myeloma; Non-Hodgkin's Lymphoma; Sarcoma; Soft Tissue
OCCU: Farmers; Health Care Workers; Horticulturists
TIME: 1989 - 1994

Sherbrooke

97 **Vobecky, J.** 04337
Univ. of Sherbrooke, Fac. Med., Human Nutrition Research Unit, Dept. Comm. Health Sciences, 3001 12e Ave. Nord, Sherbrooke Québec, Canada J1H 5N4
COLL: Echavé, L.V.; Langlois, S.P.; Abdulnour, E.; Vobecky, J.S.

Role of Vitamin A and Cholesterol in the Aetiology of Lung Cancer

To study the possible protective effect of vitamin A and the increased risk associated with a high cholesterol content of the diet in relation to lung cancer, a case-control study on 180 incident cases of primary lung carcinoma is being conducted. Two controls will be selected for each case, one among patients not suffering with lung carcinoma, the other from the neighbourhood of the cases. The hospital controls will have a lung X-ray to confirm the absence of cancer. Cases and hospital controls come from a university hospital. For the neighbourhood controls, the presence or absence of symptoms associated with lung cancer will be established by a questionnaire. If symptoms are present, the subject will be referred to a physician for confirmation of the diagnosis. Data on food intake of vitamin A, beta carotene, cholesterol and other relevant nutrients, including amino and fatty acids, will be gathered for the period prior to diagnosis of lung cancer as well as for the assessment of present nutritional status. A complete history of environmental and tobacco smoke exposure will be collected in order to control for confounding. Special attention will be paid to passive smoking. Stratified and multivariate analysis will be used for the identification of potential associations.

TYPE: Case-Control
TERM: Cholesterol; Diet; Nutrition; Passive Smoking; Vitamins
SITE: Lung
CHEM: Beta Carotene
TIME: 1987 - 1994

Toronto

98 **Chen, J.K.** 05018
Univ. of Toronto, Fac. of Dentistry, 4384 Medical Sciences Bldg, Toronto Ontario, Canada M5S 1A8 (Tel.: +1 416 9786624)
COLL: Katz, R.; Krutchkoff, D.; Eisenberg, E.

Trends in Lip Cancer in Connecticut, 1935-1985

The primary purpose of this study is to investigate secular trends in lip cancer incidence in Connecticut from 1935-1985. A secondary purpose is to determine the degree to which lip cancer alone influenced dramatic differences in trends previously reported for all oral cancer as compared to intra-oral squamous cell carcinoma in the same population (Cancer 65:2796-2802, 1990 and 66:1288-1296, 1990). The data source for this study is the Connecticut Tumor Registry (CTR). Established in 1935, it is both the oldest population-based cancer registry in the world and the largest in the United States. Three types of incidence rates (crude, age-specific and age-adjusted) of lip cancer (n = 2,219) are calculated for the 51-year study period. The study also examines age, sex, geographical distribution and lesion subsite analyses in addition to histopathological patterns of differentiation in lip cancer. According to the primary results, two major possible causal factors of lip cancer, solar radiation and smoking, among the local residents will be extensively analysed.

TYPE: Incidence
TERM: Geographic Factors; Radiation, Ultraviolet; Registry; Tobacco (Smoking); Trends
SITE: Lip
REGI: Connecticut (USA)
TIME: 1990 - 1994

99 **Choi, B.C.K.** 04325
Univ. of Toronto, Unit of Occupational and Environmental Health, 150 College St., Toronto Ontario, Canada M5S 1A8 (Tel.: +1 416 9786236; Fax: 9787262)
COLL: Connolly, J.

Factors Affecting Urinary Mutagenic Level of Workers with Bladder Cancer

The Ames/Salmonella Microsome Test (Ames Assay) is widely used to assay mutagenicity. The fact that cigarette smokers generally have mutagenic urine is now documented, but it has also been shown that mutagens can be detected in non-smoking workers in the rubber industry. In addition, there have been reports that bladder cancer patients have higher levels of excretion of tryptophan metabolites than do comparison subjects, and bladder cancer patients and normal donors differ in their urinary nitrosamines. It has been suggested that bladder cancer cases and normal controls may differ in their urinary mutagenicity in terms of potency of the mutagens and their type (i.e. base-pair or frame-shift). This provides a basis for the use of urinary Ames Test as a possible screening technique for bladder cancer cases in the workplace. Forty bladder cancer patients and 40 hospital controls will be selected for the study. Subjects will be non-smokers and will not have been under any radiation treatment or chemotherapy within the previous 60 days. An epidemiological questionnaire will be administered concerning the individual's exposure to industrial chemicals, artificial sweeteners, beverages and selected food items. Two 24-hour urine specimens, one taken while the individual is in the working environment and the other after he/she has been out of the working environment for at least 48 hours, will be extracted, frozen and tested for mutagenicity. For each specimen, Ames Test involving 2 bacterial strains TA98 and TA100, with and without S-9 liver homogenates, will be carried out. The following information will result from this study: (1) the bladder cancer status of the patients, (2) urinary mutagenicity as assayed by the Ames Test, and (3) epidemiological data obtained from the questionnaire. Multivariate analysis will be used to relate the type and potency of urinary mutagens to the bladder cancer status, simultaneously adjusted for epidemiological variables obtained in the questionnaires.

TYPE: Case-Control
TERM: Chemical Exposure; Mutagen; Occupation; Sweeteners
SITE: Bladder
TIME: 1988 - 1994

100 **Feuer, G.M.** 05035
Univ. of Toronto, Depts of Clinical Biochemistry and Pharmacology, 100 College St. , Toronto Ontario, Canada M5G 1L5 (Tel.: +1 416 9782262; Fax: 9785650)
COLL: Kerenyi, N.A.; Szalai, J.P.

Light Pollution and Increased Cancer Incidence

The objective of this study is to investigate cancer incidence among factory workers who work mostly in darkness compared with workers in another factory with normal lighting exposure, and with the general population. At present, cancer is responsible for almost half of all deaths among men in the same age group. The study will investigate the hypothesis that one of the most important aetiological factors in the rapid increase in malignant melanoma is the change in exposure to light in the last 100 years. Further studies are planned to examine by laboratory methods whether a higher level of serum melatonin, the oncostatic pineal hormone, is associated with the lower cancer rate which may occur in the group working in a dark environment.

TYPE: Incidence
TERM: Fluorescent Light; Occupation; Pigmentation
SITE: All Sites; Melanoma
TIME: 1991 - 1994

101 **Finkelstein, M.M.** 02819
Ontario Ministry of Labour, 400 University Ave., 7th Floor, Toronto Ontario, Canada M7A 1T7 (Tel.: +1 416 3267879; Fax: 3267889)

Mortality in Asbestos Cement Factory Workers

Mortality occurring among employees of an asbestos cement factory is being studied in relation to estimates of past dust exposure. Both chrysotile and crocidolite asbestos had been used at this factory and mortality rates from lung cancer and mesothelioma are known to be markedly elevated among exposed employees. The control groups consist of workers from non-asbestos areas of the factory as well as the general population of the province. Papers have been published in Br. J.Ind.Med. 40:138-144, 1983; Am.Rev.Resp.Dis. 129:754-761, 1984 and in Toxicol. Ind. Hlth 6:623-627, 1990.

TYPE: Case-Control; Cohort
TERM: Dusts; Occupation
SITE: Gastrointestinal; Mesothelioma; Respiratory
CHEM: Asbestos, Chrysotile; Asbestos, Crocidolite
OCCU: Cement Workers
TIME: 1978 - 1994

102 **Finkelstein, M.M.** 04300
Ontario Ministry of Labour, 400 University Ave., 7th Floor, Toronto Ontario, Canada M7A 1T7 (Tel.: +1 416 3267879; Fax: 3267889)
COLL: Beck, P.

Association between Radium in Drinking Water Supplies and the Risk of Bone Cancer Death in Ontario Youths

The aim of this case-control study is to measure the association between exposure to radium in domestic water supplies and the risk of death from bone sarcoma. The study subjects are the 327 Ontario born individuals, 25 years of age or less, who died of bone sarcoma between 1950 and 1983. Each has been matched with a control of the same age who died of any other disease cause. Water samples are collected at the residences of the subjects and radium-226 is measured by the radon emanation method. Statistical analysis is by Mantel-Haenszel and conditional logistic regression methods.

TYPE: Case-Control
TERM: Metals; Water
SITE: Bone
CHEM: Radium
TIME: 1985 - 1994

***103** **Finkelstein, M.M.** 05319
Ontario Ministry of Labour, 400 University Ave., 7th Floor, Toronto Ontario, Canada M7A 1T7 (Tel.: +1 416 3267879; Fax: 3267889)
COLL: Liss, G.M.

Cancer in the Pulp and Paper Industry

A cohort of some 33,000 workers at 14 Ontario mills will be enrolled in a retrospective mortality study. Workers will be traced through the Canadian Mortality Data Base. Hygiene data will be collected by questionnaire and by review of records. This is part of the IARC international study on cancer risks among workers in the pulp and paper industry.

TYPE: Cohort
TERM: Occupation
SITE: All Sites
OCCU: Paper and Pulp Workers
TIME: 1993 - 1996

104 **Howe, G.R.** 02379
Univ. of Toronto, National Cancer Inst. of Canada, Dept. of Prev. Med. and Biostatistics, 12 Queen's Park Crescent West, McMurrich Bldg, 3rd Floor, Toronto Ontario, Canada M5S 1A8 (Tel.: +1 416 9785097)
COLL: Baines, C.J.; Miller, A.B.

Cancer Incidence and Mortality in a Cohort of Diabetics

The primary objective of the present study is to monitor cancer incidence and mortality among diabetics in order to examine possible associations between (1) artificial sweetener use and bladder cancer; (2) factors relating to diabetes and cancer of the pancreas; (3) smoking patterns among diabetics and lung cancer. The secondary objective is to relate mortality from non-neoplastic diseases, such as cardiovascular diseases, to factors relating to diabetic status, such as control of disease. A self-administered questionnaire has been distributed containing questions on identifying information, details relating to onset and control of diabetes, and a number of lifestyle habits such as smoking and artificial sweetener use. 13,000 questionnaires have now been received. The subsequent cancer incidence and general mortality experience of the cohort will be monitored using computerized record linkage techniques to link the data to the national mortality data base and national cancer incidence reporting system maintained by Statistics Canada. In 1986 a supplementary questionnaire was distributed to the cohort to get data on current aspartame use and more specific data on past artificial sweetener use. A 55% response rate was achieved. A paper was published in J. Can. Dietetic Assoc. 46(4):288-291, 1985.

TYPE: Cohort; Incidence; Mortality
TERM: Diabetes; Lifestyle; Record Linkage; Registry; Sweeteners; Tobacco (Smoking)
SITE: Bladder; Lung; Pancreas
REGI: Canada (Can)
TIME: 1980 - 1994

105 **Howe, G.R.** 02908
Univ. of Toronto, National Cancer Inst. of Canada, Dept. of Prev. Med. and Biostatistics, 12 Queen's Park Crescent West, McMurrich Bldg, 3rd Floor, Toronto Ontario, Canada M5S 1A8 (Tel.: +1 416 9785097)
COLL: Baines, C.J.; Jain, M.

Diet and Cancer in a Cohort Study of Women

Participants in the National Study of Breast Cancer Screening have been asked to complete a self-administered dietary questionnaire previously validated on a sample of 200 participants in the Breast Screening Study in Toronto in 1981. Approximiately 60% of the total sample in the Breast Screening Study of 89,000 women age 40-59 completed the questionnaire. Using the follow-up proposed for the National Breast Screening Study and extending this if necessary by computerized record linkage with National Cancer Incidence and Mortality files, will enable the relationship between cancers in women and diet to be further evaluated. This dietary information will complement the information on risk factors for breast and other cancers already being collected in the National Breast Screening Study.

TYPE: Cohort
TERM: Diet; Female; Record Linkage; Screening
SITE: Breast (F); Colon; Kidney; Ovary; Pancreas; Rectum; Uterus (Corpus)
TIME: 1982 - 1996

***106** **Kusiak, R.A.** 05269
Ontario Ministry of Labour, Health & Safety Policy Branch, Health & Safety Studies Unit, 400 University Ave., 7th Floor, Toronto Ontario, Canada M7A 1T7 (Tel.: +1 416 3267884; Fax: 3267889)
COLL: Ritchie, A.C.; Sprinter, J.; Muller, J.

Ontario Miners' Study

This is an on-going study of the mortality of over 50,000 Ontario miners. Mortality is ascertained from file linkage with statistics Canada's mortality database. Exposure is obtained from chest clinics where miners were required to be certified as fit to work in dust exposure. Comparison is made with mortality of Ontario males and internal comparisons of the relative effects of exposure to radon, arsenic, silica, chromium and diesel exhausts will be calculated.

TYPE: Cohort
TERM: Dusts; Metals; Mining; Occupation; Record Linkage
SITE: Leukaemia; Lung; Stomach
CHEM: Arsenic; Chromium; Diesel Exhaust; Radon; Silica
OCCU: Miners
TIME: 1978 - 1996

***107** **L'Abbé, K.A.** 05337
Univ. of Toronto, Faculty of Medicine, Dept. of Prev. Med. Biostat., 12 Queen's Park Cres. W., 4th Floor McMurrich Bldg, Toronto Ontario, Canada M5S 1A8 (Tel.: +1 416 9784982; Fax: 9788299)
COLL: Howe, G.R.

Cohort Study of a Ten Percent Sample of the Canadian Labour Force

A follow-up study of 457,224 males and 242,196 females constituting an approximate 10 percent sample of the Canadian Labour Force between 1965 and 1971 is being conducted. Information on the occupation and industry in which each individual was working is available. The objective of the follow-up of this cohort is to act as a monitoring system to detect previously unsuspected associations between workplace exposures and mortality risk, and also to evaluate hypotheses derived from other sources. The mortality experience of the cohort to the end of 1992 will be determined by computerized record linkage to the Canadian National Mortality Data Base. Initial focus will be on associations for females in occupation and industry-specific analyses. A job exposure matrix developed in Canada will be used to translate job title information into occupational exposures.

TYPE: Cohort
TERM: Female; Occupation; Record Linkage
SITE: All Sites
TIME: 1993 - 1996

108 **Marrett, L.D.** 04551
Ontario Cancer Treatment & Research Foundation, Div. of Epidemiology & Statistics, 620 University Ave., Toronto Ontario, Canada M5G 2L7 (Tel.: + 1 416 9719800; Fax: 9716888)
COLL: Weir, H.K.

Aetiology of Malignant Germ Cell Tumours

Malignant germ cell tumours (MGCT) occur mainly in the testis but can also occur in the ovary and at extra-gonadal sites. The majority of testicular tumours and ovarian tumours in females under age 15 are of germ cell origin. Incidence rates for testicular cancer in Ontario, as well as in other parts of the world, have been rising in recent years. The childhood incidence of ovarian cancer and MGCT (all sites) also appears to be increasing. The reasons for the increases are unknown. This case-control study was designed to elucidate the aetiology of malignant germ cell tumours (MGCT) in Ontario, Canada. 556 cases were identified through the Ontario Cancer Registry and constitute all residents of Ontario, aged 59 or less, diagnosed with MGCT between 1986 and 1990. 1,122 controls were selected from the general population of Ontario and frequency-matched to the cases for age and sex. Relevant exposure information (medical, occupational, and hormonal) covering the period from pre-conception to diagnosis, were collected through mailed questionnaires to subjects and their mothers (for living subjects aged 16 and over) or to the mothers of deceased cases and those subjects under the age of 16 at time of diagnosis. In addition, this study included pathology review of all identified cases for the purpose of classifying tumours according to histological subgroups (pure seminoma, pure nonseminoma, and mixed seminoma and nonseminoma). Classifying disease state according to these histological subgroups may elucidate aetiological risk factors specific to these subgroups. Analysis is proceeding.

TYPE: Case-Control
TERM: Classification; Histology; Hormones; Occupation; Registry
SITE: Ovary; Testis
REGI: Ontario (Can)
TIME: 1988 - 1994

***109** **Marrett, L.D.** 05488
Ontario Cancer Treatment & Research Foundation, Div. of Epidemiology & Statistics, 620 University Ave., Toronto Ontario, Canada M5G 2L7 (Tel.: + 1 416 9719800; Fax: 9716888)
COLL: Sloan, M.

Incidence of Cancer in Registered Indians in Ontario, 1966-1991

The purpose of this study is to build a database which will permit description of patterns in registered Indians in Ontario. Computerized lists of registered Indians in Ontario during 1966-1991 from Indian and Northern Affairs Canada will be linked with data in the Ontario Cancer Registry records of Ontario residents treated in Manitoba from 1966 to 1991 from the Manitoba Cancer Registry. Linkage with the Manitoba registry is important since a proportion of registered Indians who live in the remote north-western part of Ontario receive cancer treatment in this province. Files have been obtained and linkage across years within the registered Indian file is in progress.

TYPE: Incidence
TERM: Data Resource; Ethnic Group; Indians; Record Linkage; Registry
SITE: All Sites
REGI: Manitoba (Can)
TIME: 1992 - 1994

***110** **Marrett, L.D.** 05489

Ontario Cancer Treatment & Research Foundation, Div. of Epidemiology & Statistics, 620 University Ave., Toronto Ontario, Canada M5G 2L7 (Tel.: +1 416 9719800; Fax: 9716888)

COLL: King, W.

Great Lakes Basin Cancer Risk Assessment

The aim of this study is to assess the risk of cancers of the colon, rectum and bladder as related to consumption of Great Lakes water. A case-control design is employed, with cases being ascertained through the Ontario Cancer Registry and controls being identified from households selected from a telephone consumer database. The questionnaire is administered by telephone, using a computer assisted telephone interview system. About 1,000 cases for each cancer site and a total of 2,000 controls will be included.

TYPE: Case-Control
TERM: Chlorination; Registry; Water
SITE: Bladder; Colon; Rectum
CHEM: Trihalomethanes
REGI: Ontario (Can)
TIME: 1992 - 1994

***111** **Marrett, L.D.** 05490

Ontario Cancer Treatment & Research Foundation, Div. of Epidemiology & Statistics, 620 University Ave., Toronto Ontario, Canada M5G 2L7 (Tel.: +1 416 9719800; Fax: 9716888)

COLL: Greenberg, M.D.; Bryan, H.; Chiarelli, A.; Darlington, G.

Reproductive Outcomes in Female Childhood Cancer Survivors

This is a province-wide retrospective cohort study to determine the risk of adverse reproductive outcomes in female survivors of childhood cancer. The association between anti-cancer therapy given during childhood for cancer and ensuing fertility in adulthood will be assessed. For those women who have had pregnancies, the risk of having an abnormal outcome (defined as spontaneous abortion, stillbirth, low birth weight or early neonatal death in offspring) will be examined. Survivors of childhood cancer will be ascertained through the Ontario Cancer Registry. Females diagnosed with any malignancy (excluding non-melanotic carcinoma of skin), diagnosed before age 20 from 1964 to 1986, who have survived for at least 5 years after diagnosis and are currently alive, have survived to at least 18 years of age and are a resident of Ontario at time of diagnosis, will be included. Subjects will be traced to current address using a variety of medical and non-medical sources. Data on anti-cancer therapy will be obtained from the cancer treatment centres, hospitals and physicians. Reproductive outcomes will be ascertained from respondent's recall of events through the use of a questionnaire administered over the telephone. The expected number of participants is 1,300.

TYPE: Cohort
TERM: Abortion; Chemotherapy; Fertility; Late Effects; Pregnancy; Radiotherapy; Registry
SITE: All Sites; Childhood Neoplasms
REGI: Ontario (Can)
TIME: 1993 - 1995

112 **Miller, A.B.** 00251

Univ. of Toronto, Dept. of Preventive Medicine & Biostatistics, McMurrich Bldg, 4th Floor, 12 Queen's Park Crescent West, Toronto Ontario, Canada M5S 1A8 (Tel.: +1 416 9785662; Fax: 9788299)

COLL: Howe, G.R.

Cancer Following Multiple Fluoroscopies

This cohort study is designed to evaluate the carcinogenic risk from ionizing radiation in the dose range 0 to 100 rad and, for the major cancer sites, to establish if there is a possible dose-response relationship. The study population consists of patients first admitted to Canadian tuberculosis sanatoria between the years 1930 and 1952. A substantial proportion (40 to 50%) were fluoroscoped regularly in conjunction with artificial pneumothorax treatment and consequently accumulated a substantial radiation dose consisting of many exposures over a period of time. Information on the admission and treatment records of these patients was extracted from the original sanitoria files. The basic patient data consist of approximately 130,000 patient records. Three computerized record linkage steps using the identifying information present on the records have been undertaken, the first internally linking records to bring together admissions referring to the same individual, the second linking these composite records to the mortality data base maintained by Statistics Canada for the years 1950 through 1975, in order to ascertain the mortality experience of the cohort, and the third to the mortality data prior to 1950 and for 1976 through 1980. A dosimetry study has been completed and published (Health Phys. 35:259-70, 1978) which determined the dose experienced by various body organs exposed to radiation from fluoroscopy. The dosimetry information has been applied to the individual patient records in order to estimate the radiation dose received by each patient. The analysis will consist of estimation of relative risks with respect to mortality for different cancer sites as a function of dose calculated relative to those members of the study population who were not fluoroscoped. A preliminary report by Howe on breast cancer mortality has been published. (Radiation Carcinogenesis: Epidemiology and Biological Significance, (eds) J.D. Boice, Jr and J. F. Fraumeni, Jr; Raven Press, New York 1984, pp. 119-129). A definitive report on this experience has been published in N. Eng. J. Med. 321:1285-1289, 1989. A fourth linkage with national cancer incidence files is now planned.

TYPE: Cohort
TERM: Clinical Records; Radiation, Ionizing; Record Linkage; Treatment
SITE: All Sites
TIME: 1976 - 1994

113 **Miller, A.B.** 01835
Univ. of Toronto, Dept. of Preventive Medicine & Biostatistics, McMurrich Bldg, 4th Floor, 12 Queen's Park Crescent West, Toronto Ontario, Canada M5S 1A8 (Tel.: +1 416 9785662; Fax: 9788299)
COLL: Thompson, D.W.; Clarke, E.A.; Wall, C.

Natural History of Dysplasia of the Uterine Cervix

In most models of the natural history of squamous carcinoma of the cervix, dysplasia is regarded as the first recognizable pathological process in an orderly progression from normal to malignant epithelium. Since not all dysplasia become malignant, there is need to clarify the extent to which regression forms part of the natural history of dysplasia. The present study aims at seeking information on the natural history of cytologically identified but untreated dysplasia in comparison with a sample of normal women with smears examined in the same laboratory. There were 22,753 women with minimal dysplasia, 23,321 with mild, 11,264 with moderate and 2,697 with severe dysplasia. The control series included 20,976 women. Information was also sought on the possible association of oral contraceptive use and the risk of progression of dysplasia. Data extraction is complete. Analyses to date show an increasing risk of invasive cancer with increasing degrees of dysplasia. The relative risks for the manifestation of a malignancy in a subsequent cervical smear are 1.48, 3.42, 20.9 and 71.5 for women with minimal, mild, moderate and severe dysplasia, respectively. The cohort has been linked to the Ontario Cancer Registry to provide further data on risk of progression, and analysis of the findings is in progress.

TYPE: Case-Control
TERM: Oral Contraceptives; Premalignant Lesion
SITE: Uterus (Cervix)
TIME: 1978 - 1994

114 **Miller, A.B.** 02001
Univ. of Toronto, Dept. of Preventive Medicine & Biostatistics, McMurrich Bldg, 4th Floor, 12 Queen's Park Crescent West, Toronto Ontario, Canada M5S 1A8 (Tel.: +1 416 9785662; Fax: 9788299)
COLL: Baines, C.J.; Howe, G.R.; Wall, C.; Boyes, D.A.; Hill, G.B.; Bowman, D.M.; Simard, A.; Fabia, J.; Bassett, A.A.; Simor, I.; Deschenes, L.; Bethune, G.; Catton, G.; Cantin, J.; Bush, H.

Breast Cancer Screening

A long-term randomized trial of screening for breast cancer has been initiated in Canada to find whether screening will reduce the mortality rate from breast cancer. Several projects have indicated that screening is of benefit in women over 50, but that more needs to be known about its benefit in women in their forties. For women over 50, the question remains whether screening should include mammography or whether physical examination alone will suffice. Fifteen centres in Canada recruited 1,000-15,000 women between the ages of 40 and 59. Women aged 40-49 either underment mammography and physical examination annually, or were followed by mail annually. Women aged 50-59 had annual screening, half by physical examination alone and half by both physical examination and mammography. The screening continued for three or four years, and all women are being followed for at least a further five years. Every participant had an initial physical examination and instruction in breast self-examination. Recruitment into the study ceased on 31 March 1985. 89,968 women were enrolled.

Annual rescreening continued to 1988. The 7-year mortality results have been published in CMAJ 147:1459-1488, 1992).

TYPE: Cohort
TERM: Mammography; Screening
SITE: Breast (F)
TIME: 1980 - 1996

115 **Miller, A.B.** 02628
Univ. of Toronto, Dept. of Preventive Medicine & Biostatistics, McMurrich Bldg, 4th Floor, 12 Queen's Park Crescent West, Toronto Ontario, Canada M5S 1A8 (Tel.: +1 416 9785662; Fax: 9788299)
COLL: Coldman, A.J.; Morrison, B.J.

Extension of the British Columbia Cohort Study

In a previous study, two cohorts of women born in 1914-18 and 1929-33 were identified and studied from the records of the British Columbia Screening programme 1949-1969. The analysis showed that regression is an important part of the natural history of carcinoma of the cervix. In the present study, the period of observation for the two cohorts has been updated to 1980 and a third cohort born in 1944-1948 identified. This will permit the answering of questions relating to the efficacy of screening and evaluate the incidence of the precursors of invasive cancer of the cervix in a presumed high risk young cohort. This will enable predictions from mathematical models of screening based on the previous analysis to be refined and will provide the opportunity to include a large body of data in that currently being collected from a number of countries by the IARC. The three cohorts comprise over 300, 000 longitudinal records of women. The basic records of the cohorts have been extracted, internal record linkage has been conducted to eliminate duplication, and the file is now undergoing analysis.

TYPE: Cohort
TERM: High-Risk Groups; Mathematical Models; Screening
SITE: Uterus (Cervix)
TIME: 1981 - 1994

Vancouver

116 **Band, P.R.** 03605
BC Cancer Control Agency, Div. of Epidemiology, Biometry and Occupational Oncology, 600 W 10th Ave., Vancouver BC, Canada V5Z 4E6 (Tel.: +1 604 8776000)
COLL: Spinelli, J.J.; Gallágher, R.P.; Threlfall, W.J.; Raynor, D.; Wong, M.

Identification of Occupational Cancer Risk Factors

The objective of this study is to identify occupational cancer risk factors. Life-time occupational histories of male cancer patients aged 20 years and over ascertained by the British Columbia Cancer Registry, and of controls from the general population are being obtained to determine whether employment in individual occupations or industries is associated with an increased risk for specific cancers. A self-administered questionnaire requesting detailed occupational histories as well as information on ethnic origin, education, smoking history and alcohol consumption will be mailed to each case and control. An estimated number of 5,000 cancer cases and 1,500 controls per year will be studied. For each individual cancer site, occupations and industry titles as well as combinations of similar occupations and industries will be examined. In addition, exposure-response and latency effects will be analysed.

TYPE: Case-Control
TERM: Alcohol; Dose-Response; Education; Environmental Factors; Ethnic Group; Latency; Occupation; Registry; Tobacco (Smoking)
SITE: All Sites
REGI: Br. Columbia (Can)
TIME: 1983 - 1994

117 **Band, P.R.** 03606
BC Cancer Control Agency, Div. of Epidemiology, Biometry and Occupational Oncology, 600 W 10th Ave., Vancouver BC, Canada V5Z 4E6 (Tel.: +1 604 8776000)
COLL: Spinelli, J.J.; Gallágher, R.P.; Bert, J.L.; Grace, J.R.; Moody, J.; Svirchev, L.M.

Detection of Carcinogens in the Workplace

The long-term objective of this project is to create a job-exposure matrix for the Province of British Columbia which can be used to establish relationships between specific types of cancer and specific chemical substances. The job exposure matrix will be used in conjunction with the identification of occupational cancer risk factors by allowing exposures to be inferred from the work history of cancer cases. To initiate the job exposure matrix, the study of a specific industry, the Trail Operations of Cominco, Limited, has been undertaken. The Trail Operations, involving zinc recovery and refining, lead smelting, sulphuric acid production and fertilizer production, is sufficiently complex to provide an excellent case study upon which the methodology of developing the exposure matrix can be tested and improved.

TYPE: Methodology
TERM: Chemical Exposure; Fertilizers; Metals; Occupation
SITE: All Sites
CHEM: Lead; Sulphuric Acid; Zinc
OCCU: Chemical Industry Workers; Fertilizer Workers; Miners, Zinc-Lead; Smelters, Lead
TIME: 1984 - 1994

118 **Horsman, D.E.** 04875
BCCA, Cytogenetic Lab., 600 W. 10th Ave., Vancouver BC, Canada V5Z 4E6 (Tel.: + 1 604 8776000)
COLL: Sparling, T.

Effects of Smoking on the Development of Myelodysplasia

This case-control study will assess the risk of myelodysplasia (MDS) associated with smoking. A smoking history will be obtained on all newly diagnosed patients with MDS, as diagnosed by haematological, marrow morphologic and cytogenetic investigations. About 200 cases will be recruited, with 200 population controls. Smoking data will be obtained by a questionnaire mailed to all identified patients' physicians.

TYPE: Case-Control
TERM: Tobacco (Smoking)
SITE: Leukaemia; Myelodysplastic Syndrome
TIME: 1989 - 1994

119 **McBride, M.L.** 04880
British Columbia Cancer Agency, Div. of Epidemiology, Biometry and Occupational Oncology, 600 W 10th Ave., Vancouver BC, Canada V5Z 4E6 (Tel.: + 1 604 8776000; Fax: 8724596)
COLL: Thériault, G.P.; Fincham, S.F.; Choi, N.W.; Robson, D.; Gallágher, R.P.

Childhood Leukaemia and Extremely Low Frequency Electromagnetic Field Exposure

Several studies have suggested that children exposed to intense electric or magnetic fields at extremely low frequency may be at higher risk of leukaemia. A population-based case-control study will be conducted of the association between exposure to electric and magnetic fields and risk of leukaemia in children aged up to 14, with particular emphasis on leukaemia subtypes. All children newly diagnosed with leukaemia from 1 January 1989, to 31 December 1993, and resident in the principal cities and surrounding areas of each of the participating provinces will be eligible. For each case, one control child will be matched on birthdate, sex and city of residence from the most appropriate available source. Data on magnetic field exposure will be collected using questionnaires, activity diaries, wiring configuration assessment, and a personal field dosimeter. Average exposure, duration of high exposure, and various components of electromagnetic field exposure can be analysed separately. The risk of all childhood leukaemia and each subtype of leukaemia in relation to 60 Hz field exposure will be examined. Associations will also be assessed with family history of cancer and congenital anomalies; parental occupational and recreational exposures; gestational factors and other exposures surrounding pregnancy; post-natal factors, including radiation, and exposure to viral infections, chemicals, and medications; and socio-economic status.

TYPE: Case-Control
TERM: Chemical Exposure; Childhood; Congenital Abnormalities; Drugs; Electromagnetic Fields; Infection; Intra-Uterine Exposure; Occupation; Radiation, Ionizing; Socio-Economic Factors
SITE: Leukaemia
TIME: 1990 - 1994

***120** **Rosin, M.P.** 05282

British Columbia Cancer Agency, Div. of Epidemiology, Biometry & Occupational Oncology, 600 West 10th Ave., Vancouver BC, Canada V5Z 4E6 (Tel.: +1 604 2914064; Fax: 2913040)

COLL: Anwar, W.A.; El Din Zaki, S.S.

Genetic and Proliferative Changes in the Urinary Bladder of Schistosoma Haematobium Patients

The primary objective of the study is to develop a better understanding of the role of inflammation and trauma in the development of urinary bladder cancer in patients with Schistosoma haematobium infections. The research focuses on the potential involvement of reactive oxygen species generated by activated inflammatory cells in the induction of genetic damage in the bladder. The approach used is to probe archival biopsies from schistosomiasis patients with cystitis, pre-malignant change, or cancer, with three groups of biomarkers: markers for genetic change (in situ hybridization with chromosome-specific probes and the micronucleus test), oxidative damage (immunohistochemical analysis of antibodies to damaged protein, lipid and DNA), and cell proliferation (immunohistochemical assessment of proliferating cell nuclear antigen). 30 biopsies of each histological type will be examined. The study will determine whether interactions occur between chromosomal breakage (levels of damage and specific changes), proliferation levels, and oxidative stress in individuals during conditions of bladder inflammation.

TYPE: Molecular Epidemiology
TERM: BMB; Biomarkers; Biopsy; Chromosome Effects; DNA; Genetic Markers; Infection; Micronuclei
SITE: Bladder
LOCA: Egypt
TIME: 1989 - 1995

121 **Yang, C.** 05097

BC Cancer Control Agency, Div. of Epidemiology, Biometry & Occupational Oncology, 600 W. 10th Ave., Vancouver BC, Canada V5Z 4E6 (Tel.: +1 604 8776000; Fax: 8724596)

COLL: Band, P.R.; Gallágher, R.P.; Spinelli, J.J.; Daling, J.R.; Weiss, N.S.

Influence of Reproductive and Occupational Exposures and Passive Smoking on the Incidence of Breast Cancer

The specific aim of this population-based case-control study is to examine the association of breast cancer with the following: prior abortion, postpartum use of lactation suppressing agents, postmenopausal use of progestin, occupation and history of passive smoking. All women in British Columbia (BC) aged 75 years or less diagnosed with breast cancer between June 1988 and November 1989 were identified through the BC Cancer Registry. The 1989 Provincial Voters' List was used to select randomly a group of controls with an age distribution similar to that of the cases. Information regarding personal characteristics, pregnancy, and medical and occupational histories was ascertained by postal questionnaire. Telephone calls were used to follow-up non-responders and to obtain missing information. At the close of recruitment, 1109 cases and 1074 controls had returned the questionnaire. A logistic regression model will be used as the primary analytic tool.

TYPE: Case-Control
TERM: Abortion; Hormones; Lactation; Occupation; Passive Smoking; Registry
SITE: Breast (F)
REGI: Br. Columbia (Can)
TIME: 1988 - 1994

Winnipeg

122 **Choi, N.W.** 04870

Manitoba Cancer Treatment & Research Foundation, Dept. of Epidemiology & Biostatistics, 100 Olivia St., Winnipeg Manitoba, Canada R3E 0V9 (Tel.: (204)7872178; Fax: 7836875)

COLL: Nelson, N.A.; Moodie, P.F.; Oh, H.C.; West, P.; Musto, R.

Cancer among Registered Indians in Manitoba

The Manitoba Cancer Registry will be linked to the Medical Services Bureau's data base, which includes some 60,000 registered Indians who have resided in Manitoba at any time during a nine-year period, 1978-1986, in order to identify Indians who have been diagnosed with cancer during this period. Expected numbers of various cancers will be calculated on the basis of cancer incidence in the non-Indian population of Manitoba. Certain cancers in excess or deficit will be further investigated by means of case-control studies. A standardized questionnaire will be used with the assistance of nursing station personnel.

TYPE: Incidence
TERM: Indians; Record Linkage; Registry
SITE: All Sites
REGI: Manitoba (Can)
TIME: 1991 - 1994

123 **Oh, H.C.** 05139
Manitoba Cancer Treatm. & Res. Found., Dept. of Epidemiology & Biostatistics, 100 Olivia St., Winnipeg Manitoba, Canada R3E 0V9 (Tel.: +1 204 7872178; Fax: 7836875)
COLL: Choi, N.W.

Passive Smoking and Lung Cancer

In this study, the effect of involuntary smoking on non-smokers and mild smokers is of interest. Lung cancer risk will be assessed in relation to both involuntary smoking volume and that of voluntary smoking. Collection of data has already started on known risk factors of lung cancer cases who come from the Manitoba Cancer Registry, including past smoking and involuntary smoking in conjuction with the study on lung cancer and home environment. This is a community based case-control study using systematic interview questionnaires. The possibility of expanding this study to other centres in Canada is being explored.

TYPE: Case-Control
TERM: Passive Smoking; Registry
SITE: Lung
REGI: Manitoba (Can)
TIME: 1988 - 1994

CHILE

Santiago

124 **Nervi, F.** 04728
Pontificia Univ. Catolica de Chile, Depto. de Gastroenterologia, Casilla 114-D , Santiago, Chile (Tel.: +56 2 397780; Fax: 2225515 ; Tlx: 240395 pucva cl)
COLL: Chianale, J.

Gallbladder Cancer: Prevalence, Early Detection, and Aetiology in a High-Risk Area.
The aim of this study is to detect subjects with gallbladder cancer in the pre-symptomatic stage in a Chilean population. The prevalence of cholesterol gallstones (the major risk factor for gallbladder cancer) is approximately 45% in Chileans older than 20 years. Clinical survey and echography will be used in a selected population of 10,000 subjects and also a small Indian village in Southern Chile (1,000). This first stage of the study will last from 1990-1995. Cases and subjects with potential risk factors (e.g. chronic carriers of S. Typhi) will be followed to estimate incidence of both gallstone disease and gallbladder cancer. Analysis of data will include univariate and multivariate logistic regression methods. A paper was published in Int. J. Cancer 46:1131-1133, 1990 (J. Chianale et al.).

TYPE: Cross-Sectional
TERM: Gallstones
SITE: Gallbladder
TIME: 1990 - 1995

125 **Serra, I.** 05221
Univ. of Chile, Faculty of Medicine, School of Public Health, Independencia 939, Santiago Correo 7, Chile (Tel.: +56 2 776560/5352; Fax: 776560/5363)
COLL: Báez, S.; Calvo, A.; Decinti, E.; Maturana, M.

Socio-Economic Status and Urban-Rural Residence as Risk Factors for Biliary Tract Cancer in Chile
The aim of this retrospective study is to investigate whether particular subgroups of the Chilean population, stratified by socio-economic levels and by residential districts (comunas, categorised as urban, rural or intermediate), have a greater risk of dying from biliary tract cancer than the general population. The hypothesis is that the poor are more prone to this cancer. A random sample of approximately 130 of the national total of 1,178 deaths from biliary tract cancer in 1986 are being analysed to establish the risk of dying associated to the socio-economic status, place of residence and occupation. Socio-economic level will be measured through years of school. The 1982 census will provide the national figures.

TYPE: Case Series
TERM: Occupation; Rural; Socio-Economic Factors; Urban
SITE: Bile Duct; Gallbladder
TIME: 1991 - 1994

Beijing

126 **Cai, S.X.** 05011
Chinese Academy of Preventive Medicine, Inst. of Occupational Medicine, 29 Nan Wei Rd, Beijing 100050, China (Tel.: 338761/260; Fax: 3011875)
COLL: Huang, M.Y.; Wang, A.R.; He, Y.P.; Su, C.Y.; Zhang, Z.Q.; Zhang, Z.; Zhou, Q.L.

Cancer Morbidity and Mortality Among Workers in the Chromate Production Industry

A cooperative group is undertaking a national survey to strengthen the prevention and treatment of occupational cancer in China, to study the morbidity of some common occupational tumours, to develope guidelines for the control of occupational tumours, and to provide scientific bases for the formulation of prevention and treatment measures. A retrospective survey is being carried out among 2,548 workers in chromate producing plants in Chongging, Shanghai, Tinghou, Qingdao, Suzhou and Tiangjin, and 5,458 control workers in the same cities (a total of 8,006 from start of operations to 31 December 1989, more than 30 years). Deaths from all causes and morbidity of all malignant neoplasms are being investigated.

TYPE: Cohort
TERM: Occupation
SITE: All Sites; Lung
CHEM: Chromium
TIME: 1982 - 1994

127 **Chen, J.S.** 03616
Chinese Academy of Preventive Medicine, Inst. of Nutrition and Food Hygiene, 29 Nan Wei Rd, Beijing 100050, China (Tel.: 3011875)
COLL: Campbell, T.C.; Li, J.; Peto, R.

Diet and Nutritional Status and Cancer Mortality in China

The second ecological study on diet, lifestyle and cancer mortality covered 69 counties and started in 1989. Data on food consumption, anthropometry, blood pressure, lung function and socio-economic status have been collected and are being tabulated. Analyses of urinary N-nitrosoamino acids, plasma retinol, carotenoids, tocopherols and minerals have been completed. Other urine and blood assays are underway. A retrospective mortality survey, including all causes of death in 1986-1988 is now being analysed. This mortality survey includes a large case-control study on smoking and disease which involves more than 10,000 cases of lung cancer. Studies on the relationship between oesophageal cancer mortality and urinary excretion of N-nitrosoamino acids in the 69 countries have been completed. Preliminary statistical analysis showed significant positive correlations between oesophageal cancer mortality rates and urinary levels of excreted N-nitrosoproline, N-nitrososarcosine and the nitrosation potential.

TYPE: Case-Control; Correlation; Mortality
TERM: Diet; Lifestyle; Minerals; Nutrition; Socio-Economic Factors; Tobacco (Smoking); Trace Elements; Vitamins
SITE: Breast (F); Colon; Leukaemia; Liver; Lung; Nasopharynx; Oesophagus; Rectum; Stomach
CHEM: Beta Carotene; N-Nitroso Compounds; Retinoids; Tocopherol
TIME: 1983 - 1995

128 **Huang, M.Y.** 05012
Chinese Academy of Preventive Medicine, Inst. of Occupational Medicine, 29 Nan Wei Rd, Beijing 100050, China (Tel.: 338761/260; Fax: 3011875)
COLL: Cai, S.X.; Zhang, X.M.; Hung, K.L.; Wu, H.C.; Hu, X.L.; Zhang, M.Q.; Liang, X.Z.; Zhou, D.N.; Wang, Z.Y.; Lian, X.X.; Ding, C.Y.; Cheng, S.M.; Yang, C.; Chen, N.H.; Xie, C.L.; Cai, X.C.

Occupational Tumours in Workers Exposed to Vinyl Chloride

In order to improve the prevention and treatment of occupational malignant tumours and to understand their prevalence among workers exposed to vinyl chloride, a cooperative group was organised and 13 polyvinyl chloride production plants in China investigated. This survey covered about 6,000 workers exposed to vinyl chloride and about 7,000 unexposed workers who served as controls. A retrospective investigation has been completed, and a prospective study will examine all causes of death, all causes of malignant tumours and liver cancer on the basis of the previous survey.

TYPE: Cohort
TERM: Occupation; Plastics
SITE: All Sites; Liver
CHEM: Vinyl Chloride
OCCU: Vinyl Chloride Workers
TIME: 1982 - 1994

129 **Liang, S.** 05013
Chinese Academy of Preventive Medicine, Inst. of Occupational Medicine, 29 Nan Wei Rd, Beijing 100050, China (Tel.: 338761)
COLL: Duan, B.R.; Chen, S.Y.

Retrospective Cohort Mortality Study of Lung Cancer in Workers Exposed to Dust

The objective of this study is to investigate whether workers in the mining industry have an increased risk of dying from lung cancer compared with members of the general population. Possible associations between (1) mining and lung cancer; (2) factors related to lung cancer such as pneumoconiosis, dust levels, chemicals; and (3) smoking patterns in relation to dust exposure and lung cancer will be examined. A paper was published in the Chinese J. Ind. Hyg. and Occup. Dis. 5:26-28, 1987.

TYPE: Cohort
TERM: Dusts; Mining; Occupation; Pneumoconiosis; Tobacco (Smoking)
SITE: Lung
OCCU: Miners
TIME: 1990 - 1994

130 **Lu, S.H.** 04681
Chinese Academy of Medical Sciences, Cancer Institute, Zuoanmen Wai, Panjiayao, Beijing 100021, China (Tel.: 781331/461)
COLL: Chui, S.X.; Luo, F.J.; Li, F.M.; Guo, L.P.; Yang, C.S.; Montesano, R.; Yang, W.X.

Relevance of N-Nitroso Compounds to Oesophageal Cancer in China

N-Nitroso compounds occur widely in the environment and most of them are carcinogenic to many species of animals. Previous work suggests that human exposure to nitroso compounds and their precursors is high in Lin-xian, a high oesophageal cancer incidence area in northern China. This project is intended to examine the role of these compounds in the aetiology of oesophageal cancer in Lin-xian with specific aims outlined below: (1) to assess the human exposure to N-nitroso compounds by studying the identities and quantities of nitroso compounds in the foods that are suspected risk factors, such as "pickled vegetables" and "mouldy grain", and determining the level of N-nitroso compounds in representative food samples before and after cooking; (2) to analyse the N-nitroso compounds in urine and gastric juice collected from three groups of individuals: early oesophageal cancer patients, those with severe oesophageal dysplasia, and normal individuals (controls); (3) to study N-nitroso compound-activated oncogenes in oesophageal epithelia of fetus and monkey; 4) to analyse DNA adducts in adjacent tissues and cancerous tissues of oesophagus collected from Lin-xian; 5) to determine whether supplementation with vitamins C and E decreases N-nitroso compounds in urine.

TYPE: Cross-Sectional
TERM: DNA; Prevention; Vitamins
SITE: Oesophagus
CHEM: N-Nitroso Compounds
TIME: 1985 - 1994

131 **Tao, S.C.** 04280
Beijing Inst. for Cancer Research, Dept. of Epidemiology, Da-Hong-Luo-Chang St., Western District, Beijing 100034, China (Tel.: 662974)
COLL: Guo, J.

Dietary Factors in the Aetiology of Breast Cancer among Postmenopausal Women

Certain epidemiological and experimental data indicate that there is a relation between diet and breast cancer incidence; relevant hypotheses concern the effects of fat, protein and total calories. The aim of this case-control study is to assess dietary factors and other characteristics of potential aetiological significance in the occurrence of breast cancer among postmenopausal women. Dietary and nutritional factors will be studied among 250-300 histologically diagnosed female breast cancer cases resident in the Beijing urban area. Nearest neighbours will be used as controls, matched for age, sex and marital status.

TYPE: Case-Control
TERM: Diet; Menopause
SITE: Breast (F)
TIME: 1988 - 1994

132 **Wei, L.** 03940
Ministry of Public Health, Laboratory of Industrial Hygiene, Dept. of Radiation Health, 2 Xinkang St., Deshengmenwai, P.O.Box 8018, Beijing 100088, China
COLL: Zha, Y.; Tao, Z.; Chen, D.; Yuan, R.; He, W.

Cancer Mortality and Mutational Diseases in a High-Background Radiation Area of Yangjiang, China

Measurements revealed that individual external exposure to the environmental radiation is 330mRmR/yr (average) in the high background radiation area (HBRA), and 114mRmR/yr (average) in a neighbouring control area (CA). About 80,000 inhabitants in each area whose families lived there for two or more generations are being observed to study whether there is any increase in cancer mortality or frequency of mutational diseases in the HBRA. Studies of environmental carcinogens and mutagens other than natural radiation by means of case - control study and commune survey showed that the factors believed to be related to mutational diseases were not statistically different between these two areas, except that the frequency of medical X-ray exposure of people in CA was a little higher than that in HBRA. About one million person-years in HBRA and as many in CA were observed for cancer mortality and sex-, age- and site-specific cancer mortality analysed. The total numbers of 31 kinds of hereditary diseases and congenital deformities in children below 12 years old between two areas (13,425 examinees in HBRA, 13,087 in CA) were almost identical. However, the frequency of Down's Syndrome in HBRA was statistically higher than that in CA, but the incidence in HBRA was within the range of spontaneous morbidity and dependent of material age. 25,258 and 21, 837 children were examined in HBRA and CA respectively.

TYPE: Cohort; Correlation; Mortality
TERM: Congenital Abnormalities; Down's Syndrome; Environmental Factors; Mutation Rate; Mutation, Germinal; Radiation, Ionizing
SITE: All Sites
TIME: 1972 - 1994

133 **Wu, M.** 03948
Chinese Academy of Medical Sciences, Cancer Inst., Dept. of Cellular Biology, Panjia Yuan, Beijing 100021, China (Tel.: +861 771 8927; Fax: 5058)
COLL: Hu, N.; Wang, X.Q.; Zhu, D.; Wang, L.; Mao, X.Z.; Jia, Y.T.; Zhang, C.L.; He, L.J.

Genetic Epidemiology of Susceptibility to Oesophageal Cancer in Yangcheng County and Yangquan City

Previous studies confirmed the existence of a familial aggregation of oesophageal cancer (EC) in Yangchen county, Shanxi Province, and indicated the presence of both an autosomal recessive gene and environmental factors in the aetiology of EC in the Yocun Commune, Linxian, Henan Province. Genetic abnormalities responsible for the susceptibility to EC are now being investigated with RFLP, using various DNA markers on chromosomal regions where cytogenetic changes were observed frequently in EC cells. Mutations in genes involved in DNA repair and carcinogenesis are being studied in samples collected from Yangchen County and Yangquan City in Shanxi Province since 1988.

TYPE: Genetic Epidemiology
TERM: Chromosome Effects; DNA Repair; High-Risk Groups; Linkage Analysis; Oncogenes; RFLP
SITE: Oesophagus
TIME: 1988 - 1995

***134** **Wu, M.** 05421

Chinese Academy of Medical Sciences, Cancer Inst., Dept. of Cellular Biology, Panjia Yuan, Beijing 100021, China (Tel.: +861 771 8927; Fax: 5058)

COLL: Hu, N.; Taylor, P.; Dawsey, S.M.; Bonney, G.E.; He, L.J.; Han, X.Y.

Oesophageal Cancer in Yangcheng County, Shanxi Province

The goal of the study is to investigate genetic factors in the aetiology of oesophageal cancer in Yangcheng County, Shanxi Province, China. All households were interviewed in 1979 to determine family history of oesophageal cancer. In 1989, vital status for all family members from three villages was determined, and families who reported a positive family history of oesophageal cancer in 1979 were interviewed a second time. The risk of oesophageal cancer was evaluated by comparing family and individual rates of oesophageal cancer during the interval 1979-1989, stratified by number of family members with oesophageal cancer prior to 1979. More families with prior oesophagel cancer history reported new oesophageal cancer deaths during the follow-up period than families without history (19% vs 5%). Oesophageal cancer rates increased with increasing positivity of family history and adjustment for other risk factors did not substantially alter this result. Comparison of risk in blood and non-blood relatives living in the same environment, as well as segregation analysis, are now being done. Molecular genetic studies (RFLP and SSCP) will continue in order to confirm the genetic hypothesis. A paper was published in Int. J. Epidemiol. 21:877-882, 1992.

TYPE: Cohort; Genetic Epidemiology
TERM: BMB; DNA; Familial Factors; Genetic Factors; RFLP; Segregation Analysis
SITE: Oesophagus
TIME:

135 **Xie, D.** 05040

Chinese Academy of Preventive Medicine, Inst. of Occupational Medicine, 29 Nan Wei Rd, Beijing 100050, China (Tel.: 338761/312)

COLL: Fang, F.

Genetic Epidemiology in the Population of a High Incidence Area for Lung Cancer

This project aims to explore the relationship between pre-malignant lesions and exposure level to benzo(a)pyrene in the population of a high incidence area for lung cancer, Xuanwei county. Levels of UDS (unscheduled DNA synthesis) and micronucleus frequency in human lymphocytes will be determined in individuals exposed to high, moderate and low levels of benzo(a)pyrene for 20 years and in a control group. The study population will be confined to women.

TYPE: Molecular Epidemiology
TERM: BMB; DNA; Environmental Factors; Female; Lymphocytes; Micronuclei; Premalignant Lesion
SITE: Lung
CHEM: Benzo(a)pyrene
TIME: 1991 - 1994

136 **Yin, S.** 04809

Chinese Academy of Preventive Medicine, Inst. of Occupational Medicine, 29 Nan Wei Rd, Beijing, China (Tel.: 3014187; Fax: 3011875)

COLL: Blot, W.J.; Li, G.

Study of Benzene Exposure in China

The aim of the study is to examine the dose-response relationship between benzene and leukaemia and other blood disorders. It is planned to carry out a retrospective cohort study. About 70,000 workers exposed to benzene and 30,000 control workers employed during 1972-1987 in 12 cities are being studied. Information on industrial hygiene including benzene concentration in the factories is collected. The benzene level for each job title is estimated. The personal information includes the details of occupational history, vital status, cause of death, and diagnostic material for leukaemia and blood disorders. The main sources of data are abstracted from the salary lists, job registry form, medical and physical exam records, and records of benzene measurement in the factories.

TYPE: Cohort
TERM: Occupation; Solvents
SITE: Leukaemia
CHEM: Benzene
TIME: 1986 - 1994

137 **You, W.C.** 04810

Beijing Inst. for Cancer Research, Da-Hong-Luo-Chang St., Beijing 100034, China (Tel.: 6037079)

COLL: Li, J.Y.; Jin, M.L.; Zhang, L.; Chang, Y.S.; Zhao, L.; Zeng, X.R.; Liu, W.D.; Yang, B.Q.; Hu, S.; Han, Z.X.; Zhao, R.X.; Liu, X.Q.; Xu, G.W.; Blot, W.J.; Kneller, R.; Keefer, L.

Precancerous Gastric Lesions in Relation to Stomach Cancer in China

The gastric cancer precursor study in Shandong province is a seven year study designed to investigate the aetiology of precancerous lesions of the stomach. A variety of possible aetiological factors, including diet and smoking are being investigated. In addition, the study attempts to quantify rates of transition in gastric mucosal status over time, including transition from chronic atrophic gastritis and dysplasia to cancer. Subjects are identified from population rosters in 14 randomly selected villages in five towns located in Linqu county, a rural area in Shandong, People's Republic of China. 3,434 subjects were interviewed and given a physical and endoscopic examination. Biopsies, blood, urine and gastric juice were collected from individuals in the study. The characteristics of prevalence rates of precancerous gastric lesions, their relationships with N-nitroso compounds in urine and gastric juice, micronutrients in serum and environmental factors are being investigated. This is a collaborative study between China and the United States. Papers appeared in J. Natl Cancer Inst. 84:1261-1266, 1992, Cancer Res. 53:1317-1321, 1993 and in Cancer Epidemiology, Biomarkers & Prevention 2:113-117, 1993

TYPE: Cohort
TERM: Atrophic Gastritis; BMB; Diet; Micronutrients; Nutrition; Premalignant Lesion; Serum; Tissue; Tobacco (Smoking); Trace Elements; Urine
SITE: Stomach
CHEM: Ferritin; N-Nitroso Compounds
TIME: 1989 - 1995

138 **Zhang, S.Z.** 04765

Lab. of Industrial Hygiene, Ministry of Public Health, Div. of Radiation Health, 2 Xinkang St., Deshengmenwai, Beijing 100088, China (Tel.: 1166-238)

COLL: Sun, Q.; Tao, Z.

Cancer Mortality in a Steel Company

Cases comprise male workers who died of liver cancer (51 cases) and stomach cancer (42 cases) in Baotou Steel-Iron Company during the period 1972-1987. Two controls were individually matched to each case on the basis of nationality, sex, age and year of death. All cases and controls were listed from the death registry in the company, and their occupational files were then reviewed; wives, parents or children were also interviewed to collect detailed information on factors such as occupational exposure to X-ray and inhaled thorium-containing dusts, as well as smoking, alcohol history, medical X-ray, diet, and socio-economic status. Odds ratios will be used as the measure of association between the occupational risk factors and cancers studied. The purpose of the study is to explore whether the occupational factors, especially radiation exposure, lead to an increased risk of liver cancer and stomach cancer.

TYPE: Case-Control
TERM: Alcohol; Diet; Dusts; Occupation; Radiation, Ionizing; Socio-Economic Factors; Tobacco (Smoking)
SITE: Liver; Stomach
CHEM: Thorium
TIME: 1989 - 1994

139 **Zhang, S.Z.** 05033

Lab. of Industrial Hygiene, Ministry of Public Health, Div. of Radiation Health, 2 Xinkang St., Deshengmenwai, Beijing 100088, China (Tel.: 1166-238)

COLL: Wang, Q.L.; Tin, D.Y.

Lung Cancer Risk from Indoor Radon and Its Daughters

A case-control study is being carried out to evaluate the lung cancer risk from indoor radon and its daughters. Approximately 400 cases of lung cancer diagnosed during the period 1981-1990 will be included in the study, and an equal number of controls living in the same areas selected at random will be matched by age, sex and calendar year of death. A personal interview will obtain information on socio-demographic and environmental factors. The level of radon, radon daughters and gamma radiation in parts of dwellings will be measured. The results of primary measurements indicated that the concentration of radon reaches 50-300 Bq/M3. Logistic regression analysis will be used to examine the relationship between lung cancer risk and radon exposure in the home.

TYPE: Case-Control
TERM: Radiation, Ionizing
SITE: Lung
CHEM: Radon
TIME: 1990 - 1995

140 **Zhang, Z.W.** 04006
Beijing Inst. for Cancer Research, Da-Hong-Luo-Chang St., Beijing 100034, China (Tel.: 6023658)
COLL: Gou, Y.R.; Zhao, L.; Song, W.Z.; Zhang, Y.

Psychosocial Factors and Cancer: A Case-Control Study

In this case-control study 200 cases with stomach cancer and 200 with breast cancer diagnosed after July 1986 in the Beijing area will be selected. 400 healthy individuals and 400 other cancer patients will be selected as controls. Each subject will be investigated using a questionnaire on personality and life events. The study aims to investigate: (1) the relationship between personality type and cancer incidence; (2) the personality and characters of cancer patients and (3) the relationship between life events and cancer incidence.

TYPE: Case-Control
TERM: Psychological Factors
SITE: All Sites; Breast (F); Stomach
TIME: 1986 - 1994

141 **Zhang, Z.W.** 04585
Beijing Inst. for Cancer Research, Da-Hong-Luo-Chang St., Beijing 100034, China (Tel.: 6023658)
COLL: Guo, Y.R.; Zhao, L.; Song, W.Z.; Zhang, Y.

Psychosocial Factors and Cancer: A Cohort Study

In this prospective study, 15,000 people living in the Beijing area will be observed for at least five years for cancer incidence. Their age is from 30 to 60 years, and they have middle or higher education degrees. Personality type and life events of every subject will be investigated and scored in the first year. Cancer incidence of each personality type (including depression, anxiety, social control etc.) will be compared to estimate the combinative action of life events. This is the first prospective study on psychosocial factors and cancer in China. This project is in cooperation with the National Institute of Psychology and Beijing Public Security Bureau. One characteristic is to use local Security Policemen as investigators. The aim is to investigate the relationship between cancer incidence and personality type, and to develop new methods of cancer prevention and treatment.

TYPE: Cohort
TERM: Psychosocial Factors
SITE: All Sites
TIME: 1989 - 1994

Changchun

***142** **Lin, X.W.** 05513
Norman Bethune Univ. of Med. Sciences, School of Preventive Medicine, 7, Xinmin St., Changchun 130021, China (Tel.: 645911-411)
COLL: Fan, H.X.; Guo, S.P.; Li, Y.H.

Investigation of Cancer Clusters in Yong Ji County, JiLin Province

Yong Ji county of JiLin is a severely polluted area along the Songhuea river, which has been polluted by industrial waste for nearly 30 years. Cancer mortality in the region has been studied retrospectively during 1984 to 1986 and revealed a cluster of cancer deaths along the river and only sporadic events in regions far from the river. The mortality data were analysed by Poisson model. The investigation is continuing.

TYPE: Mortality
TERM: Cluster; Geographic Factors; Waste Dumps; Water
SITE: All Sites
TIME: 1987 - 1994

Guangzhou

143 **Hu, M.X.** 04078
Sun Yat Sen Univ. Medical Sciences, Department of Medical Statistics & Community Medicine, Zhongshan Rd II, Guangzhou, China (Tel.: 778223; Tlx: 0074)
COLL: Liu, Q.

Nasopharyngeal Carcinoma in Sihui County
In order to study further the relationship between dietary habits, especially the consumption of salted fish, and nasopharyngeal carcinoma incidence in a high-risk population, another case-control study on nasopharyngeal carcinoma is being carried out in Sihui County. All the patients with nasopharyngeal carcinoma diagnosed with histological verification during the study period in this area are selected as cases. The controls are matched to the cases on the basis of age, sex and residence. Conditional logistic regression techniques will be applied to analyse the data.

TYPE: Case-Control
TERM: Diet
SITE: Nasopharynx
TIME: 1986 - 1994

144 **Li, C.C.** 03657
Tumor Hosp., Cancer Inst., Sun Yat Sen Univ. Medical Sciences, Head & Neck Dept., Dongfeng 4th Rd East, Guangzhou, China (Tel.: 778223/2792)
COLL: Liu, M.Z.

Serum VCA-IgA and MA-IgA Investigation for Nasopharyngeal Cancer (NPC) in a High Risk Area
In a prospective study in a high risk area of Guangdong, sera from normal individuals and from persons with nasopharyngeal mucosal hyperplastic lesions (NPHL) will be taken. VAC-IgA and MA-IgA antibody investigations will be carried out by IE or IF tests using genetically engineered antigen as a diagnostic antigen. 100 persons with positive VCA-IgA and MA-IgA antibody results and 100 persons with negative results, as well as the individuals presenting with NPHL will be followed-up for a five-year period, to determine the occurrence of NPC in different individuals.

TYPE: Cohort
TERM: Antibodies; Premalignant Lesion; Sero-Epidemiology
SITE: Nasopharynx
TIME: 1986 - 1994

Hangzhou

145 **Jiao, D.A.** 05226
Zhejiang Medical Univ., Cancer Inst., 68# Jiefang Rd, Hangzhou 310009, China (Tel.: +86 571 727427; Fax: 771571 ; Tlx: 35036 zmu cn)
COLL: Chen, K.

Assessment of Validity and Reliability of Retrospective Dietary Survey Method
Dietary factors are very important risk factors or protective factors for cancers and cardiovascular diseases. A valid and reliable survey method is an urgent requirement. A quantitative food survey method has been designed. From the quantity of food and size of portions taken by the subjects, the nutritional elements (e.g. protein, fat, crude fibre, vitamins) are calculated. Assessment of the validity and reliability of this method is being carried out. A total of 233 research subjects are included (105 males, 128 females).

TYPE: Methodology
TERM: Diet; Nutrition
SITE: All Sites
TIME: 1988 - 1994

146 **Yang, G.** 05225
Zhejiang Medical Univ., Cancer Inst., 68# Jiefang Rd, Hangzhou 310009, China (Tel.: +86 571 727427; Fax: 771571 ; Tlx: 35036 zmu cn)
COLL: Zheng, S.; Yu, H.; Ma, X.Y.; Jiao, D.A.; Yuan, Y.C.; Yao, K.Y.; Zhou, L.; Chen, K.

Family Aggregation and Genetic Factors in Common Cancers

The aim of this population-based case-control study is to assess the relationship between several common cancers in China (including stomach, colorectum, liver, breast, lung) and their genetic background, as well as the interrelation between environmental and genetic factors. 3,200 cancer cases diagnosed in the period 1987-1991 will be compared with 3,200 non-consanguineous controls matched for sex, age and residence. Information collected includes pedigree, sex, age, history of residence, residence of family, diet, smoking habits and family disease history (especially history of cancer). Information is collected from cases and their spouses, their first-degree relatives and their second-degree relatives. The analysis indicators are the cumulative incidence rate, familial aggregation, relative risk, heritability and genetic model.

TYPE: Case-Control; Genetic Epidemiology
TERM: Diet; Environmental Factors; Familial Factors; Genetic Factors; Genetic Markers; Geographic Factors; Multiple Primary; Pedigree; Segregation Analysis; Tobacco (Smoking)
SITE: Breast (F); Colon; Liver; Lung; Rectum; Stomach
TIME: 1991 - 1995

147 **Yu, H.** 05230
Zhejiang Medical Univ., Cancer Inst., 68# Jiefang Rd, Hangzhou 310009, China (Tel.: +86 571 727427; Fax: 771571 ; Tlx: 35036 zmu cn)
COLL: Zheng, S.; Jiao, D.A.; Chen, K.; Yang, G.; Zhou, L.; Yuan, Y.C.; Ma, X.Y.; Shao, Y.W.

Intervention Trial of Colorectal Cancer in a High Incidence Area

Jiashan County of Zhejiang Province has had the highest mortality rate of colorectal cancer in China (22.65/100,000 in 1974-1976, 21.30/100,000 in 1986-1989). Epidemiological studies previously conducted in the county showed that dietary habits, contaminated drinking water, parasite infestation and low selenium content in soil and grains were closely associated with colorectal cancer. The county is randomly divided into trial and control areas with attention given to avoid contamination of control areas from trial areas. The intervention measures include health education through community and mass media, safe drinking water supply, prevention and treatment of parasite diseases (schistosomiasis and hookworm infestation), provision of selenium-salt in trial area. The receptive rate of health education, prevalence rate of parasitic diseases, the consumption rate of selenium salt and the compliance with advice on healthy lifestyles are used as intermediate indicators for evaluation. The final indicator for evaluation will be the reduction in incidence rate and mortality from colorectal cancer as well as cancers of all sites. An economic evaluation (cost-benefit and cost-effectiveness analysis) will also be made.

TYPE: Intervention
TERM: Cost-Benefit Analysis; High-Risk Groups; Lifestyle; Parasitic Disease; Water
SITE: Colon; Rectum
CHEM: Selenium
TIME: 1991 - 2000

148 **Zheng, S.** 05227
Zhejiang Medical Univ., Cancer Inst., 68# Jiefang Rd, Hangzhou 310009, China (Tel.: +86 571 727427; Fax: 771571 ; Tlx: 35036 zmu cn)
COLL: Wu, J.M.; Yu, H.; Jiao, D.A.; Chen, K.; Zhou, L.; Ma, X.Y.; Shoa, Y.W.; Yuan, Y.C.; Sun, Q.R.; Shu, W.X.; Tang, H.X.

Screening for Colorectal Cancer in China

The aim of this study is to establish a practical and cost-effective screening model for colorectal cancer in China. Based on results of previous laboratory and epidemiological studies, a screening programme is being implemented in a high incidence area. More than 70,000 residents (30+ years old) are recruited for screening. In this programme, the combination of immunological faecal occult blood test and computerized risk assessment for screenees are being used as a primary screening procedure, fiberoptic sigmoidoscopy as the secondary screening and the full colonoscopy and/or air-barium enema as the diagnostic procedure. Evaluation of the efficacy of the programme will be based on whether there is a change of mortality from cancer of all sites as well as in mortality from colorectal cancer; a cost-benefit and cost-effectiveness analysis will also be done.

TYPE: Cohort
TERM: Cost-Benefit Analysis; Screening
SITE: Colon; Rectum
TIME: 1986 - 1995

149 Zheng, S. 05229
Zhejiang Medical Univ., Cancer Inst., 68# Jiefang Rd, Hangzhou 310009, China (Tel.: +86 571 727427; Fax: 771571 ; Tlx: 35036 zmu cn)
COLL: Jiao, D.A.; Yu, H.; Chen, K.; Zhou, L.; Yang, G.; Lai, K.D.; Cheng, Z.Y.; Gao, P.; Xia, L.F.; Yang, W.X.

Colorectal Cancer in Six Regions of China

The incidence and mortality of colorectal cancer varies greatly in different areas of China. The hypothesis is that this variation is due to differences in lifestyle and dietary habits. To verify the hypothesis, a population-based case-control study of colorectal cancer is conducted in six regions of China: Dalian, Shenyang (Northeast), Yingchuan (Northwest), Guiyang (Southwest), Zhengzhou (Middle) and Hangzhou (East). The study includes 200 pairs of colorectal adenocarcinoma cases and controls (1:1 match) from each region (1,200 pairs in total). Past lifestyle, including diet, food energy intake, physical activity, family history and reproductive history (females) are assessed. A comparison of risk factors will be made between different regions.

TYPE: Case-Control
TERM: Alcohol; Diet; Family History; Lifestyle; Physical Activity; Reproductive Factors; Tobacco (Smoking)
SITE: Colon; Rectum
TIME: 1988 - 1994

150 Zhou, L. 05228
Zhejiang Medical Univ., Cancer Inst., 68# Jefang Rd, Hangzhou 310009, China (Tel.: +86 571 727427; Fax: 771571 ; Tlx: 35036 zmu cn)
COLL: Yu, H.; Chen, K.; Yang, G.; Zheng, S.

Risk Factors for Colorectal Cancer

A case-control study is being conducted to study risk factors in populations at high risk for developing colorectal cancer. All patients detected in a screening programme carried out in Jia Shan County are included in the study. Information on diet, medical history, family history, etc. is being collected. Cases will be matched with controls for age, sex and residence. Conditional regression analysis will be performed.

TYPE: Case-Control
TERM: Diet; Family History; High-Risk Groups
SITE: Colon; Rectum
TIME: 1991 - 1994

Harbin

151 Dai, X.D. 03787
Harbin Medical Univ., Cancer Research Inst., Dept. of Cancer Epidemiology, Ha-Ping Rd, Harbin 150040, China (Tel.: 65003)
COLL: Wan, X.R.; Lin, C.Y.; Zhang, C.Y.; Gao, Y.J.; Jiang, J.S.

Nasopharyngeal Carcinoma in North China

This is a comparison study of epidemiological characteristics and aetiological factors related to NPC in endemic and non-endemic regions. The study relates to the low-incidence areas. (1) In a case -control study 100 patients with NPC will be compared with 200 matched controls, 100 patients with other tumours and 100 normal individuals. The EBV antibodies of serum will be determined by immunoperoxidase and immunofluorescence assay. (2) Chromosomal analysis of tumours of different histological subtypes will be undertaken. (3) Differences in patterns of smoking, exposure to potential carcinogens, etc. will be explored in patients with the more differentiated NPCs and those with undifferentiated NPCs. (4) The role of genetics in NPC will be investigated by collecting pedigrees from multiple-case NPC families. In a case-control study, 100 patients with NPC will be matched with 100 patients with tumours of the head and neck. All these patients must belong to families of which the last three generations were born in Heilongjiang Province.

TYPE: Case-Control
TERM: Antibodies; Chemical Exposure; Genetic Factors; Infection; Pedigree; Tobacco (Smoking); Virus
SITE: Nasopharynx
TIME: 1985 - 1994

152 **Dai, X.D.** 05147
Harbin Medical Univ., Cancer Research Inst., Dept. of Cancer Epidemiology, Ha-Ping Rd, Harbin 150040, China (Tel.: 65003)
COLL: Sun, X.W.; Liu, C.B.; Shi, Y.B.; Len, C.Y.

Risk Factors for Lung Cancer among Women in North-East China

A case-control study among women in Harbin is being carried out. The study aims to investigate the role of smoking and other risk factors in the aetiology of lung cancer in women. The cases include all newly diagnosed primary lung cancer in females aged less than 70 in the study areas, between 1991 and 1994. A population control will be selected for each case, matched on sex and five-year age group. A pre-coded questionnaire will be used. Information about demographic factors, active and passive smoking exposure, lifetime residential and occupational histories, diet and cooking practices, personal history of non-malignant lung diseases, history of tuberculosis and cancer of lung in first-degree relatives will be collected. Logistic regression will be used for the analysis.

TYPE: Case-Control
TERM: Coal; Cooking Methods; Diet; Female; Occupation; Passive Smoking; Tobacco (Smoking)
SITE: Lung
TIME: 1991 - 1995

153 **Hu, J.** 04770
Harbin Medical College, Dept. of Epidemiology, 41 Da Zhi St., Harbin 150001, China
COLL: Yunyuan, L.

Estimate of Latency in Stomach Cancer

This study will estimate the latency of stomach cancer caused by smoking and drinking and provide a theoretical basis for prevention. Data on smoking, drinking and diet for different time periods will be obtained by interview in about 200 cases with histologically confirmed cancer of the stomach and 400 controls. Analysis by logistic regression for risk factors will be carried out together with the estimation of latency of stomach cancer.

TYPE: Case-Control
TERM: Alcohol; Diet; Latency; Tobacco (Smoking)
SITE: Stomach
TIME: 1987 - 1994

154 **Hu, J.** 05068
Harbin Medical College, Dept. of Epidemiology, 41 Da Zhi St., Harbin 150001, China
COLL: Liu, Y.Y.

Estimation of Latency of Lung Cancer Caused by Smoking

This study will estimate the latency of lung cancer caused by different types of cigarette. Data concerning smoking were obtained by interview and questionnaire in about 170 cases with histologically confirmed cancers of the lung and about 340 controls. By means of calculating accumulated effective exposure score and excess exposure fraction, the latency of lung cancer will be estimated using a multivariate model.

TYPE: Case-Control
TERM: Latency; Tobacco (Smoking)
SITE: Lung
TIME: 1987 - 1994

***155** **Hu, J.** 05406
Harbin Medical College, Dept. of Epidemiology, 41 Da Zhi St., Harbin 150001, China
COLL: Liu, Y.Y.

Diet and Lung Cancer

A case-control study is carried out in Harbin city to investigate the role of diet in the aetiology of lung cancer. Over 200 incident cases and the same number of controls will be interviewed in hospital wards. Data on average consumption (frequency and amount) of single food items will be obtained through a dietary history questionnaire. Both univariate and multivariate analyses will be carried out, producing odds ratio estimates.

TYPE: Case-Control
TERM: Diet
SITE: Lung
TIME: 1992 - 1995

156 **Yu, Z.F.** 04118
Harbin Medical Univ., Dept. of Epidemiology, 41 Dong Da Zhi St., Harbin 150001, China (Tel.: 341157)
COLL: Li, L.; Dong, J.

Aetiological Study of Cervical Cancer in North China

Cervical cancer, despite its declining mortality in China, is still frequent. In order to study the aetiology of cervical cancer, a case-control study is being carried out at nine hospitals in North China. All new cases diagnosed over a one year period will be identified (200 expected) and controls (400) matched with cases according to age at the time of diagnosis are selected from hospitals in the same region for each case. A detailed questionnaire will elicit information on genital hygiene, sexual activity, socio-economic status, familial cancer history, cigarette smoking habits, chronic cervicitis and other risk factors. Logistic regression methods will be used to analyse the data. In addition, trends in mortality from cervical cancer over some 20 years in the Daowai district of Harbin city will be analysed by birth cohort.

TYPE: Case-Control; Mortality
TERM: Birth Cohort; Reproductive Factors; Sexual Activity; Socio-Economic Factors; Tobacco (Smoking); Trends
SITE: Uterus (Cervix)
TIME: 1982 - 1994

157 **Yu, Z.F.** 05180
Harbin Medical Univ., Dept. of Epidemiology, 41 Dong Da Zhi St., Harbin 150001, China (Tel.: 341157)
COLL: Dong, J.; Pu, L.M.; Love, E.J.

Risk Factors for Breast Cancer with Different Histological Types

A case-control study will be carried out to investigate the aetiology of breast cancer of different histological types. The cases will be in-patients with newly diagnosed and histologically confirmed breast cancer, selected from the seven major hospitals in Harbin and the major hospitals in the other large cities of the Heilongjiang. Two controls will be matched to each case on age (within 2.5 years) and usual place of residence. The interview questionnaire will include environmental, genetic, menstrual, marital, dietary, and reproductive factors, history of breast feeding, family history etc. Conditional logistic regression will be used in the analysis.

TYPE: Case-Control
TERM: Diet; Environmental Factors; Genetic Factors; Histology; Reproductive Factors
SITE: Breast (F)
TIME: 1991 - 1994

Hefei

***158** **Chen, R.L.** 05305
Anhui Medical Univ., School of Public Health, Hefei 230032, China
COLL: Wei, L.; Chen, R.; Coggon, D.

Lung Cancer in Copper Miners

In a previous cohort study an excess of lung cancer mortality was found in copper miners (O:60; E:40.75; SMR = 147, $p < 0.01$). To further explore the aetiology of lung cancer in copper miners, a case-control study is being carried out within the cohort. Lung cancer cases are obtained from the cohort, including those still alive or those deceased after 1980 (cases deceased before 1980 will be excluded to decrease recall bias). More than 80 case will be included. Each case is matched to four controls, randomly chosen from the cohort and matched for age (+/- 2 years) and date of death (<2 years). For deceased cases two of the four controls must be deceased. Information is obtained on occupational history, lifestyle (smoking and drinking habits, economic status, area of residence, cooking methods, diet, medical history and family history of cancer). Logistic regression model will be used to carry out multivariate analysis. Sampling and chemical analysis of ore dust from the mines will also be done.

TYPE: Case-Control
TERM: Dusts; Metals; Mining; Occupation; Tobacco (Smoking)
SITE: Lung
CHEM: Copper; Iron; Silica
OCCU: Miners, Copper
TIME: 1991 - 1994

Henyang

159 **Fan, J.X.** 03488
Ministry of Nuclear Industry, Hunan No 415 Hospital, Huan Henyang, China (Tel.: 22851)
COLL: Wang, L.H.; Zhou, C.T.; Nie, G.H.; Huang, Y.H.

Primary Lung Cancer in Uranium Miners

A number of retrospective epidemiological investigations aim to identify the relationship between lung cancer in uranium miners and exposure to radon decay products, to provide a basis for protection of workers in uranium mines. However several questions remain unanswered, in particular, the relationship between the risk of cancer and the cumulative radiation exposure in terms of working level-months. Uranium miners in the Middle-South of China will be grouped into cohorts. Information required includes: date of birth, date of entry into mine, work history, date of leaving mine, health and cumulative exposure doses (WLM). Those suffering from lung cancer will be ascertained with date of diagnosis, tumour histology type, X-ray appearance, presence of metastases and date of death being recorded. About 15,000 people exposed to radon and its daughters are available for study. The control group is to be made up of general regional populations. A paper has been published in Huazhong J. Ind. Hyg. 1:23, 1985.

TYPE: Cohort
TERM: Histology; Metals; Occupation; Radiation, Ionizing; Time Factors
SITE: Lung
CHEM: Radon; Uranium
OCCU: Miners, Uranium
TIME: 1985 - 1994

Kaohsiung

***160** **Chang, W.Y.** 05260
Kaohsiung Medical College, Dept. of Internal Medicine, Hepatobiliary Div., 100 Shih-Chuan 1st Rd, Kaohsiung 80708, Taiwan, China (Tel.: +886 7 3121101/6014; Fax: 3213931)
COLL: Chen, S.C.; Wang, L.Y.; Hsieh, M.Y.; Chuang, W.L.; Liu, T.Y.

Screening of Hepatocellular Carcinoma in High Risk Groups: A Family Study

The objectives of the study are: (1) to investigate whether the family members of patients with hepatocellular carcinoma (HCC) have an increased risk of developing HCC compared with age- and sex-matched controls from a community survey, and (2) to assess the possible risk factors for the developement of HCC in these subjects. Approximately 2,000 family members of HCC patients and 2,000 controls will be studied. Serum alphafoetoprotein, hepatitis B surface antigens and anti-hepatitis C antibodies will be measured. All the subject will have abdominal sonography and will be interviewed by trained interviewers with a structured questionnaire. Univariate and multivariate methods will be used in the statistical analysis.

TYPE: Cohort
TERM: BMB; HBV; HCV; Screening
SITE: Liver
TIME: 1992 - 1994

Nanjing

161 **Ding, J.H.** 04716
Jangsu Cancer Research Inst., Dept. of Epidemiology, No 42, Bei Zi Ting Rd , Nanjing 210009, China (Tel.: 713125)
COLL: Lin, Y.T.

Risk Factors for Colorectal Cancer

This case-control study aims to identify risk factors associated with large bowel cancer and to compare differences in the risk factors between colon and rectum cancer. 100 colon and 100 rectum cancer cases, newly diagnosed, will be interviewed using a specially designed questionnaire to obtain information on items such as dietary habits, lifestyle, medical history and familial cancer history. Individuals with diseases other than cancers and digestive system diseases, matched individually with cases for age and sex, will be selected from local hospitals to serve as controls. Methods of analysis will include Mantel-Haenszel analysis and conditional logistic regression analysis.

TYPE: Case-Control
TERM: Diet; Familial Factors; Lifestyle
SITE: Colon; Rectum
TIME: 1989 - 1994

162 **Jin, Z.G.** 05112
Inst. of Occupational Medicine of Jiang-Su Province, 212 He-Ban-Chun, Mia-Gao-Qiao, Nanjing 210038, China (Tel.: 651176)
COLL: Xue, S.Z.

Relationship between Bladder Cancer and Chlordimeform

The carcinogenic potential of chlordimeform (CDM) and its metabolites has been suspected since the 1970's. An increased incidence of carcinomas of the urinary bladder among workers exposed to p-chloro-o-toluidine, the main metabolite of CDM in humans, has been reported. A preliminary case-control study (55 bladder cancer cases and 55 controls) has been carried out to assess the relationship between bladder cancer and the use of CDM by pesticide operators. Preliminary results show that there is no relationship between CDM and bladder cancer after CDM has been used in the area for 10 years, but there is a significant relationship between smoking and bladder cancer. Recruitment will be continued for a further five-year period.

TYPE: Case-Control
TERM: Occupation; Pesticides
SITE: Bladder
CHEM: Chlordimeform
OCCU: Pesticide Workers
TIME: 1988 - 1995

163 **Jin, Z.G.** 05113
Inst. of Occupational Medicine of Jiang-Su Province, 212 He-Ban-Chun, Mia-Gao-Qiao, Nanjing 210038, China (Tel.: 651176)
COLL: Tao, B.; Xue, S.Z.; Lin, M.

Cancer and Polymorphism of N-Acetyltransferases in Chinese

N-acetyltransferase polymorphism is related to various types of cancer, for example, urinary bladder cancer to slow acetylator phenotype, colon cancer to rapid phenotype, etc. There is a significant difference in the gene frequency of N-acetyltransferases between Chinese and other races. This study will examine the difference between N-acetylation ratio of patients with cancer and the general population in areas of high cancer mortality in China, for example, hepatic cancer in Qin Dong county, oesophageal cancer in Huai-An county. A local, genetically balanced group of 300 students, and 50-100 patients with cancer will be assessed by chemical methods for their acetylation type. The study will test whether there is an association between specific cancer risks and the Chinese genetic make-up.

TYPE: Cross-Sectional; Molecular Epidemiology
TERM: Chinese
SITE: Liver; Oesophagus; Stomach
TIME: 1991 - 1994

CHINA

Nanning

164 **Liu, Q.F.** 04548

Cancer Institute of Guangxi, 6 Bin-hu Rd, Nanning Guangxi, 530027, China (Tel.: 21477344)
COLL: Mo, C.C.; Zhong, S.C.; Ran, R.Q.; Yeh, F.S.

Randomised Trial of Green Tea and Herbs in Prevention of Primary Hepatocellular Carcinoma in High Risk Area

PHC is closely related to aflatoxin B1 intake from food and to hepatitis B infection in Guangxi. Green tea, as well as certain Chinese traditional herbs, seems to have an inhibitory effect on hepatocarcinogenesis induced by aflatoxin B1 in rats. The aim of this follow-up study is to confirm the role of the factors suspected of involvement in PHC and the potential inhibitory effect of green tea and herbs. Since 1986, more than 15,000 serum specimens have been collected from the inhabitants of a high incidence area for PHC in Guangxi and tested for HBsAG, anti-HBs and anti-HBc. Meanwhile, urine specimens from children as well as adults have been quantitatively assayed by HPLC for aflatoxin M1. According to the results of the examinations, 3,000 individuals, considered to be at high risk for PHC were randomly divided into three subcohorts. All the members of each subcohort were routinely given tablets of green tea, radix salvia miltiorrhizae (a Chinese traditional herb) and a placebo, respectively. The risk of developing PHC between the three subcohorts will be compared in follow-up.

TYPE: Intervention
TERM: HBV; Mycotoxins; Tea
SITE: Liver
CHEM: Aflatoxin
TIME: 1986 - 1994

Shanghai

165 **Gao, R.N.** 02754

Shanghai Cancer Inst., Dept. of Epidemiology, 2200 Xie Tu Rd, Shanghai 200032, China (Tel.: +86 21 376550)
COLL: Gao, Y.T.; Guo, B.C.; Wu, X.N.; Yu, E.X.

Primary Liver Cancer in Shanghai

The study aims to determine the association of primary liver cancer with HBV infection and other known or suspected factors. 150 in-patients with primary liver cancer in the urban area of Shanghai will be interviewed. Two patient controls (other cancer and other diseases with the exception of liver diseases) will be matched for each case in the same hospital by sex, age (+/- five years) and place of residence. Sera of cases and controls will be tested for various HBV markers (HBsAG, Anti-HBs, Anti-HBc, HBeAG, Anti-HBe). Information on occupation, smoking habits, alcohol consumption, previous liver diseases, and family cancer history will also be collected by trained interviewers. The results will be expressed in the relative risk of these factors to primary liver cancer.

TYPE: Case-Control
TERM: Alcohol; Familial Factors; Infection; Sero-Epidemiology; Tobacco (Smoking); Virus
SITE: Liver
TIME: 1982 - 1994

166 **Gao, R.N.** 04436

Shanghai Cancer Inst., Dept. of Epidemiology, 2200 Xie Tu Rd, Shanghai 200032, China (Tel.: +86 21 376550)
COLL: Tu, J.T.; Yang, G.; Ji, B.T.; Gao, Y.T.; Xu, G.X.; Mo, S.Q.

Colorectal Cancer in Shanghai

This population-based case-control study of colorectal cancer aims to examine the association of colorectal cancer of different anatomical locations and histological types with specific nutrients, food consumption, physical activity, reproductive history, hormone use, family cancer history, bowel disease history, psychological factors, exposure to tobacco, alcohol, occupation, radiation, drugs, etc. About 1,600 newly diagnosed colorectal cancer cases aged 15-69 years among the permanent residents of Shanghai urban area will be identified through the Shanghai Cancer Registry during the period from September 1988 to August 1990. Controls will be randomly selected from the general population by age-sex frequency matching. All cases and controls will be interviewed by trained interviewers with a structured questionnaire. The data will be analysed by univariate and multivariate methods.

TYPE: Case-Control
TERM: Alcohol; Diet; Drugs; Familial Factors; Histology; Hormones; Nutrition; Occupation; Physical Activity; Psychosocial Factors; Radiation, Ionizing; Registry; Reproductive Factors; Tobacco (Smoking)
SITE: Colon; Rectum
REGI: Shanghai (Chi)
TIME: 1988 - 1994

167 **Gao, Y.T.** 05109
Shanghai Cancer Inst., Dept. of Epidemiology, 2200 Xie Tu Rd, Shanghai 200032, China (Tel.: +86 21 4315510; Fax: 4331428)
COLL: Ji, B.T.; Jin, F.; Blot, W.J.; McLaughlin, J.K.

Cancers of the Oesophagus, Colon, Rectum and Pancreas in Shanghai

The aim of the study is to identify risk factors for cancers of the oesophagus, colon, rectum and pancreas. Newly diagnosed cases of these cancers between 30-74 years of age among permanent residents of Shanghai urban area for the period 1 October 1990 to 30 September 1992 will be included in the study. The numbers of cases for each site to be interviewed are approximately 900 for oesophageal cancer, 900 each for colon and rectum cancers, and 460 for pancreatic cancer. Frequency matching will be used in the study according to the age and sex distribution of the cases of four cancer sites registered by the Shanghai Cancer Registry during 1984. 1500 controls will be randomly selected from the general population of Shanghai urban area. This control group will be used for each cancer site investigated in the study. A common questionnaire will be used for all study subjects, covering demographic factors, residential history, tobacco use, alcohol consumption, tea drinking, occupational history, dietary intake, medical history, family history of cancer, menstruation and reproductive history. Analysis of each cancer site and its subgroups (by histological type and anatomic site) will be conducted by logistic regression.

TYPE: Case-Control
TERM: Alcohol; Diet; Familial Factors; Menstruation; Occupation; Physical Activity; Registry; Reproductive Factors; Tea; Tobacco (Smoking)
SITE: Colon; Oesophagus; Pancreas; Rectum
REGI: Shanghai (Chi)
TIME: 1990 - 1994

168 **Liu, P.L.** 04308
Shanghai Medical Univ., School of Public Health, Dept. of Epidemiology, 138 Yi Xue Yuan Rd, Shanghai 200032, China (Tel.: +86 21 4311900/43; Fax: 4330543 ; Tlx: 33325 SFMC CN)
COLL: Jin, T.H.; Chen, G.C.

Sero-Epidemiological Study of HBV in a Population with High Incidence of PLC

PLC is one of the most common malignant neoplasms in China, approximately 100,000 people dying from the disease annually. Qidong, Nanhuei and Haimen counties are endemic areas with the highest mortality. This project is designed to investigate some 5,000 persons from these areas and to examine their HBV markers. The aim is to determine whether the highest mortality from PLC is accompanied by a high frequency of HBV markers, and to describe the distribution of HBV markers and immunity to HBV in the three counties. It is hoped that a relationship between PLC and hepatitis B infection will be established.

TYPE: Correlation
TERM: Infection; Virus
SITE: Liver
TIME: 1988 - 1994

169 **Liu, P.L.** 04879
Shanghai Medical Univ., School of Public Health, Dept. of Epidemiology, 138 Yi Xue Yuan Rd, Shanghai 200032, China (Tel.: +86 21 4311900/43; Fax: 4330543 ; Tlx: 33325 SFMC CN)
COLL: Wong, Y.F.

Role of Viruses in Cervical Cancer

The aim of the study is to evaluate a possible association between viral infections and cervical cancer. Viruses known and suspected to be involved in the aetiology of cervical cancer, including HPV, HSV-2 and CMV will be examined together with their interaction with other possible factors, such as oral contraceptive use and sexual behaviour. Blood and fresh tissue samples from biopsies or surgical specimens will be collected from about 100 untreated cases and 100 controls. Protocols will be developed to assess the presence of markers of exposure to potential carcinogens in the specimens. DNA hybridization methods will be used to assess the proportion of subjects who are positive for specific types of HPV. Serological markers will be used for HSV-2, CMV and other viruses.

TYPE: Case-Control
TERM: Biopsy; CMV; DNA; HPV; HSV; Oral Contraceptives; Sero-Epidemiology; Sexual Activity
SITE: Uterus (Cervix)
TIME: 1991 - 1994

170 **Shen, F.M.** 04314
Shanghai Medical Univ., School of Public Health, Dept. of Epidemiology, Yi Xue Yuan Rd 138, Shanghai 200032, China (Tel.: +86 21 4311900/216)
COLL: Haile, R.W.C.; Zheng, Z.S.

Genetic Epidemiology of Breast Cancer in Shanghai

The aims of the study are (1) to describe and quantify the observed family aggregation of breast cancer; (2) to carry out segregation analysis by fitting genetic models in order to examine the mode of inheritance; and (3) to perform genetic linkage analysis for those families in which cancer susceptibility is inherited as a single major gene locus. Interim matched case-control analysis with 113 cases (46 with bilateral and 67 with unilaterial involvement) and 227 controls (68 sisters of probands with breast cancer and 159 general population controls) indicated that family history, benign breast disease, obesity, early menarche and higher education increased the risk of breast cancer. Simple segregation analysis indicated that the main transmission model may be a recessive Mendelian gene model. On complex segregation analysis, while a dominant gene model could explain the distribution of synchronous premenopausal bilaterial breast cancer, a recessive Mendelian gene model predominates in all asynchronous bilateral breast cancers. The study is being continued to increase sample size and to address the molecular genetics as well.

TYPE: Case-Control; Genetic Epidemiology; Molecular Epidemiology
TERM: Age; Genetic Markers; Linkage Analysis; Menarche; Menopause; Parity; Registry; Segregation Analysis
SITE: Breast (F)
REGI: Shanghai (Chi)
TIME: 1987 - 1994

171 **Shu, X.O.** 04492
Shanghai Cancer Inst., 2200 Xie Tu Rd, Shanghai 200032, China (Tel.: +86 21 315360)
COLL: Gao, Y.T.; Zheng, W.; Yin, D.M.; Gao, R.N.; Jin, F.

Childhood Cancer in Shanghai

The general purpose of this case-control study is to determine the importance of various prenatal and childhood environmental exposures in the aetiology of childhood neoplasms such as leukaemia, brain tumour, lymphomas and cancer of liver and bone. Special emphasis will be placed on prenatal and early childhood exposures to chemicals, radiation, drugs, tobacco and alcohol, and also on the influence of high risk occupations and industries of the parents in the development of childhood cancer in their offspring. About 900 neoplasms diagnosed in children aged 0-14 between January 1980 and December 1989 will be identified through the Shanghai Cancer Registry. There will be about 300 leukaemia cases, 150 brain tumour cases, 100 lymphoma cases and 350 other cancer cases. An age-sex-frequency matched control group with 500 healthy children will be randomly selected from the general population. The mother and father of the study subjects will be interviewed in person. The information collected from the cases and controls will be compared, in order to identify factors associated with the development of different sites of childhood cancer. After 1990, the data collection will be continued for some rare childhood cancers so that the number of cases and controls will be large enough to evaluate the risk factors of these rare diseases as well. Clinical information for all cases will be abstracted from medical records to examine the prognostic factors of childhood cancer.

TYPE: Case-Control
TERM: Alcohol; Chemical Exposure; Drugs; Intra-Uterine Exposure; Occupation; Prognosis; Radiation, Ionizing; Registry; Tobacco (Smoking)
SITE: Bone; Brain; Leukaemia; Liver; Lymphoma
REGI: Shanghai (Chi)
TIME: 1988 - 1994

172 **Tao, X.G.** 04974
Shanghai Medical Univ., School of Public Health, Dept. of Environmental Health, 138 Yi Xue Yuan Rd, Shanghai 200032, China (Tel.: +86 21 4311900; Fax: 4330543 ; Tlx: 33325 sfmc cn)
COLL: Hong, C.J.; Yu, S.Z.

Risk of Male Lung Cancer Attributed to Indoor Coal Combustion

Lung cancer is now the leading cause of death from cancer among males in Shanghai. Indoor air pollution from coal combustion may make somecontribution. The purpose of this study is to explore the risk of lung cancer death in male residents who live in coal-using families. Stratified by two extreme levels of ambient sulphur dioxide and inhalable particulate concentrations, four areas were chosen in the city. In each of the areas, two neighbouring communities were chosen, one using coal and the other using coal gas as fuel. The percentage of smokers and the ambient environment of the two chosen communities in each area are similar. A total of 117,039 person-years was observed from 1 January 1978 to 31 December 1987 for all males in the eight communities. Standardized mortality rates for lung cancer among males in the eight communities are calculated. Relative risk and population attributable risk are being used to evaluate the risks.

TYPE: Cohort
TERM: Air Pollution; Coal; Male
SITE: Lung
CHEM: Sulphur Dioxide
TIME: 1987 - 1994

***173** **Tao, X.G.** 05418
Shanghai Medical Univ., School of Public Health, Dept. of Environmental Health, 138 Yi Xue Yuan Rd, Shanghai 200032, China (Tel.: +86 21 4311900; Fax: 4330543 ; Tlx: 33325 sfmc cn)
COLL: Zhu, H.G.; Yu, S.Z.

Air Pollution and Risk of Lung Cancer

The purpose of this study is to elucidate the reasons for the high incidence of lung cancer, especially among women and non-smokers, in Shanghai, and clarify the association between indoor and outdoor air pollution and lung cancer and their synergism with passive smoking on lung cancer in the city, controlling for possible confounders. All persons born before 1 January 1970 and living in 12 blocks provided with air pollutant monitoring stations, are included in the cohort (about 720,000 persons) and followed up for lung cancer from 1 January 1988 to 31 December 1993. The exposure observation period finished by the end of 1979. All lung cancer cases (all histological types) were interviewed with a questionnaire and information recorded on smoking, occupation, disease history, domestic fuel types, cooking methods, lifestyle, etc., before 1980. 5% of the total healthy population were also sampled by stratified, randomized cluster sampling methods, with Residential Group as the smallest cluster (with about 20 families each). These were interviewed, using the same questionnaire. Unconditional logistic regression model or Cox model will be used in the analysis.

TYPE: Case-Control; Cohort
TERM: Air Pollution; Cooking Methods; Passive Smoking; Tobacco (Smoking)
SITE: Lung
TIME: 1988 - 1994

174 **Tu, J.T.** 04352
Shanghai Cancer Inst., Dept. of Epidemiology, No. 25, Lane 2200, 2200 Xie Tu Rd, Shanghai 200032, China (Tel.: +86 21 375590/139)
COLL: Chen, H.Q.; Chen, X.L.

Cancer Screening Among Factory Workers

The benefit of mass screening for several cancer sites (breast, cervix, stomach, etc.) is largely accepted. The aim of this study is to evaluate the efficacy of a cancer screening programme conducted among about 100,000 factory workers. The target cancer sites are stomach, rectum, liver and lung in both sexes and breast and cervix in females. A standard questionnaire, including important risk factors for the target cancers, will be developed and used in interviewing. High risk individuals for different cancer sites will then be drawn out of the general population. Specified screening procedures will be used for early detection of different cancers. After the first mass screening, health monitoring for high risk individuals will subsequently be conducted by the trained doctors of the clinic in the factory. The length of the interval between mass screenings will be decided according to the staging of the new cases detected each year. All cancer cases at the target sites will be included in the analysis whether discovered by screening or not. The staging and survival time of cancer cases will be used as an intermediate index and changing cancer incidence and mortality as a final index for evaluating the efficacy of this cancer screening programme.

TYPE: Intervention
TERM: High-Risk Groups; Monitoring; Screening; Stage; Survival
SITE: Breast (F); Liver; Lung; Rectum; Stomach; Uterus (Cervix)
TIME: 1988 - 2003

175 **Wang, Y.** 05095
Chinese Academy of Sciences, Shanghai Inst. of Nuclear Research, P.O. Box 800204 , Shanghai 201800, China (Tel.: +86 21 9528476; Fax: 950021 ; Tlx: 30910 SINRS CN)

Trace Elements and Liver Cancer

This project on Chongming Island, a high incidence area for liver cancer with an uneven geographical distribution, investigates the relation between trace elements and liver cancer. Two communities with high liver cancer mortality (more than 40 per 100,000) and two with lower mortality ($20 per 100,000) were selected for the study. Samples of scalp hair and blood (5 ml) were collected three times, in March, July and November. Each time, 100 healthy adults aged between 20 and 40 (75 females, 25 males) were randomly selected from each community (600 samples in total). A case-control study of the association of liver cancer with trace elements is being done. 100 newly diagnosed cases of liver cancer among the permanent farmers of Chongmin Island will be identified during the study period. Controls of the same age, sex and similar economic background will be randomly selected from the general population on the island. All subjects are peasants from locations with no industrial contamination. The hair trace element content is analysed by X-ray fluorescence, with ICAP for serum analysis, and differential pulse polarography catalytic wave analysis for selenium. The data will be analysed by univariate and multivariate methods.

TYPE: Case-Control; Correlation
TERM: Trace Elements
SITE: Liver
CHEM: Cadmium; Copper; Iron; Lead; Selenium; Zinc
TIME: 1984 - 1994

176 **Xiang, Y.B.** 04577
Shanghai Cancer Inst., Shanghai Cancer Registry, 2200 Xie Tu Rd, Shanghai 200032, China (Tel.: +86 21 315360)
COLL: Gao, Y.T.; Jin, F.

Cancer in Shanghai: Relative Survival Analysis

Shanghai Cancer Registry has been collecting data on cancer patients since 1972. There are about 14,000 new cancer cases every year. Incidence, mortality and survival rates can be calculated. Survival analysis of lung cancer has been reported in China, but population-based survival analysis of all cancers has not been reported in detail, and relative survival rates have not been used. About 150,000 cancer cases were known to have died by 30 December 1987, the closing date for this analysis, but for about 40,000 cancer cases the vital status at this date is unknown, and they are being followed up to determine whether they are dead, alive or lost to follow-up. Using these data and the general population mortality data the observed and relative survival rates for cancer will be obtained to evaluate the results of activities for prevention and medical care of cancer in Shanghai. Comparisons by age and sex, and with other countries will be considered. Time trends in survival will also be analysed. Main methods of analysis are provided by a specialised statistics package of relative survival analysis.

TYPE: Incidence; Mortality
TERM: Registry; Survival; Trends
SITE: All Sites
REGI: Shanghai (Chi)
TIME: 1988 - 1994

177 **Xue, S.Z.** 04674
Shanghai Medical Univ., School of Public Health, Dept. of Occupational Health, Box 206, Shanghai 200032, China (Tel.: +86 21 4311900/214; Fax: 4330543 ; Tlx: 33325 SFMC CN)
COLL: Wang, Z.Q.; Zhou, D.H.

Lung Cancer Incidence Among Employees Exposed to Low Levels of Technical CMME

A previous study on lung cancer mortality among employees exposed to relatively high concentrations of technical CMME demonstrated an excess of lung cancer. To study whether lung cancer risk will be similarly raised under improved working conditions with much lowered CMME concentrations, a new cohort of 500 employees has been followed since the early 1970's. The results will be helpful in verifying the dose-response relationship and checking the effectiveness of preventive measures and the adequacy of the current exposure limit. Local mortality data are used as the basis for calculation of indirect SMRs. Data are collected through field survey and interview.

TYPE: Cohort
TERM: Dose-Response; Occupation; Prevention
SITE: Lung
OCCU: Chemical Industry Workers
TIME: 1988 - 1996

178 **Xue, S.Z.** 04807
Shanghai Medical Univ., School of Public Health, Dept. of Occupational Health, Box 206, Shanghai 200032, China (Tel.: +86 21 4311900/214; Fax: 4330543 ; Tlx: 33325 SFMC CN)
COLL: Jin, Z.G.; Tao, B.; Ling, M.

Susceptibility to Naphthylamine Carcinogenesis, and N-Acetyl-Transferase Polymorphism.

Thousands of tons of antioxidants A & D derived from naphthylamines (technical grade alpha- and beta-isomers) have been produced and used in rubber processing in China since the mid 1950's. Cases of urinary bladder cancer have rarely been reported. The phenotype of rapid N-acetyltransferase activity might be responsible for this. A cohort study of the incidence and mortality of urinary bladder cancer among the exposed workers and a survey on the polymorphism of N-Acetyltransferase in those people is designed. The size of the cohort is estimated at over 500 and the period of follow-up would be around 20-25 years. Data will be collected through interviewing and urine analysis. Preliminary results reveal that most of the general population subjects are rapid acetylators and the few cases of bladder cancer are all slow acetylators.

TYPE: Cohort
TERM: Analgesics; Genetic Factors; Occupation
SITE: Bladder
CHEM: 1-Naphthylamine; 2-Naphthylamine; Amines, Aromatic
TIME: 1991 - 1995

179 **Yuan, J.M.** 04581
Shanghai Cancer Inst., 2200 Xie Tu Rd, Shanghai 200032, China (Tel.: +86 21 315360)
COLL: Yu, M.C.; Gao, Y.T.; Henderson, B.E.

Nasopharyngeal Carcinoma in Shanghai

This population-based case-control study is designed to investigate the association of NPC with (1) intake of preserved food commonly used in Shanghai; (2)occupational exposure to dust, smoke and fumes; (3) active and passive exposure to tobacco smoke; (4) exposure to incense; and (5) history of chronic ear and nose diseases. Cases will be histologically confirmed incident cases of NPC among permanent residents of Shanghai Urban aged 15-74, diagnosed from 1 January 1987 to 31 December 1990 and identified through the Shanghai Cancer Registry. It is expected to interview about 800 cases. The mothers of all cases and controls under age 50 years who reside in Shanghai will be interviewed (mothers of about 150 cases and 150 controls). 800 controls will be randomly selected from the general population of Shanghai Urban by age-sex frequency-matching. All interviews will be conducted in person by four trained nurses using a structured questionnaire. All information will be analysed using univariate and multivariate methods.

TYPE: Case-Control
TERM: Diet; Dusts; Occupation; Passive Smoking; Registry; Tobacco (Smoking)
SITE: Nasopharynx
REGI: Shanghai (Chi)
TIME: 1988 - 1994

180 **Zheng, W.** 04354
Shanghai Cancer Inst., Dept. of Epidemiology, 2200 Xie Tu Rd 25, Shanghai 200032, China
COLL: Shu, X.O.; Dan, R.P.

Leukaemia in Urban Shanghai

A population-based case-control study is underway to investigate the risk factors for leukaemia in urban Shanghai. Approximately 500 new cases classified by FAB are expected to be identified through the Shanghai Cancer Registry during the period 1987-1989. A control will be selected for each case by multistage random sampling, matched on age and sex. Trained interviewers will use a standard questionnaire to obtain information on factors such as residential history, occupation and occupational exposure, drug usage, previous diseases, different sources of radiation exposure, dietary history and personal habits. The clinical information for all new cases diagnosed during 1985 to 1989 will be abstracted from medical records to examine the prognostic factors for leukaemia and to estimate the incidence rates for leukaemia of different histological cell types.

TYPE: Case-Control
TERM: Diet; Drugs; Histology; Lifestyle; Occupation; Radiation, Ionizing; Registry
SITE: Leukaemia
REGI: Shanghai (Chi)
TIME: 1987 - 1994

181 **Zheng, W.** 04586
Shanghai Cancer Inst., Dept. of Epidemiology, 2200 Xie Tu Rd 25, Shanghai 200032, China
COLL: Gao, R.N.; Shu, X.O.; Ji, B.T.; Gao, Y.T.

Case-Control Study of Cancers of the Oral Cavity, Nose, Nasal Sinus and Larynx

The objectives of this study are to (1) investigate the association of cancers of the oral cavity, pharynx (nasopharynx excluded), nose, nasal sinus and larynx with the consumption of tobacco and alcohol; (2) investigate the relationship of these cancers to dietary related variables, radiation exposure, dental hygiene habits and previous diseases; (3) identify high risk occupations, industries and occupational carcinogenic exposure. Incident cases under 75 years old for carcinoma of oral cavity, pharynx (200 cases), nose, nasal sinus (150 cases) and larynx (300 cases) will be identified through Shanghai Cancer Registry during the period from Jan. 1988 to Dec. 1989 in Urban Shanghai. An age-sex frequency-matched control group with 400 subjects will be randomly selected from the general population. Interviews will be conducted in person with a structured questionnaire.

TYPE: Case-Control
TERM: Alcohol; Diet; Hygiene; Occupation; Radiation, Ionizing; Registry; Tobacco (Smoking)
SITE: Larynx; Nasal Cavity; Oral Cavity; Pharynx
REGI: Shanghai (Chi)
TIME: 1988 - 1994

182 **Zhu, H.G.** 04977
Shanghai Medical Univ., School of Public Health, Dept. of Environmental Health, 138 Yi Xue Yuan Rd, Shanghai 200032, China (Tel.: +86 21 4311900/203; Fax: 4330543 ; Tlx: 3325 sfmc cn)
COLL: Tao, X.G.; Zhao, Q.Y.

Mutagenicity of Drinking Water and Risk of Stomach Cancer

It has been reported that the mutagenicity of drinking water increases progressively along the Huangpu River from the upper to the lower reaches. The positive samples (1983-1985) in both tap and raw water from the upper reaches are 0%, while those from the lower reaches are 100% and 96.9% respectively according to the Ames test. The purpose of this study is to assess the relationship between the mutagenicity of drinking water and the risk of male stomach and liver cancer, which now comprise about

45% of male cancer deaths among residents of Shanghai. Two communities have been chosen, one drinking tap water and another drinking river water directly from the upper reach area, and two further communities with the same practices from the lower reach area. 178,442 person-years were observed from 1 January 1984 to 31 December 1988, including all males born after 31 December 1959 in the four communities. Standardized mortality rates for stomach and liver cancer are calculated. Odds ratios from unconditional logistic regression with 14 independent variables are used to evaluate the risks of stomach and liver cancers when drinking lower reach water.

TYPE: Correlation
TERM: Mutagen; Water
SITE: Liver; Stomach
TIME: 1990 - 1994

Shenyang

***183** **Xu, Z.Y.** 05297
Liaoning Public Health and Antiepidemic Station, Environmental Epidemiology Branch, Ji-Xian St. 42-1, Shenyang 110005, China (Tel.: +86 24 364898; Fax: 363239 ; Tlx: 80363 LUCRP CN)
COLL: Blot, W.J.

Stomach and Lung Cancer and Occupational Exposure in the Iron-Steel Industry

This case-control study, carried out in an integrated iron-steel factory with 200,000 workers, will test the hypotheses that the following factors increase the risk of lung cancer: (1) co-existence of inhalable particles and organic volatile compounds in the air; (2)exposure to silica dust; and (3) working in particular areas (metal smelting, pouring etc.) or particular jobs (furnace repairing, casting etc.). It will also test the hypothesis that the following factors increase the risk of stomach cancer: (4) exposure to metal dusts and other dusts; (5) exposure to cutting fluids, benzo(a)pyrene and high temperatures and that risk varies with chemical composition (e.g. ferric oxide, iron ore) and in exposures combined with PAH. The study comprise 600 stomach cancer cases, 600 lung cancers cases and 1,000 controls. The controls were randomly selected from a computerized database (personnel data files) of active, retired and deceased workers and matched for sex and birth year. A structured questionnaire will be used to obtain information on demographic characteristics, residence, smoking habits, medical and lifetime occupational histories. Unconditional logistic models will be used to analyse the data.

TYPE: Case-Control
TERM: Coal; Dusts; Metals; Occupation; Petroleum Products
SITE: Lung; Stomach
CHEM: Benzo(a)pyrene; Iron; PAH; Steel
OCCU: Foundry Workers; Furnacemen; Metal Workers; Smelters; Steel Workers
TIME: 1991 - 1994

Taipei

***184** **Yang, C.S.** 05373
National Taiwan Univ., College of Medicine Graduate Inst. of Microbiology, 1, Sec. 1, Jen-Ai Rd, Taipei Taiwan, China (Tel.: +886 2 3562220; Fax: 3915180)
COLL: Cheng, Y.J.; Chen, C.J.; Hsu, M.M.; Chen, I.H.; Chen, J.Y.; Pan, W.H.; Chan, C.C.; Hildesheim, A.; Brinton, L.A.; Levine, P.H.; Hoover, R.N.

Nasopharyngeal Carcinoma in Taipei

This population-based case-control study of nasopharyngeal carcinoma aims to explore the role of genetic and environmental factors, such as familial aggregation, residential history, medical history, use of herbal medicine, exposure to tobacco and alcohol, occupation, food consumption, etc. About 400 newly biopsy-confirmed nasopharyngeal carcinoma patients, aged 75 years or younger and living in Taipei city/county for more than 6 months, will be recruited as the case group from the Otolaryngologic clinics of two general teaching hospitals in Taipei from July 1991 to February 1994. Controls matched for age, sex and residential area will be randomly selected from the records of local household registration offices. Cases and controls will be intereviewed by trained interviewers with a structured questionnaire. Biological samples (30 ml blood, spot urine, 12 hrs. overnight urine, toenail clippings and biopsied tissues) will also be collected and analysed for various biomarkers. The mothers of cases and controls will also be interviewed to obtain early childhood dietary history. The data will be analysed by multivariate methods, including conditional logistic regression analysis.

TYPE: Case-Control
TERM: Alcohol; BMB; Biomarkers; Blood; Diet; Familial Factors; Genetic Factors; Occupation; Plants; Tissue; Tobacco (Smoking); Toenails; Urine
SITE: Nasopharynx
TIME: 1991 - 1994

Taiyuan

185 **Sun, S.** 04803
China Inst. for Radiation Protection, Dept. of Radiation Medicine, P.O. Box 120, Taiyuan 030006, China (Tel.: +86 351 774005; Fax: 740707)
COLL: Wang, Y.; Li, W.; Yuan, L.; Li, S.

Epidemiological Studies of Workers in the Nuclear Industry in China

The nuclear industry has been operated for about 30 years in China. It is important to know the possible late effects among workers with a history of occupational exposure. Historical cohort studies were conducted in the following four sub-cohorts: uranium prospecting teams; miners in uranium mines; workers in uranium plants and nuclear power plants. The period of follow-up is 1971-1986; the total number of subjects about 40,000. SMR and RR were analysed in different dose groups. Preliminary results show that there is no significant difference of incidence of cancer between exposure and control groups in nuclear plants. The only finding was an increase of lung cancer in uranium miners.

TYPE: Cohort
TERM: Occupation; Radiation, Ionizing
SITE: All Sites
CHEM: Radon; Uranium
OCCU: Miners; Power Plant Workers
TIME: 1986 - 1994

Xilinhaote

186 **Latan, A.** 04192
Hospital of Xilinguole League, Dept. of Surgery, Xilinhaote, China

Gastric Cancer in Mongolian Herdsmen

The results of a mortality study carried out in 1977 showed an adjusted death rate for stomach cancer of 61.93 for Mongolian herdsmen compared with 15.93 for the general population. In the county of Xianghuang the adjusted death rate was as high as 107.25. The aim of the present study is to assess the relationship between dietary habits of Mongolian herdsmen and their high gastric cancer mortality. Their main diet consists of large amounts of Mongolian tea with milk, wheat porridge, dried meat, mouldy cheese and sour milk. Most of them are smokers and also heavy drinkers. Their annual consumption of green vegetables and fruit is very low. A preliminary study included 39 cases from Xianghuang and two control groups of similar size, one consisting of people who died from cardiovascular diseases, one of apparently healthy subjects. Controls were matched with cases for sex, age, socio-economic status and residence. Relatives of the deceased persons were interviewed, as well as the healthy people and medical records and life style investigated. Results of the preliminary study show an association between gastric cancer and intake of tea, dried meat, mouldy cheese, sour milk and alcohol, but not with smoked and salted food.

TYPE: Case-Control
TERM: Alcohol; Diet; Lifestyle; Tobacco (Smoking)
SITE: Stomach
TIME: 1977 - 1994

Zhengzhou

187 **Lu, J.B.** 02826
Henan Cancer Research Inst., Dept. of Epidemiology, Zhengzhou Henan Province, China (Tel.: +86 371 552149/461; Fax: 542452)
COLL: Muñoz, N.; Yang, C.S.

Epidemiological Studies of Oesophageal Cancer

The aim of this work is (1) registration of cancer and of death from all causes to enable incidence and mortality studies to be carried out in the population of Lin Xian and Yu Xian counties (1,600,000) and in 10 counties in Henan (population 7,000,000). (2) The natural history of oesophageal pre-cancerous lesions is being studied by observation of 400 cases of oesophagitis; in addition, intervention studies are carried out on 80 cases of dysplasia under treatment, 80 cases of dysplasia pending treatment, and 80 normal subjects acting as controls. Every year blood samples are collected from these subjects for determination of vitamin A, vitamin B, carotene, vitamin C, etc. (3) Studies on prevention of oesophageal cancer will be carried out in Lin Xian county. A paper has been published in Int. J. Cancer 36:645-645, 1985.

TYPE: Incidence; Mortality
TERM: High-Risk Groups; Premalignant Lesion; Prevention; Vitamins
SITE: Oesophagus
CHEM: Beta Carotene
TIME: 1975 - 1994

188 **Lu, J.B.** 04095
Henan Cancer Research Inst., Dept. of Epidemiology, Zhengzhou Henan Province, China (Tel.: +86 371 552149/461; Fax: 542452)
COLL: Sang, J.Y.; Dueng, W.Z.

Precursor Lesions of Oesophageal Cancer: A Prospective Study of Oesophageal Epithelial Dysplasia in Linxian County

In 1975, 19,250 subjects aged over 30 were examined cytologically in a screening programme for oesophageal cancer in the high-risk population of Linxian county. 958 subjects were selected from these for a prospective study. They were divided into three groups: group A, composed of 294 cases which were diagnosed through cytology as having severe dysplasia; group B, made up of 328 cases with mild dysplasia, and group C consisting of 336 subjects with a normal oesophagus. Age, sex and residence in all the three groups were matched. They will be followed-up for 15 years (1975-1990), and some of the suspected risk factors and co-factors for cancer such as smoking habits, type of water used and socio-economic status, will be investigated by interview on a continuing basis. Papers have been published in Clin. Oncol. 12:195, 1985; Chinese J. Cancer 6(4):250-253, 1987 and in Int. J. Cancer 41:805-808, 1988.

TYPE: Cohort
TERM: Cytology; Premalignant Lesion; Screening; Socio-Economic Factors; Tobacco (Smoking); Water
SITE: Oesophagus
CHEM: Calcium; Zinc
TIME: 1975 - 1994

***189** **Qiu, S.L.** 05355
Henan Medical Univ., Inst. of Medical Sciences, 40 University Rd, Zhengzhou 450052, China (Tel.: +86 371 663871; Fax: 3109)
COLL: Yang, G.R.; Zhao, L.; Lee, S.S.; Mai, K.

Intervention of Precancerous Lesions of the Oesophagus

To further determine whether administration of the Chinese herb R (CHR) affects oesophageal precancerous lesions, an intervention study of precancerous lesions was undertaken in 1993 in a high-risk area (in the villages of Zheng Tun, Meng Zhuang, Dong Xia Feng and Lee Gu in Huixian Country). 1,443 people living in these villages were examined by endoscopy and two or more biopsy specimens were taken from the middle and lower third of the oesophagus, or from macroscopic lesions,for routine histopathological diagnosis and incubation with tritiated thymidine. 200 cases suffering from moderate or severe chronic oesophagitis, alone or together with atypical hyperplasia of the epithelium, were randomly allocated to treatment with 4 gr of CHR or to a placebo daily for one year. The effect will be evaluated by histopathology and micro-autoradiography of the endoscopic biopsy mucosa.

TYPE: Intervention
TERM: BMB; High-Risk Groups; Plants
SITE: Oesophagus
TIME: 1993 - 1995

---

190 **Yang, G.R.** 04808

Henan Medical Univ., Henan Inst. of Medical Sciences, 40 University Rd , Zhengzhou Henan 450052, China (Tel.: +86 371 663817; Tlx: 3109)

COLL: Qiu, S.L.

Prevention and Treatment of Oesophageal Carcinoma in a High-Risk Population in China.

To reduce the morbidity and mortality of oesophageal cancer in a high-risk population, endoscopic surveys and intervention studies have been carried out in Huxian county, Henan Province during the last seven years in 1,793 subjects submitted to endoscopic examinations and biopsy. 36 cases have been diagnosed as oesophageal or cardial carcinoma, including 29 early and 7 advanced cancer. All of the early cancers were treated by surgery or endoscopic Nd-YAG laser therapy with good curative results. 791 cases suffering from chronic oesophagitis with basal cell hyperplasia and/or dysplasia have been assigned to a double-blind intervention study. Of these, 176 were treated with 10,000 IU retinol, 200mg riboflavine and 200mg vitamin C once a week; 60 cases received 200mg new retinoic analogue RII once a week in the period 1984-1985; and 126 cases were treated with 600mg of calcium per day between 1986-1987. Another 200 cases have been randomly allocated to treatment with 1200mg of calcium or to placebo per day since Dec. 1989. A regular cancer registry and follow-up study have been carried out in order to evaluate whether the secondary prevention mentioned above could reduce the morbidity and mortality of oesophageal cancer in these populations.

TYPE: Intervention
TERM: Vitamins
SITE: Oesophagus
CHEM: Calcium
TIME: 1983 - 1994

San José

191 **Sittenfeld, A.N.A.** 04252

Univ. de Costa Rica, Centro de Investigacion en Biologia Celular y Molecular, San José, Costa Rica (Tel.: 246749)

COLL: Arce, V.; Brenes, F.; Visona, K.; Marten, A.

Correlation between Hepatocellular Carcinoma and HBV Markers in the General Population of Costa Rica

Hepatocellular carcinoma (HCC) has an incidence of 1.21/100,000 (approximately 40 cases a year) in Costa Rica, although there are geographical variations and areas of relatively high risk, which are endemic for HBV. It has been proposed that areas of low HBV incidence have a relatively small risk for HCC. The main objective of the present study is to determine the relationship between HBV (serum and tissue markers) and aflatoxin in patients with HCC. It is also hoped to establish the usefulness of serial alpha-foetoprotein molecular variants as an early indicator of HCC in a population of 200 HBsAg carriers from one of the high HBV incidence areas. Retrospective studies will be performed on a bank of serum samples from the HBsAg carriers, collected during a period of 12 years (1974-1986). Papers have been published in "Viral Hepatitis and AIDS". International Symposium of Viral Hepatitis and AIDS-LSU-ICMRT. San José, Costa Rica, Dec. 1-5, 1986. Villarejos, V.M. (ed.). Trejos Hnos, 1987, pp. 379-382 and 383-386, and J. Immunol. Methods 106:19-26, 1988.

TYPE: Correlation
TERM: HBV; Mycotoxins
SITE: Liver
CHEM: Aflatoxin
TIME: 1974 - 1994

CROATIA

Zagreb

192 **Bauman, A.** 03176
Inst. for Medical Research and Occupational Health, Dept. for Radiation Protection, Ksaver 158, P.O.Box 291, 41000 Zagreb, Croatia (Tel.: (041)434188; Fax: 274572)
COLL: Kovac, J.; Novakovich, M.; Lokobauer, N.; Marovic, G.; Ambrosic, D.

Cancer Risk from Mutagenic Agents at a Coal Fired Power Plant

In a coal-fired power plant in the western part of Yugoslavia, extensive risk assessment has been carried out for many years. Athracite coal with high uranium content (25 ppm) and 7-9% of sulphur, meavy metals and benzo(a)pyrene is again being used. Epidemiological studies and constant monitoring are being done. A special case is a former radioactive slug waste pile site from the highly contaminated coal mine abandoned 90 years ago and on which houses are built and where two generations of the same families live. Both sites are investigated in cooperation with local medical services. Radon is being measured in houses, some of which are built with radioactive slag. At present teratogenic effects are being investigated with special emphasis on the child-bearing population. Many women are descendents of seven generations of coal miners, which were working in the radioactive coal mine. The results will be added to the existing cancer register.

TYPE: Case-Control; Incidence; Mortality
TERM: Air Pollution; Coal; Dose-Response; Environmental Factors; Metals; Occupation; Radiation, Ionizing; Record Linkage
SITE: Gastrointestinal; Leukaemia; Lung
CHEM: Benzo(a)pyrene; PAH; Radon; Sulphur Dioxide; Uranium
OCCU: Miners, Coal
TIME: 1990 - 1995

193 **Bauman, A.** 03609
Inst. for Medical Research and Occupational Health, Dept. for Radiation Protection, Ksaver 158, P.O.Box 291, 41000 Zagreb, Croatia (Tel.: (041)434188; Fax: 274572)
COLL: Kovac, J.; Cesar, D.; Marovic, G.; Horvat, D.; Konjevic, R.; Pticar, M.

Cancer Risk in the Fertilizer Industry

The aim of this study is to assess the risk of working in, or living near, a phosphate agricultural chemical plant. A uranium extraction pilot plant installed in Zagreb has now been abandoned. Carcinogens such as uranium, thorium, sulphuric and phosphoric acids, fluorides and nitrates are sources of air and soil pollution in the working environment and around the plant. At present 450 workers have been examined and chromosome aberrations, SCE and internal contamination with lead-210 have been detected. Carcinogenic effects are investigated. The awareness of the population that phosphate fertilizers are radiation sources when applied in agriculture is a psychological threat.

TYPE: Incidence; Mortality
TERM: Chemical Exposure; Environmental Factors; Fertilizers; Metals; Mutagen; Occupation; Radiation, Ionizing
SITE: Gastrointestinal; Lung
CHEM: Fluorides; Nitrates; Phosphates, Inorganic; Sulphuric Acid; Thorium; Uranium
OCCU: Agricultural Workers
TIME: 1990 - 1995

Habana

194 **Fernández, L.M.** 04621
Inst. Nacional de Oncologia, 29 y E. Vedado, Habana 10400, Cuba (Tel.: +53 7 325667; Fax: 328480 ; Tlx: 051 2662 HONCOL-CU)
COLL: Caraballoso, M.; Lence, J.; Diaz, A.

Cancer Mortality in Cuba

The main objective of this study is the analysis of cancer mortality data by factors such as person-years of life lost, geographical distribution, time trends, and the effects of age, calendar period and birth cohort. Data are provided by the statistical office of the Ministry of Public Health: these are tabulated by sex, residence and site (all sites, lung, breast), to produce age-specific and adjusted rates; these will be used for life-tables, regression models and generalised linear models. These studies will provide a description of geographic mortality patterns, for use in further analytical studies and in formulation of health policies.

TYPE: Mortality
TERM: Age; Birth Cohort; Geographic Factors; Time Factors; Trends
SITE: All Sites; Breast (F); Lung
TIME: 1987 - 1994

***195** **Fernández, L.M.** 05481
Inst. Nacional de Oncologia, 29 y E. Vedado, Habana 10400, Cuba (Tel.: +53 7 325667; Fax: 328480 ; Tlx: 051 2662 HONCOL-CU)
COLL: Caraballoso, M.; Molina, A.; Lence, J.; Martin, A.; Rodriguez, A.; Diez, M.; Caceres Diaz, C.; Gonzalez, S.; Carreras, M.V.; Sankaranarayanan, R.; Parkin, D.M.

Evaluation of the Cuban National Cancer Control Programme

Descriptive and analytical epidemiological studies are being developed to evaluate specific components of the Cuban Cancer Control Programme such as oral, breast and cervical cancer screening. Oral examination is offered to all individuals above the age of 15 years; physical examination of the breast is offered to all women above the age of 30 years and mammography to high risk women among them; cervical cytology is done for older women. The descriptive studies considered are comparisons of national mortality trends and stage distributions before and after introduction of the programme; correlation between intensity of screening and changes in mortality or stage distribution. Case-control studies are also being considered. The data sources for these studies will be the National Cancer Registry and the Statistical Office of the Ministry of Public Health of Cuba.

TYPE: Case-Control; Incidence; Mortality
TERM: Registry; Screening; Stage; Trends
SITE: Breast (F); Oral Cavity; Uterus (Cervix)
REGI: Cuba (Cub)
TIME: 1992 - 2000

CZECH REPUBLIC

Brno

196 **Augustin, J.** 04891

Czech National Cancer Registry, Dept. of Cancer Epidemiology, Tr. kpt. Jarose 8 , 602 00 Brno, Czech Republic

COLL: Bilek, O.; Kolcova, V.; Mifek, J.; Wotke, R.; Zejda, R.

Cancer Incidence and Mortality in the Czech Republic

Cancer incidence and mortality in four defined regions, each with about 100,000 inhabitants, will be compared and analysed. The areas investigated are different with respect to environment, pollution and health conditions. Standardized data from the Czech National Cancer Registry (1976-1990) will be used, as well as information about morbidity and mortality of other diseases. Information about smoking and lifestyle will also be collected. The aim of the study is to identify risk factors and to assess their role in the occurrence of cancer in the four areas.

TYPE: Incidence; Mortality
TERM: Environmental Factors; Geographic Factors; Lifestyle; Registry; Tobacco (Smoking)
SITE: All Sites
REGI: Czech (Cze)
TIME: 1991 - 1994

Prague

***197** **Kubik, A.** 05336

Inst. of Chest Diseases, Budínova 67, 18071 Prague 8, Czech Republic (Tel.: 8182675)

COLL: Bonassi, S.; Filiberti, R.; Reissigova, J.

Regional and National Variation in Lung Cancer Mortality in Italy and the Czech and the Slovak Republics

Geographical differences in cancer mortality among countries and regions reflect variation in factors such as occupation, environmental pollution, diet and lifestyle, especially smoking habits, and can be helpful in evaluating the impact of preventive measures. The objective of the present study is to review and analyse data on lung cancer mortality in three geographical areas of Italy (North, Central, South), and in the Czech and Slovak Republics in the 1970s and 1980s in relation to trends in cigarette consumption since the 1950s. The baseline data (e.g. annual numbers of death from lung cancer, estimates of the resident populations, estimates of tobacco consumption) are available from the official statistical institutions of the countries under study. Lung cancer mortality will be modelled using appropriate models as a function of sex, age at death, year of death, birth cohort, region, smoking etc. and interactions among them. Predictions will be calculated.

TYPE: Correlation; Mortality
TERM: Birth Cohort; Geographic Factors; Projection; Tobacco (Smoking); Trends
SITE: Lung
LOCA: Czech Republic; Italy; Slovak Republic
TIME: 1993 - 1994

198 **Srám, R.J.** 04222

Psychiatric Research Inst., Lab. of Pharmacogenetics, 181 03 Prague 8 - Bohnice, Czech Republic (Tel.: 8552247, 8552005)

COLL: Holá, N.

Cytogenetic Analysis in Groups Occupationally Exposed to Mutagens and Carcinogens

The effect of occupational exposure to mutagens and carcinogens on the frequency of chromosome aberrations in peripheral lymphocytes is studied in groups of workers exposed to the following agents: (1) halogenated ethers, bis(chloromethyl) ether and CMME, (2) substances released during soft-coal pressure gasification, (3) substances released during soft-coal open-cast mining, (4) mineral oils, (5) substances in the petrochemical industry. Approximately 300 workers and 50 matching controls are analysed each year. The aim of the study is to evaluate the effect of preventive hygiene measures in decreasing the frequency of chromosomal aberrations, and the possible relationship between cytogenetic data and types of biological markers.

TYPE: Cohort; Molecular Epidemiology
TERM: Chromosome Effects; Lymphocytes; Mining; Occupation; Petroleum Products; Prevalence; Prevention
SITE: Inapplicable
CHEM: BCME; Mineral Oil
OCCU: Chemical Industry Workers; Glass Workers; Miners, Coal; Petrochemical Workers
TIME: 1985 - 1994

199 **Srám, R.J.** 04223
Psychiatric Research Inst., Lab. of Pharmacogenetics, 181 03 Prague 8 - Bohnice, Czech Republic (Tel.: 8552247, 8552005)
COLL: Holá, N.

Occupational Exposure to Cytostatics in Hospitals

The effect of cytostatics in various clinical departments is followed using the cytogenetic analysis of peripheral lymphocytes in nurses and medical doctors. The aim of this study is to check working conditions and by new hygiene measures to decrease the level of cytostatics so as not to induce an increase in chromosomal aberrations. Approximately 60 nurses and medical doctors and 25 matching controls are analysed each year.

TYPE: Cohort; Molecular Epidemiology
TERM: Chemotherapy; Chromosome Effects; Lymphocytes; Occupation; Prevention
SITE: Inapplicable
OCCU: Health Care Workers
TIME: 1986 - 1994

200 **Vonka, V.** 03310
Inst. of Hematol. & Blood Transfusion, Dept. of Experimental Virology, Korunní 108, 101 03 Prague 10, Czech Republic (Tel.: +42 2 732015; Fax: 732015)
COLL: Kanka, J.; Roth, Z.; Hamsikova, E.

Role of a Virus in the Aetiology of Cervical Cancer

The prospective study on the cervical neoplasia-HSV2 relationship has been completed. In the course of the study more than 10,000 women were enrolled and subsequently followed. A total of 150 cases of moderate to severe dysplasia, 83 cases of carcinoma in situ and 21 cases of invasive carcinoma were detected. More than 60% of patients were found at enrollment, the other cases developing in originally healthy subjects. No differrence in the prevalence of HSV2 antibodies were found between the patients and the matched control subjects. These results did not provide any support for the hypothesis that HSV2 is aetiologically involved in cervical neoplasia. Presently, HPV is strongly suspected to be an aetiological agent in cervical cancer. Serological tests are being introduced (J. gen. Virol 71:2719-2724, 1990, 72:2577-2581, 1991, 73:429-432, 1992; Acta Virol. 34:433-443, 1990 and Int. J. Cancer 51:837-838, 1992). As soon as reliable tests are available, sera collected at enrollment will be tested for HPV antibodies and the risk of developing cervical cancer associated with past HPV infections will be determined.

TYPE: Cohort
TERM: Antibodies; BMB; HPV; HSV; In Situ Carcinoma; Infection; Premalignant Lesion; Sero-Epidemiology; Serum; Virus
SITE: Uterus (Cervix)
TIME: 1975 - 1994

DENMARK

Aarhus

201 **Autrup, H.** 04677
Steno Inst. of Public Health, University Park, Bldg 180, 8000 Aarhus, Denmark
COLL: Christensen, J.M.; Sorsa, M.I.; Farmer, P.B.

Biomonitoring of Styrene Exposed Individuals
The aim of the study is to develop and validate new methods to assess exposures to styrene, and to apply these methods in studies of exposed populations. Styrene has been classified by IARC as a possible carcinogen. There is, however, an urgent need for more research to establish the actual cancer risk of the more than 10,000 styrene workers in Denmark and for the more than 100,00 workers in the rest of Europe. Exposure occurs in situations where manufacturing of glassfibre-reinforced materials such as boats and windmill whips. 50 individuals with a well defined occupational exposure to styrene and a non-exposed control group (25 individuals) will be investigated. The methods to develop and validate are: (1) Identification and quantification of styrene-DNA adducts by 32P-post-labelling technique (2) Quantification of styrene hemoglobin adducts by GC-MS. (3) Detection and quantification of urinary styrene-DNA repair products by combination of HPLC and GC-MS methods. The proposed study also includes: (4) Air-sampling, and sampling of biological materials, and anlysis for styrene in air, serum and urine, and analysis of micronuclei in lymphocytes. Detection of DNA adducts in urine may be an indirect measure of the biological activity within hours prior to sampling. It is expected that the present study will provide new information on the relation between external dose, biological active dose and cytogenetic parameters. The methods developed and used in the project may also be valuable for future investigations of exposure to other potential carcinogens.

TYPE: Methodology; Molecular Epidemiology
TERM: BMB; DNA Adducts; DNA Repair; Lymphocytes; Micronuclei; Occupation; Plastics; Serum; Urine
SITE: Inapplicable
CHEM: Styrene
OCCU: Mineral Fibre Workers
TIME: 1989 - 1994

***202** **Autrup, H.** 05390
Steno Inst. of Public Health, University Park, Bldg 180, 8000 Aarhus, Denmark
COLL: Poulsen, H.E.; Wassermann, K.; Olsen, J.H.; Grandjean, P.; Dragsted, L.; Raffn, E.

Occupation and Cancer
Cancer cases notified in the years 1985-1989 will be linked to a large population series of controls from the Central Population Register. These new cases and population controls will have a continuous working history of more than 20 years. The use of information on chemicals and compounds on file in the Product register will be intensified with the dual purpose of extending the surveillance programme for work-place induced cancers in Denmark, and of testing in particular a set of animal carcinogens for potency in humans at doses seen at the work-place. The linked material will include approximately 300,000 cancer cases and 150,000 population controls. The material will be analysed as multiple case-control studies.

TYPE: Correlation
TERM: Occupation; Registry
SITE: All Sites
REGI: Denmark (Den)
TIME: 1993 - 1996

***203** **Autrup, H.** 05391
Steno Inst. of Public Health, University Park, Bldg 180, 8000 Aarhus, Denmark
COLL: Poulsen, H.E.; Wassermann, K.; Olsen, J.H.; Grandjean, P.; Dragsted, L.; Raffn, E.

Ischaemic Diseases and Cancer
Some occupations carry a high risk of both cancer and ischaemic diseases. While tobacco smoking is one well-described risk factor that is shared by both groups of patients, it is of particular interest to evaluate the relative risk of non-smoking induced cancers, including cancer of the colon, rectum, breast, prostate, brain and nervous system. Two studies are being carried out: (1) Correlation study: Since large-scale record linkages have been performed between (a) the Cancer Registry and ATP, and (b)

National Hospital Discharge Register and the ECP, it will be possible to measure the occurrence of ischaemic cardiovascular disease and cancer in workers in a large number of Danish industries, and to evaluate the occupational correlation between the diagnostic categories; and (2) Cohort study: Rosters of patients with acute myocardial infaction (ICD-8:410), other acute or chronical atherosclerotic heart diseases (411, 412), angina pectoris (413), and atherosclerotic vascular diseases (440, etc) will be defined in the files of the hospital discharge register. Linkage of the personal identification number with the files for the Central Population Register (for determination of vital status) and the cancer register will be performed. The observed number of cancers after date of discharge for cardiovascular disease will be compared with the expected frequency of cancer had the cohort members experienced the same rates as prevailed in the general population. A subsample of patients diagnosed with both diseases will be studied with respect to the two hypothesis (1) that oxidative stress and exposure to genotoxins play an important role in the aetiology, and (2) that they share a common susceptibility pheno/genotype.

TYPE: Cohort; Correlation
TERM: Antioxidants; Genotype/Phenotype; Occupation; Record Linkage; Registry
SITE: Brain; Breast (F); Colon; Nervous System; Prostate; Rectum
REGI: Denmark (Den)
TIME: 1993 - 1996

***204** **Autrup, H.** 05392
Steno Inst. of Public Health, University Park, Bldg 180, 8000 Aarhus, Denmark
COLL: Poulsen, H.E.; Wassermann, K.; Olsen, J.H.; Grandjean, P.; Dragsted, L.; Raffn, E.

Transportation Workers and Cancer

An excess risk of lung, laryngeal and bladder cancer has been reported in workers in the transportation sector. It is estimated that a total of 10 cases of lung cancer, 5 cases of bladder cancer and 1.5 cases of larynx cancer will be registered each year among workers registered in these trade unions. Based primarily on trade union membership lists, cohorts of bus drivers (6,000), mail carriers (4,000), and harbour workers (1,500) are identified. Date of first employment is also registered and cohort members will be linked to the Central Population Register. Information on cancer is obtained during the period of follow-up by linkage to the files of the Cancer Registry, and the incidence cancer rate will be obtained for groups exposed at different levels of genotoxic material. Groups of 50 non-smoking individuals will be selected for biomarker analysis. Cancer patients still alive at the time of identification and controls selected at random from the cohorts, will be interviewed and biological samples, spot urine and blood (40 ml), will be collected for analysis of susceptibility markers. The DNA adduct level will be determined in T-lymphocytes and the DNA repair capacity will be analysed in different lymphocyte subpopulations by unscheduled DNA synthesis. The level of oxidative stress will be assessed by determination of 8-hydroxydeoxyguanosine in urine and oxidative damage to serum proteins. The metabolic geno/phenotypes will be determined and the activity will be related to the level of the biomarkers. Urine samples will be analysed for the presence of PAH and for the presence of material than induce aneuploidy in yeast.

TYPE: Cohort
TERM: Antioxidants; BMB; Biomarkers; Blood; DNA Adducts; DNA Repair; Genotype/Phenotype; Occupation; Record Linkage; Registry; Urine
SITE: Bladder; Larynx; Lung
CHEM: PAH
OCCU: Bus Drivers; Dockers; Railroad Workers; Transport Workers
REGI: Denmark (Den)
TIME: 1993 - 1996

***205** **Autrup, H.** 05393
Steno Inst. of Public Health, University Park, Bldg 180, 8000 Aarhus, Denmark
COLL: Poulsen, H.E.; Wassermann, K.; Olsen, J.H.; Grandjean, P.; Dragsted, L.; Raffn, E.

Traffic and Childhood Cancer

A Danish case-control study of childhood cancer will be analysed with respect to the complete residential history of the children taking into account the role of the type and intensity of traffic nearby the homes. All cases of leukaemia, malignant lymphoma and brain tumours notified to the Danish Cancer Registry in 1968-1990, in the age-groups 0-14 (1,590) are included in the study. Population controls (4,390) are chosen at random from the Central Population Register. Parents have been identified as part of a study on parental employment at time of conception and risk of cancer in the offspring, and a full residential history has been obtained. Information on current and past traffic density and estimated air pollution from other sources will be obtained from the municipalities and central authorities.

TYPE: Case-Control
TERM: Air Pollution; Childhood; Parental Occupation; Registry
SITE: Brain; Leukaemia; Lymphoma
CHEM: Vehicle Exhaust
REGI: Denmark (Den)
TIME: 1993 - 1996

***206** **Autrup, H.** 05394
Steno Inst. of Public Health, University Park, Bldg 180, 8000 Aarhus, Denmark
COLL: Poulsen, H.E.; Wassermann, K.; Olsen, J.H.; Grandjean, P.; Dragsted, L.; Raffn, E.

Testicular Cancer

Testicular cancer has the highest incidence in Denmark and other Nordic countries compared with the rest of the world and the incidence has increased steadily for several decades. The causative agent(s) or factor(s) remain obscure in spite of considerable scientific efforts. The only identified risk factor is cryptorchidism. However, a number of studies suggest that an environmental agent may play an important role in the aetiology of testicular cancer. The individual susceptibility in relation to the activity of carcinogen metabolizing enzymes will be investigated in a case-control study of 90 testicular cancer patients from the 1947-1971 birth cohort, with no recurrence of the disease, and a random sample from the male population from the same area. Patient information will be obtained by questionnaire and blood samples will be analysed for P450 and glutathione-S-transferase genotypes. Urine samples will be analysed for CYP1A2 and CYP2A3 activity index (caffeine metabolism and 6-beta hydroxycortisol excretion respectively) and acetylator phenotype (caffeine metabolism). Oxidative stress is estimated from the activity of xanthine oxidase (caffeine metabolism) and urinary excretion of 8-hydroxydeoxyguanosine.

TYPE: Case-Control
TERM: Blood; Enzymes; Metabolism; Registry; Urine
SITE: Testis
REGI: Denmark (Den)
TIME: 1993 - 1996

207 **Bundgaard, T.** 04769
Aarhus Univ. Hosp., ENT Dept., 8000 Aarhus C, Denmark (Tel.: +45 86 125555)
COLL: Wildt, J.; Elbrond, O.

Cancer of the Oral Cavity

The purpose of this matched case-control study is to estimate the relative risks of the use of tobacco, chewing tobacco, snuff, and alcohol, of dental status and the pursuit of certain professions in the development of oral cancer. The cases are consecutively referred patients with oral cancer in a geographical area with 1.4 million inhabitants. Controls (3 per case) will be matched to cancer cases for age and sex using the Danish Central Registry of Persons. About 180 cases and 480 referents will take part in the study. Methods of analysis will include Mantel-Haenszel analysis and logistic regression.

TYPE: Case-Control
TERM: Alcohol; Hygiene; Registry; Tobacco (Chewing); Tobacco (Smoking)
SITE: Oral Cavity
REGI: Denmark (Den)
TIME: 1986 - 1994

208 **Olsen, J.** 00923
Aarhus Univ., Inst. of Social Medicine, Høegh Guldbergsgade 8, 8000 Aarhus C, Denmark (Tel.: +45 86 138822; Fax: 139919)
COLL: Sabroe, S.

Cancer among 10,000 New Members of the Carpenters' and Cabinet Makers' Trade Union

Most cohorts in occupational epidemiology are based on cross-sectional sampled active workers, thus making the studies vulnerable to the so called "healthy worker selection". In this study cohort members enter at the beginning of their work career and are followed over time to estimate cancer incidence according to occupational exposure. 10,000 carpenters and cabinet makers were recruited in 1977-1980 and data on cancer will be obtained by linkage with the Danish Cancer Registry during the period 1990 to 2000. Detailed data on the occupational exposures are based on questionnaires sent out in 1980 and updated information from samples of the entire study population (both cases and controls - "case-base" study).

TYPE: Case-Control; Cohort
TERM: Chemical Exposure; Dusts; Occupation; Registry; Wood
SITE: Leukaemia; Nasal Cavity; Respiratory
OCCU: Cabinet Makers; Carpenters, Joiners; Lacquerers
REGI: Denmark (Den)
TIME: 1977 - 2000

209 **Sabroe, S.** 04334
Aarhus Univ., Inst. of Social Medicine, Høegh-Guldbergsgade 8, 8000 Aarhus C, Denmark
COLL: Olsen, J.; Søgaard, J.; Sell, A.; Schultz, H.

Testis Cancer and Birthweight

In Denmark more than 200 new testis-cancer cases are diagnosed every year. Apparently the incidence rate in Denmark is among the highest in the world. This is probably due to both genetic and environmental causes, and the epidemiology of the disease indicates that the first stages in the cancer process take place very early in life, maybe even in utero. The aim of the study is, in a case-control design, to compare birthweight and age of gestation in a group of unselected testis cancer patients with men of the same age without testis cancer. Thus birthweight is an indirect measure of damaging influences on the foetus. Cases are the 300 youngest patients from the radiotherapy unit in Aarhus. Age-matched males born in the same municipality have been selected as controls. Birth information has been obtained from the midwifery case records. An analysis of the seasonal variation in birth month of the testis cancer patients will also be carried out.

TYPE: Case-Control
TERM: Birthweight; Seasonality
SITE: Testis
TIME: 1987 - 1994

Copenhagen

210 **Anderson, L.** 04517
Danish Cancer Society, Danish Cancer Registry, Rosenvængets Hovedvej 35, Box 839, 2100 Copenhagen, Denmark (Tel.: +45 31 268866; Fax: 260090)
COLL: Storm, H.H.; Johansen, H.

Late Effects of Thorotrast among Danish Patients 1932-1947

Information on approximately 1,100 patients injected with Thorotrast by cerebral angiography during the period 1935-1946 was collected at the Finsen Institute, and these patients have been followed for cancer occurrence. (Faber, Environ, Res. 18:37-43, 1979). The aim of this study is to re-establish the cohort with enough detail to permit record linkage with the Danish Cancer Registry, the Danish death registry and the National patient discharge registry. Cancer occurrence and mortality (all diseases) will be studied in relation to latency and radiation dose from Thorotrast, as well as other background variables. Data collection is based on hospital records, the old Thorotrast file and routine data collection from registries.

TYPE: Cohort
TERM: Dose-Response; Drugs; Late Effects; Latency; Radiation, Ionizing; Registry
SITE: All Sites
CHEM: Thorotrast
REGI: Denmark (Den)
TIME: 1989 - 1994

211 **Ewertz, M.** 04592
Danish Cancer Society, Div. of Cancer Epidemiology, (Danish Cancer Registry), Strandblvd 49, 2100 Copenhagen Ø, Denmark (Tel.: +45 31 268866)
COLL: Adami, H.O.; Holmberg, L.; Karjalainen, S.; Tretli, S.

Breast Cancer among Males

While breast cancer is common among women, it is an extremly rare disease in men: about 1% of all malignant breast neoplasms arise in males. Probably due to the rarity of the disease, little is known about the aetiology. In order to investigate the epidemiology of male breast cancer, a collaboration was agreed upon between the cancer registries of Denmark, Finland, Norway and Sweden. A first study describes the trend in male breast cancer incidence in the Nordic countries, confirming an approximately two-fold variation in the rates, while a second study includes a detailed analysis of the survival of 1,429 male patients with breast cancer. A case-control study is now underway to study aetiological factors. It includes about 150 men diagnosed with primary carcinoma of the breast between 1 August 1987 and 30 July 1991 in Denmark, Norway and Sweden. Data have been collected by self-administered (mailed) questionnaires, supplemented by telephone interviews when necessary. Papers have been published in Int. J. Cancer 43:27-31, 1989 and in Cancer 64:1177-1182, 1989 (Adami, H.O. et al.).

TYPE: Case-Control; Incidence
TERM: Registry; Survival; Trends
SITE: Breast (M)
LOCA: Denmark; Finland; Norway; Sweden
REGI: Denmark (Den); Finland (Fin); Norway (Nor); Sweden (Swe)
TIME: 1987 - 1994

212 Giwercman, A. 04439
Rigshospitalet, Univ. Dept. of Growth and Reproduction, Blegdamsvej 9 , 2100 Copenhagen Ø, Denmark (Tel.: +45 31 35455068)
COLL: Skakkebæk, N.E.

Prevalence of Carcinoma-In-Situ of the Testis among Infertile Men

Infertile men are known to have an increased risk of testicular cancer. Invasive cancer of the testis can be prevented if the malignancy is detected at the stage of carcinoma-in-situ(CIS). It is the aim of this study to assess the prevalence of CIS of the testis in the above-mentioned group of men in order to decide whether routine screening for early testicular neoplasia should be offered to these men. Testicular biopsies will be performed in 500 infertile men. Biopsy specimens will be evaluated by means of light microscopic examination. The prevalence of CIS of the testis in this group of men will be compared to the frequency of CIS in an autopsy series of gonads from 400 men who suffered sudden, unexpected death.

TYPE: Cross-Sectional
TERM: Autopsy; Biopsy; In Situ Carcinoma; Infertility
SITE: Testis
TIME: 1984 - 1994

213 Hansen, E.S. 03066
Univ. of Copenhagen, Inst. of Social Medicine, Blegdamsvej 3, 2200 Copenhagen N, Denmark (Tel.: +45 31 357900)

Role of Inhalation of Combustion Products in the Aetiology of Cancer and Atherosclerosis

Inhalation exposure to particulate and gaseous compounds from fossil fuel combustion is expected to increase cancer risk as well as the risk of developing atherosclerosis. The study will be carried out as a comparison of the follow-up experience of nine exposed occupational cohorts and three non-exposed cohorts. Comparisons will concern mortality from diseases related to atherosclerosis and incidence of carcinomas of the respiratory and gastrointestinal tracts, the urinary system, the reproductive organs, the CNS, the skin, the blood- and lymph-forming tissues and sarcomas. In this study the effects of occupational exposure will be evaluated in the light of the relative significance of particular characteristics of the exposure. Effect modification by other risk factors will be considered. Papers have been published in Scand. J. Work Environm. Hlth 15:45-46, 1989, Br. J. Ind. Med. 46:582-585, 1989 and 47:805-809, 1990.

TYPE: Cohort
TERM: Air Pollution; Chemical Exposure; Environmental Factors; Metals; Occupation; Petroleum Products; Registry
SITE: Bladder; Brain; Gastrointestinal; Haemopoietic; Male Genital; Respiratory; Sarcoma; Skin
CHEM: Arsenic; Chromium; Nickel; PAH
REGI: Denmark (Den)
TIME: 1983 - 1994

***214** **Høyer, A.P.** 05266

Danish Cancer Society Research Centre, Div. for Cancer Epidemiology, Strandblvd 49, Box 839, 2100 Copenhagen Ø, Denmark (Tel.: +45 35 268866; Fax: 260090)
COLL: Engholm, G.

Risk Factors in Breast Cancer Aetiology

The aim of this historical prospective study is to evaluate the risk of breast cancer in relation to social class, age at menarche and menopause, reproductive variables, use of exogenous hormones, smoking, alcoholic beverage and coffee consumption, body build and serum lipids, with particular emphasis on the latter. The Glostrup Population Studies were established primarily to study risk factors for cardiovascular disease and have been, since 1982, a part of the WHO Multicentre Investigations Monitoring Trends and Determinants in Cardiovascular Diseases (MONICA). Study subjects consist of all women (5,207) examined in the studies between 1964 and 1986 from 30 to 80 years of age. The participation rate varies with year of entrance into the studies, with an overall rate of 78.5 percent. Information on demographic, hormonal and reproductive variables, and life-style factors has been collected with a standardized questionnaire. Measurements of weight and height, and fasting serum lipids were obtained in connection with a clinical examination. Breast cancer cases occurring after entrance into the study are identified by linkage to the Danish Cancer Registry. Vital status at the end of follow-up is determined by linkage to the Central Population Registry using the unique 10-digit personal identification number issued to all inhabitants in Denmark. Univariate and multivariate methods are used in the analysis of data. A paper has been published in Cancer Causes and Control 3:403-408, 1992.

TYPE: Cohort
TERM: Alcohol; Coffee; Hormones; Lipids; Physical Factors; Registry; Reproductive Factors; Tobacco (Smoking)
SITE: Breast (F)
REGI: Denmark (Den)
TIME: 1989 - 1994

215 **Kruger Kjær, S.** 04593

Danish Cancer Society, Danish Cancer Registry, Rosenvængets Hovedvej 35, Box 839, 2100 Copenhagen Ø, Denmark (Tel.: +45 35 268866)
COLL: Bock, J.; Lynge, E.; Dahl, C.; Faber Vestergaard, B.; de Villiers, E.M.

Case-Control Study Focussing on "The Male Factor" for Cervical Cancer

Cervical cancer is the sixth most frequent type of cancer among women in Denmark, and the number of precancerous lesions diagnosed increased up to 1982, especially in younger women. The sexual habits of a woman are known to be associated with her risk for developing cervical cancer, but recent studies have indicated that the number of female partners that a man has had may also be of importance. The aim of the study is to test the following specific hypotheses: (1) Men with early onset of sexual activity and multiple sexual partners (including prostitutes) increase the risk for cervical cancer in their female partners; (2) Women who have first inercourse at an early age and multiple partners are at a high risk of developing cervical cancer; (3) Use of oral contraceptives increases the risk for cervical cancer; (4) Smoking is an independent risk factor for cervical cancer; (5) Early age of onset of smoking increases the risk for cervical cancer. The investigation is a population-based case-control study with a two-phase data collection. In the first phase, women living in Greater Copenhagen and diagnosed with carcinoma in situ of cervical cancer in 1985-86 at the age of 20-49 years have been identified in the Cancer Registry. In the second phase, a subgroup who report having had only one sexual partner will be investigated. Controls have been selected as a random sample of women from the Central Population Register, stratified for age and area of residence to match the age-residence distribution of the cases. The controls for the second phase will be a subgroup with the same eligibility as the cases. Examination of case and control women together with their male partners is currently under way.

TYPE: Case-Control
TERM: Age; Oral Contraceptives; Registry; Sexual Activity; Time Factors; Tobacco (Smoking)
SITE: Uterus (Cervix)
REGI: Denmark (Den)
TIME: 1985 - 1994

216 **Lynge, E.** 05183

Danish Cancer Society, Danish Cancer Registry, Rosenvængets Hovedvej 35, Box 839, 2100 Copenhagen Ø, Denmark (Tel.: +45 35 268866)
COLL: Sunde, L.; Skouv, J.; Kruse, T.; Soll-Johanning, H.

Genetic Indicators for Heredity of Colorectal Cancer

Parents of patients with colorectal cancer have a 1.6-1.9 fold risk of colorectal cancer. A study is now being undertaken where genetic indicators for heredity of colorectal cancer is investigated. Specimens from patients with both parents affected and from patients with none or one parent and at least one sibling affected will be examined for loss of heterozygosity and other genetic aberrations. Subsequently, the identified lesions will be further characterized.

TYPE: Genetic Epidemiology
TERM: Genetic Factors; Genetic Markers; Heredity; High-Risk Groups; Registry
SITE: Colon; Rectum
REGI: Denmark (Den)
TIME: 1991 - 1994

217 **Lynge, E.** 05184
Danish Cancer Society, Danish Cancer Registry, Rosenvængets Hovedvej 35, Box 839, 2100 Copenhagen Ø, Denmark (Tel.: +45 35 268866)
COLL: Engholm, G.

Lung Cancer and the Environment

The purpose is to estimate the independent contribution to lung cancer occurrence in Denmark from occupational and environmental exposures. Both the overall and histology-specific lung cancer incidence will be analysed, based on data from the Occupational Cancer Register 1970-1985. In order to control for smoking, multivariate methods for standardization of lung cancer risk in occupational groups, geographical region, etc. will be made. The standardization will use data from about 100,000 personal interviews on smoking performed by Gallup during 1969-1975.

TYPE: Incidence
TERM: Environmental Factors; Histology; Occupation; Registry; Tobacco (Smoking)
SITE: Lung
REGI: Denmark (Den)
TIME: 1990 - 1994

218 **Melbye, M.** 03040
Danish Cancer Society, Danish Cancer Registry, Rosenvængets Hovedvej 35, P.O. Box 839, 2100 Copenhagen Ø, Denmark (Tel.: +45 35 268866; Fax: 260090)
COLL: Biggar, R.J.; Ebbesen, P.

Natural History of Human Immunodeficiency Virus (HIV) Infection among Homosexual Men

A random selection of 260 homosexual men, all members of a Danish Gay organisation, has been studied approximately every year since 1981. Questionnaire information covering demographic data and sexual behaviour history has been obtained, and serum, lymphocytes, urine, saliva, and semen have been collected. The objective is to study the long-term consequences of HIV infection with respect to immunological changes, and clinical criteria (including AIDS and related cancers). Much of the information already gathered from the first year's study has been covered in a review in Br. Med. J. 292:5-12, 1986.

TYPE: Cohort
TERM: AIDS; Homosexuality; Infection; Sero-Epidemiology; Sexual Activity; Virus
SITE: All Sites; Hodgkin's Disease
CHEM: Nitrites
TIME: 1981 - 2001

***219** **Mellemgaard, A.** 05349
Danish Cancer Society, Danish Cancer Registry, Rosenvængets Hovedvej 35, Box 839, 2100 Copenhagen Ø, Denmark (Tel.: +45 35 268866)
COLL: Olsen, J.H.; Engholm, G.

Danish Case-Control Study of Renal Cell Carcinoma

In a population-based case-control study of histologically verified cases of renal cell carcinoma a number of risk factors are being investigated. Approximately 400 cases and 400 population controls, matched for age and sex, are included. Information on risk factors are obtained through a structured interview, performed in the subjects homes. The study has so far identified cigarette smoking, high relative weight, some hormonal factors and occupational exposure to hydrocarbons as possible risk factors. Multivariate regression methods are being used in the analysis.

TYPE: Case-Control
TERM: Hormones; Insecticides; Occupation; Pesticides; Registry; Tobacco (Smoking)
SITE: Kidney
CHEM: Hydrocarbons
REGI: Denmark (Den)
TIME: 1989 - 1994

220 **Nielsen, N.H.** 05076
Univ. Inst. of Forensic Pathology, Danish Cancer Registry,, Frederik den Femtes Vej, 11 , 2100 Copenhagen Ø, Denmark (Tel.: +45 35 268866; Fax: 269000)
COLL: Lanier, A.K.; Gaudette, L.A.; Nikitin, Y.P.; Storm, H.H.

Cancer in Circumpolar Inuit Eskimos

Systematically collected cancer incidence data 1969-1988 from circumpolar areas is being compiled with specific reference to Inuit (Eskimo) population groups. The data originate from population-based Inuit cancer registries in Alaska (Center for Disease Control, Anchorage), Canada (Statistics Canada, Ottawa), Greenland (Danish Cancer Registry) and USSR (Institute of Internal Medicine, Novosibirsk). In most circumpolar areas the patterns are changing rapidly, parallel to changes in lifestyle. A comprehensive description of the findings in the world Inuit population will be presented including trends and comparisons between Inuit groups and with cancer incidence in other parts of the world. The results will be published in a monograph which will also include related background information. A summary is planned in the Inuit language.

TYPE: Incidence
TERM: Eskimos; Ethnic Group; Lifestyle; Registry; Trends
SITE: All Sites
LOCA: Canada; Greenland; Russian Federation; United States of America
REGI: Canada (Can); Denmark (Den)
TIME: 1990 - 1994

221 **Olsen, J.H.** 04595
Danish Cancer Society, Danish Cancer Registry, Rosenvængets Hovedvej 35, Box 839, 2100 Copenhagen Ø, Denmark (Tel.: +45 35 268866)
COLL: Boice, J.D.; Fraumeni, J.F.; Jensen, J.

Cancer Risk in Epileptics and their Children Following Treatment with Anticonvulsive Drugs

The aims of the study are to evaluate the risk for cancer among children of epileptic mothers who were treated intensively with phenobarbital and other anticonvulsive drugs to prevent seizures. The risk for cancer among epileptics, especially liver cancer and leukaemia, will also be evaluated. Account will be taken of the fact that some patients received injections of Thorotrast for cerebral angiography. Approximately 8,000 patients admitted for epilepsy to the Danish epilepsy treatment centre in Filadelfia, Dianalund, during 1933-62 have been linked to the files of the Danish Cancer Registry. In order to identify children born after the date of admission of a patient to the centre; the patients are being traced in population registries all over the country. Similarly, information on their children will be linked to the files of the Cancer Registry in order to detect cancers among the children.

TYPE: Cohort
TERM: Drugs; Epilepsy; Radiation, Ionizing; Registry
SITE: All Sites; Childhood Neoplasms; Leukaemia; Liver
CHEM: Barbiturates; Thorotrast
REGI: Denmark (Den)
TIME: 1989 - 1994

222 **Olsen, J.H.** 04596
Danish Cancer Society, Danish Cancer Registry, Rosenvængets Hovedvej 35, Box 839, 2100 Copenhagen Ø, Denmark (Tel.: +45 35 268866)
COLL: de Nully Brown, P.; Holländer, H.C.

Parental Occupation and Risk for Childhood Cancer

Parental occupation is one suspected risk factor for the occurrence of cancer in childhood. Reports point to occupations that involve exposure to hydrocarbons, lead or chemicals, and to the occupations of higher social classes. The aim of this case-control study is to detect possible associations between the industrial employment of the parents at the time of conception and during pregnancy and the risk of the child subsequently developing a malignancy. Approximately 2,000 cases of childhood cancer and 10,000 controls are included in the study. The parents of both cases and controls are being identified in the files of the Central Population Register, and occupational histories are established through the Supplementary Pension Fund Danish Cancer Registry linkage study file, with special attention to the period around conception of the index child and the pregnancy of the mother. By applying the evaluation of chemical exposures used in other case-control studies (on brain cancer and multiple myeloma) being carried out in the Registry to the combinations of industrial activity and job-title held by the parents of children included in the present study, it may be possible to reanalyse the material according to type of chemical exposure at the work place of the parents. Included among the exposures are chemicals such as hydrocarbons, organic solvents, paints, rubber, plastics and metals.

TYPE: Case-Control
TERM: Intra-Uterine Exposure; Metals; Occupation; Paints; Plastics; Registry; Rubber; Solvents
SITE: Childhood Neoplasms
CHEM: Hydrocarbons
REGI: Denmark (Den)
TIME: 1989 - 1994

223 **Storm, H.H.** 04599
Danish Cancer Society, Div. of Cancer Epidemiology, (Danish Cancer Registry), Strandblvd 49 Box 839, 2100 Copenhagen Ø, Denmark (Tel.: +45 35 268866; Fax: 269000)
COLL: Boice, J.D.; Kleinerman, R.A.; Stovall, M.

Lung Cancer Risk Following Radiation for Cancer of the Uterine Cervix

An international cohort study followed by several case-control studies addressing the risk of new cancers after treatment for cancer of the uterine cervix is the basis of this project. Information on exposure to radiation and potential confounding factors was extracted from hospital records. An important observation from the studies is that cancer risk seems to be elevated for more than 30 years after primary treatment, and the risk seems to increase with time. In collaboration with the Radiation Epidemiology Branch, NCI, USA, data are being collected on cervical cancer patients who developed lung cancer ten or more years following treatment for first cancer. The cases will be matched to suitable controls. A separate analysis is being carried out on the Danish data. The first risk that will be examined is for second primary breast cancer. Additional potential risk factors have been taken into consideration, and the registration and abstraction of cases may be considered more uniform than in the international material. The approach taken is a case-control study, in which the matching ratio is as large as possible in order to narrow the confidence intervals of observed risks. The dose to the breast is approximately 30 rad, and no elevation of risk for breast cancer as a function of dose has been observed. Surgical and radiogenic ablation of the ovaries had an overall protective effect against the development of breast cancer.

TYPE: Case-Control; Cohort
TERM: Multiple Primary; Radiotherapy; Registry
SITE: All Sites; Breast (F); Lung; Uterus (Cervix)
LOCA: Denmark; United States of America
REGI: Denmark (Den)
TIME: 1987 - 1994

224 **Storm, H.H.** 04600
Danish Cancer Society, Div. of Cancer Epidemiology, (Danish Cancer Registry), Strandblvd 49 Box 839, 2100 Copenhagen Ø, Denmark (Tel.: +45 35 268866; Fax: 269000)
COLL: Stovall, M.; Engholm, G.; Anderson, M; Curtis, R.E.

Risk for Leukaemia Following Treatment for Breast Cancer

In a study of multiple primary cancers in Denmark occurring in 1943-82, a relative risk of 2.3 was observed for acute nonlymphocytic leukaemia following treatment for breast cancer during the entire observation period. Some 70% of the breast cancer patients had received combination treatment with surgery and radiotherapy. In order to evaluate the leukaemogenic effect of radiation treatment for breast cancer, a matched case-control study nested in the cohort of breast cancer patients was begun in 1985.

The study involved 21 cases of chronic lymphocytic leukaemia, matched to 39 controls, and 72 cases of other leukaemias, matched to 167 controls. The relative risks for non-chronic lymphocytic leukaemia were calculated by radiation dose to the bone marrow: a small, non-significant increase in risk was found, but no meaningful trend with received dose of radiation to the bone marrow was observed. A parallel study has been carried out in Connecticut, USA, by the Radiation Epidemiology Branch of NCI. The two studies will be evaluated together.

TYPE: Case-Control
TERM: Dose-Response; Multiple Primary; Radiotherapy; Registry
SITE: Breast (F); Leukaemia; Leukaemia (CLL)
REGI: Denmark (Den)
TIME: 1985 - 1994

225 **Tjønneland, A.** 04591
Danish Cancer Society, Div. for Cancer Epidemiology, (Danish Cancer Registry), Strandblvd. 49, Box 839, 2100 Copenhagen Ø, Denmark (Tel.: +45 35 268866; Fax: 260090)
COLL: Overvad, K.; Ebbesen, P.; Vuust, J.; Olsen, J.; Storm, H.H.

Diet, Cancer and Health

A large prospective cohort study is plannes to examine the association beteween specific dietray components, food or nutrients (alone or in combination) and the risks for cancer at specific sites. A further objective is to create a data bank of information on dietary factors, personal habits and biological samples, which can be used to test future hypotheses on cancer risk. The study population will consist of 60,000 persons aged 50-64 years, drawn from the Central Population Register. Information on usual adult diet and known important risk factors for cancer, such as smoking and reproductive history, will be obtained by a self-administered and interviewer-checked questionnaire. Anthropometric measurements (e.g., height, weight, body impedance) will be made, and biological samples (e.g., blood, urine, nails) stored. The biological sampling will be repeated, and follow-up questionnaires mailed to participants every five years to asses dietary changes. The study population will be followed to record cases of cancer, death and emigration by linkage to the Cancer Registry and the Central Population Register. Follow-up for other diseases will be feasible by record linkage with other disease registries. With these combined approaches, an almost complete follow-up is envisaged.

TYPE: Cohort
TERM: BMB; Blood; Diet; Nutrition; Physical Factors; Registry; Reproductive Factors; Tissue; Tobacco (Smoking); Toenails; Urine
SITE: All Sites
REGI: Denmark (Den)
TIME: 1989 - 1996

Herlev

226 **Binder, V.** 05134
Herlev Hosp., Medical Gastrointestinal Dept., Herlev Ringvej 75, 2730 Herlev, Denmark (Tel.: +45 44 535300; Fax: 535332)

Survival and Bowel Cancer Occurrence in a Cohort of Patients with Ulcerative Colitis and Crohn's Disease

All patients diagnosed with ulcerative colitis (N = 1,161) or Crohn's disease (N = 373) in 1962-1987 and residents of Copenhagen County have been followed up from diagnosis of inflammatory bowel disease to the end of 1987 (mean ll.7 years for ulcerative colitis and 8.5 years for Crohn's disease). Eight and three bowel cancers, respectively, have been reported to date, which is not more than expected in the background population.

TYPE: Cohort
TERM: Crohn's Disease; Ulcerative Colitis
SITE: Colon; Small Intestine
TIME: 1962 - 1994

Odense

227 **Birkeland, S.A.** 03399

Odense University Hospital, Dept. of Nephrology, 5000 Odense C, Denmark(Fax: +45 65 906413)

Nordic Project on Renal Transplantation and Cancer

One complication in renal transplantation with lifelong immunosuppression is development of patients with tumours. The aim of this study, which includes all patients with renal transplantation performed in the Nordic countries, is to measure cancer risk relative to that of the general population, and to evaluate factors such as living related donor vs. cadaver kidney donor, type and site of tumours, HLA-donor-recipient mismatch, donor-recipient differences in age, sex and AB0 blood group, presence of lymphocytotoxic antibodies in the recipients or not, time for tumour occurrence after transplantation, recipient kidney disease, number of transplantations performed per recipient, patient- and graft-survival, number of blood transfusions given to the recipients, type and amount of immunosuppressive drugs, period of uraemia before transplantation and occurrence or not of any viral disease. The project at present includes the period up to 1982 with about 6,000 transplanted patients observed for about 30,000 person years, in whom 400-500 tumours have been observed. The original data from the transplantation centres are registered and updated with the civil registries, the Scandiatransplant registry and the cancer registries, and risk is evaluated using observed/expected ratios.

TYPE: Cohort
TERM: Age; Blood Group; Drugs; HLA; Immunosuppression; Registry; Sex Ratio; Time Factors; Transplantation; Virus
SITE: All Sites
CHEM: Azathioprine
LOCA: Denmark; Finland; Iceland; Norway; Sweden
REGI: Denmark (Den); Finland (Fin); Iceland (Ice); Norway (Nor); Sweden (Swe)
TIME: 1971 - 1994

228 **Brok, K.E.** 04798

Odense University Hospital, Dept. of Urology, Sdr Blvd 29, 5000 Odense C, Denmark (Tel.: +45 66 143333)
COLL: Mommsen, S.

Case-Control Study of Prostatic Cancer

The aim of this study is to examine risk factors in the aetiology of prostatic cancer. Cases comprise all patients with newly diagnosed prostatic cancer in the period 1990-93 living in the island of Fyn, Denmark. About 300 cases are expected in the period. Controls are matched for race, sex and age and are randomly selected from the national population register. All cases and controls are interviewed for their lifestyle habits, dietary habits, occupational exposure, sexual history, history of general and urological diseases and urological symptoms. Biochemical epidemiology of prostatic cancer is examined by blood samples from cases and controls (androgens, oestrogens, lipids).

TYPE: Case-Control
TERM: Biochemical Markers; Diet; Hormones; Lifestyle; Occupation; Sexual Activity
SITE: Prostate
TIME: 1989 - 1994

229 **Grandjean, P.** 03723

Odense Univ., Inst. of Community Health, Dept. of Environmental Medicine, Winsløwparken 17, 5000 Odense C, Denmark (Tel.: +45 66 158600; Fax: 65918296 ; Tlx: 59964 os dk)
COLL: Juel, K.; Olsen, J.H.

Mortality and Cancer Morbidity of Cryolite Workers

Heavy occupational fluoride exposure occurred in cryolite workers in the past. In 1931, skeletal fluorosis was discovered as a disease new to medicine when cryolite workers were examined. At least 80 cases have occurred at the factory in Copenhagen where cryolite ore from Greenland is processed. Detailed personnel records are available from 1924. An excess occurrence of urinary bladder cancer was seen during 1943-1989 in male workers employed for at least six months between 1924 and 1961. A paper was published in J. Nat. Cancer Inst. 84:1903-1909, 1992.

TYPE: Cohort
TERM: Chemical Exposure; Environmental Factors; Occupation
SITE: All Sites
CHEM: Fluorides
OCCU: Cryolite Workers
TIME: 1981 - 1994

230 **Holm, N.V.** 02306
Univ. of Odense, Inst. of Clinical Genetics, Winslowsvej 17, 5000 Odense, Denmark (Tel.: +45 66 158600)

Cancer Aetiology Studies in the Danish Twin Population

The Danish Twin Registry consists of the total population of twin pairs born in Denmark during a certain period. The present study includes all same-sexed twin pairs born during the period 1881-1930 in which both partners survived the age of 14 years, in all about 11,000 pairs. The study comprises an evaluation of the importance of genetic factors in the development of breast cancer, gastro-intestinal cancer, uterus cancer, testicular cancer, malignant melanoma and leukaemia. In twin pairs with one of these neoplasms, cancer of all sites will be studied. More intensive studies are carried out in breast cancer twins. A matched case-control study of monozygotic twin pairs discordant for breast cancer will evaluate various risk factors (number of births, age at first birth, menopause and height). To evaluate genetic heterogeneity a twin family study will be performed. A report (J. Nat. Cancer Inst., 65:285-298, 1980) dealt with concordance rates and various risk factors of breast cancer and another one (Cancer Surv. 1:17-32, 1982) with concordance rates of breast cancer, colorectal cancer and leukaemia.

TYPE: Case-Control; Cohort
TERM: Age; Genetic Factors; Menopause; Parity; Physical Factors; Twins
SITE: Breast (F); Gastrointestinal; Leukaemia; Melanoma; Testis; Uterus (Cervix); Uterus (Corpus)
TIME: 1979 - 1994

231 **Jørgensen, K.E.** 04241
Univ. Hospital of Odense, Inst. of Oto-Rhino-Laryngology, ENT-Dept., Sødra Blvd 29, 5000 Odense C, Denmark (Tel.: +45 66 143333/2810)
COLL: Hansen, O.; Bastholt, L.

Occurrence of Secondary Cancers in a Population of Patients Treated for Laryngeal Cancer

From 1965 to 1987, 820 patients were treated for laryngeal cancer at the Oncological Centre and ENT Department, University Hospital of Odense. Since 1979 all patients have been prospectively registered in a computer-system, partly in order to be able to analyse the results of treatment, and partly as patients have been included in a countrywide randomized project concerning a radiosensitizer. During the whole period all cases of secondary cancer have been registered and an abnormally high frequency of pulmonary cancer has already been observed. The total population of Denmark is being used as a control group in cooperation with the Danish Cancer Registry.

TYPE: Cohort
TERM: Multiple Primary; Registry; Treatment
SITE: Larynx; Lung
REGI: Denmark (Den)
TIME: 1979 - 1994

Risskov

232 **Mortensen, P.B.** 04777
Aarhus Psychiatric Hosp., Inst. of Psychiatric Demography, Skovagervej 2, 8240 Risskov, Denmark (Tel.: +45 86 177777)

Neuroleptics and Cancer Risk in Schizophrenics and Other Psychotic Patients

The aim of these studies is to clarify the association between use of neuroleptics and other psychotropic drugs and cancer risk in psychotic patients. The study consists of a longitudinal case-register-based study of: (1) a cohort of 6,178 schizophrenic in-patients identified in 1957 and followed-up through 1984 and (2) a cohort consisting of all patients (approx. 70,000) admitted for the first time to psychiatric hospitals during the period 1970-1986 with a diagnosis of functional psychosis (ICD 295-299). These two cohorts are checked against the Danish Cancer Registry and site-specific cancer incidence rates determined. A series of case-control within cohort studies are performed in order to determine the impact of neuroleptics and other psychotropic drugs (including reserpine) on cancer risk. So far, studies of male cancer of the bladder and prostate in males and cancer of the breast and uterine cervix in females have been performed in the cohort of schizophrenic patients.

TYPE: Case-Control; Cohort
TERM: Drugs; Record Linkage; Registry
SITE: Bladder; Breast (F); Lung; Prostate; Uterus (Cervix)
CHEM: Phenothiazines; Reserpine
REGI: Denmark (Den)
TIME: 1984 - 1994

Vejle

***233 Sherson, D.** 05387
Vejle Hospital, Dept. of Occupational Medicine, Kabbeletoft 25, 7100 Vejle, Denmark (Tel.: +45 75 727233; Fax: 75831220)
COLL: Omland, Ø.; Autrup, H.; Hansen, Å. M.

Biomarkers in Foundry Workers

This study investigates the usefulness of urinary 1-hydroxypyrene (HPU) and serum benzo(a)pyrene-albumin adducts as biomarkers. It is a cross-sectional study including 70 male iron foundry workers and 70 matched controls. HPU is measured on Monday and Friday mornings using HPLC. Adducts are measured using an ELISA assay, detecting benzo(a)pyrene-diolepoxide-I. HPU has been used to monitor PAH exposure in a variety of jobs. With more experience it may be possible in the future to utilize HPU as a standard monitoring test of PAH-exposed workers. Serum benzo(a)pyrene-albumin adducts are being investigated as a possible early marker of cancer risk.

TYPE: Cross-Sectional; Methodology
TERM: Biomarkers; Occupation
SITE: Inapplicable
CHEM: PAH
OCCU: Foundry Workers
TIME: 1990 - 1994

Cairo

234 **Ebeid, N.I.** 04297
Military Medical Academy, Depts of Occupational Medicine, Gynaecology & Biochemistry, 99, 26 July St., Bonlac, Cairo, Egypt (Tel.: +20 2 3488928)
COLL: Wahba, R.; Hafez, A.H.

Schistosomiasis and Exposure to Industrial Pollutants in the Aetiology of Liver Cancer

Both Schistosomiasis and, to a lesser extent, HBV infection are common in Egypt. The aim of this study is to examine the role of concomitant exposure to welding fumes (as an industrial pollutant) as an added carcinogenic factor. Few positive HBV cases have been detected in this study, so viral hepatitis has been excluded. The study has been extended to comprise 98 individuals of whom 48 are exposed to welding fumes, while 50 are not exposed (control group). Ultrasonography of the liver is being carried out for the majority of cases. Alpha phytolic protein testing will be done shortly afterwards. Smoking habits are also being investigated.

TYPE: Cross-Sectional
TERM: Metals; Occupation; Parasitic Disease; Tobacco (Smoking); Trace Elements; Welding
SITE: Liver
CHEM: Cadmium; Copper; Lead
OCCU: Welders
TIME: 1987 - 1994

Tallinn

235 **Hint, E.** 04831
Inst. of Exp. & Clinical Medicine, Ministry of Health, Dept. of Clinical Oncology, 42 Hiiu St., Tallinn EE0016, Estonia (Tel.: 0145 514300/514381; Fax: 248260)
COLL: Purde, M.; Eomois, M.; Lilleorg, A.

Registry for Population at High Risk of Familial Breast Cancer

Epidemiological investigations in Estonia since 1968 have revealed some peculiarities in familial breast cancer. Breast cancer in relatives increases the risk for both malignant and benign breast disease. Familial breast cancer is detected earlier and survival rates are better than non-familial breast cancer. These findings have prompted creation of a registry of women including healthy relatives of breast cancer patients. Information will be collected by questionnaires and the verification of breast cancer will be checked through the Estonian Cancer Registry. Relatives will be invited to take part in mammographic screening. Of specific interest will be the study of hormonal profile (oestradiol and prolactin) and of immune status by MCA and Ca-125 markers in breast cancer patients' daughters and sisters. To date approximately 4,000 relatives have been registered. Papers have been published in Estonian Physician 50:4228-430, 1992.

TYPE: Cohort; Registry
TERM: BMB; Biochemical Markers; Familial Factors; High-Risk Groups; Hormones; Mammography; Registry; Serum
SITE: Breast (F)
REGI: Estonia (Est)
TIME: 1990 - 1994

236 **Hint, E.** 05140
Inst. of Exp. & Clinical Medicine, Ministry of Health, Dept. of Clinical Oncology, 42 Hiiu St., Tallinn EE0016, Estonia (Tel.: 0145 514300/514381; Fax: 248260)
COLL: Tekkel, M.

Identification of Precancerous Breast Disease

In Estonia a prospective study of breast cancer was started in 1974. To date information about risk factors has been collected from more than 30,000 women by self-administered questionnaire, checked by the physician and completed if necessary. The risk factors under study include: family history of breast cancer, early menarche, age at first intercourse, age at first delivery, nulliparity, prolonged breast feeding, and late menopause. Four study groups of women aged 25 to 64 have been created: (1) 6,500 women from the general female population of 13 regions, undergoing medical examination within a public health programme and considered to be representative of the adult female population of Estonia as a whole; (2) 9,000 healthy women; (3) 20,000 women with benign breast diseases; and (4) 850 women with breast cancer. Groups 2-4 are a result of regular consultations with specialists on breast disease. The list of all women under study is checked through the Estonian Cancer Registry. It is hypothesized that if the distribution (%) of one or more of the risk factors examined in the group of patients with benign breast diseases is higher than in the groups of "healthy women" or "general female population", but lower than in the group of "breast cancer patients", then the variable would be of importance for identification of precancerous conditions. On the basis of the results obtained, measures for prevention of breast cancer will be taken. A collaborative study with the Finnish Cancer Registry is being planned.

TYPE: Cohort
TERM: High-Risk Groups; Prevention; Registry
SITE: Breast (F)
REGI: Estonia (Est)
TIME: 1991 - 1994

Espoo

237 **Sipponen, P.I.** 02195

Jorvi Hosp., Dept. of Pathology, 002740 Espoo 74, Finland

COLL: Seppälä, K.; Varis, K.S.; Kekki, M.; Siurala, M.

Frequency of Intestinal Metaplasia with Colonic Type Characteristics in the Stomach of Patients with Gastric Carcinoma and Benign Lesions

The investigation is designed to clarify the frequency of intestinal metaplasia (IM) with histochemical colonic-type characteristics in the stomach of asymptomatic subjects and of patients with gastric lesions of different types. The aim is to find out whether the occurrence of IM with colonic-like characteristics is associated with gastric carcinoma or with lesions considered to be pre-malignant. The study is performed by using histochemical staining methods, known to be characteristic to the colonic mucosa and mucosubstances, on mucosa specimens obtained by gastroscopy. 125 patients with verified gastric carcinoma and 301 first-degree relatives, 62 patients with pernicious anaemia, and 183 first-degree relatives, 406 consecutive outpatients and 358 controls matched from a large Finnish population study are being examined. The investigation will continue as a follow-up study in patient groups selected according to the presence or absence of IM with colonic type characteristics in the tissue samples. Results were published in Acta path. microbiol. scand. Sect. A, 88:217-224, 1980 and Annals of Clinical Res. 13:139-143, 1981.

TYPE: Case-Control
TERM: Histology; Premalignant Lesion
SITE: Stomach
TIME: 1979 - 1994

238 **Siurala, M.** 04494

Louhentie 1 A 2, Espoo 13, Finland (Tel.: +358 90 466355)

COLL: Sipponen, P.I.; Ihamäki, T.; Kekki, M.

Benign and Malignant Gastric Tumours in Gastric Carcinoma Families and Controls

In 1972 and 1976, 301 first-degree relatives of 73 consecutive probands with gastric carcinoma were examined by endoscopy with direct vision biopsy (average 11.2 specimens per subject), penta-gastrin test, serum pepsinogen I and II etc, and in a similar way 358 first-degree relatives of 73 control probands, matched to the cancer probands for age, sex, place of birth, residence and occupation (Scand. J. Gastroent. 14:801-812, 1979 and APMIS 87:457-462, 1979). In the first part of the study the occurrence of benign and malignant stomach tumours in gastric carcinoma and control families will be evaluated using data obtained from Finnish death statistics and the Finnish Cancer Registry. The follow-up time will be on average 14.5 years. The occurrence of stomach tumours will also be referred to the morphological and functional status of the subjects at the first examination. A paper was published in Scand. J. Gastroenterol. 26:16-23, 1991.

TYPE: Cohort
TERM: Familial Factors; Registry
SITE: Stomach
REGI: Finland (Fin)
TIME: 1988 - 1994

Helsinki

239 **Husgafvel-Pursiainen, K.T.** 04833

Inst. of Occupational Health, Topeliuksenkatu 41 a A, 00250 Helsinki, Finland (Tel.: +358 0 47471; Fax: 4747208)

COLL: Anttila, S.; Hackman, P.; Heikkilä, L.; Hietanen, E.; Hirvonen, A.; Karjalainen, A.; Ridanpää, M.; Taikina-aho, O.; Vainio, H.

Lung Cancer, Asbestos and Smoking in Finland

The aims of this study are (I) to establish practical criteria based on occupational history, pulmonary fibre content and histopathology for the identification of lung cancer related to occupational asbestos exposure; (2) to study host-related factors with biochemical and molecular biology methods; and (3) to study biological aspects of lung cancer, such as activation of cellular proto-oncogenes using the same material, and DNA adduct formation in lung vs. white blood cell DNA. Lung tissue will be collected from randomly selected patients with lung cancer. On the day before surgery, patients will be interviewed on their occupational and smoking histories. Resected lung material will be used to study fibre content, macroscopical changes, histological type and extent of the tumour, microscopical fibrosis, asbestos bodies, several enzyme parameters, oncogene activation (tumour, peripheral lung tissue, white blood cells), and DNA adduct formation (tumour, peripheral lung tissue, white blood cells). So far about 65 cases have been included; the project aims to double the number of subjects. No control group is included. Various electron-microscopical, biochemical, immunohistochemical, histological and molecular methods will be used.

TYPE: Case Series; Molecular Epidemiology
TERM: BMB; Biochemical Markers; DNA Adducts; Fibrosis; Histology; Occupation; Oncogenes; Tissue; Tobacco (Smoking)
SITE: Lung
CHEM: Asbestos; Mineral Fibres
TIME: 1988 - 1994

240 **Kaprio, J.** 01228
Univ. of Helsinki, Dept. of Public Health Science, Finnish Twin Cohort Study, Mannerheimintie 96 A, P.O.Box 52, 00014 Univ. of Helsinki, Finland (Tel.: +358 0 43461; Fax: 4346671)
COLL: Koskenvuo, M.J.; Pukkala, E.I.; Teppo, L.

Cancer in Twins

The study of twins may elucidate the relative role of different environmental factors in cancer aetiology. The Finnish Twin Cohort Study contains data on some 34,500 adult twins of the same sex. Zygosity was determined by questionnaire in 1975. Record linkage with the Finnish Cancer Registry for cancer cases incident to 1989 has been carried out. Site-specific subsamples are being identified to examine how discordant identical pairs differ. The twin cohort has been extended to include all twin births between 1958 and 1986. A total of 23,500 twin pairs (of same or different sex), were found. This data base can be used for various genetic epidemiological studies.

TYPE: Case-Control; Cohort
TERM: Environmental Factors; Record Linkage; Registry; Twins
SITE: All Sites
REGI: Finland (Fin)
TIME: 1975 - 1995

241 **Knekt, P.B.** 04461
Social Insurance Institution, P.O. Box 78, 00381 Helsinki, Finland (Tel.: +358 0 4343591; Fax: 4343333 ; Tlx: 122375 KELA-SF)
COLL: Aromaa, A.; Maatela, J.; Hakulinen, T.R.; Saxén, E.A.; Teppo, L.; Aaran, R.K.; Hakama, M.K.; Nikkari, T.; Alfthan, G.; Peto, R.

Serum Antioxidants and Risk of Cancer

The aim of the present epidemiological study is to examine the role of vitamin A, vitamin E and selenium in the aetiology of cancer by studying the association between serum beta carotene, retinol, retinol-binding protein, alpha-tocopherol, and selenium levels and the incidence of cancer. During 1966-1972 the Social Insurance Institution's Mobile Clinic Health Examination Survey carried out multiphasic screening examinations in various parts of Finland. As a part of that survey, serum samples were taken and stored at -20 degrees C. Ten years later the present longitudinal study was started to investigate the association between serum micronutrient levels and the incidence of cancer. The study population comprised about 40,000 men and women, initially aged 15-99 years. Cancer incidence data from the nationwide Finnish Cancer Registry were linked to the health examination data set. During a mean follow-up of about ten years, cancer was diagnosed in 1,100 persons. The serum micronutrient levels were determined from the stored serum samples collected from these cancer patients and from 2,000 controls matched for sex, municipality and age. The beta carotene, retinol, and alpha-tocopherol measurements were performed using HPLC and the selenium determinations by electrothermal atomic absorption spectrometry. The data analyses are mainly based on the conditional logistic model.

TYPE: Case-Control; Cohort
TERM: BMB; Metals; Registry; Sero-Epidemiology; Serum; Vitamins
SITE: All Sites; Breast (F); Colon; Lung; Prostate; Rectum; Stomach; Uterus (Cervix)
CHEM: Beta Carotene; Retinoids; Selenium; Tocopherol
REGI: Finland (Fin)
TIME: 1980 - 1994

242 **Knekt, P.B.** 04539
Social Insurance Institution, P.O. Box 78, 00381 Helsinki, Finland (Tel.: +358 0 4343591; Fax: 4343333 ; Tlx: 122375 KELA-SF)
COLL: Seppänen, R.; Aromaa, A.; Teppo, L.

Diet and Cancer

The aim of the present epidemiological study is to examine the predictive role of different nutrients on the risk of cancer. During 1966-1972 the Social Insurance Institution's Mobile Clinic Health Examination Survey carried out multiphasic screening examinations in various parts of Finland. As a part of that study, a dietary survey of about 10,000 men and women, initially aged 15-99 years, was carried out. The data on the average daily food consumption during the previous year were collected by a dietary history interview. Cancer incidence data from the nationwide Finnish Cancer Registry were linked to the health examination data set. During a mean follow-up of 20 years, cancer was diagnosed in about 800 persons. The associations between the nutrients and the incidence of cancer are being estimated, mainly based on Cox's life-table regression model.

TYPE: Cohort
TERM: Diet; Registry
SITE: Breast (F); Colon; Lung; Prostate; Rectum; Stomach
REGI: Finland (Fin)
TIME: 1986 - 1995

243 **Koskinen, H.** 04542
Inst. of Occupational Health, Topeliuksenkatu 41 a A, 00250 Helsinki, Finland (Tel.: +358 90 47471)
COLL: Hernberg, S.G.; Heikkilä, P.; Kauppinen, T.P.; Kurppa, K.; Lahdensuo, A.; Liippo, K.; Nurminen, M.; Partanen, T.J.; Tala, E.; Tossavainen, A.

Lung Cancer and Occupational Exposure

There are about 2,000 cases of lung cancer annually in a population of 4.9 million in Finland. In this pilot study it is proposed to evaluate a large scale multi-centre epidemiological study on occupational causes of lung cancer in Finland. The feasibility study was conducted in the catchment area of two university lung clinics where incident, living lung cancer patients are interviewed. The exposure experience of the study cases will be supplemented with an interview of referents from the base population of the clinics (random sample of the base population as well as referents with some other cancers than lung cancer).

TYPE: Case-Control
TERM: Occupation
SITE: Lung
TIME: 1988 - 1994

244 **Partanen, T.J.** 04780
Inst. of Occupational Health, Topeliuksenkatu 41a A, 00290 Helsinki, Finland (Tel.: +358 0 47471; Tlx: 121394)
COLL: Hernberg, S.G.; Koskinen, H.; Kauppinen, T.P.; Degerth, R.; Pukkala, E.I.

Primary Pancreatic Cancer and Chemical Exposures at Work

This national case-referent study aims at revealing connections between pancreatic cancer and industries, occupations, and occupational chemical exposures previously prevalent in Finland. The case series comprised all histologically confirmed primary exocrine pancreatic cancer cases aged 40-74 years at diagnosis notified to the Finnish Cancer Registry during the period 1984-87 and deceased during the same period. The referent series consisted of the deceased cases of stomach, colon, and rectal cancers, diagnosed in Finland during the same period. The next-of-kin of both the cases and the referents received a postal questionnaire on the job history of the case or the referent as well as on smoking; alcohol, coffee and sugar consumption; and diabetes mellitus. The work histories were coded into industry, job titles and chemical exposures. The translation of work histories into exposures was done by two experienced industrial hygienists, using all the available sources of information. The data analysis concentrated on the estimation of the risk ratios (point estimates and confidence intervals) for the different industries, job titles, and exposures.

TYPE: Case-Control
TERM: Alcohol; Chemical Exposure; Coffee; Diabetes; Occupation; Registry; Tobacco (Smoking)
SITE: Pancreas
REGI: Finland (Fin)
TIME: 1989 - 1994

245 **Pukkala, E.I.** 03242
Finnish Cancer Registry, Liisankatu 21 B, 00170 Helsinki 17, Finland (Tel.: +358 0 135331; Fax: 1355378)
COLL: Kauppinen, T.P.

Occupation, Socio-Economic Status and Education as Risk Determinants of Cancer

All cancer cases diagnosed in Finland after the year 1970 are linked with the national census file from 31 December 1970. The data on occupation, socio-economic status and education are combined with data on cancer, and differences in cancer incidence are analysed by taking into account relevant information from various sources in Finland about possible risk or protective factors for the population groups in question. The results will be linked with a Finnish job exposure matrix (under preparation).

TYPE: Incidence
TERM: Alcohol; Diet; Education; High-Risk Groups; Occupation; Record Linkage; Registry; Socio-Economic Factors; Tobacco (Smoking)
SITE: All Sites
REGI: Finland (Fin)
TIME: 1988 - 1997

246 **Pukkala, E.I.** 04143
Finnish Cancer Registry, Liisankatu 21 B, 00170 Helsinki 17, Finland (Tel.: +358 0 135331; Fax: 1355378)
COLL: Lahermo, P.; Gustavsson, N.; Hakulinen, T.R.

Comparison of Cancer Incidence and Geochemical Data

All cancer incidence data of the Finnish Cancer Registry from the year 1953 and later (approximately 600,000 cases) are connected with the trace elements data obtained by the Geochemical Survey of Finland covering the whole of the area of Finland. After adjustment for known confounding factors associations between trace elements and cancer incidence will be sought.

TYPE: Correlation; Incidence
TERM: Registry; Trace Elements
SITE: All Sites
REGI: Finland (Fin)
TIME: 1993 - 1997

247 **Riihimaki, V.** 03991
Inst. of Occupational Health, Dept. of Industrial Hygiene and Toxicology, Topeliuksenkatu 41a A, 00250 Helsinki, Finland (Tel.: +358 0 47471; Tlx: 121394)
COLL: Asp, S.; Hernberg, S.G.; Pukkala, E.I.

Mortality and Cancer Morbidity among Chlorinated Phenoxyacid Applicators in Finland

Chlorinated phenoxyacids have been associated with increased risk of soft tissue sarcomas and lymphomas. This ongoing cohort study aims to assess mortality and cancer morbidity patterns in nearly 2,000 men involved from 1955 through 1972 with the spraying of phenoxyacids (2, 4-D, 2,4,5-T) to control brushwood in forest plantations, by roadsides and railways and along electric power lines. The cohort was formed in 1972 from the personnel records of four authorities/companies. All workers who, by that time, had been assigned to spraying for at least two weeks were eligible. Subsequent exposure histories and smoking habits of the workers have recently been explored. The cohort has been followed-up to date of death or to appearance in the cancer registry as a collaborative effort of the

Institute of Occupational Health and the Finnish Cancer Registry, and the mortality as well as cancer morbidity experience will be compared with that of the general population.

TYPE: Cohort
TERM: Chemical Exposure; Herbicides; Occupation; Pesticides; Registry
SITE: All Sites; Lymphoma; Sarcoma
CHEM: 2,4,5-T; 2,4-D; Phenoxy Acids
OCCU: Herbicide Manufacturers; Herbicide Sprayers
REGI: Finland (Fin)
TIME: 1972 - 1994

248 **Sankila, R.** 03246
Finnish Cancer Registry, Liisankatu 21 B, 00170 Helsinki 17, Finland (Tel.: +358 0 176290)
COLL: Hakulinen, T.R.; Pukkala, E.I.; Teppo, L.

Multiple Cancer

The data base of the Finnish Cancer Registry on multiple cancer (about 8,000 cases in 1953 to 1985) is analysed in order to detect positive and negative associations between different types of cancer. The effects of several factors (including age, histology, treatment, etc.) will be analysed in depth.

TYPE: Cohort
TERM: Age; Histology; Multiple Primary; Prevention; Registry; Treatment
SITE: All Sites
REGI: Finland (Fin)
TIME: 1988 - 1994

249 **Sorsa, M.I.** 04608
Inst. of Occupational Health, Topeliuksenkatu 41 a A, 00250 Helsinki, Finland (Tel.: +358 0 47471; Fax: 4747208)
COLL: Peltonen, K.; Osterman-Golkar, S.

Development of Biological and Chemical Monitoring Methods for Risk Evaluation in 1,3-Butadiene Exposure

1,3-Butadiene is an unusually potent indirectly genotoxic chemical which is among the top 30 chemicals in the world production list. The aims of the present study are to develop biological (cytogenetics, Hb-alkylation) and chemical (different metabolites and conjugation products) methods applicable for human exposure assessment. The study includes both experimental (rat exposures) and worker studies in butadiene manufacturing (30 workers) and in the use of butadiene in paper lamination (30 workers). All studies include both on-site and off-site controls.

TYPE: Cross-Sectional; Methodology; Molecular Epidemiology
TERM: Animal; BMB; Chromosome Effects; Monitoring; Occupation; Plastics
SITE: Inapplicable
CHEM: Butadiene
OCCU: Petrochemical Workers; Plastics Workers
LOCA: Finland; Portugal; Slovak Republic
TIME: 1988 - 1994

Imatra

***250** **Jäppinen, P.T.** 05332
Enso Gutzeit Oy, Center for Occupational Health, 55800 Imatra, Finland (Tel.: +358 54 6894331; Fax: 6894330)
COLL: Boffetta, P.; Kogevinas, M.

Cancer Risk Among Workers in the Pulp and Paper Industry (Finnish Sub-Study)

The paper and pulp industry employs hundreds of thousands of workers worldwide. The few prospective epidemiological studies in this industry indicate that cancer risk, particularly for lung cancer, gastrointestinal cancer and neoplasms of the lymphatic tissue, may be elevated, but evidence is still not convincing. The objective of this international cohort study co-ordinated by IARC is to evaluate cancer risk in relation to specific processes and specific exposures in this industry. Personnel employed in plants producing pulp, paper and paper products and in mills involved in recycling will be included.

Cohorts are currently being assembled, and it is expected that the international study will include data for more than 100,000 workers. The Finnish sub-cohort is expected to consist of approximately 5,000 pulp and paper workers.

TYPE: Cohort
TERM: Occupation; Registry
SITE: All Sites
OCCU: Paper and Pulp Workers
REGI: Finland (Fin)
TIME: 1992 - 1996

Kuopio

251 **Notkola, V.J.** 04248
Kuopio Regional Inst. of Occupational Health, P.O. Box 93, 70701 Kuopio, Finland

Occupational Cancer Mortality among Farmers in Finland 1979-1985

The aim of this study is to compare the cause-specific cancer mortality among farmers in Finland during 1979-85 to the mortality of the total economically active population. The study is based on the Finnish farm register at 31 December 1978. Male and female farmers working in the farms at registered were defined as the population at risk. Data on all deaths during 1979-1990 are obtained from death certificate data in the Finnish register on causes of death. Morbidity information is obtained from the registers of the Finnish Cancer Institute. The total study population includes about 150,000 farmers. Estimation of relative cancer risks is made by use of a log-linear model and indirect age standardization has also been used.

TYPE: Cohort
TERM: Occupation; Registry
SITE: Brain; Hodgkin's Disease; Leukaemia; Lung; Prostate
OCCU: Farmers
TIME: 1983 - 1994

252 **Syrjänen, K.J.** 03693
Univ. of Kuopio, Dept. of Pathology, Kuopio Cancer Research Centre, P.O.B. 1627, 70211 Kuopio, Finland (Tel.: +358 71 162740; Fax: 162753)
COLL: Parkkinen, S.; Castrén, O.; Saarikoski, S.; Väyrynen, M.; Syrjänen, S.; Yliskoski, M.; Kellokoski, J.; Kataja, V.; Hippeläinen, M.; Chang, F.; Hongxiu, J.; Tervahauta, A.; Mäntyjärvi, R.

Natural History of Genital HPV Infections and Cervical Cancer

A long-term prospective follow-up study was started in October 1981 to assess the natural history of human papillomavirus (HPV) infections in the uterine cervix and their associations with CIN and cervical cancer. To date a total of 532 women have been followed-up for a mean of 83 (SD: 23) months, by cervical punch biopsy or Pap smears, and colposcopy repeated at 6-monthly intervals. The clinical course of the lesions is analysed using the life-table technique, as well as the Cox-model. Cervical swabs (for C. trachomatis and HSV), and HPV, HSV and CMV serology are done. PAP smears and punch biopsies are analysed for the cytopathic changes of HPV and for concomitant CIN. Expression of HPV-encoded proteins is demonstrated by the indirect immunoperoxidase (IP-PAP) technique on frozen sections. In fresh biopsy samples (or frozen sections), the immunocompetent cells within the in-situ inflammatory infiltrate are phenotypically characterised using monoclonal antibodies to define T cell subsets, NK (natural killer) cells and Langerhans cells. In-situ hybridisation, Southern blot and PCR techniques are used to detect HPV DNA with cloned HPV types 6, 11, 16, 18, 31, 33, 42 DNA probes. Biopsy specimens will also be tested for amplification of the major cellular oncogenes (c-onc) and growth factor receptors. All patients are interviewed for their sexual and smoking habits by a detailed questionnaire. HLA-typing is now being completed. The incidence and prevalence of cervical HPV infections in an unselected Finnish female population are being assessed in the mass-screening programme of the Finnish Cancer Society. Since 1985, females have been also invited to a third (treatment) group, now comprising 530 women, and randomised to four treatments: ionization, laser, cryotherapy and interferon. Since 1986, over 300 male partners of the treated women have been examined for HPV, and treated by either laser or interferon. Over 250 papers have been published since the onset of the project.

TYPE: Cohort; Intervention
TERM: Antibodies; Biopsy; Condyloma; DNA; HPV; Infection; Prevention; Screening; Sero-Epidemiology; Sexual Activity; Tobacco (Smoking); Treatment
SITE: Female Genital; Male Genital
TIME: 1981 - 1995

253 **Syrjänen, K.J.** 05029
Univ. of Kuopio, Dept. of Pathology, Kuopio Cancer Research Centre, P.O.B. 1627, 70211 Kuopio, Finland (Tel.: +358 71 162740; Fax: 162753)
COLL: Eskelinen, M.; Johansson, R.; Tuomisto, J.; Jägeroos, H.; Jänne, J.; Alhonen, L.; Kettunen, K.; Länsimies, E.; Soimakallio, S.; Partanen, K.; Airaksinen, O.; Vehviläinen-Julkunen, K.; Meriläinen, P.; Naukkarinen, A.; Kosma, V.M.; Syrjänen, S.; Penttilä, I.; Hämäläinen, E.; Mahlamäki, E.; Heinonen, K.; Mononen, I.; Lehtonen, J.; Ahonen, R.; Mäntyjärvi, R.; Hippeläinen, M.

Female Breast Cancer

A long-term prospective follow-up study was started in March 1990 to assess the epidemiology and risk factors as well as the biology of female breast cancer in Kuopio province, Finland. Three series of women are included: (1) all women with primary breast cancer in Kuopio province; (2) all women with a benign breast lesion diagnosed in Kuopio University Hospital, and a series of age-matched healthy controls from the population registry. The study includes: (1) evaluation and assessment of risk factors; (2) the possibilities of prevention; (3) early diagnosis and (4) prognostic evaluation as well as treatment and follow-up. In addition, samples of tumours and of blood will be taken from all patients enrolled to be analysed using modern biomedical technology, including DNA-techniques, chromosomal analyses and biochemical assays. This is a multi-institutional project with workers from 15 different institutions in Kuopio University and Kuopio University Hospital. It is hoped that this multi-disciplinary approach, will permit assessment of the risk factors for breast cancer, as well as the mechanisms regulating its biological behaviour which could be used as prognostic indicators of this disease.

TYPE: Case-Control
TERM: BMB; Blood; Prevention; Prognosis; Registry; Treatment
SITE: Benign Tumours; Breast (F)
REGI: Finland (Fin)
TIME: 1990 - 2005

Tampere

254 **Hakama, M.K.** 00198
Univ. of Tampere, Dept. of Public Health, PB 607, 33101 Tampere, Finland (Tel.: +358 31 156111; Fax: 156057 ; Tlx: 22415)
COLL: Kallio, M.; Pukkala, E.I.

Effect of a Mass Screening Programme on the Risk of Cervical Cancer

The aim of the project is to evaluate the effect of an organized mass screening programme for cervical cancer on the incidence and mortality of cervical cancer and to monitor changes in efficacy by time and by population groups. The programme is nationwide and population-based and consists of essential elements such as identification of target population, quality control and information on attendance and risks. Data collection is based on the national population registry, the cancer registry and registration of deaths. The Finnish female population is about 2.5 million and the organized programmes cover the ages 30 to 60 years.

TYPE: Cohort
TERM: Registry; Screening
SITE: Uterus (Cervix)
REGI: Finland (Fin)
TIME: 1970 - 2000

***255** **Hakama, M.K.** 05403
Univ. of Tampere, Dept. of Public Health, PB 607, 33101 Tampere, Finland (Tel.: +358 31 156111; Fax: 156057 ; Tlx: 22415)
COLL: Kallio, M.; Pukkala, E.I.

Effect of a Mass Screening Programme on the Risk of Breast Cancer

The aim of the project is to evaluate the effect of an organized mass screening programme for breast cancer on the incidence and mortality of breast cancer and to monitor changes in efficacy by time and by population groups. The programme is nationwide and population-based and consists of essential elements such as identification of target population, quality control and information on attendance and risks. Data collection is based on the national population registry, cancer registry and registration of deaths. The Finnish female population is about 2.5 million and the organized programmes cover the ages 50 to 64 years.

TYPE: Cohort
TERM: Registry; Screening
SITE: Breast (F)
REGI: Finland (Fin)
TIME: 1987 - 2000

***256 Hakama, M.K.** 05533
Univ. of Tampere, Dept. of Public Health, PB 607, 33101 Tampere, Finland (Tel.: +358 31 156111; Fax: 156057 ; Tlx: 22415)
COLL: Muñoz, N.; Buiatti, E.; Parkin, D.M.

Review of Chemoprevention Trials for Cancer

Chemoprevention studies have been extensively used in recent years to determine the contribution of dietary constituents - including micronutrients and oligoelements - to cancer aetiology, and their potential for preventive interventions. The great majority of such studies use intermediate endpoints - cellular changes which have been associated with cancer (with varying degrees of certainty). Work began late in 1992 to create an inventory of all such studies, and a meeting will be held in early 1994 where the results of completed studies will be reviewed.

TYPE: Intervention
TERM: Micronutrients; Prevention; Trace Elements
SITE: All Sites
TIME: 1992 - 1994

Albi

257 **Grosclaude, P.** 05110
Registre des Cancers du Tarn, Recherche en Epidémiologie et Prévention, Chemin des 3 Tarn, 81000 Albi, France (Tel.: +33 63475951)
COLL: Roumagnac, M.; Duchene, Y.

Aetiology of Prostatic Cancer
This is a case-control study of prostatic cancer and potential risk factors such as sexual activity, various occupational risk factors, and dietary habits. Patients with prostatic cancer (histologically diagnosed from 1 October 1991) and controls living in the 'departement' of Tarn will be studied. Two controls will be obtained for each case, one from the general population and one hospital patient. The information will be collected by investigating doctors with the use of a questionnaire. The survey will be carried out on 200 patients and 400 controls. Recruitment will last one year. This study is part of a multi-centre inquiry in collaboration with two other cancer registries (Martinique and Geneva).

TYPE: Case-Control
TERM: Diet; Occupation; Registry; Sexual Activity
SITE: Prostate
REGI: Geneva (Swi); Martinique (Mar); Tarn (Fra)
TIME: 1991 - 1994

Caen

258 **Gignoux, M.** 03721
CHU Niveau 3, Registre des Tumeurs Digestives, Côte de Nacre, 14040 Caen Cedex, France (Tel.: +33 31063106/4464)
COLL: Launoy, G.; Pottier, D.; Rougereau, A.

Epidemiology of Cancers of the Digestive Tract in Calvados
The aim of the study (covering a population of over 500,000) is to evaluate the annual incidence by localization, age and sex, of carcinomas of the digestive tract. Data are collected from public and private hospitals, laboratories and physicians. Special attention is accorded to environmental and geographical patterns, profession, pre-cancerous diseases, methods of treatment and survival, particularly for carcinoma of the oesophagus for which Calvados is a high risk area. Specific studies include: (1) Identification of high-risk groups for oesophageal cancer. Areas having higher incidence of oesophagel cancer than others in the "departement", mainly rural, have been identified and important differences between socio-professional categories within these areas noted. Populations at risk from the three French digestive tract cancer registries (Cote d'Or, Haut Garonne, Calvados) will be compared. (2) Surveys on survival of patients with digestive cancer and its possible recent improvement. For several localizations, prognostic factors are studied. For instance, rural environment has been identified as a bad prognostic factor for female colorectal cancer. More than ten years (1978-1987) of data are available, and studies are carried out on the trends of the different digestive cancers: oesophagus, stomach, colorectal, liver and pancreas.

TYPE: Incidence
TERM: Environmental Factors; Geographic Factors; Occupation; Rural; Survival; Treatment; Trends
SITE: Colon; Gallbladder; Liver; Oesophagus; Pancreas; Rectum; Stomach
REGI: Caen (Fra)
TIME: 1978 - 1995

***259** **Launoy, G.** 05459
CHU Niveau 3, Registre des Tumeurs Digestives, Ave. Cäte-de-Nacre , 14040 Caen Cedex, France (Tel.: +33 31063106/6392)
COLL: Pottier, D.; Gignoux, M.; Faivre, J.; Milan, C.; Pienkowski, P.

Alcohol, Diet and Cancer of the Oesophagus in France
A multicentre case-control study is conducted in three French areas (Calvados, Cote d'Or, Haute-Garonne) to determine the effect of different types of beverages on the risk of cancer on the oesophagus and the possible protective role of vitamins. In particular, the study will enable the

assessment of possible differences between the effect of apple-derived versus other alcohol, as well as between strong and light alcoholic drinks. It is expected to include 300 cases and 600 controls in the study.

TYPE: Case-Control
TERM: Alcohol; Diet; Registry; Vitamins
SITE: Oesophagus
REGI: Côte d'Or (Fra); Calvados (Fra); Haute-Garonne (Fra)
TIME: 1991 - 1994

Dijon

260 **Faivre, J.** 03860
Registre des Cancers Digestifs, Fac. de Médecine, Registre des Tumeurs, 3, Faubourg Raines, 21000 Dijon Cedex, France (Tel.: +33 80652323)
COLL: Backley, D.; Biasco, G.; Boutron, M.C.; De Oliveira, H.; Estève, J.; Giacosa, A.; Hill, M.J.; Kasper, H.; Maskens, A.; Thomson, M.H.; Wiebecke, B.; Wilpart, M.

Case-Control Study of Patients with Adenomatous Polyps or Cancer of the Large Bowel

Two separate studies have been prepared. In an epidemiological study three groups of cases will be included in each centre: small adenomas (n=150), large adenomas (n=150), adenocarcinomas (n=150), and two groups of controls: general population controls (n=300), and polyp free controls (n=300). Most information will be obtained by means of questionnaires administered by specially trained dieticians. The following information will be collected: personal history (demographic data, medical history, anthropometric data), description of adenomas or cancers, dietary history (using a diet history questionnaire collecting dietary information by meal). The objective of the second study, a clinical study, is to obtain in each centre 15 cases with small adenomas, 15 cases with large adenomas, 15 with adenocarcinomas, and 15 polyp free controls, for whom diet history, faeces, serum and cell kinetic studies will be available.

TYPE: Case-Control
TERM: Alcohol; BMB; Diet; Faeces; Lifestyle; Premalignant Lesion; Serum
SITE: Colon
LOCA: Belgium; France; Germany; Italy; Portugal; United Kingdom
TIME: 1986 - 1994

261 **Faivre, J.** 04428
Registre des Cancers Digestifs, Fac. de Médecine, Registre des Tumeurs, 3, Faubourg Raines, 21000 Dijon Cedex, France (Tel.: +33 80652323)
COLL: Bedenne, L.; Durand, G.; Milan, C.; Arveux, P.; Boutron, M.C.

Evaluation of Mass Screening for Colorectal Cancer

This controlled trial compares the effect of systematic screening for colorectal adenomas in a well defined population of 45,000 subjects aged 45-74 with a comparable population in which there is no screening. In the test group the subjects will be offered faecal occult blood testing. Initially the whole population will be sent an explanatory letter and a document on colorectal cancer and the possibilities for screening. Information will also be given in local newspapers, municipal newsletters, local and regional radio and television programmes. The test will be prescribed by the general practitioners (GP) during the first three months of the study, then mailed to the subjects who have not received it from their GP. Tests will be sent to a reference laboratory. A total colonoscopy will be offered to individuals with a positive test. The first analyses will examine the acceptance rate and the positive predictive value of the tests for cancers and polyps. Refusal and control subjects will be followed through the records of the digestive tract cancer registry for the area. Long term evaluation of efficacy of screening wil be judged on reduction of mortality from colorectal cancer in the screened population.

TYPE: Cohort; Intervention
TERM: Faeces; Registry; Screening
SITE: Colon; Rectum
REGI: Dijon I (Fra)
TIME: 1987 - 1994

Fontenay-aux-Roses

***262** **Bard, D.** 05446
Inst. de Protection et de Sureté Nucléaire, LEADS, SEGR, DPHD, Ave. General Leclerc, BP No. 6, 92265 Fontenay-aux-Roses, France (Tel.: +33 1 46547669; Fax: 46548829)
COLL: Hubert, P.; Giraud; Gelas, J.M.

Cancer Risk in Workers of the Nuclear Industry in France

The aim of this cohort mortality study (1960-1991) is to assess cancer risks that might be linked with occupational chronic exposure to low levels of ionizing radiation. Any type of cancer death will be considered, with special attention on leukaemia. The study population is made of all workers ever monitored for ionizing radiation exposure (about 50,000) within the two major French companies in this field. Individual exposure data will be collected from corporate dosimetry records, bearing on both external and internal dose. Vital status will be assessed through the corporate social insurance system. Cause of death will be ascertained through death certificates obtained from the national registration system for causes of death. Concurrent exposure to chemicals, smoking habits and lifestyle data will be collected from corporate medical records. National vital statistics will be used for external comparisons. Sex, age, age at first exposure, yearly dose rate, cumulative lifelong dose, will be considered, along with other relevant variables. Internal comparisons will be made to yield a relative excess risk estimate per unit increase in dose.

TYPE: Cohort
TERM: Dose-Response; Occupation; Radiation, Ionizing
SITE: All Sites; Leukaemia
OCCU: Radiation Workers
TIME: 1992 - 1996

263 **Tirmarche, M.** 03473
Inst. de Protection et de Sureté Nucléaire, LEADS, SEGR, DPHD, Ave. General Leclerc, BP No. 6, 92265 Fontenay-aux-Roses, France (Tel.: +33 1 46547194; Fax: 46548829 ; Tlx: 204841)
COLL: Gelas, J.M.; Flamant, R.

Uranium Miners in France

Lung cancer has been identified as being linked to exposure to radon daughters in uranium miners in the USA, Czechoslovakia and Canada. The aim of this study is to verify this relation for French uranium miners and to study, if possible, the risk of death due to other cancers in this group of workers. The mortality of this cohort of uranium underground miners will be compared with that expected from the national statistics for the French male population of the same age groups and the same calendar periods. The vital status of these miners has been studied up to December 1985. The analysis of cancer mortality is limited to those miners having experienced more than two years of underground mining. The mean period of survey is 26 years, and the mean duration of exposure to radon and its decay products is 14 years. As mean annual exposure and working conditions were modified during the year 1956, the analysis of cancer mortality is based on two groups, separated by year of first exposure to radon and its decay products (before or since 1956). In both cohorts, a statistically significant excess of lung cancer can be observed. Initial results have been published in the Proceedings of the International Congress of IRPA VII, Sydney, 10-17 April 1988, vol. 1, p. 171-175, Pergamon Press. Intermediate analysis is described in Cancer Detection & Prevention 6:169-172, 1992 and a more detailed report of the mortality data and dose-relationship between cumulative exposure to radon and its decay proudcts and lung cancer mortality will be published in Br. J. Cancer in 1993. The cohort will now be extended to a larger group of about 3,000 uranium miners in order to estimate more precisely the risk of lung cancer linked to low annual exposures to radon (1-3 WLM per year).

TYPE: Cohort
TERM: Dose-Response; Metals; Mining; Occupation; Radiation, Ionizing
SITE: All Sites; Lung
CHEM: Radon; Uranium
OCCU: Miners, Uranium
TIME: 1981 - 1994

FRANCE

***264** **Tirmarche, M.** 05497

Inst. de Protection et de Sureté Nucléaire, LEADS, SEGR, DPHD, Ave. General Leclerc, BP No. 6, 92265 Fontenay-aux-Roses, France (Tel.: +33 1 46547194; Fax: 46548829 ; Tlx: 204841)

COLL: Tymen, G.

Indoor Radon and Lung Cancer Risk

A case-control study has been initiated in the region of Brittany and Vendee (France) in order to estimate the risk of lung cancer in subjects exposed to indoor radon. This is part of a larger European study coordinated by the Commission of the European Communities. This French study contributes 600 cases and 1,200 controls; the protocol is comparable to that adopted in Germany, Belgium and Luxemburg in the Ardennes-Eiffel study. The risk of lung cancer related to the last 30 years of exposure to indoor radon will be estimated, taking into account smoking history (active and passive) and occupational exposures.

TYPE: Case-Control
TERM: Environmental Factors; Late Effects; Radiation, Ionizing; Tobacco (Smoking)
SITE: Lung
CHEM: Radon
TIME: 1991 - 1996

Grenoble

***265** **Perdrix, A.M.** 05473

Inst. Univ. de Med. du Travail et d'Ergonomie (IUMTE), Service de Med. du Travail, 38043 Grenoble Cedex 09, France (Tel.: +33 76765442; Fax: 76544201)

COLL: Romazzini, S.; Pellet, F.; Lutz, J.-M.; Gonthier, C.; De Gaudemaris, R.

Mortality among Workers Exposed to Tungsten Carbide and Cobalt

The aim of the study is to determine whether mortality from lung cancer is higher among workers exposed to tungsten carbide and cobalt than in the general departmental population. A historical cohort study is conducted among 3,850 workers exposed between 1971 and 1981. Job history is collected for each subject. The observed numbers of deaths from different causes is compared with the expected numbers based on gender, five-year age-group and calendar-year specific rates from the departmental population.

TYPE: Cohort
TERM: Metals; Occupation
SITE: Lung
CHEM: Cobalt; Tungsten Carbide
OCCU: Metal Workers
TIME: 1992 - 1995

La Tronche

266 **Faure, J.R.** 02612

Fac. de Médecine, Lab. de Médecine Légale et de Toxicologie, Chemin Duhamel, 38700 La Tronche, France (Tel.: +33 76424888/377)

COLL: Barrett, L.; De Gaudemaris, R.; Thony, C.

Cancer Death Rates in a Population of Screw-Cutters

The role of solvents and cutting oils in the occurrence of cancers other than skin cancers has been suspected, as a result of experiments and of some epidemiological surveys. Death statistics for the population of a district where the screw-cutting industry employs one quarter of the employed population (12,000 workers, of a total population of 115,000 with 53,000 gainfully employed), are being analysed to find out whether there is an excess of cancer mortality among screw-cutting workers. This study will be complemented by subsequent establishment of a cancer registry in the district. The following information is available in death certificates: age, sex, residence, occupation at time of death or before retirement, cause of death, anatomical site of cancer. The control group is made up of the general population, for which the same information is available. A prior study established the levels of trichloroethylene in the workshop atmosphere, and in the blood, urine and exhaled air of 188 exposed workers.

TYPE: Mortality
TERM: Chemical Exposure; Environmental Factors; Occupation; Solvents
SITE: All Sites
CHEM: Hydrocarbons, Halogenated; Mineral Oil; Trichloroethylene
OCCU: Screw Cutters
TIME: 1979 - 1994

Lille

267 **Fenaux, P.** 04959
Centre Hospitalier Universitaire, Serv. des Maladies du Sang, 1, Place de Verdun , 59700 Lille, France (Tel.: +33 20444348; Fax: 20444094)
COLL: Haguenoer, J.M.Y.; Nisse, C.; Pamart, B.; Quiquandon, I.; Preudhomme, C.

Environmental Factors in the Pathogenesis of Myelodysplastic Syndromes

Myelodysplastic syndromes (MDS) are the most frequent preleukaemic states. About 65 new cases of MDS are referred to this institution every year. A case-control study on the role of environmental factors in the pathogenesis of MDS is being conducted, using the method of Siemiatycki (J. Nat. Cancer Inst. 66:217-225, 1981). The study includes one control for each patient, sex- and age-matched with the patient, and living in the same area. Interviewers, who are not informed of the diagnosis, carry out an in-depth occupational, environmental and health study at the homes of cases and controls, using a checklist of specific chemicals and substances. Results of the study will be pooled with those from similar studies, using the same methods, performed in Cardiff, Bournemouth and Leeds in the United Kingdom.

TYPE: Case-Control
TERM: Chemical Exposure; Environmental Factors; Occupation; Premalignant Lesion
SITE: Myelodysplastic Syndrome
TIME: 1991 - 1994

268 **Lefebvre, J.L.** 04838
Centre Oscar Lambret, Head and Neck Oncology Dept., Rue F. Combemale , 59020 Lille Cedex, France
COLL: Adenis, L.; Joveniaux, A.; Cambier, L.

Prospective Epidemiological Study of Head and Neck Cancers

This prospective study concerns new patients presenting with at least one cancer originating in the upper aerodigestive tract, particularly oral cavity, oropharynx, larynx and hypopharynx. The following variables are recorded and compared with for the overall population and between these four sites; sex, age, marital status, smoking (duration of exposure, type of tobacco smoked, daily consumption, total consumption before diagnosis, pack years, etc.), drinking habits and occupational exposures (heat, cold, humidity, dryness, coal dust, plaster, cement or sand dust, wood dust, metal dusts or fumes, vegetal dust, chemical dusts or fumes). Special attention is given to the comparison of smoking and drinking habits and occupational exposure.

TYPE: Case Series
TERM: Alcohol; Hygiene; Occupation; Tobacco (Smoking)
SITE: Hypopharynx; Oral Cavity; Oropharynx
TIME: 1986 - 1994

Lyon

***269** **Armstrong, B.K.** 05539
Int. Agency for Research on Cancer, 150, cours Albert Thomas, 69372 Lyon Cedex 08, France (Tel.: +33 72738485; Fax: 72738575 ; Tlx: 380 023)
COLL: English, D.R.; Gopalan, H.N.B.; Hill, D.J.; Kjellstroem, T.; Koren, H.S.; Kricker, A.; Le Marchand, L.; McMichael, A.J.; Nakazawa, H.; Weatherhead, E.; Yamasaki, H.

International Research Programme on Health, Solar UV Radiation and Environmental Change (Intersun)

The aims of the study are: to evaluate accurately the quantitative relationship between solar ultra violet radiation at the surface of the earth (UVR) and human health effects, develop reliable predictions of the health consequences of changes in UVR, provide baseline estimates of the incidence of health effects of

UVR in representative populations around the world, and develop practical ways of monitoring change in these effects over time in relation to environmental and behavioural change. UV irradiance, health effects of UV exposure, consitutional sensitivity to the sun and present and lifetime sun exposure will be measured in 12-15 population centres in different parts of the world over a 5-year period. Case-control and cross-sectional studies of the relationship of UV exposure to health effects will be carried out in these populations. The usefulness of biological markers of UV exposure to the skin as measures of UV exposure and effects will be evaluated. Standard techniques for analysis of descriptive and case-control studies will be used. Methods will be developed for adequate analysis of ecological correlations appropriately adjusted for confounding.

TYPE: Case-Control; Correlation
TERM: Biomarkers; Monitoring; Radiation, Ultraviolet
SITE: Melanoma; Skin
LOCA: Australia; France; United States of America
TIME: 1993 - 1998

270 **Boffetta, P.** 04422
Int. Agency for Research on Cancer, Unit of Analytical Epidemiology, 150, cours Albert Thomas, 69372 Lyon Cedex 08, France (Tel.: +33 72738485; Fax: 72738575 ; Tlx: 380023 circ f)
COLL: Wild, C.; Nakazawa, N.; Shuker, D.; Saracci, R.; Boyle, P.; Kyrtopoulos, S.; Henry-Amar, M.; Hemminki, K.; Cerny, T.; Dicato, M; Karjalainen, S.; Juliusson, G.; Katsouyanni, K.; Kvinnsland, S.; Levi, F.G.; Lopez, J.J.; Martin-Moreno, J.; Pangalis, G.A.; Rilke, F.; Simonato, L.; Storm, H.H.; Swerdlow, A.J.; Van Leeuwen, F.E.; Van Oosterom, A.T.; Plesko, I.; Sedkackova, E.; Walewski, J.; Zaridze, D.; Diehl, V.

Markers of DNA Damage and Risk of Second Malignancy Following Chemotherapy

A collaborative group of major hospitals treating Hodgkin's disease (HD) patients will be established to create a central register of HD patients. These patients will be followed up for response to therapy and occurrence of second malignancies, particularly leukaemia, non-Hodgkin's lymphoma and lung cancer. In parallel to the registration of index and second cancers, blood samples will be taken from each patient, immediately following diagnosis, during therapy and subsequent to therapy. The red and white blood cells will be separated and frozen. Two case-control investigations will then be carried out, on response to therapy and risk of second malignancy respectively. For patients who will not respond to theray and for a sample of patients who will respond to therapy, as well as for patients who develop a second malignancy and for matched controls, who remain free of a second malignancy, the blood samples will be analysed for markers of DNA damage. In case markers of DNA damage are shown to be linked to response to therapy or to risk of second malignancy, the frequency and the level of such markers should be increased among cases as compared to controls. A total of at least 6,000 HD patients will be included in the study, from whom 150 cases of secondary leukaemia are expected within a 10-year period. Markers of DNA damage that are currently available include DNA adducts of alkylating agents, such as procarbazine, dacarbazine, and nitrogen mustards (mechlorathamine), mutation in ras oncogenes, and activation of tumour suppressor genes such as p52. Unspecific markers of damage, such as micronuclei and SCE will be investigated in a sub-group of patients.

TYPE: Case-Control; Cohort
TERM: BMB; Biomarkers; Blood; Chemotherapy; DNA Adducts; Drugs; Micronuclei; Multiple Primary; Oncogenes; Registry; SCE
SITE: All Sites; Hodgkin's Disease
CHEM: Dacarbazine; Nitrogen Mustard; Procarbazine
LOCA: Belgium; Czech Republic; Denmark; Finland; France; Germany; Greece; Italy; Luxembourg; Netherlands; Norway; Poland; Russian Federation; Slovak Republic; Spain; Sweden; Switzerland; United Kingdom
REGI: Denmark (Den); Finland (Fin); Norway (Nor); Padova (Ita); Poland (Pol); Sweden (Swe); Vaud (Swi)
TIME: 1993 - 2004

271 **Boffetta, P.** 05193
Int. Agency for Research on Cancer, Unit of Analytical Epidemiology, 150, cours Albert Thomas, 69372 Lyon Cedex 08, France (Tel.: +33 72738485; Fax: 72738575 ; Tlx: 380023 circ f)
COLL: Saracci, R.; Kogevinas, M.; Wong, O.; Steenland, K.N.; Nordberg, G.F.; Fanning, D.; Kazantzis, G.; Davies, J.; Cocco, P.L.; Szeszenia-Dabrowska, N.

International Collaborative Study on Lead-Exposed Workers

Cohort studies of workers exposed to lead have suggested increases in lung and stomach cancers. The first phase of this project is a combination of existing cohorts, which will be reanalysed in an identical manner with respect to duration of exposure and time since first exposure. On the basis of the results of the first phase and recruitment of new cohorts, further steps of the project may include a combined update of existing and new cohorts, a comprehensive industrial hygiene survey and the conduct of nested case-control studies.

TYPE: Cohort
TERM: Metals; Occupation; Registry; Time Factors
SITE: All Sites; Lung; Stomach
CHEM: Lead
OCCU: Battery Plant Workers; Smelters, Lead
REGI: Sweden (Swe)
TIME: 1991 - 1994

***272** **Boffetta, P.** 05514
Int. Agency for Research on Cancer, Unit of Analytical Epidemiology, 150, cours Albert Thomas, 69372 Lyon Cedex 08, France (Tel.: +33 72738485; Fax: 72738575 ; Tlx: 380023 circ f)
COLL: Merler, E.; Pompe-Kirn, V.; Zaridze, D.; Garcia-Gomez, M.; Sällsten, G.

Cancer Risk Among Mercury Miners

The objective of the study is to examine cancer risk among workers employed in mercury mining and milling in Europe. Previous epidemiological studies have suggested a possible increased risk of lung cancer. A collaborative group of researchers from the four European countries with large mercury mines was formed. The total population suitable for the study is approximately 10,000 workers. All cohorts will be followed for mortality or cancer incidence. An industrial hygienist will evaluate the levels of exposure to mercury, silica, radon and other possible carcinogens.

TYPE: Cohort
TERM: Dusts; Metals; Occupation; Registry
SITE: All Sites; Brain; Kidney; Lung
CHEM: Mercury; Radon; Silica
OCCU: Millers; Miners
LOCA: Italy; Slovenia; Spain; Ukraine
REGI: Slovenia (Slvn)
TIME: 1992 - 1996

273 **Bosch, F.X.** 03857
Int. Agency for Research on Cancer, Unit of Field and Intervention Studies, 150, cours Albert Thomas, 69372 Lyon Cedex 08, France (Tel.: +33 72738485; Fax: 72738575 ; Tlx: 380023 circ f)
COLL: Muñoz, N.; Rodriguez, M.C.; Hernandez, J.M.; Castillo, R.; Grifols, R.; Lluch, A.; Plasencia, J.; Moreno, V.; Castellsague, X.

Follow-up of a Cohort of HBsAg-positive Blood Donors in Catalonia

Mortality data from the region of Catalonia in Spain suggest that it is a high-risk area for liver cancer within the context of the Western industrialized world: death rates in 1983 were 12.8 among males and 8.3 among females. A study has been initiated to assemble a cohort of about 3,000 HbsAg-positive blood donors and to link their names with local death certificate files to determine liver cancer risk. Data are being obtained from five major blood banks operating in the area. An analysis was conducted in 1987 linking the names of the HbsAg carriers to the 1981 census. No excess of liver cancer could be identified. Current work is in progress using the 1990 census data.

TYPE: Cohort
TERM: Antigens; Infection; Virus
SITE: Liver
LOCA: Spain
TIME: 1985 - 1994

274 **Bosch, F.X.** 04710

Int. Agency for Research on Cancer, Unit of Field and Intervention Studies, 150, cours Albert Thomas, 69372 Lyon Cedex 08, France (Tel.: +33 72738485; Fax: 72738575 ; Tlx: 380023 circ f)

COLL: Muñoz, N.; Peto, J.; Schiffman, M.H.; Shah, K.V.; Manos, M.; Kurman, R.; Sherman, R.; Alihonou, E.; Barry, T.M.; Bayo, S.; Crespo de Britton, R.; Daudt, A.; Gauthier, P.; Ghadirian, P.; Kitinya, J.N.; Malik, M.A.O.; Puig-Tintoré, L.M.; Ríos-Dalenz, J.L.; Rolón, P.A.; Tafur, L.; Teyssie, A.R.; Torroella, M.; Vila Tapia, A.; Wabinga, H.; Zatonski, W.A.

International Biological Study on Cervical Cancer

The International Biological Study on Cervical Cancer (IBSCC) is an international project aimed at creating a repository of cervical cancer tissue for use in studying markers of exposure to known or suspected risk factors. In the first phase, samples of invasive cervical cancer have been collected in 20 countries with varying incidence rates of cervical cancer. All cases submitted a biopsy slide to be reviewed. DNA/RNA hybridization methods will be used to assess the prevalence of specific types of HPV. All the assays will be performed in the same laboratory, using PCR band methods. A brief questionnaire has been used to assess exposure to other known risk factors for cervical cancer and serum samples will be used to measure antibody against hormone STDS.

TYPE: Cross-Sectional
TERM: BMB; Biomarkers; Biopsy; Cytology; HPV; PCR; Tissue
SITE: Uterus (Cervix)
TIME: 1989 - 1994

275 **Bosch, F.X.** 04868

Int. Agency for Research on Cancer, Unit of Field and Intervention Studies, 150, cours Albert Thomas, 69372 Lyon Cedex 08, France (Tel.: +33 72738485; Fax: 72738575 ; Tlx: 380023 circ f)

COLL: Benito, E.; Esteva, M.; Mulet, M.; Obrador, A.; Estève, J.; Muñoz, N.

Familial Study of Diet and Colorectal Cancer

The role of diet in colorectal cancer has been examined in case-control studies. Recent studies suggest inherited susceptibility to adenomatous polyps and subsequent adenocarcinoma of the colon and rectum in a segment of the population. The use of controls with similar genetical susceptibility to the cases would presumably provide a more precise estimation of the role of diet in the development of colorectal cancer. This study will evaluate the association of colorectal cancer and dietary factors among cases and controls using siblings of cases as controls. Cases were all newly diagnosed colorectal cancers in the population of Majorca from January 1990; 300 cases have been included. Controls were the traceable sibling of cases during the same period. To estimate food consumption, a semi-quantitative food frequency questionnaire was administered by trained interviewers. The study is currently bein analysed. Preliminary results indicate that familiarity is not a major confounder in dietary studies.

TYPE: Case-Control
TERM: Diet; Genetic Factors
SITE: Colon; Rectum
LOCA: Spain
TIME: 1990 - 1994

276 **Brémond, A.G.** 04709

INSERM U 265, 151, Cours Albert Thomas, 69424 Lyon Cedex 03, France (Tel.: +33 72330123; Fax: 72348784)

COLL: Coste, I.; Victoria, J.; Courtial, I.

Evaluation of a Breast Cancer Mass Screening Programme

A breast cancer screening programme began in April 1987 in the Rhone 'departement'. All women aged 50 to 69 living in this 'departement' receive a personal invitation to participate. The screening test used is a single oblique view mammogram every two years. For each positive test (abnormal mammogram) the final diagnosis and, if cancerous, stage of cancer are recorded. The objectives of the study are (1) to assess the cost/benefit of screening, and (2) to measure the effectiveness and efficiency of this type of screening programme in the context of the French health system. A breast cancer registry was established in 1988 to evaluate the screening programme. The quality of the screening test has been evaluated and cancers compared in the following three groups: new cancers, cancers appearing in the interval between two screening rounds, cancers in women not responding to the screening programme.

TYPE: Cross-Sectional; Incidence; Registry
TERM: Cost-Benefit Analysis; Mammography; Registry; Screening
SITE: Breast (F)
REGI: Rhône (Fra)
TIME: 1987 - 1994

***277** **Cardis, E.** 05527
Int. Agency for Research on Cancer, 150, cours Albert Thomas, 69372 Lyon Cedex 08, France (Tel.: +33 72738508; Fax: 72738575 ; Tlx: 380 023)
COLL: Armstrong, B.G.; Lavé, C.; Blettner, M.; Carpenter, L.M.; Cowper, G.; Fix, J.; Gilbert, E.S.; Hakama, M.K.; Hill, C.G.; Howe, G.R.; Kaneko, M.; Kendall, G.M.; Muirhead, C.R.; Salmon, L.; Yoshimura, T.; Ashmore, J.P.; Bernar, J.; Chanteur, J.; Diez Sacristan, A.; Eklöf, M.; Engels, H.; Gray, J.; Green, L.M.; Hosoda, Y.; Kaldor, J.; Malker, H.; Moser, M.; Rytömaa, T.; Schüler, G.; Seitz, G.

International Collaborative Study of Cancer Risk among Radiation Workers in the Nuclear Industry
The aim of this study is to assess directly the carcinogenic effects of low-dose protracted exposure to low LET ionising radiation, predominantly X and gamma rays. The study population consists of all workers employed for at least on year and monitored for external radiation in public and private nuclear organisations of the participating countries. The study is designed as a retrospective cohort study. Individual annual estimates of dose from X and gamma rays and neutrons are being obtained for each individual in the study cohorts. Flags identifying workers with substantial dose from radioactivity intake will be constructed. Follow-up will be for mortality in all countries and for cancer morbidity in Australia, Canada, Finland, Sweden and the UK. The study period in most countries will range from the beginning of operations in the participating facilities until 31 December 1991. Comparisons of risks will be internal, by level of radiation dose. Estimates of risk of all cancers combined (excluding leukaemia), and leukaemia (excluding CLL), will be derived and compared with those based on high dose studies.

TYPE: Cohort
TERM: Occupation; Radiation, Ionizing; Registry
SITE: All Sites; Leukaemia
OCCU: Radiation Workers
LOCA: Australia; Belgium; Canada; Finland; France; Germany; Japan; Spain; Sweden; Switzerland; United Kingdom
REGI: Alberta (Can); Br. Columbia (Can); E. Anglia (UK); Finland (Fin); Manitoba (Can); Mersey (UK); N.W. Region (UK); N.W. Territory (Can); New Brunswick (Can); Newfoundland (Can); Northern UK (UK); Nova Scotia (Can); Ontario (Can); Oxford I (UK); Quebec (Can); S. Western (UK); Saskatchewan (Can); Scotland (UK); Sweden (Swe); Thames (UK); Trent (UK); W. Midlands (UK); Wales (UK); Wessex (UK); Yorkshire (UK)
TIME: 1993 - 1998

278 **Duclos, J.C.** 03054
Inst. Univ. de Médecine du Travail, 8, Ave. Rockefeller, Domaine Rockefeller, 69373 Lyon Cedex 08, France (Tel.: +33 78777000/4503)
COLL: Pignat, J.C.; Normand, J.C.

Ethmoid Tumours and Harmful Dusts
In this study analyses of wood dust (composition and particle size) are carried out in plants where ethmoid adenocarcinomas have been reported among the workforce. The existence of specific histological types of ethmoid cancer in such plants is being investigated. Four to six cases are anticipated annually. A systematic study of histological changes in the nasal mucosa of wood workers, classified by 5-year age-groups, according to the type of wood used and the granulometry of the work places where they are exposed, is also being carried out. The aim of this histological study is to assess the architectural and cytological changes in relation to length of exposure, granulometric concentrations and type of wood used. The approximate number of exposed workers is 250. Other patients visiting the ENT department will serve as controls.

TYPE: Case Series; Cross-Sectional
TERM: Dusts; Environmental Factors; Histology; Occupation; Wood
SITE: Nasal Cavity
OCCU: Wood Workers
TIME: 1984 - 1994

279 **Hours, M.** 04445

Univ. Claude Bernard, Inst. d'Epidémiologie, 8, Ave. Rockfeller, Domaine Rockfeller, 69373 Lyon Cedex 08, France (Tel.: +33 78777000/4691)

COLL: Ayzac, L.; Bergeret, A.; Dananche, B.; Fabry, J.; Févotte, J.

Occupational Risks for Urinary Bladder and Lung Neoplasms

The main purpose of this study is to identify relative risks of bladder or lung cancer which are significantly high for some occupational activities or exposures, whether previously suspected or not. This case-control study includes 500-600 cases detected over five years in a highly industrialized area; matched controls are chosen from among other in-patients. The study involves a detailed occupational interview and a review of this interview by industrial experts in order to identify probable occupational exposures.

TYPE: Case-Control
TERM: Chemical Exposure; Occupation
SITE: Bladder; Lung
TIME: 1984 - 1994

280 **Hours, M.** 05164

Univ. Claude Bernard, Inst. d'Epidémiologie, 8, Ave. Rockfeller, Domaine Rockfeller, 69373 Lyon Cedex 08, France (Tel.: +33 78777000/4691)

COLL: Févotte, J.; Dananche, B.; Philippe, J.; Fabry, J.; Bergeret, A.; Boiron, O.; Fière, D.; Coiffier, B.; Hollard, D.; Broustet, A.; Cicollela, A.; Degos, L.; Martin, C.; Carli, P.M.; Guy,; Guyotat,

Acute Myeloid Leukaemia and Exposure to Pesticides or Glycol Ethers

The aim of this case-control study is to identify the risk of acute myeloid leukaemia associated with occupational and domestic exposure to pesticides and/or glycol ethers. 220 cases are expected over three years and will be matched to 220 hospital controls. Cases and controls will be interviewed by trained interviewers about leisure habits and past work histories. Each questionnaire will be examined by chemists and work hygienists to determine exposure to pesticides and glycol ethers, as well as confounding exposures (e.g. exhaust gas, benzene, radiation, ethylene oxide, etc.).

TYPE: Case-Control
TERM: Occupation; Pesticides; Solvents
SITE: Leukaemia (AML)
CHEM: Glycol Ethers
TIME: 1990 - 1994

281 **Kogevinas, M.** 05195

Int. Agency for Research on Cancer, Unit of Analytical Epidemiology, 150, cours Albert-Thomas, 69372 Lyon Cedex 08, France (Tel.: +33 72738485; Fax: 72738575 ; Tlx: 380023 circ f)

COLL: Saracci, R.; Ferro, G.; Boffetta, P.; Andersen, A.; Hutchings, S.; Biocca, M.; Coggon, D.; Lundberg, I.S.; Lynge, E.; Partanen, T.J.; Kolstad, H.; Astrup-Jensen, A.; Bellander, T.; Bjerk, J.E.; Pannett, B.; Pfäffli, P.

IARC International Cohort Study on Workers Exposed to Styrene

Increased risk of leukaemia and lymphoma has been suggested in studies of workers exposed to styrene in the rubber and plastics industry. A historical cohort study was conducted in Denmark, Finland, Italy, Norway, Sweden and the UK, enrolling about 40,000 workers employed in the reinforced plastics industry, where high exposure to styrene occurs. Exposure to styrene was reconstructed through job histories, environmental and biological monitoring data and production records of plants in the study. All cohorts are followed for mortality. Cancer incidence data are available in five countries and cases of lymphoma and leukaemia will also be traced in Italy. So far, the pattern indicates no increased mortality from neoplasms in general, or from common epithelial cancers. A small increased risk of leukaemias and lymphomas was observed in the subgroup of workers with more than one year of exposure. The study is being completed with the analysis of cancer incidence data, analysis by cumulative exposure and the conduct of a nested case-control study of lymphoma and leukaemia.

TYPE: Case-Control; Cohort
TERM: Occupation; Plastics; Registry
SITE: All Sites
CHEM: Styrene
OCCU: Laminators; Plastics Workers
LOCA: Denmark; Finland; Italy; Norway; Sweden; United Kingdom
REGI: Denmark (Den); Finland (Fin); Norway (Nor); OPCS (UK); Sweden (Swe)
TIME: 1988 - 1994

282 **Kogevinas, M.** 05196
Int. Agency for Research on Cancer, Unit of Analytical Epidemiology, 150, cours Albert-Thomas, 69372 Lyon Cedex 08, France (Tel.: +33 72738485; Fax: 72738575 ; Tlx: 380023 circ f)
COLL: Boffetta, P.; Saracci, R.; Vainio, H.; Winkelmann, R.; Ferro, G.; Andersen, A.; Facchini, L.; Hours, M.; Henneberger, P.; Jäppinen, P.T.; Lynge, E.; Persson, B.; Pearce, N.E.; Rodrigues, V.; Soskolne, C.L.; Boal, W.; Coggon, D.; Merletti, F.; Sunyer, J.; Heederik, D.; Wild, P.; Miyake, H.; Szadkowska-Stanczyk, I.; Kielkowski, D.; Ahrens, W.; Autier, P.; Bergeret, A.; Bethwaite, P.; Kauppinen, T.P.; Teschke, K.

IARC International Cohort Study on Cancer Risk among Workers in the Pulp and Paper Industry

The paper and pulp industry is spread world-wide, and employs hundreds of thousands of workers. The few prospective epidemiological studies in this industry indicate that cancer risk, particularly lung cancer, gastrointestinal cancer and neoplasms of the lymphatic tissue, may be elevated, but evidence is still not convincing. The objective of this study is to evaluate cancer risk in relation to specific processes and exposures in this industry. Personnel employed in plants producing pulp, paper and paper products and in mills involved in recycling will be included. Cohorts are currently being assembled, and it is expected that the international study will include data for more than 100,000 workers. Two distinct phases are planned. In a first phase, a retrospective cohort study will be conducted. Depending on the results, nested case-control studies on specific neoplasms will be considered. A parallel industrial hygiene study is being conducted.

TYPE: Cohort
TERM: Chemical Exposure; Dusts; Dyes; Occupation; Registry
SITE: All Sites
CHEM: Chlorine; Chlorophenols; Dioxins; Formaldehyde
OCCU: Paper and Pulp Workers
LOCA: Belgium; Brazil; Canada; Denmark; Finland; France; Germany; Italy; Japan; Netherlands; New Zealand; Norway; Poland; Portugal; South Africa; Spain; Sweden; United Kingdom; United States of America
REGI: Canada (Can); Denmark (Den); Finland (Fin); New Zealand (NZ); Norway (Nor); OPCS (UK); Sweden (Swe)
TIME: 1991 - 1996

283 **Møller, H.** 04558
Int. Agency for Research on Cancer, Unit of Carcinogen Identification and Evaluation, 150, cours Albert Thomas, 69732 Lyon Cedex 08, France (Tel.: +33 72738485; Fax: 72738575)
COLL: Ewertz, M.; Skakkebæk, N.E.; Giwercman, A.; Rørth, M.; Maase, H.

Risk Factors for Testicular Cancer and Cryptorchidism

Denmark has the highest registered incidence of testicular cancer (TC) in the world. The disease is most frequent in the young; 70% of cases are under 40 years of age at the time of diagnosis. The only known strong risk factor for TC is cryptorchidism, but several observations indicate that this association may be indirect, and caused by underlying aetiological factors common to both TC and cryptorchidism. Two parallel case-control studies were carried out in Denmark, one involving 514 TC and 720 controls, one involving 387 men treated for cryptorchidism and 416 controls. The men were interviewed by telephone, and their mothers were approached with a self-administered questionnaire. Factors being evaluated are: the mother's use of hormones, her height and weight, details about the pregnancy, her use of tobacco and alcohol, her occupational exposures before and during pregnancy, weight of the son at birth, occurrence of congenital malformations, infectious diseases during childhood, history of testicular trauma, the man's use of tobacco and alcohol, and occupational exposures.

TYPE: Case-Control
TERM: Alcohol; Congenital Abnormalities; Cryptorchidism; Hormones; Infection; Intra-Uterine Exposure; Occupation; Physical Factors; Pregnancy; Registry; Tobacco (Smoking); Trauma
SITE: Testis
REGI: Denmark (Den)
TIME: 1988 - 1994

284 **Muñoz, N.** 03879
Int. Agency for Research on Cancer, Unit of Field and Intervention Studies, 150, cours Albert Thomas, 69372 Lyon Cedex 08, France (Tel.: +33 72738485; Fax: 72738575 ; Tlx: 380023 circ f)
COLL: Bosch, F.X.; Aristizabal, N.; Ascunce, N.; Gili, M.; Gonzalez, L.C.; Izarzugaza, I.; Moreo, P.; Navarro, C.; Tafur, L.; Viladiu, P.; Shah, K.V.; de Sanjosé, S.; Guerrero, E.; Santamaria, M.; Alonso de Ruiz, P.

Cervical Cancer, Male Sexual Behaviour and Papilloma Virus in High-and Low-Risk Areas for Cervical Cancer

Colombia and Spain are the two countries in which one of the highest and one of the lowest incidence rates for cervical cancer have been reported, respectively. A collaborative case-control study has been conducted to determine: (1) how much of the 10-fold differential in risk between these two countries is due to female sexual behaviour; (2) how much is due to male sexual behaviour; and (3) the role of HPV in the development of this tumour in both populations. In Spain, the study included 250 in situ and 226 invasive cases, and in Colombia 276 in situ and 180 invasive cases. 636 male partners in Spain and 472 in Colombia have also participated. A one-to-one control was selected from the files of the laboratory where the in situ cases are diagnosed and from a representative sample of the female population of the area (controls for the invasive cases). Case accrual was terminated in Spain in June 1988 and in April 1989 in Colombia. The number of study subjects interviewed is 2,968. 89% of these agreed to provide a cytological sample and 93% a serum sample. HPV markers were looked for in cervical cells obtained from both cases and controls, in fresh tissue samples from biopsies or surgical specimens collected from untreated cases, and in cell specimens taken from uterine cervix and male urethras. The Virapap, Southern blot and PCR techniques for HPV-DNA hybridization were used. Serological markers for HSV-2, CMV, chlamydia, gonorrhoea and syphilis have also been measured. The study has been conducted in nine provinces in Spain (Gerona health district, Zaragoza, Sevilla, Murcia, Salamanca, Navarra, Alava, Guipuzcoa and Vizcaya) and in Cali, Colombia. Analysis of the questionnaires revealed that early age at first sexual intercourse, a high number of sexual partners, low level of education and practice of prostitution were the main risk factors among women. A high number of sexual partners of the husbands increased the risk of cervical cancer in their wives, but only in Spain. A strong association with HPV (mainly HPV16) was observed in both countries. Residual moderate associations were also found with chlamydia and hormonal factors (use of oral contraceptives and high parity). Final statistical analysis to assess the relative contribution of the various risk factors is in progress.

TYPE: Case-Control
TERM: Biomarkers; HPV; HSV; Infection; Sero-Epidemiology; Sexual Activity; Sexually Transmitted Diseases; Vitamins
SITE: Uterus (Cervix)
CHEM: Beta Carotene; Retinoids
LOCA: Colombia; Spain
TIME: 1985 - 1994

285 **Muñoz, N.** 05185
Int. Agency for Research on Cancer, Unit of Field and Intervention Studies, 150, cours Albert Thomas, 69372 Lyon Cedex 08, France (Tel.: +33 72738485; Fax: 72738575 ; Tlx: 380023 circ f)
COLL: Oliver, W.; Vivas, J.; Buiatti, E.; Lopez, G.; Peraza, S.; Cano, E.; Alvarez, N.; Castro, D.; Sanchez, V.; de Contreras, O.; de Sanjosé, S.; Benz, M.; Correa, P.; Sobala, G.

Chemoprevention Trial on Precancerous Lesions of the Stomach

A double-blind placebo-controlled randomized trial is being carried out to assess the ability of treatment for H. pylori and treatment with antioxidants (beta carotene and vitamins C and E) to induce regression of chronic gastritis (with or without atrophy or intestinal metaplasia) or to block its progression to dysplasia, as compared with a group receiving a placebo. It will include 3,000 subjects aged 35-64 years. Histological, histochemical and biochemical end-points will be used to assess the efficacy of the two treatments. Pilot studies have shown a very high prevalence of H. pylori (90%) and resistance to metronidazole was found in 82% of H. pylori strains isolated from this population. A sub-trial carried out in 220 subjects using a double blind design compared treatment with bismuth subcitrate and amoxycillin with a placebo given during two weeks. Eradication rates of 6.5% in the treatment group and of 2.0% in the placebo group were obtained. In view of these results the main trial with antioxidant vitamins has been initiated. 1,020 subjects have been recruited up to May 1993.

TYPE: Intervention
TERM: Antioxidants; BMB; Blood; H. pylori; Premalignant Lesion; Prevention; Vitamins
SITE: Stomach
CHEM: Beta Carotene
LOCA: Venezuela
TIME: 1991 - 1996

286 **Muñoz, N.** 05186
Int. Agency for Research on Cancer, Unit of Field and Intervention Studies, 150, cours Albert Thomas, 69372 Lyon Cedex 08, France (Tel.: +33 72738485; Fax: 72738575 ; Tlx: 380023 circ f)
COLL: Oliver, W.; Vivas, J.; Lopez, G.; Peraza, S.; Alvarez, N.; Parkin, D.M.; Pisani, P.; de Sanjosé, S.; Benz, M.

Case-Control Study of Stomach Cancer

The aims of this case-control study are: (a) to identify the main risk factors for stomach cancer; and (b) to evaluate the efficacy of a screening programme for stomach cancer in Tachira state. It is planned to include 300 cases of histologically confirmed stomach cancer and two groups of 300 controls each (hospital and neighbourhood controls), matched by sex and age. Exposure to the risk factors and intensity of screening will be investigated by means of a questionnaire and review of screening records. Antibodies to H. pylori will be measured in serum from cases and controls and genetic alterations in cancerous and normal gastric mucosa will be assessed. Up to May 1993, 119 cases and 119 hospital and neighbourhood controls have been recruited. A preliminary analysis will be carried out to decide on the final sample size required.

TYPE: Case-Control
TERM: BMB; H. pylori; Screening; Serum
SITE: Stomach
LOCA: Venezuela
TIME: 1991 - 1995

287 **Parkin, D.M.** 04552
Int. Agency for Research on Cancer, Unit of Descriptive Epidemiology, 150, cours Albert Thomas, 69372 Lyon Cedex 08, France (Tel.: +33 72738485; Fax: 72738575 ; Tlx: 380023 circ f)
COLL: Masuyer, E.; Augustin, J.; Barlow, L.; Bennett, B.; Bobev, D.; Coebergh, J.W.W.; Draper, G.J.; Sinnaeve, J.; Hansluwka, H.E.; Friedl, H.P.; Ivanov, E.; Karjalainen, S.; Kriauciunas, R.; Langmark, F.; Lutz, J.-M.; Merabishivili, V.; Michaelis, J.H.; Moehner, M.; Plesko, I.; Pompe-Kirn, V.; Rahu, M.A.; Raymond, L.; Schuler, D.; Storm, H.H.; Terracini, B.; Tyczynski, J.

European Childhood Leukaemia-Lymphoma Incidence Study (ECLIS)

This collaborative project was started in 1988 with the support of the Radiation Protection Programme of the European Commission, and involves the participation of representatives from cancer registries in 21 European countries. The objective is to follow geographic and temporal trends in the incidence of childhood leukaemia in Europe from 1980 until the mid 1990's, and to evaluate whether any changes can be related to exposure to radioactive material from the accident at Chernobyl in April 1986. In 1992 the study was extended to encompass almost all of the western part of the former USSR. Cancer registries are supplying data on cases of childhood leukaemia and lymphoma and on populations-at-risk, so that incidence rates by cell type may be calculated for sub-national areas. Collaboration has been established with UNSCEAR to obtain estimates of the total body radiation dose attributable to the Chernobyl accident in children under age 15. The first results covering data to the end of 1988 have been published in Eur. J. Cancer 29A:87-95, 1993. An updated analysis incorporating data to the end of 1990 will be completed during 1993.

TYPE: Incidence
TERM: Childhood; Radiation, Ionizing; Registry; Trends
SITE: Leukaemia; Lymphoma
LOCA: Austria; Belarus; Belgium; Bulgaria; Czech Republic; Denmark; Estonia; Finland; France; Germany; Hungary; Italy; Lithuania; Netherlands; Norway; Poland; Russian Federation; Slovak Republic; Slovenia; Sweden; Switzerland; United Kingdom
REGI: Berlin (Ger); Denmark (Den); Estonia (Est); Finland (Fin); Geneva (Swi); Isère (Fra); Lithuania (Lit); Mainz (Ger); Norway (Nor); Oxford I (UK); Piedmont (Ita); Poland (Pol); Scotland (UK); Slovakia (Slvk); Slovenia (Slvn); St Petersburg (Rus); Sweden (Swe); The Hague (Net)
TIME: 1988 - 1996

288 **Parkin, D.M.** 05216
Int. Agency for Research on Cancer, Unit of Descriptive Epidemiology, 150, cours Albert Thomas, 69372 Lyon Cedex 08, France (Tel.: +33 72738485; Fax: 72738575 ; Tlx: 380023 circ f)
COLL: Laudico, A.V.; Ngelangel, C.A.; Munson, M.L.; Reyes, M.G.; Robles, E.

Screening for Cancer of the Breast in the Philippines

Screening for breast cancer by mammography, with or without physical examination of the breast, has been shown to be effective in reducing mortality from breast cancer in women over 50 years. However, since the equipment is expensive, such programmes are inappropriate for developing countries, even where breast cancer incidence is moderately elevated, as in the Manial area of the Philippines. A protocol has been developed for a randomized controlled trial of screening for breast cancer in 330,000 women aged 35-64, using physical examination by trained nurses as the sole screening modality. A pilot study of 14,000 women in the age range 35-64 was undertaken in 1991-1992 to investigate various aspects of feasibility and compliance and to estimate predictive value of physical examination in this population.

TYPE: Methodology
TERM: Screening
SITE: Breast (F)
LOCA: Philippines
TIME: 1991 - 1994

289 **Parkin, D.M.** 05219
Int. Agency for Research on Cancer, Unit of Descriptive Epidemiology, 150, cours Albert Thomas, 69372 Lyon Cedex 08, France (Tel.: +33 72738485; Fax: 72738575 ; Tlx: 380023 circ f)
COLL: Masuyer, E.; Zatonski, W.A.; Tyczynski, J.; Tarkowski, W.; Matos, E.; Brancker, A.; Iscovich, J.M.; Bernstein, L.; Swerdlow, A.J.

Cancer in Polish Migrant Populations

People of Polish origin have migrated to several countries which differ in their environment and lifestyle. The overall objective of this study is to examine the risk of cancer in Polish populations in comparison to that of the host countries to which they have migrated. The countries studied include Argentina, Australia, Canada, England and Wales, France, Israel and the USA. For migrants to Australia, Israel and the USA, cancer risk will be examined according to duration of stay in the host country; for Canada, the risk in first-generation migrants (born in Poland) will be compared with that in the Canada-born population of Polish parentage. The results for migrants to France and Australia have been published (Zyczynski, J., Bull. Cancer 79:789-800, 1992).

TYPE: Incidence; Mortality
TERM: Geographic Factors; Migrants
SITE: All Sites
LOCA: Argentina; Australia; Canada; France; Israel; United Kingdom; United States of America
TIME: 1991 - 1994

***290** **Parkin, D.M.** 05528
Int. Agency for Research on Cancer, Unit of Descriptive Epidemiology, 150, cours Albert Thomas, 69372 Lyon Cedex 08, France (Tel.: +33 72738485; Fax: 72738575 ; Tlx: 380023 circ f)
COLL: Kogevinas, M.; Hung, N.C.; Anh, C.T.; Dai, L.C.; Cordier, S.E.; Rafaël, M.; Rivera-Pomar, J.M.; Stellman, S.

Soft Tissue Sarcoma and Non-Hodgkin's Lymphoma in Relation to Exposure to Herbicides in Viet Nam

During the second Indochina war, large quantities of herbicides contaminated with dioxins were sprayed onto the territory of what was, at the time, South Vietnam. Most of this activity took place in 1965-1971, but because of the relatively long biological half life of dioxins, human exposure will have been more prolonged. The objective of the study is to investigate whether any excess risk for soft tissue sarcoma and NHL exists. 150 cases of soft tissue sarcoma and 150 cases on NHL, each with two hospital controls, will be interviewed, and samples of blood and adipose tissue stored. Exposures will be estimated indirectly initially, based on a careful residential history and the known geographical location, type, and quantity of herbicide sprayed by US forces. If positive findings emerge, it will be possible to proceed to direct measurements of dioxins in adipose tissue of subjects. A pilot study was completed in early 1993 and the main study has now started.

TYPE: Case-Control
TERM: BMB; Blood; Herbicides; Tissue
SITE: Non-Hodgkin's Lymphoma; Sarcoma; Soft Tissue
CHEM: Dioxins
LOCA: Viet Nam
TIME: 1992 - 1995

***291** **Parkin, D.M.** 05529
Int. Agency for Research on Cancer, Unit of Descriptive Epidemiology, 150, cours Albert Thomas, 69372 Lyon Cedex 08, France (Tel.: +33 72738485; Fax: 72738575 ; Tlx: 380023 circ f)
COLL: Masuyer, E.; Khlat, M.; Iscovich, J.M.

Cancer in Migrant Populations in Relation to Age at Migration or Duration of Stay

The value of migrant studies is greatly enhanced when it is possible to estimate how the risk of cancer varies in migrants according to the age at which they migrated, or to the duration of their residence in the new environment. These aspects can be studied in data sets which record date of migration. The risk of melanoma in migrant populations to Australia according to these two variables has been studied (Khlat, M. et al, Am. J. Epidemiol. 135:1103-1113, 1992). In Israel, the cancer registry records date of migration for all cancer cases. The data set previously used to study cancer in migrants to Israel during 1961-1981 is being updated to 1989, and the effect of age at arrival and duration of stay on the risk of certain major cancers (stomach, large bowel, breast, prostate, melanoma) will be investigated for the principal migrant populations.

TYPE: Incidence
TERM: Age; Migrants; Registry; Time Factors
SITE: Breast (F); Colon; Melanoma; Prostate; Stomach
LOCA: Australia; Israel
REGI: Israel (Isr)
TIME: 1992-1995

***292** **Parkin, D.M.** 05530
Int. Agency for Research on Cancer, Unit of Descriptive Epidemiology, 150, cours Albert Thomas, 69372 Lyon Cedex 08, France (Tel.: +33 72738485; Fax: 72738575 ; Tlx: 380023 circ f)
COLL: Masuyer, E.; Iscovich, J.M.

Cancer in Offspring of Migrants to Israel

The objective of this study is to examine the risk of cancer in the Jewish population born in Israel according to birthplace of parents, and to compare this with the risk in migrants. The population born in Israel is still quite young, and for this and other technical reasons the analysis is confined to cancers appearing in the young (under age 30). Earlier studies of Jewish migrants (Steinitz, R. et al, IARC Scientific Publications No. 98, IARC, Lyon) showed large differences in incidence according to birthplace, and the persistence of a differential in the offspring of such migrants, when compared with individuals whose parents were born in Israel, will imply an important hereditary component in aetiology. The data comprise all records of cancer cases aged 0-29 for the period 1961-1989 (10,256 cases). Of these 2,660 are migrants, 6,554 offspring of migrants and 'third generation' (Israel-born, with parents born in Israel). The principal cancer sites to be studied are those occurring in a young population: leukaemias, Hodgkin's disease and NHL, CNS neoplasms, neuroblastoma, malignant bone tumours, soft tissue sarcoma, melanoma and carcinomas.

TYPE: Incidence
TERM: Childhood; Geographic Factors; Heredity; Migrants; Registry
SITE: Bone; Hodgkin's Disease; Leukaemia; Melanoma; Nervous System; Neuroblastoma; Non-Hodgkin's Lymphoma; Sarcoma; Soft Tissue
LOCA: Israel
REGI: Israel (Isr)
TIME: 1992-1995

***293 Parkin, D.M.** 05532
Int. Agency for Research on Cancer, Unit of Descriptive Epidemiology, 150, cours Albert Thomas, 69372 Lyon Cedex 08, France (Tel.: +33 72738485; Fax: 72738575 ; Tlx: 380023 circ f)
COLL: Pisani, P.; Sriamporn, S.; Vatanasapt, V.; Ohshima, H.

Epidemiology of Cholangiocarcinoma in Thailand

A cohort study has been started in north-eastern Thailand to clarify the role of diet as a source of carcinogens (aflatoxin, nitrate and nitrosamines), of protective agents (some vitamins and antioxidants) and as a vehicle for Opisthorchis viverrini (OV) infection in the aetiology of cholangiocarcinoma (CCA) and HCC in a populaton at very high risk of CCA. Recruitment of a cohort of 10,000 people exploits a screening programme offered to the whole population of the region. Information on exposure to betel-nut chewing, usual dietary intake, tobacco and alcohol consumption, together with socio-demographic variables, will be collected through a personal interview. Blood and urine samples will be stored for study of hepatitis B and C infection, intake of aflatoxins and antibodies to OV. During the first year of recruitment 1,300 subjects were enrolled. Selection of small groups with particular characteristics will be made for cross-sectional studies on markers of endogenous nitrosation, on the expression of some p450:s and on exposure to alkylating agents.

TYPE: Cohort; Cross-Sectional
TERM: Alcohol; Betel (Chewing); Diet; HBV; HCV; Infection; Mycotoxins; Tobacco (Chewing); Tobacco (Smoking); Vitamins
SITE: Liver
CHEM: Aflatoxin; N-Nitroso Compounds; Nitrates
LOCA: Thailand
TIME: 1993-2008

***294 Parkin, D.M.** 05534
Int. Agency for Research on Cancer, Unit of Descriptive Epidemiology, 150, cours Albert Thomas, 69372 Lyon Cedex 08, France (Tel.: +33 72738485; Fax: 72738575 ; Tlx: 380023 circ f)
COLL: Sankaranarayanan, R.; Ngelangel, C.A.; Esteban, D.; Jayant, K.; Krishnan Nair, M.; Mathew, B.

Screening for Cancer of the Cervix in Developing Countries

Detection of early invasive cancers by simple visual inspection of the cervix in asymptomatic women is being promoted by the WHO as a simpler alternative to the more resource intensive cytological screening. The sensitivity and specificity of this procedure is being tested in Barshi Tehsil, Sholapur district, Maharashtra, India and in Kazhakuttam, Panchayath, Trivandrum, Kerala, India. In a population-based survey in Barshi, more than 2,000 women,aged 35 years or over, have been subjected to unaided visual inspection by speculum examination to score the clinical appearance of cervix and Pap smear. Approximately 6,000 women in Kazhakuttam are being subjected to the same examinations. A controlled trial involving more than 70,000 women (35 years or more) in Kerala is being planned to evaluate the relative effectiveness of Pap smear/aided visual inspection/unaided visual inspection in (1) preventing invasive cancer of the cervix; (2) detecting invasive cancer at an early stage; (3) preventing late stage (III and IV) disease; and (4) preventing death from cervix cancer.

TYPE: Intervention; Methodology
TERM: Cost-Benefit Analysis; Cytology; Prevention; Screening
SITE: Uterus (Cervix)
LOCA: India; Philippines
TIME: 1994-2004

295 **Riboli, E.** 04895

Int. Agency for Research on Cancer, Unit of Analytical Epidemiology, 150, cours Albert Thomas, 69372 Lyon Cedex 08, France (Tel.: +33 72738485; Fax: 72738575 ; Tlx: 380023 circ f)

COLL: Saracci, R.; Kaaks, R.; Slimani, N.; Clavel, F.; Villelminot, S.; Berrino, F.; Pisani, P.; Vineis, P.; Gafà, L.; Tumino, R.; Day, N.E.; Khaw, K.T.; Bingham, S.; Key, T.; Forman, D.; Trichopoulou, A.; Katsouyanni, K.; Collette, H.J.A.; Kromhout, D.; Wahrendorf, J.; Boeing, H.; González, C.A.; Torrent, M.; Quiros Garcia, J.R.; Navarro, C.; Del Moral, A.; Martinez, C.

Prospective Studies on the Role of Nutrition in the Aetiology of Cancer

The objective of the project is to investigate the relationship between diet, nutritional status and the risk of cancer at several sites. The plan is to set up large prospective studies in European populations characterized by different dietary habits and different patterns of cancer incidence. In each collaborating country, cohorts of 70,000 middle-aged subjects will be recruited using appropriate sampling techniques suitable for each specific population. Blood will be collected from study participants and stored at a low temperature. Laboratory analyses will be carried out at a later stage on samples from subjects who do or do not develop cancer. Cohorts will be followed up through population cancer registries or other suitable sources until a sizeable number of cancer cases has been collected. Cancer incidence at the most common sites (stomach, colon, rectum, breast, lung, urinary tract, uterus) will be analysed in relation to information on diet and nutritional status. A pilot phase was completed in 1991 in Italy, France, Spain and the UK. Extension of the study to Greece, Germany and the Netherlands was started in 1990. Data collection for the main study was started in 1991 in France, and in 1992 in Italy, Spain and the UK, and is planned for 1993 in Greece, Germany and the Netherlands. Data collection should be completed within four years and follow-up will start in 1995-1996.

TYPE: Cohort
TERM: BMB; Blood; Diet; Environmental Factors; Nutrition; Urine
SITE: All Sites
LOCA: France; Germany; Greece; Italy; Netherlands; Spain; Sweden; United Kingdom
TIME: 1989 - 2000

296 **Saracci, R.** 02700

Int. Agency for Research on Cancer, Unit of Analytical Epidemiology, 150, cours Albert Thomas, 69372 Lyon Cedex 08, France (Tel.: +33 72738485; Fax: 72738575 ; Tlx: 380023 circ f)

COLL: Kogevinas, M.; Winkelmann, R.; Boffetta, P.; Ferro, G.; Becher, H.; Bertazzi, P.A.; Coggon, D.; Green, L.M.; Kauppinen, T.P.; Littorin, L.; Lynge, E.; Mathews, J.D.; Neuberger, M.; Pearce, N.E.; Bueno de Mesquita, H.B.; Fingerhut, M.A.; Needham, L.; Benn, T.

International Registry of Persons Exposed to Phenoxy Acid Herbicides and Contaminants

Cholophenoxy herbicides have been used extensively since the mid-1950s. These compounds may be contaminated during the production process with polychlorinated dioxins and furans, including tetrachlorodibenzo-p-dioxin (dioxin, TCDD), which is a widespread contaminant of the general environment. An international study of 18,910 production workers and sprayers exposed to chlorophenoxy herbicides, chlorophenols and contaminants (principally dioxins and dibenzofurans) is conducted. The study includes workers from Australia, Austria, Canada, Denmark, Finland, Italy, the Netherlands, New Zealand, Sweden and the United Kingdom, and has recently been enlarged with four cohorts from Germany and 12 cohorts from the USA. The first mortality follow-up has been completed. An excess risk was observed for soft-tissue sarcoma, while no excess was observed for non-Hodgkin's lymphoma. Risks appeared elevated for cancers of the thyroid, testis, other endocrine glands, nose and nasal cavity, based on small numbers of deaths. Two nested case-control studies ofsoft-tissue sarcoma and non-Hodkgin's lymphoma examining the importance of various occupational exposures are being completed. A further mortality and cancer incidence follow-up and a study to determine serum levels of dioxin and furans in a sample of workers are underway. This project is conducted through collaboration between IARC and the NIEHS of the US.

TYPE: Cohort; Incidence
TERM: Herbicides; Occupation; Pesticides; Registry
SITE: All Sites
CHEM: Chlorophenols; Dioxins; Furans; Phenoxy Acids
OCCU: Agricultural Workers; Chemical Industry Workers; Forest Workers; Herbicide Manufacturers; Herbicide Sprayers; Railroad Workers
LOCA: Australia; Austria; Canada; Denmark; Finland; Germany; Italy; Netherlands; New Zealand; Sweden; United Kingdom; United States of America
REGI: Denmark (Den); Finland (Fin); New Zealand (NZ); Sweden (Swe)
TIME: 1984 - 1994

297 **Saracci, R.** 04874

Int. Agency for Research on Cancer, Unit of Analytical Epidemiology, 150, cours Albert Thomas, 69372 Lyon Cedex 08, France (Tel.: +33 72738485; Fax: 72738575 ; Tlx: 380023 circ f)

COLL: Boffetta, P.; Riboli, E.; Bartsch, H.; Lang, M.; Ahrens, W.; Benhamou, E.; Benhamou, S.; González, C.A.; Mendes, A.; Merletti, F.; Simonato, L.; Winck, J.C.; Hirsch, A.F.; Trédaniel, J.; Pershagen, G.; Vutuc, C.; Kreienbrock, L.; Jindal, S.K.

Lung Cancer in Nonsmokers and Environmental Tobacco Smoke

A case-control study has been started in 12 collaborating centres in 8 countries to investigate the relationship between exposure to environmental tobacco smoke and to other environmental risk factors (occupational exposures, air pollution, diet) and the risk of lung cancer in subjects who have never smoked tobacco. Data are collected by personal interview. Self-reported smoking or nonsmoking status will be cross-checked by interview of spouses in a subsample of subjects. Biological samples (urine and blood) will also be collected in a subsample to obtain biochemical validation of current smoking status. Data collection will continue to the end of 1993 to reach a total of 400 cases and 600 controls. Blood samples are collected from a subgroup of patients. Individual susceptibility to lung cancer will be investigated by measuring genetic polymorphism to enzymes implicated in the metabolism of carcinogens (P450 CYP1A1 and CYP2DG; GSTM1)

TYPE: Case-Control
TERM: Air Pollution; Biochemical Markers; Diet; Occupation; Passive Smoking
SITE: Lung
LOCA: Canada; France; Germany; India; Italy; Portugal; Spain; Sweden; Switzerland
TIME: 1989 - 1994

298 **Saracci, R.** 04893

Int. Agency for Research on Cancer, Unit of Analytical Epidemiology, 150, cours Albert Thomas, 69372 Lyon Cedex 08, France (Tel.: +33 72738485; Fax: 72738575 ; Tlx: 380023 circ f)

COLL: Boffetta, P.; Andersen, A.; Bertazzi, P.A.; Frentzel-Beyme, R.R.; Olsen, J.; Simonato, L.; Teppo, L.; Westerholm, P.; Ferro, G.; Cherrie, J.; Winter, P.D.; Plato, N.

Health Effects of Man-Made Mineral Fibres in the Producer Industry

A prospective study of workers employed in 13 man-made mineral fibre (MMMF) plants from seven European countries was started in 1976. Environmental measurements of MMMF concentration have been carried out by the Institute of Occupational Medicine in Edinburgh, UK. Follow-up through 1982 (Scand. J. Work Environ. Health 12 (Suppl. 1):34-47, 1986) showed an increase in lung cancer mortality (189 observed, 151.2 expected), which was correlated with time since first exposure, but not duration of employment. The increase was concentrated among workers employed in the early technological phase, when no dust-suppressing agent was used. In order to investigate this finding, an extension of the follow-up until the end of 1991 is now in progress. A nested case-control study of lung cancer has been planned. Its aims are: (a) the investigation of the confounding effect of tobacco smoking and other occupational and non-occupational exposures to known or suspected carcinogens, (b) a detailed assessment of MMMF exposure. Feasibility of the case-control study is now under way.

TYPE: Case-Control; Cohort
TERM: Occupation; Registry; Tobacco (Smoking)
SITE: All Sites; Lung
CHEM: Mineral Fibres
OCCU: Mineral Fibre Workers
LOCA: Denmark; Finland; Germany; Italy; Norway; Sweden; United Kingdom
REGI: Denmark (Den); Finland (Fin); Norway (Nor); Sweden (Swe)
TIME: 1990 - 1994

299 **Sasco, A.J.** 04487

Int. Agency for Research on Cancer, Unit of Analytical Epidemiology, 150, cours Albert Thomas, 69372 Lyon Cedex 08, France (Tel.: +33 72738412; Fax: 72738575 ; Tlx: 380023 circ f)

COLL: Riboli, E.; Saracci, R.; Hu, M.X.; Ging, L.

Breast Cancer and Reproductive and Endocrine Factors

The aim of the study is to evaluate the relationship between hormonal profiles and breast cancer incidence in premenopausal women. The study uses a case-control approach in a population with a low incidence of the disease. Study group was selected from the population of Guangdong province in China. Incident cases, all premenopausal, were pair-matched to control women on the basis of age (within two years), and residence. Women taking contraceptive pills or any other hormonal treatment, reserpine or tranquilliser were excluded, as were women having or having had a pregnancy in the preceding 12 months, whether carried to full term or ending in a spontaneous or induced abortion, women having lactated in the preceding six months, and women having documented hormonal diseases, gynaecological conditions or chronic debilitating conditions. A detailed questionnaire is administered to cases and controls, including the following items: personal identification data, details of diagnosis, repproductive life history, age at menarche, personal history of disease, family history of cancer, diet history and other factors. Saliva and blood specimens were collected between day 20 and day 24 of the menstrual cycle. Laboratory determinations of testosterone, oestradiol, progesterone, prolactin and growth hormone will be carried out. The Chinese component of the study has now been completed and analyses of the questionnaire data were done. Preliminary results show a positive association of breast cancer with late age at first full term birth, early age at menarche, university education and some aspects of diet. Analysis of the biological samples and further statistical analysis will be carried out in 1993.

TYPE: Case-Control
TERM: Alcohol; Chinese; Diet; Drugs; Familial Factors; Hormones; Menarche; Race; Reproductive Factors; Tobacco (Smoking)
SITE: Breast (F)
CHEM: Oestradiol; Progesterone; Prolactin; Testosterone
LOCA: China
TIME: 1988 - 1994

300 **Sasco, A.J.** 04488
Int. Agency for Research on Cancer, Unit of Analytical Epidemiology, 150, cours Albert Thomas, 69372 Lyon Cedex 08, France (Tel.: +33 72738412; Fax: 72738575 ; Tlx: 380023 circ f)
COLL: Saracci, R.; Ahlbom, A.N.; Belli, S.; Benhamou, S.; Berrino, F.; Bourke, G.J.; Brown, T.; Chilvers, C.E.D.; Hatton, F.; Iversen, O.H.; Kauppinen, T.P.; Karli, M.; Maximilien, R.; Moulin, J.J.; Teissier, C.; Tirmarche, M.; Van Leeuwen, F.E.; van Barneveld, T.; Wennborg, H.

International Study of Cancer Risk in Biology Research Laboratory Workers

A review of the literature on health risks linked to work in laboratories has shown the paucity of studies on cancer risk for populations of research workers, with the exception of chemists, who have an increased risk of leukaemia/lymphoma. This underlines the need for an evaluation of cancer risk among biology laboratory workers, with an emphasis on risk linked to genetic engineering and molecular biology. A three-step approach has been decided: (1) retrospective cohort study (started in 1991), with about 70,000 subjects; (2) prospective cohort study, with periodical reassessment of exposure and outcome (to be started in 1994 and pursued until 2016); (3) case-control studies nested within the cohort. Death registries and cancer registries will be used to ascertain cases for the retrospective component, and registries (where available) and ad hoc surveillance for the prospective component. A feasibility study, aiming at a detailed assessment of exposures, present and historical, has been conducted in 1988-1989 and results will be published shortly. A paper has been published in Medecine/Sciences 5:489-498, 1989. The retrospective cohort study is presently being conducted in public research institutions of eight European countries (Finland, France, Ireland, Italy, the Netherlands, Norway, Sweden and United Kingdom), covering biomedical and agronomic research.

TYPE: Case-Control; Cohort
TERM: Chemical Exposure; Occupation; Registry
SITE: All Sites
OCCU: Laboratory Workers
LOCA: Finland; France; Ireland; Italy; Netherlands; Norway; Sweden; United Kingdom
TIME: 1988 - 2016

301 **Sasco, A.J.** 05223
Int. Agency for Research on Cancer, Unit of Analytical Epidemiology, 150, cours Albert Thomas, 69372 Lyon Cedex 08, France (Tel.: +33 72738412; Fax: 72738575 ; Tlx: 380023 circ f)
COLL: Parkin, D.M.; Chollat-Traquet, C.; Gupta, P.C.; Peto, R.

Tobacco Use and All Causes of Mortality in India

Little information is available on precise estimates of mortality and morbidity attributable to tobacco in developing countries. Few studies have been conducted on the use of tobacco other than in the form of cigarettes. It is therefore proposed to explore the feasibility of a prospective cohort study of 100,000 men, aged 35 and over, chosen from the lists of voters in Bombay. Provided the results of the feasibility study are acceptable, a five- to ten-year follow-up study will be started to assess the cause-specific mortality of the cohort, with particular attention being given to cancer and other causes of death in relation to the various forms of tobacco use (smoking, chewing, snuff, etc.).

TYPE: Cohort
TERM: Tobacco (Chewing); Tobacco (Smoking); Tobacco (Snuff)
SITE: All Sites
LOCA: India
TIME: 1990 - 2000

***302** **Sasco, A.J.** 05359
Int. Agency for Research on Cancer, Unit of Analytical Epidemiology, 150, cours Albert Thomas, 69372 Lyon Cedex 08, France (Tel.: +33 72738412; Fax: 72738575 ; Tlx: 380023 circ f)
COLL: Mignotte, H.; Brémond, A.G.; Fasquel, D.; Zlatoff, P.; Pobel, D.

Endometrial Cancer Following Breast Cancer

The aim of this case-control study is to evaluate the carcinogenic potential of various treatments for breast cancer and in particular the use of tamoxifen, which is now proposed as a preventive agent for women at high risk of breast cancer. Cases are women diagnosed with endometrial cancer, who had previously had breast cancer. Four controls will be selected for each case among women who were diagnosed with breast cancer at the same date and age as the cases, but who did not develop endometrial cancer. Information on all relevant variables is being collected from the clinical records. For the current study, conducted in Lyon, 30 to 40 cases are anticipated. If results are promising, an extension to an international setting is envisaged.

TYPE: Case-Control
TERM: Drugs; Multiple Primary; Reproductive Factors; Treatment
SITE: Breast (F); Uterus (Corpus)
CHEM: Tamoxifen
TIME: 1992 - 1994

303 **SEARCH Programme,** 03498
Int. Agency for Research on Cancer, Unit of Analytical Epidemiology, 150, cours Albert Thomas, 69372 Lyon Cedex 08, France (Tel.: +33 72738485; Fax: 72738575 ; Tlx: 380023 circ f)
COLL: Preston-Martin, S.; Little, J.; Boyle, P.; Choi, N.W.; Cordier, S.E.; Filippini, G.; Gurevicius, R; Holly, E.A.; McCredie, M.R.E.; Modan, B.; Mueller, B.A.; Peris-Bonet, R.

Brain Tumours in Children

The objectives are to evaluate the role of parental and childhood exposure to N-nitroso compounds, to quantify the importance of known risk factors (ionizing radiation and genetic predisposition), and to explore the relationship with other suggested risk factors including head trauma, barbiturates and parental occupational exposures. In utero and perinatal experiences are of particular interest. The biological mothers of 1,000 children aged 0-19 with a primary tumour of the brain or cranial meninges and of an equal or greater number of population controls will be interviewed, as will fathers of cases and controls.

TYPE: Case-Control
TERM: Childhood; Drugs; Environmental Factors; Intra-Uterine Exposure; Radiation, Ionizing; Trauma
SITE: Brain
CHEM: Barbiturates; N-Nitroso Compounds
LOCA: Australia; Canada; France; Israel; Italy; Lithuania; Spain; United States of America
TIME: 1985 - 1994

304 **SEARCH Programme** 03914
Int. Agency for Research on Cancer, Unit of Analytical Epidemiology, 150, cours Albert Thomas, 69372 Lyon Cedex 08, France (Tel.: +33 72738485; Fax: 72738575 ; Tlx: 380023 circ f)
COLL: Howe, G.R.; Little, J.; Gurevicius, R; McMichael, A.J.; McNeil, J.J.; Ménégoz, F.; Modan, B.; Wahrendorf, J.

SEARCH Study of Adult Brain Tumours

This study will investigate the aetiology of tumours of the brain and cranial meninges in adults (aged over 25). The study has been designed specifically to investigate (1) the role of exposure to dietary N-Nitroso compounds, associated chemicals and known inhibitors of nitrosation; (2) occupational exposures; and (3) other putative risk factors in the aetiology of brain cancers. The study comprises a series of population-based case-control studies conducted in a number of international centres, with the SEARCH programme of IARC acting as the central co-ordinating centre. It is anticipated that data collection will continue until the end of 1993 and that over 1,000 cases will be assembled. Data analysis is in progress.

TYPE: Case-Control
TERM: Chemical Exposure; Diet; Occupation
SITE: Brain
CHEM: N-Nitroso Compounds
LOCA: Australia; Canada; France; Germany; Israel; Lithuania; Sweden; United States of America
TIME: 1985 - 1994

305 **Vainio, H.** 05205
Int. Agency for Research on Cancer, Unit of Carcinogen Identification & Evaluation, 150, cours Albert Thomas, 69372 Lyon Cedex 08, France (Tel.: +33 72738485; Fax: 72738575 ; Tlx: 380023 circ f)
COLL: Demers, P.; Boffetta, P.; Kogevinas, M.; Matos, E.; Saracci, R.; Blair, A.E.; Bonassi, S.; Merler, E.; Miller, B.A.; Robinson, C.F.; Roscoe, R.J.; Seniori Costantini, A.R.; Stern, F.; Winter, P.D.

Re-Analysis of the Carcinogenic Risk of Exposures in the Wood and Leather Industries: Cohort Studies

Work in the furniture and cabinet making is considered to entail a carcinogenic risk to humans (IARC group 1), and carpentry and joinery to be possibly carcinogenic to humans (IARC group 2B), whereas lumber and sawmill, pulp and paper industries were not classifiable as to their carcinogenicity to humans (IARC group 3). When leather industries were evaluated, work in boot and shoe manufacture and repair was considered also as IARC group 1, whereas leather tanning and processing and other leather goods manufacture industries were included in group 3. Exposures to specific chemicals or mixtures were not evaluated on these occasions, but a number of chemicals occurring in the wood and leather industries have also been tested for carcinogenicity (e.g. formaldehyde, chlorophenols, tannins, benzene, chromium and chromium compounds). The aim of this project is to conduct a combined re-analysis of the available epidemiological studies. It will consist of: (1) the combination of relevant cohort studies of wood and leather workers into separate data sets; (2) a systematic assessment of available information on exposures, including data not used or not presented in previous reports; (3) a combined analysis according to specific exposures and job titles.

TYPE: Cohort
TERM: Dusts; Leather; Occupation; Plastics; Solvents; Wood
SITE: All Sites; Bladder; Nasal Cavity
CHEM: Benzene; Formaldehyde
OCCU: Leather Workers; Wood Workers
LOCA: France; Italy; United Kingdom; United States of America
TIME: 1991 - 1994

306 **Vainio, H.** 05388
Int. Agency for Research on Cancer, Unit of Carcinogen Identification & Evaluation, 150, cours Albert Thomas, 69372 Lyon Cedex 08, France (Tel.: +33 72738485; Fax: 72738575 ; Tlx: 380023 circ f)
COLL: Demers, P.; Boffetta, P.; Kogevinas, M.; Leclerc, A.; Luce, D.; Matos, E.; Saracci, R.; Belli, S.; Bolm-Audorff, U.; Brinton, L.A.; Comba, P.; Fukuda, K.; Hardell, L.O.; Hayes, R.B.; Gérin, M.; Magnani, C.; Merler, E.; Rodello, S.; Shibata, A.; Preston-Martin, S.; Vaughan, T.L.; Zheng, W.

Re-Analysis of the Carcinogenic Risk of Exposures in the Wood and Leather Industries: Case-Control Studies

Work in the furniture and cabinet making is considered to entail a carcinogenic risk to humans (IARC Group 1), and the carpentry and joinery to be possibly carcinogenic to humans (IARC Group 2), whereas lumber and sawmill, paper and pulp industries were not classifiable as to their carcinogenicity to humans (IARC Group 3). When leather industries were evaluated, work in boot and show manufacture and repair was considered also as IARC Group 1, whereas leather and tanning and processing in other leather goods manufacture industries were included in Group 3. Exposures to a number of chemicals or mixtures were not evaluated on these occasions, but several chemicals occurring in the wood and leather industries have also been tested for carcinogenicity (e.g. formaldehyde, chlorophenols, tannins, benzene, chromium and chromium compounds). The aim of this project is to conduct a combined re-analysis of the available epidemiological studies. It will consist of: (1) the combination of relevant case-control studies of sinonasal cancer into a single data set; (2) a systematic assessment of available information on exposures, including data not used or not presented in previous reports; (3) a combined analysis according to specific exposures and job titles. This project is being coordinated with INSERM, who will be examining the carcinogenic risk in other than wood and leather related occupations.

TYPE: Case-Control
TERM: Dusts; Leather; Occupation; Plastics; Solvents; Wood
SITE: Nasal Cavity
CHEM: Benzene; Formaldehyde
OCCU: Leather Workers; Wood Workers
LOCA: China; France; Germany; Italy; Japan; Sweden; United Kingdom
TIME: 1991 - 1994

Marseille

307 **Bernard, J.L.** 03962
Fac. de Médecine Nord, Registre des Cancers de l'Enfant des Regions Paca et Corse, 27, Blvd Jean Moulin, 13916 Marseille Cedex 20, France (Tel.: +33 91698941; Fax: 91698942)
COLL: Bernard, E.; Scheiner, C.; Perrimond, H.; Coste, D.; Raybaud, C.; Mariani, R.; Thyss, A.

Paediatric Tumour Registry

A population-based paediatric tumour registry was set up in 1984 in two South-Eastern regions of France, Provence-Alpes-Cote d'Azur and Corsica. This study surveys a population of 857,000 children under age 15. All incident cases of malignant tumours and leukaemias are registered. In addition, a case-control study is being conducted with focus on genetic risk factors. On a nationwide basis, information is being collected on cancer in sibs and twins and on unusual pathological associations in children with cancer and their families.

TYPE: Case-Control; Incidence; Registry
TERM: Childhood; Familial Factors; Genetic Factors; Registry
SITE: Childhood Neoplasms; Leukaemia
REGI: Marseille (Fra)
TIME: 1984 - 1999

Meylan

308 **Lutz, J.-M.** 04966
Registre du Cancer de l'Isère, 21 Chemin des Sources, 38240 Meylan, France (Tel.: +33 76907610; Fax: 76518667)
COLL: Romazzini, S.; Colonna, M.; Laydevant, G.

Study of Chloroprene Industry

This retrospective follow-up study investigates the possible association between exposure to chloroprene (or to its manufacturing process) and cancer mortality and morbidity, using cancer registry data. Workers will be followed from 1965 up to 1990. A continuation of follow-up is planned in case of a positive association of mortality with the exposure of interest.

TYPE: Cohort
TERM: Chemical Exposure; Occupation; Registry
SITE: All Sites
CHEM: Chloroprene
REGI: Isère (Fra)
TIME: 1990 - 1994

Montpellier

309 **Domergue, J.** 04204
Centre Paul Lamarque, Inst. du Cancer, Dépt. de Chirurgie, 2, Ave. Bertin Sans, B.P. 5054, 34000 Montpellier, France (Tel.: +33 67613100)
COLL: Pujol, H.; Corbin, A.

Role of Hormonal Therapy in Preventing Breast Cancer in Patients with Ductal or Lobular Hyperplasia

It has been proved that women with lobular and/or ductal hyperplasia are at high risk of breast cancer. While several studies are being conducted to evaluate the role of tamoxifen in reducing the incidence of breast cancer in women at familial risk, to date a method of following this high risk-group has not been evolved. The aim is to compare modifications in breast samples of patients randomly assigned to receive hormonal therapy with either progesterone or tamoxifen and patients not treated. Patients included are those with ductal or lobular "atypical hyperplasia", diagnosed by a pathologist, with or without a familial history of breast cancer. Blocks are kept for further cancer marker studies. About 200 patients recruited over 2 years will be allocated by blocked randomisation into two groups: (1) no treatment (2) progesterone for days 5-25 of cycle (women aged 30-39) plus tamoxifen (women aged 40-50). Every six months patients are seen in clinics for clinical examination and mammography. After two years of follow-up, patients will be offered new biopsies to evaluate histological changes.

TYPE: Intervention
TERM: Biopsy; Drugs; Familial Factors; High-Risk Groups; Hormones; Mammography; Premalignant Lesion; Prevention
SITE: Breast (F)
CHEM: Progesterone; Tamoxifen
TIME: 1985 - 1994

Nîmes

310 **Daurès, J.P.** 04531
Centre Hosp. Régional de Nîmes, 5, rue Hoche, 30000 Nîmes, France (Tel.: +33 66273403; Fax: 66273433)
COLL: Momas, I.

Bladder Cancer in the Département of Hérault

This case-control investigation on bladder cancer has four aims: indentification of risk factors in bladder cancer, of interactions between factors, and of dose-response relations between exposure and bladder cancer. 250 incident cases with transitional-cell carcinoma and without any other cancer and resident in Herault, who came to public and private urology departments in Herault between 1 January 1987 and 30 December 1988 are matched to two randomly selected controls, one in the Herault population, the other at Lapeyronie Hospital, free of any cancer. The inquiry deals with: (1) way of life (smoking, nutrition, water, alcohol, coffee, artificial sweeteners); (2) drugs (phenacetin); (3) occupational exposures; (4) agricultural activities. The analysis of data will be by statistical methods for matched data.

TYPE: Case-Control
TERM: Alcohol; Chemical Exposure; Coffee; Drugs; Nutrition; Occupation; Sweeteners; Tobacco (Smoking); Water
SITE: Bladder
CHEM: Phenacetin
OCCU: Agricultural Workers; Dyestuff Workers; Rubber Workers
TIME: 1987 - 1994

***311** **Daurès, J.P.** 05433
Centre Hosp. Régional de Nîmes, 5, rue Hoche, 30000 Nîmes, France (Tel.: +33 66273403; Fax: 66273433)
COLL: Schraub, S.

Case-Control Study of Lymphoma

This unmatched case-control study aims to investigate the role of ionising radiation (occupational and therapeutic), tobacco, alcohol, previous illnesses other than HIV-related, drug exposure, occupational exposures to solvents, paints, dyes and PCB, occupational and domestic exposures to electromagnetic fields and pesticides (mainly phenoxy herbicides and other herbicides used in wine-growing), in the causation of Hodgkin's and non-Hodgkin's lymphoma. It is anticipated to recruit 900 cases through the anatomophatologists in the Languedoc-Roussillon area over the period 1 January 1992 to 31 December 1994. 1,800 population controls will be randomly selected from the electoral rolls. Odds ratios will be calculated using the Mantel-Haenszel method; dose-response relationship will be tested. Potential confounders will be controlled for and potential effect modifiers will also be investigated using unconditional multiple logistic regression.

TYPE: Case-Control
TERM: Alcohol; Drugs; Dyes; Electromagnetic Fields; Herbicides; Infection; Lifestyle; Occupation; Paints; Pesticides; Radiation, Ionizing; Registry; Solvents; Tobacco (Smoking)
SITE: Hodgkin's Disease; Non-Hodgkin's Lymphoma
CHEM: Benzene; PCB; Phenoxy Acids
OCCU: Agricultural Workers; Farmers
REGI: Doubs (Fra)
TIME: 1992 - 1996

Paris

312 **Blanc, C.** 04925
Electricite de France - Gaz de France, Serv. Général de Médecine de Contrôle, 22-30 Ave. de Wagram, 75008 Paris, France (Tel.: +33 1 47647209; Fax: 47648201)
COLL: Puppinck, C.; Chevalier, A.; Godard, C.; Callet, B.; Goldberg, M.; Guenel, P.

Establishment of a Cancer Registry among Workers in the French National Electricity and Gas Company

Sick leave, invalidity and mortality among the 150,000 workers of EDF-GDF are recorded, with diagnosis, by the Social Security Department of the company. For every case of cancer detected the Department physicians complete a questionnaire, stating the date of diagnosis, the site of the tumour and its morphology. From 1978 to 1989 about 3,000 cases have been detected. New cases are recorded directly. The registry will permit descriptive morbidity and mortality studies and aetiological studies of occupational hazards. Presently a survey of exposure to electromagnetic fields is being carried out.

TYPE: Incidence; Mortality; Registry
TERM: Electromagnetic Fields; Occupation
SITE: All Sites
OCCU: Electrical Workers; Gas Workers
TIME: 1978 - 1994

***313** **Calmettes, C.** 05501
INSERM U113, 27, Rue de Chaligny, 75571 Paris Cedex 12, France (Tel.: +33 1 40011313; Fax: 40011499)
COLL: Feingold, N.; Rosenberg, M.; Lenoir, G.; Modigliani, E.; Guliana, J.M.

Epidemiology of Medullary Carcinoma of the Thyroid

1,870 cases of medullary carcinoma of the thyroid have been included in a nationwide study carried out by the Working Group of Calcitonin Tumours. Almost 30% of the cases are hereditary and grouped in 127 families (45 isolated forms, 75 MEN 2a and 7 MED 2b). A genetic study of 25 families is being carried out and geographical and clinical data are now being analysed. Endocrinological, pathological and molecular biology studies are also being carried out. Overall results to date have enabled the establishment of a detection, surveillance and treatment protocol which is largely distributed.

TYPE: Genetic Epidemiology
TERM: BMB; Genetic Factors; Heredity
SITE: Thyroid
TIME: 1982 - 1999

314 **Hubert, A.** 05165
Inst. Pasteur, Unité d'Epidémiologie des Virus Oncogènes, 28, rue du Dr Roux, 75724 Paris Cedex 15, France (Tel.: +33 1 45688930; Fax: 45688931)
COLL: Jeannel, D.; Crognier, E.; de Thé, G.; Sancho-Garnier, H.; Eschwege, E.; Slimane, B.

Dietary and Environmental Risk Factors for Nasopharyngeal Carcinoma in Morocco and among Moroccan Migrants

Nasopharyngeal carcinoma will be studied in an endemic country, Morocco, where no epidemiological research on this subject has, as yet, been performed; in addition, NPC will also be studied in a migrant population from Morocco in southern France. The programme includes three parts: (1) estimation of the incidence of NPC in Morocco and in Moroccans of the Provence-Alpes-Cote d'Azur region; (2) anthropological study of dietary changes among Moroccan migrants; (3) study of the environmental risk factors in Morocco through a case-control study in Morocco, together with genetic (HLA profiles) and virological (EBV) studies.

TYPE: Case-Control; Incidence
TERM: BMB; Diet; EBV; Environmental Factors; HLA; Maghrebians; Migrants; Serum
SITE: Nasopharynx
LOCA: France; Morocco
TIME: 1991 - 1994

315 **Leclerc, A.** 04546
INSERM, U88, 91, Blvd de l'Hôpital, 75634 Paris Cedex 13, France
COLL: Brugère, J.

Laryngeal Cancer and Occupation

In this case-control study, patients with cancer of the glottis (n = 112), supraglottis (n = 84), epilarynx (n = 104) and hypopharynx (n = 206) diagnosed in 12 specialised departments of French hospitals were compared with unmatched hospital patients (n = 300). Every case and control was interviewed between 1989 and 1991, with a detailed questionnaire on alcohol (amount and type), tobacco (amount and type), passive smoking, and questions about past occupations. For each occupation, questions are asked about the tasks performed, and materials manipulated. The analysis will be performed in two stages, based on past occupations and based on selected occupational exposures, according to occupational hygiene experts, or job-exposure matrices. Occupational exposures to asbestos, sulphuric acid, chromium, nickel, silica, and vehicle exhaust will be studied. In the two stages of the analysis, age, alcohol and tobacco consumption will be controlled for. The role of other risk factors such as type of alcohol and tobacco, passive smoking will be studied. Differences between the four sites of cancer will be stressed.

TYPE: Case-Control
TERM: Alcohol; Chemical Exposure; Dusts; Occupation; Passive Smoking; Tobacco (Smoking)
SITE: Hypopharynx; Larynx
CHEM: Asbestos; Chromium; Nickel; Silica; Sulphuric Acid; Vehicle Exhaust
TIME: 1989 - 1994

***316** **Leclerc, A.** 05293
INSERM, U88, 91, Blvd de l'Hôpital, 75634 Paris Cedex 13, France
COLL: Belli, S.; Magnani, C.; Hayes, R.B.; Gérin, M.; Bolm-Audorff, U.; Demers, P.

Occupational Risk Factors for Sino-Nasal Cancer: a Reanalysis of Several Case-Control Studies

This study is a combined reanalysis of several case-control studies. The objective is to examine the risk of nasal cancer associated with known or suspected risk factors, with more specific occupations and exposures than was previously possible. The analysis includes 650 cases and 1,412 controls from seven case-control studies. Different approaches will be used: analysis by job titles; analysis by substances, based on a job exposure matrix or exposure assessment by experts. The analysis by substances will focus on textile, formaldehyde and occupational exposures in agriculture.

TYPE: Case-Control
TERM: Chemical Exposure; Dusts; Occupation; Plastics; Textiles
SITE: Nasal Cavity
CHEM: Formaldehyde
OCCU: Agricultural Workers
LOCA: France; Germany; Italy; Netherlands; United States of America
TIME: 1992 - 1995

***317** **Luce, D.** 05340
INSERM, U88, 91, Blvd de l'Hôpital, 75634 Paris Cedex 13, France (Tel.: +33 1 45846374; Fax: 45838302)
COLL: Goldberg, P.; Goldberg, M.; Nicolau, J.; Billon-Galland, M.A.; Brochard, P.; Févotte, J.; Hours, M.; Dubourdieu, D.

Respiratory Cancer in New-Caledonia. Environmental, Occupational and Lifestyle Risk Factors

The objective of this population-based case-control study in New-Caledonia is to examine the associations between respiratory cancers and (1) occupational risk factors (carcinogenic substances used in industrial processes); (2) environmental factors (presence of mineral fibres in the environment); (3) lifestyle factors (smoking, alcohol, food consumption and specific nutrients). Cases will be identified through the New-Caledonia Cancer Registry. All incident cases of cancer of the lung, larynx, sinonasal cavities, pleura and hypopharynx, diagnosed from January 1993, will be included in the study. The expected number of cases is approximately 70 per year. Controls will be randomly selected from the general population and matched for age and sex, with a 1:2 case to control ratio. Cases and controls will be interviewed by trained interviewers. The questionnaire includes information on socio-demographic characteristics, residential history, smoking habits, alcohol and food consumption, and occupational history. The questionnaire on occupational history includes a detailed description of each job held during the working life. Specific questionnaires, designed to help the interviewer to probe with more technical questions, will also be used for occupations that occurred frequently. Exposure to carcinogenic substances will be assessed from the occupational questionnaire by industrial hygiene experts. In addition, mineral fibres will be measured and identified from air and soil samples and from biological material.

TYPE: Case-Control
TERM: Alcohol; Chemical Exposure; Diet; Nutrition; Occupation; Registry; Tobacco (Smoking)
SITE: Hypopharynx; Larynx; Lung; Nasal Cavity; Pleura
CHEM: Mineral Fibres
LOCA: New Caledonia
REGI: Nouméa (Fra)
TIME: 1993 - 1995

Reims

318 **Delisle, M.J.** 04425
Inst. Jean Godinot, Serv. de Médecine Nucléaire, Registre des Cancers de la Thyroïde, 1, rue du Général Kornig BP 171, 51056 Reims Cedex, France (Tel.: +33 26060504; Fax: 26060045 ; Tlx: 830048)
COLL: Théobald, S.; Schvartz, C.; Maès, B.; Vaudrey, C.; Pochart, J.M.

Thyroid Carcinoma in Champagne-Ardenne

Since 1967, 1,227 clinically diagnosed and histologically confirmed thyroid cancers have been registered and followed up in patients living in the following French administrative areas: Aisne (n = 290), Ardennes (n = 234), Aube (n = 128), Marne (n = 494), Haute Marne (n = 28), others (n = 53). Inclusion of all patients in the administrative areas Marne and Ardennes has been ensured since 1979 by creating a regional multidisciplinary group which manages nearly all the thyroid cancers. These two areas are used to define the population denominator for a population-based site-specific registry. By the end of 1992, 693 patients had been registered in this population-based site-specific registry. Items registered are: name, sex, date of birth, address, incidence date, histology and source of information, place of birth, age at incidence date, clinical and post surgical stagings using the TNM system, treatments, outcome and occurrence of other cancers. Quality control was carried out at the end of 1992. The main purposes of this registration are to study time and space incidence trends, to assess therapeutic strategies, genetic and therapeutic risks for the occurrence of second cancer and to identify new prognostic factors.

TYPE: Case Series; Incidence; Registry
TERM: Clinical Records; Histology; Multiple Primary; Prognosis; Stage; Survival; Treatment
SITE: Thyroid
TIME: 1967 - 1999

***319** **Delisle, M.J.** 05313
Inst. Jean Godinot, Serv. de Médecine Nucléaire, Registre des Cancers de la Thyroïde, 1, rue du Général Kornig BP 171, 51056 Reims Cedex, France (Tel.: +33 26060504; Fax: 26060045 ; Tlx: 830048)
COLL: Théobald, S.; Schvartz, C.; Maès, B.; Vaudrey, C.; Pochart, J.M.

Occurrence of Second Cancers After Thyroid Carcinoma

In a cohort of 928 thyroid cancer patients (Thyroid cancer Registry of Champagne-Ardenne) 36 developed a second cancer after treatment of the thyroid cancer. A case-control study is now in progress with the aim to study therapeutic and genetic risks. The cases will be the patients who developed a second cancer. There will be a minimum of three controls per case matched for age at diagnosis and sex. Demographic data, tumour data and treatment history are available in the registry.

Subjects will be interviewed by a trained medical secretary and information on family history and 131 radioiodine administration collected.

TYPE: Case-Control
TERM: Genetic Factors; Multiple Primary; Radiotherapy
SITE: All Sites; Thyroid
TIME: 1993 - 1994

Vandoeuvre

***320 Clavel, T.** 05448
Inst. National de Recherche et de Sécurité (INRS), Ave. de Bourgogne, BP 27, 54500 Vandoeuvre Cedex, France (Tel.: +33 83502000; Fax: 83502097 ; Tlx: INRS 850 778 F)

Mortality Study Among Workers Producing Stainless and Alloyed Steels

A mortality study will be carried out among workers producing stainless and alloyed steels to determine whether exposure to chromium nickel, iron oxide compounds, and PAH may have induced a risk of lung cancer. This study will be carried out in two phases: a cohort study and a nested case-control study, in collaboration with the Occupational Medicine and the Institute of Epidemiology of Lyon (IEL). The cohort will be defined as all men employed for at least one year between 1 January 1968 and 31 December 1991. The factory will provide individual records for all the cohort members, including names, dates, places of birth and periods of employment. The smoking habits will be known from medical records. The vital status will be recorded through the registry offices of the birthplaces by the IEL. The causes of death will be obtained by INSERM in the national registry of causes of death. A first analysis will consist in calculating SMR. External comparison is done with the French general population. A nested case-control study will also be carried out. The cases will be workers who died from lung cancer.

TYPE: Case-Control; Cohort
TERM: Dusts; Metals; Occupation
SITE: Lung
CHEM: Asbestos; Chromium; Ferric Oxide; Nickel; PAH; Silica
OCCU: Steel (Stainless) Workers; Steel Workers
TIME: 1993 - 1996

321 L'Huillier, M.C. 05235
Hôpital d'Enfants, Serv. de Médicine Infantile II, Rue du Morvan, 54511 Vandoeuvre Cedex, France (Tel.: +33 83558120; Fax: 83596069; Tlx: 960561)
COLL: Sommelet, D.

Childhood Cancer Registry and Risk Factors

Following the results of a case-control study based on cases collected by the Childhood Cancer Registry of Lorraine (population-based registry created in 1983, registering solid tumours, leukaemias and all malignant and benign brain tumours; population base: 535,236 children aged 0 to 15), a data abstract form is completed by the parents of each child entered into the registry as from 1 January 1991. The form takes into consideration many potential risk factors (genetic, personal, environmental), focussing on history of infections and familial cancers.

TYPE: Registry
TERM: Environmental Factors; Family History; Genetic Factors; Infection; Registry
SITE: Childhood Neoplasms
REGI: Lorraine (Fra)
TIME: 1991 - 1994

***322 Moulin, J.J.** 05465
Inst. National de Recherche et de Sécurité, Dépt. d'Epidémiologie, Ave. de Bourgogne, B.P. 27, 54500 Vandoeuvre Cedex, France (Tel.: +33 83502000; Fax: 83502097 ; Tlx: 850778 inrs f)

Mortality Study among Workers Producing Stainless Steel and Magnetic Steel Sheets

This study is aimed at assessing whether workers exposed to metals (chromium and nickel), PAHs and silica in a steel plant are at high risk of mortality from lung cancer when compared with a suitable regional population. The study will consist of a cohort of nearly 7,000 workers employed between 1 January 1960 and 31 May 1989. The follow-up period will be 1 January 1968 to 31 May 1989. A historical description of all the industrial processes involved in the past will be attempted in order to take into account past exposures. The job histories will be provided by the administration of the factory. In addition, all available data on smoking habits will be collected from medical records. Causes of deaths will be provided by the INSERM. Cause-specific SMRs will be computed according to exposure, taking into account duration, latency, calendar periods and tobacco smoking.

TYPE: Cohort
TERM: Dusts; Metals; Occupation
SITE: Lung
CHEM: Asbestos; Chromium; Iron; Nickel; PAH
OCCU: Steel (Stainless) Workers
TIME: 1991 - 1995

***323** **Moulin, J.J.** 05466
Inst. National de Recherche et de Sécurité, Dépt. d'Epidémiologie, Ave. de Bourgogne, B.P. 27, 54500 Vandoeuvre Cedex, France (Tel.: +33 83502000; Fax: 83502097 ; Tlx: 850778 inrs f)

Mortality Study of Steel Workers

A historical prospective mortality study is being carried out in a cohort of workers employed in a plant having produced mild steel, using open hearth furnaces and still producing sheets made of stainless and mild steel. The aim of the study is to assess the cancer risk due to the occupational exposure of chromium and nickel, PAHs and asbestos. The cohort consists of all workers employed between 1 January 1960 and 31 January 1990, i.e. about 5,000 workers. Deaths will be taken into account throughout the follow-up period from 1 January 1968 to 31 January 1990. Causes of death will be provided anonymously by the national file of the INSERM. Job histories and smoking habits of the workers will be traced through administrative and medical records of the plant. Occupational exposures will be detailed, using a historial description of the industrial processes that have been used for the last 60 years. The results will be given as SMRs, computed for various causes of death, using regional mortality rates as a reference. This analysis will take into account the duration of exposure, time elapsed since first exposure and calendar periods. As far as possible, data on smoking will be included.

TYPE: Cohort
TERM: Dusts; Metals; Occupation
SITE: Colon; Lung
CHEM: Asbestos; Chromium; Nickel; PAH; Silica
OCCU: Steel (Stainless) Workers
TIME: 1992 - 1995

***324** **Moulin, J.J.** 05467
Inst. National de Recherche et de Sécurité, Dépt. d'Epidémiologie, Ave. de Bourgogne, B.P. 27, 54500 Vandoeuvre Cedex, France (Tel.: +33 83502000; Fax: 83502097 ; Tlx: 850778 inrs f)
COLL: Soutar, C.; Auburtin, G.

Mortality Study of Workers Producing Ceramic Fibres

A historial prospective mortality study will be carried out among workers producing ceramic fibres (CF) in seven European factories located in UK (two factories), Germany (one factory) and France (four factories). The study will aim to assess possible risks of cancer of the respiratory system and of non-malignant respiratory diseases, in relation to occupational exposure to CF. The INRS will be in charge of the French study, in collaboration with the Institut de Medecine du Travail de Saint Etienne and the Institute of Occupational Health of Birmingham. The total cohort will comprise 1,500 workers, 50% being employed in the four French factories. The cohort will consist of all workers ever employed for more than one year. Deaths will be traced in the INSEE. Causes of death will be ascertained anonymously through the national file of causes of death (INSERM). The statistical analysis will compute SMRs, using an external reference (national or local populations). Principal investigator of this project is the Institute of Occupational Medicine in Edinburgh (UK). Collaborating Centre is INERIS (France).

TYPE: Cohort
TERM: Latency; Occupation; Time Factors
SITE: Mesothelioma; Respiratory
CHEM: Mineral Fibres
OCCU: Ceramic Workers
LOCA: France; Germany; United Kingdom
TIME: 1993 - 1995

***325** **Moulin, J.J.** 05468
Inst. National de Recherche et de Sécurité, Dépt. d'Epidémiologie, Ave. de Bourgogne, B.P. 27, 54500 Vandoeuvre Cedex, France (Tel.: +33 83502000; Fax: 83502097 ; Tlx: 850778 inrs f)

French Mortality Study among Hard Metal Workers

A prospective historical mortality study will be carried out among workers in factories producing hard metals. This study will aim to assess the possible occupational lung cancer risk in relation to cobalt exposure. The follow-up period will be 1968-1992. The observed mortality will be compared with that expected using national and regional rates, with adjustment for age, sex and calendar time. A job-exposure matrix will be used, together with individual job histories, to assess occupational exposures. The statistical analysis will take into account duration of exposure, time since first exposure and tobacco smoking.

TYPE: Cohort
TERM: Metals; Occupation
SITE: Lung
CHEM: Cobalt
OCCU: Metal Workers
TIME: 1993 - 1995

326 **Pham, Q.T.** 03989
INSERM U115, Santé au Travail et Santé Publique, Méthodes et Applications, Rte de la Fôret de Haye, 54505 Vandoeuvre, France (Tel.: +33 83565656)
COLL: Gabiano, P.

Mortality Among Iron Miners: A Prospective Study

Previous studies have shown an increase of lung cancer in iron miners. This study will investigate for this occupation the risk of developing cancers of other sites. A mortality study has been initiated in a cohort of 13,921 miners, active and retired, alive on 1 January 1982, in which the distribution of underground and surface workers and their length of employment are globally known. A record is established for each deceased person, containing information on name, date of birth, date of entering the mine, length of work in the mine or elsewhere, previous and subsequent main occupations and duration. Information on height, weight, tobacco and alcohol consumption, hospitalization, accidents at work, place of death, sources of death information, cause(s) of death (immediate, consecutive and contributing) and autopsy results, is also recorded. The relative frequency of causes of death will be calculated for the whole cohort and possible relationship with occupational exposures established. After five years 1,806 were deceased. Causes of death are known for 1,395. Follow-up continues.

TYPE: Cohort
TERM: Alcohol; Chemical Exposure; Metals; Mining; Occupation; Physical Factors; Time Factors; Tobacco (Smoking)
SITE: Lung; Stomach
CHEM: Iron
OCCU: Miners, Iron
TIME: 1982 - 1994

***327** **Wild, P.P.** 05499
Inst. National de Recherche et de Sécurité, Ave. de Bourgogne, BP 27, 54500 Vandoeuvre Cedex, France (Tel.: +33 83502000; Fax: 83502097 ; Tlx: inrs 850 778 f)
COLL: Bergeret, A.

Mortality Study in the Pulp and Paper Industry

IARC has initiated a worldwide multicentre cohort study of the potential cancer risks in the pulp and paper industry (see project Kogevinas, M.). The French cohort will comprise about 7,000 subjects from four pulp producing factories and will be followed up for mortality from 1968 to 1992. The exposure is assessed by individual job histories and by a standardized questionnaire on the industrial processes, completed by the factories. The mortality of the cohort will be compared both with French and regional mortality rates by standard life table methods. Internal comparisons based on Poisson regression and/or Cox models will also be attempted. The study will involve collaboration with the Institute of Epidemiology in Lyon, which will be responsible for the collection of the occupational items and the vital status of the cohort members. The INRS will be responsible for the statistical analysis. Furthermore, INRS will take part in the international study, contributing to the data base and participating in the final analysis.

TYPE: Cohort
TERM: Occupation
SITE: All Sites
OCCU: Paper and Pulp Workers
TIME: 1993 - 1995

Verneuil en Halatte

328 **Auburtin, G.** 04714
INERIS, Parc Technologique Alato, Boite Postale no. 2, 60550 Verneuil en Halatte, France (Tel.: +33 44556596; Fax: 556699 ; Tlx: 140094 F)

Lung Cancer and Pneumoconiosis in Coal Miners and Ex-Miners

The objective of this study is to determine the relationship between lung cancer and coalworker's pneumoconiosis (CWP) while controlling for smoking. It is a population-based case-control study among underground miners and ex-miners in the Nord-Pas de Calais collieries. Incident cases of primary bronchiolar and lung cancer over a period of two years among all men living in the Nord-Pas de Calais Region will be registered through a network of pneumologists and pathologists. Only cases from the defined population will be included. Controls, selected from the same population, will be individually matched to cases on age and years of experience underground. The main risk factor to be studied is the presence of CWP and its degree of severity. Definitions of CWP according to medico-legal criteria and to radiological criteria alone, will both be used in the analyses. This study should provide a better understanding of the role of CWP in the development of broncho-pulmonary lung cancer. Positive findings could provide a scientific basis for compensation for lung cancer in a person suffering from CWP, and encourage prevention strategies before diagnosis of pneumoconiosis and during follow-up of pneumoconiotic patients.

TYPE: Case-Control
TERM: Occupation; Pneumoconiosis
SITE: Lung
OCCU: Miners, Coal
TIME: 1990 - 1994

Villejuif

329 **Benhamou, S.** 04815
Inst. Gustave Roussy, INSERM U351, Rue Camille Desmoulins, 94805 Villejuif Cedex, France (Tel.: +33 1 45594139; Fax: 46787430)
COLL: Benhamou, E.; Le Mab, G.; Grosclaude, P.; Raymond, L.; Sancho-Garnier, H.; Luboinski, M.

Prostate Cancer, Sexual Behaviour and Diet

It is proposed to carry out a collaborative population-based case-control study in Geneva, Switzerland and in two counties (departements) in France for which prostate cancer represents the highest incidence rate in males. The study will also be performed in Paris in different hospitals. The aim of this study is to determine the influence of sexual activity, sexually transmitted diseases and diet (in particular fats) in prostate cancer. Cases will be taken from the Tarn, Martinique and Geneva cancer registries. Sex- and age-matched controls will be chosen from the population census list. Risk factors will be assessed by interview. In each area, 150 cases and 300 controls are expected.

TYPE: Case-Control
TERM: Diet; Fat; Registry; Sexual Activity; Sexually Transmitted Diseases
SITE: Prostate
LOCA: France; Switzerland
REGI: Geneva (Swi); Martinique (Mar); Tarn (Fra)
TIME: 1990 - 1994

330 **Benhamou, S.** 04816
Inst. Gustave Roussy, INSERM U351, Rue Camille Desmoulins, 94805 Villejuif Cedex, France (Tel.: +33 1 45594139; Fax: 46787430)
COLL: Lenfant, M.H.; Paoletti, C.; Flamant, R.

Lifestyle and Medical Drug Consumption in Renal Cell Carcinoma and Renal Pelvis Cancer.

A hospital-based case-control study has been undertaken since 1987 to study possible causes of renal cell carcinoma and renal pelvis cancer. The main environmental risk factors investigated are consumption of tobacco, alcohol, coffee and the use of analgesics and diuretics. All cases are histologically confirmed. Two controls per case, hospitalised for a non-tobacco-related disease (one for a malignant disease and one for a non-malignant disease) are matched for sex, age at interview (± 5 yrs) and hospital. 300 cases and 600 controls are expected.

TYPE: Case-Control
TERM: Alcohol; Analgesics; Coffee; Diuretics; Occupation; Tobacco (Smoking)
SITE: Kidney
TIME: 1987 - 1994

331 **Benhamou, S.** 04817
Inst. Gustave Roussy, INSERM U351, Rue Camille Desmoulins, 94805 Villejuif Cedex, France (Tel.: +33 1 45594139; Fax: 46787430)
COLL: Benhamou, E.; Schrameck, C.

Smoking and Female Lung Cancer

As part of an international case-control study on lung cancer and environmental tobacco smoke exposure, a hospital-based case-control study on lung cancer among women is being carried out. The aim is to evaluate the influence of different characteristics of smoking habits (in particular type of cigarettes and tar levels). Lung cancer will be histologically confirmed and 2 female controls hospitalised for a non-tobacco-related disease and matched for age (± 5 yrs), and hospital will be included.

TYPE: Case-Control
TERM: Female; Occupation; Passive Smoking; Tobacco (Smoking)
SITE: Lung
TIME: 1990 - 1994

***332** **Benhamou, S.** 05245
Inst. Gustave Roussy, INSERM U351, Rue Camille Desmoulins, 94805 Villejuif Cedex, France (Tel.: +33 1 45594139; Fax: 46787430)
COLL: Laplanche, A.C.; Koscielny, S.

Mortality among People Working in Research Laboratories

An international retrospective study, initiated by IARC, is being mounted to evaluate mortality from cancer and other diseases in public research institutions over the last 20 to 30 years. In France, INSERM is one institution taking part in the study. The cohort consists of all persons employed for at least one year at INSERM (about 6,500). Subjects will be classified by employment category and exposure. Cause-specific mortality will be collected for each deceased member of the cohort. SMRs will be used in the analysis. French population rates and internal comparison groups will be used as reference. The cohort is presently being constituted, using employee files. Data collection is due to be completed in 1994.

TYPE: Cohort
TERM: Chemical Exposure; Occupation
SITE: All Sites
OCCU: Laboratory Workers
TIME: 1992 - 1995

333 **Clavel, F.** 02301
Inst. Gustave Roussy, INSERM U351, Rue Camille Desmoulins, 94805 Villejuif Cedex, France
COLL: Flamant, R.; Laplanche, A.C.

Vinyl Chloride in Relation to Malignant Tumours

The purpose of this prospective study is to examine the relationship between vinyl chloride (VC) exposure and pathology (in particular malignant tumours) and causes of death. It is being conducted in factories producing VC (monomer or polymer). Only male workers, aged between 40 and 55 years, are considered. For each worker exposed, a worker never exposed to VC, of the same age (+/-two years), working in the same plant, is used as a control. Both exposed and non-exposed workers are interrogated by the same physician. The form contains some identification items (age, civil status, place of residence), information on exposure, including the whole occupational history, on tobacco and alcohol habits and health status. 2,282 workers were entered into the study and will be seen yearly for several years in order to follow their health status and note possible changes in relation to VC exposure. Preliminary analysis will now start.

TYPE: Case-Control; Cohort
TERM: Alcohol; Chemical Exposure; Male; Occupation; Plastics; Tobacco (Smoking)
SITE: All Sites
CHEM: Hydrocarbons, Halogenated; Vinyl Chloride
TIME: 1980 - 1994

334 **Clavel, F.** 04980
Inst. Gustave Roussy, INSERM U351, Rue Camille Desmoulins, 94805 Villejuif Cedex, France
COLL: Cordier, S.E.; Flandrin, G.; Conso, F.; Limasset, J.G.

Occupational Risk Factors in Hairy Cell Leukaemias

A high proportion of mechanics, farmers, and physicians has been reported in a case series of hairy cell leukaemias (HCL). This observation has given rise to the hypothesis that there is a relationship between HCL and exposure to benzene, other solvents, pesticides and radiations, already known or suspected to be involved in other types of leukaemia. A case-control study including all HCL cases diagnosed in France between 1 January 1980 and 1 January 1989 has been set up in 27 departments of haematology. Two hospital controls are matched to each case on sex, age, year and place of hospitalization, and area of residence. The study will include about 400 cases and 800 controls. A detailed interview of patients investigates the exposures of interest and results are reviewed for coding by a panel of experts in industrial hygiene and occupational medicine. For pesticides, exposure assessment is performed by specialized physicians on the farms concerned.

TYPE: Case-Control
TERM: Electromagnetic Fields; Occupation; Pesticides; Radiation, Ionizing; Solvents
SITE: Leukaemia
CHEM: Benzene
TIME: 1988 - 1994

335 **Cordier, S.E.** 05049
INSERM U170, 16 bis Ave. Paul Vaillant-Couturier, 94807 Villejuif, France (Tel.: +33 1 45595034; Fax: 45595151)

Cancer Risk among Laboratory Workers

A cohort study was set up in order to investigate a cluster of rare cancers among young scientists at the Institut Pasteur in Paris. 3,765 persons having worked at the Institute for at least six months between 1971 and 1986 were identified. Their mortality was studied up to the end of 1987. A nested case-control study has also been carried out, including 23 cases of incident and fatal cancers of the pancreas, bone, brain and haemapoeitic system and 4 controls per case, matched for sex and year of birth.

TYPE: Case-Control; Cohort
TERM: Chemical Exposure; Occupation
SITE: All Sites
OCCU: Laboratory Workers
TIME: 1986 - 1994

***336** **de Vathaire, F.** 05311
Inst. Gustave-Roussy, INSERM U351, Rue Camille-Desmoulins, 94800 Villejuif Cedex, France (Tel.: +33 45596457; Fax: 46787430)
COLL: Oberlin, O.; LeMerle, J.; Chauaudra, J.; Hardiman, C.; Rumeau, N.; Hawkins, M.M.; Bell, C.M.J.; Campbell, S.

Second Cancers After a First Cancer in Childhood

This is a cohort study of second cancers after cancer in childhood, in order to improve the knowledge of (1) the dose response-relationship between the dose of external low LET radiation received during childhood and the risk of second cancers for the following particular sites: thyroid, breast, bone, skin, brain, and lung; (2) the effects of the fractionation of the dose; (3) the carcinogenic effect of various antimitotic treatments; and (4) the preferential associations between particular types of first and second cancers. Clinical characteristics, as well as the doses and lengths of use of each chemotherapeutic agent, are collected by the paediatricians from each participating centre. The radiation doses received by all the children at different anatomical sites are evaluated with the help of a program developed at the Gustave Roussy Institute. The expected size of the cohort is 5,000 children.

TYPE: Cohort
TERM: Chemotherapy; Childhood; Dose-Response; Multiple Primary; Radiation, Ionizing; Time Factors
SITE: All Sites
LOCA: France; United Kingdom
TIME: 1989 - 1994

***337** **de Vathaire, F.** 05312
Inst. Gustave-Roussy, INSERM U351, Rue Camille-Desmoulins, 94800 Villejuif Cedex, France (Tel.: +33 45596457; Fax: 46787430)
COLL: Benhamou, S.; Benhamou, E.; Fragu, P.H.; Avril, M.F.

Thyroid and Skin Effects of Irradiation for a Haemangioma During Childhood

The objective of this cohort study is to investigate the importance of dose rate in the risk of radio-induced tumours. About 1,140 children irradiated for a haemangioma from 1946-1973 at the Gustave Roussy Institute, are being studied in order to evaluate the long-term effects of irradiation to the thyroid and skin. 226Ra, 192Ir, 90Y, 32P, 90Sr, as well as X-rays had been used for treatment. For each haemangioma in each patient, the surface skin dose and the dose at 10 mm depth has been estimated, as well as the dose to the thyroid.

TYPE: Cohort
TERM: Dose-Response; Radiotherapy
SITE: Skin; Thyroid
TIME: 1984 - 1994

***338** **de Vathaire, F.** 05399
Inst. Gustave-Roussy, INSERM U351, Rue Camille-Desmoulins, 94800 Villejuif Cedex, France (Tel.: +33 45596457; Fax: 46787430)
COLL: Schlumberger, M.; Challeton, C.; Francese, C.; de la Genardière, E.; Delisle, M.J.; Couyette; Ceccarelli, C.; Pinchera, A.

Second Cancers and Pregnancy Outcome of Patients Treated for Differentiated Thyroid Carcinoma

A cohort study is being carried out (1) to evaluate the role of radioiodine administration in the risk of congenital malformation in the offspring of patients treated for thyroid carcinoma; (2) the incidence of second cancers in these patients; and (3) to study cancer in the families of these patients. It is hoped to include about 3,000 patients treated for a thyroid cancer. Women are interviewed by a specialised intereviewer. Information about each pregnancy, and family history of cancer is collected. To be able to take into account not only the doses to various anatomical sites due to thyroid fixation, but also that due to the metastases of the thyroid cancer, appropriate software is currently being developed.

TYPE: Cohort
TERM: Congenital Abnormalities; Multiple Primary; Radiation, Ionizing; Treatment
SITE: All Sites
CHEM: Iodine
LOCA: France; Italy
TIME: 1989 - 1994

339 **Henry-Amar, M.** 03726
Inst. Gustave-Roussy, Dépt. de Biostatistique et d'Epidémiologie, 39-53, Rue Camille Desmoulins, 94805 Villejuif Cedex, France (Tel.: +33 1 45594142; Fax: 46787430)

Second Cancers after Radiotherapy and/or Chemotherapy for Hodgkin's Disease

The aim of this study is to determine whether radiotherapy or chemotherapy or their combination used in the treatment of Hodgkin's disease is a factor of increased risk for a second cancer. If an increased risk is observed, it will be related to exposure, time relationships and type of second primary. This is a retrospective study on a cohort of approximately 1,600 patients, based on medical records dating from 1960. Particular attention will be paid to the following second primaries: acute leukaemia, non-Hodgkin's lymphoma and bronchus carcinoma. 19 secondary acute leukaemias were observed among 871 adult patients followed at least for one year post initial therapy. The sole factor related to leukaemia risk was the total dose of nitrogen mustard given. Subsequent analyses will be performed taking into account other second primary cancers than leukaemia. A paper has been published in Rec. Res. Cancer Res. 17:270-283, 1989.

TYPE: Cohort
TERM: Chemotherapy; Multiple Primary; Radiation, Ionizing; Time Factors; Treatment
SITE: All Sites; Leukaemia; Lung; Non-Hodgkin's Lymphoma
TIME: 1985 - 1994

340 **Henry-Amar, M.** 04270
Inst. Gustave-Roussy, Dépt. de Biostatistique et d'Epidémiologie, 39-53, Rue Camille Desmoulins, 94805 Villejuif Cedex, France (Tel.: +33 1 45594142; Fax: 46787430)

Second Malignancies Following Radiotherapy and Chemotherapy for Hodgkin's Disease

The aim of this multi-centre case-control study is to ascertain the risk of second acute leukaemia, non-Hodgkin's lymphoma and various solid tumours in relation to the type of treatment given, radiotherapy and/or chemotherapy. The extent of radiotherapy and doses of each drug administered have been recorded. Cases were patients who developed a second malignancy at least one year after having been treated according to one of the six clinical trials conducted from 1963 to 1987 by the EORTC Lymphoma Cooperative Group or the Groupe Pierre et Marie Curie. Overall, 74 second primary cancers have been observed, 10 acute leukaemias, 11 non-Hodgkin's lymphomas, and 53 solid tumours. Three controls per case have been chosen at random using the following criteria: sex, age within 1 year of case's age, date of diagnosis of Hodgkin's disease within 3 years of case's diagnosis, duration of survival at least as long as the interval between the diagnosis of the first and second primary cancers in the case. The histological material has been reviewed, both that of the second primary for cases, and that of the initial Hodgkin's disease material for all cases and controls.

TYPE: Case-Control
TERM: Chemotherapy; Multiple Primary; Radiotherapy
SITE: Gastrointestinal; Leukaemia; Lung; Non-Hodgkin's Lymphoma
LOCA: Belgium; France; Germany; Italy; Netherlands
TIME: 1987 - 1994

***341** **Hill, C.G.** 05251
Inst. Gustave-Roussy, Dépt. de Statistique Médicale, INSERM U287, 39, rue Camille Desmoulins, 94805 Villejuif Cedex, France (Tel.: +33 1 45594116; Fax: 46787430)
COLL: Laplanche, A.C.; Hattchouel, J.M.

Cancer Mortality Around French Nuclear Sites

Cancer mortality, especially for leukaemia, is being studied in the population (under age 25) residing near six French nuclear plants operating in 1975 (Nature 347:755-757, 1990). The study will be extended to include six sites operating between 1975 and 1985. The causes of all the deaths occurring in communes located within 10 miles of each nuclear plant will be obtained from INSERM. Census data will be obtained from INSEE. The mortality observed in the exposed communes will be compared with the mortality expected from national mortality statistics.

TYPE: Mortality
TERM: Radiation, Ionizing
SITE: All Sites
TIME: 1992 - 1994

***342** **Laplanche, A.C.** 05255
Inst. Gustave Roussy, Dépt. Biostat. Epidémiol., Rue Camille Desmoulins, 94805 Villejuif, France (Tel.: +33 1 45594127; Fax: 46787430)
COLL: de Vathaire, F.

Leukaemia Mortality in French Communes with a Large and Rapid Population Increase

Kinlen suggested that the excess in leukaemia mortality observed in the population under age 25 living near nuclear installations in the UK might be attributed to a rapid increase in the population, leading to viral infections. This study will investigate populations under age 25, residing in the 43 French communes with a population increase exceeding 100% between two consecutive censuses, during the period

1968–1990. Causes of death were obtained from INSERM and census data from INSEE. The observed mortality will be compared with the mortality expected on the basis of national rates.

TYPE: Mortality
TERM: Demographic Factors; Virus
SITE: Leukaemia
TIME: 1991 - 1994

343 **Stücker, I.** 05157
INSERM U170, 16, ave. Paul Vaillant-Couturier, 94807 Villejuif Cedex, France (Tel.: +33 1 45595033; Fax: 595151)
COLL: Laurent, P.; Bechtel, P.; Brochard, P.

Cytochrome P450: Genetic Polymorphism and Environmental Risk Factors in Lung Cancer
This study aims to investigate the statistical interaction between exposure to PAH and capacity to metabolise them (i.e. inducibility of aryl hydrocarbon hydroxylase (AHH) by the regulator gene Ah). The case-control design includes 300 incident lung cancer cases and 300 hospital controls. All subjects are males aged 75 or less, born in metropolitan France from parents born in metropolitan France. One control per case is matched on age and area of residence. Investigations include a detailed questionnaire on tobacco smoking and occupational PAH exposure; Western blot analysis for the measurement of AHH inducibility; measure of glutathione transferase; activity of AHH estimated by caffeine test on urine of smokers.

TYPE: Case-Control; Molecular Epidemiology
TERM: BMB; Dusts; Occupation; Tobacco (Smoking)
SITE: Lung
CHEM: Asbestos; Chromium; Iron; Nickel; PAH
TIME: 1989 - 1994

GAMBIA

Banjul

344 **Bah, E.** 05135

Int. Agency for Reserach on Cancer, Medical Research Council Laboratories, Gambia Hepatitis Intervention Study, P.O. Box 273 , Fajara Banjul, Gambia (Tel.: +220 95229; Fax: 96117)

COLL: Jack, A.; Inskip, H.M.

The Gambia Cancer Registry

This national population-based cancer registry was started in July 1986 as part of the Gambia Hepatitis Intervention Study. The primary objective of the registry is to monitor cancer incidence in the country. Data collected by the registry for the next 30 years will be used to evaluate the protective effectiveness of infant hepatitis B immunization in presenting subsequent primary hepatocellular carcinoma in the adult. Information on all malignant tumours occurring within the country is obtained from hospitals and clinics both in the public and private sector and analysed annually by age, sex and site. For the past four years (July 1986 to June 1990), a total of 1,002 cancers have been registered. A paper describing the first two years of operation was published in Br. J. Cancer 62:647-650, 1990.

TYPE: Incidence; Intervention; Registry
TERM: Data Resource; HBV; Registry; Vaccination
SITE: All Sites
REGI: Gambia (Gam)
TIME: 1986 - 2021

345 **Jack, A.D.** 03900

Int. Agency for Research on Cancer, Medical Research Council Laboratories, Gambia Hepatitis Intervention Study, Fajara Banjul, Gambia (Tel.: +220 95229; Fax: 96117)

COLL: Maine, N.; Tomatis, L.T.; Bosch, F.X.; Greenwood, B.; Whittle, H.C.; Armstrong, B.K.; Mendy, M.; Bah, E.; George, M.; Hall, A.J.; Smith, P.G.; Kane, M.; Parkin, D.M.; Day, N.E.

Gambia Hepatitis Intervention Study

The long-term aim of this study is to assess the effectiveness of hepatitis B vaccine in the prevention of PHC. This cancer is common in The Gambia, a country in which approximately 15% of adults are carriers of the hepatitis B virus. The first phase of this project was completed at the end of 1990, by which time information on 125,000 children had been collected. Almost half of these children received hepatitis B vaccine, which had been progressively incroporated into The Gambia's Expanded Programme of Immunization during a five-year period. Subsets of this cohort have been studied in detail and have shown that the vaccine has been very effective in preventing infection and the carrier state. The effectiveness of the vaccine in the prevention of liver cancer will be determined by linkage of the database with The Gambia's cancer registry, but results are not expected for many years. Many ancillary studies have been conducted to examine the natural history of hepatitis B virus infection and response to the vaccine and its cost-effectiveness. New areas of work are concentrating on the effects of aflatoxin exposure and the examination of mutant hepatitis B viruses against which the vaccine may fail to protect.

TYPE: Cohort; Incidence; Intervention
TERM: BMB; HBV; Prevention; Registry; Vaccination
SITE: Liver
CHEM: Aflatoxin
REGI: Gambia (Gam)
TIME: 1986 - 2021

Berlin

346 **Garbe, C.** 04625
Freie Univ. Berlin, Klinikum Steiglitz, Hautklinik & Poliklinik, Hindenburgdamm 30, W 1000 Berlin 45, Germany (Tel.: +49 30 7983923)
COLL: Breibart, E.W.; Roser, M.; Burg, G.; Jung, G.; Weiss, J.; Kresbach, H.; Soyer, H.P.; Kreysel, H.W.; Weckbecker, J.; Schnyder, U.W.; Panizzon, R.; Wolff, H.H.; Zaun, H.; Bahmer, F.A.; Tilgen, W.; Orfanos, C.E.

Early Diagnosis and Risk Factors of Malignant Melanoma
A multi-centre case-control study including over 600 malignant melanoma (MM) patients and the same number of age- and sex-matched controls is underway. Persons under investigation answer a questionnaire about their sun exposure habits and sun reaction, about their awareness of pigmented moles and, for MM patients, about the reasons which underlie their decision to consult a doctor. In a whole body examination, pigmentational characteristics and all pigmented moles are carefully documented. Histological examination is performed for clinically atypical naevi. Relative risks for developing MM will be calculated in relation to potential risk factors such as sunburn, different recreational activities, occupation, hormonal factors and immunosuppression, as well as in relation to risk indicators like numbers of common and dysplastic naevi, further pigmented moles and pigmentational characteristics. Factors influencing early diagnosis will be analysed.

TYPE: Case-Control
TERM: Hormones; Immunosuppression; Lifestyle; Naevi; Occupation; Radiation, Ultraviolet
SITE: Melanoma
LOCA: Austria; Germany; Switzerland
TIME: 1989 - 1994

347 **Kohlmeier, L.** 05166
Inst. für Sozialmedizin und Epidemiologie, Bundesgesundheitsamt, General-Pape-Str. 64, W 1000 Berlin 42, Germany (Tel.: +49 30 78007197; Fax: 78007109 ; Tlx: 184016)
COLL: Thamm, M.; Mund-Hoym; Randow

Antioxidant Status and Breast Cancer in East and West Berlin
The study is designed to test the hypothesis that a dose-response relationship exists between antioxidant status (levels of beta carotene, tocopherols and selenium) and risk of breast cancer. A comparison of the risk profiles in both eastern and western German populations is also planned. In each clinic a minimum of 100 cases are to be recruited and matched 1:2 with population controls. The exposure assessment will be via analysis of toenails and adipose tissue as well as computerized interviews.

TYPE: Case-Control
TERM: Antioxidants; Dose-Response
SITE: Breast (F)
CHEM: Beta Carotene; Selenium; Tocopherol
TIME: 1990 - 1994

Berlin-Buch

348 **Geissler, E.** 04744
Max-Delbrück Centre for Molecular Medicine, Robert Rössle Str. 10 , O 1115 Berlin-Buch, Germany (Tel.: +49 30 94062328; Fax: 94063824)
COLL: Staneczeck, W.

Simian Virus (SV40) and Human Intracranial Tumours
Millions of people were vaccinated with polio vaccine contaminated with SV40 during 1955-1962. This study will try to assess whether this viral contaminant might be involved in malignant late effects. In 30 out of 110 human cerebral tumours, footprints of SV40 or of SV40-like viruses could be found by indirect immuno-fluorescence and by DNA hybridization. About 750,000 persons treated with SV40-contaminated polio vaccine between 1960-1962 are now being followed up for incidence of

malignancies, compared with a similar number of persons who were treated with SV40-free polio vaccine (Progr. Med. Virol. 37:211-22, 1990).

TYPE: Cohort
TERM: Vaccination; Virus
SITE: Brain
REGI: Berlin (Ger)
TIME: 1965 - 1994

Bremen

349 **Greiser, E.** 04986
Bremen Inst. for Prevention, Research and Social Medicine, Grünenstr. 120 , W 2800 Bremen, Germany (Tel.: +49 421 595960; Fax: 5959665)
COLL: Lotz, I.; Hoffmann, W.

Leukaemia and Malignant Lymphoma in the Vicinity of an Industrial Waste Dump

A previous study revealed an increased incidence of leukaemias and malignant lymphomas in a rural district near a disused waste-dump. The current study begins with the registration of all cases of leukaemia and malignant lymphoma (ICD-9 200-208), diagnosed between 1984 and 1990 in all municipalities neighbouring the waste dump. Data are collected from all local public and private hospitals, physicians, public health offices and laboratories. A case-control study, including all registered leukaemia patients with age- and sex-matched population controls, has been set up to identify and examine the role of risk factors for leukaemias and malignant lymphomas. 700 cases and 1400 age- and sex-matched population controls are expected. Risk factors covered will include exposure to solvents, PAH, ionizing radiation, herbicides and pesticides, certain therapeutic drugs, occupations known to be associated with an elevated risk for this group of diseases and vicinity to the dump site. Cases and controls will be interviewed by trained personnel, using a standardized questionnaire.

TYPE: Case-Control
TERM: Drugs; Environmental Factors; Herbicides; Occupation; Pesticides; Radiation, Ionizing; Solvents; Waste Dumps
SITE: Leukaemia; Lymphoma
CHEM: PAH
TIME: 1990 - 1994

***350** **Greiser, E.** 05435
Bremen Inst. for Prevention, Research and Social Medicine, Grünenstr. 120 , W 2800 Bremen, Germany (Tel.: +49 421 595960; Fax: 5959665)
COLL: Hoffmann, W.

Leukaemias and Malignant Lymphomas in Children and Adults in Three Rural Districts Adjacent to a Nuclear Power Plant

In 1991, a suspected increase in childhood leukaemia close to a nuclear power plant in the rural commune of Elbmarsch, Lower Saxony, northern Germany, raised considerable public concern. In a population of about 1,400 children under 15 years, five cases of leukaemia were diagnosed in young children between December 1989 and May 1991. A retrospective incidence study covering the operation period of the plant beginning in 1984 up to 1993 has been initiated. The aim of this study is to prepare a high-resolution temporal and spatial incidence map of leukaemia and malignant lymphoma in all ages, using multiple data sources including local hospitals, regional therapy centers and University hospitals, Public Health offices, pathological/histological laboratories and general practitioners. The study area covers the three adjacent rural districts in the Federal States of Schleswig-Holstein and Lower Saxony. Appropriate epidemiological, statistical and mathematical methods will be applied to evaluate age- and disease-specific temporal and spatial clustering.

TYPE: Incidence
TERM: Cluster; Mapping; Radiation, Ionizing
SITE: Leukaemia; Lymphoma
TIME: 1992 - 1994

351 **Jöckel, K.H.** 04306

Bremen Inst. for Prevention Research and Social Medicine, Div. of Biometry and Data Processing, Grünenstr. 120, W 2800 Bremen 1, Germany (Tel.: +49 421 5959651; Fax: 5959668)

COLL: Ahrens, W.; Bolm-Audorff, U.; Jahn, I.

Occupational Risk Factors for Lung Cancer

A case-control study with 1,000 cases and the same number of population controls will be carried out, investigating the association between risk of developing a bronchial carcinoma and serveral occupational factors. A detailed job history including description of the work place and exposure-information about known occupational carcinogens will be obtained by means of a standardized interview. Detailed information about smoking habits will also be obtained from each subject. It is intended to establish a more detailed description of occupational hazards for lung cancer from the identification of occupational risk-groups and possible associations with certain substances.

TYPE: Case-Control
TERM: Chemical Exposure; Occupation; Tobacco (Smoking)
SITE: Lung
TIME: 1987 - 1994

***352** **Jöckel, K.H.** 05333

Bremen Inst. for Prevention Research and Social Medicine, Div. of Biometry and Data Processing, Grünenstr. 120, W 2800 Bremen 1, Germany (Tel.: +49 421 5959651; Fax: 5959668)

COLL: Bolm-Audorff, U.; Pohlahelm, H.; Vilgus, B.

Lifestyle and Occupational Risk Factors for Cancer of the Lower Urinary Tract

In a hospital-based case-control study 239 male and 61 female patients with cancer of the lower urinary tract and the same number of controls (patients with non-neoplastic diseases of the lower urinary tract), matched for sex, age and area of residence, were interviewed for their job and smoking history. The study aims to assess the association between cancer risk and occupational factors, controlling for smoking, nutritional and drinking habits, and use of analgesics. Statistical analysis is by conditional logistic regression.

TYPE: Case-Control
TERM: Analgesics; Chemical Exposure; Diet; Occupation; Tobacco (Smoking)
SITE: Bladder; Ureter; Urethra
TIME: 1989 - 1994

Erfurt

353 **Schunk, W.W.** 05126

Medizinische Akademie Erfurt, Inst. für Arbeitsmedizin, Gustav-Freytag-Str. 1 , O 5082 Erfurt, Germany (Tel.: 387311)

Cancer in Rubber Workers in Thuringia

This study of workers in the rubber industry has been initiated with the aim of measuring exposure to asbestos in talc in the last 20 years. Over 200 workers were exposed. They will be compared with 100 non-exposed workers. Many exposed workers are now retired. Medical histories will be collected and clinical, functional and x-ray examinations carried out. The study is mainly aimed at assessing the relation between exposure to asbestos (quantity and type) in talc and cancer, especially cancer of the lung and mesothelioma.

TYPE: Cohort
TERM: Dusts; Occupation; Rubber
SITE: Lung; Mesothelioma
CHEM: Asbestos; Talc
OCCU: Rubber Workers
TIME: 1991 - 1994

Essen

354 **Popp, W.** 04999

Univ. of Essen (GHS), Inst. for Hygiene and Industrial Med., Hufelandstr. 55 , W 4300 Essen 1, Germany (Tel.: +49 201 7234574)

COLL: Norpoth, K.; Vahrenholz, C.; Schürfeld, C.; Müller, C.

Validity of Molecular Epidemiological Methods in Occupational Medicine

The aim of the study is to test the validity and usefulness of certain methods of molecular epidemiology (SCE, alkaline elution) in occupationally exposed groups. These methods, partly combined with methods of biological monitoring, are being applied to groups of workers who are exposed to chromium and nickel (welders), benzene and toluene (shoemakers) and ethylene oxide (hospital personnel). For every group, consisting of 20-40 persons each, control groups, matched for age, sex and smoking habits, are formed. An additional positive control group consists of multiple myeloma patients receiving therapy with the alkylating agent melphalan.

TYPE: Methodology; Molecular Epidemiology
TERM: Drugs; Occupation; SCE; Solvents
SITE: Inapplicable
CHEM: Benzene; Chromium; Ethylene Oxide; Melphalan; Nickel; Toluene
OCCU: Health Care Workers; Shoemakers-Repairers; Welders
TIME: 1990 - 1994

***355** **Popp, W.** 05354

Univ. of Essen (GHS), Inst. for Hygiene and Industrial Med., Hufelandstr. 55 , W 4300 Essen 1, Germany (Tel.: +49 201 7234574)

COLL: Norpoth, K.; Vahrenholz, C.; Schell, C.; Kraus, R.; Porschnev, M.

Biomarkers in Molecular Epidemiology

Different biomarkers (SCE, alkaline, filter elution, 32p-postlabelling of molecular epidemiology) have been applied to groups of 20 to 40 persons, occupationally exposed to carcinogens: benzene, chromium/nickel, melphalan and ethylene oxide. A control group of about 100 persons has also been tested. The investigations will be extended to include new biomarkers and new exposure groups: coke oven workers, nurses, oral cancer patients, radon exposed workers and, again, benzene and ethylene oxide exposed workers. The new biomarkers are flow cytometry (determination of lymphocyte sub-populations), chromosomal aberrations including G-banding, determination of oncogene proteins in serum and oncogene analysis by PCR and sequencing. The aim of the study is to assess the validity of different biomarkers.

TYPE: Methodology; Molecular Epidemiology
TERM: Biomarkers; Chromosome Effects; Drugs; G-Banding; Metals; Occupation; Oncogenes; PCR; SCE; Solvents
SITE: Inapplicable
CHEM: Benzene; Chromium; Ethylene Oxide; Melphalan; Nickel; Radon
OCCU: Coke-Oven Workers; Health Care Workers; Radiation Workers
LOCA: Germany; Russian Federation
TIME: 1992 - 1995

Gummersbach

356 **Schmauz, R.** 04385

Postfach 10 08 08, W 5270 Gummersbach, Germany

COLL: Okong, P.; Owor, R.

Case-Control Study of Squamous Cell Genital Cancer

Current aetiological hypotheses for genital cancers will be examined in a case-control study in Uganda, an area of high incidence. Cancer will be squamous tumours of cervix, vulva and penis. Controls will be non-cancer patients seen in the same hospital for other surgical procedures. About 300 cases are expected over 3 years. Interviews will focus on sexual habits, socio-economic factors and tribal origin. Tissue will be examined for HPV by Southern blot.

TYPE: Case-Control
TERM: Ethnic Group; HPV; Sexual Activity; Socio-Economic Factors
SITE: Penis; Uterus (Cervix); Vulva
LOCA: Uganda
TIME: 1987 - 1994

Hamburg

***357** **Manz, A.** 05256
Centre for Chemical Workers' Health, Dept. of Health, Fuhlsbüttler Str. 401 , 2000 Hamburg 60, Germany (Tel.: +49 40 63853807; Fax: 6325848)
COLL: Flesch-Janys, D.; Waltsgott, H.; Berger, J.E.; Nagel, S.; Dwyer, J.H.

Cancer Incidence and Mortality among Workers in a Chemical Plant Contaminated with Dioxin

The carcinogenicity of dioxins and furans to humans will be investigated in a cohort of 1,583 workers (1,184 men, 399 women), heavily exposed to these substances in a herbicide and insecticide producing plant in Hamburg, during 1952-1984. For the males, the mortality follow-up will be extended from 1989 to 1992. The exposure of the cohort to dioxins and furans will be investigated by measurements of blood concentrations to develop a quantitative exposure indicator. A nested case-control design is used to evaluate the potential confounding by smoking habits and exposure to other carcinogens. Cancer mortality will be analysed for a dose-response relationship by proportional hazard models using a cohort of gas workers as reference group. The objective of the study of the female cohort is to investigate breast cancer incidence. A quantitative exposure indicator for dioxins and furans will be constructed using measurement of blood concentrations. Information on main risk factors will be obtained by a structured questionnaire. A first report was published in The Lancet 338:959-964, 1991.

TYPE: Case-Control; Cohort
TERM: Occupation
SITE: All Sites; Breast (F)
CHEM: Dioxins
OCCU: Herbicide Manufacturers; Pesticide Workers
TIME: 1987 - 1994

Hannover

358 **Baltrusch, H.J.F.** 05146
Medizinische Hochschule, Immunhämatologie, Transfusionsmedizin - Blutbank 8350, Konstanty-Gutschow-Str. 8, W 3000 Hannover 61, Germany (Tel.: +49 511 5322083; Fax: 5325550)
COLL: Stangel, W.

Biopsychosocial Setting of Polycythaemia Vera

Polycythaemia vera (PV) is a rare myeloproliferative disease of largely unknown aetiology. Clinical observations suggest that the disease might be precipitated by psychosocial stress. An interdisciplinary controlled study has been initiated in order to ascertain the role of such stress in the aetiology and clinical course of PV. Consecutive PV patients and their matched pair controls are studied by a structured interview technique and a series of psychological tests, including a questionnaire measuring Type C (cancer-prone) behaviour. Family Attitudes Questionnaire, State-Trait Anxiety Inventory, Habits of Nervous Tension, Courtauld Emotional Control Scale, Rorschach and Thematic Apperception Test. Assessment of patients' coping behaviour is also included in the study as well as aspects of guidance and treatment. So far, 40 patients have been studied. Partial results were published in Psychiat. Fenn. Suppl.:133-142, 1981 and Ann. NY Acad. Sci. 521:1-15, 1988.

TYPE: Case-Control
TERM: Psychosocial Factors; Stress
SITE: Haemopoietic
TIME: 1979 - 1994

GERMANY

Heidelberg

359 **Becker, N.** 03574
German Cancer Research Centre, Div. of Epidemiology, Im Neuenheimer Feld 280 , 69120 Heidelberg, Germany (Tel.: +49 6221 422383; Fax: 422203; Tlx: 461562 dkfz d)
COLL: Abel, U.R.; Frentzel-Beyme, R.R.

Allergy and Cancer

Participation of the immune system in cancer defense mechanisms has been discussed for a long time. It is known that there is an increased cancer incidence in persons with immune suppression. Whether the reverse also applies, namely, that persons with an enhanced reaction of the immune system, such as those with various allergies, experience a reduced cancer incidence is however disputed. Studies already carried out on this subject showed a wide variety of results. This study aims to investigate again the question of the part played by the immune system in the pathogenesis of cancer in persons with a previous diagnosis of allergy. A retrospective follow-up study is being carried out of 5,000 patients attending a university dermatological hospital for the years 1973-1974. Persons with a negative allergy-test admitted to the same hospital for the purpose of excluding an allergic disease serve as controls.

TYPE: Cohort
TERM: Allergy; Immunology
SITE: All Sites
TIME: 1985 - 1994

360 **Blettner, M.** 04943
German Cancer Research Centre, Div. of Epidemiology, Im Neuenheimer Feld 280 , 69120 Heidelberg, Germany (Tel.: +49 6221 422383; Fax: 422203)
COLL: Renz, K.; Seitz, G.; Wahrendorf, J.

Historical Cohort Study among Workers in Nuclear Power Stations

The study investigates the relationship between occupational exposure to low-level ionising radiation and causes of death. A retrospective cohort study will include approximately 4000 current and past employees from several nuclear power stations in Germany. Exposure information will be obtained from personal records for all workers who had been monitored for occupational exposure to radiation. Internal comparisons will be made between groups with different duration and degree of exposure. The study will be part of a large international cohort study, in preparation at IARC.

TYPE: Cohort
TERM: Occupation; Radiation, Ionizing
SITE: All Sites
OCCU: Radiation Workers
TIME: 1989 - 1995

361 **Chang-Claude, J.** 04704
German Cancer Research Centre, Div. of Epidemiology, Im Neuenheimer Feld 280 , 69120 Heidelberg, Germany (Tel.: +49 6221 422373; Fax: 422203; Tlx: 461562 dkfz d)
COLL: Wahrendorf, J.; Qiu, S.L.; Yang, G.R.; Muñoz, N.; Crespi, M.; Raedsch, R.; Thurnham, D.; Correa, P.

Precancerous Lesions of the Oesophagus among Young Persons in China

A cross-sectional survey was conducted in 1988 among 545 persons between 15 and 26 years of age in Huixian, Henan Province, a high risk area for oesophageal cancer in the People's Republic of China. The objective was to collect information on the prevalence of precancerous lesions of the oesophagus at early ages and to identify risk factors associated with the development of such lesions. This group of young people is followed up for further investigations in the future. The observation of a familial aggregation of precancerous lesions and oesophageal cancer has led to a survey among all patients with oesophageal cancer occurring in these communities in the years 1970 to 1990, assessing the occurrence in their first and second degree relatives. These data are now analysed for familial aggregation and by segregation analysis. Papers were published in Lancet in 1989 (Wahrendorf, J.) and in Cancer Research in 1990.

TYPE: Cohort; Genetic Epidemiology
TERM: Age; Alcohol; BMB; Biopsy; Cytology; Diet; Familial Factors; Lifestyle; Micronuclei; Plasma; Premalignant Lesion; Prevalence; Tobacco (Smoking); Urine; Vitamins
SITE: Oesophagus
CHEM: N-Nitroso Compounds
LOCA: China
TIME: 1987 - 1994

***362** **Frentzel-Beyme, R.R.** 05515
German Cancer Research Centre, Div. of Epidemiology, Im Neuenheimer Feld 280 , 69120 Heidelberg, Germany (Tel.: +49 6221 422384; Fax: 422203; Tlx: 461562 dkfz d)
COLL: Adzersen, K.H.; Becker, N.; Wahrendorf, J.

Cancer Mortality in German Iron Foundries

A large scale cohort study of foundry workers is at present being carried out in the western part of Germany. A sample of 40 foundries (from over 400) is considered for inclusion, depending on adequate conditions for recruitement of the cohorts as off 1950. This will permit an observation period of 20 years and more for large parts of the cohort. Changes in the use of silica-containing materials will be documented and used for internal comparisons. The distribution of foundries by technologies applied ensures inclusion of exposure to risk factors such as nitrosamines and solvents. Cancer morbidity will also be assessed and checked against the cancer registry in one area and data from health insurance schemes elsewhere. Routine follow-up of the cohort for mortality will be carried out through the personal registration scheme of Germany.

TYPE: Cohort
TERM: Metals; Occupation; Plastics; Registry; Solvents
SITE: Brain; Larynx; Lung; Lymphoma; Oesophagus; Oropharynx; Stomach; Tongue; Tonsil
CHEM: Amines, Aromatic; Benzene; Furans; Iron; Isocyanate; N-Nitroso Compounds; Phenol
OCCU: Foundry Workers; Smelters
REGI: Saarland (Ger)
TIME: 1993 - 1995

363 **Kahn, T.M.** 05070
German Cancer Research Centre, Inst. for Applied Tumour Virology, Im Neuenheimer Feld 242, W 6900 Heidelberg, Germany (Tel.: +49 6221 424853; Fax: 424852)
COLL: Bercovich, J.; Chang-Claude, J.; Turazza, E.; Claudiani, J.; Gurucharri, C.; Sprovieri, O.; Valacco, A.; Pires Torres, C.; Ojeda, R.; Amestoy, G.; Bayo, J.; Collalto, A.; Haimovici, L.; Nishijama, S.; Gissmann, L.; Grinstein, S.; zur Hausen, H.

HPV Infection and Cancer of the Cervix and the Upper Respiratory Tract in Argentina

Mortality rates for cervical cancers in Argentina vary widely by geographical region, and mortality rates for oropharyngeal and laryngeal cancers are high. This study aims to determine the prevalence of HPV infection in women with a normal cervix and patients presenting with premalignant and malignant lesions of the genital and upper respiratory tracts from different geographical, ethnic and socio-economic origins. Patients were recruited from 10 hospitals in the Buenos Aires area, which includes persons from all over the country. Cervical smears obtained from colposcopically normal patients will be tested for HPV using filter in situ hybridization, and PCR and biopsy material from cancers and lesions will be tested by Southern blot hybridization. A case-control study for cervical cancers is being carried out to recruit 250 cases and 500 age-matched controls using a questionnaire to gain information on risk factors for cervical cancer. Analysis will be carried out to elucidate the risk factors explanatory of the large regional differences in cervical cancer mortality rates.

TYPE: Case Series; Case-Control
TERM: Alcohol; HPV; Premalignant Lesion; Reproductive Factors; Socio-Economic Factors; Tobacco (Smoking)
SITE: Oral Cavity; Oropharynx; Respiratory; Uterus (Cervix)
LOCA: Argentina
TIME: 1989 - 1994

---

364 **Maier, H.** 05169

Univ. Heidelberg, Hals-, Nasen- & Ohrenklinik, Im Neuenheimer Feld 400 , W 6900 Heidelberg, Germany (Tel.: +49 6221 566703; Fax: 565913; Tlx: 461745 unikl hd)

COLL: Lessner, K.; Dietz, A.; Gewelke, U.; Heller, W.D.

Multiple Primary Tumours in Laryngeal Cancer

The primary objective of this prospective study is to investigate the risk for multiple primary tumours associated with tumour stage, tobacco consumption, alcohol consumption, diet, occupation and social status in patients with laryngeal cancer. In the first part of the study (1988/1989), 164 patients with squamous cell carcinoma of the larynx were interviewed by a trained interviewer using a structured questionnaire. In the second part of the study, cases are followed up at monthly intervals for five years in order to detect local recurrences, metastases and secondary or multiple primary tumours. By May 1991, a total of 36 patients had died, 25 due to the occurrence of secondary or multiple tumours. A first analysis of the data is planned in 1992.

TYPE: Cohort
TERM: Alcohol; Diet; Multiple Primary; Occupation; Stage; Tobacco (Smoking)
SITE: Larynx
TIME: 1988 - 1994

365 **Raue, F.** 05000

Medizinische Universitätsklinik, Endokrinologische Ambulanz, Luisenstr. 5 , W 6900 Heidelberg, Germany (Tel.: +49 6221 568605; Fax: 563101)

Register of Medullary Thyroid Carcinoma

The objective of this retrospective and prospective register is to identify patients with hereditary medullary thyroid carcinoma (MTC). Approximately 25% of this rare type of thyroid carcinoma is genetically determined (multiple endocrine neoplasia Type II) and can be detected at an early stage by family screening. A questionnaire has been distributed containing questions on personal information, details relating to the age at diagnosis and histological findings, the present status of the patients, and about family screening and surgical therapy. Lymphocytes of members of MTC-families were stored for further genetic studies. 741 patients with MTC (25% familial cases) have been reported by 33 cooperative centers (Med. Klin. 85:113-117, 1990; Clin. Investig. 71:7-12, 1993); about 40 new cases have been registered annually since 1988. Survival data and prognostic factors have been calculated. This register will provide a basis for further collaborative diagnostic and therapeutic studies.

TYPE: Genetic Epidemiology
TERM: BMB; Genetic Factors; Lymphocytes; Registry
SITE: Thyroid
TIME: 1988 - 1994

366 **Schlaefer, K.O.H.** 05175

German Cancer Research Centre, Div. of Epidemiology, Im Neuenheimer Feld 280 , 69120 Heidelberg, Germany (Tel.: +49 6221 422367; Fax: 422203)

COLL: Wahrendorf, J.

Aetiology of Inherited Colorectal Cancer

The aim of this study is to investigate genetic and environmental factors of two types of inherited colon cancer - hereditary non polyposis colorectal cancer and hereditary flat adenoma syndrome. Both types are inherited in an autosomal dominant mode, the location of the gene is not yet known. Factors responsible for the age-of-onset are unclear and gene environment interactions may play an important role. The aim is to investigate the following factors: (1) nutrition, especially fat, vitamins, and fibres, and nutrients enhancing the production of bile acid; (2) earlier viral and bacterial infections, especially polio-, arbo- and cytomegaloviruses and streptoccocal and clostridial infections; and (3) the proliferation rate of healthy colon tissue, to determine if an increase in rate is a prognostic factor for the proximity of the onset of the disease. Collaboration with major centres for surgical treatment of colorectal cancers will be established. This will be a cohort study with a nested case-control study. The control will be the next elder sib, if not diseased, or the sib closest in age. Families will be collected according to the Amsterdam criteria, i.e. three persons affected, age at diagnosis less than 65 years, and at least two generations affected. Families will be followed up for at least 10 years. At regular clinical check-ups, an interviewer will administer a questionnaire, and blood, stool and polyp/cancer tissue samples will be collected. It is expected to identify a cohort of about 100 to 130 families (300 to 500 people).

TYPE: Case-Control; Cohort; Genetic Epidemiology
TERM: BMB; Bile Acid; Blood; Diet; Faeces; Familial Adenomatous Polyposis; Genetic Factors; Heredity; Infection; Linkage Analysis; Nutrition; Pedigree; Polyps; Premalignant Lesion; Tissue; Virus
SITE: Colon; Rectum
TIME: 1993 - 2003

367 **Schlehofer, B.** 04700
German Cancer Research Centre, Div. of Epidemiology, Im Neuenheimer Feld 280 , 69120 Heidelberg, Germany (Tel.: +49 6221 422351; Fax: 422203; Tlx: 461562 dkfz d)
COLL: Michaelis, J.H.; Kaatsch, P.; Schlehofer, J.; Blettner, M.; Wahrendorf, J.

Case-control study of Parvovirus Infections and Childhood Leukaemia

In this study protective effects of parvovirus-infections in humans will be investigated. Animal experiments have shown that incidence and growth rate of tumours of different localizations could be reduced by Parvoviruses (H1- or adeno-associated parvoviruses). For adeno-associated virus (AAV) serological findings also indicate suppressive effects in human cancer. Parvovirus B-19, a virus which causes diseases (exanthema subitum) in humans, has not yet been studied for a protective effect. The target tissue for B-19 replication is the bone marrow. In a nationwide case-control study performed jointly by the German Cancer Research Center (Dept. of Epidemiology and Dept. of Tumorvirology) and the National Registry of Childhood Malignancies at the University of Mainz, 156 incident cases of childhood leukaemia and 201 age-sex-matched hospital controls will be investigated. Serological analysis have been established to determine the presence and titres of antibodies against H1, adeno-associated, and B-19 parvoviruses. In addition to the serological tests, the parents have been asked to respond to a self-administered questionnaire about possible risk factors in the aetiology of leukaemia, including life-style, occupational and medical history 3 months before and during pregnancy as well as environmental and health history. Infectious diseases of the children prior to diagnosis have been taken into consideration. Analyses are in progress.

TYPE: Case-Control
TERM: Antibodies; Childhood; Lifestyle; Occupation; Registry; Virus
SITE: Leukaemia
REGI: Mainz (Ger)
TIME: 1989 - 1994

368 **Schlehofer, B.** 04701
German Cancer Research Centre, Div. of Epidemiology, Im Neuenheimer Feld 280 , 69120 Heidelberg, Germany (Tel.: +49 6221 422351; Fax: 422203; Tlx: 461562 dkfz d)
COLL: Boeing, H.; Wahrendorf, J.; Waldherr, R.; Heuer, C.; Niehoff, D.

Case-control Study on Renal Cell Carcinoma

The rates of renal cell cancer have been increasing in several countries. To identify aetiological risk factors for this tumour, a population-based case-control study in the Rhein-Neckar-Odenwald area is being carried out. In 10 hospitals 277 incident cases aged 20 to 75 years have been interviewed during the two years 1989-90. 286 population controls were drawn randomly from the residential list of the study are. The main aspects of the face-to-face interview are questions on drug consumption (diuretics and analgesics), medical and smoking history, but there is also a focus on exposures to occupational risk factors. Several questions are related to the risk of obesity and physical activities and a self-administered questionnaire for dietary habits is also given to cases and controls. Special attention will be paid to the possible association with exposure to pesticides and herbicides: in a substudy for a 30% sample of the cases, fat tissue of the renal capsule will be analysed for persistent chlorinated hydrocarbons and compared with results from the renal fat tissue of age- and sex-matched autopsy controls. This study is part of an international study conducted with similar protocols in Australia, Austria, China, Denmark Sweden and the USA. Analyses are in progress.

TYPE: Case-Control
TERM: Diet; Drugs; Herbicides; Obesity; Occupation; Pesticides; Physical Activity; Tobacco (Smoking)
SITE: Kidney
TIME: 1989 - 1994

369 **ter Meulen, J.** 04862
German Cancer Research Centre, Inst. for Virological Research, Im Neuenheimer Feld 280, 69120 Heidelberg, Germany (Tel.: +49 6221 484629; Fax: 401271)
COLL: Pawlita, M.; Chang-Claude, J.; Mgaya, H.N.; Luande, J.

HPV Infection and Other Sexually Transmitted Diseases in Tanzanian Women

The present study aims to identify risk groups for HPV infection which is associated with a high risk of CIN and cervical cancer. The study subjects will be some 600 female inpatients, admitted for a variety of gynaecological and obstetrical disorders, excluding those associated with cervical cancer. Investigations include: application of a questionnaire on recognized risk factors for cervical cancer, blood sampling for HBV, syphilis and HIV serology, and collection of cervical swabs for Papstaining and of cervical scrapings for detection of HPV-DNA with PCR-technology. The occurrence of specific HPV-types in Tanzanian cervical carcinomas, which is important for the determination of the appropriate primers for the PCR-reaction, is currently being investigated. The data will be analysed by log-linear models.

TYPE: Cross-Sectional
TERM: Cytology; HBV; HIV; HPV; PCR; Premalignant Lesion; Sexually Transmitted Diseases
SITE: Uterus (Cervix)
LOCA: Tanzania
TIME: 1989 - 1994

370 **van Kaick, G.** 02529
German Cancer Research Centre, Inst. of Nuclear Medicine, Im Neuenheimer Feld 280, 69120 Heidelberg, Germany (Tel.: +49 6221 484563; Fax: 401271; Tlx: 461562 DKFZ D)
COLL: Muth, H.; Wagner, G.; Kaul, A.

"Thorotrast" Research Project

Thorotrast, a 25% colloidal solution of Thorium dioxide was used as a contrast medium between 1930-1950. After intravenous injection the ThOy2-particles were stored in the reticuloendothelial system (RES). The project aims at discovering the late effects of chronic radiation exposure by studying a large number of patients injected with Thorotrast, comparing these results with a corresponding control group and examining the relationship between late effects and radiation dose. The study includes: biophysical investigations to calculate the total effective radiation dose; clinical, biochemical and radiological examinations to determine the state of health; classification of the causes of death for patients who died; statistical analysis of the results obtained. To date the names of 5,159 Thorotrast patients and 5,160 control patients have been obtained from records in different hospitals of West Germany: 901 Thorotrast patients and 669 control patients have been examined clinically. Living patients were re-examined every two years. Apart from this epidemiological study the following investigations are being carried out: animal experiments to (1) investigate the distribution of Thorium dioxide particles within the body and calculate the resulting tissue doses and (2) estimate the non-radiation effects of deposited Thorium dioxide. Papers appeared in Radiation Carcinogenesis: Epidemiology and Biological Significance, (eds) J.D. Boice, Jr and J.F. Fraumeni, Jr, Raven Press, New York, 1984, pp. 253-262 and in BIR Report 21:Risks from Radium and Thorotrast, (eds) D.M. Taylor et al., British Institute of Radiology, London, 1989, pp. 97-104.

TYPE: Cohort
TERM: Dose-Response; Drugs; Late Effects; Radiation, Ionizing
SITE: Bone; Leukaemia; Liver; Lung; Lymphoma; Spleen
CHEM: Thorotrast
TIME: 1968 - 1994

371 **Wahrendorf, J.** 04367
German Cancer Research Centre, Div. of Epidemiology, Im Neuenheimer Feld 280, 69120 Heidelberg, Germany (Tel.: +49 6221 422200; Fax: 422203; Tlx: 461562 dkfz d)
COLL: Becher, H.; Saracci, R.

Cohort Studies of Persons Exposed to Phenoxy Acid Herbicides and their Contaminants

Phenoxy acid herbicides have been widely used throughout the world since the early 1940's. A group of related compounds are the chlorophenols which are used in the manufacture of some of the above herbicides and as a preservative or a pesticide. These compounds which are of great commercial importance have been known to be contaminated with dioxins and furans which are formed during their manufacture and for which well-documented acute toxic effects in man are known. There is also concern as to whether human exposure does result in long-term effects, such as cancer. Contributing to a European IARC-coordinated study workers, employed in three chemical companies from the early 1950's are enrolled in this study. Establishment of the cohort and provisions for a mortality follow-up are underway.

TYPE: Cohort
TERM: Chemical Exposure; Herbicides; Occupation; Pesticides
SITE: All Sites
CHEM: Chlorophenols; Dioxins; Phenoxy Acids
OCCU: Chemical Industry Workers
TIME: 1987 - 1994

372 **Wahrendorf, J.** 04928
German Cancer Research Centre, Div. of Epidemiology, Im Neuenheimer Feld 280 , 69120 Heidelberg, Germany (Tel.: +49 6221 422200; Fax: 422203; Tlx: 461562 dkfz d)
COLL: Boeing, H.; Korfmann, A.; Mastrinsohn, C.; Klett, M.; Bohlscheid, S.; Wappler, G.

Cohort Study on Diet and Cancer

Following intensive pilot studies for the assessment of dietary habits and methodology of follow-up the establishment of a cohort is now commencing. A total of about 50,000 persons in the age-range 35 to 64, randomly drawn from the residential list of the populations, will be recruited in the area of Heidelberg and Potsdam. Dietary habits will be assessed by a validated self-administrable questionnaire. Further, questions will be asked on general lifestyle aspects and medical history. A sample of 30 ml blood will be drawn and processed for storage at -70 degrees C. Follow-up will be through biennial postal contact of the participants. This study is part of the EPIC project.

TYPE: Cohort
TERM: BMB; Blood; Diet; Lifestyle; Nutrition
SITE: All Sites
TIME: 1993 - 1998

373 **Weisgerber, U.** 04944
German Cancer Research Centre, Div. of Epidemiology, Im Neuenheimer Feld 280 , 69120 Heidelberg, Germany (Tel.: +49 6221 422384; Fax: 422203)
COLL: Boeing, H.; Raedsch, R.; Waldherr, R.; Rozen, P.

Randomized Double-Blind Intervention Trial with Calcium in Polypectomized Patients

In a double-blind randomized trial with 100 polypectomized patients it will be tested whether 2 g of calcium daily, given as lactate gluconate in effervescent tablets, reduces cell proliferation of the colon crypts compared to placebo treated patients. All participants will undergo a run-in phase for 3 months with placebo and are then assigned randomly to calcium or placebo treatment for 9 months. Biopsies will be taken before and at the end of the study from the recto-sigmoidal region. Two different methods are considered for measuring cell proliferation: bromodeoxyuridine (BrdU) and proliferating cell nuclear antigen (PCNA). Each participant will be visited every 3 months at home for compliance assessment and the distribution of new tablets. In connection with these visits urine will be collected, as well as a 48-h stool on dry ice. The biological material will be utilized to study the metabolic effect of calcium on the lumen of the colon and to control for compliance. Parameters of interest are bile acids and fatty acid metabolism and their binding to calcium. Aliquots of stool will be stored at -80 C for additional analyses.

TYPE: Intervention
TERM: BMB; Diet; Faeces; High-Risk Groups; Minerals; Premalignant Lesion; Prevention; Urine
SITE: Colon; Rectum
CHEM: Calcium
TIME: 1989 - 1994

Köln

374 **von Karsa, L.** 03260
Zentralinst. füur die Kassenärztliche Versorgung in der Bundesrepublik Deutschland, Herbert-Lewinstr. 5, W 5000 Köln 41, Germany (Tel.: +49 4221 4005130; Fax: 8883242)
COLL: Flatten, G.; Lang, A.

GERMANY

Continuous Evaluation of the German National Screening Programme

In the German statutory health insurance system (over 90% of the population) an annual cancer check-up is offered to women (from age 20, pap smear; from age 30, pap smear and breast physical examination with instruction in breast self examination; from age 45, faecal occult blood testing (FOBT) is added) and for men (from age 45, digital rectal examination + FOBT). Screening contacts are documented in a standard way, data are collected and analysed centrally by the Zentralinstitut. Attendance rates and detection rates are computed (by age and sex) and a number of special analyses conducted. Information obtained will be used for evaluation and further development of screening programmes (screening interval, inclusion of mammography, follow-up of positive FOBT).

TYPE: Cross-Sectional
TERM: Screening
SITE: Breast (F); Colon; Prostate; Rectum; Skin; Uterus (Cervix)
TIME: 1972 - 1994

Mainz

375 **Michaelis, J.H.** 05170
Univ. Mainz, Inst. für Medizin, Statistik & Dokumentation, Langenbeckstr. 1 , W 6500 Mainz, Germany (Tel.: +49 6131 173252; Fax: 172968)
COLL: Kaatsch, P.

Nationwide Registry of Childhood Malignancies

A nationwide registry of childhood malignancies was established in 1980 by the two German societies of paediatric oncology and haematology as a combination of a population-based and a clinical registry. Documentation of German controlled clinical trials is integrated in the system. More than 95% of all childhood malignancies diagnosed in the FRG are registered. Notification of cases by over 100 hospitals and departments of paediatric oncology and haematology is voluntary. After admission of a newly diagnosed child, a basic form is sent to the registry. In response the cooperating physician gets a tumour-specific questionnaire which contains items about epidemiological aspects, the patient's history and diagnostic procedures. Long-term follow-up data are collected periodically to calculate survival and to get information about late effects or secondary malignancies. Incidence by sex, age, site and geographical location are estimated annually, based on more than 1,000 registered children per year. Based on subgroups of the more than 12,000 registered patients, some specific epidemiological studies are being performed, e.g. on viral aetiology of leukaemia, regional clusters, low-dose ionizing radiation effects. Further information about the registry is published in annual technical reports in German, available on request.

TYPE: Incidence; Registry
TERM: Cluster; Late Effects; Multiple Primary; Radiation, Ionizing; Registry; Survival; Virus
SITE: Childhood Neoplasms
REGI: Mainz (Ger)
TIME: 1980 - 1994

376 **Michaelis, J.H.** 05171
Univ. Mainz, Inst. für Medizin, Statistik & Dokumentation, Langenbeckstr. 1 , W 6500 Mainz, Germany (Tel.: +49 6131 173252; Fax: 172968)
COLL: Kaatsch, P.

Potential Influence of thc Proximity to Nuclear Plants on the Incidence of Childhood Malignancies

Following public discussion on a possible increase in incidence of childhood malignancy in the proximity of nuclear plants, a study was started in 1989. The aim of the study is to estimate cancer incidence around all 20 nuclear plants in Western Germany in a defined region up to 15 km from each plant (divided into three zones) and to compare this with the incidence in comparable regions, matched for each plant, by distance, density of population, and other parameters of demographic structure. In order to evaluate further potential differences in incidence, a special questionnaire is used for obtaining information on lifestyle variables and occupation of the parents, possibly elevated genetic risk, course of pregnancy and childhood development, and exposure to environmental factors.

TYPE: Incidence
TERM: Environmental Factors; Genetic Factors; Lifestyle; Parental Occupation; Radiation, Ionizing; Registry
SITE: Childhood Neoplasms; Leukaemia; Lymphoma; Neuroblastoma; Wilms' Tumour
REGI: Mainz (Ger)
TIME: 1989 - 1994

Munich

377 **Spiess, H.** 03997
Univ. of Munich, Pettenkoferstr. 8A, W 8000 Munich 2, Germany (Tel.: +49 89 51603677)
COLL: Mays, C.W.

Effects of Ra-224 on Humans

The hypothesis is that radiation can induce specific diseases in humans. 899 patients who received repeated injections of Ra-224 after World War II, mostly for the treatment of tuberculosis or ankylosing spondylitis, are being followed. The diseases inducible by Ra-224 include malignant bone sarcomas, benign exostoses, growth retardation, tooth breakage, kidney diseases, liver diseases, and cataracts. At three-year intervals the patients are contacted by questionnaire. The information is used to predict the risk from other types of radiation exposure.

TYPE: Cohort
TERM: Radiation, Ionizing
SITE: Bone; Kidney; Liver
TIME: 1952 - 2000

Oldenburg

378 **Baltrusch, H.J.F.** 02866
International Psycho-Oncology Project, Bergstr. 10, W 2900 Oldenburg, Germany (Tel.: +49 541 44112147)
COLL: Bastecky, J.; Ebigbo, P.O.; Forsén, A.; Gehde, E.; Grassi, L.; Illiger, H.J.; Németh, G.; Stangel, W.; Zhang, Z.W.

Biobehavioural Precursors of Cancer

Clinical studies in cancer patients have revealed a number of personality traits, such as lack of closeness to parents, suppression of unacceptable feelings, in particular anger, avoidance of emotional conflicts, harmonizing behaviour, exaggerated altruism and negligence of own health. The use of these variables for possible screening of persons being at a high cancer risk (biobehavioural cancer risk profile or Type C behaviour pattern) is studied. To date, 1,300 European and 120 Nigerian cancer patients (both sexes and at all sites) have been investigated using the Thomas Family Attitude Questionnaire (FAQ). 1600 cancer-free individuals served as controls. Evaluation of the results showed that male cancer patients (testis, large bowel) and female cancer patients (breast, uterus, ovary, vulva) showed a statistically higher significant lack of relation to parents in the FAQ than their age-matched controls. As a second step, the Courtauld Emotional Contral Scale (CECS), measuring the suppression of anger, anxiety and depressive mood, was administered to 130 male and 80 female cancer patients and to 150 controls. As a third step, a study, measuring "Type C" behaviour, assessing anger-in, anger-out, emotional control, rationality, depression, anxiety, social support and optimism has been set up and is in progress. It is hoped to assess the possible relationships between personality factors, coping style, neuro-endocrine and psychoneuroimmunological parameters and cancer progression. Partial resulta have been published in Ann. NY Acad. Sci. 521:1-15, 1988, Int. J. Neurosci. 51:257-260, 1990 and in Ann. NY Acad. Sci. 650:355-362, 1992.

TYPE: Case-Control
TERM: Psychological Factors; Stress
SITE: All Sites
LOCA: Austria; China; Czech Republic; Finland; Germany; Greece; Hungary; Italy; Nigeria; Slovak Republic; Yugoslavia
TIME: 1981 - 1994

Wuppertal

***379** **Wichmann, H.E.** 05511
Univ. of Wuppertal Dept. of Labor Safety and Environmental Medicine, FB 14, Gauß-Str. 20, W 42097 Wuppertal, Germany (Tel.: +49 202 4392088; Fax: 4392068)
COLL: Kreienbrock, L.; Goetze, H.-J.; Dingerkus, G.; Kreuzer, M.; Heinrich, J.; Wolke, G.; Keller, G.; Muller, K.M.; Atay, Z.; Bolm-Audorff, U.

Lung Cancer and Indoor Radon in the Federal Republic of Germany

A case-control study involving more than 3,000 cases and 3,000 frequency matched controls is being conducted. Cases are selected from clinics at twelve hospitals in East Bavaria, Saarland and Eifel area and Thuringia and Saxony during the period 1990-1994. The study is limited to subjects who have lived in Germany for at least 25 years. Only pathology-confirmed lung cancer cases are included. An independent pathological confirmation is obtained. All cases are alive. Controls are obtained from the general population. Cases are matched for sex, age and latest residence area. Personal interviews are being carried out. Information is obtained on occupation, residence, cigarette smoking, passive smoking, family history of cancer, diet (Vitamin A intake), genetic issues and pet birds. Radon measurements are done in all homes occupied by the subjects during the last 35 years. Two alpha track detectors are placed in the house for a one year reading and two charcoal canisters for a 3 day reading; one of each in the living room and one in the bedroom.

TYPE: Case-Control
TERM: Animal; BMB; Diet; Occupation; Passive Smoking; Radiation, Ionizing; Tobacco (Smoking)
SITE: Lung
CHEM: Radon
TIME: 1990 - 1996

Athens

***380** **Kalapothaki, V.** 05441
Athens Univ., Mecial School, Dept. of Hygiene and Epidemiology, 75 M. Asias St. , 11527 Athens, Greece (Tel.: +30 1 7715803; Fax: 7704225)
COLL: Petridou, E.; Kogevinas, M.; Kosmidi, E.; Haidas, S.; Kalmanti, M.

Electromagnetic Fields and other Risk Factors in the Aetiology of Childhood Leukaemia in Greece
The aim of this case-control study is to explore the role of electromagnetic fields and other risk factors in the aetiology of childhood leukaemia in Greece. Information will be collected on a number of parameters: exposures prior to conception, during gestation and between birth and a reference date prior to diagnosis of the disease, residential history, socio-demographic factors, growth parameters, medical history and treatments, vaccination, history of breast feeding, family medical history, parental occupational exposures, parents' smoking habits and exposures to solvents, pesticides, diagnostic radiation, etc., as well as direct measurements, wire configuration and distance measurements. Blood samples will be collected for analysis. It is expected to enroll 150 cases and 300 hospital and population controls in the study during 1993-1994.

TYPE: Case-Control
TERM: BMB; Blood; Childhood; Electromagnetic Fields; Infection; Intra-Uterine Exposure; Parental Occupation; Pesticides; Radiation, Ionizing; Solvents
SITE: Leukaemia
TIME: 1993 - 1995

381 **Linos, D.** 03736
Athens Medical School Aretaiion Hosp., Dept. of Surgery, 76 Vasilissis Sofias Ave. , Athens, Greece
COLL: Koutras, D.

Aetiology of Thyroid Cancer
The study aims to investigate the role of several risk factors in the development of thyroid cancer in the Greek population. Among the factors being investigated are radiation, socio-economic and personal characteristics, dietary habits and prior medical conditions. The study design is that of the matched case-control study (group matching) with two groups of controls. The first group of controls will include persons with benign thyroid disease; the second will be hospital controls, seen in the hospital for emergency medical care. Matching factors will include age (+/- 5 years) and sex. The diagnosis of thyroid cancer or benign thyroid disease will be confirmed by histological examination, and absence of thyroid disease in the general control group will be verified by clinical and laboratory examinations.

TYPE: Case-Control
TERM: Diet; Histology; Physical Factors; Radiation, Ionizing; Socio-Economic Factors
SITE: Thyroid
TIME: 1985 - 1994

382 **Papaevangelou, G.J.** 04274
Athens School of Hygiene, WHO Collaborating Center on AIDS, 196 Alexandras Ave. , 115 21 Athens, Greece (Tel.: +30 1 6467941; Fax: 7781829)
COLL: Roumeliotou, A.; Kallinikos, G.; Economidou, J.

Epidemiology of AIDS in Greece
This is a study of factors responsible for Kaposi's Sarcoma in AIDS cases as well as in HTLV III negative patients. Anti-HTLV-III positive individuals are followed up to study the natural history of the infection. The sample includes 200 homosexuals, 120 haemophiliacs, 20 drug addicts and 20 other asymptomatic HTLV-III positive subjects. Demographic, socio-economic and environmental factors as well as detailed biochemical, serological and other laboratory and personal characteristics are studied and correlated to the development of Kaposi's Sarcoma and other malignancies. The epidemiological characteristics of classical Kaposi's Sarcoma are also studied in an effort to discern the difference between these two entities.

TYPE: Cohort
TERM: AIDS; Drugs; Environmental Factors; HTLV; Haemophilia; Homosexuality; Infection; Sero-Epidemiology; Socio-Economic Factors; Virus
SITE: Kaposi's Sarcoma
TIME: 1980 - 1994

383 **Papaevangelou, G.J.** 04275
Athens School of Hygiene, WHO Collaborating Center on AIDS, 196 Alexandras Ave. , 115 21 Athens, Greece (Tel.: +30 1 6467941; Fax: 7781829)
COLL: Roumeliotou, A.; Kallinikos, G.; Economidou, J.

Natural History of Viral Hepatitis B

This is a prospective study to understand the factors responsible for establishment of chronicity and development of primary liver cancer. It includes several aspects of the natural history of viral hepatitis B. More than 2,000 patients hospitalized in the Infectious Diseases Hospital of Athens have been included in this study. Patients are followed up to understand the factors responsible for the establishment of chronic hepatitis B. HBV, HDV markers, biochemical profiles, immunological, histological and immunochemical studies are performed. The relation of demographic, socio-economic, environmental and genetic factors to the development of the chronic disease and its further evolution are examined. In parallel, chronic HBsAg carriers detected through existing screening programmes, are studied in detail and followed up.

TYPE: Cohort
TERM: Environmental Factors; Familial Factors; HBV; Immunology; Infection; Socio-Economic Factors
SITE: Liver
TIME: 1980 - 1994

***384** **Petridou, E.** 05440
Univ. of Athens Medical School, Dept. of Hygiene and Epidemiology, 75 M. Asias St. , 11527 Athens, Greece (Tel.: +30 1 7773840; Fax: 9324300)
COLL: Trichopoulos, D.; Kosmidi, E.; Haidas, S.; Kalmanti, M.

Age of Exposure to Infections and Childhood Leukaemia Risk

The aim of the study is to assess the importance of age at infection and other factors in the aetiology of childhood leukaemia. Cases are 136 children with leukaemia diagnosed over a period of four years; controls are 187 children visiting out-patient clinics of the children's hospitals through which cases were identified. Information is provided by one of the parents of the child in telephone interviews. Multiple logistic regression is used in the analysis and day care attendance is used as a proxy to early infectious exposure.

TYPE: Case-Control
TERM: Age; Childhood; Infection
SITE: Leukaemia
TIME: 1987 - 1994

***385** **Petridou, E.** 05442
Univ. of Athens Medical School, Dept. of Hygiene and Epidemiology, 75 M. Asias St. , 11527 Athens, Greece (Tel.: +30 1 7773840; Fax: 9324300)
COLL: Lauzi, A.; Proukakis, C.; Trichopoulos, D.; Kosmidi, E.; Haidas, S.; Piperopoulou, F.; Koliouskas, D.; Kalmanti, M.; Kassimos, D.

Possible Consequences of the Chernobyl Nuclear Accident on the Incidence of Childhood Leukaemia

The objective of this study is to investigate whether childhood leukaemia incidence has increased in Greece after exposure to ionising radiation due to the Chernobyl nuclear accident. 1,055 children under 16 years of age with leukaemia and coming from all over Greece, were identified through the five hospitals treating childhood leukaemia during the period 1981-1992. The questionnaires contained information on calendar year of diagnosis, sex, place of birth and residence, type of leukaemia, birth order and sibship size. The collection of data has been completed and cases have been classified according to year of diagnosis in three periods: 1980-June 1986 (before the accident-reference period), July 1986-June 1988 (accident-period with possible overdiagnosis due to subjective bias) and July 1988-June 1991 (after the accident/control period). Measurements of the ionising radiation in Ra-226 and Cs-137 were provided by the Department of Medical Physics of the University of Athens.

TYPE: Case Series; Correlation
TERM: Childhood; Radiation, Ionizing
SITE: Leukaemia
TIME: 1990 - 1994

Hong Kong

386 **Koo, L.C.** 02246
Univ. of Hong Kong, Medical Faculty, Dept. of Community Medicine, 5 Sassoon Rd, Li Shu Fan Bldg, Hong Kong, Hong Kong (Tel.: +852 5 8199289; Fax: 8176528)
COLL: Ho, J.H.C.

Cultural, Environmental, and Familial Backgrounds of Female Lung Cancer Patients in Hong Kong
This retrospective study intends to identify the environmental and/ or ethnic risk factors which contribute to the unusually high rates of lung cancer among Chinese females in Hong Kong. Chinese females have among the highest world age-adjusted incidence rates and a very low proportion of ever-smokers. 200 female lung cancer patients and 200 neighbourhood female controls matched for age, socio-economic status, and residence were interviewed between 1981-1983 on their life histories. The semi-structured interviews focused on: (1) basic demographic background; (2) residential history; (3) smoking history (including passive smoking at home and/or workplace); (4) occupational history; (5) personal habits and hygiene; (6) medical history; and (7) dietary history. Emphasis was placed on recall data concerning environmental exposure to suspected initiators, promoters, or protectors 20 or more years before diagnosis. Analysis of the possible interrelationships of these factors in increasing or reducing risk among ever- or never-smokers is continuing. Papers have been published in Nutr. Cancer 11:155-172, 1988, Environm. Res. 52-23-33, 1990 and in Int. J. Epidemiol. 19:514-523, 1990.

TYPE: Case-Control
TERM: Chemical Exposure; Chinese; Diet; Environmental Factors; Female; Hygiene; Lifestyle; Occupation; Passive Smoking; Socio-Economic Factors; Tobacco (Smoking)
SITE: Lung
TIME: 1980 - 1995

Shatin

***387** **Lloyd, O.L.L.** 05482
Chinese Univ. of Hong Kong, Dept. of Community & Family Medicine, Lek Yuen Health Centre, Shatin New Territories, Hong Kong (Tel.: +852 6928786; Fax: 6063500)
COLL: Wong, T.W.; Wong, S.L.; Yu, T.S.

Cancer and the Environment in Hong Kong
The study aims to investigate mortality time trends in Hong Kong in the early 60's, the geographical mapping of mortality during 1979-1988 and incidence during 1983-1989 in the 28 census districts. The mapping will be done according to ethnicity, socio-economic status, education and environmental factors, such as radon and air pollution. Whether geographicl analytical techniques are suitable in a country like Hong Kong with a relatively unstable population will be determined, based on stability of mapping patterns for mortality between two quinquennia, evidence for biological plausibility of some of these patterns and the comparability of mortality and incidence patterns.

TYPE: Correlation; Incidence; Mortality
TERM: Air Pollution; Demographic Factors; Environmental Factors; Geographic Factors; Mapping; Registry
SITE: All Sites
CHEM: Radon
REGI: Hong Kong (HK)
TIME: 1992 - 1994

Budapest

388 **Ábráham, E.** 02058

Fövarosi Bajcsy-Zsilinszky Korhaz, Tüdögondozó Intézet, X. Köbányai ut 45. 1475, Budapest, Hungary

COLL: Czanik, P.; Dinya, E.; Karácsonyi, L; Sali, A.

Prospective Study to Determine High-Risk Groups for Lung Cancer

Screening of the population between the ages of 40-74, both sexes, in an industrial area of Budapest with 87,000 inhabitants, was started in 1968 with the aim of early detection of pulmonary diseases. Between 1975-1978, at the time of screening, supposed risk factors for lung cancer: smoking, occupational exposure, chronic respiratory complaints, fibrotic lung lesions, were determined in about 30,000 persons, 90-97% of the population group concerned. Until 31 December 1985, 343 people were diagnosed with lung cancer. Mathematical-statistical analysis (with adaptation of the log-linear model) made possible the ranking of the single risk factors for lung cancer both separately and cumulatively. The number of the possible risk factor variants is 180. On the basis of this ranking the population was classified into four groups in respect of lung cancer: risk free, moderate risk, high risk and super-high risk groups. The risk of lung cancer for people belonging to the super-high risk groups is about 20 times higher on average than for those in the risk-free groups. Immunological and bronchopathological examinations aiming to clear up the carcinogenic effects of the single risk-factors are continuing.

TYPE: Cohort
TERM: High-Risk Groups; Premalignant Lesion; Screening; Tobacco (Smoking)
SITE: Lung
TIME: 1968 - 1995

389 **Bánóczy, J.E.** 04009

Semmelweis Medical School, Dept. of Conservative Dentistry, Mikszáth Kálmán Tér 5 , 1088 Budapest VIII, Hungary (Tel.: +36 1 1131854; Fax: 1336508)

COLL: Dombi, C.

Epidemiological Investigations and Screening of Oral Carcinomas and Precancerous Lesions

Due to the high and still increasing incidence of oral carcinomas and precancerous lesions, a need for new methods of early diagnosis arose, as well as the aim to establish the prevalence and incidence rates of oral tumours and precancerous lesions, and to find the appropriate methods by which the widest range of a given population might be screened regularly. In Hungary, complex multiphasic screening in connection with compulsory lung screening has been practiced for several years. On this basis regular screening examinations were started at the lung screening stations in the 6th and 3rd districts of Budapest, including clinical, cytological and histological examinations, and introducing early treatment of detected cases. The sample size will comprise about 20,000 individuals, and it is planned to carry out a longitudinal survey, for a period of 7 to 8 years. Cost-effectiveness will be evaluated.

TYPE: Cohort
TERM: Cost-Benefit Analysis; Cytology; Histology; Premalignant Lesion; Screening; Treatment
SITE: Head and Neck; Oral Cavity
TIME: 1986 - 1996

390 **Czeizel, A.** 03932

National Inst. of Hygiene, Dept. of Human Genetics and Teratology, Gyáli ut 2-6, 1097 Budapest IX, Hungary (Tel.: +36 1 335773; Tlx: 225349 oki h)

Mutagenic Effects of Chemical Poisoning

Foetal death, congenital anomalies, birthweight, etc. in the offspring and chromosome aberrations, SCEs and other mutagenic endpoints in the peripheral lymphocyte cultures of persons surviving from suicide attempt using high doses of chemicals for self-poisoning are studied. The aim is to detect the somatic and germinal mutagenic consequences of high doses of chemicals, e.g. drugs and pesticides. In addition the teratogenic consequences of suicide attempts during pregnancy are examined in the offspring (including childhood tumours, and behavioural development). From 1990 the effects of some antioxidants (natural Vitamin E, and beta carotene) are studied on the chromosome aberrations of peripheral lymphocytes in persons who attempt suicide with high doses of chemicals. Recent papers were published in Mutat. Res. 269:35-39, 1992 and Mutat. Res. 298:131-137, 1992 (Bao et al.).

TYPE: Molecular Epidemiology
TERM: Chemical Mutagenesis; Chromosome Effects; Congenital Abnormalities; Drugs; Lymphocytes; Mutation, Germinal; Mutation, Somatic; SCE
SITE: Inapplicable
TIME: 1984 - 1995

391 **Gundy, S.** 04404
National Inst. of Hygiene, Dept. of Human Genetics & Teratology, Lab. of Human Mutagenesis, Gyáli ut 2-6, Budapest IX. 1966, Hungary
COLL: Bodrogi, I.; Baki, M.

Chromosomal Changes in Peripheral Blood Lymphocytes of Young Testicular Cancer Patients Treated with Chemotherapy

Lymphocytes of young males aged 20-30 years with primary neoplasms of the testis are examined for chromosomal aberrations and SCEs. Possible relationships will be established between chromosomal changes and the dose (for body weight), type and duration of chemotherapy treatment, including the individual sensitivity of patients, and the repair time of somatic cells and blood transfusions. Extrapolation will be made in the future from somatic cells (peripheral blood lymphocytes) to germinal cell injury. The duration of the study will be at least five years. Peripheral blood lymphocytes are used for microculture methods. Chromosomal preparations are strained with FPG technique. Chromosome and chromatid-type aberrations' distribution of I-II-III cell cycles and the SCE frequency are recorded. The chromosomal changes will be examined before, during and following the different chemotherapy programme. A paper has been published in Neoplasma 36 (4):457-464, 1989.

TYPE: Molecular Epidemiology
TERM: Chemotherapy; Lymphocytes; SCE
SITE: Testis
TIME: 1987 - 1994

Miskolc

392 **Takács, S.** 02862
Inst. of Public Health & Epidemiology, Lab. of Community Hygiene, Almos u. 10 , 3526 Miskolc, Hungary (Tel.: +36 46 54612; Fax: 58060)
COLL: Ujszászy, L.; Bokros, F.; Tatár, A.; Ferencz, T.

Environmental Exposures of Cases of Digestive and Bladder Cancer

Geographical differences between tumour incidence may be influenced by environmental effects. According to animal experiments and human observations, nitrate/nitrite, N-nitrosamines may play a role. The purpose of the present study is to examine the geographical distribution of the incidence of cancers of the digestive tracts and bladder and determine the level of these environmental exposures. Information on the lifestyle and environment of cancer patients is being obtained by personal interview and/or questionnaire and the nitrate, nitrite, and ammonium content of the drinking water usually consumed by these persons, estimated. The cases will be analysed according to histology and localisation of cancer. The control group will be chosen from the lowest incidence areas. In investigations completed in 1982-1983, levels of four trace elements were determined in foodstuffs, drinking water, sediments and human organs (liver, lung, kidney, adrenal glands). Data on 172 persons who died from cancer of various sites were analysed. Trace element concentration was measured in the tumours and in healthy tissue from the same organ as the tumour. Tissue from stomach and colon tumours and normal tissue from regions close to the tumours, taken from 49 patients during operation, was analysed. It was found that copper content was significantly higher in the tumour tissue. Further elements investigated are Al, B, Ba, Cr, Hg, Li, Mo, Ni, Pb, Sr, Se and Ti. The studied subjects are: blood, cerebrospinal- and amniotic fluid.

TYPE: Case-Control
TERM: Environmental Factors; Geographic Factors; Lifestyle; Water
SITE: Bladder; Gastrointestinal
CHEM: Chromium; N-Nitroso Compounds; Nickel; Nitrates; Nitrites; Selenium
TIME: 1979 - 1995

393 **Takács, S.** 04670
Inst. of Public Health & Epidemiology, Lab. of Community Hygiene, Almos u. 10 , 3526 Miskolc, Hungary (Tel.: +36 46 54612; Fax: 58060)
COLL: Radóczy, M.; Déri, Z.

Indoor Radon Concentrations and Lung Cancer

The goal of this investigation is to study the correlation between respiratory cancer occurrence and exposure to radon. The concentration of radon gas is measured in the flats of the diseased and of a reference group. The diseased group is selected from the data base of the oncological screening centre. The first reference group is selected at random from people living in the same area and not suffering from respiratory cancer. Members of the second reference group are selected from an area where the occurrence of respiratory diseases is very low. It is planned to carry out measurements in 200 flats with two detectors in a room and one detector in the soil. In this way the correlation between the radon emanation of the soil and the radon concentration of the living room can be investigated. The detector equipment consists of two alpha trace-detectors, which can detect the total alpha-exposure (radon and thoron) and the alpha exposure solely from radon, so that thoron exposure can be evaluated from the difference between the two. The latest results do not show any direct evidence for a causative relation between radon daughter exposure and respiratory cancer. The study of the relation between haemopoietic tumours (leukaemia) and concentration of indoor radon is now being planned.

TYPE: Case-Control
TERM: Metals; Radiation, Ionizing
SITE: Haemopoietic; Leukaemia; Lung
CHEM: Radon; Thoron
TIME: 1983 - 1995

Nyiregyháza

394 **Juhász, L.** 00573
County Hosp., Cancer Registry of the County Szabolcs-Szatmár-Bereg, Szent István u. 68, 4401 Nyiregyháza, Hungary (Tel.: +36 42 12222/1600; Fax: 14392 ; Tlx: 73426)
COLL: Dauda, G.

Gastric Cancers in the County of Szabolcs-Szatmár-Bereg

This is a study of gastric cancer in a defined county in Hungary. For each newly diagnosed case of gastric cancer two hospital controls are chosen, matched by age and residence, with a view to better characterising high-risk groups. In the second phase of the study 284 cases and 429 controls were interviewed. Questions are asked on food habits and family history. Biopsy and autopsy material from patients is typed according to the Järvi-Laurén system. In 1972-1987, 423 cases were so classified. The distribution was: intestinal 31.2%, diffuse 55.3%, mixed 13.5%, figures which differ from those obtained by others. This information is compared with data on epidemiological factors. Data show that not only stomach cancer but also ulcers occur significantly more often in families of cases than in families of controls. Papers have been published in: DAB: Development of County Cancer Registry and the Use of Data, 1989, pp 96-136.

TYPE: Case-Control
TERM: Diet; Familial Factors; High-Risk Groups; Histology; Registry
SITE: Stomach
REGI: Szabolcs-Szatmár (Hun)
TIME: 1976 - 1994

395 **Juhász, L.** 00691
County Hosp., Cancer Registry of the County Szabolcs-Szatmár-Bereg, Szent István u. 68, 4401 Nyiregyháza, Hungary (Tel.: +36 42 12222/1600; Fax: 14392 ; Tlx: 73426)

High-Risk Groups in Breast Cancer

For each newly diagnosed breast cancer case, two age- and residence-matched hospital controls are chosen. Questions are asked on demographic variables, reproduction-associated variables and family history. A series of articles relating to this study has been published in Hungarian. In the second phase of the study 531 cases and 390 controls were interviewed. The data confirm the previous finding, namely that breast cancer occurs more often in the families of patients than in controls. No differences were found in the occurrence of other cancers between families of patients and controls, or by degrees of relationship. Prostate cancer showed a clustering in families of patients with breast cancer but not in families of controls. (Familial Cancer. Ist Int. Res. Conf. Basel, 1985. (Eds: Mueller & Weber) Karger, Basel pp. 63-65). A paper has been published in Development of County Cancer Registry and the Use of Data, 1989, pp 137-152.

TYPE: Case-Control
TERM: Familial Factors; High-Risk Groups; Histology; Registry; Reproductive Factors
SITE: Breast (F)
REGI: Szabolcs-Szatmár (Hun)
TIME: 1976 - 1994

Reykjavik

***396** **Gunnarsdóttir, H.** 05346
Administration of Occupational Safety and Health, Medical Div., Biildshoefdi 16 , 112 Reykjavik, Iceland (Tel.: +354 1 672500; Fax: 674086)
COLL: Rafnsson, V.

Cancer Incidence among Nurses

The objective of this retrospective cohort study is to investigate cancer incidence among nurses, bearing in mind that nurses are exposed to potential carcinogens in their work environment. There is special focus on breast cancer, as other studies have shown an excess of this cancer among nurses. A cohort of 2,159 female nurses is being compared with the general female population. Information is obtained from the Nurses Association membership rolls, which includes all registered nurses who worked in Iceland and graduated 1920-1979. Follow-up is to 1989. 58% of the cohort graduated 1970 or later, i.e. the cohort is rather young. Most of these women had children and many of them before the age of thirty. Record linkage is being done with the Icelandic Cancer Registry. Preliminary results show an excess of total cancer. SIR is elevated for cancer of the lung, breast, kidney, bladder, brain, lymphosarcoma and leukaemia. There is a steady gradient of breast cancer the longer the latency time up to 50 years. A case-control study within the cohort is being considered.

TYPE: Cohort
TERM: Age; Chemical Exposure; Occupation; Parity; Registry
SITE: All Sites; Breast (F)
OCCU: Health Care Workers
REGI: Iceland (Ice)
TIME: 1991 - 1994

***397** **Gunnarsdóttir, H.** 05347
Administration of Occupational Safety and Health, Medical Div., Biildshoefdi 16 , 112 Reykjavik, Iceland (Tel.: +354 1 672500; Fax: 674086)
COLL: Rafnsson, V.

Cancer Incidence Among Female Manual Workers

The objective of this retrospective cohort study is to investigate whether the incidence of cancer, especially cancer of stomach, lung and cervix, is higher among female manual workers than among other Icelandic women. These types of cancer have shown to be in excess in lower socio-economic groups. 16,779 women were identified from the membership records of two pension funds for manual workers between 1970 and 1986. Follow-up was to 1989. Record-linkage was done with the Icelandic Cancer Registry. Preliminary results show that overall incidence is lower than expected. However, when ten years latency is considered total cancer is in excess. Cancer of lung and cervix is in excess both when latency time is considered and not, while the results for cancer of the stomach are more inconsistent. As no connection could be seen so far between the excess and length of employment, it is tempting to conclude that it is not the work itself, but the way of life in this socio-economic group, that could be determinant for the incidence of cancer.

TYPE: Cohort
TERM: Latency; Registry; Socio-Economic Factors; Time Factors
SITE: All Sites; Lung; Stomach; Uterus (Cervix)
REGI: Iceland (Ice)
TIME: 1991 - 1994

398 **Hallgrímsson, J.** 01818
Univ. of Iceland, Dept. of Pathology, P.O. Box 1465, 121 Reykjavik, Iceland (Tel.: +354 1 601900; Fax: 6010904)
COLL: Tulinius, H.; Bjarnason, Ó.; Magnússon, B.; Thorhallsson, P.; Geirsson, G.B.; Blondal, H.; Arnorsson, J.V.; Benediktsdottir, K.; Agnarsson, B.A.; Jónasson, J.G.; Isaksson, H.J.; Bjornsson, J.

Histological Classification of Tumours in Iceland according to the WHO Classification

This is a comprehensive study of tumours and tumour-like lesions in Iceland according to the WHO's International Histological Classifications of Tumours. All tumours occurring in the country in the periods

1955–1974 and 1955–1984 will be classified and, when desirable, tumour-like lesions at some sites will be included. This will represent a standard approach that is hoped to yield valuable data on tumours, their incidence and comparisons with other geographical areas. At the same time information will become available on the implementation of the WHO classifications. The WHO has given both scientific and financial support to this work. Results for skin, bone, kidney, upper respiratory tract, cervix, urinary bladder, lung, lymphomas, ovary, breast thyroid, endometrium, testis, soft tissue, stomach and oesophagus have been published. CNS, oral, intestinal, appendiceal, pancreatic, biliary and prostatic tumours are now being studied.

TYPE: Incidence; Methodology
TERM: Autopsy; Biopsy; Classification; Histology
SITE: Appendix; Bile Duct; Colon; Nervous System; Oesophagus; Oral Cavity; Pancreas; Prostate; Small Intestine; Soft Tissue
TIME: 1970 - 1994

399 **Tryggvadottir, L.** 05177
Icelandic Cancer Registry, Icelandic Cancer Society, Skogarhlid 8, Box 5420, 125 Reykjavik, Iceland (Tel.: +354 1 621414; Fax: 621417)
COLL: Tulinius, H.; Olafsdottir, G.

Age at Diagnosis of Familial and Sporadic Breast Cancer

Two related hypotheses are tested: (1) that familial breast cancer cases are, on average, younger at diagnosis than sporadic cases; and (2) that this relationship is obscured for recently diagnosed cases, because in this group the probability of having a first degree relative with diagnosed breast cancer will be higher with increasing age of the woman. This effect should diminish with time since diagnosis. A familial case is defined as a woman who has a first degree relative with breast cancer. Information on families of breast cancer patients comes from the Icelandic Cancer Registry, which is population-based, and contains extensive data on families of about half of all breast cancer cases diagnosed in the country since 1910, around 1000 cases. It is planned to use multiple regression analysis with age at diagnosis as the dependent variable and year of diagnosis and familiality as independent variables.

TYPE: Incidence
TERM: Age; Familial Factors; Registry
SITE: Breast (F)
REGI: Iceland (Ice)
TIME: 1983 - 1994

400 **Tryggvadottir, L.** 05178
Icelandic Cancer Registry, Icelandic Cancer Society, Skogarhlid 8, Box 5420, 125 Reykjavik, Iceland (Tel.: +354 1 621414; Fax: 621417)
COLL: Tulinius, H.; Sigurdsson, K.; Larusdottir, M.K.; Johannesson, B.

Total Number of Menstrual Cycles and the Risk of Breast Cancer

The hypothesis of this study is that: a woman's probability of being diagnosed with breast cancer increases with the total number of ovulatory menstrual cycles. Information on length of menstrual cycles, age at menarche and menopause, number of completed pregnancies and abortions, total duration of lactation and use of oral contraceptives comes from an Icelandic databank. The data has been gathered as part of population-based cervical cancer screening in Iceland, via interviewer-administered questionnaires. Over 45,000 women have answered these questions since 1979. Around 900 of these women have been diagnosed with breast cancer. The design is a nested case-control study.

TYPE: Case-Control; Cohort
TERM: Abortion; Lactation; Menarche; Menopause; Menstruation; Oral Contraceptives; Pregnancy; Registry
SITE: Breast (F)
REGI: Iceland (Ice)
TIME: 1990 - 1994

401 **Tryggvadottir, L.** 05179
Icelandic Cancer Registry, Icelandic Cancer Society, Skogarhlid 8, Box 5420, 125 Reykjavik, Iceland (Tel.: +354 1 621414; Fax: 621417)
COLL: Tulinius, H.; Olafsdottir, G.

Linkage of a Breast Cancer Susceptibility Locus to the ABO Locus

The aim of this study is to test the hypothesis that a breast cancer susceptibility gene is linked to the ABO locus. Information on families of breast cancer patients comes from the population-based Icelandic Cancer Registry, which contains extensive data on families of about half of all breast cancer cases diagnosed in the country since 1910 around 1000 cases. ABO blood group information is obtained by record linkage with data from the Icelandic Blood Bank and from hospitals. Linkage analysis will be carried out to evaluate the association of breast cancer risk with blood group.

TYPE: Genetic Epidemiology
TERM: Blood Group; Record Linkage; Registry
SITE: Breast (F)
REGI: Iceland (Ice)
TIME: 1988 - 1994

Ahmedabad

402 **Adhvaryu, S.G.** 03595
Gujarat Cancer and Research Inst., Dept. of Cancer Biology, New Civil Hospital Campus , Ahmedabad 380016, India (Tel.: 378454/378459)
COLL: Dave, B.J.; Trivedi, A.H.

Cytogenetic Studies in Individuals Chewing Tobacco and Areca Nut

The habit of chewing tobacco in combination with areca nut and/or lime is a very common practice in South East Asia. The habit has been shown to be strongly associated with cancer of the oral cavity, which is the predominant type of cancer among males in India. Since not all tobacco chewers develop oral cancer, the individual's genome might have an important role in susceptibility. The aim is to develop parameters which may prove useful in identifying individuals prone to tobacco induced genomic damage. Cytogenetic parameters such as C-band heteromorphism in chromosomes 1, 9 and 16, spontaneous and mutagen induced SCE frequencies and chromosome aberrations in lymphocytes as well as frequency of micronucleated cells in exfoliated buccal mucosa cells are being studied in the following three groups: (1) normal healthy individuals, not consuming tobacco in any form, (2) individuals chewing tobacco for the last two years at least, but without morphological alterations in the mucosa and (3) individuals with sub-mucous fibrosis, leukoplakia or oral cancer.

TYPE: Molecular Epidemiology
TERM: Areca Nut; Chromosome Effects; Genetic Factors; Mutation, Somatic; SCE; Tobacco (Chewing)
SITE: Oral Cavity
TIME: 1984 - 1994

403 **Balar, D.B.** 04394
Gujarat Cancer & Research Inst., Dept. of Pathology, New Civil Hospital Campus , Asarwa Ahmedabad 380016, India (Tel.: 377463; Tlx: 121-680 gcri in)
COLL: Patel, T.B.; Patel, R.D.

Role of HPV Infection in Cervical Cancer

HPVs have been suggested as aetiological agents in cancer of the cervix. Delineation of disease patterns has been hindered by the long disease latency: based on earlier studies of known aetiological factors, i.e. age at marriage, duration of sexual activity and multiplicity of partners, an incubation period of 20-25 years has been suggested. In 1986, 711 cases of cervical cancer, all in married women of low socio-economic status, were seen at the M.P. Shah Cancer Hospital. 14 cases were under age 30, and for the rest age at marriage was between 20 and 30 years, giving a latent period of 20 to 30 years between beginning sexual activity and onset of the disease. All cases of cervical cancer attending the hospital for a period of three years from 1988 to 1990 will be studied to test the hypothesis of a possible role of HPV. Factors to be examined include age at marriage, details of sexual habits, age at disease onset, and personal hygiene. Immunohistochemical techniques will be used to study the strain of papilloma virus present in the cervical biopsies of patients to assess the role of HPV infection.

TYPE: Case Series
TERM: HPV; Infection; Sexual Activity
SITE: Uterus (Cervix)
TIME: 1985 - 1994

404 **Bhatavdekar, J.M.** 04262
Gujarat Cancer Society, Div. of Research, New Civil Hosp. Campus, Asarwa Ahmedabad 380016, India (Tel.: +91 272 37845458; Fax: 375490 ; Tlx: 1216680 gcri in)
COLL: Balar, D.B.; Patel, T.B.

Risk Factors for Epithelial Ovarian Carcinoma and their Relationship to Oestrogen- and Progesterone-Receptor Status

Epithelial ovarian cancer is the third most frequently observed malignancy in the state of Gujarat. The risk factors for this malignancy (age at menarche, menstruation, parity, abortions, age at first full term pregnancy, age at menopause, family history of ovarian cancer) will be compared with hormonal and oestrogen- and progesterone-receptor status in approximately 150 histologically confirmed epithelial ovarian cancer patients and in 60 healthy controls matched for age and parity. The aim is to determine if any association exists between the aetiological risk factors, hormones and oestrogen- and progesterone-receptor status. Preliminary data suggest a higher occurrence of epithelial ovarian cancer in women with more than four children, abnormal levels of pituitary gonadotrophins, and lower socio-economic group (income less than Rs. 500/-per month).

TYPE: Case-Control
TERM: Abortion; Hormones; Menarche; Menopause; Menstruation; Parity; Socio-Economic Factors
SITE: Ovary
TIME: 1987 - 1994

405 **Bhatavdekar, J.M.** 04369
Gujarat Cancer Society, Div. of Research, New Civil Hosp. Campus, Asarwa Ahmedabad 380016, India (Tel.: +91 272 37845458; Fax: 375490 ; Tlx: 1216680 gcri in)
COLL: Balar, D.B.; Patel, T.B.

Risk Factors in Oropharyngeal Cancer and Sialoglycoprotein Levels

Oropharyngeal cancer is the number one malignancy in males in Gujarat. The risk factors for oropharyngeal carcinoma are e.g. ulcers in the oral cavity, tobacco chewing, snuffing, betel quid consumption, cigarette or bidi (indigenous cigarette) smoking and alcohol consumption. In women, habits of tobacco snuff inhalation and use of tobacco powder for cleaning teeth have also been found to be associated in some cases with oropharyngeal cancer. The aim of this study is to determine if any association exists between the risk factors and abnormal levels of protein-bound-sialic acid, lipid-soluble-sialic acid and free sialic acid in oropharyngeal carcinoma patients. Sialoglycoproteins will be measured in 100 oropharyngeal cancer patients and 100 age-matched healthy controls who will be followed for two years.

TYPE: Case-Control
TERM: Alcohol; Betel (Chewing); Biochemical Markers; Tobacco (Chewing); Tobacco (Smoking)
SITE: Oropharynx
TIME: 1987 - 1994

406 **Bhatavdekar, J.M.** 04955
Gujarat Cancer Society, Div. of Research, New Civil Hosp. Campus, Asarwa Ahmedabad 380016, India (Tel.: +91 272 37845458; Fax: 375490 ; Tlx: 1216680 gcri in)
COLL: Shah, N.G.; Kapadia, A.; Giri, D.D.; Balar, D.B.; Patel, N.L.

Epidemiological and Endocrinological Study of Ovarian Cancer

Only a few studies of hormone concentrations in ovarian cancer have been carried out. Results for breast cancer indicate that hormonal abnormalities may also precede and presumably favour the onset of epithelial ovarian cancer or they may be related to the evolution of the disease. Various endocrinological aspects of epithelial ovarian cancer will be examined in 100 Indian females. At the same time, known epidemiological risk factors such as age at menarche, parity, age at first full term birth, dietary habits, socio-economic status, and lifestyle, will be assessed to permit examination of the possible impact of the various risk factors on endocrinological and biological parameters. The endocrine parameters luteinizing hormone, follicle stimulating hormone, prolactin, oestradiol, progesterone, testosterone, dehydroepiandrosterone sulphate, androstenedione, etc. will be studied by RIA.

TYPE: Case Series
TERM: Diet; Hormones; Lifestyle; Menarche; Parity; Socio-Economic Factors
SITE: Ovary
TIME: 1991 - 1994

***407** **Bhatavdekar, J.M.** 05374
Gujarat Cancer Society, Div. of Research, New Civil Hosp. Campus, Asarwa Ahmedabad 380016, India (Tel.: +91 272 37845458; Fax: 375490 ; Tlx: 1216680 gcri in)
COLL: Patel, D.; Balar, D.B.

Endogenous Peptide and Steroid Hormones in Men with Colorectal Carcinoma

A possible role of peptide and steroid hormones in the risk of colorectal cancer has been reported. The current study explores in more detail the effects of risk factors and circulating peptide and steroid hormones in men with colorectal cancer. 150 patients and 50 age-matched healthy controls will be enrolled. Risk factors like, e.g. age, sex, socio-economic status, smoking and alcohol consumption, diet, family history of cancer, site of cancer, will be studied. The circulating hormones (FSH, LH, prolactin, oestradiol, progesterone, testosterone) will be estimated by RIA.

TYPE: Case-Control
TERM: Hormones
SITE: Colon; Rectum
CHEM: Oestradiol; Progesterone; Prolactin; Testosterone
TIME:

408 **Giri, D.D.** 04960
The Gujarat Cancer & Research Inst., New Civil Hosp. Campus, Asarwa Ahmedabad 380016, India (Tel.: 37845459; Tlx: 121 6680 gcri in)
COLL: Bhatavdekar, J.M.; Balar, D.B.

Biochemical Characteristics in Female Breast Cancer

Breast cancer ranks second among Indian female cancers (age standardised rate approximately 20/100,000), with indications that incidence is increasing. Hyperprolactinaemia has been found in breast cancer patients at this Institute, a finding not always reflected in studies published in the West. The aim of this case-control study is to examine various risk factors, i.e., age at menarche, parity, age at first birth, dietary habits, height, weight, socio-economic status, etc., among 200 breast cancer patients and controls belonging to the various socio-religious communities which make up the Gujarat population. The study will also investigate the steroid and peptide hormone profile, as well as the steroid receptor expression patterns in patients belonging to the various communities in an attempt to better understand the epidemiology of breast cancer in this region.

TYPE: Case-Control
TERM: Biochemical Markers; Diet; Hormones; Menarche; Parity; Physical Factors; Socio-Economic Factors
SITE: Breast (F)
TIME: 1991 - 1994

409 **Patel, T.B.** 04968
Gujarat Cancer & Research Inst., Div. Epidemiology & Biostatistics, M.P. Shah Hosp., New Civil Hosp. Campus, Asarwa Ahmedabad 380016, India (Tel.: 37845459; Tlx: 121 6680 gcri in)
COLL: Patel, N.L.; Patel, D.; Balar, D.B.; Giri, D.D.

Nasopharyngeal Cancer (Lympho-epithelioma) in West India

During 1986 to 1990, 37 cases of nasopharyngeal carcinoma were treated at the Institute. 67% were children or adolescents (under 20 years of age). 33 cases came from various districts of Gujarat and four from the province of Rajasthan. This study aims to detect the presence, if any, of regional or familial clustering of the disease among cases already registered in the Institute and normally residing in Gujarat State. A second objective is to examine the association of EBV and lympho-epithelioma in this population, and to look for expression of epithelial/lymphoid and other transfers by the tumour cells to understand the cell biology. Immunohistochemical methods will be used to detect epithelial and lymphoid cell markers. In situ hybridization techniques will be used to localise EBV DNA on archival material.

TYPE: Case Series
TERM: Cluster; EBV
SITE: Nasopharynx
TIME: 1991 - 1994

Barshi

410 **Jayant, K.** 05024
Rural Cancer Registry, Cancer Hosp., Agalgaon Rd,, Barshi 413 401 Maharashtra, India
COLL: Gulati, S.S.; Notani, P.N.; Kamat, M.R.

Role of Penile Hygiene in the Aetiology of Cancer of the Cervix Uteri

A study of differential cervical cancer rates in the various religious groups in Bombay showed that the combined effect of two risk factors, early age at first coitus and poor penile hygiene, was the sum of their separate independent effects (Br. J. Cancer 56:685, 1987). To confirm this finding, a case-control study has been initiated. Cervical cancer patients from Bombay attending the Tata Memorial Hospital, as well as age-, community- and social class-matched controls from the general population are intervivewed by a trained medical social worker. The questionnaire includes questions on tobacco habits, besides those on sexual activity and hygiene, for patients, controls and partners. To date, data on 60 cases and an equal number of controls have been collected. When about 200 cases have been accumulated, the data will be analysed using univariate and multivariate methods.

TYPE: Case-Control
TERM: Hygiene; Sexual Activity
SITE: Uterus (Cervix)
TIME: 1988 - 1994

***411 Jayant, K.** 05470
Rural Cancer Registry, Cancer Hosp., Agalgaon Rd,, Barshi 413 401 Maharashtra, India
COLL: Rao, R.S.; Nene, R.S.; Dale, P.S.

Rural Cancer Registry

The first population-based Rural Cancer Registry in India was established in 1988. Since its inception, the Registry has contributed useful data for control of cancer in this rural population. Cancer of the oesophagus is the leading cancer in males. Incidence of cancers of the oesophagus, hypopharynx and parts of the oral cavity, comprising the buccal mucosa, gum and floor of mouth, are not significantly different from the corresponding rates reported by the Bombay Cancer Registry. However, incidence of penile cancer is higher. In the females cervical cancer is the predominant cancer and has a high incidence rate. One of the positive results of the registry activity has been a significant down-staging of cervical cancer over the years, due to the interaction of the social investigators with the community and the consequent increase in cancer awareness.

TYPE: Incidence; Registry
TERM: Data Resource
SITE: All Sites
TIME: 1988 - 1999

Bombay

***412 Desai, P.B.** 05315
Tata Memorial Centre, Dr Ernest Borges Marg, Parel, Bombay 400012, India (Tel.: +91 22 4146750; Fax: 4146937 ; Tlx: 01173649 tmc in)
COLL: Deo, M.G.; Rao, R.S.; Sanghvi, L.D.

Epidemiological Studies of Radiation Workers and their Families at Nuclear Power Stations in India

The nuclear Power Corporation of India and the Tata Memorial Centre are conducting jointly a health survey of employees in nuclear power stations and their families. The main objectives of the study are (1) to obtain prevalence and incidence data on cancer, with emphasis on radiation-induced cancer, such as leukaemia and cancer of the thyroid; (2) to assess the effect of radiation on workers in nuclear facilities and in the population of neighbouring areas; and (3) to construct a database to monitor the adverse effects of radiation, if any, on a continuing basis. A pilot study has been completed at Tarapur Power Plant near Bombay. The study subjects will have physical examination, chest x-ray, blood analyses, routine cytological test and other laboratory tests. The study will mainly focus on cancer and genetic diseases. The initial study at Tarapur will be followed up with similar studies at other power stations in the country. The total cohort will consist of 13,300 radiation workers and 53,300 family members.

TYPE: Cohort
TERM: Occupation; Radiation, Ionizing
SITE: All Sites
OCCU: Radiation Workers
TIME: 1992 - 1999

413 Gupta, P.C. 03327
Tata Inst. of Fundamental Research, Basic Dental Research Unit, Homi Bhabha Rd, Bombay 400005, India (Tel.: +91 22 2152317; Fax: 2152110 ; Tlx: 011-83009 tifr in)
COLL: Mehta, F.S.; Pindborg, J.J.

Mortality Experience in Relation to Tobacco Chewing and Smoking Habits

Chewing of tobacco with lime, betel leaf, areca nut and smoking of bidis, chuttas etc. is practised widely in India. Recent studies have shown that the age-adjusted mortality among bidi smokers is greater than among tobacco chewers and mortality among betel tobacco chewers, bidi smokers and chutta smokers is significantly higher compared to non-users of tobacco. In the present study attempts are being made to determine the possible causes for this excess mortality. In a house-to-house survey, a sample of 36,000 tobacco users (age 15 years and over) in three districts of India is being interviewed annually and examined for the presence of oral cancer. Since there is no death certification system in rural India and often expert medical help is neither sought nor available, it is difficult in general to find out the precise cause of death. Therefore a list of various possible symptoms is prepared and information on symptoms before death is being collected from the next-of-kin. Similar information is also being collected from a control group of non-users of tobacco.

TYPE: Cohort
TERM: Betel (Chewing); Tobacco (Chewing); Tobacco (Smoking)
SITE: Oral Cavity
TIME: 1977 - 1994

414 **Gupta, P.C.** 04535
Tata Inst. of Fundamental Research, Basic Dental Research Unit, Homi Bhabha Rd, Bombay 400005, India (Tel.: +91 22 2152317; Fax: 2152110 ; Tlx: 011-83009 tifr in)
COLL: Mehta, F.S.; Pindborg, J.J.

Cancer Risk Following Premalignant Oral Lesions

Certain oral mucosal lesions such as leukoplakia, the palatal changes associated with reverse smoking and conditions such as submucous fibrosis are known to increase the risk of development of oral cancer. Precise relative risk estimates are however not yet available. One complicating factor is a strong association of tobacco chewing and smoking with oral cancer, as well as with precancerous lesions and conditions. In this study, cohorts consisting of 36,000 tobacco chewers and smokers are being followed annually over a period of ten years in three areas of India, by house-to-house visits. These individuals are examined every year by dentists and the clinical diagnosis is carefully documented. The diagnosis of oral cancer is histologically confirmed. This study will provide incidence rates of oral cancer among those with specific types of oral precancerous lesions and conditions and also among those without any preceding oral precancerous lesions or conditions. The incidence rates will be calculated by the method of person-years and will be adjusted for age, sex and the type of tobacco usage. Possible effects of changes in tobacco habits would also be studied.

TYPE: Cohort
TERM: Betel (Chewing); Premalignant Lesion; Tobacco (Chewing); Tobacco (Smoking)
SITE: Oral Cavity
TIME: 1977 - 1994

***415** **Gupta, P.C.** 05486
Tata Inst. of Fundamental Research, Basic Dental Research Unit, Homi Bhabha Rd, Bombay 400005, India (Tel.: +91 22 2152317; Fax: 2152110 ; Tlx: 011-83009 tifr in)
COLL: Sasco, A.J.; Parkin, D.M.; Peto, R.; Lopez, A.D.

Assessment of Cause-Specific Tobacco Attributable Mortality in Bombay, India

Current estimates of tobacco attributable mortality in India and other developing countries are mostly speculative. This study aims to address tobacco attributable mortality in a cohort of 100,000 subjects, aged >35 years, in Bombay, India, in collaboration with Tata Institute of Fundamental Research, Bombay. The enrolment into the cohort started in February 1991 and about 50,000 subjects have so far been interveiwed. The voters list is the main sampling frame. A preliminary analysis of habits in 25,000 subjects revealed that 58.8% are current chewers, 6.7% smokers and 4.9% are both smokers and chewers. Cause-specific mortality in this cohort will be assessed based on the data from the municipal death registration systems and cancer incidence via the Bombay Cancer Registry. Relative risks using person-year methods will be estimated in users in relation to tobacco use and other variables. Methodological issues related to recruitment, migration, repeatability, follow-up and death ascertainment are also being addressed.

TYPE: Cohort
TERM: Registry; Tobacco (Chewing); Tobacco (Smoking); Tobacco (Snuff)
SITE: All Sites
REGI: Bombay (Ind)
TIME: 1990 - 1995

***416** **Gupta, P.C.** 05487
Tata Inst. of Fundamental Research, Basic Dental Research Unit, Homi Bhabha Rd, Bombay 400005, India (Tel.: +91 22 2152317; Fax: 2152110 ; Tlx: 011-83009 tifr in)
COLL: Dharkar, D.K.; Mehra, S.N.

Oral Cancer Control in Ratlam District

Oral cancer is common in Ratlam district of central India and most subjects with this disease present in late stages. 115 basic health workers in one 'tehsil' (part of a district) have been trained in early detection and primary prevention of oral cancer and mouth self-examination. During their routine house-to-house visits, they examine the mouths of tobacco users, aged 35 and over. A proforma is also filled out for each individual, containing details of tobacco habits. Individuals with suspected oral lesions are referred to the district headquarters for further check-up. The physicians practicing in the area have also been trained in early detection of oral cancer and education campaigns ere conducted to promote primary prevention and mouth self-examination. Over a period of about 9 months, the health workers have interviewed and examined over 60,000 individuals. The data are now being analysed.

TYPE: Intervention
TERM: Education; Prevention; Screening; Tobacco (Chewing); Tobacco (Smoking)
SITE: Oral Cavity
TIME: 1992 - 1994

417 **Jussawalla, D.J.** 04458
Bombay Cancer Registry, Indian Cancer Society, Lady Ratan Tata Medical & Res. Center, M. Karve Rd, Bombay 400021, India (Tel.: +91 22 2047436)
COLL: Yeole, B.B.; Notani, P.N.

Survival of Breast Cancer Patients in Bombay

Data collected by the Bombay Cancer Registry on breast cancer patients for the years 1985 to 1990 and death records maintained by the Department of Vital Statistics of Bombay Municipal Corporation for the years 1985-1994 will be reviewed to study the survival rates. Firstly incident cases will be matched with the death certificates maintained by the Municipal Corporation. The vital status of the remaining unmatched patients will be ascertained first by reply postcards, then by home visits, and five-year survival rates will be calculated.

TYPE: Incidence
TERM: Registry; Survival
SITE: Breast (F)
REGI: Bombay (Ind)
TIME: 1989 - 1994

***418** **Jussawalla, D.J.** 05424
Bombay Cancer Registry, Indian Cancer Society, Lady Ratan Tata Medical & Res. Center, M. Karve Rd, Bombay 400021, India (Tel.: +91 22 2047436)
COLL: Yeole, B.B.; Natekar, M.V.; Sabnis, S.D.

Validation of the Bombay Cancer Registry Data: Quality of Data and Completion of Coverage

A population survey is being undertaken in two sections of the Bombay metropolitan area to ascertain completeness of coverage of the Bombay Cancer Registry to verify the accuracy of registrations and diagnostic details, and to assess mortality and cause of death. All households will be visited and relevant information will be collected from all the members of the family. Family members will be interviewed by specially trained persons, using a structured questionnaire. Cancer morbidity data collected in this survey will be matched to the information recorded in the cancer registry.

TYPE: Incidence; Methodology; Mortality
TERM: Registry
SITE: All Sites
REGI: Bombay (Ind)
TIME: 1993 - 1995

419 **Mehta, F.S.** 01032
Tata Inst. of Fundamental Research, Basic Dental Research Unit, Homi Bhabha Rd, Bombay 400005, India (Tel.: +91 22 219111; Tlx: 011-3009)
COLL: Pindborg, J.J.; Gupta, P.C.; Aghi, M.B.; Daftary, D.K.

Oral Cancer and Precancerous Lesions in Relation to Tobacco Use in Rural Indian Populations

Tobacco usage as practised in different parts of India is associated with oral cancer and precancerous lesions. This study aims to demonstrate a reduction in the incidence of precancerous lesions when tobacco habits are discontinued. A sample of 12,000 individuals with tobacco habits has been examined and interviewed in each of three rural areas. Villagers are being exposed to a concentrated programme of education, motivation and guidance. These individuals are re-examined annually to assess changes in tobacco habits, development of new oral precancerous lesions and changes in previously diagnosed lesions. Preliminary results showed that among the individuals who stopped or substantially reduced their tobacco habits, the reqression of leukoplakia was significantly higher and the incidence of leukoplakia significantly lower (Lancet.i: 1235-1238, 1986). The health education was of significant help in stopping the tobacco habits of the individuals (IARC Sci. Publ. 74) 307-17, 1986) and of special help to difficult subgroups like tobacco chewers (Am. Publ. Health 76:709, 1986). Results of the 10-years follow-up show e.g. that education was related to stopping or reducing smoking; that the incidence of oral leukoplakia was lower in the intervention group than in the control group and that among the inviduals in the intervention group who reported having stopped smoking, almost no oral mucosal lesions related to bidi smoking developed. Further analyses of these 10-year follow-up data continue.

TYPE: Cross-Sectional
TERM: Premalignant Lesion; Prevention; Tobacco (Chewing); Tobacco (Smoking)
SITE: Oral Cavity
TIME: 1977 - 1994

***420** **Mehta, F.S.** 05491
Tata Inst. of Fundamental Research, Basic Dental Research Unit, Homi Bhabha Rd, Bombay 400005, India (Tel.: +91 22 219111; Tlx: 011-3009)
COLL: Pindborg, J.J.; Gupta, P.C.; Hebert, J.R.; Krishnaswamy, K.; Daftary, D.K.

Population-Based Study of the Relationship of Nutrition with Oral Precancerous Lesions

Oral precancerous lesions are strongly associated with tobacco chewing and smoking habits. The aim of this study is to investigate the association of nutrition with oral precancerous lesions. A questionnaire incorporating nutrition information suitable for a large population-based study will be developed. Population samples from villages in the districts of Ernakulam, Srikakulam and Bhavnagar will be selected, so as to provide 12,000 individuals above the age of 15 years. These individuals will be interviewed for their tobacco habits by trained investigators and examined for the presence of oral cancer and oral precancerous lesions by dentists using house-to-house approach. The individuals will be re-interviewed and re-examined in about a year. A nested case-control study will be conducted for those developing new oral precancerous lesions with a detailed dietary questionnaire. Analysis will be conducted separately for the cross-sectional phase, follow-up phase and case-control phase.

TYPE: Case-Control; Cohort; Cross-Sectional
TERM: Diet; Nutrition; Premalignant Lesion; Tobacco (Chewing); Tobacco (Smoking)
SITE: Oral Cavity
TIME: 1992 - 1995

***421** **Mehta, F.S.** 05492
Tata Inst. of Fundamental Research, Basic Dental Research Unit, Homi Bhabha Rd, Bombay 400005, India (Tel.: +91 22 219111; Tlx: 011-3009)
COLL: Pindborg, J.J.; Gupta, P.C.; Aghi, M.B.; Daftary, D.K.

Primary and Secondary Prevention of Oral Cancer using Basic Health Workers

It has been demonstrated that basic health workers can detect oral cancer at an early stage during their routine work (Cancer Detection and Prevention, 9: 219-25, 1986). It has also been shown that health education against tobacco use carried out by dentists and social scientists was effective in stopping or reducing the tobacco consumption and decreasing the risk of oral precancer (Lancet, I: 1235-8, 1986) The present study was undertaken to assess whether basic health workers can undertake an entire programme of primary and secondary prevention of oral cancer. Five basic health workers were trained in primary prevention and early detection of oral cancer in Srikakulam district, Andhra Pradesh. Each one of them was stationed in a village and assigned a group of nearby villages covering an estimated 5,000 population, aged 15 years and over. These basic health workers then interviewed all the individuals in the assigned villages, examined their oral cavities and educated them against tobacco use. These individuals were re-interviewed and re-examined after one and two years by the same basic health worker. Individuals with suspected lesions were referred to a team of dentists. The dentists also performed a random check on part of the sample to reconfirm the diagnosis. Analsyis of the data is in progress.

TYPE: Cohort; Intervention
TERM: Education; Prevention; Screening; Tobacco (Chewing); Tobacco (Smoking)
SITE: Oral Cavity
TIME: 1989 - 1994

***422 Nambi, K.S.V.** 05351
Bhabha Atomic Research Centre, Div. of Environmental Assessment, Tromba , Bombay 400 085, India (Tel.: +91 22 5559369; Fax: 5560750 ; Tlx: 011-71017 barc in)
COLL: Mayya, Y.S.; Joshi, P.V.

Cancer among Nuclear Workers and the General Population

This is a historial prospective study to investigate whether workers in Indian nuclear establishments have an increased risk of dying from cancer of various sites, compared with the general population. Approximately 10,000 workers employed in Bhabha Atomic Research Centre, Bombay, 3,000 workers employed at various nuclear facilities at Tarapur and 3,000 employees at the Nuclear Fuel Complex, Hyderabad, have been studied, covering periods from 1975 to 1991. A minimum database required to obtain significant results have been prepared for occupational workers, populations living in normal background radiation areas and those living in high backgroun radiation areas. Risk coefficients obtained from ICRP recommendations are being used and extrapolated to Indian life tables. Publications appreared in Current Sci. 59:733-736, 1990, Ind. J. Cancer 28:61-69, 1991 and Arch. Environ. Hlth 47:155-157, 1992.

TYPE: Cohort
TERM: Occupation; Radiation, Ionizing
SITE: All Sites
OCCU: Radiation Workers
TIME: 1988 - 1998

423 Notani, P.N. 05077
Tata Memorial Centre, Cancer Research Inst., Epidemiology Unit., Dr E. Borges Marg , Bombay 400012, India
COLL: Jayant, K.

Role of Diet in Cancer of the Female Breast

Nutritional factors, in particular fat intake, appear to play an important role in endocrine-dependent cancers. A case-control study has been initiated to investigate the possible relationship between dietary factors and cancer of the female breast, taking into account the other risk factors related with reproductive history, obesity, personal habits (tobacco, alcohol) and family history of cancer. A questionnaire is administered to patients attending the Tata Memorial Hospital, which is the largest cancer treatment facility in the region. A first control group is obtained from patients who are diagnosed as not having cancerous or precancerous lesions. A second group is composed of individuals from the general population. To date, a little over 200 cases and around 100 each of hospital and population controls have been interviewed. Recruitment of cases and controls is in progress.

TYPE: Case-Control
TERM: Alcohol; Diet; Obesity; Reproductive Factors; Tobacco (Smoking)
SITE: Breast (F)
TIME: 1988 - 1994

424 **Rao, D.N.** 05124
Tata Memorial Hosp., Dr E. Borges Marg, Bombay 400012, India (Tel.: +91 22 4129761; Fax: 4129937 ; Tlx: 01173649 TMC IN)
COLL: Desai, P.B.

Digestive Tract Cancers in Western India

The aim of this hospital-based case-control study is to identify high-risk groups for gastrointestinal cancer. Available data indicate that cancer of the oesophagus has a high incidence in northern India, whereas stomach cancer incidence is high in the south. The incidence of colon cancer is low but it is likely to increase due to changes in lifestyle. Patients will be interviewed by social investigators for habits and customs, dietary practices and other related information before being examined by the clinician. Four major gastrointestinal gastrointestinal cancers - oesophagus, stomach, colon and rectum will be included. Over 2,000 cases and 4,000 control patients are expected. Stratified analysis will be carried out mainly to test the hypothesis of a protective effect of vegetarian diet, and the additional influence of alcohol and tobacco. It is proposed to stratify for age, sex and residence (Bombay, rest of Maharashtra state and others). Relative risk for each of the factors will be calculated by using test-based estimation procedure. Patient accrual will be completed by 1991.

TYPE: Case-Control
TERM: Alcohol; Diet; Tobacco (Smoking)
SITE: Colon; Oesophagus; Rectum; Stomach
TIME: 1987 - 1994

425 **Yeole, B.B.** 05098
Indian Cancer Society, Bombay Cancer Registry, 74 Jerbai Wadia Rd, Parel, Bombay 400012, India (Tel.: +91 22 4122351)
COLL: Jayant, K.; Jussawalla, D.J.; Natekar, M.V.

Trends in Tobacco-Related Cancers in Females of Greater Bombay

Upper alimentary and respiratory cancers are known to be aetiologically associated with the habit of chewing or smoking tobacco. Data collected by the Bombay Cancer Registry for the period 1964 to 1987 will be used to study incidence trends for cancers of the tongue, buccal mucosa, oropharynx, hypopharynx, larynx and lung. Differences in cancer incidence for these sites will be examined, taking into consideration the pattern of tobacco habits in the different population groups. Log-linear regression models will be used to examine trends in the crude, age-adjusted and age-specific incidence rates.

TYPE: Incidence
TERM: Female; Registry; Tobacco (Chewing); Tobacco (Smoking); Trends
SITE: Hypopharynx; Larynx; Lung; Oral Cavity; Oropharynx; Tongue
REGI: Bombay (Ind)
TIME: 1991 - 1994

426 **Yeole, B.B.** 05100
Indian Cancer Society, Bombay Cancer Registry, 74 Jerbai Wadia Rd, Parel, Bombay 400012, India (Tel.: +91 22 4122351)
COLL: Jayant, K.; Jussawalla, D.J.; Kadam, V.T.

Cancer in the Gujarati Population in Bombay

The Gujarati population in Bombay comprises Hindu, Muslim, Jain and Parsi religious groups. Amongst these different groups, there is wide variation in habits, customs and socio-economic status. Striking differences in the relative frequency of cancer at various sites are also observed. The site pattern of cancer in this community will be assessed by religion and the magnitude and nature of differences observed will be examined to assess the extent to which these could be ascribed to variations noted in the different lifestyles. The basic data for the study will be obtained from the Bombay Cancer Registry for the period 1986-1988. As Indian Census figures do not provide information on population by mother tongue and age, age-standardised cancer ratios (ASCAR) will be calculated.

TYPE: Incidence
TERM: Lifestyle; Registry; Religion; Socio-Economic Factors
SITE: All Sites
REGI: Bombay (Ind)
TIME: 1991 - 1994

***427 Yeole, B.B.** 05422
Indian Cancer Society, Bombay Cancer Registry, 74 Jerbai Wadia Rd, Parel, Bombay 400012, India (Tel.: +91 22 4122351)
COLL: Ascherio, A.; Kamat, M.R.; Jussawalla, D.J.

Case-Control Study of Prostate and Ovarian Cancers in Bombay

This case-control study aims to examine the hypothesis that vasectomy increases the risk of prostate cancer and that tubal ligation reduces the risk of ovarian cancer. Cases will be all patients with prostate and ovarian cancers admitted to Tata Memorial Hospital and Bombay residents only from other hospitals, reporting to Bombay Cancer Registry. Hospital controls will be selected among cases of oesophagus, rectum, colon and larynx cancer from Tata Memorial Hospital and among cancers of all sites in Bombay residents from the other hospitals. It is expected that 1,000 prostate cancer cases and 2,000 ovarian cancer cases will be identified and included in the study. Cases and controls will be interviewed by trained medical social workers.

TYPE: Case-Control
TERM: Registry; Tubal Ligation; Vasectomy
SITE: Ovary; Prostate
REGI: Bombay (Ind)
TIME: 1993 - 1996

Chandigarh

***428 Behera, D.** 05290
Postgraduate Inst. of Medical Education & Research, Dept. of Pulmonary Medicine, Sector 12, Chandigarh 160012, India (Tel.: 32351/231)

Biomass Fuel and Lung Cancer

The aim of the study is to assess a possible relationship between the use of biomass fuel and lung cancer in women. 200 women with lung cancer will be included in the study. An equal number of age-matched women without lung cancer will serve as controls. Data on smoking history, details of cooking fuel used and duration of cooking will be obtained. Comparison will also be made between different types of fuels (biomass, liquid petroleum gas, kerosene, and mixed).

TYPE: Case-Control
TERM: Cooking Methods; Female; Petroleum Products; Tobacco (Smoking)
SITE: Lung
CHEM: Kerosene
TIME: 1993 - 1995

Goa

429 Vaidya, S.G. 04572
Goa Cancer Society, Dr E. Borges Rd, Dona Paula Goa 403-004, India (Tel.: 3426/5884)
COLL: Vaidya, N.S.; Kamat, V.; Shetye, S.B.

Assessment of the Efficacy of an Anti-Tobacco Community Education Programme

A project is under way in Goa to assess the efficacy of an anti-smoking and anti-tobacco community education programme. The aim of the project is to test the effectiveness of educating the community through school children. The study population is the rural population of about 685,000 (1981 census), divided into three zones, two experimental and one control. About 10,000 individuals aged 15 and above are selected at random in each zone by a two-stage sampling procedure (villages, individuals). The selected individuals are interviewed and examined by dental surgeons and researchers for tobacco habits and oral precancerous lesions. Since the first survey was carried out (1986-1988), education programmes have been introduced in the experimental zones. In one zone, education is carried out through schools and school children, and in the other through schools, primary health centres and anganwadis (centres for health care of pre-school children and their mothers, run by the Department of Social Welfare of the Government). Baseline results among the 14,363 males interviewed were 30.2% smokers and 3.5% using smokeless tobacco, and among the 15,350 females, 7.4% smokers and 13.1% using smokeless tobacco. 21.8% of the 4,829 male tobacco habituees, have oral precancerous lesions and 12.3% of the 3,151 female habituees. Efficacy of the intervention programmes will be tested after three surveys.

TYPE: Intervention
TERM: Premalignant Lesion; Prevention; Tobacco (Chewing); Tobacco (Smoking)
SITE: Oral Cavity
TIME: 1986 - 1994

Hyderabad

***430** **Krishnaswamy, K.** 05377
National Inst. of Nutrition, Jamai-osmania, Tarnaka, Hyderabad 500 007, India (Tel.: +91 842 868909; Fax: 869074 ; Tlx: 0425-7022)
COLL: Mukundan, M.A.; Naidu, A.N.; Prasad, M.P.R.; Krishna, T.P.; Amrender Reddy, G.; Annapurna, V.V.

Dietary Intervention on Oral Precancerous Lesions

The aims are to study (1) the effect of micronutrient supplementation on oral precancerous lesions; (2) the role of nutrient intake on oral precancerous lesions (case-control study); (3) intermediary endpoints (biomarkers). In a tertiary prevention trial, four villages were selected, two to receive supplementation (case groups) and two to receive placebo (control groups). The number of subjects in each group are 150 (all smokers). Administration is done biweekly under supervision for a period of one year. Blood samples, oral epithelial cell smears and cell washings were collected at initiation of the study and at the end of the treatment period, to carry out biochemical, micronucleus and DNA adduct tests. Dietary intake by 24 hour recall and food frequency methods was also collected.

TYPE: Intervention
TERM: BMB; Biomarkers; Blood; Cytology; DNA Adducts; Micronuclei; Micronutrients; Premalignant Lesion
SITE: Oral Cavity
TIME: 1990 - 1994

Lucknow

431 **Kushwaha, M.S.** 03340
King Georges Medical College, Postgrad. Dept. Pathol. & Bacteriology, Lymphoma & Leukaemia Registry, Lucknow 226003, India (Tel.: 82554)
COLL: Misra, N.C.; Nath, P.; Misra, P.K.; Jaiswal, M.S.D.; Kumar, A.

Lymphoma-Leukaemia Registry

A Lymphoma-Leukaemia Registry was established in 1971. More than 1,000 cases of lymphomas including Hodgkin's disease, and more than 1,000 cases of all types of leukaemias have been recorded. The centre records diagnosis, clinical, haematological and histopathological data, therapy and follow-up of the cases encountered in the city of Lucknow and province of Uttar Pradesh, with a population of nearly 110 million. An attempt is being made to identify factor(s) responsible for the genesis and promotion of these diseases through family and environmental histories. Morphological findings are also being correlated with the biological behaviour of the disease. For the last four years multiple myeloma has been regularly registered and studied in a similar manner. So far 70 cases are on record. Cytochemical and immunological marker studies are done in cases of lymphoproliferative disorders. The centre also provides teaching and research facilities for undergraduates, postgraduates and young scientists, acts as a referral service for expert opinion in the field, and investigates various problems associated with these diseases. A paper has been published in Leukem. Res. 9(6):799-802, 1985.

TYPE: Registry
TERM: Biochemical Markers; Environmental Factors; Histology
SITE: Hodgkin's Disease; Leukaemia; Lymphoma; Multiple Myeloma
TIME: 1985 - 1994

Madras

432 **Bharati Arumugam, S.** 03319
Madras Medical College, Inst. of Pathology & Electron Microscopy, Park Town, Madras 600 003, India (Tel.: +91 44 30001/39181)
COLL: Narendran, P.; Kanaka, T.S.; Balakrishnan, T.S.; Logamuthukrishnan, T.; Kalyanaraman, S.; Ramamurthy, B.; Reginald, S.; Arumugam, S.; Govindan, S.

Brain Tumours in India

The main object of the study is to assess the frequency of occurrence and type of brain tumours in India. The project includes the preparation of statistics on the occurrence of these tumours in studies reported by neuropathologists in various parts of India and the establishment of a Tamil Nadu State neuropathology registry, including histopathology. Since the inception of the registry in 1972, a total of nearly 4,300 intracranial solid tumours have been diagnosed by one neuropathologist. The slides, paraffin blocks and tissue material are available for more extensive studies.

TYPE: Registry
TERM: Classification; Histology
SITE: Brain
TIME: 1972 - 1994

433 **Gajalakshmi, C.K.** 04611
Cancer Inst., Epidemiology Div. and Cancer Registry, Adyar, Madras 600020, India (Tel.: +91 44 2350131; Fax: 412185)
COLL: Shanta, V.

Breast Cancer in Madras, India

Breast cancer is the second most important cancer site in Madras. This study will examine the hypothesis that (1) premenopausal and postmenopausal women have different risk factors; (2) that age at first child birth greater than 25 years confers a higher risk in postmenopausal women; and (3) that premenopausal women with less than four children or nulliparous women a have higher risk. A case-control study will be done, using all breast cancer cases (about 1,000) registered in the population-based cancer registry from 1989 to 1992, and two groups of controls for each case. The first group will consist of cancer patients who have no disease in the breast, gynaecological organs or endocrine glands (all sites except ICD-O 174, 180-184, 193 & 194); the second group will be patients who attend the general out-patient department, without cancer. Information will be obtained on age, sex, religion, education, income, hypertension, diabetes, age at menarche, age at first child birth, number of children, age at last child birth, age at menopause, history of hysterectomy and oophorectomy and use of contraceptives.

TYPE: Case-Control
TERM: Contraception; Diabetes; Menarche; Menopause; Parity; Registry; Religion; Socio-Economic Factors
SITE: Breast (F)
REGI: Madras (Ind)
TIME: 1989 - 1994

***434** **Gajalakshmi, C.K.** 05262
Cancer Inst., Epidemiology Div. and Cancer Registry, Adyar, Madras 600020, India (Tel.: +91 44 2350131; Fax: 412185)
COLL: Ravichandran, K.

Second Cancers Following Primary Uterine Cervical Cancer

The aim of the study is to determine whether patients with cancer of the cervix are at increased risk to develop second cancers, following therapy for their first primary. Patients with histologically confirmed cervical cancer as their first primary, diagnosed between 1960-1989, form the study cohort and will be followed up till December 1991. Second cancers which develop more than 12 months after the diagnosis of the first primary are included in the analysis. Data are abstracted from the case records by a trained abstractor. Internal comparison will be made between the cases who received radiotherapy treatment and those who did not. Comparison will also be made with the general population.

TYPE: Cohort
TERM: Multiple Primary; Radiotherapy
SITE: Uterus (Cervix)
TIME: 1991 - 1995

435 **Shanta, V.** 05155
Madras Metropolitan Tumor Registry, Cancer Inst. (WIA), Canal Bank Rd, Adyar, Tamil Nadu, Madras 600020, India (Tel.: +91 44 412714; Fax: 412185)
COLL: Vasanthi, L.; Swaminathan, R.; Chacko, P.

Childhood Leukaemia and Lymphoma

This retrospective survey will establish the epidemiological profile of childhood leukaemias and lymphoma. Case records of children admitted between 1984-1990 with lymphoma (n = 141) or leukaemia (n = 174) will serve as the source of data. Demographic, environmental, socio-economic and cultural particulars of the study subjects and their families will be obtained. Medical history of the study subjects from the time of conception to presentation with clinical features will be abstracted, with particular reference to parental consanguinity and age at the time of pregnancy, birth order of the patients, growth and developmental mile-stones and childhood infections. Relevant details about the siblings of the study subjects will also be obtained. Clinical profiles of the study subjects will be collected to find out the usual modes of presentation to the physician.

TYPE: Case Series
TERM: Birth Order; Childhood; Clinical Records; Environmental Factors; Infection; Sib; Socio-Economic Factors
SITE: Leukaemia; Lymphoma
TIME: 1991 - 1994

***436** **Shanta, V.** 05284
Madras Metropolitan Tumor Registry, Cancer Inst. (WIA), Canal Bank Rd, Adyar, Tamil Nadu, Madras 600020, India (Tel.: +91 44 412714; Fax: 412185)
COLL: Sasaki, R.

Study of Breast Cancer

The aim of this hospital-based case-control study is (1) to identify the risk factors related to lifestyle, reproductive history, hormone serum levels (oestrone, oestradiol, oestriol and testosterone) and levels of oestrogen (progesterone) receptor in cancer tissue; (2) to detect differences of serum levels of beta carotene, vitamins and lipids between Indian and Japanese breast cancer patients, and (3) to assess psychological distress at the time of diagnosis of breast cancer. 200 breast cancer cases are being matched for age (+/- 5 years), sex, religion and mother tongue with 200 healthy controls (relatives of cancer patients attending the outpatient clinic of the Cancer Institute) and 200 cancer controls (patients suffering from cancers other than breast, gynaecological tract or endocrine glands). Data are collected by direct interview and will be analysed using conditional logistic regression methods.

TYPE: Case-Control
TERM: Hormones; Lifestyle; Lipids; Reproductive Factors; Vitamins
SITE: Breast (F)
CHEM: Beta Carotene; Oestradiol; Oestriol; Oestrogens; Oestrone; Progesterone; Testosterone
LOCA: India; Japan
TIME: 1992 - 1994

New Delhi

437 **Luthra, U.K.** 03737
Indian Council of Medical Research, Cytology Research Centre, Maulana Azad Medical College Campus, Bahadur Shah Zafar Marg, New Delhi 110002, India (Tel.: +91 11 3311889)
COLL: Das, D.K.; Gupta, M.M.; Seghal, A.; Murthy, N.S.; Singh, V.; Bhatnagar, P.; Sharma, B.K.; Das, B.C.; Agarwal, S.S.; Bhambhani, S.; Sodhani, P.; Menon, R.; Kumar, D.; Dutta, S.; Sharma, J.K.; Kashyap, V.; Gupta, S.; Attam, K.; Singh, K.; Singh, M.; Chadha, B.; Chadha, P.

Uterine Cervical Dysplasia

The biological behaviour of precancerous and early cancerous lesions of the cervix is not clearly understood. Identification of relevant risk factors and detection and management of such lesions are important in the prevention and management of invasive cancer of the cervix. A prospective multi-disciplinary study involving various parameters such as clinical, epidemiological, cytopathological, colposcopic, virological, cytogenetic, immunological and ultrastructural studies on precancerous and early cancerous lesions of the cervix has been underway at the Cytology Research Centre, New Delhi, India since 1976 with the aim of understanding the biological behaviour of these lesions of cervix, and of discovering relevant risk factors, particularly the delineation of possible high risk dysplasia from relatively low risk dysplasia. This prospective cohort study proposes to register 1,000 dysplasias with twice the number of controls, one group of controls matched for age and parity and the other for age only and follow them for 5-15 years. The virological parameter includes sero-epidemiology of HSV and detection of HPV in cervical smears as well as in the tissues. The cytogenetic component includes study of chromatid exchange, silver stained nucleolus organizing regions and chromosomal aberrations. At present a cohort of 950 dysplasias with matched controls is being followed with a view to determining the importance of the above factors in progression or regression of the dysplastic lesions.

TYPE: Cohort
TERM: Chromosome Effects; Cytology; HPV; HSV; Premalignant Lesion; Screening; Sero-Epidemiology
SITE: Uterus (Cervix)
TIME: 1976 - 1994

438 **Luthra, U.K.** 05168
Indian Council of Medical Research, Cytology Research Centre, Maulana Azad Medical College Campus, Bahadur Shah Zafar Marg, New Delhi 110002, India (Tel.: +91 11 3311889)
COLL: Kumar, D.; Dutta, S.; Sharma, J.K.; Kashyap, V.; Chadha, P.; Chadha, B.; Gupta, S.; Attam, K.; Sunderwa, J.; Singh, K.; Singh, M.; Padubidri

Uterine Cervical Dysplasia

A study carried out here during 1976-1988 highlighted some socio-demographic and biological parameters as important risk factors for the development of cervical dysplasia and its progression to cervical cancer. Nutritional studies and molecular techniques for analysis of HPV genotypes were introduced towards the end of the study. This study was initiated from 1988 to elucidate the role of HPV along with other possibly preventable co-factors in the process of cervical carcinogenesis and management of precancerous and early cancerous lesions of the cervix. It is proposed to register a cohort of 400 mild-to-moderate dysplasia subjects and 800 age-matched controls, along with their male spouses, for follow-up of three to six years. The end-point of the study is progression to severe dysplasia or carcinoma in situ. So far 80 women with confirmed dysplasia, along with 160 controls, have been recruited. The subjects are recruited from women attending gynaecology outpatient departments of eight collaborating hospitals from the metropolitan city of Delhi.

TYPE: Cohort
TERM: Cytology; Diet; HPV; HSV; Premalignant Lesion; Sexual Activity
SITE: Uterus (Cervix)
TIME: 1988 - 1995

Pune

439 **Vaidya, R.** 04975
Armed Forces Medical College, Dept. of Preventive and Social Medicine, Sholapur Rd, Pune 411040, India
COLL: Ghosh, M.K.

Cancer of the Upper Aerodigestive Tract

The aim of this hospital-based study is to improve understanding of the epidemiology of carcinomas of the upper aerodigestive tract among Indian armed forces personnel and their families undergoing treatment at the Malignant Diseases Treatment Centre in Pune. The role of aetiological factors such as smoking, alcohol consumption, use of snuff, dietary habits and state of oral hygiene in upper aerodigestive cancers will be examined. An attempt will be made to analyse the factors which result in delay in reporting to medical centres after onset. Approximately 100 cases and an equal number of controls will be studied. The controls will be patients admitted to the same hospital, matched for age, sex and socio-economic status, who have never suffered from any malignancy. All cases and controls will be personally interviewed, using a specially designed and tested questionnaire.

TYPE: Case-Control
TERM: Alcohol; Diet; Hygiene; Tobacco (Smoking); Tobacco (Snuff)
SITE: Hypopharynx; Larynx; Oesophagus; Oral Cavity; Oropharynx
TIME: 1991 - 1994

Trivandrum

440 **Krishnan Nair, M.** 05075
Regional Cancer Centre, Medical College Campus, Trivandrum 695011, Kerala, India (Tel.: +91 471 71904; Fax: 65347 ; Tlx: 435413 td in)
COLL: Padmanabhan, T.K.; Gangadharan, P.; Sreedevi Amma, N.; Ramachandran, T.P.; Sankaranarayanan, R.

Cancer Registry in an Area of High Natural Radiation

Radiation-emitting sands in the coastal taluk Karungappalli, Kerala, South India, have caused much concern, especially about carcinogenesis. A population-based cancer registry is monitoring cancer incidence in the taluk since 1990. The population covered is 383,326. Each individual is personally interviewed to collect socio-economic, occupational, lifestyle and dietary data. Agriculture, coir making, fishing and cashew processing are the major occupations. The area of high natural background radiation (annual average 700 millirems) involves a population of 100,000. Radiation measurements are undertaken to estimate the individual doses received by the residents. Cancer incidence and mortality in the population of the radiation exposed area will be compared with that of the population in the rest of the taluk.

TYPE: Incidence; Registry
TERM: Radiation, Ionizing
SITE: All Sites
TIME: 1990 - 2000

441 **Parukutty Amma, K.** 05079
Regional Cancer Center, Medical College Campus, P.O. Box 2417, Trivandrum 695011 Kerala, India (Tel.: +91 471 74541/71541; Fax: 165347 ; Tlx: 435413 TD IN)
COLL: Sankaranarayanan, R.; Cherian, V.; Padmakumary, G.; Rajeevkumar, S.; Krishnan Nair, M.; Gangadharan, P.

Childhood Cancer in Kerala

Every year, about 200 cases of childhood cancer are registered by the hospital registry in the Regional Cancer Centre, Trivandrum. This project aims to study the relative frequency of various cancers, sex ratio, age distribution, yearly trends and survival. Efforts are under way to organise a statewide Paediatric Cancer Regsitry to record clinical, pathological, therapeutic and survival data on all children newly diagnosed with cancer. The descriptive epidemiology of childhood cancers registered during 1983-1989 has been published (Lancet 2:455-456, 1991) and the survival experience of patients with ALL, lymphoma, Wilms' tumour, rhabdomyosarcoma and neuroblastoma have been reported (Am. J. Paediatric Haemat. Oncol. 1993).

TYPE: Relative Frequency
TERM: Sex Ratio; Survival; Trends
SITE: Childhood Neoplasms
TIME: 1982 - 1996

442 **Sankaranarayanan, R.** 05084
Regional Cancer Centre, Medical College Campus, Trivandrum 695011, Kerala, India (Tel.: +91 471 71904; Fax: 65347 ; Tlx: 435413 TD IN)
COLL: Krishnan Nair, M.; Padmakumary, G.; Cherian, V.; Duffy, S.W.; Day, N.E.; Gangadharan, P.

Risk Factors for Breast Cancer in Kerala

Breast cancer, which accounts for 21% of cancers in women in Kerala, is a major malignancy, revealing an increase in occurrence. The aim of this unmatched case-control study is to evaluate the relationship between socio-economic status, reproductive factors, diet history and menstrual status. A detailed questionnaire is administered to cases and controls concerning reproductive history, age at menarche and marriage, parity and diet history. Cases (n = 1,000) will be women diagnosed with breast cancer at the Regional Cancer Centre, Trivandrum from 1990-1992. Controls (n = 2,000) will be selected among apparently healthy visitors of patients with non-malignant conditions at the Medical College, Trivandrum.

TYPE: Case-Control
TERM: Diet; Menstruation; Parity; Reproductive Factors; Socio-Economic Factors
SITE: Breast (F)
TIME: 1990 - 1994

443 **Sankaranarayanan, R.** 05085
Regional Cancer Centre, Medical College Campus, Trivandrum 695011, Kerala, India (Tel.: +91 471 71904; Fax: 65347 ; Tlx: 435413 TD IN)
COLL: Krishnan Nair, M.; Padmakumary, G.; Cherian, V.; Duffy, S.W.; Day, N.E.; Gangadharan, P.

Cervical Cancer in Kerala

Cervical cancer is the most common cancer diagnosed among women at the Regional Cancer Centre, Trivandrum (23%). In this unmatched case-control study, a detailed questionnaire will be administered to 1,500 cervical cancer cases to elicit information on education, socio-economic status, familial cancer history, marital history, number of pregnancies, age at first coitus and pregnancy, number of sexual partners, genital hygiene, diet and promiscuity of the partner. 2,000 controls will be selected at the time of diagnosis of the case from visitors to patients with non-malignant conditions at the Medical School, Trivandrum. Unconditional logistic regression will be used to analyse the data.

TYPE: Case-Control
TERM: Diet; Education; Hygiene; Parity; Sexual Activity; Socio-Economic Factors
SITE: Uterus (Cervix)
TIME: 1990 - 1994

444 **Sankaranarayanan, R.** 05086
Regional Cancer Centre, Medical College Campus, Trivandrum 695011, Kerala, India (Tel.: +91 471 71904; Fax: 65347 ; Tlx: 435413 TD IN)
COLL: Krishnan Nair, M.; Nair, P.P.; Mathew, B.; Sudhakaran, P.R.; Menon, R.; Varghese, C.; Padmavathy Amma, B.; Sreedevi Amma, N.

Evaluation of the Chemopreventive Potential of Vitamin A, Beta Carotene, Spirulina Algae and Vitamin E in Oral Precancerous Lesions and Oral Cancer

Chemopreventive potential of vitamin A (300,000 iu/week, administered orally to 55 individuals for 6 months), beta carotene (360 mg/week orally for 6 months; n = 57) and Spirulina Algae, a blue-green alga rich in micronutrients (1 g daily for 1 year; n = 60), have been evaluated in placebo-controlled clinical trials in subjects with oral leukoplakia and submucous fibrosis in Kerala, India. Preliminary analysis revealed a complete regression rate of leukoplakia varying from 33% to 55% in the different trial arms. More than 80% of the subjects were available for re-evaluation. The complete responders are now followed up for recurrences. Vitamin E (800 iu/daily, orally for 6 months; n = 60) is currently being evaluated. These small intervention trials aim (1) to evaluate several biomarkers, and (2) to identify the most suitable active agent with the least toxicity on prolonged administration to be used in a larger trial with sufficient power and oral cancer as end point.

TYPE: Intervention
TERM: Biomarkers; Premalignant Lesion; Vitamins
SITE: Oral Cavity
CHEM: Beta Carotene
TIME: 1991 - 1994

445 **Sebastian, P.** 05028
Regional Cancer Centre, Medical College Campus, Trivandrum 695011, Kerala, India (Tel.: 72963, 76799; Fax: 65347 ; Tlx: 43413 td in)
COLL: Bhattathiri, V.N.; Sreelekha, T.T.; Ramani, P.; Vijayakumar, T.; Cherian, T.; Krishnan Nair, M.

Plasma Glutathione - A Tumour Marker in Oral Cancer

Plasma glutathione levels have been reported to be low in malignancy, returning to normal levels after successful treatment. Serial estimations of plasma glutathione may therefore show alterations before the recurrence becomes manifest clinically. The low values of plasma glutathione in cancer are probably due to high levels of gamma-glutamyl transpeptidase enzyme associated with several tumours. This study aims to evaluate the role of plasma glutathione as a tumour marker in oral cancer, the commonest cancer in India. The plasma glutathione levels will be estimated before treatment in 50 oral cancer patients undergoing radical radiotherapy, 50 patients undergoing surgery and in 50 age- and sex-matched healthy controls. The post-treatment glutathione values will be estimated two months after treatment and then every three months for two years. Plasma glutathione values before and after treatment will be correlated to clinical status of the disease and will be compared to the levels in the controls. The study is to be done over a period of three years.

TYPE: Case-Control
TERM: Tumour Markers
SITE: Oral Cavity
TIME: 1989 - 1994

Varanasi

446 **Mohapatra, S.C.** 05214
Banaras HIndu Univ., Inst. of Medical Sciences, Dept. of Preventive and Social Medicine, BHU Campus, Varanasi 221005, India (Tel.: 42668)
COLL: Meenakshi, M.; Shukla, H.S.

Nutritional Factors in Breast Cancer

This hospital-based case-control study will examine the role of nutritional factors in breast cancer. Approximately 200 cases of breast cancer will be registered between March 1989 and February 1992. Controls will be selected from other surgical units, matched by age, sex, geographical area, parity and socio-economic status. Each case will be matched with one normal control and with one woman with benign breast disease. Nutritional data will be obtained by use of standardized utensils which contain known weights of raw and cooked foods prevalent in the study area. Local nutrient conversion tables will be used. Serum levels of certain amino-acids will also be recorded.

TYPE: Case-Control
TERM: BMB; DNA; Diet; Nutrition; Serum
SITE: Breast (F)
TIME: 1989 - 1994

Semarang

447 **Sarjadi** 04034
Diponegoro Univ., Medical Faculty, Kariadi Teaching Hosp., Dept. of Pathology, Jalan dr. Soetomo no. 16, Semarang 50231, Indonesia (Tel.: +91 24 311476/31152)
COLL: Indrawijaya,; Tjahjono; Sugondo, T.

Minimum Incidence Rates of Cancer in the Semarang City Population, Based on Microscopically Diagnosed Cancers

Population-based cancer data are still difficult to obtain in Indonesia. Starting in 1970, minimum incidence rates of cancer have been calculated in the Semarang City population, based on cases microscopically diagnosed in the Department of Pathology, Diponegoro University. Despite being based only on histopathologically confirmed cases, the age-standardized rates for nasopharynx and cervix cancer were high, both exceeding those observed in Singapore Malays. Rates for cervix cancer were similar to those observed in Hongkong and Rangoon. An increase in the incidence of lung cancer from 1970 to 1981 was also observed. Data for 1981-1985 were published in 1987. This is a cooperative study between the Indonesian Cancer Society and the Dept. of Pathology, Diponegoro University.

TYPE: Incidence
TERM: Geographic Factors
SITE: Lung; Nasopharynx; Uterus (Cervix)
TIME: 1970 - 1994

448 **Sarjadi** 04758
Diponegoro Univ., Medical Faculty, Kariadi Teaching Hosp., Dept. of Pathology, Jalan dr. Soetomo no. 16, Semarang 50231, Indonesia (Tel.: +91 24 311476/31152)
COLL: Indrawijaya,; Tirtosugondo; Kasno

Cancer Incidence 1985-1994 in Semarang

This project involves the creation of the first population-based cancer registry in Indonesia, for the city of Semarang. Data will be collected from all health care facilities in the city, and incidence rates will be derived. The aim is to generate a database for epidemiological research, to identify the most frequent malignancies and to provide the information required for health care planning and cancer control activities.

TYPE: Incidence
TERM: Registry
SITE: All Sites
TIME: 1985 - 1994

Yogyakarta

449 **Soeripto** 04641
Gadjah Mada Univ., Faculty of Medicine, Dept. of Pathology, Jl. Kesehatan Sekip , Yogyakarta, Indonesia (Tel.: 902466)
COLL: Prijono Tirtoprodjo; Achmad Ghozali; Irianiwati

Cancer Incidence in Yogyakarta: Population

Population-based cancer registration at Srandakan district, Bantul county, Yogyakarta is being developed. The aim of this study is to determine cancer incidence in Yogyakarta town during the period 1994to 1996, through population-based cancer registration. This is a combination of hospital-based registration, department of pathology-based registration and population survey. The form containing identifying information, clinical diagnosis, and histopathological diagnosis will be distributed to the health centres, private practice physicians, and hospitals in both areas. Methods of analysis will be age-adjusted standardization, using as the denominator the combined population in the area.

TYPE: Incidence; Registry
TERM: Registry
SITE: All Sites
TIME: 1994 - 1996

450 **Soeripto** 04642

Gadjah Mada Univ., Faculty of Medicine, Dept. of Pathology, Jl. Kesehatan Sekip , Yogyakarta, Indonesia (Tel.: 902466)

COLL: Endang Soetristi; Irianiwati

Incidence of Childhood Cancer in Yogyakarta

The objective of the study is to determine the incidence of childhood cancer in Yogyakarta with hospital-based cancer registration. All childhood cancer patients at the hospitals in this area were collected from medical records. Patients from outside Yogyakarta were excluded. Age-adjusted standardisation will be used for analysis. The childhood population of Yogyakarta will be used for denominator. Data have been published in "International Incidence of Childhood Cancer", Parkin D.M., Stiller C.A., Draper G.J., Bieber C.A., Terracini B. and Young J.L. (Eds). IARC Sci. Publ. No 87, Lyon, France, 1988.

TYPE: Incidence
TERM: Childhood; Clinical Records; Registry
SITE: All Sites
TIME: 1989 - 1998

Dublin

451 **Bourke, G.J.** 05016
Univ. College of Dublin, Dept. of Community Medicine and Epidemiology, Earlsfort Terrace, Dublin 2, Ireland (Tel.: 2693244/6345; Fax: 1754568)
COLL: Daly, L.; Herity, B.

Total and Cancer Mortality in Research Workers in Agriculture

The aim of this historical cohort study is to ascertain if employees in an agricultural research institute have an excess all-cause or cancer mortality compared to the general population and, if so, to identify groups of workers at increased risk. The study cohort consists of approximately 1,500 administrative, research and technical staff who were employed for at least two consecutive years in the institute from 1960 to 1986. Mortality follow-up to the end of 1988 is done with personnel records, and causes of death will be determined from the Registrar General's records. Statistical analysis of the data will be based on person-years of exposure, with the use of regression techniques where appropriate. The study cohort forms part of the IARC International Study of Cancer Risk in Biology Research Laboratory Workers.

TYPE: Cohort
TERM: Occupation
SITE: All Sites
OCCU: Administrative Workers; Laboratory Workers
TIME: 1989 - 1994

ISRAEL

Beer Sheva

452 **Goldsmith, J.R.** 04746
Ben Gurion Univ. of the Negev, Faculty of Health Sciences, P.O. Box 653, 84120 Beer Sheva, Israel (Tel.: +972 57 400876; Fax: 277342; Tlx: 5253 unast il)
COLL: Sobel, S.; Arnon, L.

Childhood and Aricultural Chemicals in the Negev

The possible associations of cancer incidence with birth defects and genetic abnormalities are being investigated. Two kibbutzim (co-operative settlements) with increased incidence of cancer have been detected and a case-referent study is now being conducted, using non-cancer residents of the kibbutzim as one referent population and residents of adjacent kibbutzim of the same age group, sex, and country of origin as a second referent population. Of special concern is the possible exposures to agricultural chemicals, such as herbicides and pesticides. The area includes a large chemical industry complex and waste dump as well as a nuclear research facility.

TYPE: Case-Control
TERM: Cluster; Congenital Abnormalities; Ethnic Group; Herbicides; Occupation; Pesticides
SITE: All Sites
OCCU: Farmers
REGI: Israel (Isr)
TIME: 1989 - 1994

***453** **Goldsmith, J.R.** 05500
Ben Gurion Univ. of the Negev, Faculty of Health Sciences, P.O. Box 653, 84120 Beer Sheva, Israel (Tel.: +972 57 400876; Fax: 277342; Tlx: 5253 unast il)
COLL: Kordysh, E.; Quastel, M.; Lubin, F.

Cancer and Other Conditions among Russian Immigrants Exposed at Chernobyl

This study will evaluate the cancer risk in a group of 1,200 Russian Jewish immigrants exposed to radiation during the Chernobyl nuclear accident in April 1986 and will offer these individuals councelling. Gradients of Cs-137 have been measured and demonstrated in the body burden, depending upon the occupation and place of residence of the subjects at the time of the accident. Almost half of the children in this group have palpable thyroid swellings and about 10% have nodular goiters. This population is being monitored for cancer incidence in collaboration with the Israel Cancer Registry.

TYPE: Cohort; Incidence
TERM: Clinical Effects; Migrants; Occupation; Radiation, Ionizing; Registry
SITE: All Sites; Thyroid
OCCU: Radiation Workers
REGI: Israel (Isr)
TIME: 1991 - 1996

454 **Odes, S.** 04946
Soroka Medical Center, Gastroenterology Unit, P.O. Box 151, 84101 Beer Sheva, Israel (Tel.: +972 57 660242; Fax: 74696)

Oesophageal Carcinoma in Indian Jews in Israel

A prospective cohort study of oesophageal cancer is being carried out in the Indian Jewish (n = 8,461) population of Southern Israel. The study district has a total population of 450,700, and the cohort includes 70% of all Jewish migrants from India. Cases are obtained from clinical, radiological, endoscopic and pathology records at the hospitals and clinics and from the Central Cancer Registry in Jerusalem. Interviews with patients and/or relatives provide data about aetiological factors, including diet. Incidence data are age-adjusted to the population of Israel. A recent report (J. Clin. Gastroenterol. 12(2):222-227, 1990) for the period 1961 to 1985 shows a significantly higher mean age-adjusted incidence rate in immigrant Indian males, 6.5 (per 100,000), and Indian females, 17.2, than non-Indian males, 2.7 and non-Indian females, 2.1. The mean female to male ratio was: Indian 1.75, non-Indians 0.59. The age at diagnosis was lower in Indian females (54.6 years) than Indian males (66.2 years) Risk factors appear to be poverty in all Indians, smoking in men and spicy diets in women. The study is expected to continue for the next 20 years.

TYPE: Cohort
TERM: Diet; Indians; Lifestyle; Tobacco (Smoking)
SITE: Oesophagus
REGI: Israel (Isr)
TIME: 1977 - 1997

455 **Odes, S.** 05121
Soroka Medical Center, Gastroenterology Unit, P.O. Box 151, 84101 Beer Sheva, Israel (Tel.: +972 57 660242; Fax: 74696)
COLL: Krugliak, P.; Fraser, G.

Frequency of Colonic Adenomas and Cancer in High, Medium and Average Risk Populations in Southern Israel

A screening programme for colorectal adenomas and cancer has operated from 1983-1987 and again from 1990 in the population (age-group 40-70) in Beer Sheba in southern Israel in order to determine the frequency of positive findings and the efficacy of faecal occult blood testing, sigmoidoscopy and colonoscopy. Screenees are interviewed to determine the level of their risk according to the criteria of Rozen et al. (Front. Gastrointest. Res. 10:164, 1986). The categories are: high risk - previous colorectal tumours (I) or inflammatory bowel disease (II); moderate risk - female genital tumour (III) or family history of colorectal tumour (IV); average risk - asymptomatic persons over the age of 40 (V). In 1983-1987 the number of screenees in these categories was 22, 11, 11, 77 and 980 respectively. In the total group there were 41 cases wih polyps and 1 case with colon cancer. Faecal occult blood tests are done on the whole screened population, while total colonoscopy is offered to people in groups I and II and flexible sigmoidoscopy carried out in groups III, IV and V. The faecal occult blood test had poor sensitivity and positive predictive value compared with sigmoidoscopy and colonoscopy. The study was reactivated in 1990, including 750 new screenees.

TYPE: Case Series
TERM: High-Risk Groups; Polyps; Screening
SITE: Colon; Rectum
TIME: 1983 - 1994

Haifa

456 **Rennert, G.** 04194
National KH Center for Cancer Control, Dept. of Community Medicine & Epidem., Carmel Medical Center, 7 Michal St., Haifa 34362, Israel (Tel.: +972 4 250474; Fax: 344358)

Time Trends in Lung Cancer Incidence among Jewish Immigrants to Israel

This study is designed to evaluate the trends in incidence rates of lung cancer among Jews who have immigrated to Israel from different countries in the world. The incidence rates will be compared to rates in the country of origin (country of birth) and to the total rates in Israel according to the length of time which has elapsed since immigration. The cases will include all 9,894 Jews born outside of Israel with lung cancer diagnosed in Israel after immigration and reported to the Israel Cancer Registry between the years 1962-1982. This study will be limited by the availability of rates of lung cancer in the corresponding countries from which the cases emigrated.

TYPE: Incidence
TERM: Ethnic Group; Migrants; Registry; Time Factors
SITE: Lung
REGI: Israel (Isr)
TIME: 1986 - 1994

457 **Rennert, G.** 04911
National KH Center for Cancer Control, Dept. of Community Medicine & Epidem., Carmel Medical Center, 7 Michal St., Haifa 34362, Israel (Tel.: +972 4 250474; Fax: 344358)
COLL: Ben Harush, M.

Childhood Malignancies in Northern Israel

The study is designed to evaluate possible risk-factors for childhood malignancies among historical cases treated in the only paediatric oncology facility in Northern Israel. Age- and sex-matched hospitalized controls will be assigned to the cases. Paternal and maternal (including pregnancy-related) factors will be studied as well as various childhood exposures. The estimated number of cases is about 400. Hospital controls will be recruited prospectively and matched on same current age as cases to ensure similar number of years of recall.

TYPE: Case-Control
TERM: Intra-Uterine Exposure; Occupation
SITE: Childhood Neoplasms
TIME: 1990 - 1994

458 **Rennert, G.** 04913
National KH Center for Cancer Control, Dept. of Community Medicine & Epidem., Carmel Medical Center, 7 Michal St., Haifa 34362, Israel (Tel.: +972 4 250474; Fax: 344358)

Risk Factors and Protective Factors for Cancer - A Prevalence Study

A sample of 3,000 Israeli adults of both sexes will be studied in Haifa. Information will be obtained through home interviews on smoking, diet, sun-exposure and reproductive factors, occupational history and history of radiation exposure, as well as on family history of cancer. The study population will be identified through stratified sampling (by sex and age groups 30-44 and 45-65) of the list of all people insured by the major insurance company in Israel (covering more than 80% of the population) and will be followed up for changes in habits and occurrence of cancer. Analysis will include estimating prevalence of certain health habits with comparisons made between the studied groups in the different follow-up periods.

TYPE: Cross-Sectional
TERM: Diet; Familial Factors; Occupation; Radiation, Ionizing; Radiation, Ultraviolet; Reproductive Factors; Tobacco (Smoking)
SITE: All Sites
TIME: 1990 - 1995

***459** **Rennert, G.** 05411
National KH Center for Cancer Control, Dept. of Community Medicine & Epidem., Carmel Medical Center, 7 Michal St., Haifa 34362, Israel (Tel.: +972 4 250474; Fax: 344358)
COLL: Kitzes, R.

Debrisoquine Metabolism Phenotype Prevalence in Jewish Lung Cancer Patients and Healthy Controls

Lung cancer incidence in Israel is very low in spite of significant smoking prevalence. This study will investigate the possibility of some genetically-mediated resistence to lung cancer in the Israeli population by studying the metabolism of debrisoquine, an inducer of some relevant CYT P450 enzymes, after a single administration. The investigation will include about 200 lung cancer patients (which will also go through a detailed risk factor interview) and 100 healthy volunteers (smokers and non-smokers). Prevalence rates of extensive and poor metabolizers will be calculated in both groups and stratified according to smoking habits.

TYPE: Case-Control; Molecular Epidemiology
TERM: Drugs; Enzymes; Genetic Factors; Prevalence
SITE: Lung
TIME: 1993 - 1996

***460** **Rennert, G.** 05412
National KH Center for Cancer Control, Dept. of Community Medicine & Epidem., Carmel Medical Center, 7 Michal St., Haifa 34362, Israel (Tel.: +972 4 250474; Fax: 344358)
COLL: Almog, R.

Hepatitis as a Cause of Hepatocellular Carcinoma in Israel

The aim is to study the relative role of hepatitis B and C in the causation of HCC in Israel with a population of immigrants from developed and developing countries. It is hypothesized that the attributable risks of HCC due to HBV/HCV will be lower than those seen in southeast Asia or Africa, but higher than in the western world. A case-control study of all HCC cases diagnosed in Israel during a period of two years (about 200 cases) and a 1:2 matched population control group will be carried out. Information on hepatitis, alcohol, hormones, diet etc. will be collected via questionnaires and serum samples (hepatitis markers) and tissue from liver biopsies will be obtained in the cases. Data analysis will include conditional logistic regression.

TYPE: Case-Control
TERM: Alcohol; Diet; HBV; HCV; Hepatitis; Hormones; Migrants
SITE: Liver
TIME: 1993 - 1996

***461 Rennert, G.** 05413
National KH Center for Cancer Control, Dept. of Community Medicine & Epidem., Carmel Medical Center, 7 Michal St., Haifa 34362, Israel (Tel.: +972 4 250474; Fax: 344358)

Risk and Prognostic Factors in Breast Cancer - Data from the National Mammography Project

This cohort study will evaluate a wide range of risk factors postulated in the aetiology of breast cancer among all women participating in the national mammography project in Israel. All women are interviewed on reproductive history, family history, use of oral contraceptives and hormone replacement therapy, alcohol consumption, body mass index, smoking etc. All pertinent radiology, pathology, surgery and oncology reports are computerized for each participant in the context of the follow-up, quality control and evaluation programme of the mammography project. About 100,000 women are anticipated to enter the database in the first year. Data will be analysed to yield relative risk estimates for malignant and benign breast lesions as detected in the programme during the follow-up period.

TYPE: Cohort
TERM: Alcohol; Family History; Hormones; Mammography; Oral Contraceptives; Reproductive Factors; Surgery; Tobacco (Smoking)
SITE: Breast (F)
TIME: 1992 - 2000

***462 Rennert, G.** 05414
National KH Center for Cancer Control, Dept. of Community Medicine & Epidem., Carmel Medical Center, 7 Michal St., Haifa 34362, Israel (Tel.: +972 4 250474; Fax: 344358)
COLL: Shapiro, S.; Rennert, H.

Israeli Chernobyl Health Effects Study

This study is planned to evaluate prospectively adverse health effects, mainly malignant disease, among immigrants to Israel from areas known to have been exposed to radiation from the Chernobyl accident: Belarus, Ukraine and Russia. The study group includes all eligible Russian immigrants who registered with the study center (currently about 6,000). Two control groups of similar size are also studied; one of immigrants from the irradiated area, who did not register with our study center (to control for selection bias) and one of immigrants from Russian areas not exposed to radiation (to serve as a reference group). The study group provided demographic information, detailed residential history since the time of accident, any available exposure data and data on disease prevalence before 1986 and disease incidence thereafter. The study group and two control groups will be followed-up by computer linkage with computerized national hospitalization data and national mortality data in Israel. The primary care clinic records of the participants will also be investigated. Data analysis will measure the risk magnitude of malignant diseases following exposure to radiation (with crude dose stratification) using multivariate analysis.

TYPE: Cohort
TERM: Geographic Factors; Migrants; Radiation, Ionizing
SITE: Brain; Breast (F); Leukaemia; Lung; Thyroid
TIME: 1992 - 2000

463 Robinson, E. 04312
RAMBAM Medical Center, TECHNION-Israel Inst. of Technology, Northern Israel Oncology Center, P.O.B. 9602, Haifa 31096, Israel (Tel.: +972 4 517581; Fax: 514481)
COLL: Rennert, G.; Nasrallah, S.; Adler, Z.; Neugut, A.

Clinical Characteristics of Patients with Multiple Primary Neoplasms

Multiple primary neoplasms (MPNs) constitute almost 10% of all cancer seen in the USA. Very little is known about how they differ from first primary neoplasms. It is the purpose of the proposed investigation to explore the stage and grade distribution, and survival of various MPNs and to compare these characteristics to those of first primary neoplasms. Three population-based cancer registries will be studied. At Columbia, the Connecticut Tumor Registry and other SEER Registry data will be analysed, while the Israel Cancer Registry data will be analysed at the Technion in Israel. For each registry, second neoplasms of interest will be compared to first neoplasms at the same site for the characteristics listed above. For example, the survival of secondary acute leukaemia will be compared to the survival of de novo acute leukaemia. The use of both an American and an Israeli registry will allow generalizability of results and confirmation of important findings. Second neoplasms to be studied will be those that occur at a rate greater than expected by chance. A monograph detailing these results will be prepared. Knowledge about differences in clinical behavior may lead to changes in therapeutic approach; to more aggressive surveillance for second neoplasms; and to further clues about the biology of carcinogenesis. Recent papers appeared in Cancer Detect. Prev. 16:297-303, 1992, Cancer J. 5:245-248, 1992 (Neugut, Al) and in Cancer 71:172-176, 1993.

TYPE: Cohort
TERM: Multiple Primary; Registry; Stage; Survival
SITE: Breast (F); Colon; Head and Neck; Leukaemia; Lymphoma; Thyroid
LOCA: Israel; United States of America
REGI: Connecticut (USA); Israel (Isr)
TIME: 1987 - 1994

Jerusalem

464 **Iscovich, J.M.** 04834
Ministry of Health, Dept. of Epidemiology, Israel Cancer Registry, 107 Hebron Rd , 93480 Jerusalem, Israel (Tel.: +972 2 247172; Fax: 381772)
COLL: Steinitz, R.; Robinson, R.; Rennert, G.

Contralateral Breast Cancer: Definition and Implication for Breast Cancer and Incidence Rates.
The question of how to distinguish contralateral (bilateral) breast cancer from spread or multicentric site of the disease and its implications for incidence rates is being investigated in the Israel Cancer Registry. From a cohort of 22,653 female Jews with first breast cancer diagnosed in Israel during the period 1960-1986 and followed until August 1988, 1,395 women have subsequently developed contralateral disease. A review of the pathological and medical reports available in the Israel Cancer Registry is being undertaken, using the following criteria (separately or in combination): stage at first diagnosis, time elapsed between diagnoses, histology and appearance of metastasis in relation to the date of the second breast cancer diagnosis.

TYPE: Cohort
TERM: Histology; Multiple Primary; Registry; Stage; Time Factors
SITE: Breast (F)
REGI: Israel (Isr)
TIME: 1988 - 1994

465 **Kark, J.D.** 03915
Hebrew Univ., Faculty of Medicine, Hadassah SPH & Community Medicine, Dept. of Social Medicine, P.O. Box 1172, Ein Karem, 91000 Jerusalem, Israel (Tel.: +972 2 447113)
COLL: Goldbourt, U.; Levine, C.; Wahrendorf, J.; Martinsohn, C.

Ethnic, Dietary, Clinical, Socio-Demographic and Biochemical Risk Factors for Cancer
Cancer incidence is being studied in a cohort of 10,000 civil servants and municipal employees first examined in 1963. 1,200 cases of cancer have been identified though the Israel National Cancer Registry, matched by name and identity card number. Histology is available for over 80%. The goals of the project are to identify ethnic, dietary, clinical, socio-demographic and biochemical risk factors for cancer. The large number of cases should permit analysis of incidence by site as well as by histological sub-type. A peculiar aspect of the analysis is the examination of religious orthodoxy (assessed through type of religious education as well as through self-definition) in subjects eventually contracting cancer compared with those without cancer. A number of case-control studies are envisaged to investigate: coffee and cigarette smoking synergism in lung cancer; fruit and vegetable consumption and lung cancer; dietary fat and colon cancer. In a later stage, reexamination of some 20,000 frozen sera (from the study examination of 1963, 1965 and 1968) is envisaged, in order to correlate cancer incidence to various parameters not evaluated earlier in the study period. Given the number of cases and the availability of dates of diagnosis, factors associated with survival of cancer patients might also be elucidated. A paper on the association of blood pressure and cancer mortality has been published in J. Nat. Cancer Inst. 77:63-70, 1986.

TYPE: Case-Control; Cohort
TERM: Coffee; Diet; Ethnic Group; Fat; Fruit; Lipids; Registry; Religion; Socio-Economic Factors; Tobacco (Smoking); Vegetables
SITE: All Sites; Colon; Lung
REGI: Israel (Isr)
TIME: 1985 - 1994

466 **Steinitz, R.** 03852
Israel Cancer Registry, Ministry of Health, Dept. of Epidemiology, 20 King David St. , 91000 Jerusalem, Israel (Tel.: +972 2 247172; Fax: 381772)
COLL: Iscovich, J.M.

Multiple Primary Malignancies in the Israel Cancer Registry

The Israel Cancer Registry's system of registering each tumour separately and linking tumours in the same patient by a common accession number in the index file carrying the demographic information, offers the possibility of studying the incidence of cancer at multiple sites. Approximately 4,000 patients with multiple primary malignancies will be available for the study, not including basal cell and squamous cell carcinoma of the skin. Expected number of cases will be established by observing person-years of exposure to the risk of developing additional primary cancers, and then applying the appropriate incidence rates for each site, sex, age group and continent of origin. From this, the relative risk of developing additional primary cancers will be calculated.

TYPE: Cohort
TERM: Multiple Primary; Registry
SITE: All Sites
REGI: Israel (Isr)
TIME: 1985 - 1994

Tel Aviv

467 **Chaitchik, S.** 04190
Tel Aviv Medical Center, Int. of Oncology, Ichilov Hosp., Dept of Oncology, 6 Weizman St., Tel Aviv 64239, Israel (Tel.: +972 3 210453; Tlx: 342298 tlvmc)
COLL: Shenberg, C.; Biran, T.; Mantel, M.; Weininger, J.

Evaluation of the Levels of Selenium in a Population with Breast Cancer and a Normal Population at Risk

The purpose of the study is to evaluate levels of selenium in patients with breast cancer. The patients are referred to the Tel-Aviv Medical Center for diagnosis and surgery and will be compared with a matched normal population. 60 patients with breast cancer will be included in the study and about 120 women at risk will be evaluated using a highly sensitive XRF method. A preliminary study was carried out in order to compare the selenium concentration in breast cancer patients and healthy subjects (controls). The plans are to continue until at least 120 humans have been compared.

TYPE: Case-Control
TERM: Biochemical Markers
SITE: Breast (F)
CHEM: Selenium
TIME: 1987 - 1994

468 **Chaitchik, S.** 04686
Tel Aviv Medical Center, Int. of Oncology, Ichilov Hosp., Dept of Oncology, 6 Weizman St., Tel Aviv 64239, Israel (Tel.: +972 3 210453; Tlx: 342298 tlvmc)
COLL: Shenberg, C.; Biran, T.; Mantel, M.; Weininger, J.

Serum Selenium and Colo-Rectal Cancer

The aim of this study is to evaluate serum selenium levels in patients with colo-rectal cancer. Analyses will be performed on surgical specimens, both on tumour tissue and adjacent normal tissue. In a preliminary study of breast cancer cases and healthy controls, serum selenium was significantly lower in cases (weighted mean 0.076 ppm, SE 0.014) than in controls (0.119, SE 0.023), and selenium level was inversely related to clinical stage. In the pilot study, 20 patients with colo-rectal cancer will be studied.

TYPE: Case-Control
TERM: Biochemical Markers; Trace Elements
SITE: Breast (F); Colon; Rectum
CHEM: Selenium
TIME: 1991 - 1994

469 **Kahan, E.** 03763
Tel Aviv Univ., Sackler Sch. Med., Inst. of Occupational Health, Dept. of Epidemiology, University St., Ramat Aviv, 69978 Tel Aviv, Israel (Tel.: +972 3 425827)
COLL: Hare, C.

Survey of Latent Carcinoma of Prostate in Israel

Carcinoma of prostate (CP) is defined as latent (LCP) when detected in surgical specimens from patients clinically not suspected, or at autopsy. In post mortems and biopsies carried out during the last decade, a very high frequency of LCP was found. The Jewish population in Israel comprises people originating from over 100 countries. A previous study showed that the annual incidence of CP in Israel's population over 50 years old is 72 per 100,000, but in two groups there is a significant difference. Jews originating from Greece and Bulgaria showed a high annual rate: 103.9; Jews originating from Yemen had an extremely low rate: 31.4. The present survey aims to collect demographic information about LCP in order to compare prevalence rates of LCP by country of origin with the existing incidence rate for CP in the population. If the incidence rate differences between Jews from various origins is the expression of different exposures to risk, similar proportional differences in prevalence of LCP in Jews, by country origin, would be expected.

TYPE: Cross-Sectional
TERM: Autopsy; Ethnic Group; Prevalence
SITE: Prostate
TIME: 1986 - 1994

470 **Kahan, E.** 04459
Tel Aviv Univ., Sackler Sch. Med., Inst. of Occupational Health, Dept. of Epidemiology, University St., Ramat Aviv, 69978 Tel Aviv, Israel (Tel.: +972 3 425827)
COLL: Luria,; Derazne, E.; Peretz, H.

Physical Activities and Colon Cancer

Colon cancer is one of the most frequent malignancies in Western countries and in Israel. Studies suggest a decreased risk among high-activity occupations and among members of kibbutzim (cooperative settlements) compared to urban dwellers. This may be due to the high-activity life-style in kibbutzim. The purpose of this study is to evaluate whether there is a negative correlation between physical activity (recreation, home, work, etc.) and incidence of colon cancer. This will be a hospital-based case-control study of 200 cases and 200 matched controls. Cases will be randomly chosen among living colon cancer patients aged 25-69 years, diagnosed in the Institute of Cancer of Beilinson Medical Center, Israel, from 1986-1990. Matched controls will be chosen among living patients with cancers other than colon cancer. A questionnaire about physical activity will be performed and confounding factors, such as socio-demographic, diet components, vitamin A, family history of cancer and anthropometric characteristics will also be assessed. Data analysis will be done using logistic analysis of non-parametric data on physical activity, (moderate, intermediate and high activity), against incidence of colon cancer.

TYPE: Case-Control
TERM: Diet; Familial Factors; Physical Activity; Vitamins
SITE: Colon
TIME: 1989 - 1994

Tel Hashomer

471 **Modan, B.** 04556
Chaim Sheba Medical Center, Dept. of Clinical Epidemiology, 52621 Tel Hashomer, Israel (Tel.: +972 3 5303261; Fax: 348360)
COLL: Kleinerman, R.A.

Potential Radiocarcinogenicity of Heart Catheterisation in Childhood

This is a pilot study to determine the availability of data for a long-term follow-up study of possible carcinogenic effects among about 15,000 children who underwent cardiac catheterisation for congenital heart disease, under X-ray control, during the period 1955-1969. Dosimetry was obtained by reconstruction of old measurements and treatment plans, and cancer cases and deaths will be ascertained through registries. The objective will be to re-assess the carcinogenic effect of low-dose irradiation. Thus far +/- 12,000 records have been accrued from hospitals in the USA, United Kingdom, Netherlands and Israel. Preliminary results from Israel will be available in 1993.

TYPE: Cohort
TERM: Childhood; Radiation, Ionizing
SITE: All Sites
LOCA: Israel; United Kingdom; United States of America
TIME: 1986 - 1994

472 **Modan, B.** 05172
Chaim Sheba Medical Center, Dept. of Clinical Epidemiology, 52621 Tel Hashomer, Israel (Tel.: +972 3 5303261; Fax: 348360)
COLL: Ron, E.; Alfaudari, E.

Cancer and Benign Tumours Following Scalp Irradiation

This is a cohort study of 10,834 subjects irradiated in childhood and approximately 16,000 sibling and neighbourhood controls. The cohort is linked with the cancer registry and a central population register to identify cases and the vital status of all subjects. The aim is to estimate cancer rates and the duration of any excess risk. Recent emphasis is on low dose radiation, in view of the increased risk of thyroid and breast cancer observed thus far following low-dose exposures (Lancet 1:629-631, 1989).

TYPE: Cohort
TERM: Childhood; Latency; Radiation, Ionizing; Registry
SITE: All Sites; Benign Tumours
REGI: Israel (Isr)
TIME: 1965 - 2010

***473** **Modan, B.** 05536
Chaim Sheba Medical Center, Dept. of Clinical Epidemiology, 52621 Tel Hashomer, Israel (Tel.: +972 3 5303261; Fax: 348360)
COLL: Kaplan, S.D.; Lubin, F.; Preston-Martin, S.

Epidemiology of Brain Tumours

The main hypothesis is that nitrosamine compounds affect the development of brain tumours in conjunction with other environmentally determined factors, primarily diet. Two studies are being carried out, one in children and one in adults, based on similar questionnaires, but with different sources of cases and controls. The childhood study is based on all neurosurgery departments in the country and will include approximately 300 cases and 600 population controls. The adult study is based on one major medical centre and includes 139 cases and 278 controls. Two control groups are being used, one consisting of hospital patients from the same treatment area as the cases and one consisting of friends designated by the cases or their spouses.

TYPE: Case-Control
TERM: Childhood; Diet
SITE: Brain
CHEM: N-Nitroso Compounds
TIME: 1987 - 1994

***474** **Modan, B.** 05537
Chaim Sheba Medical Center, Dept. of Clinical Epidemiology, 52621 Tel Hashomer, Israel (Tel.: +972 3 5303261; Fax: 348360)
COLL: Lubin, F.

Epidemiology of Breast Cancer

This is a case-control study involving 1,100 newly diagnosed breast cancer cases from hospitals in central Israel and 2,200 controls (neighbourhood and hospital). Cases and hospital controls are selected by a study nurse by individual record review. Neighbourhood controls are selected through the Central Population Registry. The role of hormonal, ethnic and selected environmental factors in the aetiology of breast cancer will be studied.

TYPE: Case-Control
TERM: Environmental Factors; Ethnic Group; Hormones
SITE: Breast (F)
TIME: 1988 - 1994

***475 Modan, B.** 05538
Chaim Sheba Medical Center, Dept. of Clinical Epidemiology, 52621 Tel Hashomer, Israel (Tel.: +972 3 5303261; Fax: 348360)
COLL: Alfandry, E.; Chetrit, A.; Lusky, A.

Potential Carcinogenicity of Phototherapy in Infancy

To determine the potential carcinogenicity of phototherapy in infancy a case-control study is being carried out, based on all Israeli children <12 years of age, who developed a malignant neoplasm between 1972 and 1984. Cases are traced through the National Cancer Registry and matched to two controls for sex, birth weight, hospital and date of birth. The medical records of the cases and controls were reviewed in each of the 33 hospitals and information on phototherapy recorded. Determination of carcinogenic risk will be made and confounders assessed. Odds ratios are estimated using the Mantel-Haenszel method. Adjusted odd ratios are derived from conditional logistic analysis.

TYPE: Case-Control
TERM: Childhood; Photochemotherapy; Radiation, Ultraviolet; Registry
SITE: All Sites
REGI: Israel (Isr)
TIME: 1988 - 1994

***476 Shpilberg, O.** 05295
Chaim Sheba Medical Center, Inst. of Hematology, Bitan 27, Tel Hashomer 52621, Israel (Tel.: +972 3 5302565; Fax: 5303506)
COLL: Modan, M.; Modan, B.

Familial Aggregation of Haematological Neoplasms

This study will attempt to answer the questions: (1) is there a genetic predisposition to haematological malignancies? and (2) at what level in the haematopoietic lineage does the predisposition occur? All patients with haematological neoplasms diagnosed since 1986 will be included. Two control groups will be used: (1) all patients with non-malignant haematological neoplasms; and (2) patients with type II diabetes mellitus. A self-administered questionnaire is used, requesting a full list of 1st and 2nd degree relatives, their vital status, current age or age at death and their chronic diseases.

TYPE: Case-Control
TERM: Familial Factors; Genetic Factors
SITE: Haemopoietic
TIME: 1990 - 1994

Aviano

477 **Franceschi, S.** 05051

Aviano Cancer Centre, Epidemiology Unit, Via Pedemontana Occ., 33081 Aviano PN, Italy (Tel.: +39 434 659232; Fax: 652182)

COLL: Bidoli, E.; Guarneri, S.; Talamini, R.; Amadori, A.; Nanni, O.; Filiberti, R.; Giacosa, A.; Merlo, F.; D'Avanzo, B.; La Vecchia, C.; Negri, E.; Parazzini, F.; Decarli, A.; Ferraroni, M.; Levi, F.G.; Salvini, S.

Validation of a Dietary Questionnaire for Case-Control Studies

The aim of the present study is to evaluate the reproducibility and validity of a new food frequency questionnaire to be used in case-control studies in different parts of Italy (provinces of Pordenone, Milan, Genoa, Forli, and Naples) and the Canton of Vaud (Switzerland). The questionnaire includes 93 foods and beverages in addition to 13 questions concerning general eating patterns. This form will be administered by trained interviewers twice (once in winter and once in summer) to approximately 600 volunteers, aged 35-74, recruited in various ways (e.g. local newspapers and magazines, health organizations, open university programmes, etc.), and will be compared to the results of two one-week diet records, including detailed descriptions of each food (i.e. weight, brand, method of preparation and recipes). Comparison of nutrient intake score computed from this relatively simple questionnaire with absolute intake obtained from extensive diet record collection will be performed, both before and after adjustment for total caloric intake. Additional analysis will also be carried out in order to define further the influence of season on data collection (reproducibility), the specific foods that contribute nutrients of interest and the importance of recording portion size.

TYPE: Methodology
TERM: Diet; Nutrition
SITE: All Sites
TIME: 1991 - 1994

***478** **Franceschi, S.** 05291

Aviano Cancer Centre, Epidemiology Unit, Via Pedemontana Occ., 33081 Aviano PN, Italy (Tel.: +39 434 659232; Fax: 652182)

COLL: Decarli, A.; Ferraroni, M.; La Vecchia, C.; Negri, E.; D'Avanzo, B.; Parazzini, F.; Tavani, A.; Nanni, O.; Amadori, D.; Montella, M.; Conti, E.M.S.; Barbone, F.; Bidoli, E.; Guarneri, S.; Talamini, R.; Salvini, S.; Giacosa, A.; Filiberti, R.; Merlo, F.; Simonato, L.; Zambon, P.

Cancer of the Breast, Ovary, Colorectum and Upper Aero-Digestive Tract

The aim of this case-control study is (1) to estimate the role of dietary habits, including alcohol consumption, in the aetiology of cancer of the breast, ovary, colorectum and upper aero-digestive tract, and (2) to collect information on the independent influence of other non-dietary factors (e.g. physical activity, relevant medical history and reproductive factors). Cases will be histologically confirmed incident cases of cancer under age 75, permanent residents in the study area. It is anticipated that 2,000 breast cancer cases, 1,000 digestive tract cancer cases and 800 ovarian cancer cases will be studied. Controls will be approximately 3,000 individuals admitted to local hospitals for a wide range of acute non-neoplastic conditions. Interviews will be performed in hospital by trained interviewers. Relative risk estimates will be computed after allowance for potential confounding factors.

TYPE: Case-Control
TERM: Alcohol; Diet; Physical Activity; Registry; Reproductive Factors; Tobacco (Smoking)
SITE: Breast (F); Colon; Larynx; Oesophagus; Oral Cavity; Ovary; Rectum
REGI: Genoa (Ita); Latina (Ita)
TIME: 1992 - 1997

479 **Tirelli, U.** 04503

Centro di Riferimento Oncologico, Div. di Oncologia Medicale, Grupo Italiano Cooperat. AIDS e Tumori, Via Pedemontana Occidentale, 33081 Aviano, Italy (Tel.: +39 434 659284; Fax: 652997 ; Tlx: 652182 cerion i)

COLL: Monfardini, S.; Carbone, A.; Franceschi, S.; Diego, S.; Mauro, B.

ITALY

Malignant Tumours in Association with HIV Infection

Since 1981 AIDS has been associated with an increased number of malignancies, in particular Kaposi's Sarcoma (KS) and non-Hodgkin's lymphoma (NHL), but recently other tumours, such as Hodgkin's Disease (HD), have been observed with peculiar clinical pathological characteristics. In Italy, a Cooperative Study Group on HIV-related Tumours (GICAT) was established in 1986 (chairmen S. Monfardini and M. Moroni), and in Europe and EORTC AIDS and Tumor Study Group was established in 1992 (chairman U. Tirelli). By 1993, more than 800 HIV-related tumours had been recorded, mainly in intravenous drug abusers, in accordance with the overall epidemiology of the epidemic in Italy. In Italy, at least 15% of patients with AIDS are affected by tumours, in particular NHL and KS. Within the GICAT and the EORTC AIDS and Tummor Study Group several clinical trials have been started, while epidemiological studies, especially on HD, KS and NHL, are in progress. There is an evidence in HD of a significant increased incidence of mixed cellularity and depletion lymphocyte subtypes. Mixed cellularity subtype has also been significantly correlated with EBV.

TYPE: Incidence
TERM: AIDS; HIV; High-Risk Groups
SITE: Hodgkin's Disease; Kaposi's Sarcoma; Non-Hodgkin's Lymphoma
TIME: 1986 - 1996

Bari

480 **Assennato, G.** 04952
Univ. of Bari, Inst. of Occupational Health, Piazza Giulio Cesare, 77124 Bari, Italy (Tel.: +39 80 278216; Fax: 5575451)
COLL: Ferri, G.M.; Porro, A.; Kardhashi, A.; Misciagna, G.; Colucci, G.; Centonze, S.; Boeing, H.

Diet and Lung Cancer

The main objective of this study is to evaluate the association between specific nutrients (Vitamin A, E, D, retinol binding proteins) and incidence of lung cancer in a cohort of 20,000 subjects aged 50years or older, of both sexes and without cancer, living in five health districts of the Province of Bari in southern Italy. The cohort is selected from the Biochemical Laboratory of the Public and Private Hospitals of the area. Baseline data will be measured by food frequency questionnaires validated on a cohort sub-group. Blood samples will be stored. Outcome (incident cases of lung, breast and colon cancer) will be identified by a new integrated notification system based on general practitioner and pathology services. All cases will be histologically confirmed. Periodic linkages between the cohort database and administrative registries will be performed to keep track of migrations and deaths. A case-referent study will be nested using stored blood samples to compare levels of nutrients and other variables between lung cancer cases and controls matched on sex, age, and date of blood collection. Statistical analysis will be performed using multivariate methods.

TYPE: Case-Control; Cohort
TERM: BMB; Blood; Diet; Nutrition; Vitamins
SITE: Breast (F); Colon; Lung
TIME: 1990 - 1994

481 **Assennato, G.** 04953
Univ. of Bari, Inst. of Occupational Health, Piazza Giulio Cesare, 77124 Bari, Italy (Tel.: +39 80 278216; Fax: 5575451)
COLL: Ferri, G.M.; Porro, A.; Strickland, P.; Tockman, M.S.; Poirier, M.C.

DNA Adducts of Polycyclic Aromatic Hydrocarbons and Membrane Antigens in Occupational Groups at High Risk of Lung Cancer

The objective of the study is to evaluate the association between indicators of biological effective dose indicators (DNA adducts of PAH) and indicators of early biological effects (membrane antigens stained with immuno-assay procedures based on monoclonal antibodies against small cells lung cancer and non-small cell lung cancer antigens) among 290 coke oven workers at a steel plant in southern Italy, at high risk of lung cancer. A pilot investigation was carried out on a sub-group of 23 workers in 1988. Data collection began in May 1990and ended in June 1991. Every worker gave a blood sample, and underwent sputum induction, spirometry and a questionnaire interview (including respiratory symptoms, smoking and occupational history, and potential sources of confounding hydrocarbon exposure). To date PAH-DNA adducts have been measured in peripheral lymphocytes of 69 workers by competitive dissociation-enhanced lanthanide fluoro-immunoassay (DELFIA). Sputum cytomorphology by Papanicolaou method demonstrated 89 squamous metaplasias (36.3% of the sample) and 26 dysplasias (15.8%). Monoclonal antibodies (against small cell lung cancer and non-small cell lung cancer antigens) immuno-stained with Avidin-Biotin complex with a double bridge amplification system will be applied on sputum specimens preserved in Saccomanno fixative. Follow-up of the workers will be carried out to detect incidence and mortality for lung cancer. The data will be analysed by univariate and multivariate methods.

TYPE: Cohort; Molecular Epidemiology
TERM: Antigens; BMB; Blood; DNA Adducts; Occupation; Sputum; Tobacco (Smoking)
SITE: Lung
CHEM: PAH
OCCU: Coke-Oven Workers
TIME: 1988 - 1995

***482** **Marinaccio, L.** 05341
Oncology Inst., Div. of Gynecological Prevention, Via Amendola 209, 70126 Bari, Italy (Tel.: +39 80 5283136; Fax: 5283119)
COLL: Casamassima, A.; Traversa, A.; Simone, G.; Triantafillou, V.A.; Weisel, S.; Valentino, L.

Relationship of Peripheral Lymphocyte Subset Pattern to the Development or Progression of Cervical Intraepithelial Neoplasia

Alterations of the peripheral blood lymphocyte subset pattern (PBLSP) are among the host factors considered possibly to be involved in promotion and/or progression of human carcinogenesis. The role of PBLSP in the development of CIN is not yet clear. The aim of this study is twofold: (1) to assess whether CIN patients differ significantly from healthy age- and reproductive status-matched women without CIN as to overall or singular aspects of PBLSP. It is anticipated that 60 cases and 120 controls will be enrolled over a two year period; (2) to determine whether PBLSP alterations antedate CIN development and/or parallel its progression and possible response to therapy. From January 1992, the first 30 CIN cases and 30 matched controls, involved in the study under (1) above are being followed for a minimum of two years. Results of flow cytometry analysis of PBLSP will be correlated with data from the general and gynaecological patient history, and with the results of colposcopy, vaginal and endocervical cytology and endocervical biopsy.

TYPE: Case-Control
TERM: Cytology; Immunologic Markers; Lymphocytes; Premalignant Lesion; Promotion; Screening
SITE: Uterus (Cervix)
TIME: 1992 - 1994

Cagliari

483 **Carta, P.** 04736
Univ. di Cagliari, Ist. di Medicina del Lavoro, Via S. Giorgio 12, 09124 Cagliari, Italy (Tel.: +39 70 670481; Fax: 654350)
COLL: Cocco, P.L.; Picchiri, G.F.

Mortality Study of Lead and Zinc Miners in Sardinia

The mortaility of 5,000 Sardinian metal miners with at least one year's employment between 1960 and 1971 was followed up until 31 December 1987. Exposure to radon daughters is currently being assessed by recent measurements and estimates for the past made, using available information on ventilation devices. Exposure to respirable silica dust was assessed by the Institute and the mining company since the 1960's. SMRs for lung cancer and other groups of diseases will be calculated using the age- and year-specific mortality for the Sardinian general population as reference.

TYPE: Cohort
TERM: Mining; Occupation; Silicosis
SITE: All Sites; Lung
CHEM: PAH; Radon; Silica
OCCU: Miners, Zinc-Lead
TIME: 1984 - 1994

484 **Carta, P.** 05160
Univ. di Cagliari, Ist. di Medicina del Lavoro, Via S. Giorgio 12, 09124 Cagliari, Italy (Tel.: +39 70 670481; Fax: 654350)
COLL: Cocco, P.L.; Cherchi, P.

Cancer Mortality Among Sardinian Aluminium Workers

A preliminary cohort study of 1,148 workers of a Sardinian aluminium reduction plant with pre-baked anode process, followed from 1971 to 1990, was completed in 1992. (Med. Lav. 83:530-535, 1993). The all site mortality did not differ from the expected. Lung and bladder cancer rates were lower than expected, while an excess for pancreatic cancer was observed among workers employed in the production plant (rodding plant), but based on only three cases. An attempt to relate the mortality experience to workplace PAH exposure is currently underway with a further follow-up study and a case-control study. The cohort will be followed up until December 1995, comparing observed deaths to expected numbers derived from age- and year-specific mortality rates for the Sardinian male population. A hospital-based case-control study of all pancreatic cancer cases during 1971 and 1990 in the south of Sardinia will be conducted. Age-matched controls will be randomly selected from the same hospitals. Diet, smoking habits, residence and specific occupations will also be assessed.

TYPE: Case-Control; Cohort
TERM: Occupation
SITE: All Sites; Bladder; Lung; Lymphoma; Pancreas
CHEM: Aluminium; PAH; Tars
OCCU: Aluminium Workers; Smelters, Aluminium
TIME: 1990 - 1994

***485** **Carta, P.** 05385
Univ. di Cagliari, Ist. di Medicina del Lavoro, Via S. Giorgio 12, 09124 Cagliari, Italy (Tel.: +39 70 670481; Fax: 654350)

Cancer Mortality among Stibnite Miners and Antimony Smelter Workers in Sardinia

A cohort study of 516 Sardinian workers employed in a stibnite mine (355 males and 79 females) and in an antimony smelter plant (82 males) is being undertaken to ascertain the mortality from cancer of all sites and lung, compared with sex- and age-specific regional rates. The cohort members include all workers employed between January 1941 and June 1986. Information on the ore characteristics, the processing methods and on the environmental conditions will be available from the Regional Mining District. An attempt will be made to obtain personal data, including smoking habits, from the company medical records.

TYPE: Cohort
TERM: Metals; Occupation
SITE: All Sites; Lung
CHEM: Antimony
OCCU: Miners; Smelters
TIME: 1992 - 1995

486 **Cocco, P.L.** 04737
Univ. di Cagliari, Ist. di Medicina del Lavoro, Via S. Giorgio 12, 09124 Cagliari, Italy (Tel.: +39 70 670481; Fax: 654350)
COLL: Todde, P.F.; Manca, P.; Manca, M.B.; Formera, S.

Mortality of G6PD-Deficient Subjects

The importance of investigating the consequences of the genetic deficiency of the enzyme glucose-6-phosphate dehydrogenase (G6PD) has been stressed, particularly concerning cancer and cardiovascular diseases. To test the hypothesis of health consequences of G6PD deficiency, a cohort about 2,000 male subjects (all ages), found to be G6PD deficient during a regional screening campaign conducted in Sardinia in 1981 was collected. The mortality of the cohort will be ascertained from 1982 through 1991, evaluating the causes of death reported on the death certificates. Standardized mortality ratios will be calculated for cardiovascular diseases and all cancers, as well as for other selected causes of death, using the Sardinian male population age- and year-specific mortality rates as reference.

TYPE: Cohort
TERM: Enzymes; G6PD; Genetic Factors
SITE: All Sites
TIME: 1990 - 1994

487 **Cocco, P.L.** 05161
Univ. di Cagliari, Ist. di Medicina del Lavoro, Via S. Giorgio 12, 09124 Cagliari, Italy (Tel.: +39 70 670481; Fax: 654350)
COLL: Carta, P.; Flore, C.; Cherchi, P.

Mortality in Lead and Zinc Foundries: Follow-Up 1932-1990

The mortality experience of workers in two lead and zinc foundries in southern Sardinia for the periods 1932-1971 and 1970-1989 respectively, is currently under investigation. Members of the older cohort were 1,950 male employees with a minimum of one year of employment in the same foundry. Vital status has been ascertained for about 95% of this cohort; death certificates have been obtained for 505 out of 548 deceased subjects and coded to ICD-9. Limited exposure data are available for lead, cadmium, arsenic and silica. Standardized mortality ratios will be calculated for groups of diseases and single causes of interest deriving expected numbers from age- and year-specific mortality rates for the Sardinian male population. The ascertainment of the vital status of the members of the younger cohort is still in progress.

TYPE: Cohort
TERM: Dusts; Metals; Occupation
SITE: Bladder; Kidney; Leukaemia; Liver; Lung
CHEM: Arsenic; Cadmium; Lead; PAH; Silica; Zinc
OCCU: Smelters, Lead
TIME: 1984 - 1994

Casale Monferrato

488 **Botta, M.** 04084
Osp. Santo Spirito, Reparto Medicina Donne, Via Giolitti 2, 15033 Casale Monferrato, Italy (Tel.: +39 142 344928)
COLL: Castagneto, B.; Cocito, V.; Magnani, C.; Terracini, B.; Brusa, M.; Degiovanni, D.

Cohort Study of Workers in Asbestos-Cement Production in Casale Monferrato, Italy

A cohort of 3,370 workers in a factory producing asbestos-cement products in 1950 or after is being followed-up. The plant started the activity in 1907 and employed up to 1,600 workers (both sexes). Environmental conditions are described as rather poor in the past but no measures of asbestos concentration are available before 1978. Mortality in the cohort has been compared with Italian mortality rates and with local rates as well. A paper has been published in Med. Lav., 78(6):451-453, 1987 (Magnani, C. et al.). The study is being extended to clerks working in the same plant. Work history within the plant is being reconstructed for all workers in the cohort from administrative and medical records. The feasibility of a nested case-control study of respiratory cancers is under evaluation.

TYPE: Cohort
TERM: Chemical Exposure; Dusts; Occupation
SITE: Respiratory
CHEM: Asbestos
OCCU: Cement Workers; Office Workers
TIME: 1985 - 1994

489 **Botta, M.** 04357
Osp. Santo Spirito, Reparto Medicina Donne, Via Giolitti 2, 15033 Casale Monferrato, Italy (Tel.: +39 142 344928)
COLL: Cocito, V.; Magnani, C.; Terracini, B.; Brusa, M.; Degiovanni, D.; Castagneto, B.

Relatives of Workers in Asbestos Cement Production in Casale Monferrato, Italy

Several reports have suggested a carcinogenic risk associated with non-occupational exposure to asbestos but only a few formal epidemiological studies have been carried out. The present project is an expansion of an occupational cohort mortality study of workers in asbestos-cement production in Casale Monferrato at work in 1950 or afterwards. A large excess in mortality from respiratory cancers and asbestosis has been found (Magnani C., Med. Lav. 78(6):451-453, 1987). Wives of asbestos cement workers included in the cohort are being identified through data available from the registrar's offices of the municipalities where their husbands lived. They will be followed through 1964-1988 and the causes of death will be collected. It is estimated that number of person-years at risk will be around 37,000, which corresponds to a minimum detectable relative risk of about 3 for lung cancer (p = 0.05, power = 80%). A pilot study has been conducted with good results. Wives of clerks working in the same plant will be included as referents. The feasibility of the inclusion of the offspring in the cohort will be evaluated.

TYPE: Cohort
TERM: Chemical Exposure; Dusts; Environmental Factors; Minerals
SITE: All Sites; Lung
CHEM: Asbestos
TIME: 1988 - 1994

Cassano Murge

***490 Molinini, R.** 05464
Foundation 'Clinica del Lavoro', IRCCS, Div. of Occupational Medicine, Via per Mercadante, 70020 Cassano Murge, Italy (Tel.: +39 80 763111)
COLL: Paoletti, L.; Assennato, G.; Ferri, G.M.; Pollice, L.M.; Caruso, G.; Iacobellis, U.

Asbestos Fibres and Bladder Cancer

The difference of fibre content in samples of bladder tissue of subjects with and without bladder cancer, with different exposure to asbestos, will be studied. Fibre measurements will be performed by TEM and the quality studied by x-ray diffraction. Exposure to asbestos will be assessed using a questionnaire able to identify jobs at different levels of exposure, and validated before the beginning of the study. The preliminary phase consists of a feasibility study with 10 subjects for every study group (A: bladder tissue of subjects with bladder cancer; B: healthy bladder tissue from the same subjects; C: healthy bladder tissue of subjects with other, unrelated diseases; D: healthy bladder tissue of dead subjects). The second phase will consist of a case-control study based on the same group selection. The cases and living controls will be selected from the Urological Division and the deceased controls from the Pathology Department. A stratified analysis will be performed and logistic conditional regression used. The case-control study will have 85 subjects per group.

TYPE: Case-Control
TERM: Biopsy; Dusts
SITE: Bladder
CHEM: Asbestos
TIME: 1993 - 1997

Castellana Grotte

491 Leandro, G. 04464
Osp. Spec. Gastroenterologia, "S. De Bellis", Ist. di Ricovero e Cura a Carattere Scientifico, Viale Valente, 70013 Castellana Grotte (BA), Italy (Tel.: +39 80 735122)

Hepatocellular Carcinoma in Cirrhotic Patients. A Follow-up Study

Hepatocellular carcinoma (HCC) is one of the most common cancers in the world. Liver cirrhosis could itself be a premalignant condition acting as a promoter, since about 75% of HCC are associated with cirrhosis. This study, therefore, aims at following cirrhotic patients in order to determine which possible risk factors are associated with HCC. The clinical and serological data of cirrhotic patients admitted to our hospital will be analysed through probabilistic models (linear discriminant fuction analysis, multiple logistic regression, etc.). These models provide a powerful method of computer-assisted diagnosis. This should permit (1) determination of the risk factors for HCC in cirrhotic patients; (2) evaluation of their possible epidemiological relevance (e.g. the relation of the HBsAG carrier state to HCC); (3) making diagnostic laboratory tests more effective. In this way, sub-populations at highest risk of HCC can be identified: intensive clinical screening and periodic follow-up would make earlier diagnosis possible and consequently improve the prognosis. To date 645 cirrhotic patients have been included in the study. About 1,000 are expected.

TYPE: Case Series; Cohort
TERM: Cirrhosis; HBV; High-Risk Groups; Premalignant Lesion
SITE: Liver
TIME: 1985 - 1994

Catania

492 **Belfiore, A.** 04924
Univ. of Catania, Cattedra di Endocrinologia, Osp. Garibaldi, Piazza S. Maria di Gesu , 95124 Catania, Italy (Tel.: +39 95 2544856; Fax: 7158072)
COLL: Garofalo, R.; Giuffrida, D.; La Rosa, G.L.; Ippolito, O.; Fiumara, A.

Thyroid Cancer Associated with Graves' Disease

The aim of the study is to test the hypothesis that hyperthyroidism, in particular Graves' Disease, is a risk factor for thyroid cancer. In a previous study (J. Clin. Endocrinol. Metab. 70:830-835, 1990) 359 hyperthyroid patients and 582 euthyroid patients who underwent surgery during a 6-year period were examined retrospectively. It was found that thyroid cancer was most aggressive when associated with Graves' disease, least aggressive when associated with toxic adenoma, and with intermediate aggressiveness in euthyroid patients. It was therefore suggested that thyroid stimulating anti-bodies in Graves' disease may stimulate thyroid cancer growth and aggressiveness by acting through the TSH receptor. All patients seen in this clinic (which represent the great majority of all thyroid patients seen in an area of about one million inhabitants), with one or more thyroid nodules, will be selected for surgery and followed up for occurrence of thyroid cancer.

TYPE: Cohort
TERM: Hyperthyroidism
SITE: Thyroid
TIME: 1988 - 1994

Florence

***493** **Bechi, P.** 05395
Univ. of Florence, Clinic of Surgery 3, Viale Morgagni 85, 50134 Florence, Italy (Tel.: +39 55 412029; Fax: 4240133)
COLL: Becciolini, A.; Dei, R.; Balzi, M.; Amorosi, A.; Bruno, L.; Nesi, S.

Helicobacter Pylori Infection and Gastric Cancer

It has been suggested that H. pylori infection is related to gastric carcinogenesis. In order to test this hypothesis a cohort study is being undertaken in a high risk area for gastric cancer. Male dyspeptic subjects (aged 55-68) are being studied by upper endoscopy; presence of H. pylori infection is ascertained by classical methods and by molecular detection. The subjects will be stratified in two groups according to presence or absence of H. pylori infection. The histological findings will be compared in the two groups, especially precancerous and cancerous lesions. It is anticipated that 400 subjects will be included in the study and followed up for 10 years with endoscopy every two years. The effects of H. pylori infection on gastric mucosal proliferative activity will also be investigated. In the two groups of subjects gastric biopsy specimens will be incubated with 3H-thymidine and cell kinetic findings will be compared in H. pylori positive and H. pylori negative subjects.

TYPE: Cohort
TERM: BMB; H. pylori; Histology
SITE: Stomach
TIME: 1993 - 2005

***494** **Brandi, M.L.** 05259
Univ. of Florence Medical School, Dept. of Clinical Physiopathology, Endocrine Unit, Viale Pieraccini 6, 50139 Florence, Italy (Tel.: +39 55 4224334; Fax: 641026)

Incidence and Premorbid Genetic Test of Multiple Endocrine Neoplasia Type I and Type II Syndromes

The primary objective of this study is to evaluate the incidence of multiple endocrine neoplasia Type I (MEN I) and Type II (MEN II) syndromes in the Italian population. About 80 kindreds with MEN I and MEN II syndromes have been identified during the period March 1989 to present. All cases are interviewed by trained interviewers with a structured questionnaire, including questions on symptoms, precursor signs and family history of the disease. An Italian register of MEN I and MEN II will be created. To improve prevention of the two syndromes, preclinical genetic testing of gene carriers is also being performed in the clinically well characterized families.

TYPE: Case Series
TERM: BMB; Genetic Markers; Prevention
SITE: Benign Tumours; Endocrine Glands
TIME: 1989 - 1994

495 **Carli, P.** 04904
Univ. di Firenze, 2nda Clinica di Dermatologia, Via della Pergola 58 , 50121 Florence, Italy (Tel.: +39 55 2478356; Fax: 2478356 (2 pm-8 am))
COLL: Biggeri, A.; Bondi, R.; Urso, C.; Giannotti, B.

Dysplastic and Atypical Naevi as Risk Factors for Malignant Melanoma

The aim of this study is to investigate, in a mediterranean population, the relationship between cutaneous melanoma (CM) and the presence of atypical and dysplastic naevi. The number of common acquired naevi (and their topographic distribution) and the presence of small congenital naevi will also be evaluated. Other known risk factors will be investigated as potential confounders (acute and chronic sunlight exposure, phenotype, phototype). Physical assessment of skin colour and measurement of minimal erythemal dose will be performed. Cases are patients admitted in the period 1990-1992 to the 2nd Dermatology Clinic of the University of Florence, with a histologically confirmed diagnosis of CM, residents in the Florence area and aged 20-69 years. 150 or more cases are expected. A random sample of the resident population drawn from computerized demographic files will be utilized as the source of the control group. Two controls per case will be matched for sex, age and place of residence.

TYPE: Case-Control
TERM: Naevi; Pigmentation; Premalignant Lesion; Radiation, Ultraviolet
SITE: Melanoma
TIME: 1990 - 1994

496 **Chellini, E.** 04610
Centre for Study & Prevention of Cancer (CSPO), Via S. Salvi, 12, 50136 Florence, Italy (Tel.: +39 55 5662647; Fax: 677489)
COLL: Buiatti, E.; Merler, E.; Paci, E.; Seniori Costantini, A.R.; Zappa, M.; Biancalani, M.; Dini, S.; Pingitore, R.; Megha, T.; Tosi, P.; Comin, C.; Santucci, M.

Malignant Mesothelioma in Tuscany

All incident cases of malignant mesothelioma, histologically diagnosed in the Pathology Institutes involved in the project, are collected for the region of Tuscany. In order to obtain a reliable histological diagnosis, the histological sections of all cases are reviewed by the pathologists of the regional panel for malignant mesothelioma, and a consensus diagnosis is made, also using immunohistochemical techniques. Cases or, if deceased, their nearest relatives are interviewed by a trained interviewer in order to collect data on occupational and environmental exposure to asbestos or other risk factors. Detection of mineral fibres in pleura and lung tissue of those cases of pleural malignant mesothelioma who undergo surgical treatment is done by transmission electron microscopy and associated methods. A retrospective study was carried out on cases diagnosed in three Pathology Institutes in the period 1970-1987. This surveillance system permits the evaluation of past occupational asbestos exposure, which may also be present today, and subsequently the implementation of preventive measures. The relevance of non occupational exposures is also evaluated.

TYPE: Case Series; Incidence; Registry
TERM: Dusts; Histology; Immunologic Markers
SITE: Lung; Mesothelioma; Pleura
CHEM: Asbestos; Mineral Fibres
TIME: 1988 - 1994

497 **Geddes, M.** 04627
U.O. Epidemiologia, Via di S. Salvi 12, 50135 Florence, Italy (Tel.: +39 55 5662695)
COLL: Paci, E.; Masala, G.; Vannucchi, G.; Zappa, M.

Adequacy of Information Collected from Substitutes in Case-control Studies

The study compares information on smoking habits and occupational history collected from two sources. The first source is represented by personal interviews with 255 males with lung cancer, resident in the province of Florence, who were interviewed in the period 1980-84 during a case-control study. All the cases died before 1988. The second source is a self-administered questionnaire completed by a close relative, generally the wife, to whom a questionnaire similar to the one used in the first interview was sent by mail. Non-responders are interviewed by telephone. The aims of the study are (1) to evaluate the concordance between the original information and the information obtained from the proxy; and (2) to verify if the substitutes' compliance with the questionnaire depends on variables such as degree of relationship with the deceased, age, place of residence and level of education.

TYPE: Methodology
TERM: Occupation; Tobacco (Smoking)
SITE: Lung
TIME: 1987 - 1994

498 **Geddes, M.** 04201
U.O. Epidemiologia, Via di S. Salvi 12, 50135 Florence, Italy (Tel.: +39 55 5662695)
COLL: Balzi, D.; Biggeri, A.; Buiatti, E.; Chellini, E.; Cecconi, R.; Gaspari, R.

Cancer Risk in Italian Migrants within Italy

The aim of the study is to verify the risk for all cancers and for some specific cancer sites in the Italian population migrated from the South and North-East to some northern areas of Italy, compared with the local population. The hypothesised protective effect of being born in the South or in the North-East of Italy will be estimated in relation to the duration of the period of migration, to the age at migration and to the specific area of birth. The study is a multicentric case-control study. Cases are all cancer decendents aged 35-84 years in 1985-89 who were resident in the municipalities of Bologna, Firenze, Genova, and Torino. Two controls are matched to each case by sex, year of birth and place of residence. For each person information regarding demographic data, place of birth and year of immigration will be collected. Relative risks and mortality rates will be calculated for several cancer sites, evaluating the effect of sex, age at migration, period of immigration and duration of residence in the immigration area.

TYPE: Case-Control
TERM: Age; Geographic Factors; Migrants; Registry; Time Factors
SITE: All Sites
TIME: 1987 - 1994

499 **Palli, D.** 04273
Center for Study and Prev. of Cancer, Epidemiology Sect., Viale Volta 171 , 50131 Florence, Italy (Tel.: +39 55 578062; Fax: 578955)
COLL: Buiatti, E.; Carli, S.; Ciatto, S.; Paci, E.; Rosselli del Turco, M.

Screening for Breast Cancer: Evaluation by a Case-Control Study

A population based screening programme for breast cancer has been started in 1970 in a rural area near the city of Florence. The resident female population in the age-group 40-70 years (34,000 women at the 1981 census), is invited, every two years, to have a mammographic examination (double-view) in a mobile unit. Recently a case-control study has been carried out (Palli et al. Int. J. Cancer 38:501-4, 1986) showing a significant protective effect for participating women in the older age-group (50+ years). Monitoring of the results by means of an extension of the case-control study has been planned. Additional cases (women dead from breast cancer with a diagnosis after the start of the programme) will be identified every year on the basis of the mortality data available for the area; a matched group of controls will be periodically sampled from the general population. An average of 20 cases is expected every year.

TYPE: Case-Control
TERM: Screening
SITE: Breast (F)
TIME: 1987 - 1994

500 **Seniori Costantini, A.R.** 03510

Center for Study and Prev. of Cancer, Epidemiology Sect., Via di S. Salvi 12, 50136 Florence, Italy (Tel.: +39 55 5662647; Fax: 677489)
COLL: Chellini, E.; Miligi, L.R.; Merler, E.; Scarpelli, A.R.

Occupational Cancer Risk Assessment in Tuscany

The target of this project is 'mapping' of exposure in the main industrial activities in Tuscany. Collection of data in the leather and textile industries was carried out during the period 1984-1992, with the cooperation of the Occupational Local Health Services of the region. Information concerns number and type of departments, number of workers in different jobs, chemicals used. Job exposure matrices are being created, using a computer software developed for this purpose. A first mapping of exposure to asbestos has been completed. Annual updating of data is planned. A job exposure matrix for the shoe and leather industries has also been established.

TYPE: Methodology
TERM: Dusts; Dyes; Leather; Occupation; Solvents; Textiles; Wood
SITE: All Sites
CHEM: Asbestos
OCCU: Asbestos Workers; Cabinet Makers; Glass Workers; Leather Workers; Potters; Textile Workers; Wood Workers
TIME: 1989 - 1994

Genoa

501 **Gennaro, V.** 05211

Cancer Research Inst., Dept. of Epidemiology & Biostatistics, Viale Benedetto XV, 10 , 16132 Genoa, Italy (Tel.: +39 10 357787; Fax: 352999 ; Tlx: 286356 istex i)
COLL: Fontana, V.; Ceppi, M.; Puntoni, R.; Manti, A.

Cohort Study Among Oil Refinery Workers in Genoa

The cause-specific mortality in about 1,500 blue- and white-collar workers employed between 1949 and 1988 at an oil refinery in Genoa is being analysed, adjusting for sex, age and calendar year. Mortality is compared with the general population. Blue-collar workers are considered as potentially exposed to hydrocarbons, asbestos, nickel, chromium and other agents. 101 persons (6.7%) have been lost to follow-up. 196 deaths have been observed and evaluated by Cox regression, SMR and Kaplan-Meier analyses. Preliminary analysis show, using internal comparisons, a statistically significant excess ($p < 0.05$) in the exposed groups for all causes of death, accidents, ill-defined diseases, non-neoplastic diseases of the digestive and respiratory tract, cancer of the oesophagus, respiratory tract, mesothelioma and leukaemias.

TYPE: Cohort
TERM: Dusts; Occupation; Solvents
SITE: All Sites
CHEM: Asbestos; Benzene; Chromium; Hydrocarbons; Nickel
OCCU: Petroleum Workers
TIME: 1988 - 1994

502 **Puntoni, R.** 03818

Inst. of Oncology, Dept. Epidemiology & Biostatistics, Viale Benedetto XV, 10, 16132 Genoa, Italy (Tel.: +39 10 300767)
COLL: Merlo, F.; Ceppi, M.; Stagnaro, E.

Cancer Mortality among Genova Dockyard Workers

The objective of the present study is to evaluate the mortality experience in longshoremen employed in the shipyard of Genova. Since several substances regularly handled in this branch of the shipyard are likely to represent a hazard for the workers, the aims of the study are: (1) to identify sub-groups of the worker population at risk; (2) to identify the substances reponsible for the occurrence of higher than expected cancer specific mortality figures; (3) to define cancer as well as non-cancer mortality rates by age, length of exposure, age at entry and job title. Smoking habits will be accounted for using indirect estimates as reported by some authors. The study refers to the period 1974-1985 . Mortality data will be collected using standard techniques in order to guarantee comparability between study and referent groups. 4,500 workers have been include in the study, accounting for about 50,000 person-years of observation.

TYPE: Cohort
TERM: Age; Chemical Exposure; Occupation; Time Factors; Tobacco (Smoking)
SITE: All Sites
OCCU: Shipyard Workers
TIME: 1985 - 1994

503 **Vercelli, M.** 04511
Inst. of Clinical & Experimental Oncology, Dept. of Descriptive Epidemiology, Viale Benedetto XV, 10, 16132 Genoa, Italy (Tel.: +39 10 3534961; Fax: 352999 ; Tlx: 216353 istex i)
COLL: Puntoni, R.; Ceppi, M.; Orengo, A.; Reggiardo, G.; Orlandini, C.; Casella, C.; De Lucia, G.

Cancer Incidence in Genoa by Place of Birth and Length of Residence

Previous mortality study for breast, stomach and lung tumours in Genoa's residents coming from other areas of Italy and treated in Genoa, has pointed out interesting results which confirm the importance of migration as a risk factor for these tumours. This study proposes to analyse with the same method, 1986-1987 incidence data deviving from Genoa Cancer Registry (25,000 inhabitants, 3,500 cancer cases/year) and to verify the concordance of incidence and mortality risks. In particular, the aim is to understand the role of incidence data in giving more indications on cell morphology.

TYPE: Incidence
TERM: Environmental Factors; Geographic Factors; Latency; Migrants; Registry; Time Factors
SITE: All Sites
REGI: Genoa (Ita)
TIME: 1991 - 1994

L'Aquila

504 **Corrao, G.** 03967
Univ. of L'Aquila, School of Medicine, Dept. of Int. Medicine & Publ. Health, Via Verdi 28, 67100 L'Aquila, Italy (Tel.: +39 862 433485; Fax: 433433)
COLL: Russo, R.; Carle, F.; Lepore, A.R.; di Orio, F.; Vineis, P.; Ciccone, G.

Evaluation of Cancer Risk in Farmers with Professional Exposure to Pesticides

A follow-up study is being conducted on 40,000 men living in different regions of Italy (South Piedmont, Abruzzo, Molise) who were licensed to use potentially toxic pesticides during the period 1970-74. In comparison with the total experience of the population in the study area, the Piedmont sub-cohort showed an apparent excess of skin malignancies and lymphomas. A follow-up for mortality is now underway in order to confirm the preliminary findings. Mortality rates in the cohort will be compared with the experience of the population in the study area. A limitation of the follow-up study is the non-availability of information about exposure to specific pesticides. Therefore, a case-control study within the cohort is planned. In this study, cases with malignancies for which an excess has been suggested in the follow-up study will be compared with a sample of the total cohort. The purpose is to interpret any cancer excess in the cohort in terms of specific exposures. Information will be collected through interviews with cases and controls or their relatives.

TYPE: Case-Control; Cohort
TERM: Chemical Exposure; Occupation; Pesticides
SITE: All Sites; Brain; Haemopoietic; Skin
OCCU: Farmers
TIME: 1985 - 1994

505 **Corrao, G.** 04822
Univ. of L'Aquila, School of Medicine, Dept. of Int. Medicine & Publ. Health, Via Verdi 28, 67100 L'Aquila, Italy (Tel.: +39 862 433485; Fax: 433433)
COLL: Busellu, G.P.; Di Placido, R.; Recchia, C.; di Orio, F.

Diet and Other Risk Factors for Gastric Cancer

Excess mortality from gastric cancer has been observed in L'Aquila district. A case-control study has been designed in order to evaluate risk factors for this cancer. Cases will be 120 new histological diagnoses of gastric cancer admitted to endoscopic services of L'Aquila Province in the course of two years. In the same period 240 control patients admitted to provincial hospitals for acute, non-neoplastic and non-digestive disease will be recruited. Trained interviewers will identify and question cases and controls, using standard questionnaires to obtain information on personal characteristics, alcohol and smoking habits, dietary history and use of certain drugs. The role of these factors and their interactions will be quantitatively estimated.

TYPE: Case-Control
TERM: Alcohol; Diet; Drugs; Tobacco (Smoking)
SITE: Stomach
TIME: 1990 - 1994

Milan

506 **Berrino, F.** 03480
Ist. Nazionale per lo Studio e la Cura dei Tumori, Servizio di Epidemiologia, Via G. Venezian 1, 20133 Milan, Italy (Tel.: +39 2 2366342; Fax: 2362692 ; Tlx: 333290 tumist i)
COLL: Ferrario, F.; Macaluso, M.; Pisani, P.; Baldasseroni, A.; Axerio, M.

Occupational Exposures and the Aetiolology of Kidney Cancer and Malignant Lymphomas

Population-based case-control studies are increasingly used to test associations between chemicals and neoplasms through the use of job-histories as proxies for the history of exposure to chemicals. This study aims to apply this methodology in evaluating the role of a series of substances and occupations in the aetiology of kidney cancer and malignant lymphomas. All incident cases in the Varese Province (where the Lombardy Cancer Registry is situated) during the period 1986-1989 will be included. Expected numbers are 200 and 400 respectively for kidney cancer and lymphoma cases. Two controls per case are extracted from the electoral rolls of the province. Occupational histories are collected by means of a structured questionnaire, developed in detail for any occupational setting where exposures are likely to occur. A handbook for easy consultation by interviewers was developed. The list of exposures/occupations to be tested is the following: asbestos, polycyclic aromatic hydrocarbons, arsenic, beryllium, chromium, cadmium, lead, nickel and their compounds, chlorophenols and phenoxy-acids, BCME and CMME, chloropyrene, vinyl chloride, styrene, benzol chloride, dimethylsulphate, epichlorohydrin, formaldehyde, trichloroethylene, tetrachloroethylene; leather and allied industries, wood and allied industries, pulp and paper manufacturing, and the rubber industry. Information about known potential confounders is also collected through interview.

TYPE: Case-Control
TERM: Chemical Exposure; Dusts; Leather; Metals; Occupation; Pesticides; Plastics; Registry; Rubber; Solvents; Wood
SITE: Kidney; Lymphoma
CHEM: Arsenic; Asbestos; Benzoyl Chloride; Beryllium; Cadmium; Chlorophenols; Chloropyrenes; Chromium; Dimethylsulphate; Epichlorohydrin; Formaldehyde; Hydrocarbons; Hydrocarbons, Halogenated; Lead; Nickel; PAH; Phenoxy Acids; Styrene; Tetrachloroethylene; Trichloroethylene; Vinyl Chloride
REGI: Varese (Ita)
TIME: 1984 - 1994

507 **Berrino, F.** 04421
Ist. Nazionale per lo Studio e la Cura dei Tumori, Servizio di Epidemiologia, Via G. Venezian 1, 20133 Milan, Italy (Tel.: +39 2 2366342; Fax: 2362692 ; Tlx: 333290 tumist i)
COLL: Pisani, P.; Muti, P.; Crosignani, P.; Micheli, A.; Totis, A.; Fissi, R.; Mazzoleni, C.; Pierotti, M.; Secreto, G.

Hormones and Diet in the Aetiology of Breast Cancer

The ORDET (Ormoni e Dieta nella Eziologia dei Tumori) is a prospective study of 10,000 healthy women to be recruited in Varese Province and to be followed up through the local Cancer Registry. The major aims are to clarify the role of endogenous hormones and dietary habits in the aetiology of breast cancer, and to study the relation between these two factors. Recruitment began in 1987 and is expected to be completed by 1991. Each woman is requested to provide both standardised information and biological specimens to be stored at -80 degrees C. Information is available on: social and cultural factors, menstrual and reproductive history, diet and alcohol consumption, history of nutritional status and physical activity, various anthropometric measurements, clinical breast examination or mammography, blood pressure, pulse rate, cutaneous sebum production, hirsutism score and psychological tests. The biological bank includes aliquots of serum, plasma, red blood cell membranes, red blood cell cytoplasm, leucocytes, prepared from blood samples taken in the early morning on an empty stomach, aliquots of

12hr urine collections and a toenail sample. 135 incident cases are expected within the first ten years of follow-up.

TYPE: Cohort
TERM: Alcohol; BMB; Diet; Hormones; Mammography; Physical Activity; Physical Factors; Registry; Reproductive Factors; Socio-Economic Factors; Toenails; Urine
SITE: Breast (F)
REGI: Varese (Ita)
TIME: 1987 - 2000

508 **Bertazzi, P.A.** 03709
Univ. Milano, Inst. Occup. Health, Clinic "Luigi Devoto", Dept. of Epidemiology, Via San Barnaba 8, 20122 Milan, Italy (Tel.: +39 2 5450812; Fax: 55187172 ; Tlx: 320484 unimi i)
COLL: Pesatori, A.C.; Zocchetti, E.; Guercilena, S.; Consonni, D.; Landi, M.T.; Tironi, A.

Long-Term Investigation of the Cancer Risk of Persons Potentially Exposed to TCDD after the Seveso Accident in 1976

The aim of the project is to investigate the possible long term effects of the accident on the health experience of the population living in the TCDD contaminated area. Nearly 300,000 subjects were included in the study; ten percent of them had been living in zones definitely or probably contaminated by TCDD, and the remaining were enrolled and concurrently examined as reference group. The types of studies feasible turned out to be mortality and cancer incidence. Mortality results for the period 1976-1986 have been published (Am. J. Epidemiol. 129, 1187, 1989; Scand. J. Work Environ. Health 15, 85, 1989; Med. Lav. 80, 316, 1989); they showed an increase of cardiovascular deaths early after the accident, and suggestive increases of certain types of cancers without, however, definite and consistent patterns. The collection of cancer incidence data for the same period, 1976-1986, is nearly completed. All hospital records in the Region of Lombardy containing any kind of indication of cancer problems have been linked with the vital statistics records of people residing in the Seveso area since the accident date. After the linkage, each record is checked against the original medical documentation in the relevant hospitals. Results so far support the hypothesis of a slightly increased risk of hepatobiliary cancer, soft tissue sarcoma, and haematological neoplasms. In the same area, oestrogen dependent tumours showed a decrease. These findings are consistent with anticipated TCDD effects. The follow-up period has been extended and case-control studies have been initiated, using recently developed biological markers.

TYPE: Case-Control; Cohort
TERM: Biomarkers; Chemical Exposure; Environmental Factors; Record Linkage
SITE: All Sites; Leukaemia; Lymphoma; Soft Tissue
CHEM: Dioxins
TIME: 1985 - 1995

509 **Bertazzi, P.A.** 05220
Univ. Milano, Inst. Occup. Health, Clinic "Luigi Devoto", Dept. of Epidemiology, Via San Barnaba 8, 20122 Milan, Italy (Tel.: +39 2 5450812; Fax: 55187172 ; Tlx: 320484 unimi i)
COLL: Della Foglia, M.; Castiglione, G.; Donelli, S.; Zocchetti, C.; Pesatori, A.C.; Consonni, D.

Cancer Risk in Workers Exposed to Dyes and Formaldehyde in Textile Production

After identification of all textile dyeing and finishing shops located within the territories of three Local Health Units in northern Italy, enumeration of all persons employed during the period 1950-1979 has been completed. The production process in each shop was examined in detail and each job characterised in terms of the type of chemicals used and the time period in use. Individual work histories are available. It is anticipated that the minimum exposure categorisation will be: dyes (and possibly specific types of dyestuff); dyes and formaldehyde; formaldehyde; others. Additional exposure information is also being retrieved. Around 5,000 workers will meet the admission criteria. Their mortality will be examined for the period 1950-1990. Local and regional mortality rates are available to compute age-, sex-, year-, and cause-specific expected figures. For workers alive as of 1977, cancer morbidity will also be examined with the cooperation of the regional hospital discharge registration system of the Lombardy Tumour Registry. The focus of the study is on bladder cancer risk (18-22 cases expected) and exposure to suspected carcinogenic dyes, and formaldehyde exposure and nasal and pharyngeal cancer (6-8 cases expected).

TYPE: Cohort
TERM: Dyes; Occupation
SITE: All Sites; Bladder; Nasal Cavity; Pharynx; Respiratory
CHEM: Benzidine; Chromium; Formaldehyde
OCCU: Textile Dyers
REGI: Varese (Ita)
TIME: 1991 - 1994

510 **Boyle, P.** 04401
European Inst. of Oncology Div. of Epidemiology & Biostatistics, Via Ripamonti 332/10 , 20141 Milan, Italy (Tel.: +39 2 57408795; Fax: 57408883)
COLL: Little, J.; Bartsch, H.; Bruzzi, P.A.; Dicato, M; Franceschi, S.; George, W.O.; Ghadirian, P.; Hietanen, E.; James, W.P.T.; Katsouyanni, K.; La Vecchia, C.; Leake, R.E.; Levi, F.G.; MacMahon, B.; Mahon, G; Martin-Moreno, J.; O'Higgins, N.; Piettanen, P.; Plesko, I.; Salonen, J.T.; Trichopoulos, D.; Zaridze, D.; Zatonski, W.A.

SEARCH Study of Breast and Colorectal Cancer
The aims of this international, collaborative series of population-based case - control studies of breast cancer and colorectal cancer are to examine the relationships between intake of alcohol, specific nutrients, foods and food groups, energy intake, output and balance, reproductive and hormone use history, family history and risk of breast cancer and colorectal cancer. Among the principal hypotheses to be tested are (1) that cases of breast cancer will have a higher intake of alcoholic beverages than controls; and (2) that cases of colon and breast cancer will have higher intakes of total fat, saturated fat and protein, higher total energy intake and lower total energy output, lower vegetable intake and more first-degree relatives with a history of the same cancer than age- and sex-matched controls. Controls will be selected from the general population.

TYPE: Case-Control
TERM: Alcohol; Diet; Familial Factors; Hormones; Reproductive Factors
SITE: Breast (F); Colon; Rectum
LOCA: Canada; Estonia; Finland; France; Greece; Ireland; Italy; Luxembourg; Poland; Russian Federation; Slovak Republic; Slovenia; Spain; Switzerland; United Kingdom
TIME: 1988 - 1994

511 **Duca, P.** 04905
Univ. of Milan, Inst. of Biometry & Medical Statistics, Via Venezian 1 , 20133 Milan, Italy (Tel.: +39 2 2361302; Fax: 2362930)
COLL: Bisanti, L.; Braga, M.; Bellini, A.; Zucchi, A.; Musch, G.; Marubini, E.

Mortality Study of Workers Occupationally Exposed to Motor Vehicle Exhaust
This study is part of a larger collaborative project aimed to evaluate the cancer risk in workers (taxi-drivers, petrol station workers, etc.) occupationally exposed to environmental pollutants (motor-vehicle exhausts, heating exhausts, dusts) in the Italian metropolitan areas. In particular, this study will deal with a cohort of 5,000 traffic wardens active in Milan at any time from 1945-1950 to date. They are expected to contribute almost 75,000 person-years and 320 deaths. Exposure will be estimated based on length of employment and on direct (air sample analyses) and indirect (annual consumption of motor-vehicle and heating fuel) indicators of environmental pollution. Information on date and cause of death will be obtained from the Census Office at the place of residence. ICD-code of the cause of death will be provided by the National Bureau of Statistics.

TYPE: Cohort
TERM: Air Pollution; Dusts; Occupation
SITE: All Sites; Larynx; Lung
CHEM: Vehicle Exhaust
OCCU: Drivers; Petrol Station Attendants
TIME: 1990 - 1994

512 **La Vecchia, C.** 03370
Inst. Pharmacol. Research "Mario Negri" Dept. of Clinical Pharmacology, Unit of Epidemiology, Via Eritrea 62, 20157 Milan, Italy (Tel.: +39 2 39014.1; Tlx: 3546277)
COLL: D'Avanzo, B.; Franceschi, S.; Gentile, A.; Liberati, C.; Negri, E.; Parazzini, F.; Tavani, A.; Tognoni, G.; Decarli, A.

Case-Control Surveillance of Neoplastic Disease

Case-control studies on histologically confirmed cancers of several sites have been underway since 1979. Cases are patients admitted to University and General Hospitals in the Greater Milan area, and controls are persons admitted to the same network of hospitals with acute, non-neoplastic disease. Trained interviewers identify and question cases and controls, using standard questionnaires to obtain information on personal characteristics and habits including cigarette smoking, alcohol and methylxanthine consumption, a validated dietary history, a problem-oriented medical history, and use of selected drugs (including, for females, contraceptive practices and hormonal therapy). At present the following sites are included (approximate numbers per year given in brackets):oral cavity (20), oesophagus (40), stomach (100), intestines (colon and rectum) (150), liver (30), gallbladder and bile ducts (20), pancreas (30), larynx (20), breast (400), cervix (50), corpus uteri (50), ovary (50), prostate (80), bladder (50), kidney (30), thyroid (50), Hodgkin's disease (30), other lymphomas (30), multiple myelomas (50). All cases are histologically confirmed and diagnosed within the year prior to interview. Some 1,000 controls per year are interviewed. Recent papers appeared in Int. J. Epidemiol. 20:39-44, 1991, Cancer 69:2276-2282, 1992 and in Int. J. Cancer 50:567-571, 1992.

TYPE: Case-Control
TERM: Age; Alcohol; Diet; Drugs; Hormones; Oral Contraceptives; Premalignant Lesion; Tobacco (Smoking)
SITE: All Sites
CHEM: Methylxanthines
TIME: 1979 - 1995

513 **La Vecchia, C.** 04062
Inst. Pharmacol. Research "Mario Negri" Dept. of Clinical Pharmacology, Unit of Epidemiology, Via Eritrea 62, 20157 Milan, Italy (Tel.: +39 2 39014.1; Tlx: 3546277)
COLL: Decarli, A.; Parazzini, F.; Negri, E.

Smoking and Tobacco Related Neoplasms in Italy

Of all cancers, tobacco-related sites showed the most pronounced rises in mortality over the last decades. The present project aims to: (1) Analyse trends in total tobacco and cigarette sales (from 1900 onwards) and smoking (from 1950 onwards) in Italy as a whole and in various geographical areas (provinces or regions) for each sex and age groupand, for comparative purposes, in other Eruopean countries; (2) measure tar, nicotine and carbon monoxide yields and market shares of various cigarette brands; (3) study trends in mortality from lung and other neoplasms, in the whole of Italy from 1955 onwards, and in various regions or provinces from 1970 onwards using cross-sectional and cohort approaches; analysis of tobacco-related risks from the case-control surveillance of neoplastic diseases contracted by this study group; (4) offer wide coverage to data on smoking and tobacco-related neoplasms to ultimately reduce smoking prevalence in the population. Trends in tobacco consumption are estimated from published and unpublished data of the State Monopoly, the Central Institute of Statistics and the Departments of Health and Finance, the major problem being to give a reasonable estimate of smuggling. Death certification rates are derived from publications and copies of the original computer tapes obtained from the Central Institute of Statistics with extracts of all primary death records. Analyses of mortality rates are based both on standard cross sectional approach (age-specific and age-standardized rates, with major attention given to the younger groups) and cohort methods including age period and cohort models. Recent papers appeared in Cancer Res. 50:6502-6507, 1990, Eur. J. Cancer 27:94-104, 1991 and 28:514-599, 1992.43:784-785, 1989.

TYPE: Correlation; Mortality
TERM: Birth Cohort; Drugs; Mathematical Models; Time Factors; Tobacco (Smoking); Trends
SITE: Bladder; Kidney; Larynx; Lung; Oesophagus; Oral Cavity; Pancreas; Pharynx
CHEM: Carbon Monoxide; Nicotine; Tars
TIME: 1984 - 1995

***514** **Landi, M.T.** 05426
EPOCA, Research Center on Occupational, Clinical & Environmental Epidemiology, Università degli Studi, Via San Barnaba 8 , 20122 Milan, Italy (Tel.: +39 2 5450812; Fax: 55187172 ; Tlx: 320484 UNIMI I)
COLL: Bertazzi, P.A.; Clark, G.C.; Bell, D.; Lucier, G.W.; Cosma, G.; Garte, S.J.; Mocarelli, P.; Needham, L.; Hankinson, O.; Caporaso, N.E.

Molecular Epidemiology of Dioxin-Related Diseases in Seveso

The objective of this project is to investigate the role of genetic susceptibility factors in determining who becomes adversely affected by TCDD exposure. Ah receptor, arnt and CYP1a1 genotypes and expression, and EROD and N-oxidation activity will be used as biomarkers of susceptibility to TCDD-related diseases. Biological markers of exposure to TCDD, tobacco smoke and other potential confounding factors will be investigated. The study will be conducted within the Seveso cohort and will proceed in three phases: (1) A pilot study covering subjects randomly selected from an exposed and a non-exposed area, aimed to validate those assays which have been only partially characterized in humans; (2) A case-control study including subjects who had been diagnosed with chloracne after the accident in 1976, and controls selected from the same exposed area, to investigate whether the two groups exhibit differences in susceptibility markers. (3) A case-control study of cancer cases identified in the exposed area between 1976 and 1992, and prospectively between 1993 and 1995, and healthy controls selected from the same exposed area, to investigate whether cases and controls exhibit differences in susceptibility markers. All cases and controls will be interviewed with a structured questionnaire and a blood sample will be collected from each subject.

TYPE: Case-Control; Molecular Epidemiology
TERM: BMB; Biomarkers; Blood; Genetic Markers; PCR; RFLP; Tobacco (Smoking)
SITE: Bile Duct; Brain; Breast (F); Liver; Non-Hodgkin's Lymphoma; Respiratory; Sarcoma; Skin; Soft Tissue; Testis
CHEM: Dioxins
TIME: 1992 - 1995

***515** **Muti, P.** 05431
National Cancer Inst. of Milan, Dept. of Epidemiology, Via Venezian, 1 , Milan 20133, Italy (Tel.: +39 2 2390460; Fax: 2362692)
COLL: Micheli, A.; Krogh, V.; Fissi, R.; Sciajno, R.; Bolelli, G.; Celentano, E.; Monagle, L.; Panico, R.; Trevisan, M.; Berrino, F.

ORDET Project: Relationship between Steroid Hormones, Adiposity and other Cancer Risk Factors in Women

The aims of the present study are to describe, on a cross-sectional basis, the relationship between blood and urinary sex steroid profile, quantity and distribution of body fat, serum lipids, energy intake and physical activity indicators in a sample of 1,000 cancer-free women, consecutively enrolled for ORDET study (prospective study on hormones and diet in the aetiology of breast cancer) between March and May 1991. Further aims are to investigate the relationship between hormonal profile and environmental factors such as smoking, alcohol, saturated and unsaturated fatty acids. Data have been collected through an administered questionnaire and anthropometrical measurements have been made by trained observers according to a standardized protocol. Blood has been drawn in fasting condition between 8.00 a.m. and 9.00 a.m. The hormonal determinators are made by the same laboratory using RIA direct methods. Statistical analysis will be carried out computing Indexes of agreement between different variables adjusting for possible confounders when appropriate.

TYPE: Cross-Sectional
TERM: Alcohol; BMB; Diet; Fatty Acids; Tobacco (Smoking)
SITE: Breast (F)
CHEM: Steroids
TIME: 1992 - 1995

516 **Parazzini, F.** 03183
Ist. di Ricerche Farmacologiche "Mario Negri", Dipto di Farmacologia Clinicale, Via Eritrea, 62, 20157 Milan, Italy (Tel.: +39 2 357941; Fax: 3546277 ; Tlx: 331268 NEGRI I)
COLL: La Vecchia, C.; Fasoli, M.; Decarli, A.

Risk Factors for Gestational Trophoblastic Diseases in Italy

In order to evaluate the relation of genetic and environmental factors to the aetiology of trophoblastic diseases, a case-control study was started in June 1981. To date about 350 cases of complete and partial histologically confirmed hydatidiform mole and 60 cases of choriocarcinoma, have been collected. Controls are women admitted for normal delivery matched with cases within five year age groups. Cases and controls are identified and interviewed in a network of collaborating teaching and general hospitals in the greater Milan area. A standard questionnaire is used to obtain information on socio-demographic factors, smoking habits, alcohol and methylxanthine consumption, a few selected dietary habits, gynaecological and obstetric data, history of lifetime contraceptive practices or other female hormone use, history of diseases or other factors thought to be important in the aetiology of trophoblastic diseases, and family history of repeated spontaneous abortions or trophoblastic diseases. Papers appeared in Am. J. Obstet. Gynecol. 158:93-100, 1988, Gynecol. Oncol. 31:310-314, 1988 and in Obstet. Gynecol. 78:1039-1045, 1991.

TYPE: Case-Control
TERM: Abortion; Alcohol; Contraception; Diet; Familial Factors; Hormones; Parity; Tobacco (Smoking)
SITE: Choriocarcinoma; Hydatidiform Mole
CHEM: Methylxanthines
TIME: 1981 - 1995

517 **Parazzini, F.** 04477
Ist. di Ricerche Farmacologiche "Mario Negri", Dipto di Farmacologia Clinicale, Via Eritrea, 62, 20157 Milan, Italy (Tel.: +39 2 357941; Fax: 3546277 ; Tlx: 331268 NEGRI I)
COLL: La Vecchia, C.; Decarli, A.; Franceschi, S.; Gramenzi, A.

Case-Control Surveillance of Breast and Female Genital Tract Cancers

Although the role of hormonal, reproductive and general lifestyle habits in breast and female genital tract carcinogenesis is well understood, several questions remain unsettled, as for instance, oral contraceptive use or oestrogen replacement therapy and the risk of breast cancer, or the interaction between parity and sexual factors in cervical cancer. To identify and better quantify relative and attributable risks in female cancers, a large case-control surveillance of histologically confirmed benign and malignant breast, ovarian, endometrial, cervical and vulvar tumours has been conducted since 1982. Controls (age-matched in five-year groups with cases) are women hospitalized for conditions other than hormonal, gynaecological or neoplastic acute disorders. Cases and controls are identified in a large network of teaching and general hospitals in the Greater Milan area. A total of about 400 cases of breast cancer, 50 ovarian, endometrial, and cervical cancer, 20 vulvar cancer and 800 controls per year will be collected. Recent papers appeared in Am. J. Obstet. Gyn. 164:522-527, 1991, Am. J. Epidemiol. 135:35-40, 1992 and in Cancer 69:2276-2282, 1992.

TYPE: Case-Control
TERM: Diet; Hormones; Oral Contraceptives; Parity; Sexual Activity
SITE: Benign Tumours; Breast (F); Ovary; Uterus (Cervix); Uterus (Corpus); Vulva
TIME: 1982 - 1995

518 **Veronesi, U.** 05231
Ist. Nazionale dei Tumori, Via G. Venezian 1, 20133 Milan, Italy (Tel.: +39 2 2390; Fax: 70602991)
COLL: Costa, A.; De Palo, G.; Formelli, F.; Maltoni, C.; Marubini, E.; Coopmans de Yoldi, G.

Breast Cancer Chemoprevention with Fenretinide

The main aim of this study is to evaluate the effectiveness of the synthetic retinoid fenretinide in preventing breast cancer. Fenretinide inhibits chemically induced mammary carcinomas in different experimental animals. Due to its peculiar concentration in the mammary gland and fat (also demonstrated in human) fenretinide is here proposed as the agent of choice for a breast cancer prevention study. The design of the study is to orally administer for five years 200 mg daily of fenretinide (with a three-day drug holiday at the end of each month) to stage I-II breast cancer patients between 33-68 years of age. This intervention schedule is to be randomly assigned versus no treatment in a population of 3,500-5,000 subjects, depending on the biological activity of the retinoid. The principal endpoint of the study is the possible decrease of incidence of new primaries in the contralateral breast. Local, regional and distant recurrences of the disease as well as new primaries in organs other than the breast will also be recorded and analyzed. No placebo is foreseen for the control arm as extensive data are already available on both acute and chronic toxicity of fenretinide: moreover, ethical considerations, too, do not recommend it due to the length of the trial. However, the protocol foresees a blind, unbiased review of all mammograms; cytological and histological examinations of all biopsied lumps are also performed blindly. This study started on 1 March 1987, and by 15 April 1991, 2,308 patients had already been randomized: 1,160 in the fenretinide arm and 1,148 in the control arm. The major analysis will be a comparison between the curves for the cumulative incidence of new contralateral primaries over time, when all patients have been followed for seven years. The log-rank two-sample statistics will be used to compare the time for the first tumour to occur among control patients with that of the fenretinide treated patients.

TYPE: Intervention
TERM: BMB; Prevention; Recurrence
SITE: Breast (F)
CHEM: Retinoids
TIME: 1985 - 1994

Modena

519 **Aggazzotti, G.** 05041
Univ. di Modena, Ist. di Igiene, Via G. Campi 287, 41100 Modena, Italy (Tel.: +39 59 360084; Fax: 363057)
COLL: Fantuzzi, G.; Predieri, G.; Righi, E.

Exposure to Trihalomethanes in Indoor Swimming Pools

The study involves biological monitoring of human exposure to trihalomethanes (THM) in indoor swimming pools, particularly to chloroform, a potential carcinogen (IARC group 2B, U.S. National Toxicology Program group 2). Blood levels of chloroform have been evaluated in exposed (127 swimmers and visitors in indoor swimming pools) and 40 non exposed subjects. Samples of end-expired air (alveolar air) collected from 160 swimmers and 26 controls are also being analysed in order to assess the validity of this index when evaluating environmental exposure. Data about subjects are collected by questionnaire and also include information on possible exposure to solvents outside the swimming pool (e.g. in hobby activities); the type of activity practised in the swimming pool (visitor, swimmer, style of swimming, etc.) is recorded, as well as time spent at the pool. Samples are analysed by gas chromatography with an electron capture detector, and confirmed by mass-spectrometry.

TYPE: Methodology
TERM: Monitoring
SITE: All Sites
CHEM: Chloroform; Trihalomethanes
TIME: 1985 - 1995

***520** **Aggazzotti, G.** 05389
Univ. di Modena, Ist. di Igiene, Via G. Campi 287, 41100 Modena, Italy (Tel.: +39 59 360084; Fax: 363057)
COLL: Fantuzzi, G.; Righi, E.; Predieri, G.; Moscardelli, S.; Tartoni, P.L.

Occupational and Environmental Exposure to Perchloroethylene

Perchloroethylene (PCE) is a potential carcinogen (IARC group 2B) widely used as dry cleaning agent. This study involves biological monitoring of PCE in dry cleaners and their family members in order to evaluate the amount of exposure. Human exposure is evaluated collecting end-expired air (alveolar air) samples from dry cleaners in their shops and in their homes, where the presence of PCE is due to exhaled breath of the exposed subjects and to residuals in clothes. The level of PCE inside the homes is also evaluated in samples of alveolar air collected from family members of dry cleaners. Levels of PCE in ambient air are evaluated by spot samples and by personal passive dosimeters. Samples are analysed by gas-chromatography with an electron capture detector and confirmed by mass-spectrometry. Personal data on dry cleaners and their family members and information about the laundries and the washing procedures are collected by a structured questionnaire. A pilot study has been performed involving 18 dry cleaning shops (32 cleaners), 16 homes and 13 family members. To compare levels of PCE in dry cleaners' homes and in alveolar air samples collected from their family members in the same houses, 16 private and public areas were chosen at random in the same municipality. Inside the public areas samples of alveolar air were collected from 16 control subjects with no known exposure, matched with the family members of dry cleaners for age and sex and exposure. PCE was found in most of the houses visited; samples collected inside the homes of dry cleaners showed significantly higher levels than those in control houses.

TYPE: Cross-Sectional; Methodology
TERM: Monitoring; Occupation; Solvents
SITE: All Sites
CHEM: Perchloroethylene
OCCU: Dry Cleaners
TIME: 1992 - 1997

521 **Ponz de Leon, M.** 04480
Ist. di Patologia Medica, Policlinico, Via del Pozzo 71, 41100 Modena, Italy (Tel.: +39 59 379269)
COLL: Roncucci, L.; Pedroni, M.; Benatti, P.; Zanghieri, G.; Digregorio, C.; Fante, R.; Losi, L.; De Pietri, P.

Registry of Digestive Tract Cancers

Digestive tract cancers, in particular colorectal tumours, have been registered since 1984 in a predominantly urban population in Northern Italy. Of special interest is familial occurrence of these and other neoplasms. For this purpose a detailed genealogical tree is drawn for the registered patients, limited to first-degree relatives. Occurrence of cancer in relatives is recorded and compared with that in a suitable control group drawn from the general population, matched for age (+/- 5 years) and sex. The specific objective of this study is to determine the frequency of cancer among relatives, to ascertain the total burden of cancer in the general population attributable to hereditary tumours (Lynch syndromes I and II), and to explore the possible implication of different types of genetic transmission in colorectal cancer.

TYPE: Cohort
TERM: Familial Factors; Heredity; Registry
SITE: Colon; Rectum
REGI: Modena (Ita)
TIME: 1984 - 1994

Monfalcone

522 **Bianchi, C.** 01963
Hosp. of Monfalcone, Lab. of Pathological Anatomy, Via Galvani, 34074 Monfalcone, Italy (Tel.: +39 481 487500)
COLL: Brollo, A.; Ramani, L.; Zuch, C.

Asbestos Exposure and Malignancies in the Monfalcone Area

The objectives are: (1) to obtain data on asbestos exposure in the Monfalcone area, north-eastern Italy (about 60,000 inhabitants); (2) to identify asbestos related tumours; and (3) to prevent adverse effects in subjects exposed to asbestos in the past. Three parameters are investigated in a necropsy series (about 1,700 cases):hyaline pleural plaques, lung asbestos bodies and work history. The relationships between the above parameters and smoking and alcohol habits and malignancies are analysed. Available data indicate that severe exposure to asbestos occurred in the Monfalcone area until the 1970s. Working in shipyards was the main source of exposure. A large number of pleural and pulmonary asbestos-related tumours were observed and a possible relationship between asbestos and liver cancer emerged. A publication appeared in IARC Scientific Publications No. 112, E. Riboli and M. Delendi (eds), Lyon, 1991, pp 127-140.

TYPE: Correlation
TERM: Air Pollution; Alcohol; Autopsy; Chemical Exposure; Dusts; Environmental Factors; Occupation; Tobacco (Smoking)
SITE: All Sites
CHEM: Asbestos
TIME: 1980 - 1994

Novafeltria

***523** **Venturi, S.** 05371
Serv. d'Igiene e Santià Pubblica, USL No.1, Piazza Bramante 2, 61015 Novafeltria, Italy (Tel.: +39 541 928205; Fax: 920856)
COLL: Venturi, M.; Guidi, A.

Dietary Iodine and Gastric Cancer

The aim of this study is to test the hypothesis that iodine deficiency, or, in some cases, iodine excess is a risk factor for gastric cancer. Recent regional and district data on the epidemiology of endemic goitre and gastric cancer in Italy will be correlated. Gastric cancer trends (until 1997) will be monitored in the populations (districts of Montefeltro, Bolzano and Garfagnana) where iodine prophylaxis have been introduced only in recent years (1981-1985). Atrophic gastritis prevalence in Montefeltro district will be reconsidered, and gastric cancer mortality trends in the province of Aosta, where the iodine prophylaxis was interrupted in 1975, will be analysed. The iodine trapping ability of the stomach and the thyroid gland is inhibited by goitrogens, such as nitrate, thiocyanate and salt, previously studied as risk factors for gastric cancer. Previous studies have shown that iodine deficient individuals have more gastric cancer and atrophic gastritis than non deficient subjects; it is believed that iodine is able to regulate gastric trophism and antagonise the action of the above mentioned iodine inhibitors. A paper was published in Eur. J. Cancer Prev., 2:17-23, 1993.

TYPE: Correlation; Mortality
TERM: Atrophic Gastritis; Goitre; Trends
SITE: Stomach
CHEM: Iodine
TIME: 1992 - 1997

Padova

524 **Chieco-Bianchi, L.** 04376
Univ. degli Studi di Padova, Ist. di Oncologia, Via Gattamelata, 64 , 35128 Padova, Italy (Tel.: +39 49 8071859; Fax: 8072854)
COLL: De Rossi, A.; Amadori, A.; Del Mistro, A.; Calabro, L.; Gallegaro, L.; Miazzo, G.; Whittle, H.C.; Hall, A.J.; Bosch, F.X.

Natural History of Human Retrovirus Infections in The Gambia

The principal objective of this study is to evaluate the prevalence and the pattern of spread of retrovirus infections in Gambian children. This will be done by evaluating virus-specific serum antibodies as well as by monitoring other markers for viral infection (antigenaemia, Ig in vitro synthesis, in situ hybridization virus isolation). This study will provide information on virus transmission in paediatric populations, types of retrovirus involved and, possibly, their pathogenic spectrum. Children with signs of AIDS or related pathological conditions will be evaluated for virological and immunological parameters. Tumour samples, either frozen or paraffin embedded, will be processed for in situ hybridization using virus probes. If available, viable biological material will be studied for virus detection. Serological analysis for antibodies to HIV-1, HIV-2 and HTLV-I/II on the samples collected in 1988 and 1989 from mother-child pairs is now complete. Results on HIV-1 and HIV-2 seroprevalence rates have been published (J. Acquir. Imm. Def. Synd. 5:19-24, 1992).

TYPE: Cross-Sectional
TERM: AIDS; Childhood; HIV; Prevalence
SITE: All Sites
LOCA: Gambia
TIME: 1988 - 1994

Palermo

525 **Pinzone, F.** 05082
Sicilian Regional Health Adm., Epidemiology Unit, Via Vaccaro 5, 90145 Palermo, Italy (Tel.: +39 91 6969220)
COLL: Dardanoni, L.; Pagliaro, L.; Brignone, G.; Palazzotto, G.; Traina, A.; Romano, F.M.; Cusimano, R.; Pisa, R.

Palermo Cancer Registry

A population-based cancer registry has been created in Palermo, registering all incident cases of neoplastic disease (ICD-9 140-208) among residents in Palermo urban area (approximately 750,000 inhabitants).

TYPE: Incidence; Registry
TERM: Data Resource
SITE: All Sites
TIME: 1991 - 1999

Perugia

526 **La Rosa, F.** 04723

Univ. of Perugia, Dept. of Hygiene, Via del Giochetto, 06100 Perugia, Italy (Tel.: +39 75 5853329; Fax: 5853317)

COLL: De Bartolomeo, A.; Greco, M.; Gubbiotti, C.; Mastrandrea, V.; Minelli, L.; Patavino, V.; Petrinelli, A.M.

Risk Factors in Colon Cancer

The project aims to acquire knowledge on some biological parameters, particularly nutritional ones, to be used as markers in cancer of the colon, and to study the markers of individual susceptibility to this cancer. Different population groups will be studied: healthy people, subjects affected by ulcerative colitis or polyposis, subjects with adenomas, and patients with cancer of the colon. The project comprises the following phases: (1) compilation of anamnestic and dietary data; (2) survey of serological markers by quantifying axotemia, cholesterol, lipids, Ca++, selenium and some vitamin concentrations A, B2, C, E); (3) survey of biological markers, by isolating the T-lymphocytes in blood samples and by studying the DNA damage and the individual ability to repair them; (4) assessment of faecal bile acid (FBA) concentrations, primary FBA/secondary FBA ratio, litocholic acid/deoxycholic acid ratio both in the faecal water and in the raw faeces, the pH and Ca++ concentrations of faecal water, the cytotoxicity and mutagenicity of faecal extracts.

TYPE: Case Series; Cohort; Correlation
TERM: BMB; Bile Acid; Biomarkers; Blood; Cholesterol; DNA Repair; Diet; Faeces; Familial Factors; Lymphocytes; Serum; Trace Elements; Ulcerative Colitis; Vitamins
SITE: Colon
CHEM: Selenium
TIME: 1993 - 1997

Ragusa

***527** **Gafà, L.** 05320

Ragusa Cancer Registry, Piazza Igea, 97100 Ragusa Sicily, Italy (Tel.: +39 932 600681)

COLL: Dardanoni, L.; Tumino, R.

Frequency of Aberrant Crypt Foci in Apparently Healthy Mucosa of the Colon in Patients Suffering from Different Colon Diseases

There is still little known about the histogenesis of colorectal cancers. Morphological alterations like the so called "aberrant crypt foci" (ACF), observed both in man and in experimental animals, have been found in apparently healthy areas of the colonic mucosa in patients suffering from different pathologies (familiar polyposis, benign colonic diseases, cancer). This study aims to identify, quantify and compare the frequency of ACF in the colon of patients living in two Italian areas (Modena and Ragusa) where population-based registries are active, and where incidence of colorectal cancer is significantly different. Frequency of ACF by age, sex and familiarity of colon diseases will be investigated in patients submitted to colonoscopy (about 50 in Ragusa and 100 in Modena). Data will be periodically checked against cancer registration data to evaluate the predictive importance of the early aspects of colon diseases.

TYPE: Correlation
TERM: Disease, Other; Registry
SITE: Colon
REGI: Modena (Ita); Ragusa (Ita)
TIME: 1993 - 1995

Rome

528 **Cerimele, D.M.** 01443

Catholic Univ., A. Gemelli Polyclinic, Dept. of Dermatology, Via Pineta Sacchetti , 00168 Rome, Italy (Tel.: +39 6 33054227)

COLL: Tulli, A.; Rusciani, L.; Capizzi, R.

Environmental and Genetic Factors in Multiple Skin Cancer

The main aim of this research is to study the inter-relationships between environmental and genetic factors in the development of multiple skin cancers. A questionnaire developed for the study, is being used to collect information from skin cancer patients on colour of the skin, eyes and hair, capacity to tan, duration of sun exposure to the skin, and presence now or in the past of other skin cancers. HLA typing of patients affected by two or more skin cancers will be undertaken, to confirm the possible negative correlation between HLA-B17 and the presence of multiple skin cancers. A paper has been published (Dermatologica 176, 176-181; 1988).

TYPE: Cross-Sectional
TERM: Environmental Factors; Genetic Factors; HLA; Immunology; Multiple Primary; Pigmentation; Radiation, Ultraviolet
SITE: Skin
TIME: 1988 - 1994

529 **Cerimele, D.M.** 04601
Catholic Univ., A. Gemelli Polyclinic, Dept. of Dermatology, Via Pineta Sacchetti , 00168 Rome, Italy (Tel.: +39 6 33054227)
COLL: Rotoli, M.; Celleno, L.; Cottoni, F.; Borroni, G.; Giannetti, A.; Amerio, P.L.; Lomuto, M.

Epidemiology of Kaposi's Sarcoma in Italy

The main objective of the present study is to collect information on the incidence of classic Kaposi's Sarcoma in Italy. Southern Italy is thought to be an area of high prevalence of Kaposi's Sarcoma, but no data are available, except for a very small area, north Sardinia. A preliminary investigation addressed to all dermatologists working in southern Italy has given unsatisfactory results. A sample study has been planned, to investigate all cases of Kaposi's Sarcoma observed in four areas comparable for size and population. Two areas are thought to be areas of high prevalence (Sassari, Chieti), the other two to be areas of low prevalence (Pavia, Modena). The collaboration of the dermatologists working in these four areas has been obtained. A questionnaire, similar to that used to gather the data in the Sassari area, has been distributed to collect the main clinical, pathological and genetic features of patients affected by Kaposi's Sarcoma. HLA typing has been planned to check if the immunogenetic features previously observed in the Sassari area (Acta Med. Rom. 25, 270-278; 1987) are common also to other Italian areas. A cytogenetic investigation is in progress to expand the data previously reported (Human Genet. 72, 311-317; 1986).

TYPE: Incidence
TERM: Antigens; HLA; Immunology
SITE: Kaposi's Sarcoma
TIME: 1988 - 1994

530 **Comba, P.** 04614
Ist. Superiore di Sanità, Lab. di Igiene Ambientali, Viale Regina Elena 299, 00161 Rome, Italy (Tel.: +39 6 4990; Fax: 4440064 ; Tlx: 610071 istisan)
COLL: Belli, S.; Bruno, C.; Grignoli, M.; Maiozzi, P.; Savelli, D.

Epidemiology of Asbestos-Related Disease

This project encompasses mortality and morbidity studies concerning asbestos-related disease in Italy, including both descriptive and aetiological studies. Mortality from pleural malignant neoplasms has been studied in the Italian municipalities, leading to the detection of several previously unreported high risk areas. Occupational cohort studies are being conducted among asbestos cement workers, subjects who worked in the construction and maintenance of trains, seamen, salt rock miners and subjects compensated for asbestosis. The mortality of subjects living in the neighbourhood of an asbestos mine is being studied. Papers were published in Am. J. Ind. Med. 21:681-687 and 863-872, 1992 and in Ann. Ist. Sup. San. 27:319-324, 1991 (Vetrugno et al.) and 28:589-600, 1992 (Di Paola et al.).

TYPE: Cohort; Correlation
TERM: Asbestosis; Dusts; Occupation
SITE: Mesothelioma; Peritoneum; Pleura
CHEM: Asbestos
OCCU: Asbestos Workers
TIME: 1988 - 1994

531 **Crespi, M.** 04166

National Cancer Inst. "Regina Elena", Dept. Environmental Carcinogenesis, Epidemiology & Prevention, Viale Regina Elena 291 , 00161 Rome, Italy (Tel.: +39 6 4457086; Fax: 4457086)

COLL: Giacosa, A.; Tortorelli, A.; Castellaneta, A.; Saragoni, A.; Bonaguri, C.; Falcini, F.; Ridolfi, R.; Morettini, A.; Zampi, G.C.; Guarnieri, C.; Gabrielli, M.; D'Albasio, G.; Casale, V.; Tarquini, M.; Grassi, A.; Caperle, M.; Sacripanti, P.; Scala, S.; Franze, A.; Bordi, C.; Missale, G.

Diet and Chronic Atrophic Gastritis

This is the Italian section of the European Collaborative Study on the role of diet and other factors on the aetiology of atrophic gastritis. Using a frequency dietary questionnaire, the study will aim to define different dietary profiles for cases (those with atrophic gastritis type B) and matched controls. Every participating centre (Rome, Florence, Forli, Genoa, Parma) will collect data on 120 subjects: 40 cases, 40 endoscopic controls and 40 population or hospital non-endoscopic controls (for evaluation of the possible bias inbuilt in the endoscopic population). The endoscopic protocol, the histopathological characterisation and the modalities of dietary interviews, have been standardised among the participating centres. For the Italian section, data on about one third of the whole population should be available in mid 1989.

TYPE: Case-Control
TERM: Atrophic Gastritis; Diet; Premalignant Lesion
SITE: Stomach
TIME: 1986 - 1994

532 **Crespi, M.** 05162

National Cancer Inst. "Regina Elena", Dept. Environmental Carcinogenesis, Epidemiology & Prevention, Viale Regina Elena 291 , 00161 Rome, Italy (Tel.: +39 6 4457086; Fax: 4457086)

COLL: Conti, E.M.S.; Ramazzotti, V.; Genovese, O.

Cancer in Farmers Employed in Greenhouses

The study is conducted in the Province of Latina where many greenhouses for flower and vegetable farming exist since the 1960's. All diagnosed cases (1983-1987) of soft-tissue sarcoma, non-Hodgkin's lymphoma and malignant brain tumour have been identified by the Latina Cancer Registry. The controls are represented by two randomly selected groups, one in the general population, the other in the Latina hospital from subjects free of cancer. All cases (or their relatives) and controls are interviewed by questionnaire on their occupational histories, habits, etc. The occupational histories will be analysed using a pre-planned job-exposure matrix, specific for type of agricultural production, calendar period, geographical location and pesticide use. In the same area a cohort study of approximately 9,000 men licensed to buy and use pesticides during the period 1971-1984 is planned. Their mortality for all cancers and specific sites will be compared with the general population. Information on the pesticides actually handled and occupational history and exposure in greenhouses will be obtained from a 10% random sample of the cohort. Finally, approximately 40 farmers employed in greenhouses will be asked to participate in biological monitoring for the assessment of exposure to the pesticides.

TYPE: Case-Control; Cohort
TERM: Occupation; Pesticides; Registry
SITE: All Sites; Brain; Non-Hodgkin's Lymphoma; Soft Tissue
OCCU: Agricultural Workers; Farmers; Horticulturists
REGI: Latina (Ita)
TIME: 1990 - 1994

533 **Crespi, M.** 05163

National Cancer Inst. "Regina Elena", Dept. Environmental Carcinogenesis, Epidemiology & Prevention, Viale Regina Elena 291 , 00161 Rome, Italy (Tel.: +39 6 4457086; Fax: 4457086)

COLL: Conti, E.M.S.; Caperle, M.; Ramazzotti, V.; Ferro-Luzzi, A.; Maiani, G.; Spaziani, E.

Ecological Study on Diet, Vitamins and Cancer Frequency

The study is conducted within two areas of Latina Province: Campodimele and Roccagorga. The first (800 inhabitants) seems to be characterized by prolonged life-expectancy due also to a low frequency of diet-related cancers and cardiovascular diseases; migration may be a possible bias. The second (4,500 inhabitants) represents a comparison area. Both are covered by the Latina Cancer Registry. The population under study is composed randomly by two comparable groups (400 from Campodimele and 800 from Roccagorga). Each subject will be interviewed by a semi-quantitative food consumption questionnaire. Anthropometric measurements, blood pressure and lifestyle will also be assessed.

Vitamins (carotenoids and vitamin C), cholesterol and its fractions and triglycerides will be analysed in blood. Data analysis will be performed to evaluate a possible protective effect of dietary habits and vitamin levels.

TYPE: Cross-Sectional
TERM: BMB; Cholesterol; Diet; Registry; Vitamins
SITE: All Sites; Breast (F); Colon; Stomach
CHEM: Beta Carotene
REGI: Latina (Ita)
TIME: 1991 - 1994

534 **Greggi, S.** 04269
Catholic Univ. of San Cuore, Dept. Gynaecology & Obstetrics, EORTC, Gynaecological Oncology Group, Largo A. Gemelli 8, 00168 Rome, Italy (Tel.: +39 6 30154496; Fax: 3051343; Tlx: 611330 ucatro i)
COLL: Turnbull, A.; Bosze, P.; Franceschi, S.; Ponder, B.A.J.; Baak, J.P.A.

European Registry of Familial Ovarian Cancer

Epidemiological studies have provided little data useful for the identification of women at risk for ovarian carcinoma (OC). In addition, there are no cheap, reliable screening methods and OC is almost invariably diagnosed at an advanced stage. During the 1980s, genetic evidence of the aetiology of OC has rapidly accrued. Three carcinoma-prone conditions have been recognised in which OC shows significant familial concentration : (1) site-specific OC (risk restricted to OC); (2) breast cancer/OC syndrome (breast cancer associated with OC); and (3) cancer family syndrome (Lynch syndrome II: hereditary non-polyposis colorectal cancer associated with endometrial cancer and/or OC). Families with two or more close relatives with OC have been conventionally considered suitable for epidemiological/genetic analyses. Their OC lifetime risk has been calculated to be 11-18 times higher than in the average population. Both genetic and environmental factors could account for familial aggregation of OCs, but in a subset of families, genetic background plays a major role. The hereditary pattern seems to be consistent with the segregation of an autosomal dominant mutation with variable penetrance, but the mode of genetic transmission has not been conclusively clarified. Data on familial OC are still insufficient, most being retrospectively collected, and a larger prospective study is needed. Cases from Europe have been only occasionally reported and, if sufficient suitable families can be recruited, a bank of biological samples would be extremely important for marker and molecular genetic studies. Registry plans are to evaluate the following aspects: (1) identification and registration of OC-prone families; (2) knowledge of detailed epidemiological characteristics of familial OC; (3) establishment of a bank of biological specimens from family members; (4) identification, registration and surveillance of high-risk subjects; (5) exchange of data with the US registry.

TYPE: Case Series; Incidence; Registry
TERM: BMB; Biomarkers; Familial Factors; High-Risk Groups; Hormones; Pedigree; Reproductive Factors
SITE: Ovary
CHEM: Oestrogens; Talc
LOCA: Austria; Belgium; Bulgaria; Czech Republic; Finland; France; Germany; Greece; Hungary; Israel; Italy; Netherlands; Poland; Russian Federation; Sweden; Switzerland; United Kingdom; Yugoslavia
TIME: 1989 - 1999

***535** **Lagorio, S.** 05438
Ist. Superiore di Sanita, Lab. Igiene Ambientale, Viale Regina Elena, 299, 00161 Rome, Italy (Tel.: +39 6 4990; Fax: 4440064; Tlx: 610071 istisan)
COLL: Biocca, M.; Carere, A.; Forastiere, F.; Iavarone, I.; Miceli, M.; Pasquini, A.; Perucci, C.A.; Vanacore, N.

Cohort Mortality Study of Filling Station Attendants

Gasoline vapours and exhausts contain chemicals which are known human carcinogens, such as benzene. Both gasoline exhausts and fuels have been classified by the IARC as possibly carcinogenic to humans. A historical cohort of service station attendants, enrolled through the census of filling stations run in 1980 in every Italian region, is currently underway. It aims to evaluate possible excess of cancer risk (with emphasis on neoplasms of the lympho-haemopoietic tissue, lung, brain, genitourinary system, and digestive organs) in relation to exposure to gasoline vapours and engine exhausts. 5,671 managers of service stations from the Latium and Emilia-Romagna regions are included in the study. At entry they had worked for 15 years on average, were mainly full-time employees (69%), and 71% worked in small size stations. Information available from the 1980 survey, such as yearly sales of gasoline and length of employment at entry, will be used as indirect indexes of exposure. Refuelling is generally always provided by the attendant at Italian petrol stations, and self-service pumps operate mainly at night. An estimated average quantity of gasoline sold per year per full-time employee should then be proportional to the average intensity of exposure experienced at dispensing fuel. The feasibility of a retrospective exposure assessment, based on findings from an environmental survey conducted in 1992 on a sample of current employees, is being evaluated.

TYPE: Cohort; Mortality
TERM: Occupation; Solvents
SITE: All Sites
CHEM: Benzene; Diesel Exhaust; Gasoline; Vehicle Exhaust
OCCU: Petrol Station Attendants
TIME: 1991 - 1995

***536** **Nesti, M.** 05439
National Institute for Prevention & Occupational Safety (I.S.P.E.S.L.), Via Alessandria, 220/E, 00198 Rome, Italy (Tel.: +39 06 8841001398; Fax: 8555218)
COLL: Chellini, E.; Comba, P.; Crosignani, P.; Magnani, C.; Merler, E.

National Register of Asbestos-Related Mesothelioma Cases

The National Institute for Prevention and Occupational Safety has started to conduct a National Register of Asbestos-related Mesothelioma Cases. The purposes are to assess incidence of mesothelioma in Italy, to collect information about asbestos exposure and to estimate the impact of industrial asbestos exposure on the working population with the aim of planning occupational prevention. A pilot project which involves preparing guidelines to collect asbestos-related mesothelioma cases in a standardized way should be completed by 1993.

TYPE: Incidence; Registry
TERM: Dusts; Occupation; Prevention
SITE: Mesothelioma
CHEM: Asbestos
TIME: 1992 - 1999

537 **Settimi, L.** 04894
Ist. Superiore di Sanitá, Lab. di Igiene Ambientale, Viale Regina Elena 299, 00161 Rome, Italy (Tel.: +39 6 4990; Fax: 4040064)
COLL: Belli, S.; Comba, P.; Magnani, C.; Terracini, B.; Grignoli, M.; Maiozzi, P.

Cancer Risk and Pesticide Exposure

The study has been planned to investigate on possible associations between several cancer sites and occupational exposure to pesticides and to test previously reported associations. It is a multicentric hospital-based case-control study specifically designed to investigate the subjects' occupational chemical exposures. During a 30 month period, histologically confirmed incident cases of major cancers in adults of both sexes are identified in hospitals in six Italian rural areas (Asti, Imola, Pescia, Pistoia, Grosseto, Orbetello). Trained personnel interview patients about their life-time work experience, with special emphasis on activities in agriculture. Other information includes pathological data, smoking history and diet. A team of coders analyse the occupational histories using a pre-planned job-exposure matrix, specific with respect to type of agricultural production, calendar years, geographical location and pest. Each cancer site will be compared with a selected pool of the other sites. During the first year of the study approximately 2,400 cases have been identified.

TYPE: Case-Control
TERM: Occupation; Pesticides
SITE: All Sites
OCCU: Agricultural Workers
TIME: 1990 - 1994

Torino

538 **Ciccone, G.** 03695

Univ. of Torino, Dept. Biomed. Science & Human Oncology, Serv. Epidemiology of Tumours, Via Santena 7, 10126 Torino, Italy (Tel.: +39 11 678872)

COLL: Comba, P.; Vineis, P.

Phenoxy Herbicide Exposure and Soft-Tissue Sarcoma

The study is conducted in a rural area where large amounts of 2,4,5-T and 2,4-D have been used in rice-growing. All newly diagnosed cases of soft-tissue sarcomas which occurred in 1984-1988 in both sexes are identified (the final number will be approximately 150). Histological specimens have been collected and are read independently by two pathologists. The control group is represented by a sex-stratified random sample of the general population (approximately 300 individuals). A second control group of randomly selected deceased individuals will be used for comparison with the deceased cases. All cases and controls (or their relatives) are interviewed by mail questionnaire on their occupational histories (among other items). The questionnaires are evaluated blindly by a working group (including agronomists), in order to assess the probability of exposure to phenoxy herbicides for each individual. The relative risk of soft-tissue sarcoma following exposure to phenoxy herbicides for different jobs in agriculture will be estimated. A paper has been published in Scand. J. Work Env. Hlth 13:9-17, 1986, describing the first phase of the study (1981-83).

TYPE: Case-Control
TERM: Chemical Exposure; Environmental Factors; Herbicides; Occupation; Pesticides; Rural
SITE: Soft Tissue
CHEM: 2,4,5-T; 2,4-D; Phenoxy Acids
TIME: 1984 - 1994

539 **Ciccone, G.** 04612

Univ. of Torino, Dept. Biomed. Science & Human Oncology, Serv. Epidemiology of Tumours, Via Santena 7, 10126 Torino, Italy (Tel.: +39 11 678872)

COLL: Vineis, P.; Avanzi, G.C.; Rege Cambrin, G.; Ponzio, G.

Solvent Exposure and Cytogenetics in Leukaemias and MDS

A few previous investigations have suggested that environmental exposure to chemicals, particularly in the occupational setting, could be associated with cytogenetic abnormalities in patients affected by leukaemia. Turin is a large industrialised town, where approximately 10% of the adult population is estimated to be occupationally exposed to organic solvents. A pilot study on cytogenetic abnormalities in leukaemias and MDS has therefore been undertaken. The study consists in the enrolment of all cases of these pathologies in the Main Hospital of the town (approximately 100 leukaemias and 70 MDS in one year). Such patients are interviewed with a detailed questionnaire, including the collection of specific information on solvent exposure at work. Other items which are investigated are exposure to welding fumes and cigarette smoking. All cases (except those with chronic lymphocytic leukaemia) will have a cytogenetic characterisation, performed with standardised banding techniques. The purposes of the study are: (1) to investigate possible associations between cytogenetic abnormalities (e.g. deletions or translocations) and environmental exposure; (2) to assess the possibility of including cytogenetic tests within a larger population-based investigation on leukaemia, which is now in the early planning phase.

TYPE: Case Series; Molecular Epidemiology
TERM: Chromosome Effects; Occupation; Solvents; Tobacco (Smoking); Welding
SITE: Leukaemia; Myelodysplastic Syndrome
TIME: 1988 - 1994

540 **Terracini, B.** 02703

Univ. di Torino, Depto di Scienze Biol. & Oncol. Umana, Serv. di Epidemiologia dei Tumori, Via Santena 7, 10126 Torino, Italy (Tel.: +39 11 678872/69663)

COLL: Ceci, A.; Fossati-Bellani, F.; Lo Curto, M.; Magnani, C.; Mancini, M.; Masera, G.; Paolucci, G.; Rossi, M.; Rosso, P.

Second Primary Neoplasms among Long-Term Survivors from Childhood Cancer

The occurrence of second primary cancers is being monitored among long-term survivors after some types of childhood cancer reported to the Italian Registry of off-therapy children. Risks for second cancers have been estimated in a cohort including subjects who (1) were off therapy before 1 January 1981 (onset of the Registry) and were alive on that date, or (2) were off therapy during 1981-1988. This corresponded to a total of 3,141 subjects, 505 with Hodgkin's Disease, 223 with NHL, 299 with neuroblastoma, 347 with Wilms' Tumour, 1,630 with ALL and 137 with other acute leukaemias. 1,025 subjects were born before 1970 and 2,116 after. 1,745 were males and 1,396 females. Deaths and second primary incident cancers were identified respectively through the Registrars of the towns of residence and through the Institutions where patients are being followed up. The study includes 19,259 person-years at risk, 9,571 following ALL. A total of 28 second primary cancers were diagnosed histologically during the follow-up period vs. 3.12 expected (from the sex- and age-specific rates from the Varese Cancer Registry in 1982-1987). Among long-term survivors after ALL, a primary cancer of the CNS occurred in 9 vs 0.20 expected. This observation led to the design of a case-control study nested in the ALL sub-cohort, which is presently under way. Risk factors being investigated relate to the therapy originally used for treatment of ALL.

TYPE: Cohort
TERM: Childhood; Multiple Primary; Survival; Treatment
SITE: Hodgkin's Disease; Leukaemia (ALL); Leukaemia (ANLL); Neuroblastoma; Non-Hodgkin's Lymphoma; Wilms' Tumour
TIME: 1991 - 1994

541 **Vineis, P.** 04353
Univ. di Torino, Dipto. di Scienze Biomediche e Oncologia Umana, Via Santena 7, 10126 Torino, Italy (Tel.: +39 11 678872/69032; Fax: 6635267)
COLL: Bartsch, H.; Caporaso, N.E.; Harris, C.C.; Kadlubar, F.; Tannenbaum, S.R.; Wogan, G.N.; Estève, J.; Talaska, G.; Terracini, B.; Shields, P.G.

Biochemical Epidemiology of Smoking and Bladder Cancer

A pilot study has suggested that aromatic amines contained in tobacco smoke, particularly 4-aminobiphenyl, are associated with the excess of bladder cancer observed in smokers. 4-aminobiphenyl and other aromatic amines contained in tobacco smoke bind covalently to macromolecules, including haemoglobin. The concentration of such adducts may be influenced by the variable ability to metabolize aromatic amines shown by different individuals. In this collaborative project, biological samples are being collected from healthy subjects and bladder cancer patients, and haemoglobin and DNA adducts are being measured in exfoliated cells and bladder biopsies. The subjects are also phenotyped for the activity of N-acetyltransferase and other enzymes relevant to aromatic amine metabolism, using a method based on caffeine administration. Haemoglobin adducts of 4-aminobiphenyl and other aromatic amines are measured in different metabolic subgroups (i.e. slow vs. fast metabolizers). In addition, urinary mutagenicity and the concentration of nitrosamines and aromatic amines in the urines are measured. A paper by Bartsch, H. et al. has been published in J. Nat. Cancer Inst. 82:1826, 1991.

TYPE: Case-Control
TERM: BMB; Blood; DNA Adducts; Tobacco (Smoking); Urine
SITE: Bladder
CHEM: Amines, Aromatic; N-Nitroso Compounds
TIME: 1988 - 1994

542 **Vineis, P.** 04788
Univ. di Torino, Dipto. di Scienze Biomediche e Oncologia Umana, Via Santena 7, 10126 Torino, Italy (Tel.: +39 11 678872/69032; Fax: 6635267)
COLL: Seniori Costantini, A.R.; Crosignani, P.; Miligi, L.R.; Pisani, P.; Rodello, S.; Stagnaro, E.; Tumino, R.; Funto, I.; Fontana, A.; Demicheli, V.; Masala, G.; Nanni, O.; Ramazzotti, V.; Viganò, C.

Haematolymphopoietic Malignancies in Italy

High prevalence of exposure to solvents in industrial areas (estimated 10-16% in adult males in the provinces of Torino, Varese and Firenze), and to herbicides in agricultural areas is found in some parts of Italy. A multicenter population-based case-control study on haemato-lymphopoietic malignancies is planned in 12 Italian areas, chosen on the basis of high exposure prevalence. In three years, it is intended to interview approximately 1,100 cases of leukaemia, 1,900 cases of lymphoma (Hodgkin and non-Hodgkin) and 400 cases of myeloma; 2,000 randomly selected individuals, residents of the same areas, will serve as a population control group. Particular care will be devoted to diagnostic aspects, including B- and T-cell characterization and, in four of the areas, cytogenetics. The occupational histories will be blindly evaluated by industrial hygienists and agronomists, in order to assess the probability of exposure to the chemicals of interest. The study will have a statistical power of 90% to detect a relative risk of 2.0 at the 5% significance level for associations between organic solvent exposure and leukaemias and between herbicide exposure and non-Hodgkin's lymphomas.

TYPE: Case-Control
TERM: Herbicides; Occupation; Pesticides; Solvents; Welding
SITE: Hodgkin's Disease; Leukaemia; Multiple Myeloma; Non-Hodgkin's Lymphoma
CHEM: Phenoxy Acids
OCCU: Electrical Workers; Farmers; Leather Workers; Mechanics
TIME: 1990 - 1994

***543** **Vineis, P.** 05498
Univ. di Torino, Dipto. di Scienze Biomediche e Oncologia Umana, Via Santena 7, 10126 Torino, Italy (Tel.: +39 11 678872/69032; Fax: 6635267)
COLL: Caporaso, N.E.; Hayes, R.B.; Talaska, G.; Pierotti, M.; Airoldi, M.L.; Peluso, M.

Pathogenesis of Bladder Cancer: An Epidemiological Study

Following biochemical and molecular investigations on tobacco smoke, carcinogen-haemoglobin adducts and individual susceptibility, a case-control study on the pathogenesis of bladder cancer is being conducted. The study includes 100 patients with histologically diagnosed bladder cancer and 50 healthy subjects and aims (1) to determine the phenotype and genotype for N-acyltransferase, an enzyme involved in the metabolism of carcinogenic aromatic amines; (2) to measure mutations in ras protooncogenes and in tumour suppressor genes (p53); and (3) to measure the concentration of carcinogen-haemoglobin adducts and of DNA adducts in bladder biopsies. The purpose is to investigate whether smokers and 'slow' acetylators (i.e. genetically susceptible individuals) have a higher frequency of events which are deemed to be pathogenetically relevant.

TYPE: Case-Control; Molecular Epidemiology
TERM: BMB; DNA Adducts; Enzymes; Genetic Factors; Genotype/Phenotype; Oncogenes; Tobacco (Smoking)
SITE: Bladder
TIME: 1993 - 1995

Trento

544 **Amichetti, M.** 05224
Osp. Santa Chiara, Centro per Oncologia, USL C5, Largo Medaglie d'Oro, 38100 Trento, Italy (Tel.: +39 461 903203; Fax: 920030)
COLL: Piffer, S.; Valentini, A.; Graiff, C.; Perani, B.

Breast Self-Examination in Patients with Breast Cancer Diagnosed in the Province of Trento Between 1982-1986

The primary objective of this retrospective study is to evaluate the use of breast self-examination (BSE) among patients with breast cancer in the Province of Trento (population 500,000). The hypothesis is that women with breast cancer rarely perform BSE, a simple method of early detection of breast cancer. All cases diganosed between 1982 and 1986 have been identified by searching pathological and clinical records in local hospitals: this was necessary because there is no tumour registry. Specific questionnaires have been mailed or direct interviews performed. About 650 cases have been identified so far.

TYPE: Case Series
TERM: Prevention
SITE: Breast (F)
TIME: 1988 - 1994

Verona

545 **Ricci, P.** 04188
Univ. di Verona, Ist. di Anatomia Patologica, Strada delle Grazie, 37134 Verona, Italy (Tel.: +39 45 933323)
COLL: Merler, E.; Darra, F.; Piva, C.; Carretta, D.; Lestani, M.; Domenici, R.; Magnani, C.; Terracini, B.

Carcinogenic Risks among Railway Coach Repair Workers

The cancer mortality of an historical cohort of workers employed in coach repair at the Bologna Repair Department of the Italian Railways will be studied. Included are all workers employed between 1957 and 1977 (around 3,000 subjects), whose working histories will be reconstructed using records which include the periods during which each task was performed. Vital status will be determined for each subject, and cause of death will be assessed through death certificates. The aim of the study is to evaluate the mortality risk for exposure to asbestos (exposure to a mixture of chrysotile, amosite and crocidolite since 1957) and welding fumes. The opportunity to perform some nested case-control studies will be evaluated. Cases and controls will be identified from this and other cohorts of Italian coach repair factories already enumerated.

TYPE: Cohort
TERM: Chemical Exposure; Dusts; Occupation; Welding
SITE: All Sites
CHEM: Asbestos
OCCU: Railroad Workers
TIME: 1987 - 1994

Chiba

546 **Murata, M.** 02680

Chiba Cancer Center, Research Inst., Div. of Epidemiology, 666-2, Nitona-cho, Chuo-ku, Chiba 260, Japan (Tel.: +81 43 2645431; Fax: 2628680)

COLL: Takayama, K.; Nakagawa, R.; Fukuma, S.; Shimamura, K.; Hayashi, M.

Follow-Up of Mass Survey by Use of a Population-Based Cancer Registry

Follow-up for cancer incidence and prognosis is continuing in participants of two mass screening programmes, a gynaecological cancer survey in 1977 and a gastric cancer survey in 1984. About 40,000 women participated in the former. Information was obtained about reproductive history and family history of cancer. By the end of 1991, 174 breast cancer cases had been detected. Twice as many controls were chosen from the non-affected members of the population and a nested case-control study was conducted. Family history of breast cancer, early menarche, late marriage and low parity were all found to be risk factors for breast cancer. 65,000 individuals participated in the latter study. Information was obtained through a simple questionnaire about alcohol consumption and smoking habits. By the end of 1991, 1,189 cases of various cancers had been detected. 521 male cases were matched to twice as many controls. Alcohol was found to be a risk factor for colon cancer, tobacco for lung and oesophageal cancer. Synergistic effect of both factors was observed for oesophagus and bladder.

TYPE: Cohort
TERM: Alcohol; Familial Factors; Registry; Reproductive Factors; Screening; Tobacco (Smoking)
SITE: Bladder; Breast (F); Colon; Lung; Oesophagus; Rectum; Stomach
REGI: Chiba (Jap)
TIME: 1978 - 1995

Gifu

547 **Shimizu, H.** 04491

Gifu Univ., School of Medicine, Dept. of Public Health, 40 Tsukasamachi, Gifu 500, Japan (Tel.: +81 582 651241; Fax: 659020)

COLL: Higashiiwai, H.; Sugawara, N.; Morita, N.; Hisamichi, S.; Komatsu, S.

Vitamin A and Dysplasia of Uterine Cervix

To examine the relationship between vitamin A intake or serum vitamin A (retinol) level and the occurrence of dysplasia of uterine cervix, a case-control study is being conducted. About 300 cases of uterine dysplasia are diagnosed annually in Miyagi prefecture. Study cases comprise almost half of the patients randomly selected through a mass screening programme in 1987-1988. Control series are women who visit the Miyagi Cancer Society for general health checks, matched for age and residential area. A health-nurse is interviewing both cases and controls. All of the cases and controls are being asked to provide blood samples for the analysis of serum retinol. Conditional logistic regression analysis will be used to estimate the relative risks for several risk factors for cervical dysplasia, such as number of children and frequency of sexual intercourse, as well as of vitamin A intake.

TYPE: Case-Control
TERM: Parity; Premalignant Lesion; Sexual Activity; Vitamins
SITE: Uterus (Cervix)
CHEM: Retinoids
TIME: 1987 - 1994

***548** **Tanaka, T.** 05382

Gifu University, School of Medicine, First Dept. of Pathology, 40 Tsukasa-machi, Gifu 500, Japan (Tel.: 0582 651241; Fax: 9005)

COLL: Takeuchi, T.; Inaba, S.

Morphometric Analysis of AgNORs in Urinary Cytology

The results of previous studies indicated step-wise increases of the number and area of AgNORs during bladder carcinogenesis in both humans and rodents. The primary objective of the present study is to evaluate AgNORs as a biological marker for detecting high-risk populations of urinary bladder cancer.

The second aim is to assess AgNORs morphometry as a biomarker in cancer prevention trials of cancer of the urinary bladder. Urinary cytology will be done on approximately 3,000 out-patients at three hospitals in Gifu Prefecture and the material stained for AgNORs. Subsequently, morphometric analysis of AgNORs (number of AgNORs/nucleus and area of AgNORs/nuclear area) will be performed. In addition, immunocytochemical staining of PCNA for cell proliferation index will be performed and the possible association between AgNOR indices and PCNA index will be examined. The data will be analysed by univariate and multivariate methods.

TYPE: Methodology
TERM: BMB; Biomarkers; Cytology; Urine
SITE: Bladder
TIME: 1993 - 1994

Hamamatsu

***549** **Maeda, M.** 05271
Hamamatsu Univ., School of Medicine, Dept. of Obstetrics & Gynecology, Handa-cho 3600, Hamamatsu Shizuoka 43131, Japan (Tel.: +81 53 4353209; Fax: 4352308)
COLL: Kawashima, Y.; Terao, T.; Sugimura, M.; Kobayashi, T.

Registration and Follow-Up of Gestational Trophoblastic Disease in Shizuoka Prefecture
A registration and follow-up system of gestational trophoblastic disease has been created to improve early detection and treatment of choriocarcinoma, following hydatidiform mole. The absolute numbers of choriocarcinoma have decreased in this Prefecture since registration begun (9 cases in 1977, 2 cases in 1992).

TYPE: Registry
TERM: Registry; Treatment
SITE: Choriocarcinoma; Hydatidiform Mole
TIME: 1977 - 1999

Hiroshima

550 **Kurihara, M.** 01097
Research Inst. for Nuclear Medicine and Biology, Minami-ku, 2-3 Kasumi 1-chome, Hiroshima 734, Japan (Tel.: +81 82 2511111; Fax: 2558339)
COLL: Munaka, M.; Hayakawa, N.; Ohtaki, M.; Matsuura, M.

Cancer Mortality among Atomic Bomb Survivors in Hiroshima Prefecture
The atomic bomb survivors in Hiroshima Prefecture numbered about 173,000 in 1982. In order to observe the late effects of low dose radiation, a survey on cancer mortality among the survivors from 1968 onwards is being conducted, using the data bank of the survivors in this institute. This includes comparison with mortality among the general population of all Japan and among the non-exposed population of Hiroshima Prefecture. The mortality statistics for 1968 to 1972 was published in J. Radiat. Res. 22:456-471, 1981, and those for 1973 to 1977 in the Proceedings of the Institute No.26, 134-147, 1985 (in Japanese). The mortality for 1978 to 1982 has been investigated and is now being analysed.

TYPE: Cohort; Mortality
TERM: Atomic Bomb; Radiation, Ionizing; Survival
SITE: All Sites
TIME: 1975 - 1994

551 **Shigematsu, I.** 00665
Radiation Effects Research Found., 5-2 Hijiyama Park, Minami-ku, Hiroshima 732, Japan (Tel.: +81 82 2613131; Fax: 2637279)
COLL: Thiessen, J.W.; Trosko, J.E.; Schull, W.J.; Hasegawa, Y.; Kono, T.

Late Radiation Effects of Atomic Bomb Exposure in the Population of Hiroshima and Nagasaki, Japan
The object of the study is to determine the late medical consequences of previous atomic bomb exposure in the populations of Hiroshima and Nagasaki. The project was initiated in 1947 as the Atomic Bomb Casualty Commission in collaboration with the Japanese National Institute of Health. In 1975 the name of the project was changed to the Radiation Effects Research Foundation with equal funding from

Japan and the United States. Examination facilities, computer centres and research laboratories are maintained in both Hiroshima and Nagasaki. Mortality surveillance of a cohort of about 120,000 persons in Hiroshima and Nagasaki and about 75,000 F0l offspring including controls is being conducted. Currently, approximately 9,000 persons exposed as children or adults and about 1,000 exposed in utero, with their controls, receive regular biennial clinical examinations. The most important radiation-related effects are increased incidence rates of leukaemia, multiple myeloma, and cancers of the thyroid, female breast, lung, stomach, colon and ovary. Case-control studies are now in progress to determine possible interactions between radiation exposure and a number of environmental factors. Particular attention is being focused on early cancer detection, precancerous changes, cancer education and possible radiation-induced acceleration of aging. Major laboratory emphasis is placed on various cytogenetic, histological, immunological, hormonal and biochemical changes which may represent evidence of somatic or germinal mutation. Chromosomal aberrations, impairment of cell-mediated immunity, oncogene activation, and certain tissue mutations are being correlated with radiation dose, and the development of cancer. The frequency of biochemical or DNA mutagenesis in the children of the exposed is being actively investigated. Recent refinements of total body and specific organ radiation dose estimates provide the basis for new reports regarding dose-response relationships for all radiation-induced effects.

TYPE: Case-Control; Cohort; Molecular Epidemiology
TERM: Atomic Bomb; Chromosome Effects; DNA; Dose-Response; Histology; Late Effects; Oncogenes; Radiation, Ionizing
SITE: All Sites
TIME: 1947 - 1994

552 **Tahara, E.** 04973
Hiroshima Univ. School of Medicine, Dept. of Pathology I, 1-2-3 Kasumi, Minami-ku, Hiroshima 734, Japan (Tel.: +81 82 2511111; Fax: 2515250)
COLL: Ito, H.; Yasui, W.; Yamakido, M.

Gastric Cancer in Former Workers of a Mustard Gas Factory

The primary objective of the study is to examine the association of gastric cancer in former workers at a mustard gas factory with occupation at the factory, the type of gas exposure, and period of exposure. This study also aims to obtain macroscopic and histological characteristics of gastric cancer in these populations in comparison with control cases which will be randomly selected from gastric cancers diagnosed in the general population. 64 cases of gastric cancer aged 39-84 years among about 5,000 former workers of the factory have been collected during the period from 1974 to 1990.

TYPE: Cohort
TERM: Histology; Occupation
SITE: Stomach
CHEM: Mustard Gas
TIME: 1974 - 1994

***553** **Tahara, E.** 05381
Hiroshima Univ. School of Medicine, Dept. of Pathology I, 1-2-3 Kasumi, Minami-ku, Hiroshima 734, Japan (Tel.: +81 82 2511111; Fax: 2515250)
COLL: Yasui, W.; Yokozaki, H.; Pillai, M.R.

Genetic Analysis of Oral Cancer in India

The primary objective of the study is to examine the genetic changes in oral cancer in India and compare with those of head and neck cancers in Japan. The main target genes are hst-1, int-2 and cyclin-D. These three genes are located at chromosome 11 and frequently co-amplify in the head and neck squamous cell carcinoma in Japan. This study will analyse the co-amplification of these genes in oral cancer in India and assess the significance of the phenomenon in the carcinogenesis of head and neck cancer. This study will also analyse the multinational spectrum pattern of p53 tumour suppressor gene and estimate the effect of various carcinogens in head and neck cancers by molecular epidemiological techniques.

TYPE: Genetic Epidemiology
TERM: Genetic Factors
SITE: Head and Neck; Oral Cavity
LOCA: India; Japan
TIME: 1992 - 1994

Hyogo

554 **Hashimoto, T.** 04212
Hyogo College of Medicine, Dept. of Genetics, Mukogawa-cho, 1-1, Nishinomiya, Hyogo 663, Japan (Tel.: +81 798 456587)
COLL: Furuyama, J.

Establishment of Cell Lines from Fanconi Anaemia Patients and Heterozygote, and Assay of the Sensitivity to Carcinogens of their Cell Lines

Fanconi anaemia (FA) is an autosomal recessive disease and FA cells that show hypersensitivity to cross-linking agents. But it has been unclear that the cells of FA heterozygotes showed hypersensitivity to these agents, though FA heterozygotes were estimated to have a higher risk of developing various cancers. Lymphoblastoid cell lines from FA patients and FA family members were established, and their hypersensitivity to these agents, using cytotoxic analysis and cytogenetical studies checked. 25 cell lines of 6 Japanese FA families were established and contacts with the Department of Pediatrics are being continued in order to find new FA families.

TYPE: Case Series
TERM: Genetic Factors; High-Risk Groups; Lymphocytes; Mutagen; Premalignant Lesion
SITE: Inapplicable
LOCA: United States of America
TIME: 1984 - 1994

Kita-Kyushu

***555** **Takahashi, K.** 05417
Univ. of Occup. & Environm. Health, Dept. of Environmental Epidemiology, Orio, Yahatanishi-ku, Kita-Kyushu 807, Japan (Tel.: +81 93 6917454; Fax: 6017324)
COLL: Okubo, T.; Itoh, T.; Tsuchiya, K.

Lung Cancer among Chromium Plating Workers

The aim is to examine the mortality of chromium platers with special reference to lung cancer. A prospective cohort study of 1,193 male metal platers (626 chromium platers and 567 non-chromium platers) was initiated in 1976. Vital status is ascertained through the Japanese Family Registration System and cause of death through death certificates. Several preliminary reports have been published, the latest in Arch. Environ. Hlth 45:107-111, 1990. Follow-up continues and analyses will be updated.

TYPE: Cohort
TERM: Metals; Occupation
SITE: All Sites; Lung
CHEM: Chromium
OCCU: Chromium Plating Workers
TIME: 1976 - 1996

556 **Tsuchiya, K.** 00703
Univ. of Occupational and Environmental Health, Iseigaoka, Yahatanishi-ku, Kita-Kyushu 807, Japan (Tel.: +81 93 603-1611; Fax: 6017324)
COLL: Okubo, T.; Takahashi, K.

Lung Cancer among Chromium Plating Workers

To examine frequency of lung cancer among chromium platers, 626 chromium platers and, as a reference, 567 platers with history of plating other than chromium have been followed up from 1976. The majority of the chromium handling occurred before 1970. Follow up is carried out through the Japanese Family Registration System and information on cause of death is obtained from death certificates. A preliminary report was made, based on results until 1987, in Arch. Environm. Hlth 45:107-111, 1990. The result showed that lung cancer was the only cause of death that was significantly higher than expected for all platers (16 observed, 8.9 expected; SMR 179; 95% CI 102-290). This elevated SMR, however, was not statistically significant in either of the two plater subgroups. Follow-up will continue.

TYPE: Cohort
TERM: Metals; Occupation
SITE: Liver; Lung
CHEM: Chromium
OCCU: Chromium Plating Workers
TIME: 1977 - 1997

557 **Yoshimura, T.** 04199
Univ. of Occupational and Environmental Health, Dept. of Clinical Epidemiology, 1-1, Iseigaoka, Yahatanishi-ku, Kita-Kyushu 807, Japan (Tel.: +81 93 6031611)
COLL: Ogimoto, I.; Ikeda, M.

Gastric Cancer in the Young Adult and Reproductive History

In Japan the incidence of gastric cancer among young adults is found to be even lower in males than in females. It has been suggested by clinicians that the occurrence of female gastric cancer in young adults might be related to reproductive history. In this study, it is intended to clarify the relationship between gastric cancer in the young adult and reproductive history. 100 gastric cancer cases in young adults (less than 40 years old), 200 hospital controls and 200 population controls will be collected for a matched case-control study. Information on general risk factors, such as dietary habits drinking and smoking history and reproductive history will be obtained by self-administered questionnaire.

TYPE: Case-Control
TERM: Adolescence; Alcohol; Diet; Reproductive Factors; Tobacco (Smoking)
SITE: Stomach
TIME: 1986 - 1994

Koriyama

558 **Nomizu, T.** 04654
Hoshi General Hosp., Dept. of Surgery, 2-1-16, Omachi, Koriyama 963, Japan (Tel.: +81 249 233711; Fax: 393141)
COLL: Watanabe, F.; Fukushima, T.; Takita, K.; Sato, H.; Tsuchiya, A.; Yamaki, Y.; Abe, R.

DNA Study in Cancer-Prone Families

It is suggested that genetic factors are implicated, when cancer patients are clustered in one family. Cancer-prone families have been collected in Tohoku district in Japan. The aims of the study are to define the clinical features of familial breast, gastric or colorectal (HNPCC) cancer and the clinical and genetic biological markers for familial cancer in order to apply the results to cancer prevention. Somatic mutation is being studied and flow cytometry DNA analysis carried out, especially in HNPCC. These studies will clarify the mechanism of carcinogenesis in HNPCC and familial cancer. Chromosome analysis and linkage analysis will clarify the germline mutation. Papers have appeared in Jap. J. Gastoentrol. Surg. 25:1243-1247, 1992, Jap. J. Cancer Clin. 38:1722-1730, 1992 and ESO Monographs Familial Cancer Control, p105-111, 1992.

TYPE: Case Series; Molecular Epidemiology
TERM: Biomarkers; Chromosome Effects; DNA; Familial Factors; Genetic Factors; Linkage Analysis; Mutation, Somatic; Oncogenes
SITE: Breast (F); Colon; Rectum; Stomach
TIME: 1984 - 1995

Kurume City

559 **Fukuda, K.** 04624
Kurume Univ., School of Medicine, Dept. of Public Health, 67 Asahi-machi , Kurume City 830, Japan (Tel.: +81 942 353311)
COLL: Hirohata, I.; Shibata, A.

Primary Liver Cancer and Lifestyle

This study aims to identify any risk factors for primary liver cancer including HBV infection, HCV infection, blood transfusion, alcohol drinking, aldehyde dehydrogenase phenotype and several aspects of daily lifestyle. 200 cases and an equal number of age- and sex-matched hospital controls will be studied. Information is collected by interviewer and will be analysed by multivariate analysis.

TYPE: Case-Control; Case-Control
TERM: Alcohol; Alcohol; HBV; HBV; HCV; Lifestyle; Lifestyle
SITE: Liver; Liver
TIME: 1986 - 1994

Kyoto

560 **Watanabe, H.** 04885
Kyoto Prefectural Univ. of Medicine, Dept. of Urology, Kawaramachi-Hirokoji, Kamigyo-ku, Kyoto 602, Japan (Tel.: +81 75 2515593; Fax: 2117093)
COLL: Hirayama, T.; Hirohata, T.; Aoki, K.; Watanabe, S.; Minowa, M.; Tsuchihashi, Y.

Case-Control Study for High Risk Group of Renal Cell Carcinoma

Mortality from renal cell carcinoma is increasing in Japan. To define the risk factors for this disease, we designed a matched case-control study on 200 cases of renal cell carcinoma and 200 controls. Cases will be patients aged 40-69 with renal cell carcinoma, selected from patients visiting 15 clinics in Kyoto, Osaka, Shiga and Shizuoka in the next five years. Controls will be matched for sex, age (within three years) and residence in the same prefecture. In this study, we will use two questionnaires, with 146 questions covering past and family history, physical condition, physical constitution, eating habits, drinking, smoking, drug use, occupation, income, social environment, residence, lifestyle, marriage and sexual activity. By March 1991, 100 case-control pairs had been recruited.

TYPE: Case-Control
TERM: Alcohol; Diet; Drugs; Lifestyle; Socio-Economic Factors; Tobacco (Smoking)
SITE: Kidney
TIME: 1990 - 1994

Nagasaki

561 **Amemiya, T.** 05212
Nagasaki University, School of Medicine, Dept. of Ophthalmology, 7-1 Sakamoto-machi , Nagasaki 852, Japan
COLL: Takano, J.; Kusuki, Y.

Retinoblastoma in Nagasaki Prefecture

Between 1965 and 1986, the incidence of retinoblastoma in Nagasaki Prefecture was about 1 in 16,053 births, which is the average for Japan. In Shimabara District (Nagasaki Prefecture), the incidence was however 1 in 10,331, which is higher than in any other district in Nagasaki Prefecture. The population of Shimabara has mainly immigrated from Hyogo, Aichi and Shizuoka Prefectures. The incidence of retinoblastoma in Tamba and Tajima Districts (Hyogo Prefecture) was very close to that of Shimabara, 1/10,570 and 1/10,411 respectively. The aim of this study is to assess the role of the atomic bomb in retinoblastoma. A questionnaire will be sent to retinoblastoma patients in Nagasaki and Hiroshima if possible. Survivors will be contacted for personal details and radiation dose will be calculated on the basis of data accumulated by the Atomic Bomb Effect Research Centers in Nagasaki and Hiroshima.

TYPE: Correlation; Incidence
TERM: Atomic Bomb; Radiation, Ionizing
SITE: Retinoblastoma
TIME: 1990 - 1994

562 **Ikeda, T.** 04690
Nagasaki Univ. School of Medicine, Dept. of Pathology, 12-4 Sakamoto 1 , Nagasaki 852, Japan (Tel.: +81 958 472111; Fax: 475054)
COLL: Soda, M.; Shimokawa, I.; Ishikawa, H.; Mine, M.

District Specificity of Cancer Incidence in Nagasaki, Japan

Mortality and morbidity rates in Nagasaki prefecture were the highest in Japan in 1986, and some 35% of the population of Nagasaki City were A-bomb survivors. Moreover, the high frequency of T-cell lymphoma and ATLL in Southwestern Japan including Nagasaki Prefecture is well known. This study aims to clarify the causes of increased cancer incidence, and correlation between cancer incidence and related factors, such as radiation exposure, HBV, and HTLV. Nagasaki City Cancer Registry was established in 1957, Nagasaki Tumor Tissue Registry in 1974, and Nagasaki Prefectural Cancer Registry in 1985, covering populations of 450,000, 1,000,000, and 1,600,000 respectively. This population-based study (1973-1982) revealed that the cancer risk was definitely increased for leukaemia, multiple myeloma, cancers of the breast, thyroid, skin, liver, gall bladder and pancreas. Recent publications include GANN Monograph on Cancer Research 32:41-52, 1986, Jap. J. Cancer Clin. 33:807-814, 1987, and Nagasaki Med. J. 63:306-311, 1988.

TYPE: Correlation; Incidence
TERM: Atomic Bomb; HBV; HCV; HTLV; Radiation, Ionizing; Registry
SITE: All Sites
REGI: Nagasaki (Jap)
TIME: 1978 - 1995

563 **Mine, M.** 05238
Nagasaki Univ. School of Medicine, Scientific Data Center for the Atomic Bomb Disaster, 12-4 Saskamoto-machi, Nagasaki, Japan (Tel.: +81 958 472111/2453)
COLL: Okumura, Y.; Mori, H.; Kondo, H.; Ikeda, T.; Kondo, S.

Cancer Mortality among Atomic Bomb Survivors in Nagasaki Prefecture
A cohort of 110,000 survivors of the Nagasaki atomic bomb is being followed up for mortality. The Scientific Data Center for the Atomic Bomb Disaster at Nagasaki University has collected medical data on A-bomb survivors since 1970. This study aims to investigate the correlation between cancer mortality and related factors such as radiation exposure and health examination. The mortality statistics for 1970 to 1984 were published in Radiat. Res. 103:419-431, 1985 and those for 1970 to 1988 in Int. J. Radiat. Biol. 58:1035-1043, 1990

TYPE: Cohort; Mortality
TERM: Atomic Bomb; Radiation, Ionizing; Survival
SITE: All Sites
TIME: 1970 - 1994

Nagoya

564 **Aoki, K.** 04697
Aichi Cancer Center, 1-1 Kanokoden, Chikusa-ku, Nagoya 464, Japan (Tel.: +81 52 7626111; Fax: 7635233)
COLL: Tominaga, S.; Suzuki, R.; Nakamura, K.; Yanagawa, H.; Inaba, Y.; Ohno, Y.; Sasaki, R.; Kawai, K.; Yoshimura, T.

Evaluation of Risk Factors of Cancer - A Large Scale Cohort Study
The aim of the study is to evaluate risk factors related to cancer, taking into account changes in lifestyle and other socio-medical factors. Thirty five epidemiologists in 32 institutions are participating in this study. Basic examinations were conducted in 1988-1990. A total of 125,760 inhabitants living in 29 cities and towns throughout Japan will be followed up until year 2000, mainly by death records. Cancer incidence will be studied in 12 towns and cities. A serum bank from about 50,000 persons has been established and is kept at -80 degrees Centigrade. An interim questionnaire survey of 50% of the participants started in 1993. About 2,000 deaths were observed until the end of 1991. The distribution by major cause of death was similar to that of the general population in the same period.

TYPE: Case-Control; Cohort
TERM: Alcohol; BMB; Diet; Nutrition; Occupation; Psychological Factors; Registry; Serum; Tobacco (Smoking)
SITE: All Sites
REGI: Aichi (Jap)
TIME: 1988 - 2000

565 **Ohno, Y.** 04638
Nagoya Univ., School of Medicine, Dept. of Preventive Medicine, 65 Tsurumai-cho, Showa-ku, Nagoya 466, Japan (Tel.: +81 52 7412111; Fax: 7336729)
COLL: Kubo, N.; Sakamoto, G.; Watanabe, S.; Cornain, S.; Tjindarbumi, D.; Ramli, M.; Darwis, I.; Tjahjadi, G.; Sutrisno, E.; Sri Roostini, E.; Prihartono, J.; Budiningsih, S.

Japan-Indonesia Study on Aetiology and Clinicopathology of Breast Cancer

This is a hospital-based case-control study. Elegible cases are female breast cancer patients aged 25-69 years, histologically confirmed and newly diagnosed at the Department of Surgery, Centre Hospital, University of Indonesia, Jakarta. Controls (two per case) are female patients with diseases other than cancer admitted to or visiting the same department. Cases and controls are matched, by pair matching, for sex, age (+/- 3 years) and date of admittance/visit (+/-3 months). Controls are examined by palpation to confirm absence of lumps in the breast. Routine demographic and epidemiological information is collected by direct interview of the study subject herself, at hospital, by two academic nurses. A pilot study was already done in 1988, and the study started officially from 1 April 1989. All previously reported risk factors are to be tested. A case-control study with the same study protocols will be considered in 1990 in Tokyo, Japan. In parallel with this case-control study, clinical and histopathological studies are to be performed.

TYPE: Case-Control
TERM: Alcohol; Diet; Oral Contraceptives; Physical Factors; Reproductive Factors; Tobacco (Smoking)
SITE: Breast (F)
LOCA: Indonesia; Japan
TIME: 1989 - 1995

566 **Ohno, Y.** 04639
Nagoya Univ., School of Medicine, Dept. of Preventive Medicine, 65 Tsurumai-cho, Showa-ku, Nagoya 466, Japan (Tel.: +81 52 7412111; Fax: 7336729)
COLL: Kubo, N.; Hayashi, Y.; Nagao, K.; Genka, K.; Ohmine, K.; Aoki, K.; Fukuma, S.

Lung Cancer and its Risk Factors in Okinawa, Japan

This case-control study aims to explore the risk factors for lung cancer in Okinawa, particularly among males, since they have experienced the highest lung cancer mortality in Japan over a recent decade. Eligible cases, identified at National Okinawa Hospital, include all patients aged 40-89 years with histologically confirmed primary lung cancer, diagnosed since 1 January 1988. Controls are randomly selected from the general population, using the electoral registers; matching individually for residence, sex and age to within two years with a 2:1 allocation ratio. Routine demographic and epidemiological information, including smoking and drinking habits, occupation, and dietary practices, are collected principally at the subject's home by 7-10 trained public health nurses, using a 14-page questionnaire. Interviews started in October, 1988. A main hypothesis concerns smoking habits, but the study is intended to be hypothesis-generating, since smoking in Okinawa is lower than the national average, approximately 20% in males and 10% in females. In parallel with this case-control study, comparative clinical and pathological studies are performed between Okinawa and Chiba prefecture.

TYPE: Case-Control
TERM: Alcohol; Diet; Male; Occupation; Tobacco (Smoking)
SITE: Lung
TIME: 1988 - 1995

***567** **Tajima, K.** 05476
Aichi Cancer Center Research Institute, Division of Epidemiology, Chikusa-ku, 1-1 Kanokoden Tashiro-cho, Nagoya 464, Japan
COLL: Kuroishi, T.; Hamajima, N.; Takezaki, T.; Inoue, M.; Hirose, K.

Hospital-Based Case-Control Study on Common Cancers in Aichi, Japan

A hospital-based case-control study to investigate the risk factors for common cancers in Japan, such as stomach, lung and colon cancer, has been initiated at Aichi Cancer Centre Hospital, Japan. The study aims to assess the risk factors for different histological sub-types of cancers, as well as for those cancers occurring in anatomical sub-sites in the same organs. Information is collected by the routine interview of all patients (an average of 8,500 per year) during their first visit to the hospital, using a common questionnaire. 15% of new patients are diagnosed with cancer. About 50% of the remaining non-cancer patients constitute the reference group for case-control analysis. Results from these sequential case-control studies may contribute to further understanding of the aetiology and prevention of common cancers.

TYPE: Case-Control
TERM: Environmental Factors; Lifestyle; Prevention
SITE: Breast (F); Colon; Lung; Rectum; Stomach; Uterus (Cervix)
TIME: 1987 - 1997

568 **Tokudome, S.** 03109

Nagoya City Univ. School of Medicine, Dept. of Public Health, Mizuho-ku , Nagoya 467, Japan (Tel.: +81 52 8515511; Fax: 8423830)
COLL: Kono, S.; Ikeda, M.; Ogata, M.; Shimono, M.; Makimoto, K.; Uchimura, H.

Cancer and Other Causes of Death among Psychiatric Patients

This is a follow-up study of the mortality of about 3,000 psychiatric patients institutionalized from 1945 to 1982 in order to investigate whether they are at high risk of mortality from cancer (and other causes). The patients are divided by the psychiatric diagnosis to examine possible carcinogenic effect of drugs used in psychiatric practice. Smoking and drinking habits are also studied to evaluate their possible influence on cancer mortality of the patients.

TYPE: Cohort; Mortality
TERM: Alcohol; Drugs; Tobacco (Smoking); Treatment
SITE: All Sites
TIME: 1982 - 1994

569 **Tokudome, S.** 04112

Nagoya City Univ. School of Medicine, Dept. of Public Health, Mizuho-ku , Nagoya 467, Japan (Tel.: +81 52 8515511; Fax: 8423830)
COLL: Ikeda, M.

Follow-Up Study of Blood-Transfused Patients

About 4,000 blood-transfused patients undergoing partial gastrectomy at some institutions from about 1950 to 1970 are being followed. This study aims to examine whether patients who have received blood transfusion and have incurred non-A, non-B hepatitis (C hepatitis) are a high risk of cirrhosis and hepatocellular carcinoma.

TYPE: Cohort
TERM: Cirrhosis; Infection; Virus
SITE: Liver
TIME: 1986 - 1994

570 **Tokudome, S.** 04113

Nagoya City Univ. School of Medicine, Dept. of Public Health, Mizuho-ku , Nagoya 467, Japan (Tel.: +81 52 8515511; Fax: 8423830)
COLL: Ikeda, M.

Long-Term Follow-up Study of HTLV-I Carriers

A prospective study of about 8,000 HTLV-I carriers among blood donors at 7 Red Cross Blood Centres in Kyushu, where HTLV-I carriers are highly prevalent, is proposed. The aim of this study is to investigate the incidence rate of ATL (Adult T-cell Leukaemia/lymphoma) among HTLV-I carriers and the latency period from the infection of HTLV-I to the onset of ATL. It is also proposed to study aetiological (prognostic) factors of ATL in a case-control study.

TYPE: Cohort
TERM: HTLV; Infection; Latency; Virus
SITE: Leukaemia; Lymphoma
TIME: 1986 - 1994

571 **Tominaga, S.** 04504

Aichi Cancer Centre Research Inst., 1-1 Kanokoden, Chikusa-ku, Nagoya 464, Japan (Tel.: +81 52 7626111; Fax: 7635233)
COLL: Kato, I.; Inoue, M.; Tajima, K.

Risk Factors and Natural History of Atrophic Gastritis and Stomach Cancer

To elucidate risk factors and to study the natural history of stomach cancer and chronic atrophic gastritis (CAG), a questionnaire survey was conducted on about 6,000 patients who received a gastric endoscopic examination at the Aichi Cancer Center Hospital. Common and specific risk factors will be studied from analyses of baseline data. The study subjects free from stomach cancer at baseline will be followed for at least five years to study risk factors and natural history of stomach cancer and CAG. Preliminary analyses of the follow-up data revealed that CAG is a significant risk factor for stomach cancer. Papers have appeared in Cancer Res. 50:6559-6564, 1990, Jpn J. Cancer Res. 83:1041-1046 and 1137-1142, 1992.

TYPE: Cohort; Cross-Sectional
TERM: Atrophic Gastritis; Premalignant Lesion
SITE: Stomach
TIME: 1986 - 1995

Niigata

572 **Toyoshima, H.** 05008
Niigata Univ. School of Medicine, Dept. of Public Health, 1-757 Asahimachidori , Niigata 951, Japan (Tel.: +81 25 2236161; Fax: 2291975)

Factors Related to Carcinogenesis in an Agricultural Village in Niigata Prefecture

This is part of a large scale cohort project which aims to evaluate factors related to carcinogenesis in Japan. The total cohort population is 100,000 and members will be followed up for 10 years. Information concerning lifestyle, e.g. food intake, occupation, habits, family history, etc. have been obtained by questionnaire and serum samples are stored. In the present study, DNA samples from 1,500 individuals, in addition to the data mentioned above, are being stored to be analysed for DNA abnormalities in the future.

TYPE: Cohort; Molecular Epidemiology
TERM: BMB; DNA; Diet; Lifestyle; Occupation; Serum
SITE: All Sites
TIME: 1990 - 2000

Ohtsu

573 **Watanabe, S.** 04643
Shiga Univ. of Medical Science, Dept. of Preventive Medicine, Seta-tsukinova-cho , Ohtsu 520-21, Japan (Tel.: +81 775 482186)
COLL: Mohri, I.; Funakoshi, M.

Cancer Among Workers in Chemical and Carbon Electrode Industries

Mortality from cancer and all causes is being determined in a small cohort of 259 current and retired chromate workers who were exposed to chromates for at least 5 years before 1972, and who are being followed up from 1960. Similar studies are in progress for 1,500 workers from two carbon electrode industries and 1,000 coke-oven workers from a chemical industry, who have been followed up since 1980. Their cancer mortality by site is being compared to that of the general population of Japan.

TYPE: Cohort
TERM: Metals; Occupation
SITE: All Sites
CHEM: Chromium; Tars
OCCU: Chromate Producing Workers; Coke-Oven Workers; Electrode Manufacturers
TIME: 1973 - 1994

Okinawa

574 **Sakai, R.** 03908
Ryukyu Univ., School of Health Science, Dept. of Epidemiology, Nishihara-cho, 207 Uehara, Okinawa 903-01, Japan (Tel.: +81 9889 53331)
COLL: Mori, W.; Machinami, M.

Smoking, Alcohol Drinking and Cancer Risk

Cigarette smoking and alcohol consumption are increasing in the general population in Japan. A case-control study is being conducted to investigate the influences of cigarette-smoking and alcohol-drinking in the population. About 30,000 autopsy cases (6,000 cancer cases and 18,000 controls) in the Department of Pathology, University of Tokyo from 1930 to the present day are being analysed. Control cases are the autopsy cases which did not have malignant neoplasms, or diseases associated with malignant neoplasms, cigarette-smoking and alcohol-consumption. Cancer cases are divided by site of cancer. Information on smoking and alcohol consumption is obtained from the clinical protocol. Descriptive epidemiological and matched pair analyses are performed.

TYPE: Case-Control
TERM: Alcohol; Tobacco (Smoking)
SITE: Liver; Lung
TIME: 1978 - 1994

575 **Sakai, R.** 05125
Ryukyu Univ., School of Health Science, Dept. of Epidemiology, Nishihara-cho, 207 Uehara, Okinawa 903-01, Japan (Tel.: +81 9889 53331)

Dietary Epidemiology in Okinawa

This hospital-based case-control study is being carried out to assess the carcinogenicity and anti-tumour activity of dietary plants and herbs in Okinawa, which has the lowest cancer incidence in Japan, especially of cancers of the digestive tract. Daily ingestion of these plants is evaluated by interview of all inpatients in five major general hospitals. The controls are selected among the respondents in the interview survey, after matching for age, sex, etc. The total number of subjects will be over 10,000.

TYPE: Case-Control
TERM: Diet; Plants
SITE: All Sites
TIME: 1982 - 1995

Osaka

576 **Hanai, A.** 04689
Center for Adult Diseases, Dept. of Field Research, 1-3-3 Nakamichi, Higashinari-ku, Osaka 537, Japan (Tel.: +81 6 9721181; Fax: 9727749)
COLL: Fujimoto, I.; Taniguchi, H.; Berg, J.; Ries, L.; Young, J.; Percy, C.; Van Holten, V.; Hankey, B.

Stomach Cancer Incidence Trends among Japanese in Japan and US

The relative proportions of Lauren's two main histological types of stomach carcinoma, intestinal and diffuse, have shown that (1) the intestinal type of carcinoma is seen more in older age groups, in males and in high-risk areas; (2)the diffuse type is seen more in younger age-groups, in females, and in low-risk areas; (3) the ratio of the intestinal type to the diffuse type decreases when the total incidence decreases. The histological diagnoses reported to the Osaka Cancer Registry are being analysed to assess whether or not patterns (1) and (2) are also observed in Osaka, and to ascertain if the ratio (intestinal/diffuse) will fall with an anticipated decrease of incidence in Osaka in the future. The 60,779 stomach cancers registered at the Osaka Cancer Registry from 1973 to 1985 have already been analysed by sex and age-group. In the next two years, stomach cancer incidence by histological type in Osaka from 1977 to 1987 will be compared with the same data on whites, blacks and Japanese in the USA registered by the SEER program in the USA.

TYPE: Incidence
TERM: Age; Ethnic Group; Histology; Registry; Trends
SITE: Stomach
LOCA: Japan; United States of America
REGI: Osaka (Jap); SEER (USA)
TIME: 1989 - 1994

***577** **Hiyama, T.** 05455
Center for Adult Disease, Dept. of Field Research, 1-3-3, Nakamichi Higashinari-ku, Osaka 537, Japan (Tel.: +81 6 9721181; Fax: 9727749)
COLL: Tsukuma, H.; Tanaka, H.

Case-Control Study of Liver Cancer in Osaka, Japan

Osaka is an endemic area of liver cancer. High prevalence of HBV has been suspected to be the main cause of the high incidence of liver cancer in Japan, although it has been found that the prevalence of HBsAg is rather low (1-2%) in this area compared with that in other liver cancer endemic areas. Among liver cancer patients in Osaka, prevalence of hepatitis C antibody is high, around 70%. This may suggest that HCV is the main reason for the high occurrence of liver cancer in Osaka. The relationship between

HCV and liver cancer will be studied in a case-control study. Cases will be the incident cases admitted to the institute between 1992 and 1994. All cases will be examined for HBsAg and HCV antibody. 300 case will be enrolled in the study. Controls will be patients admitted to the institute with diseases unrelated to smoking and drinking. Information of drinking and smoking habits or other lifestyle factors is derived from self-administered questionnaires. Multiple logistic analysis will be carried out.

TYPE: Case-Control
TERM: HBV; HCV
SITE: Liver
TIME: 1992 - 1994

578 **Morimoto, K.** 04691
Osaka Univ., School of Medicine, Dept. of Hygiene and Preventive Medicine, Yamadaoka 2-2, Suita Osaka 565, Japan (Tel.: +81 6 8757460; Fax: 8757459)
COLL: Takeshita, T.; Takeuchi, T.; Shirakawa, T.; Mure, K.; Inoue, C.; Shu, K.; Ogura, H.

Genetic Epidemiology of Chromosomal Damage in Human Peripheral Lymphocytes

It has been suggested that an increase in the frequencies of chromosomal alterations such as SCEs, chromosomal aberrations and micronuclei in somatic cells may have a close relationship with cancer development. In this study, the effects of lifestyle factors such as cigarette smoking, alcohol consumption, hours of work, exposure to hazardous environmental factors such as ionizing radiation, and/or various chemical muta/carcinogens on the chromosomal alterations in peripheral blood lymphocytes will be assessed. The cells of about 200 healthy persons, several hundred cancer patients and controls, and persons with cancer-prone diseases such as Fanconi anaemia, xeroderma pigmentosum, and ataxia telangiectasia will be tested for sensitivity to these agents. The ability of the immune response to prevent cancer development, especially natural killer cell activity, is being investigated. Data are analysed to elucidate the interaction between lifestyles, hereditary predisposition and exposure to environmental hazardous agents, to obtain basic data for the primary prevention of cancer.

TYPE: Case-Control; Molecular Epidemiology
TERM: Alcohol; Chemical Exposure; Chromosome Effects; Heredity; Lifestyle; Lymphocytes; Micronuclei; SCE; Tobacco (Smoking)
SITE: All Sites
TIME: 1983 - 1994

579 **Morinaga, K.** 04967
Center for Adult Diseases, Osaka, Dept. of Field Research, 3 Nakamichi 1-chome, Higashinari-ku, Osaka 537, Japan (Tel.: +81 6 9721181; Fax: 9727749)
COLL: Kohyama, N.; Satoh, Y.

Mortality in Workers Exposed to Quartz, Kaolinite, Clinoptilolite and Mordenite

A retrospective cohort study of workers exposed to quartz, kaolinite, and zeolite is being undertaken to ascertain whether there is an excess of lung cancer and high mortality from mesothelioma. The plant has two mines: quartz and kaolinite operating since 1941, and zeolite since 1966. Analytical electron microscopic findings show that the zeolite is of two types, clinoptilolite and mordenite and that the mordenite is fibrous in form. The cohort consists of nearly 950 male and 250 female workers retired since 1965.

TYPE: Cohort
TERM: Mining; Occupation
SITE: Lung; Mesothelioma
CHEM: Mineral Fibres
OCCU: Miners
TIME: 1990 - 1995

***580** **Morinaga, K.** 05430
Center for Adult Diseases, Osaka, Dept. of Field Research, 3 Nakamichi 1-chome, Higashinari-ku, Osaka 537, Japan (Tel.: +81 6 9721181; Fax: 9727749)
COLL: Sakatani, M.

Lung Cancer Risk among Silicotics by Smoking Habit

Between 1972 and 1977, nearly 600 male patients with pneumoconiosis consulted an out-patient clinic of the National Kinki-chuo Hospital and received medical examination, including chest X-ray. 533 patients were diagnosed with silicosis. The vital status of this group has been assessed up to 1985 and an extension is now being made. Smoking history was obtained by nurses at the clinic. SMRs will be calculated for smokers, ex-smokers and non-smokers. Expected number of deaths will be calculated using the Osaka male general population.

TYPE: Cohort
TERM: Dusts; Registry; Silicosis; Tobacco (Smoking)
SITE: Lung
CHEM: Silica
REGI: Osaka (Jap)
TIME: 1992 - 1997

581 **Saji, F.** 05187
Osaka Univ. Medical School, Dept. of Obstetrics & Gynecology, 1-1-50, Fukushima Fukushima-Ku, Osaka 553, Japan (Tel.: +6 451 0051)
COLL: Tanizawa, O.; Ohashi, K.; Osumi, Y.; Okada, S.

Perinatal Infection of HTLV-I: A Prospective Study

HTLV-I is the aetiological agent of some adult T-cell leukaemia/lymphoma (ATL) in humans. Epidemiological studies of ATL in Japan have revealed a tendency to familial clustering, suggesting vertical transmission of the virus. The object of this study is to determine the mode of perinatal infection of HTLV-I and the prophylactic effect on vertical transmission by prohibiting the HTLV-I carrier mothers from breast-feeding their neonates. A serological survey was undertaken among pregnant women attending University-affiliated hospitals in Osaka (9,000 deliveries a year). Blood samples of neonates born to seropositive parturients are collected at delivery and every third month after birth, and assayed for HTLV-I infection by DNA analysis. Preliminary data (J. Clin. Immunol. 9:409-414, 1989 and Cancer 66:1933-1937, 1990) suggest that postpartum infection via breast milk is a likely major perinatal transmission route. Data obtained from the survey 1987-1992 demonstrated that only 9% of the neonates born to the HTLV-I carrier mothers were seropositive after prohibiting breast feeding.

TYPE: Case Series
TERM: HTLV; Sero-Epidemiology
SITE: Leukaemia
TIME: 1987 - 1995

Osaka-sayama city

582 **Hayakawa, K.** 03197
Kinki University School of Medicine, Dept. of Public Health, 377 Ohno-higashi, Osaka-sayama city Osaka-fu 589, Japan (Tel.: +81 723 660221; Fax: 678262)
COLL: Shimizu, T.

Ageing Twins and Cancer

This is a longitudinal prospective study of 3,000 pairs of twins born before 1935. A questionnaire survey has been conducted including questions about personal, family, and medical histories, present symptoms, food, alcohol and cigarette consumption and signs of ageing (such as presbyopia or grey hair). Comprehensive medical examinations have been done on the twin subjects, including blood chemicals and other physiological functions. Lifestyle and other environmental factors are being analysed on identical twins who developed cancer in one twin only.

TYPE: Cohort; Incidence
TERM: Lifestyle; Twins
SITE: All Sites
TIME: 1981 - 1994

Otsu

583 **Aoyama, T.** 03004
Shiga University of Medical Science, School of Medicine, Dept. of Experimental Radiology, Tsukinowa-cho, Seta, Otsu 520-21, Japan
COLL: Sugahara, T.; Yamamoto, Y.; Kato, H.; Shimizu, Y.

Mortality and Health Study of Radiological Technologists to Evaluate the Risks Involved in Exposure to Low-Dose Radiation

Japanese radiological technologists are a suitable population for the study of the risk from low-level exposure to radiation. This study aims to survey the mortality and cause of death, health status (cell-mediated immunity test, 27 serum biochemical examinations, CRP test, 4 urine examinations, pulse, blood pressure, ECG, peripheral blood cell examinations, visual acuity, chromosome aberrations and physical strength), and cumulative radiation doses of this population. The total registered number of the technologists was about 22,000 at the end of March, 1986. 9,179 members born before 1950 were traced for the mortality study from 1969 to 1982 as the first follow-up. The second follow-up of this population between 1983 to 1986 was started in 1986. The health study has so far been carried out on 1,000 members of 20 branches of the Japan Association of Radiological Technologists. Comparisons will be made with the general population. Papers appeared in "Biological Effects of Low-Level Radiation", International Atomic Energy Agency, Vienna, 1983, pp. 319-328 (IAEA-SM-266/43) and in J. UOEH II (Suppl.) pp. 432-442, 1989.

TYPE: Cohort
TERM: Occupation; Radiation, Ionizing
SITE: All Sites; Brain
OCCU: Health Care Workers
TIME: 1981 - 1994

Saitama

584 **Hoshiyama, Y.** 05067
Saitama Cancer Center Research Inst., Dept. of Epidemiology, Komuro 818, Ina-machi, Saitama 362, Japan (Tel.: +81 48 7221111; Fax: 7221739)
COLL: Tagashira, Y.; Sasaba, T.; Nakachi, K.; Imai, K.

High-Risk Groups and Primary Prevention of Stomach Cancer

A hospital-based case-control study is being conducted on 294 matched pairs. Cases are histologically confirmed in Saitama Cancer Center Hospital. Controls are selected from the general population in the same cities (or towns) as the cases, and matched for sex and age (within two years). Subjects are interviewed with a questionnaire, mainly concerned with dietary habit, smoking and drinking histories. On the basis of the frequency of consumption of the selected factors, an individual risk score will be calculated for all 588 subjects.

TYPE: Case-Control
TERM: Alcohol; Diet; Lifestyle; Prevention; Tobacco (Smoking)
SITE: Stomach
TIME: 1991 - 1994

***585** **Hoshiyama, Y.** 05330
Saitama Cancer Center Research Inst., Dept. of Epidemiology, Komuro 818, Ina-machi, Saitama 362, Japan (Tel.: +81 48 7221111; Fax: 7221739)
COLL: Sasaba, T.

Stomach Cancer and its Relation to Pepsinogen I/II, Helicobacter Pylori, and Lifestyle

This is a hospital-based follow-up study for stomach cancer and its relation to pepsinogen I/II, H. pylori, and lifestyle. Pepsinogen I/II and plasma levels of H. pylori antibody will be measured, and information on food and alcohol consumption, and cigarette smoking obtained using a dietary questionnaire. The subjects are patients who visit Saitama Cancer Center Hospital for the first time. About 20,000 subjects will be followed for at least five years. Information on incidence and mortality will be obtained from relatives, family registers in local health centres and through the local government.

TYPE: Cohort
TERM: Alcohol; Diet; H. pylori; Lifestyle; Pepsinogen; Tobacco (Smoking)
SITE: Stomach
TIME: 1993 - 1998

586 Nakachi, K. 04366
Saitama Cancer Center Research Inst., Dept. of Epidemiology, 818 Komuro, Ina-Machi, Saitama 362, Japan (Tel.: +81 48 7221111; Fax: 7221739)
COLL: Imai, K.; Kawajiri, K.

Cancer Incidence in Relation to Epidemiological, Immunological and Genetic Factors

This study aims to clarify host-environmental interactions in carcinogenesis in terms of epidemiological, psychological, immunological and genetic factors. Data have been collected in one town since 1986 on a total of 3,700 residents over the age of 40. All participants answered a self-administered questionnaire on dietary habits and history, smoking and drinking habits, psychological factors, stress, etc. Assays on peripheral blood included NK cell activity, PHA response, T-cell subsets, selenium, lipid peroxidation, vitamin A, and lipo-proteins. DNAs have also been isolated from lymphocytes and stored for a study of genetic susceptibility to cancer. Papers on cancer susceptibility and germ line CYP1A1 polymorphisms appeared in Cancer Res., 51:5177-5180, 1991; J. Biochem., 110:407-411, 1991. A paper on NK activity appeared in Jpn. J. Cancer Res., 83:798-805, 1992.

TYPE: Cohort; Molecular Epidemiology
TERM: Alcohol; BMB; DNA; Diet; Familial Factors; Genetic Factors; Immunology; Lymphocytes; Psychological Factors; Stress; Tobacco (Smoking); Trace Elements; Vitamins
SITE: All Sites
CHEM: Selenium
TIME: 1986 - 2000

587 Sasaba, T. 02128
Saitama Cancer Center Research Inst., 818 Ina, Saitama 362, Japan (Tel.: +81 48 7221111; Fax: 7221739)
COLL: Hoshiyama, Y.

Liver Cancer in Relation to Clonorchiasis and Trihalomethanes

Cancer statistics in the 1960's and 1970's revealed that people in the eastern part of Saitama had a liver cancer risk nearly three times as high as that of the prefectural average. A field survey showed no correlation between geographical distribution of HBsAg positivity and death rates from liver cancer, suggesting that HBV does not play an important role in geographical clustering of liver cancer mortality in Saitama. Stool samples from the population also revealed a cluster of Clonorchiasis, an endemic parasitic disease, in the same area. In addition, in the eastern part of Saitama the drinking water is polluted by organic chemicals (trihalomethanes) at levels of 30-100 ppb. The improved agricultural water supply system resulted in rapid decrease in the number of host snails. Cancer statistics in 1990 suggest a decrease in liver cancer death rates, and disappearance of the geographical cluster in the future. Mortality trends in relation to changes in prevalence of clonorchiasis and in levels of trihalomethanes in drinking water, are now being studied. A follow-up study of 300 individuals with clonorchiasis and 400 HBV carriers is also underway.

TYPE: Correlation; Relative Frequency
TERM: Cluster; Parasitic Disease; Water
SITE: Bile Duct; Liver
CHEM: Trihalomethanes
TIME: 1987 - 1995

588 Tagashira, Y. 03594
Saitama Cancer Center Research Inst., Komuro 818, Ina-machi, Saitama 362, Japan (Tel.: +81 48 7221111; Fax: 7221739)
COLL: Sasaba, T.; Nakachi, K.; Imai, K.; Hoshiyama, Y.

Case-Control Study in Saitama Cancer Center Hospital

This is a hospital-based case-control study. Cases are all first admission in-patients to the hospital (1,000/year), and are grouped according to cancer sites and histological types. Controls are selected from the general public living in the same cities or towns as cases and matched for sex and age.

Department staff interview cases and controls with a questionnaire, which is mainly concerned with dietary habits, smoking and drinking histories, and other environmental factors. So far, more than 4,500 cases and about 1,000 controls have been interviewed. Analysis will include single factor risk estimations and interactions between more than one factor.

TYPE: Case-Control
TERM: Alcohol; Diet; Lifestyle; Tobacco (Smoking)
SITE: All Sites
TIME: 1985 - 1994

Sapporo

589 **Miyake, H.** 03979
Sapporo Medical College, Dept. of Public Health, South 1, West 17, Chuo-ku, Sapporo 060, Japan (Tel.: +81 11 6112111; Fax: 612586)
COLL: Mori, M.; Goto, R.; Masuoka, H.; Yoshida, K.; Ohba, S.; Mitsuhashi, T.

Prospective Cohort Study of the Elderly in Hokkaido

This prospective cohort study was designed in 1983 to clarify the association between nutritional factors and adult diseases including various cancers among the elderly. The baseline surveys were performed in 1984 or in 1985 by health nurses for persons over 40 years of age who lived in Hokkaido Prefecture. In total, 3,185 persons (1,531 males and 1,654 females) have been involved in the study. The questionnaire used for the baseline survey included items concerning demographic factors, health status, lifestyle, and consumption frequency of 38 food items. A follow-up survey with regard to health status is carried out every year. As a result of the recent survey, a more frequent intake of instant noodles was associated with increased mortality. A paper has been published in Jpn. J. Cancer Clin. 37:255-260, 1991.

TYPE: Cohort
TERM: Age; Diet; Lifestyle
SITE: All Sites
TIME: 1984 - 1994

590 **Miyake, H.** 04727
Sapporo Medical College, Dept. of Public Health, South 1, West 17, Chuo-ku, Sapporo 060, Japan (Tel.: +81 11 6112111; Fax: 612586)
COLL: Takeda, T.; Nishi, M.

Incidence of Malignant Disease in Children in Hokkaido

A Registry of Childhood Malignancies has been operating in Hokkaido Prefecture, Japan since 1969. Since application for public assistance to pay medical expenses depends on registration, the Registry covers almost 100% of the incidence of childhood cancer. There are about 70,000 live births annually. There has been no report on the exact incidence of malignant diseases of children in Japan, and the aim is to describe the incidence of childhood cancer (leukaemia, malignant lymphoma, Wilms' tumour, etc.) during the 20 years of registration. A paper has been published in J. Pediatr. Surg. 25:545-546, 1990.

TYPE: Incidence
TERM: Registry
SITE: Childhood Neoplasms
TIME: 1969 - 1994

591 **Takeda, T.** 04732
Sapporo National Hosp., Clinical Research Inst. Hokkaido Cancer Center, Kikusui 4-2, Shiroishi-ku, Sapporo 003, Japan (Tel.: +81 11 8119111; Fax: 8320652)
COLL: Takasugi, N.; Hanai, J.; Miyake, H.; Nishi, M.

Mass Screening for Neuroblastoma in Sapporo City

Mass screening for neuroblastoma, aimed at six-month-old infants, has been performed in Sapporo City since 1981. There are about 20,000 live births annually. Some epidemiological problems are being studied: (1) influence of screening on the incidence by birth cohort, (2) difference in incidence between groups screened and not screened, (3) cost-benefit relations of the mass screening, (4) necessity of re-screening, (5) final survival rate of the patients screened (including both true positive cases and false negative cases). Recentpapers were published in Acta Paediatr. Scand. 80:812-817, 1991, Oncology 48:31-33, 1991 and Am. J. Pediatr. Hematol./Oncol. 14:327-331, 1992.

TYPE: Incidence
TERM: Birth Cohort; Cost-Benefit Analysis; Screening
SITE: Neuroblastoma
TIME: 1981 - 1994

Sendai

592 **Fukao, A.** 04267
Tohoku Univ., School of Medicine, Dept. of Public Health, 2-1 Seiryo-machi, Sendai 980, Japan (Tel.: +81 22 2741111; Fax: 2754877)
COLL: Minami, Y.; Hisamichi, S.; Sugawara, N.

Case-Control Study of Scirrhous Cancer of the Stomach

From many epidemiological studies, it has been clarified that the major risk factor of gastric cancer is environmental and that mortality and incidence are higher in males than in females. However, mortality and incidence under 40 years of age are higher in females than in males, and the prognosis of female patients with scirrhous gastric cancer is very poor. Hypotheses are: (1) there are different risk factors between the scirrhous (undifferentiated) and the other histological type of gastric cancer (differentiated), (2) hormonal or genetic environment are important factors. Approximately 150 cases and 300 controls (females from a mass screening programme for gastric cancer) will be identified from the member hospitals and interviewed by doctors and nurses. Of particular interest are environmental, hormonal and genetic factors. Data collection is continuing and analysis will be done by the matched-pair method.

TYPE: Case-Control
TERM: Diet; Environmental Factors; Genetic Factors; Hormones; Lifestyle; Menstruation; Occupation; Prognosis; Sexual Activity
SITE: Stomach
TIME: 1987 - 1994

593 **Fukao, A.** 05052
Tohoku Univ., School of Medicine, Dept. of Public Health, 2-1 Seiryo-machi, Sendai 980, Japan (Tel.: +81 22 2741111; Fax: 2754877)
COLL: Komatsu, S.; Hisamichi, S.; Sugawara, N.; Ida, Y.; Takano, A.

Lifestyle, Personality and Cancer Incidence

The aim of this cohort study is to investigate the influence of both lifestyle and personality on cancer incidence. The cohort, comprising about 50,000 people aged 40-64 and living in Miyagi Prefecture, Japan, was questioned about lifestyle and personality (Eysenck Personality Questionnaire) in 1990. About 10,000 serum samples were collected and beta-carotene and some substances related to cancer will be measured. The state of attendance at periodic health examinations and cancer screening programmes will be followed annually. Incidence of cancer will be measured using the Miyagi Cancer Registry.

TYPE: Cohort
TERM: BMB; Lifestyle; Psychological Factors; Registry; Serum
SITE: All Sites
CHEM: Beta Carotene
REGI: Miyagi (Jap)
TIME: 1990 - 2000

Tokorozawa

***594** **Kono, S.** 05425
National Defense Medical College, Dept. of Public Health, 3-2 Namiki, Tokorozawa 359, Japan (Tel.: +81 429 951211; Fax: 950638)
COLL: Imanishi, K.; Nishikawa, H.; Ogawa, S.; Nakagawa, T.

Colorectal Polyp Study: the Second Series

A primary study showed that the risk of colorectal adenomas, especially larger adenomas, is related to alcohol intake and the consumption of rice (low intake) and meat (high intake); (Jpn. J. Cancer Res. 83:806-11, 1992; 84:13-9, 1993). This is the second series of a cross-sectional study of middle-aged men of the Self-Defense Forces (SDF) in Japan to investigate risk factors for colorectal polyps. Men retiring from the SDF receive a comprehensive health examination, in which colonoscopy is a routine procedure. A self-administered questionnaire was newly developed to assess the quantity of selected food items consumed on average in the past year, as well as other lifestyle characteristics. About 4,000 men will be recruited from the SDF Fukuoka, Kumamoto, Sapporo and Central (Tokyo) Hospitals between January 1993 and December 1994. Since total colonoscopy is done at the Central Hospital, while sigmoidoscopy is routine at the others, risk factors for polyps by location as well as by size of polyp will be investigated.

TYPE: Cross-Sectional
TERM: Alcohol; Diet; Physical Activity; Polyps; Premalignant Lesion; Prevalence
SITE: Colon; Rectum
TIME: 1993 - 1995

Tokyo

595 **Anzai, S.** 04227
Showa Univ., School of Medicine, Dept. of Public Health, 1-5-8, Hatanodai, Shinagawa-ku, Tokyo 142, Japan (Tel.: +81 3 37848134; Fax: 37847733)
COLL: Aoki, K.; Tominaga, S.; Kurihara, N.; Inaba, Y.; Fujiki, H.; Hirohata, T.; Kuroki, T.

Evaluation of Factors Affecting Cancer Incidence

Nationwide re-evaluation of the positive or negative risk for cancer, of factors such as smoking, alcohol, animal fat, green and yellow vegetables and vitamins is being undertaken, using cohort groups. Cancer sites of interest are stomach, lung, liver, large intestine, uterus, and some other sites with relatively high incidence rates. Approximately 1,000 residents and employees aged over 40 will be examined; their biological and epidemiological characteristics will be described for the first few years, and subsequently their cancer incidence will be assessed at least for the next five to ten years. The sample size, protocols, and follow-up methods including ethical problems are now being discussed.

TYPE: Cohort
TERM: Alcohol; Diet; Fat; Tobacco (Smoking); Vitamins
SITE: Colon; Liver; Lung; Stomach; Uterus (Corpus)
TIME: 1987 - 2002

596 **Hayata, Y.** 01486
Tokyo Medical College Hosp., Dep. of Surgery, 6-7-1 Nishishinjuku, Shinjuku-ku, Tokyo 160, Japan (Tel.: +81 3 3426111)
COLL: Oho, K.; Saito, Y.; Funatsu, H.; Kato, H.

Mass Surveys of Tokyo Metropolitan Government Employees by Chest X-Ray and in High Risk Cases also by Sputum Cytology

The aim is to detect lung cancer, hopefully at an early stage. Chest X-rays have been performed since 1953 on employees of the Tokyo Metropolitan Government, and 241 lung cancer cases have been detected among a total of 2,382,518 examinations. The overall detection rate per 100,000 examinations was 10.1, 25.0 in males aged 40 years and over and 10.1 in females aged 40 years and over. Among these 41.9% were stage I, 13.7% stage II and 44.4% stage III and IV. Resectability rates were 87.3%, 66.7% and 30.2% respectively. There are currently three groups of examinees: a high-risk group examined semi-annually, a group examined annually, and the group of cases not examined the previous year. The rate of early stage lung cancer detection was 24.3% in the semi-annual group, 11.1% in the annual group and 10.3% in the third group. There were also significant differences between the three groups in resectability and five-year survival rates. Since 1974 sputum cytology surveys have been conducted on a high-risk group of 4,898 examinees to detect lung cancer and atypical squamous metaplasia cells. Cases with squamous metaplasia cells are followed up at four month intervals. Metaplastic cells are examined for DNA content. Evaluation of squamous metaplasia cells and/or DNA analysis results were based on serial findings in several canine experimental series, in which lung cancer was induced in the major bronchi by intramucosal injections of 20-methylcholanthrene. The high-risk group is defined as those aged 45 or more with a cigarette index of over 400, those involved in occupations involving exposure to pollution, and those with cough or bloody sputum.

TYPE: Cohort
TERM: Cytology; DNA; High-Risk Groups; Occupation; Screening; Sputum; Tobacco (Smoking)
SITE: Lung
TIME: 1953 - 1994

597 **Inaba, Y.** 04305
Jutendo Univ. School of Medicine, Dept. of Hygiene, 1-1 Hongo, 2-chome, Bunkyo-ku, Tokyo 113, Japan (Tel.: +81 3 8133111)
COLL: Namihisa, T.; Ichikawa, S.; Sato, N.; Kikuchi, S.

Chronic Hepatitis and Liver Cirrhosis

In order to elucidate the relationship between type of chronic hepatitis or liver cirrhosis and liver cancer, a retrospective cohort study was carried out. The cohort will consist of about 1,000 cases of chronic hepatitis or liver cirrhosis whose liver tissue was biopsied at Juntendo University Hospital from 1973 to 1982. Each case was studied for type of disease, age, sex, HBV markers, history of blood transfusion, alcohol intake and history of smoking. Information is taken from their personal hospital record. The vital status of each case as of the end of 1991 was checked by out-patient or in-patient hospital records and the 'Koseki' (permanent residence record). When the case was dead, the cause of death was confirmed by the autopsy record or death certificate. Analysis of the cohort will be done by the life table method with inner cohort comparison.

TYPE: Cohort
TERM: Alcohol; Cirrhosis; HBV; Tobacco (Smoking); Virus
SITE: Liver
TIME: 1986 - 1994

598 **Kobayashi, N.** 02244
National Children's Hosp., 3-35-31 Taishido, Setagaya-ku, Tokyo 154, Japan (Tel.: +81 3 4148121; Fax: 4190381)
COLL: Matsui, I.; Tanimura, M.

Japan Children's Cancer Registry

This is a nationwide and physician-based registry of children with cancers. The registry is controlled by the Committee of the Japan Children' Cancer Registry, part of the Children's Cancer Association of Japan. Cancer morbidity in Japanese children is being assessed. During the 17 years from 1969 to 1987, 23,746 cases (13,223 males and 10,523 females) were registered including acute lymphatic leukaemias accounting for 23%, acute myeloid leukaemias 12%, neuroblastomas 10%, retinoblastomas 8%, Wilms' tumours 4%, primary liver cell carcinomas 2.2%, lymphosarcomas 2.0%, rhabdomyosarcomas 1.8%, acute monocytic leukaemias 1.3%, medulloblastomas 1.2% of total cases. The fourth report of the registry, covering 1984-1988, will be published in December 1990.

TYPE: Incidence; Registry
TERM: Childhood
SITE: Childhood Neoplasms
TIME: 1969 - 1994

599 **Kobayashi, N.** 03892
National Children's Hosp., 3-35-31 Taishido, Setagaya-ku, Tokyo 154, Japan (Tel.: +81 3 4148121; Fax: 4190381)
COLL: Hayakawa, N.; Matsui, I.

Children's Cancer-Immunodeficiency Registry

The registry is a nationwide and physician-based registry of cases with primary immunodeficiency syndromes. From 1975 to 1989, 812 cases of immunodeficiency syndromes have been registered. The most frequent syndromes recorded are common variable immunodeficiency (15.9%), chronic granulomatous disease (12.5%), selective IgA deficiency (12.0%), infantile X-linked agammaglobulinemia (9.3%), severe combined immunodeficiency (9.2%), and ataxia-telangiectasia (7.5%). In 812 registered cases, 27 cases with malignant neoplasms were reported. The incidence of malignancies in infant cases was 3.2% (21/663). This is more than 300 times that in the general population in Japan. The rate is especially high in the Chediak-Higashi syndrome (33.3%; including possible cases of blastic crisis) and ataxia-telangiectasia (15.5%). These relationships are being studied.

TYPE: Registry
TERM: Immunodeficiency
SITE: Childhood Neoplasms
TIME: 1975 - 1994

600 **Kobayashi, N.** 04836
National Children's Hosp., 3-35-31 Taishido, Setagaya-ku, Tokyo 154, Japan (Tel.: +81 3 4148121; Fax: 4190381)
COLL: Matsui, I.

Risk Factors for Childhood Malignancy in Japan

This study will examine the hypothesis that childhood malignancy is caused by genetic factors, by intra-uterine exposure to environmental factors, and by infection. Registered cases from the Japan Children's Cancer Registry (a few hundred cases of each type of malignancy, including leukaemia) will be subjected to analysis in respect of twins, associated abnormalities including chromosomal aberration and translocation, and history of exposure to environmental factors. Controls will be healthy children, if possible, and also exposure comparisons will be made between solid tumour and leukaemia cancer.

TYPE: Case-Control
TERM: Congenital Abnormalities; Drugs; Environmental Factors; Intra-Uterine Exposure; Radiation, Ionizing; Registry; Twins
SITE: Childhood Neoplasms; Leukaemia
TIME: 1985 - 1994

601 **Nakamura, K.** 04247
Showa Univ., School of Medicine, Dept. of Hygiene and Preventive Medicine, 1-5-8, Hatanodai, Shinagawa-ku, Tokyo 142, Japan
COLL: Tadera, M.; Masaki, M.

Cancer in a Working Population in the Tokyo Area

The aim is to elucidate the effects of life style (diet, alcohol drinking, smoking, marital and socio-economic status, etc.) and several biological factors such as body build, blood pressure, serum cholesterol, etc., in the occurrence of cancer. Subjects are approximately 7,000 employees of stockbroking firms, aged 40 or more, and living in the Tokyo Area. A baseline survey will be carried out in 1988. As long as the subjects are employed in a company using the Health Insurance Society, occurrence of cancer is easily detected. After resignation, information on death can be collected by means of inquiring into their family or their 'Koseki' (family register in local government).

TYPE: Cohort
TERM: Alcohol; Cholesterol; Diet; Lifestyle; Marital Status; Physical Factors; Socio-Economic Factors; Tobacco (Smoking)
SITE: All Sites; Colon; Liver; Lung; Rectum; Stomach
TIME: 1988 - 1998

602 **Tsugane, S.** 05093
National Cancer Centre Research Inst., Epidemiology Div., 5-1-1 Tsukiji, Chuo-ku, Tokyo 104, Japan (Tel.: +81 3 35422511; Fax: 35460630)
COLL: Hamada, G.S.; Iriya, K.; Kowalski, L.P.; Torloni, H.; Watanabe, S.

Stomach Cancer among Japanese and Non-Japanese Brazilians

The aim of this case-control study is to identify epidemiological and molecular aspects of stomach cancer among Japanese and non-Japanese Brazilians that could explain the difference in mortality and incidence rates between these populations. Cases will be all those patients diagnosed with stomach cancer at of several Sao Paulo hospitals from April 1991: 150 first- or second-generation Japanese Brazilians and 150 non-Asian Brazilians of either sex, aged 40-69 and living in Sao Paulo State. Controls will be non-cancer patients matched on hospital, sex, ethnicity and age within three years. All subjects will be interviewed about their lifestyle, including dietary history, and blood samples taken; fresh tumour tissue will be obtained from cases. Plasma levels of H. pylori antibody and pepsinogen will be measured, and lymphocytes analysed for polymorphism and alkyl-DNA adducts.

TYPE: Case-Control
TERM: Biomarkers; DNA Adducts; Diet; H. pylori; Japanese; Lifestyle; Pepsinogen
SITE: Stomach
LOCA: Brazil
TIME: 1991 - 1994

603 **Watanabe, S.** 03300
National Cancer Center Research Inst., 5-1-1, Tsukiji, Chuo-ku, Tokyo 104, Japan (Tel.: +81 3 5422511; Fax: 5460630)
COLL: Fujimoto, I.; Hanai, A.; Tsunematsu, Y.; Kobayashi, Y.

Multiple Primary Malignant Neoplasms

The incidence of multiple primary cancers is being analysed by person-years methods, using hospital-based registration data in the National Cancer Center, National Children's Hospital and Center for Adult Diseases. Linkage analysis with the regional cancer registry is also being done in Osaka. Cohort analysis is carried out in cases with childhood cancer. Data on second cancers in breast and lung cancer patients have been published and a publication on second cancers in childhood cancer patients is in press in Cancer Chemotherapy.

TYPE: Cohort
TERM: Childhood; Clinical Records; Multiple Primary; Registry
SITE: All Sites
TIME: 1987 - 1994

Yokohama

604 **Okamoto, N.** 04655
Kanagawa Cancer Centre, Clinical Research Inst., Dept. of Epidemiology, 54-2, Nakao-cho, Asahi-ku, Yokohama 241, Japan (Tel.: +81 45 3915761)
COLL: Morio, S.

Lung Cancer and Industrial Dusts

The aim of this study is to examine the correlation between lung cancer and industrial dusts, especially asbestos. About 60,000 people participated a in mass-screening programme for lung cancer from 1988 to 1992. These have been divided into two groups, exposed and never exposed to the dusts, and the two groups will be followed until 1993, by means of record linkage with the population-based cancer registry in Kanagawa Prefecture. Analysis will be undertaken by the person-years method.

TYPE: Cohort
TERM: Dusts; Registry
SITE: Lung
CHEM: Asbestos
REGI: Kanagawa (Jap)
TIME: 1988 - 1994

Bishkek

***605** **Kamarli, Z.P.** 05535
Kirghiz Res. Inst. Oncology & Radiology, 92, 50-year October St., Bishkek 720064, Kyrgyzstan (Tel.: (3312) 477450; Fax: 479191)
COLL: Vasilevsky, M.G.

Factors Influencing Cancer Incidence Rate in Kirghizstan

Kirghizstan is a region with low cancer incidence rates, usually explained by the high birth rate and relatively short average lifetime. However, statistical studies suggest that the ethnic structure of the population is a more important factor. Malignant tumours occur four times less frequently in the indigenous population (Kirghizes, Uzbeks, Kazakhs) than in the European population (Russians, Ukrainians, etc.), both for individual sites and for all sites combined. The indigenous population in Kirghizstan accounts for 62.7% of the total population. The difference in the incidence rate of Kirghizes and Russians are being studied by residence, sex and age. It is suggested that social and economic conditions are important. Of particular interest are the populations in high-altitude areas and their lifestyle.

TYPE: Incidence
TERM: Ethnic Group; Lifestyle; Socio-Economic Factors
SITE: Breast (F); Hodgkin's Disease; Lung; Non-Hodgkin's Lymphoma; Oesophagus; Rectum; Stomach; Uterus (Cervix)
TIME: 1991 - 1995

MALAYSIA

Kuala Lumpur

606 **Prasad, U.** 03465
Univ. of Malaya, Dept. of Otorhinolaryngology, Lembah Pantai, Kuala Lumpur 59100, Malaysia (Tel.: +60 3 75002062; Fax: 7556554 ; Tlx: 39845 unimal ma)
COLL: Pathmanathan, R.; Sam, S.K.; Paramsothy, M; Samuel, R.

EBV Serology for Detection of Early Recurrence in Nasopharyngeal Carcinoma

From studies carried out in the context of a project on EBV serology and nasopharynx carcinoma, it has been possible to indicate the significance of EBV serology, particularly IgA/VCA as a marker in the early diagnosis of NPC. The aims is now to establish its role in detecting early recurrence, which would further suggest that activation of EBV plays a part in the malignant process and that a common environmental factor may be responsible. 200 patients with confirmed NPC for whom full clinical, radiological (including CT-scan). immunological and histopathological data are available, will be followed-up in the special NPC clinic in this hospital. As well as detailed clinical examination, EBV serology will be done to detect any rising titres. Those with rising titres will be subjected to tests to detect any recurrence or metastasis. Patients without rising titres and in remission will form a control group.

TYPE: Cohort
TERM: Antibodies; Biochemical Markers; EBV; Infection; Prognosis; Sero-Epidemiology
SITE: Nasopharynx
TIME: 1986 - 1995

607 **Prasad, U.** 03674
Univ. of Malaya, Dept. of Otorhinolaryngology, Lembah Pantai, Kuala Lumpur 59100, Malaysia (Tel.: +60 3 75002062; Fax: 7556554 ; Tlx: 39845 unimal ma)
COLL: Sam, S.K.; Aye, K.M.; Kan, K.P.

Sero-Epidemiological Studies in Sabah, East Malaysia

Sabah is one of the states of Malaysia separated from Peninsular (West) Malaysia by the South China Sea. While 28.1% of the population is comprised of indigenous Kadazan people there is a significant Chinese population living in the state. Both these races have a high incidence of NPC although their culture, life style, eating habits, occupations, religion, etc. are all widely different. EBV serology has been found to be of significance in the early diagnosis of NPC in both races. The aim of the study is to screen the population using IgA/VCA titres as a marker for the diagnosis of NPC and to investigate the epidemiological factor(s) which could have contributed to the causation of this cancer. It is estimated that 100 individuals with high IgA/VCA titres will be picked up out of every 9,000 screened. These will be examined for the presence of NPC, and those without NPC will be followed every six months for five years. Control groups will be identified out of this population for comparison.

TYPE: Case Series
TERM: Chinese; EBV; Race; Screening; Sero-Epidemiology
SITE: Nasopharynx
TIME: 1986 - 1995

608 **Yadav, M.** 05189
Univ. of Malaya, Inst. of Advanced Studies Dept. of Genetics & Cellular Biology, Jalan Lembah Pantai, 59100 Kuala Lumpur, Malaysia (Tel.: +60 3 7552744/242; Fax: 7473661 ; Tlx: 39845 UNIMAL MA)
COLL: Norhanom, W.; Nurhayati, Z.A.

Role of HPV in Cervical Neoplasia

Prevalence of genital HPV infection in Malaysia is unknown. The primary objective of this study is to elucidate the distribution of HPV types commonly associated with cervical carcinoma. Investigations include collection of data on potential risk factors through analysis of questionnaires administered at a major public hospital by the researchers. About 50-100 patients will be screened annually with data on diet, reproductive history, medical history, and exposure to tobacco and alcohol. Parallel laboratory studies will include blood antibody titres for HPV, HSV, CMV and HHV-6, determination of HPV-DNA with PCR-technology on cervical scrapes and determination of type-specific HPV by nucleic acid hybridization of cervical biopsies of histologically proven carcinomas.

TYPE: Case Series; Molecular Epidemiology
TERM: Alcohol; BMB; CMV; Diet; EBV; HPV; HSV; PCR; Reproductive Factors; Tobacco (Smoking)
SITE: Uterus (Cervix)
TIME: 1991 - 1995

Cuernavaca

***609** **Hernández-Avila, M.** 05454
National Inst. of Public Health, Center for Public Health Research, Ave. Universidad #655, Col. Sta. Ma. Ahuacatitlán, Cuernavaca Morelos, Mexico (Tel.: +6 73 175391; Fax: 111148)
COLL: López, L.

Nutritional Determinants of Breast Cancer

A population-based case-control study is being conducted to test the following specific hypothesis: high intake of alcohol and fat increases the risk of breast cancer; high intake of vitamin A (both carotenoid and preformed vitamin A), C and E reduces the risk of breast cancer. 129 cases with histologically confirmed breast cancer and 1,006 controls had been interviewed up to June 1992. Preliminary results show that the risk of breast cancer increased with age, number of years of education, early age at menarche and late age at first pregnancy. These results remained similar after adjusting for age and were similar among pre- and post-menopausal women. High consumption of calories (greater than 2,000 per day) also seemed to significantly increase the breast cancer risk. This increase seemed to be related to saturated fat consumption. Although only few women reported drinking habits, it seemed that women having more than six drinks per week were more likely to be diagnosed with breast cancer, but no significant association was found. The sample size will now be increased to 482 cases to increase the statistical power of the study.

TYPE: Case-Control
TERM: Alcohol; Diet; Fat; Nutrition; Vitamins
SITE: Breast (F)
TIME: 1991 - 1994

***610** **López-Carrillo, L.T.** 05472
Mexico National Public Health Inst., Center for Public Health Research, Ave. Universidad 655, Col. Sta. Ma. Ahuacatitlán, Cuernavaca 62508, Mexico (Tel.: +52 73 112338; Fax: 112338)
COLL: Willett, W.C.; López-Cervantes, M.; Calva, J.; Costa, R.; Romieu, I.; Hernandez, M.; Dubrow, R.; Fernandez, C.; Parra, S.

Risk Factors for Gastric Cancer in three Geographic Areas of Mexico

A population-based case-control study will be conducted in three geographic areas of Mexico, involving 300 gastric cancer cases and 600 hospital controls. The hypothesis to be tested are that chilli pepper and Helicobacter pylori infection are strong risk factors for gastric cancer. The data collection instrument will be an updated version of the semi-quantitative food frequency questionnaire validated for the Mexican population (based on W. Willett's questionnaire). This questionnaire will be administered to the cases and controls by trained interviewers to detect the frequency of chilli pepper consumption in this population. In addition, to quantify the strength of the association nd its public health impact, comparison will be made between exposed and non-exposed individuals (matched by age, sex and hospital). Other dietary and nutritional factors will also be examined. At the same time, blood samples will be drawn to determine Helicobacter pylori antibodies to define the prevalence of this bacteria among cases and controls.

TYPE: Case-Control
TERM: Diet; H. pylori; Nutrition
SITE: Stomach
TIME: 1993 - 1995

Mexico

611 **Hernandez, M.** 04829
Direccion General de Epidemiologia, Aniceto Ortega 1321, Col. del Valle Mexico D.F., Mexico (Tel.: +52 5 343263; Fax: 245600)
COLL: Romieu, I.; Meneses, F.; Rojas, R.

Determinants of Uterine Cervix Cancer in Mexico City

This is a population-based case-control study to investigate the determinants of uterine cervix cancer (UCC) among a Mexican population of women to address the following hypotheses: 1) that infection of the uterine cervix by papilloma virus and/or herpes virus type II increases the risk of UCC; 2) that deficiency of some nutrients: vitamin A (both beta carotene and preformed vitamin A), vitamin E, vitamin C, folic acid, and selenium, increases the risk of UCC, with a possible interaction with virus infection. The role of factors such as socio-economic status, use of oral contraceptives, smoking status, and obstetric and gynaecologic history, as well as sexual behaviour will also be investigated. Cases will be selected through the list of positive cytologic exam (pap smears) reported to the General Directorate of Epidemiology and will include all women diagnosed with positive cytology within one year period in Mexico D.F. Two controls for each case will be selected at random from the list of negative cytology; they will be matched by health centres and age (+/-2.5 years). Enrolment of 251 cases and 441 controls is anticipated over a one year period. Questionnaires including information on sexual activity and food frequency will be applied at home by trained interviewers. During the same visit, a blood sample will be obtained from each participant. All cytology will be reexamined by an expert pathologist and viral analysis for papilloma virus and herpes virus type II will be carried out on uterine cervix epithelial cells. After descriptive and stratified analysis, a multivariate conditional logistic regression will be performed to determine the main determinants of UCC as well as to study interactions.

TYPE: Case-Control
TERM: Blood; Diet; HPV; HSV; Nutrition; Oral Contraceptives; Sexual Activity; Socio-Economic Factors; Vitamins
SITE: Uterus (Cervix)
CHEM: Beta Carotene; Folic Acid; Selenium
TIME: 1990 - 1994

612 **Romieu, I.** 04856
Direccion General de Epidemiologia, Aniceto Ortega 1321, Col. del Valle Mexico D.F., Mexico (Tel.: +52 5 343263; Fax: 245600)
COLL: Hernandez, M.; Meneses, F.; Rojas, R.

Nutritional Determinants of Breast Cancer

The wide variation in dietary intake makes Mexico a suitable setting for a study of the role of specific nutrients and foods in the genesis of breast cancer. This study is a population-based case-control study to investigate nutritional determinants of breast cancer in a Mexico population. The following specific hypotheses will be addressed: high intake of alcohol and fat increases the risk of breast cancer; intake of vitamin A (both carotenoid and preformed vitamin A), vitamin C and E reduces the risk of breast cancer. During a one year period 500 cases of newly diagnosed localized breast cancer reported to the National Mexican Cancer Registry will be enrolled and matched to neighbourhood controls. Dietary intake will be measured using a validated food frequency questionnaire designed especially for a Mexican population. Blood samples will be obtained for subsequent analysis (retinol, beta carotene, alphatocopherol, and genetic markers), as well as toenail clippings to determine selenium levels. After descriptive and stratified analysis, conditional logistic regression will be performed to determine the major determinants of breast cancer and particularly the role of specific foods and nutrients.

TYPE: Case-Control
TERM: Alcohol; BMB; Diet; Nutrition; Toenails; Trace Elements; Vitamins
SITE: Breast (F)
CHEM: Beta Carotene; Retinoids
TIME: 1990 - 1994

NETHERLANDS

Amsterdam

613 **Kriek, E.** 04679

The Netherlands Cancer Inst., Antoni van Leeuwenhoek Huis, Div. of Chemical Carcinogenesis, Plesmanlaan 121, 1066 CX Amsterdam, Netherlands (Tel.: +31 20 5122476; Fax: 172625 ; Tlx: 11273 NKI NL)

COLL: Den Engelse, L.; Van Leeuwen, F.E.; Van Schooten, F.J.; Bos, R.P.; Jongeneelen, F.J.

DNA Adducts in Rodent and Human Tissues Exposed to Polycyclic Aromatic Hydrocarbons (PAH)

A number of sensitive methods now permit detection of carcinogen-DNA adducts in human tissues and cells. The purpose of the present study is to develop the application of immunochemical methods and 32p-postlabelling analysis for the determination of PAH-DNA adducts in human and animal tissues. Investigations in operable lung cancer patients are being conducted to measure PAH-DNA adducts in DNA isolated from peripheral blood WBC, and in DNA isolated from lung tissue of selected groups of patients with a smoking history. Occupational exposure to PAH will be investigated in workers in the aluminium industry. Particular attention will be given to persistence of adducts and for this purpose selected groups of workers will be examined over a longer period of time. The metabolite 1-hydroxypyrene will be determined in the urine and will serve as an indicator of more recent exposure. All subjects will be interviewed about possible exposure to other sources of PAH (smoking). The procedures developed and the results obtained are expected to be valuable for future epidemiological studies in the following ways: (1) identification of carcinogen-DNA adducts can provide evidence of exposure of individuals to carcinogens and the amount of a particular DNA-adduct will provide a measure of the degree of exposure; (2) identification of carcinogen-DNA adducts could be used as a marker of biological effects which might be related to the appearance of tumours at a later stage, and thus would provide a means of identifying groups at high risk. Recent publications were published by Van Schooten et al. in Carcinogenesis 11:1677-1681, 1990 and 12:427-433, 1991 and in J. Natl Cancer Inst. 82:927-933, 1990.

TYPE: Methodology
TERM: Biomarkers; DNA Adducts; Immunologic Markers; Lymphocytes; Occupation; Tobacco (Smoking)
SITE: Lung
CHEM: PAH
OCCU: Aluminium Workers
TIME: 1988 - 1994

614 **Ter Schegget, J.** 04804

Univ. of Amsterdam, Academic Medical Centre, Dept. of Virology, Meibergdreef 15 , 1105 AZ Amsterdam, Netherlands (Tel.: +31 20 5664855; Fax: 916531)

COLL: Cornelissen, M.A.; Struyk, A.P.H.B.; Briet, H.A.; Van den Tweel, J.; Lammes, F.; Jebbink, M.; Van der Hoordaa, J.

Human Papilloma Virus (HPV) in Lesions of the Cervix Uteri

The aim of the study is to determine the relative risk of progression to CIN or carcinoma in women who have HPV types 6/11 and 16/18 in the cervix uteri. Prevalence of HPV types 16/18 compared to types 6/11 cervical infection will be assessed in a retrospective longitudinal study, using archival biopsies exhibiting CIN 0, CIN I or CIN II. Evaluation will start with 100 women with CIN 0 lesions, who have already been followed up using colposcopy and histology, and PCR techniques on paraffin sections.

TYPE: Cohort
TERM: Biopsy; HPV; PCR; Premalignant Lesion
SITE: Uterus (Cervix)
TIME: 1989 - 1994

***615** **Van den Brule, A.J.C.** 05369

Free University Hospital, Dept. of Pathology, Molecular Pathology Sect., De Boelelaan 1117, 1081 HV Amsterdam, Netherlands (Tel.: +31 20 5484017/5881; Fax: 5486456)

COLL: Walboomers, J.M.; Meijer, C.J.; Kruger Kjær, S.; Muñoz, N.

Prevalence of Human Papillomavirus and Epidemiology of Cervical Cancer

The objective of this population-based case-control study of cervical cancer is to investigate the prevalence of HPV genotypes in different well defined populations having either a high incidence or a low incidence of cervical cancer. HPV genotypes will be detected and typed in cervical scrapes and biopsies using PCR. During the last year HPV prevalence rates have been determined in cytomorphologically normal (Pap 1, 2) and abnormal (Pap 3a-5) scrapes in the Netherlands (Int. J. Cancer 48: 404-408, 1991). In the present study, both asymptomatic women participating in a triennial cancer screening programme (n = 2,200) and outpatients attending the gynaecologist for various complaints (n = 1,000) are involved. Factors like age and history of cervical lesions have been analysed. To compare the data obtained in the Netherlands, representing a low-incidence area for cervical cancer, with those of high-incidence areas, HPV prevalence studies have been initiated using scrapes from Brazil and Greenland. Comparison of HPV prevalence and HPV type distribution in relation to age, sexual behaviour, clinical history, cervical cancer risk etc. will greatly enhance the knowledge about the epidemiology of HPV and cervical cancer.

TYPE: Correlation
TERM: Age; BMB; Cytology; HPV; PCR; Sexual Activity
SITE: Uterus (Cervix)
LOCA: Brazil; Denmark; Greenland; Netherlands
TIME: 1991 - 1995

616 **van Leeuwen, F.E.** 03163
Netherlands Cancer Inst., Div. of Clinical Oncology, Dept. of Epidemiology, Plesmanlaan 121, 1066 CX Amsterdam, Netherlands (Tel.: +31 20 5122453; Fax: 6172625 ; Tlx: 11273)
COLL: Kroon, B.B.R.; Peterse, J.L.; Van Dongen, J.A.; Hart, A.A.M.

Benign Proliferative Breast Lesions and Breast Cancer

The aim is to establish, in a prospective study, the risk of developing a subsequent mammary carcinoma in various groups of patients with different types of histologically defined proliferative breast lesions. A very elaborate form has been designed for this purpose. Additional information is obtained from cytology and cytophotometry. Included in the study are variables such as cancer in the family, use of contraceptives, alcohol and drugs. Another goal is to detect by continous sampling whether hormonal profiles in women with benign lesions differ from those with mammary cancer and from healthy women. For the main part of the study, 1,250 women with a histologically examined proliferative breast lesion will be followed for at least ten years. Up to now 1,100 patients have entered the study.

TYPE: Cohort
TERM: Alcohol; Drugs; Familial Factors; Genetic Factors; Hormones; Oral Contraceptives
SITE: Benign Tumours; Breast (F)
TIME: 1979 - 1995

617 **van Leeuwen, F.E.** 03775
Netherlands Cancer Inst., Div. of Clinical Oncology, Dept. of Epidemiology, Plesmanlaan 121, 1066 CX Amsterdam, Netherlands (Tel.: +31 20 5122453; Fax: 6172625 ; Tlx: 11273)
COLL: Peterse, J.L.; Kroon, B.B.R.

Breast Cancer and Oral Contraceptives in Women with Benign Breast Disease

Although most studies up to now have failed to find any significant overall relationship between use of oral contraceptives (OC) and breast cancer, some studies have indicated particular sub-categories of OC users to be at increased risk. One controversy in the literature concerns the possibility of an increased risk of breast cancer in women who have used OC for prolonged periods of time and who also have a history of benign breast disease (BBD). Studies published so far have been hindered by small numbers and most of these studies lack data on histology of BBD and information as to whether OC use preceded BBD or not. In 1985 a case-control study nested in a prospective cohort study of women with biopsy-proven benign breast disease was started. The cohort presently consists of approximately 1,100 women; entry of patients will close at N = 1,250 (a follow-up time of 10 years is required). The cohort study aims at assessing the relationship between histological features of the initial (benign) lesion and subsequent breast cancer risk. All breast cancer cases in the cohort (50 expected) will be matched to four "controls" also from the cohort, that did not (yet) develop breast cancer. Matching variables include age, duration of follow-up and histology of the benign lesion. Cases and controls will be interviewed at home as to lifetime use of OC and other hormones and known breast cancer risk factors. An album has been composed, containing colour photographs of all OCs ever marketed in the Netherlands, in order to aid women in recalling the specific OC brands they used.

TYPE: Case-Control; Cohort
TERM: Hormones; Oral Contraceptives
SITE: Benign Tumours; Breast (F)
TIME: 1985 - 1994

618 **van Leeuwen, F.E.** 04076

Netherlands Cancer Inst., Div. of Clinical Oncology, Dept. of Epidemiology, Plesmanlaan 121, 1066 CX Amsterdam, Netherlands (Tel.: +31 20 5122453; Fax: 6172625 ; Tlx: 11273)
COLL: Rookus, M.A.; Braas, P.A.M.; Chorus, A.M.

Breast Cancer and Oral Contraceptives

The study aims are: (1) to evaluate the hypothesis that breast cancer risk is related to oral contraceptive (OC) use in certain sub-categories of users, e.g. women who used OCs before their first full-term pregnancy, women who used OCs before age 25, women with a family history of breast cancer, women with a history of benign breast disease, women aged 45-54 years; (2) to re-evaluate the hypothesis that breast cancer risk is related to overall use of OC. In the Netherlands OC use has been higher than in any other Western country. Also, extensive OC use at young ages started earlier than in other countries. A population-based case-control study is conducted in four regions of the Netherlands. The study will include 1,000 incident cases under 54 years of age (500 of which will be under 45 years of age). Region- and age-matched population controls are selected from the municipal registers in a 1:1 matching design. By personal interview in the participants' homes, lifetime histories will be collected on OC use, use of other hormones, reproductive factors, family history of breast cancer and a possible history of benign breast disease. In order to facilitate recall of all OC brands ever used, subjects will be shown photographs of all OCs (pills and packages) that have been on the market in the Netherlands since 1962. OC use will be checked with the woman's physician(s). This study also enables us to examine the relationship between breast cancer risk and possible confounders, such as diet, alcohol intake, weight history and waist/hip ratio.

TYPE: Case-Control
TERM: Alcohol; Diet; Drugs; Familial Factors; Hormones; Oral Contraceptives; Reproductive Factors
SITE: Breast (F)
CHEM: Oestrogens; Progesterone
TIME: 1986 - 1994

619 **van Leeuwen, F.E.** 04508

Netherlands Cancer Inst., Div. of Clinical Oncology, Dept. of Epidemiology, Plesmanlaan 121, 1066 CX Amsterdam, Netherlands (Tel.: +31 20 5122453; Fax: 6172625 ; Tlx: 11273)
COLL: Noyon, R.; Van den Belt, A.H.W.; Somers, R.

Second Cancer Risk Following Therapy for Cancers of Breast, Testis, Hodgkin's Disease and Non-Hodgkin's Lymphoma

With the advent of improved survival in many patients with advanced cancer and the increasing use of adjuvant therapy in cancer patients with minimal tumour load, the need for quantitative data on second cancer risk following such therapy has increased. The aims of this study are: (1) to assess and quantify the separate and combined effects of radiotherapy and chemotherapy in determining the risk of second lung cancer following breast cancer, Hodgkin's disease, non-Hodgkin's lymphoma (NHL) and testis cancer; (2) to evaluate aspects of treatment (including specific cytostatic drugs) that determine the risk of second leukaemia and the MDS after testis cancer and adjuvant CT for breast cancer; (3) to assess which other factors affect lung cancer and leukaemia risk (e.g. age, sex, treatment toxicity, smoking habits). These aims will be achieved by first conducting four cohort studies of patients with breast cancer, testis cancer, Hodgkin's disease and NHL. Subsequently nested case-control studies within the cohorts will be carried out: one on second lung cancer after all three above-mentioned first primaries and another on second leukaemia after breast and testis cancer. The cohorts will include patients from both the Netherlands Cancer Institute and the Dr. Daniël den Hoed Clinic, Rotterdam. In the cohort studies a limited data set will be collected from the hospital tumour registries; this part of the study serves as a tool to identify cases for the nested case-control studies. In the latter part of the study detailed treatment data (and other second cancer risk factors) will be collected from the medical records for only a small sample of the cohorts, thus making a very efficient study design. Controls will be matched to the cases for survival time, age, sex and year of diagnosis.

TYPE: Case-Control; Cohort
TERM: Age; Chemotherapy; Multiple Primary; Radiotherapy; Sex Ratio; Tobacco (Smoking)
SITE: All Sites; Breast (F); Leukaemia; Lung; Myelodysplastic Syndrome; Non-Hodgkin's Lymphoma
TIME: 1988 - 1994

620 van Leeuwen, F.E. 05130
Netherlands Cancer Inst., Div. of Clinical Oncology, Dept. of Epidemiology, Plesmanlaan 121, 1066 CX Amsterdam, Netherlands (Tel.: +31 20 5122453; Fax: 6172625 ; Tlx: 11273)
COLL: Rookus, M.A.; Benraadt, J.; Peterse, J.L.; Lebesque, J.; Van Dongen, J.A.

Risk Factors for Contralateral Breast Cancer

Of all women with unilateral breast cancer surviving 20 years, 15-20% will develop contralateral breast cancer (CLBC). In this study of CLBC, aetiological and genetic factors, lifestyle factors and therapy for the first tumour will be taken into account simultaneously, in a design that also makes it possible to examine controversial risk factors for breast cancer. Specific aims are: (1) to assess the separate and combined effects of genetic factors, known and suspected lifestyle factors and therapy for the first tumour on CLBC risk; (2) to examine controversial risk factors (such as characteristics of the natural menstrual cycle, lactation, oestrogen use, oral contraceptive use, fat consumption and alcohol consumption) for breast cancer; (3) to gain more insight into the biological mechanism of CLBC development by studying genetic and lifestyle factors as determinants of neu-oncogene overexpression in breast tumours; (4) to estimate CLBC incidence in the Netherlands according to various standard definitions and based on reviewed histological slides. A population-based case-control study will be carried out with two matched control groups and one matched second case group. All incident CLBC cases (N = 350) diagnosed in the regions covered by five comprehensive cancer centres will be eligible for the study during an intake period of three years. By comparing CLBC cases with patients with unilateral breast cancer at the age of the first tumour of the CLBC cases, genetic factors, lifestyle factors and therapy for the first tumour can be examined simultaneously. By comparing CLBC cases with a second control group of healthy women that are at least as old as the case was at CLBC diagnosis, an enhanced expression of risk factors can be expected. Therefore, this design facilitates the study of controversial breast cancer risk factors. To be able to interpret the effects thus estimated, relative risks based on a traditional design comparing cases with unilateral breast cancer newly diagnosed at the age of the second tumour of the CLBC cases with healthy controls is necessary as well. Trained interviewers will interview all women personally at home on lifetime histories of reproductive factors, family history of breast cancer, hormone use and a possible history of benign breast disease. The treatment of the first tumour will be abstracted from medical records.

TYPE: Case-Control
TERM: Alcohol; Chemotherapy; Fat; Genetic Factors; Hormones; Lactation; Lifestyle; Menstruation; Multiple Primary; Oncogenes; Oral Contraceptives; Radiotherapy; Registry
SITE: Breast (F)
CHEM: Oestrogens
REGI: Amsterdam (Net); Eindhoven (Net); Leiden (Net); Nijmegen (Net); Rotterdam (Net)
TIME: 1991 - 1996

***621 van Leeuwen, F.E.** 05419
Netherlands Cancer Inst., Div. of Clinical Oncology, Dept. of Epidemiology, Plesmanlaan 121, 1066 CX Amsterdam, Netherlands (Tel.: +31 20 5122453; Fax: 6172625 ; Tlx: 11273)
COLL: Sasco, A.J.

International Study of Cancer Risk in Biology Research Laboratory Workers

This study is part of an international study coordinated by IARC. The study aims to assess whether there is an excess cancer risk, either for overall cancer or more particularly for leukaemia, lymphomas, cancer of the brain, pancreas or bone, for persons working in research laboratories. Whether there are specific sub-groups at higher risk, either defined by job title or by their type of scientific activity (chemistry, virology, molecular biology, etc.) will also be assessed. In the Dutch part of the study a cohort of about 5,000 persons will be followed retrospectively for at least 20 years. General descriptive information and job history information will be assembled by means of the personnel registries of the participating institutes. Job exposure information will be collected at the group level through a personal interview. Two reference groups will be used: national population rates of the diseases of interest and an internal comparison group. This group consists of persons employed by the same research institutes, but not exposed to laboratory work, like e.g. epidemiologists, administrative workers and computer personnel.

NETHERLANDS

TYPE: Cohort
TERM: Chemical Exposure; Occupation
SITE: All Sites; Bone; Brain; Leukaemia; Lymphoma; Pancreas
OCCU: Laboratory Workers
TIME: 1991 - 1995

Bilthoven

622 **Bueno de Mesquita, H.B.** 05208
Nat. Inst. of Public Health & Environ. Protection, Dept. of Chronic Diseases & Environmental Epidemiology, Antonie van Leeuwenhoeklaan 9 , 3720 BA Bilthoven, Netherlands (Tel.: +31 30 742971; Fax: 367370 ; Tlx: 47215 rivm nl)
COLL: Doornbos, G.; Van der Kuip, A.M.; Kogevinas, M.; Winkelmann, M.A.

Retrospective Cancer Mortality Study of Phenoxy Herbicide Manufacturers

As part of the "IARC International Register of Persons Exposed to Phenoxy Herbicides and Contaminants", a cohort of workers who manufacture and prepare chlorophenoxy herbicides was recruited in The Netherlands. The cohort comprised 2,310 workers from two plants, operated by different companies, who were followed during the periods 1955-1985 and 1965-1986, respectively. In 1963, there had been an industrial accident in one factory with concomitant release of dioxin into the environment. Loss to follow-up was 3%. Mortality data on 963 exposed and 1,111 non-exposed men were evaluated by external and internal comparison. Compared with national rates, total mortality and cancer mortality for exposed workers were not significantly increased. A statistically insignificant increase was observed for non-Hodgkin's lymphoma. No cases of soft-tissue sarcoma were encountered. There was no increase in cancer mortality among the 139 workers probably exposed to dioxins during the 2,4,5-trichlorophenol production accident or the subsequent clean-up operations. Compared with nonexposed workers, exposed workers did not exhibit a higher total mortality. Mortality due to all cancers and respiratory cancer was insignificantly elevated. These findings suggest that the increases in cancer mortality among workers exposed to phenoxy herbicides and chlorophenols may be attributable to chance. Lack of power prevented evaluation with respect to specific cancers (Am. J. Ind. Med. 23, 289-300, 1993). Two further studies are now underway (1) to update the mortality follow-up to 31 December, 1991 and (2) to validate exposure by determination of specific polychlorinated dibenzo-p-dioxins and dibenzofurans in blood of about 60 non-exposed subjects of the cohort.

TYPE: Cohort
TERM: Herbicides; Occupation; Pesticides; Registry
SITE: All Sites; Non-Hodgkin's Lymphoma; Sarcoma; Soft Tissue
CHEM: Dioxins; Phenoxy Acids
OCCU: Chemical Industry Workers
TIME: 1983 - 1994

Maastricht

623 **Swaen, G.M.H.** 04253
Univ. of Limburg, Dept. of Occupational Medicine, Postbox 616, 6200 MD Maastricht, Netherlands (Tel.: +31 43 882386; Fax: 618685)
COLL: Sturmans, F.; de Boorder, T.

Epidemiological Study of Workers Exposed to Acrylonitrile in the Netherlands

In a retrospective cohort study cause-specific mortality patterns between 4,000 workers exposed to acrylonitrile in the past and a comparison group comprising 4,000 non-exposed workers will be compared to test the hypothesis that exposed workers experience a higher mortality from lung cancer or cancer of the prostate than those not exposed. Special emphasis will be put on estimation of the quantity of past exposure to acrylonitrile, based on measurement if available, interviews with plant industrial hygienists and other key persons. The study will encompass nine chemical companies located in The Netherlands, where acrylonitrile has been manufactured, stored, handled or used in the production process. Follow-up of the 8,000 workers enrolled in the study will be conducted by means of the Dutch system of population-based cancer registries.

TYPE: Cohort
TERM: Chemical Exposure; Occupation; Plastics; Registry
SITE: All Sites; Lung; Prostate
CHEM: Acrylonitrile
OCCU: Acrylonitrile Workers
REGI: Leiderdorp (Net)
TIME: 1986 - 1994

624 **Van den Brandt, P.A.** 04506
Univ. of Limburg, Dept. of Epidemiology, P.O.Box 616, 6200 MD Maastricht, Netherlands (Tel.: +31 43 882361; Fax: 618685 ; Tlx: 56726 NL)
COLL: Bausch-Goldbohm, R.; Van't Veer, P.; Sturmans, F.; Hermus, R.J.J.; Bode, P.; de Bruin, M.

Toenail Selenium and the Risk of Cancer

The hypothesis that selenium reduces the risk of various forms of cancer is being tested within the framework of a prospective cohort study on diet, lifestyle and cancer started in 1986 (Int. J. Epidemiol. 17:472-, 1988 and J. Clin. Epidemiol. 43:285-295, 1990). The cohort originates from the general population and consists of over 120,000 men and women, aged 55-69 years at baseline. Toenail clippings, providing a measure of selenium status, were obtained from 80,000 participants at baseline. Repeated toenail sampling will be performed annually for random samples (n = 250) of the cohort to assess intra-individual variability. The selenium content of the nails of incident cancer cases and of a random subcohort will be determined by neutron activation analysis. Cancer follow-up consists of record linkage to cancer registries and a pathology registry, for which a linkage has been developed (Int. J. Epidemiol. 19:553-558, 1990). Statistical analysis will involve a case-cohort approach. Special attention will be given to evaluating the suggested gender-specific effect of selenium.

TYPE: Case-Control; Cohort
TERM: BMB; Diet; Lifestyle; Registry; Sex Ratio; Toenails
SITE: Breast (F); Colon; Lung; Rectum; Stomach
CHEM: Selenium
REGI: Amsterdam (Net); Eindhoven (Net); Groningen (Net); Leeuwarden (Net); Leiden (Net); Leiderdorp (Net); Maastricht (Net); Nijmegen (Net); Rotterdam (Net); Tilburg (Net); Utrecht (Net)
TIME: 1986 - 1994

625 **Van den Brandt, P.A.** 04507
Univ. of Limburg, Dept. of Epidemiology, P.O.Box 616, 6200 MD Maastricht, Netherlands (Tel.: +31 43 882361; Fax: 618685 ; Tlx: 56726 NL)
COLL: Bausch-Goldbohm, R.; Dorant, E.; Sturmans, F.; Hermus, R.J.J.

Dietary Supplements and Drugs in the Aetiology of Cancer

In 1986, a prospective cohort study on diet, other lifestyle habits and the incidence of cancer was started in the Netherlands (Int. J. Epidemiol 17:472-, 1988; J. Clin. Epidemiol. 43:285-295, 1990). Baseline data were obtained by self-administered questionnaire on 120,000 men and women, aged 55-69 years. Questions referred to diet (including supplements), smoking, long-term drug use, medical history, family history of cancer and reproductive history. The present study evaluates the various hypotheses regarding long-term drug use, associated with particular (precursor) conditions (e.g. cimetidine use and gastric ulcer) and the subsequent development of cancer. The interaction with nutritional supplements and alcohol and the effect of supplements per se will also be tested. Follow-up for cancer will be performed by record linkage to a pathology register and the cancer registries in the Netherlands (Int. J. Epidemiol. 19:553-558, 1990).

TYPE: Cohort
TERM: Alcohol; Drugs; Familial Factors; Gastric Ulcer; Registry; Reproductive Factors; Tobacco (Smoking); Vitamins
SITE: All Sites
CHEM: Ascorbic Acid; Beta Carotene; Cimetidine
REGI: Amsterdam (Net); Eindhoven (Net); Groningen (Net); Leeuwarden (Net); Leiden (Net); Leiderdorp (Net); Maastricht (Net); Nijmegen (Net); Rotterdam (Net); Tilburg (Net); Utrecht (Net)
TIME: 1988 - 1994

Nijmegen

626 **Bos, R.P.** 03596
Univ. of Nijmegen, Faculty of Medicine, Dept. of Toxicology, Kapittelweg 54 , 6525 EP Nijmegen, Netherlands (Tel.: +31 80 514203; Fax: 514090; Tlx: 484211 kunm nl)
COLL: Jongeneelen, D.J.; Sessink, P.J.M.; Scheepers, P.T.J.

Development and Validation of Methods for the Detection of Occupational Exposure to Mutagenic and Carcinogenic Agents

The objective of the project is the development and validation of biological or chemical test methods for the detection of exposure to genotoxic substances (including mutagens and carcinogens). These methods can be used in biological monitoring. After initial chemical analytical research and investigations on laboratory animals, methods can be validated in exposed workers. With respect to PAH-exposure several groups of workers exposed to coal tar and coal tar products are compared with controls (1-hydroxypyrene in urine). For diesel emissions in the first period sensitive chemical analytical methods will be developed. The occupational exposure of workers will be studied. Validations will be made with respect to several parameters, such as environmental measurements and e.g. DNA-adducts (Recent papers: IARC Scientific Publ. 89:389-395, 1988 and Ann. Occup. Hyg. 32:35-43, 1988).

TYPE: Methodology; Molecular Epidemiology
TERM: Chemotherapy; Coal; DNA; Drugs; Monitoring; Mutagen Tests; Occupation
SITE: Inapplicable
CHEM: Cyclophosphamide; Diesel Exhaust; PAH
OCCU: Chemical Industry Workers; Drivers; Health Care Workers
REGI: Amsterdam (Net); Eindhoven (Net); Groningen (Net); Leeuwarden (Net); Leiden (Net); Leiderdorp (Net); Maastricht (Net); Nijmegen (Net); Rotterdam (Net); Tilburg (Net); Utrecht (Net)
TIME: 1980 - 1994

***627** **Melchers, W.J.G.** 05257
Catholic Univ. of Nijmegen, Dept. of Medical Microbiology, P.O. Box 9101 , 6500 HB Nijmegen, Netherlands (Tel.: +31 80 617574/4356; Fax: 540216)
COLL: Figueroa, M.; Velema, J.

Prevalence of HPV Infections in Women with Cervical Dysplasias, Carcinoma of the Cervix and in a Normal Population

This population-based case-control study of cervical cancer in Honduras aims to assess the association of cervical cancer and the presence of human HPV, educational level, smoking, age at first intercourse, age at first pregnancy, frequency of spontaneous abortions, use of contraceptives, number of sexual partners, history of sexually transmitted diseases, and frequency of screening tests for cervical cancer. Data will be collected for 50 cases of invasive cancer of the cervix, 50 cases of CIS, 50 cases of severe dysplasia and 50 cases of mild dysplasia. Controls (two for each case) will be women selected among participants in a screening programme for cervical cancer, matched to the cases for age and area of residence. Controls should have negative cytology result. All cases and controls will be interviewed by trained interviewers with a structured questionnaire. Standard analytical methods for case-control studies will be used, including multiple logistic regression.

TYPE: Case-Control
TERM: Abortion; BMB; Contraception; Education; HPV; Pregnancy; Screening; Sexual Activity; Sexually Transmitted Diseases; Tobacco (Smoking)
SITE: Uterus (Cervix)
LOCA: Honduras
TIME: 1992 - 1995

628 **Nelemans, P.J.** 04886
Nijmegen Univ., Dept. of Social Medicine, Epidemiology Unit, Verlengde Groenestr. 75 , 6525 EJ Nijmegen, Netherlands (Tel.: +31 80 513125)
COLL: Verbeek, A.L.M.; Ruiter, D.J.; Rampen, F.H.J.

Malignant Melanoma and Aetiological Indicators in the Netherlands

The aim of the study is to study the role of several risk factors, including sun exposure and chemical exposures, for the development of cutaneous malignant melanoma. A case-control study is being conducted in which patients with cutaneous malignant melanoma (n=200) will be compared with a sample of patients with various other types of cancer, which are not known to be associated with the exposure variables under study (n = 400). An interview will provide information about exposure variables and, in a physical examination by a dermatologist, constitutional variables will be defined and the number of naevi will be counted. The source of patients is the registry of the Comprehensive Cancer Centre which registers the incidence of cancer in the region around Nijmegen (IKO). The independent effects of the variables under study will be examined with the use of multiple logistic regression analysis.

TYPE: Case-Control
TERM: Chemical Exposure; Naevi; Occupation; Radiation, Ultraviolet
SITE: Melanoma
REGI: Nijmegen (Net)
TIME: 1989 - 1994

629 **van der Gulden, J.W.J.** 05031
Nijmegen Univ., Dept. of Occupational Medicine, Verlengde Groenestr. 75 , 6525 EJ Nijmegen, Netherlands (Tel.: +31 80 613119)
COLL: Kolk, J.; Kiemeney, L.A.L.; Verbeek, A.L.M.

Prostate Cancer and Occupational Exposure

In 1990, a case-referent study was started on the relation between prostate cancer and occupation (job title and function) and occupational exposures. In this study cases are defined as men in whom prostate cancer was histologically confirmed between 1 January and 1 April, 1990. Referents are patients who were treated for benign prostate hyperplasia in this period, and in whom no signs of malignancy were found on histological examination. 469 cases were selected from the registry of the Comprehensive Cancer Centre IKO. 1872 controls were selected from the National Archive of Pathology Reports (PALGA). Mail questionnaires were sent to all participants to collect information on work history and occupational exposures and on confounding factors such as age, smoking, drinking and socio-economic status. The questionnaire is a slightly modified version of a validated questionnaire on occupational history, developed recently by an EEC Working Party. For farmers, metal workers, repairmen and mechanics, several specific questions were added, since for these occupational groups an increase of risk was found in a feasibility study, executed in 1989. Aetiological analyses commenced early 1991 using the case-referent approach. A paper was published in J. Occup. Med. 34:402-409, 1992.

TYPE: Case-Control
TERM: Fertilizers; Metals; Occupation; Paints; Pesticides; Registry; Solvents; Welding
SITE: Prostate
CHEM: Vehicle Exhaust
OCCU: Farmers; Mechanics; Metal Workers
REGI: Nijmegen (Net)
TIME: 1989 - 1994

630 **van Dijck, J.A.A.M.** 01886
Univ. of Nijmegen, Dept. of Medical Informatics and Epidemiology, P.O.Box 9101 , 6500 HB Nijmegen, Netherlands (Tel.: +31 80 614295; Fax: 613505)
COLL: Verbeek, A.L.M.; Straatman, H.; Peer, P.G.M.; Hendriks, J.H.C.L.; Holland, R.; Mravunac, M.

Impact of Mass Screening on Morbidity and Mortality of Breast Cancer in Nijmegen

This population-based breast cancer screening project was initiated in 1975, with single-view mammography as the only screening procedure. All women born before 1946 (n = 36,000) are invited biennially. The attendance rate have declined from 80% in round 1 to 55% in round 9, which was completed in 1992. The first results up to 1982, analysed by means of a case-control study with 62 cases of breast cancer death, suggested a 50% breast cancer mortality reduction. No effect was observed in the youngest group, aged 35-49 at entry. Using more recent data up to 1992, further analyses are being undertaken on age-specific effects, particularly under age 50 and above age 70 and the impact of the number and frequency of examinations. In addition, the breast cancer mortality in the total population of Nijmegen will be compared with that of the neighbouring city of Arnhem, where no population screening had been initiated until 1989. The effect of screening on breast cancer morbidity (distant metastases) will also be studied. The results will be incorporated into a mathematical model presented by Eddy (1980) to determine how frequently women of different age-groups should be screened, accounting for adverse effects such as false positive and false negative screening results. A study of the length of time by which diagnosis is advanced is underway. Publications have appeared in Br. J. Cancer 59:929-932, 1989 and

63:261-264, 1991 and in Int. J. Cancer 43:226-230 and 1055-1060, 1989. Lancet I: 591-593, 1222-1224, 1984 and I: 865-866, 1985.

TYPE: Case-Control; Cohort; Incidence; Intervention; Mortality
TERM: Cost-Benefit Analysis; Mammography; Mathematical Models; Screening; Survival; Time Factors
SITE: Breast (F)
TIME: 1975 - 1996

631 **Vooijs, P.G.** 04672
Univ. of Nijmegen, Dept. of Pathology, Lab. of Cytopathology, Geert Grooteplein Zuid 24, 6536 HK Nijmegen, Netherlands (Tel.: +31 80 614389; Fax: 540520)
COLL: Habbema, J.D.F.; Lubbe, J.T.N.; Pal, R.; Otto, L.P.

Information System for Cervical Cancer Screening

A nationwide screening programme for cervical cancer is being organized in the Netherlands, based on the results of three pilot projects. All pathology laboratories involved use the same protocol for coding and reporting of cytological findings and adopt the same recommendations for follow-up of abnormalities. They will be linked to a central pathology diagnoses database (PALGA). Linkage of screening results to previous and follow-up cytological and histological findings permits epidemiological studies on a regional and national level, and measurement of the impact of various screening protocols. This will provide an insight into the natural history of cervical cancer. It will offer the opportunity to evaluate quality control protocols. An information system will be developed for a uniform evaluation of registered data. The MISCAN computer program can be used for a detailed evaluation of the screening programme and for comparison of the cost effectiveness of future screening programmes. A paper appeared in Acta Cytol. 33:825-830, 1989.

TYPE: Cohort
TERM: Cost-Benefit Analysis; Cytology; Histology; Mathematical Models; Screening
SITE: Uterus (Cervix)
TIME: 1989 - 1994

Sittard

632 **Verduijn, P.G.** 05094
Hospital of Sittard, Dept. of Otolaryngology, Walramstr. 23, 6131 BK Sittard, Netherlands (Tel.: +31 46 597792)
COLL: Hayes, R.B.; Looman, C.; Habbema, J.D.F.; van der Maas, P.J.

Late Effects of Irradiation for Eustachian Tube Dysfunction

This non-concurrent prospective study aims to examine the association of nasopharyngeal radium irradiation and the origin of tumours in the head and neck region. The second question concerns the possible influence of this type of irradiation on hormone balance, as a result of the pituitary gland being in the treatment field. Further, it is investigated whether information can be obtained regarding dose-effect relationships in a dose range (0-50 cGy) for which little is known in man. In exposed (n = 2,542) and control groups, selected from the same clinical records, cause of death will be established for the deceased and the subjects who are still alive will be surveyed with questions concerning their health. First results show that some risk of tumour induction is involved with nasopharyngeal radium irradiation. No increased mortality due to cancer was observed. However, a statistically significant increase in the cumulative incidence for all cancers combined was observed. Separation of the individual tumour sites did not produce a significant excess. Preliminary results of follow-up until February 1, 1985 (mean follow-up 25.3 years) were published in Ann. Otol. Rhinol. Laryngol. 98:839-844, 1989.

TYPE: Cohort
TERM: Dose-Response; Radiotherapy
SITE: Head and Neck
LOCA: Belgium; Germany; Netherlands
TIME: 1982 - 2002

The Hague

633 **Coebergh, J.W.W.** 04164
Dutch Childhood Leukemia Study Group, Dr Van Welylaan 2, P.O. Box 60604, 2566 ER (2506 LP) The Hague, Netherlands (Tel.: +31 70 657930; Fax: 617427)
COLL: Van der Does-van den Berg, A.; van Wering, E.R.; Rammeloo, J.A.; Kamps, W.A.

Trends in Incidence, Mortality and Geographical Distribution of Childhood Leukaemia in The Netherlands (1973-1988)

Incidence, mortality and geographical distribution of childhood leukaemia in The Netherlands (1973-1987) are being examined by continuing an epidemiological register-based study of childhood leukaemia in The Netherlands (van Steensel-Moll et al., Br.J.Cancer 47:471-475, 1983). Uniform examination and review of bone marrow slides are performed in the central DCLSG laboratory. Type, age and sex specific leukaemia incidence rates in The Netherlands are compared with those in other countries, analysing rates according to year of diagnosis and year of birth. Time trends in incidence and mortality by age, sex and type of leukaemia, are also studied. Special attention is given to acute lymphocytic leukaemia through analysis of white blood cell count at diagnosis and immunophenotype since 1979.

TYPE: Incidence; Mortality
TERM: Age; Childhood; Geographic Factors; Registry; Trends
SITE: Leukaemia; Leukaemia (ALL)
REGI: The Hague (Net)
TIME: 1988 - 1994

Utrecht

634 **Baanders-van Halewijn, E.A.** 03316
Preventicon, Inst. Public Health & Epidemiology, Radboudkwartier 261-263 , 3511 CK Utrecht, Netherlands (Tel.: +31 30 313884)
COLL: de Waard, F.; Collette, H.J.A.; Slotboom, B.J.

Historical Cohort Study of Ovarian Cancer

The purpose of the study is to try and find correlations between ovarian cancer and a number of suspected risk factors, by means of a historical cohort study in the female population of Utrecht and suburbs, aged 50-73 years. The data collected during the DOM- project (a cohort study for early detection of breast cancer covering 38,000 women screened periodically from 1974 to 1980) will be used to discover new hypotheses about the aetiology and natural history of ovarian cancer which could give clues for prevention. These hypotheses will be tested in a subsequent study on a population of women born in the years 1932-1941 (the Lutine project, carried out in 1982-1983). Hypotheses will also be able to be tested using the endocrinological examinations of urinary specimens collected at intake into the study (first screening for breast cancer).

TYPE: Cohort
TERM: Hormones; Prevention; Screening
SITE: Ovary
TIME: 1984 - 1994

635 **Baanders-van Halewijn, E.A.** 03856
Preventicon, Inst. Public Health & Epidemiology, Radboudkwartier 261-263 , 3511 CK Utrecht, Netherlands (Tel.: +31 30 313884)
COLL: de Waard, F.; Thijssen, J.H.H.; Blankenstein, M.A.; Collette, H.J.A.; van Noord, P.A.H.; Slotboom, B.J.

Relation Between Urinary Hormone Excretion and Ovarian Carcinoma

This prospective study is a further development of a cohort study of correlations between a number of risk factors and ovarian carcinoma. It is now hoped to find clues for prevention by obtaining more insight into endocrinological factors possibly related to these risk factors. Hypotheses are derived from (1) earlier studies of breast and endometrial cancer, given the aetiological relationships between these and ovarian cancer, and (2) from the publications of Cuzick et al. (1983) on ovarian cancer and of Fishman et al. (1984) on breast and endometrial cancer. Correlations between some of the risk factors and the results of biochemical measurement of the concentrations in urine of the three oestrogens (oestrone, oestradiol and oestriol), of the androgenmetabolites (androsterone, aetiocholanolone and dehydroepiandrosterone (DHEA) and of the 16 alpha-hydroxylase product of oestradiol: (16 alpha-hydroxyoesterone) will be analysed by comparing the results of patients and controls. The hypothesis is that clinical evidence of ovarian cancer is preceded by increased concentrations of 16 alpha-hydroxyoestrone and lower levels of androgen metabolites, in particular DHEA in the urine. Urine samples have been collected from the participants of the DOM-project, (a cohort study for early detection of breast cancer) at time of intake. About 19,000 women born between 1911 and 1931 were screened periodically from December 1974 to March 1985 and delivered urine specimens one, two or three times over the period. This provides a unique possibility to carry out chemical analyses of samples taken long before diagnosis. In the cohort about 100 cases of ovarian cancer are expected to be diagnosed in the period December 1974 to the middle of 1987.

TYPE: Cohort
TERM: Drugs; Hormones; Urine
SITE: Ovary
CHEM: Aetiocholanolone; Androsterone; Dehydroepiandrosterone Sulphate; Oestradiol; Oestriol; Oestrone
TIME: 1986 - 1994

636 **Collette, H.J.A.** 04104
Dept. of Epidemiology, Universiteitsweg 100, 3584 CG Utrecht, Netherlands (Tel.: +31 30 539111)
COLL: de Waard, F.; Slotboom, B.J.; Fracheboud, J.; Beijerinck, D.; Mittenburg, W.H.Q.

DOM-Project for Early Diagnosis of Breast Cancer

The DOM-project (Diagnostisch Onderzoek Mammacarcinoom) is a population-based screening project in which use is made of mammography (together with inspection and palpation) to detect breast cancer at an early stage. It was started in the city of Utrecht towards the end of 1974, and a few years later it was introduced in several suburbs as well. At first women born in 1911-1925 were invited to participate, but later on women born 1926-1945 received an invitation as well. To date nearly 60,000 women have participated at least once. There are three steps of control in this study: (1) process evaluation, consisting of response percentage, pick-up rate, predictive value of the test, etc.; (2) early outcomes, such as stage distribution of detected cases, number of false negative tests, etc.; (3) late outcomes, i.e. survival and mortality rates. The last mentioned is most important: the main aim of such a screening programme is that it should result in a decrease in mortality of the disease the programme is focussed on. As the programme has been going on for more than ten years an enormous amount of data has been collected and computerised. These data are useful both for evaluation of this screening programme, for describing the natural history of breast cancer and for performing cohort nested case-control studies in other cancers.

TYPE: Mortality
TERM: BMB; Screening; Survival
SITE: Breast (F)
TIME: 1974 - 2000

637 **den Tonkelaar, I.** 05050
Univ. of Utrecht, Dept. of Public Health and Epidemiology, Radboudkwartier 261-263 , 3511 CK Utrecht, Netherlands (Tel.: +31 30 331529; Fax: 340634)
COLL: Collette, H.J.A.; Seidell, J.C.; van Noord, P.A.H.; Baanders-van Halewijn, E.A.; de Waard, F.; Wynne, H.J.A.

Fat Distribution and Breast Cancer Incidence and Survival in Postmenopausal Women

The purpose of this study is an investigation of the relation between fat distribution (as measured by central and peripheral skin folds) and breast cancer incidence and survival. The study will be performed in about 20,000 post-menopausal women who were aged 49-68 years in 1975-1979 and who are participating in a breast cancer screening project (the DOM-I project). The relationship between fat distribution and known risk factors for malignant breast disease (such as overweight, age, family, history of breast cancer, parity and age at first delivery) will be evaluated.

TYPE: Cohort
TERM: Age; Familial Factors; Fat; Obesity; Parity; Registry; Survival
SITE: Breast (F)
REGI: Utrecht (Net)
TIME: 1991 - 1994

638 **van Noord, P.A.H.** 04318
Preventicon, Inst. Public Health & Epidemiology, Div. of Epidemiology, Radboudkwartier 261-263, 3511 CK Utrecht, Netherlands (Tel.: +31 30 313884)
COLL: Maas, M.J.; de Bruin, M.; Noordhoek, J.

Prediagnostic Nail Selenium Levels in Colorectal Tumours

It has been suggested that selenium in humans may have a preventive effect in cancer occurrence. Given the observations made in animal studies, breast and colorectal cancers are the tumours best suited for studying the hypothesis in humans. Classical case-control studies in humans have not been able to exclude the possibility that decreased selenium levels found in cancer patients are not caused by the tumour. It is known that selenium can concentrate in or around cancer tissue. Several biological mechanisms have been advanced for selenium: (1) protection against free radical induced damage through the selenium containing enzyme glutathione peroxidase; (2) antioxidative properties of selenium, and (3) antagonism of heavy metals (Hg, As, Cd). The latter are relevant for their local effects in relation to colorectal cancers. This study is of the case-cohort type. Biological samples from approximately 19,000 women participating in the DOM project for breast cancer screening, have been entered in a biological bank. Selenium behaves biologically as sulphur and is accumulated in keratine, a protein polymer with high sulphur and selenium contents. The collection of toenail clippings to be used in this study was started in mid 1982. First the relation of selenium to breast cancer was addressed: no decreased nail selenium levels were found in a breast cancer case-cohort study (Int. J. Epidemiol. 16(2):318-322, 1987). Given the lower incidence of colorectal tumours a longer follow-up was needed. Selenium levels in the specimens will be determined at IRI Delft by Neutron Activation of the Se 77m isotope.

TYPE: Cohort
TERM: BMB; Toenails; Trace Elements
SITE: Colon; Rectum
CHEM: Selenium
TIME: 1987 - 1994

Wageningen

639 **Kok, F.J.** 05025
Wageningen Agricultural University Dept. of Epidemiology and Public Health, Dreijenlaan 1 P.O.Box 238, 6700 AE Wageningen, Netherlands (Tel.: +31 837082080; Fax: 837082782 ; Tlx: 45015 NL)
COLL: Huttunen, J.K.; Kohlmeier-Arab, L.; Martin-Moreno, J.; Gutzwiller, F.; Strain, J.J.; Van't Veer, P.

EURAMIC Breast Cancer Study

The EURAMIC breast cancer study is part of the EC-funded European Study on Antioxidants, Myocardial Infarction and Breast Cancer. The breast-cancer arm of this multi-centre case-control study is designed to establish an association between early stage breast cancer and the diet-related antioxidants alpha-tocopherol, beta-carotene and selenium, both separately and in combination. 70 cases and 70 controls are being enrolled from each of 5 study centres. Postmenopausal women, 50-69 years of age, with a stable dietary pattern during the past year, are eligible. Breast cancer cases (ICD-O 174, ductal carcinoma, tumour size < 5 cm, without distant metastases at diagnosis) are recruited from collaborating hospitals or screening centres. Controls are recruited from the same source population as the cases, e.g., via local population registries, health centres, general practitioners etc. Long-term exposure to the antioxidants alpha-tocopherol and beta-carotene, and selenium is assessed from aspirated subcutaneous fat and big toenails. Among cases, needle aspiration and toenail collection is conducted within one to four weeks after diagnosis. In order to guarantee standardised analysis and to minimize sources of laboratory error, biomarkers are analysed in a central laboratory. Background data from cases and controls include relevant dietary factors, breast cancer risk factors, and general lifestyle indicators. The use of biomarker data, the broad range of exposure to antioxidants in different countries and the simultaneous measurement of several antioxidants offers opportunities for a concise test of the antioxidant hypothesis for breast cancer, as one of the major age-related diseases among women in the EC.

TYPE: Case-Control
TERM: Antioxidants; BMB; Biomarkers; Diet; Fat; Nutrition; Toenails
SITE: Breast (F)
CHEM: Beta Carotene; Selenium; Tocopherol
LOCA: Germany; Netherlands; Spain; Switzerland; United Kingdom
TIME: 1990 - 1994

Zeist

640 **Hermus, R.J.J.** 03237
Toxicology and Nutrition Inst. TNO, P.O. Box 360, 3700 AJ Zeist, Netherlands (Tel.: +31 3404 44144; Fax: 57224)
COLL: Sturmans, F.; Bausch-Goldbohm, R.; Van den Brandt, P.A.; Van't Veer, P.; Dorant, E.

Diet and Cancer of the Breast, Colon, Rectum, Stomach and Lung

Results of epidemiological and animal studies suggest that there is a relation between diet and cancer incidence; relevant hypotheses concern the effects of fat, type of fat, meat, fibre (from different sources), selenium, alcohol and vitamins A, C, E and beta carotene. In the Netherlands, a prospective cohort study has been started to estimate the association between dietary habits as well as nutrient composition of the diet and onset of cancer of the breast, colon, rectum, stomach and lung. In a pilot study (1984-1985) a self-administered semi-quantitative food frequency questionnaire, which includes approximately 170 food items , was developed. In September 1986, a cohort was recruited by inviting a general population sample aged 55-69, randomly selected from 204 municipalities all over the country. Baseline questionnaire data have been collected from over 120,000 men and women. The questionnaire also asked for information on lifestyle factors, occupational, medical and family history. Furthermore, toenail clippings of the majority of participants were collected. Baseline measurements are repeated every year in a subsample. Five year follow-up of the cohort will yield about 300, 400, 200, 700 and 1,200 cases of stomach, colon, rectum, breast and lung cancer respectively. Follow-up information on total and site-specific cancer incidence, is being obtained in collaboration with the cancer registries and the National Data Base of Pathology Records (PALGA). The cohort study will be analysed using the data from the cases and a large subcohort (case-cohort design). Results will be available from 1991 onwards. Papers have been published by Van den Brandt et al. in IARC Techn. Rep. No. 4:79, 1988, J. Clin. Epidemiol. 43 (3):285, 1990 and in Int. J. Epidemiol. 19:553, 1990.

TYPE: Cohort
TERM: Alcohol; Diet; Familial Factors; Fibre; Lifestyle; Metals; Registry; Vitamins
SITE: Breast (F); Colon; Lung; Rectum; Stomach
CHEM: Beta Carotene; Selenium
REGI: Amsterdam (Net); Groningen (Net); Leiden (Net); Maastricht (Net); Nijmegen (Net); Rotterdam (Net); Tilburg (Net); Utrecht (Net)
TIME: 1984 - 1994

641 **Kampman, E.** 04787
TNO Toxicology & Nutrition Inst., Sect. of Epidemiology, P.O.Box 360 , 3700 AJ Zeist, Netherlands (Tel.: +31 34 0444144; Fax: 0457224 ; Tlx: 40022 civo nl)
COLL: Van't Veer, P.; Kok, F.J.; Schneijder, P.; Hermus, R.J.J.

Fermented Milk Products, Calcium and Colon Cancer

Dietary habits characterised by high intake of fat, energy and protein have been linked to a high incidence of colon cancer. Protective effects have been suggested for dietary fibre, vegetables, fruit, calcium and dairy products, including fermented dairy products. To investigate the association of fermented milk products and/or calcium with colon cancer risk, a case-control study will be conducted in the Netherlands from 1990 to 1993. Since the impact of diet on the aetiology of colon cancer may vary for different parts of the colon, these will be studied separately. Newly diagnosed colon cancer patients (n=250; ICD-0 153; adenocarcinoma) will be compared with 'healthy' population controls (n=250), frequency-matched for sex and age (5-year interval). Average daily intake of calcium, (fermented) milk products and other dietary factors will be assessed using a dietary history technique (for cases, referring to the year before clinical symptoms). Furthermore, data on major risk factors of colon cancer, life-style habits and medical history will be obtained. Both cases and controls will be interviewed at their homes. Mean intakes among cases and controls will be compared using a t-test of paired samples. Stratified analysis will be used to indentify potential confounders. Logistic regression will be used to evaluate confounding and adjust the odds ratio estimates.

TYPE: Case-Control
TERM: Anatomical Distribution; Dairy Products; Diet; Fat; Fibre; Registry
SITE: Colon
CHEM: Calcium
REGI: Utrecht (Net)
TIME: 1989 - 1994

642 **Van Poppel, G.** 04633
TNO-CIVO Toxicology and Nutrition Inst., P.O. Box 360, 3700 AJ Zeist, Netherlands (Tel.: (03404)44144; Fax: 57224)
COLL: Kok, F.J.; Van Bladeren, P.; Hermus, R.J.J.

Biomarkers for DNA Damage in Smokers and Non-Smokers: Effects of Beta Carotene

Beta carotene has been associated with decreased risk for lung and bladder cancer in smokers. As an antioxidant, beta carotene may protect against DNA damage and cancer. Micronuclei in bronchial mucosal cells and SCE in lymphocytes have been studied before and after three months' beta carotene supplementation (20mg/day) in 150 smokers in a double-blind placebo-controlled intervention trial (Int. J. Cancer 51:355-358, 1992; Br. J. Cancer 66:1164-1168, 1992; Carcinogenesis 13:303-305, 1992). 50 non-smokers and 50 passive smokers were studied without intervention (Mut. Res. 279:233-238, 1992). Plasma, sputum and urine samples have been stored and are now used to address additional hypotheses for application of biomarkers in molecular epidemiology (8 hydroxy deoxyguanosine, 4 aminobiphenyl-haemoglobin and thioethers in urine, DNA adducts in lymphocytes and sputum).

TYPE: Intervention; Molecular Epidemiology
TERM: Biomarkers; DNA; DNA Adducts; Lymphocytes; Micronuclei; Passive Smoking; SCE; Sputum; Tobacco (Smoking); Urine
SITE: Bladder; Lung
CHEM: Beta Carotene
TIME: 1989 - 1994

NEW CALEDONIA

Nouméa

643 **Finau, S.A.** 04355

South Pacific Commission, Pacific Islands Cancer Registry, Box D5, Nouméa Cedex, New Caledonia (Tel.: +687 262000/363; Fax: 263818 ; Tlx: sopacom 3139nm)

COLL: Henderson, B.E.; Kolonel, L.N.; Le Marchand, L.

Pacific Islands Cancer Registry

The aim of the Pacific Islands Cancer Registry is to provide information on cancer from all the Pacific Islands and compare site specific incidence rates by country and ethnic groups. Registrars visit the islands to collect information from hospitals, laboratories and medical statistics sections or from well established registries. Valuable data are also obtained from tumour registries located near the major referral centres in the United States, New Zealand and Australia. Projects will be carried out on the community level with the aim to define risk factors and reduce cancer. These projects will be evaluated for their efficiency in reducing risk behaviours.

TYPE: Incidence; Registry
TERM: Ethnic Group; Geographic Factors; Lifestyle; Registry
SITE: All Sites
REGI: Pacific Islands (Fra)
TIME: 1977 - 1994

Auckland

644 **Kay, R.G.** 01417
Univ. of Auckland, Medical School, Dept. of Surgery, Park Rd, Auckland 3, New Zealand (Tel.: +64 9 795780; Fax: 770956)
COLL: Holdaway, I.M.; Stewart, A.W.; Harvey, V.J.; Carter, J.F.; Gillman, J.C.; Mason, B.H.; Newman, P.; Bierre, A.

Auckland Breast Cancer Study Group - Breast Cancer Survey

This study is concerned with the epidemiology of breast cancer in a multi-racial society in New Zealand. It comprises the 2,706 new cases of breast cancer diagnosed in Auckland in the nine years between 1976 and 1985. Importance is attached to season of tumour detection, height-weight ratios, reproductive history, history of contraceptive practices, previous breast disease, blood group, family history and bilateral disease. Clinical data on oestrogen and progesterone receptor values, adjuvant therapy, grading, nodal status, recurrence and death dates, will be evaluated in relation to these factors. Recent papers have been published in Br. J. Cancer 61:137-141, 1990, Cancer Res. 50:5883-5886, 1990 and Breast Cancer Res. Treat. 15:103-108, 1990.

TYPE: Case Series
TERM: Familial Factors; High-Risk Groups; Hormones; Obesity; Race; Reproductive Factors
SITE: Breast (F)
CHEM: Androgens
TIME: 1976 - 1994

Dunedin

645 **Chang, A.R.** 04324
Univ. of Otago, Medical School, Dept. of Pathology, P.O. Box 913, Dunedin, New Zealand (Tel.: +64 24 740062)

Characteristics of Women Attending for Colposcopic Examination

Studies carried out overseas have confirmed that cervical malignancies are more prevalent in women of low socio-economic status, smokers and often women from minority ethnic groups. There have been little or no New Zealand studies on the characteristics of women with cervical malignant disease. This study is aiming to see if those women of low socio-economic status, Maoris and Polynesians and smokers have a higher prevalence of the precursor lesions of cervical cancer and also HPV infection. The latter is being assessed by DNA in situ hybridization methods. The study population is approximately 350 per annum and is being assessed at the University Teaching Hospital. The women all reside in the Otago Province, which has a stable population with a mixture of various racial groups including Maoris and Polynesians, as well as a dominant European population. All subjects in the study are interviewed prior to colposcopic assessment and relevant medical records are also scrutinised. Cytological and histological diagnoses are also being evaluated for each patient in the study. Information on methods of contraception and parity is also being obtained. The women in the study group will be compared with those in the general population. Papers appeared in Austr. NZ. J. Obstet. Gynaecol. 29:200-203, and 329-331, 1989.

TYPE: Cross-Sectional
TERM: Contraception; Ethnic Group; HPV; Infection; Reproductive Factors; Socio-Economic Factors; Tobacco (Smoking); Virus
SITE: Uterus (Cervix)
TIME: 1982 - 1994

646 **Dockerty, J.D.** 04982
Univ. of Otago Medical School, Dept. of Preventive & Social Medicine, Hugh Adam Cancer Epidemiology Unit, P.O.Box 913 , Dunedin, New Zealand (Tel.: +64 3 4797203; Fax: 4790529)
COLL: Sharples, K.J.; Skegg, D.C.G.; Elwood, J.M.; Heenan, L.D.B.; Borman, B.

Clustering of Childhood Cancer in New Zealand

Two geographical studies using New Zealand Cancer Registry data are being completed. The first is a test of an hypothesis put forward by Kinlen in 1988 on the basis of clustering observed in relatively isolated areas in the United Kingdom which had large population influxes. Kinlen's hypothesis relates to a suggested viral cause of childhood leukaemia. The study involves defining, on specific criteria, areas of New Zealand which have received significant population influxes. The rates of certain childhood cancers in these areas will be compared with national rates to see whether they are increased. The second study involves assessing whether or not there is spatial clustering of certain childhood cancers. Data collected by the cancer registry during 1976–1987 have been used. The degree of spatial clustering will be assessed using a method developed by Cuzick and Edwards (1990), which avoids several problems of previously used methods. It will, therefore, be possible to say whether childhood cancers cluster in space to a greater extent than could be expected by chance. The location of any clusters found will be defined, and possible reasons for them considered.

TYPE: Cross-Sectional; Incidence
TERM: Cluster; Geographic Factors; Registry
SITE: Childhood Neoplasms
REGI: New Zealand (NZ)
TIME: 1991 - 1994

647 **Dockerty, J.D.** 04983
Univ. of Otago Medical School, Dept. of Preventive & Social Medicine, Hugh Adam Cancer Epidemiology Unit, P.O.Box 913 , Dunedin, New Zealand (Tel.: +64 3 4797203; Fax: 4790529)
COLL: Elwood, J.M.; Skegg, D.C.G.; Sharples, K.J.; Lewis, M.E.

Childhood Cancer in New Zealand

A national case-control study of childhood cancers began in 1991; it aims to identify aetiological factors for specific malignancies (particularly ALL), and includes all childhood cancers. There are about 400 cases and 400 controls in the study, which involves carrying out home interviews of the parents. The cases have been identified through the New Zealand Children's Cancer Registry and the National Cancer Registry. The controls have been randomly selected from birth registration records, with matching to cases on age and sex. The exposures of interest include certain chemicals and drugs, diagnostic and therapeutic radiation, parental occupation, immune responsiveness, infectious agents, electromagnetic fields, and genetic factors. Standard analyses for case-control studies are being conducted.

TYPE: Case-Control
TERM: Chemical Exposure; Drugs; Electromagnetic Fields; Genetic Factors; Marijuana; Parental Occupation; Passive Smoking; Radiation, Ionizing; Registry
SITE: Childhood Neoplasms; Leukaemia (ALL)
CHEM: Chloramphenicol; Phenytoin
REGI: New Zealand (NZ)
TIME: 1992 - 1995

648 **Firth, H.M.** 04623
Univ. of Otago Medical School, Dept. of Preventive & Social Medicine, P.O.Box 913, Dunedin, New Zealand (Tel.: +64 3 4791200; Fax: 4797298)
COLL: Herbison, G.P.; Elwood, J.M.; Cooke, K.R.

Cancer by Occupation in New Zealand 1972-1986

The aim of the study is to describe cancer incidence and mortality by occupational group in New Zealand for the period 1972-1986. Cancer incidence and mortality data have been obtained from the New Zealand Cancer Registry. The report will discuss any apparent cancer excesses both in terms of the epidemiological information on which the excess is based and in terms of other knowledge of the aetiology of the condition.

TYPE: Incidence; Mortality
TERM: Occupation; Registry
SITE: All Sites
REGI: New Zealand (NZ)
TIME: 1989 - 1994

***649** **Firth, H.M.** 05485
Univ. of Otago Medical School, Dept. of Preventive & Social Medicine, P.O.Box 913, Dunedin, New Zealand (Tel.: +64 3 4791200; Fax: 4797298)
COLL: Herbison, G.P.

Historial Cohort Study of Workers in a Foundry and Heavy Engineering Plant

This historial cohort study (1945-1991) aims to assess the potential causal factors for cancers among foundry and heavy engineering workers. The hypotheses being tested are: (1) workers exposed to mineral oils are at increased risk of digestive cancers, and (2) foundry workers are at increased risk of lung cancer. From a pilot study already completed 3,572 men have been identified and 109,671 person-years accumulated. Vital status will be determined from electoral rolls and drivers' license registrations. Causes of death will be ascertained from death certificates. Environmental exposures have been determined from factory records and discussions with staff (past and present).

TYPE: Cohort
TERM: Chemical Exposure; Environmental Factors; Occupation; Registry
SITE: Gastrointestinal; Lung
CHEM: Mineral Oil
OCCU: Engineering Workers; Foundry Workers
REGI: New Zealand (NZ)
TIME: 1992 - 1994

650 **Paul, C.E.** 03223
Univ. of Otago Medical School, Dept. of Preventive & Social Medicine, P.O.Box 913, Dunedin, New Zealand (Tel.: +64 3 4797207; Fax: 4790529)
COLL: Skegg, D.C.G.; Spears, G.F.S.

Steroid Contraception and Breast Cancer

This is a national, population-based case-control study designed to determine whether use of the injectable contraceptive depot medroxyprogesterone acetate (DMPA, Depo Provera) affects the risk of breast cancer in women. In the study 891 women with newly diagnosed breast cancer, aged 25 to 54, were compared with 1,864 control subjects selected at random from the electoral rolls. Cases and controls were interviewed by telephone and information collected about known risk factors as well as about DMPA, oral contraceptives, therapeutic oestrogens, tonsillectomy, abortion, and alcohol use. Data collection is complete. Further analyses of the possible associations between contraceptive steriod use, reproductive factors, alcohol use, tonsillectomy, and the risk of breast cancer are being undertaken. Papers were published in Br. Med. J. 293:723-726, 1986, 299:759-762, 1989 and in Int. J. Cancer 46:366-373, 1990.

TYPE: Case-Control
TERM: Abortion; Alcohol; Contraception; Hormones; Oral Contraceptives; Tonsillectomy
SITE: Breast (F)
CHEM: Oestrogens; Progesterone; Progestogens
TIME: 1983 - 1994

651 **Richardson, A.K.** 05001
Univ. of Otago, Dept. of Preventive and Social Med., Hugh Adam Cancer Epidemiology Unit, P.O.Box 913, Dunedin, New Zealand (Tel.: +64 3 4797203; Fax: 4797298)
COLL: Elwood, J.M.; Doyle, T.; Williams, S.M.; McGee, R.O.

Evaluation of a Pilot Breast Cancer Screening Programme

This study aims at evaluating two pilot breast cancer screening programmes, in which two-view mammography is offered every two years to women aged 50-64. 40,000 women are eligible to participate in the pilot programmes. The evaluation will take place over the first three years of screening (one and a half screening rounds). A series of targets have been agreed, based on the early operating characteristics of programmes that have been successful in the longer term (i.e programmes that have shown decreased mortality from breast cancer among women who were offered a screening). Targets have been set for identification of eligible women, participation, film-reading validity and reliability, sensitivity and specificity, referral rate, biopsy rate, the benign to malignant biopsy ratio, and the stage distribution of cancers detected by screening. Data will be collected directly from the pilot programmes and also from the New Zealand cancer registry. The aim of this evaluation is to ensure the maximum benefit for women offered screening. The results will be used to help decide whether a national screening programme should be started in New Zealand.

TYPE: Cohort
TERM: Registry; Screening
SITE: Breast (F)
REGI: New Zealand (NZ)
TIME: 1990 - 1995

Porirua

652 **Bates, M.N.** 04979
Inst. of Environmental Health & Forensic Sciences, P.O.Box 50-348, Porirua, New Zealand
COLL: Smith, A.H.; Cantor, K.C.

Bladder Cancer and Arsenic in US Drinking Water Supplies

Studies of a Taiwanese population using water supplies with high levels of inorganic arsenic have shown high rates of mortality for cancers at a number of sites, including the bladder. The present study is investigating whether the generally much lower levels of arsenic found in water supplies in the United States are also associated with bladder cancer. Data on approximately 3,000 cases and 6,000 controls collected for the National Bladder Cancer Study in 1978, will be merged with arsenic exposure estimations, derived from data on arsenic levels in public water supplies, collected under the National Safe Drinking Water Act. Multiple logistic regression analysis will be the main method of analysis.

TYPE: Case-Control
TERM: Water
SITE: Bladder
CHEM: Arsenic
TIME: 1989 - 1994

Wellington

653 **Bethwaite, P.** 05043
Wellington School of Medicine, P.O. Box 7343, Wellington South, New Zealand (Tel.: +64 4 855999; Fax: 891661)
COLL: Pearce, N.E.; Carter, J.F.; Beard, M.; Varcoe, R.; May, S.

Environmental and Occupational Epidemiology of Adult Leukaemia

This is a case-control study contrasting the exposure histories of cases of adult-onset acute leukaemia with those of population controls. Cases are males and females aged 18 to 75 years presenting to the five main treatment centres between 1 January 1988 and 31 December 1991. Estimated enrolment is 130 cases with 4 controls per case. The principal research questions are: (1) is there an association between exposure to tobacco products and the development of adult-onset acute leukaemia, or its major subtypes, in persons aged 18 to 75 years? (2) is there an association between occupational exposure to extremely low frequency electromagnetic fields and the development of adult-onset acute leukaemia, or its major subtypes, in persons aged 18 to 75 years? (3) is there an association between occupational exposures involved in work in farming, chemical spraying, abattoirs, sawmilling, forestry or painting and the development of adult-onset acute leukaemia? (4) are there any differences in the estimates of risk for the various environmental factors between acute leukaemia patients with and without identified clonal chromosomal abnormalities?

TYPE: Case-Control
TERM: Chemical Exposure; Dusts; Electromagnetic Fields; Herbicides; Occupation; Paints; Pesticides; Solvents; Tobacco (Smoking); Wood
SITE: Leukaemia
OCCU: Electrical Workers; Farmers; Forest Workers; Herbicide Sprayers; Meat Workers; Painters; Sawmill Workers
TIME: 1990 - 1994

654 **Pearce, N.E.** 04249
Wellington Clinical School of Medicine, Wellington Hosp., Dept. of Community Health, Wellington, New Zealand (Tel.: +64 4 855999)

Mortality and Cancer Incidence among Phenoxy Herbicide Production Workers and Applicators in New Zealand

This ongoing cohort study will ascertain mortality and cancer incidence in 1,053 workers involved in the manufacture of phenoxy herbicides during the period 1969-1984 and 794 workers involved in the spraying of these chemicals during the period 1973-1984. The production workers cohort comprises all blue-collar workers employed at a single factory for at least one month. The sprayer cohort is based on the national registry of professional chemical applicators. Vital status is being ascertained through the national death index and the national Cancer Registry.

TYPE: Cohort
TERM: Chemical Exposure; Herbicides; Occupation; Pesticides; Registry
SITE: All Sites; Lymphoma; Soft Tissue
CHEM: Phenoxy Acids
OCCU: Herbicide Manufacturers; Herbicide Sprayers
REGI: New Zealand (NZ)
TIME: 1983 - 1995

655 **Pearce, N.E.** 04250
Wellington Clinical School of Medicine, Wellington Hosp., Dept. of Community Health, Wellington, New Zealand (Tel.: +64 4 855999)
COLL: Bowman, J.D.; Peters, J.M.; Garabrant, D.H.; Thomas, D.C.

Leukaemia in Electrical Workers

A preliminary New Zealand case-control study of 546 male cases of leukaemia registered with the New Zealand Cancer Registry during the period 1979-1983 found an excess risk of leukaemia in electrical workers. Measurements will be obtained of typical exposures to electromagnetic fields (as well as other relevant exposures) in electrical worker occupations. Further cases and controls will be collected and the expanded study will be reanalysed using the occupational hygiene data.

TYPE: Case-Control
TERM: Electromagnetic Fields; Occupation; Registry
SITE: Leukaemia
OCCU: Electrical Workers
REGI: New Zealand (NZ)
TIME: 1987 - 1994

NIGERIA

Calabar

656 **Otu, A.A.** 04932
Univ. of Calabar, College of Medical Sciences, Dept. of Surgery, P.M.B. 1115 , Calabar, Nigeria (Tel.: +234 87 222855; Tlx: 65103 unical)
COLL: Tabor, E.

Hepatocellular Carcinoma, Cirrhosis of the Liver and Hepatitis B Virus.
The association of PHC with liver cirrhosis and seropositivity to HBV is investigated by detailed examination of PHC patients and by testing their sera for HBV markers (HBsAg, anti-HBc, anti-HBs) by RIA tests. Samples which are positive for HBsAg are also tested for HBeAg and Anti-HBe. Patients with metastatic hepatic carcinoma and symptom-free, non cancer subjects serve as controls. There are two sex-and age-matched (within five years) controls per case. A communication was published in Cancer 60:2581-2585, 1987.

TYPE: Case-Control
TERM: Antibodies; Antigens; Cirrhosis; HBV; Sero-Epidemiology
SITE: Liver
LOCA: Nigeria; United States of America
TIME: 1978 - 1995

657 **Otu, A.A.** 04933
Univ. of Calabar, College of Medical Sciences, Dept. of Surgery, P.M.B. 1115 , Calabar, Nigeria (Tel.: +234 87 222855; Tlx: 65103 unical)

HIV Infection and Kaposi's Sarcoma
The present study explores whether Kaposi's sarcoma in Nigeria is associated with HIV infection. Serum samples from KS patients and patients with malignant melanoma of the foot (contemporaneous controls), and age- and sex-matched non-cancer controls are tested for anti-HIV antibody by enzyme-linked immunosorbent assay (ELISA). A communication was published in J. Surg. Oncol. 37:152-155, 1988.

TYPE: Case-Control
TERM: Antibodies; HIV; Sero-Epidemiology
SITE: Kaposi's Sarcoma
TIME: 1980 - 1995

***658** **Otu, A.A.** 05548
Univ. of Calabar, College of Medical Sciences, Dept. of Surgery, P.M.B. 1115 , Calabar, Nigeria (Tel.: +234 87 222855; Tlx: 65103 unical)

AIDS Associated Tumours and Infection in South Eastern Nigeria
This hospital-based case-control study investigates the association of HIV infection with Kaposi's sarcoma, Burkitt's lymphoma, Hodgkin's disease, non-Hodgkin's lymphoma and other cancers. The association with infections like tuberculosis, leprosy, acute bacterial infections and fungal infections is also investigated. Blood is collected for serological studies from cases as well as age- and sex-matched controls. So far, more than 300 cases and controls have been recruited into the study.

TYPE: Case-Control
TERM: AIDS; HIV; Infection; Sero-Epidemiology
SITE: Burkitt's Lymphoma; Hodgkin's Disease; Kaposi's Sarcoma; Non-Hodgkin's Lymphoma
TIME: 1992 - 1997

Irrua

659 **Okojie, C.G.** 00066
Zuma Memorial Hosp., Irrua Edo State, Nigeria (Tel.: +234 57 97300)
COLL: Akinsanya, A.

Aetiology of Breast Cancer

Every woman aged between 15 and 70 seen at any clinic in this hospital is examined, breasts carefully palpated, adenopathy tested for after taking history of parity, children alive, nursing period and lactation amenorrhoea. A first report covering 10,670 women over a period of 15 years showed that breast cancer is uncommon among Ishan women of Nigeria, with their early marriage (15 years), first baby by age of 16, grand parity (6-8), prolonged breast feeding leading to high serum prolactin levels and a consequent decrease in circulating gonadotrophins, delay of ovulation and longer lactation amenorrhea. The aims for the study are: (1) to confirm that breast cancer is uncommon in this area in a less developed country where the women are all poorly nourished, and (2) to investigate the important aetiological and/or protective factors in Ishan women.

TYPE: Relative Frequency
TERM: Age; Hormones; Lactation; Parity
SITE: Breast (F)
TIME: 1961 - 1994

NORWAY

Bergen

660 **Kvåle, G.** 03162
Univ. of Bergen, Inst. of Hygiene & Social Medicine, 5016 Haukeland Sykehus , Bergen, Norway (Tel.: +47 5 298060)
COLL: Heuch, I.; Nilssen, S.

Reproductive Experience and Cancer Risk

The aim is to examine in detail the effects of reproductive experience (parity, age at first and later births, age at menarche and age at menopause) on cancer occurrence and on mortality from non-cancer causes in a prospective study of women in Norway. In 1956-1959 74% of all women aged 20-69 in three counties in Norway attended a screening examination for breast cancer and responded to a questionnaire. In this cohort of 63,000 some 11,000 cases of cancer had been diagnosed by the end of 1989, more than 2,200 of these being breast cancer. Data on height, weight and ABO blood group are being collected for a sub-group of the cohort to evaluate possible confounding by these variables and to examine possible interactions with the reproductive variables in relation to cancer occurrence. Recent papers have appeared in Int. J Cancer 46:597-603, 1990 and 47:390-395, 1991 and in Acta Oncologica 31:187-194, 1992.

TYPE: Cohort
TERM: Blood Group; Menarche; Menopause; Parity; Physical Factors; Screening
SITE: All Sites; Breast (F); Gastrointestinal; Ovary; Uterus (Cervix); Uterus (Corpus)
TIME: 1955 - 1995

***661** **Moen, B.E.** 05274
Univ. of Bergen, Inst. for Occupational Medicine, Armauer Hansens Hus, Haukeland Sykehus, 5021 Bergen, Norway (Tel.: +47 5 974975; Fax: 974979)
COLL: Riise, T.

Mortality Among Seamen

The aim of the study is to examine a possible relationship between chemical exposure from the cargo on tankers and mortality in seamen. Chemical exposure caused by the cargo is the major factor distinguishing tankers from other ships. Mates are exposed to vapours from the cargo, while captains are not. Different carcinogenic agents, oils and oil products, may exist in the major part of the cargo. Mortality will be studied in a population of 1687 Norwegian captains and mates 1970-1987. The population has been identified by the Norwegian 1970 census, and the data will be matched to the Norwegian Register of Death Certificates. Each case will be age-matched at time of death to three individuals from the rest of the population alive at this date. The material will be analysed using multivariate conditional logistic regression.

TYPE: Case-Control; Cohort
TERM: Chemical Exposure; Occupation; Petroleum Products
SITE: All Sites
OCCU: Sailors
TIME: 1986 - 1994

Oslo

662 **Glattre, E.** 03505
Cancer Registry of Norway, Montebello, 0310 Oslo 3, Norway (Tel.: +47 2 506050)
COLL: Thoresen, S.Ø.

Aetiology of Thyroid Cancer in Norway

The Thyroid Cancer Project was established in 1985 as a multidisciplinary and multi-institutional study with its secretariat located in the Cancer Registry of Norway and the aim of investigating aetiological aspects of thyroid cancer. On-going studies are: prevalence of occult thyroid cancer in different age-groups and regions of Norway; arsenic-selenium antagonism in the thyroid gland; case-control study of marine fatty acids and thyroid cancer. Papers have been published in Eur. J. Cancer 28:491-495, 1992 (Akslen), Cancer Causes and Control 4:11-16, 1993 and Int. J. Cancer 53:183-187, 1993 (Akslen et al.).

TYPE: Case-Control; Cohort; Cross-Sectional
TERM: Diet; Prevalence; Registry; Survival; Trace Elements; Trends; Vitamins
SITE: Thyroid
CHEM: Arsenic; Iodine; Selenium
REGI: Norway (Nor)
TIME: 1985 - 1995

663 **Jellum, E.** 04678
Inst. of Clinical Biochemistry, Rikshospitalet, 0027 Oslo 1, Norway (Tel.: +47 2 867035; Fax: 867029)
COLL: Andersen, A.; Lund-Larsen, P.; Orjaseter, H.; Theodorsen, L.

The Janus Serum Bank and Early Detection of Cancer

The Janus-project was initiated by the Norwegian Cancer Society in 1973. The serum bank comprises over 500,000 serum samples consolidated from 170,000 donors. 2-14 consecutive samples are available from each donor. At regular intervals the Janus collection is matched against the files of the Norwegian Cancer Registry. From 1973 to 1993, 6,000 of the donors have developed some form of cancer. Frozen serum samples collected from a few months to 19 years prior to clinical recognition of their disease are available for research purposes. The aim of the Janus-project is to look for chemical, biochemical, immunological or other changes in these premorbid sera, that might be indicative of cancer development at early stages. The stability of the frozen sera has been evaluated and methods including monoclonal antibodies have been used to search for changes in cancer markers. Some recent findings are: CA-125 is elevated several months prior to diagnosis of ovarian cancer; serum thyroglobulin may be a preclinical tumour marker in subgroups of thyroid cancer; low level of selenium in serum reflects increased risk of thyroid cancer; raised antibodies in serum against EBV is a risk factor for development of Hodgkin's disease; prostate specific antigen is a sentitive tumour marker giving positive response several months prior to clinical diagnosis; and there is an inverse relation between the polyunsaturated fatty acid linoleic acid (18:2 n-6) and risk of breast cancer. On-going research include trace elements and cancer (method: neutron activation analysis and atomic absorption); hepatitis C and liver cancer, and H. pylori infection and cancer of the stomach.

TYPE: Incidence
TERM: BMB; Biochemical Markers; H. pylori; Hepatitis; Immunologic Markers; Registry; Serum; Trace Elements; Tumour Markers
SITE: All Sites
REGI: Norway (Nor)
TIME: 1973 - 1994

664 **Kjærheim, K.** 04991
Cancer Registry of Norway, Montebello, 0310 Oslo 3, Norway (Tel.: +47 22 451300; Fax: 451370)
COLL: Langmark, F.; Andersen, A.; Haldorsen, T.; Ravndal, E.; Fauske, S.

Cancer Risk in the Restaurant Business

The aim of the study is to analyse the risk of alcohol-associated cancers in male and female waiters and male cooks in Norway, and the association between cultural and structural work characteristics and alcohol consumption. The hypothesis is that a tolerant alcohol culture and certain job characteristics increase the risk of high alcohol consumption, which in turn increases the risk of cancers of the mouth, pharynx, oesophagus, larynx and liver. The study consists of three sub-studies: (1) A historical cohort study of skilled male waiters and cooks; (2) A historical cohort study of female waiters, based on union membership; (3) A cross-sectional study of male and female waiters and cooks, investigating the relation between work characteristics, alcohol consumption and tobacco smoking. A prospective study of this cohort will also be undertaken.

TYPE: Cohort; Cross-Sectional
TERM: Alcohol; Occupation; Registry
SITE: Larynx; Liver; Oesophagus; Oral Cavity; Pharynx
OCCU: Cooks; Waiters
REGI: Norway (Nor)
TIME: 1990 - 1995

***665** **Kristensen, P.** 05254
National Inst. of Occupational Health, P.O. Box 8149, Dep, 0033 Oslo 1, Norway (Tel.: +47 2 466850; Fax: 603276)
COLL: Andersen, A.; Irgens, L.M.; Bjerkedal, T.; Laake, P.; Norseth, T.

Perinatal Cancer in Families Working in Agriculture

In this study the relations between parental farm activity and reproductive outcome (perinatal cancer) in offspring will be studied. The hypothesis is that parental exposure to pesticides/biological factors related to animal farming, at the time of conception or during pregnancy, are risk factors for leukaemia and brain tumours in the offspring. A cohort of families will be established from 157,360 agricultural census holders, born after 1924 (agricultural census linkage with Central Population Registry). Information on exposure is obtained from the agricultural censuses 1969-1989. Perinatal information and incident cancers are collected by linkage with the Medical Birth Registry and the Cancer Registry of Norway. The study includes life-time follow-up (3.6 million person-years), of 223,813 children born 1965-1991. So far 589 incident cancers have been registered. Comparison will be made with internal (exposed vs non-exposed) and external (rural population of Norway who are non-farmers) control groups. Poisson regression analysis will be applied.

TYPE: Cohort
TERM: Childhood; Parental Occupation; Pesticides; Registry; Reproductive Factors
SITE: Brain; Leukaemia
OCCU: Agricultural Workers; Farmers
REGI: Norway (Nor)
TIME: 1991 - 1995

666 **Larsen, T.E.** 04907
Ullevål Hosp., Dept. of Pathology, 0407 Oslo 7, Norway (Tel.: +47 2 118925)

The Norwegian Melanoma Project

A rapid increase in the incidence of malignant melanoma (MM) has been seen in Norway in the last 20 years. These observations prompted the medical community to initiate a screening programme and during the next 2-3 years the population will be invited to screening for melanoma, dyplastic naevi and clinical risk factors at University Hospital skin departments. Criteria for participating in the screening programme will include mainly (l) having growing or changing naevi; (2) having had melanoma previously or having a family history of melanoma; (3) being sensitive to sunshine or having had sunburn episodes in childhood. Approximately 40 clinical and 10-15 "blind" histological features will be stored at a central data registry at Ulleval Hospital for later statistical analysis. It is also intended (1) to assess if early diagnosis of MM reduces incidence and mortality; (2) to carry out analyses of the prognostic relevance of histological diagnosis, as well as of the criteria on which the clinical and histological diagnosis are based, and to observe the behaviour of non-removed dysplastic naevi in relation to different risk factors; (3) to carry out a case-control study to assess if MM risk is lower in a group of individuals who replied favourably to the invitation to screening, compared with a group of individuals who did not reply. Approximately 10,300 patients have been screened so far and 3,100 skin lesions excised for histological examination. Among these approximately 100 melanomas were found. It is anticipated to continue the project until a minimum of 200 melanomas have been obtained.

TYPE: Intervention
TERM: Familial Factors; Naevi; Screening
SITE: Melanoma
REGI: Norway (Nor)
TIME: 1989 - 1995

Skien

***667** **Danielsen, T.E.** 05263
Telemark Central Hosp., Dept. of Occupational Medicine, 3710 Skien, Norway (Tel.: +47 3 556100; Fax: 556105)
COLL: Langård, S.; Andersen, A.

Lung Cancer among Authorized Norwegian Boiler Welders

The aim of this study is to investigate the incidence of lung cancer among 3,200 authorized boiler welders. The welders are registered in a central registry established in 1942. All electric welders authorized between 1942 and 1984 have been included in the study. Information on first exposure in electric welding on mild steel or stainless steel will be used when creating sub-cohorts in this historical cohort study.

TYPE: Cohort
TERM: Occupation; Registry; Welding
SITE: Lung
OCCU: Welders
REGI: Norway (Nor)
TIME: 1991 - 1994

***668 Danielsen, T.E.** 05264
Telemark Central Hosp., Dept. of Occupational Medicine, 3710 Skien, Norway (Tel.: +47 3 556100; Fax: 556105)
COLL: Langård, S.; Andersen, A.; Knudsen, Ø.; Vatle, A.

Lung Cancer among Welders in Two Norwegian Shipyards

The aim of these historical cohort studies is to investigate the incidence of lung cancer in about 1,500 shipyard welders, mainly welding manual metal arc on mild steel. The other shipyard production workers will serve as internal reference groups. The studies are restricted to male workers employed between 1940 to 1979 and 1945 to 1982 respectively. Limited information on smoking habits is available.

TYPE: Cohort
TERM: Occupation; Registry; Welding
SITE: Lung
OCCU: Shipyard Workers; Welders
REGI: Norway (Nor)
TIME: 1990 - 1994

***669 Hobbesland, Å** 05265
Telemark Central Hospt., Dept. of Occupational Medicine, 3710 Skien, Norway (Tel.: +47 3 556100; Fax: 556105)
COLL: Langård, S.; Andersen, A.

Cancer Incidence and Mortality among Workers in the Norwegian Ferro-Alloy Industry

Previous studies have shown an increased incidence of cancer of the lungs, colon and prostate among workers in the Norwegian ferro-alloy industry. This retrospective cohort study will re-evaluate previous results. Fifteen plants are included and the study comprises about 20,000 persons. The plants are, or were, producing different metals (ferro-silicon, silicon metal, ferro-manganese, silicon-manganese, ferro-chromium, ferro-vanadium) and calcium carbide. In cooperation with the Cancer Registry of Norway the cancer incidence and mortality rates of the cohort will be compared with those of the general population.

TYPE: Cohort
TERM: Dusts; Metals; Occupation; Registry
SITE: All Sites
CHEM: Asbestos; Calcium; Ferro-Alloys; PAH; Silica
OCCU: Ferro-Alloy Workers
REGI: Norway (Nor)
TIME: 1990 - 1994

***670 Langård, S.** 05270
Telemark Central Hosp., Dept. of Occupational Medicine, 3710 Skien, Norway (Tel.: +47 3 556100; Fax: 556105)
COLL: Andersen, A.; Solli, H.M.

Incidence of Cancer among Radon Exposed Workers in a Niobium Mine

The incidence of cancer in a cohort of 320 Norwegian niobium miners is being observed from January 1953 through 1991. The relationship between exposure to radon and lung cancer is the main purpose of the study. Crude standardized mortality ratios are also calculated. The work period for the cohort members is confined to the production period 1953 to 1964. It is a follow-up of a previous study which showed a strong relationship between radon daughters and lung cancer in the same cohort.

TYPE: Cohort
TERM: Metals; Occupation; Radiation, Ionizing; Registry
SITE: Lung
CHEM: Radon; Thorium
OCCU: Miners
REGI: Norway (Nor)
TIME: 1991 - 1994

***671** **Waage, H.P.** 05286
Telemark Central Hosp., Dept. of Occupational Medicine, 3710 Skien, Norway (Tel.: +47 3 556100; Fax: 556105)
COLL: Andersen, A.; Hilt, B.; Johnson, E.; Langård, S.; Opedal, E.; Silsand, T.; Urdal, P.

Secondary Prevention of Asbestos-Related Diseases in the County of Telemark, Norway

This study aims to contribute knowledge on how to successfully intervene on the increased lung disease risk experienced by previously asbestos-exposed smokers. The epidemiology of asbestos-related diseases is being examined through analyses of the 8-year cancer incidence, according to previously reported asbestos exposure and smoking habits, in a cross-section of the population born prior to 1942. The extent of asbestos exposure in the population is further being studied in a questionnaire survey among males and females born 1942-1951. Based on information on previous asbestos exposure and smoking histories, 419 smokers with presumed high risk of lung cancer were invited to participate in a physician-delivered smoking cessation programme, comprising risk-based information and counselling based on cognitive and behavioural principles. Participants were followed with a postal questionnaire at three, six and twelve months, and a second consultation after two years.

TYPE: Cohort; Cross-Sectional; Intervention
TERM: Dusts; High-Risk Groups; Registry; Tobacco (Smoking)
SITE: Lung
CHEM: Asbestos
REGI: Norway (Nor)
TIME: 1987 - 1994

Trondheim

672 **Nygaard, R.** 04852
Univ. Hosp., Dept. of Pediatrics, Kyrres gt. 17, 7006 Trondheim, Norway (Tel.: +47 7 998168; Fax: 997656)
COLL: Clausen, N.; Garvicz, S.; Jonmundsson, G.K.; Kristinsson, J.R.; Lanning, M.; Moe, P.J.; Siimes, M.A.

Relapses and Late Effects following Treatment for Childhood Leukaemia

A cohort of all patients who had completed treatment for childhood leukaemias before 1985 is being followed up with regard to late relapses, fertility, offspring and second malignant neoplasms. Risk and possible risk factors are evaluated by means of Cox analysis and SIR analysis. The first reports from this population-based study are already published, but due to the long-term aspect of late effects, the investigation continues, in cooperation with the Nordic Society for Paediatric Haematology and Oncology. Papers have appeared in Med. Pediatr. Oncol. 17:45-47, 1989, Acta Paediatr. Scand. 79 Suppl. 354:5-17 and 18-242, 1989.

TYPE: Cohort
TERM: Chemotherapy; Childhood; Fertility; Late Effects; Multiple Primary; Radiotherapy; Recurrence
SITE: Leukaemia
LOCA: Denmark; Finland; Iceland; Norway; Sweden
TIME: 1980 - 2000

Karachi

673 **Hassan, T.J.** 04240
Pakistan Medical Research Council, Research Centre, Jinnah Postgraduate Medical Centre, Karachi 35, Pakistan
COLL: Zuberi, S.J.; Ashraf, S.; Zaidi, S.M.H.

Risk Factors Involved in Gallbladder Cancer in Karachi

A case-control study is being conducted in Karachi to evaluate the risk factors for gallbladder cancer. One hundred biopsied cases will be studied. Two age- and sex-matched controls (one with normal gallbladder, one with gallbladder stones) are selected for each case. All cases and controls are screened for stones by ultra-sound. Risk factors to be evaluated are: gallstones, early pregnancy, high parity, oral contraceptives, diet, cholecystitis, liver disease, smoking, tobacco use, fasting and consumption of ghee.

TYPE: Case-Control
TERM: Diet; Gallstones; Oral Contraceptives; Reproductive Factors; Tobacco (Smoking)
SITE: Gallbladder
TIME: 1987 - 1994

674 **Zaidi, S.M.H.** 02197
Jinnah Postgraduate Medical Center Dept. of Radiotherapy, Karachi 35, Pakistan
COLL: Jafarey, N.A.; Ahmed, M.; Askari, A.; Masood, M.A.; Rehman, G.; Siddiqui, M.; Siddiqi, A.M.

Pakistan Medical Research Council Multicentre Tumour Study

Frequency data of tumours of various sites are collected from hospitals in six districts of Pakistan with the aim of (1) determining factors influencing cancer trends, particularly in males; (2) investigating possible environmental risk factors for cancer of the breast, cervix uteri and bone, and evaluating methods of treatment; (3) initiating incidence studies. Interim results have been published in internal Monographs in 1982 and 1991.

TYPE: Incidence; Relative Frequency
TERM: Environmental Factors; Treatment
SITE: All Sites; Bone; Breast (F); Uterus (Cervix)
TIME: 1977 - 1994

Rawalpindi

***675** **Manzoor, A.** 05298
Armed Forces Inst. of Pathology General Headquarters Medical Directorate, Rawalpindi, Pakistan
COLL: Mamoon, N.; Khan, A.H.; Mansoor, A.

Risk Factors for Colorectal Carcinoma in Northern Pakistan

Colorectal carcinoma is among the ten most frequent cancers in Northern Pakistan. This hospital-based case-control study aims to investigate the association of various factors, such as polyps, diet, ethnicity, occupation, socio-economic status, tobacco habits, etc. with colorectal cancer. It is expected to recruit 200 cases within 3 years; age- and sex-matched control subjects with non-neoplastic lesions will be recruited from the same hospital. Univariate, stratified and multivariate staistical analyses will be done.

TYPE: Case-Control
TERM: Diet; Ethnic Group; Nutrition; Occupation; Polyps; Socio-Economic Factors; Tobacco (Chewing); Tobacco (Smoking); Tobacco (Snuff)
SITE: Colon; Rectum
TIME: 1992 - 1994

***676** **Manzoor, A.** 05299
Armed Forces Inst. of Pathology General Headquarters Medical Directorate, Rawalpindi, Pakistan
COLL: Magrath, I.; Shad, A.T.; Mansoor, A.; Khan, A.H.; Saeed, S.; Khan, M.A.

PAKISTAN

Aetiology and Pathogenesis of Lymphoid Neoplasia in Pakistan

This study aims to summarise the epidemiological features of approximately 1,000 cases of primary lymphoid neoplasias (Hodgkin's disease, non-Hodgkin's lymphoma, ALL) seen in Northern Pakistan. In addition, the study aims to address (1) any correlation with malaria, (2) molecular aspects of these neoplasms, and (3) the treatment patterns in a developing country like Pakistan. Fresh tissue and paraffin-fixed tissue DNA will be analysed by PCR and Southern blot techniques for the presence of 14:18, which is the result of an abnormal expression of bcl2 oncogene, in follicular and diffuse large cell lymphoma.

TYPE: Case Series; Molecular Epidemiology
TERM: Biochemical Markers; DNA; Oncogenes; PCR; Treatment; Virus
SITE: Hodgkin's Disease; Leukaemia (ALL); Non-Hodgkin's Lymphoma
TIME: 1989 - 1995

Brzesko

677 **Moszczynski, P.** 03981
Regional Hosp., Provincial Immunology Lab., 68 Kosciuszki, 32-800 Brzesko, Poland (Tel.: 30006; Tlx: 066633)

Haematological and Immunological Disturbances Induced by Occupational Exposure to Organic Solvents

Haematological and immunological examinations have been continuously performed in 104 workers occupationally exposed to benzene, toluene and xylene. Examinations involved the basic haematological indices, histochemical examinations of neutrophils and lymphocytes, the NBT test, the immunoglobulin concentrations in the blood serum, the total T and B cell count, and skin tests of delayed hypersensitivity. A control group consisted of 50 non-exposed healthy subjects in the same industrial factory. As yet no benzene-induced haemopathies have been noted and the main finding was a lowered T lymphocyte count in the peripheral blood and the enzymatic alterations both in neutrophils and lymphocytes. A synergistic depressing effect of organic solvents with smoking was observed on certain immunological indexes. This phenomenon reflected in the changes of concentrations of IgA, IgD, IgG, IgM and lysozyme, and the subpopulations of T-cells. Papers appeared in Rev. Roum. Med. Int. 27:137-141, 1989, in Med. Prac. 60:337-341, 1989 and in Wiad. Lek. 45:180-184, 1992.

TYPE: Molecular Epidemiology
TERM: Chemical Exposure; Dusts; Occupation; Solvents
SITE: Haemopoietic
CHEM: Benzene; Toluene; Xylene
OCCU: Lacquerers; Varnishers
TIME: 1980 - 1995

Gliwice

678 **Zemla, B.** 03442
Inst. of Oncology, Armii Krajowej 15, 44-101 Gliwice, Poland (Tel.: 311061; Tlx: 036606 inonk)
COLL: Banasik, R.; Zielonka, I.; Kolosza, Z.

Lung Cancer in Relation to Environmental Risk Factors

This is a case-control study of lung cancer (about 300 cases and 600 population controls matched for age and sex) in the upper Silesian region, Poland. The objective is to determine risk in relation to tobacco use, alcohol consumption, and diet. Statistical analysis will take into account length of employment and work conditions, history of migration, and pre- and co-existing diseases. Information will be collected through questionnaire by interviewer. Papers have been published in Wiad. Lek XXXIX(14):946-956, 1986 and in Neoplasma 35(2):135-143, 1988.

TYPE: Case-Control
TERM: Alcohol; Diet; Migrants; Occupation; Time Factors; Tobacco (Smoking)
SITE: Lung
TIME: 1984 - 1994

679 **Zemla, B.** 04584
Inst. of Oncology, Armii Krajowej 15, 44-101 Gliwice, Poland (Tel.: 311061; Tlx: 036606 inonk)
COLL: Kolosza, Z.; Banasik, R.

Cervix Uteri and Breast Cancer Incidence in Upper Silesia

The aim of the study is to describe patterns in the geography and dynamics of cancer in Upper Silesia Region (Katowice voivodship, southern Poland) as part of the atlas of cancer incidence for this region. This is necessary for studies of suspected aetiological factors, as well as for monitoring changes in cancer frequency. Cancer incidence data (cervix, breast) by age and residence are being collected from 1989. Age-specific and age-adjusted rates have been computed and analysed for the whole region and for 43 urban towns and 46 rural county areas, for the period 1975-1985. A paper has been published in Gin. Pol. 3:143-148, 1989.

TYPE: Incidence
TERM: Geographic Factors; Rural; Trends; Urban
SITE: Breast (F); Uterus (Cervix); Uterus (Corpus)
TIME: 1989 - 1994

Krakow

680 **Pawlega, J.** 03745
Maria Sklodowska-Curie Memorial Inst. of Oncology, Cracow Branch, Dept. of Cancer Epidemiology, Garncarska 11, 31-115 Krakow, Poland (Tel.: +48 12 229900; Tlx: 0325437)

Role of Smoking, Drinking and Diet in Cancer Risk

In 1985, 1,895 people living in the Cracow Region and born in 1938 and later, were covered by a questionnaire survey concerning habits of smoking, drinking, diet and data on weight and height. During the next 15 years these citizens will be followed using the Cracow Cancer Registry in order to find associations between the above factors and particular sites of cancer.

TYPE: Cohort
TERM: Alcohol; Diet; Occupation; Physical Factors; Registry; Tobacco (Smoking)
SITE: All Sites; Breast (F); Lung; Stomach; Uterus (Cervix)
REGI: Cracow (Pol)
TIME: 1986 - 2000

Lódz

681 **Górski, T.** 04603
Sanitary-Epidemiological Station, Ul. Wodna 40, 90-046 Lódz, Poland (Tel.: +48 42 740846; Tlx: 886747)
COLL: Górecka, D.

Influence of Cigarette Smoking on SCE Frequencies among Nurses Handling Cytostatic Drugs

Genotoxic compounds and various cancer chemotherapeutic agents can interact with tobacco smoke synergistically. The aim of the study is to examine the hypothesis that tobacco smoking has a greater influence on SCE frequencies among nurses handling cytostatic drugs than the cytostatics themselves. The frequencies of SCE in lymphocytes will be investigated among hospital staff who handle anti-cancer drugs and in a control group which does not handle them (smokers and non-smokers).

TYPE: Cross-Sectional; Molecular Epidemiology
TERM: Chemotherapy; Chromosome Effects; Occupation; SCE; Tobacco (Smoking)
SITE: Inapplicable
OCCU: Health Care Workers
TIME: 1988 - 1994

682 **Stankiewicz, A.K.** 04860
Nofer's Inst. of Occupational Medicine, ul. Teresy 8, 90-950 Lódz, Poland (Tel.: +48 42 552250; Fax: 348331)

G6PD Activity in Cancer Patients and Occupationally Exposed Persons

The aim of the investigation is to determine G6PD activity in erythrocytes in patients with cancer. Data studied previously show that G6PD activity was markedly higher in patients before surgery, as compared to a control group, than after. Investigations will continue to determine whether G6PD activity depends on the progress of the neoplasms. To this effect, G6PD activity in erythrocytes in persons occupationally exposed to carcinogenic factors will also be examined. The study group consists of some 500 persons exposed to aromatic amines. Controls are 53 healthy people, 16-60 years of age, selected randomly from a group of 4,000 not exposed occupationally to carcinogens. G6PD activity is determined by a spectrophotometric method according to Kornberg and Horecker.

TYPE: Cross-Sectional
TERM: G6PD; Occupation; Tumour Markers
SITE: All Sites
CHEM: Amines, Aromatic
TIME: 1991 - 1994

683 **Stankiewicz, A.K.** 04180
Nofer's Inst. of Occupational Medicine, ul. Teresy 8, 90-950 Lódz, Poland (Tel.: +48 42 552250; Fax: 348331)
COLL: Hanke, J.; Lutz, W.; Krajewska, B.; Pilacik, B.

Genetic Susceptibility to Toxic Substances and its Relationship to Carcinogenesis

This study is aimed at selection of biochemical indicators of predisposition to neoplastic diseases. The following markers of genetically conditioned susceptibility to neoplastic disease were suggested: phenotypes PiZ and PiMZ, alfa 1-antitrypsin, phenotype of free acetylation (to be studied prospectively), high activity G6PD in erythrocytes, and rapid antipyrine metabolism (to be studied retrospectively). Based on examination of families of children afflicted with liver cirrhosis a cohort has been constructed, covering about 200 individuals with PiZ and PiMZ. Phenotypization is carried on with the help of cross immunoelectrophoresis. The cohort of free acetylation includes 50 workers representing this phenotype, occupationally exposed to aromatic amines. Acetylation phenotypes are being determined with the method recommended by WHO. In the retrospective studies, frequency of occurrence of G6PD deficiency, detected with methaemoglobin reduction test in a group of healthy subjects (about 10,000 persons), is compared with frequency of deficiency in patients with neoplastic disease (about 8,000 persons). Similar studies are being carried out based on the antipyrine test (Kellermann) in a group of healthy persons and in a group with lung cancer.

TYPE: Case-Control; Cohort; Molecular Epidemiology
TERM: Chemical Exposure; Cirrhosis; Enzymes; Genetic Factors; Genetic Markers
SITE: Lung
CHEM: Amines, Aromatic
LOCA: Malta; Poland
TIME: 1986 - 1994

684 **Stankiewicz, A.K.** 05156
Nofer's Inst. of Occupational Medicine, ul. Teresy 8, 90-950 Lódz, Poland (Tel.: +48 42 552250; Fax: 348331)

G6PD Activity in Erythrocytes and the Development of Neoplastic Disease

This study aims to measure the activity of G6PD in erythrocytes (a) in subjects with malignant tumours of various organs and tissues, before and after surgery; (b) in about 100 premenopausal women with benign breast tumours. Dehydroepiandrosterone sulphate will also be measured in these women; (c) in workers occupationally exposed to asbestos dust to evaluate the correlation between development of neoplastic disease and asbestosis.

TYPE: Molecular Epidemiology
TERM: Alcohol; Asbestosis; Dusts; G6PD; Occupation; Tobacco (Smoking); Tumour Markers
SITE: All Sites
CHEM: Asbestos; Dehydroepiandrosterone Sulphate
TIME: 1991 - 1994

685 **Starzynski, Z.** 05090
Inst. of Occupational Medicine, Dept. of Epidemiology, 8 Teresy St., Box 199 , 90-950 Lódz, Poland (Tel.: +48 42 314566; Fax: 348331; Tlx: 885360 imp pl)
COLL: Marek, K.; Kujawska, A.; Szeszenia-Dabrowska, N.; Wilczynska, U.; Szymczak, W.

Cancer Mortality in Silicotics

The main objective of the study is to follow up individuals who developed silicosis and coalworkers' pneumoconiosis and to analyse their causes of death. The immediate objectives are: (1) to determine survival time and cause of death in pneumoconiotics, with particular regard to cancer; (2) to evaluate health and social effects of the pneumoconioses. Data on some 12,000 individuals with pneumoconiosis diagnosed in the years 1970-1985, provided by the Central Register of Occupational Diseases, serve as the study material, and the general population of Poland will be used as the reference population.

TYPE: Cohort
TERM: Dusts; High-Risk Groups; Pneumoconiosis; Silicosis; Survival
SITE: All Sites; Lung
CHEM: Radon; Silica
TIME: 1991 - 1994

POLAND

***686** **Szeszenia-Dabrowska, N.** 05474

Inst. of Occupational Medicine, Dept. of Epidemiology & Statistics, Teresy 8, P.O.Box 199, 90-950 Lódz, Poland (Tel.: +48 42 314561; Fax: 348331 ; Tlx: 885360 imp pl)

COLL: Wilczynska, U.; Szymczak, W.

Evaluation of Cancer Risk due to Environmental Contamination with Mineral Fibers

The main aim of the study is to evaluate the cancer risk due to environmental asbestos exposure in Poland. The project consists of three parts: (1) evaluation of risk among persons occupationally exposed to asbestos, employed in asbestos-cement plants. The cohort consists of persons employed more than 3 months in the period 1945-1980, in the production of asbestos-cement material (n = 4,500). The cancer risk will be determined mainly on the basis of SMRs. The reference population will be the general population of Poland; (2) Measurements of asbestos concentration in selected areas using standard methods and latest techniques, using an FM-7400 Real Time Laser Fibre Monitor; (3) The risk for the general population will be estimated with the use of mathematical models, taking into account the dose-relationship determined in a study of occupationally exposed workers and measurements of dust concentrations in the environment.

TYPE: Cohort
TERM: Dose-Response; Dusts; Mathematical Models; Occupation
SITE: All Sites; Lung; Mesothelioma
CHEM: Asbestos; Mineral Fibres
OCCU: Cement Workers
TIME: 1992 - 1995

***687** **Szeszenia-Dabrowska, N.** 05475

Inst. of Occupational Medicine, Dept. of Epidemiology & Statistics, Teresy 8, P.O.Box 199, 90-950 Lódz, Poland (Tel.: +48 42 314561; Fax: 348331 ; Tlx: 885360 imp pl)

COLL: Wilczynska, U.; Szymczak, W.

Cancer Mortality Among Rubber Workers

This project is the continuation of a study of mortality among rubber workers. The cohort consists of 6,978 men and 7,906 women employed in a rubber plant in Lodz for at least 3 months during the period 1945-1973 and is followed up till the end of 1990. SMRs and PMRs will be calculated on the basis of data for the general population of Poland. Results of follow-up till the end of 1985 were published in Pol. J. Occ. Med. 2, 1991.

TYPE: Cohort
TERM: Occupation; Solvents
SITE: All Sites; Bladder; Leukaemia; Lung
CHEM: N-Nitroso Compounds; PAH
OCCU: Rubber Workers
TIME: 1991 - 1995

Warsaw

688 **Zatonski, W.A.** 04764

Oncological Center, Dept. Cancer Control & Epidemiology, Ul. Wawelska 15 , 00-973 Warsaw, Poland (Tel.: +48 22 233179; Fax: 222429 ; Tlx: 812704 inonk)

COLL: Estève, J.; Smans, M.; Tyczynski, J.

Atlas of Cancer Mortality in Central Europe

The aim of the study is to present the geographical distribution of cancer mortality in Austria, Czechoslovakia, Bulgaria, Romania, Hungary, Yugoslavia, Germany and Poland. Cancer mortality data for 1982-1986 will be collected from participating countries, broken down by year of death, cause of death, sex, geographical units, age at death etc. Results will be presented in the form of maps.

TYPE: Mortality
TERM: Geographic Factors; Mapping
SITE: All Sites
LOCA: Austria; Bosnia-Herzegovina; Bulgaria; Croatia; Czech Republic; Germany; Hungary; Poland; Romania; Slovak Republic; Yugoslavia
TIME: 1990 - 1994

689 **Zatonski, W.A.** 04812
Oncological Center, Dept. Cancer Control & Epidemiology, Ul. Wawelska 15 , 00-973 Warsaw, Poland (Tel.: +48 22 233179; Fax: 222429 ; Tlx: 812704 inonk)
COLL: Tyczynski, J.

Changes in Geographical Patterns of Cancer Mortality in Poland, 1975-1988

This study is designed to present changes in geographical patterns of cancer mortality in Poland for the time period 1975-1988. The SMR's for 49 districts (voivodships) will be calculated for three time periods: 1975-1979, 1980-1984 and 1985-1988. Direct age-adjustment will be used (world population). Analysis of changes in particular districts and in the whole country will be done, and results will be presented in the form of maps and tables.

TYPE: Mortality
TERM: Geographic Factors; Mapping
SITE: All Sites
TIME: 1990 - 1994

Tulcea

***690 Georgescu-Tulcea, N.** 05322
Tulcea Departmental Hosp., Unit of Cancerology, Corneliu Gavrilov, no. 148 , 8800 Tulcea, Romania (Tel.: 91513854)

Primary Liver Cancer

The study aims to determine the association of primary liver cancer with HBV infection and other known or suspected factors. 153 in-patients with primary liver cancer in the Tulcea Departmental Hospital will be interviewed. Two patient controls (other cancers and other diseases, except liver diseases) will be matched to each case for sex, age and place of residence (rural, urban). Sera of cases and controls will be tested for HBV markers. Information on occupation, smoking habits, alcohol consumption, previous liver diseases and family history of cancer will also be collected.

TYPE: Case-Control
TERM: Alcohol; HBV; Occupation; Tobacco (Smoking)
SITE: Liver
TIME: 1991 - 1995

Ecaterinburg

691 **Kogan, F.M.** 04382
Medical Research Center for Prevention and Health Protection in Industrial Workers, 30 Popov St., Ecaterinburg 620014, Russian Federation (Tel.: 83432)
COLL: Yatsenco, A.S.; Gurvich, E.B.; Kuzina, L.E.

Cancer Mortality in the Brake Lining and Asbestos Textile Industries

A study of some 300 persons in a brake lining plant showed that lung cancer mortality was surprisingly lower than in a similar group of asbestos textile workers living in the same city and having similar lifestyle. A further study was carried out on a cohort of about 3,000 workers in an asbestos friction products plant situated in another city. This cohort was devided into three sub-cohorts, exposed in the past to different kinds of dusts and other hazards: asbestos, vapours of phenol and fromaldehyde or benzene, asbestos bakelite or asbestos rubber dusts. The observed/expected mortality ratio for stomach cancer was more than 1 in the first sub-cohort only. In the second and third sub-cohorts no excess mortality was observed, as well as in the total cohort. In addition an experimental study was carried out on three groups of rats injected twice intraperitoneally with these dusts at one month interval. The asbestos bakelite and asbestos rubber dusts induced a much smaller number of tumours than chrysotile asbestos. In the last sub-group survival was significantly lower than in the two other groups. The limit exposure for asbestos bakelite and asbestos rubber dusts may of course be higher than for asbestos.

TYPE: Cohort
TERM: Dusts; Occupation; Plastics; Resins; Rubber; Textiles
SITE: Gastrointestinal; Lung
CHEM: Asbestos; Formaldehyde; Mineral Fibres; Phenol
OCCU: Asbestos Textile Workers; Asbestos Workers
TIME: 1988 - 1994

***692** **Kogan, F.M.** 05458
Medical Research Center for Prevention and Health Protection in Industrial Workers, 30 Popov St., Ecaterinburg 620014, Russian Federation (Tel.: 83432)
COLL: Nikitina, O.V.; Kashansky, S.V.

Evaluation of the Cancer Risk in the Asbestos Industry

The cancer mortality in asbestos miners and millers has been studied for 20 years and a very high cancer risk has been demonstrated. A vast antidust programme has been carried out and the asbestos exposure now does not exceed the MAC level by more than 2-3 fold. The aim of this study is to assess the present cancer risk in asbestos workers and to assess the efficiency of the antidust measures taken. The incidence of cancers of the lung, stomach, intestine, uterus and haemopoietic system will be studied in a cohort of miners and millers and compared with that of the general population.

TYPE: Cohort
TERM: Dusts; Occupation
SITE: Gastrointestinal; Haemopoietic; Lung; Uterus (Cervix)
CHEM: Asbestos
OCCU: Asbestos Workers; Miners, Asbestos
TIME: 1993 - 1994

Magnitogorsk

***693** **Coshcina, V.S.** 05376
Inst. of Biomedical and Ecological Problems, 72, Lenin St., 455044 Magnitogorsk, Russian Federation (Tel.: 7-70-94)
COLL: Cochacova, L.B.; Ocuneva, Z.V.

Epidemiology and Prophylaxis of Malignant Tumours in Workers in the Ferrous Metal Industry

The aim is to study the effect of working in ferrous metal industry on cancer risk and to identify particular high-risk groups. A cohort of 26,162 persons (17,413 men, 8,749 women) has been established. Eight exposed groups and one control group have been observed within 10 years. Cancers are registered among present and previous workers. Morbidity data are analysed by classical analytical epidemiology/person-years of observation methods. Lifestyle factors are studied simultaneously.

TYPE: Cohort
TERM: High-Risk Groups; Lifestyle; Occupation
SITE: All Sites
OCCU: Metal Workers
TIME: 1979 - 1994

Moscow

694 **Basieva, T.H.** 04892
Inst. of Carcinogenesis, All-Union Cancer Research Centre, Academy of Medical Sciences of the USSR, Kashirskoye shosse, 24, Moscow 115478, Russian Federation (Tel.: +7 095 3241470; Fax: 2302450)
COLL: Zaridze, D.; Kabulov, M.

Spatial, Temporal and Ethnic Distribution of Oesophageal Cancer in Karakalpakstan

The Soviet republic of Karakalpakstan, an area situated in the central Asian part of the Soviet Union, is thought to be endemic for oesophageal cancer. Age-standardized rates for oesophageal cancer for Karakalpakstan as a whole are 50.2 and 50.9 for males and females respectively. The incidence rates vary substantially in different regions of Karakalpakstan: the highest (120.0 males, 150.6 females) are reported from the Muinak region located in the north. Incidence of oesophageal cancer also varies between different ethnic groups and is highest among Kazakhs, followed by Karkalpaks; the lowest incidence is reported in Russian ethnics. The aim of the study is to analyse the pattern of oesophageal cancer, including geographical distribution, incidence in different ethnic groups, and time trends. Incidence maps f oesophageal cancer in 14 districts will be drawn. Time trends for the period 1973-1989 will be analysed, using age-period-cohort methods.

TYPE: Incidence
TERM: Ethnic Group; Geographic Factors; Mapping; Trends
SITE: Oesophagus
TIME: 1990 - 1994

695 **Bulbulyan, M.** 04902
Inst. of Carcinogenesis, Cancer Research Center, Russian Academy of Medical Sciences, Kashirskoye shosse, 24, Moscow 115478, Russian Federation (Tel.: +7 095 3241470; Fax: 2302450 ; Tlx: 411015 knife)
COLL: Savitskaya, T.Y.

Cancer among Workers Exposed to Aromatic Amines

This is a cohort study of about 4,000 male and female employees at a chemical plant, with initial exposure after 1940, and who were alive on 1 January 1975. The cohort consists of one group exposed to aromatic amines and three control groups. The aim of the study is to determine the risk of bladder cancer and other malignant neoplasms before and after improvement of work conditions. Information about cancer occurrence, including histological type, will be obtained from the Moscow Cancer Registry.

TYPE: Cohort
TERM: Occupation
SITE: All Sites; Bladder; Lung; Stomach
CHEM: Amines, Aromatic
OCCU: Chemical Industry Workers
REGI: Moscow (Rus)
TIME: 1987 - 1994

***696** **Bulbulyan, M.** 05504
Inst. of Carcinogenesis, Cancer Research Center, Russian Academy of Medical Sciences, Kashirskoye shosse, 24, Moscow 115478, Russian Federation (Tel.: +7 095 3241470; Fax: 2302450 ; Tlx: 411015 knife)
COLL: Kosoy, G.Kh.; Pavluchenko, A.E.; Astashevsky, S.V.

Cancer Epidemiology in Coke-Oven Workers in Novokuznetsk

A cohort of about 4,000 coke-oven workers with at least one year's employment in 1960-1989, exposed to PAH, nitrogen oxides, benzene and other chemicals, has been followed up to December 1990. The objective of the study is to investigate cancer incidence and mortality and to identify work areas, substances and processes within coke-ovens, responsible for high cancer levels. Cancer incidence and mortality by site is being compared with that of the general population.

TYPE: Cohort
TERM: Occupation; Solvents
SITE: All Sites
CHEM: Benzene; Nitrogen Oxide; PAH
OCCU: Coke-Oven Workers
TIME: 1990 - 1995

*697 **Bulbulyan, M.** 05505
Inst. of Carcinogenesis, Cancer Research Center, Russian Academy of Medical Sciences, Kashirskoye shosse, 24, Moscow 115478, Russian Federation (Tel.: +7 095 3241470; Fax: 2302450 ; Tlx: 411015 knife)
COLL: Ilichyova, S.A.; Astashevsky, S.V.

Prospective Mortality Study in Printing Industry Workers

The objective of the study is to determine the cancer experience of workers in the printing industry. The cohort consists of about 5,000 men and women, who have worked for at least two years at the Moscow printing plant during the period 1950-1990. These workers have had putative exposure to some potentially hazardous substances (e.g. benzene, PAH, oils, acids). Cause-specific cancer mortality rates will be compared with expected rates based on the general population.

TYPE: Cohort
TERM: Occupation; Solvents
SITE: All Sites
CHEM: Benzene; Mineral Oil; PAH
OCCU: Printers
TIME: 1991 - 1995

*698 **Bulbulyan, M.** 05506
Inst. of Carcinogenesis, Cancer Research Center, Russian Academy of Medical Sciences, Kashirskoye shosse, 24, Moscow 115478, Russian Federation (Tel.: +7 095 3241470; Fax: 2302450 ; Tlx: 411015 knife)
COLL: Shangina, O.V.; Astashevsky, S.V.

Prospective Mortality Study of Workers in the Shoe Industry

The objective of this prospective study is to investigate whether workers in the Moscow shoe industry have an increased risk of dying from cancer of various sites compared with the Moscow population. Approximately 4,000 men and women who worked at the Moscow shoe factory for two years or more between 1970 and 1993 have been identified from employment records. Deaths occurring in the study population between 1 January, 1970 and 31 December, 1992 have been identified. Information will also be obtained on cancer cases registered among living members of the cohort for the period 1988 to 1992.

TYPE: Cohort
TERM: Occupation
SITE: All Sites; Liver; Lung
CHEM: Benzene; Chloroprene
OCCU: Shoemakers-Repairers
TIME: 1991 - 1995

*699 **Bulbulyan, M.** 05507
Inst. of Carcinogenesis, Cancer Research Center, Russian Academy of Medical Sciences, Kashirskoye shosse, 24, Moscow 115478, Russian Federation (Tel.: +7 095 3241470; Fax: 2302450 ; Tlx: 411015 knife)
COLL: Yanovskaya, M.G.; Schaveleva, T.V.

Cancer Epidemiology in Greenhouse Workers

A cohort study is being conducted to investigate the possible association between pesticides and cancer mortality and incidence among greenhouse workers in the Moscow region. It is anticipated that the cohort will consist of 5,000 persons exposed at their working place. The cohort will serve as a basis for follow-up of possible long-term health effects. Exposure will be reconstructed through documents at the agricultural factories. Vital status or cause of death for each member of the cohort will be identified using Moscow state statistical data. Cancer incidence and mortality will be compared with that of the general population of Moscow region.

TYPE: Cohort
TERM: Occupation; Pesticides
SITE: All Sites
OCCU: Horticulturists
TIME: 1991 - 1998

***700 Bulbulyan, M.** 05508
Inst. of Carcinogenesis, Cancer Research Center, Russian Academy of Medical Sciences, Kashirskoye shosse, 24, Moscow 115478, Russian Federation (Tel.: +7 095 3241470; Fax: 2302450 ; Tlx: 411015 knife)
COLL: Uloyan, S.M.; Margaryan, A.G.; Astashevsky, S.V.

Chloroprene Cohort Study

The aim of the study is to investigate the possible association between the exposure to chloroprene and cancer mortality in exposed workers. The cohort consists of 2,700 male and female workers, employed in a chemical plant for one month or more between 1950 and 1988. The data were extracted from employment records of the plant. The cohort has been followed up from 1 January, 1950 to 31 December, 1988. Vital status or cause of death for each member of the cohort were identified using statistical data from the Armenian state. Cancer mortality in the cohort will be compared with national data.

TYPE: Cohort
TERM: Occupation
SITE: All Sites; Liver; Lung
CHEM: Chloroprene
OCCU: Chemical Industry Workers
LOCA: Armenia
TIME: 1990 - 1995

***701 Bulbulyan, M.** 05509
Inst. of Carcinogenesis, Cancer Research Center, Russian Academy of Medical Sciences, Kashirskoye shosse, 24, Moscow 115478, Russian Federation (Tel.: +7 095 3241470; Fax: 2302450 ; Tlx: 411015 knife)
COLL: Jourenkova, N.Yu.; Astashevsky, S.V.

Cancer Epidemiology in Fertilizer Workers

This study is being carried out at a plant situated in Moscow region. A cohort of about 5,000 male and female workers, employed for two years or more since 1945 to 1985, is being followed up to the end of 1990. The cohort is divided into several sub-groups according to the level of exposure to precursors of N-nitroso compounds, and includes production and maintenance workers, engineering staff at sulfuric, nitric, phosphoric acids, ammonia and complex fertilizers production departments, auxilliary workers and employees. Mortality and incidence rates of the population of Moscow region are used to calculate the expected numbers.

TYPE: Cohort
TERM: Fertilizers; Occupation
SITE: All Sites
CHEM: Nitrates; Nitrogen Oxide; Sulphur Dioxide
OCCU: Fertilizer Workers
TIME: 1988 - 1995

702 **Garkavtseva, R.F.** 04626
Academy of Medical Sciences, Inst. of Clinical Oncology Cancer Research Centre, Kashirskoye shosse 24, Moscow 115478, Russian Federation (Tel.: +7 95 3244257; Fax: 2302450 ; Tlx: 411015 knife)
COLL: Kazubskaya, T.P.; Nephedov, M.D.

Assessment of the Contribution of Hereditary Factors to Common Neoplasms

Pedigrees of 305 patients with multiple primary cancers (MPC) have been studied during 1990-1992. About 2% of the 1,817 first degree relatives of patients with MPC were found to have multiple cancers, a 7-fold excess of that observed in the general population. Increased genetic predisposition on the basis of genetic correlation analysis was demonstrated in families with MPC compared with families with one malignancy. Based on the multifactorial model of the hereditability of cancers of the breast, stomach and MPC, the probability of a member developing malignancy has been studied and a medical genetic councelling has been developed. This essentially consists of a registry of families with multiple cancers, evaluation of the risk of cancers in relatives of probands and their children and follow-up of high-risk groups.

TYPE: Case Series
TERM: Genetic Factors; Heredity; High-Risk Groups; Multiple Primary; Pedigree; Prevention
SITE: All Sites
TIME: 1985 - 1994

***703** **Garkavtseva, R.F.** 05502
Academy of Medical Sciences, Inst. of Clinical Oncology Cancer Research Centre, Kashirskoye shosse 24, Moscow 115478, Russian Federation (Tel.: +7 95 3244257; Fax: 2302450 ; Tlx: 411015 knife)
COLL: Kasubskaya, T.P.; Akulenco, L.V.; Kharkevich, G.Y.; Kirichenko, O.P.

Genetic Councelling and Follow-Up of Risk Groups with Malignant Neoplasms

This investigation aims to address the frequency of inheritable forms of ovarian, breast and endometrial cancers, malignant melanoma and childhood embryonic tumours in the population of Moscow. A familial cancer registry with data on 200 cancer families is being maintained. The genetic risks of 300 unaffected subjects from these families are being assessed by means of epidemiological, biochemical and cytogenetic studies. These subjects are under surveillance.

TYPE: Genetic Epidemiology; Registry
TERM: Genetic Councelling; Genetic Factors
SITE: Breast (F); Melanoma; Ovary; Uterus (Corpus)
TIME:

***704** **Glazkova, T.G.** 05522
Academy of Medical Sciences, Cancer Research Center, Kashirskoye shosse, 24, Moscow 115478, Russian Federation (Tel.: 95 3249485; Fax: 2302450)

Risk Factors for Leukaemia and Thyroid Cancer in Children in Russia

This population based case-control study aims to investigate the risk factors for leukaemia and thyroid cancers in children in Russia. Incident cases from regions of Russia apparently affected by radio-active fallout from the Chernobyl accident, and unaffected regions, are included in the study. The role of genetic, immunological, and environmental radiation factors are investigated.

TYPE: Case-Control
TERM: Childhood; Genetic Factors; Immunology; Radiation, Ionizing
SITE: Leukaemia; Thyroid
TIME: 1989 - 1994

705 **Levshin, V.F.** 05167
Academy of Medical Sciences, All-Union Cancer Research Centre, Kashirskoye Shosse 6, Moscow 115478, Russian Federation (Tel.: +7 095 3229134)

Smoking Cessation and Vitamins A and E for Lung Cancer Prevention

The main aim is to compare the efficacy of smoking cessation and diverse regimens of medication with vitamins A and E in the prevention of lung cancer among heavy smokers. In addition, the study aims to examine the association of lung cancer with life history factors other than smoking. Men aged 50-69 with a lifetime smoking history of more than 100,000 cigarettes will be included. Four comparable groups of such men will be recruited to receive different types of preventive intervention: (1) special help to stop smoking; (2) the combination of retinol palmitate 50,000 IU and alpha-tocopherol acetate 0.1 mg all the year round; (3) the same agents, but only for a four-month period per year; (4) placebo. It will take three years to recruit all the participants. The interventions will continue for five years. The follow-up period will last an average 10 years.

TYPE: Intervention
TERM: BMB; Hair; Prevention; Tobacco (Smoking); Vitamins
SITE: Lung
CHEM: Retinoids; Tocopherol
TIME: 1991 - 2001

***706 Nefedov, M.** 05379
National Centre of Medical Genetics, Moskvorechye 1, 115478 Moscow, Russian Federation (Tel.: +7 95 1118582; Fax: 3240702)
COLL: Garkavtseva, R.F.; Liapunova, N.A.

Analysis of Retinoblastoma Gene Deletions

The aim of this investigation is the development of methods to prevent retinoblastoma by genetic counselling, follow-up of "risk groups", cytogenetic and molecular genetics diagnosis of deletions in retinoblastoma gene, segregation analysis of RFLP, using blot-hybridisation and PCR. Cytogenetic analysis of 115 patients (51 bilateral retinoblastoma patients, 27 unilateral retinoblastoma and 37 healthy relatives) has been performed. Three "de novo" cytogenetic deletions (13q14.1) were detected. Microdeletions of retinoblastoma gene have been analysed in 64 Hind III restricted DNAs from blood of patients with inherited retinoblastoma (26 retinoblastoma patients, 38 healthy relatives from 19 families). Diagnostic probe pH3-8 and control p9D11 was kindly provided by M. Laland from the Children Clinic of Boston and Prof. W.C. Cavenee from the Ludwig Inst. for Cancer Research, McGill Univ. (Montreal) respectively. The probes were used in blot-hybridisation dose analysis. Six microdeletions of retinoblastoma gene in 5 families were detected. In summary, the frequency of deletions in families with hereditary retinoblastoma was 31%.

TYPE: Genetic Epidemiology
TERM: BMB; DNA; Genetic Councelling; Genetic Factors; PCR; Prevention; RFLP; Segregation Analysis
SITE: Retinoblastoma
LOCA: Russian Federation; Ukraine
TIME: 1991 - 1995

707 Remennick, L.I. 05154
All-Union Cancer Research Centre, Dept. of Cancer Epidemiology, Karshiskoye Shosse 6 , Moscow 113478, Russian Federation
COLL: Koshkina, V.S.; Okuneva, L.A.

Reproductive Factors and Abortion in Breast and Genital Cancer

This hospital-based case-control study will examine the interaction of various reproductive events and sexual characteristics in causation of breast and gynaecological cancers in a female population with low fertility and a high abortion rate. All new cases of cancers of the breast, cervix and corpus uteri, and ovary (about 300 over three years) arising in the city of Magnitogorsk will be compared to a similar number of non-cancer patients individually matched by age (within two years), residence and socio-economic status. The questionnaire will include items on occupation and living standards, detailed reproductive history (with special emphasis on abortion, including late, out-of-hospital, complicated and other types of terminations), history of gynaecological diseases and screening. All cases and controls will be interviewed by trained clinical nurses. The results will be analysed by univariate and multivariate methods.

TYPE: Case-Control
TERM: Abortion; Lifestyle; Reproductive Factors
SITE: Breast (F); Ovary; Uterus (Cervix); Uterus (Corpus)
TIME: 1991 - 1995

708 **Smulevich, V.B.** 04761
Russian Academy of Medical Sciences, Cancer Research Center, Lab. of Occupational Cancer, Kashirskoye Shosse 24, Moscow 115478, Russian Federation (Tel.: +7 095 3235944; Tlx: 411015 knife)
COLL: Belyakova, S.V.; Solionova, L.G.; Samojlov, D.V.

Parental Occupation as Cancer Risk Factor for Children: A Population-Based Case-Control Study

Parental exposure to occupational carcinogens and other factors related to cancer in children under 15 is being studied in a population-based case-control study in Moscow. In contrast to most previous studies, a detailed occupational history of both parents throughout the whole period between the first job and cancer diagnosis in a child is being assessed, with special emphasis on simultaneous exposure in both parents. The study will include all new cancer cases in the city of Moscow for the period 1986-90 as registered by the childhood cancer registry of the All-Union Cancer Research Center. Two control children, free of chronic diseases and congenital malformations, are matched to each cancer child by sex, year of birth and residence. Interview with the parents of diseased and control children is performed in a standard way in the medical setting (hospital or out-patient clinics). Besides items on occupational history and hazards, the questionnaire includes information on parental chronic diseases, obstetric history of the mother, smoking and drinking habits and genetic susceptibility. Results will be analysed by cancer site with application of standard statistical techniques.

TYPE: Case-Control
TERM: Alcohol; Childhood; Genetic Factors; Intra-Uterine Exposure; Occupation; Tobacco (Smoking)
SITE: All Sites
TIME: 1986 - 1995

709 **Smulevich, V.B.** 05128
Russian Academy of Medical Sciences, Cancer Research Center, Lab. of Occupational Cancer, Kashirskoye Shosse 24, Moscow 115478, Russian Federation (Tel.: +7 095 3235944; Tlx: 411015 knife)
COLL: Solionova, L.G.; Gorelikova, O.N.

Cancer Incidence and Mortality among Public Transport Drivers

There is substantial epidemiological evidence of higher cancer risks in drivers, presumably related to their occupational exposures (polluted microclimate, noise, vibration, etc.) and lifestyle (intense smoking, alcohol abuse, irregular eating, stress, etc.). In previous studies, significant increases in chronic morbidity (gastritis, ulcers, bronchitis, etc.) directly related to length of service were demonstrated. In an extension of these studies, the cohort of 3,038 public transport drivers of Moscow city is being followed from 31 December 1969 to 31 December 1989. Of these, 1,442 are bus drivers. Female drivers and retired male drivers will be analysed as separate subgroups. Cancer incidence and mortality measures (SIRs and SMRs) will be calculated, expected figures being based on the rates of Moscow population.

TYPE: Cohort
TERM: Lifestyle; Occupation; Stress
SITE: All Sites
OCCU: Bus Drivers
TIME: 1990 - 1995

710 **Zaridze, D.** 04916
Inst. of Carcinogenesis, Cancer Research Center, Russian Academy of Medical Sciences, Kashirskoye Shosse, 24, Moscow 115478, Russian Federation (Tel.: +7 095 3241470; Fax: 2302450 ; Tlx: 411015 KNIFE)
COLL: Zemlianaya, G.M.

Lung Cancer in Non-Smokers in Moscow

This case-control study aims to assess the effect of indoor and outdoor air pollution as a possible risk factor for lung cancer in lifetime non-smokers in Moscow. Cases are lung cancer patients who never used tobacco products, residents in Moscow, and admitted to the two largest oncological hospitals in Moscow. Hospital controls are other cancer patients, excluding those with respiratory cancer. Information on demographic characteristics, residential history (to assess lifetime exposure to outdoor air pollution), passive smoking, radon exposure (the direct measurement of radon concentrations in the present residence of the study subjects), occupation, dietary habits (consumption of food rich in vitamins), and alcohol intake is obtained from patients and controls. About 400 cases will be collected in the years 1991-1994. Preliminary results based on 162 cases and 285 controls suggest that lung cancer risk in non-smokers is elevated (statistically significant) in persons living in urban areas with heavy air-pollution and high levels of radon concentration in the present residence and is also related to husbands smoking habits.

TYPE: Case-Control
TERM: Air Pollution; Alcohol; Diet; Occupation; Passive Smoking
SITE: Lung
CHEM: Radon
TIME: 1990 - 1995

711 **Zaridze, D.** 04918
Inst. of Carcinogenesis, Cancer Research Center, Russian Academy of Medical Sciences, Kashirskoye Shosse, 24, Moscow 115478, Russian Federation (Tel.: +7 095 3241470; Fax: 2302450 ; Tlx: 411015 KNIFE)
COLL: Lifanova, Y.E.; Bukin, Y.V.; Babaeva, R.Y.; Levtchuk, A.A.; Maximovich, D.M.; Shevchenko, V.Y.; Bassalyk, L.; Kushlinski, V.; Duffy, S.W.

Diet and Breast Cancer

The aim of this case-control study is to examine the relationships between diet and risk of breast cancer. All consecutive patients with newly diagnosed breast cancer, resident in Moscow, and admitted to the breast cancer clinic, will be interviewed and blood samples collected. Age-matched neighbourhood controls are sampled from among the persons attending, for minor complaints, the same regional out-patient clinic from which the cases are recruited. Statistical analysis of 140 pairs of breast cancer patients and controls has shown that a decreased risk of post-menopausal breast cancer was associated with high intakes of cellulose, vitamine C, beta-carotene and also polyunsaturated fatty acids. High intakes of total fat resulted in a statistically insignificant decrease in the odds ratio, while saturated fats slightly increased risk of breast cancer. Alcohol use significantly increased risk of breast cancer (Int. J. Cancer 48:493-501, 1991). The concentrations of total oestradiols (E2) and free E2 were higher in breast cancer patients than in controls (Eur. J. Cancer Pev. 1:225-230, 1992). The levels of two principle polyunsaturated fatty acids were lower in cases than in controls (Int. J. Cancer 45:807-810, 1990). An addition 450 case and control pairs have been interviewed.

TYPE: Case-Control
TERM: Alcohol; Diet; Hormones; Nutrition; Reproductive Factors
SITE: Breast (F)
CHEM: Beta Carotene
TIME: 1987 - 1994

712 **Zaridze, D.** 04919
Inst. of Carcinogenesis, Cancer Research Center, Russian Academy of Medical Sciences, Kashirskoye Shosse, 24, Moscow 115478, Russian Federation (Tel.: +7 095 3241470; Fax: 2302450 ; Tlx: 411015 KNIFE)
COLL: Filipchenko, V.V.; Serdyuk, V.; Kustov, V.; Duffy, S.W.

Diet and Colorectal Cancer

The aim of the study is to investigate the role of diet in the aetiology of colorectal cancer. The study population consists of colorectal cancer patients, resident in Moscow, admitted to the Central Moscow Oncology Clinic, and hospital controls. Cases and controls are interviewed using a food frequency questionnaire. In addition, other lifestyle and environmental factors related to colorectal cancer risk will be assessed. A parallel study has been carried out in Chabarovsk. Preliminary results indicate a protective effect of cellulose, beta carotene and vitamin C, as well as of vegetable consumption. A protective effect was also oberved for intake of plyunsaturated fatty acids, from fish sauces. In Chabarovsk fish consumption in general was associated with a decrease in colorectal cancer risk. The increased risk was associated with high meat/vegetable, potein/cellulose, fat/cellulose ratios (Eur. J. Cancer 29A:112-115, 1993).

TYPE: Case-Control
TERM: Diet; Environmental Factors; Lifestyle
SITE: Colon; Rectum
TIME: 1988 - 1994

713 **Zaridze, D.** 04920
Inst. of Carcinogenesis, Cancer Research Center, Russian Academy of Medical Sciences, Kashirskoye Shosse, 24, Moscow 115478, Russian Federation (Tel.: +7 095 3241470; Fax: 2302450 ; Tlx: 411015 KNIFE)
COLL: Basieva, T.H.; Smans, M.; Winkelmann, R.; Duffy, S.W.

Atlas of Cancer Incidence and Mortality in the former USSR and Russia

Data on incident cases of cancer by age, sex and area of residence in the year 1989-1991 have been collected for the Russian Federation and the former USSR. Data on mortality from cancer are also available from the same areas for the years 1988-1991. The geographical area covered includes a vast range of climatic, social and industrial conditions, and a preliminary analysis of the data indicates substantial regional variation in cancer rates (Cancer Causes & Control 1:39-49, 1990). It is proposed to produce a comprehensive cancer atlas of the former USSR, incorporating incidence, mortality, relation between incidence and mortality where appropriate, measures of reliability and characteristics of the region. The atlas will contain maps and tabular data.

TYPE: Incidence; Mortality
TERM: Mapping
SITE: All Sites
TIME: 1990 - 1995

714 **Zaridze, D.** 04923
Inst. of Carcinogenesis, Cancer Research Center, Russian Academy of Medical Sciences, Kashirskoye Shosse, 24, Moscow 115478, Russian Federation (Tel.: +7 095 3241470; Fax: 2302450 ; Tlx: 411015 KNIFE)
COLL: Bulbulyan, M.; Maximovich, D.M.; Lifanova, Y.E.

Physicians Health Study

This prospective cohort study aims to evalute the relationship between cancer risk and a number of environmental factors, including diet, alcohol consumption, cigarette smoking, and family history of cancer. Information will be obtained by mail, using a self-administered questionnaire, from 50,000 physicians, males and females, aged over 40 years and living in Moscow and the Moscow region. Information about cancer occurrence, including histological type, will be obtained from the Moscow Cancer Registry. Recruitment to the study has been postponed, because of the changes taking place in the Russian society today and which have changed the dietary habits.

TYPE: Cohort
TERM: Alcohol; Diet; Familial Factors; Occupation; Tobacco (Smoking)
SITE: All Sites
OCCU: Health Care Workers
REGI: Moscow (Rus)
TIME: 1989 - 2010

***715** **Zaridze, D.** 05512
Inst. of Carcinogenesis, Cancer Research Center, Russian Academy of Medical Sciences, Kashirskoye Shosse, 24, Moscow 115478, Russian Federation (Tel.: +7 095 3241470; Fax: 2302450 ; Tlx: 411015 KNIFE)
COLL: Bulbulyan, M.; Astashevsky, S.V.; Petrova, N.I.; Ivkova, E.P.; Boffetta, P.

Cancer Mortality in Mercury Miners and Millers

The objective of this project is to estimate cancer risk in mercury miners and millers in connection with occupational exposures. The cohort consists of 1,235 workers (994 men and 241 women) who have been working in 1970 - 1990 with a minimum of 2 years of continuous employment, based on existing employment records. Industrial hygiene data will be critically reviewed to estimate exposure to mercury and dust containing silica. Cancers of all sites are of interest, with special emphasis on those of the brain and kidney. Cancer mortality in the cohort will be compared with national data.

TYPE: Cohort
TERM: Dusts; Metals; Occupation
SITE: All Sites; Brain; Kidney; Lung
CHEM: Mercury; Silica
OCCU: Millers; Miners
LOCA: Ukraine
TIME: 1992 - 1995

St Petersburg

716 **Abdulkadirov, K.M.** 04766
Inst. of Hematology & Blood Transfusion Clinic of Hematology, 2nd Sovetskaya St. 16 , St Petersburg 193024, Russian Federation (Tel.: 2775938; Fax: 812 2775938)
COLL: Samuskevich, I.G.

Haematoblastoses in Briansk, a Region Contaminated by Radioactive Fallout from the Chernobyl Accident

The aim of the present study is to estimate the possible effect of environmental radioactive contamination on the development of haemoblastoses and suppressed haematopoiesis in the population of the Briansk region, where the Chernobyl accident took place. Morbidity rates in the population exposed to irradiation in areas of this region (300,000 men) are compared with those in areas beyond the irradiated zone (1 million men). A retrospective analysis of haemoblastosis and suppressed haematopoiesis for 1979-1985 (background data) has shown that morbidity averaged 10,950 per 100,000 men. Rates from 1986 onward are being compared with those from 1979-1985. Collection of information on inhabitants first diganosed with haemoblastosis will continue during 1993-1996. Annual rates will be calculated.

TYPE: Incidence
TERM: Prevalence; Radiation, Ionizing
SITE: Haemopoietic
TIME: 1989 - 1996

717 **Semiglazov, V.F.** 04859
N.N. Petrov Research Inst. of Oncology, USSR Ministry of Health, St Petersburgskaya St., Pesochny-2, St Petersburg 189646, Russian Federation (Tel.: 2378748; Fax: 2378947)
COLL: Sagaidak, V.N.; Ebeling, K.

Randomized Controlled Trial of Breast Self Examination in Breast Cancer Mortality

The major objective of the study is to determine the effect of a breast self-examination (BSE) programme on mortality from breast cancer. The assessment will be made in a controlled trial. A population of about 100,000 women aged 40-64 will be randomized in St Petersburg. After a one-year feasibility study in 1984, the project will continue for 15 years. It consists of a four-year intensive education progamme (1985-1988) during which, and afterwards until 1994, the detected breast cancer cases will be registered and treated, and followed up for five years until 1999. The key issue of the study is the compliance of the population to the BSE programme. The frequency and competence of BSE practice will be defined in a subsample of 400 randomly selected women by means of a survey made at the sixth month and annually since the start of the project. The study is expected to result in the accrual of more than 400 new breast cancer cases and 200 deaths. It is also intended to evaluate a list of risk factors and take the opportunity to pin-point high-risk groups.

TYPE: Cohort; Intervention
TERM: Screening
SITE: Breast (F)
TIME: 1985 - 1999

Ulyanovsk

718 **Storozhuk, M.** 04694
Ulyanovsk District Oncology Dispensary, 92, 12th September St., Ulyanovsk 432700, Russian Federation
COLL: Loginov, A.P.; Yaroslavtsev, V.N.

Programme for the Fight against Cancer in Ulyanovsk District

According to the decision of the "Local Soviet" a programme of social and medical cancer prevention was adopted in 1988. The social programme includes air protection and reasonable use of soil and water resources. The medical programme includes early detection of cancer, rapid treatment of cancer patients and identification of high risk groups. The creation of a high-risk group data bank is planned. People are selected according to risk factors. Data on each person are stored until death or until a precancerous lesion becomes a cancer. In December 1988, data on 3,848 persons had been collected. It is anticipated that 10-15% of the total population will be included in the data bank. This will permit follow-up of high cancer risk populations, detection of cancer at initial stages and estimation of frequency of transition of cancerous lesions into cancer. The remaining population will be used as a control group.

TYPE: Cohort; Incidence
TERM: Data Resource; High-Risk Groups; Premalignant Lesion; Prevention; Treatment
SITE: All Sites
TIME: 1988 - 1995

SLOVAK REPUBLIC

Bratislava

719 **Dimitrova, E.** 03714
Inst. of Experimental Oncology, Slovak Academy of Sciences, Dept. of Epidemiology, Ul. Csl. armady 21., 812 32 Bratislava, Slovak Republic (Tel.: (47)57541)
COLL: Plesko, I.; Somogyi, J.; Kiss, J.; Kramárová, E.

Projections of Cancer Incidence and Mortality in Czechoslovakia
The main objective of this study is to determine projections of cancer incidence and mortality until the year 2000 and 2020 in Czechoslovakia. The projections are based on the evolution of incidence and mortality of individual cancer sites in the past 10-15 years and the population age-sex projections for the given years. For selected cancer sites the possible changes in risk factors and the effects of preventive measures will be taken into account in forecasting future incidence rates. Recent result were published in Cs. Zdrav. 34:281, 1986 and Neoplasma 35(6):635, 1988 and 36(4), 1989.

TYPE: Incidence; Mortality
TERM: Anatomical Distribution; Prevention; Prognosis; Projection; Time Factors
SITE: All Sites
TIME: 1986 - 1994

720 **Plesko, I.** 04781
Slovak Academy of Sciences Inst. of Cancer Research, Ul. Csl. armady 21 , 812 32 Bratislava, Slovak Republic (Tel.: (7)57541)
COLL: Vlasák, V.; Obsiniková, A.; Kramárová, E.

Skin Cancer in Slovakia
Incidence and mortality data for non-melanoma skin cancer in the National Cancer Registry for the territory of Slovakia since 1968 are highly complete and detailed, including those on basal cell carcinoma. The main aim of this descriptive study is the evaluation of time-trends of individual histological types of skin cancer and of malignant skin melanoma as well as their different and changing distribution in the various anatomical parts of the skin. The possible causes of geographical aggregation and the risk of developing either another basal cell skin cancer or cancer of another organ in persons with basal cell cancer will also be investigated.

TYPE: Incidence
TERM: Anatomical Distribution; Geographic Factors; Histology; Multiple Primary; Registry; Trends
SITE: All Sites; Melanoma; Skin
REGI: Slovakia (Slvk)
TIME: 1990 - 1994

Roznava

721 **Icsó, J.** 03865
Hospital Policlinic, Dept. of Occupational Medicine, Betliarska cesta , 04801 Roznava, Slovak Republic (Tel.: +42 942 71240)
COLL: Szöllösová, M.; Roda, S.; Gajdosová, D.; Dobiás, K.; Nejjari, A.; Pezerat, H.

Lung Cancer in Iron-Ore Miners in Slovakia
A significantly higher incidence of lung cancer in iron-ore miners has been observed in some parts of Slovakia. Besides alpha-radiation, other factors are suggested to play a role in the development of lung cancer in exposed miners: the chemical composition of iron-ore, underground dust from diesel-mechanisms with its inorganic and organic fraction, the mould in the water or air and others. X-ray difractional analysis showed that the surface activity of mineral particles in iron-ore, which is considered as an important carcinogenic factor, is different in two mines under study: in the mine with high incidence of lung cancer this activity is high, whereas in the mine with low incidence, this activity is very low. The aim of the study is to follow the occurrence of lung cancer in a group of approximately 1,000 iron-ore miners. It will be compared with that in approximately 25,000 non-miners, over 20 years of age, living in the same district area.

TYPE: Cohort
TERM: Chemical Exposure; Dusts; Environmental Factors; Metals; Mining; Occupation; Radiation, Ionizing
SITE: Lung
CHEM: Iron; Radon
OCCU: Miners, Iron
TIME: 1980 - 1995

SLOVENIA

Ljubljana

722 **Primic-Zakelj, M.** 04482
Inst. of Oncology, Unit of Epidemiology, Zaloska 2, 61000 Ljubljana, Slovenia (Tel.: +38 61 314344)
COLL: Ravnihar, B.; Pompe-Kirn, V.; Oblak, B.; Kosmelj, K.

Breast Cancer Risk in Relation to Oral Contraceptive Use in Slovenia
In a hospital-based case-control study (1980-83) a positive association between breast cancer and oral contraceptives was established. A population-based study is now added. Women aged 25-54, residents of Slovenia since 1965, with breast cancer newly diagnosed anywhere in the republic, are studied as cases, with one control, matched by age and municipality of residence, selected from the general population for each case. During the three years of data collection (1988-1990) 624 matched pairs were expected to be personally interviewed in hospitals or at home. Additional data regarding the disease were obtained from hospital records and confirmation of oral contraceptive use was sought from prescribing doctors. Relative risk estimates will be calculated by logistic regression. Analysis started in 1991.

TYPE: Case-Control
TERM: Oral Contraceptives; Registry
SITE: Breast (F)
REGI: Slovenia (Slvn)
TIME: 1988 - 1994

723 **Rudolf, Z.** 04888
Inst. of Oncology, Zaloska 2, 61105 Ljubljana, Slovenia (Tel.: +38 61 316490; Fax: 114180)
COLL: Vlaisavljevic, V.; Kaucic, M.; Jelinncic, V.; Us, J.; Rudolf, Z.; Novak, F.

Pilot Study of Breast Cancer Screening in Six Communes of Slovenia
A pilot randomized trial on the efficiency of an organized breast cancer screening programme has been started in three different regions of Slovenia. About 12,400 randomly selected women aged 50-64 years received a personal invitation letter. Every participant has had a physical examination, breast self examination instruction, and a single mediolateral view mammogram. In the first round the response rate was 55%-75%, better in little towns and rural populations than in urban populations. In an interval of 24 - 36 months mammography and physical examination is repeated. In the second round the response rate was 55%-81%, better in urban populations. The first analysis of mortality and incidence data on breast cancer in the screened and in the unscreened group of the study population is planned for the year 2000.

TYPE: Cohort; Intervention
TERM: Screening
SITE: Breast (F)
REGI: Slovenia (Slvn)
TIME: 1989 - 2000

Maribor

724 **Vlaisavljevic, V.** 04512
Hosp. Maribor, Dept. of Gynaecology, Ljubljanska 5, 62000 Maribor, Slovenia (Tel.: +38 62 37221)
COLL: Harper, P.

Incidence of Breast Cancer in Patients with Breast Cysts
The aim of the project is to study the incidence of breast cancer in patients followed-up because of clinically manifest breast cysts. The patients are volunteers, symptomatic patients visiting the Center for Breast Diseases. All those in which a cyst is certified are included in the study. All patients are self-selected. Recruitment began in 1980 and will terminate in 1995. The hypothesis is that the incidence of breast cancer is higher in patients with breast cysts than in the normal population. The control group for this natural incidence of breast cancer would be that published by the Cancer Registry for the observed region and the population of the same age distribution and exposure period. All patients once registered will be followed by clinical examination, mammography or ultrasound. All refusers of the study will be interviewed every year about their breast (clinically manifest breast cancer).

TYPE: Cohort
TERM: Breast Cysts; High-Risk Groups; Premalignant Lesion; Registry
SITE: Breast (F)
REGI: Slovenia (Slvn)
TIME: 1980 - 1995

Cape Town

725 **Jaskiewicz, K.** 04537
Univ. of Cape Town Medical School, Dept. of Pathology, Anzio Rd, Observatory 7925, Cape Town, South Africa (Tel.: +27 21 471250; Fax: 4171789)
COLL: Stenkop, E.; Louwrens, H.

Diet as a Risk Factor for Atrophic Gastritis

The hypothesis is that endoscopic screening of pre-selected population groups for gastric cancer is an efficient way of establishing diagnosis and prevalence of early cancer and precursor lesions of the stomach. The study of dietary and other factors will give new information concerning aetiology and could be employed in prevention of this disease. Endoscopic and histological examination of gastric mucosa will be undertaken in 600 patients with dyspepsia from a population at increased risk for gastric carcinoma. The pilot study consists of 50 patients with chronic atrophic gastritis, 50 controls with normal gastric mucosa and 50 healthy controls without endoscopic examination. Subjects will be sex- and age-matched, from the same population group. A detailed questionnaire will be employed for each of the 150 subjects, and blood will be analysed for vitamins: C, A, E, B, B6, folates, nicotinic acid and for minerals: Na, K, Ca, Mg, Cu, Zn, Fe, Se. Gastric juices will be analysed for nitrates and nitrites and PH will be estimated.

TYPE: Case-Control
TERM: Atrophic Gastritis; Diet; High-Risk Groups; Nutrition; Premalignant Lesion; Vitamins
SITE: Stomach
CHEM: Copper; Iron; Nitrates; Nitrites; Selenium
TIME: 1987 - 1994

726 **Jaskiewicz, K.** 05141
Univ. of Cape Town Medical School, Dept. of Pathology, Anzio Rd, Observatory 7925, Cape Town, South Africa (Tel.: +27 21 471250; Fax: 4171789)
COLL: Robson, S.C.; Williamson, A.L.

Evaluation of HBV and HCV Markers in Biopsy Material from Patients with Hepatocellular Carcinoma

The prevalence of hepatocellular carcinoma (HCC) is extremely high in Southern Africa. The disease is thought to be linked to HBV and HCV infection, to exposure to mycotoxins, alcohol and iron. The expression of viral antigens and anti-HCV antibodies has been studied recently in groups of patients with alcoholic cirrhosis and portal hypertension, presumed non-A non-B chronic active hepatitis, post-transfusional hepatitis, hepatoma and respective control groups. The patients with hepatomas elected for the seroprevalence study of HCV antibodies were predominantly urban, of mixed racial extraction and tended to be older than the patients reported by Kew et al. Only three of the 33 patients with hepatoma had anti-HCV antibodies in contrast to 11 with serum HBsAg positivity. Recent histological studies by immunohistochemistry have shown 52% positivity for liver tissue for HBsAg in patients with HCC. Over 300 HCC biopsies from selected rural patients, 15% younger than 25 years, are available for study. Liver tissue from sporadic urban cases is also available. Normal liver histology, benign, malignant, epithelial and mesenchymal primary liver tumours; primary biliary cirrhosis; sclerosing cholangitis; presumed non-A non-B hepatitis; hepatoportal sclerosis; haemochromatosis; haemosiderosis and cryptogenic cirrhosis are also available.

TYPE: Case Series
TERM: Biopsy; Cirrhosis; HBV; HCV; Mycotoxins
SITE: Liver
CHEM: Aflatoxin
TIME: 1991 - 1994

***727** **Myers, J.E.** 05350
Univ. of Cape Town, Med. School, Dept of Community Health, Occupational Health Research Unit, Anzio Rd, Observatory 7925, Cape Town, South Africa (Tel.: +27 21 471250; Fax: 478471)
COLL: Clapp, R.W.; Kaufman, D.W.; Close, P.; Jacobs, P.; Rösenstrauch, M.

Agrichemical Exposure as a Risk for Lymphatic and Haematopoietic Cancer in the Western Cape

The aim of this case-control study is to investigate the association between exposure to agrichemicals and the various haematological and lymphoproliferative malignancies in the western Cape. The specific objectives are (1) to test the hypothesis that the subjects exposed to chemicals used in agriculture are at increased risk for non-Hodgkin's lymphhoma and leukaemia (AML and CLL); and (2) to estimate the magnitude of the association between employment in agriculture and the above mentioned malignancies and between specific agricultural chemicals and malignancies. All incident cases who present at Tygerberg, Groote Schuur or Red Cross Children's Hospitals over a period of three years will be included. Tissue samples or slides will be collected from each case for pathology review and confirmation of diagnosis. Controls matched for age, sex and race will be obtained from the source of referral of each case. Those with malignant conditions, non-malignant haematological conditions, allergic and auto-immune conditions will be examined. Information on exposure to agrichemicals will be obtained through a questionnaire administered by a trained interviewer and from historical information on pesticide use in the area. An industrial hygienist will reconstruct exposure histories from questionnaire data on farm workers and others. Data analysis will include computation of crude and adjusted odds ratios, controlling for potential confounding by age, gender, other occupational risks, by multivariate analysis.

TYPE: Case-Control
TERM: Chemical Exposure; Drugs; Occupation; Pesticides
SITE: Haemopoietic; Leukaemia (AML); Leukaemia (CLL); Non-Hodgkin's Lymphoma
OCCU: Agricultural Workers; Farmers
TIME: 1993 - 1998

728 **Warner-Learmonth, G.** 04575
Groote Schuur Hosp., Cytopathology Unit, Anzio Rd, Observatory 7925, Cape Town, South Africa (Tel.: +27 21 471250/424; Fax: 478955)
COLL: Beck, J.

Abnormal Cervical Smears in Teenagers

This is a prospective study of cervical smears in females aged 19 years and younger. Several authors have documented CIN in sexually active teenagers. However, little attention has been paid to the problem of cervical smears reported as "inflammatory atypia". A preliminary search of the records from January 1987 to August 1988 reveals that 7% of the total number of cervical smears are from patients of 19 years and under. Of these, 78% are pregnant and 21% are using some form of contraception. 1,051 of these 8,349 teenage patients have cytological abnormalities ranging from inflammatory atypia to squamous carcinoma. Every gynaecological smear submitted for review from patients of 19 years and under will be "flagged" on registration. All atypical smears from these patients will be reviewed together with all their previous smears. In this way, it is hoped to follow the progress of inflammatory atypia and record (1) its association with sexually transmitted infections, e.g. chlamydia, herpes, human papilloma virus, trichomonas and candida; and (2) whether it is a forerunner of dysplasia.

TYPE: Case Series
TERM: Adolescence; Clinical Records; Cytology; HPV; HSV; Premalignant Lesion; Sexually Transmitted Diseases
SITE: Uterus (Cervix)
TIME: 1987 - 1997

Johannesburg

729 **Kew, M.C.** 04460
Univ. of the Witwatersrand, Dept. of Medicine, York Rd, Parktown, Johannesburg 2193, South Africa (Tel.: +27 11 4883626; Fax: 6434318)
COLL: Song, E.

Aetiology and Diagnosis of Hepatocellular Carcinoma in Southern African Blacks

The purpose of the study is to determine the cause or causes of hepatocellular carcinoma in southern African blacks, who have a very high frequency of this tumour and in whom it frequently occurs at a very young age. Information on the relation between the hepatitis B virus and the tumour has been assembled; aspects that continue to be assessed are relation to age, sex, place of birth and where childhood was spent, subsequent geographical movements, alcohol consumption, cigarette smoking, taking of oral contraceptive steroids. A surveillance programme designed to detect small early hepatocellular carcinomas is also being carried out. This involves following patients known to be chronic carriers of the hepatitis B virus (or to have chronic parenchymal hepatic disease) with serial ultrasonographic examination and serum alpha-foetoprotein estimations. The relation between hepatitis C virus and hepatocellular carcinoma will also be assessed.

TYPE: Cross-Sectional
TERM: Age; Alcohol; Blacks; Geographic Factors; HBV; HCV; Oral Contraceptives; Screening; Sex Ratio; Tobacco (Smoking)
SITE: Liver
CHEM: Steroids
TIME: 1980 - 1994

730 **MacDougall, L.G.** 04549
Univ. of the Witwatersrand, Dept. of Paediatrics (Haematology), Baragwanath Hosp., P.O. Berthsam, Johannesburg 2013, South Africa (Tel.: +27 11 9331530)
COLL: Greaves, M.F.; Bernstein, R.; Cohn, R.; Pool, J.E.

Acute Leukaemia in Black and White Children and Factors Influencing Response to Treatment and Survival

The aim is to determine if significant differences exist in the pattern of acute leukaemia between black and white children aged 0-15 years living in the Johannesburg area of South Africa, with special reference to: (1) population based incidence; (2) classification of leukaemia cell type based on morphology (FAB), immunology (cell surface markers), and cytogenetics; (3) analysis of risk factors, response to treatment and survival; (4) pharmacological evaluation of chemotherapeutic drug metabolism, and (5) assessment of drug compliance. All newly diagnosed children with acute leukaemia referred to the Paediatric Haematology/Oncology Clinics at Johannesburg and Baragwanath Hosptials are enrolled in the study categories (1), (2), and (3) above. This comprises approximately 24 children annually (8 blacks, 16 white). Papers appeared in Leuk. Res. 9:765-767, 1985; Am. J. Ped. Hem./ Oncol. 8:43-51, 1985:and in South Afr. Med. J. 75:481-484, 1989.

TYPE: Case Series; Incidence
TERM: Blacks; Caucasians; Chemotherapy; Childhood; Classification; Survival; Treatment
SITE: Leukaemia
TIME: 1975 - 1994

731 **Rees, D.** 04566
National Center for Occupational Health, P.O.Box 4788, Johannesburg 2000, South Africa (Tel.: +27 11 7241844; Tlx: 422251 sa)
COLL: Simson, I.W.; Goodman, K.

Occupational and Environmental Exposures Associated with Mesothelioma in South Africa

The two major issues to be investigated in this case-control study are (1) occupational and environmental exposures associated with mesothelioma, including diet and (2) the relatively low reported incidence rate of mesothelioma in black South Africans. Cases will be notified by pathologists, oncologists, cardio-thoracists etc. There will be two controls per case: one cancer control - a subject matched for hospital, age, race, sex with cancer of any organ except the pleura or the lung, and one non-cancer control - a subject matched as above and an in-patient of a medical ward for five days or longer. 150 cases and 300 controls are expected. Data will be collected through detailed exposure questionnaire administered to the cases and controls by trained interviewers.

TYPE: Case-Control
TERM: Blacks; Diet; Environmental Factors; Occupation
SITE: Mesothelioma
TIME: 1988 - 1994

Tygerberg

732 **Hesseling, P.B.** 04536
Univ. of Stellenbosch, Tygerberg Hosp., Dept. of Paediatrics, P.O.Box 63 , Tygerberg 7505, South Africa (Tel.: +27 21 9313131; Fax: 9317810 ; Tlx: 526226)
COLL: Wessels, G.

The Tygerberg Hospital Paediatric Tumour Registry

The aim is to record accurately clinical, pathological and therapeutic data on all children newly diagnosed with cancer at this institution. At the same time as standardizing and improving therapy, and effecting a maximum follow up of 5 years, it is intended to establish a population-based register of paediatric cancer for the period 1983-1988 for Namibia. Data will be contributed to a national paediatric cancer register, and incidence and treatment results compared with national and international groups. The data are computerized. 285 new cases have been entered since 1 January 1983. The present rate of follow-up in all patients is 98%.

TYPE: Incidence; Registry
TERM: Treatment
SITE: Childhood Neoplasms
LOCA: Namibia; South Africa
TIME: 1983 - 1994

SPAIN

Alicante

733 **Vioque, J.** 04863

Univ. de Alicante, Dpto. de Salud Comunitaria, Div. de Medicina Preventiva, Campus de San Juan, Apdo 374, 03080 Alicante, Spain (Tel.: +34 65 5659811; Fax: 5658513)

Diet and Cancers of the Oesophagus, Stomach and Pancreas

It is proposed to conduct a population-based case-control study to evaluate the role of diet in relation to cancers of oesophagus, stomach and pancreas, in several Spanish provinces during a period of about 18 months. The major hypotheses to be tested will be: (1) that a diet with a high fat content is associated with increased risk of pancreatic cancer; (2) a diet rich in fresh fruit and vegetables is associated with a lower risk of oesophagus, stomach and pancreatic cancer; (3) a low intake of vitamins C and A (and specific carotenoid fractions) is associated with a higher risk; (4) a high intake of foods with a high content of salt and carbohydrates is associated with an increased risk of stomach cancer. During a period of one year all newly diagnosed cancers in five Spanish provinces will be collected: 300-400 pancreatic, 600 stomach and 300 oesophageal cancers are expected. Controls will be randomly selected from the census, and matched by sex, age and residence to cases. Data will be collected through personal interview. A core questionnaire and a food frequency questionnaire will be used. Univariate and multivariate analysis will be performed.

TYPE: Case-Control
TERM: Diet; Fat; Fruit; Nutrition; Registry; Vegetables; Vitamins
SITE: Oesophagus; Pancreas; Stomach
REGI: Basque (Spa); Navarra (Spa); Zaragoza (Spa)
TIME: 1991 - 1994

Barcelona

***734** **Porta, M.** 05277

Inst. Municipal d'Invest. Mèdica, Univ. Autònoma de Barcelona, Passeig Marítim 25-29, 08003 Barcelona, Spain (Tel.: +34 3 4851085; Fax: 4854952)

COLL: Real, F.; Malats, N.; Rifà, J.; Guarner, L.; Carrato, A.; Andreu, M.; Salas, A.; Garcia de Herreros, A.; Corominas, J.M.; Piñol, J.L.; Fernandez, E.; Marrugat, M.; Gomez, E.; Carrillo, E.

Role of Mutations in K-ras and p53 Genes in Exocrine Pancreatic Cancer and Cancer of the Biliary Tract (PANKRAS II)

The objectives are (1) to assess the clinical usefulness of detecting mutations in K-ras and p53 genes for the diagnosis of exocrine pancreatic cancer (EPC) and cancer of the biliary tract (CBT); the study includes a large number of patients (incident cases of EPC and CBT, as well as patients with benign diseases of the pancreas); (2) to assess the prognostic value of such genetic alterations among cancer cases (the study conducts an active follow-up of the cohort); (3) to analyse the prevalence of genetic alterations according to the clinical and biological characteristics of tumours; and (4) to assess the relation between genetic alterations and history of smoking, alcohol and coffee consumption, foods and occupational exposures. Personal interviews are conducted with the patients, covering clinical history, lifestyle, etc. Fresh tissue, cytology specimens, serum, haematies, leucocytes, lymphocytes, urine, nails, hair and oral mucosa are stored. It is estimated that the study will comprise over 530 cases by early 1994, including 165 cases of EPC, 120 cases of CBT and 130 cases of chronic pancreatitis.

TYPE: Case Series; Molecular Epidemiology
TERM: Alcohol; BMB; Blood; Chromosome Effects; Coffee; Cytology; Occupation; Prognosis; Tissue; Tobacco (Smoking); Toenails; Urine
SITE: Bile Duct; Gallbladder; Pancreas
TIME: 1991 - 1995

Mataro

735 **González, C.A.** 04302

IREC, Jordi Joan 5, 08301 Mataro, Spain

COLL: Badia, A.; Cardona, T.; Verge, J.; Viver, J.; Saigi, E.; Batiste, E.; Marcos, G.; Martos, M.C.; Solanilla, P.; Brullet, E.; Badosa, E.; Vida, F.; Riboli, E.

Stomach Cancer in Catalunya and Zaragoza

Gastric cancer is the second cause of death from malignant tumours in Spain. A multicentre case-control study both in high risk and low risk areas has been initiated to provide information on the role of diet and other factors. The histopathological characterization and the modalities of interviews have been standardized among the participating hospitals. Cases will be all patients with a newly diagnosed cancer of the stomach, on the basis of histological examination in the participating centres. Controls will be chosen from the same hospitals. About 350 cases and 350 controls are expected.

TYPE: Case-Control
TERM: Cooking Methods; Diet; Histology; Occupation; Tobacco (Smoking)
SITE: Stomach
TIME: 1987 - 1994

Soria

***736 Sanz-Anquela, J.M.** 05545
Hosp. General del Insalud de Soria, Paseo Santa Barbara S/N, 42003 Soria, Spain (Tel.: +34 75 221000; Fax: 229725)
COLL: Ruiz-Liso, J.M.; Villar-del-Sordo, V.; Bajador-Andreu, E.J.; Yus-Cotor, C.; Diloy-Tejero, R.; Admella-Salvador, M.C.; Soler-Monso, M.T.

Role of Borage Intake in Premalignant Gastric Conditions

Borage, a vegetable locally consumed in some provinces in the north of Spain, e.g. Navarra, La Rioja and Zaragoza, is one of the major natural source of gamma-linolenic acid (GLA). The boiled or cooked stem and leaves of the plant have enough GLA to explain a hypothetical beneficial effect on gastric carcinogenesis. A study is planned to assess the possible protective effect of borage consumption and to study risk factors for premalignant conditions and the evolution to gastric cancer. Subjects with intestinal metaplasia types III, II and I and atrophic gastritis, with Lewis system alterations, will be drawn from Zaragoza, Soria and Barcelona, provinces with different levels of borage intake and different levels of gastric cancer, and their dietary history and endoscopic findings will be recorded. The subjects will be followed up.

TYPE: Correlation
TERM: Atrophic Gastritis; Diet; Premalignant Lesion; Vegetables
SITE: Stomach
TIME: 1993 - 1998

Valencia

737 Cortes Vizcaino, C. 05019
Univ. de Valencia, Fac. de Medicina, Depto de Medicina Preventiva y Salud Publica, Aven. Blasco Ibanez 17, 46010 Valencia, Spain (Tel.: +34 6 3864166; Fax: 3864173)
COLL: Saiz Sanchez, C.; Gimenez Fernandez, F.J.; Talamante Serrulla, S.; Sabater Pons, A.; Calatayud Sarthou, A.

Digestive Cancers in Spain: Temporal and Spatial Correlations

Mortality from digestive cancers in Spain is being studied to investigate the relationship with certain risk factors such as consumption of alcohol, coffee and tobacco, and dietary factors. Mortality and demographic data are obtained from the National Institute of Statistics, and the data concerning the risk factors are collected from several surveys and studies. The geographical distribution of several digestive cancers in Spain is being correlated with the distribution of the factors mentioned. The temporal evolution of mortality from these cancers will also be correlated with that for consumption of the risk factors in Spain over a 30-year period.

TYPE: Correlation
TERM: Alcohol; Coffee; Diet; Geographic Factors; Time Factors; Tobacco (Smoking)
SITE: Colon; Gallbladder; Liver; Oesophagus; Pancreas; Rectum; Small Intestine; Stomach
TIME: 1991 - 1994

Zaragoza

738 **Sinues, B.** 04668

Univ. de Zaragoza, Fac. Med. y Hosp. Clin. Univ., Dpto de Farmacologia, Domingo Miral, 50009 Zaragoza, Spain (Tel.: 3357854)

COLL: Izquierdo, M.; Tres, S.; Bartolome, M.; Perez Viguera, J.

Biological Indicators of Occupational Cancer Risk

The aims of the study are: (1) to establish the individual susceptibility to occupationally induced carcinogenesis, and (2), to find some biological indicators for cancer risk. Occupational cancer can be related to exposure to external mutagenic agents. Important genetically determined metabolic differences increase or reduce the occupational cancer risk. The possible relation between the following factors will be investigated: (1) biological effects (measured in blood peripheral lymphocytes, by testing SCEs, chromosome aberrations, PRI, micronuclei); (2) pharmacogenetics (by phenotyping acetylator and hydroxylator status); (3) liver microsomal activity environmentally-induced (-GT, glucaric and acid excretion); and (4) internal exposure level (by measuring thioether, mutagen and premutagen urinary excretion). A study of smokers, workers exposed to vinyl chloride and workers in the dye industries exposed to arylamine has been initiated.

TYPE: Molecular Epidemiology
TERM: BMB; Biomarkers; Chromosome Effects; Dyes; Lymphocytes; Metabolism; Micronuclei; Occupation; Plastics; SCE; Tobacco (Smoking)
SITE: Angiosarcoma; Bladder; Liver; Lung
CHEM: Vinyl Chloride
OCCU: Dyestuff Workers; Plastics Workers; Textile Workers; Vinyl Chloride Workers
TIME: 1989 - 1994

Peradeniya

739 **Warnakulasuriya, K.A.** 04338
Univ. of Peradeniya, Dept. of Oral Medicine, Augusta Rd, Peradeniya, Sri Lanka (Tel.: + 94 8 88045)

Oral Precancerous Lesions and Conditions

Case-control studies on the aetiology of oral cancer and pre-cancer conducted in India have given clues to aetiology. The current study's objective is to re-evaluate the risk factors in a different geographical setting, Sri Lanka, where chewing habits are similar in many respects. The role of betel quid chewing, particularly with and without tobacco, and smoking will be studied among pre-cancer cases (aged over 20 years). The study is population-based with 456 cases detected following a screening programme. An equal number of age- and sex-matched controls from the same villages were enlisted. All cases were interviewed at the Oral Medicine Clinic and the controls at extended clinics in the field. Interviewing and coding is complete, and analysis is underway.

TYPE: Case-Control
TERM: Betel (Chewing); Premalignant Lesion; Tobacco (Chewing); Tobacco (Smoking)
SITE: Oral Cavity
TIME: 1982 - 1994

SWEDEN

Gothenburg

740 **Bengtsson, C.B.** 04557
Dept. of Primary Health Care, Redbergsv. 6, 416 65 Gothenburg, Sweden (Tel.: +46 31 840170; Fax: 842784)

Cohort Study of Women in Gothenburg, Sweden

About 1,500 women were studied in a cross-sectional study in 1968-69. Due to the methods of data collection and a high participation rate, over 90%, the participants in the study are representative of women in Gothenburg of the ages studied (initially 38-60 years). The women were studied again in 1974-75 and 1980-81. Analyses of data with respect to cancer mainly refer to the 12-year follow-up study between 1968-69 and 1980-81. A 19-year follow-up with respect to cancer morbidity and mortality has been carried out. Premorbid characteristics (observations in 1968-69) are related to incidence of cancer during the following 12 and 19 year periods. A new cross-sectional study (which means a 24-year follow-up) is underway. Publications in relation to this study have appeared in Scan. J. Soc. Med. 6:49-54, 1978, 17:141-145, 1989 and Eur. J. Clin. Nutr. 46:501-507, 1992.

TYPE: Cohort
TERM: BMB; Plasma; Serum
SITE: All Sites
TIME: 1968 - 1994

741 **Granberg, S.B.O.** 04604
Univ. of Gothenburg, Sahlgrens Hosp., Dept. of Oncology & Gynaecology, 41345 Gothenburg, Sweden (Tel.: +46 31 601000)
COLL: Köpf, I.

Heteromorphism of Heterochromatic Segment of Chromosome 1 in 25 Families with Several Cases of Ovarian Cancer

During the last decade, evidence has been forthcoming in support of the correlation between heteromorphism of human chromosome 1 q&S'h. and the incidence of various malignancies, including ovary. The aim of this study is to investigate the degree of variability of the C- band regions of chromosome 1 in human karyotypes and to determine the incidence of heteromorphism in patients with ovarian cancer and their close relatives. 25 families with two or more cases of ovarian cancer are investigated. 192 close family members are included. Peripheral blood samples will be taken for cytogenetic investigations.

TYPE: Cross-Sectional; Genetic Epidemiology
TERM: Chromosome Effects; Genetic Factors; Heredity
SITE: Ovary
TIME: 1986 - 1994

742 **Larsson, S.** 03341
Sahlgren Hosp., Dept. of Lung Medicine, Box 17301, 402 64 Gothenburg, Sweden (Tel.: +46 31 840040)
COLL: Järvholm, B.; Sörensen, S.; Flodin, U.; Edling, C.

Attributable Risk of Environmental Factors for Lung Cancer in a Swedish City

The scope of this case-control study is to estimate the attributable risk of some known lung carcinogens, such as smoking and asbestos exposure. 147 cases of lung cancer have been interviewed by questionnaire and compared with 230 controls (109 population controls and 121 hospital controls). Preliminary results showed that the aetiological fraction for smoking/ex-smoking was 95% for men and 78% for women. Male cases were more often asbestos-exposed than controls and there was a tendency towards a dose-response relationship. Risk estimates for asbestos exposure were, however, not significantly increased. The aetiological fraction for asbestos exposure among men was 16%. Few cases and controls stated exposure to other known lung carcinogens.

TYPE: Case-Control
TERM: Dusts; Environmental Factors; Tobacco (Smoking)
SITE: Lung
CHEM: Asbestos
TIME: 1983 - 1994

743 **Torén, K.Ö.** 04100
Univ. of Gothenburg, Sahlgren Hosp., Dept. of Occupational Medicine, St Sigfridsgatan 85, 412 66 Gothenburg, Sweden (Tel.: +46 31 830615)
COLL: Järvholm, B.; Sällsten, G.

Mortality and Morbidity in Paper Mill Workers in Sweden

Studies from Sweden and the US have indicated that paper and pulp mill workers have an increased mortality of cancer, mainly stomach cancer and cancers of the blood and lymphatic system, but also an increased mortality from chronic obstructive pulmonary diseases and asthma. All these studies have been done with case referent or PMR techniques. With the aim of evaluating mortality among the workers from two Swedish paper mills a case referent study, including about 1,000 subjects has been carried out. The diseases under study were stomach cancers and respiratory diseases, including lung and pleural cancers. Data about the diseases were collected from the local registers of death and burial, the exposures are estimated according to the personal files in the paper mills. (Am. J. Ind. Med. 1991). The health risk among workers in a soft paper mill, mainly with regard to exposure of paper dust, is also being evaluated in an exposed cohort with an unexposed referent cohort. The diseases under study are all forms of cancer and diseases of the respiratory tract. The study includes about 1,000 paper mill workers and 1,500 unexposed controls. These data are collected from the Cancer Registry, from the National Mortality Registry and from a mailed questionnaire. A paper has been published in Am. J. Ind. Med. 19:729-737, 1991.

TYPE: Cohort
TERM: Chemical Exposure; Dusts; Occupation; Registry
SITE: Respiratory; Stomach
OCCU: Paper and Pulp Workers
REGI: Sweden (Swe)
TIME: 1986 - 1994

744 **Torén, K.Ö.** 05176
Univ. of Gothenburg, Sahlgren Hosp., Dept. of Occupational Medicine, St Sigfridsgatan 85, 412 66 Gothenburg, Sweden (Tel.: +46 31 830615)
COLL: Hagberg, S.; Nilsson, T.; Persson, B.; Westberg, H.; Wingren, G.B.

Cancer Incidence and Mortality among Workers in the Swedish Pulp Industry

The aim of the study is to investigate if pulp mill workers have an increased risk of cancer, and lymphatic systems. The study is designed as a cohort study conducted in the south of Sweden on sulphate mills and as a case-referent study conducted in the north of Sweden on sulphate and mechanical pulp mills. Cohort members will be identified from the personnel files of the pulp mills. Mortality and cancer morbidity will be collected from the National Cancer Registry and the National Mortality Registry. The subjects in the case-referent study will be collected from the Local Registers of Death and Burial. The exposures in both studies will be estimated according to the personnel files in the pulp mills. The size of the cohort study will be about 50,000 person-years. The case-referent study will include about 100 cases of lung and stomach cancers and about 30 cases of tumours from the blood and lymphatic system.

TYPE: Case-Control; Cohort
TERM: Dusts; Occupation; Registry; Wood
SITE: Haemopoietic; Lung; Lymphoma; Stomach
CHEM: Sulphur Dioxide; Terpenes
OCCU: Paper and Pulp Workers
REGI: Sweden (Swe)
TIME: 1990 - 1994

SWEDEN

Huddinge

745 **Bistoletti, P.** 04898
Huddinge Univ. Hosp., Karolinska Inst., Dept. of Obstetrics and Gynaecology, 141 86 Huddinge, Sweden (Tel.: +46 8 7461000)
COLL: Dillner, L.; Dillner, J.; Elfgren, K.

Serum Antibodies to Synthetic E2, E4, E7 Peptides from HPV 16

The primary objective of this investigation is to study HPV antibody responses in patients with cervical neoplasia, in pregnant women and in normal controls. The serum antibody levels after treatment will be measured. We will also investigate if any HPV antibody of the IgG group can serve as a tumour marker in cervical cancer. This is a case-control study, with several samples taken from the same patient. It is expected to study 100 cases and 100 controls. Information on sexually transmitted diseases will be obtained.

TYPE: Case-Control
TERM: HPV; Sero-Epidemiology; Sexually Transmitted Diseases; Tumour Markers
SITE: Uterus (Cervix)
REGI: Sweden (Swe)
TIME: 1989 - 1995

746 **Lambert, B.** 04544
Karolinska Inst. CNT/NOVUM, 141 57 Huddinge, Sweden (Tel.: +46 8 6089254; Fax: 6081501)
COLL: Einhorn, N.

Chromosomal Aberrations and Second Malignancy in Melphalan Treated Ovarian Carcinoma Patients

The aim of the study is to evaluate the effect of cytostatic treatment with regard to the persistence of chromosomal aberrations in peripheral T-lymphocytes and the development of second malignancy. A cohort of 50 melphalan-treated ovarian carcinoma patients has been studied for more than ten years after cessation of chemotherapy. Repeated samples for cytogenetic investigations have been taken, and the cohort is regularly followed up with regard to the occurrence of second malignancy. A paper has been published in Mutat. Res. 210:353-358, 1989.

TYPE: Cohort
TERM: Chemotherapy; Chromosome Effects; Drugs; Lymphocytes
SITE: All Sites
CHEM: Melphalan
TIME: 1977 - 1994

Linköping

747 **Arbman, G.** 04161
University Hospital, Dept. of Surgery, 581 85 Linköping, Sweden (Tel.: +46 13 191000)
COLL: Axelson, O.; Sjödahl, R.; Nilsson, E.; Fredriksson, M.; Eriksson, A.B.

Occupational Risk Factors for Colon Cancer

In view of findings in earlier studies of colonic cancer in Sweden, this study has been set up to study the risk of asbestos exposure, various chemicals and anti-hypertensive drugs. Cases of colonic cancer will be collected from patient files of the hospitals in the county of Östergöland. One set of referents will be drawn from surgical patients excluding patients with malignant diseases. Another set of referents will be drawn from the population register of the area. About 200 cases and two referent groups of individuals will be collected and included in the material. Information about occupational exposure to a number of agents, anti-hypertensive drugs, diet, smoking and heridity will be obtained through questionnaires.

TYPE: Case-Control
TERM: Chemical Exposure; Diet; Drugs; Heredity; Occupation; Tobacco (Smoking)
SITE: Colon; Rectum
TIME: 1986 - 1994

Lund

748 **Albin, M.P.** 04814

Lund University, Dept. of Occupational & Environmental Medicine, 221 85 Lund, Sweden (Tel.: +46 46 173185; Fax: 143702 ; Tlx: 32764 scanlu)

COLL: Jakobsson, K.M.; Attewell, R.; Welinder, H.; Johansson, L.G.

Mortality and Cancer Morbidity in Cohorts of Asbestos-Cement Workers

Total and cause-specific mortality and cancer morbidity are being assessed in a cohort of 1,929 male asbestos cement workers with an estimated median exposure of 1.2 fibres per ml. A local cohort of 1,233 industrial workers not exposed to asbesos is used for comparison. A nested case-control study will be carried out using referents from the exposed cohort. Dose-response relationships for lung cancer, mesothelioma, stomach cancer and colo-rectal cancer will be analysed. Data from the follow-up until 1986 are being analysed.

TYPE: Case-Control; Cohort
TERM: Dose-Response; Dusts; Occupation; Registry
SITE: Colon; Lung; Mesothelioma; Rectum; Stomach
CHEM: Asbestos
OCCU: Asbestos Workers
REGI: Lund (Swe)
TIME: 1982 - 1994

749 **Hagmar, L.E.** 05057

University Hospital Dept. of Occupational and Environmental Medicine, 221 85 Lund, Sweden (Tel.: +46 46 173171; Fax: 46143702; Tlx: 32764 SCANLU)

COLL: Svensson, B.G.; Möller, T.R.

Cancer Incidence in Cohorts of Swedish Fishermen

In Sweden, the main route of exposure to persistent polychlorinated polyaromatic compounds, such as polychlorinated dibenzo-p-dioxins and furans (PCDD/Fs) and PCBs, is a high intake of fatty fish from the Baltic Sea. Swedish fishermen have a high dietary intake of fish and are therefore a suitable group in which the carcinogenic risk can be evaluated. Preliminary data showed an increased risk for skin cancer, lip cancer, stomach cancer and possibly for multiple myeloma among 1,360 fishermen from the southeast coast of Sweden (Scand. J. Work Environ. Health 18:217-224, 1992). The total cohort comprises about 13,000 subjects and analyses of cancer incidence are now underway. As there may be sex-related differences in the biological sensitivity to these xenobiotics, a cohort of approximately 9,000 fishermen's wives will be established. So far dietary interviews has been performed in 500 subjects (exposed and referents from regional populations), and blood sampling has been done in 240 subjects. Congener specific analyses of PCDD/Fs and PCBs in pooled blood samples will be performed. Cohort based case-control studies will be performed for certain cancers in order to decrease the exposure (dietary) misclassifiction and to account for confounding factors.

TYPE: Case-Control; Cohort
TERM: Alcohol; BMB; Blood; Diet; Occupation; Registry; Tobacco (Smoking)
SITE: All Sites
CHEM: Dioxins; Furans; PCB
OCCU: Fishermen
REGI: Sweden (Swe)
TIME: 1988 - 1994

750 **Hagmar, L.E.** 05058

University Hospital Dept. of Occupational and Environmental Medicine, 221 85 Lund, Sweden (Tel.: +46 46 173171; Fax: 46143702; Tlx: 32764 SCANLU)

COLL: Gerhardsson, L.; Skerfving, S.; Schütz, A.

Cancer Incidence among Lead-Exposed Smelter Workers

This study aims to test the hypothesis of the carcinogenic potential of lead. A cohort of 848 lead-exposed workers from a secondary lead smelter, employed from 1942 and onwards is established. Since 1968, blood lead is analysed several times a year for each employee. These data will be used in dose-response calculations.

TYPE: Cohort
TERM: BMB; Blood; Dose-Response; Metals; Occupation; Registry
SITE: All Sites
CHEM: Lead
OCCU: Smelters, Lead
REGI: Lund (Swe)
TIME: 1991 - 1994

751 **Hagmar, L.E.** 05059
University Hospital Dept. of Occupational and Environmental Medicine, 221 85 Lund, Sweden (Tel.: +46 46 173171; Fax: 46143702; Tlx: 32764 SCANLU)
COLL: Welinder, H.

Cancer Incidence in a Cohort of Workers from the Polyurethane Foam Manufacturing Industry
The aim of the study is to determine whether workers exposed to toluene diisocyanate (TDI) or 4,4'-diphenylmethane diisocyanate (MDI) have increased cancer incidence. TDI is carcinogenic in rodents and both MDI and TDI are mutagenic. These substances are widely used in industry, but no data are available on their carcinogenicity to humans. A cohort of about 8,000 workers from 11 Swedish polyurethane foam manufacturing plants, employed from 1960 onwards, has been identified. Cohort analyses (Br. J. Med. 50: , 1993), as well as cohort based case-referent studies (Br. J. Med. 50: , 1993) have been performed. Follow-up continues.

TYPE: Case-Control; Cohort
TERM: Occupation; Plastics; Registry
SITE: All Sites
CHEM: Diisocyanates
OCCU: Plastics Workers
REGI: Sweden (Swe)
TIME: 1990 - 1994

752 **Hagmar, L.E.** 05060
University Hospital Dept. of Occupational and Environmental Medicine, 221 85 Lund, Sweden (Tel.: +46 46 173171; Fax: 46143702; Tlx: 32764 SCANLU)
COLL: Welinder, H.

Cancer Incidence in a Cohort of Workers Exposed to Ethylene Oxide
This is a cohort study of cancer incidence among 2,170 ethylene oxide exposed workers, employed for at least one year since 1972, in two plants producing disposable medical equipment. Present results show that the levels of hydroxyethyl adducts to N-terminal valine in haemoglobin fitted well with the values estimated for airborne exposure to ethylene oxide. There was no overall increase of cancer in the cohort during the period 1964-1985. No cases of leukaemia were found, but other haematopopietic and lymphatic tumours (SIR 1.54, 95% CI 0.32-4.5). Further follow-up, including five more years of observation, will be performed.

TYPE: Cohort
TERM: Dose-Response; Occupation; Registry
SITE: All Sites; Leukaemia; Lymphoma
CHEM: Ethylene Oxide
REGI: Sweden (Swe)
TIME: 1988 - 1995

753 **Hagmar, L.E.** 05061
University Hospital Dept. of Occupational and Environmental Medicine, 221 85 Lund, Sweden (Tel.: +46 46 173171; Fax: 46143702; Tlx: 32764 SCANLU)
COLL: Brögger, A.; Hansteen, I.L.; Heim, S.; Högstedt, B.; Knudsen, L.; Lambert, B.; Linnainmaa, K.; Mitelman, F.; Nordenson, I.; Reuterwall, C.; Salomaa, S.; Skerfving, S.; Sorsa, M.I.

Inter-Nordic Prospective Study on Cytogenetic Endpoints and Cancer Risk
To investigate whether high rates of chromosomal aberrations, SCE or micronuclei in peripheral lymphocytes indicate an increased risk for subsequent cancer, a prospective cohort study of 2,969 subjects cytogenetically examined between 1970 and 1988 in any of eight Nordic laboratories, was initiated. To reduce the effects of inter-laboratory variation, the results of the three cytogenetic endpoints were trichotomized for each laboratory (low, medium, high). In the first follow-up there was a positive

association between level of chromosome aberrations and subsequent cancer risk, which almost reached statistical significance (Cancer Genet. Cytogenet. 45:85-92. 1990). However, the follow-up period used is still too short to allow firm conclusions. Further follow-ups will be performed every five years.

TYPE: Cohort; Molecular Epidemiology
TERM: Chromosome Effects; Lymphocytes; Micronuclei; SCE
SITE: All Sites
LOCA: Finland; Norway; Sweden
TIME: 1988 - 2005

754 **Hagmar, L.E.** 05062
University Hospital Dept. of Occupational and Environmental Medicine, 221 85 Lund, Sweden (Tel.: +46 46 173171; Fax: 46143702; Tlx: 32764 SCANLU)
COLL: Åkesson, B.; Möller, T.R.

Cancer Incidence in PVC-Processing Workers Exposed to Low Levels of Vinyl Chloride Monomer, Asbestos, and Plasticizers

A significant increase in total cancer incidence, and especially in respiratory cancers, was found in a cohort of 2,031 male workers at a PVC processing plant, employed for at least three months during the period 1945-1980 (Am. J. Ind. Med. 17:553-565, 1990). However, no significant exposure-response associations between exposure estimates for PVC, asbestos, and plasticizers and cancer incidence was found. In 1993, when five more years of observation have been added, a follow-up of the cohort will be performed.

TYPE: Cohort
TERM: Dusts; Occupation; Plastics; Registry
SITE: All Sites
CHEM: Asbestos; PVC; Vinyl Chloride
OCCU: Plastics Workers
REGI: Sweden (Swe)
TIME: 1986 - 1994

755 **Hagmar, L.E.** 05063
University Hospital Dept. of Occupational and Environmental Medicine, 221 85 Lund, Sweden (Tel.: +46 46 173171; Fax: 46143702; Tlx: 32764 SCANLU)
COLL: Schütz, A.

Cancer Incidence among Leather Tanners in Three Swedish Factories

The general aim of this study is to investigate the association between cancer mortality and exposure to chemical substances in tannery factories. The study includes a cohort of about 3,500 workers from three Swedish tanneries. The cohort analysis will be completed with a nested case-referent study within the cohort, evaluating specific chemical exposures for cases and referents.

TYPE: Case-Control; Cohort
TERM: Chemical Exposure; Leather; Occupation; Pesticides; Plastics
SITE: All Sites
CHEM: Chlorophenols; Chromium; Formaldehyde
OCCU: Leather Workers
TIME: 1990 - 1994

756 **Hagmar, L.E.** 05064
University Hospital Dept. of Occupational and Environmental Medicine, 221 85 Lund, Sweden (Tel.: +46 46 173171; Fax: 46143702; Tlx: 32764 SCANLU)
COLL: Bellander, T.; Möller, T.R.

Cancer Morbidity in Nitrate Fertilizer Workers

In a cohort of 2,131 workers employed in a fertilizer manufacturing factory for at least three months from 1963 (when nitrate fertilizers were first manufactured) to 1985, no increased incidence of stomach cancer or lung cancer was found (Int. Arch. Occup. Environ. Health 63:63-67, 1991). The cohort will now be augmented with about 3,200 workers from the other fertilizer manufacturing plant in Sweden. In this plant, nitrate fertilizers have been manufactured since the 1930's.

TYPE: Cohort
TERM: Fertilizers; Occupation; Registry
SITE: Lung; Stomach
CHEM: Nitrates
OCCU: Fertilizer Workers
REGI: Sweden (Swe)
TIME: 1986 - 1994

757 **Hagmar, L.E.** 05065
University Hospital Dept. of Occupational and Environmental Medicine, 221 85 Lund, Sweden (Tel.: +46 46 173171; Fax: 46143702; Tlx: 32764 SCANLU)

Cancer Incidence in a Cohort of Piperazine-Exposed Workers in a Chemical Plant

It has previously been shown that occupational exposure to piperazine dust causes endogenous formation of N-mononitrosopiperazine which is a possible carcinogen (Int. Arch. Occup. Environ. Health 60:25-29, 1988). A previous follow-up of a cohort of 664 workers, mainly exposed to piperazine, and employed during the period 1942-1979, showed a significant increase of malignant lymphomas and myelomas (Scand. J. Work Environ. Health 12:545-551, 1986). A nested case-referent study within the cohort indicated a possible association with piperazine, although the observed numbers were too few to allow any firm conclusions. A further follow-up, including six more years of observation, will now be performed.

TYPE: Case-Control; Cohort
TERM: Chemical Exposure; Drugs; Dusts; Occupation; Plastics; Registry
SITE: All Sites; Lymphoma; Multiple Myeloma
CHEM: Ethylene Oxide; Piperazine; Urethane
OCCU: Chemical Industry Workers
REGI: Sweden (Swe)
TIME: 1982 - 1994

***758** **Jakobsson, K.M.** 05331
University Hospital Dept. of Occupational & Environmental Medicine, 221 85 Lund, Sweden
COLL: Albin, M.P.; Hagmar, L.E.

Site-Specific Colorectal Cancer Morbidity in Workers Exposed to Mineral Dust and Fibres

The object of this reanalysis of previously conducted cohort studies is to explore whether exposure to mineral fibres or mineral dust is a risk factor for cancer especially in the proximal colon. Cancer morbidity (Cancer Registry data 1958-1989) in cohorts of asbestos cement and cement workers is compared not only with the general population, but also with other cohorts of industrial workers and fishermen, as the incidence of cancer in colon and rectum is affected by factors associated with social class (e.g. physical activity, diet and body mass). A comparison between cancer morbidity and mortality data is also being performed. A total of 2,500 workers exposed to asbestos cement and cement, and 12,000 referents, are included.

TYPE: Cohort
TERM: Dusts; Occupation; Registry
SITE: Colon; Rectum
CHEM: Asbestos; Mineral Fibres
OCCU: Asbestos Workers; Cement Workers; Fishermen
REGI: Sweden (Swe)
TIME: 1992 - 1994

759 **Mitelman, F.** 02282
University Hospital Dept. of Clinical Genetics, 221 85 Lund, Sweden (Tel.: +46 46 173360; Fax: 131061)

Registry of Chromosome Aberrations in Cancer

Chromosome aberrations in human cancer and leukaemia are non-random and tend to cluster to a limited number of specific chromosomes. Experimental evidence indicates that the chromosomal pattern of malignant cells may be related to aetiological factors. The objective of this study is to gain information on this aspect of human cancer by relating the chromosomal pattern to: (1) exposure to known environmental carcinogens, (2) exposure to potential carcinogenic agents, and (3) geographical regions. The material is collected from published cases obtained by computer-based literature scans. By the end of 1992, approximately 18,000 neoplasms with chromosome aberrations identified by banding techniques studied in Africa, Asia, Europe, the USA and Japan were included in the registry. The data, including information on histopathological diagnosis, sex, age, previous neoplasms or other significant disorders, hereditary disorders, exposure to potential mutagenic and/or carcinogenic agents, chromosome preparation technique, and clinical findings, is computerized. The total organized material has been published in "Catalog of Chromosome Aberrations in Cancer" (4th edition), Wiley-Liss, Inc., New York, 1991.

TYPE: Methodology; Molecular Epidemiology
TERM: Chromosome Effects; Environmental Factors; Genetic Markers; Geographic Factors; Mutagen; Mutation, Somatic
SITE: All Sites
TIME: 1975 - 1994

Örebro

***760** **Andersson, S.O.** 05289
Örebro Medical Centre Hosp., Dept. of Urology, 701 85 Örebro, Sweden
COLL: Adami, H.O.

Epidemiology of Prostatic Cancer

This is a population-based case-control study of prostatic cancer investigating different risk factors, whith special reference to diet. 250 cases and 250 age-matched controls, randomly selected from a population registry, will be studied. All controls will undergo physical examination, including rectal palpation. Personal interviews will be held with a questionnaire focussing on diet, occupation, heredity, alcohol and tobacco consumption, medical history and education. Blood samples will be taken for hormonal assays (FSH, LH, testosterone, free testosterone, oestradiol, androstenedione)

TYPE: Case-Control
TERM: Alcohol; BMB; Diet; Education; Tobacco (Smoking)
SITE: Prostate
TIME: 1989 - 1994

***761** **Hardell, L.O.** 05405
Örebro Medical Centre, Dept. of Oncology, 701 85 Örebro, Sweden (Tel.: +46 19 151546; Fax: 183510)
COLL: Nordström, M.; Eriksson, M.H.

Case-Control Study on Non-Hodgkin's Lymphoma

The aim is to investigate occupations and occupational exposures as risk factors for non-Hodgkin's lymphoma. The study comprises 450 male cases and 900 population-based male controls. Data are assessed by mailed questionnaires and completed over the telephone, if necessary.

TYPE: Case-Control
TERM: Occupation; Registry
SITE: Non-Hodgkin's Lymphoma
REGI: Sweden (Swe)
TIME: 1992 - 1995

Örnsköldsvik

762 **Rutegård, J.N.** 04731
Örnsköldsvik Hosp., Dept. of Surgery, 891 89 Örnsköldsvik, Sweden (Tel.: +46 660 89000)
COLL: Stenling, R.; Roos, G.; Jonsson, B.

Ulcerative Colitis, DNA Abnormalities and Colorectal Cancer Risk

In an unselected population of patients with ulcerative colitis from a defined catchment area colorectal cancer risk and the reliability of a colonoscopic surveillance programme will be evaluated, and the frequency of DNA aneuploidy and its association with dysplasia and cancer studied. All patients (n = 127) from the catchment area for the period 1961-1983 are included. Surviving patients who have not been operated have been enrolled into a colonoscopic surveillance programme since 1979. DNA analysis has been performed since 1984. The programme now comprises about 110 patients. Patients are examined regularly and biopsy specimens are collected from the entire colon. Clinical data are recorded and the specimens examined histologically and by flow cytometric DNA analysis. Extra specimens for other cancer markers, such as Ag-NOR, PCNA and p53, are also obtained.

TYPE: Cohort
TERM: BMB; DNA; Screening; Tissue; Tumour Markers; Ulcerative Colitis
SITE: Colon; Rectum
REGI: Sweden (Swe)
TIME: 1979 - 1999

Solna

763 **Hogstedt, C.** 03522
National Inst. Occupational Health, Div. of Occupational Medicine, 171 84 Solna, Sweden (Tel.: +46 8 7309342; Tlx: 7309000)
COLL: Westerlund, B.

Cancer Incidence and Mortality among Lumberjacks with and without Exposure to Chlorinated Phenoxy Acid Mixtures

Payrolls for a forestry company in the middle of Sweden have been examined to identify foremen and lumberjacks who had been exposed for at least six days to mixtures of chlorinated phenoxy acids (2,4-D; 2, 4,5-T) and DDT. 141 exposed men with an average exposure of 30 days and 16 foremen with an average exposure of 176 days were identified, as well as 235 unexposed lumberjacks. The national incidence rates are being used to calculate gender, calendar year and five-year age-class specific expected numbers. The first report was published in 1979 and showed a significant increase of cancer among the foremen. The cohorts are now being followed for the period 1979-1985.

TYPE: Cohort
TERM: Chemical Exposure; Herbicides; Occupation; Pesticides
SITE: All Sites
CHEM: 2,4,5-T; 2,4-D; Phenoxy Acids; Silvex
OCCU: Forest Workers
TIME: 1984 - 1994

764 **Reuterwall, C.** 04730
National Inst. of Occupational Health, Dept. of Occupational Medicine, 171 84 Solna, Sweden (Tel.: +46 8 7309649; Fax: 7309860; Tlx: 15816 ARBSKY S)
COLL: Gustavsson, P.; Bellander, T.

Cancer Incidence and Mortality among Swedish Laboratory Technicians

Laboratory work within medical research may be associated with exposure to several physical and chemical agents with documented mutagenic or carcinogenic effects. Cancer incidence and mortality are being studied in a cohort of laboratory technicians employed at medical research laboratories (approx. 1,500 subjects) and clinical laboratories (approx. 1,000 subjects). Basic exposure information is collected from personnel records and vital status is determined from computerized population registries. Data on underlying causes of death are obtained from Statistics Sweden, data on cancer incidence from the Swedish Cancer Registry. Expected numbers of deaths and cancers are standardized for geographical region, sex, age group, and calendar year. This professional group is still rather young, and the first follow-up can only reveal relatively large increases of risk; later follow-ups, however, will have better statistical power. Depending on the results from the follow-up, an attempt to identify single risk factors will be made in a case-referent study, nested within the cohort. Technicians working at other laboratories might also be included in future studies.

TYPE: Cohort
TERM: Chemical Exposure; Drugs; Occupation; Radiation, Ionizing; Solvents
SITE: All Sites
OCCU: Laboratory Workers
REGI: Sweden (Swe)
TIME: 1989 - 1994

Stockholm

765 **Ahlbom, A.N.** 03705
Nat. Inst. of Environmental Medicine, Dept. of Epidemiology, Doktorsringen 18, Box 60208, 104 01 Stockholm, Sweden (Tel.: +46 8 7287470/6400; Fax: 313961)
COLL: Rodvall, Y.; Spännare, B.

Brain Tumours in Adults

The aetiology of different histopathological forms of brain tumours is basically unknown. Case reports, animal experiments and epidemiological studies, however, do support the theory that external factors are of importance. The objective of the present study is, within the framework of an international collaborative project (SEARCH), coordinated by the IARC, for each of the major histopathological types of brain tumours (1) to evaluate the importance of N-nitroso compounds; (2) to quantify the effects of certain well known risk-factors, and (3) to investigate the association with several other personal characteristics and exposures. Cases are all patients in the area covered by the Department of Neurosurgery at the Uppsala University Hospital with malignant primary brain tumour or meningioma during the period 1987 to 1989 between the ages of 25 and 74. 350 cases are be included in the study. One control is selected from the general population of the catchment area for each case. The data collection was performed by use of a questionnaire and subsequent telephone interviews. The analysis of data started in March 1991.

TYPE: Case-Control
TERM: Chemical Exposure; Environmental Factors; Lifestyle
SITE: Brain
CHEM: N-Nitroso Compounds
TIME: 1985 - 1994

766 **Ahlbom, A.N.** 04226
Nat. Inst. of Environmental Medicine, Dept. of Epidemiology, Doktorsringen 18, Box 60208, 104 01 Stockholm, Sweden (Tel.: +46 8 7287470/6400; Fax: 313961)
COLL: Feychting, M.

Extremely Low Frequency Electromagnetic Fields and Cancer Risk

Several independent groups have reviewed the epidemiological evidence of an association between exposure to 50 Hz electromagnetic fields and cancer and have concluded that while existing data do not warrant firm conclusions, the hypothesis of possible cancer risk should be followed up. The present study takes advantage of the Swedish census system to define a study base of those people who have lived within 300 metres of any of the 250 and 400 kV power lines in Sweden during 1960-1985. Within this study population a case-control study is being done on all cases of childhood cancer, leukaemia and brain tumours in adults, identified by record linkage to the Swedish Cancer Registry. Exposure is assessed in three different ways: (1) by direct measurements, (2) by estimation of the fields generated by the power lines, and (3) by use of distance between the power line and the homes in the study as an indirect exposure measure. Controls will be matched for sex, age, place and period of residence, and selected at random from the same study population.

TYPE: Case-Control
TERM: Electromagnetic Fields; Environmental Factors; Registry
SITE: Brain; Childhood Neoplasms; Leukaemia
REGI: Sweden (Swe)
TIME: 1987 - 1994

767 **Ahlbom, A.N.** 05042
Nat. Inst. of Environmental Medicine, Dept. of Epidemiology, Doktorsringen 18, Box 60208, 104 01 Stockholm, Sweden (Tel.: +46 8 7287470/6400; Fax: 313961)
COLL: Sasco, A.J.; Wennborg, H.

Cancer Risk in Biological Research Laboratories in Sweden

The aim of the study is to describe cancer incidence and cancer mortality among biological research laboratory workers in Sweden during the last 20 years. Another aim is to make comparisons between different laboratory occupations and, if possible, to connect any excess risk with specific kinds of laboratory work or exposure. The study will be conducted as a retrospective cohort study and will be included in an international multi-center study coordinated by IARC. Biomedical and biological research groups at seven Swedish research institutions will be included in the study. Comparisons will be made with the general population and with research institutions without biological activity. Personal data will be collected by questionnaire. Exposure assessment will be done at the research group level by interviews and questionnaires. The National Swedish Cancer Register will be used to identify the cases.

TYPE: Cohort
TERM: Occupation; Registry
SITE: All Sites
OCCU: Laboratory Workers
REGI: Sweden (Swe)
TIME: 1991 - 1994

***768 Bellander, T.** 05396
Karolinska Hospital, Dept. of Occupational Medicine, 104 01 Stockholm, Sweden (Tel.: +46 8 7293261; Fax: 334333)
COLL: Järup, L.; Elinder, C.G.; Axelson, O.; Bruhn, R.; Johansson, G.; Spång, G.

Long-term Health Effects of Occupational Exposure to Cadmium and Nickel

Since the middle of the 1940's health effects of occupational exposure to cadmium have been studied in a battery factory in southern Sweden. Earlier studies in this factory have revealed increased lung cancer risks, but no dose-response relationship. A possible reason for the lack of a dose-response relation is the rather crude exposure estimates that have been used so far. A major aim in the present study is therefore to evaluate available industrial hygiene data to get individual cumulative exposure estimates for both cadmium and nickel, the latter being a possible confounder in the study of lung cancer risks due to cadmium exposure in battery workers. The exposure assessment will be further enhanced by utilizing biological data (cadmium in blood and urine), in-vivo measurements (for some workers) and the toxico-kinetic model for cadmium. The refined exposure data will then be used in a re-analysis of previously collected data to study lung cancer risks due to cadmium exposure and possible modifying effects from nickel and smoking.

TYPE: Cohort
TERM: BMB; Blood; Metals; Occupation; Registry; Urine
SITE: Lung
CHEM: Cadmium; Nickel
OCCU: Battery Plant Workers
REGI: Sweden (Swe)
TIME: 1992 - 1995

769 Eklund, G. 04086
Karolinska Hosp., Radiumhemmet, Dept. of Cancer Epidemiology, 104 01 Stockholm, Sweden (Tel.: +46 8 7363036)
COLL: Tomenius, L.

Power Lines, 50-Hz Electromagnetic Fields and Childhood Tumours in Sweden

Wertheimer and Leeper published in the Am. J. Epidemiol. 109:273-284, 1979 an epidemiological study from the greater Denver area in Colorado, USA, that showed an excess of electrical wiring configurations suggestive of high current flow near the homes of children who had developed cancer. They proposed that the magnetic field from current in the power lines may directly or indirectly affect cancer incidence. In order to see if the relationship could be found in Sweden a similar study was undertaken in the county of Stockholm. The result showed (Tomenius, Bioelectromagnetics 7:191, 1986) that 200 kV wires and increased 50-Hz magnetic fields, respectively, were more frequent among dwellings of cases than of controls. In order to confirm the results from the statistical point of view and to be able to determine if there is a dose/response relationship between electromagnetic environment and the incidence of childhood tumours a more expanded study, including cases from more counties in Sweden, has begun. Visible electrical constructions will be registered and the 50-Hz magnetic field measured at the birth dwelling of tumour cases reported 1958-1980 to the Swedish Cancer Registry for individuals 0-18 years of age. For each tumour case two individuals matched according to age, sex and church district of birth will be used as controls. To begin with this will be accomplished in seven counties with 1,476 malignant tumours. If necessary more counties in Sweden will later be included. The municipalities where the dwellings are located will be asked to carry out the observations and measurements of the magnetic fields.

TYPE: Case-Control
TERM: Electromagnetic Fields; Environmental Factors; Registry
SITE: Childhood Neoplasms
REGI: Sweden (Swe)
TIME: 1982 - 1994

770 **Eklund, G.** 04299
Karolinska Hosp., Radiumhemmet, Dept. of Cancer Epidemiology, 104 01 Stockholm, Sweden (Tel.: +46 8 7363036)
COLL: Lindefors, B.M.; Meirik, O.; Rutqvist, L.E.

Relationship between Legal Abortions and Breast Cancer

Several retrospective epidemiological studies have claimed that early abortions in young women are followed by an increased frequency of breast cancer. This observed increased risk is contrasted to the decreased risk observed in association with a full term pregnancy at the same ages. The intention is to use the unique registrations of every legal abortion performed in Sweden for many years. By linkage of an appropriate part of the abortion register and the Swedish cancer register of women with breast cancer, it is expected that a number of cases great enough to indicate an increased frequency of at least 25% will be obtained. The applications for legal abortion from 1966 to 1974, which amount to 165,000, will be used. Women with breast cancer in the abortion cohort will be compared with women belonging to three different cohorts.

TYPE: Cohort
TERM: Abortion; Registry
SITE: Breast (F)
REGI: Sweden (Swe)
TIME: 1985 - 1994

771 **Gerhardsson, M.R.** 04172
National Inst. of Environmental Medicine Dept. of Epidemiology, Doktorsringen 18 Box 60208, 104 01 Stockholm, Sweden (Tel.: +46 8 236900; Tlx: 19530 KIENVS)
COLL: Norell, S.E.; Bylin, G.

Allergy-Related Diseases and Risk of Cancer

Several studies have suggested that allergy is negatively associated with cancer, some have reported a positive association and others have shown no association. The material has often been too small to make a division into cancer subsites. In this 19-year follow-up study of several 100,000 Swedish subjects, the cancer incidence for subjects with allergy-related diseases will be compared with a reference group. The information will be obtained from the in-patient care register, the cause-of-death register and the cancer register. The observation period will be from 1964-83.

TYPE: Cohort
TERM: Allergy; Registry
SITE: All Sites
REGI: Sweden (Swe)
TIME: 1987 - 1994

772 **Gustavsson, P.** 04747
Karolinska Hosp., Dept. of Occupational Medicine, 104 01 Stockholm, Sweden (Tel.: +46 8 7292000; Fax: 334333)
COLL: Bellander, T.; Salmonsson, S.

Cancer Risk among Graphite Electrode Workers

Mortality and cancer incidence is being investigated among 900 employees at a Swedish company manufacturing graphite electrodes. Exposure to PAH and respirable dust is under investigation. The cumulate exposures for the workers are estimated by occupational hygienists. The mortality of the cohort is compared to that of the general population, standardising for age, sex, calendar year and geographical region (county). Dose-response relationships are being investigated by internal comparisons within the cohort.

TYPE: Cohort
TERM: Dose-Response; Dusts; Occupation
SITE: All Sites; Lung
CHEM: PAH
OCCU: Electrode Manufacturers
TIME: 1989 - 1994

***773 Gustavsson, P.** 05324
Karolinska Hosp., Dept. of Occupational Medicine, 104 01 Stockholm, Sweden (Tel.: +46 8 7292000; Fax: 334333)
COLL: Jakobsson, R.; Rudengren, C.; Pershagen, G.; Norberg, S.; Nyberg, F.

Lung Cancer in Stockholm (LUCAS). A Case-Control Study of Occupational and Environmental Exposures in Stockholm County

The study is designed to investigate aetiological factors for lung cancer in Stockholm county. Occupational exposures, environmental air pollution from traffic, heating, and industry, as well as tobacco smoking and radon in buildings are considered. Cancer cases are identified from the regional cancer registry in Stockholm, and controls are selected from the general population in Stockholm. The study includes 650 cases and 1,300 controls, all males. Data on occupational history, residence and tobacco smoking are obtained by questionnaire and telephone interview with surviving cases or next-of-kin. Occupational exposure, radon exposure and exposure to general air pollution is coded from the questionnaire data. A model for calculation of the levels of general air pollution at different locations is developed.

TYPE: Case-Control
TERM: Air Pollution; Occupation; Registry; Tobacco (Smoking)
SITE: Lung
CHEM: Radon
REGI: Stockholm (Swe)
TIME: 1992 - 1996

***774 Gustavsson, P.** 05325
Karolinska Hosp., Dept. of Occupational Medicine, 104 01 Stockholm, Sweden (Tel.: +46 8 7292000; Fax: 334333)
COLL: Jakobsson, R.; Lundberg, I.S.

Lung Cancer Among Professional Drivers in Sweden

Several previous reports show an excess of lung cancer among professional drivers. This study is designed to investigate if there is an excess of lung cancer in different categories of professional drivers, and relate the findings to smoking habits. 95,000 Swedish bus, taxi and stock delivery drivers are being included in a historical cohort study. Registry information on occupation and cancer is utilized. Information on smoking habits is obtained from questionnaire data held by Statistics Sweden. The lung cancer risk is calculated with standardization for age, sex, geographic area and tobacco smoking, using all employed persons as a reference group.

TYPE: Cohort
TERM: Occupation; Registry; Tobacco (Smoking)
SITE: Lung
CHEM: Diesel Exhaust; Vehicle Exhaust
OCCU: Bus Drivers; Drivers; Taxi Drivers
REGI: Sweden (Swe)
TIME: 1993 - 1995

***775 Hall, P.** 05249
Karolinska Hosp., Radiumhemmet, Dept. of General Oncology, 104 01 Stockholm, Sweden (Tel.: +46 8 7293382; Fax: 348640)
COLL: Holm, L.E.; Lundell, G.; Mattsson, A.

Cancer Risk after Exposure to Iodine-131

The aim of this study is to evaluate the cancer risk among patients receiving 131-I. Three cohorts, consisting of (1) patients given diagnostic doses of 131-I (n=35,000); (2) patients receiving 131-I therapy for hyperthyroidism (n=10,500); and (3) patients with thyroid cancer (n=1,000), are being studied. Patient characteristics as well as administered activity of 131-I have been obtained from medical records. Organ doses for each individual have been calculated. The cohorts were matched with the Swedish Cancer Register, which was also used to calculate the expected number of cases on a person-year at risk basis.

TYPE: Cohort
TERM: Radiation, Ionizing; Registry
SITE: All Sites; Leukaemia; Thyroid
CHEM: Iodine
REGI: Sweden (Swe)
TIME: 1975

***776** **Hall, P.** 05250
Karolinska Hosp., Radiumhemmet, Dept. of General Oncology, 104 01 Stockholm, Sweden (Tel.: +46 8 7293382; Fax: 348640)
COLL: Holm, L.E.; Mattsson, A.

Cancer Risk after Exposure to Thorotrast

The aim of the study is to evaluate the cancer risk among patients receiving injections of the alpha-emitting x-ray contrast medium Thorotrast. A previously identified cohort of neurological patients receiving injections of Thorotrast for cerebral arteriography is now being re-established. The original cohort consists of approximately 1,100 patients who will now be traced through local parishes and registers using individual identification numbers. The Swedish Cancer Register will be used to determine the cancer incidence and to calculate the expected number of cancers.

TYPE: Cohort
TERM: Drugs; Metals; Radiation, Ionizing; Registry
SITE: All Sites
CHEM: Thorium; Thorotrast
REGI: Sweden (Swe)
TIME: 1992 - 1995

777 **Holm, L.E.** 03332
Karolinska Hosp., Radiumhemmet, Dept. of Cancer Prevention, 104 01 Stockholm, Sweden (Tel.: +46 8 7362266)
COLL: Fuerst, C.J.; Lundell, M.

Malignant Tumours and Leukaemia Following Radiotherapy for Haemangiomas in Childhood

The incidence of malignant tumours and leukaemia is being analysed in approximately 20,000 patients following radiotherapy for cavernous haemangiomas. The treatment was given between 1920 and 1959 and consisted mainly of application of radium needles on the skin or of X-rays. Approximately 17% of the patients did not receive any radiotherapy. More than 90% of the patients were under 15 years of age at the time of treatment and 88% were less than five years. The great majority were six months old. Dosimetry is being performed on a phantom corresponding to a six months old child. The population of patients will be matched with the Swedish Cause of Death Registry and the Swedish Cancer Registry.

TYPE: Cohort
TERM: Childhood; Radiation, Ionizing; Radiotherapy; Registry
SITE: All Sites; Leukaemia
REGI: Sweden (Swe)
TIME: 1983 - 1994

***778** **Lindblom, A.** 05429
Karolinska Hosp. Dept. of Clinical Genetics, 104 01 Stockholm 60, Sweden
COLL: Nordenskjöld, H.

Hereditary Non-Polyposis Colon Cancer

The aims of the study are to collect families with non-polyposis hereditary colon cancer and to obtain DNA from as many healthy and affected persons as possible to try to localize the gene responsible for this disease by linkage analysis. It will also be attempted to obtain tumours from affected individuals to study loss of heterozygosity, in an attempt to find candidate regions for genes involved in tumour development.

TYPE: Genetic Epidemiology
TERM: DNA; Linkage Analysis
SITE: Colon
TIME: 1993 - 1996

779 **Lundberg, I.S.** 04329
Karolinska Hosp. Dept. of Occupational Medicine, 104 01 Stockholm, Sweden
COLL: Bellander, T.

Mortality and Cancer Incidence among Styrene Exposed Workers

Styrene is a solvent with suggested carcinogenic properties in animal experiments. Some epidemiological studies of styrene exposed workers have shown excess deaths from leukaemia and lymphoma. Workers in the manufacture of glass reinforced polyester plastics have been heavily exposed to styrene. Increased frequencies of chromosomal aberrations, sister chromatid exchanges and micronuclei have been found in peripheral lymphocytes of such workers. This study examines the cancer incidence and mortality pattern of 1,500 to 2,000 workers employed in glass reinforced plastics manufacturing from the 1950's until today. The cohort contains all workers employed in around 30 factories scattered over Sweden. The observed figures from different causes of death and cancer sites will be compared with expected figures generated from the Swedish general population.

TYPE: Cohort
TERM: Chemical Exposure; Occupation; Plastics
SITE: All Sites; Leukaemia; Lymphoma
CHEM: Styrene
OCCU: Plastics Workers
TIME: 1986 - 1994

***780** **Nilsson, B.R.** 05547
Karolinska Hospital Unit of Cancer Epidemiology, Solnavägen 1, 104 01 Stockholm, Sweden (Tel.: +46 8 7294893; Fax: 326113)
COLL: Rotstein, S.; Gustavson-Kadaka, S.; Hakulinen, T.R.; Aareleid, T.; Rahu, M.A.

Cancer Incidence and Survival in Migrants to Sweden

Cancer incidence in approximately 1 million migrants to Sweden from 180 different countries will be compared with that in the Swedish population and with the incidence of the residents in the countries of origin. The study period cover the years 1968-1988. Relative survival of cancer patients of different sites will be compared as well. Data for the period 1974-1988 in 17,149 Estonian-born migrants are now being analysed.

TYPE: Cohort; Incidence
TERM: Geographic Factors; Lifestyle; Migrants; Race; Registry; Survival
SITE: All Sites
REGI: Sweden (Swe)
TIME: 1989 - 1995

***781** **Nise, G.** 05495
Karolinska Hosp., Dept. of Occupational Health, 104 01 Stockholm, Sweden (Tel.: +46 8 7292771; Fax: 334333)
COLL: Hogstedt, C.; Törnqvist, M.

Mortality and Cancer Incidence among Ethylene Oxide Exposed Workers

An increased cancer incidence was reported in a cohort of 709 ethylene oxide exposed workers (Hogstedt, 1988). This cohort will now be enlarged to include all subjects exposed during the 1970s and 1980s. Data on production and work tasks over time will be collected and a retrospective exposure assessment performed. Expected mortality and cancer morbidity rates will be calculated using calendar-year, cause, gender, and 5-year age-group specific rates from national registers. Exposure-response relationships will be assessed using individual cumulative exposure estimates for ethylene oxide. The cohort will now be followed until 1992 through the Swedish registries of mortality and cancer incidence. Further, the possibility of using specific haemoglobin adducts to estimate actual exposure will be studied.

TYPE: Cohort
TERM: Chemical Exposure; Occupation; Protein Binding; Registry
SITE: All Sites; Leukaemia; Lymphoma
CHEM: Ethylene Oxide
OCCU: Chemical Industry Workers
REGI: Sweden (Swe)
TIME: 1992 - 1995

782 Pershagen, G. 04478
Karolinska Inst., Inst. of Environmental Medicine, Box 60208, 104 01 Stockholm, Sweden (Tel.: +46 8 7287460; Fax: 313961)
COLL: Axelson, O.; Damber, L.

Domestic Radon Exposure and Lung Cancer in Sweden

Radon in dwellings is the major source of exposure in ionising radiation for the Swedish population and a few studies suggest that it is of aetiological importance for lung cancer. The objective of this study is to make a quantitative assessment of dose-response relationship. It is based on 1,500 cases lung cancer and 3,000 controls aged 35-74 from the Swedish cancer registry 1980-1984. There will be radon measurements and inspections of all houses in which study subjects lived in for two or more years between 1947 and 3 years prior to the end of follow-up. A questionnaire will also be submitted to study subjects or next-of-kin on smoking, passive smoking, occupations etc. The analysis will focus on assessment of dose-response relationship as well as on interactions between different exposures.

TYPE: Case-Control
TERM: Dose-Response; Occupation; Passive Smoking; Radiation, Ionizing; Registry; Tobacco (Smoking)
SITE: Lung
CHEM: Radon
REGI: Sweden (Swe)
TIME: 1987 - 1994

783 Reizenstein, P. 04661
Karolinska Hosp., Dept. of Haematology, Solnavägen, 104 01 Stockholm, Sweden (Tel.: +46 8 7294112; Fax: 313665)
COLL: Ringertz, H.; Beckman, M.

Radiation Exposure During Pregnancy and Childhood Malignancy

The purpose is to see whether the offspring of the 10% of Swedish mothers subjected to X-ray pelvimetry or the higher percentage subjected to other X-rays have an increased risk for malignancy. This study will exploit a computer register for the period 1971-1978 comprising a complete record of some 4 million X-ray examinations in a population base of approximately one million. About 13,000 pelvimetries and an equal number of routine chest X-rays in pregnant women will be matched against the birth register and the cancer register to examine the incidence of childhood malignancy in relation to pre-natal radiation exposure. A data set of more than twice this size on babies born to mothers not subjected to X-rays will serve as an unexposed comparison group. Simultaneously it is proposed to study children to fathers subjected to diagnostic X-ray.

TYPE: Cohort
TERM: Intra-Uterine Exposure; Pregnancy; Radiation, Ionizing; Registry
SITE: Childhood Neoplasms
REGI: Sweden (Swe)
TIME: 1988 - 1994

784 Rutqvist, L.E. 04970
Karolinska Hosp., Radiumhemmet, Dept. of Oncology, 104 01 Stockholm 60, Sweden (Tel.: +46 8 7292980; Fax: 348640)
COLL: Lewin, F.

Squamous Cancer of the Mouth, Pharynx, Larynx and Oesophagus

This is a population-based case-control study in two defined geographical regions with a population of 3 million. All cases of squamous cell cancer among males aged 40-79 years in the populations are included. A random sample of males in the general population serve as a control group. Data are collected through personal interviews (smoking and drinking habits, non-smoking tobacco use, dietary and occupational history, etc.). A total of 600 cases and 600 controls will be included.

TYPE: Case-Control
TERM: Alcohol; Diet; Occupation; Registry; Tobacco (Chewing); Tobacco (Smoking); Tobacco (Snuff)
SITE: Larynx; Oesophagus; Oral Cavity; Pharynx
REGI: Lund (Swe); Stockholm (Swe)
TIME: 1988 - 1994

***785 Svartengren, M.U.** 05368
Huddinge Hospital, Inst. of Occupational Medicine, Rehabgatan 62, 141 86 Stockholm, Sweden (Tel.: +46 8 7465362; Fax: 7113046)
COLL: Hogstedt, C.; Bellander, T.; Elinder, C.G.

Health Effects from Work in Hard Metal Industry

The aim of the study is to assess whether exposure to cobalt in hard metal industry can give rise to lung cancer. A cohort of more than 3,000 men, employed for at least 3 months, has been established, and will be followed up to 1990. Exposure over time will be estimated for all individuals based on information from cobalt in the air and urine, type of work and time period.

TYPE: Cohort
TERM: Metals; Occupation; Registry
SITE: Lung
CHEM: Cobalt
OCCU: Metal Workers
REGI: Sweden (Swe)
TIME: 1992 - 1995

786 Törnberg, S.A. 04861
Karolinska Hosp., Radiumhemmet, Dept. of Oncology and Cancer Epidemiology, 104 01 Stockholm 60, Sweden (Tel.: +46 8 7293194; Fax: 348640)
COLL: Carstensen, J.M.; Holm, L.E.

Gynaecological Cancer and Serum Cholesterol

A cohort of 46,000 women participated in a general health screening in 1963-1965. Examinations included serum cholesterol and serum beta-lipoprotein levels as well as height and weight. The data were matched with the Swedish Cancer Register and Cause of Death Register until 1987. The purposes of the present studies are to analyse the relationships between the above mentioned parameters and subsequent risk of cancers of the ovary, endometrium and cervix uteri as well as of breast cancer survival. Earlier studies of the same cohort on breast cancer, colorectal cancer, stomach cancer and total cancer incidence have been published, e.g. in JNCI 81:1917-1922, 1989.

TYPE: Cohort
TERM: Cholesterol; Physical Factors; Registry; Survival
SITE: Breast (F); Ovary; Uterus (Cervix); Uterus (Corpus)
REGI: Sweden (Swe)
TIME: 1990 - 1994

787 Wiklund, K. 04375
Karolinska Hosp., Radiumhemmet, Dept. of Cancer Epidemiology, 104 01 Stockholm, Sweden (Tel.: +46 8 7292398; Fax: 326113)
COLL: Holm, L.E.; Dich, J.; Eklund, G.

Drinking Water and Cancer

The aim is to investigate whether there is an increased risk of lung, colon, rectum and urinary bladder cancers with chlorine in the drinking water, and of stomach, urinary bladder, colon and rectum cancers with nitrate and nitrite in the drinking water. Associations between potential cancer risk for all sites and different qualities of the drinking water mutagenicity, method of cleaning, alkalinity, water hardness, aluminium, copper, chromium, fluoride, creosote, permanganate, number of coliform bacteria, conductivity, ammonium, manganese, alkalinity, nitrates, nitrites, phosphorus, carbonic acid and siliconic acid will also be examined. Cancer risks for different municipalities in Sweden will be compared with the drinking water quality. Information on water quality for the major supplies will be derived for the period since 1961. Cancer morbidity data will be obtained from the nationwide Swedish Cancer Registry, which covers almost all newly diagnosed cancer cases.

TYPE: Correlation
TERM: Registry; Water
SITE: All Sites; Bladder; Colon; Lung; Rectum; Stomach
CHEM: Chlorine; Nitrates; Nitrites
REGI: Sweden (Swe)
TIME: 1987 - 1995

Umeå

788 **Emmelin, A.** 04470
Univ. of Umeå, Dept. of Epidemiology and Health Care Research, 901 85 Umeå, Sweden (Tel.: +46 90 101213; Fax: 138977 ; Tlx: 54005)
COLL: Nyström, L.; Wall, S.

Swedish Dock Workers and Lung Cancer

For several years dock workers have suspected that occupational exposures arising from loading and unloading work by truck may be carcinogenic. In 1979 The National Swedish Board of Occupational Safety initiated a retrospective cohort study of all permanently employed dockers in Sweden who had been first employed before 1974 and for a continuous period of at least six months. The present analysis of the cohort, followed from 1 January 1961 until 1 January 1981 (6,076 workers; 97,076 person-years) showed an increasing excess of lung cancer incidence and mortality. To further analyse whether these trends could be related to the exposure from diesel exhaust fumes a case-referent study within the cohort has been performed of all cases dying of primary lung cancer between 1961 and 1982. Four referents matched by age and port were selected for each case. A questionnaire about residence, work history, special exposure in the port and smoking habits was sent to alive controls and relatives of deceased cases and controls. Work history within the port and smoking habits were validated by interviews with work superintendents and/or safety delegates. Information about the use of trucks and fuel consumption was also collected to generate an exposure index.

TYPE: Case-Control
TERM: Occupation; Tobacco (Smoking)
SITE: Lung
CHEM: Diesel Exhaust
OCCU: Dockers
TIME: 1986 - 1994

***789** **Lenner, P.H.** 05338
University Hospital of Umeå Dept. of General Oncology, 901 85 Umeå, Sweden (Tel.: +46 90 101990; Fax: 127464)
COLL: Dillner, J.; Hallmans, G.; Stendahl, U.; Wiklund, F.; Lehtinen, M.

Population-Based Sero-Epidemiological Study of Cervical Cancer

The primary aim of this study is to further elucidate the role of various microbiological agents, including HPV, in the aetiology of cervical cancer. Sera from 94 consecutive incident cervical cancer patients were matched to 188 age- and sex-matched controls, derived from a population-based serum bank including sera from the population in Vasterbotten and Norrbotten in northern Sweden. The sera were analysed for IgG- and IgA- antibodies against a panel of 12 antigens derived from HPV types 6, 11, 16 or 18 as well as against HSV type I and II, Chlamydia Trachomatis, CMV, EBV and Bovine Papilloma virus. In a preliminary analysis increased relative risks were found for IgG to various epitopes on HPV 16 or 18. An increased risk was also found for IgG to Chlamydia Irachomatis.

TYPE: Case-Control
TERM: BMB; CMV; HPV; HSV; Sero-Epidemiology; Serum
SITE: Uterus (Cervix)
TIME: 1985 - 1994

790 **Lundström, N.G.** 04993

Umeå University Dept. of Environmental Medicine, 901 87 Umeå, Sweden (Tel.: +46 90 101700; Fax: 901196)

COLL: Nordberg, G.F.; Gerhardsson, L.; Jin, T.

Lead and Cancer

Blood lead measurements among occupationally exposed people have been reported to the Swedish National Board of Occupational Health and Safety (NBOSH) since 1971. In this register 10,245 persons were registered until July 1990, when registration ended. The majority (9,000 persons) provided up to 16 samples during the period 1977-1990. Some 87,257 samples are registered. By using data on lead-exposed workers at the Roennskaer smelter and data from the Swedish lead register at the NBOHS, an investigation will be made of a possible association between lead exposure in Sweden and the frequency of cancer and other causes of death. The study concerning cancer is coordinated with a larger international multi-centre study at IARC. The study can be divided into two parts. The first is a follow-up of an earlier study concerning lead-exposed workers at the Roennskaer smelter, so including a larger number of lead-exposed persons, which improves the study capacity to detect possible correlations, for instance with the lead level in blood. The second part concerns the register at the NBOHS. Both cohorts will be compared with the Swedish Cancer Registry. Results will be standardized and compared with the causes of death for Sweden and also with a reference population of industrial workers without any specific risk factors.

TYPE: Cohort
TERM: Metals; Occupation; Registry
SITE: Kidney; Lung
CHEM: Lead
OCCU: Smelters, Lead
REGI: Sweden (Swe)
TIME: 1991 - 1994

791 **Nyström, L.** 04637

Univ. of Umeå, Dept. of Epidemiology and Health Care Research, 901 85 Umeå, Sweden (Tel.: +46 90 101213; Fax: 138977 ; Tlx: 54005)

COLL: Wall, S.; Larsson, L.G.; Andersson, I.; Bjurstam, N.; Fagerberg, G.; Frisell, J.; Tabar, L.; Rutqvist, L.E.

Breast Cancer Screening with Mammography: An Overview of the Swedish Randomised Trials

This overview, based on 282,777 women followed for 5-13 years in the Swedish randomised trials in Malmö, Kopparberg, Östergötland, Stockholm and Gothenburg, showed an overall 24% significant (95% CI 12-34%) reduction in breast cancer mortality among those invited to mammography screening compared with those who were not invited. To avoid the potentisl risk of differential misclassification causes of death were assessed by an independent endpoint committee after a blinded review of all the deceased breast cancer cases. There was a consistent risk reduction associated with screening in all studies, although the point estimate of the relative risk for all ages varied non-significantly between 0.68 and 0.84. The cumulative breast cancer mortality by time since randomisation was estimated at 1.3/1000 within six years in the invited group, compared with 1.6 in the control group. The corresponding figures after nine and twelve years were 2.6/3.3 and 3.9/5.1 respectively. The largest reduction in breast cancer mortality, 29%, was seen among women aged 50 to 69 at randomisation. Among women in the age group 40-49 when invited to screening, there was a non-significant 13% reduction in the estimated breast cancer mortality. In this age group the cumulative mortality was similar in the invited group and the control group during the first eight years of follow-up. After eight years there was a difference in favour of the invited women. There was no evident of any detrimental effect of screening in terms of breast cancer mortality in any age group. Among women aged 70-74 screening seemed to have had only a marginal impact. A paper has been published in The Lancet 341:973-978, 1993.

TYPE: Cohort; Intervention
TERM: Mammography; Registry
SITE: Breast (F)
REGI: Sweden (Swe)
TIME: 1989 - 1994

792 **Sandström, A.** 02719

Univ. of Umeå, Dept. of Epidemiology and Public Health, 901 85 Umeå, Sweden (Tel.: +46 90 102731; Fax: 138977)

COLL: Dahlgren, L.; Lönnberg, G.; Wall, S.

Occurrence of Cancer at a Smelter

Earlier studies of 3,915 workers first employed at a smelter during 1928-1966 showed an excess mortality up to 1977 of 11% compared to Swedish males, due to cancer (mostly lung) and circulatory diseases. Follow-up of this cohort for a further six years showed a decreasing trend in lung cancer incidence. The cohort has now been extended with workers employed later (up to 1980) and also with white-collar workers to further study mortality of different groups employed at the smelter. A questionnaire study comprising about 7,000 persons employed at the smelter 1928-1979 will add information on smoking habits to the entire cohort and will permit a deeper analysis of the interaction of smelter work and smoking habits. In combination with more in depth interviews with a small number of persons, strategically chosen from the cohort, it is aimed to elucidate the antagonism between the demand for a safe work environment and the demand for a well-paid employment. Papers have been published in Excerpt. Med. Int. Congr. Series 829:155-158, 1988 and in Br. J. Ind. Med. 46:82-89, 1989.

TYPE: Cohort
TERM: Occupation; Tobacco (Smoking)
SITE: All Sites; Lung
CHEM: Arsenic
OCCU: Smelters, Copper
TIME: 1987 - 1994

Uppsala

793 **Adami, H.O.** 04414

University Hosp., Cancer Epidemiology Unit, 751 85 Uppsala, Sweden (Tel.: +46 18 665045; Fax: 503431)

COLL: Andersson, S.O.; Johansson, J.E.

Risk Factors for Prostate Cancer in Örebro County

The aim is to analyse risk factors for prostate cancer with special emphasis on dietary factors. During a three year period, all patients under 80 years diagnosed as having prostate cancer in a strictly defined geographical area (Örebro county in central Sweden), and an equal number of age-matched controls from the general population will be eligible for inclusion in the study. Personal interviews have been carried out with about 250 patients and 250 controls. Blood samples were obtained from all study subjects and stored.

TYPE: Case-Control
TERM: BMB; Diet; Hormones; Occupation; Serum
SITE: Prostate
TIME: 1989 - 1994

794 **Adami, H.O.** 04790

University Hosp., Cancer Epidemiology Unit, 751 85 Uppsala, Sweden (Tel.: +46 18 665045; Fax: 503431)

COLL: Hansson, L.E.; Nyrén, O.; Baron, J.A.; Hardell, L.O.; Bruce, Å.; Bergström, R.

Risk Factors for Stomach Cancer

A population-based case-control study is being carried out in two counties in northern Sweden and three counties in southern Sweden. The study population consists of all individuals younger than 80 years, born in Sweden and living in any of these counties. The aim is to carry out face-to-face interviews with a total of 550 newly diagnosed cases and with 900 controls, comprising a stratified sample from the general population. Dietary habits are thoroughly characterised during various periods of life from adolescence and later. Detailed information is also gathered concerning socio-economic factors, smoking, use of snuff and alcohol intake and on occupational history.

TYPE: Case-Control
TERM: Alcohol; Diet; Occupation; Socio-Economic Factors; Tobacco (Smoking)
SITE: Stomach
TIME: 1989 - 1995

---

795 **Adami, H.O.** 04792
University Hosp., Cancer Epidemiology Unit, 751 85 Uppsala, Sweden (Tel.: +46 18 665045; Fax: 503431)
COLL: Lindblad, P.; McCredie, M.R.E.; McLaughlin, J.K.; Wahrendorf, J.

Case-Control Study of Renal Adenocarcinoma

The main aim of this international collaborative study is to analyse the risk of renal adenocarcinoma associated with the use of diuretics and analgesics; and to investigate the possible relationship with occupation, notably in the petroleum industry. Cases will comprise all persons aged 20-79 years diagnosed as having renal adenocarcinoma in 1989-1991 in seven counties in central Sweden. Population-based controls will be selected by stratified sampling from the general population. Face to face interviews will be carried out with an estimated 450 cases and the same number of controls using a standardized questionnaire. Questions relate to use of analgesics, diuretics, oestrogens and diet pills; tobacco use and passive smoking; consumption of alcohol and coffee; reproductive history; dietary habits (using a food frequency questionnaire); medical history relating to diseases of the kidney, heart and thyroid; diabetes and other cancers; occupation (current and usual). In addition to a separate analysis of the Swedish component of the study, a pooled analysis of the data from all centers is envisaged.

TYPE: Case-Control
TERM: Alcohol; Analgesics; Coffee; Diabetes; Diet; Diuretics; Drugs; Hormones; Obesity; Occupation; Passive Smoking; Registry; Reproductive Factors; Tobacco (Smoking)
SITE: Kidney
CHEM: Oestrogens; Paracetamol
OCCU: Petroleum Workers
LOCA: Australia; Denmark; Germany; Sweden; United States of America
REGI: Uppsala (Swe)
TIME: 1989 - 1994

796 **Adami, H.O.** 05015
University Hosp., Cancer Epidemiology Unit, 751 85 Uppsala, Sweden (Tel.: +46 18 665045; Fax: 503431)
COLL: Ekbom, A.; Galanti, R.; Inskip, P.D.

Diagnostic X-Rays and Risk of Thyroid Cancer

The aim of this study is to clarify whether diagnostic x-rays, including dental x-rays, during various periods of an individual's life, are a risk factor for the development of differentiated (papillary and follicular) thyroid cancer. A population-based case-control study will be carried out in the Uppsala health care region. Eligible patients are the 500 most recently diagnosed cases notified to the regional cancer registry. For each case, one control, matched by age and gender, will be selected from the population register. For each individual, the place of residence during various periods of life will be clarified from parish authorities. From this information follows unequivocally, due to the structure of the Swedish medical system, the hospital to which they belonged at any given time. Information concerning diagnostic x-ray procedures will then be sought in the rosters of the departments of radiology. It is believed that by this method nearly complete lifetime exposure information will be retrieved for all individuals in the study. A sample of surviving cases and controls will be contacted for a short telephone interview to clarify the importance of differential and non-differential misclassification when information on exposure to diagnostic x-rays is obtained by interview rather than by a review of radiology records.

TYPE: Case-Control
TERM: Radiation, Ionizing; Registry
SITE: Thyroid
REGI: Uppsala (Swe)
TIME: 1990 - 1994

797 **Bruce, Å.** 04424
Swedish National Food Administration, Box 622, 751 26 Uppsala, Sweden (Tel.: +46 18 175500; Fax: 105848 ; Tlx: 76121 SLVUPS S)
COLL: Adami, H.O.; Holmberg, L.; Bergkvist, L.; Bergström, R.; Becker, W.; Ohlander, E.-M.

Diet and Breast Cancer

The aim is to study the importance of diet for the risk of developing breast cancer in middle-aged Swedish women. A case-control study is being coordinated within a general mammographic screening programme in two counties in Sweden. Some 45,000 women aged 40-70 were screened in 1986-1988 in the county of Västmanland. The same number of women will be screened in 1988-1990 in the county of Uppsala. A questionnaire is distributed with the invitation to the mammography. It includes questions on social conditions (marital status, education, children, familial cancer), a few quantitative questions on food items and a food frequency section inquiring about 60 food items. The completed form is brought by the women to the screening and filed at the clinic. In a second phase women with diagnosed breast cancer (500 expected) and an equal number of age-stratified controls are interviewed on the telephone by a nutritionist. The interview starts by identifying the woman's "dietary periods" during which her diet has been virtually constant. These periods are separated e.g. by life events causing changes in her diet, such as moving, marriage, childbirth. The last period is checked against the questionnaire previously sent in. For each period structured questions are asked on the consumption of foods listed in the questionnaire. Portion sizes are estimated. The diets of the cases and the controls will be compared as to the intake of individual foods and nutrients such as fat, fibre, retinol, carotenes and selenium.

TYPE: Case-Control
TERM: Diet; Familial Factors; Fat; Fibre; Mammography; Metals; Nutrition; Socio-Economic Factors; Vitamins
SITE: Breast (F)
CHEM: Beta Carotene; Retinoids; Selenium
TIME: 1986 - 1994

798 **Edling, C.** 04426
University Hosp., Dept. of Occupational Medicine, 751 85 Uppsala, Sweden (Tel.: +46 18 663644)

Mining, Radon and Lung Cancer

Lung cancer due to radon exposure is a well-known health hazard among miners. However, there is still debate about the "acceptable" exposure level and the interaction with smoking. In this cohort study 600 miners, who had their first underground experience in 1970 or later, i.e. when the exposure levels had been reduced to about 200-400 Bq/m3, are followed. Data on health status and smoking habits are collected every third year throughout the study. The lung cancer incidence is compared to that expected computed from local statistics.

TYPE: Cohort
TERM: Mining; Occupation; Tobacco (Smoking)
SITE: Lung
CHEM: Radon
OCCU: Miners
TIME: 1986 - 1994

799 **Edling, C.** 04427
University Hosp., Dept. of Occupational Medicine, 751 85 Uppsala, Sweden (Tel.: +46 18 663644)
COLL: Friis, L.

Cancer Incidence among Sewage Workers

Sewage workers are exposed to several possible mutagens and carcinogens. There is one study reporting urinary mutagens in municipal sewage workers and case reports on cancer in the lung, stomach and kidneys. The aim of this cohort study is to assess whether there is an increased risk of cancer among sewage workers. The cohort consists of about 800 men employed for at least one year during 1965-1986. The incidence is compared to expected numbers computed from national and local statistics.

TYPE: Cohort
TERM: Occupation
SITE: All Sites
OCCU: Sewage Workers
TIME: 1989 - 1994

800 **Ekbom, A.** 04039
Univ. Hospital of Uppsala, Cancer Epidemiology Unit, 751 85 Uppsala, Sweden (Tel.: +46 18 665047; Fax: 503431)
COLL: Adami, H.O.; Pinczowski, D.; Helmick, C.; Zack, M.M.

Cancer Risk in Patients with Inflammatory Bowel Disease

The aim of this historical cohort study is to estimate the risk of malignant disease, particularly of the large bowel, in patients with ulcerative colitis (UC) and Crohn's disease (CD). Two sources of information are being used in order to identify all patients who fell ill with an inflammatory bowel disease during the period 1965-1983 in a defined geographical area - the Uppsala health care region with about 1.2 million inhabitants. A computerized in-patient register covering all discharge diagnoses in the study area during the pertinent period and the diagnostic ledgers at all departments of clinical pathology in the area have been scrutinized. In a second phase, which is now ongoing, additional information about clinical characteristics, extent of bowel involvement, disease activity, pharmacological treatment and histopathological features will be retrieved from hospital records and from histopathological reports for a "nested" case-control study. By these means, a cohort of about 3,000 patients with UC and 1,500 patients with CD has now been developed. After computerized record linkages to the death registry and the cancer registry, the observed and expected number of malignant diseases in the cohort will be calculated.

TYPE: Cohort
TERM: Crohn's Disease; Record Linkage; Registry; Ulcerative Colitis
SITE: Colon; Rectum
REGI: Sweden (Swe)
TIME: 1986 - 1994

***801 Ekbom, A.** 05317
Univ. Hospital of Uppsala, Cancer Epidemiology Unit, 751 85 Uppsala, Sweden (Tel.: +46 18 665047; Fax: 503431)
COLL: Adami, H.O.; Inskip, P.D.; Galante, R.; Ron, E.

Risk Factors for Thyroid Cancer

This case-control study aims to investigate diagnostic x-ray exposure as a risk factor for papillary and follicular carcinoma of the thyroid in Sweden. 484 cases diagnosed in central Sweden in 1980-1992 are matched for sex, age, county of residence and living status at the time of diagnosis to 484 controls, randomly selected from the population registry. The place of residence for all the subjects up to the date of diagnosis of the case will be established. The hospital to which the subjects belonged at any given time could be identified from this information, which will help in obtaining data on diagnostic exposures from the ledgers of x-ray departments. The extent of misclassification of exposure will be assessed by telephone interview of a sample of surviving cases and controls.

TYPE: Case-Control
TERM: Radiation, Ionizing
SITE: Thyroid
TIME: 1991 - 1994

802 Holmberg, L. 04443
University Hospital, Cancer Epidemiology Unit, 756 54 Uppsala, Sweden (Tel.: +46 18 663000)
COLL: Adami, H.O.; Sparén, P.

Breast Cancer in Women Treated for Cancer in Situ of the Breast

The aim is to analyse the importance of cancer in situ of the female breast as a risk factor for development of invasive breast cancer and death from breast cancer. The natural history of cancer in situ of the breast is largely unknown and available evidence has emerged mainly from small retrospective series. Further knowledge about the possible importance of detecting and treating preinvasive cancers has become a major concern when a large number of such lesions are being detected within mammographic screening programmes. All cases of cancer in situ notified to the Swedish Cancer Registry from 1958 through 1985 have therefore been identified. This cohort, total 1,454, will be followed up by means of computerised linkages - based on the individually unique national registration number - to the National Cancer and Death Registries. By these means it will be possible to calculate person-years at risk and to identify all incident cases and all women deceased with breast cancer as an underlying cause of death. Based on these data, SMRs will be analysed using age of cancer in situ, years of follow-up and calender year of diagnosis (a proxy variable for diagnosiis within or outside screening programmes) as determinants of risk. This approach will be followed by nested case-control studies in order to adjust for possible confounding factors and to enable more detailed studies of histopathological characteristics as determinants of risk.

TYPE: Case-Control; Cohort
TERM: High-Risk Groups; In Situ Carcinoma; Registry
SITE: Breast (F)
REGI: Sweden (Swe)
TIME: 1988 - 1994

803 **Persson, I.R.** 02084
University Hospital, Unit of Cancer Epidemiology, 751 85 Uppsala 14, Sweden (Tel.: +46 18 663745; Fax: 503431)
COLL: Adami, H.O.; Bergkvist, L.; Hoover, R.N.; Shairer, C.

Hormone Replacement Therapy and Risk of Breast and Reproductive Cancers

A prospective cohort study was commenced in 1977, including all women within the Uppsala health care region, who had been prescribed replacement hormones. Prescriptions were collected during a three-year period from 1977 through 1980, including both oestrogens and oestrogen-progestogen combinations, which had been prescribed because of menopausal symptoms. The established cohort comprises about 23,200 women, whose characteristics and detailed exposure, i.e. compliance and lifetime duration of intake, have been studied through questionnaires sent on several occasions. Follow-up has been made through linkage with the Swedish Cancer Registry in order to ascertain the occurrence of breast and various gynaecological cancers prospectively. These outcomes have hitherto been analysed for observation periods through 1983-1984 and through 1987. Int. J. Epidemiol. 17:732-737, 1988; Am. J. Epidemiol. 130:503-510, 1989; N. Engl. J. Med. 321:293-297, 1989. In 1987-1988 a comprehensive questionnaire was sent to 13,000 of the women in the cohort, providing detailed data on co-variates in exposure of all these individuals. This subcohort will be subject to forthcoming analyses on both incidence and mortality of cancer diseases.

TYPE: Cohort
TERM: Hormones; Menopause; Record Linkage; Registry; Treatment
SITE: Breast (F); Female Genital
CHEM: Oestrogens; Progestogens
REGI: Sweden (Swe)
TIME: 1977 - 1994

804 **Persson, I.R.** 04789
University Hospital, Unit of Cancer Epidemiology, 751 85 Uppsala 14, Sweden (Tel.: +46 18 663745; Fax: 503431)
COLL: Adami, H.O.; Naéssen, T.; Falkeborn, M.; Brinton, L.A.

Oophorectomy and Hysterectomy and the Risk of Breast or Ovarian Cancer

Aim: To investigate whether bilateral oophorectomy or hysterectomy affects the risk of subsequent development of breast or other cancers.A cohort of women who have undergone a gynaecological operation with a hysterectomy and/or bilateral oophorectomy was defined from the Inpatient Registry, holding data on all hospital admissions during the period 1965 through 1983 in the Uppsala health care region. About 17,000 women were enrolled. The individual women were followed up through linkage to the cancer registry, to ascertain the observed number of cases. The relative risk estimates were formed as the ratio of the observed and expected number of cases, the latter calculated from the person-years of observation in the cohort and the age-specific incidence rates of these two cancers in the background population. The risk is being analysed in the cohort as a whole and among subgroups of it, according to ovarian and uterine status through cohort and case-cohort designs. Data in the Inpatient Registry are supplemented with information from the patients' medical records regarding the exact type of operation, the indication for the operation and other relevant factors.

TYPE: Cohort
TERM: Hysterectomy; Oophorectomy; Registry
SITE: Breast (F); Ovary
REGI: Uppsala (Swe)
TIME: 1990 - 1994

SWITZERLAND

Basel

805 **Müller, H.J.** 03985
Kantonsspital Basel, Dept. of Research, Lab. of Human Genetics, Hebelstr. 20 , 4031 Basel, Switzerland (Tel.: +41 61 2652362; Fax: 2611500)
COLL: Weber, W.; Stalder, G.A.; Torhorst, J.K.H.; Obrecht, J.P.; Speck, B.; Signer, E.

Familial Cancer in Switzerland

This register contains detailed pedigrees of over 3,500 families of cancer patients. Special emphasis is laid on families with (1) breast cancer aggregation in families site-specific or in association with the neoplasms of the Li-Fraumeni/SBLA syndrome; (2) colorectal cancer including FAP and HNPCC, Crohn's disease and ulcerative colitis; (3) Fanconi's anaemia and related disorders; and (4) cancer of the skin. Cancer diagnoses of patients and relatives are verified through standard pathology or cytology reports. The incidence and spectrum of neoplasms in relatives are compared with those in the cancer registry of Basel. DNA from a subset of breast cancer patients and their relatives is being analysed for mutations in the tumour suppressor gene p53 for identifying cancer-prone individuals. Similar DNA studies are performed for colorectal cancer families concentrating on polymorphisms on chromosome 5 (FAP) and 17/18 (HNPCC). Cytogenetic studies and estimations of DNA repair capacities are carried out in families with an assumed increased genomic instability.

TYPE: Genetic Epidemiology; Registry
TERM: DNA Repair; Familial Factors; Genetic Factors; Genetic Markers; Heredity; Multiple Primary; Pedigree; Radiation, Ultraviolet; Registry; Screening
SITE: All Sites; Bone; Breast (F); Childhood Neoplasms; Gastrointestinal; Leukaemia; Lymphoma; Ovary; Testis
REGI: Basel (Swi)
TIME: 1982 - 1994

806 **Stähelin, H.B.** 04179
Kantonsspital, Medizinisch-geriatrische Klinik, Hebelstr. 10, 4031 Basel, Switzerland (Tel.: +41 61 252525)
COLL: Ulrich, J.; Torhorst, J.K.H.

Basle Study: Mortality Follow-Up

This is an on-going prospective study with the aim of analysing risk factors for cardiovascular disease (CVD), cancer and senile dementia. The influence of life style, cardiovascular risk factors, and plasma vitamins on cancer and CVD is investigated. 1960 Basle Study I: Baseline data collection, plasma lipids and clinical data, 6, 000 participants. 1965 Basle Study II: Follow-up. Clinical data, plamsa lipids, 5,000 participants. 1971 Basle Study III: Follow-up. Plasma lipids, plasma vitamins, life style, approximately 4,500 participants. 1980 Mortality follow-up 1965-1980, 600 cases. 1986 Mortality follow-up: 450 additional cases. 1990 Mortality follow-up: 360 additional cases. A mailed survey was sent to approximately 3,600 survivors, asking for information on life style, nutrition, activities of daily life, and medical history. A paper has been published in Am. J. Clin. Nutr. 53:265S-269S, 1991.

TYPE: Cohort
TERM: Diet; Lifestyle; Lipids; Nutrition; Physical Activity; Vitamins
SITE: All Sites; Colon; Lung; Stomach
TIME: 1985 - 1994

Geneva

807 **Bouchardy, C.** 04931
Geneva Tumour Registry, La Cluse 55, Geneva, Switzerland (Tel.: +41 22 202291)
COLL: Benhamou, S.; Paoletti, C.; Sancho-Garnier, H.; Dayer, P.; Galteau, M.M.

Cancer Risk Related to Tobacco according to Individual Capacity to Metabolize Probe Drugs

The activation of many of the carcinogens identified in cigarette smoke is catalysed by the cytochrome P450-dependent mixed function oxidase system. Some of these enzymes are genetically controlled and there is great variation in activity between individuals. The aim of the study is to identify smokers at very high risk of cancer by testing some of their enzymatic capacity possibly involved in the metabolism of carcinogens in cigarette smoke. This study, involving smokers (or ex-smokers), is a multicentric hospital-based case-control investigation. Specific enzymatic capacity will be investigated in 500 histologically confirmed cancers of lung, larynx or buccal cavity and in 500 non-cancer controls admitted to a network of public hospitals in the Paris area and in Caen. In particular, phenotypic tests using dextromethorphane and mephenytoin will be done in all patients and other probe drugs will be added in a sample of the studied population.

TYPE: Case-Control
TERM: Drugs; Enzymes; Registry; Tobacco (Smoking)
SITE: Larynx; Lung; Oral Cavity; Pharynx
LOCA: France
REGI: Geneva (Swi)
TIME: 1988 - 1994

***808** **Morabia, A.** 05344
Hôpital Cantonal Universitaire Unité d'Epidémiol. Clinique, 25, rue Micheli-du-Crest, 1211 Geneva 14, Switzerland (Tel.: +41 22 3729552; Fax: 3476486)

Smoking and Breast Cancer

The aim of the study is to determine whether there is an association between breast cancer and either active or passive smoking. In contrast to previous reports on this association, unexposed subjects will be women who are neither active nor passive smokers. Eligible cases are all women younger than 75 years, resident in the Canton of Geneva, and diagnosed with a first invasive breast cancer in 1992 or 1993 (approximately 250). Controls will be a random sample of the general population of the same age (approximately 750). All subjects will be interviewed in a mobile epidemiology unit.

TYPE: Case-Control
TERM: Passive Smoking; Tobacco (Smoking)
SITE: Breast (F)
TIME: 1992 - 1994

Lausanne

809 **Levi, F.G.** 03921
Registre Vaudois des Tumeurs, Inst. Univ. de Médecine Sociale et Préventive, CHUV Falaises 1, 1011 Lausanne, Switzerland (Tel.: +41 21 314908; Fax: 230303)
COLL: La Vecchia, C.; Decarli, A.; Cislaghi, C.; Randriamiharisoa, A.; Negri, E.; Boyle, P.

Analyses of Cancer Death Certification Data in Switzerland and Other European Countries

Swiss cancer death certifications from 1950 will be utilized (1) to carry out in a comprehensive analysis of the number of certified deaths; age-specific and age-standardized rates; proportional mortality by site; in graphical and numerical form, and (2) to develop statistical modelling techniques for disentangling the effects of cohort of birth, period of death and age on cancer mortality rates. Similar analyses are being made for 26 major European countries, using the WHO mortality database. Recent papers have been published in Wld Hlth Statist. Quart. 45:117-164, 1992, J. Cancer Res. Clin. Oncol. 119:165-171, 1993 and in Eur. J. Cancer 29:431-470, 1993.

TYPE: Methodology; Mortality
TERM: Age; Birth Cohort; Geographic Factors; Mathematical Models; Time Factors
SITE: All Sites
TIME: 1987 - 1995

810 **Levi, F.G.** 04465
Registre Vaudois des Tumeurs, Inst. Univ. de Médecine Sociale et Préventive, CHUV Falaises 1, 1011 Lausanne, Switzerland (Tel.: +41 21 314908; Fax: 230303)
COLL: La Vecchia, C.; Delaloye, J.F.; de Grandi, P.; Gulie, C.; Negri, E.

Risk Factors for Endometrial Cancer in the Canton of Vaud, Switzerland

Endometrial cancer is the most frequent invasive malignancy of the female genital tract in Switzerland and the third most common neoplasm with an estimated number of 800 incident cases per year. It is well known that the risk is higher among women with elevated levels or availability of serum oestrogens, such as obese women or women on menopausal replacement treatment. Most studies, however, have concentrated upon the isolated effect of each single factor, in the absence of an integrated approach to the relative importance of various risk factors and their interactions. Further, there are factors, such as diet or exercise, whose relation with endometrial carcinogenesis is plausible, but far from being proven or quantified. The present proposal is based on a multi-national case-control study of endometrial cancer, originally designed at the Environmental Epidemiology Branch of the American National Cancer Institute to re-assess the role of major identified risk factors for endometrial cancer and their interactions and to explore newer hypotheses, chiefly in the field of diet. The research is based on a case-control design. Cases will be patients with histologically confirmed endometrial cancer registered by the Cancer Registry of the Canton of Vaud. Controls will be subjects admitted for conditions not relating to neoplastic, gynecologic or hormone-related diseases to the same hospitals where cases had been identified, matched by sex, age and region of residence. It is planned to obtain data on 200 cases and 600 controls over a five-year period. The population base provided by the Cancer Registry will allow reliable estimation not only of relative, but also of absolute excess risks, which is a further original aspect of the study, at least in relation to European populations. Papers or reports from this study have been published in Cancer Causes and Control 2:99-103, 1991 and Brit. J. Cancer 66:7202-722, 1992.

TYPE: Case-Control
TERM: Diet; Hormones; Obesity; Physical Activity; Registry; Reproductive Factors
SITE: Uterus (Corpus)
REGI: Vaud (Swi)
TIME: 1988 - 1995

Bangkok

811 **Srivatanakul, P.** 04497
National Cancer Inst., Research Div., Rama VI Rd, Bangkok 10400, Thailand (Tel.: +66 2 2461294/302; Tlx: INSNACA)
COLL: Muñoz, N.; Bosch, F.X.; Estève, J.; Khuhaprema, T.; Rinsurongkawong, S.; Cheirsilpa, A.; Tanprasert, S.

Cohort Study of HBsAg Carriers in Bangkok
The objectives of the present study are (1) to determine the role of other risk factors such as aflatoxin, HCV, tobacco, alcohol, and N-nitroso compounds in the development of PHC and other chronic liver diseases; (2) to determine the risk of developing PHC and other chronic liver diseases among a cohort of HBsAg carriers as compared to that of the general population; (3) to evaluate the value of AFP levels and ultrasonography in the detection of early PHC. (These data will provide the necessary information for the formulation of cost-effective screening programmes of high-risk groups for PHC in Thailand and other developing countries); (4) to assess the benefits of early treatment of PHC. A Group of 2,000 HBsAg positive male blood donors over 30 years of age have been selected from among blood donors, Army personnel and other people who come for check up at the Out-Patient Clinic. At the end of the five-year follow-up period a case-control study within the cohort will be carried out to compare both PHC cases and cases of chronic hepatitis and cirrhosis with appropriate controls in relation to the risk factors.

TYPE: Case-Control; Cohort
TERM: Alcohol; Antigens; Cost-Benefit Analysis; HBV; High-Risk Groups; Hormones; Liver Disease; Mycotoxins; Screening; Tobacco (Smoking)
SITE: Liver
CHEM: Aflatoxin; N-Nitroso Compounds; Steroids
TIME: 1987 - 1994

***812** **Srivatanakul, P.** 05416
National Cancer Inst., Research Div., Rama VI Rd, Bangkok 10400, Thailand (Tel.: +66 2 2461294/302; Tlx: INSNACA)
COLL: Ianijan, P.; Lalitwingsa, S.; Vipasrinimitr, S.; Tipayadarapanit, V.; Unpanye, P.; Martin, N.C.; Suriya, P.; Saengsingaew, V.; Parkin, D.M.; Boffetta, P.

Studies of Respiratory Cancer in Lampang Province, Thailand
Lampang province includes several large open-cast lignite mines, and in proximity to these are nine coal-fired power stations. This project is designed to ascertain whether there is any increased risk of respiratory cancer in association with air pollution from these electricity generating stations. (1) A population-based cancer registry is being established to better quantify the reportedly high risk of respiratory cancers; at present no incidence data are available. Incidence rates will be calculated for districts (or groups of districts) within the province. (2) A case-control study will recruit 200 cases of lung cancer from residents of Lampang. Three control groups of the same size will be selected from hospital patients (excluding respiratory diseases/other relevant cancers), blood donors and the general population. Exposure to pollution will be estimated from residential history, and from measurements of metals (cadmium, chromium, arsenic), DNA adducts of PAH and their varying metabolistes. Tobacco smoking, occupational history and exposure to domestic pollution will be taken into account.

TYPE: Case-Control; Incidence; Registry
TERM: Air Pollution; DNA Adducts; Mining; Occupation; Registry
SITE: Lung
REGI: Lampang (Tha)
TIME: 1993 - 1995

Hat Yai

813 **Chongsuvivatwong, V.** 05181
Prince Songkla Univ., Fac. of Medicine, Epidemiology Unit, Hat Yai 90112, Thailand (Tel.: +66 74 245677; Fax: 235174 ; Tlx: 62168 UNISONE TH)
COLL: Muñoz, N.; Bosch, F.X.; Chichareon, S.; Vithsupakorn, K.

Case-Control Study on Cervical Carcinoma

The hypotheses to be tested are whether squamous cell carcinoma is related to sexually transmitted agents (HPV, HSV-2, chlamydia and CMV), sexual behaviours of the patient and of her partner, use of oral contraceptives, tobacco smoking and the effect of screening for cervical carcinoma. Cases are incident cases in the University Hospital, with histological confirmation. Controls are female patients admitted to the surgical ward of the same hospital, who are in the same age range as the cases and whose admission diagnoses are not related to the above exposure variables. Current husbands of the case or control women are ascertained. 300 cases, 300 controls and their husbands are needed. Information on past sexual behaviour of the subject is obtained at interview in the hospital. Serum is obtained from all subjects. Exfoliated cervical cells from the cases and the controls and exfoliated cells of the glans penis and from the urethra are obtained from the husbands. Analysis will be done using unconditional logistic regression techniques.

TYPE: Case-Control
TERM: BMB; CMV; HPV; HSV; Infection; Oral Contraceptives; Screening; Serum; Sexual Activity; Sexually Transmitted Diseases
SITE: Uterus (Cervix)
TIME: 1991 - 1994

Ankara

***814** **Bilir, N.** 05246
Hacettepe Univ. Fac. of Medicine Dept. of Public Health, Ankara 06100, Turkey
COLL: Yilmaz, G.

Cancer Incidence in Diyarbakir

Cancer has been a notifiable disease in Turkey since 1983. A population-based cancer registration system has been established in the province of Diyarbakir, with a population of 1.2 million. Active data collection will start by 1 January 1993. In addition to identifiable and basic demographic data, some information on the personal characteristics of cancer patients will be obtained by trained registry personnel. The study will be extended to other provinces in different geographical regions of Turkey.

TYPE: Incidence; Registry
TERM: Data Resource
SITE: All Sites
TIME: 1993 - 1999

Istanbul

815 **Akdas, A.** 04260
Marmara Univ., School of Medicine, Dept. of Urology, Tophanelioglu cad., No. 13-15 Altunizade, Istanbul 81190, Turkey (Tel.: +90 1 3255865; Fax: 3250323)
COLL: Simsek, F.; Ersev, D.; Türkeri, L.; Bilir, N.

Aetiology of Testicular Tumours

In this study the possible role of some factors (exposure to X-ray, heat, smoke, dust and chemical substances, testicular trauma, history of cryptorchidism or mumps, wearing tight underwear and pants, sexual activity) in the aetiology of testicular tumours will be investigated. The cases will be all patients with a histologically proven testicular tumour seen in the department of urology. The approximate number is estimated as 15 patients per year. Age-matched male controls will be taken from orthopaedics and general surgery wards. Patients with a diagnosis of malignant disease will be excluded. Necessary information for the cases and controls will be collected by questionnaire.

TYPE: Case-Control
TERM: Chemical Exposure; Radiation, Ionizing; Sexual Activity; Tobacco (Smoking); Trauma
SITE: Testis
TIME: 1985 - 1994

816 **Akdas, A.** 05158
Marmara Univ., School of Medicine, Dept. of Urology, Tophanelioglu cad., No. 13-15 Altunizade, Istanbul 81190, Turkey (Tel.: +90 1 3255865; Fax: 3250323)
COLL: Simsek, F.; Ilker, Y.; Alican, Y.; Türkeri, L.

Prostate Cancer: Genetic and Epidemiological Factors

In this study the HLA complexes of patients with proven adenocarcinoma of the prostate will be determined in order to find out any possible relationship between genotype and the disease. The possible role of occupation, exposure to drugs and x-ray, heat, chemical substances, medical history of venereal disease, sexual activity and its spectrum in the aetiology of prostate cancer will also be investigated in these patients. The approximate number of cases is estimated at 25 per year. Age-matched controls will be recruited from patients with histologically proven benign prostatic hypertrophy and the information will be obtained by questionnaire.

TYPE: Case-Control
TERM: Chemical Exposure; Drugs; Genetic Factors; HLA; Occupation; Sexual Activity; Sexually Transmitted Diseases
SITE: Prostate
TIME: 1992 - 1996

Aberdeen

***817** **Collins, A.R.** 05248
Rowett Research Inst., Greenburn Rd, Bucksburn, Aberdeen AB2 9SB, Scotland, United Kingdom (Tel.: +44 224 712751; Fax: 715349 ; Tlx: 739988 Rowett G.)
COLL: Arthur, J.; Duthie, G.; Beattie, J.

Modification of Antioxidant Status and Indices of Free Radical Mediated Damage in Relation to Cancer and Heart Disease

The hypothesis being tested is that dietary antioxidants reduce DNA damage caused by free radicals and may protect against cancer. The link between nutrition and cancer will be investigated in a population known to be at relatively high risk of cancer. 100 healthy males (50 smokers, 50 non-smokers), aged 50-59, will be recruited through local health centres. Initial measurements of plasma vitamin E will be done to allow matching of pairs (non-smoker/smoker) with the same vitamin E levels. Baseline levels will be established for plasma vitamin E, carotenoids, vitamin C, lipid peroxidation, cholesterol, DNA-damage in lymphocytes, excreted oxidised nucleotide (8-OH-dG) in urine, and HPRT-mutations in lymphocytes (the last three markers represent indices of cancer risk). Half of each group will then be given daily supplement of antioxidant vitamins (C, E, beta carotene) and, over a period of 20 weeks, changes in the biomarkers will be investigated.

TYPE: Intervention; Molecular Epidemiology
TERM: Antioxidants; BMB; Biomarkers; Cholesterol; DNA Repair; Diet; HPRT-Mutants; Nutrition
SITE: Inapplicable
CHEM: Beta Carotene; Retinoids
TIME: 1992 - 1994

***818** **Russell, E.** 05357
Univ. of Aberdeen, Medical School, Dept. of Public Health, Foresterhill, Polwarth Bldg, Aberdeen AB9 2ZD, Scotland, United Kingdom (Tel.: +44 224 681818; Fax: 662994)
COLL: Wilson, B.J.; Russell, D.; Haites, N.E.; Ewen, S.W.B.; Collins, A.R.; Brown, K.; Brunt, P.

Antioxidant Status, DNA Damage, and the Development of Colorectal Neoplasia

The primary objective of this study is to examine the influence of dietary antioxidants on the progression towards colorectal cancer and is based on the assumption of an adenoma-carcinoma sequence. A cohort of 100-200 individuals at higher than average risk of developing colorectal cancer (by virtue of a first degree family history or the already detected presence of colorectal adenomas) will be invited to attend for genetic assessment and counselling as well as for colonoscopy. Other initial data collected will include blood antioxidant levels and gene mutations in epithelial cells shed in faeces. The subjects will be followed up for 12-18 months, with repeat measurements of antioxidants and gene mutations and a repeat colonoscopy at the end of the follow-up. The relationship between the change in the size and number of adenomas at colonoscopy and (1) genetic risk and antioxidant levels, as putative aetiological factors, and (2) gene mutations in faeces, as a potential marker of neoplastic progression, will be examined using univariate and multivariate techniques. Long term outcome will be monitored using the Scottish Cancer Registration Scheme.

TYPE: Cohort
TERM: Antioxidants; Diet; Genetic Factors; High-Risk Groups; Registry
SITE: Colon; Rectum
REGI: Scotland N.E. (UK)
TIME: 1993 - 1997

Barnett

819 **Harte, G.A.** 04987
Nuclear Electric plc, Health and Safety Dept.,, Barnett Way, Barnett Gloucester GL4 7RS, United Kingdom (Tel.: +44 452 652222; Fax: 652776 ; Tlx: 43501)
COLL: Taylor, R.H.; Davies, N.F.; Brodie, P.; Turp, J.E.

Mortality Study of Employees of Nuclear Electric PLC

Occupational histories and radiation exposure data are being collected for approximately 20,000 employees of Nuclear Electric plc (formerly the Central Electricity Generating Board) who have been classified as radiation workers at any time since the company started operating nuclear power stations in 1962. Mortality in this cohort will be studied in relation to occupational exposure to ionizing radiation and, possibly, other physical and chemical agents.

TYPE: Cohort
TERM: Chemical Exposure; Occupation; Radiation, Ionizing
SITE: All Sites
OCCU: Radiation Workers
TIME: 1990 - 1994

Belfast

820 **MacKenzie, G.** 05202
Queens Univ. of Belfast, Dept. of Epidemiology and Public Health, Mulhouse Bldg, Grosvenor Rd, Belfast BT12 6BJ, N. Ireland, United Kingdom (Tel.: +44 232 240503; Fax: 236298)
COLL: McNair, R.

Atlas of Cancer Mortality in Northern Ireland

The purpose of this project is to construct a colour atlas showing the temporo-spatial distribution of the major cancers in Northern Ireland during the period 1981-1989. The atlas will also be analytical and use 3-D imaging techniques to illustrate statistical modelling and to identify outliers. An interpretative commentary based on the detailed statistical analyses undertaken will be included. The study has been extended to include some new incidence data.

TYPE: Incidence; Mortality
TERM: Mapping; Mathematical Models
SITE: All Sites
TIME: 1990 - 1994

Birmingham

821 **Allan, R.N.** 04516
General Hosp., Gastroenterology Unit, Steelhouse Lane, Birmingham B4 6NH, United Kingdom (Tel.: +44 21 2368611/5024; Fax: 2367626)
COLL: Jewell, D.P.; Hellers, G.

Incidence of Cancer in Crohn's Disease

The aim of this study is to establish the pattern and incidence of malignant disease in patients with Crohn's Disease. Patients will be recruited from three centres (Birmingham, Oxford and Stockholm). Only primary referrals within five years of onset of disease between 1954 and 1965 and followed for a minimum of 20 years will be included, which will eliminate any secondary or tertiary referral bias in relation to specialist hospital practice. Some 250 patients will be collected from each centre. The morbidity rates for cancer in patients with Crohn's Disease will be compared with the morbidity rates for cancer in each geographical area in the study. The cancer risk will be calculated for all sites within the digestive system and for extra-intestinal cancer.

TYPE: Cohort
TERM: Crohn's Disease; Ulcerative Colitis
SITE: All Sites; Colon; Rectum
LOCA: Sweden; United Kingdom
TIME: 1989 - 1994

***822** **Griffiths, R.K.** 05323
Univ. of Central Birmingham, Medical School, Health Serv. Res. Cent. Inst. of Public and Environm. Health, Edgbaston , Birmingham B15 2TJ, United Kingdom (Tel.: +44 21 4146774)
COLL: Gilman, E.A.; Alexander, F.E.; Cartwright, R.A.; McKinney, P.A.; Day, N.E.; Beral, V.; Birch, J.M.; Chilvers, C.E.D.; Peto, J.

UK Childhood Cancer Study - West Midlands Region

This population-based case-control study of the causes of cancer in children will gather detailed information on each of the around 1,000 children diagnosed with cancer every year in England and Wales, and on twice as many healthy children. A similar investigation is going on in Scotland. The study has been designed to investigate five hypotheses relating to increased childhood cancer risk: (1) child's exposure to natural or man-made radiation, either during the pregnancy or after birth; (2) similar exposure to potentially hazardous chemicals; (3) occupational exposure of the father to either radiation or chemicals; (4) child's exposure to electro-magnetic fields; (5) child's exposure and response to infections in the early years of life. Controls will be randomly selected from the general population and matched for age and sex. All cases and controls will be interviewed by trained interviewers with a structured questionnaire. In collaboration with the National Radiological Protection Board, measurements of background gamma radiation and radon levels will be made in the houses in which the study families have lived. Blood samples collected from the case families will be examined to determine genetic and immunogical characteristics.

TYPE: Case-Control
TERM: BMB; Blood; Chemical Exposure; Electromagnetic Fields; Infection; Parental Occupation; Radiation, Ionizing
SITE: Childhood Neoplasms; Leukaemia
TIME: 1992 - 1997

823 **Kneale, G.W.** 04540
Univ. of Birmingham, Dept. of Public Health and Epidemiology, Edgbaston, Birmingham B15 2TJ, United Kingdom
COLL: Stewart, A.M.; Gilman, E.A.; Knox, E.G.

Spatial and Temporal Distributions of Childhood Cancers in Great Britain

The study aims to characterise and interpret the spatial, temporal and spatio-temporal distributions of childhood cancers in Britain. Cases include all children who have died of cancer or leukaemia, in England, Scotland and Wales, under the age of 16 years, in the period 1953 to 1981 (who form part of the Oxford Survey of Childhood Cancers, an on-going national case-control study), together with all childhood registrations of leukaemia and non-Hodgkin lymphoma diagnosed in the years 1966 to 1983. Spatial distributions will be mapped by local authority area and area rates will be examined for spatial heterogeneity. Proximity analyses will also be used to examine case distributions with respect to local environmental features which may represent a source of risk. Analyses of the temporal distribution will include examination of cohort childhood cancer rates, specifically looking for age, period and cohort effects. Data will be examined for spatio-temporal clustering, i.e. excesses of cases occurring close together in space and in time, using the method of Knox and Mantel. Identified "clusters" will be examined for associations with local geographical circumstances and temporal events, to test hypotheses on changes in case ascertainment, survival, or migration; on local toxic or radiation hazards; or on transmission of a viral agent.

TYPE: Case Series; Incidence; Mortality
TERM: Birth Cohort; Childhood; Cluster; Geographic Factors; Mapping; Radiation, Ionizing; Time Factors; Virus
SITE: Childhood Neoplasms; Leukaemia; Non-Hodgkin's Lymphoma
TIME: 1986 - 1994

824 **Mann, J.R.** 04245
Birmingham Children's Hosp., Dept. of Oncology, Ladywood Middleway, Birmingham B16 8ET, United Kingdom (Tel.: +44 21 4544851; Fax: 4564697)
COLL: Cameron, A.H.; Powell, J.E.; Parkes, S.E.; Raafat, F.; Stevens, M.C.G.; Griffiths, R.K.

Childhood Cancer: Epidemiological and Aetiological Investigations

Clinical, epidemiological, treatment and survival data on all children diagnosed with malignancies in the West Midlands region since 1957 (5,000 cases), and prospectively since 1984, have been assembled in the West Midlands Regional Children's Tumour Research Group. The purpose of the project is to use these data for epidemiological studies, such as determining changes in incidence in relation to time, variations in lifestyle, geographical and other environmental factors. In addition, retrospective studies on selected diagnostic groups (pathology is reviewed) in relation to clinical features, results of treatment and tumour characteristics (flow cytometry, gene probes) are being undertaken.

TYPE: Incidence; Registry
TERM: BMB; Environmental Factors; Ethnic Group; Geographic Factors; Lifestyle; Parental Occupation; Time Factors
SITE: Childhood Neoplasms
REGI: Birmingham (UK)
TIME: 1984 - 1996

825 **Sorahan, T.M.** 04493
Univ. of Birmingham, Dept. of Public Health & Epidemiology, Edgbaston , Birmingham B15 2TT, United Kingdom (Tel.: +44 21 4143985; Fax: 4144036)
COLL: Davies, P.

Mortality and Cancer Morbidity of Production Workers in the Polyurethane Foam Industry

Study subjects are all 8,500 past and present workforce employees, from 11 participating factories in England and Wales, with a minimum period of employment of six months and with some employment in the period 1958-1979. The nature of the study is an historical prospective cohort study. Detailed job histories have been collected. Cancer rates among those employees with "higher" levels of exposure to toluene diisocyanate will be compared with cancer rates of all other employees.

TYPE: Cohort
TERM: Occupation; Plastics
SITE: All Sites
CHEM: Diisocyanates
OCCU: Plastics Workers
TIME: 1985 - 2000

826 **Stewart, A.M.** 04929
Univ. of Birmingham, Faculty of Medicine, Dept. of Social Medicine, Birmingham B15 2TJ, United Kingdom (Tel.: +44 21 4721311; Fax: 4144036)
COLL: Cummins, C.W.; Nussbaum, R.H.

Late Effects of Low-Level Radiation

The aim is to determine whether military personnel who were stationed during 1943-1962 at Hanford Nuclear Reservation, USA, where there were many radioactive releases, are at greater risk of dying of cancer than a control group at Fort Lewis, also in Washington State. It is expected that 25,000 veterans in each camp will be traced from army records. Those deceased will be identified, as well as their cause of death, and a retrospective cohort study carried out. Exposure data will be available from a Department of Energy study which should be completed by 1994. Some preliminary exposure data and dose reconstructions are available from the State of Washington. Cooperation has been obtained from the US Veterans Administration.

TYPE: Cohort
TERM: Radiation, Ionizing
SITE: All Sites
LOCA: United States of America
TIME: 1988 - 1994

827 **Stewart, A.M.** 00858
Univ. of Birmingham, Faculty of Medicine, Dept. of Social Medicine, Birmingham B15 2TJ, United Kingdom (Tel.: +44 21 4721311; Fax: 4144036)
COLL: Kneale, G.W.; Sorahan, T.M.; Gilman, E.A.

Oxford Survey of Childhood Cancers

This long-standing retrospective survey was tarted in 1953 to study the aetiology of childhood cancers. It consists of interviews with the parents of children under 16 dying of cancer or leukaemia in the United Kingdom and with parents of control children matched for age, sex and district; antenatal information is confirmed by application to clinics and doctors, and the case child's medical history is followed up through general practitioner and hospital notes. Surviving children have similarly been interviewed and followed up since 1962 by means of the regional cancer registries. Present projects include: (1) A study of the effects of drugs in pregnancy and of ultrasound during pregnancy (2) A study of parental occupations and their possible relationship to the aetiology of childhood cancers; (3) A study of congenital defects found in children who later develop cancer or leukaemia and congenital defects in their siblings, with a similar study in matched controls and their siblings; (4) A study of second primary tumours occurring in children after cure of their initial cancer with special attention to the effect of radiation and genetics; (5) A study of cancers among siblings of children with cancer.

TYPE: Case-Control; Cohort
TERM: Clinical Effects; Congenital Abnormalities; Drugs; Intra-Uterine Exposure; Multiple Primary; Mutation, Somatic; Occupation; Pregnancy; Radiation, Ionizing; Radiotherapy; Registry; Sib; Ultrasound; Virus
SITE: Childhood Neoplasms; Leukaemia
TIME: 1953 - 1994

828 **Wallace, D.M.A.** 04976
Queen Elizabeth Hosp., Department of Urology, Clinical Trials Unit, Edgbaston, Birmingham B15 2TH, United Kingdom (Tel.: +44 21 6272288; Fax: 6272289)
COLL: Harrington, J.M.; Bathers, S.

Feasibility Study for Surveillance of Occupational Urothelial Tumours

The primary objective of this study is to assess the feasibility of establishing a surveillance procedure for occupational urothelial tumours. Further aims will be to identify the areas and occupations where more detailed epidemiological investigations should be carried out, to establish a database of the occupations of workers exposed to the recognised urothelial carcinogens, to establish a register of cases with urothelial cancer exposed at work to urothelial carcinogens, to advise the clinicians, to co-ordinate the further investigation of cases where occupational exposure is suspected and to provide an information service to urologists and other clinicians managing urothelial cancer cases. The cases will be all newly diagnosed patients with urothelial cancer in the West Midlands area over a two-year period. All patients will be screened using an occupational history questionnaire. Suspected exposure will be fully investigated by a trained interviewer. Controls will be selected randomly from patients in the same hospital in the same period of time. To test the reproducibility of interview information a proportion of the interviews will be conducted again by a second interviewer. The assessment of occupational exposure will be retrospective, based on information collected by hygienists in the Institute of Occupational Health, producing a job-exposure matrix for estimation of occupational exposure to certain and suspect carcinogens. The final analysis of the data will concentrate on identifying and investigating unusual clusters of cases that appear to have a common aetiology. The feasibility of further epidemiological studies on occupational urothelial cancer will be examined where the data suggest aetiological hypotheses.

TYPE: Case-Control; Methodology
TERM: Chemical Exposure; Data Resource; Occupation; Registry
SITE: Urinary Tract
REGI: W. Midlands (UK)
TIME: 1990 - 1994

Bridgend

829 **Powell, D.E.** 03906
Princess of Wales Hosp., Dept. of Pathology, Coity Rd, Bridgend Mid-Glam. CF32 0PT, United Kingdom

Epidemiology of Leukaemia and Lymphoma

In this longitudinal study of the space and time distribution of leukaemia and lymphoma, a continuous 25-year period is being monitored in a very stable population of 140,000 served by one diagnostic centre. All cases are referred to the one laboratory. Environmental factors studied include housing, schools and water. Banked sera are studied for viruses. Contacts of cases are traced.

TYPE: Correlation
TERM: Cluster; Environmental Factors; Time Factors; Virus
SITE: Leukaemia; Lymphoma
TIME: 1962 - 1994

Brighton

830 **Bridges, B.A.** 04900
Univ. of Sussex, MRC Cell Mutation Unit, Falmer, Brighton BN1 9RR, United Kingdom (Tel.: +44 273 678123; Fax: 678121)
COLL: Arlett, C.F.; Cole, J.

Mutation Frequency of Lymphocytes in Individuals Exposed to Radon
The aim is to see whether individuals living in dwellings with high levels of radon gas have a higher frequency of mutations in their peripheral lymphocytes than those living in dwellings with low levels of radon. The frequency of T-lymphocytes resistant to 6-thioguanine will be measured using a cloning assay. Results of a pilot study are now available as a result of which a further, more extensive study is in progress.

TYPE: Molecular Epidemiology
TERM: Lymphocytes; Radiation, Ionizing
SITE: Inapplicable
CHEM: Radon
TIME: 1990 - 1994

Bristol

831 **Golding, J.** 05148
Univ. of Bristol, Hosp. for Sick Children, Inst. of Child Health, St. Michael's Hill, Bristol BS2 8BJ, United Kingdom (Tel.: +44 272 225967; Fax: 255051)
COLL: Adam, H.; Parks, S.; Verd, S.; Stejskal, J.; Ignatyeva, R.

Childhood Cancer in Six European Centres
This cohort study will involve prospective collection of information on a geographically-based population from pregnancy through until the children are seven years of age. The study cohort comprises 38,000 children. As the data collection progresses, identification of all children with cancer will be made. Prospective information concerning features of the environment, including type of housing, maternal smoking and alcohol consumption, use of illicit drugs and features of the diet, together with maternal illnesses and drugs taken, will be available for analysis. Features of childhood including exposure to passive smoke and other possible mutagens will also be available. The overall wealth of data should enable features of the background of children who develop cancer to be identified.

TYPE: Cohort
TERM: Alcohol; BMB; Blood; Diet; Drugs; Environmental Factors; Mutagen; Passive Smoking; Tissue; Tobacco (Smoking); Urine
SITE: Childhood Neoplasms
CHEM: Nicotine
LOCA: Czech Republic; Greece; Russian Federation; Spain; United Kingdom
TIME: 1991 - 2000

832 **Golding, J.** 05149
Univ. of Bristol, Hosp. for Sick Children, Inst. of Child Health, St. Michael's Hill, Bristol BS2 8BJ, United Kingdom (Tel.: +44 272 225967; Fax: 255051)
COLL: Mott, M.; von Kries, R.

Vitamin K and Pethidine in the Aetiology of Childhood Cancer
An earlier unexpected finding implicated pethidine and vitamin K in the aetiology of childhood cancer. This study is specifically designed to test the hypothesis that these two substances are implicated in childhood cancer. Information is being abstracted from all case-notes of children born in Bristol since 1965 who have developed cancer. The control group concerns the maternal records of a random sample of all other births from 1965. Abstraction of data is done by clerks who are blind as to whether it is a case or control that is under review. Extensive information on potential confounders is also being abstracted.

TYPE: Case-Control
TERM: Clinical Records; Drugs; Vitamins
SITE: Childhood Neoplasms
CHEM: Pethidine
TIME: 1990 - 1994

833 **Harvey, I.M.** 05066
Health Care Evaluation Unit, Canynge Hall Whiteladies Rd, Bristol BS8 2PR, United Kingdom
COLL: Marks, R.; Frankel, S.

Solar Keratosis and Non-Melanoma Skin Cancer in the Elderly Population of South Wales

This descriptive epidemiological study aims to describe the prevalence and incidence of solar keratoses and non-melanoma skin cancer amongst persons over 60 years of age in South Wales. A random sample (n = 1,000) of individuals has been selected and visited by a research dermatologist. Sun exposed areas of the skin have been examined and lesions photographed for validation purposes. Data have been collected on skin type, ancestry and estimated lifetime sun exposure in order to investigate the influence of these factors on prevalence of the lesions.

TYPE: Cross-Sectional
TERM: Pigmentation; Premalignant Lesion; Radiation, Ultraviolet
SITE: Skin
TIME: 1988 - 1994

Cambridge

834 **Davies, T.W.** 05138
Inst. of Public Health, Forrie Site, Robinson Way, Cambridge CB2 2SR, United Kingdom (Tel.: +44 223 330318; Fax: 330330)

East Anglian Diet and Health Study 1992

This case-control study will examine the hypothesis that consumption of dairy products in adolescence is positively correlated with the risk of testicular cancer. Cases will consist of living men who have had testicular cancer. Two cancer controls matched for age (within 2.5 years) are taken from the same cancer registry, as well as two population controls selected by general practitioners and matched for age. Data on current diet will be collected using a self-completed questionnaire and diary, and estimates of past diet will be obtained from the mothers. A pilot study is being conducted first using 20 cases and 80 controls. The main study will include some 200 cases and 800 controls. Date collection is now almost complete and analysis will start.

TYPE: Case-Control
TERM: Adolescence; Dairy Products; Diet; Registry
SITE: Testis
REGI: E. Anglia (UK)
TIME: 1990 - 1994

Cardiff

835 **Jacobs, A.** 04989
Univ. of Wales College of Medicine, Dept. of Epidemiology & Community Med., Heath Park, Cardiff CF4 4XN, Wales, United Kingdom (Tel.: +44 222 747747/2318)
COLL: West, R.R.; Fenaux, P.; Lowenthal, R.M.

Chemical and Environmental Exposures in Myelodysplasia

The main aim of this case-control study is to investigate the lifetime occupational and personal history of patients with myelodysplasia (MDS) with respect to exposure to chemicals and radiation. The study seeks 420 case-control pairs. A secondary aim will be to estimate the incidence of MDS in defined populations. All cases of MDS aged 15+ diagnosed since 1 October 1988, resident in defined areas, will be included. The controls will be matched for sex, age, district of residence and year of presentation. Each control will be randomly selected from patients attending out-patient clinics of the hospital which ascertained the case. Cases and controls will be interviewed by trained interviewers using a structured questionnaire and analysis will be carried out using the methods of Breslow and Day for conditional logistic regression and multivariate analysis of matched data.

TYPE: Case-Control
TERM: Chemical Exposure; Occupation; Radiation, Ionizing
SITE: Myelodysplastic Syndrome
TIME: 1989 - 1994

836 **Mansel, R.E.** 04994
Univ. of Wales College of Medicine, Dept. of Surgery, Heath Park, Cardiff CF4 4XN, Wales, United Kingdom (Tel.: +44 222 747747)
COLL: Gravelle, I.H.; West, R.R.; Miers, M.E.

Wolfe Mammographic Patterns as Predictors of Breast Cancer Risk

Mammographic parenchymal patterns have been proposed as a method of identifying women at high risk of developing breast cancer. In 1976, Wolfe described four breast patterns with reported relative risks as great as 37:1 from the highest (DY) to the lowest risk class (N1). Subsequent studies have produced much smaller quantitative estimates of risk, and several have failed to support Wolfe's hypothesis. This study aims to determine if Wolfe patterns are predictive of breast cancer risk. It is a prospective study of subsequent development of breast cancer among a cohort of 5,000 women who attended the symptomatic breast clinic at the University Hospital of Wales from 1980-1985, and underwent mammography. Each woman was allocated a Wolfe code of N1, P1, P2 or DY. Those with no known breast cancer have been mailed annually and asked to report any breast symptoms. Those known to have a breast cancer have been followed through breast clinic attendance and general practitioner contacts. Data concerning other risk factors have been collected through questionnaires and at clinic attendance. The study will determine the rate of subsequent breast cancer for each of the Wolfe codes allocated at the time of the initial mammogram and will review the role of breast parenchymal patterns in breast cancer risk.

TYPE: Cohort
TERM: High-Risk Groups; Mammography
SITE: Breast (F)
TIME: 1982 - 1994

Cumbria

837 **Berry, R.J.** 02679
Westlakes Research Inst., Ingwell Hall, Westlakes Science & Technology Park, Moor Row, Cumbria CA24 3JZ, United Kingdom (Tel.: +44 946 591147; Fax: 514042)
COLL: Slovak, A.; Binks, K.; McElvenny, D.

Cancer Risk in Radiation Workers

Mortality and morbidity studies of the British Nuclear Fuels workforce, begun in 1976, have been extended to include both current and, as far as practicable, all former employees. Mortality patterns for the Sellafield workforce have been examined with respect to external ionizing radiation (Smith and Douglas, Br. Med. J. 293:845, 1986). Both of these studies are expected to include cancer morbidity and further follow-up to be completed during 1993. A further extension of the Sellafield study will incorporate organ-specific radiation doses derived from data on plutonium in urine. The total BNFL workforce study concerns approximately 40,000 present and past employees.

TYPE: Cohort
TERM: Late Effects; Occupation; Radiation, Ionizing; Registry
SITE: All Sites
OCCU: Radiation Workers
TIME: 1976 - 1994

Didcot

***838** **Duncan, K.P.** 05314
National Radiological Protection Board, Chilton Didcot Oxfordshire, OX11 0RQ, United Kingdom (Tel.: +44 235 831600; Fax: 833891 ; Tlx: 837124)
COLL: Kendall, G.M.; Muirhead, C.R.

National Registry for Radiation Workers

A long-term follow-up study is being conducted of occupational radiation exposure and mortality, in particular from cancer. The registry consists of all workers in the UK, who are monitored regularly for exposure to radiation, and whose dose records are kept and can (with consent) be transferred. Follow-up information is obtained principally from the National Health Service Central Registers, as well as from organisations such as the Department of Social Security. The first analysis has recently been published (Kendall et al., Br. Med. J. 304:220-225, 1992), based on a cohort of 95,217 workers having a mean lifetime dose of 33.6 mSv. The cohort is still fairly young, and the number of deaths from all causes was about 6,600. There was a clear healthy worker effect relative to national rates: SMRs of 85 for all causes and 86 for all malignant neoplasms, based on a 10 year lag. Within the cohort there was evidence of an increasing trend in risk with radiation dose for leukaemia, excluding chronic lymphatic and, to a lesser extent, for all cancers. However, the confidence intervals for the associated risk estimates were wide, and corresponding risk factors recommended by the International Commission on Radiological Protection fall well within these ranges. Follow-up of the cohort will continue and additional groups of workers will be included, in order to increase the power of the study.

TYPE: Cohort; Registry
TERM: Monitoring; Occupation; Radiation, Ionizing
SITE: All Sites
OCCU: Radiation Workers
TIME: 1976 - 1999

839 **MacGibbon, B.H.** 01116
National Radiological Protection Board, Chilton, Didcot Oxon OX11 ORQ, United Kingdom (Tel.: +44 235 831600; Fax: 833891 ; Tlx: 837124 RADPRO G)
COLL: Kendall, G.M.; Muirhead, C.R.

National Registry for Radiation Workers

A long-term follow-up study is being conducted to investigate any relationship between occupational exposure to radiation and mortality, in particular from cancer. Any worker in the U.K. who is monitored regularly for exposure to radiation, and for whom dose records are kept and can (with consent) be transferred, is entered in the registry. Currently about 100,000 workers are included, having a collective dose in excess of 3000 man Sv. Follow-up information is being obtained principally from the National Health Service Central Registers, but also from the Department of Social Security. Data validation includes checks of health physics records and sample checks of the follow-up mechanisms. The main analysis will consist of a test for trend in mortality rates across different dose groups. The estimate and confidence limits for the trend in cancer rates with dose will be compared with results from studies of populations exposed to high radiation doses, such as the Japanese atomic bomb survivors.

TYPE: Cohort; Incidence; Registry
TERM: Dose-Response; Monitoring; Occupation; Radiation, Ionizing; Trends
SITE: All Sites
OCCU: Radiation Workers
TIME: 1976 - 1994

840 **MacGibbon, B.H.** 03502
National Radiological Protection Board, Chilton, Didcot Oxon OX11 ORQ, United Kingdom (Tel.: +44 235 831600; Fax: 833891 ; Tlx: 837124 RADPRO G)
COLL: Darby, S.C.; Kendall, G.M.; Muirhead, C.R.; Doll, R.

Health of Participants in the UK Atmospheric Nuclear Weapon Tests

At the request of the Ministry of Defence (MOD), a study is being carried out of mortality and cancer incidence among participants in the UK programme of atmospheric nuclear weapons tests, held in Australia and the Pacific. A group of 22,347 participants and a control group of 22,326 servicemen and civilians who did not participate were identified from MOD archives. The National Health Service Central Registers provide the principal mechanism for tracing the study subjects and acquiring death certificates and cancer registration details. However, additional follow-up via the Department of Social Security increased the number of deaths fully identified by 6.5% for the period to the beginning of 1984 (J. Epidemiol. Comm. Hlth 45(1):65-70, 1991). Based on the follow-up over this period, it was concluded that test participation does not have a detectable effect on life expectancy or on the risk of cancer, apart possibly from small hazards of leukaemia and multiple myeloma (Br. Med. J. 296:332-338, 1988). A further five years' follow-up is planned. Analyses based on the combined participant and control cohorts have also been used to examine mortality among UK servicemen who served abroad in the 1950s and

1960s (Br. J. Ind. Med. 47:793-794); mortality from non-violent causes was less than that expected from rates for all men in England and Wales.

TYPE: Cohort
TERM: Radiation, Ionizing
SITE: All Sites
TIME: 1983 - 1994

Dundee

841 **Williams, F.L.R.** 05009
Ninewells Hosp. and Medical School, Dept. of Epidemiology and Public Health, Dundee DD1 9SY, Scotland, United Kingdom (Tel.: +44 382 60111/2083; Fax: 644197 ; Tlx: 76293 uldund g)
COLL: Lloyd, O.L.L.

Lung Cancer and Sources of Air-Borne Pollution

It has been suggested that abnormal sex ratios of births may signal the presence of environmental stresses. If so, the detection of such abnormalities would constitute a simple screening procedure to alert medical and environmental health authorities to hazards to health. Previous work has demonstrated a link between abnormal sex ratios of births, a high incidence of lung cancer, and air-borne pollution from foundries. The purpose of this (pilot) study is to evaluate the hypotheses that the sex ratios of births are abnormal (either low or high) in areas where the parents have been residentially exposed to air pollution from foundries or incinerators; and that mortality from all causes and/or from lung cancer in these areas is abnormally high. The sex ratios and mortality will be investigated around 12 sources of air pollution; 12 comparison areas will also be investigated.

TYPE: Correlation; Mortality
TERM: Air Pollution; Sex Ratio
SITE: Lung
TIME: 1990 - 1994

Edinburgh

***842** **Dunlop, M.G.** 05469
Western General Hosp., MRC Human Genetics Unit, Crewe Rd, Edinburgh EH4 2XU Scotland, United Kingdom (Tel.: +44 31 3322471; Fax: 3432620)
COLL: Prosser, J.; Wyllie, A.; Bird, C.C.; Carter, D.

Family Studies in Colorectal Cancer

To delineate the prevalence of genetic susceptibility to colorectal cancer in Scotland, case-control studies of a consecutive unselected series of Scottish patients with colorectal cancer are being carried out. Full ascertainment is made of all 1st and 2nd relatives, including all causes of morbidity and mortality. Controls are from family doctor registers. 228 cases and 142 controls have been interviewed by a research nurse. Family data confirmation has been carried out from hospital records, cancer registration and central population registers of births and deaths on 107 individuals to date. The aim is to (1) assess the genetic effect on the incidence of cancer in Scotland; (2) allow analysis of the risk of developing colorectal cancer in the relatives of colorectal cancer cases; (3) carry out segregation analysis to assess the mode of genetic inheritance; and (4) obtain blood samples for genetic linkage analysis.

TYPE: Case-Control; Genetic Epidemiology
TERM: BMB; Blood; Linkage Analysis; Segregation Analysis
SITE: Colon; Rectum
TIME: 1990 - 1994

***843** **Dunlop, M.G.** 05480
Western General Hosp., MRC Human Genetics Unit, Crewe Rd, Edinburgh EH4 2XU Scotland, United Kingdom (Tel.: +44 31 3322471; Fax: 3432620)
COLL: Prosser, J.; Wyllie, A.; Bird, C.C.; Carter, D.

Extracolonic Features of Polyposis Coli in Cases of Non-Polyposis Colorectal Cancer

This is a case-control study of the prevalence of retinal pigmentation and mandibular oeseomas (features associated with polyposis coli) in consecutive series of non-polyposis colorectal patients. A sub-analysis, depending on extent of family history and age at diagnosis, will be carried out. To determine whether there is an association of retinal pigmentation +/- osteomas in non-polyposis colorectal cancer, 250 cases and 150 controls have undergone x-ray examination of the mandible and retinal examination by indirect ophthalmoscopy +/- photography.

TYPE: Case-Control
TERM: Age; Disease, Other; Familial Factors
SITE: Colon; Rectum
TIME: 1991 - 1994

844 **MacLaren, W.M.** 04550
Inst. of Occupational Medicine, 8 Roxburgh Place, Edinburgh EH8 9SU, United Kingdom (Tel.: +44 31 6675131)
COLL: Agius, R.; Crawford, N.

Exposure to Low Levels of Radon and Thoron Daughters

The aim of the project is to study relationships between exposure to low levels of radon and thoron daughters and mortality, particularly that due to lung cancer, at 10 British coal mines. The study population consists of approximately 15,000 British coal miners employed at 10 mines in Scotland, the North and Midlands of England, and South Wales. All of these men attended either the first round of medical surveys (1953-1958), or the second (1958-1963), of the British National Coal Board's Pneumoconiosis Field Research. Miners involved in this project were medically examined at five-yearly intervals over a period of approximately 27 years. A systematical attendance record system has ensured that information on where the men worked within the mines and for how long, is stored on computer file. The intention is to combine these data with underground measurements of radon and thoron daughter activities, gathered during the 1970's, in order to produce estimates of cumulative exposure. The facilities of the National Health Service Central Register are being used to collect mortality data, and the statistical analysis will compare exposure histories of miners who have died with those of survivors.

TYPE: Cohort
TERM: Metals; Occupation; Radiation, Ionizing; Registry
SITE: Lung
CHEM: Radon; Thoron
OCCU: Miners, Coal
TIME: 1988 - 1994

Glasgow

***845** **Reid, R.** 05356
Univ. of Glasgow, Western Infirmary, Dept. of Pathology, Scottish Bone Tumour Registry, Glasgow G11 6NT, United Kingdom
COLL: Hamblen, D.L.

Scottish Bone Tumour Registry

A prospective registry of bone tumours diagnosed in Scotland was established in 1962, with long term follow-up of malignant lesions indefinitely, and of benign lesions for a minimum of five years. Data collected include histological sections, radiographs and comprehensive clinical data. The registry is used to validate cancer registration notifications and to provide a database for pathological and orthopaedic research.

TYPE: Incidence; Registry
TERM: BMB; Histology; Registry
SITE: Bone
REGI: Scotland (UK)
TIME: 1962 - 1999

Hartshill

846 **Knight, T.** 04964
Univ. of Keele, School of Postgrad. Medicine, Thornburrow Drive, Hartshill Stoke-on-Trent ST4 7QB, United Kingdom (Tel.: +44 782 716699; Fax: 747319)
COLL: Forman, D.; Newell, D.; Hengels, K.J.; Wyatt, J.; Buchanan, D.

Biochemical Markers for Detection of Gastric Precancerous Disease

The aim of this study is to assess the feasibility of using serum levels of pepsinogen I and II to detect atrophic gastritis (AG) and intestinal metaplasia (IM) in a high risk population of 1,000 men aged 25-60 years, employed in local industries. Subjects with Pepsinogen I level <20 ng/ml or Pepsinogen I:II <1.5 will be designated cases of AG with/without IM (estimated prevalence 2-4%, i.e. 20-40 cases). All cases of AG and IM and age-matched controls per case, will be endoscoped and biopsied (2 prepyloric, 2 at the incisura, 2 fundic) to confirm the diagnosis. Serum gastrin levels, the prevalence of intragastric infection with H. pylori and current and dietary patterns will be assessed using a three-day semi-quantitative dietary record and a food frequency questionnaire. Serum pepsinogen and gastrin levels will be determined by RIA. Infection with H. pylori will be determined by measurement of antibodies to the organism in serum. Dietary intake data will be analysed by computer software developed at the MRC Dunn Nutrition Unit, Cambridge, UK.

TYPE: Case-Control; Cohort
TERM: Atrophic Gastritis; BMB; Biochemical Markers; Diet; H. pylori; Premalignant Lesion; Serum
SITE: Stomach
TIME: 1990 - 1994

Lancaster

847 **Gorst, D.W.** 04985
Royal Lancaster Infirmary, Ashton Rd, Lancaster LA1 4RP, United Kingdom (Tel.: +44 524 65944)

Northwest Adult Leukaemia Register

This study is a prospective collection of demographic, medical and follow-up data on all cases of acute myeloid, acute lymphoblastic and chronic granulocytic leukaemia occurring in adults (aged >15) in a geographically predefined area. Data are used to test hypotheses of incidence (e.g. coastal excess) and to establish true outcome and survival figures in unselected patients. At present, there are over 1,000 patients on file. Data are obtained from the haematologist on diagnosis using a proforma, stored and analysed using microcomputer and standard statistical techniques.

TYPE: Incidence; Registry
TERM: Geographic Factors; Survival
SITE: Leukaemia (ALL); Leukaemia (AML); Leukaemia (CML)
TIME: 1987 - 1994

Leeds

848 **Cade, J.E.** 04956
Dept. of Public Health Medicine, 20 Hyde Terrace, Leeds LS2 9LN, United Kingdom (Tel.: +44 532 334862; Fax: 334852)
COLL: Taylor, I.; Waters, W.E.; Perry, M.; Jackson, A.; Campbell, M.

Diet in Women with Asymptomatic Breast Cancer

The aim of the study is to compare nutrient intake (especially fat intake) between women with asymptomatic breast cancer, benign breast disease and normal, control women. Over a two-year period, women attending the assessment clinics of the breast screening programmes in Portsmouth and Southampton will be included in the study. All women are interviewed prior to diagnosis. The interview includes questions on diet and other possible risk factors and women are given a more detailed dietary questionnaire for completion unaided. Height, weight, waist and hip girth are also measured. It is anticipated that about 300 women with breast cancer, 400 women with benign breast disease and 1,400 control women will be included in the study. The control group will be randomly selected from women who are found at the assessment clinic to have no breast disease. Due to the advent of breast screening in the UK for women aged 50-64 years, this is the first opportunity to study women with asymptomatic breast cancer. The opportunity for bias is reduced since at the time of interview the diagnosis is not known to the women or to the interviewer.

TYPE: Case-Control
TERM: Diet; Fat; Nutrition; Physical Factors
SITE: Breast (F)
TIME: 1989 - 1994

849 **Cartwright, R.A.** 05215
Univ. of Leeds, Leukaemia Research Fund, Centre for Clinical Epidemiology, 17 Springfield Mount, Leeds LS2 9NG, United Kingdom (Tel.: +44 532 443517; Fax: 426065)
COLL: Alexander, F.E.; Staines, A.; McNally, R.J.Q.

Case-Control Study of Adult Acute Leukaemia

The hypotheses to be tested are: (1) household exposure to radon gas is greater in past case-houses than in control-houses; (2) cases have more occupations related to 'electrical' work; (3) aplasia-forming drugs have been used more frequently by cases; and (4) family history of similar conditions is more frequent in case families. 800 or more cases, and double the number of controls obtained from general practitioners' records, will be interviewed. Data on exposures will be obtained from medical records and from measurements of radon gas.

TYPE: Case-Control
TERM: Chemical Exposure; Drugs; Family History; Occupation
SITE: Leukaemia
CHEM: Radon
OCCU: Electrical Workers
TIME: 1991 - 1996

***850** **Cartwright, R.A.** 05520
Univ. of Leeds, Leukaemia Research Fund, Centre for Clinical Epidemiology, 17 Springfield Mount, Leeds LS2 9NG, United Kingdom (Tel.: +44 532 443517; Fax: 426065)
COLL: Alexander, F.E.; McNally, R.J.Q.; Staines, A.

Leukaemia and Lymphoma Registry

This registry collects data on cases of leukaemia and lymphoma diagnosed in all hospitals in the areas of the UK covered by this investigation, which was initiated in 1984. The registered cases are later cross-checked with the regional registries for completion and elimination of multiple registrations. All the case are diagnostically reviewed by haematologists and pathologists. So far, 45,033 cases have been registered. Analyses carried out include geographical distribution and time trends.

TYPE: Incidence; Registry
TERM: Clinical Records; Data Resource; Geographic Factors; Trends
SITE: Leukaemia; Lymphoma
TIME: 1984 - 1994

851 **Howel, D.** 04988
Univ. of Leeds, Academic Unit of Public Health Med., 30 Hyde Terrace , Leeds LS2 9LN, United Kingdom (Tel.: +44 532 334856; Fax: 334852)
COLL: Arblaster, L.; Swinburne, L.; Schweiger, M.; Gibbs, A.R.

Asbestos Exposure and Mesothelioma: A Community Study

This is a case-control study aiming to compare asbestos exposure in two groups, and in particular the relative contributions of occupational, household, neighbourhood and incidental contact. Information will be sought from the surviving relatives of recent cases (approximately 200) and matched controls in Yorkshire. The data collected will be related to the results of a quantitative mineral fibre analysis by electron microscopy of lung samples of both cases and controls.

TYPE: Case-Control
TERM: Air Pollution; Chemical Exposure; Dusts; Environmental Factors; Occupation; Registry
SITE: Mesothelioma
CHEM: Mineral Fibres
REGI: Yorkshire (UK)
TIME: 1990 - 1994

***852** **Staines, A.** 05540
Univ. of Leeds, Leukaemia Research Fund Centre for Clinical Epidemiology, 17 Springfield Mount, Leeds LS2 9NG, United Kingdom (Tel.: +44 532 443517; Fax: 426065)
COLL: Cartwright, R.A.; Bailey, C.; Lewis, I.; Kinsey, S.; Proctor, L.

Yorkshire Regional Childhood Tumour Registry

This is a children's cancer registry covering the entire population (approximately 3.5 million people) in the Yorkshire Regional Health Authority, from 1974 to the present. Cases are ascertained from the regional treatment centres, neurosurgical centres, the regional cancer registry, the national children's tumour registry and the UK Childhood Cancer Study Group records. All children with malignant disease, intracranial tumours, and a number of related conditions, such as histiocytosis and aplastic anaemia are registered. The cases without histological confirmation are also registered, but this is clearly recorded. The vital status of all cases is ascertained every two years. The registry is used for a number of studies, including descriptive epidemiology, case-control studies, age-period cohort analyses, clustering and survival analysis. A special interest of this registry includes epidemiology, histological review and survival analysis of childhood intracranial tumours registered between 1974 and 1988.

TYPE: Case-Control; Incidence; Registry
TERM: Cluster; Histology; Registry; Survival
SITE: Childhood Neoplasms
REGI: Oxford II (UK); Yorkshire (UK)
TIME: 1982 - 1994

Leicester

***853** **Farmer, P.B.** 05318
Univ. of Leicester, MRC Toxicology Unit, Hodgkin Bldg, Lancaster Rd , Leicester LE1, United Kingdom
COLL: Autrup, H.; Waters, R.; Kyrtopoulos, S.; Srám, R.J.

Human Exposure to Urban Environmental Carcinogenic Pollutants

The objective of the project is the development of methods for assessing the genetic burden, resulting from exposure to chemical carcinogens, in the general population. This will be achieved by (1) the design of novel biomonitoring methods for detecting adducts of carcinogens with DNA and protein; (2) a comparison of adduct levels in DNA from different tissues which will indicate the most appropriate source to use for biomonitoring purposes, and (3) the establishment of baseline data for adduct levels, for later epidemiological studies.Carcinogen DNA adducts will be determined by 32P-postlabelling, GC-MS and competitive repair assays and oxidative damage to DNA by MS or HPLC. The level of damage will be determined in tissues from smokers and non-smokers and from rural/non-rural population.

TYPE: Methodology
TERM: Air Pollution; Biomarkers; DNA Adducts; Lymphocytes; Transplacental; Urban
SITE: Inapplicable
LOCA: Czech Republic; Denmark; Greece; United Kingdom
TIME: 1993 - 1995

854 **Jones, D.R.** 03837
Univ. of Leicester, Dept. of Epidemiology & Public Health, Leicester Royal Infirmary, Clinical Sciences Bldg, P.O.Box 65, Leicester LE2 7LX, United Kingdom (Tel.: +44 533 523196; Fax: 523272)
COLL: Goldblatt, P.O.

Cancer Mortality Following Widowhood in the OPCS Longitudinal Study

This study makes use of the OPCS Longitudinal study, in which routinely collected data on deaths, and deaths of a spouse: occurring in a 1% sample of the population of England and Wales in the period 1971-1985, are linked together, and with 1971 and 1981 census records of sample members. The timing and causes of death following the potentially very stressful event of conjugal bereavement may thus be analysed. The effects of several measures of socio-economic status, including social class are being investigated, as are potential effects of.social or familial support, measured by household structure and numbers of children. Papers appeared in Stress Med. 2:129-140, 1986 and J. Biosociol. Sci. 19:107-121, 1987.

TYPE: Cohort
TERM: Psychosocial Factors; Socio-Economic Factors; Stress
SITE: All Sites; Breast (F); Lung; Stomach
TIME: 1982 - 1994

855 **Jones, D.R.** 04448

Univ. of Leicester, Dept. of Epidemiology & Public Health, Leicester Royal Infirmary, Clinical Sciences Bldg, P.O.Box 65, Leicester LE2 7LX, United Kingdom (Tel.: +44 533 523196; Fax: 523272)

COLL: Ellman, R.; Thomas, B.A.

Bereavement, Coping Strategies, Suppression of Emotion and Breast Cancer Risk

This study aims to test the hypotheses that (1) death of a spouse and (2) a tendency to suppress emotions (particularly anger) are associated with a raised risk of breast cancer. The roles played by coping strategies adopted in response to stressors, and by the availability of social support will be investigated. More than 12,000women screened at Guildford have provided brief details of any widowhood suffered, and have completed a questionnaire about their emotional and coping responses. Follow-up to detect new cases of or deaths from breast cancer, and hence allow the risk associated with bereavement, etc. to be estimated, is in progress. Other risk factors which are already measured in evaluating the screening programme will be used as covariates in the analyses.

TYPE: Cohort
TERM: Psychosocial Factors; Registry; Screening; Stress
SITE: Breast (F)
TIME: 1989 - 1994

Liverpool

856 **Osman, J.** 02170

Health and Safety Executive, Epidemiology & Medical Statistics Unit, Stanley Precinct, Magdalen House, Bootle, Liverpool Merseyside L20 3QZ, United Kingdom (Tel.: +44 51 9514535; Fax: 951331)

COLL: Hodgson, J.T.

Asbestos Survey

All asbestos workers in the UK are seen at 2-yearly intervals and a morbidity questionnaire completed with clinical examination of lung fields. The mortality experience of the same population is also recorded. Hygiene measurements from some factories are available from 1972. Smoking histories are recorded. A mortality report on male asbestos workers in England and Wales between 1971-1981 was published in Br. J. Ind. Med. 43:158-164, 1986. This study provides continuous surveillance of illness and death in employees known to be exposed to asbestos. An analysis of mortality up to 1990 is planned for 1993.

TYPE: Cohort
TERM: Dusts; Minerals; Occupation; Tobacco (Smoking)
SITE: All Sites; Respiratory
CHEM: Asbestos
OCCU: Asbestos Workers
TIME: 1971 - 1994

857 **Osman, J.** 04473

Health and Safety Executive, Epidemiology & Medical Statistics Unit, Stanley Precinct, Magdalen House, Bootle, Liverpool Merseyside L20 3QZ, United Kingdom (Tel.: +44 51 9514535; Fax: 951331)

COLL: Agius, R.; Hodgson, J.T.

Mortality Study of Scottish Hard Rock Quarry Workers

Current and past workers of about 30 hard rock quarries in Scotland has been enrolled in a prospective mortality study as part of a combined morbidity and mortality study. The total numbers are about 1,500 current and 1,500 past workers. Dust levels and mineralogical composition has been measured at all quarries, and for different quarrying processes. The principal research question for the study was the exposure/response relationship for silicosis, and for any related excesses in respiratory disease mortality. Lung cancer and mortality will also be examined in relation to silica exposure, and compared with national mortality rates.

TYPE: Cohort
TERM: Dusts; Minerals; Occupation
SITE: Lung
CHEM: Silica
OCCU: Quarry Workers
TIME: 1988 - 1994

858 **Osman, J.** 04475
Health and Safety Executive, Epidemiology & Medical Statistics Unit, Stanley Precinct, Magdalen House, Bootle, Liverpool Merseyside L20 3QZ, United Kingdom (Tel.: +44 51 9514535; Fax: 951331)
COLL: Leon, D.A.

Print Workers Lung Cancer Case-Control Study

A mortality study of 10,791 males in the printing industry found a significant excess of lung cancer amongst machine assistants. In order to make a more definite assessment of the possible lung cancer risk associated with the printing industry, a case-control study was set up to investigate whether there were any relationships with duration of exposure, time period of exposure and place of employment. The study design is case-control, nested within the main cohort study. There are three controls for each of the 121 cases. The controls are selected at random from the population of men of the same age as the case at death and who were at risk in the period that the death occurred.

TYPE: Case-Control
TERM: Dose-Response; Latency; Occupation
SITE: Lung
OCCU: Machinists; Printers
TIME: 1987 - 1994

859 **Osman, J.** 05122
Health and Safety Executive, Epidemiology & Medical Statistics Unit, Stanley Precinct, Magdalen House, Bootle, Liverpool Merseyside L20 3QZ, United Kingdom (Tel.: +44 51 9514535; Fax: 951331)
COLL: Hodgson, J.T.; McCaig, R.H.

Childhood Leukaemias in Offspring of Radiation Workers

This is a case-control study based on a population drawn from the past and present workforce of the nuclear fuel reprocessing plant at Sellafield. Information will be gained from official records and other sources. The aim of the study is to evaluate the effects on risk of the occupational histories of the fathers of children with leukaemia and non-Hodgkin's lymphoma. Factors taken into account will include internal and external exposure to ionising radiation, exposure to known carcinogenic chemicals and involvement in contamination incidents.

TYPE: Case-Control
TERM: Chemical Exposure; Childhood; Parental Occupation; Radiation, Ionizing; Registry
SITE: Leukaemia; Non-Hodgkin's Lymphoma
OCCU: Power Plant Workers
TIME: 1990 - 1994

860 **Pharoah, P.** 04479
Univ. of Liverpool, Dept. of Public Health, P.O. Box 147, Liverpool L69 3BX, United Kingdom (Tel.: +44 51 7945593; Fax: 7945588 ; Tlx: 627095 unilpl g)
COLL: Ashby, D.; Blettner, M.; Greaves, J.; Roberts, R.J.

Cancer Risk after Exposure to Cutting Oil Mist and Cutting Fluids

A historical prospective mortality study is being undertaken to assess if long term exposure to cutting fluids, during the course of machining, grinding or other cutting operations, is associated with excess mortality, mainly from digestive and respiratory cancers. Data from 25,000 current and past employees, from two automobile factories on Merseyside, form the cohort for the study. Data on the workers are being obtained from personnel records and followed up via OPCS. Site-specific mortality rates will be computed for each exposure category and compared to the rates expected based on the general UK population. Internal comparisons will be made between groups with and without exposure to particular types of cutting fluids. Dose-response models for length and level of exposure will be investigated.

TYPE: Cohort
TERM: Chemical Exposure; Dose-Response; Latency; Occupation; Petroleum Products; Registry
SITE: Colon; Lung; Rectum; Stomach
CHEM: Mineral Oil; PAH
OCCU: Automobile Workers; Machinists
TIME: 1988 - 1994

861 **Pharoah, P.** 04998
Univ. of Liverpool, Dept. of Public Health, P.O. Box 147, Liverpool L69 3BX, United Kingdom (Tel.: +44 51 7945593; Fax: 7945588 ; Tlx: 627095 unilpl g)
COLL: Ashby, D.; Blettner, M.

Effects of Exposure to Oil Mist

The aim of this cohort study is to determine whether there is an excess of cancer mortality following exposure to oil mist. A group of workers exposed to oil mist in two car manufacturing industries and a control group of non-exposed persons employed in the same industries are being compared. The cohort, which will involve 25,000 men, is being traced through the National Health Service Central Register, which provides information on cause of death.

TYPE: Cohort
TERM: Occupation; Petroleum Products; Registry
SITE: Lung; Stomach
CHEM: Oil Mist
OCCU: Automobile Workers
TIME: 1988 - 1994

London

***862** **Atkin, W.S.** 05444
Imperial Cancer Research Fund St Mark's Hosp., City Road, London EC1V 2PS, United Kingdom (Tel.: +44 71 6082323; Fax: 250)
COLL: Cuzick, J.M.; Northover, J.M.A.

Long-term Risk of Colorectal Cancer in Symptomatic Patients Aged between 50 and 60 years with a Negative Sigmoidoscopy

65% of colorectal cancers are located within the reach of the 65 cm flexible sigmoidoscope. A single flexible sigmoidoscopy screening at some time between 55 and 60 years would probably prevent almost 50% of colorectal cancers occurring up to age 80. This hypothesis is based on (1) previous findings that after removal of adenomas at sigmoidoscopy, the risk of subsequent rectal cancer becomes low for many years (New Engl. J. Med., 326:658-662, 1993; Atkin et al.), and (2) the prevalence of adenomas in average-risk persons undergoing flexible sigmoidoscopy screening increases with age, but peaks before age 60 at about 10%. This suggests that a single screening should identify the majority of persons with distal adenomas. Conversely, persons with a negative sigmoidoscopy should be at low risk of developing distal bowel cancer. This study aims to test this hypothesis and to determine the age at which screening sigmoidoscopy would be most effective. Considered for inclusion were all patients aged between 50 and 60 years who presented at St Mark's Hospital between 1958 and 1970, who underwent rigid sigmoidoscopy and in whom no adenomas or cancers were found. A total of 6,500 patients are being traced through the National Health Service Central Register of the OPCS. The incidence and mortality of colorectal cancer in this cohort will be compared with that of the local population (the region covered by the Thames Cancer Registry).

TYPE: Cohort
TERM: Registry; Screening
SITE: Colon; Rectum
REGI: Thames (UK)
TIME: 1992 - 1994

863 **Barreto, S.M.** 05210
London School of Hygiene and Tropical Medicine, Dept. of Epidemiology, Keppel St., London WC1E 7HT, United Kingdom (Tel.: +44 71 6368636; Fax: 4365389)
COLL: Swerdlow, A.J.; Smith, P.G.; Boffetta, P.; Kogevinas, M.; Andrade, A.

Historical Cohort Study of Brazilian Steelworkers

A retrospective cohort study is being conducted to assess the all-cause mortality of 12,000 Brazilian steelworkers between 1 January 1979 and 31 December 1990. National and regional age-specific death rates will be used to generate expected numbers of deaths for this cohort. Work and health histories are being obtained from company records and cause of death will be taken from death certificates.

TYPE: Cohort
TERM: Metals; Occupation
SITE: All Sites
CHEM: Steel
OCCU: Steel Workers
LOCA: Brazil
TIME: 1989 - 1994

***864** **Barreto, S.M.** 05302
London School of Hygiene and Tropical Medicine, Dept. of Epidemiology, Keppel St., London WC1E 7HT, United Kingdom (Tel.: +44 71 6368636; Fax: 4365389)
COLL: Swerdlow, A.J.; Smith, P.G.; Boffetta, P.; Kogevinas, M.; Andrade, A.

14 Years Follow-Up Study of Brazilian Steel Workers

A retrospective cohort study is being conducted to assess the cause-specific mortality of 17,000 Brazilian steel workers between 1 January 1979 to 31 December 1992. Follow-up is complete for 80% of cohort members. Several national databases are being used to trace cohort members. Local and regional age- and cause-specific death rates will be used to generate expected numbers of deaths. Work and exposure histories are being obtained from company records. Major health hazards are benzene, PAH and silica dust. Causes of deaths will be taken from the death certificate, and coded to ICD-9. A sample will be compared with hospital records in order to estimate the validity of death certificates.

TYPE: Cohort
TERM: Dusts; Occupation; Solvents
SITE: Leukaemia; Respiratory
CHEM: Benzene; PAH; Silica
OCCU: Steel Workers
TIME: 1991 - 1994

***865** **Bourne, T.H.** 05247
King's College Hosp., Dept. of Obstetrics & Gynaecology, Ovarian Gynaecol. Ultrasound Unit, Denmark Hill, London SE5 8RX, United Kingdom (Tel.: +44 71 3263168; Fax: 7372546)
COLL: Collins, W.; Campbell, S.; Hampson, J.; Reynolds, K.

Prevalence and Incidence of Ovarian Cancer in Women with a Family History of the Disease

The aim of the study is (1) to determine the prevalence and incidence of ovarian and other cancers in women with a family history of the disease; (2) to investigate the use of vaginal ultrasound, colour doppler and serum tumour markers as screening methods for the early stages of the disease; (3) to collect blood and tissue from the cohort for genetic analyses. 2,000 women with one or more first degree relatives with ovarian cancer will be recruited. Blood samples will be taken and stored and vaginal ultrasound performed. All patients with cysts will have surgery. Patients will be followed-up every 6 months for repeat scan and blood samples.

TYPE: Cohort; Cross-Sectional; Incidence
TERM: BMB; Blood; Familial Factors; Screening; Tissue; Tumour Markers
SITE: Ovary
TIME: 1998 - 1995

866 **Cuzick, J.M.** 04589
Imperial Cancer Research Fund, Dept. of Mathematics, Statistics and Epidemiology, Lincoln's Inn Fields P.O.Box 123, London WC2A 3PX, United Kingdom (Tel.: +44 71 2420200; Tlx: 265107 ICRF G)
COLL: DeStavola, B.; Cartwright, R.A.; Glashan, P.

Genetic Basis of Palmar Keratoses in Bladder Cancer

A very high prevalence of palmar keratoses in patients with bladder cancer has been reported. A relationship was found with smoking, but the relation was not strong enough to explain the findings. The hypothesis is that palmar keratoses might be a marker for the way carcinogens are metabolised. The present case-control study examines the genetic component of palmar keratoses in bladder cancer. Cases are first-degree blood relatives of bladder cancer patients. Two control groups are used; the spouse and his/her first-degree relatives and general hospital controls. Cases and controls are interviewed and their palms inspected for keratoses. 200 will be studied, along with their spouses, and family members. 400 hospital controls will be sought.

TYPE: Case-Control
TERM: Genetic Factors; Premalignant Lesion; Tobacco (Smoking)
SITE: Bladder
TIME: 1986 - 1994

867 **Cuzick, J.M.** 04053
Imperial Cancer Research Fund, Dept. of Mathematics, Statistics and Epidemiology, Lincoln's Inn Fields P.O.Box 123, London WC2A 3PX, United Kingdom (Tel.: +44 71 2420200; Tlx: 265107 ICRF G)
COLL: Singer, A.; Tessy, G.; Hollingworth, T.

Cervix Cancer and Papilloma Virus

The aim is to determine the incidence and progressive potential of papilloma viruses in the causation of cervix cancer, and to examine interaction with other possible factors including smoking, oral contraceptive use, and sexual behaviour. 100 patients with invasive cervix cancer under the age of 40 will be examined in a case-control study, with controls matched by age and family doctor. A prospective study of 2,000 women attending Family Planning Clinics, who are HPV16 positive but with negative cytologies, will also be carried out. Controls will also be selected from this cohort. A quantitative PCR assay for HPV is being used to relate disease state to the level of HPV16.

TYPE: Case-Control; Cohort
TERM: HPV; Oral Contraceptives; PCR; Sexual Activity; Tobacco (Smoking)
SITE: Uterus (Cervix)
TIME: 1985 - 1994

868 **Davies, A.P.** 04823
The London Hosp., Dept. of Obstetrics and Gynaecology, Ovarian Cancer Screening Unit, Whitechapel, London E1 1BB, United Kingdom (Tel.: +44 71 3777674)

Multimodal Screening Programme for Familial Ovarian Cancer

It is planned to create a register of 5,000 women aged 35 years and over who are at increased risk of developing ovarian cancer in view of their family history. An at-risk woman will be defined as having: (1) one first-degree relative with ovarian cancer; or (2) two second-degree relatives with ovarian cancer; or (3) two or more relatives who have developed other cancers before the age of 50 (Lynch II families). Details of their family history and exposure to other suspected risk factors will be collected. All volunteers will undergo annual screening for 3 years using a combination of serum CA-125 and real-time pelvic ultrasonography. It is hoped to: a) assess the specificities and sensitivities of the screening methods individually and in combination for pre-clinical ovarian cancer; b) assess whether screening has had an effect upon the stage distribution of the disease at presentation and upon mortality from ovarian cancer; c) identify the genetic basis for the increased cancer risk.

TYPE: Cohort; Genetic Epidemiology
TERM: Familial Factors; Genetic Factors; Screening; Stage
SITE: Ovary
TIME: 1990 - 1994

869 **Elliott, P.** 05194
London School of Hygiene and Tropical Medicine, Keppel St. (Gower St.), London WC1E 7HT, United Kingdom (Tel.: +44 71 9272415; Fax: 5804524)

Small Area Health Statistics Unit (SAHSU)

This is a new government-funded project set up in reponse to a recommendation of a state enquiry into the raised incidence of childhood leukaemia near the nuclear reprocessing plant at Sellafield. The main objective is to study the risk of disease, especially cancer, near industrial sources of environmental pollution. The project is exploiting British mortality data from 1981 and cancer incidence data from 1974 (1975 in Scotland) as well as small-area population data from the decennial census. Geographical data retrieval is by postcode of residence, there being some 1.6 million postcodes in the UK with an average of 15 households per postcode. Health and population data can be rapidly assembled for circles of arbitrary size around any point in Britain, and the observed number of cases related to those expected from national or regional rates, after standardisation for age, sex and socio-economic classification of area. Current projects include a study of haemangiosarcoma of the liver around vinyl chloride works, leukaemia around benzene works and respiratory cancers around coke works. Studies of larynx cancer around incinerators of oils and solvents, and the distribution of childhood leukaemia in Britain (in collaboration with the Childhood Cancer Research Group in Oxford) have recently been published.

TYPE: Incidence; Mortality
TERM: Air Pollution; Cluster; Data Resource; Environmental Factors; Geographic Factors; Petroleum Products; Plastics; Registry; Solvents
SITE: All Sites; Larynx; Leukaemia; Liver; Lung; Lymphoma
CHEM: Benzene; Vinyl Chloride
REGI: OPCS (UK)
TIME: 1987 - 1994

870 **Filakti, H.H.** 02178
OPCS, LS Medical Analysis Sect., LS Unit, 10 Kingsway, London WC2B 6JP, United Kingdom
COLL: Macdonald-Davies, I.; Bethune, A.; Harding, S.

Cancer Incidence and Survival in a 1% Census Sample

Data on cancer incidence and survival from the OPCS Longitudinal Study are being analysed to separate mortality differentials into those associated with differential incidence and those associated with differential survival. Major innovations include the systematic analysis of national data by survival periods within one year and by cause of death. Factors considered in the analysis of differentials include area of residence, social class, housing, household and family circumstances and marriage and fertility history.

TYPE: Cohort
TERM: Lifestyle; Marital Status; Reproductive Factors; Socio-Economic Factors; Survival
SITE: All Sites
TIME: 1980 - 1994

871 **Fraser, P.M.** 04435
London Sch. Hygiene & Tropical Medicin, Dept. Epidemiol. & Population Science, Epidemiological Monitoring Unit, Keppel St. (Gower St.), London WC1E 7HT, United Kingdom (Tel.: +44 71 6368638)
COLL: Day, N.E.

Leukaemogenicity of Cytostatic Drugs Used in Cancer Therapy

The aim of this study is to estimate the relative leukaemogenicity of different cytostatic regimes and to establish dose-response curves for agents shown to induce leukaemia. The relevance is two-fold. First, leukaemia is probably the most serious long-term side effect of cytostatic therapy and quantitative information on the degree of effect should be of value to clinicans in choosing treatment regimes. Second, exposure to cytostatic agents represents one of the few occasions on which humans are deliberately exposed to known carcinogens in a situation where both the dose is closely monitored and long-term follow-up ensured. It is thus the most favourable circumstance in which to estimate in quantitative terms the relationship between dose and duration of treatment, and risk of leukaemia development. Case-control studies will be carried out within cohorts of individuals with one of the following malignancies -Hodgkin's disease, non-Hodgkin's lymphoma, ovarian cancer, testicular cancer and breast cancer. The cohorts will be identified from the records of six cancer registries, covering a population of 22 million. Cases of acute and non-lymphocytic leukaemia occurring in these cohorts during 1960-1986 subsequent to the diagnosis of the initial cancer will be identified. The treatment given for the index cancer of these leukaemia cases and a corresponding set of matched controls will be abstracted from the patients' hospital records at the treating centre. Over 250 leukaemia cases will be available for study. The methods of statistical analysis will be based on conditional logistic regression with general relative risk models.

TYPE: Case-Control; Cohort
TERM: Chemotherapy; Dose-Response; Multiple Primary; Radiotherapy; Registry
SITE: Leukaemia
REGI: E. Anglia (UK); Mersey (UK); Oxford I (UK); Thames (UK); W. Midlands (UK); Yorkshire (UK)
TIME: 1988 - 1994

872 **Kazantzis, G.** 01490
Imperial College of Science, Technology & Medicine, Environm. Geochemistry Research Dept., Prince Consort Rd, London SW7 2BP, United Kingdom (Tel.: +44 71 5895111; Fax: 2258544 ; Tlx: 929484)
COLL: Sullivan, K.R.; Ades, A.E.; Armstrong, B.G.; Blanks, R.G.

Case-Control Studies of Cadmium Exposed Workers

Case-control sets have been taken from a cohort of 6,995 cadmium exposed workers in England (Lancet i: 1424-1427, 1983) and from two other cohorts of approximately 3,000 men, for more detailed investigation of past cadmium and other environmental exposures. With regard to prostatic cancer, marginally increased risks were observed after "high" or "medium " exposure, but these were not statistically significant (Br. J. Ind. Med. 42:540-545, 1985). With regard to lung cancer, mortality was examined in a cohort of 4,393 men employed in a zinc-lead-cadmium-smelter. An increased lung cancer risk was found, particularly evident in those employed for more than 20 years, with a statistically significant trend in SMRs with increasing duration of employment. Matched logistic regression was used to compare the cumulative exposures of cases of lung cancer to those of controls. The increasing lung cancer risk could not be accounted for by cadmium, but was associated with estimates of cumulative exposure to arsenic and to lead, although it was not possible to determine whether this increased risk was due to arsenic, lead or to other contaminants in the smelter (Br. J. Ind. Med. 45:435-442, 1988). Further case-control sets from the cohorts are now being examined. (Heavy Metals in the Environment, J.P. Vernet (Ed.) Vol. 1, pp 304-307, Geneva). An international case-control study on lung and prostatic cancer has been proposed.

TYPE: Case-Control
TERM: Metals; Occupation
SITE: All Sites; Lung; Prostate
CHEM: Arsenic; Cadmium; Lead; PAH
TIME: 1983 - 1994

873 **Kazantzis, G.** 03492
Imperial College of Science, Technology & Medicine, Environm. Geochemistry Research Dept., Prince Consort Rd, London SW7 2BP, United Kingdom (Tel.: +44 71 5895111; Fax: 2258544 ; Tlx: 929484)
COLL: Lam, T.H.; Sullivan, K.R.; Blanks, R.G.

Cohort Mortality Study of Cadmium Exposed Workers

A cohort of 6,995 men born before 1940 and exposed to cadmium for more than one year between 1942 and 1970 was initially followed up until the end of 1979, updated for a five year period 1984 and now updated for a further five year period to the end of 1989. In the initial study no excess of deaths due to prostatic cancer, cerebrovascular disease or renal disease was found. A significant excess of lung cancer was observed in men employed for more than ten years but this was not related to intensity of exposure (Lancet i: 1424-1427, 1983). Results of the five year update to include mortality experience to the end of 1984 were published in Scand. J. Work Environ. Health 14:220-223, 1988 and Environ. Chem. 27:113-122, 1990. In the second five-year update to the end of 1989, the mortality from prostate cancer, low throughout the study, had fallen still further. Lung cancer mortality was borderline significantly increased with some evidence of a trend across exposure categories but which does not attain significance. There was suggestive evidence of a relationship with both intensity and duration of employment in workers employed before 1940. Stomach cancer mortality was significantly increased in the cohort as a whole, but this was not related to intensity of exposure. The study showed no evidence that cadmium may act as a prostatic carcinogen. The excess lung cancer risk cannot at present be attributed to cadmium, owing to the presence of multiple confounding factors in the working environment. Data collection is continuing, and lung cancer risk is being examined in relation to estimates of cumulative cadmium exposure.

TYPE: Cohort
TERM: Chemical Exposure; Environmental Factors; Metals; Occupation
SITE: Kidney; Lung; Prostate
CHEM: Cadmium
TIME: 1979 - 1994

874 **Lund, V.J.** 05073
Univ. College of London, Inst. of Laryngology and Otology, 330 Gray's Inn Rd, London WC1X 8DA, United Kingdom (Tel.: +44 71 8378855; Fax: 8339480)

Malignant Melanoma of the Nasal Cavity and Paranasal Sinuses

The aim of the study is (1) to assess environmental and industrial factors which may be of aetiological importance; and (2) to correlate prognosis with histopathology. A core group of patients have been identified as a result of a previous study and will be available for a case-control study with particular emphasis on formaldehyde exposure and work in the tyre manufacturing industry, where several cases have occurred. Histological patterns in a retrospective group of 60 patients are being examined using immunocytochemistry and DNA flow cytometry, to determine possible prognostic criteria in a condition where conventional classifications of melanoma (Clark, Breslow) cannot be applied.

TYPE: Case Series; Case-Control
TERM: BMB; Environmental Factors; Histology; Occupation; Prognosis
SITE: Melanoma; Nasal Cavity
CHEM: Formaldehyde
OCCU: Rubber Workers
TIME: 1990 - 1994

875 **Marmot, M.G.** 04774
Univ. College London, Dept. of Epidemiology & Public Health, Gower St. , London WC1E 6EA, United Kingdom (Tel.: +44 71 3807602; Fax: 3807608)
COLL: Davey-Smith, G.; Stansfeld, S.

Stress and Health Study

Social class differences in mortality and morbidity from a wide range of diseases persist in the UK, USA and other industrialized countries. In a previous study of the British Civil Service, an unexplained threefold higher mortality from cardiovascular and other disease was found in the lowest compared to the highest employment grade. In a new study set up to investigate this, 10,314 British Civil Servants have been enrolled in a longitudinal study of health and disease. The overall aim is to study the effect on health and disease both of the work environment - psychological workload, control over work (both its pacing and content), opportunity for use of skills and social support at work - and of social supports; and the interaction between these psychosocial factors and other established risk factors in the aetiology of chronic disease. To date, a cross-sectional questionnaire survey and medical examination have been completed and a questionnaire re-survey is in progress. Longitudinal data on incidence of cardiovascular and other diseases will be collected by: (1) repeat medical screening and questionnaire administration to the 10,314 participants; (2) collection of sickness absence data from the Civil Service Central Monitoring System; (3) obtaining medical diagnostic data from General Practitioners on prolonged sickness absences; (4) cancer registration and mortality from the National Health Service Central Register.

TYPE: Incidence
TERM: Occupation; Psychological Factors; Psychosocial Factors; Stress
SITE: All Sites
REGI: OPCS (UK)
TIME: 1985 - 1994

876 **Marmot, M.G.** 04995
Univ. College London, Dept. of Epidemiology & Public Health, Gower St. , London WC1E 6EA, United Kingdom (Tel.: +44 71 3807602; Fax: 3807608)
COLL: Swerdlow, A.J.; Grulich, A.; Head, J.

Cancer Mortality in Migrants to England and Wales

Migrant studies have traditionally been of use in separating the effects of the environment and genetics in cancer aetiology. By linking routinely collected country of birth information from death certificates and the 1971 and 1981 censuses, it is possible to calculate site specific cancer mortality rates, as well as mortality rates from other fatal diseases, for immigrant populations. A previous study of migrant mortality based on routinely collected statistics around the 1971 census has been published by OPCS. This study will cover a longer time period, 1970-1985, and contain additional information. In particular, time trends of disease will be analysed, and possible interacting effects of social class and region of residence examined. So far work has been completed on cancer mortality in Italian migrants and a study of cancer mortality in migrants from East and West Africa and the Caribbean has been published (Grulich et al.). Planned work includes a study of mortality in migrants from Hong Kong and China and a study of cancer and cardiovascular mortality in migrants from South Asia.

TYPE: Incidence; Mortality
TERM: Linkage Analysis; Migrants; Socio-Economic Factors; Trends
SITE: All Sites
TIME: 1990 - 1994

877 **McCartney, A.C.E.** 04847
St Thomas's Hosp., U.M.D.S., Depts of Ophthalmology & Pathology, Lambeth Palace Rd, London SE1, United Kingdom (Tel.: +44 71 9289292; Fax: 4013661, 4019062)
COLL: Foster, A.; Thomson, J.

Incidence and Spread of Retinoblastoma in Rural Ghana

An apparently abnormally high frequency of retinoblastoma is being observed in the hospital of Agogo, with a marked preponderance of spread into the orbit and optic nerve. The frequency, tribal patterns and methods of prevention are being investigated and correlated with histopathological data, especially the type of spread from the globe of the eye. The possibility of gene amplification by extensive intermarriage is being investigated. Histopathological data include morphometric analysis and immunohistochemical analysis.

TYPE: Case Series
TERM: Ethnic Group; Genetic Factors; Prevention
SITE: Retinoblastoma
LOCA: Ghana; United Kingdom
TIME: 1989 - 1994

***878** **McGregor, J.M.** 05273
Guy's Hosp., Dept. of Dermatology, St Thomas's St., London SE1 9RT, United Kingdom (Tel.: +44 71 9554142; Fax: 9554584)
COLL: Levison, D.A.; MacDonald, D.M.

Skin Cancer in Renal Allograft Recipients

The aim of this study is to examine the incidence and morbidity of non-melanoma skin cancer in a population of 500 renal allograft recipients. Possible associations with cumulative UV exposure, duration of immunosuppression, age at transplantation and current age, will be examined by use of a questionnaire. A possible association with papillomavirus will be examined including molecular techniques.

TYPE: Cohort
TERM: Age; BMB; Immunosuppression; PCR; Radiation, Ultraviolet; Transplantation
SITE: Skin
TIME: 1991 - 1994

879 **McPherson, C.K.** 04555
London School of Hygiene and Tropical Medicine, Dept. of Public Health and Policy, Keppel St. (Gower St.), London WC1E 7HT, United Kingdom (Tel.: +44 71 6368636; Fax: 4363611 ; Tlx: 8953474)

Early Use of Oral Contraceptives and Breast Cancer in Women Aged 35-44

The proportion of women who are at risk of developing breast cancer and who have had prolonged exposure to oral contraceptives (OC) has only recently become large enough to test for possible delayed effects of prolonged OC use before first pregnancy on breast cancer risk. Since the literature on this association is so conflicting a case-control study of women aged 35-44 in three regions is proposed. Around 15% of these women will have been exposed to OCs for more than four years before first pregnancy. 400 cases and 400 controls matched by age and general practitioner (GP) will be recruited and interviewed to obtain accurate OC use data with specially-designed calendars and a detailed schedule. OC histories will be checked by reference to GP records.

TYPE: Case-Control
TERM: Late Effects; Latency; Oral Contraceptives; Registry
SITE: Breast (F)
REGI: Scotland N. (UK)
TIME: 1988 - 1994

880 **Phillips, R.** 05174
St. Marks' Hosp. for Diseases of the Colon and Rectum, City Rd, London EC1V 2PS, United Kingdom (Tel.: +44 71 2531050)
COLL: Spigelman, A.D.; Farmer, K.C.R.; Williams, C.B.; Talbot, I.

Gastrointestinal Polyps and Cancer in Familial Adenomatous Polyposis

The aim of this study is to examine the natural history of upper gastrointestinal polyps in familial adenomatous polyposis and to assess the effect of the non-steroidal anti-inflammatory drug sulindac in the management of these polyps. Approximately 150 patients attend the endoscopy unit for surveillance of upper gastrointestinal polyps and 40 patients have been selected for a placebo-controlled trial of sulindac. Oesophago-gastro-duodenoscopy is performed by a single endoscopist using a side-viewing duodenoscope. A videotape record is kept for future comparison in addition to a diagram. Tissue biopsies are taken for histology and cell proliferation studies (bromodeoxyuridine DNA labelling). The interval between surveillance is either one or three years depending on the severity of the polyps. The role of bile in the aetiology of these polyps is being examined.

TYPE: Case Series; Intervention
TERM: BMB; Bile Acid; DNA; Drugs; Familial Adenomatous Polyposis; Histology; Polyps; Premalignant Lesion; Prevention; Tissue
SITE: Gastrointestinal
TIME: 1988 - 1998

881 **Smith, P.G.** 04128
London School of Hygiene and Tropical Medicine, Dept. of Epidemiol. and Popul. Sciences, Keppel St., London WC1E 7HT, United Kingdom (Tel.: +44 71 9272246; Fax: 4364230 ; Tlx: 8953474)
COLL: Douglas, A.; Omar, R.

Mortality of Workers in a Nuclear Reprocessing Plant

A retrospective-prospective cohort study of the 14,000 workers who worked at the Sellafield nuclear reprocessing plant at any time prior to 1976 is being carried out. The objective is to identify any hazards associated with this form of occupational exposure to ionising radiation. Analyses will include comparison of mortality and cancer incidence rates with the general population and between workers exposed to different doses of internal and external radiation. Cancer incidence data will also be examined. A paper appeared in Br. Med. J. 293:845-854, 1986.

TYPE: Cohort
TERM: Occupation; Radiation, Ionizing
SITE: All Sites
TIME: 1976 - 1994

882 **Szarewski, A.M.** 05005
Imperial Cancer Research Fund, P.O.Box 123, Lincoln's Inn Fields, London WC2A 3PX, United Kingdom (Tel.: +44 71 2693006)
COLL: Cuzick, J.M.; Jenkins, D.; Singer, A.; Turk, J.

Effect of Vitamin Supplementation on CIN-I in Cervical Epithelium

Epidemiological studies have suggested that deficiencies in the consumption of beta carotene, vitamin C and vitamin E may increase tne risk of cervical cancer and cervical intraepithelial neoplasia. This appears to be an independent risk factor after controlling for variables such as age at first intercourse and number of sexual partners. The effects of vitamin deficiency or supplementation on Langerhans cells in the cervix have not been previously studied. It is thought that vitamin supplementation is likely to be most effective at the earliest stages of disease and CIN1 offers a unique opportunity for studying the effects of dietary supplementation over a relatively short period of time. 400 women volunteers whose most recent smear shows mild dyskaryosis will be randomised (double blind) to take either a vitamin supplement (containing 18 mg beta carotene, 150 mg vitamin C and 75 mg vitamin E) or placebo for 12 months. Blood samples will be taken at each visit to check levels of these vitamins. The women will be colposcoped at the beginning and end of the study. At the last visit, biopsies will be taken from both the lesion and a normal area for histology and immunocytochemistry, principally regarding Langerhans cells. A vaginal infection screen will be carried out at the same time. Ethical approval has been obtained from both the Middlesex Hospital Ethics Committee and Islington Ethics Committee.

TYPE: Intervention
TERM: BMB; Diet; Vitamins
SITE: Uterus (Cervix)
TIME: 1990 - 1994

883 **Szarewski, A.M.** 05006
Imperial Cancer Research Fund, P.O.Box 123, Lincoln's Inn Fields, London WC2A 3PX, United Kingdom (Tel.: +44 71 2693006)
COLL: Cuzick, J.M.; Jenkins, D.; Singer, A.; Turk, J.

Assessment of the Effect of Smoking Cessation on Langerhans Cells in Cervical Epithelium and CIN-I

Epidemiological studies have established that cigarette smoking is a risk factor for CIN. No study has yet looked at the effects of smoking cessation on Langerhans (antigen presenting) cells in cervical epithelium. 300 women volunteers who have mild dyskaryosis on their most recent cervical smear will attempt to stop smoking for six months. It is estimated that about 10% will succeed. These women will have a colposcopy at the beginning of the study to confirm that they do not have more severe disease than CIN-I. Only a biopsy of a normal area will be taken so as not to affect the natural history of the lesion. At the six month visit, both another normal biopsy and one from the abnormal area will be taken. All biopsies will be analysed histologically and for a number of immunological parameters, of which the most important will be Langerhans cells. The women's smoking histories will be checked using salivary cotinine assays at each visit. There will also be about 30 women who smoke but have a normal smear, and 30 non-smokers, with or without an abnormal smear to act as controls. Whenever a biopsy for immunological testing is taken, an infection screen will be performed, since vaginal/cervical infection could affect the local immune response. Ethical approval has been obtained from both the Middlesex Hospital Ethics Committee and Islington Ethics Committee.

TYPE: Cohort
TERM: BMB; Cytology; Histology; Tobacco (Smoking)
SITE: Uterus (Cervix)
TIME: 1990 - 1994

884 **Wald, N.J.** 02088
St Bartholomew's Hosp., Med. Coll., Wolfson Inst. of Preventive Med., Charterhouse Square, London EC1M 6BQ, United Kingdom (Tel.: +44 71)2530661/8373)
COLL: Bailey, A.

The Oxford Study of Men in a Private Health Insurance Scheme

In this prospective study about 22,000 men, members of the British United Provident Association (BUPA), have been recruited. Follow-up is by notification of death (1,136) or cancer (285) from the National Health Service Central Register. The main aim is to investigate factors which may predispose to cancer, in particular tobacco smoking and blood micronutrient levels such as vitamin A, beta carotene and vitamin E. Carboxyhaemoglobin levels and serum or urinary cotinine levels have been used as an index of tobacco smoke absorption. Both vitamin A (retinol) and vitamin E were significantly associated with the risk of cancer and the most likely explanation is that both these low levels were metabolic consequences, rather than precursors, of the cancer. In contrast, the significantly lower levels of beta carotene found in the subjects who subsequently developed cancer, affecting the risk directly or indirectly. Since the last report in this Directory further work has been carried out on passive smoking, serum cholesterol and cancer, vitamin E and colorectal cancer, Helicobacter pylori and stomach cancer and blood pressure and stroke.

TYPE: Cohort
TERM: Cholesterol; H. pylori; Passive Smoking; Registry; Tobacco (Smoking); Vitamins
SITE: All Sites
CHEM: Beta Carotene; Carbon Monoxide; PAH; Retinoids; Tars
REGI: OPCS (UK)
TIME: 1975 - 1994

885 **Williams, C.B.** 02966
St. Mark's Hosp., City Rd, London EC1 V2PS, United Kingdom (Tel.: +44 71 2531050)
COLL: Atkin, W.A.; Nicholls, R.J.; Macrae, F.A.

Colorectal Adenoma Follow-Up Study

This is a prospective, randomised, long-term follow-up study of 1,000 colorectal adenoma patients based on colonoscopy, supplemented where necessary by double contrast barium enema. Patients accruing to the study were allocated to high or low-risk status. Statistical analysis is now underway to make recommendations on follow-up protocols.

TYPE: Cohort
TERM: Polyps
SITE: Colon; Rectum
TIME: 1982 - 1994

***886** **Wolff, S.P.** 05383
Univ. of London, Univ. College, Medical School, 5 University St., London WC1E 6JJ, United Kingdom (Tel.: +44 71 3809678; Fax: 3809837)
COLL: Ben-Shlomo, Y.

Leukaemia Incidence in Relation to Environmental Benzene Exposure

The incidence of childhood leukaemia is examined in this geographical study, in relation to differential exposure to benzene as a result of different patterns of motor vehicle exposure. Levels of benzene within automobiles can approach occupational limits. Population risk for childhood leukaemia (based on published rates for lymphoid leukaemia in different areas in the UK is assessed in the context of levels of car ownership, mobility and lifestyle patterns, as well as other socio-economic variables. Individual benzene doses are estimated on the basis of socio-economic status and lifestyle and corroborated by individual sampling. A paper was published in Experientia 48:301-304, 1992.

TYPE: Correlation
TERM: Childhood; Cluster; Lifestyle; Socio-Economic Factors; Solvents
SITE: Leukaemia; Lymphoma
CHEM: Benzene; Vehicle Exhaust
TIME: 1992 - 1995

Macclesfield

887 **Tomenson, J.A.** 01145
ICI Epidemiology Unit, Alderley Park, Macclesfield Cheshire SK10 4TJ, United Kingdom (Tel.: +44 625 515409)
COLL: Rose, F.G.; Wood, A.

Applications of a Computerised Occupational Health Records System

For several years, ICI has had computerised records of the mortality and serious morbidity experience of all its employees in the United Kingdom. These data are analysed annually for geographical or product group correlations by calculating of proportional ratios. The data are also used as an invaluable source for ad hoc studies of specific compounds or occupational groups, particularly those described in the literature as presenting an excess risk. If an in-depth investigation is justified, the health and industrial hygiene records held at the production sites on a distributed computer system can be examined. Sufficient demographic and job histories have now been collected for historical prospective cohort studies to be conducted electronically, and for SMR to be calculated. A paper has been published in Banbury Report 9:177-189, 1981.

TYPE: Cohort
TERM: Chemical Exposure; Occupation
SITE: All Sites
TIME: 1975 - 1994

Manchester

888 **Kay, C.R.** 01090
Royal College of General Practitioners, Manchester Research Unit, 8 Barlow Moor Rd , Manchester M20 OTR, United Kingdom (Tel.: +44 61 4457771; Fax: 4459650)
COLL: Hannaford, P.C.; Beral, V.

Cancer in Oral Contraceptive Users

This is a cohort study of 23,000 oral contraceptive users and 23,000 non-users, recruited over 14 months starting in May 1968 by 1,400 general practitioners in the UK. The cohort now consists of approximately 15,000 women. Full details of newly reported morbidity, operations, pregnancy and, when appropriate, cause of death are collected at regular intervals. Patients no longer under observation in the main study are 'flagged' at the National Health Service Central Registry, for notification of death or cancer registration. Comprehensive details of all oral contraceptives used are collected, as well as hormonal replacement treatments. The aim is to investigate all aspects of health with regard to oral contraceptive effects; special interests include cardiovascular effects and genital and breast cancers. Particular emphasis is now placed on past use of oral contraceptives.

TYPE: Cohort
TERM: Oral Contraceptives
SITE: All Sites; Breast (F); Female Genital
TIME: 1968 - 1994

Newcastle-upon-Tyne

889 **Parker, L.** 04564
Newcastle Univ., Medical School, Children's Cancer Unit, Dept. of Child Health, Framlington Place, Newcastle-upon-Tyne NE2 4HH, United Kingdom (Tel.: +44 91 2226000/6958; Fax: 2226222 ; Tlx: 53654 uninew g)
COLL: Craft, A.W.; Robinson, A.

Maternal Malignancies and Soft Tissue Sarcoma in Children

The Northern Children's Malignant Disease Registry holds information on all children and young adults diagnosed with cancer in the Northern Region since 1968. This study aims to investigate further the relationship between soft tissue sarcoma and maternal malignancies, particularly of breast (Birch et al., Br. J. Cancer 49:325-331, 1984). The families of 400 patients diagnosed with soft tissue sarcoma before age 25, since 1954 will be traced. Questionnaires will be sent to general practitioners and any cases of maternal cancers will be investigated using medical records and pathological specimens whenever possible. Incidence of malignancies in the case mothers will be compared with that of a control population.

TYPE: Incidence
TERM: Childhood; Familial Factors; High-Risk Groups; Registry
SITE: All Sites; Breast (F); Sarcoma; Soft Tissue
REGI: Newcastle (UK)
TIME: 1988 - 1994

890 **Parker, L.** 05038
Newcastle Univ., Medical School, Children's Cancer Unit, Dept. of Child Health, Framlington Place, Newcastle-upon-Tyne NE2 4HH, United Kingdom (Tel.: +44 91 2226000/6958; Fax: 2226222 ; Tlx: 53654 uninew g)
COLL: Craft, A.W.; Smith, J.; Dickinson, H.

Radiation Exposure in the Parents of Children with Cancer, Congenital Malformation and Down's Syndrome

The aim of the study is to determine whether workers at the Sellafield nuclear fuel reprocessing plant who are parents of children with cancer or congenital malformation or whose children died in infancy have been exposed to an unusually high dose of internal or external radiation when compared with their colleagues whose children do not have these conditions. The children born in Cumbria to the 27,000 men and women employed at the Nuclear Installation, Sellafield (NIS), since 1950 and to around 6,000 contract radiation workers will be identified by linking the workforce file with that of the Cumbrian birth register. It is anticipated that up to 20,000 children will be so identified, 20-30 of whom will have developed a malignant disease and up to 100 will have died. Case children will be identified from cancer registration, congenital malformation registration and death certificates. Radiation records held by the plant will be obtained. Matched control parents will be selected from the workforce file, and radiation dose in the control group will be compared with that in the case-parent group. Matching of parents will be based on age, sex, year of first employment and year of birth of child.

TYPE: Case-Control
TERM: Congenital Abnormalities; Down's Syndrome; Intra-Uterine Exposure; Occupation; Radiation, Ionizing; Registry
SITE: Childhood Neoplasms
OCCU: Radiation Workers
REGI: Newcastle (UK)
TIME: 1990 - 1994

891 **Parker, L.** 05039
Newcastle Univ., Medical School, Children's Cancer Unit, Dept. of Child Health, Framlington Place, Newcastle-upon-Tyne NE2 4HH, United Kingdom (Tel.: +44 91 2226000/6958; Fax: 2226222 ; Tlx: 53654 uninew g)
COLL: Craft, A.W.; Smith, J.; Dickinson, H.

Incidence of Cancer, Congenital Malformation (Including Stillbirth) and Down's Syndrome in Cumbria

The purpose of the proposed research is to determine whether (1) children born to workers employed at the nuclear fuel reprocessing plant at Sellafield are at increased risk of cancer, congenital malformation (including stillbirth) or Down's Syndrome, in comparison with the incidence of these conditions in the rest of Cumbria; and (2) there is any increased risk for these conditions with increasing radiation exposure in parents. The incidence of cancer, congenital malformation and infant mortality rate will be determined for Cumbrian children of the plant workforce and for all Cumbrian children born over a similar time period. Any increased risk in the children of the Sellafield workers will be determined. There have been 27,000 employees since 1950, as well as 6,000 contract radiation workers. It is anticipated that up to 20,000 offspring of these workers will be identified and the incidence of ill-health in these children will be compared with the incidence in the remaining 240,000 Cumbrian children born since 1950.

TYPE: Cohort
TERM: Congenital Abnormalities; Dose-Response; Down's Syndrome; Radiation, Ionizing; Registry
SITE: Childhood Neoplasms
REGI: Newcastle (UK)
TIME: 1990 - 1994

Nottingham

892 **Hardcastle, J.D.** 03556
Univ. of Nottingham, Univ. Hospital, Queens Medical Centre, Dept. of Surgery, Floor E, West Block, Nottingham NG7 2UH, United Kingdom (Tel.: +44 602 709245)
COLL: Chamberlain, J.

Randomised Controlled Trial of Faecal Occult Blood Screening for Colorectal Cancer

The purpose of this trial is to measure the effectiveness of faecal occult blood testing in reducing the mortality from colorectal cancer and to measure the economic effects on health service resources of such a screening service. Data will also be obtained that could be used in the long-term to measure the effectiveness of faecal blood occult screening in reducing the incidence of colorectal cancer by removing adenomas. The effect of different methods of education will also be evaluated in order to achieve the highest compliance. 156,000 persons aged 50-74 years will be identified from General Practitioners' lists and randomly allocated by household to test or control group. The test group will be offered faecal occult blood screening at 2-yearly intervals and each individual will be followed for a minimum of 7 years. The control group are monitored for development of colorectal neoplasia. Persons leaving the trial area have their records flagged at the National Health Service Central Register to permit follow-up for development of colorectal neoplasia and date/cause of death.

TYPE: Cohort; Intervention
TERM: Screening
SITE: Colon; Rectum
TIME: 1981 - 1994

893 **Logan, R.F.** 04965
Univ. of Nottingham Medical School, Dept. Publ. Health Med. & Epidem., Queen's Medical Centre, Clifton Blvd, Nottingham NG7 2UH, United Kingdom (Tel.: +44 602 709308; Fax: 709316 ; Tlx: 37346 uninot g)
COLL: Little, J.; Hardcastle, J.D.

Diet and Colorectal Cancer

This case-control study takes advantage of a current colorectal cancer screening trial, which allows the study of asymptomatic as well as symptomatic colorectal cancer patients. Cases and controls are being interviewed in their homes about their past dietary habits using a food frequency approach. Enquiry is also being made about past occupations, leisure activities, past illnesses and drug intakes and family history. Because of difficulties interviewing patients with advanced cancers only patients with Duke's stage A or B cancers are being approached. Two series of control subjects are being assembled: one series who have had recent surgery and a second from family practitioner lists and the records of the screening trial. By February 1991, 80 patients with asymptomatic cancers and 145 with symptomatic cancers had been interviewed. The reponse rate has been 88% in the cancer patients and 78% in the controls. From comparisons with an earlier study of diet in patients with colorectal polyps, it is hoped to identify what dietary constituents or combinations thereof are associated with the development of colorectal cancer. Constituents of particular interest are animal fat and protein intakes and cereal and total fibre intakes. Data collection is now completed case series of 240. Analysis is in progress.

TYPE: Case-Control
TERM: Diet; Drugs; Family History; Fat; Fibre; Nutrition; Occupation; Physical Activity; Protein
SITE: Colon; Rectum
TIME: 1988 - 1994

894 **Logan, R.F.** 05182
Univ. of Nottingham Medical School, Dept. Publ. Health Med. & Epidem., Queen's Medical Centre, Clifton Blvd, Nottingham NG7 2UH, United Kingdom (Tel.: +44 602 709308; Fax: 709316 ; Tlx: 37346 uninot g)
COLL: Little, J.; Hardcastle, J.D.

Colorectal Adenomatous Polyps, Vitamins and Cholesterol

The aim of the study is to determine whether the risk of asymptomatic colorectal adenomas increases with decreasing serum levels of vitamin A and carotene, vitamin E and cholesterol. Blood samples are being obtained at outpatient clinics for the investigation of patients who are found to be faecal occult blood (FOB) positive in the Department of Surgery screening trial and for the follow-up of patients in whom colorectal adenomas or carcinoma were detected as a result. It is anticipated that serum concentrations of these vitamins will be available for about 200 subjects with adenomas, 50 with carcinomas, and 160 FOB-positive subjects in whom no adenoma or carcinoma was found.

TYPE: Case-Control
TERM: BMB; Biomarkers; Blood; Cholesterol; Polyps; Premalignant Lesion; Vitamins
SITE: Colon; Rectum
CHEM: Beta Carotene; Retinoids
TIME: 1989 - 1994

Oxford

895 **Beral, V.** 04291

Imperial Cancer Research Fund, Cancer Epidemiology Unit, Radcliffe Infirmary, Gibson Labs, Oxford OX2 6HE, United Kingdom

COLL: Fraser, P.M.; Rooney, C.; MacOnochie, N.

Prostatic Cancer in Employees of the United Kingdom Atomic Energy Authority

In the United Kingdom Atomic Energy Authority Mortality Study a statistically significant increase in mortality from prostatic cancer was found in certain groups of radiation workers: men with the highest recorded cumulative exposures to external radiation, and men who had been monitored for exposure to radionuclides, especially tritium. The availability and validity of the information required for a case -control study based on UKAEA records has been assessed and found satisfactory, and record forms have been developed and piloted. A case - control study involving 137 cases of prostatic cancer and three controls per case is now underway to determine if prostatic cancer risk can be linked to any specific aspect or aspects of the employment history of the individuals concerned.

TYPE: Case-Control
TERM: Chemical Exposure; Occupation; Radiation, Ionizing
SITE: Prostate
TIME: 1986 - 1994

896 **Beral, V.** 04420

Imperial Cancer Research Fund, Cancer Epidemiology Unit, Radcliffe Infirmary, Gibson Labs, Oxford OX2 6HE, United Kingdom

COLL: Fraser, P.M.; Smith, P.G.; Douglas, A.; Carpenter, L.M.; Booth, M.

Nuclear Industry Combined Epidemiological Analysis

The data from three separate studies of employees in the nuclear industry are being combined so that the relationship between exposure to low-level ionising radiation and mortality can be described with greater precision than is possible in any of the studies individually. The combined study population will be of the order of 75,000, including employees of the United Kingdom Atomic Energy Authority (Brit. Med. J. 291:440-447, 1985) of the Atomic Weapons Establishment (Brit. Med. J. 297:757-770, 1988), and of the Sellafield plant of British Nuclear Fuels (Smith, P., Douglas, A.: Br. Med. J. 293:845-854, 1986). All workers will be followed-up to the end of 1986, and approximately 12,000 deaths are expected of which about 3,000 will be attributed to cancer. Analyses will include comparisons within the workforce of the mortality of employees exposed to external and internal sources of radiation in relation to other employees, and of mortality in relation to the recorded level of exposure to external radiation. Comparisons of mortality rates in the combined workforce with national rates will also be made.

TYPE: Cohort
TERM: Dose-Response; Occupation; Radiation, Ionizing
SITE: All Sites
OCCU: Radiation Workers
TIME: 1987 - 1994

897 **Beral, V.** 02340

Imperial Cancer Research Fund, Cancer Epidemiology Unit, Radcliffe Infirmary, Gibson Labs, Oxford OX2 6HE, United Kingdom

COLL: Fraser, P.M.; Carpenter, L.M.; Booth, M.; MacOnochie, N.

United Kingdom Atomic Energy Authority Mortality Study

In 1985 and 1987 the mortality in 40,000 employees of the United Kingdom Atomic Energy Authority during 1946-1976 was reported (Brit. Med. J. 291:435-439 and 440-447,1985; Inskip et al, Brit. J. Ind. Med. 44:149-160, 1987). The average duration of follow-up was 16 years, the numbers of deaths from cancers at many sites were small, and wide confidence limits surrounded the estimated risks associated with exposure to specific doses of external radiation. Since that report, the follow-up of the entire study population has been placed on a continuous basis and a new analysis covering the years up to 1986 is planned giving an average duration of follow-up of 23 years with numbers of deaths about two thirds greater than in the previous analysis. Site-specific mortality and latency will thus be examined in greater detail than before, and the risks associated with specific doses of external radiation estimated with greater precision. Ascertainment of registered cancers will permit the examination of non-fatal as well as fatal cancers.

TYPE: Cohort
TERM: Dose-Response; Latency; Occupation; Radiation, Ionizing; Registry; Time Factors
SITE: All Sites
OCCU: Radiation Workers
REGI: OPCS (UK)
TIME: 1980 - 1994

***898** **Beral, V.** 05531
Imperial Cancer Research Fund, Cancer Epidemiology Unit, Radcliffe Infirmary, Gibson Labs, Oxford OX2 6HE, United Kingdom
COLL: Parkin, D.M.; Ngilimana, P.J.; Sindikubwabo, B.; Ngendahayo, L.; Bigirimana, V.; Newton, R.

Kaposi's Sarcoma, Non-Hodgkin's Lymphoma and Carcinoma of the Cervix in Africa in Relation to Infection with HIV

The first study was started in Butare, Rwanda, in 1992 and is being extended to Bujumbura, Burundi, during 1993. All cancer cases recorded by the cancer registry are interviewed with a specially tested questionnaire, and blood taken for antibody studies. Recruitement of non-cancer controls (hospital patients without malignancy and hospital visitors) is also planned. The cancers primarily of interest are Kaposi's sarcoma, non-Hodgkin's lymphoma and cervix cancer. The main emphasis of the study is on likely exposure to infectious diseases through sexual or faeco-oral spread.

TYPE: Case-Control
TERM: Antibodies; HIV; Sexually Transmitted Diseases
SITE: Kaposi's Sarcoma; Non-Hodgkin's Lymphoma; Uterus (Cervix)
LOCA: Rwanda
TIME:

899 **Cook-Mozaffari, P.J.** 04529
Imperial Cancer Research Fund, Cancer Epidemiology Unit, Gibson Labs, Radcliffe Infirmary, Woodstock Rd, Oxford OX2 6HE, United Kingdom (Tel.: +44 865 56337)
COLL: Forman, D.; Darby, S.C.; Doll, R.

Data Bank for Local and Temporal Variations in Cancer Mortality

Tabulations of cancer deaths in England and Wales by pre-1974 Local Authority Areas have been obtained from OPCS for the years 1959-1980 and for the sites of cancer listed in the IARC "Cancer Incidence in Five Continents" series. Aggregations of areas into post-1974 Country Districts or approximations of Country Districts have been made to facilitate the comparison of death rates across the 1974 reorganization of boundaries and to overcome the lack of census information for the year 1981 by pre-1974 Local Authority Areas. Data have so far been analysed for the period 1969-1978 as a basis for studies of geographical variation in the occurrence of leukaemia and cancer of the bladder. It is planned to mobilise the data for the remaining years and to extend investigations to other sites of cancer.

TYPE: Methodology; Mortality
TERM: Geographic Factors; Registry
SITE: All Sites; Bladder; Leukaemia
REGI: OPCS (UK)
TIME: 1988 - 1994

900 **Cook-Mozaffari, P.J.** 04706
Imperial Cancer Research Fund, Cancer Epidemiology Unit, Gibson Labs, Radcliffe Infirmary, Woodstock Rd, Oxford OX2 6HE, United Kingdom (Tel.: +44 865 56337)
COLL: Peto, J.; Davies, J.M.

Occurrence of Cancer of the Bladder in Relation to Suspected High-Risk Occupations in England and Wales

Mapping of the death for cancer of the bladder for men aged 25-64 at the level of County District has shown a concentration of elevated rates in certain port and estuarine areas. Multiple regression analyses show a strong geographical association with the chemical industry and with employment in "sea transport and port services". A death-certificate case-control study is being undertaken in areas of elevated risk to explore further the pattern of occupational risk.

TYPE: Case-Control
TERM: High-Risk Groups; Occupation
SITE: Bladder
OCCU: Chemical Industry Workers
TIME: 1988 - 1994

901 **Darby, S.C.** 02036
Imperial Cancer Research Fund, Cancer Epidemiology Unit, Radcliffe Infirmary, Gibson Labs., Oxford OX2 6HE, United Kingdom (Tel.: +44 865 53762)
COLL: Weiss, H.; Doll, R.

Mortality of Patients Treated with X-irradiation for Ankylosing Spondylitis

The follow-up of approximately 14,000 patients treated with X-rays for ankylosing spondylitis between 1935 and 1950 has shown, in conjunction with many other studies, that ionising radiation can cause cancer in nearly every organ in the body. The study, which has shown that the cancer risk is increased for at least 20 years after exposure, has been continued, to find out how long the effect lasts and what total risk is associated with a given level of exposure. Results from the extended follow-up of patients who received only a single course of treatment indicate that the increased risk has substantially disappeared by approximately 30 years after exposure. Beyond this the cancer risk in this population returns approximately to normal levels. Detailed dosimetry calculations for a sample of one in every 15 patients in the study have been carried out. Dose-response relationships are now being examined. Follow-up of patients who received more than one course of treatment is also under way. Papers have been published in Br. J. Cancer 55:179-190, 1987 and in Br. J. Radiol. 61:212-220, 1988.

TYPE: Cohort
TERM: Dose-Response; Radiation, Ionizing; Time Factors
SITE: All Sites; Leukaemia
TIME: 1955 - 2000

902 **Darby, S.C.** 03621
Imperial Cancer Research Fund, Cancer Epidemiology Unit, Radcliffe Infirmary, Gibson Labs., Oxford OX2 6HE, United Kingdom (Tel.: +44 865 53762)
COLL: Reeves, G.; Doll, R.

Mortality of Women Treated with X-Radiation for Metropathia Haemorrhagica

Previous follow-up of approximately 2,100 patients treated with X-irradiation for metropathia haemorrhagica in Scotland between 1940 and 1960 has shown an excess of deaths from leukaemia and from cancers of the heavily irradiated sites between five and 20 years after treatment. Over the same period the number of deaths from cancer of the breast was below expectation. Preliminary results from an extended follow-up of this group confirm previous findings and indicate that the excess rate of leukaemia is about 1.0 per million women per year per rad in the first 20 years after treatment. This figure is in accord with estimates derived from the survivors of the atomic bomb explosions in Hiroshima and Nagasaki if a simple linear dose-response relationship is assumed. However, studies of women treated with radiation for cancer of the cervix and patients treated with X-rays for ankylosing spondylitis have suggested that this assumption is incorrect when high doses of radiation are delivered to a small volume of marrow. In these studies a lower risk of leukaemia has been observed which may be explained by cell-sterilization at high doses.

TYPE: Cohort
TERM: Dose-Response; Radiation, Ionizing
SITE: All Sites; Leukaemia
TIME: 1965 - 2000

903 **Darby, S.C.** 04370
Imperial Cancer Research Fund, Cancer Epidemiology Unit, Radcliffe Infirmary, Gibson Labs., Oxford OX2 6HE, United Kingdom (Tel.: +44 865 53762)
COLL: Silcocks, P.B.; Doll, R.

Lung Cancer in South West England

Radon in houses is one of the most important sources of radiation exposure to which the general public is subject. The principal parts of the body liable to be affected are the bronchi, where radionuclides emitting alpha particles tend to be deposited. The experience of men working in mines where the radon content of the air is unusually high provides good evidence that this type of exposure is a cause of lung cancer. Several studies of the possible effect of radon in houses have already been carried out in other countries. These have confirmed that a small effect is likely to be produced, but further information is needed about the size of the risk. The present study is being carried out in South West England where the highest radon concentrations in houses in Britain are found, and is designed to provide more precise estimates of the size of the risk. It is hoped to collect data for about 600 patients with lung cancer in the course of three years, and at least twice that number of control subjects. Lung cancer patient are interviewed as soon as possible after a presumptive diagnosis of lung cancer has been made. The control group is made up of two parts: first an age-sex-matched sample of patients admitted to the same hospitals with a wide variety of conditions unrelated to smoking, and second an age-sex-matched sample of community controls chosen from the general population of the region. Radon levels in the past houses of both cases and controls are being measured. A paper has been published in Nature 344:824, 1990.

TYPE: Case-Control
TERM: Environmental Factors; Radiation, Ionizing
SITE: Lung
CHEM: Radon
TIME: 1987 - 1994

904 **Draper, G.J.** 02452
Univ. of Oxford, Radcliffe Infirmary, Childhood Cancer Research Group, 57 Woodstock Rd, Oxford OX2 6HJ, United Kingdom (Tel.: +44 865 310030; Fax: 514254)
COLL: Stiller, C.A.; Brownbill, P.A.; Bunch, K.J.

Aetiological Studies of Childhood Cancer

A number of related studies are being carried out to investigate possible environmental and genetic factors in the aetiology of childhood cancer. These include (1) analyses of geographical variations internationally; (2) investigation of the occurrence of cancer in the relatives (parents, sibs, twins, children) of children with cancer, particularly children with retinoblastoma; (3) record linkage studies of cohorts of children who may have an increased risk of cancer or for whom possible aetiological factors can be studied; (4) the effects of parity and parental age; (5) a study of perinatal factors in relation to childhood cancer, in particular investigating the recent suggestion that prophylactic intramuscular vitamin K may be a risk factor. Data from the national registry of childhood tumours maintained by the Group (about 1,200 cases per year) are used for these and other investigations. Recent papers were published in Br. J. Cancer 64:543-548, 1991, and 66:211-219, 1992 and in Int. J. Cancer 52:538-543, 1992 and 53:371-376, 1993.

TYPE: Genetic Epidemiology; Incidence
TERM: Familial Factors; Geographic Factors; Intra-Uterine Exposure; Mutation, Germinal; Record Linkage; Registry; Sib; Twins
SITE: Childhood Neoplasms; Retinoblastoma
REGI: Oxford II (UK)
TIME: 1975 - 1994

905 **Draper, G.J.** 04688
Univ. of Oxford, Radcliffe Infirmary, Childhood Cancer Research Group, 57 Woodstock Rd, Oxford OX2 6HJ, United Kingdom (Tel.: +44 865 310030; Fax: 514254)
COLL: O'Connor, C.M.; Stiller, C.A.; Vincent, T.J.; Bithell, J.F.; Cook-Mozaffari, P.J.; Muirhead, C.R.

Geographical Studies of Childhood Leukaemia and Cancer

Studies are carried out based on the National Registry of Childhood Tumours which covers virtually all cases of childhood cancer and leukaemia (i.e. those diagnosed at ages 0-14) in England, Scotland and Wales from 1962 onwards, i.e. about 30,000 cases, with addresses, postcodes, wards, (approximate) grid reference and other geographical variables. For children born in 1962 onwards cancer registrations are being linked to birth records and the district of birth added. The objectives of the analyses are to investigate variations in incidence according to both place of diagnosis and place of birth. Plans for the study include analyses of the effect on incidence rates of various geographical and socio-economic factors and of environmental factors such as radon and gamma radiation. Rates around nuclear installations (or other suspected high-incidence areas) will be compared with those for other areas. A particularly appropriate method for investigating the type of hypothesis of most interest for the present study is the 'Poisson maximum' method proposed by Stone (1988). A series of analyses is presented in 'The Geographical Epidemiology of Childhood Leukaemia and non-Hodgkin Lymphomas in Great Britain 1966-83'. G.J. Draper (ed). HMSO, London, 1991. A paper has also been published in Br. Med. J. 306:89-94, 1993.

TYPE: Correlation; Incidence
TERM: Cluster; Geographic Factors; Radiation, Ionizing; Record Linkage; Registry; Socio-Economic Factors
SITE: Childhood Neoplasms; Leukaemia
REGI: Oxford II (UK)
TIME: 1986 - 1994

***906** **Draper, G.J.** 05451
Univ. of Oxford, Radcliffe Infirmary, Childhood Cancer Research Group, 57 Woodstock Rd, Oxford OX2 6HJ, United Kingdom (Tel.: +44 865 310030; Fax: 514254)
COLL: Vincent, T.J.; Bunch, K.J.; Kendall, G.M.; Muirhead, C.R.; Sorahan, T.M.

Incidence of Cancer in the Children of Workers Occupationally Exposed to Ionising Radiation

It has been suggested that pre-conception exposure to ionising radiation is a possible cause of leukaemia in the offspring. This hypothesis is being tested by linking radiation records for 100,000 individuals in the UK National Registry of Radiation Workers with records of childhood cancer covering nearly all cases born in Great Britain from 1952 onwards. Controls are chosen from birth registers and are matched to the cases by year of birth and (usually) area of birth. The registers of radiation workers and childhood cancer already exist, but in order to carry out the record linkage it is necessary also to obtain records relating to the births of the children with cancer. The study is described in the National Radiological Protection Board, Radiological Protection Bulletin 129:10-14, 1992.

TYPE: Case-Control
TERM: Radiation, Ionizing; Record Linkage; Registry
SITE: Childhood Neoplasms
REGI: Oxford II (UK)
TIME: 1992 - 1995

907 **Forman, D.** 04430
Imperial Cancer Research Fund, Cancer Epidemiol. & Clin. Trials Unit, Radcliffe Infirmary, Gibson Lab., Gibson Bldg , Oxford OX2 6HE, United Kingdom (Tel.: +44 865 53951; Fax: 310545)
COLL: Pike, M.C.; Chilvers, C.E.D.; Oliver, R.T.D.

Case-Control Study of Testicular Cancer

800 men, aged 15-49 years, with histologically confirmed germ-cell tumours of testis diagnosed between January 1984 and September 1986 have been interviewed in eight different regions of England and Wales. They were asked about their exposure to a number of potential risk factors suggested as being involved in the disease, and their responses are being compared with those from matched population controls. As particular emphasis is being placed on possible in utero hormonal exposures, the mothers of cases and controls have been sent a postal questionnaire concerned with exposures during pregnancy.

TYPE: Case-Control
TERM: Dusts; Hormones; Intra-Uterine Exposure; Occupation; Radiation, Ionizing; Sexual Activity; Solvents
SITE: Testis
CHEM: Asbestos; Mineral Oil
TIME: 1984 - 1994

908 **Forman, D.** 04431
Imperial Cancer Research Fund, Cancer Epidemiol. & Clin. Trials Unit, Radcliffe Infirmary, Gibson Lab., Gibson Bldg , Oxford OX2 6HE, United Kingdom (Tel.: +44 865 53951; Fax: 310545)
COLL: Key, T.; Beral, V.; Hannaford, P.C.; Kay, L.

Prospective Study of Men and Women Aged 45-59 Years

It is proposed to establish a prospective cohort of 50,000 men and women recruited through general practitioners and through breast cancer screening. Subjects will be asked to provide a blood sample and to complete a questionnaire mainly concerned with diet and exercise. As cancers become identified, relevant blood variables and questionnaire information for case groups can be compared with those for disease free subjects. In this way, many fundamental questions concerned with the aetiology of cancer can be addressed. Pilot studies are being undertaken with the Aylesbury District Breast Screening Unit and with some general practitioners. Assessments will be made of compliance, ease of sample collection and validity of the questionnaire. This study will be part of the European Prospective Study of Diet and Cancer, coordinated by IARC.

TYPE: Cohort
TERM: BMB; Blood; Diet; Physical Activity
SITE: Breast (F); Colon; Prostate; Stomach
TIME: 1988 - 1994

909 **Forman, D.** 04432
Imperial Cancer Research Fund, Cancer Epidemiol. & Clin. Trials Unit, Radcliffe Infirmary, Gibson Lab., Gibson Bldg , Oxford OX2 6HE, United Kingdom (Tel.: +44 865 53951; Fax: 310545)
COLL: Oliver, R.T.D.; Bodmer, J.

Register of Familial Testicular Tumours

A register has been established to systematically record familial cases of testicular cancer. Such a register should serve firstly to generate more reliable statistics about the familial prevalence of the disease and, secondly, as a resource for biological material for genetic studies. Over 40 confirmed cases of testicular tumours in first degree relatives had been reported to the registry, mainly from oncologists around the country. Worldwide, only about 100 such families have been reported in the literature, and it is clear that the register will add considerably to the number of known cases. In addition, families are also recorded where the index case has a testicular tumour and the relative an undescended testis. To date, 38 such families are registered. At present, blood samples are collected from families reported to the registry in order to examine in more detail the genetic associations with this disease. A sib-pair analysis has been conducted to look at the relationship between tumour and HLA haptotype.

TYPE: Case Series
TERM: BMB; Genetic Factors; HLA
SITE: Testis
TIME: 1985 - 1994

910 **Forman, D.** 04433
Imperial Cancer Research Fund, Cancer Epidemiol. & Clin. Trials Unit, Radcliffe Infirmary, Gibson Lab., Gibson Bldg , Oxford OX2 6HE, United Kingdom (Tel.: +44 865 53951; Fax: 310545)
COLL: Doll, R.; Ferguson-Smith, J.; Bryson, D.; Leech, S.; Packer, P.

Study of Workers Exposed to Nitrate Fertilizers

This study was designed to establish whether employees of a factory producing inorganic fertilizers experience excess mortality from cancer as a result of their exposure to high concentrations of nitrate dust. The cohort, comprising some 1,400 workers, employed for one year or more since 1946, is particularly important in assessing the role of nitrates in cancer aetiology, as many workers have been heavily exposed for several years. The original analysis of these data, covering the period 1946-1981 (Al-Dabbagh et al., Br. J. Ind. Med. 43:507-515, 1986) showed no significant excess for any major form of cancer. The study is currently being updated to include deaths up until 1989. This will allow increased statistical power in analysing the results and a more detailed consideration of rare types of cancer. Also urine samples are being obtained from a representative sample of the employees in order to monitor current exposure levels.

TYPE: Cohort
TERM: BMB; Dusts; Fertilizers; Occupation; Urine
SITE: Bladder; Lung; Oesophagus; Stomach
CHEM: Nitrates
OCCU: Fertilizer Workers
TIME: 1982 - 1994

911 **Forman, D.** 04453
Imperial Cancer Research Fund, Cancer Epidemiol. & Clin. Trials Unit, Radcliffe Infirmary, Gibson Lab., Gibson Bldg , Oxford OX2 6HE, United Kingdom (Tel.: +44 865 53951; Fax: 310545)
COLL: Goss, B.; Swerdlow, A.J.

Epidemiology of Naevi in Childhood

The presence of naevi (moles) is a major risk factor for malignant melanoma. Very little is known about the natural history of naevi, yet there is evidence that congenital naevi and those that appear early in childhood are of special aetiological significance. The aims of this study are to obtain reliable figures for the prevalence of naevi in a sample of Oxford children at birth and 4 years old. A proportion of these children will then be followed up in order to quantify the appearance of naevi during childhood. As well as making naevus counts by anatomical site, the distribution of naevi by size and type will be assessed. The relationship between the appearance of naevi and various risk factors, genetic and environmental, will be investigated. 1,000 new-born babies and 250 4-year old children had been examined and 250 of the babies have been re-examined at the age of one year.

TYPE: Cohort
TERM: Childhood; Environmental Factors; Genetic Factors; Naevi
SITE: Melanoma
TIME: 1988 - 1994

912 **Forman, D.** 04454
Imperial Cancer Research Fund, Cancer Epidemiol. & Clin. Trials Unit, Radcliffe Infirmary, Gibson Lab., Gibson Bldg , Oxford OX2 6HE, United Kingdom (Tel.: +44 865 53951; Fax: 310545)
COLL: Sitas, F.; Newell, F.; Lachlan, G.; Jass, J.; Wild, C.; Wyatt, J.

Epidemiology of Chronic Atrophic Gastritis in Kenya

Parts of rural Kenya are experiencing an epidemic of acute gastritis in young people. Possibly related to this are anecdotal reports of increases in gastric cancer also affecting young people. Two systematic dyspepsia investigations of patients and asymptomatic volunteers in the age range 18-25 years have been carried out. These have shown a high prevalence of atrophic gastritis and a high rate of gastric infection with Heliobacter pylori. 57% of the volunteer group had atrophic gastritis, 70% had evidence of current H. pylori infection and all of them had antibodies of past infection. The occurrence of these findings in such young people is extremely unusual and contrasts strongly with the pattern in developed countries. A survey of about 100 young children has also been carried out to determine the age at which H. pylori infections became established and to examine other factors that may be involved in the aetiology. The role of mycotoxins, particularly aflatoxin, in the aetiology of gastritis has been investigated by examining blood samples from the same groups. Aflatoxin exposure was significantly higher in the dyspeptic patients.

TYPE: Cross-Sectional
TERM: Antibodies; Childhood; Infection; Mycotoxins; Premalignant Lesion; Urine
SITE: Stomach
CHEM: Aflatoxin
LOCA: Kenya
TIME: 1987 - 1994

913 **Forman, D.** 04455
Imperial Cancer Research Fund, Cancer Epidemiol. & Clin. Trials Unit, Radcliffe Infirmary, Gibson Lab., Gibson Bldg , Oxford OX2 6HE, United Kingdom (Tel.: +44 865 53951; Fax: 310545)
COLL: Sitas, F.; Newell, D.; Jewell, D.P.; Buiatti, E.; Palli, D.; Elwood, P.C.; Webb, P.

Epidemiology of Chronic Atrophic Gastritis

Most gastric cancers are preceded by a long period of chronic atrophic gastritis, which may be considered as a premalignant lesion. The role of two serological markers is being evaluated as predictors of gastritis, for eventual use in identifying populations at high risk of gastric cancer. The markers under examination are the serum level of the enzyme pepsinogen I and the presence of antibodies to the bacterium Heliobacter pylori. Parallel serum and gastric biopsy samples have been obtained from 100 patients attending endoscopy clinics in Oxford. The biopsies exhibited a range of gastric histopathology from normal to severely gastritic. Using the markers one can discriminate between those with and without gastritis with a high degree of sensitivity. It is now planned to investigate the serologically defined prevalence of gastritis in a cross-section of populations in Italy (n=800) and Wales (n=4,000).

Case-control studies will then be conducted, comparing dietary and other factors in persons with and without gastritis.

TYPE: Case-Control; Cross-Sectional
TERM: Antibodies; Biochemical Markers; Biopsy; Diet; Enzymes; H. pylori; Pepsinogen; Premalignant Lesion
SITE: Stomach
LOCA: Italy; United Kingdom
TIME: 1987 - 1994

914 **Forman, D.** 04926
Imperial Cancer Research Fund, Cancer Epidemiol. & Clin. Trials Unit, Radcliffe Infirmary, Gibson Lab., Gibson Bldg , Oxford OX2 6HE, United Kingdom (Tel.: +44 865 53951; Fax: 310545)
COLL: Møller, H.; Tjønneland, A.; Palli, D.; Zakelj, M.P.; Zatonski, W.A.; de Backer, G.; Boeing, H.; Tulinius, H.; Calheiros, J.M.; Abid, L.; Manousos, O.; Kyrtopoulos, S.; Coleman, M.P.; Roy, P.; Wild, C.; Newall, D.; Elder, J.; Knight, T.; Webb, P.; Haubrich, T.; Hengels, K.J.; Fukao, A.; Tsugane, S.; Kaye, S.; Potter, J.

European Correlation Study of Biological Markers for Gastritis and Gastric Cancer (EUROGAST)

The primary aim of this international collaborative study is to investigate putative biological correlates of gastric cancer. Centres have been chosen on the basis of contrasting gastric cancer rates and within each centre 100 men and women from each of the two age groups 25-34 and 55-64 years will be randomly selected from the general population. Each participant will provide a blood sample and complete a brief questionnaire. Sera will be assayed for H. pylori antibodies, pepsinogen, and a number of antioxidant micronutrients. All of these parameters, which have been related to the development of gastritis, will be correlated with gastric cancer incidence. Lymphocyte DNA preparations will be assayed for the presence of alkylating agent-DNA adducts, formed by the activity of N-nitroso compounds.

TYPE: Correlation
TERM: BMB; DNA Adducts; H. pylori; Micronutrients; Pepsinogen; Serum
SITE: Stomach
CHEM: N-Nitroso Compounds
LOCA: Algeria; Belgium; Denmark; Germany; Greece; Iceland; Italy; Japan; Poland; Portugal; Slovenia; United Kingdom; United States of America
TIME: 1989 - 1994

915 **Hawkins, M.M.** 04799
Univ. of Oxford, Childhood Cancer Research Group, 57 Woodstock Rd, Oxford OX2 6HJ, United Kingdom (Tel.: +44 865 310030; Fax: 514254)
COLL: Kinnier Wilson, L.M.

Second Primary Neoplasms after Childhood Cancer

Survivors of childhood cancer are at a greater risk of developing a subsequent cancer than the corresponding general populations. The purpose of our work is to monitor the incidence of cancers occurring subsequent to childhood cancer diagnosed in Britain, and to investigate the aetiology of these second primary cancers. The study is based on the national population-based register for childhood neoplasms and utilizes information from a national register for multiple primary tumours which has been established. The incidence of secondary leukaemia is being investigated in a cohort of approximately 16,500 one-year survivors of childhood cancer diagnosed between 1962 and 1983. The incidence of solid tumours as second primaries is being studied among about 13,400 three-year survivors of childhood cancer diagnosed between 1940 and 1983. A case-control study of secondary leukaemia is being carried out to investigate the relationship between the relative risk of subsequent leukaemia and the types and doses of chemotherapy and radiotherapy received by patients. Twenty seven secondary leukaemias have so far been included. It is planned to study the aetiology of other types of second primary tumours using a case-control design. Papers have been published in Br. J. Cancer 56:331-338 and 339-347, 1987, Int. J. Radiat. Oncol. Biol. Phys. 19:1297-1301 and in Br. Med. J. 304:951-958, 1992.

TYPE: Case-Control; Cohort
TERM: Chemotherapy; Childhood; High-Risk Groups; Multiple Primary; Radiotherapy; Survival
SITE: Leukaemia
REGI: Oxford II (UK)
TIME: 1981 - 1994

916 **Hawkins, M.M.** 04906
Univ. of Oxford, Childhood Cancer Research Group, 57 Woodstock Rd, Oxford OX2 6HJ, United Kingdom (Tel.: +44 865 310030; Fax: 514254)

Pregnancies and Offspring of Childhood Cancer Survivors

As a result of the substantial increases in survival following childhood cancer, an increasing number of survivors reach reproductive age and consider having a family. The outcome of such pregnancies and the health of all offspring are of considerable interest because the therapy which many survivors received is potentially mutagenic to germ cells, and because of the heritable component of childhood cancer and the association between some childhood cancers and congenital abnormalities, and in order to provide reliable estimates of various outcomes for those who counsel survivors and their families. Survivors of reproductive age are identified from the National Registry of Childhood Tumours and information is obtained from general practitioners' records and hospital consultants. Questions include those relating to infertility, spontaneous abortion, termination or stillbirths experienced by survivors, and also any congenital abnormalities, cancer or other genetic disease among offspring born to survivors. There are 2,900 survivors born in 1962 or before, and 2,650 survivors born between 1963 and 1968. A cohort of about 1,500 offspring of survivors has been established and will be followed up to compare mortality and cancer observed with that expected from general population rates. Recent papers include Int. J. Cancer 43:399-402 and 975-978, 1989, Br. J. Obstet. Gynaecol. 96:378-380, 1989 and in J. Natl Cancer Inst. 83:1643-1650, 1991.

TYPE: Cohort
TERM: Abortion; Congenital Abnormalities; Familial Factors; Genetic Factors; Infertility; Mutation, Germinal; Pregnancy; Survival
SITE: Childhood Neoplasms
REGI: Oxford II (UK)
TIME: 1981 - 1994

917 **Hawkins, M.M.** 04927
Univ. of Oxford, Childhood Cancer Research Group, 57 Woodstock Rd, Oxford OX2 6HJ, United Kingdom (Tel.: +44 865 310030; Fax: 514254)
COLL: Robertson, C.M.

Long-Term Survival, Causes of Late Deaths and Cure after Childhood Cancer

There have been great improvements in survival following childhood cancer in recent decades. The purpose of our work is to monitor the mortality occurring after patients have survived several years from diagnosis of childhood cancer in Britain, and to examine in detail the causes of such deaths. The study is based on the National Registry for Childhood Tumours. There are about 11,500 patients who were diagnosed before 1982 and are known to have survived at least three years. Comparison of the numbers of deaths observed with those expected from mortality occurring among the general population identifies departures from expected mortality and provides information relating to cure and late adverse effects of childhood neoplastic disease and its treatment. This work has been of benefit to survivors obtaining life insurance. The number of deaths occurring among individuals who have survived at least five years are sufficiently small (750 among 4,100 five-year survivors diagnosed before 1971, in Britain) to investigate in each case the course of events leading to death from hospital notes, post-mortem reports and death certificates. This detailed look at the sequence of events leading to death has identified some deaths that might have been preventable. Papers were published in Arch. Dis. Child. 64:798-897, 1989 and 65:1356-1363, 1990.

TYPE: Cohort
TERM: Late Effects; Survival; Treatment
SITE: Childhood Neoplasms
REGI: Oxford II (UK)
TIME: 1981 - 1994

918 **Key, T.** 04876
Imperial Cancer Research Fund, Cancer Epidemiology Unit, Radcliffe Infirmary, Gibson Bldg, Oxford OX2 6HE, United Kingdom (Tel.: +44 865 311933; Fax: 310545)
COLL: Bishop, D.T.; Silcocks, P.B.

Case-Control Study of Prostate Cancer

This is a case-control study of men with prostate cancer, matched withhealthy controls of the same age selected from the same general practice as the case. Cases and controls are interviewed at home to collect information on diet, exposure to metals, and sexual behaviour, and to collect samples of blood, hair and nails. Preliminary analysis will be done on 300 case-control pairs, to test the hypothesis that risk is directly related to dietary fat, to exposure to cadmium or nickel, and to a venereally transmitted agent.

TYPE: Case-Control
TERM: BMB; Diet; Fat; Hair; Metals; Sexual Activity; Sexually Transmitted Diseases; Toenails
SITE: Prostate
CHEM: Cadmium; Nickel
TIME: 1990 - 1994

***919 Key, T.** 05334
Imperial Cancer Research Fund, Cancer Epidemiology Unit, Radcliffe Infirmary, Gibson Bldg, Oxford OX2 6HE, United Kingdom (Tel.: +44 865 311933; Fax: 310545)
COLL: Forman, D.; Moore, J.W.; Savage, P.; English, R.; Shepstone, B.; Gravelle, H.

Diet and Mammographic Patterns

Wolfe classified the mammographic appearance of breast parenchyma into four patterns which are related to the risk of developing breast cancer. The hypothesis being tested in this study is that high risk mammographic patterns are associated with a low intake of carotene and fibre and a high intake of saturated fat and alcohol. Mammographic patterns and diets have been assessed in 500 women attending for breast cancer screening in Aylesbury, UK. Measurements of serum concentrations of carotenoids are in progress.

TYPE: Cross-Sectional
TERM: Alcohol; BMB; Diet; Fat; Fibre; High-Risk Groups; Mammography; Serum
SITE: Breast (F)
CHEM: Beta Carotene
TIME: 1990 - 1994

920 Mant, D. 03839
University of Oxford, Department of Community Medicine, Gibson Labs., Radcliffe Infirmary, Woodstock Rd, Oxford, United Kingdom (Tel.: +44 865 511293)
COLL: Pike, M.C.; Vessey, M.P.; Chilvers, C.E.D.

Adenocarcinoma of the Cervix

The objective of the study is to determine the contribution of oral contraception, among other factors, to the aetiology of adenocarcinoma of the cervix in women under 45 years. It is intended to study 150 cases (diagnosed between 1 January 1984 and 31 December 1988) and 450 age-matched controls selected from general practice lists. A comparison group of 300 cases of squamous carcinoma of the cervix in women under 40 years of age and 900 age-matched controls will also be sought. A formal structured questionnaire will be administered to both cases and controls by a trained interviewer. Pathological specimens will be reviewed independently to confirm the diagnosis and evidence of HPV infection will also be sought by DNA-DNA hybridisation.

TYPE: Case-Control
TERM: DNA; HPV; Histology; Infection; Oral Contraceptives; Virus
SITE: Uterus (Cervix)
TIME: 1985 - 1994

921 Neil, H.A.W. 04849
Univ. of Oxford, Dept. of Public Health & Primary Care, Radcliffe Infirmary, Woodstock Rd, Oxford OX2 6HE, United Kingdom (Tel.: +44 865 511293)
COLL: Thorogood, M.; Cohen, D.L.; Mann, J.I.

Cholesterol, Cancer and Diabetes

294 diabetic patients identified in a population-based survey in 1982 in Oxford agreed to participate in a clinical study. Various clinical and biochemical variables were measured, including lipoprotein profile. The records of patients were flagged by the National Health Service Central Registry so that deaths could be identified. By 31 December 1990, 93 deaths had occurred. The aim of the study is to examine the predictors of mortality, particularly in type II diabetes, and to examine the relationship between cholesterol concentration and cancer mortality in diabetes. Survival analyses will be undertaken using Cox's proportional hazard model.

TYPE: Cohort
TERM: Biochemical Markers; Cholesterol; Diabetes; Survival
SITE: All Sites
TIME: 1982 - 1999

922 **Stiller, C.A.** 04930
Univ. of Oxford, Childhood Cancer Research Group, 57 Woodstock Rd, Oxford OX2 6HJ, United Kingdom (Tel.: +44 865 310030; Fax: 514254 ; Tlx: 83147 ccrg)
COLL: Bunch, K.J.; Draper, G.J.; Eatock, E.M.; Loach, M.J.; Passmore, S.J.

Incidence of Childhood Cancer

The National Registry of Childhood Tumours includes virtually all cases of cancer and leukaemia in children aged 0-14 in England, Scotland and Wales from 1962 onwards. Incidence rates are analysed for all childhood cancers and for particular diagnostic groups. Analyses in progress or planned include (1) trends in incidence rates; (2) completeness of ascertainment of cases by national record systems; (3) national registry of childhood myelodysplasia; (5) incidence rates in relation to residence in new towns and other indicators of population mixing which might be related to infectious factors in the aetiology of leukaemia. International variations in incidence rates are studied using data from the IARC study of international childhood cancer incidence. Recent papers were published in Br. J. Cancer 63:993-999, 1991; 64:543-548 and 549-554, 1991; and in Int. J. Cancer 52:538-543, 1992.

TYPE: Incidence; Registry
TERM: Childhood; Ethnic Group; Geographic Factors; Infection; Migrants; Registry; Trends
SITE: Childhood Neoplasms; Leukaemia; Myelodysplastic Syndrome
REGI: Oxford II (UK)
TIME: 1975 - 1995

923 **Thorogood, M.** 04671
Univ. of Oxford, Dept. Commun. Medicine & Gen. Practice, Radcliffe Infirmary, Woodstock Rd, Oxford OX2 6HE, United Kingdom (Tel.: +44 865 511293; Fax: 310545)
COLL: Vessey, M.P.; Mann, J.R.; McPherson, C.K.

Cancer and Other Diseases in Vegetarians

This is a prospective study of people eating different diets, including some 300 vegans, 800 fish eaters who do not eat meat, 5,000 lacto-ovo-vegetarians and 5,000 omnivorous controls (the friends and relatives of the non-meat-eaters). Subjects have been flagged with the National Health Service Central Register, and information on both deaths and cancer registrations is being provided. Some 3,000 subjects have provided blood samples, and 5,000 have completed 4-day estimated-weight dietary records. The aim of the study is to examine the different mortality experience of non-meat-eaters and meat-eaters, especially with reference to digestive tract cancers, breast cancer, and cardiovascular disease.

TYPE: Cohort
TERM: BMB; Blood; Diet; Registry; Vegetarian
SITE: Breast (F); Gastrointestinal
REGI: OPCS (UK)
TIME: 1980 - 1994

924 **Vessey, M.P.** 00762
Univ. of Oxford, Dept. of Public Health & Primary Care, Gibson Labs., Radcliffe Infirmary, Gibson Bldg, Oxford OX2 6HE, United Kingdom (Tel.: +44 865 319107)
COLL: Painter, R.; Villard-Mackintosh, L.

Long-Term Follow-Up Study of Women Using Different Methods of Contraception

Seventeen family planning clinics are participating in a prospective study of the beneficial and adverse effects of different methods of contraception. Over 17,000 women are under observation of whom 57% were using oral contraceptives on admission to the study; 25% were using a diaphragm and 18% were using an intrauterine device. During follow-up, each subject is questioned at return visits to the clinic and a record of pregnancies and their outcome, hospital visits, changes in contraceptive methods and the results of cervical smears, is accumulated. Women who default are sent a postal questionnaire and if this is not returned, are telephoned or visited in their homes to collect the necessary information. So far, data representing 290,000 woman-years of follow-up are available for analysis. Losses to follow-up have been fewer than 3 per 1,000 per annum. Analyses completed to date concern the entry characteristics of the subjects, patterns of morbidity and mortality, outcome of pregnancy, fertility after discontinuation of contraception, and efficacy of different methods. Most of the findings fit in well with those of retrospective studies and with those of the Royal College of General Practitioners prospective study. A detailed interim report of the study was published in the Journal of Biosocial Science in October 1976. About 70 publications have now emerged from this study.

TYPE: Cohort
TERM: Contraception; Cytology; Fertility; Oral Contraceptives; Parity; Pregnancy
SITE: All Sites; Breast (F); Uterus (Cervix)
TIME: 1968 - 1994

925 **Vessey, M.P.** 01289
Univ. of Oxford, Dept. of Public Health & Primary Care, Gibson Labs., Radcliffe Infirmary, Gibson Bldg, Oxford OX2 6HE, United Kingdom (Tel.: +44 865 319107)
COLL: Hunt, K.

Mortality and Cancer Incidence in Women Receiving Hormone Replacement Therapy

In this investigation women seen at 20 menopause clinics in different parts of the country who were currently using hormone replacement treatment and had used it continuously for at least one year, were recruited and are being followed up indefinitely using the facilities of the National Health Service Central Registries. Mortality rates and cancer incidence rates are computed and comparisons are made with national mortality and cancer incidence data. Cancer of the endometrium and cancer of the breast are sites of particular interest, as well as cardiovascular disease. About 5,000 women are taking part in the study. Reports were published by Hunt et al. in Br. J. Obstet. Gynaecol. 94:620-635, 1987 and 97:1080-1086, 1990.

TYPE: Cohort
TERM: Hormones; Menopause; Registry
SITE: Breast (F); Uterus (Corpus)
REGI: OPCS (UK)
TIME: 1978 - 1994

926 **Vessey, M.P.** 01430
Univ. of Oxford, Dept. of Public Health & Primary Care, Gibson Labs., Radcliffe Infirmary, Gibson Bldg, Oxford OX2 6HE, United Kingdom (Tel.: +44 865 319107)
COLL: Fairweather, D.; Meara, J.

Follow-Up Study of Women Exposed to DES in Pregnancy and their Offspring

In the early 1950s, a randomised double-blind controlled trial of the prophylactic use of stilboestrol in pregnancy was conducted at a London teaching hospital. The study involved over 1,000 primigravidae and the treatment group received stilboestrol according to the Smith and Smith regimen. The participants in the study and their offspring have been identified and traced and death certificates obtained for those who have died. Information about the health of survivors has also been sought from general practitioners and the analysis of the results has been published in Brit. J. Obstet. and Gynaecol. 90:1007-1017, 1983. Long-term follow-up will continue indefinitely. A recent publication appeared in Brit. J. Obstet. and Gynaecol. 90:620-622, 1989.

TYPE: Cohort
TERM: Drugs; Hormones; Intra-Uterine Exposure
SITE: Breast (F); Female Genital
CHEM: DES
TIME: 1978 - 1994

927 **Vessey, M.P.** 03000
Univ. of Oxford, Dept. of Public Health & Primary Care, Gibson Labs., Radcliffe Infirmary, Gibson Bldg, Oxford OX2 6HE, United Kingdom (Tel.: +44 865 319107)
COLL: Colin Jones, D.G.; Langman, M.J.S.; Lawson, D.H.

Post-Marketing Surveillance of the Safety of Cimetidine

A total of 10,000 takers of cimetidine and a like number of controls were recruited at Oxford, Portsmouth, Nottingham and Glasgow and followed up for one year. A large excess of gastric cancer was seen in the takers group, but the great majority of the cancers were the result of cimetidine being given to patients with symptoms of undiagnosed malignancy. The takers have been 'labelled' in the National Health Service Central Registries and are being followed up. Papers have been published in Br. J. Med. 291:1084-1088, 1985, Alimentary Pharmacology & Therapeutics 1:167-177, 1987, The Lancet i:1453, 1989 and in Gut 33:1280-1284, 1992.

TYPE: Cohort
TERM: Drugs; Registry
SITE: Gastrointestinal
CHEM: Cimetidine
REGI: OPCS (UK)
TIME: 1978 - 1994

Salisbury

928 **Thomas, H.F.** 01915
Wessex Regional Health Authority, 9078 Campbell Rd, Salisbury SP1 3BG, United Kingdom (Tel.: +44 722 332322)
COLL: Winter, P.D.; Donaldson, L.

Mortality Study of Pest Control Officers

A study is being conducted to examine the mortality of pest control officers employed by local authorities in England and Wales. At first an interest was taken in the use of 'antu' (alpha-naphthylthiourea), a rodenticide which chemical tests suggested could be associated with bladder cancer, and which was generally withdrawn from use in the UK in 1967. A report on sub-groups exposed to 'antu' has been published (Br. Med. J. 285:927-931, 1982) suggesting that there was an increased incidence of bladder tumours. A population of 1,500 men presently employed is now being followed-up to examine their mortality experience over the next 10 to 20 years. Information on pesticides handled and personal habits, e.g., smoking and alcohol consumption, has been obtained from a 10% random sample of men (J.R.S.H. 6:204-206, 1986). 115 deaths have been notified to date. An analysis of causes is being undertaken.

TYPE: Cohort
TERM: Alcohol; Chemical Exposure; Drugs; Insecticides; Occupation; Pesticides; Tobacco (Smoking)
SITE: All Sites
CHEM: Lindane; Naphthylthiourea, Alpha; Warfarin
OCCU: Pesticide Workers
TIME: 1979 - 2012

Southampton

***929** **Alexander, F.E.** 05445
Royal South Hants Hosp., Leukaemia Research Fund Centre for Clinical Epidemiology, SW Study Group, Graham Rd, Southampton SO9 4PE, United Kingdom (Tel.: +44 703 825836; Fax: 825836)
COLL: Jarrett, R.; Taylor, M.; Cartwright, R.A.; Staines, A.

Case-Control Study of Lymphoid Malignancies in Young Adults

This is a population-based case-control study to test the following hypotheses: (1) HD (and possibly also ALL and NHL) in young adults arises as an abnormal response to late first exposure to a herpes virus or other infectious agents; predisposing host factors include HLA class II type; (2) HD, NHL and ALL cases have lived unusually close together at some time prior to diagnosis. Cases are all newly diagnosed cases, aged 16-24 years, resident at diagnosis in defined geographical areas (anticipated number: 350). Two controls per case are matched for age, sex and health authority of residence. Data are being collected in face to face interviews and through abstraction from medical records. Blood (for HLA classification and serological analysis for anti-bodies to EBV, HHV6, HHV7) and saliva samples (for detecting anti-bodies to EBV, CMV, VZV) are obtained from cases and controls, and blocks from lymphoma cases. Conditional logistic regression and methods for spatial clustering will be used in the analysis.

TYPE: Case-Control
TERM: BMB; Blood; Cluster; EBV; HLA; Saliva; Virus
SITE: All Sites; Hodgkin's Disease; Leukaemia; Non-Hodgkin's Lymphoma
TIME: 1991 - 1996

930 **Coggon, D.** 04795
Univ. of Southampton, Southampton General Hosp., MRC Environmental Epidemiology Unit, Tremona Rd, Southampton SO9 4XY, United Kingdom (Tel.: +44 703 777624; Fax: 704021)
COLL: Barker, D.J.P.; Winter, P.D.

Food Storage and Domestic Crowding in Childhood and Stomach Cancer

This study tests the hypothesis that domestic crowding and poor food storage facilities in childhood are risk factors for stomach cancer. Information about food storage facilities and the number and size of rooms in 6,000 houses have been abstracted from the records of a housing survey carried out in Chesterfield in 1936. Occupants of these houses will be identified from the 1939 census and followed up for mortality. Risk of stomach cancer will be related to type of food storage facility and measures of crowding in 1939.

TYPE: Cohort
TERM: Childhood; Diet; Socio-Economic Factors
SITE: Stomach
TIME: 1989 - 1994

931 **Coggon, D.** 04796
Univ. of Southampton, Southampton General Hosp., MRC Environmental Epidemiology Unit, Tremona Rd, Southampton SO9 4XY, United Kingdom (Tel.: +44 703 777624; Fax: 704021)
COLL: Pannett, B.; Winter, P.D.; Wright, H.

Study of Workers Exposed to Mineral Acid Mists

This study tests the hypothesis that occupational exposure to mineral acid mists is a cause of laryngeal cancer. A retrospective cohort is being assembled to include exposed and unexposed workers from the steel and battery manufacturing industries. Subjects will be followed for mortality and cancer incidence through the National Health Service Central Register. The association between laryngeal cancer and acid mists will be examined principally by a nested case-control approach.

TYPE: Case-Control; Cohort
TERM: Occupation
SITE: Larynx
CHEM: Sulphuric Acid
OCCU: Battery Plant Workers
TIME: 1989 - 1994

932 **Coggon, D.** 05048
Univ. of Southampton, Southampton General Hosp., MRC Environmental Epidemiology Unit, Tremona Rd, Southampton SO9 4XY, United Kingdom (Tel.: +44 703 777624; Fax: 704021)
COLL: Maitland, N.

Lung Cancer in Butchers

This study tests the hypothesis that an apparent excess of lung cancer among meat workers in routine statistics of occupational mortality and cancer incidence is explained by a carcinogenic papilloma virus. Lung cancer deaths in butchers in England and Wales during 1987 and 1988 have been identified from death certificates, and histological material is being sought from the tumours where it is available. It is hoped to obtain approximately 50 specimens together with material from a similar number of lung tumours in controls of the same age and sex who were not butchers. Evidence of papillomavirus DNA in the tumours will be sought by a PCR method.

TYPE: Case-Control
TERM: Autopsy; HPV; Occupation; PCR
SITE: Lung
OCCU: Meat Workers
TIME: 1990 - 1994

***933** **Houlston, R.S.** 05252
Southampton General Hosp., Dept. of Clinical Genetics, Tremona Rd, Southampton S09 4XY Hants, United Kingdom (Tel.: +44 703 796166)
COLL: Kee, F.; Norton, N.E.

Genetic Epidemiology of Colorectal Cancer in Northern Ireland

The aim is to study the genetic epidemiology of colorectal cancer in Northern Ireland. Family histories of cancer have been obtained from over 300 probands with colorectal cancer ascertained through a cancer registry. These pedigrees will be analysed by complex segregation techniques to determine the most likely model underlying the familial basis of this cancer.

TYPE: Genetic Epidemiology
TERM: Genetic Factors; Segregation Analysis
SITE: Colon; Rectum
TIME: 1992 - 1994

***934** **Inskip, H.M.** 05407
Univ. of Southampton, Southempton General Hosp., MRC Environmental Epidemiology Unit, Tremona Rd, Southampton S09 4XY, United Kingdom (Tel.: +44 703 777624; Fax: 704021)
COLL: Winter, P.D.; Wield, G.

Seascale Birth and School Cohort Studies

These studies were set up in 1984 following concern about raised levels of childhood leukaemia in the area surrounding the Sellafield nuclear site. Children born or attending schools in Seascale, a village close to the site, have been followed up to obtain information about mortality and cancer incidence. Studies of children born during 1950-1983 or attending schools in Seascale from 1955-1984 have been published (Gardner et al., British Medical Journal 1987; 295:819-827) and they indicated that the raised level of leukaemia was concentrated in those born in the area. These cohort studies are being extended to include children born as early as 1930 and as late as 1987 and attempts are being made to link the fathers of the birth cohort to the Sellafield workforce, so that their occupational radiation exposure prior to conception of the child can be examined in relation to the child's subsequent risk of leukaemia.

TYPE: Cohort
TERM: Childhood; Occupation; Parental Occupation; Radiation, Ionizing
SITE: Childhood Neoplasms; Leukaemia; Lymphoma
OCCU: Power Plant Workers
TIME: 1984 - 1994

***935** **Inskip, H.M.** 05408
Univ. of Southampton, Southempton General Hosp., MRC Environmental Epidemiology Unit, Tremona Rd, Southampton S09 4XY, United Kingdom (Tel.: +44 703 777624; Fax: 704021)
COLL: Styles, L.

Study of Royal Naval Submariners

Submariners live in an artificial environment and are exposed to potentially hazardous factors such as diesel fumes, recycled air and, on nuclear submarines, ionising radiation. The aim of the study is to examine the long-term effects on mortality and morbidity of working in a submarine environment. A cohort study of some 15,000 submariners who first undertook submarine training between 1960 and 1979 are being followed up until the end of 1989. The cohort has been identified from records held by the Navy and information on deaths occurring in-service has been extracted. Information on those who have left the Navy is being obtained from the National Health Service Central Register. Cancer incidence data are also being collected. Initial comparisons will be made with mortality in the civilian population. It is planned that nested case-control studies focussing on diseases of interest will be conducted.

TYPE: Cohort
TERM: Environmental Factors; Occupation; Radiation, Ionizing
SITE: All Sites
CHEM: Diesel Exhaust
OCCU: Military Servicemen
TIME: 1986 - 1994

***936** **Inskip, H.M.** 05423
Univ. of Southampton, Southempton General Hosp., MRC Environmental Epidemiology Unit, Tremona Rd, Southampton S09 4XY, United Kingdom (Tel.: +44 703 777624; Fax: 704021)
COLL: Winter, P.D.

West Cumbria Leukaemia and Lymphoma Case-Control Study

Following a government inquiry into childhood cancer levels in West Cumbria near the Sellafield nuclear site, a case-control study of lymphomas and leukaemias was conducted. It included 97 cases diagnosed in West Cumbria between 1950 and 1985 in persons under the age of 25 years and 1,001 controls matched on age, sex and area of residence at birth. The specific objectives were to identify factors that might explain the high rates of childhood leukaemia in the area. The main finding of this study was a strong statistical association between the child's risk of leukaemia and paternal occupational exposure to external radiation before the child's conception (Gardner et al., British Medical Journal 1990; 300:423-434). This study is being extended to examine cases diagnosed after 1985 and to examine occupational radiation exposure in greater detail, including internal exposure to radionuclides.

TYPE: Case-Control
TERM: Childhood; Occupation; Parental Occupation; Radiation, Ionizing
SITE: Leukaemia; Lymphoma
OCCU: Radiation Workers
TIME: 1984 - 1994

937 **Morton, N.E.** 04680
Univ. of Southampton, Southampton General Hosp., Dept. of Community Medicine, South Block, Southampton SO9 4XY, United Kingdom (Tel.: +44 703 777222; Tlx: 47661)
COLL: Iselius, E.L.

Research Programme in Genetic Epidemiology

Two epidemiology projects are conducted. The leukaemia project tests the hypothesis that susceptibility in Down's syndrome depends on mode of origin and that development of acute leukaemia following transient leukaemia of the newborn requires a mutation at an unlinked locus. Mode of origin is being determined from DNA markers. The control group consists of Down's cases without leukaemia. The neurofibromatosis project tests the hypothesis that sporadic cases are mostly germinal mutations, with a high frequency determined by a large gene size, consistently estimated directly from segregation analysis of pedigrees, and indirectly from reproductive performance assuming mutation-selection equilibrum. The linkage project is creating recombination maps of chromosomes 10 and X by multiple pairwise analysis, using DNA markers in families and XY trisomies.

TYPE: Genetic Epidemiology
TERM: DNA; Down's Syndrome; Mutation, Germinal; Pedigree; Segregation Analysis
SITE: Leukaemia
TIME: 1988 - 1994

938 **Pippard, C.** 04909
Univ. of Southampton, MRC Environmental Epidemiology Unit, Tremona Rd , Southampton, United Kingdom (Tel.: +44 703 777624)
COLL: Barker, D.J.P.; Bridges, B.A.

Cohort Study of Ataxia-Telangiectasia Patients

The study aims to determine whether the incidence of any cancer is elevated in homozygotes and heterozygotes. Heterozygotes comprise a significant proportion of the population and may be an at-risk group. The patients and their parents, grandparents, siblings and children are being registered. They will be traced at the National Health Service Central Register and copies of cancer registration and death certificates obtained. Observed numbers of cancers in the cohort will be compared with expected numbers derived from national statistics allowing for age, sex, time period and region of residence. Allowance will be made for the probability of heterozygosity in grandparents.

TYPE: Cohort
TERM: Ataxia Telangiectasia; High-Risk Groups; Registry
SITE: All Sites
REGI: OPCS (UK)
TIME: 1982 - 1994

939 **Winter, P.D.** 03632
Univ. of Southampton, Southampton General Hosp., MRC Environmental Epidemiology Unit, Tremona Rd, Southampton SO9 4XY, United Kingdom (Tel.: +44 703 777624; Tlx: 47661 sotonu g)
COLL: Coggon, D.; Pannett, B.; Jones, R.D.

Mortality Follow-Up of Pottery Workers

This is a historical cohort mortality follow-up of 6,000 pottery workers who were previously surveyed for respiratory morbidity in 1970 during a cross-sectional survey. The National Health Service Central Register and the Department of Health and Social Security records have been used to identify the vital status and cause of death of these workers during the ensuing years. The primary objective is to examine the consequences of low exposure to silica dust in a large occupational cohort, particularly levels of lung cancer, which has been shown to be raised among silicotics. Exposure will be assessed by occupational histories, job titles and dust measurements. Cigarette smoking habits were collected in 1970 and can be adjusted for in the analysis. A publication by Winter et al. appeared in IARC Scientific Publication No. 97, 1990.

TYPE: Cohort
TERM: Dusts; Occupation; Registry
SITE: All Sites; Lung
CHEM: Silica
OCCU: Potters
TIME: 1985 - 1994

Sutton

940 **Bell, C.M.J.** 04954
Thames Cancer Registry, 15 Cotswold Rd, Sutton Surrey SM2 5PY, United Kingdom (Tel.: +44 81 6427692)
COLL: Horwich, A.

Treatment for Early Stage Seminoma of the Testis

This is a population-based study of a cohort of patients with testicular cancer (N = 831) diagnosed at an early stage between 1961-1985 in the two South Thames Health Regions of England. The patients are being followed up to 1989 to determine the excess risk of second cancers and any excess of causes of death. The aim is to test findings in other studies of an excess of leukaemia and urinary tract cancer following radiotherapy to the pelvis.

TYPE: Cohort
TERM: Multiple Primary; Radiotherapy; Registry
SITE: All Sites; Testis
REGI: Thames (UK)
TIME: 1988 - 1994

941 **Chamberlain, J.** 01301
Inst. of Cancer Research, Cancer Screening Evaluation Unit, Sect. of Epidemiology, Block D, Cotswold Rd, Sutton Surrey SM2 5NG, United Kingdom (Tel.: +44 81 6427692)
COLL: Price, J.L.; Joslin, C.A.; Blamey, R.W.; Forrest, A.P.M.

Trial of Early Detection of Breast Cancer

This is an observational trial which aims to determine the effectiveness of alternative early detection programmes in reducing mortality from breast cancer. Two populations of approximately 25,000 women aged 45-64 have been offered screening by clinical examination and mammography. Two others of the same size and age distribution have been offered education in breast self-examination, and four additional similar populations have served as comparison groups, offered no additional detection services above what is normally provided as good current practice. Field-work was completed in 1988 but all the women are flagged so that mortality from breast cancer can be monitored until at least 1992. The main analysis concerns population-based mortality from breast cancer. Other analyses include use of the case-control method regarding screening history in women who have died or have not died from breast cancer. Papers have been published in Br. J. Cancer 44:618-627, 1981, and the Lancet 11:411-416, 1988.

TYPE: Cohort; Mortality
TERM: Mammography; Screening
SITE: Breast (F)
TIME: 1979 - 1994

942 **Moss, S.M.** 04882
Inst. of Cancer Research, Dept. of Health, Cancer Screening Evaluation Unit, Block D, 15 Cotswold Rd, Sutton Surrey SM2 5NG, United Kingdom (Tel.: +44 81 6438901; Fax: 7707876)
COLL: Chamberlain, J.; Thomas, B.A.; Kirkpatrick, A.

Effect on Breast Cancer Mortality of Annual Mammographic Screening Starting at Age 40

This is a multicentre randomized controlled trial, designed to provide a definitive answer to the open question of the proportion of breast cancer deaths which can be avoided by mammographic screening of women in their 40's. One hundred ninety-five thousand women aged 40 to 41 will be identified from Family Practitioner Committee Registers and randomly allocated to a group of 65,000 in the study group and 130,000 in the control group. Those in the study group will be offered mammographic screening every year for seven years; those in the control group will not be offered screening until they are 50. Both groups will then join the routine 3-yearly breast screening service. During the course of the study all breast cancer cases (no matter how diagnosed) and all deaths among study and control group women will be recorded. Mortality from breast cancer will be compared between the two groups.

TYPE: Intervention
TERM: Mammography; Screening
SITE: Breast (F)
TIME: 1990 - 2000

Taunton

943 **Bowie, C.** 04947
Somerset Health Authority, Dept. of Public Health, District Headquarters, Wellsprings Rd, Taunton TA2 7PQ, United Kingdom (Tel.: +44 823 333491; Fax: 272710)
COLL: O'Boyle, P.; Ewings, P.; Pocock, R.

Cancer of the Prostate

This study aims to investigate a possible link between prostate cancer and the following: alcohol, tobacco, sexual activity, diet, family history of cancer, occupation (farming in particular), water hardness, use of pesticides, fertilizers and other chemicals, radiation exposure, vitamin supplements, past illness, zinc and cadmium exposure. It is anticipated that approximately 180 cases of prostate cancer will be identified by the Hospitals' Pathology Service during the period May 1989 to June 1991. There will be two control groups, one consisting of patients with benign prostatic conditions and one of hospital patients whose reason for admission was not for urinary or prostatic problems. The controls will be age-matched with the cases. The cases and controls will be interviewed by trained interviewers with a structured questionnaire. Multivariate and univariate methods will be used.

TYPE: Case-Control
TERM: Alcohol; Diet; Fertilizers; Occupation; Pesticides; Radiation, Ionizing; Sexual Activity; Tobacco (Smoking); Vitamins; Water
SITE: Prostate
CHEM: Cadmium; Zinc
TIME: 1989 - 1994

944 **Bowie, C.** 05101
Somerset Health Authority, Dept. of Public Health, District Headquarters, Wellsprings Rd, Taunton TA2 7PQ, United Kingdom (Tel.: +44 823 333491; Fax: 272710)
COLL: Bowie, S.H.U.

Radon in Dwellings and Lung Cancer

The hypothesis is that radon found in the dwellings in the U.K. causes lung cancer and that there is a geographical association between districts in England and Wales with high radon levels in dwellings and both local geological features and lung cancer mortality adjusted for smoking and other known causes of lung cancer. The study uses OPCS data on lung cancer mortality for district council areas, radon measurements supplied by NRPB, smoking histories from lifestyle surveys and air pollution data from the

Warren Springs Laboratory. This ecological study is designed to correlate lung cancer rates with these population-level data.

TYPE: Correlation
TERM: Air Pollution; Environmental Factors; Radiation, Ionizing; Tobacco (Smoking)
SITE: Lung
CHEM: Radon
TIME: 1990 - 1994

Tyne & Wear

945 **Johnson, S.** 00956
Cookson Ceramics & Minerals Ltd, Cookson House, Willington Quay, Wallsend Tyne & Wear NE28 6UQ, United Kingdom (Tel.: +44 91 2622211; Fax: 2950119; Tlx: 537357)
COLL: Billington, N.A.

Lung Cancer in the Antimony Industry

A prospective epidemiological survey of workers engaged in the production of antimony metal, antimony oxide, and antimony sulphide is being undertaken to ascertain whether there is an excess of lung cancer by comparison with local death rates and death rates in other occupations. The survey includes all persons working at the plant on 1 January 1961 or joining after that date, and is likely to continue until 1990 or later. Persons leaving employment are traced. The calculations of expected deaths are age-specific and related to man-years of exposure. Other factors studied are smoking, sulphur, arsenic and area of residence.

TYPE: Cohort
TERM: Chemical Exposure; Environmental Factors; Metals; Occupation; Tobacco (Smoking)
SITE: Lung
CHEM: Antimony; Arsenic; Zircon
OCCU: Antimony Process Workers
TIME: 1961 - 1994

UNITED STATES OF AMERICA

Albuquerque

946 **Becker, T.M.** 04450

Univ. of New Mexico, School of Medicine, Dept. of Medicine, 900 Camino de Salud , Albuquerque NM 87131, United States of America (Tel.: +1 505 2775541)

COLL: Helgerson, S.; Wheeler, C.M.; Jordan, S.W.; Rosenfeld, P.; Mertz, G.; Icenogle, J.; Stone, K.M.

Natural History of Human Papillomavirus Infection

This prospective cohort study will evaluate the role of cervical HPV infection as a risk factor for cervical dysplasia. Cohorts totalling 600 Native American, Hispanic, and White women with and without cytologic/ colposcopic evidence of mild cervical dysplasia will be followed for 3-5 years. Presence of type-specific HPV DNA will be determined by nucleic acid hybridization. Multiple logistic regression will be used to explore risk factors for progression of dysplasia, including HPV infection, other sexually transmitted infections, cigarette smoking, and contraceptive use.

TYPE: Cohort
TERM: Contraception; DNA; HPV; Premalignant Lesion; Sexually Transmitted Diseases; Socio-Economic Factors; Tobacco (Smoking)
SITE: Uterus (Cervix)
TIME: 1988 - 1994

947 **Samet, J.M.** 02430

Univ. of New Mexico School of Medicine, Cancer Centre, New Mexico Tumor Registry, 900 Camino de Salud NE, Albuquerque NM 87131, United States of America (Tel.: +1 505 2775541; Fax: 2778572)

COLL: Morgan, M.V.; Key, C.R.; Pathak, D.R.; Valdivia, A.A.

Lung Cancer Mortality in New Mexico's Uranium Miners

Underground uranium miners are at increased risk of lung cancer because of exposure to radon daughters. In New Mexico, a large population of uranium miners have been exposed to radon daughters at concentrations lower than most previously studied groups. A cohort study of approximately 4,000 male underground miners has been implemented to assess the exposure-response relationship between radon daughter exposure and lung cancer mortality and to descibe the interaction between cigarette smoking and radon daughters. Exposures are documented with multiple sources and information on cigarette smoking is available from records of mining-related clinical examinations. Mortality has been evaluated until 1985 and conventional cohort and multivariate analyses are in progress. Papers were published in Hlth Phys. 56:415-421, 1989 and 61:745-752, 1991.

TYPE: Cohort
TERM: Metals; Mining; Occupation; Radiation, Ionizing; Tobacco (Smoking)
SITE: Lung
CHEM: Radon; Uranium
TIME: 1977 - 1994

***948** **Samet, J.M.** 05483

Univ. of New Mexico School of Medicine, Cancer Centre, New Mexico Tumor Registry, 900 Camino de Salud NE, Albuquerque NM 87131, United States of America (Tel.: +1 505 2775541; Fax: 2778572)

COLL: Bartow, S.A.; McFeeley, P.; Pathak, D.R.; McPherson, R.S.

Breast Cancer Epidemiology in New Mexico Hispanic Women

This case-control study aims to address: (1) risk factors for breast cancer in Hispanic and non-Hispanic white women; (2) changes in risk factor distributions across birth cohorts of Hispanic women; and (3) the effect of Hispanic ethnicity on breast cancer risk, after controlling for established breast cancer risk factors. Complementary investigations include an autopsy study and the development of a bank of malignant and non-malignant breast tissue for future molecular and cellular studies. A total of 1,600 subjects under the age of 75 and resident in New Mexico will be interviewed. Cases include 400 Hispanic and 400 non-Hispanic women. Controls are frequency-matched for age-group, ethnicity and health planning district and are identified through random digit dialling procedures. Personal interviews are conducted to collect information on medical history, medication usage, use of cigarettes and alcohol and physical activity.

TYPE: Case-Control
TERM: Alcohol; Autopsy; BMB; Diet; Drugs; Ethnic Group; Physical Activity; Registry; Tissue; Tobacco (Smoking)
SITE: Breast (F)
REGI: New Mexico (USA)
TIME: 1991 - 1996

Ann Arbor

949 **Sheikh, K.** 04203
Univ. of Michigan, School of Public Health, Dept. of Epidemiology, 109 Observatory St. , Ann Arbor MI 48109, United States of America (Tel.: + 1 313 7642159)
COLL: Wolfe, R.A.; Higgins, I.T.T.; Carman, W.J.; Harlan, L.

Obesity and Cancer in Women

The objective of the study is to determine whether obesity is associated with common cancers in the female population of Tecumseh, Michigan. The relationship between the type of obesity, body frame, the patterns of change in obesity over time, reproductive factors and risk of cancer of the breast, gastrointestinal tract and reproductive organs will be investigated. Other known or suspected risk factors for cancer to be included in the study are physical inactivity, family medical history, other morbid conditions, use of exogenous hormones, serum cholesterol and uric acid, smoking and alcohol consumption. For a subset of the population, dietary energy and intake of nutrients will also be examined. Since 1959, the Tecumseh population (3,523 women over age 16) have participated in an on-going prospective study. Data have been gathered on demographic and social characteristics, dietary intake, biochemical parameters, exposure to female hormones, morbidity and mortality. This 2-year project will utilize the existing data and newly collected data. Cancer cases have been identified through 1979, a morbidity and mortality follow-up will be carried out to ascertain cancers diagnosed in the female population between 1979 and 1986. 104 cases of breast cancer, 78 cases of uterine cancer and 40 cases of colon cancer are expected to be available for this investigation. Living cases and the relatives of deceased cases will be interviewed and medical records of the cases reviewed to confirm the diagnosis and to determine the approximate time of onset of cancer. In univariate and multivariate analysis, the risk of cancer will be assessed according to risk factor status.

TYPE: Cohort
TERM: Alcohol; Cholesterol; Diet; Familial Factors; Hormones; Obesity; Physical Activity; Physical Factors; Tobacco (Smoking); Uric Acid
SITE: Breast (F); Colon; Female Genital; Ovary; Uterus (Cervix); Uterus (Corpus)
TIME: 1986 - 1994

Arlington

950 **Brett, S.M.** 05045
Environ Corporation, 4350 N. Fairfax Drive, Arlington VA 22203, United States of America (Tel.: + 1 703 5162300; Fax: 5162345)
COLL: Rodricks, J.V.; Wilcock, K.; Starr, T.; Chinchilli, V.

Leukaemia Risk Associated with Occupational Benzene Exposure

A re-analysis of the leukaemia risk associated with occupational benzene exposure will be conducted as an update of an earlier study of Pliofilm workers. The risk of leukaemia will be estimated by applying independently developed estimates of cumulative benzene exposure to individuals in this cohort and the use of conditional logistic regression techniques.

TYPE: Cohort
TERM: Occupation; Solvents
SITE: Leukaemia
CHEM: Benzene
TIME: 1990 - 1994

UNITED STATES OF AMERICA

Atlanta

951 **Anderson, M.S.** 04518
Center for Chronic Disease Prevention & Health Promotion, Div. Chronic Dis. Control & Community Intervention, 1600 Clifton Road, NE, Atlanta GA 30333, United States of America (Tel.: +1 404 4884390)
COLL: Smith, R.A.; Liff, J.; Steiner, C.; Campolucci, S.

Georgia Cervical Cancer Prevention and Control Demonstration Project

The goals of this study are to identify the cultural, psychological, economic, and medical care system barriers to early detection and prevention of invasive cervical cancer; and to design, implement, and evaluate public health interventions to overcome identified barriers. The study includes a population-based, case-control study of black and white women residing in two metropolitan Atlanta counties. Approximately 500 women diagnosed with cervical carcinoma in situ and invasive cervical cancer will be interviewed. Control subjects (n = 1,000) are women in the same age group selected from the same geographical area by random digit dialling techniques. Specific projects include: (1) Interviews eliciting information on demographic characteristics, reproductive and sexual history, cancer screening, and medical care history; (2) Review of prior cytology and histology specimens of diagnosed cases conducted by consultant pathologists and cytotechnologists; (3) Review of medical records of diagnosed cases for information on prior Pap smears, treatment, and follow-up of cervical neoplasia. Intervention will be designed based on information obtained from interviews, and review of slides and medical records. These interventions will include evaluation components so that model cervical cancer prevention programmes can be implemented in other areas.

TYPE: Case-Control
TERM: Cytology; Histology; In Situ Carcinoma; Prevention; Reproductive Factors; Screening; Sexual Activity; Treatment
SITE: Uterus (Cervix)
TIME: 1986 - 1994

***952** **Drotman, D.P.** 05521
Centers for Disease Control, Nat. Center for Infectious Disease, Div. of HIV/AIDS, 1600 Clifton Rd, Atlanta GA 30333, United States of America (Tel.: +1 404 6392003; Fax: 6392029)
COLL: Haverkos, H.W.

AIDS Related Kaposi's Sarcoma

AIDS cases reported to the Centres for Disease Control and Prevention from the entire US since 1983 form the database for this study. A standard set of demographic, behavioural and medical variables are reported for each case. The attempt is to identify the co-factors that contribute to the development of Kaposi's sarcoma in persons infected with HIV. Since more than 90% of AIDS related Kaposi's sarcoma is reported in homosexual men, the co-factors are probably associated with homosexual lifestyle.

TYPE: Cohort
TERM: AIDS; Clinical Records; HIV; Homosexuality
SITE: Kaposi's Sarcoma
TIME: 1983 - 1999

953 **Heath, C.W.** 03096
American Cancer Society, 1599 Clifton Rd NE, Atlanta GA 30329, United States of America (Tel.: +1 404 3297686; Fax: 3251467)
COLL: Thun, M.J.; Calle, E.E.

Cancer Prevention Study II

Confidential questionnaires were collected in 1982 from 1.2 million persons over age 30 by 77,000 volunteers in a prospective study relating lifestyles and exposures to the development of cancer and other diseases. Mortality follow-up uses linkage to the National Death Index. Death certificates are obtained from state health departments. Mortality analyses are being conducted regarding cancer risk in relation to such factors as diet, drugs and medicines, occupational exposures, history of diseases, low tar/nicotine cigarette smoking, passive smoking, drinking habits, etc. Recent publications appeared in New Engl. J. Med. 325:1593-1596, 1991, Lancet 339:1268-1278, 1992 and in J. Nat. Cancer Inst. 84:1491-1500, 1992.

TYPE: Cohort
TERM: Alcohol; Diet; Drugs; Lifestyle; Occupation; Passive Smoking; Prevention; Tobacco (Smoking)
SITE: All Sites
TIME: 1982 - 1994

Baltimore

954 **Giardiello, F.M.** 00773
Johns Hopkins Univ., School of Medicine, 600 N. Wolfe St., Blalock Bldg 942, Baltimore MD 21205, United States of America (Tel.: +1 301 9552635)
COLL: Levin, L.S.; Hamilton, S.R.; Zamcheck, N.; Danes, B.S.; Yardley, J.; Owens, A.H.; Baylin, S.B.; Traboulsi, E.I.; Maumenee, I.H.; Offerhaus, G.J.A.; Vogelstein, B.; Booker, S.V.

Families at High Risk of Colorectal Cancer

Families with hereditary polyposis and hereditary cancer of the large bowel from pedigrees ascertained in six adjacent states of the United States are being studied. Subjects are invited to undergo periodic gastrointestinal, ocular and dental examinations. The areas of interest are: natural history of the disease, detailed formal genetic analysis of age of onset, genetic fitness and evidence of heterogeneity, anomalies such as aneuploidies and polyploidies in tissue culture, histological analysis of biopsy specimen, dermatoglyphics, biochemical and molecular markers, dermatological aspects, longitudinal studies of CEA, development of a regional polyposis registry, supporting experimental viral studies in animals, editing of a quarterly Polyposis Newsletter, and organising a hereditary and colon cancer group, IMPACC (Intestinal Multiple Polyposis and Colorectal Cancer). Papers were published in Am. J. Med. Genet. 29:323-332, 1988; Ophthalmology 95:964-969, 1988; and Semin. Surg. Oncol. 3:137-139, 126-133, 1987.

TYPE: Genetic Epidemiology
TERM: Genetic Factors; Genetic Markers; High-Risk Groups; Histology; Pedigree; Premalignant Lesion; Registry
SITE: Colon; Gastrointestinal; Rectum
TIME: 1973 - 1994

955 **Gold, E.B.** 01893
Johns Hopkins Univ., Dept. of Epidemiology, 615 N. Wolfe St., Baltimore MD 21205, United States of America
COLL: Leviton, A.; Austin, D.F.; Child, M.A.; Kolonel, L.N.; Aschenbrener, C.; Key, C.R.; Lyon, J.L.; Weiss, N.S.; West, D.W.; Gilles, F.H.; Satariano, W.A.; West, C.; Mueller, N.; Lopez, R.

Aetiology of Childhood Brain Tumours

In this case-control study of childhood brain tumours, data have been collected on 361 cases occurring in children under 16 years of age and 1,083 controls selected by random digit dialling. Data collection ceased in 1981. Most of the analyses to date have been concerned with parental occupational exposures. In one component of the data set, employment in the aerospace industry was not associated with increased risk of brain tumour in an offspring (J. Nat. Cancer Inst. 77:17-19, 1986). Employment of either parent in the medicine and science industries was associated with an increased risk of tumours. Employment in the agriculture industry and white collar employment in general were also associated with childhood brain tumours. More recently, investigations have included analyses of risk of childhood brain tumours associated with familial aggregation of birth defects or malignancies and with risk associated with parental smoking and alcohol and caffeine consumption. A recent publication from this study reported that tumours in grandparents and great-grandparents were associated with a younger age at presentation of brain tumours in children (Cancer Causes and Contr. 1:75-79, 1990).

TYPE: Case-Control
TERM: Alcohol; Childhood; Coffee; Congenital Abnormalities; Parental Occupation; Registry; Tobacco (Smoking)
SITE: Brain
REGI: SEER (USA)
TIME: 1977 - 1994

956 **Groopman, J.D.** 04605
Johns Hopkins Univ., School of Public Health, 615 N. Wolfe St., Baltimore MD 21117, United States of America (Tel.: (301)9550029; Fax: 9550617)
COLL: Montesano, R.; Hall, A.J.; Wogan, G.N.; Wild, C.; Bosch, F.X.

Monitoring Human Exposure to Aflatoxins in the Gambia

Aflatoxin B1 (AFB1) is one of the most potent liver carcinogens known for experimental animals. This chemical is also present in the environment as a result of mould related spoilage of foods and commodities. Consumption of these foods results in high intake of AFB1 by people living in many regions of Asia and Africa. A number of researchers have used classic epidemiological methods to examine the possible causal relationship between AFB1 in the diet and liver cancer. While strong associations between AFB1 exposure and liver cancer have been found, new technologies have only recently become available to monitor individual exposure and metabolism of aflatoxin B1 in people. The development of biomarkers to assess the exposure to aflatoxin will help in efforts to limit the potential risk in human populations. In this study it is proposed to examine both seasonal and annual variation in aflatoxin intake, metabolism, urinary and albumin disposition in people living in the Gambia, who are at high risk of developing liver cancer. These studies will assess the correspondence between dietary intake of aflatoxins with biological markers, such as excreted AFB-DNA adducts, oxidative metabolites in urine, aflatoxin M1 in human milk and covalently bound aflatoxin to serum albumin, to determine which of these markers have utility for non-invasively assessing the exposure status of people at high risk for liver cancer.

TYPE: Cross-Sectional; Molecular Epidemiology
TERM: Biomarkers; Diet; HBV; Metabolism; Monitoring; Mycotoxins
SITE: Liver
CHEM: Aflatoxin
LOCA: Gambia
TIME: 1988 - 1994

957 **Jacobson, L.P.** 04962
Johns Hopkins Univ., School of Hygiene & Public Health, Dept. of Epidemiology, 624 N. Broadway, Baltimore MD 21205, United States of America (Tel.: +1 301 9554320; Fax: 9557587)
COLL: Detels, R.; Munoz, A.; Phair, J.; Rinaldo, C.; Saah, A.

Aids-Related Kaposi's Sarcoma in the Multi-Centre Aids Cohort Study

The overall objective of this prospective study is to describe the natural history of HIV-I among homosexual/bisexual men in the US. In 1984-1985, 4,954 homosexual men were enrolled in four collaborative centres - Baltimore, Chicago, Los Angeles and Pittsburgh; in 1987 enrollment was re-opened. At semi-annual follow-up visits, these men answer a standardized questionnaire, undergo a physical examination and provide specimens for laboratory analyses. Clinical outcomes, including the development of AIDS, malignancies and mortality, are ascertained continuously. Several investigations are being performed within the study to describe the epidemiology of KS in this population. Initial analyses have demonstrated that incidence is not decreasing. Comparing men with KS to men who first developed a non-KS AIDS diagnosis, this study showed that men with KS had more sexual encounters prior to their diagnosis, thereby supporting the theory that a sexually-transmitted agent may be related to the development of KS. Further investigations to explore the timing of such factors are underway. Other research on KS of the Multi-center AIDS Cohort Study includes examination of the effects of prophylaxis and markers in the progression and prognosis of Kaposi's sarcoma. The study provides a unique opportunity to describe the changes in incidence of KS according to maturity of infection, immunosupression and prophylactic interventions.

TYPE: Cohort
TERM: AIDS; BMB; HIV; Homosexuality; Sexual Activity
SITE: Kaposi's Sarcoma
TIME: 1984 - 1995

958 **Offerhaus, G.J.A.** 05037
Johns Hopkins Medical Institutions, Dept. of Pathology, Carnegie Room 214 , Baltimore MD 21205, United States of America (Tel.: +1 301 9553511; Fax: 9553438)
COLL: Tersmette, A.C.; Giardiello, F.M.; Goodman, S.N.; Vandenbroucke, J.P.; Tytgat, G.N.; Hoedemaeker, P.J.

Longterm Follow-Up of Gastrectomy Patients for Gastric Stump Cancer

This is a long-term follow-up study of a large post-gastrectomy cohort of patients who had surgery for ulcer between 1931-1960, in order to estimate the risk of gastric stump cancer; to evaluate the impact of smoking; to assess risk of other gastro-intestinal cancer in this cohort; and to evaluate the effect of screening on mortality due to gastric cancer in a subset of patients of this cohort who underwent endoscopy plus biopsy. The cohort has 2,633 patients; follow-up is more than 99% complete; 504 asymptomatic patients participated in the endoscopic screening study.

TYPE: Cohort
TERM: Screening; Surgery; Tobacco (Smoking)
SITE: Bile Duct; Colon; Pancreas; Rectum; Stomach
LOCA: Netherlands
TIME: 1975 - 1994

959 **Strickland, P.** 05099
Johns Hopkins Sch. Hyg. & Public Hlth, Dept. of Environmental Health Sciences, 615 N. Wolfe St., Baltimore MD 21205, United States of America
COLL: West, S.; Taylor, H.; Nethercott, J.; Whitmore, S.E.; Grossman, L.

Susceptibility to Skin Cancer in Maryland Watermen

The overall goal of this project is to understand the mechanisms involved in determining individual susceptibility to skin cancer. This goal will be addressed by examining a previously identified population of over 800 Maryland watermen who are at high risk for non-melanoma skin cancer (basal cell carcinoma and squamous cell carcinoma). Overall prevalence of non-melanoma skin cancer is 10-15% and 20-25% in older age-groups (70-80 years). In this population, accurate estimates of individual lifetime UV exposures have been calculated from a combination of interview data, working characteristics, and field measurements. Other phenotypic characteristics that influence skin cancer risk such as skin pigmentation, ease of sunburning (skin type), and hair colour, have been recorded for all individuals in this population. This nested case-control study, including approximately 80 cases and 80 controls, will investigate the role of DNA repair in determining individual susceptibility to skin cancer. DNA repair efficiency of cancer cases and age-matched controls will be assessed by two DNA repair assays: a lymphocyte-plasmid reactivation assay and an immunoassay for DNA photoproduct repair in lymphocyte and fibroblasts. The results of this study will (1) contribute to overall understanding of the mechanisms that determine susceptibility to skin (and other) cancers in man, (2) aid in the identification of individuals at elevated risk for skin cancer, and (3) be useful in estimating increased risks to the general population resulting from potential increases in solar UV at the earth's surface.

TYPE: Case-Control; Cohort
TERM: DNA Repair; Occupation; Pigmentation; Radiation, Ultraviolet
SITE: Skin
OCCU: Dockers; Fishermen
TIME: 1990 - 1995

960 **Strickland, P.** 05233
Johns Hopkins Sch. Hyg. & Public Hlth, Dept. of Environmental Health Sciences, 615 N. Wolfe St., Baltimore MD 21205, United States of America
COLL: Rothman, N.; Poirier, M.C.

Molecular Biomonitoring in Wildland Firefighters

This study was designed to characterize the personal exposure of a representative sample of wildland firefighters to a variety of PAHs, and to examine the usefulness of several molecular biomarkers of PAH exposure (PAH-DNA adducts, PAH-haemoglobin adducts, and PAH urine metabolites). In addition, seasonal changes in pulmonary function and respiratory symptoms were examined. Fieldwork was completed during the 1988 fire season (June to September) in Northern California. Approximately 100 firefighters took part in two study designs. The first study examined a group of 60 firefighters before the fire season began and again after the season was over. The second study examined a group of 20 firefighters with very heavy combustion product exposure at the Yellowstone forest fire on their return to California. Twenty age-, sex-, and race-matched controls were also examined. Blood and urine samples were collected from each individual. Spirometry measurements were taken to assess pulmonary functions. Each volunteer filled out a questionnaire regarding demographics, dietary history, previous exposure to PAH, and recent firefighting activity. Only non-smokers were admitted to the study.

TYPE: Methodology; Molecular Epidemiology
TERM: BMB; Biomarkers; DNA Adducts; Occupation; Tissue
SITE: Lung
CHEM: PAH
OCCU: Firemen
TIME: 1988 - 1994

Berkeley

961 **Buffler, P.A.** 05102
Univ. of California School of Public Health, 19 Earl Warren Hall, Berkeley CA 94720, United States of America (Tel.: +1 510 6422082; Fax: 6435676)
COLL: Cooper, S.P.; Burau, K.; Delclos, G.; Downs, T.D.; Key, M.; Tucker, S.; Whitehead, L.

University of Texas Skin Health Study
The specific aim of this research is to evaluate the contribution of occupational exposure to the prevalence of keratosis, a usually benign skin lesion that can be a precursor to skin cancer, among workers in a Paraquat production plant. The study design is a cross-sectional study which will compare the prevalence of keratoses between the current workers of the plant and an age, race, sex frequency-matched group of their non-family friends who have never worked at the plant. Approximately two friends were selected for each worker yielding a total of 364 study participants. Three major sources of data for this study include: (1) an interview questionnaire to ascertain demographic and medical information and sunlight exposure history; (2) a full body dermatology examination by a board of certified dermatologists to ascertain skin conditions; and (3) company records to ascertain a complete work history at the plant from which an exposure score will be developed.

TYPE: Cross-Sectional
TERM: Herbicides; Occupation; Premalignant Lesion; Radiation, Ultraviolet
SITE: Skin
OCCU: Miners, Copper
TIME: 1990 - 1994

962 **King, M.C.** 02275
Univ. of California School of Public Health, Dept. Biomed. & Env. Health Sciences, Berkeley CA 94720, United States of America

Genetic Epidemiology of Breast Cancer in Families
The objective of the study is to determine the genetic, cultural, and environmental factors that cause breast cancer to cluster in some families. Large, extended families with high incidence of breast cancer are investigated using tools from genetic analysis and epidemiology. The existence of possible susceptibility alleles, greatly increasing cancer risk in women of the high-risk genotypes, is tested by genetic analysis of DNA polymorphisms that may "track" the inheritance of the susceptibility gene within a specific family. Investigation of other genetic mechanisms of inheritance of susceptibility to breast cancer is in progress. In addition, the interaction of genetic and environmental risk factors influencing breast cancer risk among female relatives in high-risk families is being studied using epidemiological techniques. The prevalence of genetically-influenced breast cancer among American Caucasian women has been estimated. Recent papers were published in Genet. Epidemiol. 3 (Suppl. 1):3-14, 15-36, 87-92, 1987; Am. J. Epidemiol. 125:308-318, 1987 and in Hum. Genet. Mapping 9:157, 1987.

TYPE: Cross-Sectional; Genetic Epidemiology
TERM: Caucasians; Cluster; DNA; Environmental Factors; Familial Factors; Genetic Factors; Genetic Markers; High-Risk Groups; Lifestyle; Pedigree; Prevalence
SITE: Breast (F)
TIME: 1974 - 1994

Bethesda

963 **Alavanja, M.C.R.** 04261
NCI/NIH, Epidemiology & Biostatistics Program, Div. of Cancer Etiology, Executive Plaza North, Room 543B, Bethesda MD 20892, United States of America (Tel.: +1 301 4961611)
COLL: Brownson, R.C.; Boice, J.D.; Lubin, J.; Hrubec, Z.

Lung Cancer Among Non-Smoking Missouri Women with Residential Exposure to Radon

This study will attempt to determine the risk of lung cancer among non-smoking females with domestic radon exposure. A population-based case-control study of all non-smoking female lung cancer cases in the state of Missouri will be conducted. Cases, controls and next-of-kin will be made in their current and previous homes. Approximately 600 cases will be accrued from January 1987 through July 1991. Approximately 1,400 age-matched will be collected during the same time intervals. Controls in age-group 30 to 64 years will be randomly selected from Missouri motor vehicles registration records. Controls who are 65 years old or older will be randomly selected from Health Care Finance Administration files of Missouri residents. Diet, occupations, pregnancy history, and family cancer history will also be assessed.

TYPE: Case-Control
TERM: Abortion; Diet; Familial Factors; Female; Occupation; Parity; Registry; Reproductive Factors
SITE: Lung
CHEM: Radon
REGI: Missouri (USA)
TIME: 1988 - 1994

964 **Biggar, R.J.** 04523

NCI/NIH, Environmental Epidemiology Branch, Viral Epidemiology Sect., HNC 3348, Room 434D, Executive Plaza Bldg North, Bethesda MD 20892-4200, United States of America (Tel.: +1 301 4968115; Fax: 4969146)

HIV Seroconverters Registry

The natural history of HIV infection can be evaluated best when the time since infection with HIV is documented. Because most institutions have relatively few subjects with documented HIV seroconversion, a registry has been prepared of such cases. To be eligible, last negative and first positive serology must be available, along with birth day, sex, risk group and follow-up history of AIDS-related conditions (upon request, questionnaires will be sent to groups interested). Time to AIDS and risk of different manifestations of AIDS by risk group, age and sex, will be evaluated. So far over 1,000 seroconverters are enrolled. Follow-up questions will be sent annually. New participants are welcome.

TYPE: Cohort
TERM: AIDS; HIV; Sero-Epidemiology
SITE: Kaposi's Sarcoma; Lymphoma
TIME: 1987 - 1994

965 **Blair, A.E.** 04292

National Cancer Institute, NIH, Occupational Studies Sect., Executive Plaza North, Room 418, Bethesda MD 20892, United States of America (Tel.: +1 301 4969093; Fax: 4021819)

COLL: Alavanja, M.C.R.; Cantor, K.C.; Dosemeci, M.; Fraumeni, J.F.; Figgs, L.; Hayes, R.B.; Heineman, E.; Hoover, R.N.; Pottern, L.M.; Rothman, N.; Stewart, P.A.; Ward, E.M.; Zahm, S.H.

Occupational Studies of Cancer

The Occupational Studies Section conducts a comprehensive programme of occupational studies to evaluate the role workplace exposures play in the origin of cancer. To accomplish this goal the Section conducts: (1) descriptive and hypothesis-generating studies to identify promising new areas of research and to sharpen hypotheses, (2) analytical studies to test hypotheses regarding specific exposures and specific cancers, and (3) methodological studies to develop and improve procedures used to assess exposures and conduct epidemiological studies. Proportionate mortality, cohort mortality, and case-control designs are employed. Study populations are assembled from companies, labour unions, professional associations, government agencies, hospitals, and tumour registries. Studies underway are designed to investigate cancer and a variety of exposures including wood dusts (furniture workers), Chinese workers exposed to benzene, industrial workers exposed to phenol, aircraft maintenance workers, cutting oils (automotive workers), metals and solvents (jewellery manufacturers), pesticides (applicators, employees of a lawn care company, grain millers, noxious weed applicators and Department of Agriculture employees), acrylonitrile (manufacturers and plastic producers), silica (miners, brickmakers, and others), talc (ceramic fixture manufacturers), workers exposed to benzidine and formaldehyde (embalmers, resin producers, and others). Studies of occupations with a variety of exposures include plumbers, shipyard workers, professional chemists, and firefighters. Case-control studies of mesothelioma, leukaemia, non-Hodgkin's lymphoma, multiple myeloma and cancers of the brain, lung, stomach, soft-tissue sarcoma, and bladder have been designed to evaluate a variety of occupational exposures as well as other factors that may be associated with the development of these cancers. Methodological projects to develop and improve techniques used to assess exposures in occupational studies include reliability of exposure estimation by industrial hygienists, comparison of estimates of exposure to specific substances using varying amounts of industrial hygiene information, experimental modification of workplace parameters to evaluate their effects on airborne concentrations of formaldehyde, comparison of the potential for exposure as estimated by industrial hygienists with that reported by workers, effects of exposure misclassification on risk estimates, and comparison of historical estimates of exposure made by industrial hygienists with actual measured levels.

TYPE: Case-Control; Cohort; Cross-Sectional; Methodology
TERM: Chemical Exposure; Dusts; Metals; Occupation; Pesticides; Petroleum Products; Plastics; Registry; Solvents; Wood
SITE: Bladder; Brain; Haemopoietic; Lung; Lymphoma; Mesothelioma; Sarcoma; Stomach
CHEM: Acrylonitrile; Asbestos; Benzene; Formaldehyde; Silica; Talc
OCCU: Acrylonitrile Workers; Agricultural Workers; Automobile Workers; Cabinet Makers; Chemists; Dry Cleaners; Farmers; Firemen; Heavy Equipment Operators; Herbicide Sprayers; Jewellery Manufacturers; Miners; Morticians
LOCA: China; Denmark; Turkey; United States of America
TIME: 1978 - 1994

966 **Blair, A.E.** 04820
National Cancer Institute, NIH, Occupational Studies Sect., Executive Plaza North, Room 418, Bethesda MD 20892, United States of America (Tel.: +1 301 4969093; Fax: 4021819)
COLL: Stewart, P.A.; Zahm, S.H.; Falk, R.T.; Ward, M.; Kross, B.; Popendorf, W.; Burmeister, L.F.

Methodologic Study to Enhance Assessment of Pesticide Exposure in Epidemiological Studies of Cancer

Methods to assess exposure to pesticides in case-control studies of cancer must be improved if further progress in this important area is to be made. This project is designed to obtain information on the detail and quality of information regarding pesticide use that can be obtained from farmers or their surrogates, particularly for exposures many years in the past. Information on pesticide use will be obtained from a detailed questionnaire, from records kept by farmers, from records kept by suppliers, from exposure measured by dermal deposition, and from biochemical monitoring of the blood and urine. Two hundred farmers in Iowa and their next-of-kin (wives, sibs, and children) will participate in the study. Approximately 100 of these farmers will be asked to participate in the biologic monitoring portion of the project. Information from the various sources will be compared to identify combinations of data that appear most promising for use in epidemiologic investigations of pesticides and cancer.

TYPE: Methodology
TERM: Biochemical Markers; Occupation; Pesticides
SITE: Inapplicable
CHEM: 2,4-D; Alachlor; Atrazine; Malathion
OCCU: Farmers
TIME: 1989 - 1995

***967** **Blair, A.E.** 05397
National Cancer Institute, NIH, Occupational Studies Sect., Executive Plaza North, Room 418, Bethesda MD 20892, United States of America (Tel.: +1 301 4969093; Fax: 4021819)
COLL: Stewart, P.A.

Extended Follow-Up of Workers Employed at an Aircraft Maintenance Facility

The mortality experience of 14,457 workers employed at an Air Force Base between 1952 and 1956 is being evaluated to assess cancer risks associated with exposure to organic solvents. Excesses of non-Hodgkin's lymphoma and multiple myeloma, particularly among women, were found in the initial follow-up through 1981. The mortality follow-up is now being extended through 1992. The cohort will also be linked to the Utah Tumor Registry (the state where the facility is located) to obtain data on cancer incidence. Mortality and morbidity rates among the exposed workers will be compared with rates in the general population and among unexposed members of the cohort. The original investigation included a detailed assessment of exposures by industrial hygienists. The current effort includes interviews with approximately 200 workers to obtain more details on workplace exposures.

TYPE: Cohort
TERM: Occupation; Paints; Registry; Solvents
SITE: All Sites; Multiple Myeloma; Non-Hodgkin's Lymphoma
OCCU: Aircraft Workers
REGI: Utah (USA)
TIME: 1993 - 1997

968 **Blot, W.J.** 04279
National Cancer Institute, NIH, Biostatistics Branch, EPN 431, 6130 Executive Blvd, Bethesda MD 20892-4200, United States of America (Tel.: +1 301 4964153; Fax: 4020081)
COLL: Dosemeci, M.; McLaughlin, J.K.; Blot, W.J.; Glen, R.; Reger, R.; Hearl, F.; McCawley, M.; Wu, Z.B.; Chen, J.

Retrospective Morbidity and Mortality Study of Silica Exposed Workers in Wuhan and Other Areas
Workers in mines and factories where silica exposure is high will be enrolled in a retrospective mortality study to test the hypothesis that silica exposure causes lung cancer. The cohort will consist of all persons who worked for at least one year during the period 1 January 1960-31 December 1974. Approximately 80,000 workers are expected to be enrolled. Levels of radon, silica, arsenic, nickel and cadmium will be measured in selected work sites. Since about 1963, the government has required that work sites with silica exposure conduct regular industrial hygiene measures of silica levels at the work site. These historical data will form the basis for exposure assessment. For each member of the cohort, a minimum basic data set of information will be collected for identification and follow-up. For persons included in a silicosis registry, additional information will be collected on medical history and complete work history. For selected silicotis, lung cancer deaths and appropriate controls, a nested case - control study will be conducted which will collect information on cigarette smoking, detailed employment, information and other personal characteristics. Rates from the Chinese national mortality survey of 1973-1975 will be used in generating expected numbers of deaths. Internal comparisons of workers with high vs low silica exposure will also be made. Special attention will be paid to assessing mortality among the sub-cohort with silicosis. Data from the nested case -control study will be examined using stratified and logistic regression analyses to evaluate risks of lung cancer and silicosis by intensity and duration of silica exposure, controlling for smoking.

TYPE: Case-Control; Cohort
TERM: Chemical Exposure; Dose-Response; Dusts; Metals; Occupation; Tobacco (Smoking)
SITE: Lung
CHEM: Arsenic; Cadmium; Nickel; Radon; Silica
LOCA: China
TIME: 1986 - 1994

***969** **Blot, W.J.** 05303
National Cancer Institute, NIH, Biostatistics Branch, EPN 431, 6130 Executive Blvd, Bethesda MD 20892-4200, United States of America (Tel.: +1 301 4964153; Fax: 4020081)
COLL: Chow, W.H.; Gammon, M.D.; Lo, S.H.; Schoenberg, J.; Risch, H.A.; Dubrow, R.; Mayne, S.T.; West, B.; Vaughan, T.L.; Stanford, J.L.; Hantsen, P.

Adenocarcinomas of the Oesophagus and Gastric Cardia
The primary objective of this collaborative study is to identify risk factors for the rapidly increasing adenocarcinomas of the oesophagus and gastric cardia, and compare them with risk factors for other cancers of the oesophagus and stomach. This population-based case-control study will be conducted by three centres across the US (Yale University, Columbia University/New Jersey State Health Department, and Fred Hutchinson Cancer Research Center/University of Washington), with the NCI serving as the data co-ordinating centre. Over a two-year period starting April 1993, about 300 cases each of adenocarcinoma of the oesophagus and non-cardia stomach, 400 cases of gastric cardia cancer, and 300 cases of squamous cell carcinomas of the oesophagus will be recruited. About 700 randomly selected population controls will be included for comparison. Information on medication history (such as H2-receptor antagonists), and medical conditions (such as Barrett's oesophagus and oesophagitis), diet, smoking and alcohol consumption, family history of cancer and occupational history will be obtained by personal interview. A blood sample will be collected from selected cases and controls for assays of antibodies to Helicobacter pylori, and stored for future analysis of genetic polymorphisms. Formalin-fixed tissue blocks will be collected from a subset of cases.

TYPE: Case-Control
TERM: Alcohol; BMB; Blood; Diet; Disease, Other; Drugs; H. pylori; Occupation; Tobacco (Smoking)
SITE: Oesophagus; Stomach
TIME: 1993 - 1995

970 **Boice, J.D.** 02363
NCI/NIH, Radiation Epidemiology Branch, 6130 Executive Blvd, Executive Plaza North, R408, Bethesda MD 20892-4200, United States of America (Tel.: +1 301 4966600; Fax: 4020207)
COLL: Land, C.E.; Beebe, G.W.; Kleinerman, R.A.; Chen, K.W.; Curtis, R.E.; Ron, E.; Lubin, J.; Jablon, S.; Inskip, P.D.; Travis, L.B.; Linet, M.S.; Alavanja, M.C.R.; Friedman, D.; Hatch, E.; Doody, M.M.; Shriner, D.

Radiation-Induced Cancer

The Radiation Epidemiology Branch (REB) plans and conducts independent and cooperative epidemiological research to identify and quantify the risk of cancer in populations exposed to ionising and non-ionising radiation, especially at low-dose levels (populations with documented therapeutic, diagnostic, occupational, environmental or military exposures are studied). The risk of radiation-induced cancer in terms of tissues at risk, dose-response, radiation quality, fractionation of dose, time since exposure, sex, age at exposure and at observation are characterised, as well as possible modifying influences of other environmental and host factors. The REB develops statistical and epidemiological methodologies to facilitate epidemiological research; explores and formulates models of radiation carcinogenesis that may help define basic mechanisms of cancer induction, including the integration of experimental findings with epidemiological observations; conducts case-control and cohort studies of cancer risk in patient populations given diagnostic or therapeutic radiation alone or in combination with cytotoxic drugs and other forms of treatment; conducts population studies to examine possible analogues of radiation carcinogenesis in man, such as the induction of cytogenetic abnormalities in circulating lymphocytes, and integrates laboratory markers of radiation exposure and tissue response into epidemiological studies designed to clarify the patterns of cancer risk and the mechanisms of action. It also advises and collaborates with other agencies and individuals involved in radiation research and regulatory activities. Recent publications appeared in N Engl. J. Med. 326:1745-1751, 1992, (Curtis et al), Nature 36:113, 1992 (Ron et al) and in Health Physics 63:1-100, 1992 (Guilmette & Mays).

TYPE: Case-Control; Cohort; Methodology
TERM: Age; Chemotherapy; Dose-Response; Mathematical Models; Multiple Primary; Physical Factors; Radiation, Ionizing; Radiation, Non-Ionizing; Radiotherapy
SITE: All Sites; Breast (F); Leukaemia; Lung; Thyroid
LOCA: Canada; China; Czech Republic; Denmark; Estonia; Finland; Germany; Israel; Japan; Norway; Slovenia; Sweden; United Kingdom; United States of America
TIME: 1977 - 1994

971 **Brinton, L.A.** 04938
NCI/NIH, Environmental Epidemiology Branch, Executive Plaza North, HNC 3342, Room 443, Bethesda MD 20892-4200, United States of America (Tel.: +1 301 4961691)
COLL: Schairer, C.; Schiffman, M.H.; Nasca, P.C.; Mallin, K.; Richart, R.M.

Case-Control Study of Cancers of the Vagina and Vulva

A case-control study of cancers of the vagina and vulva has been completed in two US areas (Chicago and Upstate New York). Included were women with newly diagnosed invasive or in situ cancers of these two sites. The study began in February 1986 and included 12 months of retrospectively and 18 months of prospectively ascertained cases. Personal interviews focusing on medical, sexual, reproductive, and hygiene factors were conducted with 209 vulvar cancer patients, 41 vaginal cancer patients, and 445 community controls (identified either through random digit dialling or Health Care Financing lists). Results pertaining to vulvar cancer risk were recently published (Obstet. Gynecol. 75:859-866, 1990) and showed significant relationships of risk with sexual behaviour, smoking, histories of genital warts, and previous abnormal Pap smears. The results for vaginal cancer were published in Gynecol. Oncol. 38:49-54, 1990) and showed that the primary predictors of risk were low socio-economic status, a history of genital warts and previous genial abnormalities. Blood specimens drawn at the time of interview are currently being analysed for antibodies to several sexually transmitted agents, and future analyses will attempt to assess relationships with sexual behaviour in detail.

TYPE: Case-Control
TERM: BMB; Blood; Hygiene; Reproductive Factors; Screening; Sexual Activity; Sexually Transmitted Diseases; Tobacco (Smoking)
SITE: Vagina; Vulva
TIME: 1986 - 1994

972 **Brinton, L.A.** 04936
NCI/NIH, Environmental Epidemiology Branch, Executive Plaza North, HNC 3342, Room 443, Bethesda MD 20892-4200, United States of America (Tel.: +1 301 4961691)
COLL: Potischman, N.A.; Barrett, R.J.; Berman, M.L.; Mortel, R.; Twiggs, L.B.; Wilbanks, G.D.; Swanson, C.A.

Case-Control Study of Endometrial Cancer

A multi-disciplinary study of endometrial cancer has been conducted in five centres in the US (Chicago, IL; Hersey, PA; Los Angeles, CA; Minneapolis, MN; Winston-Salem, NC). Approximately 400 cases, 300 community controls and 200 hospital controls being treated for benign endometrial conditions were enrolled. The major aim is to evaluate the interrelationships with risk of endometrial cancer. Results related to exogenous oestrogen usage, cigarette smoking, reproductive factors, fat distribution patterns, physical activity, alcohol intake and dietary factors have been published. Blood samples are being analysed for hormone levels and blood lipids, and will be correlated to risk factors and to risk of disease. A methodological study has been completed at one centre with data on fatty acid composition of plasma, red blood cell membranes, adipose depots and cheek cells. These fatty acid profiles will be compared to each other and to estimates of dietary fatty acid intakes. Although the major focus of this study is on nutritional determinants, the study will also provide opportunity to examine several other relevant issues, including risks related to stage of disease, histological type of tumour, and the determinants of benign disease.

TYPE: Case-Control
TERM: Alcohol; Blood; Diet; Fat; Hormones; Micronutrients; Nutrition; Reproductive Factors; Tobacco (Smoking)
SITE: Uterus (Corpus)
TIME: 1988 - 1994

973 **Brinton, L.A.** 04939
NCI/NIH, Environmental Epidemiology Branch, Executive Plaza North, HNC 3342, Room 443, Bethesda MD 20892-4200, United States of America (Tel.: +1 301 4961691)
COLL: Schairer, C.; Storm, H.H.; Persson, I.R.

Follow-up Study of Patients with Gynaecological Operations

The purpose of this project is to evaluate the risk of cancer associated with varying indications for and types of gynaecological operations. Cohorts of approximately 60,000 patients with over 500,000 person years of observation and 15,000 patients with over 150,000 person years will be assembled by the cancer registries of Denmark and Sweden, respectively. Demographic and exposure information, as well as clinical reasons for the operation, will be abstracted from individual patient records. The rosters of patients will then be linked to cancer registry records so that patients who have developed malignancies can be identified. The observed number of cancers will be compared to the expected number, derived from sex-, age-, and calendar-year specific population based rates.

TYPE: Cohort; Incidence
TERM: Record Linkage; Registry; Surgery
SITE: Breast (F); Colon; Lung; Ovary; Uterus (Corpus)
LOCA: Denmark; Sweden
REGI: Denmark (Den); Sweden (Swe)
TIME: 1990 - 1994

974 **Brinton, L.A.** 04940
NCI/NIH, Environmental Epidemiology Branch, Executive Plaza North, HNC 3342, Room 443, Bethesda MD 20892-4200, United States of America (Tel.: +1 301 4961691)
COLL: Potischman, N.A.; Daling, J.R.; Liff, J.; Schoenberg, J.; Swanson, C.A.

Case-Control Study of Breast Cancer with Emphasis on Women Under the Age of 45 Years

The aim of this multi-centre study (Atlanta, GA; Seattle, WA; Trenton, NJ) is to evaluate breast cancer risk in relation to oral contraceptive use, alcohol consumption, dietary patterns and other factors. Cases are comprised of newly diagnosed cases under the age of 45, with the age range extended to 54 years in Atlanta. Controls in all areas will be identified through random digit dialling techniques, but in Atlanta an additional group of controls will be identified through a household survey, to allow comparative analyses using different control groups. It is anticipated that there will be 2,200 interviewed cases, 2,000 random digit dialling controls and 640 household survey controls. Personal interviews are being conducted and a variety of anthropometric measurements obtained. The interview includes a food frequency questionnaire designed to elicit information about usual adult diet as well as adolescent diet. A calibration study among a sample of controls will assess the measurement error and biases in the food questionnaire. Information on usage of oral contraceptives and previous operations will be validated against review of selected medical records. In Seattle, blood samples are being collected from a sample of subjects to enable a variety of hormonal and micronutrient analyses.

TYPE: Case-Control
TERM: Alcohol; Blood; Diet; Genetic Factors; Hormones; Micronutrients; Nutrition; Oral Contraceptives; Reproductive Factors
SITE: Breast (F)
TIME: 1990 - 1994

***975** **Chow, W.H.** 05306
National Cancer Institute, NIH, Biostatistics Branch, EPN 403, 6130 Executive Blvd, Bethesda MD 20892, United States of America (Tel.: +1 301 4963344; Fax: 4020081)
COLL: Blot, W.J.; Swanson, C.A.; Rothman, N.; Mallin, K.; Armstrong, R.W.; Sobin, L.

Stomach Cancer Among Polish Americans

This case-control study will examine reasons for the high risk of stomach cancer among Polish Americans, and the changes in their lifestyle, particularly in their diet, that may contribute to their reduced rates compared with those of their home country. 200 Polish and 200 non-Polish cases and an equal number of controls matched for age, sex, race and Polish ethnicity, will be recruited from residents of Cook County, Illinois. Interviews will be conducted using a pre-coded questionnaire to obtain information on demographic background, ethnic origin, diet, tobacco and alcohol consumption, occupational history and history of selected medical conditions and medication use. A 30 ml blood sample will be collected from a random 25% sample of subjects from each case and control group. Blood specimens will be used for assay of antibodies to Helicobacter pylori, serum pepsinogen and selected nutrient levels among controls, and stored for future analysis of genetic polymorphisms. Diagnostic slides will be reviewed by a pathologist for reclassification, using the WHO and Lauren classification. Formalin-fixed tissue blocks will be collected from as many cases as possible for future analysis of genetic mutations.

TYPE: Case-Control
TERM: Alcohol; BMB; Blood; Diet; Drugs; Ethnic Group; H. pylori; Migrants; Nutrition; Occupation; Pepsinogen; Tobacco (Smoking)
SITE: Stomach
TIME: 1993 - 1996

***976** **Chow, W.H.** 05307
National Cancer Institute, NIH, Biostatistics Branch, EPN 403, 6130 Executive Blvd, Bethesda MD 20892, United States of America (Tel.: +1 301 4963344; Fax: 4020081)
COLL: Rothman, N.; Sinha, R.; Hoover, R.N.; Blot, W.J.; Denobile, J.W.; Sweeney, W.B.; Blankenship, C.

Case-Control Study of Colon Adenomas

The primary objective of this case-control study of an asymptomatic clinic population is to investigate colon adenomas in relation to P450IA2 and N-acetyltransferase activities, enzymes that metabolise the carcinogens heterocyclic aromatic amines, which are produced by cooking meat at high temperatures. Data on selected risk factors and 50 ml of blood will be collected from 150 cases and 600 controls before sigmoidoscopy screening. All cases and 150 randomly selected controls will participate in a more in-depth study one month after diagnosis. This will involve a detailed interview, collection of three overnight urine samples, accompanied by three 24-hour food intake diaries, and a urine sample collected between four and five hours after ingestion of a 100mg dose of caffeine. The interview will elicit information on diet, cooking practice, physical activity, sunlight exposure, medications for pain and inflammation, smoking and alcohol consumption, occupation and family history. Tumour tissue blocks will be collected from as many cases as possible for future analysis of genetic mutations.

TYPE: Case-Control
TERM: Alcohol; Analgesics; BMB; Cooking Methods; Diet; Genetic Factors; Nutrition; Occupation; Physical Activity; Tobacco (Smoking)
SITE: Colon
CHEM: Amines, Aromatic; PAH
TIME: 1993 - 1995

977 **Kirsch, I.R.** 04990
National Cancer Institute, NIH, Navy Medical Oncology Branch, Naval Hosp., Bldg 8, Room 5101, 8901 Wisconson Ave., Bethesda MD 20889-5105, United States of America (Tel.: +1 301 4960909; Fax: 4960047)
COLL: Garry, V.F.; Blair, A.E.

Hybrid Antigen Receptor Genes, Genomic Instability, and Risk of Leukaemia and Lymphoma

Using PCR, it can be demonstrated that the increased frequency of lymphocyte-specific cytogenetic abnormalities found previously in workers exposed to pesticides and herbicides results from an increased frequency of interlocus V(D)J recombination. The mechanism underlying formation of hybrid antigen receptor genes in lymphocytes appears related to that underlying rearrangements between putative oncogenes and antigen receptor genes in lymphoid malignancies, also associated with exposure to pesticides. In a pilot study some 100 grain workers and other farmers are being examined before and after seasonal exposure to pesticides and herbicides for evidence of genomic instability as manifested by an increased frequency of interlocus V(D)J recombination. The control population will be age- and sex-matched unexposed individuals. If, using the PCR-based assay, significant differences can be detected in the exposed population, the study will be expanded and individuals and populations with an increased incidence of lymphocyte-specific interlocus recombination will be followed to determine comparative risk of development of leukaemia and lymphoma.

TYPE: Cohort; Molecular Epidemiology
TERM: Herbicides; Lymphocytes; PCR; Pesticides
SITE: Leukaemia; Lymphoma
OCCU: Farmers; Grain Millers
TIME: 1990 - 1994

978 **Osborne III, J.S.** 05232
National Inst. of Health, Div. of Research Grants, Epidemiology and Disease Control-1, 5333 Westbard, Room 203-C, Bethesda MD 20892-4200, United States of America (Tel.: +1 301 4967246; Fax: 4021279)
COLL: Kaplan, B.H.; Shy, C.M.

Analysis of Reported Cancer Clusters in Small Populations

The objective of this project is to develop a reliable method for assessing reports of cancer clusters in small stable populations which lack reliable denominator data for rates. The method has been successfully applied in a rural community (N=300) in North Carolina, USA (Am. J. Epidemiol. 132(1):87-95, 1990). It involves a three-step process of determining (1) secular trends in proportionate cancer mortality; (2) standardized proportionate mortality ratios (SPMR) stratified by age, sex, race, calendar period, and any other appropriate variables, and (3) 95% Poisson confidence intervals (CI) for the expected distribution of deaths. Secular trend data serve to address and validate concerns arising in the community, SPMR data assess deviation from expected mortality, and, if SPMR data indicate occurrence has differed from expected, the CI enable the likelihood of this difference arising from random case clustering to be evaluated. If mortality exceeds the upper limit of the CI, then it is not likely to represent random case clustering and an aetiological investigation is appropriate. Evaluation of this approach for morbidity and its application in another community reporting excess cancer mortality are in progress.

TYPE: Methodology
TERM: Cluster
SITE: All Sites
TIME: 1986 - 1995

979 **Rapoport, S.I.** 04969

National Inst. of Aging, NIH, Lab. of Neurosciences, Bldg 10, Room 6C103 , Bethesda MD 20892-4200, United States of America (Tel.: +1 301 4968970; Fax: 4020074)

COLL: Greig, N.H.; Ries, L.

Primary Brain Tumours in the Elderly

Data from NCI's SEER Program are analysed for annual incidence of primary brain tumours. Data are based on registries representing approximately 10% of the US population. In 1985, incidence rates of primary malignant brain tumours for persons aged 75-79, 80-84, and 85 years of age and over were 187%, 394% and 501% respectively of rates in 1973/1974. Reported incidence in younger persons varied little over the same period of time. In the future, the contribution of new diagnostic procedures introduced or augmented since 1973, including brain imaging techniques (e.g. x-ray computerized tomography), will be assessed and distinguished from a true increase in annual incidence of malignant brain tumours in the elderly.

TYPE: Incidence
TERM: Age; Registry
SITE: Brain
REGI: SEER (USA)
TIME: 1988 - 1994

980 **Scherrer, C.** 04937

NCI/NIH, Environmental Epidemiology Branch, EXecutive Plaza North, HNC 3342, Room 443, Bethesda MD 20817, United States of America (Tel.: +1 301 4961691)

COLL: Hoover, R.N.; Brinton, L.A.; Sturgeon, S.; Schatzkin, A.; Taylor, P.

Follow-Up Study of Participants in the Breast Cancer Detection Demonstration Project

A five year multi-centre breast cancer screening project, the Breast Cancer Detection Demonstration Project, was completed in 1980. At this time, a cohort of approximately 60,000 of the original 280,000 participants was chosen for further follow-up; included were all of the project-detected breast cancer cases, all those receiving a recommendation for breast biopsy, and a sample of normal screenees. For a five year period, annual telephone interviews were conducted with these subjects to elicit information on vital status, cancer incidence, and changes in breast cancer risk factors. Follow-up is now being performed on a bi-annual basis through a mailed questionnaire; this has enabled an expansion of data collection to include information on dietary factors, anthropometry, and physical activity levels. Analyses are currently underway to examine the effects of menopausal hormones on the risk of breast cancer as well as on all-cause mortality, and mortality due to cardiovascular diseases, other cancers, fractures, accidents, and suicide.

TYPE: Cohort
TERM: Diet; Hormones; Menopause; Physical Activity; Reproductive Factors
SITE: Breast (F); Colon; Lung; Ovary; Uterus (Corpus)
TIME: 1980 - 1994

981 **Sondik, E.J.** 03888

National Cancer Inst., NIH Div. of Cancer Prevention and Contol, Bldg 31, 10A49 9000 Rockville Pike, Bethesda MD 20892, United States of America (Tel.: +1 301 4968506)

COLL: Janerich, D.T.; Flannery, J.T.; Lerchen, M.L.; Swanson, G.M.; Kolonel, L.N.; Isaacson, P.; Platz, C.; Key, C.R.; Thomas, D.B.; West, D.W.; Greenberg, R.S.; Meinert, L.; Martinez, I.; Zippin, C.

Surveillance, Epidemiology, and End Results (SEER) Program

A cancer registration system is conducted in 12% of the United States population through contracts with non-profit medically oriented institutions. The system is designed to report incidence data within 11 months of the close of the calendar year and survival within 18 months. The goals of the SEER Program are to: (1) Determine periodically incidence of cancer in selected geographical areas of the US with respect to demographic and social characteristics of the population. (2) Estimate cancer incidence for the US on an annual basis (3) Monitor trends in incidence of specific forms of cancer with respect to geographical area and demographic and social characteristics of the population. (4) Determine periodically survival experience for cancer patients diagnosed among residents of selected geographical areas of the US. (5) Monitor trends in cancer patient survival with respect to form of cancer, extent of disease, therapy, and demographic, socio-economic, and other parameters of prognostic importance. (6) Identify cancer aetiological factors by conducting special studies which disclose groups of the population at high or low cancer risks. These groups may be defined by social, occupational, environmental, dietary, or other characteristics, and by drug history. (7) Identify factors related to patient survival through special studies of referral patterns, diagnostic procedures, treatment methods, and other aspects of medical care. (8) Promote specialist training in epidemiology, biostatistics, and tumour registry methodology, operation, and management.

TYPE: Incidence; Registry
TERM: Diet; Drugs; Education; Environmental Factors; Occupation; Prognosis; Socio-Economic Factors; Survival; Treatment
SITE: All Sites
LOCA: Puerto Rico; United States of America
REGI: SEER (USA)
TIME: 1973 - 1994

Birmingham

982 **Austin, H.D.** 04418
Univ. of Alabama at Birmingham, School of Public Health, Dept. of Epidemiology, Univ. Station, 202 Tidwell Hall, Birmingham AL 35294, United States of America (Tel.: +1 205 9347129)
COLL: Delzell, E.; Macaluso, M.; Rotimi, C.

Mortality from Lung and Stomach Cancer at Three Car Plants

The purpose of this study is to determine whether employees of a foundry and two engine plants experienced an elevated mortality rate from lung and stomach cancer. This will be accomplished through the use of a retrospective follow-up study and two nested case-control studies. The follow-up study will include 20,000 hourly-paid workers active at any time between 1 January 1973 and 31 December 1986, and former workers who retired before 1973, but were still active as of 1 January 1970. Cohort members will be followed through 31 December 1986 and their mortality experience compared with that of the general US and Ohio populations. Vital status will be ascertained through plant records, Ohio or Michigan death registries, the Social Security Administration, and the National Death Index. Exposures of potential interest in the nested case-control studies, in addition to smoking, include polycyclic aromatic hydrocarbons, machining oils (mineral and synthetic), silica, and welding fumes. About 150 lung cancer cases and 20 stomach cancer cases are expected. Two controls will be selected for each lung cancer decedent and four for each stomach cancer decedent. Controls will be matched to cases by race and year of birth and must have been under observation at the time of death of the corresponding case. Poisson regression and conditional logistic regression will be used for the analyses of the cohort and case-control studies, respectively.

TYPE: Case-Control; Cohort
TERM: Chemical Exposure; Dusts; Occupation; Tobacco (Smoking); Welding
SITE: All Sites; Lung; Stomach
CHEM: PAH
OCCU: Automobile Workers; Foundry Workers
TIME: 1988 - 1994

***983** **Sathiakumar, N.** 05360
Univ. of Alabama, School of Public Health, Dept. of Epidemiology, 720 South 2th St. , Birmingham AL 35294, United States of America (Tel.: +1 205 9343719; Fax: 9757058)
COLL: Delzell, E.; Cole, P.

Follow-Up Study of Mortality among Workers at the Shell Denver Plant

The purpose of this project is to study the mortality experience among workers at the Shell Denver Chemical Plant near Denver, Colorado. Attention will focus on the pattern of variation of mortality by age, duration of employment, period of hire, work area within the plant and payroll status. Specifically, the hypothesis that the chemicals aldrin and dieldrin are associated with hepato-cellular cancers will be tested. All men ever employed at the plant between 1952 and 1984, 2,500-3,000 workers, will be followed up. Identification of study subjects will be made from plant records, data tapes and hard copies of previous studies of the same plant workers. Ascertainment of vital status will be based on the records of the National Death Index, the Social Security Administration, the Colorado Division of Motor Vehicles and the plant records. Death certificates will be requested from state bureaus of vital statistics. A nosologist will code the underlying cause of death and a secondary cause of death, if that condition is a cancer. The overall and cause-specific mortality will be computed and compared with the mortality rates of the US general population, the Colorado general population and the regional county (Adams, Arapaho, Denver,

Douglas, Jefferson) populations, using the SMR as the measure of association. SMRs will be computed for the overall cohort of chemical plant workers and for sub-cohorts, specified on the basis of work area, duration of employment and time since hire in each area. Epidemiological procedures will consits of stratified analysis, using Mantel-Haenszel summary rate ratios or directly standardized rate ratios and appropriate confidence intervals.

TYPE: Cohort
TERM: Occupation; Pesticides; Time Factors
SITE: Liver
CHEM: Aldrin; DBCP; Dieldrin; Endrin; Organophosphates
OCCU: Chemical Industry Workers
TIME: 1992 - 1994

***984 Sathiakumar, N.** 05361
Univ. of Alabama, School of Public Health, Dept. of Epidemiology, 720 South 2th St. , Birmingham AL 35294, United States of America (Tel.: +1 205 9343719; Fax: 9757058)
COLL: Delzell, E.; Cole, P.

Follow-Up Study of Synthetic Rubber Workers

This is a study of the styrene-butadiene rubber industry in the US and Canada to determine if there is an increased risk of lymphatic and haematopoietic tissue cancer. The purpose is to evaluate the mortality experience of synthetic rubber workers relative to that of the US and Canadian general populations; to assess the cancer incidence experience of Canadian synthetic rubber workers relative to that of the general population of Ontario; to determine if overall and cause-specific mortality patterns vary by subject characteristics such as age, calendar time, plant, period of hire, duration of employment, time since hire and payroll status; to examine the relation between work areas within the synthetic rubber industry and cause-specific mortality patterns; to evaluate the possible relation between exposure to butadiene and styrene and the occurrence of leukaemia and other lymphatic haematopoietic tissue cancers among synthetic rubber workers. Identification of study subjects will be made from plant records and microfilmed personnel records of previous studies of the same plant workers. Ascertainment of vital status will be based on the records of the National Death Index, the Social Security Administration and several states' Division of Motor Vehicles. Death certificates will be requested from state bureaus and provinces of vital statistics. A nosologist will code the underlying cause of death and a secondary cause of death, if that condition is a cancer. A Job Exposure Matrix will be developed for butadiene, styrene and benzene. The matrix will measure exposure by using information on subjects' history of employment by production area, job title, task, duration of employment and combination of these factors. The overall and cause-specific mortality of the study group will be assessed by comparing their rates with general population mortality rates (SMRs). The cohort's mortality rates will be compared with the rates of the US, Texas, Kentucky, Louisiana or Ontario general population as appropriate. SMRs will be computed for the overall cohort of synthetic rubber workers and for sub-cohorts, specified on the basis of plant, payroll classification, duration of employment, period of hire and years since hire.

TYPE: Cohort
TERM: Occupation; Plastics; Registry; Solvents; Time Factors
SITE: Haemopoietic; Leukaemia; Lymphoma; Soft Tissue
CHEM: Benzene; Butadiene; Styrene
OCCU: Rubber Workers
LOCA: Canada; United States of America
REGI: Ontario (Can)
TIME: 1991 - 1994

Boston

985 Clapp, R.W. 05213
John Snow Research & Training, JSI/World Education, 210 Lincoln St. , Boston MA 02111, United States of America (Tel.: +1 617 4829485; Fax: 4820617 ; Tlx: 200178)
COLL: Spiegelman, D.; Davis, L.; Martin, T.

Occupational Cancer in African Americans

Mortality due to respiratory cancer and other selected cancers by occupation and race in Massachusetts has been analysed. A larger study, involving a comprehensive analysis of 11 states with large African American populations is proposed. The primary hypothesis is that when an occupational risk is present,

African Americans are disproportionately affected. Indirect adjustment for cigarette smoking will be carried out to control for confounding of specific occupation-cancer associations. Public use databases, containing data on cancer mortality, occupation and cigarette smoking prevalence, will be used and results will have relevance to attempts to prevent lung cancer, especially in new occupations and industries. The results will be presented as standardized mortality odds ratios by occupation and industry.

TYPE: Mortality
TERM: Occupation; Race
SITE: Bladder; Leukaemia; Lung
TIME: 1988 - 1994

986 **Hennekens, C.H.** 03491
Harvard Medical School, 900 Commonwealth Ave. East, Boston MA 02215, United States of America (Tel.: +1 617 7324965; Fax: 7341437)
COLL: Buring, J.E.; Gaziano, M.; Manson, J.; Ridker, P.; Stampfer, M.J.

The Physicians' Health Study (PHS)

This is a randomized double-blind placebo controlled trial of 50 mg of beta-carotene on alternate days on risks of total epithelial cancer, including lung and colon as well as cardiovascular disease in 22,071 apparently healthy male physicians aged 40-84 years in 1982. The study will include data from 12 years of treatment and follow-up in a well-nourished population of men at normal risk of cancer and low cardiovascular disease.

TYPE: Intervention
TERM: Prevention
SITE: All Sites; Colon; Lung
CHEM: Beta Carotene
OCCU: Health Care Workers
TIME: 1982 - 1996

***987** **Hennekens, C.H.** 05327
Harvard Medical School, 900 Commonwealth Ave. East, Boston MA 02215, United States of America (Tel.: +1 617 7324965; Fax: 7341437)
COLL: Buring, J.E.; Gaziano, M.; Manson, J.; Ridker, P.

Women's Health Study (WHS)

This is a randomized, double-blind placebo controlled 2x2x2 factorial trial of alternate day aspirin beta-carotene and vitamin E supplement and risks of total epithelial cancer, including lung, colon and breast, as well as cardiovascular disease in over 40,000 apparently healthy post-menopausal female nurses aged 45 and over in 1992. This study will provide data from a well- nourished population of women at normal risk of cancer and cardiovascular disease.

TYPE: Cohort; Intervention
TERM: Antioxidants; Drugs; Prevention; Vitamins
SITE: All Sites; Breast (F); Colon; Lung
CHEM: Aspirin; Beta Carotene
TIME: 1992 - 1996

***988** **Koh, H.K.** 05292
Boston Univ. School of Medicine, 80 E. Concord St., C-3, Boston MA 02118, United States of America (Tel.: +1 617 6387131; Fax: 6388551)
COLL: Norton, L.; Geller, A.C.; Miller, D.R.; Lew, R.A.; Sikes, R.

Confirmed Melanomas in National Skin Cancer Screening Programmes

The object of this study is to evaluate the American Academy of Dermatology annual, free melanoma/skin cancer screening programmes. National screening has been taking place since 1985 and almost 560,000 Americans in all 50 states have received free examinations. Persons with suspected melanoma found at screenings are encouraged to consult their own physicians for treatment. From 1989-1991, confirmed diagnoses of persons with suspected melanoma were sought through contact with the screening physician. Final diagnoses were available for only 22% of all positive screenees. In contrast, in 1992, 94% of participants were successfully contacted through direct letters and telephone calls to each screenee with suspected melanoma. From 1989-1992 data, it was found that 98.6% (283/287) of screen-detected, confirmed melanomas were Stage 1 lesions and more than 90% of all lesions were in-situ or less than 1.50 mm. Comparisons with SEER registry population-based data indicate that screening detects a slightly higher proportion of Stage 1 lesions, but little difference in thickness distribution of Stage 1 lesions. Future efforts must determine if screening for melanoma can reduce mortality.

TYPE: Cohort
TERM: Screening
SITE: Melanoma; Skin
TIME: 1989 - 1999

***989** **Lee, I.M.** 05294
Harvard University School of Public Health Dept. of Epidemiology, 677 Huntington Ave. , Boston MA 02115, United States of America (Tel.: +1 617 4321050; Fax: 5667805)
COLL: Paffenbarger Jr, R.S.

College Alumni Health Study

The College Alumni Health Study is a prospective cohort study of 35,000 male alumni from Harvard University and 22,500 male and female alumni from the University of Pennsylvania, to measure relationships between physical activity, body size, other lifestyle characteristics and chronic disease (including cancer) incidence. Data have been obtained from (1) university archives - Alumni had attended a compulsory medical examination during their tenure at both institutions. Also available from archives are data on sports played by alumni at both institutions; and (2) mailed questionnaires - Harvard alumni have been contacted on six different occasions between 1962 and 1992; Pennsylvania alumni, on four different occasions between 1962 and 1992. In December 1992, 18,800 alumni from Harvard (aged 59-98) and 15,400 from the University of Pennsylvania (aged 69-84) were alive. Non-fatal cancer in alumni has been ascertained via self-reports of physician-diagnosed disease on these same questionnaires. Such self-reports have been shown to be valid. Fatal cancer has been determined from copies of death certificates. In the past, loss to mortality follow-up has been less than 1%.

TYPE: Cohort
TERM: Lifestyle; Obesity; Physical Activity
SITE: All Sites
TIME: 1960 - 1999

990 **Poole, C.** 05026
Boston Univ. School of Public Health Epidemiology & Biostatistics Sect., 80 E Concord St., Boston MA 02118, United States of America
COLL: Rothman, K.J.; Dreyer, N.A.; Loughlin, J.E.

Cohort Mortality Study of Styrene Workers

This is a retrospective cohort mortality study of employees at ARCO Chemical Company's plant at Beaver Valley, Pennsylvania, USA. Since the mid-1940s, the plant has been engaged in the production of styrene, polystyrene and a variety of other chemicals. Cause-specific mortality rates will be compared with rates for the populations of the USA and Pennsylvania. The cohort will contain 6,000-7,000 persons. A much smaller, preliminary study of recently employed males suggested a possible increase in mortality from cancers of the central nervous system (three observed, 0.7 expected).

TYPE: Cohort
TERM: Occupation; Plastics
SITE: All Sites
CHEM: Polystyrene; Styrene
OCCU: Chemical Industry Workers
TIME: 1989 - 1994

991 **Speizer, F.E.** 00821
Harvard Medical School, Brigham & Women's Hosp., Dept. of Medicine, Channing Lab., 180 Longwood Ave., Boston MA 02115, United States of America (Tel.: +1 617 4322276)
COLL: Rosner, B.R.; Willett, W.C.; Stampfer, M.J.; Colditz, G.A.; Hennekens, C.H.; Peto, R.; Hunter, D.J.; Manson, J.E.

The Nurses' Health Study

The determinants of cancer and cardiovascular disease among women have not been well described. This prospective cohort study aims to evaluate cancer morbidity and mortality for a number of exposures including oral contraceptives, cigarette smoking, post-menopausal oestrogens, hair dyes, and family history of selected diseases. Information has been obtained by mail questionnaire for 120,700 married female registered nurses, aged 30-55 residing in 11 larger US states and listed in the Directory of the American Nurses' Association in 1972. Biennial follow-up mailings have been performed beginning in 1976. In 1980 the questionnaire was expanded to include an extensive dietary component as well as sections on drug usage and exercise. Full dietary assessment was repeated in 1984, 1986 and 1990. Continued emphasis is placed on risk factors associated with breast and other cancers. Related studies include assessment of cardiovascular outcomes and diabetes, both as a risk factor for cardiovascular events and as an independent outcome and the collection of toenail specimens to be used prospectively in a case-control fashion to assess selenium as a potential risk factor in breast and other cancers. In 1989 blood specimens were obtained on approximately 35,000 members of the cohort, and will be used for nested case-control analyses.

TYPE: Cohort
TERM: BMB; Blood; Cosmetics; Diet; Drugs; Dyes; Familial Factors; Female; Hormones; Lipids; Menopause; Metals; Occupation; Oral Contraceptives; Physical Activity; Tobacco (Smoking); Toenails
SITE: All Sites; Breast (F)
CHEM: Oestrogens; Selenium
OCCU: Health Care Workers
TIME: 1977 - 1994

992 **Stampfer, M.J.** 04570
Brigham & Women's Hosp., Dept. of Medicine, Channing Labs, 180 Longwood Ave., Boston MA 02115, United States of America (Tel.: +1 617 4322279; Fax: 7311541 ; Tlx: 4974739)
COLL: Hennekens, C.H.; Willett, W.C.; Eberlein, K.A.; Rosner, B.R.; Salvini, S.

Nutritional Biochemical Markers of Cancer and Myocardial Infartion

Pre-diagnostic levels of nutritional biochemical markers of cancer and myocardial infarction are studied in the Physicians' Health Study (PHS) cohort. The PHS is a randomized trial of aspirin and beta carotene in the prevention of cardiovascular diseases and cancer incidence, conducted among 22,071 US male physicians aged 40-84 years. As part of the trial, blood samples were collected prior to randomisation from 14,916 participants, and have been stored at -82 C. All subjects developing cancer or myocardial infarction are matched by age and smoking status to a control from the same population. Objectives are to define if any of the biochemical markers assayed is a predictor of cancer or myocardial infarction. Using a nested case-control study design, the possible effects of plasma levels of retinol, lycopene, alpha and beta carotene, alpha and gamma tocopherol, vitamin B6, selenium and specific fatty acids are being studied. Participants have been followed annually since the beginning of the study and additional information on vitamin supplementation, exercise, alcohol consumption, smoking habits and other risk factors has been collected.

TYPE: Case-Control; Cohort
TERM: Alcohol; BMB; Biochemical Markers; Drugs; Metals; Physical Activity; Tobacco (Smoking); Vitamins
SITE: All Sites
CHEM: Aspirin; Beta Carotene; Retinoids; Selenium
TIME: 1986 - 1994

993 **Stern, R.S.** 01988
Beth Israel Hosp., Harvard Medical School, Dept. of Dermatology, 330 Brookline Ave., Boston MA 02215, United States of America (Tel.: +1 617 7354591; Fax: 7354948)
COLL: Abel, E.; Epstein, J.H.; Wintroub, B.; Wolf, J.; Nigra, T.P.; Voorhees, J.; Anderson, T.F.; Harber, L.; Prystowsky, J.; Muller, S.; Taylor, J.R.; Frost, P.; Urbach, F.; Arndt, K.A.; Baughman, R.D.; Braverman, I.M.; Murray, J.; Petrozzi, J.; Fitzpatrick, T.B.; Parrish, J.A.; Gonzalez, E.

Photochemotherapy Follow-Up Study

This 16-centre prospective cohort study will assess the risk and efficacy of oral methoxsalen photochemotherapy, introduced in 1974 for the treatment of psoriasis, and now widely used for this common skin disease. The incidence of adverse events, especially skin cancers and other cancers are being documented. Data are collected on patients prior to entry. At annual intervals, each of the 1,380 patients is interviewed and periodically invited to be examined by a dermatologist. The cohort has been followed for an average of 13 years. Historical controls are being used. Since side effects are expected to be dose-related, the incidence of adverse events in patients receiving varying amounts of life-time exposures to photochemotherapy will be compared. A dose-dependent increase in the risk of squamous cell cancer has been documented. Male genitals are especially susceptible to these carcinogenic effects. Follow-up continues.

TYPE: Cohort
TERM: Drugs; Photochemotherapy; Radiation, Non-Ionizing
SITE: All Sites; Eye; Skin
CHEM: Methoxsalen
TIME: 1976 - 1994

Bronx

994 **Romney, S.L.** 05027
Univ. of Yeshiva, Albert Einstein College of Medicine, Dept. of Gynecology & Obstetrics, 1300 Morris Park Ave., Bronx NY 10461, United States of America (Tel.: +1 718 4302691; Fax: 8630599)
COLL: Burk, R.; Kadish, A.; Ho, G.; Basu, J.; Palan, P.R.; Klein, S.

Clinical Trial of Beta Carotene in Women with Moderate Cervical Dysplasia
The efficacy of oral supplementation with beta-carotene (BC) in increasing the regression rate of cervical neoplasia is being investigated in a clinical trial of women with moderate cervical dysplasia. Approximately 138 women will be recruited from the colposcopy clinic of a public hospital and randomized into either the treatment group receiving 30 mg of beta-carotene a day or the active placebo group. Subjects will be followed up by colposcopy, repeat Pap smear and HPV DNA hybridization studies at three-month intervals. The outcome, regression from moderate to mild dysplasia or to normal, will be determined by biopsy at the ninth month. Women with endpoint regression will be followed up for another six months for recurrence, and blood and cervico-vaginal lavage samples will be obtained at baseline and at each follow-up visit. These samples will be used to assess levels of BC, ascorbic acid and alpha-tocopherol. Presence of HPV infection will also be determined from the lavage sample by southern blot. Subjects' dietary patterns and exposure to risk factors for cervical dysplasia will be monitored via questionnaires. Regression and recurrence rates will be compared between treatment groups in the whole population as well as in strata defined by HPV positivity or by different levels of vitamin C and E. The question of synergistic interaction of various antioxidant micronutrients will be evaluated.

TYPE: Intervention
TERM: BMB; Biochemical Markers; Blood; Cytology; DNA; Diet; Drugs; HPV; Micronutrients; Nutrition; Premalignant Lesion; Tissue; Vitamins
SITE: Uterus (Cervix)
CHEM: Ascorbic Acid; Beta Carotene; Tocopherol
TIME: 1991 - 1995

Brookline

995 **Buring, J.E.** 05159
Harvard Medical School, Brigham & Women's Hosp., 55 Pond Ave., Brookline MA 02146, United States of America (Tel.: +1 617 7324965; Fax: 7324970)
COLL: Hennekens, C.H.; Belanger, C.; Cook, N.; Eberlein, K.A.; Hebert, P.; Satterfield, S.; Rosner, B.R.; Gordon, D.; LaMotte, F.; Manson, J.E.; Peto, R.; Cohen, L.; Prentice, R.; Goodman, D.

Randomized Trial of Beta Carotene, Vitamin E and Aspirin in Cancer and Cardiovascular Disease (Women's Health Study)
This is a randomized, placebo-controlled, double-blind trial of beta carotene and vitamin E in the prevention of cancer and cardiovascular disease, and of low-dose aspirin in the prevention of cardiovascular disease. The trial will be conducted among 40,000 post-menopausal female US nurses, aged 50 years or older, with no previous history of cancer (except non-melanoma skin cancer) or cardiovascular disease. The design of the trial is 2x2x3 factorial, with participants randomly assigned to 15mg beta carotene daily or placebo; to 200 IU vitamin E daily or placebo; and to taking either 300mg aspirin, 100mg aspirin, or a placebo every other day. Primary aims will be to investigate whether beta

carotene or vitamin E decreases total cancers of epithelial cell origin, and whether low dose aspirin, beta carotene, or vitamin E reduces the risk of important vascular endpoints. Following a run-in phase and randomization, participants will be followed by mailed questionnaires every six months, and all endpoints will be documented by medical records.

TYPE: Intervention
TERM: Antioxidants; Drugs; Female; Prevention; Vitamins
SITE: All Sites
CHEM: Aspirin; Beta Carotene; Retinoids
TIME: 1991 - 1996

996 **Rosenberg, L.** 04949
Boston Univ. Medical School, Slone Epidemiology Unit, 1371 Beacon St. , Brookline MA 02146, United States of America (Tel.: +1 617 7346006; Fax: 7385119)
COLL: Shapiro, S.; Palmer, J.R.; Strom, B.L.; Stolley, P.D.

Surveillance of Serious Illnesses and Drugs

The Slone Epidemiology Unit obtains comprehensive medication histories and data on demographic factors, reproductive history, medical history, cigarette smoking, and other exposures from hospital patients with any of a wide range of cancers and other illnesses. Special emphasis has been placed on the effect of female hormone preparations on the risk of several cancers, including breast and ovary. Large bowel cancer and other rarer cancers are also studied. The large data file is used for hypothesis generation and for the testing of hypotheses. Papers have been published in Am. J. Epidemiol. 134:1-13 and 1375-1385, 1991, J. Natl Cancer Inst. 83:355-358, 1991 and in Cancer Causes and Control 3:547-554, 1992.

TYPE: Correlation
TERM: Data Resource; Drugs; Hormones
SITE: Breast (F); Colon; Lung; Ovary; Prostate; Uterus (Corpus)
TIME: 1976 - 1994

997 **Siguel, E.N.** 03685
Boston Univ. Medical Center Hosp., Clinical Nutrition Unit, Fatty Acid Lab., P.O. Box 5, Brookline MA 02146-0001, United States of America (Tel.: +1 617 7394887; Fax: 6388603)
COLL: Lerman, R.H.

Fatty Acid Patterns in Patients with Cancer

Several projects are either planned or underway to relate the nature and extent of cancer, including the likelihood of having cancer, with the pattern of fatty acids in plasma and adipose tissue. Emphasis is on essential fatty acids and their derivatives, saturated fat and trans-fatty acids. Controls, matched by sex, and age, are drawn from individuals with similar characteristics to the patients with cancer but without any known cancer. An additional purpose is to identify fatty acid patterns or substances that behave chemically like fatty acids which distinguish plasma of subjects with cancer from normal subjects.

TYPE: Case-Control
TERM: Fat
SITE: Breast (F); Colon; Ovary; Prostate
TIME: 1985 - 1994

Brooklyn

***998** **Fruchter, R.G.** 05296
State Univ. of New York, Health Science Center, Dept. of Obstetrics & Gynecology, P.O. Box 24, 450 Clarkson Ave., Brooklyn NY 11203, United States of America (Tel.: +1 718 2701403; Fax: 2704173)
COLL: Maiman, M.; Minkoff, H.; De Hovitz, J.

Women's AIDS Cohort Study

In a study of the natural history of HIV disease in women, an evaluation of the prevalence, characteristics and progression of cervical intraepithelial neoplasia is being done. 350 HIV infected women will be followed for 3-5 years. They will have a clinical examination at entry into the study, cytology every 6 months and colposcopy every year.

TYPE: Cohort
TERM: AIDS; BMB; Cytology; Female; HIV
SITE: Uterus (Cervix)
TIME: 1992 - 1996

Buffalo

***999 Bandera, E.V.** 05301
State Univ. of New York at Buffalo, Dept. of Social and Preventive Medicine, 270 Farber Hall, Buffalo NY 14214, United States of America (Tel.: +1 716 8292975; Fax: 8292979)
COLL: Freudenheim, J.; Marshall, J.; Zielezny, M.; Priore, R.L.; Baptiste, M.S.; Brasure, J.; Graham, S.

Diet and Alcohol Consumption and Subsequent Risk of Lung Cancer in the New York State Cohort
The main objective of this study is to evaluate the association between diet and alcohol consumption and the risk of lung cancer. The New York State (NYS) Cohort comprises 32,689 men and 25,279 women, who responded to a questionnaire in 1980. They have been followed up with the assistance of the NYS Department of Vital Statistics and the Division of Cancer Control Tumor Registry. The magnitude of the cohort precludes further efforts in locating every participant due to financial and time constraints. Therefore, a nested case-control study is being conducted with the 525 lung cancer cases identified in the cohort (395 males and 130 females) and 525 controls (395 males and 130 females), randomly selected among cohort participants and matched to cases for age (+ /- 2 years), sex, race, and county of residence. Efforts to locate these subjects include contacts by mail and telephone and record linkage with the NYS Departments of Vital Statistics and Motor Vehicles and with the National Death Index. Statistical analysis will include logistic regression for the nested case-control study and survival analysis for the whole cohort.

TYPE: Case-Control
TERM: Alcohol; Diet; Fat; Registry
SITE: Lung
CHEM: Beta Carotene
REGI: New York (USA)
TIME: 1980 - 1994

1000 Graham, S. 04057
State University of New York, Dept. Social & Preventive Medicine, 2211 Main St., Buffalo NY 14214, United States of America (Tel.: +1 716 8312981)
COLL: Hreschchyshyn, M.; Sufrin, G.; Nemoto, T.; Nolan, J.; Rawls, W.E.; Campbell, C.; Marshall, J.; Vena, J.E.; Zielezny, M.; Freudenheim, J.; Hellmann, R.

Western New York Study of Cancer of Ovary, Endometrium, Breast, Prostate and Pancreas
This case-control study investigates the possible relationships between dietary habits, smoking, alcohol ingestion, occupation and reproductive history with cancers of the male and female reproductive tract and pancreas. For those cases with cancer of the breast and prostate, a blood sample will be obtained and analysed to determine vitamin, lipid and mineral levels, which will be correlated with dietary intake. To avoid hospital-related bias and to obtain detailed information on each case, all new cancer cases in Erie and Niagara Counties in New York State are interviewed, along with 500 controls of each sex, 40 years of age and over, in the counties from which the cases are selected. A two percent random sample of Health Care Finance Administration registrants is being used to obtain controls, age 65 and older, and a one percent sample of Department of Motor Vehicle registrants for those aged 40 to 65. In addition to interviewing patients, spouses receive a self-administered questionnaire to substantiate the participants' dietary history. The use of spouse interviews is primarily to ascertain validity of dietary history data.

TYPE: Case-Control
TERM: Alcohol; Diet; Infection; Occupation; Reproductive Factors; Tobacco (Smoking); Virus; Vitamins
SITE: Breast (F); Ovary; Pancreas; Prostate; Uterus (Corpus)
TIME: 1986 - 1994

Cambridge

1001 Thilly, W.G. 03938
Massachusetts Inst. of Technology, Center for Environm. Health Sciences, Div. of Toxicology, Room E18-666, 77 Massachusetts Ave., Cambridge MA 02139, United States of America (Tel.: +1 617 2536220; Fax: 2585424 ; Tlx: 921473 MITCAM)
COLL: Morley, A.A.; Albertini, R.J.; Lambert, B.; Ehrenberg, L.; Hemminki, K.; Wogan, G.N.; Tannenbaum, S.R.

Mutational Spectra as a Means of Diagnosing the Causes of Genetic Change in Man

Based on the observation that individual mutagens produce individual patterns of mutation, it is hypothesized that knowledge of mutational spectra in cell populations from the human body will yield important information about what environmental agents, if any, are causing somatic and/or germinal mutations in man. To date the study has shown (1) that the mutational spectra in human cells varies among mutagens, and (2) that all point mutations within exon 3 of hgprt can be detected in human genomic DNA by the use of a gradient denaturing gel electrophoresis. Means to amplify human DNA sequences with high fidelity have been developed. This provides means to study spectra without growth of cells. It is planned to apply the protocols, when ready, to studies of normal subjects, patients receiving chemotherapy involving known mutagens and cigarette smokers. The first phase of the human studies would involve about 100 patients. At the level of cell culture, the mutational spectra of spontaneous change and, in long term low dose protocols, those of polycyclic aromatic hydrocarbons, active oxygen species, alkylating agents, mycotoxins and halodeoxyuridines are being characterized. Papers appeared in Genome 31:590-593, 1989; Mutat. Res. 231:165-176, 1990, and in Ann. Rev. Pharmacol. Toxicol. 30:369-385, 1990.

TYPE: Methodology; Molecular Epidemiology
TERM: Chemical Mutagenesis; Drugs; Mutagen; Mutation, Germinal; Mutation, Somatic; Mycotoxins; PCR; Tobacco (Smoking)
SITE: Inapplicable
CHEM: Alkylating Agents; Oxygen; PAH
LOCA: Australia; Finland; Sweden; United States of America
TIME: 1980 - 1994

Chapel Hill

***1002 Godley, P.A.** 05434
Univ. of North Carolina, Div. of Medical Oncology, 3009 Old Clinic Bldg, CB 7305 , Chapel Hill NC 27599-7305, United States of America
COLL: Mohler, J.; Martinson, F.; Gallagher, F.; Robertson, C.M.; Campbell, M.

Biochemical Markers of Essential Fatty Acid Consumption and Risk of Prostate Cancer

The hypothesis is that the excess incidence of prostate cancer in blacks compared with whites is due, in part, to lower exposure to omega-3 fatty acids and higher exposure to omega-6 fatty acids in black men. In this case-control study, cases will be 150 patients with pathologically confirmed, recently diagnosed prostate cancer recruited from the urology clinics of the University of North Carolina Hospitals and the Durham Veterans Administration Medical Center. Controls will be 150 patients with benign prostatic hyperplasia who are free of prostate cancer on pathological examination of the prostate, frequency matched to the cases on race, age, hospital and study entry date. 40% of the subjects are expected to be black. Subcutaneous fat specimens will be obtained from the upper arm of the subjects. Venous blood from erythrocyte membranes will also be obtained. Adipose tissue and erythrocyte membrane fatty acid composition will be used as estimates of long- and short-term exposure to fatty acids. A dietary history instrument will serve as an indicator of fatty acid consumption. Data analysis will include crude comparisons of the mean omega-3 fatty acid proportions of cases and controls as well as blacks and whites. Further analyses will explore the differences in mean fatty acid proportions controlled by age, body mass and cholesterol level, and examine the risks of prostate cancer associated with omega-3 fatty acids and omega-6 fatty acids.

TYPE: Case-Control
TERM: BMB; Biochemical Markers; Blood; Diet; Fat; Fatty Acids; Nutrition; Tissue
SITE: Prostate
TIME: 1992 - 1997

1003 **Hulka, B.S.** 04631
Univ. of North Carolina, School of Public Health, Dept. of Epidemiology, CB# 7400 Rosenau 207, Chapel Hill NC 27599-7400, United States of America (Tel.: +1 919 665731)
COLL: Everson, R.; Griffith, J.; Margolin, B.; Wilcosky, T.; Vine, M.F.

Biological Markers in Epidemiological Research

This three-phase project is designed to study the use of biological markers in epidemiological research. Phase I is the development and publication of a book entitled "Biological Markers in Epidemiology" (OUP, 1990) which will serve as a textbook both for epidemiologists new to this field and for toxicologists with an interest in marker applications. Phase II is the development of pilot studies to assess the use of DNA adducts in epidemiological research. Studies currently in progress are 1) an autopsy study designed to address the issue of tissue specificity in the formation of DNA adducts among persons exposed to environmental agents such as tobacco smoke, 2) an analysis of DNA adduct formation in sperm to determine the potential for heritable mutations upon exposure to tobacco smoke, and 3) a study of DNA adduct formation in different white blood cell populations to assess whether certain populations are better than others for studying markers of environmental exposures. Phase III, now in the planning stage, involves the development of large scale epidemiological field studies integrating the use of laboratory assays of biological markers.

TYPE: Molecular Epidemiology
TERM: Autopsy; Biomarkers; DNA Adducts; Mutagen; Passive Smoking
SITE: Inapplicable
TIME: 1987 - 1994

***1004** **Loomis, D.P.** 05339
Univ. of North Carolina, School of Public Health, Dept. of Epidemiology, CB 7400 McGavran Hall, Chapel Hill NC 27599-7400, United States of America (Tel.: +1 919 9667433; Fax: 9662089)
COLL: Savitz, D.A.; Stevens, R.G.

Prostate Cancer, Malignant Melanoma and Breast Cancer among Men in Electrical Occupations

The goals of this study are (1) to determine if an excess of prostate cancer, malignant melanoma, or breast cancer occurs among men in electrical jobs; and (2) to assess the hypothesis that exposure to power frequency electric and magnetic fields causes cancer by reducing melatonin production. A case-control study is being conducted, including 250 breast cancer cases, 26,100 prostate cancer cases, 3,237 melanoma cases and 27,509 controls, who died from other causes. Cases and controls are deceased male residents of 24 US states that reported death certificate occupational data 1985-1988 and are matched for year of death. A job exposure matrix is used to identify electrical workers.

TYPE: Case-Control
TERM: Electromagnetic Fields; Occupation; Radiation, Non-Ionizing
SITE: Breast (M); Melanoma; Prostate
OCCU: Electrical Workers
TIME: 1992 - 1994

***1005** **Moorman, P.G.** 05343
Univ. of North Carolina, Dept. of Epidemiology, CB #7400, McGavran-Greenberg Hall, Chapel Hill NC 27599, United States of America (Tel.: +1 919 5445438; Fax: 9662089)
COLL: Hulka, B.S.; Hiatt, R.A.; Krieger, N.; Vogelman, J.; Orentreich, N.

Breast Cancer and HLD Levels

This nested case-control study tests three hypotheses: (1) prediagnostic levels of HDL cholesterol are higher in women who develop breast cancer than in women who remain cancer free; (2) levels of HDL are a stronger predictor of breast cancer risk in premenopausal women than in postmenopausal women; and (3) HDL cholesterol represents a common biological pathway through which the effects of some of the known risk factors for breast cancer are mediated. Analyses will be conducted using frozen serum obtained from members of the Kaiser Permanente Medical Care Program (Oakland, CA) at the time of a multiphasic health examination between 1964 and 1972. Follow-up to detect all incident cases of breast cancer extended through December 1990. Cases include 200 randomly selected white women who developed breast cancer (100 premenopausal, 100 postmenopausal) after the multiphasic examination. Each case was matched to one cancer-free control for race/ethnicity and both age at and year of the multiphasic examination; controls were required to remain in the health plan up to the time of the case's diagnosis of breast cancer. Analyses will compare HDL serum levels of cases and controls, stratified by menopausal status and adjusting for numerous potential confounding breast cancer risk factors, based on questionnaire data obtained at the examination and supplemental medical chart review.

TYPE: Case-Control
TERM: BMB; Cholesterol; Lipids; Serum
SITE: Breast (F)
TIME: 1992 - 1994

***1006 Newman, B.** 05523
Univ. of North Carolina at Chapel Hill, School of Public Health, Dept. of Epidemiology (CB # 7400), Chapel Hill NC 27599-7400, United States of America (Tel.: +1 919 9667435; Fax: 9662089)
COLL: Hulka, B.S.; Liu, E.T.; Aldrich, T.E.; Lannin, D.

Carolina Breast Cancer Study

This is a population-based, case-control study of breast cancer in white and black women. Over three years, 900 new breast cancer cases and an equal number of women without cancer, all from the age range 20-74, will be enrolled from a 24-county region, encompassing central and eastern North Carolina. Case finding will use an innovative technique for 'rapid' reporting through the Central Cancer Registry. A home visit will be conducted to complete an interview and to take a blood specimen. The suspected risk factors evaluated will be reproductive and menstrual history, hormone use, alcohol consumption, occupational exposures and socio-demographic characteristics. DNA will be extracted from lymphocytes and tumour specimens and analysed for certain genetic alterations, e.g. amplification of the oncogenes HER-2/neu and int-2, mutations of the tumour suppressor gene p53, and variations at the breast cancer susceptibility gene BRCA1 (when it is cloned). Statistical analyses will integrate information obtained from the interview with tumour characteristics; particular attention will be given to genetic alterations which may be associated with identifiable lifestyle behaviours and/or environmental exposures. Particular attention will be paid to exploring differences between black and white women.

TYPE: Case-Control
TERM: Alcohol; DNA; Hormones; Occupation; Oncogenes; Pesticides; Race; Registry; Reproductive Factors; Rural
SITE: Breast (F)
REGI: N. Carolina (USA)
TIME: 1993 - 1996

1007 Olshan, A.F. 03770
Univ. of North Carolina, Dept. of Epidemiology, McGavran-Greenberg Hall, Chapel Hill NC 27599, United States of America (Tel.: +1 919 9667424; Fax: 9662089)
COLL: Breslow, N.E.; Falletta, J.M.; Grufferman, S.; Pendergrass, T.W.; Robison, L.L.; Strong, L.C.; Waskerwitz, M.; Woods, W.G.

Risk Factors for Wilms' Tumour

The purpose of this study is to examine the relationship between specific environmental exposure and the risk of Wilms' tumour. These exposures include paternal occupational hydrocarbon and lead exposure and maternal exposure to progestins during pregnancy. Information will be collected on additional exposures which may be used to develop new aetiological hypotheses. The environmental factors will be examined within hereditary and non-hereditary sub-groups of Wilms' tumour patients to better understand the role that genetic-environmental interactions play in the development of Wilms' tumour. The nationwide case-control study will identify its cases from the National Wilms' Tumour Study. Approximately 200 cases and 200 matched controls will be obtained for one year of accrual. Controls will be identified using random digit dialling. Information collected from the questionnaire will include: socio-demographic factors, parental occupational and environmental exposures, mothers' exposures during pregnancy, and family medical history.

TYPE: Case-Control
TERM: Chemical Exposure; Environmental Factors; Familial Factors; Heredity; Intra-Uterine Exposure; Occupation; Socio-Economic Factors
SITE: Wilms' Tumour
LOCA: Canada; United States of America
TIME: 1984 - 1994

1008 **Olshan, A.F.** 05173
Univ. of North Carolina, Dept. of Epidemiology, McGavran-Greenberg Hall, Chapel Hill NC 27599, United States of America (Tel.: +1 919 9667424; Fax: 9662089)
COLL: Grufferman, S.; Pollock, B.H.; Stram, D.; Neglia, J.P.; Seeger, R.; Shah, N.

Risk Factors for Neuroblastoma

This will be the first large case-control interview study of risk factors for childhood neuroblastoma. About 640 cases (under age 19 years at diagnosis) will be identified through the Childrens Cancer Study Group and the Pediatric Oncology Group. Parents of cases and of 640 controls matched for age, sex, and race will be interviewed by telephone. Controls will be selected by random digit dialling. The study will evaluate the relative importance of risk factors for neuroblastoma reported from previous studies. These include: maternal use of medications during pregnancy (specifically alcohol, diuretics and a group of drugs that includes amphetamines, antidepressants, antipsychotics and tranquillizers); parental employment in the electronics industry and occupational exposures to electromagnetic fields; maternal age at birth and the subject's length of gestation and birth weight. The study will also evaluate drugs used by the mother during pregnancy for their potential to cause transplacental exposure of the foetus to N-nitroso compounds. Information on other potential risk factors such as family medical, gestational and delivery history will also be collected. The study will examine the possibility of neuroblastoma having a similar set of risk factors to that of other childhood cancers. Since most patients will have data collected on clinical, biological and genetic markers, potential environmental risk factors will be analysed separately for subgroups of patients defined by such characteristics as disease stage and N-myc oncogene amplification.

TYPE: Case-Control
TERM: Age; Alcohol; Diuretics; Drugs; Electromagnetic Fields; Genetic Markers; Intra-Uterine Exposure; Oncogenes; Parental Occupation; Stage; Time Factors
SITE: Neuroblastoma
CHEM: N-Nitroso Compounds
LOCA: Canada; United States of America
TIME: 1991 - 1996

1009 **Rappaport, S.M.** 04712
Univ. of South Carolina, School of Public Health, Rosenau Hall, CB 7400, Chapel Hill NC 27599, United States of America (Tel.: +1 919 9661023; Fax: 9667141)
COLL: Yager, J.W.; Smith, A.H.

Dose-Response for Occupational Styrene Exposures

This study investigates the linkages between exposure to styrene, absorbed dose, and cytogenetic response resulting from occupational exposures in the reinforced plastics industry. A group of 4-10 workers in a reinforced-plastics facility in Southern California provided air and blood samples for initial testing of measurement methods. A larger group of 30 exposed and 15 non-exposed workers was assembled at a reinforced-plastics facility in Southern Washington. A new method has been developed for measuring styrene in mixed-exhaled air which involves forced expiration through a commercial charcoal tube. Exhaled-air levels measured with this device correlated well with blood levels; thus, blood styrene concentrations can be estimated by breath concentrations. A toxicokinetic model (two-compartment open model) of styrene disposition associated with realistic exposure scenarios has been developed. Blood and exhaled-air concentrations of styrene which have been measured thus far are consistent with the results of the model. In addition to the measurement of SCEs, other cytogenetic endpoints have been added to the protocol for blood analysis, i.e., micronuclei, DNA adducts, and alkylated hemoglobin. Preliminary work has demonstrated the presence of SCEs, micronuclei, and DNA adducts in the workers' lymphocytes. A method to measure styrene-oxide adducts of haemoglobin is being developed. The major objectives are now to: (1) complete the toxicokinetic modelling (with the 2-compartment open model) and correlate styrene exposure with levels of styrene in blood and breath; (2) correlate the numbers of SCEs and micronuclei in workers' lymphocytes with the intensity and pattern of styrene exposure; (3) isolate DNA from blood samples and measure DNA adducts; (4) continue work on a HPLC method for measuring alkylated hemoglobin in the samples and possibly styrene-oxide hg adducts.

TYPE: Methodology; Molecular Epidemiology
TERM: BMB; DNA Adducts; Dose-Response; Lymphocytes; Micronuclei; Occupation; Plastics; SCE
SITE: Inapplicable
CHEM: Styrene
OCCU: Plastics Workers
TIME: 1986 - 1994

***1010 Sandler, R.S.** 05358
Univ. of North Carolina, CB 7080, 423 Burnett-Womack Bldg, Manning Drive, Chapel Hill NC 27599-7080, United States of America (Tel.: +1 919 9662511; Fax: 9661757)
COLL: Kupper, L.L.; Ulshan, M.D.; Woosley, J.T.; Campbell, M.

Epidemiology of Rectal Mucosal Proliferation

Increased rectal mucosal proliferation appears to be an early step in the progression to colorectal cancer. This study is designed to examine demographic and environmental factors associated with increased mucosal proliferative index. The specific aims of the proposed study are (1) to characterize rectal mucosal proliferative labelling index, using proliferating cell nuclear antigen, and mitotic index, using whole crypt dissection; (2) to determine the association between rectal mucosal proliferative index and the presence of colon adenomas and cancer; and (3) to determine the association beteween rectal mucosal proliferative index and specific dietary and lifestyle factors that have been associated with colorectal cancer. Study subjects will be 500 male and female patients at the University of North Carolina Hospitals. Rectal mucosal punch biopsies are taken from subjects during routine colonoscopy. Patients will be interviewed by a trained nutritionist about dietary intake, using a standard semi-quantitative food frequency questonnaire, and about a number of lifestyle and environmental factors. Pathological material are reviewed by a single pathologist to determine type of polyp, degree of atypia and presence of cancer. Data analysis will compare proliferative indices in various demographic and exposure strata.

TYPE: Cross-Sectional
TERM: BMB; Diet; Drugs; Environmental Factors; Lifestyle; Obesity; Occupation; Polyps; Tobacco (Smoking)
SITE: Colon; Rectum
TIME: 1992 - 1996

Cheltenham

***1011 Daly, M.B.** 05309
Fox Chase Cancer Center, 510 Township Line Rd, Cheltenham PA 19118, United States of America (Tel.: +1 215 7282705; Fax: 7282707)
COLL: Gallion, H.; Whittemore, A.S.; Trock, B.

Epidemiology of Ovarian Cancer

The aim of the study is to identify the familial patterns of ovarian cancer, the role of reproductive risk factors in familial cases and to study the involved gene(s) by linkage analysis. Family history and reproductive data will be collected on 1,200 ovarian cancer cases seen at 606 institutions. Blood samples will be stored for genetic marker studies. Segregation analysis will be used to describe the genetic patterns. Statistical analysis will compare risk profiles in sporadic vs familial cases.

TYPE: Case Series; Genetic Epidemiology
TERM: BMB; Blood; Genetic Factors; Genetic Markers; Reproductive Factors; Segregation Analysis
SITE: Ovary
TIME: 1991 - 1998

Chestnut Hill

1012 Rothman, K.J. 04757
Epidemiology Resources, Inc., 826 Boylston St., Chestnut Hill MA 02166, United States of America (Tel.: +1 617 7349100; Fax: 2770335)
COLL: Wilkinson, G.; Franco, S.; Manzo, N.; Dreyer, N.A.

Brain Tumours among Offspring of Electronics Workers

A follow-up study of 4,700 workers at two plants engaged in the manufacture of color television picture tubes is being conducted to examine the health of workers and their offspring. Workers employed between 1982 and 1987 are being surveyed to gather information about their offspring. Of particular interest are any occurrences of brain or other cancers in childhood. Medical records and death certificates ar being obtained for children with illnesses of interest so that a mortality analysis may be conducted. Death certificates are also being collected for workers in the cohorts. Mortality among the workers will be compared with national and state mortality rates; the mortality rates among workers at each facility will be compared.

TYPE: Cohort
TERM: Occupation
SITE: Brain; Childhood Neoplasms
OCCU: Electronics Workers
TIME: 1988 - 1994

Chicago

1013 **Chmiel, J.S.** 02229
Northwestern Univ. Medical School, Cancer Center, Biometry Sect., 680 North Lake Shore Drive, Suite 1104 , Chicago IL 60611, United States of America (Tel.: +1 312 9088655)
COLL: Berlin, N.I.

Cancer Incidence and Mortality in an Electric Utility Company

In cooperation with the Commonwealth Edison Company, the Cancer Center developed a cancer surveillance programme involving all Commonwealth Edison employees. The primary objective of the Program was to assess over a period of years certain aspects of disease and related risks in workers in an electric utility company. Primary concerns were: (1) mortaltiy from all causes; and (2) incidence of cancer. Particular attention was devoted to radiation induced cancers, with emphasis on long-term exposure to low dose radiation. From 1983 to 1988, all employees were asked to complete an initial (baseline) questionnaire on cancer history, smoking history, and demographics. Participants were contacted every two years to collect information on events occurring since the time of the last questionnaire. Job history and radiation exposure data were also obtained. Although the collection of questionnaire and medical data has been discontinued, the existing data will continue to be analysed. Employees who will be included in analyses are those who have held full-time jobs at Commonwealth Edison for one year or more. All eligible employees were followed for mortality. 20,754 were being followed for mortality and cancer incidence and an additional 5,800 employees who worked at Commonwealth Edison Company between 1975 and 1983 were being followed for mortality only. Additional follow-up for mortality is anticipated.

TYPE: Cohort
TERM: Monitoring; Occupation; Radiation, Ionizing; Tobacco (Smoking)
SITE: All Sites
OCCU: Radiation Workers
TIME: 1979 - 1994

***1014** **Davis, F.G.** 05310
Univ. of Illinois School of Publ. Hlth, Dept. of Epidemiology & Biostatistics, 2121 W. Taylor St., Chicago IL 60612, United States of America (Tel.: +1 312 9965019; Fax: 9960064)
COLL: Fisher, S.; Golberg, J.; Haenszel, W.M.; Weber, L.

Effect of Gastric Surgery on Cancer Risk

This is a retrospective cohort study involving some 15,000 subjects, to test the hypothesis that operative procedures for peptic ulcer disease increase the risk of subsequent cancer of the stomach and colorectum. All patients having operative procedures for peptic ulcer disease in the US Veterans' Administration Hospitals in 1970 or 1971 (n=7,609) were compared with 8,374 hospital controls (patients operated on for other conditions than ulcer) and followed up till 1988.

TYPE: Cohort
TERM: Gastrectomy; Operation, Previous
SITE: Colon; Rectum; Stomach
TIME: 1991 - 1994

1015 **Mallin, K.** 04725
Univ. of Illinois at Chicago School of Public Health Epidemiology-Biometry M/C 922, P.O. Box 6998, Chicago IL 60680-6998, United States of America

Mortality of Steelworkers in Northwestern Illinois

Deaths among steelworkers in a plant employing approximately 2,500 workers are being analysed. A previous investigation of bladder cancer incidence in northwestern Illinois detected a number of cases who had worked in this steel mill. Chemical exposures include solvents, degreasers, metals, paints, cutting oils, lubricants, and sealants. A previous inspection by OSHA examined trichloroethylene, methyl ethyl ketone, diisobutyl ketone, and methyl isobutyl ketone. Between 1977 and 1988 there were 583 deaths among workers in this plant. The proportional distribution of cancer deaths will be examined using standard techniques, and appropiate follow-up will be undertaken.

TYPE: Cohort
TERM: Metals; Occupation; Paints; Petroleum Products; Solvents
SITE: All Sites
CHEM: Diisobutyl Ketone; Methyl Ethyl Ketone; Methyl Isobutyl Ketone; Trichloroethylene
OCCU: Steel Workers
REGI: Illinois (USA)
TIME: 1990 - 1994

1016 **Mallin, K.** 04726
Univ. of Illinois at Chicago School of Public Health Epidemiology-Biometry M/C 922, P.O. Box 6998, Chicago IL 60680-6998, United States of America

Investigation of a Bladder Cancer Cluster in Northwestern Illinois

An excess of incident bladder cancer cases was detected in a small community near Rockford, Illinois (Amer. J. Epidemiol.). SIRs for the period 1978-1985 were 1.7 in males, and 2.6 in females, based on 21 and 10 cases, respectively. Two drinking water wells in this community are within half a mile of a landfill site that was closed in 1972. One well was closed due to contamination with trichloroethylene and other solvents. Further investigations of this cluster are under way. Known risk factors for bladder cancer are being compared for the cluster cases and cases in the surrounding towns. Interviews with cases or their next of kin have been completed and include information on smoking, occupational histories, residential histories, fluid intake, and other factors. Predominant occupations among all cases include machinists. Additional analyses using 1976-1978 data from the Illinois State Cancer Registry will also be undertaken. A paper was published in Am. J. Epidemiol. 132:S96-S106, 1990.

TYPE: Cross-Sectional
TERM: Cluster; Occupation; Solvents; Tobacco (Smoking); Water
SITE: Bladder
CHEM: Benzene; Chlorobenzene; Trichloroethylene
REGI: Illinois (USA); Iowa (USA); Wisconsin (USA)
TIME: 1987 - 1994

***1017** **Nelson, R.L.** 05353
University of Illinois Hospital, Dept. of Surgery, 1740 S. Taylor St., Rm 2204 UIH, M/C 957, Chicago IL 60612-7322, United States of America (Tel.: +1 312 9665162; Fax: 9962704)
COLL: Davis, F.G.; Kikendall, J.W.; Bowen, P.; Sutter, E.; Milner, J.

Body Iron Stores and Risk for Colon Cancer

The hypothesis is that body iron stores are positively associated with colonic adenoma risk. Individuals having colonoscopy for normal indications had blood drawn prior to colonoscopy. Patients were then classified as either being normal, having adenoma or colon cancer. Serum ferritin was measured as an index of body iron stores. Selenium and folate were also measured.

TYPE: Cross-Sectional
TERM: Trace Elements
SITE: Colon
CHEM: Ferritin; Folic Acid; Iron; Selenium
TIME: 1992 - 1994

1018 **Sharp, D.S.** 03999
Amoco Corporation, Medical Dept., 200 Randolph Drive, P.O. Box 87703, Chicago IL 60680-0703, United States of America (Tel.: (312)8563901)
COLL: Hornstra, M.H.

Relationship of Exposure to Refinery Processes and Mortality from Cancer
The objective of this follow-up study is to examine mortality patterns in order to assess any increased risk for a particular cause of death. Vital status for over 26,000 employees from ten Amoco refineries has been updated and collection of death certificates for those the Social Security Administration, National Death Index and Bureau of Motor Vehicles identify as deceased. Standardized mortality ratios relative to the US and to the specific state's population are being calculated.

TYPE: Cohort
TERM: Chemical Exposure; Occupation; Petroleum Products
SITE: All Sites
OCCU: Petroleum Workers
TIME: 1985 - 1994

Cincinnati

1019 **Brown, D.P.** 03007
NIOSH, Industrywide Studies Branch, Div. Surveillance, Hazard Evaluation & Field Studies, Biometry Sect., 4676 Columbia Parkway, Cincinnati OH 45226, United States of America (Tel.: +1 513 6842694)

Update of Completed Cohort Mortality Studies
The Industrywide Studies Branch has completed a number of retrospective cohort mortality studies during the past few years, many of which lacked conclusive results. The primary reasons for inconclusive results are too few deaths (low statistical power) and a short follow-up period (resulting in a short latency period). Thus, many of the original hypotheses tested by these studies remain unresolved questions. Examples of such studies include workers exposed to PCBs, perchloroethylene, chlorinated hydrocarbon pesticides, and styrene butadiene rubber. These cohorts are being followed through the latest possible date. The additional death information will be added to the file and the life-table analysis updated. Additional studies will be selected based on length of time since last follow-up, adequacy of the current data to address the hypothesis, and the current interest or priority of the study. The updates have been completed for the studies of workers exposed to perchloroethylene, PCBs, benzene and cadmium. Other studies currently being updated include workers exposed to attapulgite (a clay fibre), vinyl chloride and mineral wool.

TYPE: Cohort
TERM: Chemical Exposure; Dusts; Metals; Occupation; Pesticides; Plastics; Solvents
SITE: All Sites
CHEM: Attapulgite; Benzene; Butadiene; Cadmium; Hydrocarbons; Hydrocarbons, Halogenated; Mineral Fibres; PCB; Styrene; Vinyl Chloride
TIME: 1982 - 1994

***1020** **Calvert, G.M.** 05516
NIOSH, Div. of Surveillance, Hazard Evaluation and Field Studies, Industrywide Studies Branch, 4676 Columbia Parkway, R16, Cincinnati OH 45226, United States of America (Tel.: +1 513 8414481; Fax: 8414486)
COLL: Fajen, J.M.; Fleming, L.; Briggle, T.

Cytogenetic Studies among Fumigant Applicators
Animal studies involving short term gavage exposure to methyl bromide show a definite trend toward neoplastic response. Fumigation workers using methyl bromide are being studied to determine if there is an association between methyl bromide exposure and genotoxicity. Blood lymphocytes were obtained from 31 structural fumigation workers and 27 non-exposed individuals, who were recruited by the workers. The lymphocytes will be analysed for HPRT mutations and micronuclei. In addition, oropharyngeal cells were collected and will be analysed for the formation of micronuclei.

TYPE: Cross-Sectional; Molecular Epidemiology
TERM: Biomarkers; HPRT-Mutants; Lymphocytes; Micronuclei; Occupation
SITE: Inapplicable
CHEM: Methyl Bromide
OCCU: Fumigation Workers
TIME: 1992 - 1995

***1021 Davis-King, K.** 05517
NIOSH, Div. of Surveillance, Hazard Evaluation and Field Studies, Industrywide Studies Branch, 4676 Columbia Parkway, R16, Cincinnati OH 45226, United States of America (Tel.: +1 513 8414481; Fax: 8414486)
COLL: Calvert, G.M.; Schulte, P.A.

Brain Cancer among Farmers

Using a case-control design, this study will evaluate the association between rural and farm exposures (occupational and non-occupational) and development of primary malignant brain gliomas among rural residents in four states in the upper midwest of the USA. The primary hypothesis of interest is that pesticides are associated with the increased risk of brain cancer that has been noted in the rural community. In addition, the study will examine if genetic effects, such as oncogene mutations, or if alternative hypotheses presented in the literature, are associated with increased risk for brain cancer and if any of these factors acts synergistically with pesticides. A total of 600-700 cases (half male, half female) and 1,200-1,400 population controls will be interviewed.

TYPE: Case-Control
TERM: Occupation; Pesticides; Rural; Solvents
SITE: All Sites; Brain
CHEM: Nitrates
OCCU: Farmers
TIME: 1992 - 1998

1022 Hornung, R.W. 04444
NIOSH, Centres for Disease Control, 4676 Columbia Parkway, Mailstop R-4, Cincinnati OH 45226, United States of America (Tel.: +1 613 8414211; Fax: 8414500)
COLL: Roscoe, R.J.

Risk Relationship between Radon Daughter Exposure and Cigarette Smoking in Colorado Plateau Uranium Miners (1950-1987)

This study involves an update of lung cancer mortality through 1990 in the Colorado Plateau cohort of 3,346 uranium miners. Previous analyses of this cohort involved the vital status of white male miners through 1982. In addition to eight more years of follow-up, the cigarette smoking history of both living and deceased miners has been updated to include the period 1969 to 1985. One of the prime hypotheses to be tested using the new data will be the existence of a possible trend in lung cancer risk attributed to joint exposure to radon and cigarette smoking from multiplicative towards additive as a function of age of the cohort. Changes in risk coefficients compared to previous analyses will also be examined, with special attention to the decline in relative risk with time since last exposure.

TYPE: Cohort
TERM: Dose-Response; Occupation; Time Factors; Tobacco (Smoking)
SITE: Lung
CHEM: Radon
OCCU: Miners, Uranium
TIME: 1991 - 1994

1023 Penn, I. 00777
Univ. of Cincinnati Medical Center, Dept. of Surgery, 231 Bethesda Ave. (ML 558), Cincinnati OH 45267, United States of America (Tel.: +1 513 5586867; Fax: 5583580)

Cancer in Organ Transplant Recipients and Other Immunosuppressed Patients

The aim is to study the frequency and types of malignancies and their behaviour in organ transplant recipients receiving immunosuppressive therapy. The patients are subdivided into three groups: (1) those who had tumours before transplantation; (2) those in whom tumours were accidentally transplanted with the allograft; (3) those who developed tumours de novo after transplantation. Data on more than 9,000 patients have been collected up till recently. The study continues. Numerous reports have been published, the most recent appearing in Transplant. International 6:1-3, 1993 and Hematol. Oncol. Clin. N. Amer. 7(2)5:1-15, 1993.

TYPE: Cohort
TERM: Chemotherapy; Immunosuppression; Transplantation
SITE: All Sites
TIME: 1968 - 1994

1024 **Roscoe, R.J.** 00031
NIOSH, Div. of Surveillance Hazard Evaluation and Field Studies, Robert A. Taft Labs, 4676 Columbia Parkway, Cincinnati OH 45226, United States of America (Tel.: + 1 513 8414411; Fax: 8414486)
COLL: Salvan, A.; Deddens, J.; Schnorr, T.M.; Prorok, P.C.

Uranium Miners - Low Dose Study

This project aims: (1) to help evaluate the current standard for radon daughter exposures; and (2) to help assess health implications of energy industries. Since the early 1950s, NIOSH and its predecessors have been studying a cohort of 4,000 uranium miners (also known as the Colorado Plateau cohort) to determine the adverse health effects of radon daughters and other exposures. Dose-response relationships between exposures to radon daughters and lung cancer have been established and used to help set federal regulations. Data points for the dose-response curve in the area of low lifetime doses (i.e. less than 120 WLM) are, however, still needed for this cohort and for radiation carcinogenesis in general. NIOSH, in collaboration with NCI, has updated the work and smoking histories of this cohort through 1985 (last update was through 1969) and extended the vital status follow-up through 1990 (last follow-up through 1984). This updating and follow-up will allow NIOSH to finalize the occupational exposures for virtually all of the cohort and to extend the median period of observation to about 32 years. A report on lung cancer mortality among the non-smoking members of this cohort was published in J. Am. Med. Ass. 262 (5):pp 629-633, 1989. This project is associated with the project by R. W. Hornung (ID:4444).

TYPE: Cohort
TERM: Dose-Response; Dusts; Latency; Metals; Mining; Occupation; Radiation, Ionizing; Tobacco (Smoking)
SITE: Lung; Lymphoma
CHEM: Radon; Silica; Uranium
OCCU: Miners, Uranium
TIME: 1950 - 1994

1025 **Schulte, P.A.** 03821
NIOSH, Div. of Surveillance, Hazard Evaluation and Field Studies, Industrywide Studies Branch, 4676 Columbia Parkway, Mail Stop R-13, Cincinnati OH 45226, United States of America (Tel.: + 1 513 8414507; Fax: 8414550/8414540)
COLL: Hayes, R.B.; Friedland, J.M; Mazzuckelli, L.F.; Ward, E.M.; Caporaso, N.E.; Aldefer, R.

Biological Markers of Occupational Bladder Cancer

This is a case-control study, involving 150 cases and 150 controls, to identify early markers of bladder cancer to assess whether or not there are metabolic phenotypes that increase the risk of occupational bladder cancer. The markers to be utilized in the study include various oncogenes, tumour suppressor genes, growth factors and their receptors and other cellular markers.

TYPE: Case-Control
TERM: Biomarkers; Chemical Exposure; Genetic Factors; Occupation; Oncogenes
SITE: Bladder
CHEM: Amines, Aromatic
TIME: 1990 - 1996

1026 **Schulte, P.A.** 04219
NIOSH, Div. of Surveillance, Hazard Evaluation and Field Studies, Industrywide Studies Branch, 4676 Columbia Parkway, Mail Stop R-13, Cincinnati OH 45226, United States of America (Tel.: + 1 513 8414507; Fax: 8414550/8414540)
COLL: Boeniger, M.; Walker, J.; Herrick, R.; Halperin, W.E.; Griffith, J.

Biological Markers in Hospital Workers Exposed to Low Levels of Ethylene Oxides

Hospital sterilizer operators using ethylene oxide were studied to determine if there was a relationship between exposure and a panel of biological markers. A total of 74 workders from the United States and Mexico will be assessed for haemoglobin and DNA adducts, SCE, micronuclei, and HPRT mutations.

TYPE: Case Series
TERM: BMB; Biomarkers; DNA Adducts; HPRT-Mutants; Micronuclei; Occupation; SCE
SITE: Leukaemia
CHEM: Ethylene Oxide
OCCU: Health Care Workers
LOCA: Mexico; United States of America
TIME: 1988 - 1995

1027 **Ward, E.M.** 03034
NIOSH, Div. of Surveillance, Hazard Evaluation & Field Studies, 4676 Columbia Parkway, Alice Hamilton Bldg, Cincinnati OH 45226, United States of America (Tel.: +1 513 8414481; Fax: 8414486)

Investigation of Workers Exposed to MOCA

MOCA has been shown to be a potent carcinogen in animal studies and is structurally similar to aromatic amines, such as benzidine, known to cause bladder cancer in occupationally exposed workers. No adequate epidemiological studies have examined the carcinogenicity of MOCA in humans. MOCA is used as a curing agent for isocyanate containing polymers. About 33,000 US workers are currently exposed, most in workplaces where 10 or fewer employees work directly with the substance. A registry of workers exposed to MOCA at a US production facility has been established. The registry will be used for prospective surveillance of cancer incidence and mortality. Cancer incidence will be determined by notification of the cohort and telephone interview. In addition, workers from the facility will be offered a screening examination which will utilize conventional techniques to examine urine samples for abnormalities indicative of bladder cancer.

TYPE: Cohort
TERM: Occupation; Screening; Urine
SITE: All Sites; Bladder
CHEM: 4,4'-Methylene-bis(2-chloroaniline)
TIME: 1981 - 1994

1028 **Ward, E.M.** 03212
NIOSH, Div. of Surveillance, Hazard Evaluation & Field Studies, 4676 Columbia Parkway, Alice Hamilton Bldg, Cincinnati OH 45226, United States of America (Tel.: +1 513 8414481; Fax: 8414486)
COLL: Zahm, S.H.

Mortality Study of Workers Exposed to Chlorinated Naphthalenes

The aim of the study is to determine whether Halowax exposure is associated with an excess risk of soft tissue sarcoma, lymphoma or liver cancer. Chlorinated naphthalenes, sold under the trade name Halowax, are known chloracnegens. Because chlorinated naphthalenes are structurally similar to, and share common toxic effects with other chlorinated hydrocarbons that are chloracnegens (polychlorinated biphenyls (PCBs), chlorinated dibenzodioxins (dioxins, TCDDs) and chlorinated dibenzofurans (furans)), it is hypothesized that they might share common chronic health effects as well. These include an excess risk of soft tissue sarcoma and lymphoma associated with TCDDs in humans and liver tumours associated with PCBs and TCDDs in animal studies. The study population will be 8,000 workers exposed to Halowax in the early 1940's at a plant manufacturing Navy cable. Phase I will include 740 workers identified from medical records as having possible or probable chloracne and should be completed in 1.5 years. Phase II will include all 8,000 workers employed at the plant during the time when Halowax was used, and will be completed in three years. Standard NIOSH methods will be used for coding the masterfile, vital status follow-up, nosology of death certificates and life table analysis for computing standardized mortality ratios. Although the current number of workers exposed to chlorinated naphthalenes is under 5,000, the study will be of value in investigating the ultimate health outcomes of other worker groups exposed to potential chloracnegens.

TYPE: Cohort
TERM: Chemical Exposure; Occupation
SITE: Liver; Lymphoma; Soft Tissue
CHEM: Naphthalenes, Chlorinated; PCB
OCCU: Cable Manufacturers
TIME: 1983 - 1994

Columbia

1029 **Coker, A.** 05104
Univ. of South Carolina, Dept. of Epidemiology and Biostatistic, School of Public Health, Sumter and Greene, Columbia SC 29208, United States of America (Tel.: +1 803 7717353; Fax: 7774783)
COLL: Pirisi, L.

Human Papilloma Viruses and CIN among Low-Income Black Women in South Carolina

This nested case-control study will investigate the relationship between HPV, by DNA subtype, and CIN 1, CIN 2, and CIN 3, in a low-income, largely minority population. The approximately 15,000 women receiving Pap smears each year at SC DHEC clinics in the Trident Health District will form the sampling frame. From December 1990 through November 1991, the spatula used to collect each smear will be saved in individually labeled test tubes for subsequent HPV DNA typing of selected subjects. Smear results will be used to select 450 CIN cases and 200 controls with normal cervical cytology. Adjusted odds ratios and 95% confidence intervals will be used to test the following study hypotheses: (1) women with any CIN will be more likely to have HPV 16/18/33 compared with women having normal cervical cytology when controlling for confounders; (2) cases will be more likely to have HPV 6/11 compared with controls when controlling for confounders; (3) the adjusted odds ratio for the HPV 16/18/33 (or HPV 6/11) and CIN relationship will be largest for the comparison of CIN 3 cases with controls relative to the comparison of CIN 2 or CIN 1 cases with controls.

TYPE: Case-Control
TERM: BMB; Blacks; Cytology; HPV; Socio-Economic Factors
SITE: Uterus (Cervix)
TIME: 1991 - 1994

Dayton

***1030** **Chuang, T.Y.** 05386
Wrigth State Univ., Dept. of Dermatology, P.O.Box 927, Dayton OH 45401, United States of America (Tel.: +1 513 2020214; Fax: 2620211)
COLL: Reizner, G.T.; Elpern, D.J.

Skin Cancer Incidence in Kauai, Hawaii

The aim of the study is to assess the incidence of skin cancer in Kauai, Hawaii through a prospective incidence study (patients diagnosed between 1983 and 1996). A case-control study of risk factors will also be carried out. 1,200 patients will be included, as well as 1,200 matched controls (dermatological patients without skin cancer). Data will be collected through clinical visits, medical records, interview and questionnaire and will cover exposure to ultraviolet light, race, skin, eye and hair colour, freckling, moles, etc. An attempt will be made to confirm that place of birth, age at arrival in Hawaii and duration of residence are the major risk factors. Cox Multiple Regression Analysis will be used.

TYPE: Case-Control; Incidence
TERM: Geographic Factors; Pigmentation; Race; Radiation, Ultraviolet
SITE: Melanoma; Skin
TIME: 1983 - 1996

Des Moines

1031 **Young, D.C.** 04004
University of Iowa, Dept. Preventive Medicine/Radiology, 974 73rd St. , Des Moines IA 50312, United States of America
COLL: Young, C.S.; Wallace, R.B.

Women's Health Trial

Using previously identified risk factors women at high risk for breast cancer will be solicited to volunteer as participants in a study to evaluate what role, if any, dietary fat may play in the induction of breast cancer. A total of 30,000 women will be divided into a control group as well as an intervention group. The intervention group will be motivated to reduce dietary fat intake to a level of 25-30% of total calories. The end point will be breast cancer in any form, with an annual mammography being provided.

TYPE: Cohort; Intervention
TERM: Diet; Fat; Mammography; Prevention
SITE: Breast (F)
TIME: 1985 - 1994

Detroit

1032 Demets, R.Y. 04265
Wayne State Univ., Dept. of Family Medicine, 4201 St. Antoine (NHL-4J), Detroit MI, United States of America (Tel.: +1 313 5775074)

Colorectal Neoplasia among Pattern and Model Makers

This is a case-control study to test the hypothesis that work with cutting oils leads to an increased incidence of colorectal adenomas. Cases are patients with histologically proven adenomas, matched 2:1 by age with controls in the same trade. Exposure data, smoking habits and trade years are independent variables. The study subjects are participants in a longitudinal cancer screening programme. Cases are reviewed and examined every three years for 15 years. The total number of participants is 2,000, from which approximately 300 cases will be recruited. All polyps discovered are biopsied and reviewed by a single university pathologist.

TYPE: Case-Control
TERM: Chemical Exposure; Metals; Occupation; Plastics; Screening; Tobacco (Smoking)
SITE: Colon; Rectum
CHEM: Formaldehyde; Mineral Oil; Nickel
OCCU: Model and Pattern Makers
TIME: 1982 - 1997

1033 Haas, G.P. 05022
Wayne State Univ., Harper Hosp., 4160 John R. Rd, Suite 1017, Detroit MI 48201, United States of America (Tel.: +1 313 7457381; Fax: 7450464)
COLL: Edson Pontes, J.; Sakr, W.; Crissman, J.; Heilbrun, L.K.; Cassin, B.

Age- and Race- Related Trends in the Prevalence of Prostate Cancer

Carcinoma of the prostate is the most frequent cancer diagnosed and the third cause of cancer related deaths. While the clinical presentation is most common in elderly males, the disease is known to have a long latency period and very little data are available on the prevalence of prostate cancer in the younger age groups. In addition, there are marked differences between various racial groups. The aim of the study is to identify the prevalence of prostate cancer in black and white males with particular attention paid to the younger age groups (less than 50 years of age). Autopsy specimens will be obtained from men who die from unrelated causes and will be histologically step-sectioned to identify the presence of malignant or pre-malignant changes. The accumulated data will demonstrate the prevalence of prostate cancer in black and white males of various age groups and will serve as a data base for further studies of potential predisposing factors, molecular changes and the targeting of populations to be screened.

TYPE: Cross-Sectional
TERM: Age; Autopsy; BMB; Race; Screening
SITE: Prostate
TIME: 1991 - 1995

1034 Park, R. 04657
Int. Union of United Auto Workers, Health and Safety Dept., 8000 E. Jefferson St., Detroit MI 48214, United States of America (Tel.: +1 313 9265563)
COLL: Mirer, F.E.

Occupational Cancer Mortality in the Automobile Industry

Several studies are under way to assess work-related mortality in: (1) seven automotive machining plants, with primary attention to lung cancer and non-Hodgkin's lymphoma in transmission and chassis workers (approximately 5,000 deaths in total); (2) five automotive stamping plants, with emphasis on gastrointestinal cancer (approximately 4,000 deaths in total); (3) two engine plants (1,800 deaths) and (4) approximately 10 stamping and assembly plants. PMRs and MORs will be calculated using US deaths as the reference population.

TYPE: Mortality
TERM: Occupation; Petroleum Products; Welding
SITE: Gastrointestinal; Lung; Non-Hodgkin's Lymphoma
OCCU: Automobile Workers; Machinists; Metal Workers; Tool Makers; Welders
TIME: 1985 - 1994

East Lansing

1035 **Swanson, G.M.** 03431
Michigan State Univ., Comprehensive Breast Cancer Center, A211 East Fee Hall, East Lansing MI 48824, United States of America (Tel.: +1 517 3538828; Fax: 3361798)
COLL: Satariano, E.; Brissette, P.; Iwamamoto, K.

Occupational Cancer Surveillance: New Approaches

This project is operating an occupational cancer monitoring system. The system complements the Metropolitan Detroit Cancer Surveillance System (MDCSS), which is a participant in the SEER programme of the NCI. Occupational histories and related information are obtained by telephone interview. A total of approximately 20,000 case interviews are projected during the course of the study. Controls are population controls selected by random digit dialling procedures. Cases selected for the study will include all newly diagnosed cases of the specified sites between the ages of 40 and 84. New leads about occupational risks associated with cancer incidence are expected from the proposed study, particularly for women and blacks. These leads will be disseminated to local public health officials, NIOSH, and appropriate groups of workers and industries. Analytical investigation suggested by study results will follow. Monitoring occupation in terms of cancer incidence permits earlier detection of potential hazards than can be achieved by mortality studies. Evaluation of death certificate occupational information for study subjects who are deceased is underway. Papers have been published in J. Occup. Med. 27:439-444, 1985; Am. J. Publ. Hlth 77:1532-1534, 1987; and in Am. J. Med. 14:121-136, 1988.

TYPE: Case-Control; Incidence; Methodology
TERM: Blacks; Environmental Factors; Female; Lifestyle; Occupation; Registry
SITE: Bladder; Colon; Eye; Liver; Lung; Melanoma; Mesothelioma; Oesophagus; Rectum; Salivary Gland; Stomach
REGI: Michigan (USA)
TIME: 1984 - 1994

1036 **Swanson, G.M.** 04335
Michigan State Univ., Comprehensive Breast Cancer Center, A211 East Fee Hall, East Lansing MI 48824, United States of America (Tel.: +1 517 3538828; Fax: 3361798)
COLL: Schwartz, A.G.

Investigation of Tumours that Occur Excessively among Blacks

This collaborative study is being conducted by the Detroit, Atlanta and New Jersey registries in the SEER program and the National Cancer Institute. Its objectives are: (1) to identify race-specific risk factors for prostate, pancreas, oesophagus, and multiple myeloma; (2) to estimate the extent to which the risk factors may explain the black/white differences in the incidence rates of these cancers; and (3) to obtain biochemical measures of certain risk factors, such as hormones. The study design is that of a case referent study. The metropolitan Detroit component will enrol 302 male and female cases of multiple myeloma, 325 cases of cancer of the prostate; 422 male and female cases of cancer of the pancreas, and 250 male cases of cancer of the oesophagus over four years of data collection. Interviews will also be conducted with 967 controls. Cases in the age group 30-79 at diagnosis are included in the study. The interview obtains data regarding dietary factors, sexual development and behaviour, occupational history, family history of cancer, medical history, hormonal status, socio-economic status, and tobacco use. For cancers of the pancreas, extensive questions regarding coffee use are included. Data are obtained through in-person interviews, conducted in the home for most cases and controls. For cancers of the pancreas, many interviews are being conducted in hospital. Data collection is projected to continue through 1989. Analyses will be conducted separately for each type of cancer; multiple logistic regression analysis will be performed to control for potential confounders.

TYPE: Case-Control
TERM: Blacks; Diet; Familial Factors; Hormones; Occupation; Premalignant Lesion; Registry; Socio-Economic Factors
SITE: Multiple Myeloma; Oesophagus; Pancreas; Prostate
REGI: Atlanta (USA); Michigan (USA); New Jersey (USA)
TIME: 1986 - 1994

East Millstone

1037 **Gamble, J.F.** 05108
Exxon Biomedical Sciences, Inc., Div. of Occupational Health and Epidemiology, Mettlers Rd, East Millstone NJ 08875-2350, United States of America (Tel.: +1 908 8736004; Fax: 8736009)
COLL: Pearlman, E.D.

Kidney Cancer among Petrochemical Workers

This nested case-control study aims to investigate the association between gasoline hydrocarbons and kidney cancer among petrochemical workers. Cases with confirmed kidney cancers and four controls per case, frequency-matched for sex, race, date of birth and date of employment will be identified from a cohort of refinery and chemical plant employees. Detailed work and smoking history will be obtained from medical and personnel records. Quantitative exposure scores for hydrocarbons will be estimated by score years (frequency of exposure x intensity of exposure x years of exposure). The exposure-response relationship will be studied for causality.

TYPE: Case-Control
TERM: Occupation; Petroleum Products; Tobacco (Smoking)
SITE: Kidney
CHEM: Hydrocarbons
OCCU: Petrochemical Workers
TIME: 1989 - 1994

***1038** **Gamble, J.F.** 05321
Exxon Biomedical Sciences, Inc., Div. of Occupational Health and Epidemiology, Mettlers Rd, East Millstone NJ 08875-2350, United States of America (Tel.: +1 908 8736004; Fax: 8736009)
COLL: Lerman, S.E.; Holder, W.

Patient-Based Case-Control Study of Non-Melanoma Skin Cancer

This case-control study of non-melanoma skin cancer aims to examine the role of employment in the petro-chemical industry and non-occupational risk factors. Diagnosed and confirmed cases of basal cell carcinoma (BCC) and squamous cell carcinoma (SCC) will be selected from out-patient files from 1989 to the present. Cases will be restricted to white males. One control per case will be frequency matched for age. Approximately 500 cases each of BCC and SCC, and controls will be sent questionnaires with questions on non-occupational risk factors, including cumulative sun exposure, eye, skin and hair colour, sunburns, ethnic group, skin reaction to the sun, as well as a complete work history. Analysis of the risk of occupational exposure will evaluate ever employed vs. never employed and exposure-response trends using years worked as the exposure variable. Non-occupational risk factors will be evaluated using logistic regression in the calculation of both adjusted and unadjusted odds ratios.

TYPE: Case-Control
TERM: Ethnic Group; Naevi; Occupation; Pigmentation; Radiation, Ultraviolet
SITE: Skin
OCCU: Petrochemical Workers
TIME: 1989 - 1994

Edinboro

1039 **Miller, G.H.** 04140
Studies on Smoking, Inc., 125 High St., Edinboro PA 16412, United States of America (Tel.: +1 814 7345538)
COLL: Schneiderman, M.A.; De Pue, R.H.

Comparison of Cancer Incidence in Women Exposed and Non-Exposed to Passive Smoking

The study will compare the specific types of cancer contracted by non-smoking women with no known exposure and by non-smoking women who have had some lifetime exposure. The differences, if any, in cancer incidence between these two levels of exposure, will be examined. Later studies will compare these two groups with smoking women. The present study is based on data from interviews of close relatives of the deceased in Erie County, Pennsylvania during the years 1975-1976 and 1979-1980. A total of 3,295 interviews out of a total population of 10,131 deceased (including males and females) were completed though 1986. Only interviews containing all necessary data on passive smoking were included in the study, resulting in a data base of 934 non-smoking wives (199 had no known exposure and 733 had some lifetime exposure). Preliminary analysis appears to show significant differences in cancer incidence between exposed and non-exposed women, as well as differences in the types of cancer contracted. In order to increase the sample size, additional data on decedents in 1975-1976 and 1979-1980 will be obtained in 1987-1988. A study of breast cancer and passive smoking has been completed. A paper was published in Cancer Prevention and Detection 14:497-503, 1989.

TYPE: Cohort
TERM: Female; Passive Smoking
SITE: All Sites
TIME: 1974 - 1994

1040 **Miller, G.H.** 05198
Studies on Smoking, Inc., 125 High St., Edinboro PA 16412, United States of America (Tel.: +1 814 7345538)
COLL: Cox, C.E.; Depue, R.H.; Gambrill, R.D.; Hawthorne, V.; Novotny, T.; Pierce, J.P.; Schneiderman, M.A.; Wells, A.J.

Can Active or Passive Smoking Cause Breast Cancer?

In order to show whether or not active and passive smoking are associated with breast cancer a large sample from a complete population in Erie County was used to determine the impact of both active and passive smoking on the risk of breast cancer. Close relatives of the deceased were interviewed to determine three categories of exposure: (1) no known or minimal exposure; (2) passive smoking exposure, and (3) active smoking. The sample included 371, 2,116 and 1,123 women respectively in these categories. The proportion of women with breast cancer in each category were 0.5%, 4.4% and 5.5%. Results of age-adjusting provided contradictory results. The results of the pilot phase appear to show that if both active and passive smoking were eliminated, the incidence of breast cancer in women should be reduced significantly. Additional research is in progress to clarify the results and to recruit additional subjects.

TYPE: Cohort
TERM: Passive Smoking; Tobacco (Smoking)
SITE: Breast (F)
TIME: 1986 - 1994

Emeryville

1041 **Reynolds, P.** 04854
California Dept. of Health Services, Environmental Epidemiology Sect., 5900 Hollis St., Suite E, Emeryville CA 94608, United States of America (Tel.: +1 415 5403657; Fax: 5402673)
COLL: Lemp, G.F.; Saunders, L.D.

AIDS and Cancer: A Record Linkage Study

The overall objective of this project is to provide baseline information on the epidemiology of human immunodeficiency virus (HIV)-associated malignancies. This is being accomplished via record linkage between the population-based cancer registry serving the San Francisco Bay Area since 1969 (with over a quarter of a million cancer records) and the San Francisco AIDS Surveillance Program serving San Francisco since 1980 (with over 6,000 case records). A variety of analytical techniques are being used to examine the occurrence and characteristics of HIV-associated cancers.

TYPE: Incidence
TERM: AIDS; HIV; Record Linkage; Registry
SITE: All Sites; Kaposi's Sarcoma
REGI: Bay Area (USA)
TIME: 1988 - 1994

Gainesville

1042 **Pollock, B.H.** 04934
Univ. of Florida, College of Medicine, Div. of Epidemiology & Biostatistics, Dept. of Pediatrics, Suite 22, 4110 SW 34th St. , Gainesville FL 32608-2516, United States of America (Tel.: +1 904 3925198; Fax: 3928162)
COLL: Mulhern, R.K.; Armstrong, F.D.; Ryan, B.R.

Development of Intervention Strategies to Reduce Delay in Diagnosis of Childhood Cancer

Early detection of childhood cancer usually optimizes response to therapy and thus reduces subsequent mortality and morbidity. As asymptomatic screening is often impractical, the long-range goal is to improve the early symptomatic detection of childhood cancer. The specific aims are: to describe the constellation of signs and symptoms which occur prior to diagnosis; to evaluate the determinants of lagtime (time from the first onset of symptoms until diagnosis); and to assess the relationship between lagtime and prognosis. This cohort study will utilize a self-administered questionnaire, completed by parents of 1,200 children with newly diagnosed malignancies accrued to a Pediatric Oncology Group therapeutic protocol during the period 1990-1992. To determine which factors are associated with lagtime, information will be collected on: demographic factors, socio-economic status, family health-seeking behaviour and access to health services. This questionnaire data along with other data will make it possible to test the hypothesis that lagtime is independently associated with prognosis. Follow-up for survival and disease-free survival will continue up to 1993. Cox proportional hazards regression will be used to assess the association of lagtime with long-term survival with adjustement for potential confounders such as stage at diagnosis and treatment. It is anticipated that these findings will lead to the development of cancer control interventions.

TYPE: Cohort
TERM: Prognosis; Socio-Economic Factors; Survival; Time Factors
SITE: Childhood Neoplasms
LOCA: Canada; United States of America
TIME: 1990 - 1994

Greenfield

***1043** **Falck, F.** 05471
, 75 Wells St., Greenfield MA, United States of America (Tel.: +1 413 7720389; Fax: 7737596)
COLL: Deckers, P.; Ricci, A.; Wolff, M.S.; Godbold, J.

Pesticides and Polychlorinated Biphenyl Residues in Human Breast Lipids and their Relation to Breast Cancer

Various halogenated hydrocarbons, acting as either co-carcinogens or promoting agents, derived from the environment and concentrated in human fatty stores, may play a role in breast cancer risk. A pilot study was undertaken to measure and compare levels of chemical residues in mammary adipose tissue from women with malignant and non-malignant breast disease. Elevated levels of polychlorinated biphenyls, bis(4-chlorophenyl)-1, 1 dichloroethene, and bis(4-chlorophenyl)-1, 1,1-trichloroethane were found in fat samples from women with cancer, compared with those who had benign breast disease. A larger study is now underway. Cellular mechanisms by which these compounds may contribute to the aetiology of breast cancer are being examined.

TYPE: Case-Control
TERM: BMB; Biopsy; Chemical Exposure; Pesticides
SITE: Breast (F)
CHEM: DDE; DDT; PCB
TIME: 1987 - 1995

Hagerstown

1044 **Comstock, G.W.** 00283
Johns Hopkins Univ., School of Hygiene and Public Health, Training Center for Public Health Res., Box 2067, Hagerstown MD 21742-2067, United States of America (Tel.: +1 301 7913230)
COLL: Bush, T.L.

Serological Precursors for Cancer Risk Factors

In 1974, 15 ml of blood were collected from approximately 25,600 persons in Washington County, Maryland. The serum was separated from the blood cells and stored at -70 degrees C. In 1989 20 ml of blood were collected from 32,941 persons in the same county. Plasma and white blood cells were also stored at -70 degrees C. In addition to basic identification and personal information obtained when blood was drawn, participants in the 1989 study were asked to complete a dietary questionnaire at their leisure and to return them along with a toenail clipping for trace metal analysis. Dietary histories and information from plasma, white blood cells, and toenails will be analysed for limited nested case-control studies of specific cancers as these develop among participants and are identified by the cancer registry. Examples of potential cancer risk factors include vitamins (A, D, E, carotenoids), trace elements (selenium, zinc), hormones, lipids, white blood cell components (DNA adducts), and antibodies (herpes, cytomegalovirus, papillomovirus, Epstein-Barr virus). Comparison between estimated intake of various nutrients can be made. The stability of plasma components on prolonged storage or on repeated thawing and freezing will be tested in participants who do not live in the immediate area of Washington County, MD.

TYPE: Case-Control; Cohort
TERM: BMB; DNA Adducts; Diet; EBV; HPV; HSV; Hormones; Lipids; Plasma; Serum; Toenails; Trace Elements; Virus; Vitamins; White Cells
SITE: All Sites
CHEM: Beta Carotene; Retinoids
REGI: Maryland (USA)
TIME: 1974 - 1994

Hanover

1045 **Baron, J.A.** 04978
Dartmouth Medical School, 2 Maynard St., Hanover NH 03756, United States of America (Tel.: +1 603 6465542; Fax: 6466313)
COLL: Rothstein, R.; Beck, G.; Summers, R.; Haile, R.W.C.; Mandel, J.S.; Sandler, R.S.

Calcium in the Prevention of Neoplastic Polyps

This is a randomized, double-blind, placebo-controlled clinical trial of the efficacy of nutritional supplementation with calcium carbonate in preventing recurrence of neoplastic polyps of the large bowel. It is a collaborative investigation conducted at six clinical centres (Cleveland Clinic Foundation, Dartmouth-Hitchcock Medical Center, Univ. of California at Los Angeles, Univ. of Iowa, Univ. of Minnesota and Univ. of North Carolina) with a central data coordinating centre at Dartmouth. The study will involve 850 patients newly treated for neoplastic polyps, who have had at least one neoplastic polyp removed within three months of recruitment, have no known polyps remaining in the colon, and no contra-indications to calcium supplementation. After completing a three-month placebo run-in period, subjects are allocated at random to receive either placebo or calcium carbonate, three grams per day in divided doses with meals. They are monitored by questionnaire for toxicity and compliance to the drug regimen at six-month intervals throughout the study. One year and four years after the qualifying colonoscopy, each subject will receive a complete colonoscopic examination, with removal and histological examination of all polyps. The primary analysis will compare the polyp recurrence rate in the placebo group with that in the calcium group. The study will have at least 80% power to detect a 35% reduction in polyp recurrence from a 10% per year recurrence rate. It will also permit investigation of the relationship between baseline dietary calcium and polyp recurrence.

TYPE: Intervention
TERM: BMB; Polyps; Prevention; Serum
SITE: Colon
CHEM: Calcium
TIME: 1990 - 1994

Hawthorne

1046 **Swift, M.R.** 04317
New York Medical College, Div. of Human Molecular Genetics, 4 Skyline Drive , Hawthorne NY 10532, United States of America (Tel.: +1 914 3472592; Fax: 5922275)

Cancer in Families with Ataxia Telangiectasia

Ataxia telangiectasia (A-T) is a rare autosomal recessive syndrome that predisposes homozygotes to cancer. Retrospective studies have shown that A-T heterozygotes, about 1% of the general population, also have an excess risk of cancer, particularly breast cancer in women. To examine the hypothesis that A-T heterozygotes have an excess risk of cancer, a study of cancer in 161 families affected by A-T has been carried out. Cancer incidence and mortality, mortality from ischemic heart disease, and mortality from all causes were compared prospectively in adult blood relatives of patients with A-T and their spouses, who served as controls. In a case-control sub-study, occupational and fluoroscopic diagnostic exposures to radiation in female blood relatives, in whom breast cancer was first diagnosed during the prospective observation, were compared with the exposures in matched blood relatives without breast cancer. Cancer rates were significantly higher in the group of blood relatives than in their spouses, specifically in the sub-group of blood relatives who were known A-T heterozygotes. Also, female blood relatives with breast cancer were more likely to have been exposed to ionizing radiation than controls without cancer. Prospective follow-up of surviving patients is continuing and new families added to the study. Recent papers appeared in N. Engl. J. Med. 325:1831-1836, 1991, Cancer Genet. Cytogenet. 46:21-27, 1990 and 50:119-123, 1990 and in Immunodeficiency Rev. 2:67-81, 1990.

TYPE: Case-Control; Cohort
TERM: Ataxia Telangiectasia; Familial Factors; Genetic Factors; Genetic Markers; High-Risk Groups; Premalignant Lesion
SITE: All Sites; Breast (F); Leukaemia
TIME: 1980 - 1996

Honolulu

1047 **Goodman, M.T.** 04440
Univ. of Hawaii, Cancer Research Center of Hawaii, 1236 Lauhala St., Suite 407 , Honolulu HI 96813, United States of America (Tel.: +1 808 5862987; Fax: 5862982)
COLL: Kolonel, L.N.; Nomura, A.M.Y.; Hankin, J.H.; Yoshizawa, C.N.

Dietary Influences on Endometrial Cancer Risk

The primary objective of this 5-year matched case-control study is to examine the role of dietary fat, vitamin A and its precursors, and vitamin C in the development of cancer of the endometrium. A total of 391 cases, consisting of 158 prevalent cancers and 233 incident cancers, are expected to be interviewed during the study period. Eligible cases will be defined as all endometrial cancer patients admitted to the seven largest Oahu hospitals who have a histologically confirmed diagnosis of primary cancer. Two population-based controls will be matched to each case on the basis of ethnicity and age (within 5 years). In addition to the dietary data, the questionnaire will include the following subjects: demographic information, occupational history, anthropometric measurements, reproductive history, contraceptive and non-contraceptive hormone use, medical history, history of cigarette use, and physical activity. Analysis of data will consist of an assessment of the dose-response relation of total, saturated and unsaturated dietary fat, vitamin A, beta carotene and other carotenes, and vitamin C to endometrial cancer while adjusting for the effects of potential confounders.

TYPE: Case-Control
TERM: Contraception; Diet; Dose-Response; Ethnic Group; Fat; Hormones; Nutrition; Obesity; Occupation; Physical Activity; Reproductive Factors; Tobacco (Smoking); Vitamins
SITE: Uterus (Corpus)
CHEM: Beta Carotene
TIME: 1988 - 1994

***1048 Goodman, M.T.** 05453
Univ. of Hawaii, Cancer Research Center of Hawaii, 1236 Lauhala St., Suite 407 , Honolulu HI 96813, United States of America (Tel.: +1 808 5862987; Fax: 5862982)
COLL: Kolonel, L.N.; Hankin, J.H.; Wilkens, L.; Cooney, R.V.

Multiethnic Study of Pre-Invasive Cervical Lesions in Hawaii

The objective of this multi-ethnic case-control study is to examine the associations of dietary and serum nutrient levels, HPV and tobacco smoking on the risk of cervical dysplasia. Subjects will be identified in several clinics during the period 1992-1995. Cases will include 600 women with histologically confirmed squamous intraepitelial lesions. Controls will be matched 1:1 for age and clinic. A personal interview includes demographic information, diet history, sexual and reproductive history, hormone use, medical history, lifetime history of tobacco use, environmental tobacco smoke exposure and personal hygiene. Plasma or erythrocyte levels of retinol, retinol esters, alpha and gamma tocopherol, alpha and beta carotene, lycopene, canthaxanthin, beta cryptoxanthin, lutein, folate, ascorbic acid and cholesterol will be assessed in blood samples. The presence of HPV sub-types is determined in exfoliated cell samples. Data analysis will examine the independant associations and interactive effects of dietary and biochemicl measures of nutritional status, HPV, tobacco smoke exposure and other variables on the risk of cervical dysplasia, using multiple conditional logistic regression. The study sample should provide sufficient power to detect relatively weak exposure-disease relationships.

TYPE: Case-Control
TERM: BMB; Blood; Cholesterol; Contraception; Diet; HPV; Hormones; Nutrition; Physical Activity; Reproductive Factors; Sexual Activity; Tobacco (Smoking); Vitamins
SITE: Uterus (Cervix)
CHEM: Ascorbic Acid; Beta Carotene; Retinoids; Tocopherol
TIME: 1992 - 1996

1049 Hankin, J.H. 04630
Univ. of Hawaii, Cancer Research Center, 1236 Lauhala St., Suite 407, Honolulu HI 96813, United States of America (Tel.: +1 808 5488452; Fax: 5483411)
COLL: Zhao, L.P.; Kolonel, L.N.; Wilkens, L.

Development of a Telephone Interview Method for Quantitative Diet History Assessment

The aim of this study is to design a method for assessing dietary intakes for population-based cancer case-control studies among persons living in widely dispersed geographical areas. The current diet history method, administered in face-to-face interviews, will be modified for administration by telephone, and the comparability of the two methods will be tested among the five major ethnic groups of Hawaii. The current method consists of obtaining frequencies and amounts of selected food items for estimating the usual monthly intakes of calories, nutrients, and other dietary components. Amounts are selected from coloured photographs showing three portion sizes based on the eating patterns of the population. This method will be modified to provide appropriate visual aids that can be mailed to subjects for subsequent use in telephone interviews. Three hundred persons representative of the major ethnic groups will be randomly selected to receive either telephone or face-to-face interviews initially and the alternate method six months later. Dietary intakes from the two interviews will be analysed to assess intra-individual agreement, based on the estimating equation technique. Differences in agreement by sex, age, and ethnicity will be examined.

TYPE: Methodology
TERM: Diet; Ethnic Group; Nutrition
SITE: Inapplicable
TIME: 1989 - 1994

1050 **Kolonel, L.N.** 00057
Univ. of Hawaii, Cancer Research Center, Epidemiology Program, 1236 Lauhala St., Suite 407A, Honolulu HI 96813, United States of America (Tel.: +1 808 5862986; Fax: 5862982)
COLL: Wilkens, L.; Hankin, J.H.; Le Marchand, L.; Goodman, M.T.

Lifestyle and Cancer Risk

This is a long-term follow-up study of more than 40,000 Hawaii residents of both sexes, representing a random sample of households throughout the State. The purpose is to test a variety of hypotheses relating lifestyle (diet, smoking, alcohol consumption, etc.) and other factors to cancer risk in Hawaii's multi-ethnic population. Information on demographic and other risk factors was collected by personal interview during the period 1975-1980. Follow-up is achieved largely by linkage of the cohort to the population-based Hawaii Tumor Registry, since out-migration rates from Hawaii are very low.

TYPE: Cohort
TERM: Alcohol; Diet; Ethnic Group; Lifestyle; Tobacco (Smoking)
SITE: All Sites
TIME: 1975 - 1998

***1051** **Le Marchand, L.** 05428
Univ. of Hawaii, Cancer Research Center, Epidemiology Program, 1236 Lauhala St., Suite 407, Honolulu HI 96813, United States of America (Tel.: +1 808 5862988; Fax: 5862982)
COLL: Sivaraman, L.; Hankin, J.H.; Lau, A.; Kolonel, L.N.; Zhao, L.P.; Cooney, R.V.

Dietary and Genetic Susceptibility Factors for Lung Cancer

Previous work in Hawaii has shown that, after adjusting for lifetime use of tobacco and other risk factors, including diet, the lung cancer risk of Hawaiian and Caucasian smokers are 121% and 46% greater than that of Japanese smokers (Cancer Epidemiol. Biomarkers Prev. 1:103-107, 1992). A population-based case-control study was initiated to investigate the roles of diet (antioxidants and fat) and of several genetic polymorphisms involved in carcinogen activation or detoxification (CYP2D6, CYP1A1, CYP2E1, GST1, NAT2, and H-ras) in these ethnic differences in lung cancer risk. 221 Hawaiians, 239 Japanese and 320 Caucasians with lung cancer, as well as 780 population controls matched on sex, age and race, will be interviewed. A quantitative dietary history assessing complete dietary intake and a detailed smoking and occupational history are administered by trained interviewers in the subjects' homes. DNA purified from leukocytes is analysed by RFLP and PCR techniques to characterize the geno-types of cases and controls.

TYPE: Case-Control; Molecular Epidemiology
TERM: Antioxidants; BMB; DNA; Diet; Ethnic Group; Fat; Genetic Markers; Lymphocytes; PCR; Plasma; RFLP; Registry; Serum
SITE: Lung
REGI: Hawaii (USA)
TIME: 1991 - 1996

***1052** **Zhao, L.P.** 05288
Univ. of Hawaii, Cancer Research Center of Hawaii, 1236 Lauhala St. , Honolulu HI 96813, United States of America (Tel.: +1 808 5862987; Fax: 5862982)

Statistical Methods for Genetic Epidemiology of Cancer

The objective of this study is (1) to develop three estimating equation-based methods for assessing familial aggregation of cancer in a population-based cohort of pedigrees. As part of this development, the usual sampling plans based on cases only, as well as on both cases and controls will be evaluated; (2) to assemble a population-based cohort of pedigrees by linking several population registries including the Birth, Marriage and Tumor Registry. The final cohort will comprise about 3 million individuals. Using the newly developed methods, this study will assess the familial aggregation of cancer at various sites and quantify their variability across the five major ethnic groups in Hawaii (Caucasian, Japanese, Hawaiian, Chinese and Filipino).

TYPE: Cohort; Genetic Epidemiology; Methodology
TERM: Cluster; Ethnic Group; Familial Factors; Mathematical Models; Registry
SITE: All Sites
REGI: Hawaii (USA)
TIME: 1992 - 1996

Houston

1053 **Anderson, D.E.** 04715

Univ. of Texas, M.D. Anderson Cancer Center, Sect. of Human Genetics, 1515 Holcombe Blvd, Houston TX 77030, United States of America (Tel.: +1 713 7922586; Fax: 7928149)

COLL: Morton, N.E.

Genetic Epidemiology of Breast Cancer

The aim is to determine whether familial aggregation in an unselected series of families is best explained by the transmission of a major gene, a transmissible multifactorial component acting alone or in concert with a major gene, or by an independent environmental contribution. Segregation analysis under a mixed model will be applied to three data sets: the families of 199 unselected female breast cancer patients, the families of 124 consecutive breast cancer patients with bilateral disease, and the families of 115 unselected male breast cancer patients. The use of three independent data sets will increase the power to resolve the question of the cause(s) underlying the familial aggregation of breast cancer, and whether the cause(s) are the same or different in the three familial groups.

TYPE: Case Series; Genetic Epidemiology
TERM: Familial Factors; Genetic Factors; Segregation Analysis
SITE: Breast (F); Breast (M)
TIME: 1986 - 1994

1054 **Anderson, D.E.** 04941

Univ. of Texas, M.D. Anderson Cancer Center, Sect. of Human Genetics, 1515 Holcombe Blvd, Houston TX 77030, United States of America (Tel.: +1 713 7922586; Fax: 7928149)

COLL: Ferrell, R.E.

Genetics of Breast Cancer

Families with inherited forms of breast cancer, namely, Li-Fraumeni syndrome, site-specific breast cancer, breast-ovarian cancer, Cowden's disease, and the Torre-Muir syndrome are being investigated in the study. Each of these is being evaluated for evidence of genetic linkage between the appropriate disease susceptibility allele and an array of polymorphic genetic markers in an attempt to localize the susceptibility allele to a specific chromosomal site. Initial effort has been directed to the breast-ovarian cancer and site-specific families. Twenty families comprise the breast-ovarian cancer series and 30 families the site-specific series. The data are being analysed by the lod score method using the computer program LIPED which incorporates options for dealing with age-dependent expression of clinical phenotype penetrance.

TYPE: Genetic Epidemiology
TERM: Chromosome Effects; Genetic Factors; High-Risk Groups
SITE: Breast (F); Ovary
TIME: 1982 - 1994

1055 **Black, H.S.** 05034

Baylor College of Medicine, Dept. of Dermatology, Baylor Plaza, Houston TX 77030, United States of America (Tel.: +1 713 7947637; Fax: 7947638)

COLL: Wolf, J.; Herd, A.J.; Goldberg, L.; Rosen, T.; Bruce, S.; Thornby, J.I.; Foreyt, J.P.; Tschen, J.

Skin Cancer Prophylaxis by Low Fat Dietary Intervention

The specific aim of this protocol is to determine, in a closely-controlled clinical setting, whether dietary modification of lipid intake can alter the course of skin cancer development. 700 patients will be randomized into two groups of 350 each. One will be a non-intervention (NI) group in which no change in dietary habits will be initiated and from which the control rate of carcinoma occurrence will be derived. The dietary intervention (DI) group will, after an intensive dietary modification training programme, adopt a diet characterized by reduced fat intake (20% of total caloric intake). Both groups are to be examined at four-monthly intervals, over 24 months, for carcinoma incidence. Based upon a 24-month 25% rate of carcinoma occurrence in the NI group, a 35% expected improvement in the DI group, and a 35% loss to follow-up, 70% of 20 computed simulations yielded significant differences between the NI and DI groups. This intervention design will not only address the question of whether dietary intervention may be an effective tool for prevention and/or management of a common malignancy induced by ultraviolet radiation, but will also provide direct and definitive evidence for involvement of dietary lipid in carcinogenic development.

TYPE: Intervention
TERM: Diet; Lipids; Nutrition
SITE: Skin
TIME: 1990 - 1994

***1056 Bondy, M.L.** 05304
Univ. of Texas, M.D. Anderson Cancer Center, Dept. of Epidemiology, 1515 Holcombe Blvd, Box 189, Houston TX 77030, United States of America (Tel.: +1 713 7923020; Fax: 7920807)
COLL: Levin, V.; Saya, H.; Steck, P.; Lustbader, E.; Strong, L.C.

Genetic Epidemiology of Brain Tumours

Specific genetic factors associated with brain tumours are largely unknown, but certain genetic syndromes have been identified as predisposing to brain tumours. Epidemiological studies and case reports reveal clustering in families and brain tumours are the third most frequent tumour type described in the Li-Fraumeni cancer family syndrome. The purpose of study is to determine the extent to which cancer aggregates in kindreds of glioma patients and to find the most likely aetiological model to explain the cancer aggregation. Family history of cancers and hereditary syndromes will be collected on approximately 200 glioma patients per year (for 3.5 years). SIRs will be applied to determine if there is excess cancer in the glioma families compared with that expected in the general population. Segregation analysis will be conducted to determine if cancer aggregates among relatives of glioma probands with no known hereditary syndromes and if this aggregation represents unrecognized familial syndromes. In addition, evidence for genetic heterogeneity in kindreds will be evaluated by specific proband characteristics, such as p53 tumour specific and/dor germline mutations, expression of NF1-GRD transcipts, alterations of a tumour suppressor gene on chromosome 10, glioma subtypes, second malignant neoplasms, and age at diagnosis. Results may determine whether there are certain clinical and epidemiological parameters that can discriminate between the hereditary and non hereditary cases and could be used to screen for hereditary cases in the future.

TYPE: Genetic Epidemiology
TERM: BMB; Blood; Family History; Genetic Factors; Segregation Analysis; Tissue
SITE: Brain
LOCA: Canada; United States of America
TIME: 1992 - 1995

1057 Jansson, B. 04135
Univ. of Texas, M.D. Anderson Cancer Center, Dept. of Biomathematics, 1515 Holcombe Blvd, Box 273, Houston TX 77030, United States of America (Tel.: +1 713 7923392; Fax: 7924262)
COLL: Becker, F.F.

Dietary, Total Body, and Intracellular Potassium/Sodium Ratio and its Influence on Cancer

The hypothesis that the ratio K/Na is negatively associated to cancer rates has been tested in a number of different ways: geopathological, dietary and clinical. It has also been found that certain agents or conditions which may reduce cancer risk increase the intracellular K/Na ratio, while other exposures or conditions which may increase cancer risk decrease the intracellular K/Na ratio. This study uses data from various sources, such as "Cancer Mortality by County in USA", "Third National Cancer Survey", published data, to test the hypothesis in two ways: (1) phenomena known to affect K/Na are studied regarding associations with cancer, and (2) phenomena known to affect cancer rates are studied regarding associations with the K/Na ratios. The hypothesis is now also being confirmed in mice by measuring the cancer incidence rates by DMBH (mammary cancer), for groups with high, normal or low dietary K/Na ratios. A paper was published in Cancer Detect. Prev. 14:563-565, 1990.

TYPE: Correlation
TERM: Diet; Fibre; Metals; Minerals; Vitamins
SITE: All Sites; Colon
CHEM: Potassium; Sodium Chloride
TIME: 1986 - 1994

1058 **McPherson, S.** 05203

Univ. of Texas Health Science Center, School of Public Health, 1200 Herman Pressler St., Houston TX 77030, United States of America (Tel.: +1 713 7924660; Fax: 7944876)
COLL: Winn, R.J.; Nichaman, M.Z.; Levin, B.

Dietary Risk Factors for Colorectal Polyps

Colorectal adenomatous polyps (APs) and hyperplastic polyps (HPs) are precursor lesions to colorectal carcinoma. This case-control study aims to investigate the association between dietary intake of calcium, total fat and fibre and the presence of APs and HPs. Individuals with an incident diagnosis of colorectal polyps (CPs) after sigmoidoscopy/colonscopy from gastric intestinal clinic sites in Houston, Texas, are the cases. Controls will be selected from those subjects who are negative for CPs after the same diagnostic procedures. Information on physical activity, family history of cancer, smoking history, use of non-steroidal anti-inflammatory drugs and diet are being collected. Univariate and multivariate analyses will be done to calculate odds ratios and to determine interaction and confounding by other variables.

TYPE: Case-Control
TERM: Diet; Drugs; Fat; Fibre; Nutrition; Physical Activity; Polyps; Premalignant Lesion; Tobacco (Smoking)
SITE: Colon; Rectum
CHEM: Calcium
TIME: 1991 - 1994

1059 **Vogel, V.G.** 04734

Univ. of Texas, M.D. Anderson Cancer Center, 1515 Holcombe Blvd, Box 501, Houston TX 77030, United States of America (Tel.: +1 713 7928515; Fax: 7969155)
COLL: Peters, G.N.; Winn, R.J.; Bondy, M.D.

Stage Distribution and Risk Factor Survey in Women Screened for Breast Cancer

The goal of the American Cancer Society 1987 Texas Breast Screening Project was to attract women for low-cost screening mammography. The aim of this study is to ascertain whether screening for breast cancer in community centres can yield the same proportion of early-stage breast cancer as that reported in controlled trials. The project yielded more than 64,000 examinations at 306 participating facilities. Findings were radiographically classified as negative, indeterminate (further evaluation required), or suspicious for malignancy. Among the project mammograms were 2,172 suspicious examinations. A study is being conducted to determine: (1) the stage at diagnosis and the clinico-pathological outcomes of the malignancies detected at screening; and (2) the ratio of benign to malignant biopsies. Follow-up data are available for 75 % of the screened women in whom 1,122 biopsies were done; 214 cancers have been documented to date. In addition, questionnaires were given to all screenees, and 32,000 have been coded and entered into a computer database. Data collected included demographic variables, family history of breast cancer, history of benign breast disease, health beliefs and prior health practices, and exposures such as cigarette smoking and oral contraceptive use. Names and addresses are available, and prospective studies of women with positive family histories are planned to improve rates of early detection of breast malignancy in this cohort. Papers have been published in Cancer 66:1613-1620, 1990 and Cancer Detect. & Prev. 14:573-576, 1990.

TYPE: Cohort
TERM: Familial Factors; Mammography; Oral Contraceptives; Screening; Tobacco (Smoking)
SITE: Benign Tumours; Breast (F)
TIME: 1986 - 1994

1060 **Wargovich, M.J.** 04866

Univ. of Texas, M.D. Anderson Cancer Center, 1515 Holcombe Blvd, Box 501, Houston TX 77030, United States of America (Tel.: +1 713 7922828; Fax: 7911536; Tlx: 9108811556)
COLL: Lynch, P.; Levin, B.; Winn, R.J.

Calcium Intervention in Subjects with Sporadic Colon Adenoma

The hypothesis under study is that calcium in daily doses of 1500-4000 mg modifies colonic epithelial proliferation in subjects with resected adenoma. Biomarkers of cellular proliferation in this study include 3H-TdR, bromo-deoxyuridine incorporation, and proliferating cell nuclear antigen. 105 subjects are randomized to placebo or one of three graded doses of calcium and maintained at their dosage level for three months. Colon biopsies will be assessed at baseline, on study, and two months post-study to determine the degree of calcium inhibition of epithelial proliferation. These data will be used to implement a secondary phase III chemoprevention trial in the same population with the additional endpoint of adenoma recurrence.

TYPE: Intervention
TERM: Biomarkers
SITE: Colon
CHEM: Calcium
TIME: 1990 - 1994

Irvine

1061 **Anton-Culver, H.** 04415
Univ. of California at Irvine, Dept. of Medicine, Epidemiology Program, Irvine Hall, Room 244, Irvine CA 92717-7550, United States of America (Tel.: +1 714 85607401; Fax: 85604773)

Smoking and other Risk Factors for Lung Cancer by Cell Type

Few studies have examined the importance of smoking and other risk factors for the different histological types of lung cancer. The histological distribution differs dramatically between men and women, with squamous cell carcinoma being the most common cell type in men, and adenocarcinoma predominating in women. The proposed case-control study will examine the importance of smoking, occupation, and family history of cancer for specific histological types of lung cancer in men and women using data from the population-based registry of the Cancer Surveillance Program of Orange County. Cases will include all cases of small cell carcinoma and adenocarcinoma of the lung diagnosed in Orange County during 1984-1986, a total of approximately 2,500 cases. Each histological type will be compared with all other cancers excluding those thought to be associated with smoking, occupation or family history. Data available include demographic information, industry/occupation, smoking history, family history of cancer, pathology and stage of disease, treatment and survival status. The relative odds associated with risk factors under study will be compared between different histological types of lung cancer and between men and women. A multiple logistic model will be used to evaluate the independent contributions to risk of environmental and host factors and to study possible interactions between smoking and other risk factors.

TYPE: Case-Control
TERM: Familial Factors; Histology; Occupation; Registry; Tobacco (Smoking)
SITE: Lung
REGI: Orange County (USA)
TIME: 1988 - 1994

***1062** **Anton-Culver, H.** 05518
Univ. of California at Irvine, Dept. of Medicine, Epidemiology Program, Irvine Hall, Room 244, Irvine CA 92717-7550, United States of America (Tel.: +1 714 85607401; Fax: 85604773)
COLL: Samoszuk, M.

Epstein-Barr Virus in AIDS-Related Lymphomas

There has been speculation that EBV may be responsible for malignant lymphoma which frequently develops in AIDS patients. The objective of this project is to determine if EBV is present in a significant proportion of AIDS-related lymphomas. A large number of lymphoma tissue specimens is obtained from AIDS patients and from non-AIDS control patients. The tissues are analysed for the presence of EBV DNA by two state-of-the-art sensitive molecular techniques, PCR and in-situ hybridization. To date, 12 cases of AIDS-related lymphomas and approximately 85 cases of non-AIDS-related lymphomas have been accessioned. The presence of Epstein-Barr has been identified in two of the cases of AIDS-related lymphomas which also contain an extensive infiltration with an unusual white blood cell called an eosinophil. This finding is the first report of eosinophilia associated with AIDS-related lymphomas. It is significant because it confirms a recent report that Epstein-Barr infection of B-lymphocytes can induce the production of a substance (interleukin-5) that promotes eosinophil growth and migration. The presence of eosinophils in AIDS-related lymphomas with EBV will provide an interesting and valuable new approach to the biological therapy of this otherwise incurable cancer.

TYPE: Case Series
TERM: AIDS; EBV; PCR; Registry
SITE: Lymphoma
REGI: Orange County (USA)
TIME: 1991 - 1994

***1063 Anton-Culver, H.** 05526
Univ. of California at Irvine, Dept. of Medicine, Epidemiology Program, Irvine Hall, Room 244, Irvine CA 92717-7550, United States of America (Tel.: +1 714 85607401; Fax: 85604773)
COLL: Casey, G.; Fain, P.R.; Manetta, A.; Liao, S.Y.; Barker, D.

Resource Development for Genetic Studies for Breast Cancer

Family resources and specific laboratory techniques for studying hereditary breast cancer will be developed. Breast cancer patients aged 20-29 will be ascertained through the Cancer Surveillance Program of Orange County database of cancer cases for 1984-1992. Previous studies indicate that 36% of these patients will have hereditary breast cancer. Family history information through first cousins will be obtained from each proband. Segregation analysis will be applied to these data to estimate the frequency of hereditary breast cancer, and the penetrance in gene carriers. Selected pedigrees will be extended as far as possible, in order to study gene linkage of breast cancer to loci and regions on chromosome 17 which have been implicated in breast cancer including p53, D17S74, D17S5 and D17S4. The laboratory techniques needed to combine the linkage studies with loss of heterozygosity studies in tumour tissue from affected family members will be developed. Many of these procedures have already been developed in one or both of the two collaborating laboratories; however, special issues must be addressed in order to apply the techniques to studies of hereditary breast cancer. It is expected to obtain family history data from about 80 cases and to identify at least 10 families for gene linkage studies. The family resource and laboratory techniques are critical for future studies of hereditary breast cancer which are currently being planned. Specifically, the family resource and laboratory techniques will be used in future studies in order to (1) map the location of inherited breast cancer predisposing genes; (2) characterize specific mutations; (3) study the interrelationships between genotype, environmental and clinical variables.

TYPE: Genetic Epidemiology; Methodology
TERM: BMB; Genotype/Phenotype; Heredity; Metabolism; Registry; Segregation Analysis
SITE: Breast (F)
REGI: Orange County (USA)
TIME: 1992 - 1994

Ithaca

1064 Turnbull, B. 05188
Cornell University, School of Operational Research & Industrial Engineering, 227 ETC, Ithaca NY 14853, United States of America (Tel.: +1 607 2559131; Fax: 2559129)

Statistical Methodology for Monitoring Clusters of Disease

The aim is to develop and evaluate statistical methodology to monitor the occurrence of clusters of chronic disease. Geographical incidence patterns are examined and the 'randomness' hyothesis is tested. For statistical tests with a general 'omnibus' alternative, a paper was published in Am. J. Epidemiol. 132:S136-S143, 1990, using leukaemia incidence data from the New York State Cancer Registry and population data from the US Census. Tests now being developed are designed to be sensitive to alternatives in which the incidence rates are elevated in the vicinity of (multiple) putative sources of hazard. Applications will be made with toxic dump site locations and leukaemia in New York State and with nuclear installations and childhood leukaemia in Sweden.

TYPE: Methodology
TERM: Cluster; Radiation, Ionizing; Waste Dumps
SITE: Leukaemia
LOCA: Sweden; United States of America
TIME: 1987 - 1994

La Jolla

1065 **Barrett-Connor, E.L.** 03913
Univ. of California at San Diego, Dept. Community & Family Medicine, 9500 Gilman Drive, La Jolla CA 92093-0628, United States of America (Tel.: +1 619 5343511; Fax: 5347517)
COLL: Garland, C.F.; Friedlander, N.

Lipids, Diabetes and Cancer

This study was designed to look at diabetes and heart disease prospectively in subjects attending a lipid research clinic. Extensive behavioural and dietary data have also allowed cancer to be examined. Data on mortality from cancer (from death certificates) for all years and incidence according to clinical records (for the ninth and tenth years only) have been collected. There are over 4,000 adult subjects in the geographically defined population, but not all participants have all data, e.g., 24-hour diet recall is available for approximately 1,000 subjects. Data are now available on sex hormones measured in plasma obtained from approximately 1,100 men and 550 women in 1972-1974.

TYPE: Cohort
TERM: Diabetes; Diet; Lipids; Prevalence
SITE: All Sites; Breast (F); Colon; Lung; Prostate
TIME: 1972 - 1994

1066 **Garland, C.F.** 03630
Univ. of California at San Diego, Dept. of Family and Preventive Medicine, 9500 Gilman Drive, #0631C, La Jolla CA 92093-0607, United States of America (Tel.: +1 619 5340520; Fax: 5580797)
COLL: Shekelle, R.B.; Barrett-Connor, E.L.; Criqui, M.H.; Rossof, A.H.; Paul, O.

Colo-Rectal Cancer Risk in a Cohort of Men

1,954 men employed by the Western Electric Company and enrolled in the Western Electric Heart Study in 1957-1958 are being assessed for colo-rectal cancer risk. The men were aged 40-55 years at entry. They completed dietary histories (2-8 day) at entry and one year later, and 99.9% of the men have been followed until the twentieth anniversary of entry. Incidence of colo-rectal cancer will be ascertained in the population, and a prospective design will be used to assess risk according to nutritional characteristics at entry.

TYPE: Cohort
TERM: Diet; Vitamins
SITE: Colon; Rectum
TIME: 1983 - 1994

***1067** **Garland, C.F.** 05510
Univ. of California at San Diego, Dept. of Family and Preventive Medicine, 9500 Gilman Drive, #0631C, La Jolla CA 92093-0607, United States of America (Tel.: +1 619 5340520; Fax: 5580797)
COLL: Barrett-Connor, E.L.; Garland, F.C.; Gorham, E.D.; Davies, R.J.

Diet, Vitamin D, and Cancer of the Female Breast

This is a prospective study of incidence of breast cancer in 2,424 women according to baseline anthrometric characteristics, diet and sex hormones. This study will assess the role of dietary vitamin D in reducing the risk of breast cancer. Dietary sources include a 24-hour recall and a diet assessment (food frequency questionnaire).

TYPE: Cohort
TERM: Diet; Hormones; Registry; Vitamins
SITE: Breast (F)
REGI: San Diego (USA)
TIME: 1991 - 1995

Lansing

1068 **Johnson, B.C.** 03381

Michigan Dept. of Public Health, Environmental Epidemiology Div., Center for Environmental Health Science, 3500 N. Logan St., Lansing MI 48909, United States of America (Tel.: +1 517 3358350)

Case-Control Study of Soft Tissue Sarcoma

The study will assess associations between soft-tissue sarcoma cases and environmental variables. Case identification and confirmation methods will be developed to both prevent mis-classification and yield complete case finding. Although the case -control study only depends upon unbiased case finding and not complete case ascertainment, complete case finding will be accomplished to assess incidence rates within the geographical area. Cases of soft-tissue sarcoma, occurring over the past years within an eight county geographical area of eastern-central Michigan, will be identified by pathology report review. The diagnosis will be confirmed by a pathological panel review of histological slides. Controls will be selected both from hospital discharges and the resident population. Detailed occupational, personal habit and place of residence data will be obtained. Within this geographical area resides a group of workers who synthesized 2,4,5-T. Environmental exposure data will be utilized in the study analysis.

TYPE: Case-Control
TERM: Chemical Exposure; Environmental Factors; Herbicides; Histology; Lifestyle; Occupation; Pesticides
SITE: Soft Tissue
CHEM: 2,4,5-T; Phenoxy Acids
TIME: 1984 - 1994

Livermore

1069 **Moore II, D.H.** 02630

Lawrence Livermore National Lab., Biomedical Sciences Div. L-452, 700 E. Ave., P.O. Box 5507, Livermore CA 94550, United States of America (Tel.: +1 415 4225631; Fax: 4222282 ; Tlx: 9103868339)
COLL: Schneider, J.S.

Melanoma among Employees at the Lawrence Livermore National Laboratory

Causes for a two- to three-fold increase in melanoma incidence among the 10,000 employees of the Lawrence Livermore National Laboratory (LLNL), a high energy research facility located in a sunny valley in northern California, are being investigated. The study includes (1) a study of mortality among the approximately 20,000 former employees to determine whether melanoma mortality is also high; (2) interviews of all cases and a matched control for each case to determine work and sun exposure history. Papers were published in Arch. Dermatol. 126:767-769, 1990, Cancer Causes and Control 13:191-197, 1992 and in Epidemiology 4:43-47, 1993.

TYPE: Case-Control; Cohort
TERM: Chemical Exposure; Occupation; Radiation, Ionizing; Radiation, Ultraviolet
SITE: Melanoma
OCCU: Chemists; Laboratory Workers; Physicists
REGI: Bay Area (USA)
TIME: 1981 - 1994

Loma Linda

1070 **Fraser, G.E.** 04344

Loma Linda Univ., Center for Health Research, Nichol Hall, Loma Linda CA 92350, United States of America (Tel.: +1 909 8244546; Fax: 8244087)
COLL: Beeson, W.L.; Lindsted, K.D.; Sabaté, J.; Knutsen, S.

Adventist Health Study

The purpose of the study is to test the effect of consumption of foods on risk of malignant tumours. Special emphasis has been put on fruits, vegetables and meat. This cohort study enrolled 34,198

Californian Seventh-Day Adventists in 1974-1976. A follow-up study has been carried out for six years since 1982. Information on exposure has been gathered by mailed questionnaire. All hospitalisations of study subjects have been ascertained by annual mail contact. Study field representatives have visited relevant hospital and screened charts for new cancer diagnoses. The Center has collaborated with two California tumour registries to ascertain cases. Cases have been matched with California State Death Tapes. A number of papers on methods or cancer results have been published.

TYPE: Cohort
TERM: Diet; Fruit; Registry; Religion; Vegetables
SITE: All Sites
REGI: Bay Area (USA); Los Angeles (USA)
TIME: 1974 - 1995

Los Alamos

1071 **Wiggs, L.D.** 00347
Los Alamos National Lab., Epidemiology Sect., Health, Safety and Environment Div., P.O.Box 1663, Los Alamos NM 87545, United States of America
COLL: Voelz, G.L.; Galke, W.A.

Health Study of Plutonium Workers

This epidemiological programme investigates the risk of adverse human health effects, especially cancer risks, associated with internal desposition of plutonium and associated external radiation doses. Retrospective cohort mortality analyses are being conducted at the Rocky Flats, Los Alamos, and Mound facilities by comparing mortality rates of exposed versus unexposed persons at the same facility and by determining standardised mortality ratios based on US death rates. These cohorts include over 50,000 workers. Nested case-control studies are used to investigate significant mortality excesses. Upon completion of mortality studies for these cohorts, a pooled analysis of data from all facilities will be used for more statistical power.

TYPE: Case-Control; Cohort
TERM: Dose-Response; Metals; Occupation; Radiation, Ionizing
SITE: All Sites
CHEM: Plutonium
OCCU: Plutonium Workers
TIME: 1975 - 1994

Los Angeles

1072 **Bernstein, L.** 04563
Univ. of Southern California, School of Medicine, Dept. of Preventive Medicine, 1420 San Pablo St., PMB A-202, Los Angeles CA 90033-9987, United States of America (Tel.: +1 213 3421270; Fax: 3421237)
COLL: Ross, R.K.; Henderson, B.E.

Breast Cancer in Young Women

Although there is no strong evidence that oral contraceptives (OC) have an effect on breast cancer risk when used during the middle reproductive years, women who use OCs for long periods of time in the post-menarcheal period may be at increased risk of breast cancer. Previous studies that have examined this relationship disagree as to whether there is a significantly increased risk with early long-term OC use or no risk. The specific aims of this study of breast cancer in young women are (1) to examine the hypothesis that early long-term OC use is associated with an increased risk of breast cancer and to determine whether any increased risk is modified by reproductive factors and other breast cancer risk factors, the particular formulation of OC used or temporal factors such as calendar period of use, age at first use, time since first use and duration of use; (2) to establish the reliability of retrospective recall of OC use by comparing interview histories with medical records; and (3) to investigate the relationship between breast cancer and diagnostic and therapeutic radiation, abortion, regularity of menstrual cycle, physical activity, alcohol and caffeine consumption and smoking habits. A matched case-control design is being used. Cases are white, English-speaking women aged 40 and under, diagnosed with breast cancer between January 1985 and December 1990, and identified by the population-based tumour registry. Controls are individually matched to cases by age, race, parity (ever vs. never had a full-term pregnancy) and neighbourhood of residence. Structured personal interviews are being conducted. To facilitate recall, a photograph album of all OCs sold in the United States and a comprehensive calendar to record life events is used during the interview. 720 case-control pairs will be interviewed.

TYPE: Case-Control
TERM: Abortion; Alcohol; Menstruation; Oral Contraceptives; Physical Activity; Radiation, Ionizing; Radiotherapy; Registry; Reproductive Factors; Time Factors; Tobacco (Smoking)
SITE: Breast (F)
CHEM: Caffeine
REGI: Los Angeles (USA)
TIME: 1987 - 1994

1073 **Bernstein, L.** 04768
Univ. of Southern California, School of Medicine, Dept. of Preventive Medicine, 1420 San Pablo St., PMB A-202, Los Angeles CA 90033-9987, United States of America (Tel.: +1 213 3421270; Fax: 3421237)
COLL: Levine, A.M.

Epidemiology of HIV-Related Lymphoma

This study is designed to define the aetiology of high-grade lymphoma in the setting of HIV-infection by looking at the specific hypothesis that EBV-induced chronic B-lymphocyte stimulation, in the setting of altered immune function leads to unregulated B-cell proliferation, specific chromosomal translocation, c-myc activation, and subsequent development of monoclonal, B-cell lymphoma. In this population-based case-control study, cases with HIV-positive high-grade lymphoma (n = 200) will be compared with controls who are HIV-positive individuals without lymphoma and to HIV-negative lymphoma patients. Population-based controls will be matched to this latter group of patients as well. In addition to interviews, all patients will have HIV and EBV antibody testing and HLA typing. A subgroup of patients (cases and controls) will be studied at the tissue level to determine (1) if EBV genome is present and its extent; (2) if HIV provirus is present and expressed within lymphoma tissues (PCR and in situ hybridization; (3) the molecular genotype of HIV+ and HIV-negative lymphomas by cytogenetic and molecular genetic analyses and (4) if monoclonality of lymphoma tissue will differentiate HIV+ lymphoma cases from HIV+ individuals without lymphoma (kappa/lambda immunophenotype and immunoglobulin gene rearrangement). A serum and tissue bank will be established for use in future studies.

TYPE: Case-Control
TERM: BMB; EBV; HIV; HLA; PCR; Registry; Serum; Tissue
SITE: Lymphoma
REGI: Los Angeles (USA)
TIME: 1989 - 1994

***1074** **Bernstein, L.** 05503
Univ. of Southern California, School of Medicine, Dept. of Preventive Medicine, 1420 San Pablo St., PMB A-202, Los Angeles CA 90033-9987, United States of America (Tel.: +1 213 3421270; Fax: 3421237)
COLL: Wu-Williams, A.; Ross, R.K.

Aetiology of Gastric Cardia/Oesophageal Adenocarcinoma

Overall gastric cancer incidence has declined in the US during the past 50 years, but the incidence of adenocarcinoma of the gastric cardia and lower oesophagus has recently substantially increased. Compared with gastric and oesophageal cancer overall, adenocarcinoma of the gastric cardia and lower oesophagus is characterized by a high male to female ratio and is relatively more common in whites. Experimental and clinical evidence suggest that duodenal gastric or oesophageal reflux may play an aetiological role in proximal gastric carcinogenesis. Histamine2 (H2)-receptor antagonists (H2-blockers) were introduced in the mid-1970s and have become the treatment of choice for peptic ulcer disease and gastro-oesophageal reflux disorders. H2-blockers are potentially involved in gastric carcinogenesis. These drugs alter gastric acid production, increasing gastric pH and allowing reflux of alkaline duodenal refluxate into the proximal stomach and lower oesophagus. Furthermore these drugs can be nitrosated to form mutagenic N-nitroso compounds. A population-based case-control study is proposed to evaluate the role of this category of drugs (including possible effects of age at first use, dose, duration of use, latency and condition for which H2-blockers were prescribed) as well as other possible risk factors (e.g. use of other categories of drugs such as non-steroidal anti-inflammatory drugs, including aspirin, prior benign conditions of the stomach and oesophagus, smoking and alcohol use, and dietary factors in the aetiology of adenocarcinoma of the gastric cardia/lower oesophagus. Consecutively diagnosed cases will be identified by the population-based cancer registry in Los Angeles County over 4.5 years. Controls will be neighbourhood controls matched 2:1 with cases for age, race, and sex. A total of 440 case-control triplets will be interviewed during the course of this study.

TYPE: Case-Control
TERM: Age; Alcohol; Diet; Drugs; Latency; Registry; Time Factors; Tobacco (Smoking)
SITE: Oesophagus; Stomach
CHEM: Aspirin
REGI: Los Angeles (USA)
TIME: 1993 - 1998

1075 Enstrom, J.E. 00527
Univ. of California, School of Public Health, 405 Hilgard Ave., Los Angeles CA 90024-1772, United States of America (Tel.: +1 310 8252048)

Epidemiology of Cancer among Mormons

The SMR for total cancer in 350,000 California Mormons during 1960-1987 and 750,000 Utah Mormons during 1970 and 1975 compared with US whites is about 50% for "active" males, about 65% for all males, and about 75% for all females. For active males the SMR is about 25% for smoking-related cancer sites and about 65% for all other sites. Mormons are a health-conscious religious group, who appear to be similar to US whites in many respects, such as diet, socio-economic status, and urbanization, but who use about one half as much tobacco, alcohol, coffee, and tea as the general population. "Active" Mormons, defined to be church leaders known as High Priests, abstain almost entirely from tobacco, alcohol, and caffeine. In late 1979 a questionnaire survey and prospective mortality follow-up of 10,000 "active" California Mormon adults, specifically 5,400 High Priests and 4,600 wives was initiated. The survey obtained detailed demographic, lifestyle, dietary, occupational, and medical history data. Church-based and California State mortality data have been collected on the 10,000 cohort members and on all other California Mormons during the years 1980-1987. The data on lifestyle and health-related characteristics of "active" California Mormons will be compared with their detailed cancer mortality rates. Distinctive differences between these Mormons and other comparison groups will be highlighted. Using variation in key health practices (physical activity and sleep), a subgroup of middle-aged high priests has been identified with an SMR of 22% for all causes and 34% for all cancers. These analyses have provided significant understanding as to why Mormons have low cancer mortality rates. A paper has been published in J. Nat. Cancer Inst. 81:1807-1814, 1989.

TYPE: Cohort
TERM: Alcohol; Diet; Lifestyle; Occupation; Religion; Socio-Economic Factors; Tobacco (Smoking)
SITE: All Sites
TIME: 1973 - 1994

1076 Enstrom, J.E. 03009
Univ. of California, School of Public Health, 405 Hilgard Ave., Los Angeles CA 90024-1772, United States of America (Tel.: +1 310 8252048)

Cancer among Low-Risk Populations

This is a prospective epidemiological investigation of several populations that are at low risk of cancer relative to the general US population. The populations include: California physicians, whose cancer mortality trends are being examined with respect to their substantial degree of smoking cessation since 1950; Alameda County (California) residents and NHANES I Epidemiologic Follow-up Study (NHEFS) cohort members, whose cancer rates are being examined with respect to several good health practices; and various cohorts of health conscious Californians, whose cancer rates are being examined with respect to vitamin supplement usage and other lifestyle variables. Papers appeared in Am. J. Public Health 76:1124-1130, 1986 and Epidemiology 3:194-202, 1992.

TYPE: Cohort
TERM: Lifestyle; Occupation; Vitamins
SITE: All Sites
OCCU: Health Care Workers
TIME: 1976 - 1994

***1077** **Enstrom, J.E.** 05541

Univ. of California, School of Public Health, 405 Hilgard Ave., Los Angeles CA 90024-1772, United States of America (Tel.: +1 310 8252048)

COLL: Heath, C.W.; Garfinkel, L.

Tobacco-Related Disease Trends Among 118,000 Californians

This is a historical prospective epidemiological study of 118,000 California subjects (51,000 males and 67,000 females) enrolled in late 1959 in the original American Cancer Society (ACS) Cancer Prevention Study (CPS I). The study is being done collaboratively with ACS. Smoking histories and other lifestyle data were collected in 1959, 1961, 1963, 1965 and 1972, and mortality follow-up has been completed through 1972. The smoking history data indicate that about half of those who had smoked in 1959 had quit as of 1972, with substantial cessation among both men and women. The first objective is to extend follow-up of this cohort through 1991 using several passive follow-up techniques that have been shown to be highly effective among a sample of these CPS I subjects. The second objective is to reinterview a sample of 1,000 survivors for current smoking and lifestyle characteristics. The third objective is to measure the long-term preventive impact of smoking cessation on tobacco related disease mortality among various categories of CPS I subjects who smoked in 1959. Comparison will be made with CPS I subjects who never smoked and the general population.

TYPE: Cohort
TERM: Lifestyle; Prevention; Tobacco (Smoking)
SITE: All Sites
TIME: 1991 - 1995

1078 **Haile, R.W.C.** 03639

UCLA School of Public Health, Div. of Epidemiology, 405 Hilgard Ave. , Los Angeles CA 90024, United States of America (Tel.: +1 213 8258193)

COLL: Sparkes, R.S.; Paganini-Hill, A.; Thomas, D.C.; Thompson, D.W.; Siemiatycki, J.; Gatti, R.A.; Greenland, S.

Genetic-Epidemiological Study of Bilateral Breast Cancer

The objective is to identify a gene or genes involved in the aetiology of breast cancer and to investigate gene-environment interactions. Data on family history and exposure to risk factors for breast cancer will be available for members of 442 families where the index case has bilateral breast cancer diagnosed before 50 years of age. Blood samples will be available from 68 multiple-case families potentially informative for linkage. The following analyses have been completed: case-control analyses of environmental risk factors, familial risk calculations, complex segregation analyses, and linkage analyses of 28 phenotypic genetic markers. The linkage work will be extended by using selected RFLPs and will focus on segments of chromosomes 1, 11 and 13, based on preliminary suggestions of linkage from the data and results published in the literature. A sequential approach will be taken to decide which RFLPs to investigate with the full data set of 68 families. Candidate RFLPs will be first tested on 10 of the more informative families, determined by a new computer program that estimates power while allowing for reduced penetrance and age of onset corrections. Those RFLP's that yield lod scores suggestive of linkage will be considered for further analyses with more families. Linkage analyses will be conducted using LIPED and an affected-pedigree-member method that analyses all pairs of affected relatives. In addition, a regression method of linkage analysis will be used, as well as further case-control and proportional hazards analyses.

TYPE: Case Series; Genetic Epidemiology
TERM: BMB; Blood; DNA; Environmental Factors; Familial Factors; Genetic Factors; Genetic Markers; Linkage Analysis; RFLP; Segregation Analysis
SITE: Breast (F)
LOCA: Canada; United States of America
TIME: 1984 - 1994

1079 **London, S.J.** 05072

Univ. of Southern California, School of Medicine, Dept. of Preventive Medicine, Parkview Medical Bldg, B-306, 1420 San Pablo St., Los Angeles CA 90033, United States of America (Tel.: +1 213 3421092; Fax: 3423272)

COLL: Idle, J.; Adams, J.

Genetic Markers of Lung Cancer

The aims of this case-control study are (1) to examine the hypothesis that the risk of lung cancer is related to genetic polymorphism of CYP2D6, CYP1A1, CYP2E1, CYP2A6, and glutathione transferase 1; and (2) to examine whether these genetic polymorphisms differ between African Americans and Caucasians and contribute to differences in lung cancer rates that do not appear to be explained by smoking patterns alone. Cases are newly diagnosed (within 7 months of enrollment) lung cancer cases in Los Angeles County. Controls are randomly selected (frequency-matched for age and race) from driver's license lists (under age 65) and lists of Medicare beneficiaries (over age 65). It is hoped to enrol 320 cases (160 from each ethnic group) and 640 controls (320 from each ethnic group). Blood and urine samples are taken from cases and controls and each subject also completes a lung cancer risk factor and semiquantitative food frequency questionnaire. CYP2D6 eligible subjects are also phenotyped. Analysis will use standard case-control methods.

TYPE: Case-Control; Genetic Epidemiology
TERM: BMB; Blood; Diet; Ethnic Group; Genetic Factors; Genetic Markers; Urine
SITE: Lung
TIME: 1990 - 1995

1080 **Mack, T.M.** 04081
Univ. of Southern California, School of Medicine, Dept. of Preventive Medicine, 2025 Zonal Ave., PMB-B105, Los Angeles CA 90033, United States of America (Tel.: +1 213 2247255; Tlx: 9103212434)
COLL: Deapen, D.

Case-Control Studies of Cancer in Twins

Twins who have had cancer diagnoses are ascertained by newspaper advertising. Diagnoses are validated by review of pathology reports and histological sections. Mail questionnaires are distributed to affected and unaffected twins. Returned questionnaires are edited, coded and analysed using standard univariate and multivariate methods. Hypotheses under test depend upon site. In all, it is expected that completed questionnaires will be received from more than 5,000 twin pairs. Studies in tandem with these are being carried out in twins in Denmark, Finland and Sweden.

TYPE: Case-Control
TERM: Twins
SITE: All Sites
LOCA: Canada; United States of America
TIME: 1982 - 1994

1081 **Menck, H.R.** 04097
Univ. of Southern California, Cancer Surveillance Program, 1721 Griffin Ave. , Los Angeles CA 90031, United States of America (Tel.: +1 213 2247641)
COLL: Mack, T.M.; McLaughlin, J.K.

Case-Control Study of Gallbladder Cancer

The objective is to perform a case-control study of gallbladder cancer in Los Angeles County Hispanics, to evaluate the role of reproductive history, oestrogen activity, Hispanic diet, Amerindian inheritance, past gastrointestinal and metabolic disease, and environmental exposures. An attempt will be made to distinguish between the causes of gallbladder cancer and the causes of cholelithiasis. The cases will consist of all female cases of gallbladder cancer occurring in Los Angeles County in five years (n = 200); controls will consist of matched persons chosen from the neighbourhood of the case and matched cases undergoing cholecystectomy at the hospital in which the cancer case was diagnosed. A personal interview will elicit information about reproductive history, medical history and drug use, Hispanic diet, and environmental exposures.

TYPE: Case-Control
TERM: Diet; Drugs; Environmental Factors; Ethnic Group; Female; Hormones; Reproductive Factors
SITE: Gallbladder
TIME: 1984 - 1994

1082 **Paganini-Hill, A.** 04027

Univ. of Southern California, School of Medicine, Dept. of Preventive Medicine, 1721 Griffin Ave., #103, Los Angeles CA 90031, United States of America (Tel.: +1 213 2247941)
COLL: Henderson, V.

Role of Oestrogen and Vitamin A in Disease Prevention

This cohort study is designed to measure the risks and benefits of menopausal oestrogen replacement therapy in terms of incident disease and mortality. Detailed health surveys have been collected on 13,986 residents of Leisure World, Laguna Hills, CA. Follow-up of this cohortis maintained by abstraction of hospital discharge diagnoses and cancer pathology reports (from three local hospitals) and death certificates. Follow-up questionnaires were also sent to all living cohort members in 1983, 1985 and 1992. Statistical analysis will include the use of Poisson regression models and the proportional hazards model of Cox. These multivariate methods will be used to sort out possible confounding and modifying effects. Recent papers appeared in Arch. Intern. Med. 151:75-78, 1991 and in Epidemiology 2:16-25, 1991.

TYPE: Cohort
TERM: Age; Hormones
SITE: All Sites; Breast (F); Uterus (Corpus)
CHEM: Oestrogens
TIME: 1981 - 1994

1083 **Paganini-Hill, A.** 04562

Univ. of Southern California, School of Medicine, Dept. of Preventive Medicine, 1721 Griffin Ave., #103, Los Angeles CA 90031, United States of America (Tel.: +1 213 2247941)
COLL: Ross, R.K.

Hormone Replacement Therapy and Breast Cancer

This case-control study is designed to determine the effect on breast cancer risk both of cyclic oestrogen-progestogen hormonal replacement therapy and of unopposed oestrogen replacement therapy. Cases are white English-speaking women aged 55-64 years at diagnosis of breast cancer and identified in the population-based tumour registry between 1 March 1987 and 1 March 1990 and white or black women aged 55 years or older at diagnosis of breast cancer, born in 1923 or later and identifed between 1 January 1992 and 1 January 1995. Controls will be individually matched to cases by age, race and neighbourhood of residence. A structured personal interview will be conducted with validation of hormone therapy by review of physician records. A total of 3,000 case-control pairs will be analysed, which will allow for the evaluation of the effects of hormone replacement therapy in the presence of possible confounding variables such as age at and type of menopause.

TYPE: Case-Control
TERM: Hormones; Registry
SITE: Breast (F)
CHEM: Oestrogens; Progestogens
REGI: Los Angeles (USA)
TIME: 1987 - 1995

1084 **Preston-Martin, S.** 02914

Univ. of Southern California, School of Medicine, Dept. of Preventive Medicine, Parkview Medical Bldg B301, 1420 San Pablo St., Los Angeles CA 90033, United States of America (Tel.: +1 213 3421310; Fax: 3421237)
COLL: Sarkar, S.

Case-Control Study of Cancers of the Nose, Sinuses and Nasopharynx

The aims of this study are to (1) investigate the association between cancers of the nose, sinuses, and nasopharynx and occupational exposure to dust and fumes and to identify high risk occupations and industries; (2) investigate the relationship of these cancers to tobacco use, to a history of upper respiratory infection or irritation, and to radiation exposure. The responses of cases and controls to a questionnaire will be compared in order to define factors associated with the development of these diseases. Cases will be identified by the Los Angeles County, University of Southern California Cancer Surveillance Program and will include non-orientals aged 25-69, diagnosed with nose, sinus or nasopharyngeal carcinoma between 1979 and 1985. It is estimated that it will be impossible to interview 20% of the eligible cases because of deaths and refusals. A person of the same sex, race and birth year (within five years) will be identified for each case from among those living in the neighbourhood where the case lived at the time of his diagnosis. Telephone interviews with cases and controls will be conducted. The questionnaire elicits a detailed occupational history as well as information on tobacco use and radiation exposure. Additional questions elicit a history of infectious, allergic, and other conditions which caused irritation or anatomical alteration of the nasal passages. Standard matched pair methods will be used in the statistical analysis.

TYPE: Case-Control
TERM: Chemical Exposure; Dusts; Environmental Factors; Infection; Occupation; Radiation, Ionizing; Tobacco (Smoking)
SITE: Nasal Cavity; Nasopharynx
TIME: 1982 - 1994

1085 **Preston-Martin, S.** 03104
Univ. of Southern California, School of Medicine, Dept. of Preventive Medicine, Parkview Medical Bldg B301, 1420 San Pablo St., Los Angeles CA 90033, United States of America (Tel.: +1 213 3421310; Fax: 3421237)
COLL: Thomas, D.C.; Sarkar, S.

Lip Cancer in Women

The aims of this case-control study are to investigate: (1) whether use of lip coverings protects against cancer of the lip; (2) the relationship of lip cancer to sun exposure; and (3) the association of lip cancer to other suggested risk factors such as use of tobacco. 80 female cases of lip cancer under age 75 diagnosed in Los Angeles County from 1972-1985 will be included. Controls will be selected by random digit dialling. Cases and controls will be interviewed by telephone, using a standard questionnaire which obtains a lifetime history of outdoor jobs, recreational exposure to sunlight, use of various types of lip coverings, and use of tobacco products. The analysis will use unmatched case-control comparisons.

TYPE: Case-Control
TERM: Cosmetics; Female; Radiation, Ultraviolet; Tobacco (Smoking)
SITE: Lip
TIME: 1986 - 1994

1086 **Preston-Martin, S.** 04069
Univ. of Southern California, School of Medicine, Dept. of Preventive Medicine, Parkview Medical Bldg B301, 1420 San Pablo St., Los Angeles CA 90033, United States of America (Tel.: +1 213 3421310; Fax: 3421237)
COLL: Thomas, D.C.

X-Rays and Drugs in the Aetiology of Acute Myelogenous Leukaemia

This study aims to investigate whether adult onset of acute myelogenous leukaemia is related to (1) diagnostic X-rays in the ten years prior to diagnosis, (2) use of phenylbutazone and other bone marrow depressant drugs, or (3) other suggested leukaemia risk factors such as occupational exposure to chemicals or electromagnetic fields. Information from interviews and from medical charts will be collected on 400 leukaemia cases and 400 matched neighbourhood controls.

TYPE: Case-Control
TERM: Chemical Exposure; Drugs; Electromagnetic Fields; Immunosuppression; Radiation, Ionizing
SITE: Leukaemia (AML)
CHEM: Phenylbutazone
TIME: 1987 - 1996

1087 **Preston-Martin, S.** 04565
Univ. of Southern California, School of Medicine, Dept. of Preventive Medicine, Parkview Medical Bldg B301, 1420 San Pablo St., Los Angeles CA 90033, United States of America (Tel.: +1 213 3421310; Fax: 3421237)
COLL: Mack, W.S.; Daling, J.R.; Holly, E.A.

Childhood Brain Tumours and N-Nitroso Exposures

Childhood brain tumours (CBT) are the most common solid tumour in children aged 0-19 years. Epidemiological studies to date suggest that suspected risk factors such as ionising radiation, head trauma and known genetic predispositions account for only a small proportion of incident cases. A compelling experimental model suggests that N-nitroso compounds (NOC) may cause brain tumours in humans, particularly when exposure is transplacental. Since the tumours that are induced by transplacental NOC exposure of non-human primates occur in young adults as well as in immature animals it seems reasonable to include all cases diagnosed to age 20. It is proposed to study all cases of primary CBT diagnosed from 1984-1990 among children aged 0-19 living in all areas of the West Coast that are covered by population-based tumour registries; this includes 19 counties in the Los Angeles (LA), San Francisco (SF) and Seattle areas. In both SF and Seattle the biological mothers of 105 children with CBT diagnosed during this 7 year period will be interviewed; in LA 300 case mothers will be interviewed. In each region random-digit-dial controls, stratified by age and sex, will be selected with a control: case ratio of 2:1 for SF and Seattle and 1:1 in LA. It is expected to complete interviews with mothers of 510 CBT cases, 720 controls and 1,000 fathers. The questionnaire asks about sources of NOC exposure such as diet, drinking water, drugs, cosmetics, rubber products and parental occupational exposures. Pathology slides of all cases will be reviewed by a single paediatric neuropathologist. Data for each of the three West Coast regions will be analysed separately and will also be pooled. The combined West Coast regions will have sufficient statistical power to conduct separate analyses for the two most common histological types of CBT - astrocytoma and medulloblastoma. Data will also be comparable with data from a concurrent international study.

TYPE: Case-Control
TERM: Childhood; Cosmetics; Diet; Drugs; Histology; Intra-Uterine Exposure; Occupation; Registry; Rubber; Water
SITE: Brain
CHEM: N-Nitroso Compounds
REGI: Bay Area (USA); Los Angeles (USA); Seattle (USA)
TIME: 1988 - 1994

1088 **Preston-Martin, S.** 04801
Univ. of Southern California, School of Medicine, Dept. of Preventive Medicine, Parkview Medical Bldg B301, 1420 San Pablo St., Los Angeles CA 90033, United States of America (Tel.: +1 213 3421310; Fax: 3421237)
COLL: Navidi, W.; Bowman, J.D.; Thomas, D.C.; Kaune, W.

Brain Tumours in Children and Exposure to Magnetic Fields

The objective of this study is to test epidemiologically whether magnetic field exposure is a risk factor for brain tumour in children. Subjects are taken from an on-going study in Los Angeles County. Cases include all children under age 20 with a primary tumour of the brain, cranial nerves or cranial meninges diagnosed from 1984-1990. Controls will be children of similar ages selected by random digit dialling. Subjects will total 300 cases and 300 controls. Wiring configurations and exterior magnetic fields will be measured at all subject residences within Los Angeles County from birth until diagnosis; a total of 1,200 such residences is expected. Interior exposure will be modeled from exterior measurements, and validated from interior measurements taken in a subset of 150 residences.

TYPE: Case-Control
TERM: Childhood; Electromagnetic Fields; Registry
SITE: Brain
REGI: Los Angeles (USA)
TIME: 1990 - 1994

1089 **Preston-Martin, S.** 05083
Univ. of Southern California, School of Medicine, Dept. of Preventive Medicine, Parkview Medical Bldg B301, 1420 San Pablo St., Los Angeles CA 90033, United States of America (Tel.: +1 213 3421310; Fax: 3421237)
COLL: Blot, W.J.; Sarkar, S.

Follow-Up Study of Oral and Pharyngeal Cancer

The aims of this study are to examine the influence of a variety of possible risk factors such as tobacco and alcohol use and diet on the development of a second cancer in persons with a cancer of the oral cavity or pharynx (ICD-O codes 141, 143-146, 148-149). Of the 595 cases of oral or pharyngeal cancer identified in 1984-1985, 41 cases subsequently developed a second cancer up to June 30, 1989. Three matched controls have been identified for each case. A control is defined as a person who had the initial oral or pharyngeal cancer but had not developed a second cancer by the time the case developed the second cancer. The cases and controls will be interviewed with regard to their use of tobacco and alcohol and diet since the development of their oral or pharyngeal cancer.

TYPE: Case-Control
TERM: Alcohol; Diet; Multiple Primary; Registry; Tobacco (Smoking)
SITE: Oral Cavity; Pharynx
REGI: Los Angeles (USA)
TIME: 1990 - 1994

1090 **Sarkar, S.** 05087
Univ. of Southern California, School of Medicine, Dept. of Preventive Medicine, Parkview Medical Bldg, B301, 1420 San Pablo St., Los Angeles CA 90033, United States of America (Tel.: +1 213 3421663; Fax: 3421237)
COLL: Preston-Martin, S.; Blot, W.J.

Feasibility Study of Human Papilloma Virus and Oral and Pharyngeal Cancer

The objective of the study is to determine whether HPV is associated with oral and pharyngeal cancers. At present, the oral and pharyngeal biopsy tissues of 20 cases of oral and pharyngeal cancer are being studied, using the PCR technique to detect the presence of DNA sequences related to HPV. If a higher prevalence of HPV in the tumour tissue than that reported in the general population is found, it is proposed to do a large case-control study using 550 cases and 550 controls identified from an earlier study of oral and pharyngeal cancer.

TYPE: Case Series
TERM: BMB; Biopsy; HPV; Registry
SITE: Oral Cavity; Pharynx
REGI: Los Angeles (USA)
TIME: 1991 - 1994

Louisville

1091 **Tamburro, C.H.** 02475
Univ. of Louisville, Sch. of Medicine, Depts of Medicine, Pharm. & Toxicology, Div. Occup. Toxicol., Liver Res. Center, HSC 119A, 500 S. Jackson St., Louisville KY 40292, United States of America (Tel.: +1 502 5885252; Fax: 5886867)
COLL: Greenberg, R.S.; Wong, J.

Medical Surveillance of Vinyl Monomer Exposure in Chemical Industry Workers

The programme is the long term continuation of the "medical surveillance of occupational environments" which has become the prototype of prospective comprehensive medical surveillance systems for chemical industries. The original surveillance system was directed at cancer control and involves collection and storage of basic core data on job classification, work history records, chemical exposure rating, and medical illness and disease record data. Subsequently each cancer case and other organ diseases as identified are studied, with special emphasis on the development of liver cancer. Research is directed to the development of molecular biological monitoring. The vinyl monomer programme continues the original data base composed of 5,500 previous employees, 1,200 previously active employees and 600 presently active employees. Annually over 70% of the work force is re-examined and new information added to the data base. All vinyl chloride and related chemicals continue to be monitored. Data collection and storage methods, and subsequent study of disease, have already been shown to be cost effective. Recent analysis of the medical surveillance of vinyl monomers data has identified the human hepatotoxic, carcinogenic, and biological threshold level for vinyl chloride induced hepatic angiosarcoma and hepatic fibrosis. Study continues regarding gastrointestinal lung, brain and haematopoietic cancers. Papers have been published in Hepatol. 4:413-418, 1984; and in Am. J. Med. 78:68-76, 1985 (Liss, G.M. et al.).

TYPE: Cohort; Methodology
TERM: Chemical Exposure; Environmental Factors; Metals; Occupation; Petroleum Products; Plastics; Solvents
SITE: Brain; Gastrointestinal; Haemopoietic; Liver; Lung
CHEM: Acrylamide; Acrylonitrile; Butadiene; Hexane; Hydrocarbons; Hydrocarbons, Halogenated; Mercury; PVC; Phenol; Styrene; Toluene; Vinyl Acetate; Vinyl Chloride; Vinylidene Chloride
OCCU: Chemical Industry Workers
TIME: 1974 - 1994

1092 **Tamburro, C.H.** 03526
Univ. of Louisville, Sch. of Medicine, Depts of Medicine, Pharm. & Toxicology, Div. Occup. Toxicol., Liver Res. Center, HSC 119A, 500 S. Jackson St., Louisville KY 40292, United States of America (Tel.: +1 502 5885252; Fax: 5886867)
COLL: Greenberg, R.S.; Wong, I.

Assessment of Health Effects of Acrylonitrile Exposure in Chemical Industry Workers

This program is an off-shoot of the medical surveillance of occupational environments program. It applies the health surveillance system to the assessment of possible health effects of acrylonitrile related compounds in exposed worker populations. The study covers a population of approximately 600 to 800 previous workers and 200 to 300 active employees. The retrospective assessment data include a work exposure history, and a rank ordered chemical exposure rating of 10-14 chemicals. These include acrylonitrile, acrylamides, acrylic acid, butadiene, hydrocarbons, halogenated hydrocarbons, methanol, phenol, toluene and caprylyl chloride. Medical data and cancer occurrence data for the retrospective cohort are available. The prospective programme includes annual data on medical examinations and laboratory screening studies for major organ systems. A sub-cohort population has, in addition, radiological screening evaluations for the lung and liver. Environmentally validated job specific determinations of exposure for acrylonitrile are also available for the years 1978, 1979, and 1980. Evaluation is underway to determine the biological threshold for hepatotoxicity and carcinogenicity of acrylonitrile and related compounds. Hepatol. 2:692, 1982 (Liss, G.M.); Seminars in Liver Disease 4:159-169, 1984 (Tamburro, C.H.); Am. J. M. 78:68-76, 1985.

TYPE: Cohort
TERM: Chemical Exposure; Occupation; Plastics; Rubber; Solvents
SITE: Colon; Liver; Lung
CHEM: Acrylamide; Acrylic Acid; Acrylonitrile; Butadiene; Caprylyl Chloride; Hydrocarbons; Hydrocarbons, Halogenated; Methanol; Phenol; Toluene; Vinyl Chloride
OCCU: Acrylonitrile Workers
TIME: 1974 - 1994

1093 **Tamburro, C.H.** 04762
Univ. of Louisville, Sch. of Medicine, Depts of Medicine, Pharm. & Toxicology, Div. Occup. Toxicol., Liver Res. Center, HSC 119A, 500 S. Jackson St., Louisville KY 40292, United States of America (Tel.: +1 502 5885252; Fax: 5886867)
COLL: Wong, J.

Health Effects of Heavy Metal Exposure in Chemical Catalyst Workers.

This is a prospective medical surveillance program of heavy metal catalyst workers utilising the health surveillance system to assess and identify any possible health effects of heavy metal and metallic compounds. These include aluminium, arsenic, antimony, barium, beryllium, cadium, chromium, cobalt, copper, iron, lead, magnesium, manganese, nickel, molybdenum, selenium, silver, uranium and zinc. The programme involves approximately 600 workers from two separate plants dealing with complex processes of different composition. Retrospective work histories and ranked chemical exposure ratings are available on all 83 catalyst processes. Exposure data will include past work history, chemical exposure assessment for each year, job classification, and work exposure rank, based on building and process. Annual and biannual medical examination including laboratory screening for major organ systems and radiological, electrocardiographic and pulmonary function studies are performed regularly. Environmental exposure data is validated by atomic absorption determination of environmental levels of certain heavy metals. Medical disorders with a potential relationship to heavy metal exposure, including cancer, are being studied in relation to these exposures.

TYPE: Cohort
TERM: Chemical Exposure; Metals; Occupation
SITE: Colon; Kidney; Liver; Lung; Skin
OCCU: Metal Workers
TIME: 1983 - 1994

1094 **Tamburro, C.H.** 04846
Univ. of Louisville, Sch. of Medicine, Depts of Medicine, Pharm. & Toxicology, Div. Occup. Toxicol., Liver Res. Center, HSC 119A, 500 S. Jackson St., Louisville KY 40292, United States of America (Tel.: +1 502 5885252; Fax: 5886867)
COLL: Fortwengler, P.; Wong, J.

Health Effects of Polychlorinated Hydrocarbon Exposure

Retrospective analysis of various populations of individuals exposed to polychlorinated hydrocarbons through accidental environmental exposure or through occupational contact is being carried out. Populations with documented environmental exposure to various polychlorinated hydrocarbons and a work or environmental exposure history will be collected. Exposures will be rank ordered, and environmental (water, air or soil) measurements of the index chemical(s) taken in addition to either blood and/or fat biopsy level determinations of the toxic materials under study. In most cases, detailed medical examination with comprehensive histories and laboratory screening studies for major organ systems are also included. Methodology is being developed to determine if any causal relationships exist between disease(s) (including cancer), and environmental exposure to polychlorinated hydrocarbons. It is hoped to provide data on actual hazardous levels, duration of body burden and biological threshold for disease development (including cancer).

TYPE: Cohort
TERM: Chemical Exposure; Environmental Factors; Occupation
SITE: All Sites
CHEM: Dioxins; Furans; PCB
TIME: 1980 - 1994

Lowell

1095 **Eisen, E.A.** 03769
Univ. of Lowell, Dept. of Work Environment, College of Engineering, One University Ave., Lowell MA 01854-2881, United States of America (Tel.: +1 508 9343278; Fax: 4525711)
COLL: Monson, R.R.; Tolbert, P.E.; Smith, T.

Cancer Mortality among Autoworkers exposed to Machining Fluids

NIOSH has estimated that over one million workers in the US are currently exposed to machining fluids. Previous studies have suggested an association between this exposure and respiratory, digestive and skin cancers. The present cohort study was initiated to assess whether long/term exposure to machining fluids in the course of machining, grinding, or other cutting operations, is associated with excess cancer mortality. The cohort includes all hourly workers who worked at least three years at one of three US automobile production plants. A cohort of 46,384 workers was defined based on company personnel records, and verified by Social Security employment records. An extensive exposure assessment was carried out at each plant. Lifetime exposures to straight, soluble and synthetic cutting oils, as well as to specific components and additives are currently being estimated for each subject in the cohort. The study population is 79% white, 90% male, and, on average, worked for 15 years. The average date of birth varies across the three plants from 1920 to 1939. The follow-up period extends from 1943 to 1984, by the end of which time 21% (10,142) of the cohort had died according to Social Security and National Death Index records. To date, cause of death information has been obtained for 93% of those known to be dead. Based on these data, SMRs have been estimated for each plant, using both the US and local county populations as reference. The overall SMR for death from all causes and from all cancer was 0.9. Slight excesses of 1.1 to 2.0, have been observed for some of the respiratory and digestive cancers. Dose-response models for specific cancers are being fitted, and the observed excess mortality from cancers of larynx and brain is being investigated with a nested case-control study.

TYPE: Case-Control; Cohort
TERM: Chemical Exposure; Occupation
SITE: All Sites; Brain; Larynx
CHEM: Mineral Oil
OCCU: Automobile Workers
TIME: 1985 - 1994

Madison

1096 **Messing, E.** 04208
Univ. of Wisconsin, School of Medicine, Div. of Urology, 600 Highland Ave., G5/347, Madison WI 53792, United States of America (Tel.: +1 608 2631359)
COLL: Young, T.B.; Hunt, V.B.; Roecker, E.B.

Early Detection of Urological Cancer by Home-Screening for Haematuria

To determine whether haematuria screening with urinary dipsticks can detect early-stage transitional cell carcinoma (TCC) of the bladder and renal cell carcinoma (RCC), a home testing programme was designed. A total of 2,982 healthy men age 50 or older, were recruited from five primary health care clinics. Subjects were solicited through their primary care physicians and asked to test their urine once a day for two weeks during month one and nine. Subjects were also asked to complete a questionnaire to gather information on socio-demographic factors, occupational exposures, smoking, medical histories, and other factors. If any dipstick test is at least "trace" positive, a urological evaluation is conducted, which consists of a physical examination and routine laboratory tests, intravenous urogram, cystoscopy, and bladder wash cytology. A total of 1,737 men have agreed to participate and were eligible to do so. Of these 1,340 men actually completed the testing procedure. After the first testing period, 283 had tested positive for haematuria and 192 received a complete urological work-up. Of those completed worked up, 16 had urological malignancies (9 transitional cell carcinoma, 1 renal cell carcinoma, 6 cancer of the prostate) and 47 had other serious urological diseases.

TYPE: Cohort
TERM: Occupation; Screening; Socio-Economic Factors; Tobacco (Smoking); Urine
SITE: Bladder; Kidney
TIME: 1985 - 1994

1097 **Newcomb, P.A.** 04469
Univ. of Wisconsin, Comprehensive Cancer Center, 420 N. Charter St., Room 6795, Madison WI 53706, United States of America (Tel.: +1 608 2637890)
COLL: MacMahon, B.; Willett, W.C.; Longnecker, M.

Alcohol Consumption, Lactation, and Breast Cancer Risk

A population-based case-control study is underway to evaluate the relationship between age-specific alcohol consumption and lactation and the occurrence of breast cancer. Recent evidence suggests that both factors are determinants of risk with modest, but potentially important, magnitudes of effect. Specific hypotheses to be addressed include: (1) cessation of alcohol intake is associated with a lower risk of breast cancer than is continued consumption, (2) alcohol intake before age 30 is associated more strongly with increased breast cancer risk than is consumption at later ages, and (3) increasing duration of lactation is associated with decreased risk of breast cancer among premenopausal women. To test these and other hypotheses, telephone interviews will be conducted with approximately 6,000 women with newly diagnosed breast cancer and 6,000 randomly selected community members. Cases will be identified from tumour registries in Wisconsin, Massachusetts, Maine, and New Hampshire. Controls will be selected from general population lists (driver's license lists and Medicare beneficiary lists). Cases and controls will be queried about past and current alcohol consumption and lactation history, as well as other medical and demographic factors known to influence breast cancer occurrence. The University of Wisconsin Clinical Cancer Center and the Harvard University School of Public Health both serve as coordinating centers for this multi-state study.

TYPE: Case-Control
TERM: Age; Alcohol; Lactation
SITE: Breast (F)
REGI: Maine (USA); Massachusetts (USA); New Hampshire (USA); Wisconsin (USA)
TIME: 1988 - 1994

1098 **Newcomb, P.A.** 04850
Univ. of Wisconsin, Comprehensive Cancer Center, 420 N. Charter St., Room 6795, Madison WI 53706, United States of America (Tel.: +1 608 2637890)

Alcohol Consumption and Risk of Large Bowel Cancer in Women

The relationship between alcohol consumption and risk of cancers of the colon and rectum is currently being evaluated in a population-based case-control study. Approximately 1,000 women newly diagnosed with colorectal cancer during 1990-1992 will be interviewed by telephone. The interview elicits past and current alcohol consumption history as well as reproductive experiences and other medical and demographic factors known to influence the occurence of colorectal cancer. In addition, subjects will complete a food frequency questionnaire regarding dietary practices prior to diagnosis. For comparison, women without colorectal cancer will be selected from the general population using drivers' license and medicare lists and interviewed concurrently with the case group.

TYPE: Case-Control
TERM: Alcohol; Diet; Female; Registry; Reproductive Factors
SITE: Colon; Rectum
REGI: Wisconsin (USA)
TIME: 1990 - 1994

Marshfield

1099 **Banerjee, T.K.** 04767
Marshfield Clinic, 1000 N. Oak Ave., Marshfield WI 54449, United States of America (Tel.: +1 715 3875134)
COLL: Layde, P.; Nordstrom, D.; Weber, J.

Pilot Study of Familial Factors in Cancer in a Defined Population Area

It is uncertain whether the aetiological significance of familial clustering of cancer represents inherited susceptibility or common familial exposure to some cultural, dietary, or other environmental stimulus. A case-control study is being conducted in a defined population area to describe cancer patterns and to compare the prevalence of a history of cancer among relatives of a sample of 100 new cancer cases to that among relatives of a sample of 100 age-and sex-matched controls randomly selected from the same postal code areas as the cases and to analyse the validity of information obtained from cases and controls about cancer in their relatives. Cases will be chosen prospectively from those which are newly diagnosed and reported to the population-based Marshfield Tumor Registry from the time when the study begins. Odds ratios for the association of family history of various cancers and the risk of cancer will be calculated using matched pair and conditional logistic regression techniques.

TYPE: Case-Control
TERM: Familial Factors; Genetic Factors
SITE: All Sites
TIME: 1990 - 1994

Maywood

1100 **Swinnen, L.J.** 05004
Loyola Univ. of Chicago, Fac. of Medicine, 2160 S. First Ave., Maywood IL 60153, United States of America (Tel.: +1 708 2168539; Fax: 2169335)
COLL: Costanzo-Nordin, M.R.; Fisher, S.G.

Lymphoproliferative Disorder Following Cardiac Transplantation

In a prior study, a significant increase in the incidence of post- transplant lymphoproliferative disorder (PTLD) was found to be associated with the use of OKT 3. The current study examines the effect of discontinuing the prophylactic use of OKT 3. All patients transplanted at Loyola University will be considered and compared with existing historical control groups of previously transplanted patients with or without anti-T-cell therapy. The total patient number is expected to be approximately 250. Multivariate analyses will be performed to assess the influence of other immunosuppressive agents.

TYPE: Cohort
TERM: BMB; Immunosuppression; Transplantation; Treatment
SITE: Lymphoma
TIME: 1990 - 1994

UNITED STATES OF AMERICA

Memphis

***1101 Kritchevsky, S.B.** 05268

Univ. of Tennessee, Center for Health Sciences Dept. of Biostatistics & Epidemiology, Lamar Alexander Bldg 877 Madison Ave., Memphis TN 38163, United States of America (Tel.: +1 901 5778757; Fax: 5287235)

Fluctuations of Plasma Cholesterol and Cancer Occurrence in the Lipid Research Clinics Primary Prevention Trial

The objective of the study is to determine whether the lowering of plasma cholesterol promote tumour growth. Data from the Lipid Research Clinics Primary Prevention Trial will be used. The trial randomized 3,906 men to either a placebo or to the bile acid sequestrant cholestyramine. Lipids were measured every two months during the average of 7.4 years of follow-up. An additional 6 years post-trial follow-up is also available. The analysis will relate the frequency of cholesterol decreases to the rate of cancer incidence during the post trial follow-up period.

TYPE: Cohort
TERM: Cholesterol
SITE: All Sites
LOCA: Canada; United States of America
TIME: 1993 - 1995

1102 Sharp, G.B. 05003

Univ. of Tennessee, College of Graduate Health Sciences, Dept. of Biostatistics & Epidemiology, 877 Madison Ave., Suite 330, Memphis TN 38163, United States of America (Tel.: +1 901 5778220; Fax: 5286517)

COLL: Cole, P.

Epidemiology of Vaginal Clear Cell Adenocarcinoma

The major objective of this case-control study is to describe risk factors for this cancer among women who were exposed in utero to DES. Self-administered questionnaires were returned by 108 DES-positive cases and by 456 DES-positive controls. Initial studies have focused on risk factors for endometrial cancer, including height and body mass level during adolescence. Future studies will determine if there are case/control differences in age at menarche and in nutritional intakes during adolescence. Papers have been published in the Am. J. Obstet. Gynecol. 162:994-1001, 1990 and Cancer 66:2215-2220, 1990.

TYPE: Case-Control
TERM: Diet; Drugs; Intra-Uterine Exposure; Menarche; Nutrition
SITE: Vagina
CHEM: DES
TIME: 1990 - 1994

***1103 Sharp, G.B.** 05285

Univ. of Tennessee, College of Graduate Health Sciences, Dept. of Biostatistics & Epidemiology, 877 Madison Ave., Suite 330, Memphis TN 38163, United States of America (Tel.: +1 901 5778220; Fax: 5286517)

Oesophageal and Gastric Cancer

This case-control study will examine the relationship of oesophageal and gastric cancers of two histological types with intake of micronutrients, nitrosamines, nitrites and nitrates, with exposure to alcohol and tobacco, and with exposure to viruses. About 200 oesophageal and gastric cancer cases will be identified through the Tennessee Cancer Reporting Service from 1993 to 1995. Controls will be drawn from a computerised list of licensed drivers. Cases and controls will be interviewed by mail or telephone, using a revised version of the nutritional assessment questionnaire developed by G. Block (NCI). Univariate and multivariate methods will be used to analyse the data.

TYPE: Case-Control
TERM: Alcohol; Diet; Histology; Nutrition; Registry; Tobacco (Smoking); Virus
SITE: Oesophagus; Stomach
CHEM: N-Nitroso Compounds; Nitrates; Nitrites
REGI: Tennessee (USA)
TIME: 1992 - 1995

Midland

1104 **Olsen, G.W.** 04295
Dow Chemical Company, Epidemiology Dept., Health & Environmental Sciences Div., 1803 Bldg, Midland MI 48674, United States of America (Tel.: +1 517 6366923; Fax: 6361875)
COLL: Bodner, K.M.; Ramlow, J.M.; Kemp, M.; Spadacene, N.W.; Bloemen, L.J.

Epidemiological Surveillance of a Chemical Industry Workforce

Cause-specific mortality surveillance is being carried out on approximately 100,000 present and former employees of The Dow Chemical Company employed since 1940. Nested case-control and chemical-specific cohort mortality studies are conducted. Among the chemicals produced at the major production sites are herbicides and insecticides, epoxy resins, benzene, styrene, magnesium, chlorinated solvents, vinyl chloride, polyethylene and various consumer products.

TYPE: Case-Control; Cohort
TERM: Chemical Exposure; Herbicides; Insecticides; Metals; Occupation; Pesticides; Plastics; Solvents
SITE: Brain; Kidney; Liver; Lung
CHEM: Benzene; Epoxy Resins; Magnesium; Polyethylenes; Styrene; Vinyl Chloride
OCCU: Chemical Industry Workers
LOCA: Australia; Canada; Netherlands; United States of America
TIME: 1979 - 1994

Minneapolis

1105 **Folsom, A.R.** 03627
Univ. of Minnesota, Div. of Epidemiology, School of Public Health, 515 Delaware St. S.E., 1-210 Moos Tower, Minneapolis MN, United States of America (Tel.: +1 612 6249950; Fax: 6258950)
COLL: Sellers, T.A.; Potter, J.D.; Munger, R.G.; Wallace, R.B.; Prineas, R.; Kushi, L.; Bostick, R.

Iowa Women's Health Study

A population-based cohort of 41,837 women aged 55-69 years was recruited to test the hypothesis that increased abdominal fat distribution, independent of body mass, is associated with increased incidence of endometrial cancer, breast cancer, and total mortality. Considerable exposure data on the cohort were obtained in a baseline survey in 1986, and cancer incidence and mortality data are now being obtained by linkage to the Iowa Health (SEER) Registry and National Death Index. Published five-year results indicated that abdominal adiposity was independently associated with increased risk of breast cancer, total mortality and several other nonfatal cancer endpoints ascertained by self report. However, the association of abdominal adiposity with endometrial cancer risk was not independent of body mass. Follow-up for cancer incidence and mortality is being continued through at least eight years. Further follow-up will allow verification of early results and extend them to rarer neoplasms which also may be related to fat distribution. Extended follow-up will also permit testing of dietary hypotheses for cancer, including possible interactions of diet with fat distribution. Further follow-up will enable comparison of cancer rates among women living in farms versus those not living in farms. A nested case-control study of families will also be conducted to test whether body size or shape explains the familial clustering of breast and endometrial cancer.

TYPE: Cohort
TERM: Cluster; Diet; Fat; Obesity; Physical Factors; Registry
SITE: All Sites; Breast (F); Uterus (Corpus)
REGI: Iowa (USA)
TIME: 1985 - 1995

1106 Gilbertsen, V. 03971
Univ. of Minnesota, Dept. of Surgery, 420 Delaware St. SE, Box 195 UHMC, Minneapolis MN 55455, United States of America (Tel.: +1 612 6274151)
COLL: Mandel, J.S.; Bond, J.; Snover, D.; Bradley, M.; Ederer, F.; Schuman, L.M.; Williams, S.

Trial of Faecal Occult Blood Screening for Colorectal Cancer Detection

This is a randomized clinical trial to determine whether screening for occult blood using the haemoccult test will result in a reduction in mortality from colorectal cancer. Over 45,000 volunteers who are residents of Minnesota between the ages of 50 and 80 have been randomized to three groups. One group is screened by annual faecal occult blood tests for five years, the second is screened every two years over the same period, while the controls are followed-up by annual questionnaire. Follow-up of almost the entire population has been maintained over several years and information on all deaths is carefully assessed. First results were published in New Engl. J. Med. 328:1365-1371, 1993 (Mandel et al.).

TYPE: Mortality
TERM: BMB; Screening
SITE: Colon; Rectum
TIME: 1975 - 1995

1107 Mandel, J.S. 03975
Univ. of Minnesota, School of Public Health, Div. Environm. & Occup. Health, 420 Delaware St. SE, Box 197, UHMC, Minneapolis MN 55455, United States of America (Tel.: +1 612 6274200)

Cancer Risk in X-Ray Technologists

The objective of this study is to evaluate the health status, primarily with respect to cancer, of all current and former members of the American Registry of Radiologic Technologists (ARRT). About 180,000 individuals whose location is known are being sent a detailed questionnaire to obtain data on exposure to radiation, a history of cancer and potential confounding factors. In addition, dosimetry data will be obtained. All cancers identified from the questionnaire will be validated by sending a diagnostic form to the person's physician. In addition to evaluating morbidity data, death certificates are being obtained on all deceased ARRT subjects. These data will be coded and analysed to examine relationships between mortality and morbidity patterns and radiation exposure. This study is important for three reasons: first, the large, well defined cohort readily available for study of a variety of cancers associated with radiation exposure, second, the varying levels of radiation exposure and the availability of considerable dosimetry data and third, the large proportion of females (70%) provides an opportunity to assess radiation effects in this group.

TYPE: Cohort
TERM: Chemical Exposure; Dose-Response; Occupation; Radiation, Ionizing
SITE: All Sites; Breast (F); Leukaemia; Lung; Thyroid
OCCU: Health Care Workers
TIME: 1982 - 1994

1108 Robison, L.L. 04756
Univ. of Minnesota, Div. of Pediatric Hematology/Oncology, Box 422 UMHC , Minneapolis MN 55455, United States of America (Tel.: +1 612 6262778; Fax: 6262815)
COLL: Buckley, J.D.; Kersey, J.; Neglia, J.P.; Reaman, G.H.; Severson, R.K.; Hammond, G.D.; Heerema, N.A.; Miller, D.; Potter, J.D.; Sather, H.N.; Trigg, M.

Acute Lymphoblastic Leukaemia in Children

The Children's Cancer Study Group is conducting a case-control study of the role of risk factors for childhood ALL within biologically-defined subgroups. An estimated total of 2,000 newly diagnosed cases of ALL ascertained from January 1989 to December 1992 will be included. Patient subgroups are defined on the basis of FAB morphology, immunophenotyping, and cytogenetics. One control per case is selected by random digit dialling and is individually matched to the case on age, sex, and race. Telephone interviews are being conducted with the parents of both cases and controls. Postulated risk factors under investigation include maternal reproductive history, exposure during the index pregnancy, exposure to ionizing radiation, exposure to pesticides and solvents, timing and extent of infectious exposures, and familial/ genetic components.

TYPE: Case-Control
TERM: Childhood; Familial Factors; Genetic Factors; Intra-Uterine Exposure; Pesticides; Radiation, Ionizing; Reproductive Factors; Solvents
SITE: Leukaemia (ALL)
TIME: 1988 - 1995

1109 **Robison, L.L.** 04782
Univ. of Minnesota, Div. of Pediatric Hematology/Oncology, Box 422 UMHC , Minneapolis MN 55455, United States of America (Tel.: +1 612 6262778; Fax: 6262815)
COLL: Hammond, G.D.; Buckley, J.D.; Arthur, D.C.; Woods, W.G.; Pentz, M.A.; Wilkins, J.R.; London, S.J.

Acute Nonlymphoblastic Leukaemia in Children

The Children's Cancer Study Group (CCSG) has recently completed an analytic study designed to assess specific environmental exposures as potential risk factors for ANLL in children. The findings from this study provide the direction for the proposed epidemiological investigation of childhood ANLL. The study will include approximately 630 cases of ANLL diagnosed during a 4-year period in persons up to 18 years of age. The primary objectives of the study are (1) to confirm the association of pesticide exposure and ANLL risk and to identify the substances or class of pesticides responsible; (2) to confirm the association of solvents and petroleum products and ANLL risk and determine the class of solvent responsible; (3) to confirm the association of maternal marijuana use and ANLL risk; and (4) to identify subgroups, defined by age at diagnosis, FAB morphology, cytogenetic abnormality, or clinical features of the leukaemia, in which the associations with pesticides, solvent, or marijuana exposure are strongest. Cases will be ascertained through the member institutions of the Children's Cancer Study Group. Cases of ANLL will be compared with a series of age-, sex-, and race-matched regional controls selected using random digit dialling. Parents of cases and controls will be interviewed by telephone. Within most of the morphologic subgroups of ANLL (FAB morphology classification: M0-M7), the study will have a sample size that provides sufficient power for assessing the study hypotheses. Classification of cases according to FAB morphology and cytogenetics wil be made by centralised review.

TYPE: Case-Control
TERM: Childhood; Classification; Marijuana; Pesticides; Petroleum Products; Solvents
SITE: Leukaemia (ANLL)
TIME: 1989 - 1994

***1110** **Robison, L.L.** 05415
Univ. of Minnesota, Div. of Pediatric Hematology/Oncology, Box 422 UMHC , Minneapolis MN 55455, United States of America (Tel.: +1 612 6262778; Fax: 6262815)
COLL: Boice, J.D.; Breslow, N.E.; Donaldson, S.; Green, D.; Hays, D.; Li, F.P.; Meadows, A.T.; Mulvihill, J.J.; Neglia, J.P.; Nesbit, M.; Ochs, J.; Packer, R.; Potter, J.; Sklar, C.; Strong, L.C.; Zeltzer, L.

Childhood Cancer Survivor Study

The Childhood Cancer Survivor Study is designed to investigate the long-term effects of cancer and its associated therapies. A collaborative retrospective cohort study is being conducted, involving the identification and active follow-up of approximately 25,000 survivors of cancer diagnosed before 21 years of age, between 1970 and 1986. For comparison purposes, a variety of data sources will be utilized, including siblings, within cohort, and vital/health statistics. The specific aims of the study are to expand on known late effects, as well as to investigate specific hypotheses relating to new risk factors for adverse events and provide information which will facilitate the development of primary and secondary prevention strategies. Topics to be investigated include the occurrence of second neoplasms, organ dysfunction (e.g. cardiac, gonadal), early death, decreased fertility, offspring with adverse events, and family history of cancer.

TYPE: Cohort
TERM: Disease, Other; Family History; Fertility; Late Effects; Multiple Primary; Survival; Treatment
SITE: All Sites; Childhood Neoplasms
TIME: 1993 - 1998

1111 **Sellers, T.A.** 05143
Univ. of Minnesota, School of Public Health, Div. of Epidemiology, 1300 South Second St., Suite 300, Minneapolis MN 55455, United States of America (Tel.: +1 612 6261733; Fax: 6240315)
COLL: Potter, J.D.; Rich, S.S.; King, R.A.; Anderson, V.E.; Kuni, C.C.; Kushi, L.; Bartow, S.; Wiesner, G.; McGovern, P.G.

Epidemiological and Genetic Studies of Breast Cancer

A historical cohort study of families identified by the University of Minnesota over 40 years ago is being conducted to identify and quantify gene-environment interactions in the pathogenesis of breast cancer. The exposed group is defined as the first- and second-degree relatives of breast cancer patients treated at the University of Minnesota between 1944 and 1952; unexposed women will be the spouses of first- and second-degree male relatives. Women in both groups will be administered questionnaires on diet, reproductive history, etc., and will be referred for a mammogram to confirm disease status. Collection of blood samples on selected families will permit linkage analysis to confirm the presence of a breast cancer susceptibility gene(s).

TYPE: Cohort; Genetic Epidemiology
TERM: BMB; Blood; Diet; Genetic Factors; Linkage Analysis; Reproductive Factors
SITE: Breast (F)
TIME: 1990 - 1996

***1112 Sellers, T.A.** 05363
Univ. of Minnesota, School of Public Health, Div. of Epidemiology, 1300 South Second St., Suite 300, Minneapolis MN 55455, United States of America (Tel.: +1 612 6261733; Fax: 6240315)
COLL: Folsom, A.R.; Potter, J.D.; Rich, S.S.

Cancer Epidemiology of Postmenopausal Women in the Iowa Women's Health Study

The aim of the study is to examine whether the familial aggregation of postmenopausal breast cancer can be partly explained by obesity and body fat distribution. Incident breast cancer cases from an on-going prospective study of 41,837 Iowa women are age-matched to non-cases. A total of 265 cases and 265 controls were identified and mailed a questionnaire to identify first degree female relatives. Approximately 2,000 family members were identified. Over 90% have returned mailed questionnaires regarding cancer history, reproductive history, and anthropomoterics. Blood samples are being collected on 16 multiplex families for genetic linkage analysis. Other analyses include case-control comparisons of family members and segregation analysis of case families, including lifestyle co-variates.

TYPE: Case-Control; Cohort; Genetic Epidemiology
TERM: BMB; Blood; Lifestyle; Linkage Analysis; Obesity; Segregation Analysis
SITE: Breast (F)
TIME:

1113 Severson, R.K. 05002
Univ. of Minnesota, Div. of Hematology/Oncology, Box 422, UMHC, 420 Delaware St. SE, Minneapolis MN 55455, United States of America (Tel.: +1 612 6274438)
COLL: Robison, L.L.; Pollock, B.H.; Hammond, G.D.; Ross, J.A.

Paediatric Cancer Distribution in the US: A Geographical Analysis

The primary objective of this ecological study is to determine areas within the US where the observed to expected ratios of paediatric cancer differ notably from 1. It has been estimated that almost 10,000 incident cancers are diagnosed annually in the US in persons under the age of 20. During 1989, over 6,000 new cases were diagnosed by member institutions of the Children's Cancer Study Group (CCSG) and the Paediatric Oncology Group (POG). These observed cases have been coded geographically to specific regional areas within the US. Using census data and SEER program rates, expected numbers for the age groups of 0-4, 5-9, 10-14 and 15-19 have been generated. Regional maps of age-, sex-, and race-specific observed/expected numbers are being created for specific cancer sites including leukaemia, lymphoma and brain tumours. Preliminary analyses have indicated several geographical patterns where the observed number of childhood cancer patients is appreciably less than the expected number. Further data will be collected for 1990 incident cases.

TYPE: Incidence
TERM: Mapping
SITE: Childhood Neoplasms
TIME: 1990 - 1994

1114 **Woods, W.G.** 05096
Univ. of Minnesota Hospital & Clinic, Box 454 UMHC, 420 Delaware St. SE, Minneapolis MN 55455, United States of America (Tel.: +1 612 6262778; Fax: 6262815)
COLL: Lemieux, B.; Tuchman, M.

Screening for Neuroblastoma in Infants

The Quebec Neuroblastoma (NB) Screening Project was initiated to assess the clinical and biological aspects of screening infants for the presence of NB in North America. The experimental cohort, all infants born in the province of Quebec during the five-year period 1989-1994, are eligible for screening at three weeks and six months of age. Parents mail urine-saturated and dried filter papers to the screening laboratories, where the catecholamine metabolites VMA and HVA are assessed by a two-stage screening assay, thin-layer chromatography followed by gas chromatography-mass spectrometry. Infants with elevated levels after repeat testing are offered a thorough non-invasive diagnostic evaluation looking for the presence of NB. Patients diagnosed with NB are then uniformly staged, treated, and followed, similarly to clinically detected patients either never screened or missed by screening. Mortality from NB of all children diagnosed as part of the cohort will be compared to results in several population-based, non-screened control groups. Secondary endpoints include NB incidence, clinical staging, and survival. In the process, specific biological parameters will be studied to determine (1) if known prognostic factors in NB detected clinically are of equal prognostic importance in screen-detected tumours; and, (2) if such studies shed light on the hypothesis that NB is more than one distinct disease. These parameters include tumour N-myc oncogene amplification and expression, DNA content, cytogenetics, gangliosides, and histology; serum ferritin, neuron-specific enolase, and gangliosides; and urinary catecholamine metabolites. Finally, as secondary aims, the effect of preclinical detection of NB on health care costs and the maternal psychosocial consequences of cancer screening in infants will be determined. The results obtained in this study will help determine whether screening of infants for the detection of NB will lead to a reduction in population-based mortality, and to shed light on the biology of this malignancy.

TYPE: Cohort
TERM: BMB; Registry; Screening
SITE: Neuroblastoma
LOCA: Canada; United States of America
REGI: Canada (Can); Delaware (USA); Florida (USA); Minnesota (USA)
TIME: 1989 - 1997

Nashville

1115 **Dupont, W.D.** 04015
Vanderbilt Univ., School of Medicine, Dept. of Preventive Medicine, Div. of Biostatistics, A-1124 Medical Center North, Nashville TN 37232-2637, United States of America (Tel.: +1 615 3222001; Fax: 3438722)
COLL: Page, D.L.; Parl, F.F.; Cole, P.

Breast Cancer Risk after Atrophic or Apocrine Changes

Preliminary studies suggest that the presence of atrophy in benign breast biopsies is an important risk factor for subsequent breast cancer. This risk may be as low as one-third or as high as five times that of the general population, depending on other prognostic factors such as age at biopsy or a family history of breast cancer. This wide variation may result from different types of atrophy, similar in morphological appearance but with different aetiologies and associated with different cancer risks. In this case there may be a positive correlation between the type of atrophy associated with high cancer risk and these other factors. Papillary apocrine change is a distinctive lesion seen in 27% of benign breast biopsies, and has never been rigorously evaluated as a breast cancer risk factor. In a retrospective cohort study of approximately 8,100 consecutive women who underwent benign breast biopsy at one of three Nashville hospitals between 1952-1968 (Vanderbilt or Baptist hospitals) or 1950-1968 (St. Thomas Hospital), all biopsies will be reviewed and reclassified for the presence and degree of atrophic or papillary apocrine changes. Information on subsequent breast cancer outcome and epidemiological risk factors will be obtained from interviews with patients or their next of kin. This information has already been obtained on 60% of the study subjects who have participated in prior studies. 90% follow-up is expected. Variation in risk will be assessed by means of intra-study control groups of biopsied patients, as well as external control groups from the Third National Cancer Survey in Atlanta and the Cancer in Connecticut data base. Data will be analysed using hazard regression methods and other statistical techniques to elucidate cancer risks associated with various constellations of risk factors from a longitudinal data base.

This study will substantially increase the precision with which breast cancer risk can be predicted on the basis of histological and non-anatomical findings.

TYPE: Cohort; Methodology
TERM: Biopsy; Premalignant Lesion
SITE: Benign Tumours; Breast (F)
TIME: 1986 - 1994

1116 **Dupont, W.D.** 05020
Vanderbilt Univ., School of Medicine, Dept. of Preventive Medicine, Div. of Biostatistics, A-1124 Medical Center North, Nashville TN 37232-2637, United States of America (Tel.: +1 615 3222001; Fax: 3438722)
COLL: Vnencak-Jones, C.L.; Parl, F.F.; Page, D.L.; King, M.C.

Epidemiology of Molecular Risk Factors for Breast Cancer

A nested case-control study of women with benign breast disease is being conducted to determine the effect of oncogenic and histological factors on breast cancer risk. The study cohort will consist of all women without a prior history of breast cancer who underwent benign breast biopsies at Vanderbilt Hospital between 1959 and 1988, at St. Thomas Hospital between 1964 and 1988 or at Baptist Hospital between 1982 and 1988. Biopsy slides and paraffin-embedded biopsy tissue are available for all these women. The slides of the entry biopsies will be reviewed blindly and classified according to Page's histological criteria (New Engl. J. Med. 312:146-151, 1985). The study cohort contains approximately 10,913 women. Cases will be cohort members who develop breast cancer during follow-up (approximately 586 women). Two controls matched for age, year, benign histological diagnosis and hospital of their entry biopsy at risk for developing breast cancer will be chosen from the cohort when their matched case patient develops this disease. In situ hybridization techniques will be used to look for altered expression, amplification or transcription of specific oncogenes within different types of breast lesions in the entry biopsies of cases and controls. The oncogenes which will be evaluated include the HER-2/neu, c-myc, c-myb, c-rasHa and rb oncogenes. Other oncogenes will be evaluated as evidence accrues suggesting that they may have a role in the pathogenesis of breast cancer.

TYPE: Case-Control; Cohort; Molecular Epidemiology
TERM: BMB; Biopsy; Histology; Oncogenes
SITE: Breast (F)
TIME: 1990 - 1995

New Haven

***1117** **Mayne, S.T.** 05272
Yale Univ. School of Medicine, Dept. of Epidemiology & Public Health, 60 College St. Box 3333, New Haven CT 06510, United States of America (Tel.: +1 203 7586274; Fax: 7856980)
COLL: Zheng, T.Z.; Goodwin, W.J.; Fallon, B.; Cooper, D.; Friedman, C.

Beta Carotene Chemoprevention of Early Stage Head and Neck Cancer

This double blind placebo controlled study was initiated to determine whether supplemental beta carotene inhibits second primary cancers and/or tumour recurrences in patients curatively treated for early stage cancers of the oral cavity, pharynx, and larynx. Patients with incident cancer of the oral cavity, pharynx, or larynx are identified throughout Connecticut. Following treatment of the initial cancer, eligible patients are invited to participate in the study. Those who are judged compliant after a brief placebo run-in period are randomized to receive 50 mg beta carotene or placebo as a daily supplement. Data collection includes an extensive baseline intereview for risk factor information, update interviews yearly, collection of blood samples three times in the first year and yearly thereafter, and clinical follow-up by the patient's regular physician. This study will attempt to enroll 600 patients. The design of this trial allows for an examination of the efficacy of beta carotene in this clinical setting, the efficacy of community-based chemoprevention research, and a detailed evaluation of risk factors for second primary cancers, especially in the group of patients randomized to placebo.

TYPE: Intervention
TERM: BMB; Blood; Multiple Primary; Registry
SITE: All Sites; Larynx; Oral Cavity; Pharynx
CHEM: Beta Carotene
REGI: Connecticut (USA)
TIME: 1991 - 1999

***1118 Risch, H.A.** 05279
Yale Univ. School of Medicine, Dept. of Epidemiology & Public Health, 60 College St., P.O. Box 3333, New Haven CT 06510, United States of America (Tel.: +1 203 7852848; Fax: 7854497)
COLL: Mayne, S.T.; Dubrow, R.; West, A.B.

Oesophageal and Gastric Adenocarcinoma

A population-based case-control study within the state of Connecticut is being carried out to ascertain dietary, lifestyle and other risk factors for adenocarcinoma of the oesophagus and gastric cardia, and to compare these risk factors with those for other cancers of the oesophagus and stomach. In total, about 175 newly-diagnosed cases will be identified by the Rapid Case Ascertainment System. An approximately equal number of randomly selected population controls will be frequency matched to the cases for age and sex. Controls will be identified using random-digit dialling for those under 65 years of age, and those 65 years and over will be randomly chosen from rosters provided by the Health Care Financing Administration. In addition, the Rapid Case Ascertainment System will be used to obtain an age-sex frequency-matched sample of about 175 newly-diagnosed cases of squamous-cell carcinoma of the oesophagus and adenocarcinoma of other parts of the stomach for comparison with the cases. Cases and controls will be interviewed in their homes by a trained interviewer, using a standardised structured questionnaire, to ascertain dietary history, tobacco use, alcohol consumption, usage of a number of medications, and various occupational exposures and predisposing medical conditions. Average daily consumption of a number of dietary nutrients will be calculated from the diet histories, and univariate and multivariate logistic regression analyses will be used to estimate relative risks for comparison of the oesophageal and gastric-cardia adenocarcinoma cases with the controls and with the other cancer cases. This is part of a three-centre collaborative study, involving Connecticut, the state of New Jersey and western Washington state.

TYPE: Case-Control
TERM: Alcohol; Analgesics; Coffee; Diet; Drugs; Lifestyle; Occupation; Registry; Tobacco (Smoking)
SITE: Oesophagus; Stomach
REGI: Connecticut (USA)
TIME: 1992 - 1995

***1119 Risch, H.A.** 05280
Yale Univ. School of Medicine, Dept. of Epidemiology & Public Health, 60 College St., P.O. Box 3333, New Haven CT 06510, United States of America (Tel.: +1 203 7852848; Fax: 7854497)
COLL: Howe, G.R.; Strand, L.; Malcolm, E.

Menopausal Hormone Usage and Cancer in Saskatchewan

The association between the occurrence of cancers of various sites, and previous usage of menopausal oestrogens, progestogens and oral contraceptives, will be examined in a record-linkage cohort study using the Saskatchewan Health Prescription Drug Plan Database. For this study, the cohort of all women aged 43-49 in 1976 (n = 33,003), resident in Saskatchewan, will be identified from the Saskatchewan Health Subscription Plan (SHSP) Master Registration File. The cohort will be linked by SHSP registration number to the Drug Plan Database for the period January 1976 to June 1987, and to the Provincial Cancer Registry Database for the period March 1960 to December 1990. SHSP annually updates its information on whereabouts and vital status of residents, and cause and date of death or emigration from the province will be included in the cohort data. For each cancer site analysed, women having a cancer at this site before 1976 will be omitted from the analysis; analysis will be by Cox (survival) regression, with time-dependent covariates for the drug exposures.

TYPE: Cohort
TERM: Hormones; Menopause; Oral Contraceptives; Record Linkage; Registry
SITE: Breast (F); Colon; Rectum; Thyroid; Uterus (Corpus)
CHEM: Oestrogens; Progestogens
LOCA: Canada
REGI: Saskatchewan (Can)
TIME: 1990 - 1994

***1120** **Risch, H.A.** 05281
Yale Univ. School of Medicine, Dept. of Epidemiology & Public Health, 60 College St., P.O. Box 3333, New Haven CT 06510, United States of America (Tel.: +1 203 7852848; Fax: 7854497)
COLL: Marrett, L.D.; Howe, G.R.; Jain, M.

Dietary Factors and Epithelial Ovarian Cancer

This project will examine the relationship between dietary and reproductive factors and ovarian cancer occurrence in southern Ontario, Canada. Quantitative diet histories from about 450 women diagnosed with ovarian cancer over a three-year interval will be compared with the same information obtained from a random sample of about 550 women of similar ages in the general population. Each study subject will be interviewed at home by a trained dietary interviewer. From the information recorded about frequency and portion size of usual foods consumed, average daily intake of a number of nutrients, such as fat, protein, carbohydrate, and various vitamins will be calculated. Differences in nutrient intake between cases and controls will be evaluated by logistic regression. In particular consumption of types of fat, and usage of or intolerance to dairy products will be examined.

TYPE: Case-Control
TERM: Dairy Products; Diet; Fat; Nutrition; Registry; Reproductive Factors
SITE: Ovary
LOCA: Canada
REGI: Ontario (Can)
TIME: 1990 - 1994

New Orleans

1121 **Fontham, E.T.** 04052
Louisians State Univ., Medical Center, Dept. of Pathology, 1901 Perdido St. , New Orleans LA 70112, United States of America (Tel.: +1 504 5686094; Fax: 5686037)
COLL: Correa, P.; Chen, V.W.; Greenberg, R.S.; Liff, J.; Coates, R.J.; Buffler, P.A.; Alterman, T.; Wu-Williams, A.; Reynolds, P.; Boyd, P.; Austin, D.F.

Lung Cancer in Non-Smoking Women

This is a population-based case-control study of women who have never used tobacco products and develop lung cancer. The study is designed to test the hypothesis that involuntary exposure to environmental tobacco smoke (ETS) increases the risk of lung cancer in non-smoking women. Cases are ascertained in five major US metropolitan areas: Atlanta, Houston, Los Angeles, New Orleans, and San Francisco Bay Area. Interview data will be compared with two control groups: a population group selected by random digit dialling (ages 20-64) and from the Health Care Finance Administration roles (ages 65+) and a group of women diagnosed with colon cancer. Both control groups are composed of women who report themselves to be lifetime non-smokers. Detailed information is obtained on involuntary exposure to ETS during childhood and adult life. Information is also sought on other lung cancer risk factors such as occupation, diet, residential factors, family history of cancer, etc. Urine specimens and hair samples are collected from study subjects for analysis of cotinine content to validate subjects' reported status with respect to recent smoking habits. Radon monitors were placed in the current residence of a sample of cases and controls. Histopathological review of diagnostic material is evaluated by one pulmonary pathologist. Stratified analyses and logistic regression techniques will be used to examine risk of lung cancer by histological type associated with exposures to ETS while controlling for confounding variables.

TYPE: Case-Control
TERM: Biochemical Markers; Passive Smoking
SITE: Lung
CHEM: Radon
TIME: 1985 - 1994

New York

1122 **Berwick, M.** 04313
Cancer Prevention Research Inst. Cancer Prevention Research Unit, 11 East 22nd St., 8th Floor, New York NY 10010, United States of America (Tel.: +1 212 5330555; Fax: 5330798)
COLL: Barnhill, R.; Roush, G.C.; Dubin, N.

Lethal Melanoma and Skin Examination

Prevention of mortality from malignant melanoma is based on the untested concept that visual inspection of the skin will lead to early detection and effective treatment. This case-control study will test the hypothesis that lethal cutaneous melanoma is inversely associated with skin examination. All incident cases of cutaneous melanoma (CM; expected number of cases 1,150) newly diagnosed in Connecticut residents from September 1986 to May 1989 will be identified by rapid ascertainment. About 1,200 population-based controls, frequency-matched to the age-sex distribution of cases, will be obtained by random digit dialling. Cases and controls will be interviewed in person regarding their skin examination practices and risk factors for CM. From the case group, lethal CM (n = 220 to 250) will be compared with the controls with respect to skin examination practices, particularly self-examination. Adjustment will be made for factors that explain the majority of CM in the general population. Non-lethal, invasive CM and in situ CM will also be compared with the general population after adjustment for aetiological factors. Two sources indicate that the prevalence of self-examination in healthy persons at risk for melanoma from the general population is 32% or greater. With 32% as the prevalence of self-examination the study has a statistical power of 80% to detect a relative risk of 1.7 for lethal CM in those not practising self-examination.

TYPE: Case-Control
TERM: Prevention
SITE: Melanoma
TIME: 1986 - 1994

1123 **Brandt-Rauf, P.W.** 04602
Columbia University, School of Public Health, Division of Environmental Sciences, 60 Haven Ave., New York NY 10032, United States of America (Tel.: +1 212 3053464; Fax: 3058119)
COLL: Niman, H.L.; Neugut, A.I.; Perera, F.P.; Vainio, H.; Hemminki, K.

Serum Oncogene Protein Screening

The purpose of this research is to develop a new molecular epidemiological method for the surveillance of occupational cancer. This method is based on the use of monoclonal antibody immuno-blotting assays for the detection of oncogene protein products in serum that can be used to screen for early detection of neoplastic changes in occupational cohorts at risk for malignant disease due to workplace exposures. This approach is currently being applied to several such cohorts (including hazardous waste workers and foundry workers) and matched normal controls (total numbers of approximately 100), and further studies are planned on other occupational cohorts and on cohorts of cancer patients to follow their response to therapy.

TYPE: Methodology; Molecular Epidemiology
TERM: Antibodies; Monitoring; Mutagen Tests; Occupation; Screening
SITE: All Sites
OCCU: Foundry Workers; Waste Incineration Workers
TIME: 1987 - 1994

***1124** **Breuer, B.** 05447
Strang Cancer Prevention Center, 428 East 72 St., New York NY 10021, United States of America (Tel.: +1 212 7944900; Fax: 7944958)
COLL: Brandt-Rauf, P.W.; Pincus, M.; Miller, D.G.; Minick, R.

erbB-2, myc, and ras Proteins in the Sera and Tumours of Breast Cancer Patients

The aim of this study is to determine the prevalence of elevated levels of erbB-2, myc, and ras proteins in the pre-operative sera of women with breast cancer. In order to confirm that the tumour is the source of the protein, whenever possible, matching tumour specimens are procured, as well as second blood specimens that are drawn approximately nine months after surgery. Oncoprotein levels in the tumour are measured with immunohistochemical techniques. Serologocial levels are assayed by immunoblotting with monoclonal antibodies to specific peptides within each oncoprotein, respectively. this is an on-going study, and so far 36 cases have been recruited. The controls are female patients of the Strang Cancer Prevention Center who have no evidence of breast cancer. The long-term aim of the project is to determine the utility of these oncoproteins as biological markers of very early breast cancer.

TYPE: Case-Control
TERM: Biomarkers
SITE: Breast (F)
TIME: 1991 - 1995

***1125 Cuttner, J.** 05449
Mount Sinai Medical Center Dept. of Medicine Div. of Hematology, One Gustave L. Levy Place, Box 1079, New York NY 10029, United States of America (Tel.: +1 212 2416481; Fax: 9961029)
COLL: Spivack, M.; Tabrizi, D.; Troy, K.; Rosina, O.; Wallenstein, S.; Rubenstein, P.

HLA Class I Antigens: Chronic Lymphocytic Leukaemia

The specific aim of this study is to perform HLA typing on patients with haematological malignancies, including CLL, hairy cell leukaemia, NHL and HD. HLA Class I typing, using standard lymphocyte microcytoxicity assays, will be performed. Epidemiological information will be obtained, including ethnic background, family history of haematological malignancy or solid tumours. In addition, information on patient's exposure to drugs, radiation or smoking will be obtained. The control population will consist of persons of the same ethnic background seen at the New York Blood Center. Class I HLA typing have already been performed on 50 patients with CLL. Initial results, published in Proc. ASCO 10:131, 1991, show an increased incidence of the Class I antigen B35 and antigen group A24(9),B35,Cw4 in Ashkenazi Jews and European Caucasians.

TYPE: Case-Control
TERM: BMB; Drugs; Ethnic Group; Family History; HLA; Radiation, Ionizing; Tobacco (Smoking)
SITE: Hodgkin's Disease; Leukaemia (CLL); Non-Hodgkin's Lymphoma
TIME: 1991 - 1994

1126 German III, J.L. 02265
The New York Blood Center, Lab. of Human Genetics, 310 E. 67th St. , New York, United States of America (Tel.: +1 212 5703075; Fax: 5703195)
COLL: Passarge, E.

Bloom's Syndrome Registry

Bloom's syndrome is a rare, recessively transmitted disorder which predisposes to cancer of the wide variety of sites and types that affects the general population. Diagnosed cases throughout the world comprise the Bloom's Syndrome Registry (166 accessions to date), and contact with most affected families is maintained, either directly or through their physician. 60 of the persons included in the registry have developed at least one cancer. 83 cancers have been diagnosed at mean age 24.7 (range 2-46). 19 were acute leukaemia (mean age at diagnosis 17 years), and 40 were carcinoma (mean age at diagnosis 32 years). The others were lymphoma and Hodgkin's disease (20), osteosarcoma, Wilms' tumour, and medulloblastoma. Progress reports are made in the journal Clinical Genetics (latest in volume 25:57-69 and 93-110, 1989), particular emphasis being laid on cancer occurrence in homozygous affected persons and their parents (obligate heterozygotes).

TYPE: Registry
TERM: DNA; Genetic Markers; High-Risk Groups; Premalignant Lesion
SITE: All Sites
TIME: 1963 - 1999

***1127 Muscat, J.E.** 05275
American Health Foundation Div. of Epidemiology, 320 E. 43rd St., New York NY 10017, United States of America (Tel.: +1 212 5512530; Fax: 6872339)
COLL: Wynder, E.L.

Acetaminophen and Nonsteroidal Anti-Inflammatory Drugs in Relation to Colorectal Cancer Risk

The purpose of this case-control study is to determine whether regular intake of pain control medication reduces the risk of colorectal cancer. 500 patients with colorectal cancer and 500 control subjects have been interviewed with respect to dietary history, occupation, tobacco, alcohol, family history of cancer and medication use. Preliminary results show a significant protective effect with regular aspirin consumption, although no trend was found with duration. Acetaminophen appears to have a protective effect although few patients took this on a regular basis. The mechanism of nonsteroidal anti-inflammatory drugs may be inhibition of prostaglandins, which are associated with tumour promotion.

TYPE: Case-Control
TERM: Analgesics; Drugs
SITE: Colon; Rectum
CHEM: Aspirin; Paracetamol
TIME: 1990 - 1994

1128 **Neugut, A.I.** 03164
Columbia Univ. College of Physicians & Surgeons, 630 West 168 St., New York NY 10032, United States of America (Tel.: +1 212 3059414; Fax: 3059413)
COLL: Garbowski, G.; Forde, K.A.; Treat, M.R.; Waye, J.; Fenoglio-Preiser, C.

Colorectal Polyps

The objective is to identify potential risk factors for adenomatous polyps of the colon. Questionnaires were administered to patients undergoing colonoscopic polypectomy. The control group consisted of patients in whom colonoscopic examination showed no neoplasia. Currently data are being analysed from telephone interviews with 303 adenomatous polyp patients (with no history of colon cancer), 509 normal controls (with no history of colon cancer or polyps), and 113 colon cancer patients. 198 patients with recurrent adeomatous polyps and 348 recurrent controls (postpolypectomy but normal on index colonoscopy) are also being studied.

TYPE: Case-Control
TERM: Premalignant Lesion
SITE: Colon; Rectum
TIME: 1983 - 1994

1129 **Neugut, A.I.** 05014
Columbia Univ. College of Physicians & Surgeons, 630 West 168 St., New York NY 10032, United States of America (Tel.: +1 212 3059414; Fax: 3059413)
COLL: Garbowski, G.; Benson, M.; Buttyan, R.; Romas, N.; Rosner, W.; Brandt-Rauf, P.W.

Prostate Cancer Among Ethnic Groups in New York

The objective is to identify risk factors and biological markers which distinguish prostate cancer among different ethnic groups with varying risk in New York City, specifically whites, blacks and Hispanics. Cancer cases are collected at Columbia-Presbyterian Medical Center and its affiliates and data collected on various risk factors, as well as serum and tissue specimens. Controls are other hospitalised men. To date 100 cases and 180 controls have been collected. Case-control comparisons are done overall and by ethnic group.

TYPE: Case-Control
TERM: Ethnic Group; Race
SITE: Prostate
TIME: 1986 - 1995

1130 **Pasternack, B.S.** 01874
New York Univ. Medical Center, Dept. of Environmental Medicine, 550 1st Ave. , New York NY 10016, United States of America (Tel.: +1 212 2635289; Fax: 2635019)
COLL: Toniolo, P.G.; Shore, R.E.; Levitz, M.; Bruning, P.F.; Riboli, E.; Strax, P.

Endocrine and Environmental Factors in Breast Cancer

This study comprises a prospective investigation of endocrinological factors and dietary components involved in initiation or promotion of breast cancer. Some 16,000 women, both premenopausal and postmenopausal will have had a complete breast examination, including questionnaire, clinical examination and mammography. Annual blood samples taken from these women have been frozen and stored. Pre-menopausal women were given menstrual cycle recording calendars, which will allow each blood sample to be dated relatively to the end of the current menstrual cycle and provide evidence of extra-short or extra-long cycles. In a nested case-control study, telephone interviews are conducted on women who develop breast cancer and their matched controls to obtain more detailed medical, reproductive and socio-economic information. Subsequently, stored blood samples for the matched sets are retrieved for laboratory analyses. To date, approximately 300 invasive cases have been identified, 90 of which were diagnosed at the time of entry into the cohort and the remaining at or between subsequent screening. 185 cases are postmenopausal and 115 premenopausal.

TYPE: Case-Control; Cohort
TERM: BMB; Diet; Drugs; Hormones; Nutrition; Screening
SITE: Breast (F)
CHEM: Androgens; Oestrogens; Prolactin
TIME: 1984 - 1994

1131 **Perera, F.P.** 03588
Columbia Univ., School of Public Health, Div. Environmental Health Sciences, 60 Haven Ave. B-109, New York NY 10032, United States of America (Tel.: +1 212 3053465)
COLL: Santella, R.M; Jeffrey, A.M.; Brandt-Rauf, P.W.; Tsai, W.Y.; Pero, R.W.; Hemminki, K.; Walles, S.A.S.; Mayer, J.

Molecular Epidemiology - Chemical Carcinogenesis: Adducts, Oncogenes and Cytogenetic Markers.
The objective of these studies is to develop, validate and apply a battery of biologic markers of carcinogenesis in human subjects. Several related studies of model populations (lung cancer patients, exposed workers, smokers, groups with community exposure, tobacco smoke exposed children, chemotherapy patients) and appropriate controls are being carried out. A battery of measurements (DNA and protein adducts, sister chromatid exchange, chromosomal aberrations, micronuclei, oncogene activation, tumour suppressor gene activation and gene mutation) is being obtained from the same biological samples along with detailed health and exposure histories. Markers of genetic susceptibility and nutritional status are also being evaluated. Recent papers have been published in Nature 360:256-258, 1992 and Cancer Research 52:3558-3565, 1992.

TYPE: Correlation; Methodology; Molecular Epidemiology
TERM: Biomarkers; Chemical Exposure; Chromosome Effects; DNA Adducts; DNA Binding; Genetic Markers; Micronuclei; Passive Smoking; Protein Binding; SCE; Tobacco (Smoking)
SITE: All Sites; Lung
TIME: 1983 - 1994

1132 **Toniolo, P.G.** 05129
New York Univ. Medical Center, Dept. of Environmental Medicine, 341 E. 25th St. , New York NY 10010, United States of America (Tel.: +1 212 2636499; Fax: 2638570)
COLL: Pasternack, B.S.; Shore, R.E.

Mammographic Parenchymal Patterns and Breast Cancer Risk
This case-control study nested within a cohort aims to explore the relationship between mammographic parenchymal patterns, the dietary intake of fat, serum levels of endogenous reproductive hormones (oestrogens, androgens and prolactin) and the subsequent risk of breast cancer. The study is prospective in design, in that parenchymal patterns, dietary data, and endocrine function are assessed prior to breast cancer detection, often years in advance. Information is being collected on approximately 250 histologically confirmed breast cancer cases, diagnosed between 1985 and 1992, and on 650 individually matched controls, selected at random within risk sets defined by matching criteria of age, duration of storage of serum specimens, number of blood donations, and menopausal status from among the members of the same cohort from which the cases arose. The study requires consensus classification of mammograms according to Wolfe's patterns and percent densities by two radiologists. Extensive background information on potential confounders is available.

TYPE: Case-Control; Cohort
TERM: BMB; Diet; Fat; Hormones; Mammography; Serum
SITE: Breast (F)
TIME: 1991 - 1994

1133 **Wallach, R.C.** 01194
New York Univ. School of Medicine, 530 First Ave., New York NY 10016, United States of America (Tel.: +1 212 2635199)

Risk Factors for Ovarian Cancer
Failure to develop techniques for early diagnosis of ovarian cancer leads to the late presentation of most cases at a time when cure is unlikely. No significant major risk factors have been identified in the genesis of ovarian cancer. Patients with ovarian cancer are identified and their records are studied in an attempt to correlate factors in personal, medical, social, occupational and other environmental conditions that may exert a carcinogenic effect related to the development of ovarian cancer.

TYPE: Case Series
TERM: Clinical Records; Occupation; Reproductive Factors; Socio-Economic Factors
SITE: Ovary
TIME: 1975 - 1995

1134 **Wynder, E.L.** 01900
American Health Foundation, Div. of Epidemiology, 320 E. 43rd St., New York NY 10017, United States of America (Tel.: +1 212 9531900)
COLL: Stellman, S.; Ng, S.K.C.; Moore, M.

Epidemiology of Smoking-Related Diseases

The objective of this case-control study is to identify and examine the effect of risk factors associated with cigarette smoking, plus confounding factors, on lung, bladder, pancreatic and kidney cancer, myocardial infarction and cancers of the upper respiratory and upper alimentary tracts. The effects of smoking cigarettes with varying levels of tar and nicotine on the risk of disease will be measured and the use of the newer "less hazardous" cigarettes monitored. Questionnaires are administered to hospitalized patients with an index disease, and to hospitalized controls in 18 US hospitals. A total of 1,250 cases and an equal number of hospital controls are interviewed annually. The major parameters of tobacco usage investigated include: (1) type, duration, and amount of tobacco used by brand, (2) tar and nicotine yields of cigarette smoked, (3) inhalation practices and (4) length of cigarette smoked. These data are analysed in relation to various demographic characteristics, and risk ratios are calculated. One of the fundamental purposes of the study is to determine the risk of the various tobacco-related diseases for individuals with long histories of smoking filter cigarettes in general and low tar/low nicotine yielding cigarettes in particular. A secondary purpose is to investigate the role of nutritional factors in the aetiology of these diseases. Recent papers appeared in Cancer 62:1223-1230, 1988; Int. J. Cancer 43:190-194 and 42:325-328, 1988 and Cancer Res. 48:4405-4408, 1988.

TYPE: Case-Control
TERM: Drugs; Tobacco (Smoking)
SITE: Bladder; Kidney; Larynx; Lung; Oesophagus; Oral Cavity; Pancreas
CHEM: Carbon Monoxide; Nicotine; PAH; Tars
TIME: 1982 - 1994

1135 **Wynder, E.L.** 03924
American Health Foundation, Div. of Epidemiology, 320 E. 43rd St., New York NY 10017, United States of America (Tel.: +1 212 9531900)
COLL: Stellman, S.

Environmental Tobacco Smoke in Relation to Lung Cancer in Non-Smokers

The purpose of this case-control study is to determine whether lung cancer in non-smokers is associated with exposure to second hand tobacco smoke. Detailed information will be collected on lifetime exposure to other people's tobacco smoke from 180 newly diagnosed lung cancer patients who have never smoked and from a comparison group of 540 lifetime non-smokers. Major components of environmental tobacco smoke (ETS) exposure addressed in the questionnaire include exposure in childhood, in adulthood, at home, at work, in transportation, and in public places. For each member of the household (both in childhood and adulthood) who smoked, the average number of hours per day of exposure, the number of years of exposure, and the quality (intensity) of exposure are asked. Lifetime non-smoking status of study subjects as well as the smoking habits of the current spouse will be validated by reviewing the hospital chart and contacting the spouse. The effect of ETS exposure will be assessed individually for the major components of childhood and adulthood. In addition, a cumulative exposure index will be devised which takes into account exposure from all sources. Information will be collected on other factors which may affect the risk of lung cancer in non-smokers, i.e., occupation, diet, residence, and past exposure to therapeutic radiation. These factors will be controlled for in the analysis. The effect of ETS exposure will be assessed for all histological types of lung cancer combined, as well as for the major specific types (i.e. squamous cell and adenocarcinoma). An effort will be made to obtain information on the location of the lesion within the lung (i.e. central/peripheral) to assess whether it is associated with ETS exposure.

TYPE: Case-Control
TERM: Diet; Environmental Factors; Histology; Occupation; Passive Smoking; Radiation, Ionizing
SITE: Lung
TIME: 1985 - 1994

UNITED STATES OF AMERICA

Newark

1136 **Mashberg, A.** 02582
Univ. of Medicine and Dentistry, New Jersey Medical School, Div. of Surgical Oncology, 185 S. Orange Ave., Newark NJ 07103, United States of America (Tel.: +1 201 6761000/1627; Fax: 6767752)
COLL: Boffetta, P.; Garfinkel, L.

Alcohol and Oral Squamous Carcinoma

The object of the final segment of this study is to evaluate the carcinogenic effects of alcohol and tobacco on specific anatomical sites of the oral cavity/oropharynx in a population of heavy drinkers and cigarette smokers. Series of 353 oral cavity cancer male patients with 347 invasive cancers and 64 carcinomas in situ, collected at a New Jersey Veteran Administration Medical Center, were interviewed with respect to tobacco smoking and alcohol drinking. Sites of origin of cancers are classified as floor of the mouth (N=144), oral tongue (N=51), anterior tonsillar pillar (N=46), soft palate (N=41), lingual aspect of retromolar trigone (N=10), alveolar ridge, buccal mucosa and hard palate (less than 10 cases). The remaining 51 cases have cancers from multiple sites. Odds ratios of tobacco and alcohol consumption will be calculated by comparing each site with floor of the mouth. Preliminary analysis suggests that tobacco smoking appears more strongly associated with cancers of soft palate, retromolar trigone and oral tongue, whereas alcohol seems to exert a stronger carcinogenic effect on the floor of the mouth and oral tongue. A paper has been published in Int. J. Cancer 52:530-533, 1992 (Boffetta, P.).

TYPE: Case-Control
TERM: Alcohol; Anatomical Distribution; High-Risk Groups; Tobacco (Smoking)
SITE: Oral Cavity; Oropharynx
TIME: 1981 - 1994

Oakland

1137 **Friedman, G.D.** 00851
Kaiser Permanente Medical Care Program, Div. of Research, 3451 Piedmont Ave. , Oakland CA 94611, United States of America (Tel.: +1 510 4502129; Fax: 4502073)
COLL: Selby, J.V.

Surveillance for Drugs that May be Carcinogenic

The objective is to evaluate medicaments for possible carcinogenic effects. To seek hypotheses, 143,000 persons will be classified according to use and non-use of various medicaments by means of computer-stored records collected from 1969-1973 by the San Francisco Kaiser-Permanente Clinic (1.3 million prescriptions for 3,400 drugs). The development of various kinds of cancer will be followed up by surveillance of hospital records. Drug-cancer associations found would be subject to further study. More detailed evaluation of hypotheses are by case-control studies using the Kaiser-Permanente records, with attention to potential confounding variables. Papers were published in Int. J. Cancer 41:677-682, 1988, Cancer Res. 49:5736-5747, 1989 and in Am. J. Epidemiol. 136:1415-1416, 1992.

TYPE: Cohort
TERM: Clinical Records; Drugs; Monitoring; Record Linkage
SITE: All Sites
TIME: 1977 - 1999

1138 **Friedman, G.D.** 01977
Kaiser Permanente Medical Care Program, Div. of Research, 3451 Piedmont Ave. , Oakland CA 94611, United States of America (Tel.: +1 510 4502129; Fax: 4502073)
COLL: Sidney, S.

Surveillance of Health Effects of Potentially Less Hazardous Cigarettes

This is a prospective epidemiological study designed to investigate the health effects of smoking, with particular interest in the recently introduced smoking products which have reduced yields of selected toxic constituents. The differences between health effects of the older and newer smoking products are to be evaluated. A self-administered questionnaire was completed by most Kaiser health plan subjects given a multiphasic health checkup (MHC) at the KFRI Oakland-San Francisco facilities, 1979 - 1986. Approximately 115,000 subjects are involved. The questionnaire asks for details of smoking habits, as well as including more general questions on occupation, demographic/socio-economic characteristics, and dietary practices. Questionnaire responses, in conjunction with mortality and morbidity data (hospitalization, physician visits, days of incapacity) and MHC data, are to be analysed for relationship between smoking habits and adverse health effects. Long-term follow-up is in progress. Publications appeared in West. J. Med. 145:651-656, 1986; Am. J. Epidemiol. 129:1305-1309, 1989, and in Am. J. Public Health 79:1415-1416, 1989.

TYPE: Cohort
TERM: Diet; Occupation; Socio-Economic Factors; Tobacco (Smoking)
SITE: All Sites
TIME: 1979 - 1999

1139 **Goldhaber, M.K.** 04793
Kaiser Permanente Medical Care Program of Northern California, Div. of Research, 3451 Piedmont Ave., Oakland CA 94611, United States of America (Tel.: +1 415 9872188)
COLL: Friedman, G.D.; Armstrong, M.A.; Golditch, I.

Retrospective Follow-up Study of the Sequelae of Tubal Sterilization

This is a study to determine if certain forms of female sterilization are associated with increased incidences of serious disease, including cancer, and death. Subjects for the study (n = 42,922) were all women sterilized in the Kaiser Permanente Medical Care Program in Northern California during 1971-1984, who had given birth to their most recent child in California. Data on age, race, interval since last birth, and parity at the time of the sterilization were extracted from the children's birth certificates and used to match sterilized women to controls (n = 48,215) within the Kaiser membership. Study and control cohorts were computer-linked to cancer registry and hospitalization files at Kaiser Permanente. Relative risks will be calculated by the Cox proportional hazards model for blocked data.

TYPE: Cohort
TERM: Tubal Ligation
SITE: Breast (F); Ovary; Uterus (Cervix); Uterus (Corpus)
TIME: 1986 - 1994

1140 **Hiatt, R.A.** 04131
Kaiser Permanente Medical Care Program, Div. of Research, 3451 Piedmont Ave. , Oakland CA 94611, United States of America (Tel.: +1 510 4502109; Fax: 4502072)
COLL: Krieger, N.

Benign Breast Disease, Exogenous Hormone Use and Breast Cancer

In order to clarify the relationship between benign breast disease (BBD), exogenous hormones and breast cancer (BC), a three part study has been undertaken using a cohort of approximately 2,500 women with biopsy documented benign breast disease. BC risk in this cohort was previously found to be directly related to the degree of atypia (Cancer 39(6): 2603-2607, 1977). This previous study has been updated by determining all additional BCs that have occurred in this group through 1985 and making some corrections in the dataset. In a nested case-control study, BC cases will be compared with women with BBD and without cancer for hormone use after the diagnosis of BBD as documented in their medical records. Third, in a second nested case-control study, hormone use before BBD will be examined among a sub-sample of women with BBD and matched health plan controls. (Am. J. Epidemiol. 135:619-631, 1992).

TYPE: Case-Control; Cohort
TERM: Drugs; Hormones; Premalignant Lesion
SITE: Benign Tumours; Breast (F)
CHEM: Oestrogens
TIME: 1986 - 1994

1141 **Hiatt, R.A.** 05234
Kaiser Permanente Medical Care Program, Div. of Research, 3451 Piedmont Ave. , Oakland CA 94611, United States of America (Tel.: +1 510 4502109; Fax: 4502072)

Renal Cell Cancer and Use of Thiazide

Previous studies have suggested that thiazide use may increase the risk of renal cell carcinoma. These studies need to be confirmed and, in addition, data are needed to assess duration, dose and any effect of the indication for thiazide use. The current case-control study selected all cases of renal cancer among persons who had a multiphasic health check-up while members of a large pre-paid health plan from 1964-1988. Controls also took such an examination and were matched on age, sex, and year of examination. Data on body mass index, smoking and alcohol use, coffee and tea consumption, race/ethnicity and family history of cancer will be taken from the answers to questions recorded at the time of examination. Medical record reviewers will abstract information on thiazide use.

TYPE: Case-Control
TERM: Alcohol; Clinical Records; Coffee; Diuretics; Race; Registry; Tea; Tobacco (Smoking)
SITE: Kidney
REGI: Bay Area (USA); California (USA)
TIME: 1991 - 1994

***1142 Hiatt, R.A.** 05328
Kaiser Permanente Medical Care Program, Div. of Research, 3451 Piedmont Ave. , Oakland CA 94611, United States of America (Tel.: +1 510 4502109; Fax: 4502072)
COLL: Vogelman, J.; Orentreich, N.; Krieger, N.

Breast Cancer and Hormone Levels

This nested case-control study tests three hypotheses: (1) prediagnostic serum DHEA, DHEAS, and oestrone sulfate (ES) levels are higher among breast cancer patients than controls; (2) postdiagnostic serum DHEA, DHEAS, and ES levels are higher among breast cancer patients than controls, and (3) the prediagnostic and postdiagnostic serum levels remain the same among breast cancer patients. Analyses will be conducted using frozen serum obtained from members of the Kaiser Permanente Medical Care Program (Oakland, CA) at the time of a multiphasic health examination between 1964 and 1972. Cases are all women aged 55 and older in this cohort who had two multiphasic examinations during this interval and who were diagnosed with breast cancer between these (n = 35). Each case is matched to two controls for age at and year of the first multiphasic examination. Analyses will compare serum levels of cases and controls, and also among cases (pre- versus post-diagnostic levels), adjusting for numerous potential confounding breast cancer risk factors, based on questionnaire data obtained at the examination and supplemental medical chart review.

TYPE: Case-Control
TERM: BMB; Hormones; Serum
SITE: Breast (F)
TIME: 1992 - 1994

***1143 Hiatt, R.A.** 05329
Kaiser Permanente Medical Care Program, Div. of Research, 3451 Piedmont Ave. , Oakland CA 94611, United States of America (Tel.: +1 510 4502109; Fax: 4502072)
COLL: Drezner, M.; Vogelman, J.; Orentreich, N.; Krieger, N.

Breast Cancer and Vitamin D Levels

This nested case-control study tests the hypothesis that risk for breast cancer is inversely related to level of Vitamin D. Analyses will be conducted using frozen serum obtained from members of the Kaiser Permanente Medical Care Program (Oakland, CA) at the time of a multiphasic health examination between 1964 and 1972. Follow-up to detect all incident cases extended through December 1990. Cases include 100 randomly selected white women who developed post-menopausal breast cancer after the multiphasic examination. Each case was matched to one cancer-free control for race/ethnicity and both age at and year of the multiphasic examination; controls were also required to remain in the health plan until the time of the case's diagnosis with breast cancer. Analyses will compare serum levels of Vitamin D among cases and controls, adjusting for numerous potential confounding breast cancer risk factors, based on questionnaire data obtained at the examination and supplemental medical chart review.

TYPE: Case-Control
TERM: BMB; Serum; Vitamins
SITE: Breast (F)
TIME: 1992 - 1994

***1144 Krieger, N.** 05335
Kaiser Foundation Research Inst. Div. of Research, 3451 Piedmont Ave. , Oakland CA 94611, United States of America (Tel.: +1 510 4502157; Fax: 4502072)
COLL: Wolff, M.S.; Hiatt, R.A.; Vogelman, J.; Orentreich, N.

Breast Cancer and Environmental Carcinogens

This nested case-control study tests the hypothesis that serum levels of DDE (a metabolite of DDT) and PCBs are higher among women who develop breast cancer than among women who remain cancer-free. Analyses will be conducted using frozen serum obtained from members of the Kaiser Permanente Medical Care Program at the time of a multiphasic health examination carried out between 1964 and 1972. Follow-up to detect all incident cases of breast cancer extended through December 1990. Cases include 150 randomly selected women (50 white, 50 black, 50 Asian), who developed breast cancer after the multiphasic examination. Each case was matched to one cancer-free control for race/ethnicity and both age at and year of the multiphasic examination; controls were required to remain in the health plan up to the time of the case's diagnosis with breast cancer. Analyses will compare serum levels of DDE and PCBs among cases and controls, adjusting for numerous potential confounding breast cancer risk factors, based on questionnaire data obtained at the examination and supplemental medical chart review.

TYPE: Case-Control
TERM: BMB; Pesticides; Serum
SITE: Breast (F)
CHEM: DDE; DDT; PCB
TIME: 1992 - 1994

1145 Selby, J.V. 04784
Kaiser Permanente Medical Care Program, Div. of Research, 3451 Piedmont Ave. , Oakland CA 94611, United States of America (Tel.: +1 415 9872106; Fax: 9873027)
COLL: Friedman, G.D.; Quesenberry, C.P.

Case-Control Evaluations of Screening Tests for Colorectal Cancer

This three-year study evaluates the efficacy of digital rectal examination, sigmoidoscopy, and faecal occult blood testing for preventing mortality from colorectal cancer. In separate studies, cases of fatal colorectal cancer within reach of digital examination will be compared with controls who did not die by the date of death of the case, matched to the case for age, sex and length of Program membership. Exposures to the screening test of interest, other screening tests, and a history of prior colorectal neoplasms or a family history of colorectal cancer will be sought by review of medical records. Estimates of efficacy for the most recent screen will be obtained from the odds ratios for exposure to the test, adjusted for the number of prior tests, for other screens, and for other risk factors. Estimates of the most efficacious screening interval will also be calculated. The number of cases in each study is: sigmoidoscopy, 300 cases, two controls per case; digital rectal examination, 250 cases, one control per case; faecal occult blood, 650 cases, one control per case.

TYPE: Case-Control
TERM: Familial Factors; Prevention; Screening; Time Factors
SITE: Colon; Rectum
TIME: 1989 - 1994

Olympia

1146 Milham Jr, S. 04908
Washington State Dept. of Health, ED-13, Olympia WA 98504, United States of America (Tel.: +1 206 1536408)

Washington State Occupational Surveillance System

The occupational and cause of death information for 600,000 male deaths 1950-1989 and 100,000 female deaths 1974-1989 will be analysed using an age- and year of death- standardized proportionate mortality ratio programme. Detailed cause of death analysis (160 causes) will be published for 220 occupational categories for males and 50 categories for females. Previous analyses of this data set detected a multiple myeloma excess in nuclear workers, a lung cancer excess in copper smelter workers and an excess of leukaemia and non-Hodgkin's lymphoma in electrical workers.

TYPE: Mortality
TERM: Occupation
SITE: All Sites
TIME: 1968 - 1994

Philadelphia

1147 **Bunin, G.R.** 05046
Children's Hosp. of Philadelphia, 34th and Civic Center Blvd, Philadelphia PA 19104, United States of America (Tel.: +1 215 5901445)
COLL: Buckley, J.D.; Hammond, D.; Ruccione, K.; Sather, H.N.; Woods, W.G.

Parental Occupation and Childhood Cancer

The Children's Cancer Study Group is conducting a hypothesis-generating case-control study of parental occupation and all major types of childhood cancer. The study uses data on parental occupation from a self-administered questionnaire. The cancers to be studied and the approximate number of cases to be included are: acute lymphocytic leukaemia 1500, acute non-lymphocytic leukaemia 200, Hodgkin's disease 200, non-Hodgkin's lymphoma 200, neuroblastoma 300, Wilm's tumour 200, osteosarcoma 100, astrocytoma 100, primitive neuroectodermal tumour 100, rhabdomyosarcoma 100, and germ cell and gonadal tumour 90. Data on 800 controls selected by random-digit dialling are also available. The occupational exposures of interest include exposure to paints, solvents, metals, and pesticides.

TYPE: Case-Control
TERM: Metals; Occupation; Paints; Parental Occupation; Pesticides; Solvents
SITE: Childhood Neoplasms
OCCU: Farmers; Metal Workers; Welders
LOCA: Canada; United States of America
TIME: 1990 - 1994

1148 **Bunin, G.R.** 05047
Children's Hosp. of Philadelphia, 34th and Civic Center Blvd, Philadelphia PA 19104, United States of America (Tel.: +1 215 5901445)
COLL: Yandell, D.; Meadows, A.T.

Correlation of Mutations in the Retinoblastoma Gene with Parents' Exposures

This pilot study will investigate the possible association of specific exposures that may increase the risk of retinoblastoma with specific types of alterations in the retinoblastoma gene. 10-50 cases from an epidemiological study of retinoblastoma will be studied. The cases to be included are those without a family history who had an exposure that was associated with increased risk. DNA from the case, the tumour, and the parents of the case will be analysed by molecular biological techniques to locate and characterise the mutation in the retinoblastoma gene. The type and location of mutations in exposed cases will be compared to those in unexposed cases.

TYPE: Molecular Epidemiology
TERM: DNA; Mutation, Germinal; Parental Occupation
SITE: Retinoblastoma
OCCU: Metal Workers; Welders
LOCA: Canada; United States of America
TIME: 1991 - 1994

1149 **London, W.T.** 02480
Fox Chase Cancer Center, 7701 Burholme Ave., Philadelphia PA 19111, United States of America (Tel.: +1 215 7282204; Fax: 7283574)
COLL: Buelow, K.B.; MacMahon, B.; Redeker, A.; Smith, M.

Relationship of HBV and Genetic Changes to Aetiopathogenesis of Hepatocellular Carcinoma

Chronic infection with HBV is associated with 80% of hepatocellular carcinoma (HCC) cases worldwide. The precise role of HBV in pathogenesis of HCC is not understood. The aim of this study is to examine the relationship of HBV DNA integration to genetic changes in HCC cells. Tumours will be identified prospectively in a cohort of 5,000 HBV carriers in the Philadelphia area, in a cohort of 1,500 carriers in Alaska and in referred cases in Los Angeles. Status of HBV DNA in tumour and non-tumour tissues will be compared as will the pattern on southern blots of DNA probes for chromosomal regions suspected of containing genes critical to HCC development (i.e. 4, 11, 13). Specifically, loss of heterozygosity in tumour tissues compared with normal tissues at a rate significantly above background will be considered as genetic changes in the pathogenetic pathway.

TYPE: Cohort
TERM: DNA; HBV; Oncogenes
SITE: Liver
TIME: 1987 - 1994

1150 **Meadows, A.T.** 00318
Children's Hosp. of Philadelphia, Div. of Oncology, 34th and Civic Center Blvd, Philadelphia PA 19104, United States of America (Tel.: +1 215 5902804)
COLL: Banfi, A.; Baum, E.; d'Angio, G.J.; Green, D.; LeMerle, J.; Morris-Jones, P.H.; Nesbit, M.; Newton, W.A.; Obringer, A.; Sallan, S.; Siegel, S.; Voûte, P.; Woods, W.G.; Zipursky, A.

Second Malignant Neoplasms after Childhood Cancer
The Late Effects Study Group registers second malignant neoplasms (SMN) occurring in individuals treated for childhood cancer at any of the 13 member institutions and surviving for at least two years after diagnosis. The registry, which now includes over 300 individuals, serves as a resource for studying many aspects of SMN: (1) incidence; (2) histopathology; (3) identification of risk factors such as treatment and genetic factors; (4) unusual tumour associations and (5) congenital anomalies associated with SMN. A recent paper described 91 patients who developed bone sarcomas as SMNs (Cancer 67:193-201, 1991).

TYPE: Registry
TERM: Chemotherapy; Congenital Abnormalities; Genetic Factors; Late Effects; Multiple Primary; Mutation, Somatic; Radiotherapy
SITE: Childhood Neoplasms
LOCA: Canada; France; Italy; Netherlands; United Kingdom; United States of America
TIME: 1972 - 1995

1151 **Meadows, A.T.** 04310
Children's Hosp. of Philadelphia, Div. of Oncology, 34th and Civic Center Blvd, Philadelphia PA 19104, United States of America (Tel.: +1 215 5902804)
COLL: Bunin, G.R.; Albright, L.A.; Allen, J.C.; Boessel, J.; Buckley, J.D.; Neuberg, R.W.; Rorke, L.B.; Shiminski, T.; Smithson, W.A.; Deutsch, M.

Brain Tumours in Young Children
The role of genetic predisposition, preconception exposure to mutagens, and gestational exposure to carcinogens in the aetiology of astrocytoma and primitive neuroectodermal tumour medulloblastoma is being investigated. As these factors are likely to influence the development of brain tumour early in life, the study is restricted to patients diagnosed at ages 0-5. Specifically, family history of seizures and cancer, preconception exposures through parents occupation, gestational exposure to alcohol, medication, nitrosamines and maternal diet are being examined. Cases were identified through the Childrens Cancer Study Group, a multi-institution cooperative group conducting clinical trials. The group consists of 34 member institutions in the US and Canada. Controls matched to cases on race and birthdate (+ 1 year) were obtained by thelphone random digit dialling. Parents of cases and controls were interviewed by telephone. Data on 332 case-control pairs were collected. Separate analyses forastrocytoma and primitive neuroectodermal tumour/medulloblastoma are being conducted.

TYPE: Case-Control
TERM: Alcohol; Childhood; Familial Factors; Intra-Uterine Exposure; Metals; Occupation; Solvents
SITE: Brain; Childhood Neoplasms
CHEM: N-Nitroso Compounds
LOCA: Canada; United States of America
TIME: 1987 - 1994

1152 **Strom, B.L.** 03382
Univ. Pennsylvania, School of Medicine, Clinical Epidemiology Unit, Department of Medicine, Room 225L NEB/S2, Philadelphia PA 19104, United States of America (Tel.: +1 215 8984623)
COLL: Soloway, R.D.; Stolley, P.D.; West, S.; Litvak, J.; Ríos-Dalenz, J.L.; Rodriguez-Martinez, H.

Biochemical Epidemiology of Biliary Tract Cancer

Biliary tract cancer, uncommon in the US but of considerable international importance, provides a unique opportunity to combine biochemistry and epidemiology to shed light on the possible aetiology. The marked differences in incidence between countries have led to many hypotheses about aetiology, including biochemical hypotheses arising out of preliminary work by the investigators: marked differences have been demonstrated in the bile and gallstone constituents of normal, cholelithiasis, and cancer patients in Bolivia and US. In order to further investigate the risk factors and aetiology of biliary tract cancers, a case-control study will examine subjects collected from two collaborating Latin American centres: La Paz, Bolivia and Mexico City, Mexico. Three case groups will be recruited: patients undergoing abdominal surgery who have newly-diagnosed (1) cancer of the gallbladder, (2) cancer of the extrahepatic biliary tract, and (3) cancer of the ampulla of Vater. Each group of cases will be compared to two set of controls: (1) abdominal surgery patients with cholelithiasis and (2) abdominal surgery patients who have no biliary tract disease. Personal interviews will be used to obtain demographic characteristics, prior medical history, family history, and exposure to agents presumed to be risk factors for biliary cancer. Bile, blood, and, where appropriate, gallstone specimens will be obtained from cases and controls during surgery and analysed for various biochemical parameters. 55 cases and 440 controls are expected annually.

TYPE: Case-Control
TERM: Biochemical Markers; Gallstones
SITE: Bile Duct; Gallbladder
LOCA: Bolivia; Mexico; United States of America
TIME: 1984 - 1994

1153 **Weiss, W.** 00021
Hahnemann Univ., 3912 Netherfield Rd, Philadelphia PA 19129, United States of America (Tel.: +1 215 8494971)

Respiratory Effects of Chloromethyl Ethers

A prospective study of 125 male workers at a chemical plant by periodic chest X-rays and questionnaires was conducted from 1963 to 1968. In 1965 a spirogram was done in 103 men. Additional follow-up has been continued to the present. 94 men exposed to chloromethyl methyl ether containing bis(chloromethyl)ether as a contaminant were divided into three exposed groups according to the magnitude of individual exposure indexes. There was a dose-response relationship between cumulative chemical exposure and the prevalence of chronic cough and expectoration, and the incidence of bronchogenic carcinoma. There was an inverse relationship between smoking habits and the risk of lung cancer. Of 94 exposed men, 22 have developed lung cancer (15 small cell). Only 1 of 31 unexposed men has developed lung cancer, not small cell. The shape of the epidemic curve has been investigated graphically and by calculating period SMRs. Since 1975-1979 the curve has been descending. Recent analysis shows that the way the cohort was assembled in 1963 resulted in a biased sample due primarily to inadvertent selection of men with higher exposures. Follow-up continues. Papers have been published in J. Occup. Med. 22:527-529, 1980; J. Nat. Cancer Inst. 69:1265-1270, 1982 and in J. Occup. Med. 31:102-105, 1989. Final follow-up as of December 1992 is now in progress.

TYPE: Cohort
TERM: Chemical Exposure; Dose-Response; Histology; Occupation; Tobacco (Smoking)
SITE: Lung
CHEM: BCME; CMME
TIME: 1963 - 1994

Pittsburgh

1154 **Grufferman, S.** 04136
Univ. of Pittsburgh, School of Medicine, Dept. Clin. Epidemiol. & Family Med., 33550 Terrace St., M-200 Scaife Hall, Pittsburgh PA 15261, United States of America (Tel.: +1 412 6488933; Fax: 6489114)
COLL: Ambinder, R.F.; Schwartz, A.G.; Yang, P.

Case-Control Study of Hodgkin's Disease in Children

A case-control study of childhood HD has been initiated with cases obtained from the Children's Cancer Group and Pediatric OncologyGroup. Controls, individually matched to cases for age, sex, and race, are selected by random digit dialling. Data on cases and controls are collected by telephone interviews of the parents. So far, 333 cases and 328 controls have been interviewed. A total of 415 cases and 675 controls is expected. The major focus of the study is on the risk of HD in relation to early childhood infectious diseases. Approximately 36% of childhood cases from a pilot study were found to have EBV in their tumour tissues. Other risk factors to be evaluated include childhood environmental exposures, breast feeding, familial factors, socio-economic status, parental occupation, exposures and use of recreational drugs during pregnancy. A repository of plasma and white blood cell samples from cases, and of sections of their Hodgkin's disease tumour tissue, will be maintained for future investigation.

TYPE: Case-Control
TERM: BMB; Childhood; Drugs; EBV; Environmental Factors; Infection; PCR; Parental Occupation; Plasma; Segregation Analysis; Socio-Economic Factors; White Cells
SITE: Hodgkin's Disease
LOCA: Canada; United States of America
TIME: 1991 - 1996

***1155 Grufferman, S.** 05401
Univ. of Pittsburgh, School of Medicine, Dept. Clin. Epidemiol. & Family Med., 33550 Terrace St., M-200 Scaife Hall, Pittsburgh PA 15261, United States of America (Tel.: +1 412 6488933; Fax: 6489114)
COLL: Maurer, H.M.; Ruymann, F.B.; Schwartz, A.G.; Yang, P.

Case-Control Study of Childhood Rhabdomyosarcoma

This is an interview case-control study of childhood rhabdomyosarcoma (RMS). The major goal is to investigate potential aetiological and environmental factors. The following risk factors will be studied: father's cigarette smoking, immunization history, preventable infectious diseases, environmental exposure to chemicals, mother's age at subject's birth, family history of asthma and socio-economic status. In addition, the possible role of heredity will be investigated as suggested by reports of familial aggregation of RMS with other soft tissues sarcomas and breast cancer. The Intergroup Rhabdomyosarcoma Study (IRS) which serves to identify cases, obtains subjects from the Pediatric Oncology Group and the Children's Cancer Group. Approximately 85% of all new cases in the US are entered into IRS protocols each year and all diagnoses are subjected to central expert pathology review. A total of 322 cases and 322 matched community controls have been accrued from 69 medical centres in 42 states. Controls are from the same communities as cases and are of the same race and sex and of similar age. Controls are identified by random-digit dialling, and all subjects' parents are interviewed by telephone. Preliminary findings show that parents' use of cocaine and marijuana during the year before the child's birth is associated with a significantly increased risk of RMS. Routine childhood immunizations were found to be protective, but this appears to be an artifact resulting from immunizations being stopped once a diagnosis of RMS is made. It was also found that in utero radiation exposure was a risk factor for RMS. The first study's findings that fathers', but not mother's, cigarette smoking was a strong risk factor for the disease in their children was not confirmed. An excess of major congenital malformations and neurofibromatosis was observed in RMS cases. Analysis continues.

TYPE: Case-Control
TERM: Age; Asthma; Chemical Exposure; Childhood; Cocaine; Heredity; Immunology; Infection; Intra-Uterine Exposure; Marijuana; Radiation, Ionizing; Socio-Economic Factors; Tobacco (Smoking)
SITE: Rhabdomyosarcoma
TIME: 1987 - 1994

1156 Marsh, G.M. 03562
Univ. of Pittsburgh, Graduate School of Public Health, Dept. Biostat. & Environm. Epidemiology, 130 De Soto St., Pittsburgh PA 15261, United States of America (Tel.: +1 412 6243032)
COLL: Enterline, P.E.

Mortality Surveillance in Fibreglass and Mineral Wool Workers

This is an update and extension of a previous historical-prospective mortality study by the same investigators, of 15,016 fibreglass production/maintenance workers fromm 11 plants and 1,874 mineral wool workers from six plants, who were employed a year or more (six months for two plants) between 1 January 1945 and 31 December 1963 (1 January 1940-1963 for one plant). The current study will include male and female workers hired until 1978 and will examine the mortality experience of the cohort ultimately until 1989. The primary objective is to determine whether exposures associated with the production of man-made mineral fibres are associated with excess mortality risks due to malignant or non-malignant respiratory disease. The current study will also include the collection of data on smoking habits and an expanded assessment of historical environment exposures in the plants.

TYPE: Cohort
TERM: Chemical Exposure; Dusts; Environmental Factors; Occupation; Tobacco (Smoking)
SITE: Respiratory
CHEM: Glass Fibres; Mineral Fibres
OCCU: Mineral Fibre Workers
TIME: 1987 - 1994

1157 **Marsh, G.M.** 04138
Univ. of Pittsburgh, Graduate School of Public Health, Dept. Biostat. & Environm. Epidemiology, 130 De Soto St., Pittsburgh PA 15261, United States of America (Tel.: +1 412 6243032)
COLL: Leviton, L.C.; Talbott, E.O.

Health Registry Study of Drake (Kilsdonk) Chemical Company

The Drake "Superfund" waste site is located in Lock Haven, Pennsylvania, on the site of the former Drake Chemical Company. Drake and its predecessors produced, used or stored many chemicals including beta-naphthylamine, benzidine and benzene. The EPA has been involved in the clean-up of the waste site since February 1982. Several studies have revealed excess bladder cancer rates among males in the Lock Haven area and among former employees of the Drake and predecessor companies. The purpose of the current research is to register and undertake medical surveillance of a cohort of workers known to be at increased risk of bladder cancer due to occupational exposure to beta-naphthylamine. The registry has been designed to: (1) determine the vital status of all former Drake workers (2) determine the current addresses of all living cohort members, (3) notify and enrol as many cohort members as possible, (4) establish and maintain a programme of medical surveillance for bladder cancer, (5) analyse the results of the surveillance programme, (6) evaluate the bladder risk in the cohort and (7) evaluate the total and other cause-specific mortality risks in the cohort.

TYPE: Cohort
TERM: Analgesics; Chemical Exposure; Occupation; Solvents
SITE: All Sites; Bladder
CHEM: 2-Naphthylamine; Benzene; Benzidine
TIME: 1986 - 1994

1158 **Marsh, G.M.** 03660
Univ. of Pittsburgh, Graduate School of Public Health, Dept. Biostat. & Environm. Epidemiology, 130 De Soto St., Pittsburgh PA 15261, United States of America (Tel.: +1 412 6243032)

Update of the Indian Orchard Plant Mortality Study

This is an update of a previous historical prospective mortality study of 2,479 male hourly employees who worked one or more years between 1 January 1949 and 31 December 1966 at a plastics producing plant in Massachusetts. Vital status was determined for the cohort through 1976. In the original study, comparisons with the local county white male mortality experience revealed a slight excess in digestive system cancer and a statistically significant excess ($p<0.05$) in genito-urinary cancer. A secondary nested matched case-control study revealed possible associations between rectal cancer and cellulose nitrate production and between prostatic cancer and polystyrene processing. The original cohort also served as a data base for a proportional mortality study designed to examine the long term health effects of formaldehyde exposure. Reports of the original findings have been published (Br. J. Industr. Med. 39:313-322, 1982; J. Occup. Med. 25:219-230, 1983). The updating of this cohort through 1983 will add to the number of deaths and the duration of follow-up and will include complete work history information for the entire study population. These new data will allow an historical prospective analysis of total and cause specific mortality patterns relative to specific factors in the occupational environment of these men. The updated study will also include a matched case-control analysis of all respiratory cancer, buccal/pharyngeal cancer deaths, and a prospective assessment of mortality patterns among workers with previous employment in the vinyl chloride polymerization and processing areas.

TYPE: Case-Control; Cohort
TERM: Chemical Exposure; Environmental Factors; Occupation; Plastics
SITE: All Sites
CHEM: Formaldehyde; Hydrocarbons; Hydrocarbons, Halogenated; Nitrates; Styrene; Vinyl Chloride
OCCU: Chemical Industry Workers; Plastics Workers
TIME: 1985 - 1994

1159 **Marsh, G.M.** 04139
Univ. of Pittsburgh, Graduate School of Public Health, Dept. Biostat. & Environm. Epidemiology, 130 De Soto St., Pittsburgh PA 15261, United States of America (Tel.: +1 412 6243032)
COLL: Esmen, N.A.

Mortality Patterns among Chemical Plant Workers Exposed to Formaldehyde and Other Substances
In 1986 the National Cancer Institute and Formaldehyde Institute reported on a cohort mortality study of 26,651 workers employed at some time before 1966 in one or more of 10 facilities using formaldehyde in the United States. This study revealed a statistically significant excess of nasopharyngeal cancer (4 observed, less than 1 expected) in a sub-cohort of 4,389 workers from one plant using formaldehyde. It was not, however, possible to draw definitive conclusions concerning this excess due to the small number of deaths and to difficulties in assessing the contribution of extraneous or confounding factors. The aim of the present study is to perform an enhanced, extended, and independent historial-prospective cause-specific mortality study of the Connecticut plant workforce. The cohort will include about 6000 employees hired between 1941 and 1986. Using both external and internal comparisons the mortality experience of the cohort will be examined from 1 January 1945 through 31 December 1984 with emphasis on mortality from malignant neoplasms and particularly nasopharyngeal cancer. This study will test the hypothesis that workers exposed not to formaldehyde alone but to formaldehyde in conjunction with particulates are at an increased risk of developing cancers in the nasopharynx.

TYPE: Cohort
TERM: Chemical Exposure; Occupation; Plastics
SITE: All Sites; Nasopharynx
CHEM: Formaldehyde
TIME: 1986 - 1994

1160 **Redmond, C.K.** 00126
Univ. of Pittsburgh, School of Public Health, Dept. of Biostatistics, 130 De Soto St., Pittsburgh PA 15261, United States of America
COLL: Costantino, J.P.; Rockette, H.E.; Mazumdar, S.; LeGasse, A.A.; Bass, G.I.

Cancer Mortality in Steel Workers
The objective of this study is the identification of work areas, occupations, and processes within the steel industry that are associated with unusual mortality from specific diseases indicative of health hazards in the work environment. Two cohorts of steel workers consisting of (1) all male workers employed at seven US Allegheny County steel plants in 1953 and (2) all oven workers employed at twelve coke plants, plus controls, in the USA and Canada in 1951-1955 have been followed for mortality, 1953-1975. Mortality rates for specific causes of death among men employed in various jobs and work areas are compared with rates for the total steel worker population. Reports have been prepared on coke oven workers, crane operators, and open hearth workers, sheet and tin mill workers, and masons. Excess cancer mortality was observed in several work areas including coke ovens (respiratory), foundry (genito-urinary), electric furnace (urinary), sheet and tin mill (lymphatic and haemopoietic) and blacksmith shop (respiratory) (Env. Hlth Perspect. 52:67-73, 1983). Efforts to update until 1982 work histories and vital status of all coke-oven workers and selected portions of the remaining steelworkers cohort until 1982 have been completed. Currently this updated information is being used to develop modelling methodology for carcinogenic risk assessment of exposure to environmental pollutants. Some preliminary work has been published (Am. J. Epidemiol. 128:860-873, 1988 and Environm. Hlth Perspect. 90:271-277, 1991).

TYPE: Cohort
TERM: Metals; Occupation
SITE: All Sites; Genitourinary; Haemopoietic; Lymphoma; Respiratory
CHEM: Steel
OCCU: Coke-Oven Workers; Steel Workers
LOCA: Canada; United States of America
TIME: 1971 - 1994

1161 **Redmond, C.K.** 01962
Univ. of Pittsburgh, School of Public Health, Dept. of Biostatistics, 130 De Soto St., Pittsburgh PA 15261, United States of America
COLL: Costantino, J.P.; Arena, V.; LeGasse, A.A.; Bass, G.I.

Mortality Study of High Nickel Alloy Workers

There is evidence of increased incidence of respiratory cancers, including cancers of the nasal sinuses, among workers in the nickel refining industry. In order to determine whether any such relationship might exist among workers in the high nickel alloy (more than 20% Ni) industry, cause-specific mortality and especially site-specific cancer mortality among workers in 11 study plants across the USA were investigated. The mortality experience of nickel-exposed workers in the study plants was compared to the population of workers involved in all phases of operation in the study plants, and to the total US mortality experience. The study was a historical prospective follow-up on a cohort of approximately 25,000 workers with at least one year of job experience during the period 1956-1960. A report on cause specific mortality patterns as of 1977 has been prepared (IARC Scient. Publ. No 53, 1984). Overall, no increased risk was observed for cancers of the nasal sinuses, larynx or kidney. Cancer of the liver and large intestine did demonstrate statistically significant standard US mortality ratios. Observed increases were found primarily among longer term workers, but were not concentrated in a particular work area or job category. In addition, excess mortality from lung cancer was found among men employed in maintenance jobs, but it is unclear whether the greater risk is directly associated with nickel exposure since such an excess was not evident in other job categories where nickel exposure was also present. No conclusions regarding a causal association with nickel exposure have been drawn at this time. Continued observation of the cohort is recommended, since the cohort is still relatively young and additional follow-up would result in a substantial increase in expected cancer deaths. Currently, efforts are underway to update the vital status follow-up of the cohort through 1988.

TYPE: Cohort
TERM: Metals; Occupation
SITE: All Sites; Colon; Liver; Nasal Cavity; Rectum; Respiratory
CHEM: Nickel
OCCU: Nickel Workers
TIME: 1978 - 1994

1162 **Schwartz, A.G.** 05088
Univ. of Pittsburgh School of Med., Dept. of Clinical Epidemiology & Preventive Medicine, M200 Scaife Hall, Pittsburgh PA 15261, United States of America
COLL: Moll, P.P.; Swanson, G.M.

Familial Risk of Lung Cancer

A study of the familial risk of lung cancer and other respiratory disease among families identified through a population-based series of 401 non-smoking lung cancer cases aged 40-84, 135 lung cancer cases less than 40, and age-, sex-, race-matched population-based controls is being conducted in the Detroit Metropolitan area. Telephone interviews are used to collect data on cancer history, respiratory disease history, smoking, passive smoking, occupation, and demographic variables for each of the estimated 7,000 first-degree relatives of the cases and controls. Various measures of familial risk will be used in the determination of disease risk among relatives after adjustment for individual risk factors. The modification of a new statistical method, the family risk index method, appropriate for the analysis of family data, is being undertaken to measure familial heterogeneity of cancer risk. In addition, logistic regression and proportional hazards modelling will be used for data analysis.

TYPE: Case-Control; Methodology
TERM: Familial Factors; Occupation; Passive Smoking; Tobacco (Smoking)
SITE: Lung
TIME: 1990 - 1995

Portland

1163 **Nussbaum, R.H.** 04996
Portland State Univ., Depts of Physics and Environmental Sciences and Research, P.O.Box 751, Portland OR 97207-0751, United States of America (Tel.: +1 503 2225643; Fax: 7254882)
COLL: Stewart, A.M.; Cummins, H.W.

Late Effects of Low-Level Radiation

A cohort of about 18,000 military personnel who were stationed at Camp Hanford (Washington State) during several years of large releases of radioactive substances has been identified. In cooperation with the US Veterans Administration Environmental Epidemiology Branch, death certificates will be traced. As a first step, cancer mortality among Hanford personnel will be compared with that of a matched cohort of up to 100,000 military personnel, stationed during the same period at Fort Lewis (Washington State) Camp. A ten-year follow-up is planned and other end-points might be added. Statistical analysis will be carried out at Birmingham University (United Kingdom). The hypothesis to be tested is that cancer mortality among a radiation-exposed cohort cannot be distinguished from that of an unexposed cohort.

TYPE: Cohort
TERM: Occupation; Radiation, Ionizing
SITE: All Sites
OCCU: Military Servicemen
TIME: 1987 - 1994

1164 **Weinmann, S.A.** 05131
Univ. of Washington, School of Public Health and Community Medicine, Kaiser Permanente, Northwest Region, 3715 N. Interstate Ave. , Portland OR 97227, United States of America (Tel.: + 1 503 2493318; Fax: 2493320)
COLL: Weiss, N.S.; Psaty, B.; White, E.; Siscovick, D.; Ragunathan, T.E.; Glass, A.G.

Risk Factors for Renal Cell Carcinoma
The primary objective of this case-control study is to examine the role of diuretic drugs in the aetiology of renal cell carcinoma. Other prescription medications, including anti-hypertensives, sex hormones, and diet pills will also be investigated, as will personal and family medical history, patterns of adult weight, and tobacco use. The study will involve 212 cases (127 males and 85 females). Controls will be health plan members of the Kaiser Permanente, Northwest Region, individually matched to cases on sex, age, months of health plan membership, date of entry into plan, and presence in plan on case diagnosis date. Exposure information will be obtained from out- and in-patient medical records with a possible interview component, to be determined from the results of a pilot study comparing chart data with interview responses. Results will be analysed using standard statistical methods for case-control studies including logistic regression analysis.

TYPE: Case-Control
TERM: Diuretics; Drugs; Hormones; Tobacco (Smoking)
SITE: Kidney
TIME: 1989 - 1994

Raleigh

1165 **Aldrich, T.E.** 05190
North Carolina Cancer Registry, North Carolina Div. of Statistics & Information Serv., P.O. Box 29538, Raleigh NC 27626-0538, United States of America (Tel.: + 1 919 7334728; Fax: 7338485)

Monitoring of Rare Cancers as Sentinel Events
The Cancer Surveillance Section routinely evaluates spatial and temporal patterns of selected rare cancers, the hypothesis being that shifts in the occurrence of these events can serve as sentinels of potential public health significance. Currently studies of leukaemia, paediatric cancers and cancers of liver, pancreas, soft tissue and non-Hodgkin's lymphoma are under way. The leukaemia study is being conducted in several communities and evaluates incidence in relation to recognized point sources of potential leukaemogens (e.g. benzene). The paediatric study is a statewide investigation of rates for 1975-1990, focussing on geographical variation of specific sites. Liver cancer, soft tissue sarcoma and non-Hodgkin's lymphoma are being studied in counties with paper mills. One collaborative study has been performed with a tumour registry at a veterinary hospital to search for compatible patterns between animal and human cancers. The pancreas study is a two-county study evaluating potential pesticide exposure and disease incidence between 1985-1990. Sample sizes range from 30 to 300 cases, depending on the time period and geographical area involved. Statistical studies focus on geographical, temporal and personal characteristics (age, race, sex) and disease traits (cell type) to evaluate evidence of increased risk.

TYPE: Case Series; Correlation; Incidence
TERM: Animal; Cluster; Geographic Factors; Monitoring; Pesticides; Registry; Solvents; Time Factors; Trends
SITE: Childhood Neoplasms; Leukaemia; Liver; Non-Hodgkin's Lymphoma; Pancreas; Soft Tissue
CHEM: Benzene; Dioxins
REGI: N. Carolina (USA)
TIME: 1990 - 1995

1166 **Aldrich, T.E.** 05192
North Carolina Cancer Registry, North Carolina Div. of Statistics & Information Serv., P.O. Box 29538, Raleigh NC 27626-0538, United States of America (Tel.: +1 919 7334728; Fax: 7338485)
COLL: Morris, P.; McLawhorn, K.

Evaluation of Cancer Cluster Reports

Reports of increased cancer occurrence from members of the public are systematically evaluated using the computer program CLUSTER. Spatial characteristics of case occurrence, temporal trends and personal traits (age, race, sex, occupation) are evaluated. In all cases, the hypothesis being tested is: is there an increased occurrence of cancer, and if so, does the pattern of increase suggest an environmental risk factor? Most studies are very small (3 to 30 cases). No control groups are involved: increases are evaluated based on national incidence and evidence of environmental patterns is evaluated based on randomness of case occurrence. Surveillance will be carried out, in most cases for a minimum of five years. Special follow-up studies are also being performed. Poisson distribution tests are used for many techniques, extra-Poisson for two of them. Chi-square techniques are also used with likelihood ratio solutions for two of the methods. The computer program CLUSTER is routinely used with these analyses (Agency for Toxic Substances and Disease Registries, Atlanta, GA, 1993). These studies all involve local incidence data from North Carolina communities.

TYPE: Incidence; Methodology
TERM: Cluster; Environmental Factors; Registry; Trends
SITE: All Sites
REGI: N. Carolina (USA)
TIME: 1989 - 1995

Richland

1167 **Kathren, R.L.** 02842
Hanford Environmental Health Found., Dept. of Research, P.O.Box 100 , Richland WA 99352, United States of America (Tel.: +1 509 3768650)
COLL: Omohunndro, E.L.; Mahaffey, J.; Gilbert, E.S.

Department of Energy Hanford Health and Mortality Study

The health effects of occupational low-level ionizing radiation on nuclear industry workers have been of concern since the association of leukaemia, osteosarcoma, and aplastic anaemia with high-level exposure to ionizing radiation was noted early in this century. This prospective cohort study was designed originally to evaluate the mortality patterns of Hanford Site employees as a function of their radiation exposure. Its scope has been expanded to include reproductive outcomes and assessment of occupational non-radiation exposures and ongoing monitoring of major medical problems among currently active workers. These objectives require: (1) the maintenance and updating of a personnel roster representing 15,000 current employees; (2) active mortality follow-up of translocated and retired workers; (3) periodic medical examination of active employees; (4) collection of individual internal and external ionizing radiation exposure data; and (5) interval data analysis for detection of evolving detrimental health effects. To avoid the underassessment of risk inherent in a "healthy worker" population, controls consist of internal cohort risk sets with comparable demographic and work history characteristics. The potential effects of radiation exposure are assessed with a modified Cox proportional hazards model. Hanford site data also have been submitted to external investigators for corroboration of conclusions regarding the health effects of low-level radiation exposure. A paper has been published in Health Phys. 55:1945-1981, 1988.

TYPE: Cohort
TERM: Occupation; Radiation, Ionizing; Radiation, Non-Ionizing
SITE: All Sites
OCCU: Radiation Workers
TIME: 1944 - 1994

1168 **Stevens, R.G.** 04571
Battelle Pacific Northeast Labs, P.O.Box 999, Richland WA 99352, United States of America (Tel.: +1 509 3756941)

Iron and Risk of Cancer

The first National Health and Nutrition Examination Survey (NHANES I) cohort of 14,000 subjects drawn as a sample of the United States population is being followed for mortality and morbidity outcome. During the period 1971-75, the subjects were identifid and given a medical examination, detailed dietary questionnaire, and a series of blood tests. The project is concerned with determining whether and to what extent body iron stores, and dietary intake influence cancer risk. The hypothesis is based on two possible biological mechanisms: (1) iron can catalyse the production of oxygen radicals, and these may be proximate carcinogens or promoters, and (2) iron may be a limiting nutrient to the survival and growth of an existing cancer cell. Of the original 14,000 adult subjects aged 25-79 identified in 1971-1975, 3,355 men survived at least four years after a blood test for transferrin saturation. Of these, 242 developed cancer by 1981-1984. Transferrin saturation was significantly higher in the men who developed cancer than in those who did not. Among women, only those 8% with the highest transferrin saturation showed any suggestion of increased cancer risk. These results were reported in N. Engl. J. Med. 319(16):1047-1052, 1988. Future studies are directed toward the effect of iron in prognosis after cancer diagnosis, effect of iron in radiation sensitivity, and effect of iron on risk of a second cancer after cancer therapy.

TYPE: Cohort
TERM: Diet; Nutrition; Promotion; Trace Elements
SITE: All Sites
CHEM: Iron
TIME: 1985 - 1994

Rochester

1169 **Talley, N.J.** 05007
Mayo Clinic and Foundation, 200 First St., Rochester MN 55905, United States of America (Tel.: +1 507 2556027; Fax: 2556318)
COLL: DiMagno, E.P.; Blaser, M.J.

Association of Helicobacter Pylori with Gastric Adenocarcinoma

The aim is to determine whether H. pylori is a risk factor for gastric adenocarcinoma at different sites. A case-control study is being undertaken including patients with gastric adenocarcinoma, patients with cancers elsewhere (colon, oesophagus, lung) and cancer-free controls (asymptomatic persons, benign non-gastric diseases). Demographic and clinical data have been collected and serum obtained for analysis.

TYPE: Case-Control
TERM: BMB; H. pylori
SITE: Stomach
TIME: 1990 - 1994

Rockville

1170 **Devesa, S.S.** 03968
NCI, Div. of Cancer Etiology, Epidemiology & Biostatistics Program, Biostatistics Branch, Executive Plaza Bldg North, Room 415F, Rockville MD 20852, United States of America (Tel.: (301)4964153)

Analysis of Cancer Incidence and Mortality Patterns in the United States

Utilizing data derived from several population-based NCI surveys, the Connecticut Tumor Registry, and the SEER program, incidence patterns are being assessed among residents of several geographical areas. Mortality data provided by the National Center for Health Statistics are also being analysed. Current analysis focuses on the patterns for more specific forms of cancer among various racial/ethnic groups and by histologic category.

TYPE: Incidence; Mortality
TERM: Blacks; Caucasians; Registry; Trends
SITE: All Sites
REGI: Connecticut (USA); SEER (USA)
TIME: 1987 - 1994

1171 **Dosemeci, M.** 04618
National Cancer Inst., NIH, Environmental Epidemiology Branch, Occupational Studies Sect., Executive Plaza North, Room 418H Bldg EPN, Rockville MD 20892-4200, United States of America (Tel.: +1 301 4969093; Fax: 4021819)
COLL: Hayes, R.B.; Blair, A.E.; Unsal, M.; Fraumeni, J.F.

Occupational Exposure and Cancer Risk in Turkey

The objective of this study is to identify the possible associations between various occupational exposures and cancer risk in Turkey. Between 1978 and 1984 about 7,500 cases of cancer (excluding haematopoietic malignancies) were treated among industrial workers in SSK, Okmeydani Hospital, in Istanbul, Turkey. In addition to the general demographic information, full employment histories, including workplace name, job title, duration of employment, tobacco and alcohol use were systematically collected for all admissions to the cancer clinic. Abstraction of data was started in November 1988. Exposure assessment will be carried out semi-quantitatively (none, low, medium an high) using a specially developed job-industry exposure matrix for about 20 different chemicals or chemical groups. Cancer risk analyses for various sites will be carried out using standard epidemiological methods for case-control studies. Selected cases of cancer sites other than the one under consideration will form the control group. In the analyses, standardized odds ratios by industry, occupation, duration of employment, level of exposure, cumulative exposures and probability of exposure will be presented as measures of risks for various sites of cancer.

TYPE: Case-Control
TERM: Chemical Exposure; Clinical Records; Dose-Response; Dusts; Metals; Occupation; Pesticides; Physical Activity; Plastics; Solvents; Wood
SITE: All Sites
CHEM: Acrylonitrile; Arsenic; Asbestos; Benzene; Benzidine; Chromium; Formaldehyde; Hydrocarbons, Chlorinated; Nickel; PAH; Phenol; Silica; Styrene
OCCU: Construction and Maintenance Workers; Engineering Workers; Miners; Petroleum Workers; Potters; Rubber Workers; Shipyard Workers; Textile Workers
LOCA: Turkey
TIME: 1988 - 1996

***1172** **Dosemeci, M.** 05316
National Cancer Inst., NIH, Environmental Epidemiology Branch, Occupational Studies Sect., Executive Plaza North, Room 418H Bldg EPN, Rockville MD 20892-4200, United States of America (Tel.: +1 301 4969093; Fax: 4021819)
COLL: Blair, A.E.; Fraumeni, J.F.; Hayes, R.B.; Figgs, L.; Hoover, R.N.; Heineman, E.

Cancer Mortality by Occupation and Industry

The objective of this project is to generate hypothesis for associations between occupational risk factors and all cancer sites, using mortality files from 24 states of the USA. Since 1984, 24 states started coding (1980 US census occupational codes) industry and occupation information from death certificates.

Approximately 3.1 million death certificates from 1984-1989 have been coded, edited and entered in the computer. A user-friendly risk evaluation programme has been developed to carry out PMR analyses by occupation, industry or industry/occupation combination. A monograph that covers the results of these analyses for cancers (637,770 cancer deaths) will be published. In addition, the data will be analysed by occupation exposures (physical and chemical risk factors), using job exposure matrices especially developed for this data. Another user-friendly programme is in the developing stage to carry out case-control studies of 36 sites against non-cancers from the same data for detailed analyses by occupational exposures and confounding factors.

TYPE: Methodology; Mortality
TERM: Dose-Response; Occupation
SITE: All Sites
TIME: 1992 - 1996

1173 **Pottern, L.M.** 03350
NCI, NIH, Environmental Epidemiology Branch, Occupational Studies Sect., Executive Plaza North 418, Rockville MD 20892, United States of America (Tel.: +1 301 4969093)
COLL: Miller, B.A.; Blair, A.E.; Stewart, P.A.; McCammon, C.; Zey, J.

Mortality Study of Workers Exposed to Acrylonitrile

A cohort mortality study will be conducted of approximately 7,000 workers occupationally exposed to acrylonitrile (AN) from plants that began producing or using AN between 1952 and 1962. Workers will be identified from historical company records and traced to the present to determine vital status. Job histories, industrial hygiene measurements, limited biological monitoring data, smoking information, and death certificates will be used to evaluate associations between AN exposure and cause-specific mortality. The mortality of the cohort will be compared with that of the general US population, regional populations, and an unexposed worker population, if feasible.

TYPE: Cohort
TERM: Chemical Exposure; Environmental Factors; Occupation; Plastics; Tobacco (Smoking)
SITE: All Sites; Lung
CHEM: Acrylonitrile
OCCU: Acrylonitrile Workers
TIME: 1985 - 1994

***1174** **Rabkin, C.S.** 05278
National Cancer Inst., NIH Viral Epidemiology Branch, 6130 Executive Blvd EPN, Room 434, Rockville MD 20852, United States of America (Tel.: +1 301 4968115; Fax: 4020817)
COLL: Blattner, W.A.; Goedert, J.J.; Royal, S.

National Cancer Institute Registry of HIV Infected Haemophiliacs

Since the early 1980s, the Viral Epidemiology Branch of the NCI has been collaborating with 16 haemophilia treatment centres in a study of 1,100 HIV-infected and 700 HIV-uninfected haemophiliacs. To further investigate the effects of HIV on cancer, all HIV-infected haemophiliacs notified at other centres are being registered for a prospective study. The study has three major objectives: (1) to determine the incidence and histological subtypes of HIV-associated NHL and estimate the effects of duration of HIV infection, age, CD4+ lymphopenia and other risk factors; (2) to investigate pathogenetic mechanisms such as EBV activation, oncogene rearrangements, and cytokine stimulation; and (3) to determine incidence and aetiological factors of other tumours which may occur in excess. The design includes a cohort study (with optional storage of serum and lymphocytes) and a nested case-control study of incident cases of cancer. Treatment centres are invited to participate.

TYPE: Case-Control; Cohort
TERM: Age; BMB; EBV; HIV; Haemophilia
SITE: All Sites; Lymphoma; Non-Hodgkin's Lymphoma
LOCA: Brazil; Canada; Italy; Spain; Sweden; United Kingdom; United States of America
TIME: 1992 - 1999

***1175** **Shaw, G.L.** 05364
NCI, NIH, Biomarkers and Prevention Research Branch, 9610-C Medical Center Drive, Suite 300, Rockville MD 20850, United States of America (Tel.: +1 301 4023128; Fax: 4024422)
COLL: Tucker, M.A.; Blessing, J.

Risk of Second Malignancy after Treatment for Ovarian Cancer

This study in collaboration with the Gynecologic Oncology Group (GOG) will evaluate the carcinogenicity of cisplatin and adriamycin and update the previously studied cohorts treated with Melphalan. Sixteen clinical trials entering 4,354 women, dating back to 1971, were identified for this study. Approximately 2,200 women were estimated to have survived at least one year after diagnosis and these files in the GOG Statistical Offices have been manually reviewed for diagnosis of a second malignant neoplasm or excessive/protracted haematological toxicity from the treatment. Approximately 10% of the patients, entered on 12 trials still under active follow-up at GOG have no follow-up since January 1990 and tracking is proceeding by using the National Death Index and contacting directly the institutions where the patient was treated. GOG is pursuing the tracking of patients still on active protocols. If excess of second malignancies is observed, controls from the same cohort will have dosage information abstracted for a nested case-control analysis.

TYPE: Case-Control; Cohort
TERM: Chemotherapy; Drugs; Multiple Primary
SITE: All Sites; Ovary
CHEM: Adriamycin; Cisplatin; Melphalan
TIME: 1990 - 1994

1176 **Tucker, M.A.** 04950
National Cancer Institute, NIH, Genetic Epidemiology Branch, Family Studies Sect., 6130 Executive Blvd, Rockville MD 20892, United States of America (Tel.: +1 301 4964375; Fax: 4024489)
COLL: Hartge, P.; Holly, E.A.; Sagebiel, R.W.; Halpern, A.; Clark, W.

Case-Control Study of Malignant Melanoma

The purpose is to estimate the effects of host and environmental factors on malignant melanoma aetiology. 600 cases will be identified over a 2-year period at two pigmented lesion clinics; 1,200 controls will be selected from clinics within the same university hospital system. Whole body skin examinations, photography, biopsy data, personal interviews, and self-administered diet questionnaires will be used. The study is a collaborative project of the NCI, the University of California, and the University of Pennsylvania. The separate effects of normal and dysplastic naevi and the interaction of sunlight and moles will be estimated with this large dataset.

TYPE: Case-Control
TERM: Biopsy; Diet; Naevi; Radiation, Ultraviolet
SITE: Melanoma
TIME: 1990 - 1994

1177 **Ward, M.** 05032
NCI, Environmental Epidemiology Branch, Occupational Studies Sect., 6130 Executive Blvd, Rockville MD 20892, United States of America (Tel.: +1 301 49699093; Fax: 4969146)
COLL: Zahm, S.H.; Heineman, E.; Cantor, K.P.; Stewart, P.A.; Blair, A.E.

Gastric Cancer in Eastern Nebraska

A case-control study of stomach cancer, using the histological classification of Lauren, will be conducted in the 66 counties of eastern Nebraska. The main objective is to evaluate the association of agricultural exposures with the two types of stomach cancer. Agricultural exposures, which are also hypothesized risk factors for stomach cancer, include fertilizer and pesticide use, dust exposure and ingestion of nitrate from contaminated drinking water. About 300 cases from the period 1988-1993 will be identified from the Nebraska Tumour Registry. Population controls are frequency-matched to the age, sex and year of birth of the cases. Interviews will be conducted with subjects and next-of-kin in 1991.

TYPE: Case-Control
TERM: Dusts; Fertilizers; Histology; Occupation; Pesticides; Registry; Water
SITE: Stomach
CHEM: Nitrates; Nitrites
REGI: Nebraska (USA)
TIME: 1991 - 1994

1178 **Zahm, S.H.** 03192
NCI, Environmental Epidemiology Branch, Occupational Studies Sect., 6130 Executive Blvd, R418K, HNC 3344, Executive Plaza Bldg North, Rockville MD 20892-4200, United States of America (Tel.: +1 301 4969093; Fax: 4021819 ; Tlx: 248232 NIH UE)
COLL: Roscoe, R.J.

Calculation of Mortality Rates for Blue-Collar Workers

Mortality studies conducted by epidemiological research organizations usually use United States or other general population death statistics for comparison. While general population statistics are good because usually based on very large numbers, their validity in worker comparisons is questionable because of differences between workers and the general population. Such differences include the following: (1) workers appear to smoke more (on the average) than the general population, (2) they are economically more homogeneous, and (3) they are generally healthier than the general population because of selection at the time of hiring and the physical activity associated with many jobs. These drawbacks to general population comparisons are well documented in the occupational health literature. Epidemiologists have little alternative to general population comparisons because comparison rates based upon sufficiently large worker populations do not exist. NCI and NIOSH are collaborating to develop a computerized data base from completed studies of worker populations. About 280,000 workers from NCI cohort studies and 230,000 workers from NIOSH studies will be included in the data base. The pooled worker population will be used as a comparison group in future mortality studies. Updating will include vital status follow-up and obtaining death certificates for the deceased. Calculation of death rates will be accomplished using programs written for the purpose. A paper has been published in Am. J. Epidemiol. 123:918-919, 1986.

TYPE: Methodology; Mortality
TERM: Occupation
SITE: All Sites
TIME: 1983 - 1994

1179 **Zahm, S.H.** 03646
NCI, Environmental Epidemiology Branch, Occupational Studies Sect., 6130 Executive Blvd, R418K, HNC 3344, Executive Plaza Bldg North, Rockville MD 20892-4200, United States of America (Tel.: +1 301 4969093; Fax: 4021819 ; Tlx: 248232 NIH UE)
COLL: Fraumeni, J.F.

Cancer Mortality among Members of the American Chemical Society

A retrospective cohort study of men who were members of the American Chemical Society will be conducted to provide evidence on whether cancer mortality among chemists differs from that expected. Approximately 50,000 men were members prior to 1955 and were active members in 1965. They will be followed to December 1978. SMRs will be calculated. Particular attention will be given to malignant lymphoma, leukaemia, and cancer of the pancreas. Length of membership, age at entry into the Society, and type of employment will be considered.

TYPE: Cohort
TERM: Chemical Exposure; Occupation
SITE: Leukaemia; Lymphoma; Pancreas
OCCU: Chemists
TIME: 1980 - 1994

1180 **Zahm, S.H.** 04288
NCI, Environmental Epidemiology Branch, Occupational Studies Sect., 6130 Executive Blvd, R418K, HNC 3344, Executive Plaza Bldg North, Rockville MD 20892-4200, United States of America (Tel.: +1 301 4969093; Fax: 4021819 ; Tlx: 248232 NIH UE)

Mortality Study of Lawn Care Workers: A Retrospective and Prospective Study

Phenoxy acetic acid herbicides have been associated with lymphoma and soft tissue sarcoma. Pesticide-exposed groups may be at higher risk of these tumours as well as leukaemia, lung cancer, and multiple myeloma. To investigate these hypotheses, a study of over 20,000 former and current lawn care service workers in one company is being conducted. The applicators have been exposed to 2,4-D and other pesticides for up to 90 days per year. A retrospective study of workers employed between 1969-1980 will be done to compare their mortality to that of the US general population. A prospective component will allow future study of these workers and newly-hired workers.

TYPE: Cohort
TERM: Chemical Exposure; Herbicides; Occupation; Pesticides
SITE: All Sites; Brain; Leukaemia; Lung; Multiple Myeloma; Non-Hodgkin's Lymphoma; Soft Tissue
CHEM: 2,4-D; Phenoxy Acids
OCCU: Herbicide Manufacturers; Herbicide Sprayers
TIME: 1987 - 1994

1181 **Zahm, S.H.** 04289
NCI, Environmental Epidemiology Branch, Occupational Studies Sect., 6130 Executive Blvd, R418K, HNC 3344, Executive Plaza Bldg North, Rockville MD 20892-4200, United States of America (Tel.: +1 301 4969093; Fax: 4021819 ; Tlx: 248232 NIH UE)
COLL: Garrity, T.J.; Telles, J.L.; Heineman, E.

Mortality Study of Philadelphia Firefighters

A mortality study of firefighters is being conducted to investigate the hypothesis that the introduction of synthetic building materials and exposure to other products of combustion may result in increased cancer risk. The cohort will consist of approximately 10,000 firefighters hired since 1910. Approximately 5,600 have been actively employed since 1945, when development of new building materials began. Mortality will be examined according to employment characteristics such as calendar time, job title, duration of employment, assignment to engine or ladder companies, number of runs per year for assigned company, area of city, and age and year of first employment. Introduction of diesel equipment and potential for exposure in five houses will be assessed.

TYPE: Cohort
TERM: Age; Chemical Exposure; Occupation; Time Factors
SITE: All Sites; Brain; Colon; Leukaemia; Rectum
OCCU: Firemen
TIME: 1987 - 1994

Salt Lake City

1182 **Lyon, J.L.** 03896
Univ. of Utah, School of Medicine, Dept. of Family & Preventive Medicine, 50 N Medical Dr., Salt Lake City UT 84112, United States of America (Tel.: +1 801 5817234)
COLL: Archer, V.E.; Schiager, K.J.

Cancer Risk from Radon in Homes

A case-control study is being done to evaluate the risk of radon in homes among smokers and non-smokers separately. Lung cancer cases will be collected over a three year period: approximately 450 smokers and 350 non-smokers. Controls will be selected by random digit dialling supplemented by records of health financing for the elderly. Controls will be matched on age, sex, race and cigarette usage. A telephone interview will obtain information an all residences the person has lived in, plus supplementary information on lifestyle and diet. Radon is measured in as many of the residences as practical by a year-long track etch technique. Adjustments will be made for changes in the houses over time. For houses not measured, estimates will be made based on geology and house features. Lifetime radon exposures are calculated. Odds ratios will be obtained for different lifetime radon and cigarette exposures. Regression analyses will be used. An attempt will be made to determine at which age the radon exposures were most important.

TYPE: Case-Control
TERM: Age; Radiation, Ionizing; Time Factors; Tobacco (Smoking)
SITE: Lung
CHEM: Radon
REGI: Idaho (USA); Utah (USA)
TIME: 1989 - 1994

1183 **Slattery, M.L.** 05089
Univ. of Utah, School of Medicine, Dept. of Family and Preventive Medicine, 50 N Medical Drive, Room 1C26, Salt Lake City UT 84132, United States of America (Tel.: +1 801 5817234)
COLL: Potter, J.D.; Friedman, G.D.; Caan, B.

Diet, Physical Activity and Reproduction as Risk Factors for Colon Cancer

The purpose of this study is to assess the interaction between dietary intake, physical activity, reproductive history, body size, and genetic factors as they relate to the development of colon cancer. This is a large multi-centre case-control study of 2,400 cases and 2,400 controls. Cases and controls will be population-based in Utah and Minnesota; they will come from the membership records at the Kaiser Health Plan. Detailed dietary intake data will be ascertained using a previously validated diet history questionnaire. Detailed information on physical activity, reproductive history, and family history of cancer will also be ascertained in an interviewer-administered questionnaire. Cases should be interviewed within three months of diagnosis since a rapid-reporting system will be used to identify cases. Data will be analysed to determine how diet, physical activity, reproductive history, age, sex, and tumour site interact in the development of colon cancer.

TYPE: Case-Control
TERM: Diet; Genetic Factors; Physical Activity; Registry; Reproductive Factors
SITE: Colon
REGI: Bay Area (USA); Minnesota (USA); Utah (USA)
TIME: 1991 - 1996

San Francisco

1184 Cleaver, J.E. 03930
Univ. of California, Lab. of Radiobiology and Environmental Health, Box 0750, San Francisco CA 94143-0750, United States of America (Tel.: +1 415 6664563)

Molecular Studies of DNA, Mutagenicity, and Carcinogenicity

Human DNA repair deficient diseases, xeroderma pigmentosum, and Cockayne syndrome, which involves hypersensitivity to ultraviolet light and other clinical symptoms, are being investigated at molecular and genetic levels. Particular emphasis is placed on rapid diagnostic methods for patient and prenatal diagnosis and studies of the fine structure of regulation and control of DNA repair and cloning of relevant genes. Additional studies are being made into mechanisms of reversion and detailed rates of repair of thymine dimers and (6-4) photoproducts. Recent publications include : Exp. Cell Research 22, 513-520, 1989; Teratogenesis Carcinogenesis and Mutagenesis 9, 147-155, 1989; Carcinogenesis, 10, 1691-1696, 1989.

TYPE: Molecular Epidemiology
TERM: Biopsy; Congenital Abnormalities; DNA Repair; Premalignant Lesion; Radiation, Ultraviolet; Xeroderma Pigmentosum
SITE: Skin
TIME: 1968 - 1994

1185 Holly, E.A. 04363
Cancer Epidemiology Studies, 1388 Sutter St. Suite 920, San Francisco CA 94109, United States of America (Tel.: +1 415 4763345; Fax: 4763350)
COLL: Whittemore, A.S.; Felton, J.S.

Mutagenic Mucus in the Uterine Cervix of Smokers

An association has been noted between cancer of the uterine cervix and cigarette smoking after controlling for other known risk factors. In this study, the Ames/Salmonella microsomal test was performed on cervical fluids of 333 non-smokers and 364 smokers to determine whether smokers are more likely to have a positive outcome for mutagenicity. Subjects for this study were accrued by physicians at 11 clinics who collected the specimens themselves to assure data-collection uniformity. To collect the specimen, the cervix was washed with 1 ml of sterile nonpyrogenic electrolyte solution, the fluid was drawn into a pipette, the wash was repeated and the fluid was then again drawn into the pipette, placed in a test tube and frozen. Patients were interviewed at the time of fluid collection by a professional interviewer. The structured questionnaire covered smoking habits, passive smoking environment, sexual history, contraceptive use, history of sexually transmitted diseases, and recent diet. The relationship between laboratory results and exposure characteristics will be examined using a multiple-logistic model to asess possible interaction, to control for potentially confounding variables, and to obtain point estimates and confidence limits for the odds ratios for each variable under study.

TYPE: Correlation
TERM: Diet; High-Risk Groups; Mathematical Models; Mutagen; Oral Contraceptives; Passive Smoking; Sexual Activity; Sexually Transmitted Diseases; Tobacco (Smoking)
SITE: Uterus (Cervix)
TIME: 1987 - 1994

1186 Holly, E.A. 04456
Cancer Epidemiology Studies, 1388 Sutter St. Suite 920, San Francisco CA 94109, United States of America (Tel.: +1 415 4763345; Fax: 4763350)
COLL: Whittemore, A.S.; McGrath, M.S.; Dorfman, R.R.

Non-Hodgkin's Lymphoma and Retroviral Tests

A population-based case-control study of non-Hodgkin's lymphoma is under way to include more than 1,500 newly diagnosed patients and more than 2,000 controls in the San Francisco region. A rapid case-finding system is used to identify cases newly diagnosed in all hospitals in the six-county San Francisco Bay Area region. Controls are identified using random digit-dial, and are frequency-matched to cases for age and sex. Structured personal interviews are conducted by trained interviewers in study subjects' homes. Data are collected on occupational factors, common allergies, vaccinations, drugs used for treatment of allergies, other disturbances of the immune system, viral infections, travel to foreign countries, and among men, history of homosexual experiences. A sample of lymphoma patients with AIDS-related diseases, a sample of lymphoma patients without these diseases and a sample of controls will be tested for anitibodies to HIV and other viruses. The laboratory results will be used with the questionnaire response data to assess risk factors for lymphoma by histological type.

TYPE: Case-Control
TERM: AIDS; Allergy; Drugs; HIV; Histology; Homosexuality; Immunology; Occupation; Vaccination
SITE: Non-Hodgkin's Lymphoma
TIME: 1988 - 1994

1187 Moscicki, A.B. 05036
Univ. of California at San Francisco, 400 Parnassus Ave., AC-01 Box 0374, San Francisco CA 94143, United States of America (Tel.: +1 415 4762184; Fax: 4766106)
COLL: Palefsky, J.; Bolan, G.; Darraugh, T.; Brescia, B.; King, E.; Schoolnik, G.; Winkelstein, W.; Schachter, J.; Daniels, T.; Benowitz, N.

Natural History of Human Papillomavirus

The aims are: (1) to observe the natural history of HPV and (2) to determine risk factors related to the development of cervical neoplasia in adolescents. This population was chosen for several reasons: adolescents with HPV and limited years of sexual activity demonstrate more clearly an early infection (short exposure time) than most adult populations who have been sexually active for many years. Since rates of neoplasia are increasing in younger women, co-factors identified in this group may represent important factors for the acceleration of neoplastic development. Finally, adolescence is a time of abundant metaplastic activity, so presenting an ideal population to study the role of cervical immaturity in neoplasia. 1,568 females aged 13-19 years and positive for HPV DNA will be asked to undergo a colposcopic examination and face-to-face interview. Females sexually active for less than two years with a positive HPV DNA test and no evidence of cervical neoplasia (N = 960) will be asked to enroll in a longitudinal study. Patients will be examined at the initial visit and every four months or until CIN develops. Colposcopic examinations, cytology, test for sexually transmitted diseases and face-to-face interviews for information on sexual behaviour, contraceptive use, and cigarette use will be performed at scheduled intervals. 400 HPV negative control subjects will be followed similarly for comparison. This study will describe the natural history of latent HPV infection and calculate risk factors for the accelerated development of CIN in a case-control and cohort population. The results from this study will be related to prevention and education in young women as well as to identifying a clinical model for the role of viruses in abnormal cellular development.

TYPE: Case-Control; Cohort
TERM: Adolescence; Contraception; Cytology; HPV; Sexual Activity; Sexually Transmitted Diseases; Tobacco (Smoking)
SITE: Uterus (Cervix)
TIME: 1990 - 1995

San Mateo

1188 **Wong, O.** 03549
Applied Health Sciences, 181 2nd Ave., Suite 628 P.O.Box 2078, San Mateo CA 94401, United States of America

Prospective Mortality Registry of Refinery Employees in the Petroleum Industry

The objective is to develop and implement a mortality registry of refinery employees in the petroleum industry. In its first year of operation, the registry will consist of 43,000 employees from 53 refineries (13 companies). By its fifth year, it will likely have 55, 000 participants. This prospective data base will be used to monitor the cause-specific mortality pattern of petroleum refinery employees.

TYPE: Cohort; Mortality; Registry
TERM: Chemical Exposure; Occupation; Petroleum Products; Registry
SITE: All Sites
OCCU: Petroleum Workers
TIME: 1984 - 1994

1189 **Wong, O.** 03704
Applied Health Sciences, 181 2nd Ave., Suite 628 P.O.Box 2078, San Mateo CA 94401, United States of America

Mortality of Petroleum Industry Employees Exposed to Downstream Gasoline

The objective of this study is to determine whether employees exposed to downstream gasoline in the petroleum industry are at higher risk of mortality from malignant and non-malignant diseases, in particular kidney cancer, when compared to a suitable control population. The study will consist of approximately 50,000 employees from five major US petroleum companies, and the study period will cover 1941-1985. Cause-specific SMRs, relative risks and other appropriate statistics will be calculated by exposure, latency, etc.

TYPE: Cohort
TERM: Chemical Exposure; Latency; Occupation; Petroleum Products
SITE: Kidney
CHEM: PAH
OCCU: Petroleum Workers
TIME: 1986 - 1994

Seattle

1190 **Beresford, S.A.A.** 04818
Univ. of Washington, Sch. Publ. Health & Community Medicine, Dept. of Epidemiology, SC-36, Seattle WA 98195, United States of America (Tel.: +1 206 5439512; Fax: 5438525)
COLL: Weiss, N.S.; McKnight, B.; Wilbur, D.

Endometrial Cancer Risk and Postmenopausal Hormone Use

This case-control study is designed to increase knowledge on the role of exogenous oestrogens and progestins in altering the risk for endometrial carcinoma. Although the adverse effects of oestrogens alone are well established, the effect of adding progestins has not been adequately quantified. The specific aims are: (1) to estimate the risk of endometrial cancer associated with combined oestrogen and progestogen therapy, relative both to no therapy, and to oestrogen-only therapy in post-menopausal women; (2) to determine whether this risk is further influenced by the number of days in the cycle on which progestogen is given in addition to oestrogen; (3) to determine the length of the oestrogen-free interval after which the risk of endometrial cancer approaches background rates, for different durations of prior oestrogen use. The cases for this study are women aged 45-74 years, residing in King, Pierce or Snohomish counties, Washington, USA, who were diagnosed with endometrial cancer between January 1987 and December 1990. Cases are being identified through the Cancer Surveillance System, a population-based cancer registry that has served the area since 1974. All endometrial cancer cases are subject to independent histological review. It is anticipated that approximately 660 incident endometrial carcinoma cases will be enrolled. About 660 controls are being recruited through random digit dialling. Identical in-person interviews are administered to cases and controls to collect information on hormone use, reproductive history and other variables of interest. Color photographs of both contraceptive and noncontraceptive hormones are used to facilitate recall. A self-administered dietary questionnaire (developed by NCI) is also collected from each respondent.

TYPE: Case-Control
TERM: Diet; Hormones; Oral Contraceptives; Registry; Reproductive Factors
SITE: Uterus (Corpus)
CHEM: Oestrogens; Progestogens
REGI: Seattle (USA)
TIME: 1991 - 1994

***1191 Cook, L.S.** 05398
Univ. of Washington, School of Public Health & Community Medicine, Dept. of Epidemiology, SC-36, 1959 Pacific Ave., Health Sciences Bldg, Seattle WA 98195, United States of America (Tel.: +1 206 5431065; Fax: 5438525)
COLL: Weiss, N.S.; Daling, J.R.; McKnight, B.; Moore, D.E.; Schwartz, S.M.; White, E.

Tamoxifen Therapy and Risk of Reproductive Cancers

The success of tamoxifen in prolonging survival and recurrence-free survival among breast cancer patients has initiated the study of its use in the primary prevention of breast cancer. However, the long-term risks associated with tamoxifen therapy have not been adequately evaluated, particularly with respect to subsequent cancer occurrence. This population-based, nested case-control study will examine whether tamoxifen use is a risk factor for cancer of the uterine corpus and/or ovary among women diagnosed with unilateral breast cancer. Additionally, the variation in the duration of tamoxifen use will be examined as it relates to the subsequent occurrence of primary contralateral breast cancer. A cohort of women diagnosed with unilateral breast cancer between 1 January, 1978 and 31 December, 1990 will be identified from the population-based cancer registry of western Washington. Women who subsequently developed a new primary malignancy of the contralateral breast (n = 235), ovary (n = 39), or uterine corpus (n = 45) as of 31 December, 1991 will be selected as cases. A sample of women in this cohort who did not develop a second primary malignancy will be selected as controls (n = 756). They will be matched at initial breast cancer diagnosis for age, menopausal status, stage of cancer, and year of diagnosis. Tamoxifen therapy and other treatment information for the initial unilateral breast cancer, along with information on other potential risk factors will be obtained by abstracting hospital medical records of cases and controls and through a self-administered questionnaire sent to physicians who were involved in each patients' care. Analyses will compare the relative risk of each second primary cancer in relation to use of tamoxifen, adjusted for potentially confounding characteristics.

TYPE: Case-Control; Cohort
TERM: Drugs; Prevention; Registry; Time Factors
SITE: Breast (F); Ovary; Uterus (Corpus)
CHEM: Tamoxifen
REGI: Seattle (USA)
TIME: 1992 - 1994

1192 Davis, S. 04377
Fred Hutchinson Cancer Research Center, Public Health Science Div., Program in Epidemiology (MP474), 1124 Columbia St. , Seattle WA 98104, United States of America (Tel.: +1 206 6675134; Fax: 6674787)

Identification and Characterization of Case Aggregation Patterns in Hodgkin's Disease

The study aims primarily to investigate the potential role of an infectious process in the aetiology of Hodgkin's disease (HD). Case aggregations in the past will be determined and aetiological relationships characterized and quantified. Data from a population-based case-control study conducted in north-western Washington are used. Cases were Caucasian residents of King, Pierce, and Snohomish counties diagnosed with HD between January 1974 and December 1982, who were 15-74 years of age at diagnosis. Of 313 cases eligible, 281 (90%) were located and personally interviewed. For each case, one control was selected at random from the same population, matched according to sex, approximate age, and general socio-economic status. The research program aims to: (1) determine whether cases aggregate through prior interpersonal contact in school, employment or residential settings more than would be expected to occur by chance in the same population; (2) improve previously developed measures of interpersonal contact and develop additional indicators among study subjects; (3) characterize and quantify any patterns of case aggregation observed in terms of complexity, first-order interdependence and various postulated periods of biological activity (i.e., susceptibility and infectivity) and (4) evaluate the role of host factors related to immunocompetence and antigenic stimulation in any case aggregation patterns observed.

TYPE: Case-Control; Methodology
TERM: Caucasians; Cluster; Infection
SITE: Hodgkin's Disease
TIME: 1988 - 1994

1193 **Davis, S.** 04451
Fred Hutchinson Cancer Research Center, Public Health Science Div., Program in Epidemiology (MP474), 1124 Columbia St. , Seattle WA 98104, United States of America (Tel.: +1 206 6675134; Fax: 6674787)
COLL: Bley, L.D.

Nutritional Factors and Non-Hodgkin's Lymphoma

This study is a population-based case-control study of non-Hodgkin's lymphoma (NHL) in adults. The primary purpose is to investigate the role of animal protein, fats, and Vitamin A in the development of NHL. Cases will consist of all persons aged 20-64 newly diagnosed with NHL in King, Pierce, and Snohomish Counties in western Washington State between 1 June 1988 and 31 May 1989 (140 cases anticipated). An equal number of controls will be randomly selected from the populations of the same three counties, matched to the cases regarding age and sex. Detailed dietary data will be collected using a self-administered questionnaire developed at the National Cancer Institute, and a short telephone interview will be conducted with study subjects to elicit information regarding lifestyle and demographic factors. Results from this investigation are expected to provide the first epidemiological evidence regarding a possible role of specific dietary constituents in the development of NHL.

TYPE: Case-Control
TERM: Diet; Fat; Lifestyle; Nutrition; Protein; Vitamins
SITE: Non-Hodgkin's Lymphoma
TIME: 1988 - 1994

1194 **Davis, S.** 04824
Fred Hutchinson Cancer Research Center, Public Health Science Div., Program in Epidemiology (MP474), 1124 Columbia St. , Seattle WA 98104, United States of America (Tel.: +1 206 6675134; Fax: 6674787)
COLL: Kopecky, K.J.; Hamilton, T.E.; Amundson, B.; Garbe, P.

Hanford Thyroid Disease Study

This cohort study is designed to determine whether thyroid morbidity (including hypothyroidism and benign and malignant neoplasms) is increased among persons exposed to atmospheric releases of I-131 between 1944 and 1957 from the Hanford Nuclear Site in eastern Washington. A pilot phase will be conducted initially, to include up to 1,590 persons who represent both those most likely to have been exposed to I-131 releases as well as those presumed to be unexposed to such releases. Attempts will be made to locate all study subjects and to conduct personal interviews with each (or their next-of-kin, if deceased) regarding residential, medical, and personal habit histories. Each subject's mother or other close relative will also be located and interviewed regarding aspects of the subject's childhood that are relevant to I-131 exposure. Based upon interview data and in collaboration with the Hanford Environmental Dose Reconstruction Project, estimates of thyroid radiation dose will be obtained for each individual. Medical and pathology records will be sought to verify self-reported thyroid disease. Each study subject will be examined independently by two endocrinologists, and will undergo a thyroid ultrasound examination. Blood specimens will be obtained for thyroid function tests. For those found to have a thyroid nodule, permission will be sought to conduct a fine needle aspiration. Data obtained in the pilot study will be used to perform power calculations and to design a full study to estimate the risk of benign and malignant forms of thyroid disease among exposed subjects.

TYPE: Cohort
TERM: Childhood; Radiation, Ionizing
SITE: Thyroid
CHEM: Iodine
TIME: 1989 - 1996

***1195 Davis, S.** 05450

Fred Hutchinson Cancer Research Center, Public Health Science Div., Program in Epidemiology (MP474), 1124 Columbia St. , Seattle WA 98104, United States of America (Tel.: +1 206 6675134; Fax: 6674787)

COLL: Thomas, D.B.; Moolgavkar, S.H.; Stevens, R.G.; Kaune, W.; Sever, L.

Electric Power and the Risk of Breast Cancer

This population-based case-control study is designed to investigate whether exposure to extremely low frequency magnetic fields and light-at-night in residential settings increase the risk of breast cancer in women. Eligible cases are all women between the ages of 20 and 74 in a two-county area of northwestern Washington State, diagnosed with invasive breast cancer between November 1992 and April 1995. A random sample of 800 women will be enrolled in the study. Control women, equal in number and comparable in age and race, will be chosen from the same two-county area by random digit dialling. A personal interview will be conducted with each subject. Magnetic field and ambient light intensity will be measured in each subject's current residence continuously for 48 hours at 10-second intervals. A scaled diagram will be drawn of all electrical hardware within 140 feet of each subject's current residence, and all residences within northwestern Washington State up to ten years prior to the reference date. Measurements and wiring configuration data will be used to derive cumulative exposure classifications. Analyses will focus on associations between each of the two primary exposures of interest, and their interaction, and the risk of breast cancer, allowing for the combined influence of other established and suspected risk factors for breast cancer.

TYPE: Case-Control
TERM: Alcohol; Diet; Electromagnetic Fields; Fat; Nutrition; Occupation; Radiation, Non-Ionizing; Registry
SITE: Breast (F)
CHEM: Oestrogens
REGI: Seattle (USA)
TIME: 1992 - 1996

***1196 Habel, L.A.** 05402

Univ. of Washington, School of Public Health & Community Medicine, Dept. of Epidemiology, SC-36, 1959 Pacific Ave., Health Sciences Bldg, Seattle WA 98195, United States of America (Tel.: +1 206 5431065; Fax: 5438525)

COLL: Daling, J.R.; Weiss, N.S.; Stanford, J.L.; Self, S.G.; Porter, P.

Risk of Recurrence Following Breast Conserving Surgery for Ductal Carcinoma In Situ

There has been a substantial increase in the diagnosis of ductal carcinoma in situ (DCIS) over the last decade. Limited information is available regarding the clinical outcomes of women treated with breast-conserving therapies for this disease. A case-control study is being conducted to estimate the rates of recurrence following breast-conserving surgery for DCIS and to evaluate tumour and patient characteristics that may influence these rates. A cohort of 750 women diagnosed with DCIS from January 1980 - June 1992, treated by tumour excision, has been identified from a population-based cancer registry. Each member of the study cohort will be sent a questionnaire to obtain information about recurrences and selected patient characteristics. Cases will consist of all women in the cohort with a reported recurrence of their breast cancer. They will be compared with a randomly selected subset of women from the total cohort. Tumour characteristics will only be collected on cases and members of the sub-cohort. This information will be obtained by pathological evaluation of fixed tissue from the initial and recurrent tumours. Cox proportional hazards modelling will be used to estimate the relative risk of recurrent breast cancer for various factors of interest.

TYPE: Case-Control; Cohort
TERM: In Situ Carcinoma; Recurrence; Registry; Surgery
SITE: Breast (F)
REGI: Seattle (USA)
TIME: 1992 - 1994

1197 Henderson, M.M. 03725

Fred Hutchinson Cancer Research Center, Cancer Prevention Research Unit, 1124 Columbia St., Seattle WA 98104, United States of America (Tel.: +1 206 4674678)

COLL: Anderson, G.; Baker, M.; Bowen, D.; Chu, J.; Goodman, G.; Grizzle, J.; Kinne, S.; Kristal, A.; Omenn, G.; Peterson, A.; Sarason, B.; Sarason, I.; Thompson, B.; Thornquist, M.; Urban, N.; White, E.; Beresford, S.A.A.; Bankson, D.; Feng, Z.; Kestin, M.; Koepsell, T.; Taplin, S.; Wagner, E.

Cancer Prevention Research Unit

The Cancer Prevention Research Unit is a federally funded programme with a core of cancer research scientists and multiple interrelated research studies in cancer prevention and control. Studies include the efficacy of vitamin A/beta carotene in preventing lung cancer in high risk groups (smokers/asbestos workers); the efficacy of folic acid in reversing cervical dysplasia; whether a school-based prevention programme can deter children from using smokeless tobacco; whether women trained to adopt a low-fat diet will maintain the change over time and whether their husbands adopt a similar eating pattern; whether behavioural changes brought about through a community organisation (the Seattle Archdiocese) will motivate large numbers of people to make dietary (low fat/high fibre) changes; whether breast cancer screening rates among women over 50 can be improved , utilising a community organisation intervention strategy; which methods of communication are most effective in encouraging women to take part in mammography screening programmes; whether worksite quit-smoking rates can be improved with a comprehensive multi-level, or "stepped" approach; and, in a multi-institutional statistical/methodological study, development of new strategies for community-based studies. Pilot studies investigate whether precursors of colon cancer can be predicted through new screening tests, and the effect of worksite smoking policies on workers' smoking habits.

TYPE: Cohort; Correlation; Cross-Sectional; Incidence; Intervention; Methodology
TERM: Biochemical Markers; Diet; Dusts; Fat; Fibre; High-Risk Groups; Mammography; Nutrition; Occupation; Prevalence; Prevention; Screening; Tobacco (Smoking); Vitamins
SITE: All Sites; Breast (F); Colon; Lung; Uterus (Cervix)
CHEM: Asbestos; Beta Carotene
OCCU: Asbestos Workers
TIME: 1982 - 1994

1198 Herrinton, L. 05133
Univ. of Washington, Sch. of Public Health & Comm. Med., Dept. of Epidemiology, SC-36, Seattle WA 98195, United States of America (Tel.: +1 202 6674630; Fax: 5438525)
COLL: Stanford, J.L.; Beresford, S.A.A.; Weiss, N.S.; Scott, C.R.

Consumption and Metabolism of Lactose and Galactose and the Occurrence of Epithelial Ovarian Cancer

The aims of the study are to test the hypothesis that ovarian cancer risk is increased by (1) consuming relatively high levels of dietary lactose and galactose; (2) having relatively high levels of urinary galactose following ingestion of lactose; and (3) having relatively high erythrocyte levels of metabolic intermediates of galactose following ingestion of lactose, or being heterozygous for a low-activity variant of a specific enzyme of galactose metabolism. Caucasian residents of western Washington who are diagnosed with Stage I epithelial ovarian cancer during the period 1989 to 1991, will be identified through the Cancer Surveillance System, a population-based tumour registry. Each case will be asked to identify a friend control, matched to the case on age. Study subjects will complete a self-administered food frequency questionnaire and an in-person interview, and will undergo an oral lactose challenge followed by collection of blood and urine samples. The samples will be assayed for enzymes and intermediates of lactose and galactose metabolism. 121 cases are expected.

TYPE: Case-Control
TERM: BMB; Blood; Diet; Enzymes; Metabolism; Registry; Urine
SITE: Ovary
REGI: Seattle (USA)
TIME: 1991 - 1994

***1199 Lazovich, D.** 05427
Univ. of Washington, Dept. of Epidemiology, SC-36, Seattle WA 98195, United States of America
COLL: Weiss, N.S.; White, E.; McKnight, B.; Wagner, E.; Stevens, N.G.

Evaluation of Efficacy of the Faecal Occult Blood Test

Although widely used as a screening test for the early detection of colorectal cancer, the efficacy of faecal occult blood testing (FOBT) to prevent death from colorectal cancer is unknown. A population-based case-control study is being conducted to evaluate the efficacy of FOBT. Cases (n = 239) are all deaths from colorectal cancer occurring during the years 1986-1991 among Group Health Cooperative (GHC) members; controls (n = 478) are randomly sampled from the GHC membership list and matched for age, gender and duration of membership. Medical chart review is used to ascertain a history of FOBT. Analysis will consist of a comparison of a history of screening in defined intervals up to the diagnosis date of the case to identify the interval during which screening is most likely to be efficacious. Results from this study will be combined with two recently completed case-control studies to increase the precision with which the efficacy, if any, of FOBT can be estimated.

TYPE: Case-Control
TERM: Registry; Screening
SITE: Colon; Rectum
REGI: Seattle (USA)
TIME: 1992 - 1994

1200 **Omenn, G.** 05078
Univ. of Washington, School of Public Health and Community Medicine, SC-30, Seattle WA 98195, United States of America (Tel.: +1 206 5431144; Fax: 5433813)
COLL: Goodman, G.; Grizzle, J.; Thornquist, M.; Rosenstock, L.; Barnhart, S.; Anderson, G.; Hammar, S.; Balmes, J.; Valanis, B.; Glass, A.G.; Cullen, M.; Cherniack, M.; Keogh, J.; Meyskens, F.L.; Williams, J.

Chemoprevention of Lung Cancer in High-Risk Populations

CARET, the beta-carotene and retinol efficacy trial, is a two-armed, double-blind, randomized chemoprevention trial to test the hypothesis that oral administration of beta-carotene 30 mg/day plus retinol (retinyl palmitate) 25,000 IU/day will decrease the incidence of lung cancer in two high-risk groups: heavy smokers and asbestos-exposed workers who have smoked. The two agents are thought to contribute complementary and potentially beneficial mechanisms of action: anti-oxidant role of beta-carotene and nuclear ligand stimulation of maintenance of differentiated state of epithelial cells by retinol. Smokers are recruited from health insurance rolls; asbestos-exposed workers are recruited from workers' compensation systems, lawyers, physicians, and unions. The study is designed to detect a "maximal potential chemopreventive effect" of 33% reduction in lung cancer incidence, allowing a two-year time-lag to full effect (linear with time), accrual over five years, overall adherence to medication schedule 79% for smokers and 85% for asbestos-exposed workers at three years, 5% use of beta carotene in placebo groups (due to secular promotion of multivitamins containing active agents of the Trial), loss to competing causes of death 2%/year, and loss to follow-up 2% over full study. Statistical testing procedures involve weighted log-rank test, 80% power, 0.05 level, two-sided. The projected sample is 4,000 asbestos-exposed workers and 13,700 smokers, for a total of 17,700. Pilot studies in 1985-1988 randomized 1,845 individuals, all of whom are followed for endpoints, and 1,521 of whom are Vanguard population for investigation of potential toxicity related to cumulative dose.

TYPE: Intervention
TERM: BMB; Dusts; High-Risk Groups; Prevention; Registry; Tissue; Tobacco (Smoking)
SITE: Lung; Mesothelioma
CHEM: Asbestos; Beta Carotene; Retinoids
REGI: SEER (USA)
TIME: 1985 - 1999

1201 **Rossing, M.A.** 05132
Fred Hutchinson Cancer Research Center Div. of Public Health Sciences, 1124 Columbia St. MP 474, Seattle WA 98104, United States of America
COLL: Daling, J.R.; Weiss, N.S.; Self, S.G.; Moore, D.E.

Risk of Cancer in a Cohort of Infertile Women

Infertile women, particularly women with ovulatory abnormalities, may provide a unique opportunity to assess the role of endogenous hormones in cancer aetiology. In addition, concerns have been raised regarding the potential neoplastic effects of the powerful ovulatory stimulants currently used in the treatment of infertility. This study is a retrospective case-cohort study of cancer incidence among women who sought evaluation and treatment for infertility at any one of several infertility clinics in King County, Washington from 1974-1985. It is anticipated that approximately 4,000 women will be enrolled. Cancer incidence will be ascertained by linking cohort members to the records of the population-based tumour registry of western Washington. Medical records of a subset of women selected for the case-cohort study will be reviewed in order to assess the type of infertility, hormonal treatments received and other factors of interest. Both within-cohort comparisons and comparisons with population rates will be performed in order to assess the risk of various cancers among infertile women according to type of infertility and infertility treatments received.

TYPE: Cohort
TERM: Hormones; Infertility; Registry
SITE: All Sites
REGI: Seattle (USA)
TIME: 1990 - 1994

***1202 Self, S.G.** 05283
Fred Hutchinson Cancer Research Center, Div. of Public Health Sciences, Dept. of Biostatistics, 1124 Columbia St., MP-665, Seattle WA 98104, United States of America (Tel.: +1 206 6674267; Fax: 6675530)
COLL: Thomas, D.B.; Porter, P.

Cell Proliferation in the Epidemiology of Breast Cancer

The hypothesis is that a number of well known risk factors for breast cancer operate by modulation of the proliferation and differentiation of normal breast epithelial cells and, in many instances, this is realised through modulation of levels of steroidal hormones. The specific goals are (1): to measure cell proliferation by immunocytochemistry in normal breast epithelial cells and to relate this measure to serum levels of steroidal hormones, to immunocytochemical expression of hormone receptors, growth factors, growth factor receptors and oncogenes in normal breast epithelial, to age, menopausal status and day of menstrual cycle (for premenopausal women), and to epidemiological risk factors for breast cancer; (2) to compare measures of cell proliferation in normal breast epithelial cells of women diagnosed with breast cancer to those in women diagnosed with benign breast disease; (3) to compare serum levels of steroidal hormones among women diagnosed with breast cancer to those among control women; and (4) to compare serum levels of steroidal hormones and immunocytochemical expression of hormone receptors, growth factors, growth factor receptors and oncogenes in normal breast epithelium among women diagnosed with breast cancer to that among women diagnosed with benign breast disease. A nested case-control study within the cohort of 300,000 women participating in the Shanghai Breast Self-Examination Trial will be included. Breast biopsy specimens will be obtained from 330 incident cancer cases and from 1,000 incident cases of benign breast disease. Immunocytochemical methods will be used to measure proliferation, hormone receptors, growth factors, growth factor receptors and oncogenes in epithelial cells within the breast lesion and in normal cells adjacent to the lesion. Serum levels of steriodal hormones and questionnaire data will be obtained from these individuals and from 660 control women who will be individually matched to the cancer cases for age, menopausal status and, in premenopausal women, day of menstrual cycle on which the serum specimen is collected.

TYPE: Case-Control
TERM: Age; Hormones; Menopause; Menstruation; Oncogenes
SITE: Benign Tumours; Breast (F)
CHEM: Steroids
LOCA: China
TIME: 1992 - 1996

1203 Sherman, K.J. 04972
Fred Hutchinson Cancer Research Center, Div. of Public Health Sciences MP-381, 1124 Columbia St., Seattle WA 98104, United States of America (Tel.: +1 206 6674630; Fax: 6675948)
COLL: Beckmann, A.M.; Daling, J.R.; McDougall, J.K.; Schubert, M.; Ashley, R.L.

Oral Cancer: Epidemiology, Biochemistry, and Immunology

This case-control study will test the hypothesis that infection with HPV and/or HSV type 1 or 2 is related to an increased risk of oral cancer. The study will be conducted in three counties of western Washington. Using the local population-based registry, all men and women aged 18-65, diagnosed with squamous cell carcinoma of the tongue, gum, floor of the mouth, other and unspecified parts of the mouth, and oropharynx, who are residents of three metropolitan counties in western Washington, will be identified and invited to participate. Eligible subjects will be all incident cases (180 men, 120 women) from December 1988 to November 1994. General population controls (360 men, 240 women) will be selected by random digit dialling. Cases and controls will be interviewed regarding history of sexually transmitted diseases, sexual practices, as well as known or suspected risk factors for oral cancer. Tissue specimens and exfoliated oral cavity cells will be collected from each case. A biopsy specimen will be obtained from a segment of the interviewed controls, and exfoliated oral cavity cells will be obtained from all controls. These tissue samples will be examined using molecular hybridisation techniques for HPV. Blood samples will be collected for all cases and controls and analysed for evidence of prior exposure to HSV types 1 and 2 and several HPVs.

TYPE: Case-Control
TERM: Biomarkers; HPV; HSV; Immunology; Registry; Sexual Activity; Sexually Transmitted Diseases
SITE: Oral Cavity; Oropharynx
REGI: Seattle (USA)
TIME: 1990 - 1995

1204 **Stanford, J.L.** 04315
Fred Hutchinson Cancer Research Center, Univ. of Washington, Div. of Public Health Sci., Program in Epidemiology, 1124 Columbia St., MP-474 , Seattle WA 98104, United States of America (Tel.: +1 206 6675137; Fax: 6674787)
COLL: Weiss, N.S.; Daling, J.R.

Breast Cancer in Middle-Aged Women

The objective is to test the hypothesis that oral progestin use, with or without exogenous oestrogens, reduces the risk of breast cancer in menopausal women. Secondary aims include assessment of prescribing patterns of exogenous hormones in postmenopausal women, and the potential interrelationships between hormone replacement therapy and other hormonally related risk factors for breast cancer, e.g. age at first live birth, parity. The study is a population-based case-control investigation of incident breast cancer in women 50-64 years of age, residing in King County, Washington, USA. Cases will consist of about 500 white women diagnosed with primary breast cancer between 1 January 1988 and 30 June 1990. A population control group of 500 white women, residing in the same county, and who do not have a prior history of breast cancer will be selected by random digit dialling. Controls will be frequency matched to cases by five-year age groups. Cases and controls will be interviewed in person regarding use of exogenous hormones and other recognized risk factors for breast cancer including: detailed aspects of menstrual and reproductive history, use of exogenous hormones (oral contraceptives and hormone replacement therapy), medical history, family history of cancer, and socio-demographic factors. A standard case-control analysis will be performed, including descriptive, stratified, and multivariate analyses. Odds ratios and 95% confidence intervals will be estimated by stratified techniques and logistic regression. The risk of breast cancer will be estimated in relation to "ever use" of progestins and whether the progestin was used with or without oestrogens. Potential confounding variables will be controlled in the analysis. Possible effect modification will also be examined to assess whether specific subgroups of women experience different risk patterns.

TYPE: Case-Control
TERM: Age; Familial Factors; Hormones; Oral Contraceptives; Reproductive Factors; Socio-Economic Factors
SITE: Breast (F)
CHEM: Progesterone
TIME: 1987 - 1994

***1205** **Stanford, J.L.** 05525
Fred Hutchinson Cancer Research Center, Univ. of Washington, Div. of Public Health Sci., Program in Epidemiology, 1124 Columbia St., MP-474 , Seattle WA 98104, United States of America (Tel.: +1 206 6675137; Fax: 6674787)
COLL: Daling, J.R.; Brawer, M.K.; McKnight, B.; Kristal, A.

Prostate Cancer in Middle-Aged Men

Recent studies of prostate cancer suggest an enhanced risk associated with prior vasectomy. The incidence of prostate cancer in middle-aged men in western Washington has increased by about 60% over the past decade and vasectomy is a widely used method of contraception in this geographic area. If findings from recent studies are confirmed, this could represent a major public health problem. The primary aims of the prosposed study are to determine whether or not vasectomy is related to an increased risk of prostate cancer, and to assess whether there is evidence of a possible latent effect of vasectomy. To study the relationship between vasectomy and prostate cancer, a popualtion-based case-control study of 705 cases and 705 controls will be conducted. Eligible cases will be men 40-64 years of age who are diagnosed with incident prostate cancer between 1 January, 1993, and 31 December, 1996, and who are residents of King County in northwestern Washington State. Cases will be ascertained through the Seattle-Puget Sound SEER Cancer Registry. Controls will be identified through random digit dialling among residents of King County and will be frequency matched for age (same 5-year group). Study subjects will be intereviewed in person. Information will be obtained on history of vasectomy, reproductive and sexual history, medical history, demographic variables, and health behaviour, such as smoking and alcohol use and the prevalence of use of screening procedures for the early detection of prostate cancer. Study subjects will also complete a food frequency questionnaire. All cases and controls will be asked to provide a blood sample, and tumour tissue blocks will be requested for all cases. Data analysis will include descriptive, stratified, and multivariate analyses. The relative risk of prostate cancer in relation to vasectomy will be estimated, after accounting for the potential confounding effects of other exposure variables.

TYPE: Case-Control
TERM: BMB; Blood; Registry; Sexual Activity; Tissue; Vasectomy
SITE: Prostate
REGI: Seattle (USA)
TIME: 1993 - 1998

***1206 Thomas, D.B.** 05477
Fred Hutchinson Cancer Research Center, Program in Epidemiology, 1124 Columbia St. , Seattle WA 98104, United States of America (Tel.: +1 206 4675134; Fax: 6674787)
COLL: Gao, D.-L.; Self, S.G.

Randomised Trial of Breast Self-Examination in Shanghai

This study is a randomised trial of breast examination currently underway in 520 factories of the Shanghai Textile Industry Bureau. The purpose of the study is to determine whether regular practice of BSE reduces mortality from breast cancer. Nearly 300,000 current and retired female workers between the ages of 30 and 64 have been randomly allocated by factory to an intervention or control group. Women in the intervention group receive intensive and repeated BSE instruction, as well as motivation to continue to practice BSE on a regular basis. Their compliance with requests to practice BSE, and competency in BSE performance are monitored. All women will be followed for 10 years from initial inclusion in the study for the occurrence of breast cancer and mortality from breast cancer and all other causes using multiple active and passive reporting systems.

TYPE: Cohort; Intervention
TERM: Screening
SITE: Breast (F)
TIME: 1988 - 1998

***1207 Thomas, D.B.** 05478
Fred Hutchinson Cancer Research Center, Program in Epidemiology, 1124 Columbia St. , Seattle WA 98104, United States of America (Tel.: +1 206 4675134; Fax: 6674787)
COLL: Koetsawang, S.; Beckmann, A.M.; Kiviat, N.; Corey, L.

Papilloma Virus and Cervical Cancer in Bangkok

The objectives of this hospital-based case-control study are: (1) to determine whether specific types of HPV are causally related to intraepithelial and invasive cervical neoplasia; and if so (2) to identify circumstances under which this carcinogenic effect occurs; and (3) to clarify the role of male sexual behaviour and the role of prostitutes in the aetiology of cervical cancer and in transmission of the specific types of HPV that may be of aetiological importance in this disease. Approximately 300 cases, 300 hospital controls, and 300 clinic controls will be studied, plus 80 controls from a family planning clinic and 50 controls from among women who had a hysterectomy for non-neoplastic conditions. These women's current husbands will also be studied, as will 200 prostitutes. All subjects will be interviewed to ascertain information on sexual behaviour, contraceptive practices, smoking, and other factors that may confound or modify associations between HPV and cervical cancer. Serum specimens from all subjects will be obtained for determination of antibodies against HIV, HBV and HSV type 2. Fresh tissue specimens from the tumours of the cases and the normal cervices of the hysterectomized controls, cervical scrapings from all women, and penile swabs from the husbands will be obtained and analysed for specific types of HPV DNA using PCR techniques. Relative risks of invasive and intraepithelial cervical neoplasia will be estimated in relation to various indices of male and female sexual behaviour, and specific types of HPV, and interactions between HPV infections and such potentially modifying factors as age, steroid contraceptives, smoking, and other viral infections will be investigated. Also, the prevalence of specific types of HPV DNA will be compared in the husbands of cases and controls, the husbands and their wives, and the husbands and the prostitutes, to determine whether the same types of HPV that are related to cervical cancer are prevalent in prostitutes, and sexually transmitted.

TYPE: Case-Control
TERM: BMB; Contraception; Cytology; HBV; HIV; HPV; HSV; PCR; Serum; Sexual Activity; Tissue; Tobacco (Smoking)
SITE: Uterus (Cervix)
LOCA: Thailand
TIME: 1990 - 1994

1208 Vaughan, T.L. 04283
Fred Hutchinson Cancer Research Center, Program in Epidemiology, MP-474, 1124 Columbia St. (W404), Seattle WA 98104, United States of America (Tel.: +1 206 6674741; Fax: 6674787)
COLL: Berwick, M.; Swanson, G.M.; Isaacson, P.; Lyon, J.L.; Morgan, M.

Epidemiological Study of Nasopharyngeal Cancer

A population-based case-control study of nasopharyngeal cancer is being carried out (a) to determine if occupational and residential exposure to formaldehyde increases the risk of nasopharyngeal cancer and (b) to identify other medical, environmental and lifestyle factors associated with the disease in a low-incidence population. Eligible cases (estimated number 245) will include all persons aged 18-74 years who develop nasopharyngeal cancer between 1 April 1987 and 30 June 1991, and who reside in areas covered by six population-based cancer registries in the United States funded by the SEER Program of the NCI. A random digit dialling technique is being used to select one control per case from among residents of the same areas as the cases. Subjects are interviewed by telephone to determine their occupational and residential histories, along with other factors suspected to be associated with risk of nasopharyngeal cancer, including medical, tobacco, alcohol and dietary histories. Using exposure assessment methods already developed in a preliminary study, indices of formaldehyde exposure will be calculated both from home and workplace sources. Blood samples are being collected and stored for future analysis of HLA type and EBV antibody response. Data processing and analysis are being performed at the coordinating centre in Seattle.

TYPE: Case-Control
TERM: Alcohol; Chemical Exposure; Diet; Environmental Factors; Lifestyle; Tobacco (Smoking); Wood
SITE: Nasopharynx
CHEM: Formaldehyde
TIME: 1988 - 1994

***1209 Vaughan, T.L.** 05370
Fred Hutchinson Cancer Research Center, Program in Epidemiology, MP-474, 1124 Columbia St. (W404), Seattle WA 98104, United States of America (Tel.: +1 206 6674741; Fax: 6674787)
COLL: Camp, J.; Morgan, M.; Teschke, K.

Wood Dust Exposure and Lung Cancer

This population-based case-control study will attempt to determine if employment in wood-related occupations is associated with increased risk of lung cancer, and identify specific exposures which could explain an observed increased risk. Specific aims are: (a) to test the hypothesis that persons employed in operational jobs in any of the four major segments of the wood industry are at increased risk of lung cancer; (b) to test the hypothesis that persons exposed to respirable wood dust in particular, as well as several other potential carcinogens present in wood-related occupations, are at increased risk; and (c) to determine whether excess risk associated with any of the above factors varies by histological type, or is modified by other known risk factors, such as tobacco use, dietary intake of beta-carotene and vitamine C, or family history of lung cancer. 769 male cases, aged 18-74 years and resident in an 11-county area of western Washington, where wood-related occupations are particularly common, will be identified from the Cancer Surveillance System of the Fred Hutchinson Cancer Research Center. An equal number of controls of similar age from the same geographic area will be selected by random digit dialling. Subjects will be interviewed by telephone, using a structured questionnaire and experienced interviewers. Exposure to wood dust will be assessed by two industrial hygienists, taking into account job title and typical duties, type of manufactured product, and participation in activities known to produce measurable concentrations of respirable dust. For industries in western Washington, in which total respirable dust levels are not well-characterized, random full-shift sampling will also be accomplished during the first year. Approximately 680 samples will be taken at a minimum of eight industrial sites over two seasons. Stratified analysis and logistic regression will be used to control for the effect of major risk factors, to identify any important effect modification by these factors, and to investigate differences in risk among specific histological types of lung cancer.

TYPE: Case-Control
TERM: Dusts; Histology; Occupation; Plastics; Registry; Solvents; Tobacco (Smoking); Vitamins; Wood
SITE: Lung
CHEM: Beta Carotene; Formaldehyde; PAH; Terpenes
OCCU: Carpenters, Joiners; Paper and Pulp Workers; Sawmill Workers; Wood Workers
REGI: Seattle (USA)
TIME: 1992 - 1997

1210 **White, E.** 04576
Fred Hutchinson Cancer Research Center, 1124 Columbia St. W202, Seattle WA 98104, United States of America (Tel.: +1 202 4674678)
COLL: Daling, J.R.; Weiss, N.S.

Physical Activity and Colon Cancer

Studies of job activity and colon cancer suggest a decreased risk associated with high activity jobs. In addition, there has been a decline in colon cancer incidence among men and women under age 50 in western Washington, which may be related to an increase in leisure time activity. The specific aims of this study are: (1) to determine if physical activity (job activity plus leisure activity) is related to reduced risk of colon cancer, and (2) to determine whether leisure time activity affects colon cancer incidence independently of job related physical activity. The first aim addresses an aetiological question, the second will help answer the cancer control issue of whether an exercise programme during leisure time could be an intervention to prevent colon cancer. To study the relationship between physical activity and colon cancer, a population-based case-control study of 400 cases and 400 controls is proposed. The case group will be incident colon cancer cases age 25-59, diagnosed between January 1986 - December 1990 in three counties in Washington State identified through the Seattle-Puget Sound SEER Cancer Registry. Controls will be an age and sex stratified sample identified through random digit dialling. Cases and controls will be interviewed by telephone. Questions will concern leisure time physical activity, job activity, household activity, diet, weight, and demographic characteristics. Data analysis will focus on the effect of physical activity on colon cancer, using quantitative estimates of energy expenditure categorised by type of activity (e.g., leisure activities) and by intensity of activity. Various components of diet including calories, fat, fibre, calcium and vitamin A will also be considered in the analyses.

TYPE: Case-Control
TERM: Diet; Fat; Fibre; Nutrition; Occupation; Physical Activity; Prevention; Registry; Vitamins
SITE: Colon
REGI: Seattle (USA)
TIME: 1987 - 1994

Southfield

1211 **Reid, R.I.** 04484
The Reid Inst. for Research and Education, 29355 Northwestern Highway, Suite 215, Southfield MI 48034, United States of America (Tel.: + 1 313 3544338; Fax: 3500851)
COLL: Greenberg, M.D.; Campion, M.; Omoto, K.H.; Rutledge, L.H.; Lovincz, A.T.; Rothrock, R.

Polyploid Papillomaviral Infections and Cervical Cancer

This multi-center cohort study will evaluate the role of cervical HPV infection as a risk factor for progression of dysplasia. Women with abnormal Pap smears will undergo colposcopy and cervicography. A cohort of 300 women with biopsy-confirmed CIN 2 or lower will be followed for three years. Presence of type-specific HPV will be determined by nucleic acid hybridization. Other factors being studied include nicotine/cotinine levels in serum and cervical mucus and DNA ploidy. Differences in progression among women with various types of HPV will be analysed.

TYPE: Cohort
TERM: Cytology; DNA; HPV; Premalignant Lesion; Sexually Transmitted Diseases
SITE: Uterus (Cervix)
TIME: 1987 - 1994

1212 **Tilley, B.C.** 05030
Henry Ford Health System, Div. of Biostatistics & Research Epidemiology, 23725 Northwestern Highway, Southfield MI, United States of America (Tel.: +1 313 3548060; Fax: 3544812)
COLL: Zold-Kilbourn, P.; Vernon, S.W.; Myers, R.; Sower, M.F.; Glanz, K.

Colorectal Cancer Screening and Nutrition Intervention

This project involves evaluation of a nutrition and screening enhancement intervention programme among pattern and model automotive workers of the General Motors Corporation (GM). Pattern and model makers have been shown to have an elevated risk of colorectal cancer mortality. Consequently a colorectal cancer screening programme has been administered to eligible employees since 1980. For this intervention trial, 26 worksites, with more than 6000 employees, will be randomly selected to receive either both interventions (screening enhancement and nutrition), or the usual GM screening and any health promotion programmes currently available to employees. The nutrition intervention will combine education, environmental change, and regular contact to achieve a reduction in total fat intake, percent calories from fat, and an increase in average dietary fibre intake. The screening intervention will use constructs of the Health Belief model and Social Learning Theory to promote employee participation in the corporate screening programme. To evaluate the interventions, intervention and control worksites will be compared using a generalization of the paired t-test that accounts for cluster design.

TYPE: Cohort; Intervention
TERM: Diet; Fat; Fibre; Nutrition; Occupation; Screening
SITE: Colon; Rectum
OCCU: Model and Pattern Makers
TIME: 1991 - 1996

Stanford

1213 **Paffenbarger Jr, R.S.** 04350
Stanford Univ., School of Medicine, Dept. of Health Research and Policy, HRP Bldg, Room 113, Stanford CA 94305-5092, United States of America (Tel.: +1 415 7236417; Fax: 7256951 ; Tlx: 3731148 sumc/348402)
COLL: Whittemore, A.S.; Wing, A.L.

Physical Activity, Body Size and Cancer Incidence

Recent epidemiological evidence has associated sedentary work and leisure-time activities with increased risk of colon and breast cancer; overweight and endocrine patterns have for some time been linked with colon, breast, prostate and other cancers. A data base on 25,000 US college alumni will be constructed using retrospectively collected data for 1916-1950 (physical examination, social, and athletic records), contemporary alumni data since 1962 on the same study subjects (mail questionnaire responses pertaining to exercise, other lifestyle elements, personal health, and family disease patterns), and mortality certification. Data on 6,000 San Francisco Bay Area longshoremen are available from multiphasic health examinations, annual job classifications, and death certificate assessments. A projected questionnaire will update the alumni follow-up to 23 years, while survey of annual job assignments will extend the longshoremen follow-up to 34 years. Multivariate analyses will examine absolute measurements and changing patterns of physical activity, body size, and other lifestyle characteristics. Subjects will be classified as to physical activity by type, frequency, intensity, duration, kilocalorie expenditure, and postural nature (sitting, standing, and locomotion). Expected numbers of cancer cases offer statistical power sufficient to test hypotheses of physical activity and body size as causation for cancers of the colon, rectum, and prostate, and perhaps others such as pancreas, lung and breast. Within these two diverse populations, differences in physical activity, body size, cigarette habits, blood pressure levels, etc. significantly altered incidence of cardiovascular-respiratory diseases and estimates of longevity. Those findings invite parallel study of cancer risk, especially since much of the information gained thus far has been applicable to public health.

TYPE: Cohort
TERM: Familial Factors; Lifestyle; Physical Activity; Physical Factors; Tobacco (Smoking)
SITE: Breast (F); Colon; Lung; Pancreas; Prostate; Rectum
TIME: 1987 - 1994

1214 **Paffenbarger Jr, R.S.** 04997
Stanford Univ., School of Medicine, Dept. of Health Research and Policy, HRP Bldg, Room 113, Stanford CA 94305-5092, United States of America (Tel.: +1 415 7236417; Fax: 7256951 ; Tlx: 3731148 sumc/348402)
COLL: Whittemore, A.S.; Wu, A.; Gallágher, R.P.; Howe, G.R.

Prostate Cancer in High, Medium and Low Risk Populations

This study is a population-based case-control study of modifiable risk factors for prostate cancer among blacks, whites and Asians in Los Angeles, San Francisco and Hawaii. Investigators will use a common protocol and questionnaire to administer personal interviews to approximately 500 black patients, 500 white patients, and 500 Asian patients with histologically confirmed carcinoma of the prostate, and 1,500 black, white and Asian population-based controls. Controls will be matched to cases on age, ethnicity and region of residence. A Canadian component (Vancouver and Toronto) will contribute Asian cases and controls only. The study will examine within each ethnic group and each of the two age groups (less than 70 years, 70+ years) how prostate cancer risk varies with diet, physical activity, and body size. Serum collected from controls will be analysed for hormones and prostate-specific antigen.

TYPE: Case-Control
TERM: BMB; Diet; Physical Activity; Physical Factors; Registry; Serum
SITE: Prostate
LOCA: Canada; United States of America
REGI: Bay Area (USA); Br. Columbia (Can); Hawaii (USA); Los Angeles (USA); Ontario (Can)
TIME: 1989 - 1994

Tucson

1215 **Clark, L.C.** 05136
Univ. of Arizona, College of Medicine, Dept. of Family and Community Medicine, Epidemiology Sect., 2504 E. Elm St., Tucson AZ 85716, United States of America (Tel.: +1 602 6264890; Fax: 3217774)
COLL: Combs, G.; Turnbull, B.

Colon Cancer Screening of a Defined Population

The aim of this study is to test the hypothesis that low selenium status is related to the presence of colonic neoplasms in a defined population. This is a cross-sectional study of 1,700 patients enrolled in a pair of double-blind, placebo-controlled clinical trials on the prevention of non-melanoma skin cancers with a supplement of selenium. The supplementation levels are 200 mcg and 400 mcg per day in the form of high selenium Brewer's yeasts as the treatment and a Brewer's yeast placebo. The patients will be screened according to the ACS-NCI colon cancer screening guidelines. This study will generate two cohorts of individuals that can be followed in a prospective cohort study to evaluate the relationship of selenium status and the formation of colonic neoplasms. Patients are from seven clinical centres in the Eastern coastal plain of the US.

TYPE: Cohort; Cross-Sectional
TERM: Screening
SITE: Colon
CHEM: Selenium
TIME: 1990 - 1994

1216 **Clark, L.C.** 05144
Univ. of Arizona, College of Medicine, Dept. of Family and Community Medicine, Epidemiology Sect., 2504 E. Elm St., Tucson AZ 85716, United States of America (Tel.: +1 602 6264890; Fax: 3217774)
COLL: Combs, G.; Turnbull, B.

Nutritional Prevention of Non-Melanoma Skin Cancer

The aim of this study is to test the hypothesis that improved selenium status will prevent the formation of non-melanoma skin cancers. Treatment consists of a supplement of 200 mcg of selenium, in the form of high selenium Brewer's yeast; the placebo is an identical tablet of Brewer's yeast. The patients are drawn from seven dermatology practices in the Eastern coastal plain of the US. Patients must be under the age of 80, white, have a 5-year life expectancy, and have a negative history of hepatic or renal disease. 1,310 patients have been randomized into this study; over 6,500 person-years of observation have been collected in the dermatology clinic sites.

TYPE: Intervention
TERM: BMB; Prevention
SITE: Skin
CHEM: Selenium
TIME: 1983 - 1994

1217 **Clark, L.C.** 05145
Univ. of Arizona, College of Medicine, Dept. of Family and Community Medicine, Epidemiology Sect., 2504 E. Elm St., Tucson AZ 85716, United States of America (Tel.: +1 602 6264890; Fax: 3217774)
COLL: Combs, G.; Turnbull, B.

Prevention of Squamous Cell Carcinoma of the Skin with a Nutritional Supplement of Selenium
The aim of this study is to test the hypothesis that improved selenium status will prevent the formation of squamous cell carcinomas of the skin. A supplement of 400 mcg of selenium, in the form of high selenium Brewer's yeast is the treatment; the placebo is an identical tablet of Brewer's yeast. The patients are drawn from a private dermatology practice in Macon, Georgia. 425 patients have been randomized into the study. Patients must be under the age of 80, white, have a 5-year life expectancy, and have a negative history of hepatic or renal disease.

TYPE: Intervention
TERM: BMB; Prevention
SITE: Skin
CHEM: Selenium
TIME: 1989 - 1994

1218 **Garcia, F.** 05053
Univ. of Arizona, Cancer Center, Sect. of Epidemiology and Biometry, 1515 N Campbell St., Room 2942, Tucson AZ 85724, United States of America (Tel.: +1 602 6264010)
COLL: Stark, A.; Moon, T.E.; Cartmel, B.; Villar, H.; Giordano, G.; McNamara, D.

Tissue Lipids and Dietary Markers for Breast Carcinogenesis
A case-control study is in progress in south-eastern Arizona to assess the relationship between breast carcinogenesis and nutrient status as reflected by tissue lipids and food frequency questionnaire. Forty-eight Hispanic and non-Hispanic caucasian women with pathologically confirmed primary, non-metastatic breast carcinoma have been enrolled in the study. Using a random-digit dialling procedure, 135 community controls have been selected. Adipose tissue biopsies, and a sample of buccal epithelial cells were collected from each consenting subject. The specimens will be analysed using gas liquid and thin layer chromatography in order to identify the relative proportions of 13 individual fatty acids. A food frequency questionnaire concerning over 250 food items was completed by all subjects. Additionally a detailed questionnaire on established risk factors, menstrual history, anthropometric data, and exposure to hormones, tobacco and ethanol was completed by all cases and most controls.

TYPE: Case-Control
TERM: BMB; Diet; Lipids; Nutrition; Registry
SITE: Breast (F)
REGI: Arizona (USA)
TIME: 1988 - 1994

1219 **Lebowitz, M.D.** 03312
Univ. Arizona, College of Medicine, Arizona Medical Center, Div. of Respiratory Sciences, Tucson AZ 85724, United States of America (Tel.: +1 602 6266379; Fax: 6266970)
COLL: Burrows, B.

Longitudinal Study of Obstructive Airways Disease
A 25-year longitudinal study of obstructive airways diseases in a community population of 5,200 persons is in its 22nd year. The population is age-stratified, so that many older cohorts are included. It is family-oriented for family concordance studies. It is believed that sufficient cases of lung cancer will be

seen to test hypotheses on smoking, the interrelationship between lung cancer and obstructive airways diseases, immunological states and other environmental factors that may be related to the development of lung cancer. Death certificates will be examined. Chest X-rays, medical history, immunological and physiological tests will also be conducted. Questions will be specifically asked about occupational and other environmental exposures. At the end of the first 12 years, 12% had died; they were older, included more smokers, males, and subjects with chronic disease. Lung cancer was the underlying cause in 36/547 (6.6%). Of these, 40% had moderate or severe obstructive airways disease, but only 11% had such reported on the death certificate. There are common risk factors for both causes of death.

TYPE: Cohort
TERM: Environmental Factors; Occupation; Premalignant Lesion; Tobacco (Smoking)
SITE: Respiratory
TIME: 1972 - 1996

1220 **Moon, T.E.** 03567
Univ. Arizona Disease Prevention Cent., Epidemiology & Biometry Section, Dept. Family & Community Medicine, 1501 North Campbell St., Tucson AZ 85724, United States of America (Tel.: +1 602 6264010; Fax: 6264089 ; Tlx: 9109521238)
COLL: Levine, N.; Alberts, D.; Schreiber, M.M.; Cartmel, B.

Chemoprevention of Skin Cancer by Vitamin A

The objective of the first study is to evaluate the role of retinol (Vitamin A) in the prevention of skin cancer. Persons with a clinical history of at least ten actinic keratoses and no prior history of cancer are enrolled in a prospective randomized double blind placebo-controlled intervention trial. A total of 2,800 such subjects identified by all dermatologists in Arizona will be included. The study will evaluate the efficacy of continued pharmacological doses of retinol in the reduction of risk of skin cancer, the effects of medication compliance, motivation methods, etc. The objective of the second study is to evaluate the role of retinol (Vitamin A) or 13-cis retinoic acid (synthetic Vitamin A) in the prevention of new skin cancers. Persons with a diagnosis of 4 or more non-melanoma skin cancers are enrolled in this prospective double blind placebo controlled intervention trial. A total of 711 eligible subjects identified by community dermatologists or use of newspaper advertisements are included.

TYPE: Intervention
TERM: Diet; Metabolism; Prevention; Vitamins
SITE: Skin
CHEM: Retinoids
TIME: 1983 - 1994

1221 **Moon, T.E.** 04141
Univ. Arizona Disease Prevention Cent., Epidemiology & Biometry Section, Dept. Family & Community Medicine, 1501 North Campbell St., Tucson AZ 85724, United States of America (Tel.: +1 602 6264010; Fax: 6264089 ; Tlx: 9109521238)
COLL: Levine, N.; Schreiber, M.M.; McNamara, N.

Epidemiology of Cutaneous Cancers

The objective of these two studies is to evaluate the relative and attributable risk of skin cancer for demographic, phenotypic, environmental (including diet) and biochemical factors. The first case - control study includes at least 400 subjects with newly diagnosed primary melanoma and 800 age, sex, race, and geographical region matched controls. The second case-control study includes at least 400 subjects with newly diagnosed basal or squamous cell skin cancer with no history of cancer. Controls for this second study include 800 age, sex, race and geographical region matched subjects. Controls will be identified by a random phone digit procedure. Biochemical measurements include blood retinyl palmitate, beta carotene and selenium and tissue levels of ornithine decarboxylase, and fatty acids.

TYPE: Case-Control
TERM: Diet; Environmental Factors; Enzymes; Lipids; Metals; Vitamins
SITE: Melanoma; Skin
CHEM: Beta Carotene; Retinoids; Selenium
TIME: 1983 - 1997

1222 **Moon, T.E.** 04653
Univ. Arizona Disease Prevention Cent., Epidemiology & Biometry Section, Dept. Family & Community Medicine, 1501 North Campbell St., Tucson AZ 85724, United States of America (Tel.: +1 602 6264010; Fax: 6264089 ; Tlx: 9109521238)
COLL: Garcia, F.; Cartmel, B.; Villar, H.; King, D.; Micozzi, M.; McNamara, D.

Case-Control Study of Breast Cancer

This research applies complementary methods to quantitate macro-and micro-nutrient intake and nutrient-mediated factors to evaluate the aetiology and predict breast cancer risk in a defined white Hispanic and white non-Hispanic population in Arizona. Nutrients of primary focus include dietary lipids, retinoids, carotenoids, selenium and total calories. Nutrient-mediated factors include body size, body fat, tissue and blood fatty acids, plasma retinoids, carotenoids and selenium. Exposure estimates are based on interview, biochemical, anthropometric, and bioelectrical impedance of 600 post menopausal cases of breast cancer and 600 controls randomly selected from Arizona residents.

TYPE: Case-Control
TERM: Biochemical Markers; Diet; Fat; Lipids; Metals; Nutrition
SITE: Breast (F)
CHEM: Beta Carotene; Retinoids; Selenium
TIME: 1988 - 1994

***1223** **Moon, T.E.** 05493
Univ. Arizona Disease Prevention Cent., Epidemiology & Biometry Section, Dept. Family & Community Medicine, 1501 North Campbell St., Tucson AZ 85724, United States of America (Tel.: +1 602 6264010; Fax: 6264089 ; Tlx: 9109521238)
COLL: Dalen, J.; Bassford, T.

Arizona Women's and Adult Health Study

This randomised intervention trial aims to evaluate oestrogen replacement therapy, low fat diet or vitamin D/calcium vs placebo in the control of cardiovascular diseases, cancer and bone fractures. 3,500 women aged 50-79 years from white and native American communities throughout Arizona will be recruited. The end points of evaluation include incidence of and mortality from cardiovascular diseases, cancers of the breast, endometrium, ovary, colon and rectum and bone fractures.

TYPE: Intervention
TERM: Diet; Hormones; Registry; Vitamins
SITE: Breast (F); Colon; Ovary; Rectum; Uterus (Corpus)
CHEM: Calcium; Oestrogens
REGI: Arizona (USA)
TIME: 1993 - 2005

Tyler

1224 **McLarty, J.W.** 03768
Univ. of Texas Health Center, P.O. Box 2003, Tyler TX 75710, United States of America (Tel.: +1 214 8773451)
COLL: Kummet, T.D.

Beta Carotene, Retinol and Lung Cancer Chemoprevention

A randomized clinical trial of beta carotene and retinol versus placebo as lung cancer preventive agents is being conducted. This five-year study will enrol at least 600 persons occupationally exposed to asbestos. Bronchial epithelial changes, as evaluated by repeated sputum cytology, will be the primary measure of response. Serum levels of beta carotene will be studied with respect to the degree of sputum metaplasia or dysplasia found at initial examination and with respect to bronchial epithelial changes throughout the clinical trial.

TYPE: Intervention
TERM: Cytology; Dusts; Occupation; Premalignant Lesion; Prevention; Sputum; Vitamins
SITE: Lung
CHEM: Asbestos; Beta Carotene; Retinoids
TIME: 1984 - 1994

Washington

1225 **Byrne, J.** 02138
Children's National Medical Center Dept. of Hematology/Oncology Clinical Cancer Epidemiology Program, 111 Michigan Ave., NW, Washington DC 20010-2970, United States of America (Tel.: +1 202 7458954; Fax: 7455685)
COLL: Nicholson, H.S.; Fears, T.R.; Gail, M.G.; Haupt, R.

Late Effects in Long-Term Survivors of Childhood and Adolescent Cancer

The objective of this study is to document the ways in which different treatments for cancer in childhood or adolescence affect the fertility and quality of life of adult surviviors. Current investigations include studies of menarche, fertility, foetal deaths and menopause, psychosocial effects, late mortality and morbidity. A cohort study design is used most often; cancer survivors are matched to sibling controls and interviewed about relevant experiences. Clinical measures such as hormone assays and imaging techniques are incorporated as needed. A paper by Mostow et al. has been published in J. Clin. Oncol. 9:592-599, 1991.

TYPE: Case Series; Cohort
TERM: Adolescence; Chemotherapy; Childhood; Congenital Abnormalities; Drugs; Fertility; Intra-Uterine Exposure; Menarche; Menopause; Psychosocial Factors; Radiotherapy; Sib
SITE: Childhood Neoplasms
CHEM: Cyclophosphamide
TIME: 1978 - 1995

***1226** **Chiazze, L. Jr** 05261
Georgetown Univ. School of Medicine, Div. of Biostatistics & Epidemiology, 3750 Reservoir Rd NW, Washington DC 20007, United States of America (Tel.: +1 202 6874758; Fax: 6877230)
COLL: Watkins, D.K.

Risk Factors Associated with Lung Cancer Deaths among Employees of a Fibreglass Manufacturing Plant

This is a case-control study within a cohort of employees of a fibreglass manufacturing facility to determine the extent to which exposures to substances in the plant environment, to non-workplace factors, or to a combination of the two, may play a role in respiratory disease mortality risk. Worker profiles for exposure to respirable fibres, fine fibres, asbestos, talc, formaldehyde, silica, and asphalt fumes were developed. Employment histories provided information on length of employment, year of hire, age at first hire and an extensive interview survey obtained information on demographic characteristics, lifetime residence, occupational and smoking histories, hobbies, and personal and family medical history. Analyses using matched odds ratios are being carried out to assess the association between lung cancer and cumulative exposure history, demographic characteristics and employment variables. 164 lung cancer cases and 379 matched controls are drawn from a cohort of those with one or more years of employment between 1 January 1948 and 31 December 1963 and will be followed up until 1982.

TYPE: Case-Control; Cohort
TERM: Demographic Factors; Dusts; Occupation; Plastics; Time Factors; Tobacco (Smoking)
SITE: Lung
CHEM: Asbestos; Formaldehyde; Glass Fibres; Silica; Talc
OCCU: Mineral Fibre Workers; Plastics Workers
TIME: 1988 - 1994

1227 **Gibb, H.** 04438
US Environmental Protection Agency, Off. Health & Environmental Assessment, RD-689, 401 Main St. SW, Washington DC 20460, United States of America (Tel.: +1 202 2607315; Fax: 2600393)
COLL: Lees, P.; Stewart, W.

Derivation of Exposure-Response Relationships from Analysis of Retrospective Industrial Hygiene and Epidemiological Data: Chromium Compounds and Lung

Hexavalent chromium compounds are acknowledged to be human carcinogens. This conclusion is based on the results of occupational studies of chromate production workers and trivalent chromium and animal bioassays of hexavalent chromium compounds. Data from occupational studies of chromate production workers to date have not been detailed enough to examine the carcinogenic risk of exposure to trivalent and hexavalent chromium compounds separately. This study will examine the lung cancer mortality risk for a cohort of chromate production workers by exposure to both hexavalent and trivalent chromium compounds. Lung cancer response by solubility of the chromium compounds will also be examined, and a lung cancer dose-response analysis for the various chromium compounds will be attempted. The cohort contains over 2,500 workers. Over 200,000 industrial hygiene measurements, some of which date back to the first date of entry (1950) of workers into the cohort, will be used to estimate exposure. Smoking data on part of the cohort is available.

TYPE: Cohort
TERM: Dose-Response; Metals; Occupation
SITE: All Sites; Lung
CHEM: Chromium
OCCU: Chromate Producing Workers
TIME: 1988 - 1994

1228 **Kikendall, J.W.** 04963
Walter Reed Army Medical Center, Gastroenterology Serv., Washington DC 20307-5001, United States of America (Tel.: +1 202 5761768; Fax: 5762478)
COLL: Bowen, P.; Burgess, M.; Magnetti, C.; Woodward, J.; Sobin, L.; Langenberg, P.

Risk Factors for Colonic Neoplasia

361 subjects undergoing colonoscopy because of occult bleeding or barium enema suggestive of polyp are being studied. All were in good general health at recruitment. Each subject underwent a dietary assessment and evaluation of serum nutrients and growth factors, as well as of other cancer risk factors. Subjects were classified as controls, adenomas, or cancer according to the results of colonoscopy. The three groups are compared using univariate and multivariate analysis to define potential risk factors. To date, the data have shown that cigarettes and alcohol are independently associated with colonic adenomas and that serum gastrin is not elevated in subjects with colonic neoplasia. Assessment of the following risk factors is in progress: carotenoids, retinol, tocopherols, zinc, calcium, vitamin D, cruciferous vegetables, insulin-like growth factor 1.

TYPE: Case-Control
TERM: Alcohol; Diet; Nutrition; Polyps; Tobacco (Smoking); Vegetables; Vitamins
SITE: Colon
CHEM: Calcium; Retinoids; Tocopherol; Zinc
TIME: 1984 - 1994

1229 **McGowan, L.** 01419
George Washington Univ. Medical Center, Div. Gynecologic Oncology, Dept. of Obstetrics & Gynecology, 2150 Pennsylvania Ave. NW, Washington DC 20037, United States of America (Tel.: +1 202 9944218; Fax: 9940815)
COLL: Hoover, R.N.; Lesher, L.; Hartge, P.; Norris, H.J.

Ovarian Cancer in Greater Washington

Ovarian cancer risk has been estimated according to various reproductive and other suggested risk factors, using interview and medical records data from a case-control study. Identification was attempted of all women aged 20-79 years residing in the Washington, DC, metropolitan area who were first diagnosed at surgery with microscopically confirmed primary epithelial ovarian cancer from August 1978 to June 1981. The discharge lists of all 33 area hospitals that treated ovarian cancer were regularly checked. Cases included women with tumours of low malignant potential and frankly malignant tumours. For all potential cases, microscopic slides made from the tumour tissue were obtained. Those found not to have definite primary ovarian cancer of the epithelial type by clinical and microscopic evaluation were excluded from the study. 400 cases aged 20-79 years with histologically confirmed epithelial cancer were identified, of whom 296 were interviewed. Controls were identified from hospital discharge lists and matched to cases according to age, race, hospital, and date of discharge. A woman was not eligible to be a control if her discharge diagnosis was potentially related to the exposures under study, and it was ascertained from physician's records that the woman had at least one ovary intact. 439 women aged 20-79 years, eligible to be controls, were identified, of whom 343 were interviewed. Trained, experienced medical interviewers administered a standardized questionnaire in the patient's home. The interviewers obtained a detailed history of all gynaecological surgery and elicited which organs were involved in each procedure. Effects on ovarian cancer risk were measured by the estimated rate ratio, the ratio of ovarian cancer incidence in the exposed group to that in the unexposed. The estimates were adjusted for the effects of confounding variables by stratified contingency table analysis and by logistic regression models. Recent papers appeared in Am. J. Epidemiol. 136:1175-1183,

1184-1203 and 1212-1220, 1992 (Whittemore et al), Am. J. Epidemiol. 136:1204-1211, 1992 (Harris et al) and in J. Natl Cancer Inst. 85:142-147, 1993 (John et al).

TYPE: Case-Control
TERM: Drugs; Hormones; Infertility; Virus
SITE: Ovary
CHEM: Oestrogens; Steroids
TIME: 1978 - 1994

West Paterson

1230 **Lucas, L.J.** 03966
CYTEC Industries, 5 Garret Mountain Plaza, West Paterson NJ 07424, United States of America (Tel.: +1 201 3573351; Fax: 3573052)
COLL: Utidijian, H.M.D.; Caporossi, J.C.; Swaen, G.M.H.

Mortality of Employees Exposed to Acrylamide, Acrylonitrile and Methyl Methacrylate

Retrospective cohort mortality studies have been undertaken to determine if there is an excess cancer risk among employees exposed to the following chemicals: acrylonitrile, acrylamide and methyl methacrylate. The numbers of persons exposed are, respectively: 1,700, 3,000 and 1,300. In each study the mortality experience of exposed workers will be compared with that of an internal plant population, and with that of the general regional and national populations. Reports on cause specific mortality patterns as of 1983 have been prepared. Overall, no increased risk was observed for cancer of all combined sites. Efforts are currently underway to update work histories and vital status of all cohort members. These data will be integrated with exposure data to evaluate exposure-response relationships. Papers have been published in J. Nat. Cancer Inst. 78:192-193, 1987; J. Occup. Med. 31:41-46 and 31:614-617, 1989 (Collins, J.J. et al.).

TYPE: Cohort
TERM: Chemical Exposure; Occupation; Plastics
SITE: All Sites
CHEM: Acrylamide; Acrylonitrile; Methyl Methacrylate
OCCU: Chemical Industry Workers
LOCA: Netherlands; United States of America
TIME: 1983 - 1994

1231 **Lucas, L.J.** 04842
CYTEC Industries, 5 Garret Mountain Plaza, West Paterson NJ 07424, United States of America (Tel.: +1 201 3573351; Fax: 3573052)
COLL: Utidijian, H.M.D.; Caporossi, J.C.

Mortality Study of Employees Exposed to Ethylene Oxide

A retrospective cohort mortality study will be conducted to test the hypothesis that workers exposed to ethylene oxide during the industrial production of ethylene oxide sterilized medical products are at increased risk of mortality from malignant neoplasms, in particular leukaemia. The cohort will include about 4,600 employees identified from company records who were hired between 1952 and 1988. Using national and regional external comparisons the mortality experience of the cohort will be examined from 1 January 1957 through 31 December 1990. Exposure-response relationships will be assessed from work history data and industrial hygiene measurements.

TYPE: Cohort
TERM: Occupation
SITE: All Sites; Leukaemia
CHEM: Ethylene Oxide
TIME: 1990 - 1994

Worcester

1232 **Hebert, J.R.** 05023

Univ. of Massachusetts, Medical Center, 55 Lake Ave. North, Worcester MA 01655, United States of America (Tel.: +1 508 8564129; Fax: 8563840)

COLL: Reale, F.R.; Jederlinic, P.; Lew, R.; Rogers, E.

Relationship Between Precancerous Conditions of the Lung and Dietary Factors in High Risk Populations

This is a multi-phase study of the relationship between dietary factors, especially dietary lipids and lipid-soluble vitamins, and cytological conditions and DNA ploidy characteristics that precede frank squamous cell carcinoma of the lung. The first phase of establishing the utility of the cell markers is currently under way in a study of 30 non-smokers, 30 current smokers of 20 or more cigarettes per day, and 30 patients newly diagnosed with squamous cell carcinoma of the lung. The follow-up studies planned include an antioxidant vitamin intervention trial of current smokers and a cross-sectional study of black and white Americans from a Boston inner city neighbourhood. The latter is motivated by concern at the large and growing divergence in lung cancer rates between blacks and whites that is not attributable to cigarette smoking.

TYPE: Cross-Sectional; Intervention
TERM: Biomarkers; DNA; Diet; High-Risk Groups; Lipids; Nutrition; Premalignant Lesion; Vitamins
SITE: Lung
TIME: 1990 - 1996

Montevideo

1233 **De Stéfani, E.** 03622
Hosp. de Clinicas "Dr. Manuel Quintela, Depto de Patologia, Ave. Italia y Las Heras , Montevideo, Uruguay (Tel.: +598 2 781045; Tlx: 22348 urexpor uy)
COLL: Pellegrini, H.D.; Carzoglio, J.; Cendan, M.; Olivera, L.

Gastric Cancer in Uruguay

Although rates from stomach cancer in Uruguay have been declining since 1964, counties in the north display high rates and adjusted rates for the whole country are in the order of 23/100,000 in males and 10/100,000 in females. Ingestion of salted meat and exposure to a local infusion known as "mate" have been suspected as aetiological factors. This project attempts to evaluate the possible association of these variables with gastric cancer in a case-control study. Only incident and histologically proven cases will be eligible. Lauren's classification in intestinal and diffuse types will be employed. According to admission rates at the University Hospital, an average of 70 cases per year is expected. In the study period a total of 200 cases will be assembled and interviewed. In order to increase the statistical power of the study, four controls per case will be selected randomly from patients admitted to the same hospital. Patients with oesophageal, oropharyngeal, laryngeal and lung cancer will not be eligible, in order to assess tobacco smoking and alcohol drinking effects.

TYPE: Case-Control
TERM: Alcohol; Cooking Methods; Diet; Mate; Metals; Tobacco (Smoking)
SITE: Stomach
CHEM: Sodium Chloride
TIME: 1985 - 1994

1234 **De Stéfani, E.** 03624
Hosp. de Clinicas "Dr. Manuel Quintela, Depto de Patologia, Ave. Italia y Las Heras , Montevideo, Uruguay (Tel.: +598 2 781045; Tlx: 22348 urexpor uy)
COLL: Olivera, L.; Pellegrini, H.D.; Carzoglio, J.; Ordoqui, G.

Dark Tobacco Exposure and Lung Cancer

It has been suggested that dark tobacco exposure could be associated with a greater risk of developing lung cancer than smoking the Virginia brands. Uruguay is particularly well suited for such assessment due to the rather similar prevalence of both kinds of exposures. This project aims to assess relative risk (RR) estimates for dark and Virginia tobacco. Only incident cases of lung cancer occurring in males aged 30-79 will be selected and classified histologically according to the WHO system. In order to detect an RR of 2.5 with a significance level of 5% and a power of 0.80, at least 80 cases of each major histological type will be assembled. The control group (one control per case) will be selected from the same hospital and will include males of the same age range. Patients with respiratory conditions and neoplasms with well documented associations with smoking will be excluded. Usual residence, occupation, type of tobacco smoked, dose (in cigarettes per day) duration of smoking, years since quitting, use of hand rolled cigarettes and filter use are the variables included in the study. Preliminary analysis of data collected on 804 cases and 680 controls shows a RR 1.6 higher for dark tobacco compared with Virginia tobacco.

TYPE: Case-Control
TERM: Occupation; Tobacco (Smoking)
SITE: Lung
TIME: 1985 - 1994

***1235** **De Stéfani, E.** 05484
Hosp. de Clinicas "Dr. Manuel Quintela, Depto de Patologia, Ave. Italia y Las Heras , Montevideo, Uruguay (Tel.: +598 2 781045; Tlx: 22348 urexpor uy)
COLL: Ronco, A.L.; Mendilamarsu, S.; Kogevinas, M.; Boffetta, P.

Cancer and Occupation in Uruguay

Uruguay has the highest age-adjusted death rate from cancer of all sites in males in the Americas. The objectives of this study are to identify the possible associations between occupational exposures and cancer risk in Uruguay, estimate the number of deaths from cancer due to working in hazardous occupations, and examine the synergistic effect of certain occupational exposures and tobacco smoking. The study will provide estimates on the prevalence of exposure to pesticides and asbestos. About 2,000 histologically confirmed male cancer cases first diagnosed between 1993 and 1994 in four major hospitals in Montevideo will be included in the study. The catchment area of these studies covers the whole country. A pilot tested questionnaire will be used to request information on demographic factors, full occupational history, including a detailed evaluation of exposure to pesticides and asbestos, blond and black tobacco use and other lifestyle factors. Cancer risk for various cancer sites will be examined using standard methods of analysis of case-control studies. Cases of a cancer under consideration will be compared with selected cases of other cancer sites.

TYPE: Case-Control
TERM: Dusts; Occupation; Pesticides; Registry; Tobacco (Smoking)
SITE: All Sites
CHEM: Asbestos
REGI: Uruguay (Uru)
TIME: 1992 - 1994

1236 **Oreggia Coppetti, F.V.** 04142
Univ. of Montevideo, Hosp. de Clínicas, Dept. of Otolaryngology, French 2013 - Carrasco, Montevideo, Uruguay (Tel.: +598 2 604648)
COLL: De Stéfani, E.; Rivero, S.; Fierro, L.

Risk Factors in Cancer of the Oral Cavity, Pharynx and Larynx

An earlier study has been extended in order to gain further insight into tobacco and alcohol carcinogenesis in the oral cavity, pharynx and larynx. The effect of smoking cessation will be examined according to duration, intensity of habit, and type of tobacco smoked. Data will be collected according to a structured questionnaire at the University Hospital and Oncology Institute. The number of cases expected per year is 200; each case will be matched on age, sex, and place of residence to two controls, selected from the same hospitals, with diseases not related to tobacco and alcohol.

TYPE: Case-Control
TERM: Alcohol; Time Factors; Tobacco (Smoking)
SITE: Larynx; Oral Cavity; Pharynx
TIME: 1991 - 1994

1237 **Ronco, A.L.** 05241
Registro Nacional de Cáncer, Eduardo Acevedo 1530, 11200 Montevideo, Uruguay (Tel.: +598 2 486594)
COLL: Vassallo, J.A.

Atlas of Cancer Mortality in Uruguay 1989-1992

The aim of the study is to present the patterns of geographical distribution of cancer mortality in Uruguay for the period 1989-1992. The source is death certificates provided by the Statistics Bureau of the Ministry of Health, collected and coded (ICD) in the National Cancer Registry. The data are being processed to establish crude and age-adjusted rates for the most important types of cancer in the 19 Departments of Uruguay. The patterns of cancer distribution are also compared with those of the main non-cancer causes of death (cardiovascular, respiratory) for the same period.

TYPE: Mortality
TERM: Geographic Factors; Mapping; Registry
SITE: All Sites
REGI: Uruguay (Uru)
TIME: 1990 - 1994

1238 **Vassallo, J.A.** 04510
Registro Nacional de Cáncer, Eduardo Acevedo 1530, 11200 Montevideo, Uruguay (Tel.: +598 2 486594)
COLL: Leibovici, S.; Ronco, A.L.

Cancer Incidence and Mortality in a Department of Uruguay

Uruguay is administratively subdivided into 19 Departments. One of them, Cerro Largo, which has the highest cancer rates in the country, and adequate medical resources, is being used for detailed study of the high cancer mortality rates in Uruguay. This correlation study attempts to identify the pattern of cancer mortality in this chosen geographical region, and compare it with the whole country. The risk factors to which the population of Cerro Largo is exposed (consumption of tobacco, meat, the infusion known as "mate", etc.), will be analysed, with the object of correlating them with the magnitude of the tumour incidence and mortality rates.

TYPE: Correlation; Incidence; Mortality
TERM: Diet; Fat; Mate; Registry; Tobacco (Smoking)
SITE: All Sites
REGI: Uruguay (Uru)
TIME: 1988 - 1994

1239 **Vassallo, J.A.** 05243
Registro Nacional de Cáncer, Eduardo Acevedo 1530, 11200 Montevideo, Uruguay (Tel.: +598 2 486594)
COLL: Ronco, A.L.

Cancer in Frontier Inhabitants of Uruguay

Uruguay has long (500 Km) frontiers with Argentina and Brazil. It is well known that the lifestyle and habits in frontier regions are different from the rest of the countries involved. The cancer rates for Uruguay will be compared with those for Argentina and Brazil. The study will try to assess if the influence from the neighbouring countries on the habits of the frontier cities modifies the cancer rates. Data will be obtained from the National Cancer Registry.

TYPE: Incidence
TERM: Environmental Factors; Geographic Factors; Lifestyle; Registry
SITE: All Sites
REGI: Uruguay (Uru)
TIME: 1991 - 1994

1240 **Vassallo, J.A.** 05244
Registro Nacional de Cáncer, Eduardo Acevedo 1530, 11200 Montevideo, Uruguay (Tel.: +598 2 486594)
COLL: Ronco, A.L.

Geographic Pathology of Cancer in Uruguay and Its Correlation with Risk Factors

A study of the distribution of cancer in Uruguay is being carried out. The main goals are (1) to establish correlations with risk factors (lifestyle, habits, diet) in different geographical areas; (2) to assess the magnitude of cancer in the different areas; and (3) to obtain the necessary information to develop local and/or regional strategies for cancer prevention.

TYPE: Correlation; Incidence; Mortality
TERM: Diet; Environmental Factors; Geographic Factors; Lifestyle; Prevention; Registry
SITE: All Sites
REGI: Uruguay (Uru)
TIME: 1989 - 1994

Belgrade

1241 **Djordjevic, M.** 03078
Inst. of Oncology & Radiology, Pasterova 14, 11000 Belgrade, Yugoslavia (Tel.: +38 11 685755; Fax: 685300 ; Tlx: 11077 onkos yu)
COLL: Gec, M.

Adolescent Obesity or Malnutrition as Risk Factors for Cancer

The hypotheses being tested are that adolescent obesity or, alternatively, malnutrition in the adolescent period, are risk factors in cancer aetiology. Cancer of the genitourinary organs and the digestive system are now being studied, while a study of breast cancer has been completed. It is planned to interview 500 patients with a corresponding control group matched by age, sex, occupation, demography, etc.

TYPE: Case-Control
TERM: Adolescence; Diet; Nutrition; Obesity; Physical Factors
SITE: Gastrointestinal; Genitourinary
TIME: 1982 - 1994

1242 **Djordjevic, M.** 04740
Inst. of Oncology & Radiology, Pasterova 14, 11000 Belgrade, Yugoslavia (Tel.: +38 11 685755; Fax: 685300 ; Tlx: 11077 onkos yu)
COLL: Vuletic, L.; Jovicevic-Bekic, A.; Kanjuh, V.

Follow-up of Risk Groups for Breast Cancer

The aim of this prospective study is to justify the use of selective screening in breast cancer. Through a case-control study of breast cancer completed in 1988, 27 risk factors for breast cancer have been identified, including age, occupation, parity, oral contraception, previous diseases and injuries. Based on these data, a table of graduated risk factors by age has been created. In this prospective study, randomly selected women from the general population will be interviewed with a questionnaire to elicit these risk factors. According to the presence, combination or absence of various risk factors, women will be included in three groups with low, medium and high risk for breast cancer (at least 1000 women each group) and followed for 10 years in order to determine the incidence of breast cancer in each group. If the high-risk group should reveal significantly higher breast cancer incidence, this might open up the possibility of selective screening in a high-risk population. So far screening of 3,000 women led to the identification of a risk group of 39 women (among whom six developed breast cancer), compared with 2,961 women not at risk (four breast cancers).

TYPE: Cohort; Methodology
TERM: High-Risk Groups; Screening
SITE: Breast (F)
TIME: 1990 - 2000

1243 **Mitrovic, N.** 04754
Inst. of Oncology & Radiology, Pasterova 14, 11000 Belgrade, Yugoslavia (Tel.: +38 11 685755; Tlx: 11077 onkos yu)
COLL: Djordjevic, M.; Vuletic, L.; Kanjuh, V.; Jovicevic-Bekic, A.; Matijasevic, A.

Ovarian Cancer in Serbia

In the first part of the investigation, the aim is to determine standardized mortality and incidence rates for ovarian cancer, trend over a five-year period and demographic differences concerning the ethnic and religious characteristics of the female population. This analysis will be by histological type. In a case-control study with 300 cases and a control group of 900 disease-free women matched for age, occupation and residence the following factors will be investigated: positive family history of cancer (using Chase's life-table method in order to obtain cumulative incidence of family cancer), links with other sites (cancer of the endometrium, breast cancer and cancer of the colon), parity, hormonal impairments and oral contraception.

TYPE: Case-Control
TERM: Ethnic Group; Familial Factors; Histology; Oral Contraceptives; Parity; Religion; Trends
SITE: Ovary
TIME: 1990 - 1995

***1244** **Sokic, S.** 05365
Univ. of Belgrade, School of Medicine Inst. of Epidemiology, Visegradska 26 , 11000 Belgrade, Yugoslavia (Tel.: +38 11 641550)
COLL: Benko, A.

Case-Control Study of Hodgkin's Disease

The aim of this study is to investigate the risk factors for Hodgkin's disease. It involves 100 cases and 100 controls (patients treated for injuries or rheumatism), matched for sex, age and place of residence. In addition to demographic characteristics, factors of interest are weight and height, blood transfusions, natural diet, childhood diseases, birth order, twinning, etc. Personal case histories focus on acute and chronic diseases, allergies, congenital malformations, blood diseases, variola and tuberculosis inoculations, HGsAg bacilli carriers, appendectomy and tonsillectomy. Family case histories comprise parental diseases, notably diseases during pregnancy (including irradiation, alcohol and tobacco consumption, exposure to chemicals). Parents' exposure to ultraviolet radiation, electromagnetic waves, chemical compounds, medical drugs, tobacco, alcohol, coffee, sports, diet, housing, and other potential risk factors are also being studied. Multiple logistic regression and McNemar's analyses are being applied and odds ratios calculated.

TYPE: Case-Control
TERM: Alcohol; Allergy; Appendectomy; Blood; Chemical Exposure; Congenital Abnormalities; Diet; Intra-Uterine Exposure; Physical Factors; Radiation, Ionizing; Tobacco (Smoking); Tonsillectomy
SITE: Hodgkin's Disease
TIME: 1992 - 1994

***1245** **Sokic, S.** 05366
Univ. of Belgrade, School of Medicine Inst. of Epidemiology, Visegradska 26 , 11000 Belgrade, Yugoslavia (Tel.: +38 11 641550)
COLL: Benko, A.

Case-Control Study of Thyroid Cancer

The purpose of this case-control study is to investigate risk factors for cancer of the thyroid. It involves 100 incident thyroid cancer cases, 100 patients treated for minor injuries (control group I) and 100 goitre patients (control group II). Cases and controls are matched for sex, age (+/- 5 years) and place of residence (Serbia). Risk factors of interest are demographic characteristics, cigarette smoking, alcohol and coffee consumption, family history of goitre or hyperthyroidism, reproductive factors, history of thyroid tumours in the cases and in their families, history of malignant tumours, mumps and other diseases, chemical exposure, residence in endemic areas and other suggested risk factors. Multiple logistic regression and McNemar's test are applied. The results will be presented as odds ratios.

TYPE: Case-Control
TERM: Alcohol; Chemical Exposure; Coffee; Family History; Goitre; Hyperthyroidism; Physical Activity; Reproductive Factors; Tobacco (Smoking)
SITE: Thyroid
TIME: 1992 - 1994

***1246** **Sokic, S.** 05367
Univ. of Belgrade, School of Medicine Inst. of Epidemiology, Visegradska 26 , 11000 Belgrade, Yugoslavia (Tel.: +38 11 641550)
COLL: Benko, A.

Case-Control Study of Laryngeal Cancer

A matched case-control study has been initiated to address the potential risk factors for laryngeal cancer. The study involves 100 incident laryngeal cancer patients and 100 control subjects chosen from patients treated for minor injuries at the University Clinical Centre in Belgrade, matched for age (+/-5 years), sex and place of residence (Serbia). The risk factors studied are socio-economic status, occupational factors, previous surgical interventions, cigarette smoking, alcohol and diet. McNemar's test and multiple logistic regression will be used to analyse the data. The measure association will be odds ratio (OR) with its 95% confidence interval.

TYPE: Case-Control
TERM: Alcohol; Coffee; Diet; Surgery; Tobacco (Smoking)
SITE: Larynx
TIME: 1992 - 1994

INDEX OF INVESTIGATORS

This index identifies projects by investigator. It includes both principal investigators and collaborators.

Each project is identified by its serial number, not the page number.

For studies in which the person indexed is the principal investigator, the serial number is shown in **bold type**.

For studies included in the Directory for the first time, the serial number is preceded by an asterisk (*).

Aaran, R.K. 241
Aareleid, T. *780
Abdulkadirov, K.M. **716**
Abdulnour, E. 97
Abe, R. 558
Abel, E. 993
Abel, U.R. 359
Abid, L. 914
Ábrahám, E. **388**
Achmad Ghozali 449
Adam, H. 831
Adami, H.O. 45, 211, *760, **793**, **794**, **795**, **796**, 797, 800, *801, 802, 803, 804
Adams, J. 1079
Adenis, L. 268
Ades, A.E. 872
Adhvaryu, S.G. **402**
Adler, Z. 463
Admella-Salvador, M.C. *736
Adzersen, K.H. *362
Agarwal, S.S. 437
Aggazzotti, G. **519**, ***520**
Aghi, M.B. 419, *421
Agius, R. 844, 857
Agnarsson, B.A. 398
Ahlbom, A.N. 300, **765**, **766**, **767**
Ahmed, M. 674
Ahmed, Q.A. *54
Ahonen, R 253
Ahrens, W. 282, 297, 351
Airaksinen, O. 253
Airoldi, M.L. *543
Aït Hamouda, R. *2
Aitken, J.F. 9
Akdas, A. **815**, **816**
Åkesson, B. 754
Akinsanya, A. 659
Akulenco, L.V. *703
Alavanja, M.C.R. **963**, 965, 970
Albertini, R.J. 1001
Alberts, D. 1220
Albin, M.P. **748**, *758
Albright, L.A. 1151
Aldefer, R. 1025
Aldrich, T.E. *1006, **1165**, **1166**
Alexander, F.E. *822, 849, *850, ***929**
Alexeyeff, M. 28
Alfandry, E. *475
Alfaudari, E. 472
Alfthan, G. 241
Alhonen, L. 253
Alican, Y. 816
Alihonou, E. 274
Allan, R.N. **821**
Allen, J.C. 1151
Allouache, A. 1
Almog, R. *460
Alonso de Ruiz, P. 284
Alterman, T. 1121
Alvarez, N. 285, 286
Amadori, A. 477, 524
Amadori, D. *478
Ambinder, R.F. 1154
Ambrosic, D. 192
Amemiya, T. **561**
Amerio, P.L. 529
Amestoy, G. 363
Amichetti, M. **544**
Amorosi, A. *493
Amrender Reddy, G. *430
Amundson, B. 1194
Andersen, A. 281, 282, 298, 663, 664, *665, *667, *668, *669, *670, *671
Anderson, D.E. **1053**, **1054**
Anderson, G. 1197, 1200
Anderson, L. **210**
Anderson, M 224
Anderson, M.S. **951**
Anderson, T.F. 993
Anderson, V.E. 1111
Andersson, I. 791
Andersson, S.O. ***760**, 793
Andrade, A. 863, *864
Andreu, M. *734
Anh, C.T. *290
Annapurna, V.V. *430
Anton-Culver, H. **1061**, ***1062**, ***1063**
Anttila, S. 239
Anwar, W.A. *120
Anzai, S. **595**
Aoki, K. 560, **564**, 566, 595
Aoyama, T. **583**
Arblaster, L. 851
Arbman, G. **747**
Arce, V. 191
Archer, V.E. 1182
Arena, V. 1161
Aristizabal, N. 284
Arlett, C.F. 830
Armstrong, B.G. *277, 872
Armstrong, B.K. 31, 32, ***269**, 345
Armstrong, F.D. 1042
Armstrong, M.A. 1139
Armstrong, R.W. *975
Arndt, K.A. 993
Arnon, L. 452
Arnorsson, J.V. 398
Aromaa, A. 241, 242
Arsenault, A. 77
Arthur, D.C. 1109
Arthur, J. *817
Arumugam, S. 432
Arveux, P. 261
Aschenbrener, C. 955
Ascherio, A. *427
Ascunce, N. 284
Ashby, D. 860, 861
Ashley, R.L. 1203
Ashmore, J.P. **81**, *277
Ashraf, S. 673
Askari, A. 674
Asp, S. 247
Assennato, G. **480**, **481**, *490
Astashevsky, S.V. *696, *697, *698, *700, *701, *715
Astrup-Jensen, A. 281

Atay, Z. *379
Atkin, W.A. 885
Atkin, W.S. ***862**
Attam, K. 437, 438
Attewell, R. 748
Auburtin, G. *324, **328**
Augustin, J. **196**, 287
Austin, D.F. 955, 1121
Austin, H.D. **982**
Autier, P. 282
Autrup, H. **201**, ***202**, ***203**, ***204**, ***205**, ***206**, *233, *853
Avanzi, G.C. 539
Avril, M.F. *337
Axelson, O. 747, *768, 782
Axerio, M. 506
Aye, K.M. 607
Ayme, S. 58
Ayzac, L. 279
Baak, J.P.A. 534
Baanders-van Halewijn, E.A. **634**, **635**, 637
Babaeva, R.Y. 711
Backley, D. 260
Badia, A. 735
Badosa, E. 735
Báez, S. 125
Baghurst, P.A. *37
Bah, E. **344**, 345
Bahmer, F.A. 346
Bailey, A. 884
Bailey, C. *852
Bain, C.J. 8
Baines, C.J. 104, 105, 114
Bairati, I. 92
Bajador-Andreu, E.J. *736
Baker, G. *43
Baker, M. 1197
Baki, M. 391
Balakrishnan, T.S. 432
Balanda, K.P. 10, 11
Balar, D.B. **403**, 404, 405, 406, *407, 408, 409
Baldasseroni, A. 506
Ball, D. 82
Balmes, J. 1200
Baltrusch, H.J.F. **378**
Balzi, D. 498
Balzi, M. *493
Banasik, R. 678, 679
Band, P.R. **116**, **117**, 121
Bandera, E.V. ***999**
Banerjee, T.K. **1099**
Banfi, A. 1150
Bankier, A. 24
Bankson, D. 1197
Bánóczy, J.E. **389**
Baptiste, M.S. *999
Barbone, F. *478
Bard, D. ***262**
Barker, D. *1063
Barker, D.J.P. 930, 938
Barlow, L. 287
Barnhart, S. 1200
Barnhill, R. 1122
Baron, J.A. 794, **1045**
Barreto, J.H.S. 71
Barreto, S.M. **863**, ***864**
Barrett, L. 266
Barrett, R.J. 972
Barrett-Connor, E.L. **1065**, 1066, *1067
Barry, T.M. 274
Bartlett, S. 86
Bartolome, M. 738
Bartow, S. **1111**
Bartow, S.A. *948
Bartsch, H. 297, 510, 541
Basieva, T.H. **694**, 713
Bass, G.I. 1160, 1161
Bassalyk, L. 711
Bassett, A.A. 114
Bassford, T. *1223
Bastecky, J. 378
Bastholt, L. 231
Basu, J. 994
Bates, M.N. **652**
Bathers, S. 828
Batiste, E. 735
Battistutta, D. **5**, 8
Baughman, R.D. 993
Baum, E. 1150
Bauman, A. **192**, **193**
Bausch-Goldbohm, R. 624, 625, 640
Baylin, S.B. 954
Bayo, J. 363
Bayo, S. 274
Beamish, M. 17
Beard, M. 653
Beattie, J. *817
Becciolini, A. *493
Becher, H. 296, 371
Bechi, P. ***493**
Bechtel, P. 343
Beck, G. 1045
Beck, J. 728
Beck, P. 102
Becker, F.F. 1057
Becker, N. **359**, *362
Becker, T.M. **946**
Becker, W. 797
Beckers, R. 58
Beckman, M. 783
Beckmann, A.M. 1203, *1207
Bedenne, L. 261
Beebe, G.W. 970
Beeson, W.L. 1070
Behera, D. ***428**
Beijerinck, D. 636
Belanger, C. 995
Belfiore, A. **492**
Bell, C.M.J. *336, **940**
Bell, D. *514
Bellander, T. 281, 756, 764, ***768**, 772, 779, *785
Belli, S. 300, 306, *316, 530, 537
Bellini, A. 511
Belyakova, S.V. 708
Ben Harush, M. 457

Ben-Shlomo, Y. *886
Benatti, P. 521
Benediktsdottir, K. 398
Bengtsson, C.B. **740**
Benhamou, E. 297, 329, 331, *337
Benhamou, S. 297, 300, **329**, **330**, **331**, ***332**, *337, 807
Benito, E. 275
Benko, A. *1244, *1245, *1246
Benlatreche, K. 1
Benn, T. 296
Bennett, B. 287
Benowitz, N. 1187
Benraadt, J. 620
Benson, M. 1129
Benz, M. 285, 286
Beral, V. *822, 888, **895**, **896**, **897**, ***898**, 908
Bercovich, J. 363
Beresford, S.A.A. **1190**, 1197, 1198
Berg, J. 576
Berger, J.E. *357
Bergeret, A. 279, 280, 282, *327
Bergkvist, L. 797, 803
Bergström, R. 794, 797
Berlin, N.I. 1013
Berman, M.L. 972
Bernar, J. *277
Bernard, E. 307
Bernard, J.L. **307**
Bernstein, L. 289, **1072**, **1073**, ***1074**
Bernstein, R. 730
Berrino, F. 295, 300, **506**, **507**, *515
Berry, R.J. **837**
Bert, J.L. 117
Bertazzi, P.A. 296, 298, **508**, **509**, *514
Berwick, M. 16, **1122**, 1208
Bethune, A. 870
Bethune, G. 114
Bethwaite, P. 282, **653**
Bhambhani, S. 437
Bharati Arumugam, S. **432**
Bhat, B. 74
Bhatavdekar, J.M. **404**, **405**, **406**, ***407**, 408
Bhatnagar, P. 437
Bhattathiri, V.N. 445
Biancalani, M. 496
Bianchi, C. **522**
Biasco, G. 260
Bicevskis, M. 17
Bidoli, E. 477, *478
Bierre, A. 644
Biggar, R.J. 218, **964**
Biggeri, A. 495, 498
Bigirimana, V. *898
Bilek, O. 196
Bilir, N. ***814**, 815
Billington, N.A. 945
Billon-Galland, M.A. *317
Binder, V. **226**
Bingham, S. 295
Binks, K. 837
Biocca, M. 281, *535
Biran, T. 467, 468
Birch, J.M. *46, *822
Bird, C.C. *842, *843
Birkeland, S.A. **227**
Bisanti, L. 511
Bishop, D.T. 918
Bistoletti, P. **745**
Bithell, J.F. 905
Bjarnason, Ó. 398
Bjerk, J.E. 281
Bjerkedal, T. *665
Bjornsson, J. 398
Bjurstam, N. 791
Black, H.S. **1055**
Blair, A.E. 305, **965**, **966**, ***967**, 977, 1171, *1172, 1173, 1177
Blamey, R.W. 941
Blanc, C. **312**
Blankenship, C. *976
Blankenstein, M.A. 635
Blanks, R.G. 872, 873
Blaser, M.J. 1169
Blattner, W.A. *1174
Blessing, J. *1175
Blettner, M. *277, **360**, 367, 860, 861
Bley, L.D. 1193
Bleyen, L.J. **60**
Blizzard, C.L. *18
Bloemen, L.J. 1104
Blondal, H. 398
Blot, W.J. 136, 137, 167, *183, **968**, **968**, ***969**, *975, *976, 1089, 1090
Boal, W. 282
Bobev, D. 287
Bock, J. 215
Bode, P. 624
Bodmer, J. 909
Bodner, K.M. 1104
Bodrogi, I. 391
Boeing, H. 295, 368, 372, 373, 480, 914
Boeniger, M. 1026
Boessel, J. 1151
Boffetta, P. *250, **270**, **271**, ***272**, 281, 282, 296, 297, 298, 305, 306, *715, *812, 863, *864, 1136, *1235
Bohlscheid, S. 372
Boice, J.D. 221, 223, 963, **970**, *1110
Boiron, O. 280
Bokey, L. 8
Bokros, F. 392
Bolan, G. 1187
Bolelli, G. *515
Bolm-Audorff, U. 306, *316, 351, *352, *379
Bonaguri, C. 531
Bonassi, S. *197, 305
Bond, J. 1106
Bondi, R. 495
Bondy, M.D. 1059
Bondy, M.L. ***1056**
Bonin, A. 48
Bonney, G.E. *134
Booker, S.V. 954

Booth, M. 896, 897
Bordi, C. 531
Borlee, I. 58
Borman, B. 646
Borroni, G. 529
Bos, R.P. 613, **626**
Bosch, F.X. **273**, **274**, **275**, 284, 345, 524, 811, 813, 956
Bostick, R. 1105
Bosze, P. 534
Botta, M. **488**, **489**
Bouchardy, C. **807**
Bourke, G.J. 300, **451**
Bourne, T.H. ***865**
Boutron, M.C. 260, 261
Bowen, D. 1197
Bowen, P. *1017, 1228
Bowie, C. **943**, **944**
Bowie, S.H.U. 944
Bowman, D.M. 114
Bowman, J.D. 655, 1088
Boyd, P. 1121
Boyes, D.A. 114
Boyle, P. *46, 270, 303, **510**, 809
Braas, P.A.M. 618
Bradley, M. 1106
Bradley, W.E.C. 77
Braga, M. 511
Brancker, A. 90, 289
Brandi, M.L. ***494**
Brandt-Rauf, P.W. **1123**, *1124, 1129, 1131
Brasilino de Carvalho, M. **66**
Brasure, J. *999
Braverman, I.M. 993
Brawer, M.K. *1205
Breibart, E.W. 346
Brémond, A.G. **276**, *302
Brenes, F. 191
Brescia, B. 1187
Breslow, N.E. 1007, *1110
Brett, S.M. **950**
Breuer, B. ***1124**
Bridges, B.A. **830**, 938
Briet, H.A. 614
Briggle, T. *1020
Brignone, G. 525
Brinton, L.A. *184, 306, 804, **971**, **972**, **973**, **974**, 980
Brissette, P. 1035
Brochard, P. *317, 343
Brodie, P. 819
Brögger, A. 753
Brok, K.E. **228**
Brollo, A. 522
Broustet, A. 280
Brown, A.M. *29
Brown, D.P. **1019**
Brown, K. *818
Brown, T. 300
Brownbill, P.A. 904
Brownson, R.C. 963
Bruce, Å. 794, **797**
Bruce, S. 1055
Brugère, J. 315
Bruhn, R. *768
Brullet, E. 735
Bruning, P.F. 1130
Bruno, C. 530
Bruno, L. *493
Brunt, P. *818
Brusa, M. 488, 489
Bruzzi, P.A. 510
Bryan, H. *111
Bryson, D. 910
Buchanan, D. 846
Buckley, J.D. 1108, 1109, 1147, 1151
Budiningsih, S. 565
Buelow, K.B. 1149
Bueno de Mesquita, H.B 296, **622**
Buffler, P.A. **961**, 1121
Buiatti, E. *256, 285, 496, 498, 499, 913
Bukin, Y.V. 711
Bulbulyan, M. **695**, ***696**, ***697**, ***698**, ***699**, ***700**, ***701**, 714, *715
Bunch, K.J. 904, *906, 922
Bundgaard, T. **207**
Bunin, G.R. **1147**, **1148**, 1151
Burau, K. 961
Burg, G. 346
Burgess, M. 1228
Buring, J.E. 986, *987, **995**
Burk, R. 994
Burmeister, L.F. 966
Burrows, B. 1219
Busellu, G.P. 505
Bush, H. 114
Bush, T.L. 1044
Buttyan, R. 1129
Bylin, G. 771
Byrne, J. **1225**
Caan, B. **1183**
Caceres Diaz, C. *195
Cade, J.E. **848**
Cai, S.X. **126**, 128
Cai, X.C. 128
Calabro, A. 58
Calabro, L. 524
Calatayud Sarthou, A. 737
Calheiros, J.M. 914
Calle, E.E. 953
Callet, B. 312
Calmettes, C. ***313**
Calva, J. *610
Calvert, G.M. ***1020**, *1021
Calvo, A. 125
Calzolari, E. 58
Camargo, B. 71
Cambier, L. 268
Cameron, A.H. 824
Camp, J. *1209
Campbell, C. 1000
Campbell, M. 848, *1002, *1010
Campbell, S. *336, *865
Campbell, T.C. 127
Campion, M. 1211

Campolucci, S. 951
Camus, M. 72
Cannon, L. 10, 11
Cano, E. 285
Cantin, J. 114
Cantor, K.C. 652, 965
Cantor, K.P. 1177
Caperle, M. 531, 533
Capizzi, R. 528
Caporaso, N.E. *514, 541, *543, 1025
Caporossi, J.C. 1230, 1231
Caraballoso, M. 194, *195
Carbone, A. 479
Carden, A. 24
Cardis, E. ***277**
Cardona, T. 735
Carere, A. *535
Carle, F. 504
Carli, P. **495**
Carli, P.M. 280
Carli, S. 499
Carlin, J. 25
Carman, W.J. 949
Carpenter, L.M. *277, 896, 897
Carpenter, M. 90
Carrato, A. *734
Carreras, M.V. *195
Carretta, D. 545
Carrillo, E. *734
Carstensen, J.M. 786
Carta, P. **483**, **484**, ***485**, 487
Carter, D. *842, *843
Carter, J.F. 644, 653
Cartmel, B. 1218, 1220, 1222
Cartwright, R.A. *822, **849**, ***850**, *852, 866, *929
Caruso, G. *490
Carzoglio, J. 1233, 1234
Casale, V. 531
Casamassima, A. *482
Casella, C. 503
Casey, G. *1063
Cassiman, J.J. 62
Cassin, B. 1033
Castagneto, B. 488, 489
Castellaneta, A. 531
Castellsague, X. 273
Castiglione, G. 509
Castilho, E.A. 64, 65
Castillo, R. 273
Castrén, O. 252
Castro, D. 285
Catton, G. 114
Ceccarelli, C. *338
Cecconi, R. 498
Ceci, A. 540
Celentano, E. *515
Celleno, L. 529
Cendan, M. 1233
Centonze, S. 480
Ceppi, M. 501, 502, 503
Cerimele, D.M. **528**, **529**
Cerny, T. 270
Cesar, D. 193
Chacko, P. 435
Chadha, B. 437, 438
Chadha, P. 437, 438
Chagas, J.F.S. 66
Chaitchik, S. **467**, **468**
Challeton, C. *338
Chamberlain, J. 892, **941**, 942
Chan, C.C. *184
Chang, A.R. **645**
Chang, F. 252
Chang, W.Y. ***160**
Chang, Y.S. 137
Chang-Claude, J. **361**, 363, 369
Chanteur, J. *277
Chapuis, P. 8
Chauaudra, J. *336
Cheirsilpa, A. 811
Chellini, E. **496**, 498, 500, *536
Chen, C.J. *184
Chen, D. 132
Chen, G.C. 168
Chen, H.Q. 174
Chen, I.H. *184
Chen, J. 968
Chen, J.K. **98**
Chen, J.S. **127**
Chen, J.Y. *184
Chen, K. 145, 146, 147, 148, 149, 150
Chen, K.W. 970
Chen, N.H. 128
Chen, R. *158
Chen, R.L. ***158**
Chen, S.C. *160
Chen, S.Y. 129
Chen, V.W. 1121
Chen, X.L. 174
Cheng, S.M. 128
Cheng, Y.J. *184
Cheng, Z.Y. 149
Cherchi, P. 484, 487
Cherian, T. 445
Cherian, V. 441, 442, 443
Cherniack, M. 1200
Cherrie, J. 298
Chetrit, A. *475
Chevalier, A. 312
Chianale, J. 124
Chiarelli, A. *111
Chiazze, L. Jr ***1226**
Chichareon, S. 813
Chick, J. 12
Chieco-Bianchi, L. **524**
Child, M.A. 955
Chilvers, C.E.D. 300, *822, 907, 920
Chinchilli, V. 950
Chmiel, J.S. **1013**
Choi, B.C.K. **99**
Choi, N.W. 93, 119, **122**, 123, 303
Chollat-Traquet, C. 301
Chongsuvivatwong, V. **813**
Chorus, A.M. 618
Chow, W.H. *969, ***975**, ***976**
Christensen, J.M. 201

Christie, D.G. **19 *29**
Chu, J. 1197
Chuang, T.Y. ***10, 30**
Chuang, W.L. *160
Chui, S.X. 130
Ciatto, S. 499
Ciccone, G. 504, **538**, **539**
Cicollela, A. 280
Cislaghi, C. 809
Clapp, R.W. *727, **985**
Clark, G.C. *514
Clark, L.C. **1215**, **1216**, **1217**
Clark, W. 1176
Clarke, E.A. 113
Claudiani, J. 363
Clausen, N. 672
Clavel, F. 295, **333**, **334**
Clavel, T. ***320**
Cleaver, J.E. **1184**
Clifford, C. 14
Close, P. *727
Clutton, S. 10, 11
Coates, M.S. **39**
Coates, R.J. 1121
Cocco, P.L. 271, 483, 484, **486**, **487**
Cochacova, L.B. *693
Cocito, V. 488, 489
Coebergh, J.W.W. 287, **633**
Coggon, D. *158, 281, 282, 296, **930**, **931**, **932**, 939
Cohen, D.L. 921
Cohen, L. 995
Cohn, R. 730
Coiffier, B. 280
Coker, A. **1029**
Colditz, G.A. 991
Coldman, A.J. 115
Cole, J. 830
Cole, P. *983, *984, 1102, 1115
Coleman, M.P. 1, 914
Colin Jones, D.G. 927
Collalto, A. 363
Collette, H.J.A. 295, 634, 635, **636**, 637
Collins, A.R. ***817**, *818
Collins, J. 25, *27
Collins, W. *865
Colonna, M. 308
Colucci, G. 480
Comba, P. 306, **530**, *536, 537, 538
Combs, G. 1215, 1216, 1217
Comin, C. 496
Comstock, G.W. **1044**
Connolly, J. 99
Conso, F. 334
Consonni, D. 508, 509
Conti, E.M.S. *478, 532, 533
Cook, L.S. ***1191**
Cook, N. 995
Cook-Mozaffari, P.J. **899**, **900**, 905
Cooke, K.R. 648
Cooney, R.V. *1048, *1051
Cooper, D. 41, *1117
Cooper, S.P. 961
Coopmans de Yoldi, G. 518
Copland, P.I. 42
Coppock, E.A. 90
Corbett, S.J. **40**
Corbin, A. 309
Cordier, S.E. *290, 303, 334, **335**
Corey, L. *1207
Cornain, S. 565
Cornelissen, M.A. 614
Corominas, J.M. *734
Corrao, G. **504**, **505**
Correa, P. 285, 361, 1121
Cortes Vizcaino, C. **737**
Corvalan, C.F. 42, *43
Coshcina, V.S. ***693**
Cosma, G. *514
Costa, A. 518
Costa, R. *610
Costantino, J.P. 1160, 1161
Costanzo-Nordin, M.R. 1100
Coste, D. 307
Coste, I. 276
Cottoni, F. 529
Courtial, I. 276
Couyette *338
Cowen, A. 8
Cowper, G. *277
Cox, C.E. 1040
Craft, A.W. 889, 890, 891
Crawford, N. 844
Crespi, M. 361, **531**, **532**, **533**
Crespo de Britton, R. 274
Criqui, M.H. 1066
Crissman, J. 1033
Crognier, E. 314
Crosignani, P. 507, *536, 542
Crouch, P. 3
Cullen, M. 1200
Cummins, C.W. 826
Cummins, H.W. 1163
Curado, M.P. 70
Curtis, R.E. 224, 970
Cuschieri, A. 58
Cusimano, R. 525
Cuttner, J. ***1125**
Cuzick, J.M. *862, **866**, **867**, 882, 883
Czanik, P. 388
Czeizel, A. **390**
D'Albasio, G. 531
d'Angio, G.J. 1150
D'Avanzo, B. 477, *478, 512
Daftary, D.K. 419, *420, *421
Dahl, C. 215
Dahlgren, L. 792
Dai, L.C. *290
Dai, X.D. **151**, **152**
Dalcin, P. 62
Dale, P.S. *411
Dalen, J. *1223
Daling, J.R. 121, 974, 1087, *1191, *1196, 1201, 1203, 1204, *1205, 1210
Daly, L. 451
Daly, M.B. ***1011**
Damber, L. 782

Dan, R.P. 180
Dananche, B. 279, 280
Danes, B.S. 954
Daniels, T. 1187
Danielsen, T.E. ***667**, ***668**
Darby, S.C. 840, 899, **901**, **902**, **903**
Dardanoni, L. 525, *527
Darlington, G. *111
Darra, F. 545
Darraugh, T. 1187
Darwis, I. 565
Das, B.C. 437
Das, D.K. 437
Dauda, G. 394
Daudt, A. 274
Daurès, J.P. **310**, ***311**
Dave, B.J. 402
Davey-Smith, G. 875
David, G. 62
Davies, A.P. **868**
Davies, J. 271
Davies, J.A. 69
Davies, J.M. 900
Davies, N.F. 819
Davies, P. 825
Davies, R.J. *1067
Davies, T.W. **834**
Davis, F.G. ***1014**, *1017
Davis, L. 985
Davis, S. **1192**, **1193**, **1194**, ***1195**
Davis-King, K. ***1021**
Dawsey, S.M. *134
Day, N.E. 295, 345, 442, 443, *822, 871
Dayer, P. 807
de Backer, G. 60, 914
De Bartolomeo, A. 526
de Boorder, T. 623
de Bruin, M. 624, 638
de Contreras, O. 285
De Gaudemaris, R. *265, 266
de Grandi, P. 810
De Hovitz, J. *998
de Klerk, N.H. **30**, 34, 35, 36
de la Genardière, E. *338
De Lucia, G. 503
De Maeseneer, J. 60
de Nully Brown, P. 222
De Oliveira, H. 260
De Palo, G. 518
De Pietri, P. 521
De Pue, R.H. 1039
De Quint, P. **57**
De Rossi, A. 524
de Sanjosé, S. 284, 285, 286
De Stéfani, E. **1233**, **1234**, ***1235**, 1236
de Thé, G. *2, 314
de Vathaire, F. ***336**, ***337**, ***338**, *342
de Villiers, E.M. 215
de Waard, F. 634, 635, 636, 637
De Wals, P. 58
Deacon, M.C. 28
Deapen, D. 1080
Decarli, A. 477, *478, 512, 513, 516, 517, 809
Decinti, E. 125
Deckers, P. *1043
Deddens, J. 1024
Degerth, R. 244
Degiovanni, D. 488, 489
Degos, L. 280
Dei, R. *493
Del Mistro, A. 524
Del Moral, A. 295
Delaloye, J.F. 810
Delclos, G. 961
Delisle, M.J. **318**, ***319**, *338
Della Foglia, M. 509
Delzell, E. 982, *983, *984
Demers, P. 305, 306, *316
Demets, R.Y. **1032**
Demicheli, V. 542
Den Engelse, L. 613
Den Tonkelaar, I. **637**
Denobile, J.W. *976
Deo, M.G. *412
Depoorter, A.M. 57
Depue, R.H. 1040
Derazne, E. 470
Déri, Z. 393
Desai, P.B. ***412**, 424
Deschenes, L. 114
Desmeules, M. 85
DeStavola, B. 866
Detels, R. 957
Deutsch, M. 1151
Devesa, S.S. **1170**
Dharkar, D.K. *416
di Orio, F. 504, 505
Di Placido, R. 505
Diaz, A. 194
Dicato, M 270, 510
Dich, J. 787
Dickinson, H. 890, 891
Diego, S. 479
Diehl, V. 270
Dietz, A. 364
Diez Sacristan, A. *277
Diez, M. *195
Digregorio, C. 521
Dillner, J. 745, *789
Dillner, L. 745
Diloy-Tejero, R. *736
DiMagno, E.P. 1169
Dimitrova, E. **719**
Ding, C.Y. 128
Ding, J.H. **161**
Dingerkus, G. *379
Dini, S. 496
Dinya, E. 388
Djordjevic, M. **1241**, **1242**, 1243
Dobiás, K. 721
Dockerty, J.D. **646**, **647**
Doll, R. 840, 899, 901, 902, 903, 910
Dombi, C. 389

Domenici, R. 545
Domergue, J. **309**
Donaldson, L. 928
Donaldson, S. *1110
Donelli, S. 509
Dong, J. 156, 157
Donnan, G.A. 38
Doody, M.M. 970
Doornbos, G. 622
Dorant, E. 625, 640
Dorfman, R.R. 1186
Dosemeci, M. 965, 968, **1171**, ***1172**
Dosman, J. **93**, 94
Douglas, A. 881, 896
Downey, R. 85
Downs, T.D. 961
Doyle, T. 651
Dragsted, L. *202, *203, *204, *205, *206
Drake, J.J. 69
Draper, G.J. 287, **904**, **905**, ***906**, 922
Dreyer, N.A. 990, 1012
Drezner, M. *1143
Driscoll, T.R. 42
Drotman, D.P. ***952**
Duan, B.R. 129
Dubeau, H. 77
Dubin, N. 1122
Dubourdieu, D. *317
Dubrow, R. *610, *969, *1118
Duca, P. **511**
Duchene, Y. 257
Duclos, J.C. **278**
Dueng, W.Z. 188
Duffy, S.W. 442, 443, 711, 712, 713
Dufour, R. 82
Duncan, K.P. ***838**
Dunlop, M.G. ***842**, ***843**
Dupont, W.D. **1115**, **1116**
Dupras, G. 77
Durand, G. 261
Duthie, G. *817
Dutta, S. 437, 438
Dwyer, J.H. *357
Dwyer, T. **16**
Easton, D. *78
Eatock, E.M. 922
Eaves, E.R. 8
Ebbesen, P. 218, 225
Ebeid, N.I. **234**
Ebeling, K. 717
Eberlein, K.A. 992, 995
Ebigbo, P.O. 378
Eccles, J. 30, 34, 36
Echavé, L.V. 97
Economidou, J. 382, 383
Ederer, F. 1106
Edling, C. 742, **798**, **799**
Edson Pontes, J. 1033
Ehrenberg, L. 1001
Einhorn, N. 746
Eisen, E.A. **1095**
Eisenberg, E. 98
Ekbom, A. 796, **800**, ***801**
Eklöf, M. *277
Eklund, G. **769**, **770**, 787
El Din Zaki, S.S. *120
Elahi, M.F. *54
Elbrond, O. 207
Elder, J. 914
Elfgren, K. 745
Eliasziw, M. 84
Elinder, C.G. *768, *785
Elliott, P. **869**
Ellman, R. 855
Elpern, D.J. *1030
Elwood, J.M. 646, 647, 648, 651
Elwood, P.C. 913
Emmelin, A. **788**
Endang Soetristi 450
Engels, H. **55**, **56**, *277
Engholm, G. *214, 217, *219, 224
English, D.R. **31**, **32**, *269
English, R. *919
Enstrom, J.E. **1075**, **1076**, ***1077**
Enterline, P.E. 1156
Eomois, M. 235
Epstein, J.H. 993
Eriksson, A.B. 747
Eriksson, M.H. *761
Ersev, D. 815
Eschwege, E. 314
Eskelinen, M. 253
Esmen, N.A. 1159
Esteban, D. *294
Esteva, M. 275
Estève, J. 260, 275, 541, 688, 811
Evans, G. 28
Everhart, J. 63
Everson, R. 1003
Ewen, S.W.B. *818
Ewertz, M. **211**, 283
Ewings, P. 943
Eylenbosch, W.J. 55
Faber Vestergaard, B. 215
Fabia, J. 114
Fabry, J. 279, 280
Facchini, L. 282
Fagerberg, G. 791
Fain, P.R. *1063
Fair, M.E. 81, 89, 90
Fairley, C.K. ***37**
Fairweather, D. 926
Faivre, J. *259, **260**, **261**
Fajen, J.M. *1020
Falcini, F. 531
Falck, F. ***1043**
Falk, R.T. 966
Falkeborn, M. 804
Falletta, J.M. 1007
Fallon, B. *1117
Fan, H.X. *142
Fan, J.X. **159**
Fang, F. 135
Fanning, D. 271
Fante, R. 521
Fantuzzi, G. 519, *520
Farmer, K.C.R. 880

Farmer, P.B. 201, ***853**
Fasoli, M. 516
Fasquel, D. *302
Faure, J.R. **266**
Fauske, S. 664
Fava, A.S. 66, 70
Fears, T.R. 1225
Feingold, N. *313
Felton, J.S. 1185
Fenaux, P. **267**, 835
Feng, Z. 1197
Fenoglio-Preiser, C. 1128
Fenwick, M. 28
Ferencz, T. 392
Ferguson-Smith, J. 910
Fernandez, C. *610
Fernandez, E. *734
Fernández, L.M. **194**, ***195**
Ferrario, F. 506
Ferraroni, M. 477, *478
Ferrell, R.E. 1054
Ferri, G.M. 480, 481, *490
Ferro, G. 281, 282, 296, 298
Ferro-Luzzi, A. 533
Feuer, G.M. **100**
Févotte, J. 279, 280, *317
Feychting, M. 766
Fière, D. 280
Fierro, L. 1236
Figgs, L. 965, *1172
Figueroa, M. *627
Filakti, H.H. **870**
Filiberti, R. *197, 477, *478
Filipchenko, V.V. 712
Filippini, G. 303
Finau, S.A. **643**
Fincham, S.F. **67**, 93, 119
Fincham, S.M. **68**
Fingerhut, M.A. 296
Finkelstein, M.M. **101**, **102**, ***103**
Firth, H.M. **648**, ***649**
Fisher, S. *1014
Fisher, S.G. 1100
Fissi, R. 507, *515
Fitzpatrick, T.B. 993
Fiumara, A. 492
Fix, J. *277
Flamant, R. 263, 330, 333
Flander, L. 25
Flandrin, G. 334
Flannery, J.T. 981
Flatten, G. 374
Fleming, L. *1020
Flesch-Janys, D. *357
Flodin, U. 742
Flore, C. 487
Folsom, A.R. **1105**, *1112
Fontana, A. 542
Fontana, V. 501
Fontham, E.T. **1121**
Forastiere, F. *535
Ford, J.M. 44
Forde, K.A. 1128
Foreyt, J.P. 1055
Forman, D. 295, 846, 899, **907**, **908**, **909**, **910**, **911**, **912**, **913**, **914**, *919
Formelli, F. 518
Formera, S. 486
Forrest, A.P.M. 941
Forsén, A. 378
Fortwengler, P. 1094
Fossati-Bellani, F. 540
Foster, A. 877
Fracheboud, J. 636
Fradet, Y. 92
Fragu, P.H. *337
Franceschi, S. **477**, ***478**, 479, 510, 512, 517, 534
Francese, C. *338
Franco, E.L.F. 66, **70**, **71**
Franco, S. 1012
Frankel, S. 833
Franze, A. 531
Fraser, G. 455
Fraser, G.E. **1070**
Fraser, P.M. **871**, 895, 896, 897
Fraumeni, J.F. 221, 965, 1171, *1172, 1179
Fredriksson, M. 747
Freitag, S. 82
Frentzel-Beyme, R.R. 298, 359, ***362**
Freudenheim, J. *999, 1000
Friedl, H.P. 287
Friedland, J.M 1025
Friedlander, N. 1065
Friedman, C. *1117
Friedman, D. 970
Friedman, G.D. **1137**, **1138**, 1139, 1145, 1183
Friis, L. 799
Frisell, J. 791
Frost, P. 993
Fruchter, R.G. ***998**
Fryns, J.P. 62
Fuerst, C.J. 777
Fujiki, H. 595
Fujimoto, I. 576, 603
Fukao, A. **592**, **593**, 914
Fukuda, K. 306, **559**
Fukuma, S. 546, 566
Fukushima, T. 558
Funakoshi, M. 573
Funatsu, H. 596
Funto, I. 542
Furuyama, J. 554
Gabiano, P. 326
Gabrielli, M. 531
Gafà, L. 295, ***527**
Gail, M.G. 1225
Gajalakshmi, C.K. **433**, ***434**
Gajdosová, D. 721
Galante, R. *801
Galanti, C. 58
Galanti, R. 796
Galke, W.A. 1071
Gallagher, F. *1002
Gallágher, R.P. 116, 117, 119, 121, 1214

Gallegaro, L. 524
Gallion, H. *1011
Galteau, M.M. 807
Gamble, J.F. **1037**, ***1038**
Gambrill, R.D. 1040
Gammon, M.D. *969
Gangadharan, P. 440, 441, 442, 443
Gange, D. 28
Gao, D.-L. *1206
Gao, P. 149
Gao, R.N. **165**, **166**, 171, 181
Gao, Y.J. 151
Gao, Y.T. 165, 166, **167**, 171, 176, 179, 181
Garabrant, D.H. 655
Garbe, C. **346**
Garbe, P. 1194
Garbowski, G. 1128, 1129
Garcia de Herreros, A. *734
Garcia, F. **1218**, 1222
Garcia-Gomez, M. *272
Garfinkel, L. *1077, 1136
Garkavtseva, R.F. **702**, ***703**, *706
Garland, C.F. 1065, **1066**, ***1067**
Garland, F.C. *1067
Garland, S. *37
Garofalo, R. 492
Garrity, T.J. 1181
Garry, V.F. 977
Garte, S.J. *514
Garvicz, S. 672
Gaspari, R. 498
Gatti, R.A. 1078
Gaudette, L.A. **82**, 85, 90, 220
Gauthier, P. 274
Gavino, V. 74
Gaziano, M. 986, *987
Gec, M. 1241
Geddes, M. **497**, **498**
Gehde, E. 378
Geirsson, G.B. 398
Geissler, E. **348**
Gelas, J.M. *262, 263
Geller, A.C. *988
Genka, K. 566
Gennaro, V. **501**
Genovese, O. 532
Gentile, A. 512
George, M. 345
George, W.O. 510
Georgescu-Tulcea, N. ***690**
Gerhardsson, L. 750, 790
Gerhardsson, M.R. **771**
Gérin, M. 73, 306, *316
German III, J.L. **1126**
Gewelke, U. 364
Ghadirian, P. **74**, *79, 274, 510
Ghosh, M.K. 439
Giacosa, A. 260, 477, *478, 531
Giannetti, A. 529
Giannotti, B. 495
Giardiello, F.M. **954**, 958
Gibb, H. **1227**
Gibbs, A.R. 851
Gies, H.P. 16
Gignoux, M. **258**, *259
Gilbert, E.S. *277, 1167
Gilbertsen, V. **1106**
Giles, G.G. 13, 14, **20**, **21**, **22**, **23**, **24**, 25, *26, *27, 38
Gili, M. 284
Gilles, F.H. 955
Gillman, J.C. 644
Gilman, E.A. *822, 823, 827
Gimenez Fernandez, F.J. 737
Ging, L. 299
Giordano, G. 1218
Giraud *262
Giri, D.D. 406, **408**, 409
Gissmann, L. 363
Giuffrida, D. 492
Giwercman, A. **212**, 283
Glanz, K. 1212
Glashan, P. 866
Glass, A.G. 1164, 1200
Glattre, E. **662**
Glazkova, T.G. ***704**
Glen, R. 968
Godard, C. 312
Godbold, J. *1043
Godley, P.A. ***1002**
Goedert, J.J. *1174
Goetze, H.-J. *379
Gois Filho, J.F. 66
Golberg, J. *1014
Gold, E.B. **955**
Goldberg, L. 1055
Goldberg, M. 80, 312, *317
Goldberg, P. *317
Goldblatt, P.O. 854
Goldbourt, U. 465
Goldgar, D. *78
Goldhaber, M.K. **1139**
Golding, J. **831**, **832**
Golditch, I. 1139
Goldsmith, J.R. **452**, ***453**
Gomez, E. *734
Gonthier, C. *265
Gonzales, M. 22
González, C.A. 295, 297, **735**
Gonzalez, E. 993
Gonzalez, L.C. 284
Gonzalez, S. *195
Goodman, D. 995
Goodman, G. 1197, 1200
Goodman, K. 731
Goodman, M.T. **1047**, ***1048**, 1050
Goodman, S.N. 958
Goodwin, W.J. *1117
Gopalan, H.N.B. *269
Gordon, D. 995
Gordon, I. 19
Górecka, D. 681
Gorelikova, O.N. 709
Gorham, E.D. *1067
Górski, T. **681**
Gorst, D.W. **847**
Goss, B. 911

Goto, R. 589
Gou, Y.R. 140
Goujard, J. 58
Goulston, K. 8
Govindan, S. 432
Grace, J.R. 117
Graham, S. *999, **1000**
Graiff, C. 544
Gramenzi, A. 517
Granberg, S.B.O. **741**
Grandjean, P. *202, *203, *204, *205, *206, **229**
Grassi, A. 531
Grassi, L. 378
Grattan, H. 8
Gravelle, H. *919
Gravelle, I.H. 836
Gray, J. *277
Greaves, J. 860
Greaves, M.F. 730
Greco, M. 526
Green, A.C. 9
Green, D. *1110, 1150
Green, L.M. *277, 296
Green, M. 25, *27
Greenberg, M.D. *111, 1211
Greenberg, R.S. 981, 1091, 1092, 1121
Greenland, S. 1078
Greenwood, B. 345
Greggi, S. **534**
Greig, N.H. 979
Greiser, E. **349**, ***350**
Griffith, J. 1003, 1026
Griffiths, R.K. ***822**, 824
Grifols, R. 273
Grignoli, M. 530, 537
Grinstein, S. 363
Grizzle, J. **1197**, 1200
Grogan, D. 81
Groopman, J.D. **956**
Grosclaude, P. **257**, 329
Grossman, L. 959
Grufferman, S. 1007, 1008, **1154**, ***1155**
Grulich, A. 876
Guarner, L. *734
Guarneri, S. 477, *478
Guarnieri, C. 531
Gubbiotti, C. 526
Guenel, P. 312
Guercilena, S. 508
Guerrero, E. 284
Guidi, A. *523
Gulati, S.S. 410
Guliana, J.M. *313
Gulie, C. 810
Gundy, S. **391**
Gunnarsdóttir, H. ***396**, ***397**
Guo, B.C. 165
Guo, J. 131
Guo, L.P. 130
Guo, S.P. *142
Guo, Y.R. 141
Gupta, M.M. 437
Gupta, P.C. 301, **413**, **414**, ***415**, ***416**, 419, *420, *421
Gupta, S. 437, 438
Gurevicius, R 303, 304
Gurucharri, C. 363
Gurvich, E.B. 691
Gustavson-Kadaka, S. *780
Gustavsson, N. 246
Gustavsson, P. 764, **772**, ***773**, ***774**
Gutzwiller, F. 639
Guy, 280
Guyotat, 280
Haas, G.P. **1033**
Habbema, J.D.F. 631, 632
Habel, L.A. ***1196**
Hackman, P. 239
Haenszel, W.M. *1014
Hafez, A.H. 234
Hagberg, S. 744
Hagmar, L.E. **749**, **750**, **751**, **752**, **753**, **754**, **755**, **756**, **757**, *758
Haguenoer, J.M.Y. 267
Haidas, S. *380, *384, *385
Haider, M. 52
Haidinger, G. ***50**
Haile, R.W.C. 170, 1045, **1078**
Haimovici, L. 363
Haites, N.E. *818
Hakama, M.K. 241, **254**, ***255**, ***256**, *277
Hakulinen, T.R. 241, 246, 248, *780
Haldorsen, T. 664
Hall, A.J. 345, 524, 956
Hall, P. ***775**, ***776**
Hallgrímsson, J. **398**
Hallmans, G. *789
Halperin, W.E. 1026
Halpern, A. 1176
Hamada, G.S. 602
Hamajima, N. *567
Hämäläinen, E. 253
Hamblen, D.L. *845
Hamdi-Chérif, M. **1**, ***2**
Hamilton, S.R. 954
Hamilton, T.E. 1194
Hammar, S. 1200
Hammond, D. 1147
Hammond, G.D. 1108, 1109, 1113
Hampson, J. *865
Hamsikova, E. 200
Han, X.Y. *134
Han, Z.X. 137
Hanai, A. **576**, 603
Hanai, J. 591
Hanke, J. 683
Hankey, B. 576
Hankin, J.H. 1047, *1048, **1049**, 1050, *1051
Hankinson, D. 28
Hankinson, O. *514
Hannaford, P.C. 888, 908
Hansen, Å. M. *233
Hansen, E.S. **213**

Hansen, J. 34
Hansen, O. 231
Hansen-Koenig, D. 58
Hansluwka, H.E. 287
Hanson, J. 68
Hansson, L.E. 794
Hansteen, I.L. 753
Hantsen, P. *969
Harber, L. 993
Hardcastle, J.D. **892**, 893, 894
Hardell, L.O. 306, ***761**, 794
Hardiman, C. *336
Harding, S. 870
Hare, C. 469
Harlan, L. 949
Harper, P. 724
Harrington, J.M. 828
Harris, C.C. 541
Harris, F. 58
Hart, A.A.M. 616
Harte, G.A. **819**
Hartge, P. 1176, 1229
Harvey, I.M. **833**
Harvey, V.J. 644
Hasegawa, Y. 551
Hashimoto, T. **554**
Hassan, T.J. **673**
Hatch, E. 970
Hattchouel, J.M. *341
Hatton, F. 300
Haubrich, T. 914
Haupt, R. 1225
Haverkos, H.W. *952
Hawkins, M.M. *336, **915**, **916**, **917**
Hawthorne, V. 1040
Hayakawa, K. **582**
Hayakawa, N. 550, 599
Hayashi, M. 546
Hayashi, Y. 566
Hayata, Y. **596**
Hayes, M.V. 69
Hayes, R.B. 306, *316, *543, 632, 965, 1025, 1171, *1172
Hays, D. *1110
He, L.J. 133, *134
He, W. 132
He, Y.P. 126
Head, J. 876
Hearl, F. 968
Heath, C.W. **953**, *1077
Hebert, J.R. *420, **1232**
Hebert, P. 995
Heederik, D. 282
Heenan, L.D.B. 646
Heenan, P. 32
Heerema, N.A. 1108
Heikkilä, L. 239
Heikkilä, P. 243
Heilbrun, L.K. 1033
Heim, S. 753
Heineman, E. 965, *1172, 1177, 1181
Heinonen, K. 253
Heinrich, J. *379
Helgerson, S. 946
Heller, W.D. 364
Hellers, G. 821
Hellmann, R. 1000
Helmick, C. 800
Hemminki, K. 270, **1001**, 1123, **1131**
Henderson, B.E. 179, 643, 1072
Henderson, M.M. **1197**
Henderson, V. 1082
Hendriks, J.H.C.L. 630
Hengels, K.J. 846, 914
Henneberger, P. 282
Hennekens, C.H. **986**, ***987**, 991, 992, 995
Henry-Amar, M. 270, **339**, **340**
Herbison, G.P. 648, *649
Herd, A.J. 1055
Herity, B. 451
Hermus, R.J.J. 624, 625, **640**, 641, 642
Hernandez, J.M. 273
Hernandez, M. *610, **611**, 612
Hernández-Avila, M. ***609**
Hernberg, S.G. 243, 244, 247
Herrick, R. 1026
Hesseling, P.B. **732**
Heuch, I. 660
Heuer, C. 368
Hiatt, R.A. *1005, **1140**, **1141**, ***1142**, ***1143**, *1144
Hietanen, E. 239, 510
Higashiiwai, H. 547
Higgins, I.T.T. 949
Hildesheim, A. *184
Hill, C.G. *277, ***341**
Hill, D.J. *269
Hill, G.B. 85, 93, 114
Hill, M.J. 260
Hilt, B. *671
Hint, E. **235**, **236**
Hippeläinen, M. 252, 253
Hirayama, T. 560
Hirohata, I. 559
Hirohata, T. 560, 595
Hirose, K. *567
Hirsch, A.F. 297
Hirvonen, A. 239
Hisamichi, S. 547, 592, 593
Hiyama, T. ***577**
Ho, G. 994
Ho, J.H.C. 386
Hobbesland, Å ***669**
Hobbs, M.S.T. 30, 34, 35, 36
Hodgson, J.T. 856, 857, 859
Hoedemaeker, P.J. 958
Hoffmann, W. 349, *350
Högstedt, B. 753
Hogstedt, C. **763**, *781, *785
Holá, N. 198, 199
Holdaway, I.M. 644
Holder, W. *1038
Holländer, H.C. 222
Holland, R. 630
Hollard, D. 280
Hollingworth, T. 867
Holly, E.A. 303, 1087, 1176, **1185**, **1186**

Holm, L.E. *775, *776, **777**, 786, 787
Holm, N.V. **230**
Holmberg, L. 211, 797, **802**
Holmstock, L. 56
Hong, C.J. 172
Hongxiu, J. 252
Hoover, R.N. *184, 803, 965, *976, 980, *1172, 1229
Hopper, J.L. 14, 21, **25**, ***26**, ***27**, *46
Hornstra, M.H. 1018
Hornung, R.W. **1022**
Horsman, D.E. **118**
Horvat, D. 193
Horwich, A. 940
Hoshiyama, Y. **584**, ***585**, 587, 588
Hosoda, Y. *277
Hossain, M.A. *54
Houlston, R.S. ***933**
Hours, M. **279**, **280**, 282, *317
Howe, G.R. **104**, **105**, *107, 112, 114, *277, 304, *1119, *1120, 1214
Howel, D. **851**
Høyer, A.P. ***214**
Hreschchyshyn, M. 1000
Hrubec, Z. 963
Hsieh, M.Y. *160
Hsu, M.M. *184
Hu, J. **153**, **154**, ***155**
Hu, M.X. **143**, 299
Hu, N. 133, *134
Hu, S. 137
Hu, X.L. 128
Huang, M.Y. 126, **128**
Huang, Y.H. 159
Hubert, A. *2, **314**
Hubert, P. *262
Hufnagl, A. 51
Hulka, B.S. **1003**, *1005, *1006
Hung, K.L. 128
Hung, N.C. *290
Hunt, K. 925
Hunt, V.B. 1096
Hunter, D.J. 991
Hurley, S. **38**
Husgafvel-Pursiainen, K.T. **239**
Hutchings, S. 281
Huttunen, J.K. 639
Iacobellis, U. *490
Ianijan, P. *812
Iavarone, I. *535
Icenogle, J. 946
Ichikawa, S. 597
Icsó, J. **721**
Ida, Y. 593
Idle, J. 1079
Ignatyeva, R. 831
Ihamäki, T. 238
Ikeda, M. 557, 568, 569, 570
Ikeda, T. **562**, 563
Ilichyova, S.A. *697
Ilker, Y. 816
Illiger, H.J. 378
Imai, K. 584, 586, 588
Imanishi, K. *594
Inaba, S. *548
Inaba, Y. 564, 595, **597**
Indrawijaya, 447, 448
Infante-Rivard, C. **75**
Inoue, C. 578
Inoue, M. *567, 571
Inskip, H.M. 344, ***934**, ***935**, ***936**
Inskip, P.D. 796, *801, 970
Ippolito, O. 492
Ireland, P. 21
Irgens, L.M. *665
Irianiwati 449, 450
Iriya, K. 602
Isaacson, P. 981, 1208
Isaksson, H.J. 398
Iscovich, J.M. 289, *291, *292, **464**, 466
Iselius, E.L. 937
Ishikawa, H. 562
Ito, H. 552
Itoh, T. *555
Ivanov, E. 287
Iversen, O.H. 300
Ivkova, E.P. *715
Iwamamoto, K. 1035
Izarzugaza, I. 284
Izquierdo, M. 738
Jablon, S. 970
Jack, A. 344
Jack, A.D. **345**
Jackson, A. 848
Jacobs, A. **835**
Jacobs, I. **33**
Jacobs, P. *727
Jacobson, L.P. **957**
Jägeroos, H. 253
Jänne, J. 253
Jäppinen, P.T. ***250**, 282
Järup, L. *768
Järvholm, B. 742, 743
Jafarey, N.A. 674
Jahn, I. 351
Jain, M. 105, *1120
Jaiswal, M.S.D. 431
Jakobsson, K.M. 748, ***758**
Jakobsson, R. *773, *774
James, W.P.T. 510
Janerich, D.T. 981
Jansson, B. **1057**
Jarrett, R. *929
Jaskiewicz, K. **725**, **726**
Jass, J. 912
Jayant, K. *294, **410**, ***411**, 423, 425, 426
Jeannel, D. *2, 314
Jebbink, M. 614
Jederlinic, P. 1232
Jeffrey, A.M. **1131**
Jelinncic, V. 723
Jellum, E. **663**
Jenkins, D. 882, 883
Jensen, J. 221
Jewell, D.P. 821, 913

Ji, B.T. 166, 167, 181
Jia, Y.T. 133
Jiang, J.S. 151
Jiao, D.A. **145**, 146, 147, 148, 149
Jin, F. 167, 171, 176
Jin, M.L. 137
Jin, T. 790
Jin, T.H. 168
Jin, Z.G. **162**, **163**, 178
Jindal, S.K. 297
Jöckel, K.H. **351**, ***352**
Johannesson, B. 400
Johansen, H. 210
Johansson, G. *768
Johansson, J.E. 793
Johansson, L.G. 748
Johansson, R. 253
Johnson, B.C. **1068**
Johnson, E. *671
Johnson, S. **945**
Johnsson, R.R. 71
Jónasson, J.G. 398
Jones, D.R. **854**, **855**
Jones, R.D. 939
Jongeneelen, D.J. 626
Jongeneelen, F.J. 613
Jonmundsson, G.K. 672
Jonsson, B. 762
Joossens, J.V. 61
Jordan, S.W. 946
Jordan-Simpson, D. 90
Jørgensen, K.E. **231**
Joshi, P.V. *422
Joslin, C.A. 941
Jourenkova, N.Yu. *701
Joveniaux, A. 268
Jovicevic-Bekic, A. 1242, 1243
Juel, K. 229
Juhász, L. **394**, **395**
Juliusson, G. 270
Jung, G. 346
Jupe, D.M.L. 17
Jussawalla, D.J. **417**, ***418**, 425, 426, *427
Kaaks, R. 295
Kaatsch, P. 367, 375, 376
Kabulov, M. 694
Kadam, V.T. 426
Kadish, A. 994
Kadlubar, F. 541
Kahan, E. **469**, **470**
Kahn, T.M. **363**
Kalapothaki, V. ***380**
Kaldor, J. **41**, *277
Kallinikos, G. 382, 383
Kallio, M. 254, *255
Kalmanti, M. *380, *384, *385
Kalyanaraman, S. 432
Kamarli, Z.P. ***605**
Kamat, M.R. 410, *427
Kamat, V. 429
Kamps, W.A. 633
Kan, K.P. 607
Kanaka, T.S. 432
Kanda, J.L. 66, 70
Kane, M. 345
Kaneko, M. *277
Kanjuh, V. 1242, 1243
Kanka, J. 200
Kapadia, A. 406
Kaplan, B.H. 978
Kaplan, S.D. *473
Kaprio, J. **240**
Karácsonyi, L 388
Kardhashi, A. 480
Karjalainen, A. 239
Karjalainen, S. 211, 270, 287
Kark, J.D. **465**
Karkut, G. 58
Karli, M. 300
Karner-Hanusch, J. **51**
Kashansky, S.V. *692
Kashyap, V. 437, 438
Kasno 448
Kasper, H. 260
Kassimos, D. *385
Kasubskaya, T.P. *703
Kataja, V. 252
Kathren, R.L. **1167**
Kato, H. 583, 596
Kato, I. 571
Katsouyanni, K. 270, 295, 510
Katz, R. 98
Kaucic, M. 723
Kaufman, D.W. *727
Kaul, A. 370
Kaune, W. 1088, *1195
Kauppinen, T.P. 243, 244, 245, 282, 296, 300
Kawai, K. 564
Kawajiri, K. 586
Kawashima, Y. *549
Kay, C.R. **888**
Kay, L. 908
Kay, R.G. **644**
Kaye, A. 22
Kaye, S. 914
Kazantzis, G. 271, **872**, **873**
Kazubskaya, T.P. 702
Kee, F. *933
Keefer, L. 137
Kefford, R.F. 47
Kekki, M. 237, 238
Keller, G. *379
Kellokoski, J. 252
Kemp, M. 1104
Kendall, G.M. *277, *838, 839, 840, *906
Keogh, J. 1200
Kerenyi, N.A. 100
Kersey, J. 1108
Kesteloot, H. **61**
Kestin, M. 1197
Kettunen, K. 253
Kew, M.C. **729**
Key, C.R. 947, 955, 981
Key, M. 961
Key, T. 295, 908, **918**, ***919**
Khan, A.H. *675, *676
Khan, M.A. *676

Kharkevich, G.Y. *703
Khaw, K.T. 295
Khlat, M. *291
Khoo, S.K. **6**
Khuhaprema, T. 811
Kielkowski, D. 282
Kiemeney, L.A.L. 629
Kikendall, J.W. *1017, **1228**
Kikuchi, S. 597
Killingback, M. 8
King, D. 1222
King, E. 1187
King, M.C. **962**, 1116
King, R.A. **1111**
King, W. *110
Kinne, S. 1197
Kinnier Wilson, L.M. 915
Kinsey, S. *852
Kirichenko, O.P. *703
Kirkpatrick, A. 942
Kirsch, I.R. **977**
Kiss, J. 719
Kitinya, J.N. 274
Kitzes, R. *459
Kiviat, N. *1207
Kjaerheim, K. **664**
Kjellstroem, T. *269
Klaassen, D.J. 94
Klein, S. 994
Kleinerman, R.A. 223, 471, 970
Klett, M. 372
Kneale, G. **823**
Kneale, G.W. 827
Knekt, P.B. **241**, **242**
Kneller, R. 137
Knight, N. 5
Knight, T. **846**, 914
Knox, E.G. 823
Knudsen, L. 753
Knudsen, Ø. *668
Knutsen, S. 1070
Kobayashi, N. **598**, **599**, **600**
Kobayashi, T. *549
Kobayashi, Y. 603
Köck, M. 49
Koemeyer, H. 48
König, K. *53
Köpf, I. 741
Koepsell, T. 1197
Koetsawang, S. *1207
Kogan, F.M. **691**, ***692**
Kogevinas, M. *250, 271, **281**, **282**, *290, 296, 305, 306, *380, 622, 863, *864, *1235
Koh, H.K. ***988**
Kohlmeier, L. **347**
Kohlmeier-Arab, L. 639
Kohyama, N. 579
Kok, F.J. **639**, 641, 642
Kolcova, V. 196
Koliouskas, D. *385
Kolk, J. 629
Kolonel, L.N. 643, 955, 981, 1047, *1048, 1049, **1050**, *1051
Kolosza, Z. 678, 679
Kolstad, H. 281
Komatsu, S. 547, 593
Kondo, H. 563
Kondo, S. 563
Konjevic, R. 193
Kono, S. 568, ***594**
Kono, T. 551
Koo, L.C. **386**
Kopecky, K.J. 1194
Kordysh, E. *453
Koren, H.S. *269
Korfmann, A. 372
Korman, M. 8
Koscielny, S. *332
Koshkina, V.S. 707
Koskenvuo, M.J. 240
Koskinen, H. **243**, 244
Kosma, V.M. 253
Kosmelj, K. 722
Kosmidi, E. *380, *384, *385
Kosoy, G.Kh. *696
Koutras, D. 381
Kovac, J. 192, 193
Kowalski, L.P. 66, 70, 602
Krajewska, B. 683
Kramárová, E. 719, 720
Kraus, R. *355
Kreienbrock, L. 297, *379
Kresbach, H. 346
Kreuzer, M. *379
Krewski, D. 81
Kreysel, H.W. 346
Kriauciunas, R. 287
Kricker, A. 32, *269
Krieger, N. *1005, 1140, *1142, *1143, ***1144**
Kriek, E. **613**
Krishna, T.P. *430
Krishnan Nair, M. *294, **440**, 441, 442, 443, 444, 445
Krishnaswamy, K. *420, ***430**
Kristal, A. 1197, *1205
Kristensen, P. ***665**
Kristinsson, J.R. 672
Kritchevsky, S.B. ***1101**
Krogh, V. *515
Kromhout, D. 295
Kroon, B.B.R. 616, 617
Kross, B. 966
Kruger Kjær, S. **215**, *615
Krugliak, P. 455
Kruse, T. 216
Krutchkoff, D. 98
Ktenas, D. 21
Kubik, A. ***197**
Kubo, N. 565, 566
Kujawska, A. 685
Kumar, A. 431
Kumar, D. 437, 438
Kummet, T.D. 1224
Kundi, M. 52
Kuni, C.C. **1111**
Kunze, M.J. *50

Kupper, L.L. *1010
Kurihara, M. **550**
Kurihara, N. 595
Kurman, R. 274
Kuroishi, T. *567
Kuroki, T. 595
Kurppa, K. 243
Kushi, L. 1105, 1111
Kushlinski, V. 711
Kushwaha, M.S. **431**
Kusiak, R.A. ***106**
Kustov, V. 712
Kusuki, Y. 561
Kuzina, L.E. 691
Kvåle, G. **660**
Kvinnsland, S. 270
Kyrtopoulos, S. 270, *853, 914
L'Abbé, K.A. ***107**
L'Huillier, M.C. **321**
La Rosa, F. **526**
La Rosa, G.L. 492
La Vecchia, C. 477, *478, 510, **512**, **513**, 516, 517, 809, 810
Laake, P. *665
Lach, B. 85
Lachlan, G. 912
Lacroix, A. 74, *79
Länsimies, E. 253
Lagorio, S. ***535**
Lahaye, D. 59
Lahdensuo, A. 243
Lahermo, P. 246
Lai, K.D. 149
Lakhani, R. 73
Laleman, G.R. 56
Lalitwingsa, S. *812
Lam, T.H. 873
Lambert, B. **746**, 753, 1001
Lambert, J. 8
Lammes, F. 614
LaMotte, F. 995
Land, C.E. 970
Landi, M.T. 508, ***514**
Lang, A. 374
Lang, M. 297
Langård, S. *667, *668, *669, ***670**, *671
Langenberg, P. 1228
Langlois, S.P. 97
Langman, M.J.S. 927
Langmark, F. 287, 664
Lanier, A.K. 220
Lannin, D. *1006
Lanning, M. 672
Laouamri, S. *2
Laplanche, A.C. *332, 333, *341, ***342**
Larkins, R. 21
Larsen, T.E. **666**
Larsson, L.G. 791
Larsson, S. **742**
Larusdottir, M.K. 400
Latan, A. **186**
Lau, A. *1051
Laudico, A.V. 288
Launoy, G. 258, ***259**
Laurence, K. 58
Laurent, P. 343
Lauzi, A. *385
Lavé, C. *277
Lawson, D.H. 927
Layde, P. 1099
Laydevant, G. 308
Lazovich, D. ***1199**
Le Mab, G. 329
Le Marchand, L. *269, 643, 1050, ***1051**
Leake, R.E. 510
Leandro, G. **491**
Lebesque, J. 620
Lebowitz, M.D. **1219**
Lechat, M.F. **58**
Leclerc, A. 306, **315**, ***316**
Lee, I.M. ***989**
Lee, S.S. *189
Leech, S. 910
Lees, P. 1227
Lefebvre, J.L. **268**
LeGasse, A.A. 1160, 1161
Lehtinen, M. *789
Lehtonen, J. 253
Leibovici, S. 1238
Leigh, J. **42**, ***43**, 48
LeMerle, J. *336, 1150
Lemieux, B. 1114
Lemp, G.F. 1041
Len, C.Y. 152
Lence, J. 194, *195
Lenfant, M.H. 330
Lenner, P.H. ***789**
Lenoir, G. *78, *313
Leon, D.A. 858
Lepore, A.R. 504
Lerchen, M.L. 981
Lerman, R.H. 997
Lerman, S.E. *1038
Lesaffre, E. 61
Lesher, L. 1229
Lessner, K. 364
Lestani, M. 545
Levi, F.G. 270, 477, 510, **809**, **810**
Levin, B. 1058, 1060
Levin, L.S. 954
Levin, V. *1056
Levine, A.M. 1073
Levine, C. 465
Levine, N. 1220, 1221
Levine, P.H. *184
Levison, D.A. *878
Leviton, A. 955
Leviton, L.C. 1157
Levitz, M. 1130
Levshin, V.F. **705**
Levtchuk, A.A. 711
Lew, R. 1232
Lew, R.A. *988
Lewin, F. 784
Lewis, I. *852
Lewis, M.E. 647
Li, C.C. **144**

Li, F.M. 130
Li, F.P. *1110
Li, G. 136
Li, J. 127
Li, J.Y. 137
Li, L. 156
Li, S. 185
Li, W. 185
Li, Y.H. *142
Lian, X.X. 128
Liang, S. **129**
Liang, X.Z. 128
Liao, S.Y. *1063
Liapunova, N.A. *706
Liberati, C. 512
Liddell, F.D.K. 76
Lifanova, Y.E. 711, 714
Liff, J. 951, 974, 1121
Liippo, K. 243
Lilleorg, A. 235
Lillis, D.F. 58
Limasset, J.G. 334
Lin, C.Y. 151
Lin, M. 163
Lin, X.W. ***142**
Lin, Y.T. 161
Lindblad, P. 795
Lindblom, A. ***778**
Lindefors, B.M. 770
Lindsted, K.D. 1070
Linet, M.S. 970
Ling, M. 178
Linnainmaa, K. 753
Linos, D. **381**
Liss, G.M. *103
Little, J. 7, 303, 304, 510, 893, 894
Little, J.H. 9
Littorin, L. 296
Litvak, J. 1152
Liu, C.B. 152
Liu, E.T. *1006
Liu, M.Z. 144
Liu, P.L. **168**, **169**
Liu, Q. 143
Liu, Q.F. **164**
Liu, T.Y. *160
Liu, W.D. 137
Liu, X.Q. 137
Liu, Y.Y. 154, *155
Lloyd, O.L.L. ***387**, 841
Lluch, A. 273
Lo Curto, M. 540
Lo, S.H. *969
Loach, M.J. 922
Logamuthukrishnan, T. 432
Logan, R.F. **893**, **894**
Loginov, A.P. 718
Lokobauer, N. 192
Lomuto, M. 529
London, S.J. **1079**, 1109
London, W.T. **1149**
Longnecker, M. 1097
Lönnberg, G. 792
Looman, C. 632
Loomis, D.P. ***1004**
Lopes, L.F. 71
Lopez, A.D. *415
Lopez, G. 285, 286
Lopez, J.J. 270
López, L. *609
Lopez, R. 955
López-Carrillo, L.T. ***610**
López-Cervantes, M. *610
Losi, L. 521
Lotz, I. 349
Loughlin, J.E. 990
Louwrens, H. 725
Love, E.J. 157
Lovincz, A.T. 1211
Lowenthal, R.M. **17**, ***18**, 835
Lu, J.B. **187**, **188**
Lu, S.H. **130**
Luande, J. 369
Lubbe, J.T.N. 631
Lubin, F. *453, *473, *474
Lubin, J. 963, 970
Luboinski, M. 329
Lucas, L.J. **1230**, **1231**
Luce, D. 306, ***317**
Lucier, G.W. *514
Lund, V.J. **874**
Lund-Larsen, P. 663
Lundberg, I.S. 281, *774, **779**
Lundell, G. *775
Lundell, M. 777
Lundström, N.G. **790**
Luo, F.J. 130
Luria, 470
Lusky, A. *475
Lustbader, E. *1056
Luthra, U.K. **437**, **438**
Lutz, J.-M. *265, 287, **308**
Lutz, W. 683
Lynch, H.T. *78
Lynch, P. 1060
Lynge, E. 215, **216**, **217**, 281, 282, 296
Lyon, J.L. 955, **1182**, 1208
Ma, X.Y. 146, 147, 148
Maas, M.J. 638
Maase, H. 283
Maatela, J. 241
Macaluso, M. 506, 982
MacDonald, D.M. *878
Macdonald-Davies, I. 870
MacDougall, L.G. **730**
MacGibbon, B.H. **839**, **840**
Machinami, M. 574
Mack, T.M. **1080**, 1081
Mack, W.S. 1087
MacKenzie, G. **820**
MacLaren, W.M. **844**
MacLennan, R. 5, **7**, **8**, 9, 12, 39
MacMahon, B. 510, 1097, 1149
MacOnochie, N. 895, 897
Macrae, F.A. 8, 24, *27, 28, 885
Maeda, M. ***549**
Maès, B. 318, *319

Magnani, C. 306, *316, 488, 489, *536, 537, 540, 545
Magnetti, C. 1228
Magnússon, B. 398
Magrath, I. *676
Mahaffey, J. 1167
Mahlamäki, E. 253
Mahon, G 510
Mai, K. *189
Maiani, G. 533
Maier, H. **364**
Maiman, M. *998
Maine, N. 345
Maiozzi, P. 530, 537
Maitland, N. 932
Makimoto, K. 568
Malats, N. *734
Malcolm, E. *1119
Malik, M.A.O. 274
Malker, H. *277
Mallin, K. 971, *975, **1015**, **1016**
Maltoni, C. 518
Mamoon, N. *675
Manca, M.B. 486
Manca, P. 486
Mancini, M. 540
Mandel, J.S. 1045, 1106, **1107**
Manetta, A. *1063
Mann, J.I. 921
Mann, J.R. **824**, 923
Manos, M. 274
Manousos, O. 914
Mansel, R.E. **836**
Manson, J. 986, *987
Manson, J.E. 991, 995
Mansoor, A. *675, *676
Mant, D. **920**
Mantel, M. 467, 468
Manti, A. 501
Mäntyjärvi, R. 252, 253
Manz, A. ***357**
Manzo, N. 1012
Manzoor, A. ***675**, ***676**
Mao, X.Z. 133
Mao, Y.S. 81, **83**, **84**, **85**, 87, 88, 89, 91
Marcos, G. 735
Marek, K. 685
Margaryan, A.G. *700
Margolin, B. 1003
Mariani, R. 307
Marinaccio, L. ***482**
Marks, R. 20, 833
Marmot, M.G. **875**, **876**
Marovic, G. 192, 193
Marrett, L.D. 69, 84, **108**, ***109**, ***110**, ***111**, *1120
Marrugat, M. *734
Marsden, K. 17
Marsh, G.M. **1156**, **1157**, **1158**, **1159**
Marshall, J. *999, 1000
Marten, A. 191
Marth, E. **49**
Martin, A. *195
Martin, C. 280
Martin, N.C. *812
Martin, N.G. **9**
Martin, T. 985
Martin-Moreno, J. 270, 510, 639
Martinez, C. 295
Martinez, I. 981
Martinsohn, C. 465
Martinson, F. *1002
Martos, M.C. 735
Marubini, E. 511, 518
Marynen, P. 62
Masaki, M. 601
Masala, G. 497, 542
Masera, G. 540
Mashberg, A. **1136**
Maskens, A. 260
Mason, B.H. 644
Masood, M.A. 674
Mastrandrea, V. 526
Mastrinsohn, C. 372
Mastroiacovo, P. 58
Masuoka, H. 589
Masuyer, E. 287, 289, *291, *292
Mathew, B. *294, 444
Mathews, J.D. **13**, **14**, 296
Matijasevic, A. 1243
Matos, E. 289, 305, 306
Matsui, I. 598, 599, 600
Matsuura, M. 550
Mattsson, A. *775, *776
Maturana, M. 125
Mauad, M.A. 71
Maumenee, I.H. 954
Maurer, H.M. *1155
Mauro, B. 479
Maximilien, R. 300
Maximovich, D.M. 711, 714
May, S. 653
Mayer, J. **1131**
Mayne, S.T. *969, ***1117**, *1118
Mays, C.W. 377
Mayya, Y.S. *422
Mazumdar, S. 1160
Mazzoleni, C. 507
Mazzuckelli, L.F. 1025
McBride, M.L. **119**
McCaig, R.H. 859
McCammon, C. **1173**
McCarthy, W.H. 47
McCartney, A.C.E. **877**
McCawley, M. 968
McCredie, M.R.E. *26, **44**, **45**, ***46**, 303, 795
McDonald, A.D. 76
McDonald, J.C. **76**
McDougall, J.K. 1203
McDuffie, H.H. 93, **94**, **95**, **96**
McElvenny, D. 837
McFeeley, P. *948
McGee, R.O. 651
McGovern, P.G. **1111**
McGowan, L. **1229**
McGrath, M.S. 1186
McGregor, J.M. ***878**

McIntyre, O.R. 8
McKinney, P.A. *822
McKnight, B. 1190, *1191, *1199, *1205
McLarty, J.W. **1224**
McLaughlin, J.K. 45, 93, 167, 795, 968, 1081
McLawhorn, K. 1166
McLeish, J. 8
McLeod, G.R.C. 7, 9
McMichael, A.J. 3, *4, *269, 304
McNair, R. 820
McNally, R.J.Q. 849, *850
McNamara, D. 1218, 1222
McNamara, N. 1221
McNeil, J.J. *37, 38, 304
McPherson, C.K. **879**, 923
McPherson, R.S. *948
McPherson, S. **1058**
Meadows, A.T. *1110, 1148, **1150**, **1151**
Meara, J. 926
Mecucci, C. 62
Meenakshi, M. 446
Megha, T. 496
Mehra, S.N. *416
Mehta, F.S. 413, 414, **419**, ***420**, ***421**
Meijer, C.J. *615
Meinert, L. 981
Meirik, O. 770
Melbye, M. **218**
Melchers, W.J.G. ***627**
Mellemgaard, A. ***219**
Menck, H.R. **1081**
Mendes, A. 297
Mendilamarsu, S. *1235
Mendy, M. 345
Ménégoz, F. 304
Meneses, F. 611, 612
Menon, R. 437, 444
Merabishivili, V. 287
Meriläinen, P. 253
Merler, E. *272, 305, 306, 496, 500, *536, 545
Merletti, F. 282, 297
Merlo, F. 477, *478, 502
Mertz, G. 946
Messing, E. **1096**
Messing, K. **77**
Meyer, F. **92**
Meyskens, F.L. 1200
Mgaya, H.N. 369
Miah, M.N.I. *54
Miazzo, G. 524
Miceli, M. *535
Michaelis, J.H. 287, 367, **375**, **376**
Micheli, A. 507, *515
Micozzi, M. 1222
Miers, M.E. 836
Mifek, J. 196
Mignotte, H. *302
Mikkelsen, T. 85
Milan, C. *259, 261
Milham Jr, S. **1146**
Miligi, L.R. 500, 542
Miller, A.B. 80, 82, 104, **112**, **113**, **114**, **115**
Miller, B.A. 305, 1173
Miller, D. 1108
Miller, D.G. *1124
Miller, D.R. *988
Miller, G.H. **1039**, **1040**
Mills, C. 84
Milner, J. *1017
Minami, Y. 592
Mine, M. 562, **563**
Minelli, L. 526
Minick, R. *1124
Minkoff, H. *998
Minowa, M. 560
Mirer, F.E. 1034
Misciagna, G. 480
Misra, N.C. 431
Misra, P.K. 431
Missale, G. 531
Mitelman, F. 753, **759**
Mitsuhashi, T. 589
Mittenburg, W.H.Q. 636
Miyake, H. 282, **589**, **590**, 591
Mo, C.C. 164
Mo, S.Q. 166
Mocarelli, P. *514
Modan, B. 303, 304, **471**, **472**, ***473**, ***474**, ***475**, *476
Modan, M. *476
Modigliani, E. *313
Moe, P.J. 672
Moehner, M. 287
Moen, B.E. ***661**
Mohammad, G. *54
Mohapatra, S.C. **446**
Mohler, J. *1002
Mohri, I. 573
Molina, A. *195
Molinini, R. ***490**
Moll, P.P. 1162
Møller, H. **283**, 914
Möller, T.R. 749, 754, 756
Momas, I. 310
Mommsen, S. 228
Monagle, L. *515
Monfardini, S. 479
Mononen, I. 253
Monson, R.R. 1095
Montella, M. *478
Montesano, R. 130, 956
Montpetit, V. 85
Moodie, P.F. 122
Moody, J. 117
Moolgavkar, S.H. *1195
Moon, T.E. 1218, **1220**, **1221**, **1222**, ***1223**
Moore II, D.H. **1069**
Moore, D.E. *1191, 1201
Moore, J.W. *919
Moore, M. 1134
Moorman, P.G. ***1005**
Morabia, A. ***808**
Moreno, V. 273
Moreo, P. 284
Morettini, A. 531

Morgan, M. 1208, *1209
Morgan, M.V. 947
Mori, H. 563
Mori, M. 589
Mori, W. 574
Morimoto, K. **578**
Morinaga, K. **579**, ***580**
Morio, S. 604
Morison, D. 86
Morita, N. 547
Morley, A.A. *4, 1001
Morris, P. 1166
Morris-Jones, P.H. 1150
Morrison, B.J. 115
Morrison, H. **86**, **87**, 88, 89
Mortel, R. 972
Mortensen, P.B. **232**
Morton, N.E. **937**, 1053
Moscardelli, S. *520
Moscicki, A.B. **1187**
Möse, J.R. 49
Moser, M. *277
Moss, S.M. **942**
Moszczynski, P. **677**
Mott, M. 832
Moulin, J.J. 300, ***322**, ***323**, ***324**, ***325**
Mravunac, M. 630
Mueller, B.A. 303
Müller, C. 354
Müller, H.J. **805**
Mueller, N. 955
Muirhead, C.R. *277, *838, 839, 840, 905, *906
Mukundan, M.A. *430
Mulet, M. 275
Mulhern, R.K. 1042
Muller, J. *106
Muller, J.M. **10**, **11**
Muller, K.M. *379
Muller, S. 993
Mulvihill, J.J. *1110
Munaka, M. 550
Mund-Hoym 347
Munger, R.G. 1105
Munoz, A. 957
Muñoz, N. 187, *256, 273, 274, 275, **284**, **285**, **286**, 361, *615, 811, 813
Munson, M.L. 288
Murata, M. **546**
Mure, K. 578
Murray, J. 993
Murthy, N.S. 437
Muscat, J.E. ***1127**
Musch, G. 511
Musk, A.W. 30, **34**, **35**, **36**
Musto, R. 122
Muth, H. 370
Muti, P. 507, ***515**
Myers, J.E. ***727**
Myers, R. 1212
Mylvaganam, A. 3
Nadon, L. 73
Naéssen, T. 804
Nagao, K. 566
Nagel, S. *357
Naidu, A.N. *430
Nair, P.P. 444
Nakachi, K. 584, **586**, 588
Nakagawa, R. 546
Nakagawa, T. *594
Nakamura, K. 564, **601**
Nakazawa, H. *269
Nakazawa, N. 270
Nambi, K.S.V. ***422**
Namihisa, T. 597
Nanni, O. 477, *478, 542
Narendran, P. 432
Narod, S.A. ***78**
Nasca, P.C. 971
Nasrallah, S. 463
Natekar, M.V. *418, 425
Nath, P. 431
Naukkarinen, A. 253
Navarro, C. 284, 295
Navidi, W. 1088
Needham, L. 296, *514
Nefedov, M. ***706**
Neglia, J.P. 1008, 1108, *1110
Negri, E. 477, *478, 512, 513, 809, 810
Neil, H.A.W. **921**
Nejjari, A. 721
Nelemans, P.J **628**
Nelemans, P.J. **628**
Nelson, N.A. 122
Nelson, R.L. ***1017**
Németh, G. 378
Nemoto, T. 1000
Nene, R.S. *411
Nephedov, M.D. 702
Nervi, F. **124**
Nesbit, M. *1110, 1150
Nesi, S. *493
Nesti, M. ***536**
Nethercott, J. 959
Neuberg, R.W. **1151**
Neuberger, M. **52**, 296
Neugut, A. 463
Neugut, A.I. 1123, **1128**, **1129**
Nevin, N. 58
Newall, D. 914
Newcomb, P.A. **1097**, **1098**
Newell, D. 846, 913
Newell, F. 912
Newman, B. ***1006**
Newman, P. 644
Newton, R. *898
Newton, W.A. 1150
Ng, S.K.C. 1134
Ngelangel, C.A. 288, *294
Ngendahayo, L. *898
Ngilimana, P.J. *898
Nichaman, M.Z. 1058
Nicholls, R.J. 885
Nicholson, H.S. 1225
Nicolau, J. *317

Nie, G.H. 159
Niehoff, D. 368
Nielsen, N.H. **220**
Niessner, H. ***53**
Nigra, T.P. 993
Nikitin, Y.P. 220
Nikitina, O.V. *692
Nikkari, T. 241
Nilssen, S. 660
Nilsson, B.R. ***780**
Nilsson, E. 747
Nilsson, T. 744
Niman, H.L. 1123
Nise, G. ***781**
Nishi, M. 590, 591
Nishijama, S. 363
Nishikawa, H. *594
Nisse, C. 267
Nolan, J. 1000
Nomizu, T. **558**
Nomura, A.M.Y. 1047
Noordhoek, J. 638
Norberg, S. *773
Nordberg, G.F. 271, 790
Nordenskjöld, H. *778
Nordenson, I. 753
Nordström, M. *761
Nordstrom, D. 1099
Norell, S.E. 771
Norhanom, W. 608
Normand, J.C. 278
Norpoth, K. 354, *355
Norris, H.J. 1229
Norseth, T. *665
Northover, J.M.A. *862
Norton, L. *988
Norton, N.E. *933
Notani, P.N. 410, 417, **423**
Notkola, V.J. **251**
Nouasria-Sekfali, N. *2
Novak, F. 723
Novakovich, M. 192
Novotny, T. 1040
Noyon, R. 619
Nurhayati, Z.A. 608
Nurminen, M. *43, 243
Nussbaum, R.H. 826, **1163**
Nyberg, F. *773
Nygaard, R. **672**
Nyong'o, A. 55
Nyrén, O. 794
Nyström, L. 788, **791**
O'Boyle, P. 943
O'Connor, C.M. 905
O'Dea, K. 21
O'Higgins, N. 510
O'Neill, B.J. 40
Oberlin, O. *336
Oblak, B. 722
Obrador, A. 275
Obrecht, J.P. 805
Obringer, A. 1150
Obsiniková, A. 720
Ochs, J. *1110
Ocuneva, Z.V. *693
Odes, S. **454**, **455**
Offerhaus, G.J.A. 954, **958**
Ogata, M. 568
Ogawa, S. *594
Ogimoto, I. 557
Ogura, H. 578
Oh, H.C. 122, **123**
Ohashi, K. 581
Ohba, S. 589
Ohlander, E.-M. 797
Ohmine, K. 566
Ohno, Y. 564, **565**, **566**
Oho, K. 596
Ohshima, H. *293
Ohtaki, M. 550
Ojeda, R. 363
Okada, S. 581
Okamoto, N. **604**
Okojie, C.G. **659**
Okong, P. 356
Okubo, T. *555, 556
Okumura, Y. 563
Okuneva, L.A. 707
Olafsdottir, G. 399, 401
Oliveira, B.V. 70
Oliver, R.T.D. 907, 909
Oliver, W. 285, 286
Olivera, L. 1233, 1234
Olsen, G.W. **1104**
Olsen, J. **208**, 209, 225, 298
Olsen, J.H. *202, *203, *204, *205, *206, *219, **221**, **222**, 229
Olshan, A.F. **1007**, **1008**
Omar, R. 881
Omenn, G. 1197, **1200**
Omland, Ø. *233
Omohunndro, E.L. 1167
Omoto, K.H. 1211
Oosterlinck, A. 62
Opedal, E. *671
Ordoqui, G. 1234
Oreggia Coppetti, F.V. **1236**
Orengo, A. 503
Orentreich, N. *1005, *1142, *1143, *1144
Orfanos, C.E. 346
Orjaseter, H. 663
Orlandini, C. 503
Osborne III, J.S. **978**
Osman, J. **856**, **857**, **858**, **859**
Osterman-Golkar, S. 249
Osumi, Y. 581
Otto, L.P. 631
Otu, A.A. **656**, **657**, ***658**
Overvad, K. 225
Owens, A.H. 954
Owor, R. 356
Paci, E. 496, 497, 499
Packer, P. 910
Packer, R. *1110
Padmakumary, G. 441, 442, 443
Padmanabhan, T.K. 440
Padmavathy Amma, B. 444

Padubidri 438
Paffenbarger Jr, R.S. *989, **1213**, **1214**
Paganini-Hill, A. 1078, **1082**, **1083**
Page, D.L. **1115**, 1116
Pagliaro, L. 525
Pahwa, P. 93
Painter, R. 924
Pal, R. 631
Palan, P.R. 994
Palazzotto, G. 525
Palefsky, J. 1187
Palli, D. **499**, 913, 914
Palmer, J.R. 996
Pamart, B. 267
Pan, W.H. *184
Pangalis, G.A. 270
Panico, R. *515
Panizzon, R. 346
Pannett, B. 281, 931, 939
Paoletti, C. 330, 807
Paoletti, L. *490
Paolucci, G. 540
Papaevangelou, G.J. **382**, **383**
Paramsothy, M 606
Parazzini, F. 477, *478, 512, 513, **516**, **517**
Parent, M.E. ***79**
Park, R. **1034**
Parker, L. **889**, **890**, **891**
Parkes, S.E. 824
Parkin, D.M. *195, *256, 286, **287**, **288**, **289**, ***290**, ***291**, ***292**, ***293**, ***294**, 301, 345, *415, *812, *898
Parkkinen, S. 252
Parks, S. 831
Parl, F.F. 1115, 1116
Parra, S. *610
Parrish, J.A. 993
Parsons, W.J. 13
Partanen, K. 253
Partanen, T.J. 243, **244**, 281
Parukutty Amma, K. **441**
Pasquini, A. *535
Passarge, E. 1126
Passmore, S.J. 922
Pasternack, B.S. **1130**, 1132
Patavino, V. 526
Patel, D. *407, 409
Patel, N.L. 406, 409
Patel, R.D. 403
Patel, T.B. 403, 404, 405, **409**
Pathak, D.R. 947, *948
Pathmanathan, R. 606
Paul, C.E. **650**
Paul, O. 1066
Pavluchenko, A.E. *696
Pawlega, J. **680**
Pawlita, M. 369
Pearce, N.E. 282, 296, 653, **654**, **655**
Pearlman, E.D. 1037
Pedroni, M. 521
Peer, P.G.M. 630
Pellegrini, H.D. 1233, 1234
Pellet, F. *265
Peltonen, K. 249
Peluso, M. *543
Pendergrass, T.W. 1007
Penfold, J.C.B. 8
Penn, I. **1023**
Penttilä, I. 253
Pentz, M.A. **1109**
Perani, B. 544
Peraza, S. 285, 286
Percy, C. 576
Perdrix, A.M. ***265**
Perera, F.P. 1123, **1131**
Peretz, H. 470
Perez Viguera, J. 738
Peris-Bonet, R. 303
Pero, R.W. **1131**
Perrimond, H. 307
Perry, M. 848
Pershagen, G. 297, *773, **782**
Persson, B. 282, 744
Persson, I.R. **803**, **804**, 973
Perucci, C.A. *535
Pesatori, A.C. 508, 509
Peters, G.N. 1059
Peters, J.M. 655
Peterse, J.L. 616, 617, 620
Peterson, A. **1197**
Peto, J. 274, *822, 900
Peto, R. 127, 241, 301, *415, 991, 995
Petridou, E. *380, ***384**, ***385**
Petrinelli, A.M. 526
Petrova, N.I. *715
Petrozzi, J. 993
Pexieder, T. 58
Pezerat, H. 721
Pfäffli, P. 281
Pfeiffer, K.P. 49
Phair, J. 957
Pham, Q.T. **326**
Pharoah, P. **860**, **861**
Philippe, J. 280
Phillips, M. 33
Phillips, R. **880**
Picchiri, G.F. 483
Pichler-Semmelrock, F. 49
Pienkowski, P. *259
Pierce, J.P. 1040
Pierotti, M. 507, *543
Piettanen, P. 510
Piffer, S. 544
Pignat, J.C. 278
Pike, M.C. 907, 920
Pilacik, B. 683
Pillai, M.R. *553
Pinchera, A. *338
Pincus, M. *1124
Pinczowski, D. 800
Pindborg, J.J. 413, 414, 419, *420, *421
Pingitore, R. 496
Piñol, J.L. *734
Pinto, C.B. **64**, **65**
Pinzone, F. **525**

Piperopoulou, F. *385
Pippard, C. **938**
Pires Torres, C. 363
Pirisi, L. 1029
Pisa, R. 525
Pisani, P. 286, *293, 295, 506, 507, 542
Piva, C. 545
Plasencia, J. 273
Plato, N. 298
Platz, C. 981
Plesko, I. 270, 287, 510, 719, **720**
Pobel, D. *302
Pochart, J.M. 318, *319
Pocock, R. 943
Pohlahelm, H. *352
Poirier, M.C. 481, 960
Pollice, L.M. *490
Pollock, B.H. 1008, **1042**, 1113
Pompe-Kirn, V. *272, 287, 722
Ponder, B.A.J. 534
Ponz de Leon, M. **521**
Ponzio, G. 539
Pool, J.E. 730
Poole, C. **990**
Popendorf, W. 966
Popp, W. **354**, ***355**
Porro, A. 480, 481
Porschnev, M. *355
Porta, M. ***734**
Porter, P. *1196, *1202
Potischman, N.A. 972, 974
Potter, A. 19
Potter, J. 21, 914, *1110
Potter, J.D. 1105, 1108, 1111, *1112, 1183
Pottern, L.M. 965, **1173**
Pottier, D. 258, *259
Poulsen, H.E. *202, *203, *204, *205, *206
Powell, D.E. **829**
Powell, J.E. 824
Powles, J.W. 21
Prasad, M.P.R. *430
Prasad, U. **606**, **607**
Predieri, G. 519, *520
Prentice, R. 995
Preston-Martin, S. 303, 306, *473, **1084**, **1085**, **1086**, **1087**, **1088**, **1089**, 1090
Preudhomme, C. 267
Price, J.L. 941
Prihartono, J. 565
Prijono Tirtoprodjo 449
Primic-Zakelj, M. **722**
Prineas, R. 1105
Priore, R.L. *999
Proctor, L. *852
Proietto, J. 21
Prorok, P.C. 1024
Prosser, J. *842, *843
Proukakis, C. *385
Prystowsky, J. 993
Psaty, B. 1164
Pticar, M. 193
Pu, L.M. 157
Puig-Tintoré, L.M. 274
Pujol, H. 309
Pukkala, E.I. 240, 244, **245**, **246**, 247, 248, 254, *255
Puntoni, R. 501, **502**, 503
Puppinck, C. 312
Purde, M. 235
Qiu, S.L. ***189**, 190, 361
Qu, S. 48
Quastel, M. *453
Quesenberry, C.P. 1145
Quiquandon, I. 267
Quiros Garcia, J.R. 295
Raafat, F. 824
Rabkin, C.S. ***1174**
Radic, A. 58
Radóczy, M. 393
Raedsch, R. 361, 373
Rafaël, M. *290
Raffn, E. *202, *203, *204, *205, *206
Rafnsson, V. *396, *397
Ragunathan, T.E. 1164
Rahal, D. 1
Rahim, M.A. ***54**
Rahim, M.M. *54
Rahu, M.A. 287, *780
Rajeevkumar, S. 441
Ramachandran, T.P. 440
Ramael, M. 55
Ramamurthy, B. 432
Ramani, L. 522
Ramani, P. 445
Ramazzotti, V. 532, 533, 542
Ramli, M. 565
Ramlow, J.M. 1104
Rammeloo, J.A. 633
Rampen, F.H.J. 628
Ran, R.Q. 164
Randell, P. 32
Randow 347
Randriamiharisoa, A. 809
Rao, N.D. **424**
Rao, R.S. *411, *412
Rapoport, A. 66
Rapoport, S.I. **979**
Rappaport, S.M. **1009**
Raue, F. **365**
Ravichandran, K. *434
Ravndal, E. 664
Ravnihar, B. 722
Rawls, W.E. 1000
Raybaud, C. 307
Raymond, L. 287, 329
Raynor, D. **116**
Real, F. *734
Reale, F.R. 1232
Reaman, G.H. 1108
Recchia, C. 505
Redeker, A. 1149
Redmond, C.K. **1160**, **1161**
Reed, R. 23

Rees, D. **731**
Reeves, G. 902
Rege Cambrin, G. 539
Reger, R. 968
Reggiardo, G. 503
Reginald, S. 432
Rehman, G. 674
Reid, R. ***845**
Reid, R.I. **1211**
Reissigova, J. *197
Reizenstein, P. **783**
Reizner, G.T. *1030
Remennick, L.I. **707**
Rennert, G. **456**, **457**, **458**, ***459**, ***460**, ***461**, ***462**, 463, 464
Rennert, H. *462
Renz, K. 360
Reuterwall, C. 753, **764**
Reyes, M.G. 288
Reynolds, K. *865
Reynolds, P. **1041**, 1121
Riboli, E. **295**, 297, 299, 735, 1130
Ricci, A. *1043
Ricci, P. **545**
Rich, S.S. 1111, *1112
Richardson, A.K. **651**
Richardson, L. 73
Richart, R.M. 971
Ridanpää, M. 239
Ridker, P. 986, *987
Ridolfi, R. 531
Ries, L. 576, 979
Rifà, J. *734
Righi, E. 519, *520
Riihimaki, V. **247**
Riise, T. *661
Rilke, F. 270
Rinaldo, C. 957
Ring, I. 9, 10, 11, **12**
Ringertz, H. 783
Rinsurongkawong, S. 811
Ríos-Dalenz, J.L. 274, 1152
Risch, H.A. *969, ***1118**, ***1119**, ***1120**
Rist, M. *53
Ritchie, A.C. *106
Rivera-Pomar, J.M. *290
Rivero, S. 1236
Roberts, R.J. 860
Robertson, C.M. 917, *1002
Robinson, A. 889
Robinson, C.F. 305
Robinson, E. **463**
Robinson, K. 19
Robinson, R. 464
Robison, L.L. 1007, **1108**, **1109**, ***1110**, 1113
Robles, E. 288
Robson, D. 85, 93, 119
Robson, S.C. 726
Rockette, H.E. 1160
Roda, S. 721
Rodello, S. 306, 542
Roder, D.M. 3
Rodricks, J.V. 950
Rodrigues, V. 282
Rodriguez, A. *195
Rodriguez, M.C. 273
Rodriguez-Martinez, H. 1152
Rodvall, Y. 765
Roecker, E.B. 1096
Roels, H. 60
Rogers, E. 1232
Rojas, R. 611, 612
Rolón, P.A. 274
Romano, F.M. 525
Romas, N. 1129
Romazzini, S. *265, 308
Romieu, I. *610, 611, **612**
Romney, S.L. **994**
Ron, E. 472, *801, 970
Ronco, A.L. *1235, **1237**, 1238, 1239, 1240
Roncucci, L. 521
Rookus, M.A. 618, 620
Rooney, C. 895
Roos, G. 762
Rorke, L.B. 1151
Rørth, M. 283
Roscoe, R.J. 305, 1022, **1024**, 1178
Rose, F.G. 887
Rosen, T. 1055
Rosenberg, L. **996**
Rosenberg, M. *313
Rosenfeld, P. 946
Rosenstock, L. 1200
Rösenstrauch, M. *727
Roser, M. 346
Rosin, M.P. ***120**
Rosina, O. *1125
Rosner, B.R. 991, 992, 995
Rosner, W. 1129
Ross, J.A. 1113
Ross, R.K. 1072, *1074, 1083
Rosselli del Turco, M. 499
Rossi, M. 540
Rossing, M.A. **1201**
Rosso, P. 540
Rossof, A.H. 1066
Roth, E. 51
Roth, Z. 200
Rothman, K.J. 990, **1012**
Rothman, N. 960, 965, *975, *976
Rothrock, R. 1211
Rothstein, R. 1045
Rotimi, C. 982
Rotoli, M. 529
Rotstein, S. *780
Rougereau, A. 258
Roumagnac, M. 257
Roumeliotou, A. 382, 383
Roush, G.C. 1122
Roy, C.R. 16
Roy, P. 914
Royal, S. *1174
Rozen, P. 373
Rubenstein, P. *1125

Ruccione, K. 1147
Rudengren, C. *773
Rudolf, Z. **723**
Ruiter, D.J. 628
Ruiz-Liso, J.M. *736
Rumeau, N. *336
Rusciani, L. 528
Russell, D. *818
Russell, E. ***818**
Russell, I. 23, 25
Russo, R. 504
Rutegård, J.N. **762**
Rutledge, L.H. 1211
Rutqvist, L.E. 770, **784**, 791
Ruymann, F.B. *1155
Ryan, B.R. 1042
Rytömaa, T. *277
Saah, A. 957
Saarikoski, S. 252
Saba, L.M.B. 71
Sabaté, J. 1070
Sabater Pons, A. 737
Sabnis, S.D. *418
Sabroe, S. 208, **209**
Sacripanti, P. 531
Saeed, S. *676
Saengsingaew, V. *812
Sagaidak, V.N. 717
Sagebiel, R.W. 1176
Saigi, E. 735
Saito, Y. 596
Saiz Sanchez, C. 737
Saji, F. **581**
Sakai, R. **574**, **575**
Sakamoto, G. 565
Sakatani, M. *580
Sakr, W. 1033
Salas, A. *734
Sali, A. 388
Sallan, S. 1150
Sällsten, G. *272, 743
Salmon, L. *277
Salmonsson, S. 772
Salomaa, S. 753
Salonen, J.T. 510
Salvan, A. 1024
Salvini, S. 477, *478, 992
Sam, S.K. 606, 607
Samet, J.M. **947**, ***948**
Samojlov, D.V. 708
Samoszuk, M. *1062
Samuel, R. 606
Samuskevich, I.G. 716
Sanchez, V. 285
Sancho-Garnier, H. 314, 329, 807
Sandler, R.S. ***1010**, 1045
Sandström, A. **792**
Sang, J.Y. 188
Sanghvi, L.D. *412
Sankaranarayanan, R. *195, *294, 440, 441, **442**, **443**, **444**
Sankila, R. **248**
Santamaria, M. 284
Santella, R.M 1131
Santucci, M. 496
Sanz-Anquela, J.M. ***736**
Saracci, R. 270, 271, 281, 282, 295, **296**, **297**, **298**, 299, 300, 305, 306, 371
Saragoni, A. 531
Sarason, B. 1197
Sarason, I. 1197
Sarjadi **447**, **448**
Sarkar, S. 1084, 1085, 1089, **1090**
Sasaba, T. 584, *585, **587**, 588
Sasaki, R. *436, 564
Sasco, A.J. **299**, **300**, **301**, ***302**, *415, *621, 767
Satariano, E. 1035
Satariano, W.A. 955
Sather, H.N. 1108, 1147
Sathiakumar, N. ***983**, ***984**
Sato, H. 558
Sato, N. 597
Satoh, Y. 579
Satterfield, S. 995
Saunders, L.D. 1041
Savage, P. *919
Savelli, D. 530
Savitskaya, T.Y. 695
Savitz, D.A. *1004
Saxén, E.A. 241
Saya, H. *1056
Scala, S. 531
Scarpelli, A.R. 500
Schachter, J. 1187
Schairer, C. 971, 973
Schatzkin, A. 980
Schaveleva, T.V. *699
Scheepers, P.T.J. 626
Scheiner, C. 307
Schell, C. *355
Scherrer, C. **980**
Schiager, K.J. 1182
Schiffman, M.H. 274, 971
Schlaefer, K.O.H. **366**
Schlehofer, B. **367**, **368**
Schlehofer, J. 367
Schlumberger, M. *338
Schmauz, R. **356**
Schneider, J.S. 1069
Schneiderman, M.A. 1039, 1040
Schneijder, P. 641
Schnorr, T.M. 1024
Schnyder, U.W. 346
Schoenberg, J. *969, 974
Schoolnik, G. 1187
Schrameck, C. 331
Schraub, S. *311
Schreiber, M.M. 1220, 1221
Schubert, M. 1203
Schüler, G. *277
Schürfeld, C. 354
Schütz, A. 750, 755
Schuler, D. 287
Schull, W.J. 551
Schulte, P.A. *1021, **1025**, **1026**
Schulte-Hermann, R. *53

Schultz, H. 209
Schuman, L.M. 1106
Schunk, W.W. **353**
Schvartz, C. 318, *319
Schwartz, A.G. 1036, 1154, *1155, **1162**
Schwartz, S.M. *1191
Schweiger, M. 851
Sciajno, R. *515
Scott, C.R. 1198
SEARCH Programme **303**, **304**
Sebastian, P. **445**
Secreto, G. 507
Sedkackova, E. 270
Seeger, R. 1008
Seghal, A. 437
Seidell, J.C. 637
Seitz, G. *277, 360
Sekfali, N. 1
Selby, J.V. 1137, **1145**
Self, S.G. *1196, 1201, ***1202**, *1206
Sell, A. 209
Sellers, T.A. 1105, **1111**, ***1112**
Semenciw, R. 81, 83, 84, 86, 87, **88**, **89**
Semiglazov, V.F. **717**
Seniori Costantini, A.R. 305, 496, **500**, 542
Seppälä, K. 237
Seppänen, R. 242
Serdyuk, V. 712
Serra, I. **125**
Sessink, P.J.M. 626
Settimi, L. **537**
Sever, L. *1195
Severson, R.K. 1108, **1113**
Shad, A.T. *676
Shah, K.V. 274, 284
Shah, N. 1008
Shah, N.G. 406
Shairer, C. 803
Shanahan, E.M. *4
Shangina, O.V. *698
Shanta, V. 433, **435**, ***436**
Shao, Y.W. 147
Shapiro, S. *462, 996
Sharma, B.K. 437
Sharma, J.K. 437, 438
Sharp, D.S. **1018**
Sharp, G.B. **1102**, ***1103**
Sharpe, C. 71
Sharples, K.J. 646, 647
Shaw, G.L. ***1175**
Shaw, H.M. **47**
Sheikh, K. **949**
Shekelle, R.B. 1066
Shen, F.M. **170**
Shenberg, C. 467, 468
Shepstone, B. *919
Sherman, K.J. **1203**
Sherman, R. 274
Sherson, D. ***233**
Shetye, S.B. 429
Shevchenko, V.Y. 711
Shi, Y.B. 152
Shibata, A. 306, 559
Shields, P.G. 541
Shigematsu, I. **551**
Shimamura, K. 546
Shiminski, T. **1151**
Shimizu, H. **547**
Shimizu, T. 582
Shimizu, Y. 583
Shimokawa, I. 562
Shimono, M. 568
Shirakawa, T. 578
Shoa, Y.W. 148
Shore, R.E. 1130, 1132
Shpilberg, O. ***476**
Shriner, D. 970
Shu, K. 578
Shu, W.X. 148
Shu, X.O. **171**, 180, 181
Shuker, D. 270
Shukla, H.S. 446
Shy, C.M. 978
Sichieri, R. **63**
Siddiqi, A.M. 674
Siddiqui, M. 674
Sidney, S. 1138
Siegel, S. 1150
Siemiatycki, J. **72**, **73**, 75, 1078
Signer, E. 805
Siguel, E.N. **997**
Sigurdsson, K. 400
Siimes, M.A. 672
Sikes, R. *988
Silcocks, P.B. 903, 918
Silins, J. **90**
Silsand, T. *671
Simard, A. 114
Simonato, L. 270, 297, 298, *478
Simone, G. *482
Simor, I. 114
Simsek, F. 815, 816
Simson, I.W. 731
Sindikubwabo, B. *898
Singer, A. 867, 882, 883
Singh, K. 437, 438
Singh, M. 437, 438
Singh, V. 437
Sinha, R. *976
Sinnaeve, J. 287
Sinues, B. **738**
Sipponen, P.I. **237**, 238
Siscovick, D. 1164
Sitas, F. 912, 913
Sittenfeld, A.N.A. **191**
Siurala, M. 237, **238**
Sivaraman, L. *1051
Sjödahl, R. 747
Skakkebæk, N.E. 212, 283
Skegg, D.C.G. 646, 647, 650
Skerfving, S. 750, 753
Skinnider, L. 93
Sklar, C. *1110
Skouv, J. 216
Slattery, M.L. **1183**
Slimane, B.. 314
Slimani, N. 295

Sloan, M. *109
Slotboom, B.J. 634, 635, 636
Slovak, A. 837
Smans, M. 688, 713
Smith, A.H. 652, 1009
Smith, J. 890, 891
Smith, M. 1149
Smith, P.G. 345, 863, *864, **881**, 896
Smith, P.J. ***15**
Smith, R.A. 951
Smith, T. 1095
Smithson, W.A. 1151
Smulevich, V.B. **708**, **709**
Snover, D. 1106
Sobala, G. 285
Sobel, S. 452
Sobin, L. *975, 1228
Sobrinho, J.A. 66
Soda, M. 562
Sodhani, P. 437
Søgaard, J. 209
Sörensen, S. 742
Soeripto **449**, **450**
Soimakallio, S. 253
Sokic, S. ***1244**, ***1245**, ***1246**
Solanilla, P. 735
Soler-Monso, M.T. *736
Solionova, L.G. 708, 709
Soll-Johanning, H. 216
Solli, H.M. *670
Soloway, R.D. 1152
Somers, R. 619
Sommelet, D. 321
Somogyi, J. 719
Sondik, E.J. **981**
Song, E. 729
Song, W.Z. 140, 141
Sorahan, T.M. **825**, 827, *906
Sorsa, M.I. 201, **249**, 753
Soskolne, C.L. 282
Soutar, C. *324
Sower, M.F. 1212
Soyer, H.P. 346
Spadacene, N.W. 1104
Spännare, B. 765
Spång, G. *768
Sparén, P. 802
Sparkes, R.S. 1078
Sparling, T. 118
Spaziani, E. 533
Spears, G.F.S. 650
Speck, B. 805
Speizer, F.E. **991**
Spickett, J.T. 33
Spiegelman, D. 985
Spiess, H. **377**
Spigelman, A.D. 880
Spinelli, J.J. 93, 116, 117, 121
Spivack, M. *1125
Sprinter, J. *106
Sprovieri, O. 363
Srám, R.J. **198**, **199**, *853
Sreedevi Amma, N. 440, 444
Sreelekha, T.T. 445
Sri Roostini, E. 565
Sriamporn, S. *293
Srivatanakul, P. **811**, ***812**
St John, D.J.B. 8, 24, *27, **28**
Stacey, N. **48**
Stähelin, H.B. **806**
Stagnaro, E. 502, 542
Staines, A. 849, *850, ***852**, *929
Stalder, G.A. 805
Stampfer, M.J. 986, 991, **992**
Staneczeck, W. 348
Stanford, J.L. *969, *1196, 1198, **1204**, ***1205**
Stangel, W. 358, 378
Stankiewicz, A.K. **682**, **683**, **684**
Stansfeld, S. 875
Staples, M. 20
Stark, A. 1218
Starr, T. 950
Starzynski, Z. **685**
Steck, P. *1056
Steenland, K.N. 271
Steiner, C. 951
Steinitz, R. 464, **466**
Stejskal, J. 831
Stellman, S. *290, 1134, 1135
Stendahl, U. *789
Stenkop, E. 725
Stenling, R. 762
Stern, F. 305
Stern, R.S. **993**
Stevens, M.C.G. 824
Stevens, N.G. *1199
Stevens, R.G. *1004, **1168**, *1195
Stewart, A.M. 823, **826**, **827**, 1163
Stewart, A.W. 644
Stewart, P.A. 965, 966, *967, 1173, 1177
Stewart, W. 1227
Stiller, C.A. 904, 905, **922**
Stocker, H. 87
Stoll, C. 58
Stolley, P.D. 996, 1152
Stone, D. 58
Stone, K.M. 946
Storm, H.H. 210, 220, **223**, **224**, 225, 270, 287, 973
Storozhuk, M. **718**
Stovall, M. 223, 224
Straatman, H. 630
Strain, J.J. 639
Stram, D. 1008
Strand, L. *1119
Strax, P. 1130
Strickland, P. 481, **959**, **960**
Strom, B.L. 996, **1152**
Strong, L.C. 1007, *1056, *1110
Struyk, A.P.H.B. 614
Stücker, I. **343**
Sturgeon, S. 980
Sturmans, F. 623, 624, 625, 640
Styles, L. *935
Su, C.Y. 126
Sudhakaran, P.R. 444
Sufrin, G. 1000

Sugahara, T. 583
Sugawara, N. 547, 592, 593
Sugimura, M. *549
Sugondo, T. 447
Sullivan, K.R. 872, 873
Summers, R. 1045
Sun, Q. 138
Sun, Q.R. 148
Sun, S. **185**
Sun, X.W. 152
Sunde, L. 216
Sunderwa, J. 438
Sunyer, J. 282
Suriya, P. *812
Sutrisno, E. 565
Sutter, E. *1017
Suzuki, R. 564
Svartengren, M.U. ***785**
Svel, I. 58
Svensson, B.G. 749
Svirchev, L.M. 117
Swaen, G.M.H. **623**, 1230
Swaminathan, R. 435
Swanson, C.A. 972, 974, *975
Swanson, G.M. 981, **1035**, **1036**, 1162, 1208
Sweeney, W.B. *976
Swerdlow, A.J. 270, 289, 863, *864, 876, 911
Swift, M.R. **1046**
Swinburne, L. 851
Swinnen, L.J. **1100**
Syrjänen, K.J. **252**, **253**
Syrjänen, S. 252, 253
Szadkowska-Stanczyk, I. 282
Szalai, J.P. 100
Szarewski, A.M. **882**, **883**
Szeszenia-Dabrowska, N. 271, 685, ***686**, ***687**
Szöllösová, M. 721
Szwarcwald, C.L. 64, 65
Szymczak, W. 685, *686, *687
Tabar, L. 791
Tabor, E. 656
Tabrizi, D. *1125
Tadera, M. 601
Tafur, L. 274, 284
Tagashira, Y. 584, **588**
Tahara, E. **552**, ***553**
Taikina-aho, O. 239
Tajima, K. ***567**, 571
Takács, S. **392**, **393**
Takahashi, K. ***555**, 556
Takano, A. 593
Takano, J. 561
Takasugi, N. 591
Takayama, K. 546
Takeda, T. 590, **591**
Takeshita, T. 578
Takeuchi, T. *548, 578
Takezaki, T. *567
Takita, K. 558
Tala, E. 243
Talamante Serrulla, S. 737
Talamini, R. 477, *478
Talaska, G. 541, *543
Talbot, I. 880
Talbott, E.O. **1157**
Talley, N.J. **1169**
Tamburro, C.H. **1091**, **1092**, **1093**, **1094**
Tanaka, H. *577
Tanaka, T. ***548**
Tang, H.X. 148
Taniguchi, H. 576
Tanimura, M. 598
Tanizawa, O. 581
Tannenbaum, S.R. 541, 1001
Tanprasert, S. 811
Tao, B. 163, 178
Tao, S.C. **131**
Tao, X.G. **172**, ***173**, 182
Tao, Z. 132, 138
Taplin, S. 1197
Tarkowski, W. 289
Tarquini, M. 531
Tartoni, P.L. *520
Tatár, A. 392
Tavani, A. *478, 512
Taylor, H. 959
Taylor, I. 848
Taylor, J.R. 993
Taylor, M. *929
Taylor, P. *134, 980
Taylor, R. *29
Taylor, R.H. 819
Taylor, S.M. 69
Teissier, C. 300
Tekkel, M. 236
Telles, J.L. **1181**
Temmerman 55
Ten Kate, L.P. 58
Tenconi, R. 58
Teppo, L. 240, 241, 242, 248, 298
ter Meulen, J. **369**
Ter Schegget, J. **614**
Terao, T. *549
Terracini, B. 287, 488, 489, 537, **540**, 541, 545
Tersmette, A.C. 958
Tervahauta, A. 252
Teschke, K. 282, *1209
Tessy, G. 867
Teyssie, A.R. 274
Thamm, M. 347
Théobald, S. 318, *319
Theodorsen, L. 663
Thériault, G.P. **80**, 93, 119
Thiessen, J.W. 551
Thijssen, J.H.H. 635
Thilly, W.G. **1001**
Thomas, B.A. 855, 942
Thomas, D.B. 981, *1195, *1202, ***1206**, ***1207**
Thomas, D.C. 655, 1078, 1085, 1086, 1088
Thomas, H.F. **928**
Thomas, J. 85

Thompson, B. 1197
Thompson, D.W. 113, 1078
Thomson, J. 877
Thomson, M.H. 260
Thony, C. 266
Thoresen, S.Ø. 662
Thorhallsson, P. 398
Thornby, J.I. 1055
Thornquist, M. 1197, 1200
Thorogood, M. 921, **923**
Threlfall, W.J. 116
Thun, M.J. 953
Thurnham, D. 361
Thyss, A. 307
Tilgen, W. 346
Tilley, B.C. **1212**
Tin, D.Y. 139
Tipayadarapanit, V. *812
Tirelli, U. **479**
Tirmarche, M. **263**, ***264**, 300
Tironi, A. 508
Tirtosugondo 448
Tjahjadi, G. 565
Tjahjono 447
Tjindarbumi, D. 565
Tjønneland, A. **225**, 914
To, T. 93
Tockman, M.S. 481
Todde, P.F. 486
Törnberg, S.A. **786**
Törnqvist, M. *781
Tognoni, G. 512
Tokudome, S. **568**, **569**, **570**
Tolbert, P.E. 1095
Tomatis, L.T. 345
Tomenius, L. 769
Tomenson, J.A. **887**
Tominaga, S. 564, **571**, 595
Toniolo, P.G. 1130, **1132**
Torén, K.Ö. **743**, **744**
Torhorst, J.K.H. 805, 806
Torloni, H. 602
Torrent, M. 295
Torroella, M. 274
Tortorelli, A. 531
Tosi, P. 496
Tossavainen, A. 243
Totis, A. 507
Touabti, A. *2
Toyoshima, H. **572**
Traboulsi, E.I. 954
Traina, A. 525
Traversa, A. *482
Travis, L.B. 970
Treat, M.R. 1128
Trédaniel, J. 297
Tres, S. 738
Tretli, S. 211
Trevisan, M. *515
Triantafillou, V.A. *482
Trichopoulos, D. *384, *385, 510
Trichopoulou, A. 295
Trigg, M. 1108
Trivedi, A.H. 402
Trock, B. *1011
Trosko, J.E. 551
Troy, K. *1125
Tryggvadottir, L. **399**, **400**, **401**
Tsai, W.Y. 1131
Tschen, J. 1055
Tsuchihashi, Y. 560
Tsuchiya, A. 558
Tsuchiya, K. *555, **556**
Tsugane, S. **602**, 914
Tsukuma, H. *577
Tsunematsu, Y. 603
Tu, J.T. 166, **174**
Tuchman, M. 1114
Tuck, D.M. *18
Tucker, M.A. *1175, **1176**
Tucker, S. 961
Türkeri, L. 815, 816
Tulinius, H. 398, 399, 400, 401, 914
Tulli, A. 528
Tumino, R. 295, *527, 542
Tuomisto, J. 253
Turazza, E. 363
Turk, J. 882, 883
Turnbull, A. 534
Turnbull, B. **1064**, 1215, 1216, 1217
Turp, J.E. 819
Twiggs, L.B. 972
Tyczynski, J. 287, 289, 688, 689
Tymen, G. *264
Tytgat, G.N. 958
Uchimura, H. 568
Ujszászy, L. 392
Uloyan, S.M. *700
Ulrich, J. 806
Ulshan, M.D. *1010
Unpanye, P. *812
Unsal, M. 1171
Urbach, F. 993
Urban, N. 1197
Urdal, P. *671
Urso, C. 495
Us, J. 723
Utidijian, H.M.D. 1230, 1231
Väyrynen, M. 252
Vahrenholz, C. 354, *355
Vaidya, N.S. 429
Vaidya, R. **439**
Vaidya, S.G. **429**
Vainio, H. 239, 282, **305**, **306**, 1123
Valacco, A. 363
Valanis, B. 1200
Valdivia, A.A. 947
Valentini, A. 544
Valentino, L. *482
van Barneveld, T. 300
Van Bladeren, P. 642
van de Ven, W. 62
Van den Belt, A.H.W. 619
Van den Berghe, H. 59, **62**
Van den Brandt, P.A. **624**, **625**, 640
Van den Brule, A.J.C. ***615**
Van den Oever, R. **59**
Van den Tweel, J. 614

Van der Does-van den Berg, A. 633
van der Gulden, J.W.J. **629**
Van der Hoordaa, J. 614
Van der Kuip, A.M. 622
van der Maas, P.J. 632
van Dijck, J.A.A.M. **630**
Van Dongen, J.A. 616, 620
Van Holten, V. 576
Van Kaick, G. **370**
Van Leeuwen, F.E. 62, 270, 300, 613, **616**, **617**, **618**, **619**, **620**, ***621**
Van Marck, E. 55
Van Mieghem, E. 56
van Noord, P.A.H. 635, 637, **638**
Van Oosterom, A.T. 270
van Poppel, G. **642**
Van Schooten, F.J. 613
van Wering, E.R. 633
Van't Veer, P. 624, 639, 640, 641
Vanacore, N. *535
Vandenbroucke, J.P. 958
Vandevelde, E. 60
Vannucchi, G. 497
Varcoe, R. 653
Varghese, C. 444
Varis, K.S. 237
Vasanthi, L. 435
Vasilevsky, M.G. *605
Vassallo, J.A. 1237, **1238**, **1239**, **1240**
Vatanasapt, V. *293
Vatle, A. *668
Vaudrey, C. 318, *319
Vaughan, T.L. 306, *969, **1208**, ***1209**
Vehviläinen-Julkunen, K. 253
Velema, J. *627
Vena, J.E. 1000
Venturi, M. *523
Venturi, S. ***523**
Verbeek, A.L.M. 628, 629, 630
Vercauteren, P. 62
Vercelli, M. **503**
Verd, S. 831
Verdier, P. 89
Verduijn, P.G. **632**
Verge, J. 735
Vernon, S.W. 1212
Veronesi, U. **518**
Vessey, M.P. 920, 923, **924**, **925**, **926**, **927**
Victoria, J. 276
Vida, F. 735
Viganò, C. 542
Vijayakumar, T. 445
Vila Tapia, A. 274
Viladiu, P. 284
Vilgus, B. *352
Villar, H. 1218, 1222
Villar-del-Sordo, V. *736
Villard-Mackintosh, L. 924
Villelminot, S. 295
Villeneuve, P. 87
Vincent, T.J. 905, *906
Vine, M.F. 1003
Vineis, P. 295, 504, 538, 539, **541**, **542**, ***543**
Vioque, J. **733**
Vipasrinimitr, S. *812
Visona, K. 191
Vithsupakorn, K. 813
Vivas, J. 285, 286
Viver, J. 735
Vlaisavljevic, V. 723, **724**
Vlasák, V. 720
Vlietinck, R. 62
Vnencak-Jones, C.L. 1116
Vobecky, J. **97**
Vobecky, J.S. 97
Voelz, G.L. 1071
Vogel, V.G. **1059**
Vogelman, J. *1005, *1142, *1143, *1144
Vogelstein, B. 954
von Karsa, L. **374**
von Kries, R. 832
Vonka, V. **200**
Vooijs, P.G. **631**
Voorhees, J. 993
Voûte, P. 1150
Vuletic, L. 1242, 1243
Vutuc, C. *50, *53, 297
Vuust, J. 225
Waage, H.P. ***671**
Wabinga, H. 274
Wagner, E. 1197, *1199
Wagner, G. 370
Wahba, R. 234
Wahlqvist, M.L. 8, 21
Wahrendorf, J. 45, 295, 304, 360, 361, *362, 366, 367, 368, **371**, **372**, 465, 795
Walboomers, J.M. *615
Wald, N.J. **884**
Waldherr, R. 368, 373
Waldhör, T. *53
Walewski, J. 270
Walker, J. 1026
Wall, C. 113, 114
Wall, S. 788, 791, 792
Wallace, D.M.A. **828**
Wallace, R.B. 1031, 1105
Wallach, R.C. **1133**
Wallenstein, S. *1125
Walles, S.A.S. 1131
Walter, S.D. **69**
Waltsgott, H. *357
Wan, X.R. 151
Wang, A.R. 126
Wang, L. 133
Wang, L.H. 159
Wang, L.Y. *160
Wang, Q.L. 139
Wang, X.Q. 133
Wang, Y. **175**, 185
Wang, Z.Q. 177
Wang, Z.Y. 128
Wappler, G. 372

Ward, E.M. 965, 1025, **1027**, **1028**
Ward, M. 8, 966, **1177**
Wargovich, M.J. **1060**
Warnakulasuriya, K.A. **739**
Warner-Learmonth, G. **728**
Waskerwitz, M. 1007
Wassermann, K. *202, *203, *204, *205, *206
Watanabe, F. 558
Watanabe, H. **560**
Watanabe, S. 560, 565, **573**, 602
Waters, R. *853
Waters, W.E. 848
Watkins, D.K. *1226
Watson, P. *78
Watts, C. 24
Waye, J. 1128
Weatherall, J.W. 58
Weatherhead, E. *269
Webb, P. 913, 914
Weber, J. 1099
Weber, L. *1014
Weber, W. 805
Weckbecker, J. 346
Wei, L. **132**, *158
Weininger, J. 467, 468
Weinmann, S.A. **1164**
Weir, H.K. 108
Weisel, S. *482
Weisgerber, U. **373**
Weiss, H. 901
Weiss, J. 346
Weiss, N.S. 121, 955, 1164, 1190, *1191, *1196, 1198, *1199, 1201, 1204, 1210
Weiss, W. **1153**
Welinder, H. 748, 751, 752
Wells, A.J. 1040
Wennborg, H. 300, 767
Werner, M. 81
Wessels, G. 732
West, A.B. *1118
West, B. *969
West, C. 955
West, D.W. 955, 981
West, P. 122
West, R. 93
West, R.R. 835, 836
West, S. 959, 1152
Westberg, H. 744
Westerholm, P. 298
Westerlund, B. 763
Weyler, J. 55
Wheeler, C.M. 946
White, D. 93
White, E. 1164, *1191, 1197, *1199, **1210**
White, M.D 96
Whitehead, L. 961
Whitmore, S.E. 959
Whittemore, A.S. *1011, 1185, 1186, 1213, 1214
Whittle, H.C. 345, 524
Wichmann, H.E. ***379**
Wiebecke, B. 260
Wield, G. *934
Wiesner, G. 1111
Wiggs, L.D. **1071**
Wigle, D.T. 81, 83, 86, 88, **91**
Wiklund, F. *789
Wiklund, K. **787**
Wilbanks, G.D. 972
Wilbur, D. 1190
Wilcock, K. 950
Wilcosky, T. 1003
Wilczynska, U. 685, *686, *687
Wild, C. 270, 912, 914, 956
Wild, P. 282
Wild, P.P. ***327**
Wildt, J. 207
Wilkens, L. *1048, 1049, 1050
Wilkins, J.R. 1109
Wilkinson, G. 1012
Willett, W.C. *610, 991, 992, 1097
Williams, C.B. 880, **885**
Williams, F.L.R. **841**
Williams, J. 21, 1200
Williams, M. 12
Williams, S. 1106
Williams, S.M. 651
Williamson, A.L. 726
Wilpart, M. 260
Wilson, B.J. *818
Winck, J.C. 297
Wing, A.L. 1213
Wingren, G.B. 744
Winkelmann, M.A. 622
Winkelmann, R. 282, 296, 713
Winkelstein, W. 1187
Winn, R.J. 1058, 1059, 1060
Winter, P.D. 298, 305, 928, 930, 931, *934, *936, **939**
Wintroub, B. 993
Wogan, G.N. 541, 956, 1001
Wolf, J. 993, 1055
Wolfe, R.A. 949
Wolff, H.H. 346
Wolff, M.S. *1043, *1144
Wolff, S.P. ***886**
Wolke, G. *379
Wong, I. 1092
Wong, J. 1091, 1093, 1094
Wong, M. 116
Wong, O. 271, **1188**, **1189**
Wong, S.L. *387
Wong, T.W. *387
Wong, Y.F. 169
Wood, A. 887
Woods, W.G. 1007, 1109, **1114**, 1147, 1150
Woodward, A. **3**, ***4**
Woodward, J. 1228
Woosley, J.T. *1010
Wotke, R. 196
Wright, H. 931
Wu, A. 1214
Wu, H.C. 128
Wu, J.M. 148

Wu, M. **133**, ***134**
Wu, X.N. 165
Wu, Z.B. 968
Wu-Williams, A. *1074, 1121
Wyatt, J. 846, 912
Wyllie, A. *842, *843
Wynder, E.L. *1127, **1134**, **1135**
Wynne, H.J.A. 637
Xia, L.F. 149
Xiang, Y.B. **176**
Xie, C.L. 128
Xie, D. **135**
Xu, G.W. 137
Xu, G.X. 166
Xu, Z.Y. ***183**
Xue, S.Z. 162, 163, **177**, **178**
Yadav, M. **608**
Yager, J.W. 1009
Yamaki, Y. 558
Yamakido, M. 552
Yamamoto, Y. 583
Yamasaki, H. *269
Yanagawa, H. 564
Yandell, D. 1148
Yang, B.Q. 137
Yang, C. **121**, 128
Yang, C.S. 130, ***184**, 187
Yang, G. **146**, 147, 149, 150, 166
Yang, G.R. *189, **190**, 361
Yang, P. 1154, *1155
Yang, W.X. 130, 149
Yanovskaya, M.G. *699
Yao, K.Y. 146
Yardley, J. 954
Yaroslavtsev, V.N. 718
Yasui, W. 552, *553
Yatsenco, A.S. 691
Yeh, F.S. 164
Yeole, B.B. 417, *418, **425**, **426**, ***427**
Yilmaz, G. *814
Yin, D.M. 171
Yin, S. **136**
Yliskoski, M. 252
Yokozaki, H. *553
Yoshida, K. 589
Yoshimura, T. *277, **557**, 564
Yoshizawa, C.N. 1047
You, W.C. **137**
Young, C.S. 1031
Young, D.C. **1031**
Young, G.P. 28
Young, J. 576
Young, T.B. 1096
Yu, E.X. 165
Yu, H. 146, **147**, 148, 149, 150
Yu, M.C. 179
Yu, S.Z. 172, *173
Yu, T.S. *387
Yu, Z.F. **156**, **157**
Yuan, J.M. **179**
Yuan, L. 185
Yuan, R. 132
Yuan, Y.C. 146, 147, 148
Yunyuan, L. 153
Yus-Cotor, C. *736
Zack, M.M. 800
Zahm, S.H. 965, 966, 1028, 1177, **1178**, **1179**, **1180**, **1181**
Zaidi, S.M.H. 673, **674**
Zakelj, M.P. 914
Zambon, P. *478
Zamcheck, N. 954
Zampi, G.C. 531
Zanghieri, G. 521
Zappa, M. 496, 497
Zaridze, D. 270, *272, 510, 694, **710**, **711**, **712**, **713**, **714**, ***715**
Zatonski, W.A. 274, 289, 510, **688**, **689**, 914
Zaun, H. 346
Zejda, R. 196
Zeltzer, L. *1110
Zemla, B. **678**, **679**
Zemlianaya, G.M. 710
Zeng, X.R. 137
Zey, J. 1173
Zha, Y. 132
Zhang, C.L. 133
Zhang, C.Y. 151
Zhang, L. 137
Zhang, M.Q. 128
Zhang, S.Z. **138**, **139**
Zhang, X.M. 128
Zhang, Y. 140, 141
Zhang, Z. 126
Zhang, Z.Q. 126
Zhang, Z.W. **140**, **141**, 378
Zhao, L. 137, 140, 141, *189
Zhao, L.P. 1049, *1051, ***1052**
Zhao, Q.Y. 182
Zhao, R.X. 137
Zheng, S. 146, 147, **148**, **149**, 150
Zheng, T.Z. *1117
Zheng, W. 171, **180**, **181**, 306
Zheng, Z.S. 170
Zhong, S.C. 164
Zhou, C.T. 159
Zhou, D.H. 177
Zhou, D.N. 128
Zhou, L. 146, 147, 148, 149, **150**
Zhou, Q.L. 126
Zhu, D. 133
Zhu, H.G. *173, **182**
Zielezny, M. *999, 1000
Zielonka, I. 678
Zippin, C. 981
Zipursky, A. 1150
Zlatoff, P. *302
Zocchetti, C. 509
Zocchetti, E. 508
Zold-Kilbourn, P. 1212
Zuberi, S.J. 673
Zucchi, A. 511
Zuch, C. 522
zur Hausen, H. 363

INDEX OF TERMS

New key–words in this issue are:

Biomarkers

Cocaine

Genetic Councelling

Genotype/Phenotype

Hepatitis

Micronutrients

Protein Adducts

"Iodine Deficiency" has been deleted. Studies investigating iodine deficiency are now key–worded to the CHEMICAL "Iodine" only. Similarly, studies of iron deficiency or iron excess are key–worded to "Iron".

The list of index entries below identifies all key–words in use in this Directory. General terms such as "Reproductive Factors" are only used when more specific terms cannot easily be assigned: such specific terms are shown beneath each general term. Cross–references to the OCCUPATIONS and CHEMICALS indexes are also shown. Thus a study indexed to "Occupation" here will usually also be indexed to one or more specific occupations, and a study indexed to "Drugs" or "Metals" will usually also be indexed to one or more members of these groups in the CHEMICALS index.

Abortion
Adolescence
Age
AIDS
Air Pollution
Alcohol
Allergy
Analgesics
Anatomical Distribution
Animal
Antibodies
Antigens
Antioxidants
Appendectomy
Areca Nut
Asbestosis
Asthma
Ataxia Telangiectasia
Atomic Bomb
Atrophic Gastritis
Autopsy

Betel (Chewing)
Bile Acid
Biochemical Markers
See also: Lipids; Trace Elements; Vitamins
Biomarkers
Biopsy
Birth Cohort
Birth Order
Birthweight
Blacks
Blood
Blood Group
BMB
Breast Cysts

Caucasians
Chemical Exposure
See also specific entries in CHEMICALS index
Chemical Mutagenesis
Chemotherapy
Childhood
Chinese
Chlorination
Cholesterol
Chromosome Effects
Cirrhosis
Classification
Clinical Effects
Clinical Records
Cluster
CMV
Coal
Cocaine
Coffee
Condyloma
Congenital Abnormalities
Contraception
Cooking Methods
Cosmetics
Cost–Benefit Analysis
Crohn's Disease
Cryptorchidism
Cytology

Dairy Products
Data Resource
Demographic Factors
Diabetes
Diet
Disease, Other
Diuretics

DNA
DNA Adducts
DNA Binding
DNA Repair
Dose–Response
Down's Syndrome
Drugs
See also: Analgesics; Chemotherapy; Hormones; Oral contraceptives
See also specific entries in CHEMICALS index
Dusts
See also specific entries in CHEMICALS index
Dyes

EBV
Education
Electromagnetic Fields
Environmental Factors
Enzymes
Epilepsy
Eskimos
Ethnic Group

Faeces
Familial Adenomatous Polyposis
Familial Factors
Family History
Fat
Fatty Acids
Female
Fertility
Fertilizers
Fibre
Fibrosis
Fluorescent Light
Fruit
Fungicides

Gallstones
Gastrectomy
Gastric Ulcer
G–Banding
Genetic Councelling
Genetic Factors
Genetic Markers
Genotype/Phenotype
Geographic Factors
Goitre
G6PD

H. pylori
Haemophilia
Hair
HBV
HCV
Hepatitis
Herbicides
Heredity
High–Risk Groups
Histology
HIV
HLA
Homosexuality
Hormones
HPRT–Mutants
HPV
HSV
HTLV
Hygiene
Hyperthyroidism
Hysterectomy

Immunodeficiency
See also: AIDS
Immunologic Markers
Immunology
Immunosuppression
In Situ Carcinoma
Indians
Infection
Infertility
Insecticides
Intra–Uterine Exposure

Japanese

Lactation
Late Effects
Latency
Leather
Lifestyle
Linkage Analysis
Lipids
Liver Disease
Lymphocytes

Maghrebians
Male
Mammography
Mapping
Marijuana
Marital Status
Mate
Mathematical Models
Menarche
Menopause
Menstruation
Metabolism
Metals
See also: trace elements, specific metals and metal compounds in CHEMICAL index
Micronuclei
Micronutrients
Migrants
Minerals
Mining
Monitoring
Multiple Primary
Mutagen
Mutagen Tests
Mutation Rate
Mutation, Germinal
Mutation, Somatic

Mycotoxins
See also specific entries in CHEMICALS index

Naevi
Nutrition

Obesity
Occupation
See also specific entries in OCCUPATIONS index
Oncogenes
Oophorectomy
Operation, Previous
Oral Contraceptives

Paints
Parasitic Disease
Parental Occupation
Parity
Passive Smoking
PCR
Pedigree
Pepsinogen
Pesticides
See also: Fungicides; Herbicides; Insecticides
See also specific entries in CHEMICALS index
Petroleum Products
Photochemotherapy
Physical Activity
Physical Factors
Pigmentation
Plants
Plasma
Plastics
See also specific plastics, resins and monomers in CHEMICALS index
Pneumoconiosis
Polyps
Pregnancy
Premalignant Lesion
See also: Atrophic Gastritis; Polyps
Prevalence
Prevention
Prognosis
Projection
Promotion
Protein
Protein Binding
Psychological Factors
Psychosocial Factors

Race
See also: Asians; Blacks; Caucasians; Chinese; Eskimos; Ethnic Group; Indians; Japanese; Maghrebians
Radiation, Ionizing
Radiation, Non–Ionizing
Radiation, Ultraviolet
Radiotherapy
Record Linkage
Recurrence
Registry
Religion
Reproductive Factors
See also: Abortion; Birthweight; Contraception; Oral contraceptives; Fertility; Infertility; Lactation; Menarche; Menopause; Menstruation; Parity; Pregnancy
Resins
RFLP
Rubber
Rural

Saliva
SCE
Screening
See also: Mammography
Seasonality
Segregation Analysis
Sero–Epidemiology
Serum
Sex Ratio
Sexual Activity
Sexually Transmitted Diseases
Sib
Silicosis
Socio–Economic Factors
Solvents
See also specific solvents in CHEMICALS index
Sputum
Stage
Stress
Surgery
See also: specific operations in this index
Survival
Sweeteners
See also specific entries in CHEMICALS index

Tea
See also: Mate
Textiles
Time Factors
Tissue
Tobacco (Chewing)
Tobacco (Smoking)
See also: Passive Smoking
Tobacco (Snuff)
Toenails
Tonsillectomy
Trace Elements
Transplantation
Trauma
Treatment
See also: Chemotherapy; Drugs; Radiotherapy; Surgery
Trends
Tubal Ligation

Abortion

All Sites *111
Breast (F) 121, 400, 650, 707, 770, 1072
Childhood Neoplasms *111, 916
Choriocarcinoma 516
Hydatidiform Mole 516
Lung 963
Ovary 404, 707
Uterus (Cervix) *627, 707
Uterus (Corpus) 707

Adolescence

Childhood Neoplasms 1225
Gastrointestinal 1241
Genitourinary 1241
Stomach 557
Testis 834
Uterus (Cervix) 728, 1187

Age

All Sites 194, 227, 248, *396, 498, 502, 512, 589, 619, 809, 970, 1082, *1174, 1181
Benign Tumours *1202
Brain 979, 1181
Breast (F) 170, 194, 230, *291, *396, 399, 619, 637, 659, 970, 1082, 1097, *1202, 1204
Colon *291, *843, 1181
Gastrointestinal 230
Leukaemia 230, *384, 619, 633, 970, 1181
Leukaemia (ALL) 633
Liver 729
Lung 194, 619, 970, 1182
Lymphoma *1174
Melanoma 230, *291
Myelodysplastic Syndrome 619
Neuroblastoma 1008
Non–Hodgkin's Lymphoma 619, *1174
Oesophagus 361, *1074
Prostate *291, 1033
Rectum *843, 1181
Rhabdomyosarcoma *1155
Skin *878
Stomach *291, 576, *1074
Testis 230
Thyroid 970
Uterus (Cervix) 215, 230, *615
Uterus (Corpus) 230, 1082
Wilms' Tumour 71

AIDS

All Sites 218, 524, 1041
Burkitt's Lymphoma *658
Hodgkin's Disease 218, 479, *658
Kaposi's Sarcoma 41, 382, 479, *658, *952, 957, 964, 1041
Lymphoma 41, 964, *1062
Non–Hodgkin's Lymphoma 479, *658, 1186
Uterus (Cervix) *998

Air Pollution

All Sites 49, *387, 511, 522, 869
Bladder 213

Brain *205, 213
Gastrointestinal 192, 213
Haemopoietic 213
Inapplicable *853
Larynx 511, 869
Leukaemia 192, *205, 869
Liver 869
Lung 172, *173, 192, 297, 511, 710, *773, *812, 841, 869, 944
Lymphoma *205, 869
Male Genital 213
Mesothelioma 851
Respiratory 213
Sarcoma 213
Skin 213

Alcohol
All Sites 67, 89, 91, 116, 245, 333, 512, 522, 564, 568, 578, 586, 588, 601, 625, 680, 684, 708, 714, 749, 928, 953, 992, 1050, 1075
Benign Tumours 616
Bile Duct *734
Bladder 310, 546
Bone 171
Brain 171, 955, 1151
Breast (F) 25, *46, 63, 89, *214, 299, 423, *461, *478, 507, 510, *515, 546, 565, *609, 612, 616, 618, 620, 640, 650, 680, 711, *919, *948, 949, 974, 1000, *1006, 1072, 1097, *1195
Childhood Neoplasms 831, 1151
Choriocarcinoma 516
Colon *53, 63, 89, 149, 166, 167, 260, 424, *478, 510, 546, *594, 595, 601, 640, 737, 949, *976, 1098, 1228
Female Genital 949
Gallbladder *734, 737
Gastrointestinal 76
Hodgkin's Disease *311, *1244
Hydatidiform Mole 516
Hypopharynx 268, 315, *317, 439
Kidney 45, 76, 330, 560, 795, 1141
Larynx 59, 70, 76, 181, 315, *317, 364, 439, *478, 664, 784, 1236, *1246
Leukaemia 171
Lip 70
Liver 138, 165, 171, *293, *460, 559, 559, 559, 559, 574, 595, 597, 601, 664, *690, 729, 737, 811
Lung *50, 63, 76, *317, 326, 546, 566, 574, 595, 601, 640, 678, 680, 710, *999
Lymphoma 171
Mesothelioma 59
Nasal Cavity 181, *317
Nasopharynx 59, *184, 1208
Neuroblastoma 1008
Non–Hodgkin's Lymphoma *311
Oesophagus 63, 167, *259, 361, 424, 439, *478, 546, 664, 737, 784, *969, *1074, *1103, *1118
Oral Cavity 66, 70, 181, 207, 268, 363, 439, *478, 664, 784, 1089, 1136, 1236
Oropharynx 66, 268, 363, 405, 439, 1136
Ovary *478, 949, 1000
Pancreas 167, 244, *734, 737, 1000
Peritoneum 76
Pharynx 70, 181, 664, 784, 1089, 1236
Pleura 76, *317
Prostate *760, 943, 1000
Rectum *53, 149, 166, 167, 424, *478, 510, 546, *594, 601, 640, 737, 1098
Respiratory 363
Small Intestine 737
Stomach 63, 138, 153, 186, 326, 424, 505, 546, 557, 584, *585, 595, 601, 640, 680, 737, 794, *969, *975, *1074, *1103, *1118, 1233
Testis 283

Thyroid	*1245
Uterus (Cervix)	363, 608, 680, 949
Uterus (Corpus)	595, 949, 972, 1000
Wilms' Tumour	71

Allergy

All Sites	359, 771
Hodgkin's Disease	*1244
Non–Hodgkin's Lymphoma	1186

Analgesics

All Sites	1157
Bladder	178, *352, 1157
Colon	*976, *1127
Kidney	45, 330, 795
Oesophagus	*1118
Rectum	*1127
Stomach	*1118
Ureter	*352
Urethra	*352

Anatomical Distribution

All Sites	719, 720
Colon	641
Melanoma	720
Oral Cavity	1136
Oropharynx	1136
Skin	720

Animal

Childhood Neoplasms	1165
Inapplicable	249
Leukaemia	1165
Liver	1165
Lung	*379
Non–Hodgkin's Lymphoma	1165
Pancreas	1165
Soft Tissue	1165

Antibodies

All Sites	1123
Female Genital	252
Kaposi's Sarcoma	657, *898
Leukaemia	367
Liver	656
Male Genital	252
Nasopharynx	*2, 144, 151, 606
Non–Hodgkin's Lymphoma	*898
Stomach	912, 913
Uterus (Cervix)	200, *898

Antigens

Kaposi's Sarcoma	529
Liver	273, 656, 811
Lung	481

Antioxidants

All Sites	*987, 995

Bladder *204
Brain *203
Breast (F) *203, 347, 639, *987
Colon *203, *818, *987
Inapplicable *817
Larynx *204
Lung *204, *987, *1051
Nervous System *203
Prostate *203
Rectum *203, *818
Stomach 285

Appendectomy
Hodgkin's Disease *1244

Areca Nut
Oral Cavity 402

Asbestosis
All Sites 684
Mesothelioma 530
Peritoneum 530
Pleura 530

Asthma
Rhabdomyosarcoma *1155

Ataxia Telangiectasia
All Sites 938, 1046
Breast (F) 1046
Leukaemia 1046

Atomic Bomb
All Sites 550, 551, 562, 563
Retinoblastoma 561

Atrophic Gastritis
Stomach 137, *523, 531, 571, 725, *736, 846

Autopsy
All Sites 522
Appendix 398
Bile Duct 398
Breast (F) *948
Colon 398
Inapplicable 1003
Lung 932
Nervous System 398
Oesophagus 398
Oral Cavity 398
Pancreas 398
Prostate 398, 469, 1033
Small Intestine 398
Soft Tissue 398
Testis 212

Betel (Chewing)

Liver	*293
Oral Cavity	413, 414, 739
Oropharynx	405

Bile Acid

Colon	366, 526
Gastrointestinal	880
Rectum	366

Biochemical Markers

All Sites	663, 921, 992, 1197
Bile Duct	1152
Breast (F)	235, 408, 467, 468, 1197, 1222
Colon	468, 1197
Gallbladder	1152
Hodgkin's Disease	431, *676
Inapplicable	966
Leukaemia	431
Leukaemia (ALL)	*676
Lung	239, 297, 1121, 1197
Lymphoma	431
Multiple Myeloma	431
Nasopharynx	606
Non–Hodgkin's Lymphoma	*676
Oropharynx	405
Prostate	228, *1002
Rectum	468
Stomach	846, 913
Uterus (Cervix)	994, 1197

Biomarkers

All Sites	270, 508, 1131
Angiosarcoma	738
Bile Duct	*514
Bladder	*120, *204, *548, 642, 738, 1025
Brain	*514
Breast (F)	74, *514, 558, 639, *1124
Colon	74, 526, 558, 894, 1060
Hodgkin's Disease	270
Inapplicable	*4, *233, *355, *817, *853, 1003, *1020
Larynx	*204
Leukaemia	508, 1026
Liver	*514, 738, 956
Lung	*204, 613, 642, 738, 960, 1131, 1232
Lymphoma	508
Melanoma	*269
Nasopharynx	*184
Non–Hodgkin's Lymphoma	*514
Oral Cavity	*430, 444, 1203
Oropharynx	1203
Ovary	534
Prostate	74
Rectum	558, 894
Respiratory	*514
Sarcoma	*514
Skin	*269, *514
Soft Tissue	508, *514
Stomach	558, 602
Testis	*514
Uterus (Cervix)	274, 284

Biopsy

Appendix	398
Benign Tumours	1115
Bile Duct	398
Bladder	*120, *490
Breast (F)	309, *1043, 1115, 1116
Colon	398
Female Genital	252
Liver	726
Male Genital	252
Melanoma	1176
Nervous System	398
Oesophagus	361, 398
Oral Cavity	398, 1090
Pancreas	398
Pharynx	1090
Prostate	398
Skin	1184
Small Intestine	398
Soft Tissue	398
Stomach	913
Testis	212
Uterus (Cervix)	169, 274, 614

Birth Cohort

All Sites	194, 809
Bladder	513
Breast (F)	194
Childhood Neoplasms	823
Kidney	513
Larynx	513
Leukaemia	823
Lung	194, *197, 513
Neuroblastoma	591
Non–Hodgkin's Lymphoma	823
Oesophagus	513
Oral Cavity	513
Pancreas	513
Pharynx	513
Uterus (Cervix)	156

Birth Order

Leukaemia	435
Lymphoma	435

Birthweight

Testis	209

Blacks

All Sites	1170
Bladder	1035
Colon	1035
Eye	1035
Leukaemia	730
Liver	729, 1035
Lung	1035
Melanoma	1035
Mesothelioma	731, 1035
Multiple Myeloma	1036
Oesophagus	1035, 1036
Pancreas	1036

Prostate	1036
Rectum	1035
Salivary Gland	1035
Stomach	1035
Uterus (Cervix)	1029

Blood

All Sites	225, 270, 295, 372, 749, 750, *929, 991, *1117
Benign Tumours	253
Bile Duct	*514, *734
Bladder	*204, 541
Brain	*514, *1056
Breast (F)	21, *26, *27, *46, 74, 253, 480, *514, 908, 923, 974, 991, 1078, 1111, *1112
Childhood Neoplasms	*822, 831
Colon	21, *27, 74, 366, 480, 526, *842, 894, 908
Gallbladder	*734
Gastrointestinal	923
Hodgkin's Disease	270, *929, *1244
Larynx	*204, *1117
Leukaemia	*380, *822, *929
Liver	*514
Lung	*204, 480, 481, *768, 1079
Nasopharynx	*184
Non–Hodgkin's Lymphoma	*290, *514, *929
Oesophagus	*969
Oral Cavity	*430, *1117
Ovary	*865, *1011, 1198
Pancreas	*734
Pharynx	*1117
Prostate	21, 74, 908, *1002, *1205
Rectum	21, *27, 366, *842, 894
Respiratory	*514
Sarcoma	*290, *514
Skin	*514
Soft Tissue	*290, *514
Stomach	285, 908, *969, *975
Testis	*206, *514
Uterus (Cervix)	611, 994, *1048
Uterus (Corpus)	972
Vagina	971
Vulva	971

Blood Group

All Sites	227, 660
Breast (F)	401, 660
Gastrointestinal	660
Ovary	660
Uterus (Cervix)	660
Uterus (Corpus)	660

BMB

All Sites	225, 241, 270, 295, 372, 533, 564, 572, 586, 593, 663, 740, 749, 750, *929, 991, 992, 1044, *1117, *1174
Angiosarcoma	738
Benign Tumours	253, *494
Bile Duct	*514, *734
Bladder	*120, *204, 541, *543, *548, 738, 910
Bone	*845
Brain	*514, *1056

Breast (F) 21, 25, *26, *27, *46, 74, 235, 241, 253, 446, 480, 507, *514, *515, 518, 533, 612, 624, 636, 639, 908, *919, 923, *948, 991, *1005, *1043, *1063, 1078, 1111, *1112, 1116, 1130, 1132, *1142, *1143, *1144, 1218
Childhood Neoplasms *822, 824, 831
Colon 21, *27, 74, 241, 260, 366, 373, 480, 526, 533, 624, 638, 762, *842, 894, 908, *976, *1010, 1045, 1106
Endocrine Glands *494
Gallbladder *734
Gastrointestinal 880, 923
Hodgkin's Disease 270, *929, *1125, 1154
Inapplicable *4, 201, 249, *817, 1009
Kaposi's Sarcoma 41, 957
Larynx *204, *1117
Leukaemia *380, *822, *929, 1026
Leukaemia (CLL) *1125
Liver *160, 345, *514, 738
Lung 135, *204, 239, 241, 343, *379, 480, 481, 624, 705, 738, *768, 910, 960, *1051, 1079, 1200
Lymphoma 41, 1073, 1100, *1174
Melanoma 874
Mesothelioma 1200
Nasal Cavity 874
Nasopharynx *184, 314
Neuroblastoma 1114
Non-Hodgkin's Lymphoma *290, *514, *929, *1125, *1174
Oesophagus *134, *189, 361, 910, *969
Oral Cavity *430, 1090, *1117
Ovary 534, *865, *1011, 1198
Pancreas *734
Pharynx 1090, *1117
Prostate 21, 74, 241, *760, 793, 908, 918, *1002, 1033, *1205, 1214
Rectum 21, *27, 241, 366, 373, 624, 638, 762, *842, 894, *1010, 1106
Respiratory *514
Retinoblastoma *706
Sarcoma *290, *514
Skin *514, *878, 1216, 1217
Soft Tissue *290, *514
Stomach 137, 241, 285, 286, *493, 533, 624, 846, 908, 910, 914, *969, *975, 1169
Testis *514, 909
Thyroid *313, 365
Uterus (Cervix) *37, 200, 241, 274, 608, *615, *627, *789, 813, 882, 883, 994, *998, 1029, *1048, *1207
Vagina 971
Vulva 971
Wilms' Tumour *15

Breast Cysts

Breast (F) 724

Caucasians

All Sites 1170
Breast (F) 962
Hodgkin's Disease 1192
Leukaemia 730

Chemical Exposure

All Sites 19, 33, 42, 68, 84, 90, 117, 229, 247, 266, 282, 300, 308, *332, 333, 335, 371, *396, 489, 502, 504, 508, 522, 545, 578, *621, 623, 654, *661, 755, 757, 763, 764, 779, *781, 819, 887, 928, 982, 1018, 1019, 1094, 1095, 1107, 1131, 1157, 1158, 1159, 1171, 1173, 1180, 1181, 1188, 1230
Bladder 73, 84, 99, 213, 279, 310, *352, 965, 1025, 1157
Bone 171, *621

Brain 84, 171, 213, 304, 504, *621, 765, 965, 1091, 1095, 1104, 1180, 1181
Breast (F) *396, *1043, 1107
Childhood Neoplasms 647, *822
Colon 73, 84, 747, 860, 1032, 1092, 1093, 1181
Gastrointestinal 193, 213, *649, 1091
Haemopoietic 19, 213, 504, 677, *727, 965, 1091
Hodgkin's Disease *1244
Hypopharynx 315, *317
Inapplicable 48, 58
Kidney 73, 84, 506, 873, 1093, 1104, 1189
Larynx 315, *317, 1095
Leukaemia 33, 84, 119, 171, 208, 508, *621, 653, 779, *781, *822, 849, 859, 1107, 1179, 1180, 1181
Leukaemia (ALL) 75, 647
Leukaemia (AML) *727, 1086
Leukaemia (CLL) *727
Liver 171, 1028, 1091, 1092, 1093, 1104
Lung 73, 94, 193, 279, *317, 326, 351, 386, 489, 623, *649, 683, 721, 860, 873, 945, 965, 968, 982, 1091, 1092, 1093, 1104, 1107, 1131, 1153, 1173, 1180
Lymphoma 13, 19, 73, 84, 171, 247, 506, 508, *621, 654, 757, 779, *781, 965, 1028, 1179
Male Genital 213
Melanoma 73, 628, 1069
Mesothelioma 42, 851, 965
Multiple Myeloma 757, 1180
Myelodysplastic Syndrome 17, 267, 835
Nasal Cavity 208, *316, *317, 1084
Nasopharynx 151, 1084, 1159, 1208
Non–Hodgkin's Lymphoma *727, 859, 1180
Oesophagus 73
Pancreas 73, 244, *621, 1179
Peritoneum 42
Pleura 42, *317
Prostate 73, 84, 623, 816, 873, 895
Rectum 73, 84, 747, 860, 1032, 1181
Respiratory 208, 213, 488, 743, 1156
Rhabdomyosarcoma *1155
Sarcoma 13, 213, 247, 965
Skin 213, 504, 1093
Soft Tissue 33, 508, 538, 654, 1028, 1068, 1180
Stomach 73, 84, 326, 743, 860, 965, 982
Testis 815
Thyroid 1107, *1245
Ureter *352
Urethra *352
Urinary Tract 828
Wilms' Tumour 71, 1007

Chemical Mutagenesis

Inapplicable 390, 1001

Chemotherapy

All Sites *111, 270, *336, 339, 619, 746, 970, 1023, *1175
Breast (F) 619, 620, 970
Childhood Neoplasms *111, 1150, 1225
Choriocarcinoma 6
Gastrointestinal 340
Hodgkin's Disease 270
Hydatidiform Mole 6
Inapplicable 199, 626, 681
Leukaemia 339, 340, 619, 672, 730, 871, 915, 970
Lung 339, 340, 619, 970
Myelodysplastic Syndrome 619

Non–Hodgkin's Lymphoma 339, 340, 619
Ovary *1175
Testis 391
Thyroid 970

Childhood

All Sites 56, *336, 450, 471, 472, *475, 524, 603, 708, 777, 889
Benign Tumours 472
Bone *292
Brain 44, 57, *205, 303, *473, *665, 955, 1087, 1088, 1151
Breast (F) 889
Childhood Neoplasms 307, 598, 823, 922, *934, 1151, 1225
Hodgkin's Disease *292, 540, 1154
Leukaemia 56, 57, 119, *205, 287, *292, 307, 367, *380, *384, *385, 435, 633, *665, 672, *704, 730, 777, 823, 859, *886, 915, 922, *934, *936
Leukaemia (ALL) 540, 633, 1108
Leukaemia (ANLL) 540, 1109
Lymphoma 56, 57, *205, 287, 435, *886, *934, *936
Melanoma *292, 911
Myelodysplastic Syndrome 922
Nervous System *292
Neuroblastoma 57, *292, 540
Non–Hodgkin's Lymphoma *292, 540, 823, 859
Rhabdomyosarcoma *1155
Sarcoma *292, 889
Soft Tissue *292, 889
Stomach 912, 930
Thyroid *704, 1194
Wilms' Tumour 540

Chinese

Breast (F) 299
Liver 163
Lung 386
Nasopharynx 607
Oesophagus 163
Stomach 163

Chlorination

All Sites 84
Bladder 84, *110
Brain 84
Colon 84, *110
Kidney 84
Leukaemia 84
Lymphoma 84
Prostate 84
Rectum 84, *110
Stomach 84

Cholesterol

All Sites 533, 601, 884, 921, *1101
Breast (F) 533, 786, 949, *1005
Colon 526, 533, 601, 894, 949
Female Genital 949
Inapplicable *817
Liver 601
Lung 97, 601
Ovary 786, 949
Rectum 601, 894
Stomach 533, 601
Uterus (Cervix) 786, 949, *1048

Uterus (Corpus) 786, 949

Chromosome Effects

All Sites 62, 551, 578, 746, 753, 759, 1131
Angiosarcoma 738
Bile Duct *734
Bladder *120, 738
Breast (F) *78, 558, 1054
Colon 51, 558
Gallbladder *734
Inapplicable 198, 199, 249, *355, 390, 681
Leukaemia 62, 539
Liver 738
Lung 738, 1131
Lymphoma 62
Myelodysplastic Syndrome 539
Oesophagus 133
Oral Cavity 402
Ovary *78, 741, 1054
Pancreas *734
Rectum 51, 558
Stomach 558
Uterus (Cervix) 437
Wilms' Tumour *15

Cirrhosis

Liver 491, 569, 597, 656, 726
Lung 683

Classification

Appendix 398
Bile Duct 398
Brain 432
Colon 398
Leukaemia 730
Leukaemia (ANLL) 1109
Nervous System 398
Oesophagus 398
Oral Cavity 398
Ovary 108
Pancreas 398
Prostate 398
Small Intestine 398
Soft Tissue 398
Testis 108

Clinical Effects

All Sites *453
Childhood Neoplasms 827
Leukaemia 827
Thyroid *453

Clinical Records

All Sites 52, 112, 450, 603, 1137, 1171
Brain 85
Childhood Neoplasms 832
Kaposi's Sarcoma *952
Kidney 1141
Leukaemia 435, *850
Lung 52
Lymphoma 435, *850

Mesothelioma	52
Ovary	1133
Thyroid	318
Uterus (Cervix)	728

Cluster

All Sites	*54, *142, 452, 869, *929, 978, *1052, 1105, 1166
Bile Duct	587
Bladder	1016
Breast (F)	*78, 962, 1105
Breast (M)	*54
Childhood Neoplasms	*54, 375, 646, 823, *852, 905, 1165
Gastrointestinal	64
Hodgkin's Disease	*929, 1192
Larynx	869
Leukaemia	*350, 823, 829, 869, *886, 905, *929, 1064, 1165
Liver	587, 869, 1165
Lung	869
Lymphoma	*350, 829, 869, *886
Nasopharynx	409
Non–Hodgkin's Lymphoma	823, *929, 1165
Oropharynx	*54
Ovary	*78
Pancreas	1165
Soft Tissue	1165
Uterus (Corpus)	1105

CMV

Uterus (Cervix)	169, 608, *789, 813

Coal

Gastrointestinal	192
Hodgkin's Disease	40
Inapplicable	48, 626
Leukaemia	192
Leukaemia (CLL)	40
Lung	152, 172, *183, 192
Non–Hodgkin's Lymphoma	40
Stomach	*183

Cocaine

Rhabdomyosarcoma	*1155

Coffee

All Sites	465
Bile Duct	*734
Bladder	310
Brain	955
Breast (F)	*214
Colon	465, 737
Gallbladder	*734, 737
Kidney	45, 330, 795, 1141
Larynx	*1246
Liver	737
Lung	465
Oesophagus	737, *1118
Pancreas	244, *734, 737
Rectum	737
Small Intestine	737
Stomach	737, *1118
Thyroid	*1245
Wilms' Tumour	71

Condyloma

Female Genital	252
Male Genital	252

Congenital Abnormalities

All Sites	132, *338, 452
Brain	955
Childhood Neoplasms	600, 827, 890, 891, 916, 1150, 1225
Hodgkin's Disease	*1244
Inapplicable	58, 390
Leukaemia	119, 600, 827
Skin	1184
Testis	283

Contraception

All Sites	924
Breast (F)	25, 433, 650, 924
Choriocarcinoma	516
Hydatidiform Mole	516
Uterus (Cervix)	*627, 645, 924, 946, *1048, 1187, *1207
Uterus (Corpus)	1047

Cooking Methods

Colon	*976
Lung	152, *173, *428
Stomach	735, 1233

Cosmetics

All Sites	991
Brain	1087
Breast (F)	991
Lip	1085

Cost–Benefit Analysis

Breast (F)	276, 630
Colon	147, 148
Head and Neck	389
Liver	811
Neuroblastoma	591
Oral Cavity	389
Rectum	147, 148
Uterus (Cervix)	*294, 631

Crohn's Disease

All Sites	821
Colon	226, 800, 821
Rectum	800, 821
Small Intestine	226

Cryptorchidism

Testis	283

Cytology

All Sites	924
Bile Duct	*734
Bladder	*548
Breast (F)	924

Gallbladder *734
Head and Neck 389
Lung 596, 1224
Oesophagus 188, 361
Oral Cavity 66, 389, *430
Oropharynx 66
Pancreas *734
Uterus (Cervix) 10, 55, 274, *294, 369, 437, 438, *482, *615, 631, 728, 883, 924, 951, 994, *998, 1029, 1187, *1207, 1211

Dairy Products
Colon 641
Ovary *1120
Testis 834

Data Resource
All Sites *109, 344, *411, 525, 718, *814, 869
Breast (F) 996
Colon 996
Larynx 869
Leukaemia *850, 869
Liver 869
Lung 869, 996
Lymphoma *850, 869
Ovary 996
Prostate 996
Urinary Tract 828
Uterus (Corpus) 996

Demographic Factors
All Sites *387
Brain 22
Leukaemia *342
Lung *1226
Spinal Cord 22

Diabetes
All Sites 921, 1065
Bladder 104
Breast (F) 433, 1065
Colon 1065
Kidney 45, 795
Lung 104, 1065
Pancreas 104, 244
Prostate 1065

Diet
All Sites 49, 61, 88, 89, 145, 225, 245, 295, 372, 458, 465, 477, 512, 533, 564, 572, 575, 586, 588, 589, 601, 680, 714, 749, 806, 953, 981, 991, 1044, 1050, 1057, 1065, 1070, 1075, 1105, 1138, 1168, 1197, 1238, 1240
Benign Tumours 517
Bladder *352
Brain 304, *473, 1087
Breast (F) 21, 63, 74, 89, 105, 127, 131, 146, 157, 242, 299, 408, 423, 442, 446, *478, 480, 507, 510, *515, 517, 533, 565, *609, 612, 618, 624, 639, 640, 680, 711, 797, 848, 908, *919, 923, *948, 949, 974, 980, 991, 1000, 1031, 1065, *1067, 1105, 1111, 1130, 1132, *1195, 1197, 1218, 1222, *1223
Childhood Neoplasms 831
Choriocarcinoma 516

Colon	21, 63, 74, 89, 105, 127, 146, 149, 150, 161, 166, 167, 242, 260, 275, 366, 373, 424, 465, 470, *478, 480, 510, 526, 533, *594, 595, 601, 624, 640, 641, *675, 712, 737, 747, 806, *818, 893, 908, 949, *976, 980, *1010, 1057, 1058, 1065, 1066, 1098, 1183, 1197, 1210, 1212, *1223, 1228
Female Genital	949
Gallbladder	673, 737, 1081
Gastrointestinal	923, 1241
Genitourinary	1241
Hodgkin's Disease	*1244
Hydatidiform Mole	516
Hypopharynx	*317, 439
Inapplicable	*817, 1049
Kidney	105, 368, 560, 795
Larynx	70, 181, *317, 364, 439, *478, 784, *1246
Leukaemia	127, 180
Lip	70
Liver	127, 138, 146, *293, *460, 595, 601, 737, 956
Lung	35, *50, 63, 97, 127, 146, 152, *155, 242, 297, *317, *379, 386, 465, 480, 566, 595, 601, 624, 640, 678, 680, 710, 806, 963, 980, *999, *1051, 1065, 1079, 1135, 1197, 1232
Melanoma	1176, 1221
Mesothelioma	35, 731
Multiple Myeloma	1036
Nasal Cavity	181, *317
Nasopharynx	*2, 127, 143, 179, *184, 314, 1208
Non–Hodgkin's Lymphoma	1193
Oesophagus	63, 127, 167, *259, 361, 424, 439, 454, *478, 733, 737, 784, *969, 1036, *1074, *1103, *1118
Oral Cavity	70, 181, *420, 439, *478, 784, 1089
Oropharynx	439
Ovary	105, 406, *478, 517, 949, 980, 1000, *1120, 1198, *1223
Pancreas	105, 167, 733, 737, 1000, 1036
Pharynx	70, 181, 784, 1089
Pleura	*317
Prostate	21, 74, 92, 228, 242, 257, 329, *760, 793, 908, 918, 943, 1000, *1002, 1036, 1065, 1214
Rectum	8, 21, 105, 127, 146, 149, 150, 161, 166, 167, 242, 275, 366, 373, 424, *478, 510, *594, 601, 624, 640, *675, 712, 737, 747, *818, 893, *1010, 1058, 1066, 1098, 1212, *1223
Skin	1055, 1220, 1221
Small Intestine	737
Stomach	63, 127, 137, 138, 146, 153, 186, 242, 394, 424, 505, 531, 533, 557, 584, *585, 592, 595, 601, 602, *610, 624, 640, 680, 725, 733, 735, *736, 737, 794, 806, 846, 908, 913, 930, *969, *975, *1074, *1103, *1118, 1233
Testis	834
Thyroid	381, 662
Ureter	*352
Urethra	*352
Uterus (Cervix)	438, 443, 517, 608, 611, 680, 882, 949, 994, *1048, 1185, 1197
Uterus (Corpus)	105, 517, 595, 810, 949, 972, 980, 1000, 1047, 1105, 1190, *1223
Vagina	1102
Vulva	517
Wilms' Tumour	71

Disease, Other

All Sites	*1110
Childhood Neoplasms	*1110
Colon	*527, *843
Oesophagus	*969
Rectum	*843
Stomach	*969

Diuretics

Kidney	330, 795, 1141, 1164
Neuroblastoma	1008

DNA

All Sites 551, 572, 586, 1126
Bladder *120, 642
Breast (F) 25, 446, 558, 962, *1006, 1078
Colon 51, 558, 762, *778
Female Genital 252
Gastrointestinal 880
Hodgkin's Disease *676
Inapplicable 77, 626
Leukaemia 937
Leukaemia (ALL) *676
Liver 1149
Lung 135, 596, 642, *1051, 1232
Male Genital 252
Non-Hodgkin's Lymphoma *676
Oesophagus 130, *134
Rectum 51, 558, 762
Retinoblastoma *706, 1148
Stomach 558
Uterus (Cervix) 169, 920, 946, 994, 1211
Wilms' Tumour *15

DNA Adducts

All Sites 270, 1044, 1131
Bladder *204, 541, *543, 642
Hodgkin's Disease 270
Inapplicable 48, 201, *853, 1003, 1009
Larynx *204
Leukaemia 1026
Lung *204, 239, 481, 613, 642, *812, 960, 1131
Oral Cavity *430
Stomach 602, 914

DNA Binding

All Sites 1131
Lung 1131

DNA Repair

All Sites 805
Bladder *204
Bone 805
Breast (F) 805
Childhood Neoplasms 805
Colon 526
Gastrointestinal 805
Inapplicable 201, *817
Larynx *204
Leukaemia 805
Lung *204
Lymphoma 805
Oesophagus 133
Ovary 805
Skin 959, 1184
Testis 805

Dose–Response

All Sites 42, 116, 210, *262, 263, *336, 551, *686, 750, 752, 772, 839, 896, 897, 901, 902, 970, 1071, 1107, 1171, *1172, 1227
Bone 370
Breast (F) 224, 347, 970, 1107
Childhood Neoplasms 891

Colon 748, 860
Gastrointestinal 76, 192
Head and Neck 632
Inapplicable 1009
Kidney 76
Larynx 76
Leukaemia 192, 224, *262, 370, 752, 871, 901, 902, 970, 1107
Leukaemia (CLL) 224
Liver 370
Lung 76, 177, 192, 263, 370, *686, 748, 772, 782, 858, 860, 968, 970, 1022, 1024, 1107, 1153, 1227
Lymphoma 370, 752, 1024
Mesothelioma 42, *686, 748
Peritoneum 42, 76
Pleura 42, 76
Rectum 748, 860
Skin *337
Spleen 370
Stomach 748, 860
Thyroid *337, 970, 1107
Uterus (Corpus) 1047

Down's Syndrome

All Sites 132
Childhood Neoplasms 890, 891
Leukaemia 937

Drugs

All Sites 210, 221, 227, 270, 512, 568, 625, 746, 757, 764, *776, 928, 953, 981, *987, 991, 992, 993, 995, 1137, *1175
Benign Tumours 616, 1140
Bladder 232, 310, 513, 1134
Bone 171, 370
Brain 44, 85, 171, 303, 1087
Breast (F) *46, 232, 299, *302, 309, 616, 618, 926, *948, *987, 991, 996, 1130, 1140, *1191
Childhood Neoplasms 221, 600, 647, 827, 831, 832, 1225
Colon 166, 747, 893, *987, 996, *1010, 1058, *1127
Eye 993
Female Genital 926
Gallbladder 1081
Gastrointestinal 880, 927
Haemopoietic *727
Hodgkin's Disease 270, *311, *1125, 1154
Inapplicable 354, *355, 390, 626, 1001
Kaposi's Sarcoma 41, 382
Kidney 45, 368, 513, 560, 795, 1134, 1164
Larynx 513, 807, 1134
Leukaemia 119, 171, 180, 221, 349, 370, 600, 827, 849
Leukaemia (ALL) 647
Leukaemia (AML) *727, 1086
Leukaemia (CLL) *727, *1125
Liver 171, 221, 370
Lung 232, 370, *459, 513, 807, *987, 996, 1134
Lymphoma 41, 171, 349, 370, 757
Multiple Myeloma 757
Neuroblastoma 1008
Non-Hodgkin's Lymphoma *311, *727, *1125, 1186
Oesophagus 513, *969, *1074, *1118, 1134
Oral Cavity 513, 807, 1134
Ovary 635, 996, *1175, *1191, 1229
Pancreas 513, 1134
Pharynx 513, 807
Prostate 232, 816, 996

Rectum 166, 747, 893, *1010, 1058, *1127
Skin 993
Spleen 370
Stomach 505, *969, *975, *1074, *1118
Uterus (Cervix) 232, 994
Uterus (Corpus) *302, 996, *1191
Vagina 1102
Wilms' Tumour 71

Dusts

All Sites 42, 52, 90, *272, 282, 305, 489, 500, 501, 511, 522, 545, *669, 684, 685, *686, *715, 754, 757, 772, 856, 939, 982, 1019, 1171, 1197, *1235
Bladder 73, 305, 487, *490, 910, 965
Brain *272, *715, 965
Breast (F) 1197
Colon 34, 73, *323, 748, *758, 1197
Gastrointestinal 72, 76, 101, 691, *692
Haemopoietic 677, *692, 744, 965
Hypopharynx 315
Kidney 73, 76, *272, 487, 506, *715
Larynx 59, 76, 315, 511
Leukaemia *106, 208, 487, 653, *864
Liver 138, 487
Lung 34, 35, 36, *43, 52, 72, 73, 76, 94, *106, 129, *158, *183, *272, *320, *322, *323, 343, 353, 487, 489, 496, 511, *580, 604, *671, 685, *686, 691, *692, *715, 721, 742, 744, 748, 772, 857, 910, 939, 965, 968, 982, 1024, 1197, 1200, *1209, 1224, *1226
Lymphoma 73, 506, 744, 757, 965, 1024
Melanoma 73
Mesothelioma 34, 35, 42, 52, 59, 72, 101, 353, 496, 530, *536, *686, 748, 851, 965, 1200
Multiple Myeloma 757
Nasal Cavity 208, 278, 305, 306, *316, 1084
Nasopharynx 59, 179, 1084
Oesophagus 73, 910
Pancreas 73
Peritoneum 34, 42, 76, 530
Pleura 42, 76, 496, 530
Prostate 73
Rectum 73, 748, *758
Respiratory 101, 208, 488, 743, 856, *864, 1156
Sarcoma 965
Stomach 34, 73, *106, 138, *183, 743, 744, 748, 910, 965, 982, 1177
Testis 907
Uterus (Cervix) *692, 1197

Dyes

All Sites 282, 500, 509, 991
Angiosarcoma 738
Bladder 509, 738
Breast (F) 991
Hodgkin's Disease *311
Liver 738
Lung 738
Nasal Cavity 509
Non–Hodgkin's Lymphoma *311
Pharynx 509
Respiratory 509

EBV

All Sites *929, 1044, *1174
Hodgkin's Disease *929, 1154
Leukaemia *929

Lymphoma *1062, 1073, *1174
Nasopharynx *2, 314, 409, 606, 607
Non–Hodgkin's Lymphoma *929, *1174
Uterus (Cervix) 608

Education

All Sites 116, 245, 981
Oral Cavity *416, *421
Prostate *760
Uterus (Cervix) 443, *627

Electromagnetic Fields

All Sites 80, 312
Brain 38, 766, 1088
Breast (F) *1195
Breast (M) *1004
Childhood Neoplasms 647, 766, 769, *822
Hodgkin's Disease *311
Leukaemia *18, 119, 334, *380, 653, 655, 766, *822
Leukaemia (ALL) 647
Leukaemia (AML) 1086
Lymphoma *18
Melanoma *1004
Multiple Myeloma *18
Neuroblastoma 1008
Non–Hodgkin's Lymphoma *311
Prostate *1004

Environmental Factors

All Sites 14, 42, 49, 88, 90, 116, 132, 196, 229, 240, 266, 295, *387, 489, 503, 508, 522, 674, 759, 869, *935, 981, 1094, 1158, 1166, 1173, 1239, 1240
Bladder 213, 392, 1035
Bone 674
Brain 213, 303, 765, 766, 1091
Breast (F) *26, *27, 146, 157, *474, *567, 674, 962, 1078
Childhood Neoplasms 321, 376, 600, 766, 769, 824, 831
Colon 5, 24, *27, 34, 146, 258, *567, 712, *1010, 1035
Eye 1035
Gallbladder 258, 1081
Gastrointestinal 72, 192, 193, 213, 392, *649, 1091
Haemopoietic 213, 1091
Hodgkin's Disease 431, 1154
Inapplicable 48
Kaposi's Sarcoma 382
Kidney 873
Larynx 869
Leukaemia 192, 349, 376, 431, 435, 508, 600, 766, 829, 869
Liver 146, 258, 383, 869, 1035, 1091
Lung 34, 72, 135, 146, 192, 193, 217, *264, 386, 489, *567, *649, 721, 742, 869, 873, 903, 944, 945, 1035, 1091, 1135, 1173
Lymphoma 349, 376, 431, 435, 508, 829, 869
Male Genital 213
Melanoma 39, 874, 911, 1035, 1221
Mesothelioma 30, 34, 42, 72, 731, 851, 1035
Multiple Myeloma 431
Myelodysplastic Syndrome 17, 267
Nasal Cavity 278, 874, 1084
Nasopharynx 314, 1084, 1208
Neuroblastoma 376
Oesophagus 258, 1035
Pancreas 258
Peritoneum 34, 42
Pleura 42

Prostate	873
Rectum	5, 24, *27, 146, 258, *567, 712, *1010, 1035
Respiratory	213, 1156, 1219
Salivary Gland	1035
Sarcoma	213
Skin	39, 213, 528, 1221
Soft Tissue	508, 538, 1068
Stomach	34, 146, 258, *567, 592, 1035
Uterus (Cervix)	*567, 674
Wilms' Tumour	376, 1007

Enzymes

All Sites	486
Bladder	*543
Larynx	807
Lung	94, *459, 683, 807
Melanoma	1221
Oral Cavity	807
Ovary	1198
Pharynx	807
Skin	1221
Stomach	913
Testis	*206

Epilepsy

All Sites	221
Childhood Neoplasms	221
Leukaemia	221
Liver	221

Eskimos

All Sites	82, 220

Ethnic Group

All Sites	82, *109, 116, 220, 452, 465, 643, 1050, *1052
Breast (F)	*474, *605, *948
Childhood Neoplasms	824, 922
Colon	465, *675
Gallbladder	1081
Hodgkin's Disease	*605, *1125
Inapplicable	1049
Leukaemia	922
Leukaemia (CLL)	*1125
Lung	456, 465, *605, *1051, 1079
Myelodysplastic Syndrome	922
Non–Hodgkin's Lymphoma	*605, *1125
Oesophagus	*605, 694
Ovary	1243
Penis	356
Prostate	469, 1129
Rectum	*605, *675
Retinoblastoma	877
Skin	*1038
Stomach	576, *605, *975
Uterus (Cervix)	356, *605, 645
Uterus (Corpus)	1047
Vulva	356

Faeces

Colon	260, 261, 366, 373, 526
Rectum	261, 366, 373

Familial Adenomatous Polyposis

Colon 5, 24, 28, 366
Gastrointestinal 880
Rectum 5, 24, 28, 366

Familial Factors

All Sites 458, 586, 625, 714, 805, 889, 991, 1046, *1052, 1099
Benign Tumours 616, 1059
Bladder 546
Bone 805
Brain 1151
Breast (F) *26, *27, *46, *78, *79, 146, 235, 299, 309, 395, 399, 510, 546, 558, 616, 618, 637, 640, 644, 797, 805, 889, 949, 962, 991, 1046, 1053, 1059, 1078, 1204, 1213
Breast (M) 1053
Childhood Neoplasms 307, 805, 904, 916, 1151
Choriocarcinoma 516
Colon 5, 24, *27, 28, 51, 146, 161, 166, 167, 470, 510, 521, 526, 546, 558, 640, *843, 949, 1145, 1213
Female Genital 949
Gastrointestinal 805
Haemopoietic *476
Hydatidiform Mole 516
Leukaemia 307, 805, 1046
Leukaemia (ALL) 1108
Liver 146, 165, 383
Lung 94, 146, 546, 640, 963, 1061, 1162, 1213
Lymphoma 805
Melanoma 9, 39, 666
Multiple Myeloma 1036
Nasopharynx *184
Oesophagus *134, 167, 361, 546, 1036
Ovary *78, 534, 805, *865, 868, 949, 1243
Pancreas 167, 1036, 1213
Prostate 1036, 1213
Rectum 5, 24, *27, 28, 51, 146, 161, 166, 167, 510, 521, 546, 558, 640, *843, 1145, 1213
Retinoblastoma 904
Sarcoma 889
Skin 39
Soft Tissue 889
Stomach 146, 238, 394, 546, 558, 640
Testis 805
Uterus (Cervix) 949
Uterus (Corpus) 949
Wilms' Tumour 1007

Family History

All Sites *1110
Brain *1056
Breast (F) *26, *27, *79, *461
Childhood Neoplasms 321, *1110
Colon *27, 149, 150, 893
Hodgkin's Disease *1125
Leukaemia 849
Leukaemia (CLL) *1125
Non–Hodgkin's Lymphoma *1125
Rectum *27, 149, 150, 893
Thyroid *1245

Fat

All Sites 61, 465, 1105, 1197, 1238

Breast (F) 63, *609, 620, 637, 639, 797, 848, *919, 997, 1031, 1105, 1132, *1195, 1197, 1222
Colon 63, 465, 595, 641, 893, 997, 1058, 1197, 1210, 1212
Liver 595
Lung 63, 465, 595, *999, *1051, 1197
Non–Hodgkin's Lymphoma 1193
Oesophagus 63, 733
Ovary 997, *1120
Pancreas 733
Prostate 92, 329, 918, 997, *1002
Rectum 8, 893, 1058, 1212
Stomach 63, 595, 733
Uterus (Cervix) 1197
Uterus (Corpus) 595, 972, 1047, 1105

Fatty Acids

Breast (F) *515
Prostate *1002

Female

All Sites *107, 991, 995, 1039
Bladder 1035
Brain 96
Breast (F) 105, 991
Colon 105, 1035, 1098
Eye 1035
Gallbladder 1081
Gastrointestinal 72
Hypopharynx 425
Kidney 105
Larynx 425
Lip 1085
Liver 1035
Lung 72, 95, 135, 152, 331, 386, 425, *428, 963, 1035
Melanoma 1035
Mesothelioma 72, 1035
Multiple Myeloma 96
Non–Hodgkin's Lymphoma 96
Oesophagus 1035
Oral Cavity 425
Oropharynx 425
Ovary 105
Pancreas 105
Rectum 105, 1035, 1098
Salivary Gland 1035
Sarcoma 96
Soft Tissue 96
Stomach 1035
Tongue 425
Uterus (Cervix) *998
Uterus (Corpus) 105

Fertility

All Sites *111, 924, *1110
Breast (F) 924
Childhood Neoplasms *111, *1110, 1225
Choriocarcinoma 6
Hydatidiform Mole 6
Leukaemia 672
Uterus (Cervix) 924

Fertilizers

All Sites	83, 86, 117, *701
Bladder	910
Brain	86
Gastrointestinal	193
Leukaemia	86
Lung	193, 756, 910
Multiple Myeloma	86
Non–Hodgkin's Lymphoma	86
Oesophagus	910
Prostate	86, 629, 943
Sarcoma	83
Stomach	756, 910, 1177

Fibre

All Sites	1057, 1197
Breast (F)	640, 797, *919, 1197
Colon	640, 641, 893, 1057, 1058, 1197, 1210, 1212
Lung	640, 1197
Rectum	8, 640, 893, 1058, 1212
Stomach	640
Uterus (Cervix)	1197

Fibrosis

Lung	239

Fluorescent Light

All Sites	100
Melanoma	7, 100

Fruit

All Sites	465, 1070
Colon	465
Lung	465
Oesophagus	733
Pancreas	733
Stomach	733
Wilms' Tumour	71

Fungicides

Lung	94

G–Banding

Inapplicable	*355

G6PD

All Sites	486, 682, 684

Gallstones

Bile Duct	1152
Gallbladder	124, 673, 1152

Gastrectomy

Colon	*1014
Rectum	*1014

Stomach *1014

Gastric Ulcer

All Sites 625

Genetic Councelling

Breast (F) *703
Melanoma *703
Ovary *703
Retinoblastoma *706
Uterus (Corpus) *703

Genetic Factors

All Sites 14, *319, 486, 586, 702, 708, 805, 1046, 1099
Benign Tumours 616
Bladder 178, *543, 866, 1025
Bone 805
Brain *1056
Breast (F) 25, *26, *27, 146, 157, 230, 558, 616, 620, *703, 805, 962, 974, 1046, 1053, 1054, 1078, 1111
Breast (M) 1053
Childhood Neoplasms 307, 321, 376, 647, 805, 916, 1150
Colon *27, 28, 146, 216, 275, 366, 558, *818, *933, 954, *976, 1183
Gastrointestinal 230, 805, 954
Haemopoietic *476
Head and Neck *553
Inapplicable 554
Leukaemia 230, 307, 376, *704, 805, 1046
Leukaemia (ALL) 647, 1108
Liver 146
Lung 94, 146, *459, 683, 1079
Lymphoma 376, 805
Melanoma 9, 230, *703, 911
Nasopharynx 151, *184
Neuroblastoma 376
Oesophagus *134
Oral Cavity 402, *553
Ovary *703, 741, 805, 868, *1011, 1054
Prostate 816
Rectum *27, 28, 146, 216, 275, 366, 558, *818, *933, 954
Retinoblastoma *706, 877
Skin 528
Stomach 146, 558, 592
Testis 230, 805, 909
Thyroid *313, *319, 365, *704
Uterus (Cervix) 230
Uterus (Corpus) 230, *703
Wilms' Tumour 376

Genetic Markers

All Sites 759, 805, 1046, 1126, 1131
Benign Tumours *494
Bile Duct *514
Bladder *120
Bone 805
Brain *514
Breast (F) 25, *26, *27, *78, 146, 170, *514, 805, 962, 1046, 1078
Childhood Neoplasms 805
Colon *27, 146, 216, 954
Endocrine Glands *494
Gastrointestinal 805, 954
Leukaemia 805, 1046

Liver 146, *514
Lung 146, 683, *1051, 1079, 1131
Lymphoma 805
Melanoma 9, 47
Neuroblastoma 1008
Non–Hodgkin's Lymphoma *514
Ovary *78, 805, *1011
Rectum *27, 146, 216, 954
Respiratory *514
Sarcoma *514
Skin *514
Soft Tissue *514
Stomach 146
Testis *514, 805

Genotype/Phenotype

Bladder *204, *543
Brain *203
Breast (F) *203, *1063
Colon *203
Larynx *204
Lung *204
Nervous System *203
Prostate *203
Rectum *203

Geographic Factors

All Sites 1, *54, 69, 83, 84, 90, *142, 194, 196, 289, *387, 498, 503, 643, 688, 689, 720, 759, *780, 809, 869, 899, 1237, 1239, 1240
Bladder 84, 392, 899
Bone *292
Brain 84, *462
Breast (F) 146, 194, *462, 679
Breast (M) *54
Childhood Neoplasms *54, 646, 823, 824, 904, 905, 922, 1165
Colon 84, 146, 258, 737
Gallbladder 258, 737
Gastrointestinal 64, 392
Hodgkin's Disease *292
Kidney 84
Larynx 869
Leukaemia 84, *292, *462, 633, 823, *850, 869, 899, 905, 922, 1165
Leukaemia (ALL) 633, 847
Leukaemia (AML) 847
Leukaemia (CML) 847
Lip 98
Liver 146, 258, 729, 737, 869, 1165
Lung 146, 194, *197, 447, *462, 869
Lymphoma 84, *850, 869
Melanoma 20, *292, 720, *1030
Myelodysplastic Syndrome 922
Nasopharynx 447
Nervous System *292
Neuroblastoma *292
Non–Hodgkin's Lymphoma *292, 823, 1165
Oesophagus 258, 694, 737
Oropharynx *54
Pancreas 258, 737, 1165
Prostate 84
Rectum 84, 146, 258, 737
Retinoblastoma 904
Sarcoma 83, *292
Skin 20, 720, *1030
Small Intestine 737

Soft Tissue *292, 1165
Stomach 84, 146, 258, 737
Thyroid *462
Uterus (Cervix) 447, 679
Uterus (Corpus) 679

Goitre
Stomach *523
Thyroid *1245

H. pylori
All Sites 663, 884
Oesophagus *969
Stomach 285, 286, *493, *585, 602, *610, 846, 913, 914, *969, *975, 1169

Haemophilia
All Sites *1174
Kaposi's Sarcoma 382
Lymphoma *1174
Non–Hodgkin's Lymphoma *1174

Hair
Lung 705
Prostate 918

HBV
All Sites 344, 562
Liver *160, 164, 191, *293, 345, 383, *460, 491, 559, 559, 559, 559, *577, 597, 656, *690, 726, 729, 811, 956, 1149
Uterus (Cervix) 369, *1207

HCV
All Sites 562
Liver *160, *293, *460, 559, 559, *577, 726, 729

Hepatitis
All Sites 663
Liver *460

Herbicides
All Sites 33, 83, 86, 247, 296, 371, 452, 622, 654, 763, 1180
Brain 86, 1104, 1180
Hodgkin's Disease *311, 542
Kidney 368, 1104
Leukaemia 33, 86, 349, 542, 653, 977, 1180
Liver 1104
Lung 94, 1104, 1180
Lymphoma 13, 247, 349, 654, 977
Multiple Myeloma 86, 542, 1180
Non–Hodgkin's Lymphoma 86, *290, *311, 542, 622, 1180
Prostate 86
Sarcoma 13, 83, 247, *290, 622
Skin 961
Soft Tissue 33, *290, 538, 622, 654, 1068, 1180
Wilms' Tumour 71

Heredity

All Sites	578, 702, 805
Bone	*292, 805
Breast (F)	*78, 805, *1063
Childhood Neoplasms	805
Colon	*53, 216, 366, 521, 747
Gastrointestinal	805
Hodgkin's Disease	*292
Leukaemia	*292, 805
Lymphoma	805
Melanoma	47, *292
Nervous System	*292
Neuroblastoma	*292
Non–Hodgkin's Lymphoma	*292
Ovary	*78, 741, 805
Rectum	*53, 216, 366, 521, 747
Rhabdomyosarcoma	*1155
Sarcoma	*292
Soft Tissue	*292
Testis	805
Thyroid	*313
Wilms' Tumour	1007

High–Risk Groups

All Sites	88, 245, 685, *693, 702, 718, 889, 938, 1046, 1126, 1197
Bladder	900
Breast (F)	*78, *79, 174, 235, 236, 309, 395, 644, 724, 802, 836, 889, *919, 962, 1046, 1054, 1197, 1242
Colon	5, 24, 28, 147, 150, 216, 373, 455, *818, 954, 1197
Gastrointestinal	954
Hodgkin's Disease	479
Inapplicable	554
Kaposi's Sarcoma	479
Leukaemia	915, 1046
Liver	174, 491, 811
Lung	174, 388, 596, *671, 685, 1197, 1200, 1232
Mesothelioma	1200
Non–Hodgkin's Lymphoma	479
Oesophagus	133, 187, *189
Oral Cavity	66, 1136
Oropharynx	66, 1136
Ovary	*78, 534, 1054
Rectum	5, 24, 28, 147, 150, 174, 216, 373, 455, *818, 954
Sarcoma	889
Soft Tissue	889
Stomach	174, 394, 725
Uterus (Cervix)	115, 174, 1185, 1197

Histology

All Sites	248, 551, 720
Appendix	398
Bile Duct	398
Bone	*845
Brain	85, 432, 1087
Breast (F)	157, 395, 464, 1116
Childhood Neoplasms	*852
Colon	166, 398, 954
Gastrointestinal	880, 954
Head and Neck	389
Hodgkin's Disease	431
Larynx	59
Leukaemia	180, 431
Lung	159, 217, 239, 496, 1061, 1135, 1153, *1209
Lymphoma	431

Melanoma 720, 874
Mesothelioma 59, 496
Multiple Myeloma 431
Nasal Cavity 278, 874
Nasopharynx 59
Nervous System 398
Non-Hodgkin's Lymphoma 1186
Oesophagus 398, *1103
Oral Cavity 389, 398
Ovary 108, 1243
Pancreas 398
Pleura 496
Prostate 398
Rectum 166, 954
Skin 12, 720
Small Intestine 398
Soft Tissue 398, 1068
Stomach 237, 394, *493, 552, 576, 735, *1103, 1177
Testis 108
Thyroid 318, 381
Uterus (Cervix) 631, 883, 920, 951

HIV

All Sites 524, 1041, *1174
Burkitt's Lymphoma *658
Hodgkin's Disease 479, *658
Kaposi's Sarcoma 41, 479, 657, *658, *898, *952, 957, 964, 1041
Lymphoma 41, 964, 1073, *1174
Non-Hodgkin's Lymphoma 479, *658, *898, *1174, 1186
Uterus (Cervix) 369, *898, *998, *1207

HLA

All Sites 227, *929
Hodgkin's Disease *929, *1125
Kaposi's Sarcoma 529
Leukaemia *929
Leukaemia (CLL) *1125
Lymphoma 1073
Nasopharynx 314
Non-Hodgkin's Lymphoma *929, *1125
Prostate 816
Skin 528
Testis 909

Homosexuality

All Sites 218
Hodgkin's Disease 218
Kaposi's Sarcoma 382, *952, 957
Non-Hodgkin's Lymphoma 1186

Hormones

All Sites 512, 991, 1044, 1082, 1201
Benign Tumours 517, 616, 617, 1140, *1202
Breast (F) 121, *214, 235, 299, 309, 408, *436, *461, *474, 507, 510, 517, 616, 617, 618, 620, 644, 650, 659, 711, 803, 925, 926, 949, 974, 980, 991, 996, *1006, *1067, 1082, 1083, *1119, 1130, 1132, 1140, *1142, *1202, 1204, *1223
Choriocarcinoma 516
Colon 166, *407, 510, 949, 980, 996, *1119, *1223
Female Genital 803, 926, 949
Gallbladder 1081
Hydatidiform Mole 516

Kidney 45, *219, 795, 1164
Liver *460, 811
Lung 980, 996
Melanoma 346
Multiple Myeloma 1036
Oesophagus 1036
Ovary 108, 404, 406, 517, 534, 634, 635, 949, 980, 996, *1223, 1229
Pancreas 1036
Prostate 228, 793, 996, 1036
Rectum 166, *407, 510, *1119, *1223
Stomach 592
Testis 108, 283, 907
Thyroid *1119
Uterus (Cervix) 517, 949, *1048
Uterus (Corpus) 517, 810, 925, 949, 972, 980, 996, 1047, 1082, *1119, 1190, *1223
Vulva 517

HPRT – Mutants

Inapplicable 77, *817, *1020
Leukaemia 1026

HPV

All Sites 1044
Female Genital 252
Lung 932
Male Genital 252
Oral Cavity 363, 1090, 1203
Oropharynx 363, 1203
Penis 356
Pharynx 1090
Respiratory 363
Uterus (Cervix) *37, 55, 169, 200, 274, 284, 356, 363, 369, 403, 437, 438, 608, 611, 614, *615, *627, 645, 728, 745, *789, 813, 867, 920, 946, 994, 1029, *1048, 1187, *1207, 1211
Vulva 356

HSV

All Sites 1044
Oral Cavity 1203
Oropharynx 1203
Uterus (Cervix) 55, 169, 200, 284, 437, 438, 608, 611, 728, *789, 813, *1207

HTLV

All Sites 562
Kaposi's Sarcoma 382
Leukaemia 570, 581
Lymphoma 570

Hygiene

Hypopharynx 268, 439
Larynx 181, 439
Lung 386
Nasal Cavity 181
Oesophagus 439
Oral Cavity 181, 207, 268, 439
Oropharynx 268, 439
Pharynx 181
Uterus (Cervix) 410, 443
Vagina 971
Vulva 971

Hyperthyroidism
Thyroid 492, *1245

Hysterectomy
Breast (F) 804
Ovary 804

Immunodeficiency
Childhood Neoplasms 599

Immunologic Markers
All Sites 663
Lung 496, 613
Mesothelioma 496
Pleura 496
Uterus (Cervix) *482

Immunology
All Sites 359, 586
Kaposi's Sarcoma 529
Leukaemia *704
Liver 383
Non–Hodgkin's Lymphoma 1186
Oral Cavity 1203
Oropharynx 1203
Rhabdomyosarcoma *1155
Skin 528
Thyroid *704

Immunosuppression
All Sites 227, 1023
Leukaemia (AML) 1086
Lymphoma 1100
Melanoma 346
Skin *878

In Situ Carcinoma
Breast (F) 23, 802, *1196
Testis 212
Uterus (Cervix) 200, 951

Indians
All Sites *109, 122
Oesophagus 454

Infection
All Sites 218
Bladder *120
Breast (F) 1000
Burkitt's Lymphoma *658
Childhood Neoplasms 321, *822, 922
Colon 366
Female Genital 252
Hodgkin's Disease 218, *311, *658, 1154, 1192
Kaposi's Sarcoma 382, *658
Leukaemia 119, *380, *384, 435, 570, *822, 922
Liver 165, 168, 273, *293, 383, 569

Lymphoma	435, 570
Male Genital	252
Myelodysplastic Syndrome	922
Nasal Cavity	1084
Nasopharynx	151, 606, 1084
Non–Hodgkin's Lymphoma	*311, *658
Ovary	1000
Pancreas	1000
Prostate	1000
Rectum	366
Rhabdomyosarcoma	*1155
Stomach	912
Testis	283
Uterus (Cervix)	200, 284, 403, 645, 813, 920
Uterus (Corpus)	1000
Wilms' Tumour	71

Infertility

All Sites	1201
Childhood Neoplasms	916
Ovary	1229
Testis	212

Insecticides

All Sites	83, 86, 928
Brain	86, 1104
Kidney	*219, 1104
Leukaemia	86
Liver	1104
Lung	94, 1104
Multiple Myeloma	86
Non–Hodgkin's Lymphoma	86
Prostate	86
Sarcoma	83

Intra–Uterine Exposure

All Sites	708
Bone	171
Brain	44, 171, 303, 1087, 1151
Breast (F)	926
Childhood Neoplasms	222, 457, 600, 783, 827, 890, 904, 1151, 1225
Female Genital	926
Hodgkin's Disease	*1244
Leukaemia	119, 171, *380, 600, 827
Leukaemia (ALL)	75, 1108
Liver	171
Lymphoma	171
Neuroblastoma	1008
Retinoblastoma	904
Rhabdomyosarcoma	*1155
Testis	283, 907
Vagina	1102
Wilms' Tumour	1007

Japanese

Stomach	602

Lactation

Breast (F)	60, 121, 400, 620, 659, 1097

Late Effects

All Sites	*111, 210, 551, 837, *1110
Bone	370
Breast (F)	879
Childhood Neoplasms	*111, 375, 917, *1110, 1150
Leukaemia	370, 672
Liver	370
Lung	*264, 370
Lymphoma	370
Spleen	370

Latency

All Sites	116, 210, *397, 472, 503, 897
Benign Tumours	472
Breast (F)	879
Colon	860
Kidney	1189
Leukaemia	570
Lung	154, *397, 858, 860, 1024
Lymphoma	570, 1024
Mesothelioma	*324
Oesophagus	*1074
Rectum	860
Respiratory	*324
Stomach	153, *397, 860, *1074
Uterus (Cervix)	*397

Leather

All Sites	305, 500, 755
Bladder	305
Kidney	506
Lymphoma	506
Nasal Cavity	305, 306

Lifestyle

All Sites	88, 90, 196, 220, 372, 426, 572, 578, 582, 588, 589, 593, 601, 643, *693, 709, *780, 806, 870, 953, *989, 1050, 1075, 1076, *1077, 1239, 1240
Benign Tumours	31
Bladder	104, 392, 1035
Brain	765
Breast (F)	21, *27, 127, *436, *567, *605, 620, 624, 640, 707, 962, *1112, 1213
Childhood Neoplasms	376, 824
Colon	21, *27, 127, 147, 149, 161, 260, *567, 601, 624, 640, 712, 806, *1010, 1035, 1213
Eye	1035
Gastrointestinal	392
Hodgkin's Disease	*311, *605
Kidney	560
Leukaemia	127, 180, 367, 376, *886
Liver	127, 559, 559, 559, 559, 601, 1035
Lung	104, 127, 386, *567, 601, *605, 624, 640, 806, 1035, 1213
Lymphoma	376, *886
Melanoma	20, 31, 346, 1035
Mesothelioma	1035
Nasopharynx	*2, 127, 1208
Neuroblastoma	376
Non-Hodgkin's Lymphoma	*311, *605, 1193
Oesophagus	127, 361, 454, *605, 1035, *1118
Ovary	406, 707
Pancreas	104, 1213
Prostate	21, 228, 1213
Rectum	21, *27, 127, 147, 149, 161, *567, 601, *605, 624, 640, 712, *1010, 1035, 1213

Salivary Gland 1035
Skin 20
Soft Tissue 1068
Stomach 127, 186, *567, 584, *585, 592, 601, 602, *605, 624, 640, 806, 1035, *1118
Uterus (Cervix) *567, *605, 707
Uterus (Corpus) 707
Wilms' Tumour 376

Linkage Analysis

All Sites 876
Breast (F) 170, 558, 1078, 1111, *1112
Colon 366, 558, *778, *842
Oesophagus 133
Rectum 366, 558, *842
Stomach 558

Lipids

All Sites 465, 806, 991, 1044, 1065
Breast (F) 21, *214, *436, 991, *1005, 1065, 1218, 1222
Colon 21, 465, 806, 1065
Lung 465, 806, 1065, 1232
Melanoma 1221
Prostate 21, 1065
Rectum 21
Skin 1055, 1221
Stomach 806

Liver Disease

Liver 811

Lymphocytes

All Sites 578, 586, 746, 753
Angiosarcoma 738
Bladder 642, 738
Colon 526
Inapplicable 77, 198, 199, 201, 390, 554, 830, *853, 1009, *1020
Leukaemia 977
Liver 738
Lung 135, 613, 642, 738, *1051
Lymphoma 977
Testis 391
Thyroid 365
Uterus (Cervix) *482

Maghrebians

Nasopharynx 314

Male

All Sites 333
Lung 172, 566

Mammography

All Sites 1197
Benign Tumours 1059
Breast (F) 11, 23, 114, 235, 276, 309, *461, 507, 630, 791, 797, 836, *919, 941, 942, 1031, 1059, 1132, 1197
Colon 1197

Lung 1197
Uterus (Cervix) 1197

Mapping

All Sites 90, *387, 688, 689, 713, 820, 1237
Childhood Neoplasms 823, 1113
Leukaemia *350, 823
Lymphoma *350
Non–Hodgkin's Lymphoma 823
Oesophagus 694

Marijuana

Childhood Neoplasms 647
Leukaemia (ALL) 647
Leukaemia (ANLL) 1109
Rhabdomyosarcoma *1155

Marital Status

All Sites 601, 870
Colon 601
Liver 601
Lung 601
Rectum 601
Stomach 601

Mate

All Sites 1238
Stomach 1233

Mathematical Models

All Sites 14, *686, 809, 820, 970, *1052
Bladder 513
Breast (F) 630, 970
Kidney 513
Larynx 513
Leukaemia 970
Lung *43, 513, *686, 970
Mesothelioma *686
Oesophagus 513
Oral Cavity 513
Pancreas 513
Pharynx 513
Thyroid 970
Uterus (Cervix) 115, 631, 1185

Menarche

All Sites 660
Breast (F) 60, 170, 299, 400, 408, 433, 660
Childhood Neoplasms 1225
Gastrointestinal 660
Ovary 404, 406, 660
Uterus (Cervix) 660
Uterus (Corpus) 660
Vagina 1102

Menopause

All Sites 660, 991
Benign Tumours *1202

Breast (F)	60, 131, 170, 230, 400, 433, 660, 803, 925, 980, 991, *1119, *1202
Childhood Neoplasms	1225
Colon	980, *1119
Female Genital	803
Gastrointestinal	230, 660
Leukaemia	230
Lung	980
Melanoma	230
Ovary	404, 660, 980
Rectum	*1119
Testis	230
Thyroid	*1119
Uterus (Cervix)	230, 660
Uterus (Corpus)	230, 660, 925, 980, *1119

Menstruation

Benign Tumours	*1202
Breast (F)	25, 400, 442, 620, 1072, *1202
Colon	167
Oesophagus	167
Ovary	404
Pancreas	167
Rectum	167
Stomach	592

Metabolism

Angiosarcoma	738
Bladder	738
Breast (F)	*1063
Liver	738, 956
Lung	738
Ovary	1198
Skin	1220
Testis	*206

Metals

All Sites	90, 117, 241, 263, 271, *272, *485, *555, 573, *669, *715, 750, *776, 863, 872, 991, 992, 1015, 1019, 1057, 1071, 1160, 1161, 1171, 1227
Bladder	73, 213, 487, 965
Bone	102
Brain	213, *272, *362, *715, 965, 1091, 1104, 1151
Breast (F)	74, 241, 640, 797, 991, 1222
Childhood Neoplasms	222, 1147, 1151
Colon	73, 74, 241, *323, 640, 1032, 1057, 1093, 1161
Gastrointestinal	192, 193, 213, 1091
Genitourinary	1160
Haemopoietic	213, 393, 965, 1091, 1160
Inapplicable	77, *355
Kidney	73, *272, 487, 506, *715, 790, 873, 1093, 1104
Larynx	59, *362
Leukaemia	*106, 192, 393, 487
Liver	234, 487, 556, 1091, 1093, 1104, 1161
Lung	3, 73, *106, *158, 159, *183, 192, 193, 241, 263, *265, 271, *272, *320, *322, *323, *325, 326, *362, 393, *485, 487, *555, 556, 640, *670, *715, 721, *768, *785, 790, 844, 872, 873, 945, 947, 965, 968, 1024, 1091, 1093, 1104, 1227
Lymphoma	73, *362, 506, 965, 1024, 1160
Male Genital	213
Melanoma	73, 1221
Mesothelioma	59, 965
Nasal Cavity	1161
Nasopharynx	59
Oesophagus	73, *362

Oropharynx *362
Pancreas 73
Prostate 73, 74, 241, 629, 872, 873, 918
Rectum 73, 241, 640, 1032, 1161
Respiratory 213, 1160, 1161
Sarcoma 213, 965
Skin 213, 1093, 1221
Stomach 73, *106, *183, 241, 271, 326, *362, 640, 965, 1233
Tongue *362
Tonsil *362
Uterus (Cervix) 241

Micronuclei
All Sites 270, 578, 753, 1131
Angiosarcoma 738
Bladder *120, 642, 738
Hodgkin's Disease 270
Inapplicable 201, 1009, *1020
Leukaemia 1026
Liver 738
Lung 135, 642, 738, 1131
Oesophagus 361
Oral Cavity *430

Micronutrients
All Sites *256
Breast (F) 974
Oral Cavity *430
Stomach 137, 914
Uterus (Cervix) 994
Uterus (Corpus) 972

Migrants
All Sites 289, *453, 498, 503, *780, 876
Bone *292
Brain *462
Breast (F) 21, *291, *462
Childhood Neoplasms 922
Colon 21, *291
Hodgkin's Disease *292
Leukaemia *292, *462, 922
Liver *460
Lung 456, *462, 678
Melanoma *291, *292
Myelodysplastic Syndrome 922
Nasopharynx 314
Nervous System *292
Neuroblastoma *292
Non-Hodgkin's Lymphoma *292
Prostate 21, *291
Rectum 21
Sarcoma *292
Soft Tissue *292
Stomach *291, *975
Thyroid *453, *462

Minerals
All Sites 489, 856, 1057
Breast (F) 127
Colon 127, 373, 1057
Leukaemia 127
Liver 127

Lung 127, 489, 857
Nasopharynx 127
Oesophagus 127
Rectum 127, 373
Respiratory 856
Stomach 127

Mining

All Sites *29, 42, 90, 263, 483
Colon 34
Hodgkin's Disease 40
Inapplicable 48, 198
Leukaemia *106
Leukaemia (CLL) 40
Lung 3, 34, 36, *106, 129, *158, 263, 326, 483, 579, 721, 798, *812, 947, 1024
Lymphoma 1024
Mesothelioma 34, 42, 579
Non–Hodgkin's Lymphoma 40
Peritoneum 34, 42
Pleura 42
Stomach 34, *106, 326

Monitoring

All Sites 62, 67, 519, *520, *838, 839, 1013, 1123, 1137
Breast (F) 174
Childhood Neoplasms 1165
Inapplicable 249, 626
Leukaemia 62, 1165
Liver 174, 956, 1165
Lung 174
Lymphoma 62
Melanoma *269
Non–Hodgkin's Lymphoma 1165
Pancreas 1165
Rectum 174
Skin *269
Soft Tissue 1165
Stomach 174
Uterus (Cervix) 174

Multiple Primary

All Sites 223, 248, 270, *319, *336, *338, 339, 466, 603, 619, 702, 720, 805, 940, 970, *1110, *1117, *1175
Bone 805
Breast (F) 146, 223, 224, *302, 463, 464, 619, 620, 805, 970
Childhood Neoplasms 375, 805, 827, *1110, 1150
Colon 146, 463
Gastrointestinal 340, 805
Head and Neck 463
Hodgkin's Disease 270, 540
Larynx 70, 231, 364, *1117
Leukaemia 224, 339, 340, 463, 619, 672, 805, 827, 871, 915, 970
Leukaemia (ALL) 540
Leukaemia (ANLL) 540
Leukaemia (CLL) 224
Lip 70
Liver 146
Lung 146, 223, 231, 339, 340, 619, 970
Lymphoma 463, 805
Melanoma 720
Myelodysplastic Syndrome 619
Neuroblastoma 540
Non–Hodgkin's Lymphoma 339, 340, 540, 619

Oral Cavity 70, 1089, *1117
Ovary 805, *1175
Pharynx 70, 1089, *1117
Rectum 146
Skin 528, 720
Stomach 146
Testis 805, 940
Thyroid 318, *319, 463, 970
Uterus (Cervix) 223, *434
Uterus (Corpus) *302
Wilms' Tumour 540

Mutagen
All Sites 759
Bladder 99
Childhood Neoplasms 831
Gastrointestinal 193
Inapplicable 554, 1001, 1003
Liver 182
Lung 193
Stomach 182
Uterus (Cervix) 1185

Mutagen Tests
All Sites 1123
Inapplicable 626

Mutation Rate
All Sites 132

Mutation, Germinal
All Sites 132
Childhood Neoplasms 904, 916
Inapplicable 58, 390, 1001
Leukaemia 937
Retinoblastoma 904, 1148

Mutation, Somatic
All Sites 62, 759
Breast (F) 558
Childhood Neoplasms 827, 1150
Colon 558
Inapplicable *4, 58, 390, 1001
Leukaemia 62, 827
Lymphoma 62
Oral Cavity 402
Rectum 558
Stomach 558

Mycotoxins
Inapplicable 1001
Liver 164, 191, *293, 726, 811, 956
Stomach 912

Naevi
Benign Tumours 31
Melanoma 31, 346, 495, 628, 666, 911, 1176
Skin *1038

Nutrition

Site	References
All Sites	61, 89, 145, 225, 295, 372, 477, 564, 806, 1168, 1197
Bladder	310
Breast (F)	89, 127, 446, 480, *609, 612, 639, 711, 797, 848, 974, 1130, *1195, 1197, 1218, 1222
Colon	*53, 89, 127, 166, 366, 480, *675, 806, 893, *976, 1058, 1197, 1210, 1212, 1228
Gastrointestinal	1241
Genitourinary	1241
Hypopharynx	*317
Inapplicable	*817, 1049
Larynx	*317
Leukaemia	127
Liver	127
Lung	97, 127, *317, 480, 806, 1197, 1232
Nasal Cavity	*317
Nasopharynx	127
Non–Hodgkin's Lymphoma	1193
Oesophagus	127, 733, *1103
Oral Cavity	*420
Ovary	*1120
Pancreas	733
Pleura	*317
Prostate	*1002
Rectum	*53, 127, 166, 366, *675, 893, 1058, 1212
Skin	1055
Stomach	127, 137, *610, 725, 733, 806, *975, *1103
Uterus (Cervix)	611, 994, *1048, 1197
Uterus (Corpus)	972, 1047
Vagina	1102

Obesity

Site	References
All Sites	89, *989, 1105
Breast (F)	*46, *79, 89, 423, 637, 644, 949, 1105, *1112
Colon	89, 949, *1010
Female Genital	949
Gastrointestinal	1241
Genitourinary	1241
Kidney	45, 368, 795
Ovary	949
Rectum	*1010
Uterus (Cervix)	949
Uterus (Corpus)	810, 949, 1047, 1105

Occupation

Site	References
All Sites	19, *29, 33, 42, 52, 56, 62, 67, 68, 80, 81, 86, 87, 88, 90, 100, *103, *107, 116, 117, 126, 128, 185, *202, 229, 245, 247, *250, *262, 263, 266, 271, *272, *277, 281, 282, 296, 298, 300, 305, 308, 312, *327, *332, 333, 335, *357, 360, 371, *396, *412, *422, 451, 452, *453, 458, 483, 484, *485, 500, 501, 502, 504, 509, 511, *520, 522, 532, *535, 537, 545, *555, 564, 572, 573, 583, *621, 622, 623, 648, 654, *661, *669, 680, 682, 684, *686, *687, *693, 695, *696, *697, *698, *699, *700, *701, 708, 709, 714, *715, 749, 750, 751, 752, 754, 755, 757, 763, 764, 767, 772, 779, *781, 792, 799, 819, 825, 837, *838, 839, 856, 863, 872, 875, 881, 887, 896, 897, 928, *935, 939, 953, *967, 981, 982, 990, 991, 1013, 1015, 1018, 1019, *1021, 1027, 1071, 1075, 1076, 1094, 1095, 1107, 1123, 1138, 1146, 1157, 1158, 1159, 1160, 1161, 1163, 1167, 1171, *1172, 1173, 1178, 1180, 1181, 1188, 1197, 1227, 1230, 1231, *1235
Angiosarcoma	738
Bile Duct	125, *734
Bladder	73, 99, 162, 178, *204, 213, 279, 305, 310, *352, 484, 487, 509, *687, 695, 738, 900, 910, 965, 985, 1016, 1025, 1027, 1035, 1096, 1157
Bone	171, *621

TERM

Brain	38, 86, 96, 171, *203, 213, 251, *272, 304, *362, 504, 532, 583, *621, *715, 965, 1012, *1021, 1087, 1091, 1095, 1104, 1151, 1180, 1181
Breast (F)	121, *203, *357, *396, 680, 991, 1000, *1006, 1107, *1195, 1197
Breast (M)	*1004
Childhood Neoplasms	222, 457, 827, 890, *934, 1012, 1147, 1151
Colon	34, 73, 166, 167, *203, 258, *323, *675, 747, 748, *758, 860, 893, *976, *1010, 1032, 1035, 1092, 1093, 1161, 1181, 1197, 1210, 1212
Eye	1035
Gallbladder	125, 258, *734
Gastrointestinal	76, 101, 192, 193, 213, *649, 691, *692, 1034, 1091
Genitourinary	1160
Haemopoietic	19, 213, 504, 677, *692, *727, 744, 965, *984, 1091, 1160
Hodgkin's Disease	40, 93, 251, *311, 542
Hypopharynx	268, 315, *317
Inapplicable	*4, 48, 77, 198, 199, 201, *233, 249, 354, *355, 626, 681, 966, 1009, *1020
Kidney	45, 73, 76, *219, *272, 330, 368, 487, 506, *715, 790, 795, 873, 1037, 1093, 1096, 1104, 1189
Larynx	59, 76, 181, *204, 315, *317, *362, 364, 511, 664, 784, 931, 1095
Leukaemia	33, 56, 62, 86, *106, 119, 136, 171, 180, 192, 208, 251, *262, *277, 334, 349, 367, 487, 539, 542, *621, 653, 655, *687, 752, 779, *781, 827, 849, *864, *934, *936, 950, *984, 985, 1026, 1107, 1179, 1180, 1181, 1231
Leukaemia (AML)	280, *727
Leukaemia (CLL)	40, *727
Liver	128, 138, 171, 234, 258, 487, 556, 664, *690, *698, *700, 738, *983, 1028, 1035, 1091, 1092, 1093, 1104, 1161
Lung	3, 34, 35, 36, *43, *50, 52, 73, 76, 87, 94, *106, 126, 129, 152, *158, 159, 177, *183, 192, 193, *204, 217, 239, 243, 251, 263, *265, 271, *272, 279, 297, 298, *317, *320, *322, *323, *325, 326, 328, 331, 343, 351, 353, *362, *379, 386, 481, 483, 484, *485, 487, 497, 511, *555, 556, 566, 579, 596, 613, 623, *649, *667, *668, *670, 678, 680, *686, *687, 691, *692, 695, *698, *700, 710, *715, 721, 738, 744, 748, 756, *768, 772, *773, *774, 782, *785, 788, 790, 792, 798, *812, 844, 857, 858, 860, 861, 872, 873, 910, 932, 939, 945, 947, 960, 963, 965, 968, 982, 985, 1022, 1024, 1034, 1035, 1061, 1091, 1092, 1093, 1104, 1107, 1135, 1153, 1162, 1173, 1180, 1197, *1209, 1224, *1226, 1227, 1234
Lymphoma	13, 19, 56, 62, 73, 171, 247, 349, *362, 484, 506, *621, 654, 744, 752, 757, 779, *781, *934, *936, 965, *984, 1024, 1028, 1160, 1179
Male Genital	213
Melanoma	20, 73, 100, 346, 628, 874, *1004, 1035, 1069
Mesothelioma	30, 34, 35, 42, 52, 59, 101, *324, 353, 530, *536, 579, *686, 731, 748, 851, 965, 1035
Multiple Myeloma	86, 93, 96, 542, 757, *967, 1036, 1180
Myelodysplastic Syndrome	17, 267, 539, 835
Nasal Cavity	181, 208, 278, 305, 306, *316, *317, 509, 874, 1084, 1161
Nasopharynx	59, 179, *184, 1084, 1159
Nervous System	*203
Non–Hodgkin's Lymphoma	40, 86, 93, 96, *311, 532, 542, 622, *727, *761, *967, 1034, 1180, 1186
Oesophagus	73, 167, 258, *362, 664, 784, 910, *969, 1035, 1036, *1118
Oral Cavity	181, 268, 664, 784
Oropharynx	268, *362
Ovary	108, 1000, 1133
Pancreas	73, 167, 244, 258, 484, *621, *734, 1000, 1036, 1179
Peritoneum	34, 42, 76, 530
Pharynx	181, 509, 664, 784
Pleura	42, 76, *317, 530
Prostate	73, 86, *203, 228, 251, 257, 623, 629, 793, 816, 872, 873, 895, 943, 1000, *1004, 1036
Rectum	73, 166, 167, *203, 258, *675, 747, 748, *758, 860, 893, *1010, 1032, 1035, 1161, 1181, 1212
Respiratory	101, 208, 213, *324, 488, 509, 743, 856, *864, 1156, 1160, 1161, 1219
Salivary Gland	1035
Sarcoma	13, 93, 96, 213, 247, 622, 965
Skin	20, 213, 504, 959, 961, *1038, 1093
Soft Tissue	33, 93, 96, 532, 538, 622, 654, *984, 1028, 1068, 1180

Stomach 34, 73, *106, 138, *183, 258, 271, 326, *362, 552, 592, 680, 695, 735, 743, 744, 748, 756, 794, 860, 861, 910, 965, *969, *975, 982, 1035, *1118, 1177
Testis 108, 283, 907
Thyroid *453, 1107
Tongue *362
Tonsil *362
Ureter *352
Urethra *352
Urinary Tract 828
Uterus (Cervix) 680, *692, 1197
Uterus (Corpus) 1000, 1047
Wilms' Tumour 71, 1007

Oncogenes
All Sites 270, 551
Benign Tumours *1202
Bladder *543, 1025
Breast (F) 558, 620, *1006, 1116, *1202
Colon 558
Hodgkin's Disease 270, *676
Inapplicable *355
Leukaemia (ALL) *676
Liver 1149
Lung 239
Neuroblastoma 1008
Non–Hodgkin's Lymphoma *676
Oesophagus 133
Rectum 558
Stomach 558

Oophorectomy
Breast (F) 804
Ovary 804

Operation, Previous
Colon *1014
Rectum *1014
Stomach *1014

Oral Contraceptives
All Sites 512, 888, 924, 991
Benign Tumours 517, 616, 617, 1059
Breast (F) *26, *46, 400, *461, 517, 565, 616, 617, 618, 620, 650, 722, 879, 888, 924, 974, 991, 1059, 1072, *1119, 1204
Colon *1119
Female Genital 888
Gallbladder 673
Liver 729
Ovary 517, 1243
Rectum *1119
Thyroid *1119
Uterus (Cervix) 113, 169, 215, 517, 611, 813, 867, 920, 924, 1185
Uterus (Corpus) 517, *1119, 1190
Vulva 517

Paints
All Sites *967, 1015
Childhood Neoplasms 222, 1147
Hodgkin's Disease *311
Leukaemia 653

Multiple Myeloma *967
Non–Hodgkin's Lymphoma *311, *967
Prostate 629

Parasitic Disease
Bile Duct 587
Colon 147
Liver 234, 587
Rectum 147

Parental Occupation
Brain 57, *205, *665, 955
Childhood Neoplasms 376, 647, *822, 824, *934, 1147
Hodgkin's Disease 1154
Leukaemia 57, *205, 376, *380, *665, *822, 859, *934, *936
Leukaemia (ALL) 75, 647
Lymphoma 57, *205, 376, *934, *936
Neuroblastoma 57, 376, 1008
Non–Hodgkin's Lymphoma 859
Retinoblastoma 1148
Wilms' Tumour 376

Parity
All Sites *396, 660, 924
Benign Tumours 517
Breast (F) 60, 170, 230, *396, 408, 433, 442, 517, 637, 659, 660, 924
Choriocarcinoma 516
Gastrointestinal 230, 660
Hydatidiform Mole 516
Leukaemia 230
Lung 963
Melanoma 230
Ovary 404, 406, 517, 660, 1243
Testis 230
Uterus (Cervix) 230, 443, 517, 547, 660, 924
Uterus (Corpus) 230, 517, 660
Vulva 517

Passive Smoking
All Sites 91, 884, 953, 1039, 1131
Bladder 642
Breast (F) 121, *808, 1040
Childhood Neoplasms 647, 831
Hypopharynx 315
Inapplicable 1003
Kidney 45, 795
Larynx 315
Leukaemia (ALL) 75, 647
Lung 97, 123, 152, *173, 297, 331, *379, 386, 642, 710, 782, 1121, 1131, 1135, 1162
Nasopharynx 179
Uterus (Cervix) 1185

PCR
Bile Duct *514
Brain *514
Breast (F) *514
Hodgkin's Disease *676, 1154
Inapplicable *355, 1001
Leukaemia 977
Leukaemia (ALL) *676

Liver	*514
Lung	932, *1051
Lymphoma	977, *1062, 1073
Non–Hodgkin's Lymphoma	*514, *676
Respiratory	*514
Retinoblastoma	*706
Sarcoma	*514
Skin	*514, *878
Soft Tissue	*514
Testis	*514
Uterus (Cervix)	55, 274, 369, 608, 614, *615, 867, *1207

Pedigree

All Sites	702, 805
Bone	805
Breast (F)	*26, *27, 146, 805, 962
Childhood Neoplasms	805
Colon	*27, 146, 366, 954
Gastrointestinal	805, 954
Leukaemia	805, 937
Liver	146
Lung	146
Lymphoma	805
Nasopharynx	151
Ovary	534, 805
Rectum	*27, 146, 366, 954
Stomach	146
Testis	805

Pepsinogen

Stomach	*585, 602, 913, 914, *975

Pesticides

All Sites	33, 86, 90, 247, 296, 371, 452, 504, 532, 537, 622, 654, *699, 755, 763, 928, 1019, *1021, 1171, 1180, *1235
Bladder	162, 965
Brain	38, 86, 96, 504, 532, *665, 965, *1021, 1104, 1180
Breast (F)	*1006, *1043, *1144
Childhood Neoplasms	1147, 1165
Haemopoietic	504, *727, 965
Hodgkin's Disease	93, *311, 542
Inapplicable	966
Kidney	*219, 368, 506, 1104
Leukaemia	33, 86, 334, 349, *380, 542, 653, *665, 977, 1165, 1180
Leukaemia (ALL)	75, 1108
Leukaemia (AML)	280, *727
Leukaemia (ANLL)	1109
Leukaemia (CLL)	*727
Liver	*983, 1104, 1165
Lung	965, 1104, 1180
Lymphoma	13, 247, 349, 506, 654, 965, 977
Mesothelioma	965
Multiple Myeloma	86, 93, 96, 542, 1180
Non–Hodgkin's Lymphoma	86, 93, 96, *311, 532, 542, 622, *727, 1165, 1180
Pancreas	1165
Prostate	86, 629, 943
Sarcoma	13, 93, 96, 247, 622, 965
Skin	504
Soft Tissue	33, 93, 96, 532, 538, 622, 654, 1068, 1165, 1180
Stomach	965, 1177

Petroleum Products

All Sites	19, 86, 90, *661, 869, 1015, 1018, 1188
Bladder	73, 213, 965
Brain	86, 213, 965, 1091
Colon	73, 860
Gastrointestinal	213, 1034, 1091
Haemopoietic	19, 213, 965, 1091
Inapplicable	198
Kidney	73, 1037, 1189
Larynx	869
Leukaemia	86, 869
Leukaemia (ANLL)	1109
Liver	869, 1091
Lung	73, *183, *428, 860, 861, 869, 965, 1034, 1091
Lymphoma	19, 73, 869, 965
Male Genital	213
Melanoma	73
Mesothelioma	965
Multiple Myeloma	86
Non-Hodgkin's Lymphoma	86, 1034
Oesophagus	73
Pancreas	73
Prostate	73, 86
Rectum	73, 860
Respiratory	213
Sarcoma	213, 965
Skin	213
Stomach	73, *183, 860, 861, 965

Photochemotherapy

All Sites	*475, 993
Eye	993
Skin	993

Physical Activity

All Sites	806, *989, 991, 992, 1171
Breast (F)	*46, *478, 507, 908, *948, 949, 980, 991, 1072, 1213
Colon	149, 166, 167, 470, *478, *594, 806, 893, 908, 949, *976, 980, 1058, 1183, 1210, 1213
Female Genital	949
Kidney	368
Larynx	*478
Lung	806, 980, 1213
Oesophagus	167, *478
Oral Cavity	*478
Ovary	*478, 949, 980
Pancreas	167, 1213
Prostate	908, 1213, 1214
Rectum	149, 166, 167, *478, *594, 893, 1058, 1213
Stomach	806, 908
Thyroid	*1245
Uterus (Cervix)	949, *1048
Uterus (Corpus)	810, 949, 980, 1047

Physical Factors

All Sites	225, 601, 660, 680, 970, 1105
Breast (F)	25, 60, 63, *214, 230, 408, 507, 565, 660, 680, 786, 848, 949, 970, 1105, 1213
Colon	63, 601, 949, 1213
Female Genital	949
Gastrointestinal	230, 660, 1241
Genitourinary	1241
Hodgkin's Disease	*1244

Leukaemia 230, 970
Liver 601
Lung 63, 326, 601, 680, 970, 1213
Melanoma 230
Oesophagus 63
Ovary 660, 786, 949
Pancreas 1213
Prostate 1213, 1214
Rectum 601, 1213
Stomach 63, 326, 601, 680
Testis 230, 283
Thyroid 381, 970
Uterus (Cervix) 230, 660, 680, 786, 949
Uterus (Corpus) 230, 660, 786, 949, 1105

Pigmentation

All Sites 100
Benign Tumours 31
Melanoma 31, 100, 495, *1030
Skin 528, 833, 959, *1030, *1038

Plants

All Sites 575
Nasopharynx *184
Oesophagus *189

Plasma

All Sites 740, 1044
Breast (F) 21
Colon 21
Hodgkin's Disease 1154
Lung *1051
Oesophagus 361
Prostate 21
Rectum 21

Plastics

All Sites 90, 128, 281, 305, 333, 623, 751, 754, 755, 757, 779, 825, 869, 990, 1019, 1158, 1159, 1171, 1173, 1230
Angiosarcoma 738
Bladder 73, 305, 738, 965
Brain 38, *362, 965, 1091, 1104
Childhood Neoplasms 222
Colon 73, 1032, 1092
Gastrointestinal 691, 1091
Haemopoietic 965, *984, 1091
Inapplicable 201, 249, 1009
Kidney 73, 506, 1104
Larynx *362, 869
Leukaemia 779, 869, *984
Liver 128, 738, 869, 1091, 1092, 1104
Lung 73, *362, 623, 691, 738, 869, 965, 1091, 1092, 1104, 1173, *1209, *1226
Lymphoma 73, *362, 506, 757, 779, 869, 965, *984
Melanoma 73
Mesothelioma 965
Multiple Myeloma 757
Nasal Cavity 305, 306, *316
Nasopharynx 1159
Oesophagus 73, *362
Oropharynx *362
Pancreas 73
Prostate 73, 623

Rectum	73, 1032
Sarcoma	965
Soft Tissue	*984
Stomach	73, *362, 965
Tongue	*362
Tonsil	*362

Pneumoconiosis

All Sites	685
Lung	129, 328, 685

Polyps

Colon	366, 455, *594, *675, 885, 894, *1010, 1045, 1058, 1228
Gastrointestinal	880
Rectum	366, 455, *594, *675, 885, 894, *1010, 1058

Pregnancy

All Sites	*111, 924
Breast (F)	400, 924
Childhood Neoplasms	*111, 783, 827, 916
Choriocarcinoma	6
Hydatidiform Mole	6
Leukaemia	827
Testis	283
Uterus (Cervix)	*627, 924

Premalignant Lesion

All Sites	512, 718, 1046, 1126
Benign Tumours	31, 1115, 1140
Bladder	866
Breast (F)	309, 724, 1046, 1115, 1140
Choriocarcinoma	6
Colon	5, 260, 366, 373, *594, 894, 954, 1058, 1128
Gastrointestinal	880, 954
Head and Neck	389
Hydatidiform Mole	6
Inapplicable	554
Leukaemia	1046
Liver	491
Lung	135, 388, 1224, 1232
Melanoma	20, 31, 495
Multiple Myeloma	1036
Myelodysplastic Syndrome	17, 267
Nasopharynx	144
Oesophagus	187, 188, 361, 1036
Oral Cavity	363, 389, 414, 419, *420, 429, *430, 444, 739
Oropharynx	363
Pancreas	1036
Prostate	92, 1036
Rectum	5, 8, 366, 373, *594, 894, 954, 1058, 1128
Respiratory	363, 1219
Skin	20, 833, 961, 1184
Stomach	137, 237, 285, 531, 571, 725, *736, 846, 912, 913
Uterus (Cervix)	113, 200, 363, 369, 437, 438, *482, 547, 614, 728, 946, 994, 1211

Prevalence

All Sites	524, 1065, 1197
Breast (F)	962, 1065, 1197
Colon	*594, 1065, 1197
Haemopoietic	716
Inapplicable	198

Lung *459, 1065, 1197
Melanoma 20
Oesophagus 361
Prostate 469, 1065
Rectum *594
Skin 20
Thyroid 662
Uterus (Cervix) 55, 1197

Prevention

All Sites 248, *256, 702, 718, 719, 953, 986, *987, 995, *1077, 1197, 1240
Benign Tumours 253, *494
Breast (F) *78, 236, 253, 309, 518, 544, *567, *987, 1031, *1191, 1197
Colon 373, *567, 986, *987, 1045, 1145, 1197, 1210
Endocrine Glands *494
Female Genital 252
Gastrointestinal 880
Inapplicable 58, 198, 199
Liver 345
Lung 35, 177, *567, 705, 986, *987, 1197, 1200, 1224
Male Genital 252
Melanoma 1122
Mesothelioma 35, *536, 1200
Oesophagus 130, 187
Oral Cavity *416, 419, *421, 429
Ovary *78, 634, *1191
Rectum 8, 373, *567, 1145
Retinoblastoma *706, 877
Skin 1216, 1217, 1220
Stomach 285, *567, 584
Uterus (Cervix) *294, *567, 951, 1197
Uterus (Corpus) *1191

Prognosis

All Sites 719, 981
Benign Tumours 253
Bile Duct *734
Bone 171
Brain 171
Breast (F) *27, 253
Childhood Neoplasms 1042
Choriocarcinoma 6
Colon *27
Gallbladder *734
Hydatidiform Mole 6
Leukaemia 171
Liver 171
Lymphoma 171
Melanoma 874
Nasal Cavity 874
Nasopharynx 606
Pancreas *734
Rectum *27
Stomach 592
Thyroid 318

Projection

All Sites 719
Lung *197

Promotion

All Sites 1168

Uterus (Cervix) *482

Protein

Colon 893
Non–Hodgkin's Lymphoma 1193
Rectum 893

Protein Binding

All Sites *781, 1131
Leukaemia *781
Lung 1131
Lymphoma *781

Psychological Factors

All Sites 140, 378, 564, 586, 593, 875
Breast (F) 140
Stomach 140

Psychosocial Factors

All Sites 141, 854, 875
Breast (F) 854, 855
Childhood Neoplasms 1225
Colon 166
Haemopoietic 358
Lung 854
Rectum 166
Stomach 854

Race

All Sites *780
Bladder 985
Breast (F) 299, 644, *1006
Kidney 1141
Leukaemia 985
Lung 985
Melanoma *1030
Nasopharynx 607
Prostate 1033, 1129
Skin *1030

Radiation, Ionizing

All Sites 56, 62, 81, 87, 90, 112, 132, 185, 210, 221, *262, 263, *277, *336, *338, 339, *341, 360, *412, *422, 440, *453, 458, 471, 472, 550, 551, 562, 563, 583, 764, *775, *776, 777, 819, 826, 837, *838, 839, 840, 881, 896, 897, 901, 902, *935, 970, 1013, 1071, 1107, 1163, 1167
Benign Tumours 472
Bone 171, 370, 377
Brain 44, 171, 303, *462, 583
Breast (F) *462, 970, 1072, 1107
Childhood Neoplasms 221, 375, 376, 600, 647, 783, *822, 823, 827, 890, 891, 905, *906, *934
Colon 166
Gastrointestinal 192, 193
Haemopoietic 393, 716
Hodgkin's Disease *311, *1125, *1244
Inapplicable *4, 77, 830
Kidney 377
Larynx 181

Leukaemia 56, 62, 119, 171, 180, 192, 221, *262, *277, 287, 334, 339, 349, *350, 370, 376, *380, *385, 393, *462, 600, *704, *775, 777, *822, 823, 827, 859, 901, 902, 905, *934, *936, 970, 1064, 1107
Leukaemia (ALL) 647, 1108
Leukaemia (AML) 1086
Leukaemia (CLL) *1125
Liver 138, 171, 221, 370, 377
Lung 3, 87, 95, 139, 159, 192, 193, 263, *264, 339, 370, *379, 393, *462, *670, 721, 782, 844, 903, 944, 947, 970, 1024, 1107, 1135, 1182
Lymphoma 56, 62, 171, 287, 349, *350, 370, 376, *934, *936, 1024
Melanoma 1069
Myelodysplastic Syndrome 835
Nasal Cavity 181, 1084
Nasopharynx 1084
Neuroblastoma 376
Non–Hodgkin's Lymphoma *311, 339, 823, 859, *1125
Oral Cavity 181
Pharynx 181
Prostate 895, 943
Rectum 166
Retinoblastoma 561
Rhabdomyosarcoma *1155
Spleen 370
Stomach 138
Testis 815, 907
Thyroid 381, *453, *462, *704, *775, 796, *801, 970, 1107, 1194
Wilms' Tumour 376

Radiation, Non–Ionizing

All Sites 970, 993, 1167
Breast (F) 970, *1195
Breast (M) *1004
Eye 993
Leukaemia 970
Lung 970
Melanoma *1004
Prostate *1004
Skin 993
Thyroid 970

Radiation, Ultraviolet

All Sites 458, *475, 805
Benign Tumours 31
Bone 805
Breast (F) 805
Childhood Neoplasms 805
Gastrointestinal 805
Leukaemia 805
Lip 98, 1085
Lymphoma 805
Melanoma 7, 16, 20, 31, 39, *269, 346, 495, 628, *1030, 1069, 1176
Ovary 805
Skin 20, 32, 39, *269, 528, 833, *878, 959, 961, *1030, *1038, 1184
Testis 805

Radiotherapy

All Sites *111, 223, *319, 619, 777, 940, 970
Breast (F) 223, 224, 619, 620, 970, 1072
Childhood Neoplasms *111, 827, 1150, 1225
Gastrointestinal 340
Head and Neck 632
Leukaemia 224, 340, 619, 672, 777, 827, 871, 915, 970
Leukaemia (CLL) 224

Lung 223, 340, 619, 970
Myelodysplastic Syndrome 619
Non–Hodgkin's Lymphoma 340, 619
Skin *337
Testis 940
Thyroid *319, *337, 970
Uterus (Cervix) 223, *434

Record Linkage

All Sites 14, 68, 83, 86, 90, 91, *107, *109, 112, 122, 240, 245, 508, 1041, 1137
Bladder 104, *204, 232
Brain 86, *203
Breast (F) 105, *203, 232, 401, 803, 973, *1119
Childhood Neoplasms 904, 905, *906
Colon 105, *203, 800, 973, *1119
Female Genital 803
Gastrointestinal 192
Kaposi's Sarcoma 1041
Kidney 105
Larynx *204
Leukaemia 86, *106, 192, 508, 905
Lung 104, *106, 192, *204, 232, 973
Lymphoma 508
Multiple Myeloma 86
Nervous System *203
Non–Hodgkin's Lymphoma 86
Ovary 105, 973
Pancreas 104, 105
Prostate 86, *203, 232
Rectum 105, *203, 800, *1119
Retinoblastoma 904
Sarcoma 83
Soft Tissue 508
Stomach *106
Thyroid *1119
Uterus (Cervix) 232
Uterus (Corpus) 105, 973, *1119

Recurrence

Breast (F) 518, *1196
Leukaemia 672

Registry

All Sites 1, 14, *29, 33, 56, 67, 68, 69, 81, 82, 89, 90, *109, *111, 116, 122, 176, 196, *202, 210, 220, 221, 223, 225, 227, 240, 241, 245, 246, 247, 248, *250, 270, 271, *272, *277, 281, 282, 296, 298, 300, 308, 344, *387, *396, *397, *415, *418, 426, 448, 449, 450, *453, 465, 466, 472, *475, 498, 503, 532, 533, 562, 564, 593, 603, 622, 623, 625, 643, 648, 654, 663, *669, 680, 720, 749, 750, 751, 752, 754, 757, 767, 771, *775, *776, 777, *780, *781, 787, 805, 837, 869, 884, 889, 897, 899, 938, 939, 940, *967, 1041, *1052, 1070, 1105, *1117, 1166, 1170, 1188, 1201, *1235, 1237, 1238, 1239, 1240
Benign Tumours 253, 472
Bladder 104, *110, *204, 213, 232, 546, 787, 899, 965, 1035
Bone 171, *292, 805, *845
Brain 22, 44, 57, 171, *203, *205, 213, 251, *272, *362, 532, *665, 766, 955, 965, 979, 1087, 1088
Breast (F) 23, 25, *26, *27, *46, 60, 89, 121, 170, *195, *203, *214, 223, 224, 232, 235, 236, 241, 242, 253, *255, 276, *291, 395, *396, 399, 400, 401, 417, 433, 463, 464, *478, 507, 533, 546, 620, 624, 637, 640, 651, 680, 722, 724, 770, 786, 791, 802, 803, 804, 805, 855, 879, 889, 923, 925, *948, 973, *1006, *1063, *1067, 1072, 1083, 1105, *1119, *1191, *1195, *1196, 1218, *1223

Breast (M)	211
Childhood Neoplasms	*111, 221, 222, 307, 321, 375, 376, 590, 600, 646, 647, 766, 769, 783, 805, 827, *852, 890, 891, 904, 905, *906, 922, 1165
Choriocarcinoma	*549
Colon	24, *27, 28, 89, *110, 166, 167, *203, 216, 241, 242, 261, *291, 463, 465, *478, 521, *527, 533, 546, 624, 640, 641, 748, *758, 787, 800, *818, 860, *862, 954, 973, 1035, 1098, *1119, 1183, *1199, 1210, *1223
Eye	1035
Female Genital	803
Gastrointestinal	213, *649, 805, 923, 927, 954
Haemopoietic	213, 744, 965, *984
Head and Neck	463
Hodgkin's Disease	40, 251, 270, *292, *311
Hydatidiform Mole	*549
Hypopharynx	*317, 425
Inapplicable	58
Kaposi's Sarcoma	1041
Kidney	45, *219, *272, 506, 790, 795, 1141
Larynx	59, 181, *204, 231, *317, *362, 425, *478, 664, 784, 807, 869, *1117
Leukaemia	33, 56, 57, 171, 180, *205, 208, 221, 224, 251, *277, 287, *292, 307, 367, 376, 463, 600, 633, 655, *665, 752, 766, *775, 777, *781, 805, 827, 859, 869, 871, 899, 905, 922, *984, 1165
Leukaemia (ALL)	633, 647
Leukaemia (CLL)	40, 224
Lip	98
Liver	171, 221, 345, 664, 869, 1035, 1165
Lung	3, 36, 104, 123, *204, 217, 223, 231, 232, 241, 242, 251, 271, *272, 298, *317, *362, *397, 425, 456, 465, 546, *580, 604, 623, 624, 640, *649, *667, *668, *670, *671, 680, 744, 748, 756, *768, *773, *774, 782, *785, 787, 790, 807, *812, 844, 860, 861, 869, 939, 963, 965, 973, *999, 1035, *1051, 1061, 1200, *1209
Lymphoma	13, 56, 57, 171, *205, 247, 287, *362, 376, 463, 506, 654, 744, 752, 757, *781, 805, 869, 965, *984, *1062, 1073
Male Genital	213
Melanoma	9, 20, 39, *291, *292, 720, 1035
Mesothelioma	30, 59, 748, 851, 965, 1035, 1200
Multiple Myeloma	757, *967, 1036
Myelodysplastic Syndrome	922
Nasal Cavity	181, 208, *317
Nasopharynx	*2, 59, 179
Nervous System	*203, *292
Neuroblastoma	57, *292, 376, 1114
Non–Hodgkin's Lymphoma	40, *292, *311, 532, 622, *761, 859, *967, 1165
Oesophagus	167, *259, *362, *478, 546, 664, 733, 784, 1035, 1036, *1074, *1103, *1118
Oral Cavity	181, *195, 207, 425, *478, 664, 784, 807, 1089, 1090, *1117, 1203
Oropharynx	*362, 425, 1203
Ovary	108, *427, *478, 786, 804, 805, 973, *1120, *1191, 1198, *1223
Pancreas	104, 167, 244, 733, 1036, 1165
Pharynx	181, 664, 784, 807, 1089, 1090, *1117
Pleura	*317
Prostate	*203, 232, 241, 242, 251, 257, *291, 329, *427, 623, 629, 1036, *1205, 1214
Rectum	24, *27, 28, *110, 166, 167, *203, 216, 241, 242, 261, *478, 521, 546, 624, 640, 748, *758, 787, 800, *818, 860, *862, 954, 1035, 1098, *1119, *1199, *1223
Respiratory	208, 213, 743
Retinoblastoma	904
Salivary Gland	1035
Sarcoma	13, 213, 247, *292, 622, 889, 965
Skin	12, 20, 39, 213, 720
Soft Tissue	33, *292, 532, 622, 654, 889, *984, 1165
Spinal Cord	22
Stomach	238, 241, 242, 271, *291, *362, 394, *397, 533, 546, 576, 624, 640, 680, 733, 743, 744, 748, 756, 787, 860, 861, 965, 1035, *1074, *1103, *1118, 1177
Testis	108, *206, 283, 805, 834, 940

Thyroid 365, *453, 463, 662, *775, 796, *1119
Tongue *362, 425
Tonsil *362
Urinary Tract 828
Uterus (Cervix) *195, 215, 223, 232, 241, 254, *397, 680, 786
Uterus (Corpus) 786, 810, 925, 973, 1105, *1119, 1190, *1191, *1223
Wilms' Tumour 376

Religion
All Sites 426, 465, 1070, 1075
Breast (F) 433
Colon 465
Lung 465
Ovary 1243

Reproductive Factors
All Sites 90, 225, 458, 625, 870
Bladder 546
Brain 96, *665
Breast (F) 25, *46, 157, *214, 299, *302, 395, 423, *436, 442, *461, *478, 507, 510, 546, 565, 618, 644, 707, 711, 974, 980, 1000, *1006, 1072, 1111, 1204
Colon 149, 166, 167, *478, 510, 546, 980, 1098, 1183
Gallbladder 673, 1081
Kidney 45, 795
Larynx *478
Leukaemia *665
Leukaemia (ALL) 1108
Lung 546, 963, 980
Multiple Myeloma 96
Non–Hodgkin's Lymphoma 96
Oesophagus 167, *478, 546
Oral Cavity 363, *478
Oropharynx 363
Ovary *478, 534, 707, 980, 1000, *1011, *1120, 1133
Pancreas 167, 1000
Prostate 1000
Rectum 149, 166, 167, *478, 510, 546, 1098
Respiratory 363
Sarcoma 96
Soft Tissue 96
Stomach 546, 557
Thyroid *1245
Uterus (Cervix) 156, 363, 608, 645, 707, 951, *1048
Uterus (Corpus) *302, 707, 810, 972, 980, 1000, 1047, 1190
Vagina 971
Vulva 971

Resins
Gastrointestinal 691
Lung 691

RFLP
Bile Duct *514
Brain *514
Breast (F) *514, 1078
Colon 51
Liver *514
Lung *1051
Melanoma 47
Non–Hodgkin's Lymphoma *514
Oesophagus 133, *134
Rectum 51

Respiratory *514
Retinoblastoma *706
Sarcoma *514
Skin *514
Soft Tissue *514
Testis *514

Rubber

Brain 1087
Childhood Neoplasms 222
Colon 1092
Gastrointestinal 691
Kidney 506
Liver 1092
Lung 353, 691, 1092
Lymphoma 506
Mesothelioma 353

Rural

All Sites *1021
Bile Duct 125
Brain *1021
Breast (F) 679, *1006
Colon 258
Gallbladder 125, 258
Liver 258
Oesophagus 258
Pancreas 258
Rectum 258
Soft Tissue 538
Stomach 258
Uterus (Cervix) 679
Uterus (Corpus) 679

Saliva

All Sites *929
Hodgkin's Disease *929
Leukaemia *929
Non–Hodgkin's Lymphoma *929

SCE

All Sites 62, 270, 578, 753, 1131
Angiosarcoma 738
Bladder 642, 738
Hodgkin's Disease 270
Inapplicable 354, *355, 390, 681, 1009
Leukaemia 62, 1026
Liver 738
Lung 642, 738, 1131
Lymphoma 62
Oral Cavity 402
Testis 391

Screening

All Sites 660, 805, 1027, 1123, 1197
Benign Tumours 1059
Bile Duct 958
Bladder 546, 1027, 1096
Bone 805
Breast (F) 11, 60, *78, 105, 114, 174, *195, *255, 276, 288, 374, 499, 546, 630, 636, 651, 660, 717, 723, 805, 855, 941, 942, 1059, 1130, 1197, *1206, 1242

Childhood Neoplasms 805
Colon 28, 105, 148, 261, 374, 455, 546, 762, *862, 892, 958, 1032, 1106, 1145, 1197, *1199, 1212, 1215
Female Genital 252
Gastrointestinal 660, 805
Head and Neck 389
Inapplicable 58
Kidney 105, 1096
Leukaemia 805
Liver *160, 174, 729, 811
Lung 174, 388, 546, 596, 1197
Lymphoma 805
Male Genital 252
Melanoma 666, *988
Nasopharynx *2, 607
Neuroblastoma 591, 1114
Oesophagus 188, 546
Oral Cavity *195, 389, *416, *421
Ovary *78, 105, 634, 660, 805, *865, 868
Pancreas 105, 958
Prostate 374, 1033
Rectum 28, 105, 148, 174, 261, 374, 455, 546, 762, *862, 892, 958, 1032, 1106, 1145, *1199, 1212
Skin 374, *988
Stomach 174, 286, 546, 958
Testis 805
Uterus (Cervix) 10, 55, 115, 174, *195, 254, *294, 374, 437, *482, *627, 631, 660, 813, 951, 1197
Uterus (Corpus) 105, 660
Vagina 971
Vulva 971

Seasonality

Testis 209

Segregation Analysis

Brain *1056
Breast (F) 146, 170, 1053, *1063, 1078, *1112
Breast (M) 1053
Colon 146, *842, *933
Hodgkin's Disease 1154
Leukaemia 937
Liver 146
Lung 146
Oesophagus *134
Ovary *1011
Rectum 146, *842, *933
Retinoblastoma *706
Stomach 146

Sero–Epidemiology

All Sites 218, 241
Breast (F) 241
Burkitt's Lymphoma *658
Colon 241
Female Genital 252
Hodgkin's Disease 218, *658
Kaposi's Sarcoma 382, 657, *658, 964
Leukaemia 581
Liver 165, 656
Lung 241
Lymphoma 964
Male Genital 252

Nasopharynx 144, 606, 607
Non–Hodgkin's Lymphoma *658
Prostate 241
Rectum 241
Stomach 241
Uterus (Cervix) 169, 200, 241, 284, 437, 745, *789

Serum

All Sites 241, 564, 572, 593, 663, 740, 1044
Breast (F) 235, 241, 446, *919, *1005, 1132, *1142, *1143, *1144
Colon 241, 260, 526, 1045
Inapplicable 201
Lung 241, *1051
Lymphoma 1073
Nasopharynx 314
Prostate 241, 793, 1214
Rectum 241
Stomach 137, 241, 286, 846, 914
Uterus (Cervix) 200, 241, *789, 813, *1207

Sex Ratio

All Sites 227, 619
Breast (F) 619, 624
Childhood Neoplasms 441
Colon 624
Leukaemia 619
Liver 729
Lung 619, 624, 841
Myelodysplastic Syndrome 619
Non–Hodgkin's Lymphoma 619
Rectum 624
Stomach 624

Sexual Activity

All Sites 218
Benign Tumours 517
Breast (F) 517
Female Genital 252
Hodgkin's Disease 218
Kaposi's Sarcoma 41, 957
Lymphoma 41
Male Genital 252
Oral Cavity 1203
Oropharynx 1203
Ovary 517
Penis 356
Prostate 228, 257, 329, 816, 918, 943, *1205
Stomach 592
Testis 815, 907
Uterus (Cervix) 156, 169, 215, 284, 356, 403, 410, 438, 443, 517, 547, 611, *615, *627, 813, 867, 951, *1048, 1185, 1187, *1207
Uterus (Corpus) 517
Vagina 971
Vulva 356, 517, 971

Sexually Transmitted Diseases

Kaposi's Sarcoma 41, *898
Lymphoma 41
Non–Hodgkin's Lymphoma *898
Oral Cavity 1203
Oropharynx 1203
Prostate 329, 816, 918

Uterus (Cervix) 55, 284, 369, *627, 728, 745, 813, *898, 946, 1185, 1187, 1211
Vagina 971
Vulva 971

Sib

Childhood Neoplasms 827, 904, 1225
Leukaemia 435, 827
Lymphoma 435
Retinoblastoma 904

Silicosis

All Sites 483, 685
Lung 483, *580, 685

Socio–Economic Factors

All Sites 49, 91, 245, *397, 426, 465, 601, 854, 870, 876, 981, 1075, 1138
Bile Duct 125
Bladder 1096
Breast (F) 127, 408, 433, 442, 507, *605, 797, 854, 1204
Childhood Neoplasms 905, 1042
Colon 127, 465, 601, *675
Gallbladder 125
Hodgkin's Disease *605, 1154
Kaposi's Sarcoma 382
Kidney 560, 1096
Leukaemia 119, 127, 435, *886, 905
Liver 127, 138, 383, 601
Lung 127, 386, *397, 465, 601, *605, 854
Lymphoma 435, *886
Melanoma 20
Multiple Myeloma 1036
Nasopharynx *2, 127
Non–Hodgkin's Lymphoma *605
Oesophagus 127, 188, *605, 1036
Oral Cavity 363
Oropharynx 363
Ovary 404, 406, 1133
Pancreas 1036
Penis 356
Prostate 1036
Rectum 127, 601, *605, *675
Respiratory 363
Rhabdomyosarcoma *1155
Skin 20
Stomach 127, 138, *397, 601, *605, 794, 854, 930
Thyroid 381
Uterus (Cervix) 156, 356, 363, *397, 443, *605, 611, 645, 946, 1029
Vulva 356
Wilms' Tumour 71, 1007

Solvents

All Sites 266, 305, 500, 501, *520, *535, *687, *696, *697, 764, 869, *967, 1015, 1019, *1021, 1157, 1171
Bladder 73, 305, *687, 965, 1016, 1157
Brain 38, *362, 965, *1021, 1091, 1104, 1151
Childhood Neoplasms 222, 1147, 1151, 1165
Colon 73, 1092
Gastrointestinal 1091
Haemopoietic 677, 965, *984, 1091
Hodgkin's Disease *311, 542
Inapplicable 354, *355
Kidney 73, 506, 1104

Larynx *362, 869
Leukaemia 136, 334, 349, *380, 539, 542, 653, *687, *864, 869, *886, 950, *984, 1165
Leukaemia (ALL) 75, 1108
Leukaemia (AML) 280
Leukaemia (ANLL) 1109
Liver 869, 1091, 1092, 1104, 1165
Lung 73, *362, *687, 869, 965, 1091, 1092, 1104, *1209
Lymphoma 73, 349, *362, 506, 869, *886, 965, *984
Melanoma 73
Mesothelioma 965
Multiple Myeloma 542, *967
Myelodysplastic Syndrome 539
Nasal Cavity 305, 306
Non–Hodgkin's Lymphoma *311, 542, *967, 1165
Oesophagus 73, *362
Oropharynx *362
Pancreas 73, 1165
Prostate 73, 629
Rectum 73
Respiratory *864
Sarcoma 965
Soft Tissue *984, 1165
Stomach 73, *362, 965
Testis 907
Tongue *362
Tonsil *362

Sputum
Bladder 642
Lung 481, 596, 642, 1224

Stage
Breast (F) 174, *195, 463, 464
Colon 463
Head and Neck 463
Larynx 59, 364
Leukaemia 463
Liver 174
Lung 174
Lymphoma 463
Mesothelioma 59
Nasopharynx 59
Neuroblastoma 1008
Oral Cavity *195
Ovary 868
Prostate 92
Rectum 174
Stomach 174
Thyroid 318, 463
Uterus (Cervix) 174, *195

Stress
All Sites 378, 586, 709, 854, 875
Breast (F) 854, 855
Haemopoietic 358
Lung 854
Stomach 854

Surgery
Bile Duct 958
Breast (F) *461, 973, *1196

Colon 958, 973
Larynx *1246
Lung 973
Ovary 973
Pancreas 958
Rectum 958
Stomach 958
Uterus (Corpus) 973

Survival
All Sites 176, 550, 563, 685, *780, 870, 921, 981, *1110
Brain 22
Breast (F) 23, 174, 417, 463, 630, 636, 637, 786
Breast (M) 211
Childhood Neoplasms 375, 441, *852, 916, 917, 1042, *1110
Colon 258, 463
Gallbladder 258
Head and Neck 463
Hodgkin's Disease 540
Leukaemia 463, 730, 915
Leukaemia (ALL) 540, 847
Leukaemia (AML) 847
Leukaemia (ANLL) 540
Leukaemia (CML) 847
Liver 174, 258
Lung 174, 685
Lymphoma 463
Neuroblastoma 540
Non–Hodgkin's Lymphoma 540
Oesophagus 258
Ovary 786
Pancreas 258
Rectum 174, 258
Spinal Cord 22
Stomach 174, 258
Thyroid 318, 463, 662
Uterus (Cervix) 174, 786
Uterus (Corpus) 786
Wilms' Tumour 540

Sweeteners
Bladder 99, 104, 310
Lung 104
Pancreas 104

Tea
Colon 167
Kidney 1141
Liver 164
Oesophagus 167
Pancreas 167
Rectum 167
Wilms' Tumour 71

Textiles
All Sites 500
Gastrointestinal 691
Lung 691
Nasal Cavity *316

Time Factors

All Sites 194, 227, 271, *336, 339, *397, 498, 502, 503, 719, 809, 897, 901, 1181
Bladder 513
Brain 1181
Breast (F) 194, *291, 464, 630, 1072, *1191
Childhood Neoplasms 823, 824, 1042, 1165
Colon *291, 737, 1145, 1181
Gallbladder 737
Gastrointestinal 64, 76
Haemopoietic *984
Kidney 76, 513
Larynx 76, 513, 1236
Leukaemia 339, 823, 829, 901, *984, 1165, 1181
Liver 737, *983, 1165
Lung 76, 159, 194, 271, 326, 339, *397, 456, 513, 678, 1022, 1182, *1226
Lymphoma 829, *984
Melanoma *291
Mesothelioma *324
Neuroblastoma 1008
Non–Hodgkin's Lymphoma 339, 823, 1165
Oesophagus 513, 737, *1074
Oral Cavity 513, 1236
Ovary *1191
Pancreas 513, 737, 1165
Peritoneum 76
Pharynx 513, 1236
Pleura 76
Prostate *291
Rectum 737, 1145, 1181
Respiratory *324
Small Intestine 737
Soft Tissue *984, 1165
Stomach 271, *291, 326, *397, 737, *1074
Uterus (Cervix) 215, *397
Uterus (Corpus) *1191

Tissue

All Sites 225
Bile Duct *734
Brain *1056
Breast (F) 74, *948
Childhood Neoplasms 831
Colon 74, 366, 762
Gallbladder *734
Gastrointestinal 880
Lung 239, 960, 1200
Lymphoma 1073
Mesothelioma 1200
Nasopharynx *184
Non–Hodgkin's Lymphoma *290
Ovary *865
Pancreas *734
Prostate 74, *1002, *1205
Rectum 366, 762
Sarcoma *290
Soft Tissue *290
Stomach 137
Uterus (Cervix) 274, 994, *1207
Wilms' Tumour *15

Tobacco (Chewing)

All Sites 301, *415
Colon *675
Hypopharynx 425

Larynx	425, 784
Liver	*293
Lung	425
Oesophagus	784
Oral Cavity	207, 402, 413, 414, *416, 419, *420, *421, 425, 429, 739, 784
Oropharynx	405, 425
Pharynx	784
Rectum	*675
Tongue	425

Tobacco (Smoking)

All Sites	52, 67, 89, 91, 116, 196, 225, 245, 298, 301, 333, *415, 458, 465, 502, 512, 522, 564, 568, 578, 586, 588, 601, 619, 625, 680, 684, 708, 714, 749, 792, 856, 884, 928, 953, 982, 991, 992, 1013, 1050, 1075, *1077, 1131, 1138, 1173, 1197, *1235, 1238
Angiosarcoma	738
Benign Tumours	1059
Bile Duct	*514, *734, 958
Bladder	104, 310, *352, 513, 541, *543, 546, 642, 738, 866, 1016, 1096, 1134
Bone	171
Brain	96, 171, *514, 955
Breast (F)	*46, 89, 127, 146, *214, 299, 423, *461, *478, *514, *515, 546, 565, 619, 680, *808, *948, 949, 991, 1000, 1040, 1059, 1072, 1197, 1213
Childhood Neoplasms	831
Choriocarcinoma	516
Colon	34, *53, 89, 127, 146, 149, 166, 167, 424, 465, *478, 546, 595, 601, *675, 737, 747, 949, 958, *976, *1010, 1032, 1058, 1197, 1213, 1228
Female Genital	252, 949
Gallbladder	673, *734, 737
Gastrointestinal	76
Hodgkin's Disease	*311, *1125, *1244
Hydatidiform Mole	516
Hypopharynx	268, 315, *317, 425, 439
Inapplicable	681, 1001
Kidney	45, 76, *219, 330, 368, 513, 560, 795, 1037, 1096, 1134, 1141, 1164
Larynx	59, 70, 76, 181, 315, *317, 364, 425, 439, *478, 513, 784, 807, 1134, 1236, *1246
Leukaemia	118, 127, 171, 539, 619, 653
Leukaemia (CLL)	*1125
Lip	70, 98, 1085
Liver	127, 138, 146, 165, 171, 234, *293, *514, 574, 595, 597, 601, *690, 729, 737, 738, 811
Lung	3, 34, 35, *50, 52, 76, 94, 104, 127, 129, 146, 152, 154, *158, *173, *197, 217, 239, *264, 298, *317, 326, 331, 343, 351, *379, 386, 388, 425, *428, 465, 481, 497, 513, 546, 566, 574, *580, 595, 596, 601, 613, 619, 642, *671, 678, 680, 705, 738, 742, *773, *774, 782, 788, 792, 798, 807, 944, 945, 947, 968, 982, 1022, 1024, 1061, 1131, 1134, 1153, 1162, 1173, 1182, 1197, 1200, *1209, 1213, *1226, 1234
Lymphoma	171, 1024
Male Genital	252
Melanoma	20
Mesothelioma	34, 35, 52, 59, 1200
Multiple Myeloma	96
Myelodysplastic Syndrome	118, 539, 619
Nasal Cavity	181, *317, 1084
Nasopharynx	59, 127, 151, 179, *184, 1084, 1208
Non-Hodgkin's Lymphoma	96, *311, *514, 619, *1125
Oesophagus	127, 167, 188, 361, 424, 439, 454, *478, 513, 546, 737, 784, *969, *1074, *1103, *1118, 1134
Oral Cavity	66, 70, 181, 207, 268, 363, 413, 414, *416, 419, *420, *421, 425, 429, 439, *478, 513, 739, 784, 807, 1089, 1134, 1136, 1236
Oropharynx	66, 268, 363, 405, 425, 439, 1136
Ovary	*478, 949, 1000
Pancreas	104, 167, 244, 513, *734, 737, 958, 1000, 1134, 1213
Peritoneum	34, 76

Pharynx 70, 181, 513, 784, 807, 1089, 1236
Pleura 76, *317
Prostate *760, 943, 1000, 1213
Rectum *53, 127, 146, 149, 166, 167, 424, *478, 546, 601, *675, 737, 747, 958, *1010, 1032, 1058, 1213
Respiratory 363, *514, 856, 1156, 1219
Rhabdomyosarcoma *1155
Sarcoma 96, *514
Skin 20, *514
Small Intestine 737
Soft Tissue 96, *514
Stomach 34, 127, 137, 138, 146, 153, 186, 326, 424, 505, 546, 557, 584, *585, 595, 601, 680, 735, 737, 794, 958, *969, *975, 982, *1074, *1103, *1118, 1233
Testis 283, *514, 815
Thyroid *1245
Tongue 425
Ureter *352
Urethra *352
Uterus (Cervix) 156, 215, 363, 608, *627, 645, 680, 867, 883, 946, 949, *1048, 1185, 1187, 1197, *1207
Uterus (Corpus) 595, 949, 972, 1000, 1047
Vagina 971
Vulva 971
Wilms' Tumour 71

Tobacco (Snuff)

All Sites 301, *415
Colon *675
Hypopharynx 439
Larynx 439, 784
Oesophagus 439, 784
Oral Cavity 439, 784
Oropharynx 439
Pharynx 784
Rectum *675

Toenails

All Sites 225, 991, 1044
Bile Duct *734
Breast (F) 74, 507, 612, 624, 639, 991
Colon 74, 624, 638
Gallbladder *734
Lung 624
Nasopharynx *184
Pancreas *734
Prostate 74, 918
Rectum 624, 638
Stomach 624

Tonsillectomy

Breast (F) 650
Hodgkin's Disease *1244

Trace Elements

All Sites 246, *256, 586, 663, 1044, 1168
Breast (F) 127, 468, 612
Colon 127, 468, 526, 638, *1017
Leukaemia 127
Liver 127, 175, 234
Lung 127
Nasopharynx 127

Oesophagus 127
Rectum 127, 468, 638
Stomach 127, 137
Thyroid 662

Transplacental
Inapplicable *853

Transplantation
All Sites 227, 1023
Lymphoma 1100
Skin *878

Trauma
Brain 44, 303
Testis 283, 815

Treatment
All Sites 112, 248, *338, 339, 568, 674, 718, 981, *1110
Benign Tumours 253
Bone 674
Breast (F) 253, *302, 674, 803
Childhood Neoplasms 732, 917, *1110
Choriocarcinoma *549
Colon 258
Female Genital 252, 803
Gallbladder 258
Head and Neck 389
Hodgkin's Disease 540, *676
Hydatidiform Mole *549
Kaposi's Sarcoma 41
Larynx 59, 231
Leukaemia 339, 730
Leukaemia (ALL) 540, *676
Leukaemia (ANLL) 540
Liver 258
Lung 231, 339
Lymphoma 41, 1100
Male Genital 252
Mesothelioma 59
Nasopharynx 59
Neuroblastoma 540
Non–Hodgkin's Lymphoma 339, 540, *676
Oesophagus 258
Oral Cavity 389
Pancreas 258
Rectum 258
Stomach 258
Thyroid 318
Uterus (Cervix) 674, 951
Uterus (Corpus) *302
Wilms' Tumour 540

Trends
All Sites 176, 194, 220, 720, 839, 876, 1166, 1170
Bladder 513
Brain 22, 85
Breast (F) 65, 194, *195, 679
Breast (M) 211
Childhood Neoplasms 441, 922, 1165
Colon 65, 258

Gallbladder 258
Gastrointestinal 64
Hypopharynx 425
Kidney 513
Larynx 425, 513
Leukaemia 287, 633, *850, 922, 1165
Leukaemia (ALL) 633
Lip 98
Liver 65, 258, 1165
Lung 65, 194, *197, 425, 513
Lymphoma 287, *850
Melanoma 720
Myelodysplastic Syndrome 922
Non–Hodgkin's Lymphoma 1165
Oesophagus 65, 258, 513, 694
Oral Cavity *195, 425, 513
Oropharynx 425
Ovary 1243
Pancreas 258, 513, 1165
Pharynx 513
Prostate 65
Rectum 65, 258
Skin 720
Soft Tissue 1165
Spinal Cord 22
Stomach 65, 258, *523, 576
Thyroid 662
Tongue 425
Uterus (Cervix) 65, 156, *195, 679
Uterus (Corpus) 679

Tubal Ligation
Breast (F) 1139
Ovary *427, 1139
Prostate *427
Uterus (Cervix) 1139
Uterus (Corpus) 1139

Tumour Markers
All Sites 663, 682, 684
Colon 762
Oral Cavity 445
Ovary *865
Rectum 762
Uterus (Cervix) 745

Twins
All Sites 14, 240, 582, 1080
Breast (F) 230
Childhood Neoplasms 600, 904
Gastrointestinal 230
Leukaemia 230, 600
Melanoma 9, 39, 230
Retinoblastoma 904
Skin 39
Testis 230
Uterus (Cervix) 230
Uterus (Corpus) 230

Ulcerative Colitis
All Sites 821
Colon 226, 526, 762, 800, 821

Rectum 762, 800, 821
Small Intestine 226

Ultrasound
Childhood Neoplasms 827
Leukaemia 827

Urban
Bile Duct 125
Breast (F) 679
Gallbladder 125
Inapplicable *853
Uterus (Cervix) 679
Uterus (Corpus) 679

Uric Acid
Breast (F) 949
Colon 949
Female Genital 949
Ovary 949
Uterus (Cervix) 949
Uterus (Corpus) 949

Urine
All Sites 225, 295, 1027
Bile Duct *734
Bladder *204, 541, *548, 642, 910, 1027, 1096
Breast (F) 507
Childhood Neoplasms 831
Colon 373
Gallbladder *734
Inapplicable 201
Kidney 1096
Larynx *204
Lung *204, 642, *768, 910, 1079
Nasopharynx *184
Oesophagus 361, 910
Ovary 635, 1198
Pancreas *734
Rectum 373
Stomach 137, 910, 912
Testis *206

Vaccination
All Sites 344
Brain 348
Liver 345
Non–Hodgkin's Lymphoma 1186

Vasectomy
Ovary *427
Prostate *427, *1205

Vegetables
All Sites 465, 1070
Colon 465, 1228
Lung 465
Oesophagus 733

Pancreas	733
Stomach	733, *736
Wilms' Tumour	71

Vegetarian

Breast (F)	923
Gastrointestinal	923

Virus

All Sites	218, 227, *929, 1044
Brain	348
Breast (F)	1000
Childhood Neoplasms	375, 823, 827
Colon	366
Hodgkin's Disease	218, *676, *929
Kaposi's Sarcoma	382
Leukaemia	*342, 367, 570, 823, 827, 829, *929
Leukaemia (ALL)	*676
Liver	165, 168, 273, 569, 597
Lymphoma	570, 829
Nasopharynx	151
Non–Hodgkin's Lymphoma	*676, 823, *929
Oesophagus	*1103
Ovary	1000, 1229
Pancreas	1000
Prostate	1000
Rectum	366
Stomach	*1103
Uterus (Cervix)	200, 645, 920
Uterus (Corpus)	1000

Vitamins

All Sites	241, 533, 586, 625, 806, 884, *987, 992, 995, 1044, 1057, 1076, 1197
Breast (F)	63, 127, 241, *436, 480, 533, *609, 612, 640, 797, *987, 1000, *1067, *1143, 1197, *1223
Childhood Neoplasms	832
Colon	63, 127, 241, 470, 480, 526, 533, 595, 640, 806, 894, *987, 1057, 1066, 1197, 1210, *1223, 1228
Leukaemia	127
Liver	127, *293, 595
Lung	35, 63, 97, 127, 241, 480, 595, 640, 705, 806, *987, 1197, *1209, 1224, 1232
Melanoma	1221
Mesothelioma	35
Nasopharynx	127
Non–Hodgkin's Lymphoma	1193
Oesophagus	63, 127, 130, 187, 190, *259, 361, 733
Oral Cavity	444
Ovary	1000, *1223
Pancreas	733, 1000
Prostate	92, 241, 943, 1000
Rectum	8, 127, 241, 640, 894, 1066, *1223
Skin	1220, 1221
Stomach	63, 127, 241, 285, 533, 595, 640, 725, 733, 806
Thyroid	662
Uterus (Cervix)	241, 284, 547, 611, 882, 994, *1048, 1197
Uterus (Corpus)	595, 1000, 1047, *1223

Waste Dumps

All Sites	*142
Leukaemia	349, 1064
Lymphoma	349

Water

All Sites	49, 84, *142, 787
Bile Duct	587
Bladder	84, *110, 310, 392, 652, 787, 1016
Bone	102
Brain	84, 1087
Colon	*53, 84, *110, 147, 787
Gastrointestinal	392
Kidney	84
Leukaemia	84
Liver	182, 587
Lung	787
Lymphoma	84
Oesophagus	188
Prostate	84, 943
Rectum	*53, 84, *110, 147, 787
Stomach	84, 182, 787, 1177

Welding

All Sites	545, 982
Gastrointestinal	1034
Hodgkin's Disease	542
Leukaemia	539, 542
Liver	234
Lung	*667, *668, 982, 1034
Multiple Myeloma	542
Myelodysplastic Syndrome	539
Non–Hodgkin's Lymphoma	542, 1034
Prostate	629
Stomach	982

White Cells

All Sites	1044
Hodgkin's Disease	1154

Wood

All Sites	305, 500, 1171
Bladder	305, 965
Brain	965
Haemopoietic	744, 965
Kidney	506
Larynx	59
Leukaemia	208, 653
Lung	744, 965, *1209
Lymphoma	506, 744, 965
Mesothelioma	59, 965
Nasal Cavity	208, 278, 305, 306
Nasopharynx	59, 1208
Respiratory	208
Sarcoma	965
Stomach	744, 965

Xeroderma Pigmentosum

Skin	1184

INDEX OF SITES

Entries in this index refer mainly to sites defined at the three–digit level of the ninth revision of the International Classification of Diseases, such as "Lip" (ICD–9 140) etc. A number of other useful headings have been retained or added, however, such as "Head and Neck", "Female Genital", "Childhood Neoplasms", etc., because many projects address such groups ofmalignancies, even though they do not fit a precise topographicalcategory. Certain morphological groups have been retained or introduced into this index for the same reason, such as "Sarcoma", "Mesothelioma", "Neuroblastoma", etc. Where possible, abstracts coded to "Leukaemia" have been recoded to the main subtype(s) of leukaemia.

General terms (e.g. "Female Genital") are used only when specific terms cannot be assigned: the specific terms are shown beneath each general term. The entry "Inapplicable" is used for some genetic and molecular epidemiology projects which involve subcellular phenomena rather than study of a particular cancer site.
New site in 1994 is "Rhabdomyosarcoma".

The list of index entries below identifies all the cancer sites, types or groups in use in this Directory.

All Sites
Angiosarcoma
Appendix

Benign Tumours
Bile Duct
Bladder
Bone
Brain
Breast (F)
Breast (M)
Burkitt's Lymphoma

Childhood Neoplasms
Choriocarcinoma
Colon

Endocrine Glands
Eye

Female Genital
 See also: Ovary; Placenta; Uterus (Cervix); Uterus (Corpus); Vagina; Vulva

Gallbladder
Gastrointestinal
 See also: Anus; Appendix; Bile Duct; Colon; Gallbladder; Liver; Oesophagus; Pancreas; Rectum; Small Intestine; Stomach
Genitourinary
 See also: Female Genital; Male Genital; Urinary Tract

Haemopoietic
 See also: Leukaemia; Lymphoma
Head and Neck
 See also specific sites
Hodgkin's Disease
Hydatidiform Mole
Hypopharynx

Inapplicable

Kaposi's Sarcoma
Kidney

Larynx
Leukaemia
Leukaemia (ALL)
Leukaemia (AML)
Leukaemia (ANLL)
Leukaemia (CLL)
Leukaemia (CML)
Lip
Liver
Lung
Lymphoma
 See also: Hodgkin's Disease; Lymphosarcoma; Non–Hodgkin's Lymphoma

Male Genital
 See also: Penis; Prostate; Testis
Melanoma
Mesothelioma
Multiple Myeloma
Myelodysplastic Syndrome

Nasal Cavity
Nasopharynx
Nervous System
 See also: Brain; Cranial Nerve; Spinal Cord
Neuroblastoma
Non–Hodgkin's Lymphoma

Oesophagus
Oral Cavity
 See also: Lip; Parotid; Salivary Gland; Tongue
Oropharynx
Ovary

Pancreas
Penis
Penis
Peritoneum
Pharynx
Pleura
Prostate

Rectum

SITE

Respiratory
    See also: Larynx; Lung; Mesothelioma; Pleura
Retinoblastoma
Rhabdomyosarcoma

Salivary Gland
Sarcoma
Skin
    See also: Melanoma
Small Intestine
Soft Tissue
Spinal Cord
Spleen
Stomach

Testis
Thyroid
Tongue
Tonsil

Ureter
Urethra
Urinary Tract
    See also: Bladder; Kidney; Ureter; Urethra
Uterus (Cervix)
Uterus (Corpus)

Vagina
Vulva

Wilm's Tumour

All Sites

Abortion *111
Age 194, 227, 248, *396, 498, 502, 512, 589, 619, 809, 970, 1082, *1174, 1181
AIDS 218, 524, 1041
Air Pollution 49, *387, 511, 522, 869
Alcohol 67, 89, 91, 116, 245, 333, 512, 522, 564, 568, 578, 586, 588, 601, 625, 680, 684, 708, 714, 749, 928, 953, 992, 1050, 1075
Allergy 359, 771
Analgesics 1157
Anatomical Distribution 719, 720
Antibodies 1123
Antioxidants *987, 995
Asbestosis 684
Ataxia Telangiectasia 938, 1046
Atomic Bomb 550, 551, 562, 563
Autopsy 522
Biochemical Markers 663, 921, 992, 1197
Biomarkers 270, 508, 1131
Birth Cohort 194, 809
Blacks 1170
Blood 225, 270, 295, 372, 749, 750, *929, 991, *1117
Blood Group 227, 660
BMB 225, 241, 270, 295, 372, 533, 564, 572, 586, 593, 663, 740, 749, 750, *929, 991, 992, 1044, *1117, *1174
Caucasians 1170
Chemical Exposure 19, 33, 42, 68, 84, 90, 117, 229, 247, 266, 282, 300, 308, *332, 333, 335, 371, *396, 489, 502, 504, 508, 522, 545, 578, *621, 623, 654, *661, 755, 757, 763, 764, 779, *781, 819, 887, 928, 982, 1018, 1019, 1094, 1095, 1107, 1131, 1157, 1158, 1159, 1171, 1173, 1180, 1181, 1188, 1230
Chemotherapy *111, 270, *336, 339, 619, 746, 970, 1023, *1175
Childhood 56, *336, 450, 471, 472, *475, 524, 603, 708, 777, 889
Chlorination 84
Cholesterol 533, 601, 884, 921, *1101
Chromosome Effects 62, 551, 578, 746, 753, 759, 1131
Clinical Effects *453
Clinical Records 52, 112, 450, 603, 1137, 1171
Cluster *54, *142, 452, 869, *929, 978, *1052, 1105, 1166
Coffee 465
Congenital Abnormalities 132, *338, 452
Contraception 924
Cosmetics 991
Crohn's Disease 821
Cytology 924
Data Resource *109, 344, *411, 525, 718, *814, 869
Demographic Factors *387
Diabetes 921, 1065
Diet 49, 61, 88, 89, 145, 225, 245, 295, 372, 458, 465, 477, 512, 533, 564, 572, 575, 586, 588, 589, 601, 680, 714, 749, 806, 953, 981, 991, 1044, 1050, 1057, 1065, 1070, 1075, 1105, 1138, 1168, 1197, 1238, 1240
Disease, Other *1110
DNA 551, 572, 586, 1126
DNA Adducts 270, 1044, 1131
DNA Binding 1131
DNA Repair 805
Dose–Response 42, 116, 210, *262, 263, *336, 551, *686, 750, 752, 772, 839, 896, 897, 901, 902, 970, 1071, 1107, 1171, *1172, 1227
Down's Syndrome 132
Drugs 210, 221, 227, 270, 512, 568, 625, 746, 757, 764, *776, 928, 953, 981, *987, 991, 992, 993, 995, 1137, *1175
Dusts 42, 52, 90, *272, 282, 305, 489, 500, 501, 511, 522, 545, *669, 684, 685, *686, *715, 754, 757, 772, 856, 939, 982, 1019, 1171, 1197, *1235
Dyes 282, 500, 509, 991
EBV *929, 1044, *1174
Education 116, 245, 981
Electromagnetic Fields 80, 312

Environmental Factors	14, 42, 49, 88, 90, 116, 132, 196, 229, 240, 266, 295, *387, 489, 503, 508, 522, 674, 759, 869, *935, 981, 1094, 1158, 1166, 1173, 1239, 1240
Enzymes	486
Epilepsy	221
Eskimos	82, 220
Ethnic Group	82, *109, 116, 220, 452, 465, 643, 1050, *1052
Familial Factors	458, 586, 625, 714, 805, 889, 991, 1046, *1052, 1099
Family History	*1110
Fat	61, 465, 1105, 1197, 1238
Female	*107, 991, 995, 1039
Fertility	*111, 924, *1110
Fertilizers	83, 86, 117, *701
Fibre	1057, 1197
Fluorescent Light	100
Fruit	465, 1070
G6PD	486, 682, 684
Gastric Ulcer	625
Genetic Factors	14, *319, 486, 586, 702, 708, 805, 1046, 1099
Genetic Markers	759, 805, 1046, 1126, 1131
Geographic Factors	1, *54, 69, 83, 84, 90, *142, 194, 196, 289, *387, 498, 503, 643, 688, 689, 720, 759, *780, 809, 869, 899, 1237, 1239, 1240
H. pylori	663, 884
Haemophilia	*1174
HBV	344, 562
HCV	562
Hepatitis	663
Herbicides	33, 83, 86, 247, 296, 371, 452, 622, 654, 763, 1180
Heredity	578, 702, 805
High-Risk Groups	88, 245, 685, *693, 702, 718, 889, 938, 1046, 1126, 1197
Histology	248, 551, 720
HIV	524, 1041, *1174
HLA	227, *929
Homosexuality	218
Hormones	512, 991, 1044, 1082, 1201
HPV	1044
HSV	1044
HTLV	562
Immunologic Markers	663
Immunology	359, 586
Immunosuppression	227, 1023
Indians	*109, 122
Infection	218
Infertility	1201
Insecticides	83, 86, 928
Intra-Uterine Exposure	708
Late Effects	*111, 210, 551, 837, *1110
Latency	116, 210, *397, 472, 503, 897
Leather	305, 500, 755
Lifestyle	88, 90, 196, 220, 372, 426, 572, 578, 582, 588, 589, 593, 601, 643, *693, 709, *780, 806, 870, 953, *989, 1050, 1075, 1076, *1077, 1239, 1240
Linkage Analysis	876
Lipids	465, 806, 991, 1044, 1065
Lymphocytes	578, 586, 746, 753
Male	333
Mammography	1197
Mapping	90, *387, 688, 689, 713, 820, 1237
Marital Status	601, 870
Mate	1238
Mathematical Models	14, *686, 809, 820, 970, *1052
Menarche	660
Menopause	660, 991
Metals	90, 117, 241, 263, 271, *272, *485, *555, 573, *669, *715, 750, *776, 863, 872, 991, 992, 1015, 1019, 1057, 1071, 1160, 1161, 1171, 1227
Micronuclei	270, 578, 753, 1131
Micronutrients	*256
Migrants	289, *453, 498, 503, *780, 876
Minerals	489, 856, 1057

Mining	*29, 42, 90, 263, 483
Monitoring	62, 67, 519, *520, *838, 839, 1013, 1123, 1137
Multiple Primary	223, 248, 270, *319, *336, *338, 339, 466, 603, 619, 702, 720, 805, 940, 970, *1110, *1117, *1175
Mutagen	759
Mutagen Tests	1123
Mutation Rate	132
Mutation, Germinal	132
Mutation, Somatic	62, 759
Nutrition	61, 89, 145, 225, 295, 372, 477, 564, 806, 1168, 1197
Obesity	89, *989, 1105
Occupation	19, *29, 33, 42, 52, 56, 62, 67, 68, 80, 81, 86, 87, 88, 90, 100, *103, *107, 116, 117, 126, 128, 185, *202, 229, 245, 247, *250, *262, 263, 266, 271, *272, *277, 281, 282, 296, 298, 300, 305, 308, 312, *327, *332, 333, 335, *357, 360, 371, *396, *412, *422, 451, 452, *453, 458, 483, 484, *485, 500, 501, 502, 504, 509, 511, *520, 522, 532, *535, 537, 545, *555, 564, 572, 573, 583, *621, 622, 623, 648, 654, *661, *669, 680, 682, 684, *686, *687, *693, 695, *696, *697, *698, *699, *700, *701, 708, 709, 714, *715, 749, 750, 751, 752, 754, 755, 757, 763, 764, 767, 772, 779, *781, 792, 799, 819, 825, 837, *838, 839, 856, 863, 872, 875, 881, 887, 896, 897, 928, *935, 939, 953, *967, 981, 982, 990, 991, 1013, 1015, 1018, 1019, *1021, 1027, 1071, 1075, 1076, 1094, 1095, 1107, 1123, 1138, 1146, 1157, 1158, 1159, 1160, 1161, 1163, 1167, 1171, *1172, 1173, 1178, 1180, 1181, 1188, 1197, 1227, 1230, 1231, *1235
Oncogenes	270, 551
Oral Contraceptives	512, 888, 924, 991
Paints	*967, 1015
Parity	*396, 660, 924
Passive Smoking	91, 884, 953, 1039, 1131
Pedigree	702, 805
Pesticides	33, 86, 90, 247, 296, 371, 452, 504, 532, 537, 622, 654, *699, 755, 763, 928, 1019, *1021, 1171, 1180, *1235
Petroleum Products	19, 86, 90, *661, 869, 1015, 1018, 1188
Photochemotherapy	*475, 993
Physical Activity	806, *989, 991, 992, 1171
Physical Factors	225, 601, 660, 680, 970, 1105
Pigmentation	100
Plants	575
Plasma	740, 1044
Plastics	90, 128, 281, 305, 333, 623, 751, 754, 755, 757, 779, 825, 869, 990, 1019, 1158, 1159, 1171, 1173, 1230
Pneumoconiosis	685
Pregnancy	*111, 924
Premalignant Lesion	512, 718, 1046, 1126
Prevalence	524, 1065, 1197
Prevention	248, *256, 702, 718, 719, 953, 986, *987, 995, *1077, 1197, 1240
Prognosis	719, 981
Projection	719
Promotion	1168
Protein Binding	*781, 1131
Psychological Factors	140, 378, 564, 586, 593, 875
Psychosocial Factors	141, 854, 875
Race	*780
Radiation, Ionizing	56, 62, 81, 87, 90, 112, 132, 185, 210, 221, *262, 263, *277, *336, *338, 339, *341, 360, *412, *422, 440, *453, 458, 471, 472, 550, 551, 562, 563, 583, 764, *775, *776, 777, 819, 826, 837, *838, 839, 840, 881, 896, 897, 901, 902, *935, 970, 1013, 1071, 1107, 1163, 1167
Radiation, Non–Ionizing	970, 993, 1167
Radiation, Ultraviolet	458, *475, 805
Radiotherapy	*111, 223, *319, 619, 777, 940, 970
Record Linkage	14, 68, 83, 86, 90, 91, *107, *109, 112, 122, 240, 245, 508, 1041, 1137

Registry 1, 14, *29, 33, 56, 67, 68, 69, 81, 82, 89, 90, *109, *111, 116, 122, 176, 196, *202, 210, 220, 221, 223, 225, 227, 240, 241, 245, 246, 247, 248, *250, 270, 271, *272, *277, 281, 282, 296, 298, 300, 308, 344, *387, *396, *397, *415, *418, 426, 448, 449, 450, *453, 465, 466, 472, *475, 498, 503, 532, 533, 562, 564, 593, 603, 622, 623, 625, 643, 648, 654, 663, *669, 680, 720, 749, 750, 751, 752, 754, 757, 767, 771, *775, *776, 777, *780, *781, 787, 805, 837, 869, 884, 889, 897, 899, 938, 939, 940, *967, 1041, *1052, 1070, 1105, *1117, 1166, 1170, 1188, 1201, *1235, 1237, 1238, 1239, 1240
Religion 426, 465, 1070, 1075
Reproductive Factors 90, 225, 458, 625, 870
Rural *1021
Saliva *929
SCE 62, 270, 578, 753, 1131
Screening 660, 805, 1027, 1123, 1197
Sero–Epidemiology 218, 241
Serum 241, 564, 572, 593, 663, 740, 1044
Sex Ratio 227, 619
Sexual Activity 218
Silicosis 483, 685
Socio–Economic Factors 49, 91, 245, *397, 426, 465, 601, 854, 870, 876, 981, 1075, 1138
Solvents 266, 305, 500, 501, *520, *535, *687, *696, *697, 764, 869, *967, 1015, 1019, *1021, 1157, 1171
Stress 378, 586, 709, 854, 875
Survival 176, 550, 563, 685, *780, 870, 921, 981, *1110
Textiles 500
Time Factors 194, 227, 271, *336, 339, *397, 498, 502, 503, 719, 809, 897, 901, 1181
Tissue 225
Tobacco (Chewing) 301, *415
Tobacco (Smoking) 52, 67, 89, 91, 116, 196, 225, 245, 298, 301, 333, *415, 458, 465, 502, 512, 522, 564, 568, 578, 586, 588, 601, 619, 625, 680, 684, 708, 714, 749, 792, 856, 884, 928, 953, 982, 991, 992, 1013, 1050, 1075, *1077, 1131, 1138, 1173, 1197, *1235, 1238
Tobacco (Snuff) 301, *415
Toenails 225, 991, 1044
Trace Elements 246, *256, 586, 663, 1044, 1168
Transplantation 227, 1023
Treatment 112, 248, *338, 339, 568, 674, 718, 981, *1110
Trends 176, 194, 220, 720, 839, 876, 1166, 1170
Tumour Markers 663, 682, 684
Twins 14, 240, 582, 1080
Ulcerative Colitis 821
Urine 225, 295, 1027
Vaccination 344
Vegetables 465, 1070
Virus 218, 227, *929, 1044
Vitamins 241, 533, 586, 625, 806, 884, *987, 992, 995, 1044, 1057, 1076, 1197
Waste Dumps *142
Water 49, 84, *142, 787
Welding 545, 982
White Cells 1044
Wood 305, 500, 1171

Angiosarcoma

Biomarkers 738
BMB 738
Chromosome Effects 738
Dyes 738
Lymphocytes 738
Metabolism 738
Micronuclei 738
Occupation 738
Plastics 738
SCE 738
Tobacco (Smoking) 738

Appendix

Autopsy 398
Biopsy 398
Classification 398
Histology 398

Benign Tumours

Age *1202
Alcohol 616
Biopsy 1115
Blood 253
BMB 253, *494
Childhood 472
Diet 517
Drugs 616, 1140
Familial Factors 616, 1059
Genetic Factors 616
Genetic Markers *494
Hormones 517, 616, 617, 1140, *1202
Latency 472
Lifestyle 31
Mammography 1059
Menopause *1202
Menstruation *1202
Naevi 31
Oncogenes *1202
Oral Contraceptives 517, 616, 617, 1059
Parity 517
Pigmentation 31
Premalignant Lesion 31, 1115, 1140
Prevention 253, *494
Prognosis 253
Radiation, Ionizing 472
Radiation, Ultraviolet 31
Registry 253, 472
Screening 1059
Sexual Activity 517
Tobacco (Smoking) 1059
Treatment 253

Bile Duct

Alcohol *734
Autopsy 398
Biochemical Markers 1152
Biomarkers *514
Biopsy 398
Blood *514, *734
BMB *514, *734
Chromosome Effects *734
Classification 398
Cluster 587
Coffee *734
Cytology *734
Gallstones 1152
Genetic Markers *514
Histology 398
Occupation 125, *734
Parasitic Disease 587
PCR *514
Prognosis *734
RFLP *514
Rural 125
Screening 958
Socio-Economic Factors 125

Surgery	958
Tissue	*734
Tobacco (Smoking)	*514, *734, 958
Toenails	*734
Urban	125
Urine	*734
Water	587

Bladder

Air Pollution	213
Alcohol	310, 546
Analgesics	178, *352, 1157
Antioxidants	*204
Biomarkers	*120, *204, *548, 642, 738, 1025
Biopsy	*120, *490
Birth Cohort	513
Blacks	1035
Blood	*204, 541
BMB	*120, *204, 541, *543, *548, 738, 910
Chemical Exposure	73, 84, 99, 213, 279, 310, *352, 965, 1025, 1157
Chlorination	84, *110
Chromosome Effects	*120, 738
Cluster	1016
Coffee	310
Cytology	*548
Diabetes	104
Diet	*352
DNA	*120, 642
DNA Adducts	*204, 541, *543, 642
DNA Repair	*204
Drugs	232, 310, 513, 1134
Dusts	73, 305, 487, *490, 910, 965
Dyes	509, 738
Environmental Factors	213, 392, 1035
Enzymes	*543
Familial Factors	546
Female	1035
Fertilizers	910
Genetic Factors	178, *543, 866, 1025
Genetic Markers	*120
Genotype/Phenotype	*204, *543
Geographic Factors	84, 392, 899
High-Risk Groups	900
Infection	*120
Leather	305
Lifestyle	104, 392, 1035
Lymphocytes	642, 738
Mathematical Models	513
Metabolism	738
Metals	73, 213, 487, 965
Micronuclei	*120, 642, 738
Mutagen	99
Nutrition	310
Occupation	73, 99, 162, 178, *204, 213, 279, 305, 310, *352, 484, 487, 509, *687, 695, 738, 900, 910, 965, 985, 1016, 1025, 1027, 1035, 1096, 1157
Oncogenes	*543, 1025
Passive Smoking	642
Pesticides	162, 965
Petroleum Products	73, 213, 965
Plastics	73, 305, 738, 965
Premalignant Lesion	866
Race	985
Record Linkage	104, *204, 232
Registry	104, *110, *204, 213, 232, 546, 787, 899, 965, 1035
Reproductive Factors	546

SCE	642, 738
Screening	546, 1027, 1096
Socio–Economic Factors	1096
Solvents	73, 305, *687, 965, 1016, 1157
Sputum	642
Sweeteners	99, 104, 310
Time Factors	513
Tobacco (Smoking)	104, 310, *352, 513, 541, *543, 546, 642, 738, 866, 1016, 1096, 1134
Trends	513
Urine	*204, 541, *548, 642, 910, 1027, 1096
Water	84, *110, 310, 392, 652, 787, 1016
Wood	305, 965

Bone

Alcohol	171
BMB	*845
Chemical Exposure	171, *621
Childhood	*292
DNA Repair	805
Dose–Response	370
Drugs	171, 370
Environmental Factors	674
Familial Factors	805
Genetic Factors	805
Genetic Markers	805
Geographic Factors	*292
Heredity	*292, 805
Histology	*845
Intra–Uterine Exposure	171
Late Effects	370
Metals	102
Migrants	*292
Multiple Primary	805
Occupation	171, *621
Pedigree	805
Prognosis	171
Radiation, Ionizing	171, 370, 377
Radiation, Ultraviolet	805
Registry	171, *292, 805, *845
Screening	805
Tobacco (Smoking)	171
Treatment	674
Water	102

Brain

Age	979, 1181
Air Pollution	*205, 213
Alcohol	171, 955, 1151
Antioxidants	*203
Biomarkers	*514
Blood	*514, *1056
BMB	*514, *1056
Chemical Exposure	84, 171, 213, 304, 504, *621, 765, 965, 1091, 1095, 1104, 1180, 1181
Childhood	44, 57, *205, 303, *473, *665, 955, 1087, 1088, 1151
Chlorination	84
Classification	432
Clinical Records	85
Coffee	955
Congenital Abnormalities	955
Cosmetics	1087
Demographic Factors	22
Diet	304, *473, 1087
Drugs	44, 85, 171, 303, 1087
Dusts	*272, *715, 965

Electromagnetic Fields 38, 766, 1088
Environmental Factors 213, 303, 765, 766, 1091
Familial Factors 1151
Family History *1056
Female 96
Fertilizers 86
Genetic Factors *1056
Genetic Markers *514
Genotype/Phenotype *203
Geographic Factors 84, *462
Herbicides 86, 1104, 1180
Histology 85, 432, 1087
Insecticides 86, 1104
Intra-Uterine Exposure 44, 171, 303, 1087, 1151
Lifestyle 765
Metals 213, *272, *362, *715, 965, 1091, 1104, 1151
Migrants *462
Occupation 38, 86, 96, 171, *203, 213, 251, *272, 304, *362, 504, 532, 583, *621, *715, 965, 1012, *1021, 1087, 1091, 1095, 1104, 1151, 1180, 1181
Parental Occupation 57, *205, *665, 955
PCR *514
Pesticides 38, 86, 96, 504, 532, *665, 965, *1021, 1104, 1180
Petroleum Products 86, 213, 965, 1091
Plastics 38, *362, 965, 1091, 1104
Prognosis 171
Radiation, Ionizing 44, 171, 303, *462, 583
Record Linkage 86, *203
Registry 22, 44, 57, 171, *203, *205, 213, 251, *272, *362, 532, *665, 766, 955, 965, 979, 1087, 1088
Reproductive Factors 96, *665
RFLP *514
Rubber 1087
Rural *1021
Segregation Analysis *1056
Solvents 38, *362, 965, *1021, 1091, 1104, 1151
Survival 22
Time Factors 1181
Tissue *1056
Tobacco (Smoking) 96, 171, *514, 955
Trauma 44, 303
Trends 22, 85
Vaccination 348
Virus 348
Water 84, 1087
Wood 965

Breast (F)

Abortion 121, 400, 650, 707, 770, 1072
Age 170, 194, 230, *291, *396, 399, 619, 637, 659, 970, 1082, 1097, *1202, 1204
Alcohol 25, *46, 63, 89, *214, 299, 423, *461, *478, 507, 510, *515, 546, 565, *609, 612, 616, 618, 620, 640, 650, 680, 711, *919, *948, 949, 974, 1000, *1006, 1072, 1097, *1195
Antioxidants *203, 347, 639, *987
Ataxia Telangiectasia 1046
Autopsy *948
Biochemical Markers 235, 408, 467, 468, 1197, 1222
Biomarkers 74, *514, 558, 639, *1124
Biopsy 309, *1043, 1115, 1116
Birth Cohort 194
Blood 21, *26, *27, *46, 74, 253, 480, *514, 908, 923, 974, 991, 1078, 1111, *1112
Blood Group 401, 660

BMB 21, 25, *26, *27, *46, 74, 235, 241, 253, 446, 480, 507, *514, *515, 518, 533, 612, 624, 636, 639, 908, *919, 923, *948, 991, *1005, *1043, *1063, 1078, 1111, *1112, 1116, 1130, 1132, *1142, *1143, *1144, 1218
Breast Cysts 724
Caucasians 962
Chemical Exposure *396, *1043, 1107
Chemotherapy 619, 620, 970
Childhood 889
Chinese 299
Cholesterol 533, 786, 949, *1005
Chromosome Effects *78, 558, 1054
Cluster *78, 962, 1105
Coffee *214
Contraception 25, 433, 650, 924
Cosmetics 991
Cost–Benefit Analysis 276, 630
Cytology 924
Data Resource 996
Diabetes 433, 1065
Diet 21, 63, 74, 89, 105, 127, 131, 146, 157, 242, 299, 408, 423, 442, 446, *478, 480, 507, 510, *515, 517, 533, 565, *609, 612, 618, 624, 639, 640, 680, 711, 797, 848, 908, *919, 923, *948, 949, 974, 980, 991, 1000, 1031, 1065, *1067, 1105, 1111, 1130, 1132, *1195, 1197, 1218, 1222, *1223
DNA 25, 446, 558, 962, *1006, 1078
DNA Repair 805
Dose–Response 224, 347, 970, 1107
Drugs *46, 232, 299, *302, 309, 616, 618, 926, *948, *987, 991, 996, 1130, 1140, *1191
Dusts 1197
Dyes 991
Electromagnetic Fields *1195
Environmental Factors *26, *27, 146, 157, *474, *567, 674, 962, 1078
Ethnic Group *474, *605, *948
Familial Factors *26, *27, *46, *78, *79, 146, 235, 299, 309, 395, 399, 510, 546, 558, 616, 618, 637, 640, 644, 797, 805, 889, 949, 962, 991, 1046, 1053, 1059, 1078, 1204, 1213
Family History *26, *27, *79, *461
Fat 63, *609, 620, 637, 639, 797, 848, *919, 997, 1031, 1105, 1132, *1195, 1197, 1222
Fatty Acids *515
Female 105, 991
Fertility 924
Fibre 640, 797, *919, 1197
Genetic Councelling *703
Genetic Factors 25, *26, *27, 146, 157, 230, 558, 616, 620, *703, 805, 962, 974, 1046, 1053, 1054, 1078, 1111
Genetic Markers 25, *26, *27, *78, 146, 170, *514, 805, 962, 1046, 1078
Genotype/Phenotype *203, *1063
Geographic Factors 146, 194, *462, 679
Heredity *78, 805, *1063
High–Risk Groups *78, *79, 174, 235, 236, 309, 395, 644, 724, 802, 836, 889, *919, 962, 1046, 1054, 1197, 1242
Histology 157, 395, 464, 1116
Hormones 121, *214, 235, 299, 309, 408, *436, *461, *474, 507, 510, 517, 616, 617, 618, 620, 644, 650, 659, 711, 803, 925, 926, 949, 974, 980, 991, 996, *1006, *1067, 1082, 1083, *1119, 1130, 1132, 1140, *1142, *1202, 1204, *1223
Hysterectomy 804
In Situ Carcinoma 23, 802, *1196
Infection 1000
Intra–Uterine Exposure 926
Lactation 60, 121, 400, 620, 659, 1097
Late Effects 879
Latency 879
Lifestyle 21, *27, 127, *436, *567, *605, 620, 624, 640, 707, 962, *1112, 1213

Linkage Analysis 170, 558, 1078, 1111, *1112
Lipids 21, *214, *436, 991, *1005, 1065, 1218, 1222
Mammography 11, 23, 114, 235, 276, 309, *461, 507, 630, 791, 797, 836, *919, 941, 942, 1031, 1059, 1132, 1197
Mathematical Models 630, 970
Menarche 60, 170, 299, 400, 408, 433, 660
Menopause 60, 131, 170, 230, 400, 433, 660, 803, 925, 980, 991, *1119, *1202
Menstruation 25, 400, 442, 620, 1072, *1202
Metabolism *1063
Metals 74, 241, 640, 797, 991, 1222
Micronutrients 974
Migrants 21, *291, *462
Minerals 127
Monitoring 174
Multiple Primary 146, 223, 224, *302, 463, 464, 619, 620, 805, 970
Mutation, Somatic 558
Nutrition 89, 127, 446, 480, *609, 612, 639, 711, 797, 848, 974, 1130, *1195, 1197, 1218, 1222
Obesity *46, *79, 89, 423, 637, 644, 949, 1105, *1112
Occupation 121, *203, *357, *396, 680, 991, 1000, *1006, 1107, *1195, 1197
Oncogenes 558, 620, *1006, 1116, *1202
Oophorectomy 804
Oral Contraceptives *26, *46, 400, *461, 517, 565, 616, 617, 618, 620, 650, 722, 879, 888, 924, 974, 991, 1059, 1072, *1119, 1204
Parity 60, 170, 230, *396, 408, 433, 442, 517, 637, 659, 660, 924
Passive Smoking 121, *808, 1040
PCR *514
Pedigree *26, *27, 146, 805, 962
Pesticides *1006, *1043, *1144
Physical Activity *46, *478, 507, 908, *948, 949, 980, 991, 1072, 1213
Physical Factors 25, 60, 63, *214, 230, 408, 507, 565, 660, 680, 786, 848, 949, 970, 1105, 1213
Plasma 21
Pregnancy 400, 924
Premalignant Lesion 309, 724, 1046, 1115, 1140
Prevalence 962, 1065, 1197
Prevention *78, 236, 253, 309, 518, 544, *567, *987, 1031, *1191, 1197
Prognosis *27, 253
Psychological Factors 140
Psychosocial Factors 854, 855
Race 299, 644, *1006
Radiation, Ionizing *462, 970, 1072, 1107
Radiation, Non-Ionizing 970, *1195
Radiation, Ultraviolet 805
Radiotherapy 223, 224, 619, 620, 970, 1072
Record Linkage 105, *203, 232, 401, 803, 973, *1119
Recurrence 518, *1196
Registry 23, 25, *26, *27, *46, 60, 89, 121, 170, *195, *203, *214, 223, 224, 232, 235, 236, 241, 242, 253, *255, 276, *291, 395, *396, 399, 400, 401, 417, 433, 463, 464, *478, 507, 533, 546, 620, 624, 637, 640, 651, 680, 722, 724, 770, 786, 791, 802, 803, 804, 805, 855, 879, 889, 923, 925, *948, 973, *1006, *1063, *1067, 1072, 1083, 1105, *1119, *1191, *1195, *1196, 1218, *1223
Religion 433
Reproductive Factors 25, *46, 157, *214, 299, *302, 395, 423, *436, 442, *461, *478, 507, 510, 546, 565, 618, 644, 707, 711, 974, 980, 1000, *1006, 1072, 1111, 1204
RFLP *514, 1078
Rural 679, *1006
Screening 11, 60, *78, 105, 114, 174, *195, *255, 276, 288, 374, 499, 546, 630, 636, 651, 660, 717, 723, 805, 855, 941, 942, 1059, 1130, 1197, *1206, 1242
Segregation Analysis 146, 170, 1053, *1063, 1078, *1112
Sero-Epidemiology 241
Serum 235, 241, 446, *919, *1005, 1132, *1142, *1143, *1144
Sex Ratio 619, 624
Sexual Activity 517
Socio-Economic Factors 127, 408, 433, 442, 507, *605, 797, 854, 1204
Stage 174, *195, 463, 464

Stress 854, 855
Surgery *461, 973, *1196
Survival 23, 174, 417, 463, 630, 636, 637, 786
Time Factors 194, *291, 464, 630, 1072, *1191
Tissue 74, *948
Tobacco (Smoking) *46, 89, 127, 146, *214, 299, 423, *461, *478, *514, *515, 546, 565, 619, 680, *808, *948, 949, 991, 1000, 1040, 1059, 1072, 1197, 1213
Toenails 74, 507, 612, 624, 639, 991
Tonsillectomy 650
Trace Elements 127, 468, 612
Treatment 253, *302, 674, 803
Trends 65, 194, *195, 679
Tubal Ligation 1139
Twins 230
Urban 679
Uric Acid 949
Urine 507
Vegetarian 923
Virus 1000
Vitamins 63, 127, 241, *436, 480, 533, *609, 612, 640, 797, *987, 1000, *1067, *1143, 1197, *1223

Breast (M)

Cluster *54
Electromagnetic Fields *1004
Familial Factors 1053
Genetic Factors 1053
Geographic Factors *54
Occupation *1004
Radiation, Non–Ionizing *1004
Registry 211
Segregation Analysis 1053
Survival 211
Trends 211

Burkitt's Lymphoma

AIDS *658
HIV *658
Infection *658
Sero–Epidemiology *658

Childhood Neoplasms

Abortion *111, 916
Adolescence 1225
Alcohol 831, 1151
Animal 1165
Birth Cohort 823
Blood *822, 831
BMB *822, 824, 831
Chemical Exposure 647, *822
Chemotherapy *111, 1150, 1225
Childhood 307, 598, 823, 922, *934, 1151, 1225
Clinical Effects 827
Clinical Records 832
Cluster *54, 375, 646, 823, *852, 905, 1165
Congenital Abnormalities 600, 827, 890, 891, 916, 1150, 1225
Diet 831
Disease, Other *1110
DNA Repair 805
Dose–Response 891
Down's Syndrome 890, 891
Drugs 221, 600, 647, 827, 831, 832, 1225
Electromagnetic Fields 647, 766, 769, *822

Environmental Factors	321, 376, 600, 766, 769, 824, 831
Epilepsy	221
Ethnic Group	824, 922
Familial Factors	307, 805, 904, 916, 1151
Family History	321, *1110
Fertility	*111, *1110, 1225
Genetic Factors	307, 321, 376, 647, 805, 916, 1150
Genetic Markers	805
Geographic Factors	*54, 646, 823, 824, 904, 905, 922, 1165
Heredity	805
Histology	*852
Immunodeficiency	599
Infection	321, *822, 922
Infertility	916
Intra–Uterine Exposure	222, 457, 600, 783, 827, 890, 904, 1151, 1225
Late Effects	*111, 375, 917, *1110, 1150
Lifestyle	376, 824
Mapping	823, 1113
Marijuana	647
Menarche	1225
Menopause	1225
Metals	222, 1147, 1151
Migrants	922
Monitoring	1165
Multiple Primary	375, 805, 827, *1110, 1150
Mutagen	831
Mutation, Germinal	904, 916
Mutation, Somatic	827, 1150
Occupation	222, 457, 827, 890, *934, 1012, 1147, 1151
Paints	222, 1147
Parental Occupation	376, 647, *822, 824, *934, 1147
Passive Smoking	647, 831
Pedigree	805
Pesticides	1147, 1165
Plastics	222
Pregnancy	*111, 783, 827, 916
Prognosis	1042
Psychosocial Factors	1225
Radiation, Ionizing	221, 375, 376, 600, 647, 783, *822, 823, 827, 890, 891, 905, *906, *934
Radiation, Ultraviolet	805
Radiotherapy	*111, 827, 1150, 1225
Record Linkage	904, 905, *906
Registry	*111, 221, 222, 307, 321, 375, 376, 590, 600, 646, 647, 766, 769, 783, 805, 827, *852, 890, 891, 904, 905, *906, 922, 1165
Rubber	222
Screening	805
Sex Ratio	441
Sib	827, 904, 1225
Socio–Economic Factors	905, 1042
Solvents	222, 1147, 1151, 1165
Survival	375, 441, *852, 916, 917, 1042, *1110
Time Factors	823, 824, 1042, 1165
Tissue	831
Tobacco (Smoking)	831
Treatment	732, 917, *1110
Trends	441, 922, 1165
Twins	600, 904
Ultrasound	827
Urine	831
Virus	375, 823, 827
Vitamins	832

Choriocarcinoma

Abortion	516
Alcohol	516

Chemotherapy 6
Contraception 516
Diet 516
Familial Factors 516
Fertility 6
Hormones 516
Parity 516
Pregnancy 6
Premalignant Lesion 6
Prognosis 6
Registry *549
Tobacco (Smoking) 516
Treatment *549

Colon

Age *291, *843, 1181
Alcohol *53, 63, 89, 149, 166, 167, 260, 424, *478, 510, 546, *594, 595, 601, 640, 737, 949, *976, 1098, 1228
Analgesics *976, *1127
Anatomical Distribution 641
Antioxidants *203, *818, *987
Autopsy 398
Bile Acid 366, 526
Biochemical Markers 468, 1197
Biomarkers 74, 526, 558, 894, 1060
Biopsy 398
Blacks 1035
Blood 21, *27, 74, 366, 480, 526, *842, 894, 908
BMB 21, *27, 74, 241, 260, 366, 373, 480, 526, 533, 624, 638, 762, *842, 894, 908, *976, *1010, 1045, 1106
Chemical Exposure 73, 84, 747, 860, 1032, 1092, 1093, 1181
Chlorination 84, *110
Cholesterol 526, 533, 601, 894, 949
Chromosome Effects 51, 558
Classification 398
Coffee 465, 737
Cooking Methods *976
Cost–Benefit Analysis 147, 148
Crohn's Disease 226, 800, 821
Dairy Products 641
Data Resource 996
Diabetes 1065
Diet 21, 63, 74, 89, 105, 127, 146, 149, 150, 161, 166, 167, 242, 260, 275, 366, 373, 424, 465, 470, *478, 480, 510, 526, 533, *594, 595, 601, 624, 640, 641, *675, 712, 737, 747, 806, *818, 893, 908, 949, *976, 980, *1010, 1057, 1058, 1065, 1066, 1098, 1183, 1197, 1210, 1212, *1223, 1228
Disease, Other *527, *843
DNA 51, 558, 762, *778
DNA Repair 526
Dose–Response 748, 860
Drugs 166, 747, 893, *987, 996, *1010, 1058, *1127
Dusts 34, 73, *323, 748, *758, 1197
Environmental Factors 5, 24, *27, 34, 146, 258, *567, 712, *1010, 1035
Ethnic Group 465, *675
Faeces 260, 261, 366, 373, 526
Familial Adenomatous Polyposis 5, 24, 28, 366
Familial Factors 5, 24, *27, 28, 51, 146, 161, 166, 167, 470, 510, 521, 526, 546, 558, 640, *843, 949, 1145, 1213
Family History *27, 149, 150, 893
Fat 63, 465, 595, 641, 893, 997, 1058, 1197, 1210, 1212
Female 105, 1035, 1098
Fibre 640, 641, 893, 1057, 1058, 1197, 1210, 1212
Fruit 465
Gastrectomy *1014

Term	References
Genetic Factors	*27, 28, 146, 216, 275, 366, 558, *818, *933, 954, *976, 1183
Genetic Markers	*27, 146, 216, 954
Genotype/Phenotype	*203
Geographic Factors	84, 146, 258, 737
Heredity	*53, 216, 366, 521, 747
High-Risk Groups	5, 24, 28, 147, 150, 216, 373, 455, *818, 954, 1197
Histology	166, 398, 954
Hormones	166, *407, 510, 949, 980, 996, *1119, *1223
Infection	366
Latency	860
Lifestyle	21, *27, 127, 147, 149, 161, 260, *567, 601, 624, 640, 712, 806, *1010, 1035, 1213
Linkage Analysis	366, 558, *778, *842
Lipids	21, 465, 806, 1065
Lymphocytes	526
Mammography	1197
Marital Status	601
Menopause	980, *1119
Menstruation	167
Metals	73, 74, 241, *323, 640, 1032, 1057, 1093, 1161
Migrants	21, *291
Minerals	127, 373, 1057
Mining	34
Multiple Primary	146, 463
Mutation, Somatic	558
Nutrition	*53, 89, 127, 166, 366, 480, *675, 806, 893, *976, 1058, 1197, 1210, 1212, 1228
Obesity	89, 949, *1010
Occupation	34, 73, 166, 167, *203, 258, *323, *675, 747, 748, *758, 860, 893, *976, *1010, 1032, 1035, 1092, 1093, 1161, 1181, 1197, 1210, 1212
Oncogenes	558
Operation, Previous	*1014
Oral Contraceptives	*1119
Parasitic Disease	147
Pedigree	*27, 146, 366, 954
Petroleum Products	73, 860
Physical Activity	149, 166, 167, 470, *478, *594, 806, 893, 908, 949, *976, 980, 1058, 1183, 1210, 1213
Physical Factors	63, 601, 949, 1213
Plasma	21
Plastics	73, 1032, 1092
Polyps	366, 455, *594, *675, 885, 894, *1010, 1045, 1058, 1228
Premalignant Lesion	5, 260, 366, 373, *594, 894, 954, 1058, 1128
Prevalence	*594, 1065, 1197
Prevention	373, *567, 986, *987, 1045, 1145, 1197, 1210
Prognosis	*27
Protein	893
Psychosocial Factors	166
Radiation, Ionizing	166
Record Linkage	105, *203, 800, 973, *1119
Registry	24, *27, 28, 89, *110, 166, 167, *203, 216, 241, 242, 261, *291, 463, 465, *478, 521, *527, 533, 546, 624, 640, 641, 748, *758, 787, 800, *818, 860, *862, 954, 973, 1035, 1098, *1119, 1183, *1199, 1210, *1223
Religion	465
Reproductive Factors	149, 166, 167, *478, 510, 546, 980, 1098, 1183
RFLP	51
Rubber	1092
Rural	258
Screening	28, 105, 148, 261, 374, 455, 546, 762, *862, 892, 958, 1032, 1106, 1145, 1197, *1199, 1212, 1215
Segregation Analysis	146, *842, *933
Sero-Epidemiology	241
Serum	241, 260, 526, 1045
Sex Ratio	624
Socio-Economic Factors	127, 465, 601, *675
Solvents	73, 1092
Stage	463

Surgery 958, 973
Survival 258, 463
Tea 167
Time Factors *291, 737, 1145, 1181
Tissue 74, 366, 762
Tobacco (Chewing) *675
Tobacco (Smoking) 34, *53, 89, 127, 146, 149, 166, 167, 424, 465, *478, 546, 595, 601, *675, 737, 747, 949, 958, *976, *1010, 1032, 1058, 1197, 1213, 1228
Tobacco (Snuff) *675
Toenails 74, 624, 638
Trace Elements 127, 468, 526, 638, *1017
Treatment 258
Trends 65, 258
Tumour Markers 762
Ulcerative Colitis 226, 526, 762, 800, 821
Uric Acid 949
Urine 373
Vegetables 465, 1228
Virus 366
Vitamins 63, 127, 241, 470, 480, 526, 533, 595, 640, 806, 894, *987, 1057, 1066, 1197, 1210, *1223, 1228
Water *53, 84, *110, 147, 787

Endocrine Glands

BMB *494
Genetic Markers *494
Prevention *494

Eye

Blacks 1035
Drugs 993
Environmental Factors 1035
Female 1035
Lifestyle 1035
Occupation 1035
Photochemotherapy 993
Radiation, Non–Ionizing 993
Registry 1035

Female Genital

Alcohol 949
Antibodies 252
Biopsy 252
Cholesterol 949
Condyloma 252
Diet 949
DNA 252
Drugs 926
Familial Factors 949
Hormones 803, 926, 949
HPV 252
Infection 252
Intra–Uterine Exposure 926
Menopause 803
Obesity 949
Oral Contraceptives 888
Physical Activity 949
Physical Factors 949
Prevention 252
Record Linkage 803
Registry 803
Screening 252
Sero–Epidemiology 252

Sexual Activity 252
Tobacco (Smoking) 252, 949
Treatment 252, 803
Uric Acid 949

Gallbladder

Alcohol *734, 737
Biochemical Markers 1152
Blood *734
BMB *734
Chromosome Effects *734
Coffee *734, 737
Cytology *734
Diet 673, 737, 1081
Drugs 1081
Environmental Factors 258, 1081
Ethnic Group 1081
Female 1081
Gallstones 124, 673, 1152
Geographic Factors 258, 737
Hormones 1081
Occupation 125, 258, *734
Oral Contraceptives 673
Prognosis *734
Reproductive Factors 673, 1081
Rural 125, 258
Socio–Economic Factors 125
Survival 258
Time Factors 737
Tissue *734
Tobacco (Smoking) 673, *734, 737
Toenails *734
Treatment 258
Trends 258
Urban 125
Urine *734

Gastrointestinal

Adolescence 1241
Age 230
Air Pollution 192, 213
Alcohol 76
Bile Acid 880
Blood 923
Blood Group 660
BMB 880, 923
Chemical Exposure 193, 213, *649, 1091
Chemotherapy 340
Cluster 64
Coal 192
Diet 923, 1241
DNA 880
DNA Repair 805
Dose–Response 76, 192
Drugs 880, 927
Dusts 72, 76, 101, 691, *692
Environmental Factors 72, 192, 193, 213, 392, *649, 1091
Familial Adenomatous Polyposis 880
Familial Factors 805
Female 72
Fertilizers 193
Genetic Factors 230, 805, 954
Genetic Markers 805, 954

Geographic Factors 64, 392
Heredity 805
High–Risk Groups 954
Histology 880, 954
Lifestyle 392
Menarche 660
Menopause 230, 660
Metals 192, 193, 213, 1091
Multiple Primary 340, 805
Mutagen 193
Nutrition 1241
Obesity 1241
Occupation 76, 101, 192, 193, 213, *649, 691, *692, 1034, 1091
Parity 230, 660
Pedigree 805, 954
Petroleum Products 213, 1034, 1091
Physical Factors 230, 660, 1241
Plastics 691, 1091
Polyps 880
Premalignant Lesion 880, 954
Prevention 880
Radiation, Ionizing 192, 193
Radiation, Ultraviolet 805
Radiotherapy 340
Record Linkage 192
Registry 213, *649, 805, 923, 927, 954
Resins 691
Rubber 691
Screening 660, 805
Solvents 1091
Textiles 691
Time Factors 64, 76
Tissue 880
Tobacco (Smoking) 76
Trends 64
Twins 230
Vegetarian 923
Water 392
Welding 1034

Genitourinary

Adolescence 1241
Diet 1241
Metals 1160
Nutrition 1241
Obesity 1241
Occupation 1160
Physical Factors 1241

Haemopoietic

Air Pollution 213
Chemical Exposure 19, 213, 504, 677, *727, 965, 1091
Drugs *727
Dusts 677, *692, 744, 965
Environmental Factors 213, 1091
Familial Factors *476
Genetic Factors *476
Metals 213, 393, 965, 1091, 1160
Occupation 19, 213, 504, 677, *692, *727, 744, 965, *984, 1091, 1160
Pesticides 504, *727, 965
Petroleum Products 19, 213, 965, 1091
Plastics 965, *984, 1091
Prevalence 716
Psychosocial Factors 358

Radiation, Ionizing 393, 716
Registry 213, 744, 965, *984
Solvents 677, 965, *984, 1091
Stress 358
Time Factors *984
Wood 744, 965

Head and Neck

Cost–Benefit Analysis 389
Cytology 389
Dose–Response 632
Genetic Factors *553
Histology 389
Multiple Primary 463
Premalignant Lesion 389
Radiotherapy 632
Registry 463
Screening 389
Stage 463
Survival 463
Treatment 389

Hodgkin's Disease

AIDS 218, 479, *658
Alcohol *311, *1244
Allergy *1244
Appendectomy *1244
Biochemical Markers 431, *676
Biomarkers 270
Blood 270, *929, *1244
BMB 270, *929, *1125, 1154
Caucasians 1192
Chemical Exposure *1244
Chemotherapy 270
Childhood *292, 540, 1154
Cluster *929, 1192
Coal 40
Congenital Abnormalities *1244
Diet *1244
DNA *676
DNA Adducts 270
Drugs 270, *311, *1125, 1154
Dyes *311
EBV *929, 1154
Electromagnetic Fields *311
Environmental Factors 431, 1154
Ethnic Group *605, *1125
Family History *1125
Geographic Factors *292
Herbicides *311, 542
Heredity *292
High–Risk Groups 479
Histology 431
HIV 479, *658
HLA *929, *1125
Homosexuality 218
Infection 218, *311, *658, 1154, 1192
Intra–Uterine Exposure *1244
Lifestyle *311, *605
Micronuclei 270
Migrants *292
Mining 40
Multiple Primary 270, 540
Occupation 40, 93, 251, *311, 542

Oncogenes	270, *676
Paints	*311
Parental Occupation	1154
PCR	*676, 1154
Pesticides	93, *311, 542
Physical Factors	*1244
Plasma	1154
Radiation, Ionizing	*311, *1125, *1244
Registry	40, 251, 270, *292, *311
Saliva	*929
SCE	270
Segregation Analysis	1154
Sero–Epidemiology	218, *658
Sexual Activity	218
Socio–Economic Factors	*605, 1154
Solvents	*311, 542
Survival	540
Tobacco (Smoking)	*311, *1125, *1244
Tonsillectomy	*1244
Treatment	540, *676
Virus	218, *676, *929
Welding	542
White Cells	1154

Hydatidiform Mole

Abortion	516
Alcohol	516
Chemotherapy	6
Contraception	516
Diet	516
Familial Factors	516
Fertility	6
Hormones	516
Parity	516
Pregnancy	6
Premalignant Lesion	6
Prognosis	6
Registry	*549
Tobacco (Smoking)	516
Treatment	*549

Hypopharynx

Alcohol	268, 315, *317, 439
Chemical Exposure	315, *317
Diet	*317, 439
Dusts	315
Female	425
Hygiene	268, 439
Nutrition	*317
Occupation	268, 315, *317
Passive Smoking	315
Registry	*317, 425
Tobacco (Chewing)	425
Tobacco (Smoking)	268, 315, *317, 425, 439
Tobacco (Snuff)	439
Trends	425

Inapplicable

Air Pollution	*853
Animal	249
Antioxidants	*817
Autopsy	1003
Biochemical Markers	966

Biomarkers *4, *233, *355, *817, *853, 1003, *1020
BMB *4, 201, 249, *817, 1009
Chemical Exposure 48, 58
Chemical Mutagenesis 390, 1001
Chemotherapy 199, 626, 681
Cholesterol *817
Chromosome Effects 198, 199, 249, *355, 390, 681
Coal 48, 626
Congenital Abnormalities 58, 390
Diet *817, 1049
DNA 77, 626
DNA Adducts 48, 201, *853, 1003, 1009
DNA Repair 201, *817
Dose–Response 1009
Drugs 354, *355, 390, 626, 1001
Environmental Factors 48
Ethnic Group 1049
G–Banding *355
Genetic Factors 554
High–Risk Groups 554
HPRT–Mutants 77, *817, *1020
Lymphocytes 77, 198, 199, 201, 390, 554, 830, *853, 1009, *1020
Metals 77, *355
Micronuclei 201, 1009, *1020
Mining 48, 198
Monitoring 249, 626
Mutagen 554, 1001, 1003
Mutagen Tests 626
Mutation, Germinal 58, 390, 1001
Mutation, Somatic *4, 58, 390, 1001
Mycotoxins 1001
Nutrition *817, 1049
Occupation *4, 48, 77, 198, 199, 201, *233, 249, 354, *355, 626, 681, 966, 1009, *1020
Oncogenes *355
Passive Smoking 1003
PCR *355, 1001
Pesticides 966
Petroleum Products 198
Plastics 201, 249, 1009
Premalignant Lesion 554
Prevalence 198
Prevention 58, 198, 199
Radiation, Ionizing *4, 77, 830
Registry 58
SCE 354, *355, 390, 681, 1009
Screening 58
Serum 201
Solvents 354, *355
Tobacco (Smoking) 681, 1001
Transplacental *853
Urban *853
Urine 201

Kaposi's Sarcoma

AIDS 41, 382, 479, *658, *952, 957, 964, 1041
Antibodies 657, *898
Antigens 529
BMB 41, 957
Clinical Records *952
Drugs 41, 382
Environmental Factors 382
Haemophilia 382
High–Risk Groups 479
HIV 41, 479, 657, *658, *898, *952, 957, 964, 1041

HLA 529
Homosexuality 382, *952, 957
HTLV 382
Immunology 529
Infection 382, *658
Record Linkage 1041
Registry 1041
Sero–Epidemiology 382, 657, *658, 964
Sexual Activity 41, 957
Sexually Transmitted Diseases 41, *898
Socio–Economic Factors 382
Treatment 41
Virus 382

Kidney

Alcohol 45, 76, 330, 560, 795, 1141
Analgesics 45, 330, 795
Birth Cohort 513
Chemical Exposure 73, 84, 506, 873, 1093, 1104, 1189
Chlorination 84
Clinical Records 1141
Coffee 45, 330, 795, 1141
Diabetes 45, 795
Diet 105, 368, 560, 795
Diuretics 330, 795, 1141, 1164
Dose–Response 76
Drugs 45, 368, 513, 560, 795, 1134, 1164
Dusts 73, 76, *272, 487, 506, *715
Environmental Factors 873
Female 105
Geographic Factors 84
Herbicides 368, 1104
Hormones 45, *219, 795, 1164
Insecticides *219, 1104
Latency 1189
Leather 506
Lifestyle 560
Mathematical Models 513
Metals 73, *272, 487, 506, *715, 790, 873, 1093, 1104
Obesity 45, 368, 795
Occupation 45, 73, 76, *219, *272, 330, 368, 487, 506, *715, 790, 795, 873, 1037, 1093, 1096, 1104, 1189
Passive Smoking 45, 795
Pesticides *219, 368, 506, 1104
Petroleum Products 73, 1037, 1189
Physical Activity 368
Plastics 73, 506, 1104
Race 1141
Radiation, Ionizing 377
Record Linkage 105
Registry 45, *219, *272, 506, 790, 795, 1141
Reproductive Factors 45, 795
Rubber 506
Screening 105, 1096
Socio–Economic Factors 560, 1096
Solvents 73, 506, 1104
Tea 1141
Time Factors 76, 513
Tobacco (Smoking) 45, 76, *219, 330, 368, 513, 560, 795, 1037, 1096, 1134, 1141, 1164
Trends 513
Urine 1096
Water 84
Wood 506

Larynx

Air Pollution 511, 869
Alcohol 59, 70, 76, 181, 315, *317, 364, 439, *478, 664, 784, 1236, *1246
Antioxidants *204
Biomarkers *204
Birth Cohort 513
Blood *204, *1117
BMB *204, *1117
Chemical Exposure 315, *317, 1095
Cluster 869
Coffee *1246
Data Resource 869
Diet 70, 181, *317, 364, 439, *478, 784, *1246
DNA Adducts *204
DNA Repair *204
Dose–Response 76
Drugs 513, 807, 1134
Dusts 59, 76, 315, 511
Environmental Factors 869
Enzymes 807
Female 425
Genotype/Phenotype *204
Geographic Factors 869
Histology 59
Hygiene 181, 439
Mathematical Models 513
Metals 59, *362
Multiple Primary 70, 231, 364, *1117
Nutrition *317
Occupation 59, 76, 181, *204, 315, *317, *362, 364, 511, 664, 784, 931, 1095
Passive Smoking 315
Petroleum Products 869
Physical Activity *478
Plastics *362, 869
Radiation, Ionizing 181
Record Linkage *204
Registry 59, 181, *204, 231, *317, *362, 425, *478, 664, 784, 807, 869, *1117
Reproductive Factors *478
Solvents *362, 869
Stage 59, 364
Surgery *1246
Time Factors 76, 513, 1236
Tobacco (Chewing) 425, 784
Tobacco (Smoking) 59, 70, 76, 181, 315, *317, 364, 425, 439, *478, 513, 784, 807, 1134, 1236, *1246
Tobacco (Snuff) 439, 784
Treatment 59, 231
Trends 425, 513
Urine *204
Wood 59

Leukaemia

Age 230, *384, 619, 633, 970, 1181
Air Pollution 192, *205, 869
Alcohol 171
Animal 1165
Antibodies 367
Ataxia Telangiectasia 1046
Biochemical Markers 431
Biomarkers 508, 1026
Birth Cohort 823
Birth Order 435
Blacks 730
Blood *380, *822, *929
BMB *380, *822, *929, 1026

Caucasians	730
Chemical Exposure	33, 84, 119, 171, 208, 508, *621, 653, 779, *781, *822, 849, 859, 1107, 1179, 1180, 1181
Chemotherapy	339, 340, 619, 672, 730, 871, 915, 970
Childhood	56, 57, 119, *205, 287, *292, 307, 367, *380, *384, *385, 435, 633, *665, 672, *704, 730, 777, 823, 859, *886, 915, 922, *934, *936
Chlorination	84
Chromosome Effects	62, 539
Classification	730
Clinical Effects	827
Clinical Records	435, *850
Cluster	*350, 823, 829, 869, *886, 905, *929, 1064, 1165
Coal	192
Congenital Abnormalities	119, 600, 827
Data Resource	*850, 869
Demographic Factors	*342
Diet	127, 180
DNA	937
DNA Adducts	1026
DNA Repair	805
Dose–Response	192, 224, *262, 370, 752, 871, 901, 902, 970, 1107
Down's Syndrome	937
Drugs	119, 171, 180, 221, 349, 370, 600, 827, 849
Dusts	*106, 208, 487, 653, *864
EBV	*929
Electromagnetic Fields	*18, 119, 334, *380, 653, 655, 766, *822
Environmental Factors	192, 349, 376, 431, 435, 508, 600, 766, 829, 869
Epilepsy	221
Ethnic Group	922
Familial Factors	307, 805, 1046
Family History	849
Fertility	672
Fertilizers	86
Genetic Factors	230, 307, 376, *704, 805, 1046
Genetic Markers	805, 1046
Geographic Factors	84, *292, *462, 633, 823, *850, 869, 899, 905, 922, 1165
Herbicides	33, 86, 349, 542, 653, 977, 1180
Heredity	*292, 805
High–Risk Groups	915, 1046
Histology	180, 431
HLA	*929
HPRT–Mutants	1026
HTLV	570, 581
Immunology	*704
Infection	119, *380, *384, 435, 570, *822, 922
Insecticides	86
Intra–Uterine Exposure	119, 171, *380, 600, 827
Late Effects	370, 672
Latency	570
Lifestyle	127, 180, 367, 376, *886
Lymphocytes	977
Mapping	*350, 823
Mathematical Models	970
Menopause	230
Metals	*106, 192, 393, 487
Micronuclei	1026
Migrants	*292, *462, 922
Minerals	127
Mining	*106
Monitoring	62, 1165
Multiple Primary	224, 339, 340, 463, 619, 672, 805, 827, 871, 915, 970
Mutation, Germinal	937
Mutation, Somatic	62, 827
Nutrition	127
Occupation	33, 56, 62, 86, *106, 119, 136, 171, 180, 192, 208, 251, *262, *277, 334, 349, 367, 487, 539, 542, *621, 653, 655, *687, 752, 779, *781, 827, 849, *864, *934, *936, 950, *984, 985, 1026, 1107, 1179, 1180, 1181, 1231

Paints 653
Parental Occupation 57, *205, 376, *380, *665, *822, 859, *934, *936
Parity 230
PCR 977
Pedigree 805, 937
Pesticides 33, 86, 334, 349, *380, 542, 653, *665, 977, 1165, 1180
Petroleum Products 86, 869
Physical Factors 230, 970
Plastics 779, 869, *984
Pregnancy 827
Premalignant Lesion 1046
Prognosis 171
Protein Binding *781
Race 985
Radiation, Ionizing 56, 62, 119, 171, 180, 192, 221, *262, *277, 287, 334, 339, 349, *350, 370, 376, *380, *385, 393, *462, 600, *704, *775, 777, *822, 823, 827, 859, 901, 902, 905, *934, *936, 970, 1064, 1107
Radiation, Non-Ionizing 970
Radiation, Ultraviolet 805
Radiotherapy 224, 340, 619, 672, 777, 827, 871, 915, 970
Record Linkage 86, *106, 192, 508, 905
Recurrence 672
Registry 33, 56, 57, 171, 180, *205, 208, 221, 224, 251, *277, 287, *292, 307, 367, 376, 463, 600, 633, 655, *665, 752, 766, *775, 777, *781, 805, 827, 859, 869, 871, 899, 905, 922, *984, 1165
Reproductive Factors *665
Saliva *929
SCE 62, 1026
Screening 805
Segregation Analysis 937
Sero-Epidemiology 581
Sex Ratio 619
Sib 435, 827
Socio-Economic Factors 119, 127, 435, *886, 905
Solvents 136, 334, 349, *380, 539, 542, 653, *687, *864, 869, *886, 950, *984, 1165
Stage 463
Survival 463, 730, 915
Time Factors 339, 823, 829, 901, *984, 1165, 1181
Tobacco (Smoking) 118, 127, 171, 539, 619, 653
Trace Elements 127
Treatment 339, 730
Trends 287, 633, *850, 922, 1165
Twins 230, 600
Ultrasound 827
Virus *342, 367, 570, 823, 827, 829, *929
Vitamins 127
Waste Dumps 349, 1064
Water 84
Welding 539, 542
Wood 208, 653

Leukaemia (ALL)

Age 633
Biochemical Markers *676
Chemical Exposure 75, 647
Childhood 540, 633, 1108
DNA *676
Drugs 647
Electromagnetic Fields 647
Familial Factors 1108
Genetic Factors 647, 1108
Geographic Factors 633, 847
Intra-Uterine Exposure 75, 1108
Marijuana 647

Multiple Primary 540
Oncogenes *676
Parental Occupation 75, 647
Passive Smoking 75, 647
PCR *676
Pesticides 75, 1108
Radiation, Ionizing 647, 1108
Registry 633, 647
Reproductive Factors 1108
Solvents 75, 1108
Survival 540, 847
Treatment 540, *676
Trends 633
Virus *676

Leukaemia (AML)

Chemical Exposure *727, 1086
Drugs *727, 1086
Electromagnetic Fields 1086
Geographic Factors 847
Immunosuppression 1086
Occupation 280, *727
Pesticides 280, *727
Radiation, Ionizing 1086
Solvents 280
Survival 847

Leukaemia (ANLL)

Childhood 540, 1109
Classification 1109
Marijuana 1109
Multiple Primary 540
Pesticides 1109
Petroleum Products 1109
Solvents 1109
Survival 540
Treatment 540

Leukaemia (CLL)

BMB *1125
Chemical Exposure *727
Coal 40
Dose–Response 224
Drugs *727, *1125
Ethnic Group *1125
Family History *1125
HLA *1125
Mining 40
Multiple Primary 224
Occupation 40, *727
Pesticides *727
Radiation, Ionizing *1125
Radiotherapy 224
Registry 40, 224
Tobacco (Smoking) *1125

Leukaemia (CML)

Geographic Factors 847
Survival 847

Lip

Alcohol	70
Cosmetics	1085
Diet	70
Female	1085
Geographic Factors	98
Multiple Primary	70
Radiation, Ultraviolet	98, 1085
Registry	98
Tobacco (Smoking)	70, 98, 1085
Trends	98

Liver

Age	729
Air Pollution	869
Alcohol	138, 165, 171, *293, *460, 559, 559, 559, 559, 574, 595, 597, 601, 664, *690, 729, 737, 811
Animal	1165
Antibodies	656
Antigens	273, 656, 811
Betel (Chewing)	*293
Biomarkers	*514, 738, 956
Biopsy	726
Blacks	729, 1035
Blood	*514
BMB	*160, 345, *514, 738
Chemical Exposure	171, 1028, 1091, 1092, 1093, 1104
Chinese	163
Cholesterol	601
Chromosome Effects	738
Cirrhosis	491, 569, 597, 656, 726
Cluster	587, 869, 1165
Coffee	737
Cost–Benefit Analysis	811
Data Resource	869
Diet	127, 138, 146, *293, *460, 595, 601, 737, 956
DNA	1149
Dose–Response	370
Drugs	171, 221, 370
Dusts	138, 487
Dyes	738
Environmental Factors	146, 258, 383, 869, 1035, 1091
Epilepsy	221
Familial Factors	146, 165, 383
Fat	595
Female	1035
Genetic Factors	146
Genetic Markers	146, *514
Geographic Factors	146, 258, 729, 737, 869, 1165
HBV	*160, 164, 191, *293, 345, 383, *460, 491, 559, 559, 559, 559, *577, 597, 656, *690, 726, 729, 811, 956, 1149
HCV	*160, *293, *460, 559, 559, *577, 726, 729
Hepatitis	*460
Herbicides	1104
High–Risk Groups	174, 491, 811
Hormones	*460, 811
Immunology	383
Infection	165, 168, 273, *293, 383, 569
Insecticides	1104
Intra–Uterine Exposure	171
Late Effects	370
Lifestyle	127, 559, 559, 559, 559, 601, 1035
Liver Disease	811
Lymphocytes	738
Marital Status	601

Metabolism 738, 956
Metals 234, 487, 556, 1091, 1093, 1104, 1161
Micronuclei 738
Migrants *460
Minerals 127
Monitoring 174, 956, 1165
Multiple Primary 146
Mutagen 182
Mycotoxins 164, 191, *293, 726, 811, 956
Nutrition 127
Occupation 128, 138, 171, 234, 258, 487, 556, 664, *690, *698, *700, 738, *983, 1028, 1035, 1091, 1092, 1093, 1104, 1161
Oncogenes 1149
Oral Contraceptives 729
Parasitic Disease 234, 587
PCR *514
Pedigree 146
Pesticides *983, 1104, 1165
Petroleum Products 869, 1091
Physical Factors 601
Plastics 128, 738, 869, 1091, 1092, 1104
Premalignant Lesion 491
Prevention 345
Prognosis 171
Radiation, Ionizing 138, 171, 221, 370, 377
Registry 171, 221, 345, 664, 869, 1035, 1165
RFLP *514
Rubber 1092
Rural 258
SCE 738
Screening *160, 174, 729, 811
Segregation Analysis 146
Sero–Epidemiology 165, 656
Sex Ratio 729
Socio–Economic Factors 127, 138, 383, 601
Solvents 869, 1091, 1092, 1104, 1165
Stage 174
Survival 174, 258
Tea 164
Time Factors 737, *983, 1165
Tobacco (Chewing) *293
Tobacco (Smoking) 127, 138, 146, 165, 171, 234, *293, *514, 574, 595, 597, 601, *690, 729, 737, 738, 811
Trace Elements 127, 175, 234
Treatment 258
Trends 65, 258, 1165
Vaccination 345
Virus 165, 168, 273, 569, 597
Vitamins 127, *293, 595
Water 182, 587
Welding 234

Lung

Abortion 963
Age 194, 619, 970, 1182
Air Pollution 172, *173, 192, 297, 511, 710, *773, *812, 841, 869, 944
Alcohol *50, 63, 76, *317, 326, 546, 566, 574, 595, 601, 640, 678, 680, 710, *999
Animal *379
Antigens 481
Antioxidants *204, *987, *1051
Autopsy 932
Biochemical Markers 239, 297, 1121, 1197
Biomarkers *204, 613, 642, 738, 960, 1131, 1232
Birth Cohort 194, *197, 513
Blacks 1035

Blood *204, 480, 481, *768, 1079
BMB 135, *204, 239, 241, 343, *379, 480, 481, 624, 705, 738, *768, 910, 960, *1051, 1079, 1200
Chemical Exposure 73, 94, 193, 279, *317, 326, 351, 386, 489, 623, *649, 683, 721, 860, 873, 945, 965, 968, 982, 1091, 1092, 1093, 1104, 1107, 1131, 1153, 1173, 1180
Chemotherapy 339, 340, 619, 970
Chinese 386
Cholesterol 97, 601
Chromosome Effects 738, 1131
Cirrhosis 683
Clinical Records 52
Cluster 869
Coal 152, 172, *183, 192
Coffee 465
Cooking Methods 152, *173, *428
Cytology 596, 1224
Data Resource 869, 996
Demographic Factors *1226
Diabetes 104, 1065
Diet 35, *50, 63, 97, 127, 146, 152, *155, 242, 297, *317, *379, 386, 465, 480, 566, 595, 601, 624, 640, 678, 680, 710, 806, 963, 980, *999, *1051, 1065, 1079, 1135, 1197, 1232
DNA 135, 596, 642, *1051, 1232
DNA Adducts *204, 239, 481, 613, 642, *812, 960, 1131
DNA Binding 1131
DNA Repair *204
Dose–Response 76, 177, 192, 263, 370, *686, 748, 772, 782, 858, 860, 968, 970, 1022, 1024, 1107, 1153, 1227
Drugs 232, 370, *459, 513, 807, *987, 996, 1134
Dusts 34, 35, 36, *43, 52, 72, 73, 76, 94, *106, 129, *158, *183, *272, *320, *322, *323, 343, 353, 487, 489, 496, 511, *580, 604, *671, 685, *686, 691, *692, *715, 721, 742, 744, 748, 772, 857, 910, 939, 965, 968, 982, 1024, 1197, 1200, *1209, 1224, *1226
Dyes 738
Environmental Factors 34, 72, 135, 146, 192, 193, 217, *264, 386, 489, *567, *649, 721, 742, 869, 873, 903, 944, 945, 1035, 1091, 1135, 1173
Enzymes 94, *459, 683, 807
Ethnic Group 456, 465, *605, *1051, 1079
Familial Factors 94, 146, 546, 640, 963, 1061, 1162, 1213
Fat 63, 465, 595, *999, *1051, 1197
Female 72, 95, 135, 152, 331, 386, 425, *428, 963, 1035
Fertilizers 193, 756, 910
Fibre 640, 1197
Fibrosis 239
Fruit 465
Fungicides 94
Genetic Factors 94, 146, *459, 683, 1079
Genetic Markers 146, 683, *1051, 1079, 1131
Genotype/Phenotype *204
Geographic Factors 146, 194, *197, 447, *462, 869
Hair 705
Herbicides 94, 1104, 1180
High–Risk Groups 174, 388, 596, *671, 685, 1197, 1200, 1232
Histology 159, 217, 239, 496, 1061, 1135, 1153, *1209
Hormones 980, 996
HPV 932
Hygiene 386
Immunologic Markers 496, 613
Insecticides 94, 1104
Late Effects *264, 370
Latency 154, *397, 858, 860, 1024
Lifestyle 104, 127, 386, *567, 601, *605, 624, 640, 806, 1035, 1213
Lipids 465, 806, 1065, 1232
Lymphocytes 135, 613, 642, 738, *1051
Male 172, 566
Mammography 1197

Marital Status	601
Mathematical Models	*43, 513, *686, 970
Menopause	980
Metabolism	738
Metals	3, 73, *106, *158, 159, *183, 192, 193, 241, 263, *265, 271, *272, *320, *322, *323, *325, 326, *362, 393, *485, 487, *555, 556, 640, *670, *715, 721, *768, *785, 790, 844, 872, 873, 945, 947, 965, 968, 1024, 1091, 1093, 1104, 1227
Micronuclei	135, 642, 738, 1131
Migrants	456, *462, 678
Minerals	127, 489, 857
Mining	3, 34, 36, *106, 129, *158, 263, 326, 483, 579, 721, 798, *812, 947, 1024
Monitoring	174
Multiple Primary	146, 223, 231, 339, 340, 619, 970
Mutagen	193
Nutrition	97, 127, *317, 480, 806, 1197, 1232
Occupation	3, 34, 35, 36, *43, *50, 52, 73, 76, 87, 94, *106, 126, 129, 152, *158, 159, 177, *183, 192, 193, *204, 217, 239, 243, 251, 263, *265, 271, *272, 279, 297, 298, *317, *320, *322, *323, *325, 326, 328, 331, 343, 351, 353, *362, *379, 386, 481, 483, 484, *485, 487, 497, 511, *555, 556, 566, 579, 596, 613, 623, *649, *667, *668, *670, 678, 680, *686, *687, 691, *692, 695, *698, *700, 710, *715, 721, 738, 744, 748, 756, *768, 772, *773, *774, 782, *785, 788, 790, 792, 798, *812, 844, 857, 858, 860, 861, 872, 873, 910, 932, 939, 945, 947, 960, 963, 965, 968, 982, 985, 1022, 1024, 1034, 1035, 1061, 1091, 1092, 1093, 1104, 1107, 1135, 1153, 1162, 1173, 1180, 1197, *1209, 1224, *1226, 1227, 1234
Oncogenes	239
Parity	963
Passive Smoking	97, 123, 152, *173, 297, 331, *379, 386, 642, 710, 782, 1121, 1131, 1135, 1162
PCR	932, *1051
Pedigree	146
Pesticides	965, 1104, 1180
Petroleum Products	73, *183, *428, 860, 861, 869, 965, 1034, 1091
Physical Activity	806, 980, 1213
Physical Factors	63, 326, 601, 680, 970, 1213
Plasma	*1051
Plastics	73, *362, 623, 691, 738, 869, 965, 1091, 1092, 1104, 1173, *1209, *1226
Pneumoconiosis	129, 328, 685
Premalignant Lesion	135, 388, 1224, 1232
Prevalence	*459, 1065, 1197
Prevention	35, 177, *567, 705, 986, *987, 1197, 1200, 1224
Projection	*197
Protein Binding	1131
Psychosocial Factors	854
Race	985
Radiation, Ionizing	3, 87, 95, 139, 159, 192, 193, 263, *264, 339, 370, *379, 393, *462, *670, 721, 782, 844, 903, 944, 947, 970, 1024, 1107, 1135, 1182
Radiation, Non−Ionizing	970
Radiotherapy	223, 340, 619, 970
Record Linkage	104, *106, 192, *204, 232, 973
Registry	3, 36, 104, 123, *204, 217, 223, 231, 232, 241, 242, 251, 271, *272, 298, *317, *362, *397, 425, 456, 465, 546, *580, 604, 623, 624, 640, *649, *667, *668, *670, *671, 680, 744, 748, 756, *768, *773, *774, 782, *785, 787, 790, 807, *812, 844, 860, 861, 869, 939, 963, 965, 973, *999, 1035, *1051, 1061, 1200, *1209
Religion	465
Reproductive Factors	546, 963, 980
Resins	691
RFLP	*1051
Rubber	353, 691, 1092
SCE	642, 738, 1131
Screening	174, 388, 546, 596, 1197
Segregation Analysis	146
Sero−Epidemiology	241
Serum	241, *1051
Sex Ratio	619, 624, 841

Silicosis 483, *580, 685
Socio–Economic Factors 127, 386, *397, 465, 601, *605, 854
Solvents 73, *362, *687, 869, 965, 1091, 1092, 1104, *1209
Sputum 481, 596, 642, 1224
Stage 174
Stress 854
Surgery 973
Survival 174, 685
Sweeteners 104
Textiles 691
Time Factors 76, 159, 194, 271, 326, 339, *397, 456, 513, 678, 1022, 1182, *1226
Tissue 239, 960, 1200
Tobacco (Chewing) 425
Tobacco (Smoking) 3, 34, 35, *50, 52, 76, 94, 104, 127, 129, 146, 152, 154, *158, *173, *197, 217, 239, *264, 298, *317, 326, 331, 343, 351, *379, 386, 388, 425, *428, 465, 481, 497, 513, 546, 566, 574, *580, 595, 596, 601, 613, 619, 642, *671, 678, 680, 705, 738, 742, *773, *774, 782, 788, 792, 798, 807, 944, 945, 947, 968, 982, 1022, 1024, 1061, 1131, 1134, 1153, 1162, 1173, 1182, 1197, 1200, *1209, 1213, *1226, 1234
Toenails 624
Trace Elements 127
Treatment 231, 339
Trends 65, 194, *197, 425, 513
Urine *204, 642, *768, 910, 1079
Vegetables 465
Vitamins 35, 63, 97, 127, 241, 480, 595, 640, 705, 806, *987, 1197, *1209, 1224, 1232
Water 787
Welding *667, *668, 982, 1034
Wood 744, 965, *1209

Lymphoma

Age *1174
AIDS 41, 964, *1062
Air Pollution *205, 869
Alcohol 171
Biochemical Markers 431
Biomarkers 508
Birth Order 435
BMB 41, 1073, 1100, *1174
Chemical Exposure 13, 19, 73, 84, 171, 247, 506, 508, *621, 654, 757, 779, *781, 965, 1028, 1179
Childhood 56, 57, *205, 287, 435, *886, *934, *936
Chlorination 84
Chromosome Effects 62
Clinical Records 435, *850
Cluster *350, 829, 869, *886
Data Resource *850, 869
DNA Repair 805
Dose–Response 370, 752, 1024
Drugs 41, 171, 349, 370, 757
Dusts 73, 506, 744, 757, 965, 1024
EBV *1062, 1073, *1174
Electromagnetic Fields *18
Environmental Factors 349, 376, 431, 435, 508, 829, 869
Familial Factors 805
Genetic Factors 376, 805
Genetic Markers 805
Geographic Factors 84, *850, 869
Haemophilia *1174
Herbicides 13, 247, 349, 654, 977
Heredity 805
Histology 431
HIV 41, 964, 1073, *1174
HLA 1073

HTLV 570
Immunosuppression 1100
Infection 435, 570
Intra-Uterine Exposure 171
Late Effects 370
Latency 570, 1024
Leather 506
Lifestyle 376, *886
Lymphocytes 977
Mapping *350
Metals 73, *362, 506, 965, 1024, 1160
Mining 1024
Monitoring 62
Multiple Primary 463, 805
Mutation, Somatic 62
Occupation 13, 19, 56, 62, 73, 171, 247, 349, *362, 484, 506, *621, 654, 744, 752, 757, 779, *781, *934, *936, 965, *984, 1024, 1028, 1160, 1179
Parental Occupation 57, *205, 376, *934, *936
PCR 977, *1062, 1073
Pedigree 805
Pesticides 13, 247, 349, 506, 654, 965, 977
Petroleum Products 19, 73, 869, 965
Plastics 73, *362, 506, 757, 779, 869, 965, *984
Prognosis 171
Protein Binding *781
Radiation, Ionizing 56, 62, 171, 287, 349, *350, 370, 376, *934, *936, 1024
Radiation, Ultraviolet 805
Record Linkage 508
Registry 13, 56, 57, 171, *205, 247, 287, *362, 376, 463, 506, 654, 744, 752, 757, *781, 805, 869, 965, *984, *1062, 1073
Rubber 506
SCE 62
Screening 805
Sero-Epidemiology 964
Serum 1073
Sexual Activity 41
Sexually Transmitted Diseases 41
Sib 435
Socio-Economic Factors 435, *886
Solvents 73, 349, *362, 506, 869, *886, 965, *984
Stage 463
Survival 463
Time Factors 829, *984
Tissue 1073
Tobacco (Smoking) 171, 1024
Transplantation 1100
Treatment 41, 1100
Trends 287, *850
Virus 570, 829
Waste Dumps 349
Water 84
Wood 506, 744, 965

Male Genital

Air Pollution 213
Antibodies 252
Biopsy 252
Chemical Exposure 213
Condyloma 252
DNA 252
Environmental Factors 213
HPV 252
Infection 252
Metals 213

Occupation 213
Petroleum Products 213
Prevention 252
Registry 213
Screening 252
Sero–Epidemiology 252
Sexual Activity 252
Tobacco (Smoking) 252
Treatment 252

Melanoma

Age 230, *291
Anatomical Distribution 720
Biomarkers *269
Biopsy 1176
Blacks 1035
BMB 874
Chemical Exposure 73, 628, 1069
Childhood *292, 911
Diet 1176, 1221
Dusts 73
Electromagnetic Fields *1004
Environmental Factors 39, 874, 911, 1035, 1221
Enzymes 1221
Familial Factors 9, 39, 666
Female 1035
Fluorescent Light 7, 100
Genetic Councelling *703
Genetic Factors 9, 230, *703, 911
Genetic Markers 9, 47
Geographic Factors 20, *292, 720, *1030
Heredity 47, *292
Histology 720, 874
Hormones 346
Immunosuppression 346
Lifestyle 20, 31, 346, 1035
Lipids 1221
Menopause 230
Metals 73, 1221
Migrants *291, *292
Monitoring *269
Multiple Primary 720
Naevi 31, 346, 495, 628, 666, 911, 1176
Occupation 20, 73, 100, 346, 628, 874, *1004, 1035, 1069
Parity 230
Petroleum Products 73
Physical Factors 230
Pigmentation 31, 100, 495, *1030
Plastics 73
Premalignant Lesion 20, 31, 495
Prevalence 20
Prevention 1122
Prognosis 874
Race *1030
Radiation, Ionizing 1069
Radiation, Non–Ionizing *1004
Radiation, Ultraviolet 7, 16, 20, 31, 39, *269, 346, 495, 628, *1030, 1069, 1176
Registry 9, 20, 39, *291, *292, 720, 1035
RFLP 47
Screening 666, *988
Socio–Economic Factors 20
Solvents 73
Time Factors *291
Tobacco (Smoking) 20
Trends 720

Twins 9, 39, 230
Vitamins 1221

Mesothelioma

Air Pollution 851
Alcohol 59
Asbestosis 530
Blacks 731, 1035
BMB 1200
Chemical Exposure 42, 851, 965
Clinical Records 52
Diet 35, 731
Dose–Response 42, *686, 748
Dusts 34, 35, 42, 52, 59, 72, 101, 353, 496, 530, *536, *686, 748, 851, 965, 1200
Environmental Factors 30, 34, 42, 72, 731, 851, 1035
Female 72, 1035
High–Risk Groups 1200
Histology 59, 496
Immunologic Markers 496
Latency *324
Lifestyle 1035
Mathematical Models *686
Metals 59, 965
Mining 34, 42, 579
Occupation 30, 34, 35, 42, 52, 59, 101, *324, 353, 530, *536, 579, *686, 731, 748, 851, 965, 1035
Pesticides 965
Petroleum Products 965
Plastics 965
Prevention 35, *536, 1200
Registry 30, 59, 748, 851, 965, 1035, 1200
Rubber 353
Solvents 965
Stage 59
Time Factors *324
Tissue 1200
Tobacco (Smoking) 34, 35, 52, 59, 1200
Treatment 59
Vitamins 35
Wood 59, 965

Multiple Myeloma

Biochemical Markers 431
Blacks 1036
Chemical Exposure 757, 1180
Diet 1036
Drugs 757
Dusts 757
Electromagnetic Fields *18
Environmental Factors 431
Familial Factors 1036
Female 96
Fertilizers 86
Herbicides 86, 542, 1180
Histology 431
Hormones 1036
Insecticides 86
Occupation 86, 93, 96, 542, 757, *967, 1036, 1180
Paints *967
Pesticides 86, 93, 96, 542, 1180
Petroleum Products 86
Plastics 757
Premalignant Lesion 1036

Record Linkage 86
Registry 757, *967, 1036
Reproductive Factors 96
Socio-Economic Factors 1036
Solvents 542, *967
Tobacco (Smoking) 96
Welding 542

Myelodysplastic Syndrome

Age 619
Chemical Exposure 17, 267, 835
Chemotherapy 619
Childhood 922
Chromosome Effects 539
Environmental Factors 17, 267
Ethnic Group 922
Geographic Factors 922
Infection 922
Migrants 922
Multiple Primary 619
Occupation 17, 267, 539, 835
Premalignant Lesion 17, 267
Radiation, Ionizing 835
Radiotherapy 619
Registry 922
Sex Ratio 619
Solvents 539
Tobacco (Smoking) 118, 539, 619
Trends 922
Welding 539

Nasal Cavity

Alcohol 181, *317
BMB 874
Chemical Exposure 208, *316, *317, 1084
Diet 181, *317
Dusts 208, 278, 305, 306, *316, 1084
Dyes 509
Environmental Factors 278, 874, 1084
Histology 278, 874
Hygiene 181
Infection 1084
Leather 305, 306
Metals 1161
Nutrition *317
Occupation 181, 208, 278, 305, 306, *316, *317, 509, 874, 1084, 1161
Plastics 305, 306, *316
Prognosis 874
Radiation, Ionizing 181, 1084
Registry 181, 208, *317
Solvents 305, 306
Textiles *316
Tobacco (Smoking) 181, *317, 1084
Wood 208, 278, 305, 306

Nasopharynx

Alcohol 59, *184, 1208
Antibodies *2, 144, 151, 606
Biochemical Markers 606
Biomarkers *184
Blood *184
BMB *184, 314
Chemical Exposure 151, 1084, 1159, 1208

Chinese 607
Cluster 409
Diet *2, 127, 143, 179, *184, 314, 1208
Dusts 59, 179, 1084
EBV *2, 314, 409, 606, 607
Environmental Factors 314, 1084, 1208
Familial Factors *184
Genetic Factors 151, *184
Geographic Factors 447
Histology 59
HLA 314
Infection 151, 606, 1084
Lifestyle *2, 127, 1208
Maghrebians 314
Metals 59
Migrants 314
Minerals 127
Nutrition 127
Occupation 59, 179, *184, 1084, 1159
Passive Smoking 179
Pedigree 151
Plants *184
Plastics 1159
Premalignant Lesion 144
Prognosis 606
Race 607
Radiation, Ionizing 1084
Registry *2, 59, 179
Screening *2, 607
Sero–Epidemiology 144, 606, 607
Serum 314
Socio–Economic Factors *2, 127
Stage 59
Tissue *184
Tobacco (Smoking) 59, 127, 151, 179, *184, 1084, 1208
Toenails *184
Trace Elements 127
Treatment 59
Urine *184
Virus 151
Vitamins 127
Wood 59, 1208

Nervous System

Antioxidants *203
Autopsy 398
Biopsy 398
Childhood *292
Classification 398
Genotype/Phenotype *203
Geographic Factors *292
Heredity *292
Histology 398
Migrants *292
Occupation *203
Record Linkage *203
Registry *203, *292

Neuroblastoma

Age 1008
Alcohol 1008
Birth Cohort 591
BMB 1114
Childhood 57, *292, 540

Cost–Benefit Analysis 591
Diuretics 1008
Drugs 1008
Electromagnetic Fields 1008
Environmental Factors 376
Genetic Factors 376
Genetic Markers 1008
Geographic Factors *292
Heredity *292
Intra–Uterine Exposure 1008
Lifestyle 376
Migrants *292
Multiple Primary 540
Oncogenes 1008
Parental Occupation 57, 376, 1008
Radiation, Ionizing 376
Registry 57, *292, 376, 1114
Screening 591, 1114
Stage 1008
Survival 540
Time Factors 1008
Treatment 540

Non–Hodgkin's Lymphoma

Age 619, *1174
AIDS 479, *658, 1186
Alcohol *311
Allergy 1186
Animal 1165
Antibodies *898
Biochemical Markers *676
Biomarkers *514
Birth Cohort 823
Blood *290, *514, *929
BMB *290, *514, *929, *1125, *1174
Chemical Exposure *727, 859, 1180
Chemotherapy 339, 340, 619
Childhood *292, 540, 823, 859
Cluster 823, *929, 1165
Coal 40
Diet 1193
DNA *676
Drugs *311, *727, *1125, 1186
Dyes *311
EBV *929, *1174
Electromagnetic Fields *311
Ethnic Group *605, *1125
Family History *1125
Fat 1193
Female 96
Fertilizers 86
Genetic Markers *514
Geographic Factors *292, 823, 1165
Haemophilia *1174
Herbicides 86, *290, *311, 542, 622, 1180
Heredity *292
High–Risk Groups 479
Histology 1186
HIV 479, *658, *898, *1174, 1186
HLA *929, *1125
Homosexuality 1186
Immunology 1186
Infection *311, *658
Insecticides 86
Lifestyle *311, *605, 1193

Mapping 823
Migrants *292
Mining 40
Monitoring 1165
Multiple Primary 339, 340, 540, 619
Nutrition 1193
Occupation 40, 86, 93, 96, *311, 532, 542, 622, *727, *761, *967, 1034, 1180, 1186
Oncogenes *676
Paints *311, *967
Parental Occupation 859
PCR *514, *676
Pesticides 86, 93, 96, *311, 532, 542, 622, *727, 1165, 1180
Petroleum Products 86, 1034
Protein 1193
Radiation, Ionizing *311, 339, 823, 859, *1125
Radiotherapy 340, 619
Record Linkage 86
Registry 40, *292, *311, 532, 622, *761, 859, *967, 1165
Reproductive Factors 96
RFLP *514
Saliva *929
Sero–Epidemiology *658
Sex Ratio 619
Sexually Transmitted Diseases *898
Socio–Economic Factors *605
Solvents *311, 542, *967, 1165
Survival 540
Time Factors 339, 823, 1165
Tissue *290
Tobacco (Smoking) 96, *311, *514, 619, *1125
Treatment 339, 540, *676
Trends 1165
Vaccination 1186
Virus *676, 823, *929
Vitamins 1193
Welding 542, 1034

Oesophagus

Age 361, *1074
Alcohol 63, 167, *259, 361, 424, 439, *478, 546, 664, 737, 784, *969, *1074, *1103, *1118
Analgesics *1118
Autopsy 398
Biopsy 361, 398
Birth Cohort 513
Blacks 1035, 1036
Blood *969
BMB *134, *189, 361, 910, *969
Chemical Exposure 73
Chinese 163
Chromosome Effects 133
Classification 398
Coffee 737, *1118
Cytology 188, 361
Diet 63, 127, 167, *259, 361, 424, 439, 454, *478, 733, 737, 784, *969, 1036, *1074, *1103, *1118
Disease, Other *969
DNA 130, *134
DNA Repair 133
Drugs 513, *969, *1074, *1118, 1134
Dusts 73, 910
Environmental Factors 258, 1035
Ethnic Group *605, 694
Familial Factors *134, 167, 361, 546, 1036

Fat 63, 733
Female 1035
Fertilizers 910
Fruit 733
Genetic Factors *134
Geographic Factors 258, 694, 737
H. pylori *969
High-Risk Groups 133, 187, *189
Histology 398, *1103
Hormones 1036
Hygiene 439
Indians 454
Latency *1074
Lifestyle 127, 361, 454, *605, 1035, *1118
Linkage Analysis 133
Mapping 694
Mathematical Models 513
Menstruation 167
Metals 73, *362
Micronuclei 361
Minerals 127
Nutrition 127, 733, *1103
Occupation 73, 167, 258, *362, 664, 784, 910, *969, 1035, 1036, *1118
Oncogenes 133
Petroleum Products 73
Physical Activity 167, *478
Physical Factors 63
Plants *189
Plasma 361
Plastics 73, *362
Premalignant Lesion 187, 188, 361, 1036
Prevalence 361
Prevention 130, 187
Registry 167, *259, *362, *478, 546, 664, 733, 784, 1035, 1036, *1074, *1103, *1118
Reproductive Factors 167, *478, 546
RFLP 133, *134
Rural 258
Screening 188, 546
Segregation Analysis *134
Socio-Economic Factors 127, 188, *605, 1036
Solvents 73, *362
Survival 258
Tea 167
Time Factors 513, 737, *1074
Tobacco (Chewing) 784
Tobacco (Smoking) 127, 167, 188, 361, 424, 439, 454, *478, 513, 546, 737, 784, *969, *1074, *1103, *1118, 1134
Tobacco (Snuff) 439, 784
Trace Elements 127
Treatment 258
Trends 65, 258, 513, 694
Urine 361, 910
Vegetables 733
Virus *1103
Vitamins 63, 127, 130, 187, 190, *259, 361, 733
Water 188

Oral Cavity

Alcohol 66, 70, 181, 207, 268, 363, 439, *478, 664, 784, 1089, 1136, 1236
Anatomical Distribution 1136
Areca Nut 402
Autopsy 398
Betel (Chewing) 413, 414, 739
Biomarkers *430, 444, 1203

Biopsy 398, 1090
Birth Cohort 513
Blood *430, *1117
BMB *430, 1090, *1117
Chromosome Effects 402
Classification 398
Cost–Benefit Analysis 389
Cytology 66, 389, *430
Diet 70, 181, *420, 439, *478, 784, 1089
DNA Adducts *430
Drugs 513, 807, 1134
Education *416, *421
Enzymes 807
Female 425
Genetic Factors 402, *553
High–Risk Groups 66, 1136
Histology 389, 398
HPV 363, 1090, 1203
HSV 1203
Hygiene 181, 207, 268, 439
Immunology 1203
Mathematical Models 513
Micronuclei *430
Micronutrients *430
Multiple Primary 70, 1089, *1117
Mutation, Somatic 402
Nutrition *420
Occupation 181, 268, 664, 784
Physical Activity *478
Premalignant Lesion 363, 389, 414, 419, *420, 429, *430, 444, 739
Prevention *416, 419, *421, 429
Radiation, Ionizing 181
Registry 181, *195, 207, 425, *478, 664, 784, 807, 1089, 1090, *1117, 1203
Reproductive Factors 363, *478
SCE 402
Screening *195, 389, *416, *421
Sexual Activity 1203
Sexually Transmitted Diseases 1203
Socio–Economic Factors 363
Stage *195
Time Factors 513, 1236
Tobacco (Chewing) 207, 402, 413, 414, *416, 419, *420, *421, 425, 429, 739, 784
Tobacco (Smoking) 66, 70, 181, 207, 268, 363, 413, 414, *416, 419, *420, *421, 425, 429, 439, *478, 513, 739, 784, 807, 1089, 1134, 1136, 1236
Tobacco (Snuff) 439, 784
Treatment 389
Trends *195, 425, 513
Tumour Markers 445
Vitamins 444

Oropharynx

Alcohol 66, 268, 363, 405, 439, 1136
Anatomical Distribution 1136
Betel (Chewing) 405
Biochemical Markers 405
Biomarkers 1203
Cluster *54
Cytology 66
Diet 439
Female 425
Geographic Factors *54
High–Risk Groups 66, 1136
HPV 363, 1203
HSV 1203

Hygiene 268, 439
Immunology 1203
Metals *362
Occupation 268, *362
Plastics *362
Premalignant Lesion 363
Registry *362, 425, 1203
Reproductive Factors 363
Sexual Activity 1203
Sexually Transmitted Diseases 1203
Socio-Economic Factors 363
Solvents *362
Tobacco (Chewing) 405, 425
Tobacco (Smoking) 66, 268, 363, 405, 425, 439, 1136
Tobacco (Snuff) 439
Trends 425

Ovary

Abortion 404, 707
Alcohol *478, 949, 1000
Biomarkers 534
Blood *865, *1011, 1198
Blood Group 660
BMB 534, *865, *1011, 1198
Chemotherapy *1175
Cholesterol 786, 949
Chromosome Effects *78, 741, 1054
Classification 108
Clinical Records 1133
Cluster *78
Dairy Products *1120
Data Resource 996
Diet 105, 406, *478, 517, 949, 980, 1000, *1120, 1198, *1223
DNA Repair 805
Drugs 635, 996, *1175, *1191, 1229
Enzymes 1198
Ethnic Group 1243
Familial Factors *78, 534, 805, *865, 868, 949, 1243
Fat 997, *1120
Female 105
Genetic Councelling *703
Genetic Factors *703, 741, 805, 868, *1011, 1054
Genetic Markers *78, 805, *1011
Heredity *78, 741, 805
High-Risk Groups *78, 534, 1054
Histology 108, 1243
Hormones 108, 404, 406, 517, 534, 634, 635, 949, 980, 996, *1223, 1229
Hysterectomy 804
Infection 1000
Infertility 1229
Lifestyle 406, 707
Menarche 404, 406, 660
Menopause 404, 660, 980
Menstruation 404
Metabolism 1198
Multiple Primary 805, *1175
Nutrition *1120
Obesity 949
Occupation 108, 1000, 1133
Oophorectomy 804
Oral Contraceptives 517, 1243
Parity 404, 406, 517, 660, 1243
Pedigree 534, 805
Physical Activity *478, 949, 980

Physical Factors	660, 786, 949
Prevention	*78, 634, *1191
Radiation, Ultraviolet	805
Record Linkage	105, 973
Registry	108, *427, *478, 786, 804, 805, 973, *1120, *1191, 1198, *1223
Religion	1243
Reproductive Factors	*478, 534, 707, 980, 1000, *1011, *1120, 1133
Screening	*78, 105, 634, 660, 805, *865, 868
Segregation Analysis	*1011
Sexual Activity	517
Socio–Economic Factors	404, 406, 1133
Stage	868
Surgery	973
Survival	786
Time Factors	*1191
Tissue	*865
Tobacco (Smoking)	*478, 949, 1000
Trends	1243
Tubal Ligation	*427, 1139
Tumour Markers	*865
Uric Acid	949
Urine	635, 1198
Vasectomy	*427
Virus	1000, 1229
Vitamins	1000, *1223

Pancreas

Alcohol	167, 244, *734, 737, 1000
Animal	1165
Autopsy	398
Biopsy	398
Birth Cohort	513
Blacks	1036
Blood	*734
BMB	*734
Chemical Exposure	73, 244, *621, 1179
Chromosome Effects	*734
Classification	398
Cluster	1165
Coffee	244, *734, 737
Cytology	*734
Diabetes	104, 244
Diet	105, 167, 733, 737, 1000, 1036
Drugs	513, 1134
Dusts	73
Environmental Factors	258
Familial Factors	167, 1036, 1213
Fat	733
Female	105
Fruit	733
Geographic Factors	258, 737, 1165
Histology	398
Hormones	1036
Infection	1000
Lifestyle	104, 1213
Mathematical Models	513
Menstruation	167
Metals	73
Monitoring	1165
Nutrition	733
Occupation	73, 167, 244, 258, 484, *621, *734, 1000, 1036, 1179
Pesticides	1165
Petroleum Products	73
Physical Activity	167, 1213
Physical Factors	1213

Plastics	73
Premalignant Lesion	1036
Prognosis	*734
Record Linkage	104, 105
Registry	104, 167, 244, 733, 1036, 1165
Reproductive Factors	167, 1000
Rural	258
Screening	105, 958
Socio–Economic Factors	1036
Solvents	73, 1165
Surgery	958
Survival	258
Sweeteners	104
Tea	167
Time Factors	513, 737, 1165
Tissue	*734
Tobacco (Smoking)	104, 167, 244, 513, *734, 737, 958, 1000, 1134, 1213
Toenails	*734
Treatment	258
Trends	258, 513, 1165
Urine	*734
Vegetables	733
Virus	1000
Vitamins	733, 1000

Penis

Ethnic Group	356
HPV	356
Sexual Activity	356
Socio–Economic Factors	356

Peritoneum

Alcohol	76
Asbestosis	530
Chemical Exposure	42
Dose–Response	42, 76
Dusts	34, 42, 76, 530
Environmental Factors	34, 42
Mining	34, 42
Occupation	34, 42, 76, 530
Time Factors	76
Tobacco (Smoking)	34, 76

Pharynx

Alcohol	70, 181, 664, 784, 1089, 1236
Biopsy	1090
Birth Cohort	513
Blood	*1117
BMB	1090, *1117
Diet	70, 181, 784, 1089
Drugs	513, 807
Dyes	509
Enzymes	807
HPV	1090
Hygiene	181
Mathematical Models	513
Multiple Primary	70, 1089, *1117
Occupation	181, 509, 664, 784
Radiation, Ionizing	181
Registry	181, 664, 784, 807, 1089, 1090, *1117
Time Factors	513, 1236
Tobacco (Chewing)	784
Tobacco (Smoking)	70, 181, 513, 784, 807, 1089, 1236

Tobacco (Snuff) 784
Trends 513

Pleura

Alcohol 76, *317
Asbestosis 530
Chemical Exposure 42, *317
Diet *317
Dose–Response 42, 76
Dusts 42, 76, 496, 530
Environmental Factors 42
Histology 496
Immunologic Markers 496
Mining 42
Nutrition *317
Occupation 42, 76, *317, 530
Registry *317
Time Factors 76
Tobacco (Smoking) 76, *317

Prostate

Age *291, 1033
Alcohol *760, 943, 1000
Antioxidants *203
Autopsy 398, 469, 1033
Biochemical Markers 228, *1002
Biomarkers 74
Biopsy 398
Blacks 1036
Blood 21, 74, 908, *1002, *1205
BMB 21, 74, 241, *760, 793, 908, 918, *1002, 1033, *1205, 1214
Chemical Exposure 73, 84, 623, 816, 873, 895
Chlorination 84
Classification 398
Data Resource 996
Diabetes 1065
Diet 21, 74, 92, 228, 242, 257, 329, *760, 793, 908, 918, 943, 1000, *1002, 1036, 1065, 1214
Drugs 232, 816, 996
Dusts 73
Education *760
Electromagnetic Fields *1004
Environmental Factors 873
Ethnic Group 469, 1129
Familial Factors 1036, 1213
Fat 92, 329, 918, 997, *1002
Fatty Acids *1002
Fertilizers 86, 629, 943
Genetic Factors 816
Genotype/Phenotype *203
Geographic Factors 84
Hair 918
Herbicides 86
Histology 398
HLA 816
Hormones 228, 793, 996, 1036
Infection 1000
Insecticides 86
Lifestyle 21, 228, 1213
Lipids 21, 1065
Metals 73, 74, 241, 629, 872, 873, 918
Migrants 21, *291
Nutrition *1002

Occupation 73, 86, *203, 228, 251, 257, 623, 629, 793, 816, 872, 873, 895, 943, 1000, *1004, 1036
Paints 629
Pesticides 86, 629, 943
Petroleum Products 73, 86
Physical Activity 908, 1213, 1214
Physical Factors 1213, 1214
Plasma 21
Plastics 73, 623
Premalignant Lesion 92, 1036
Prevalence 469, 1065
Race 1033, 1129
Radiation, Ionizing 895, 943
Radiation, Non-Ionizing *1004
Record Linkage 86, *203, 232
Registry *203, 232, 241, 242, 251, 257, *291, 329, *427, 623, 629, 1036, *1205, 1214
Reproductive Factors 1000
Screening 374, 1033
Sero-Epidemiology 241
Serum 241, 793, 1214
Sexual Activity 228, 257, 329, 816, 918, 943, *1205
Sexually Transmitted Diseases 329, 816, 918
Socio-Economic Factors 1036
Solvents 73, 629
Stage 92
Time Factors *291
Tissue 74, *1002, *1205
Tobacco (Smoking) *760, 943, 1000, 1213
Toenails 74, 918
Trends 65
Tubal Ligation *427
Vasectomy *427, *1205
Virus 1000
Vitamins 92, 241, 943, 1000
Water 84, 943
Welding 629

Rectum

Age *843, 1181
Alcohol *53, 149, 166, 167, 424, *478, 510, 546, *594, 601, 640, 737, 1098
Analgesics *1127
Antioxidants *203, *818
Bile Acid 366
Biochemical Markers 468
Biomarkers 558, 894
Blacks 1035
Blood 21, *27, 366, *842, 894
BMB 21, *27, 241, 366, 373, 624, 638, 762, *842, 894, *1010, 1106
Chemical Exposure 73, 84, 747, 860, 1032, 1181
Chlorination 84, *110
Cholesterol 601, 894
Chromosome Effects 51, 558
Coffee 737
Cost-Benefit Analysis 147, 148
Crohn's Disease 800, 821
Diet 8, 21, 105, 127, 146, 149, 150, 161, 166, 167, 242, 275, 366, 373, 424, *478, 510, *594, 601, 624, 640, *675, 712, 737, 747, *818, 893, *1010, 1058, 1066, 1098, 1212, *1223
Disease, Other *843
DNA 51, 558, 762
Dose-Response 748, 860
Drugs 166, 747, 893, *1010, 1058, *1127
Dusts 73, 748, *758
Environmental Factors 5, 24, *27, 146, 258, *567, 712, *1010, 1035

Ethnic Group	*605, *675
Faeces	261, 366, 373
Familial Adenomatous Polyposis	5, 24, 28, 366
Familial Factors	5, 24, *27, 28, 51, 146, 161, 166, 167, 510, 521, 546, 558, 640, *843, 1145, 1213
Family History	*27, 149, 150, 893
Fat	8, 893, 1058, 1212
Female	105, 1035, 1098
Fibre	8, 640, 893, 1058, 1212
Gastrectomy	*1014
Genetic Factors	*27, 28, 146, 216, 275, 366, 558, *818, *933, 954
Genetic Markers	*27, 146, 216, 954
Genotype/Phenotype	*203
Geographic Factors	84, 146, 258, 737
Heredity	*53, 216, 366, 521, 747
High–Risk Groups	5, 24, 28, 147, 150, 174, 216, 373, 455, *818, 954
Histology	166, 954
Hormones	166, *407, 510, *1119, *1223
Infection	366
Latency	860
Lifestyle	21, *27, 127, 147, 149, 161, *567, 601, *605, 624, 640, 712, *1010, 1035, 1213
Linkage Analysis	366, 558, *842
Lipids	21
Marital Status	601
Menopause	*1119
Menstruation	167
Metals	73, 241, 640, 1032, 1161
Migrants	21
Minerals	127, 373
Monitoring	174
Multiple Primary	146
Mutation, Somatic	558
Nutrition	*53, 127, 166, 366, *675, 893, 1058, 1212
Obesity	*1010
Occupation	73, 166, 167, *203, 258, *675, 747, 748, *758, 860, 893, *1010, 1032, 1035, 1161, 1181, 1212
Oncogenes	558
Operation, Previous	*1014
Oral Contraceptives	*1119
Parasitic Disease	147
Pedigree	*27, 146, 366, 954
Petroleum Products	73, 860
Physical Activity	149, 166, 167, *478, *594, 893, 1058, 1213
Physical Factors	601, 1213
Plasma	21
Plastics	73, 1032
Polyps	366, 455, *594, *675, 885, 894, *1010, 1058
Premalignant Lesion	5, 8, 366, 373, *594, 894, 954, 1058, 1128
Prevalence	*594
Prevention	8, 373, *567, 1145
Prognosis	*27
Protein	893
Psychosocial Factors	166
Radiation, Ionizing	166
Record Linkage	105, *203, 800, *1119
Registry	24, *27, 28, *110, 166, 167, *203, 216, 241, 242, 261, *478, 521, 546, 624, 640, 748, *758, 787, 800, *818, 860, *862, 954, 1035, 1098, *1119, *1199, *1223
Reproductive Factors	149, 166, 167, *478, 510, 546, 1098
RFLP	51
Rural	258
Screening	28, 105, 148, 174, 261, 374, 455, 546, 762, *862, 892, 958, 1032, 1106, 1145, *1199, 1212
Segregation Analysis	146, *842, *933
Sero–Epidemiology	241

Serum 241
Sex Ratio 624
Socio–Economic Factors 127, 601, *605, *675
Solvents 73
Stage 174
Surgery 958
Survival 174, 258
Tea 167
Time Factors 737, 1145, 1181
Tissue 366, 762
Tobacco (Chewing) *675
Tobacco (Smoking) *53, 127, 146, 149, 166, 167, 424, *478, 546, 601, *675, 737, 747, 958, *1010, 1032, 1058, 1213
Tobacco (Snuff) *675
Toenails 624, 638
Trace Elements 127, 468, 638
Treatment 258
Trends 65, 258
Tumour Markers 762
Ulcerative Colitis 762, 800, 821
Urine 373
Virus 366
Vitamins 8, 127, 241, 640, 894, 1066, *1223
Water *53, 84, *110, 147, 787

Respiratory

Air Pollution 213
Alcohol 363
Biomarkers *514
Blood *514
BMB *514
Chemical Exposure 208, 213, 488, 743, 1156
Dusts 101, 208, 488, 743, 856, *864, 1156
Dyes 509
Environmental Factors 213, 1156, 1219
Genetic Markers *514
HPV 363
Latency *324
Metals 213, 1160, 1161
Minerals 856
Occupation 101, 208, 213, *324, 488, 509, 743, 856, *864, 1156, 1160, 1161, 1219
PCR *514
Petroleum Products 213
Premalignant Lesion 363, 1219
Registry 208, 213, 743
Reproductive Factors 363
RFLP *514
Socio–Economic Factors 363
Solvents *864
Time Factors *324
Tobacco (Smoking) 363, *514, 856, 1156, 1219
Wood 208

Retinoblastoma

Atomic Bomb 561
BMB *706
DNA *706, 1148
Ethnic Group 877
Familial Factors 904
Genetic Councelling *706
Genetic Factors *706, 877
Geographic Factors 904
Intra–Uterine Exposure 904
Mutation, Germinal 904, 1148

Parental Occupation 1148
PCR *706
Prevention *706, 877
Radiation, Ionizing 561
Record Linkage 904
Registry 904
RFLP *706
Segregation Analysis *706
Sib 904
Twins 904

Rhabdomyosarcoma

Age *1155
Asthma *1155
Chemical Exposure *1155
Childhood *1155
Cocaine *1155
Heredity *1155
Immunology *1155
Infection *1155
Intra–Uterine Exposure *1155
Marijuana *1155
Radiation, Ionizing *1155
Socio–Economic Factors *1155
Tobacco (Smoking) *1155

Salivary Gland

Blacks 1035
Environmental Factors 1035
Female 1035
Lifestyle 1035
Occupation 1035
Registry 1035

Sarcoma

Air Pollution 213
Biomarkers *514
Blood *290, *514
BMB *290, *514
Chemical Exposure 13, 213, 247, 965
Childhood *292, 889
Dusts 965
Environmental Factors 213
Familial Factors 889
Female 96
Fertilizers 83
Genetic Markers *514
Geographic Factors 83, *292
Herbicides 13, 83, 247, *290, 622
Heredity *292
High–Risk Groups 889
Insecticides 83
Metals 213, 965
Migrants *292
Occupation 13, 93, 96, 213, 247, 622, 965
PCR *514
Pesticides 13, 93, 96, 247, 622, 965
Petroleum Products 213, 965
Plastics 965
Record Linkage 83
Registry 13, 213, 247, *292, 622, 889, 965
Reproductive Factors 96
RFLP *514

Solvents 965
Tissue *290
Tobacco (Smoking) 96, *514
Wood 965

Skin

Age *878
Air Pollution 213
Anatomical Distribution 720
Biomarkers *269, *514
Biopsy 1184
Blood *514
BMB *514, *878, 1216, 1217
Chemical Exposure 213, 504, 1093
Congenital Abnormalities 1184
Diet 1055, 1220, 1221
DNA Repair 959, 1184
Dose–Response *337
Drugs 993
Environmental Factors 39, 213, 528, 1221
Enzymes 1221
Ethnic Group *1038
Familial Factors 39
Genetic Factors 528
Genetic Markers *514
Geographic Factors 20, 720, *1030
Herbicides 961
Histology 12, 720
HLA 528
Immunology 528
Immunosuppression *878
Lifestyle 20
Lipids 1055, 1221
Metabolism 1220
Metals 213, 1093, 1221
Monitoring *269
Multiple Primary 528, 720
Naevi *1038
Nutrition 1055
Occupation 20, 213, 504, 959, 961, *1038, 1093
PCR *514, *878
Pesticides 504
Petroleum Products 213
Photochemotherapy 993
Pigmentation 528, 833, 959, *1030, *1038
Premalignant Lesion 20, 833, 961, 1184
Prevalence 20
Prevention 1216, 1217, 1220
Race *1030
Radiation, Non–Ionizing 993
Radiation, Ultraviolet 20, 32, 39, *269, 528, 833, *878, 959, 961, *1030, *1038, 1184
Radiotherapy *337
Registry 12, 20, 39, 213, 720
RFLP *514
Screening 374, *988
Socio–Economic Factors 20
Tobacco (Smoking) 20, *514
Transplantation *878
Trends 720
Twins 39
Vitamins 1220, 1221
Xeroderma Pigmentosum 1184

Small Intestine

Alcohol 737
Autopsy 398
Biopsy 398
Classification 398
Coffee 737
Crohn's Disease 226
Diet 737
Geographic Factors 737
Histology 398
Time Factors 737
Tobacco (Smoking) 737
Ulcerative Colitis 226

Soft Tissue

Animal 1165
Autopsy 398
Biomarkers 508, *514
Biopsy 398
Blood *290, *514
BMB *290, *514
Chemical Exposure 33, 508, 538, 654, 1028, 1068, 1180
Childhood *292, 889
Classification 398
Cluster 1165
Environmental Factors 508, 538, 1068
Familial Factors 889
Female 96
Genetic Markers *514
Geographic Factors *292, 1165
Herbicides 33, *290, 538, 622, 654, 1068, 1180
Heredity *292
High–Risk Groups 889
Histology 398, 1068
Lifestyle 1068
Migrants *292
Monitoring 1165
Occupation 33, 93, 96, 532, 538, 622, 654, *984, 1028, 1068, 1180
PCR *514
Pesticides 33, 93, 96, 532, 538, 622, 654, 1068, 1165, 1180
Plastics *984
Record Linkage 508
Registry 33, *292, 532, 622, 654, 889, *984, 1165
Reproductive Factors 96
RFLP *514
Rural 538
Solvents *984, 1165
Time Factors *984, 1165
Tissue *290
Tobacco (Smoking) 96, *514
Trends 1165

Spinal Cord

Demographic Factors 22
Registry 22
Survival 22
Trends 22

Spleen

Dose–Response 370
Drugs 370
Late Effects 370
Radiation, Ionizing 370

Stomach

Adolescence 557
Age *291, 576, *1074
Alcohol 63, 138, 153, 186, 326, 424, 505, 546, 557, 584, *585, 595, 601, 640, 680, 737, 794, *969, *975, *1074, *1103, *1118, 1233
Analgesics *1118
Antibodies 912, 913
Antioxidants 285
Atrophic Gastritis 137, *523, 531, 571, 725, *736, 846
Biochemical Markers 846, 913
Biomarkers 558, 602
Biopsy 913
Blacks 1035
Blood 285, 908, *969, *975
BMB 137, 241, 285, 286, *493, 533, 624, 846, 908, 910, 914, *969, *975, 1169
Chemical Exposure 73, 84, 326, 743, 860, 965, 982
Childhood 912, 930
Chinese 163
Chlorination 84
Cholesterol 533, 601
Chromosome Effects 558
Coal *183
Coffee 737, *1118
Cooking Methods 735, 1233
Diet 63, 127, 137, 138, 146, 153, 186, 242, 394, 424, 505, 531, 533, 557, 584, *585, 592, 595, 601, 602, *610, 624, 640, 680, 725, 733, 735, *736, 737, 794, 806, 846, 908, 913, 930, *969, *975, *1074, *1103, *1118, 1233
Disease, Other *969
DNA 558
DNA Adducts 602, 914
Dose–Response 748, 860
Drugs 505, *969, *975, *1074, *1118
Dusts 34, 73, *106, 138, *183, 743, 744, 748, 910, 965, 982, 1177
Environmental Factors 34, 146, 258, *567, 592, 1035
Enzymes 913
Ethnic Group 576, *605, *975
Familial Factors 146, 238, 394, 546, 558, 640
Fat 63, 595, 733
Female 1035
Fertilizers 756, 910, 1177
Fibre 640
Fruit 733
Gastrectomy *1014
Genetic Factors 146, 558, 592
Genetic Markers 146
Geographic Factors 84, 146, 258, 737
Goitre *523
H. pylori 285, 286, *493, *585, 602, *610, 846, 913, 914, *969, *975, 1169
High–Risk Groups 174, 394, 725
Histology 237, 394, *493, 552, 576, 735, *1103, 1177
Hormones 592
Infection 912
Japanese 602
Latency 153, *397, 860, *1074
Lifestyle 127, 186, *567, 584, *585, 592, 601, 602, *605, 624, 640, 806, 1035, *1118
Linkage Analysis 558
Lipids 806
Marital Status 601
Mate 1233
Menstruation 592
Metals 73, *106, *183, 241, 271, 326, *362, 640, 965, 1233
Micronutrients 137, 914
Migrants *291, *975
Minerals 127
Mining 34, *106, 326

Monitoring 174
Multiple Primary 146
Mutagen 182
Mutation, Somatic 558
Mycotoxins 912
Nutrition 127, 137, *610, 725, 733, 806, *975, *1103
Occupation 34, 73, *106, 138, *183, 258, 271, 326, *362, 552, 592, 680, 695, 735, 743, 744, 748, 756, 794, 860, 861, 910, 965, *969, *975, 982, 1035, *1118, 1177
Oncogenes 558
Operation, Previous *1014
Pedigree 146
Pepsinogen *585, 602, 913, 914, *975
Pesticides 965, 1177
Petroleum Products 73, *183, 860, 861, 965
Physical Activity 806, 908
Physical Factors 63, 326, 601, 680
Plastics 73, *362, 965
Premalignant Lesion 137, 237, 285, 531, 571, 725, *736, 846, 912, 913
Prevention 285, *567, 584
Prognosis 592
Psychological Factors 140
Psychosocial Factors 854
Radiation, Ionizing 138
Record Linkage *106
Registry 238, 241, 242, 271, *291, *362, 394, *397, 533, 546, 576, 624, 640, 680, 733, 743, 744, 748, 756, 787, 860, 861, 965, 1035, *1074, *1103, *1118, 1177
Reproductive Factors 546, 557
Rural 258
Screening 174, 286, 546, 958
Segregation Analysis 146
Sero–Epidemiology 241
Serum 137, 241, 286, 846, 914
Sex Ratio 624
Sexual Activity 592
Socio–Economic Factors 127, 138, *397, 601, *605, 794, 854, 930
Solvents 73, *362, 965
Stage 174
Stress 854
Surgery 958
Survival 174, 258
Time Factors 271, *291, 326, *397, 737, *1074
Tissue 137
Tobacco (Smoking) 34, 127, 137, 138, 146, 153, 186, 326, 424, 505, 546, 557, 584, *585, 595, 601, 680, 735, 737, 794, 958, *969, *975, 982, *1074, *1103, *1118, 1233
Toenails 624
Trace Elements 127, 137
Treatment 258
Trends 65, 258, *523, 576
Urine 137, 910, 912
Vegetables 733, *736
Virus *1103
Vitamins 63, 127, 241, 285, 533, 595, 640, 725, 733, 806
Water 84, 182, 787, 1177
Welding 982
Wood 744, 965

Testis

Adolescence 834
Age 230
Alcohol 283
Autopsy 212
Biomarkers *514

Biopsy 212
Birthweight 209
Blood *206, *514
BMB *514, 909
Chemical Exposure 815
Chemotherapy 391
Classification 108
Congenital Abnormalities 283
Cryptorchidism 283
Dairy Products 834
Diet 834
DNA Repair 805
Dusts 907
Enzymes *206
Familial Factors 805
Genetic Factors 230, 805, 909
Genetic Markers *514, 805
Heredity 805
Histology 108
HLA 909
Hormones 108, 283, 907
In Situ Carcinoma 212
Infection 283
Infertility 212
Intra–Uterine Exposure 283, 907
Lymphocytes 391
Menopause 230
Metabolism *206
Multiple Primary 805, 940
Occupation 108, 283, 907
Parity 230
PCR *514
Pedigree 805
Physical Factors 230, 283
Pregnancy 283
Radiation, Ionizing 815, 907
Radiation, Ultraviolet 805
Radiotherapy 940
Registry 108, *206, 283, 805, 834, 940
RFLP *514
SCE 391
Screening 805
Seasonality 209
Sexual Activity 815, 907
Solvents 907
Tobacco (Smoking) 283, *514, 815
Trauma 283, 815
Twins 230
Urine *206

Thyroid

Age 970
Alcohol *1245
BMB *313, 365
Chemical Exposure 1107, *1245
Chemotherapy 970
Childhood *704, 1194
Clinical Effects *453
Clinical Records 318
Coffee *1245
Diet 381, 662
Dose–Response *337, 970, 1107
Family History *1245
Genetic Factors *313, *319, 365, *704
Geographic Factors *462

Goitre *1245
Heredity *313
Histology 318, 381
Hormones *1119
Hyperthyroidism 492, *1245
Immunology *704
Lymphocytes 365
Mathematical Models 970
Menopause *1119
Migrants *453, *462
Multiple Primary 318, *319, 463, 970
Occupation *453, 1107
Oral Contraceptives *1119
Physical Activity *1245
Physical Factors 381, 970
Prevalence 662
Prognosis 318
Radiation, Ionizing 381, *453, *462, *704, *775, 796, *801, 970, 1107, 1194
Radiation, Non-Ionizing 970
Radiotherapy *319, *337, 970
Record Linkage *1119
Registry 365, *453, 463, 662, *775, 796, *1119
Reproductive Factors *1245
Socio-Economic Factors 381
Stage 318, 463
Survival 318, 463, 662
Tobacco (Smoking) *1245
Trace Elements 662
Treatment 318
Trends 662
Vitamins 662

Tongue

Female 425
Metals *362
Occupation *362
Plastics *362
Registry *362, 425
Solvents *362
Tobacco (Chewing) 425
Tobacco (Smoking) 425
Trends 425

Tonsil

Metals *362
Occupation *362
Plastics *362
Registry *362
Solvents *362

Ureter

Analgesics *352
Chemical Exposure *352
Diet *352
Occupation *352
Tobacco (Smoking) *352

Urethra

Analgesics *352
Chemical Exposure *352
Diet *352
Occupation *352
Tobacco (Smoking) *352

Urinary Tract

Chemical Exposure 828
Data Resource 828
Occupation 828
Registry 828

Uterus (Cervix)

Abortion *627, 707
Adolescence 728, 1187
Age 215, 230, *615
AIDS *998
Alcohol 363, 608, 680, 949
Antibodies 200, *898
Biochemical Markers 994, 1197
Biomarkers 274, 284
Biopsy 169, 274, 614
Birth Cohort 156
Blacks 1029
Blood 611, 994, *1048
Blood Group 660
BMB *37, 200, 241, 274, 608, *615, *627, *789, 813, 882, 883, 994, *998, 1029, *1048, *1207
Cholesterol 786, 949, *1048
Chromosome Effects 437
Clinical Records 728
CMV 169, 608, *789, 813
Contraception *627, 645, 924, 946, *1048, 1187, *1207
Cost–Benefit Analysis *294, 631
Cytology 10, 55, 274, *294, 369, 437, 438, *482, *615, 631, 728, 883, 924, 951, 994, *998, 1029, 1187, *1207, 1211
Diet 438, 443, 517, 608, 611, 680, 882, 949, 994, *1048, 1185, 1197
DNA 169, 920, 946, 994, 1211
Drugs 232, 994
Dusts *692, 1197
EBV 608
Education 443, *627
Environmental Factors *567, 674
Ethnic Group 356, *605, 645
Familial Factors 949
Fat 1197
Female *998
Fertility 924
Fibre 1197
Genetic Factors 230
Geographic Factors 447, 679
HBV 369, *1207
High–Risk Groups 115, 174, 1185, 1197
Histology 631, 883, 920, 951
HIV 369, *898, *998, *1207
Hormones 517, 949, *1048
HPV *37, 55, 169, 200, 274, 284, 356, 363, 369, 403, 437, 438, 608, 611, 614, *615, *627, 645, 728, 745, *789, 813, 867, 920, 946, 994, 1029, *1048, 1187, *1207, 1211
HSV 55, 169, 200, 284, 437, 438, 608, 611, 728, *789, 813, *1207
Hygiene 410, 443
Immunologic Markers *482
In Situ Carcinoma 200, 951
Infection 200, 284, 403, 645, 813, 920
Latency *397
Lifestyle *567, *605, 707
Lymphocytes *482
Mammography 1197
Mathematical Models 115, 631, 1185
Menarche 660
Menopause 230, 660

Metals 241
Micronutrients 994
Monitoring 174
Multiple Primary 223, *434
Mutagen 1185
Nutrition 611, 994, *1048, 1197
Obesity 949
Occupation 680, *692, 1197
Oral Contraceptives 113, 169, 215, 517, 611, 813, 867, 920, 924, 1185
Parity 230, 443, 517, 547, 660, 924
Passive Smoking 1185
PCR 55, 274, 369, 608, 614, *615, 867, *1207
Physical Activity 949, *1048
Physical Factors 230, 660, 680, 786, 949
Pregnancy *627, 924
Premalignant Lesion 113, 200, 363, 369, 437, 438, *482, 547, 614, 728, 946, 994, 1211
Prevalence 55, 1197
Prevention *294, *567, 951, 1197
Promotion *482
Radiotherapy 223, *434
Record Linkage 232
Registry *195, 215, 223, 232, 241, 254, *397, 680, 786
Reproductive Factors 156, 363, 608, 645, 707, 951, *1048
Rural 679
Screening 10, 55, 115, 174, *195, 254, *294, 374, 437, *482, *627, 631, 660, 813, 951, 1197
Sero-Epidemiology 169, 200, 241, 284, 437, 745, *789
Serum 200, 241, *789, 813, *1207
Sexual Activity 156, 169, 215, 284, 356, 403, 410, 438, 443, 517, 547, 611, *615, *627, 813, 867, 951, *1048, 1185, 1187, *1207
Sexually Transmitted Diseases 55, 284, 369, *627, 728, 745, 813, *898, 946, 1185, 1187, 1211
Socio-Economic Factors 156, 356, 363, *397, 443, *605, 611, 645, 946, 1029
Stage 174, *195
Survival 174, 786
Time Factors 215, *397
Tissue 274, 994, *1207
Tobacco (Smoking) 156, 215, 363, 608, *627, 645, 680, 867, 883, 946, 949, *1048, 1185, 1187, 1197, *1207
Treatment 674, 951
Trends 65, 156, *195, 679
Tubal Ligation 1139
Tumour Markers 745
Twins 230
Urban 679
Uric Acid 949
Virus 200, 645, 920
Vitamins 241, 284, 547, 611, 882, 994, *1048, 1197

Uterus (Corpus)

Abortion 707
Age 230, 1082
Alcohol 595, 949, 972, 1000
Blood 972
Blood Group 660
Cholesterol 786, 949
Cluster 1105
Contraception 1047
Data Resource 996
Diet 105, 517, 595, 810, 949, 972, 980, 1000, 1047, 1105, 1190, *1223
Dose-Response 1047
Drugs *302, 996, *1191
Ethnic Group 1047
Familial Factors 949
Fat 595, 972, 1047, 1105

Female 105
Genetic Councelling *703
Genetic Factors 230, *703
Geographic Factors 679
Hormones 517, 810, 925, 949, 972, 980, 996, 1047, 1082, *1119, 1190, *1223
Infection 1000
Lifestyle 707
Menarche 660
Menopause 230, 660, 925, 980, *1119
Micronutrients 972
Multiple Primary *302
Nutrition 972, 1047
Obesity 810, 949, 1047, 1105
Occupation 1000, 1047
Oral Contraceptives 517, *1119, 1190
Parity 230, 517, 660
Physical Activity 810, 949, 980, 1047
Physical Factors 230, 660, 786, 949, 1105
Prevention *1191
Record Linkage 105, 973, *1119
Registry 786, 810, 925, 973, 1105, *1119, 1190, *1191, *1223
Reproductive Factors *302, 707, 810, 972, 980, 1000, 1047, 1190
Rural 679
Screening 105, 660
Sexual Activity 517
Surgery 973
Survival 786
Time Factors *1191
Tobacco (Smoking) 595, 949, 972, 1000, 1047
Treatment *302
Trends 679
Tubal Ligation 1139
Twins 230
Urban 679
Uric Acid 949
Virus 1000
Vitamins 595, 1000, 1047, *1223

Vagina
Blood 971
BMB 971
Diet 1102
Drugs 1102
Hygiene 971
Intra–Uterine Exposure 1102
Menarche 1102
Nutrition 1102
Reproductive Factors 971
Screening 971
Sexual Activity 971
Sexually Transmitted Diseases 971
Tobacco (Smoking) 971

Vulva
Blood 971
BMB 971
Diet 517
Ethnic Group 356
Hormones 517
HPV 356
Hygiene 971
Oral Contraceptives 517
Parity 517

Reproductive Factors	971
Screening	971
Sexual Activity	356, 517, 971
Sexually Transmitted Diseases	971
Socio-Economic Factors	356
Tobacco (Smoking)	971

Wilms' Tumour

Age	71
Alcohol	71
BMB	*15
Chemical Exposure	71, 1007
Childhood	540
Chromosome Effects	*15
Coffee	71
Diet	71
DNA	*15
Drugs	71
Environmental Factors	376, 1007
Familial Factors	1007
Fruit	71
Genetic Factors	376
Herbicides	71
Heredity	1007
Infection	71
Intra-Uterine Exposure	1007
Lifestyle	376
Multiple Primary	540
Occupation	71, 1007
Parental Occupation	376
Radiation, Ionizing	376
Registry	376
Socio-Economic Factors	71, 1007
Survival	540
Tea	71
Tissue	*15
Tobacco (Smoking)	71
Treatment	540
Vegetables	71

INDEX OF STUDY TYPES

"Study type" is a very useful (and widely used) descriptive term for epidemiological projects, and this Index enables rapid identification of, say, case–control studies of bladder cancer. Epidemiological studies can, however, be difficult to classify into conventional designs such as cohort, case–control, etc, and there may therefore be a few errors. The study type of each project has been assigned by the editors for indexing purposes, but was only shown beneath each abstract for the first time in the 1988 edition. Investigators who notice errors should submit corrections when updating their entry for the next edition.

Projects which involve more than one study type are classified under each type.

The study type "Registry" is used only for projects in which a new population–based cancer registry is being described. These abstracts are retained in the Directory for two years at most. Studies involving collaboration with a cancer registry are indexed both to the general keyword "Registry" in the TERMS index, and to the specific cancer registry in the REGISTRY index.

Case Series

All Sites 702
Benign Tumours *494
Bile Duct 125, *734
Breast (F) 25, 544, 558, 644, 1053, 1078
Breast (M) 1053
Childhood Neoplasms 823, 1165, 1225
Colon 455, 526, 558
Endocrine Glands *494
Gallbladder 125, *734
Gastrointestinal 880
Hodgkin's Disease *676
Hypopharynx 268
Inapplicable 554
Leukaemia *385, 435, 539, 581, 730, 823, 1026, 1165
Leukaemia (ALL) *676
Liver 491, 726, 1165
Lung 239, 496
Lymphoma 435, *1062
Melanoma 9, 874
Mesothelioma 496
Myelodysplastic Syndrome 539
Nasal Cavity 278, 874
Nasopharynx 409, 607
Non–Hodgkin's Lymphoma *676, 823, 1165
Oral Cavity 268, 363, 1090
Oropharynx 268, 363
Ovary 406, 534, *1011, 1133
Pancreas *734, 1165
Pharynx 1090
Pleura 496
Rectum 455, 558
Respiratory 363
Retinoblastoma 877
Soft Tissue 1165
Stomach 558
Testis 909
Thyroid 318
Uterus (Cervix) 363, 403, 608, 728

Case–Control

All Sites 19, 52, 80, 83, 84, 86, 116, 140, 223, 240, 241, 270, 281, 298, 300, *319, 333, 335, *357, 378, 452, 465, *475, 484, 498, 504, 508, 512, 532, 537, 551, 564, 575, 578, 588, 619, *661, 708, 749, 751, 755, 757, 872, *929, 970, 982, 992, *1021, 1044, 1046, 1071, 1080, 1095, 1099, 1158, 1171, *1174, *1175, *1235
Benign Tumours 253, 517, 617, 1140, *1202
Bile Duct *514, 1152
Bladder 73, 84, 99, *110, 162, 232, 279, 310, *352, 392, 484, *490, 541, *543, 652, 866, 900, 965, 1025, 1035, 1134
Bone 102, 171
Brain 38, 44, 57, 84, 85, 86, 96, 171, *205, 303, 304, *473, 504, *514, 532, 765, 766, 955, 965, *1021, 1087, 1088, 1095, 1104, 1151
Breast (F) *26, *27, *46, 74, *79, 121, 127, 131, 140, 146, 157, 170, *195, 223, 224, 230, 232, 241, 253, 299, *302, 347, *357, 395, 400, 408, 423, 433, *436, 442, 446, 467, 468, *474, *478, 480, 499, 510, *514, 517, 565, *567, *609, 612, 617, 618, 619, 620, 624, 630, 639, 650, 707, 711, 722, 797, 802, *808, 848, 879, *948, 970, 974, 997, 1000, *1005, *1006, *1043, 1046, 1072, 1083, 1097, *1112, 1116, *1124, 1130, 1132, 1140, *1142, *1143, *1144, *1191, *1195, *1196, *1202, 1204, 1218, 1222
Breast (M) 211, *1004
Burkitt's Lymphoma *658
Childhood Neoplasms 222, 307, 457, 600, 647, 766, 769, *822, 827, 832, *852, 890, *906, 1147, 1151
Choriocarcinoma 516

TYPE

Colon	*27, 28, *53, 73, 74, 84, *110, 127, 146, 149, 150, 161, 166, 167, 241, 260, 275, 366, *407, 424, 465, 468, 470, *478, 480, 510, *567, 624, 641, *675, 712, 747, 748, *842, *843, 893, 894, *976, 997, 1032, 1035, 1058, 1098, *1127, 1128, 1145, 1183, *1199, 1210, 1228
Eye	1035
Gallbladder	673, 1081, 1152
Gastrointestinal	76, 101, 192, 230, 340, 392, 1241
Genitourinary	1241
Haemopoietic	19, 358, 393, *476, 504, *727, 744, 965
Hodgkin's Disease	40, 93, 270, *311, 542, *658, *929, *1125, 1154, 1192, *1244
Hydatidiform Mole	516
Hypopharynx	315, *317, 439
Kaposi's Sarcoma	41, 657, *658, *898
Kidney	45, 73, 76, 84, *219, 330, 368, 506, 560, 795, 1037, 1104, 1134, 1141, 1164
Larynx	70, 76, 181, 315, *317, 439, *478, 784, 807, 931, 1095, 1134, 1236, *1246
Leukaemia	*18, 57, 84, 86, 118, 119, 127, 171, 180, 192, *205, 208, 224, 230, 307, 334, 340, 349, 367, *380, *384, 393, 508, 542, 600, 619, 653, 655, *704, 766, *822, 827, 849, 859, 871, 915, *929, *936, 970, 1046
Leukaemia (ALL)	75, 647, 1108
Leukaemia (AML)	280, *727, 1086
Leukaemia (ANLL)	1109
Leukaemia (CLL)	40, 224, *727, *1125
Lip	70, 1085
Liver	127, 138, 146, 165, 171, 175, *460, *514, 559, 559, 559, 559, 574, *577, 656, *690, 811, 1035, 1104
Lung	3, *50, 52, 73, 76, 94, 95, 97, 123, 127, 139, 146, 152, 154, *155, *158, *173, *183, 192, 223, 232, 241, 243, *264, 279, 297, 298, *317, *320, 328, 331, 340, 343, 351, *379, 386, 393, *428, *459, 465, 480, 484, 566, *567, 574, 619, 624, 678, 683, 710, 742, 744, 748, *773, 782, 788, 807, *812, 858, 872, 903, 932, 963, 965, 968, 970, 982, *999, 1035, *1051, 1061, 1079, 1104, 1121, 1134, 1135, 1162, 1182, *1209, *1226, 1234
Lymphoma	*18, 19, 41, 57, 73, 84, 171, *205, 349, 484, 506, 508, 744, 757, *936, 965, 1073, *1174
Melanoma	7, 73, 230, *269, 346, 495, 628, 874, *1004, *1030, 1035, 1069, 1122, 1176, 1221
Mesothelioma	30, 52, 101, 731, 748, 851, 965, 1035
Multiple Myeloma	*18, 86, 93, 96, 542, 757, 1036
Myelodysplastic Syndrome	17, 118, 267, 619, 835
Nasal Cavity	181, 208, 306, *316, *317, 874, 1084
Nasopharynx	*2, 127, 143, 151, 179, *184, 314, 1084, 1208
Neuroblastoma	57, 1008
Non–Hodgkin's Lymphoma	40, 86, 93, 96, *290, *311, 340, *514, 532, 542, 619, *658, *727, *761, 859, *898, *929, *1125, *1174, 1186, 1193
Oesophagus	73, 127, 167, *259, 424, 439, *478, 733, 784, *969, 1035, 1036, *1074, *1103, *1118, 1134
Oral Cavity	70, 181, *195, 207, 363, *420, 439, 445, *478, 739, 784, 807, 1089, 1134, 1136, 1203, 1236
Oropharynx	363, 405, 439, 1136, 1203
Ovary	108, 404, *427, *478, 517, 707, 997, 1000, *1120, *1175, *1191, 1198, 1229, 1243
Pancreas	73, 167, 244, 484, 733, 1000, 1036, 1134
Penis	356
Peritoneum	76
Pharynx	70, 181, 784, 807, 1089, 1236
Pleura	76, *317
Prostate	73, 74, 84, 86, 92, 228, 232, 241, 257, 329, *427, 629, *760, 793, 816, 872, 895, 918, 943, 997, 1000, *1002, *1004, 1036, 1129, *1205, 1214
Rectum	*27, 28, *53, 73, 84, *110, 127, 146, 149, 150, 161, 166, 167, 241, 275, 366, *407, 424, 468, *478, 510, *567, 624, *675, 712, 747, 748, *842, *843, 893, 894, 1032, 1035, 1058, 1098, *1127, 1128, 1145, *1199
Respiratory	101, 208, 363, *514
Rhabdomyosarcoma	*1155
Salivary Gland	1035
Sarcoma	83, 93, 96, *290, *514, 965
Skin	*269, 504, *514, 959, *1030, *1038, 1221

Soft Tissue 93, 96, *290, 508, *514, 532, 538, 1068
Stomach 73, 84, 127, 138, 140, 146, 153, *183, 186, 237, 241, 286, 394, 424, 505, 531, 557, *567, 584, 592, 602, *610, 624, 725, 733, 735, 744, 748, 794, 846, 913, 965, *969, *975, 982, 1035, *1074, *1103, *1118, 1169, 1177, 1233
Testis 108, *206, 209, 230, 283, *514, 815, 834, 907
Thyroid *319, 381, 662, *704, 796, *801, 970, *1245
Ureter *352
Urethra *352
Urinary Tract 828
Uterus (Cervix) 55, 113, 156, 169, *195, 215, 223, 230, 232, 241, 284, 356, 363, 410, 443, *482, 517, 547, *567, 611, *627, 707, 745, *789, 813, 867, *898, 920, 951, 1029, *1048, 1187, *1207
Uterus (Corpus) 230, *302, 517, 707, 810, 972, 1000, 1047, 1190, *1191
Vagina 971, 1102
Vulva 356, 517, 971
Wilms' Tumour 71, 1007

Cohort

All Sites 19, *29, 33, 52, 56, 68, 80, 81, 83, 86, 87, 89, 91, *103, *107, *111, 112, 126, 128, 132, 141, 185, 210, 218, 221, 223, 225, 227, 229, 240, 241, 247, 248, *250, *262, 263, 270, 271, *272, *277, 281, 282, 295, 296, 298, 300, 301, 305, 308, *327, *332, 333, 335, *336, *338, 339, *357, 359, 360, 371, 372, *396, *397, *412, *415, *422, 451, *453, 465, 466, 471, 472, 483, 484, *485, 486, 489, 501, 502, 504, 508, 509, 511, 532, *535, 545, 550, 551, *555, 563, 564, 568, 572, 573, 582, 583, 586, 589, 593, 601, 603, 619, *621, 622, 623, 625, 654, 660, *661, *669, 680, 685, *686, *687, *693, 695, *696, *697, *698, *699, *700, *701, 709, 714, *715, 718, 740, 746, 749, 750, 751, 752, 753, 754, 755, 757, 763, 764, 767, 771, 772, *775, *776, 777, 779, *780, *781, 792, 799, 806, 819, 821, 825, 826, 837, *838, 839, 840, 854, 856, 863, 870, 881, 884, 887, 888, 896, 897, 901, 902, 921, 924, 928, *935, 938, 939, 940, 953, *967, 970, 982, *987, *989, 990, 991, 992, 993, 1013, 1015, 1018, 1019, 1023, 1027, 1039, 1044, 1046, 1050, *1052, 1065, 1070, 1071, 1075, 1076, *1077, 1082, 1094, 1095, *1101, 1105, 1107, *1110, 1137, 1138, 1157, 1158, 1159, 1160, 1161, 1163, 1167, 1168, 1173, *1174, *1175, 1180, 1181, 1188, 1197, 1201, 1227, 1230, 1231
Benign Tumours 31, 472, 616, 617, 1059, 1115, 1140
Bile Duct 958
Bladder 104, 178, *204, 213, 232, 305, 484, 487, 509, 546, *687, 695, 910, 965, 1027, 1096, 1157
Bone 370, 377, *621
Brain 86, *203, 213, 251, *272, 348, *362, *462, 504, 532, 583, *621, *665, *715, 965, 1012, 1091, 1095, 1104, 1180, 1181
Breast (F) 21, *27, 60, *78, 89, 105, 114, *203, *214, 223, 230, 232, 235, 236, 241, 242, *255, *357, *396, 400, *461, *462, 463, 464, 480, 507, 546, 616, 617, 619, 624, 630, 637, 640, 651, 660, 680, 717, 723, 724, 770, 786, 791, 802, 803, 804, 836, 854, 855, 888, 908, 923, 924, 925, 926, 941, 949, 970, 973, 980, *987, 991, 1031, 1040, 1046, 1059, 1065, *1067, 1082, 1105, 1107, 1111, *1112, 1115, 1116, *1119, 1130, 1132, 1139, 1140, *1191, *1196, 1197, *1206, 1213, 1242
Childhood Neoplasms *111, 221, 783, 827, 831, 891, 916, 917, *934, 1012, 1042, *1110, 1225
Colon 21, *27, 34, 89, 105, 148, *203, 226, 241, 242, 261, *323, 366, 463, 465, 480, 521, 526, 546, 595, 601, 624, 638, 640, 748, *758, 762, 800, 806, *818, 821, 860, *862, 885, 892, 908, 949, 958, 973, 980, *987, *1014, 1065, 1066, 1092, 1093, *1119, 1161, 1181, 1197, 1212, 1213, 1215
Eye 993
Female Genital 252, 803, 888, 926, 949
Gastrointestinal 76, 101, 213, 230, *649, 660, 691, *692, 923, 927, 1091
Genitourinary 1160
Haemopoietic 19, 213, 504, *692, 744, 965, *984, 1091, 1160
Head and Neck 389, 463, 632
Hodgkin's Disease 218, 251, 270, 540
Inapplicable *4, 198, 199
Kaposi's Sarcoma 41, 382, *952, 957, 964

TYPE

Kidney	76, 105, *272, 377, 487, *715, 790, 873, 1093, 1096, 1104, 1189
Larynx	76, *204, 231, *362, 364, 511, 664, 931, 1095
Leukaemia	33, 56, 86, *106, 136, 208, 221, 230, 251, *262, *277, 339, 370, *462, 463, 487, 508, 570, 619, *621, *665, 672, *687, 752, *775, 777, 779, *781, 827, *864, 871, 901, 902, 915, *934, 950, 970, 977, *984, 1046, 1107, 1179, 1180, 1181, 1231
Leukaemia (ALL)	540
Leukaemia (ANLL)	540
Liver	128, *160, 221, 273, *293, 345, 370, 377, 383, 487, 491, 556, 569, 595, 597, 601, 664, *698, *700, 811, *983, 1028, 1091, 1092, 1093, 1104, 1149, 1161
Lung	3, 34, 36, 52, 76, 87, 104, *106, 126, 129, 159, 172, *173, 177, *204, 223, 231, 232, 241, 242, 251, 263, *265, 271, *272, 298, *320, *322, *323, *325, 326, 339, 353, *362, 370, 388, *397, *462, 465, 480, 481, 483, 484, *485, 487, 489, 511, 546, *555, 556, 579, *580, 595, 596, 601, 604, 619, 623, 624, 640, *649, *667, *668, *670, *671, 680, 683, 685, *686, *687, 691, *692, 695, *698, *700, *715, 721, 744, 748, 756, *768, 772, *774, *785, 790, 792, 798, 806, 844, 854, 857, 860, 861, 873, 910, 939, 945, 947, 965, 968, 970, 973, 980, 982, *987, 1022, 1024, 1065, 1091, 1092, 1093, 1104, 1107, 1153, 1173, 1180, 1197, 1213, *1226, 1227
Lymphoma	13, 19, 41, 56, 247, *362, 370, 463, 484, 508, 570, *621, 654, 744, 752, 757, 779, *781, *934, 964, 965, 977, *984, 1024, 1028, 1100, 1160, *1174, 1179
Male Genital	213, 252
Melanoma	31, 230, 911, *988, 1069
Mesothelioma	34, 52, 101, *324, 353, 530, 579, *686, 748, 965
Multiple Myeloma	86, 757, *967, 1180
Myelodysplastic Syndrome	619
Nasal Cavity	208, 305, 509, 1161
Nasopharynx	144, 606, 1159
Nervous System	*203
Neuroblastoma	540, 1114
Non–Hodgkin's Lymphoma	86, 339, 532, 540, 619, 622, *967, *1174, 1180
Oesophagus	*134, 188, 361, *362, 454, 546, 664, 910
Oral Cavity	66, 389, 413, 414, *420, *421, 664
Oropharynx	66, *362
Ovary	*78, 105, 634, 635, 660, 786, 804, *865, 868, 949, 973, 980, 1139, *1175, *1191
Pancreas	104, 105, 484, *621, 958, 1179, 1213
Peritoneum	34, 76, 530
Pharynx	509, 664
Pleura	76, 530
Prostate	21, 86, *203, 232, 241, 242, 251, 623, 873, 908, 1065, 1213
Rectum	21, *27, 105, 148, *203, 241, 242, 261, 366, 521, 546, 601, 624, 638, 640, 748, *758, 762, 800, *818, 821, 860, *862, 885, 892, 958, *1014, 1066, *1119, 1161, 1181, 1212, 1213
Respiratory	101, 208, 213, *324, 488, 509, 743, 856, *864, 1156, 1160, 1161, 1219
Sarcoma	13, 83, 213, 247, 622, 965
Skin	32, 213, *337, 504, *878, 959, *988, 993, 1093
Small Intestine	226
Soft Tissue	33, 508, 532, 622, 654, *984, 1028, 1180
Spleen	370
Stomach	34, *106, 137, 238, 241, 242, 271, 326, *362, *397, *493, 546, 552, 571, *585, 595, 601, 624, 640, 680, 695, 743, 744, 748, 756, 806, 846, 854, 860, 861, 908, 910, 930, 958, 965, 982, *1014
Testis	230, 940
Thyroid	*337, *453, *462, 463, 492, 662, *775, 970, 1107, *1119, 1194
Tongue	*362
Tonsil	*362
Uterus (Cervix)	10, 115, 200, 223, 230, 232, 241, 254, *397, *434, 437, 438, 614, 631, 660, 680, *692, 786, 867, 883, 924, 946, 949, *998, 1139, 1187, 1197, 1211
Uterus (Corpus)	105, 230, 595, 660, 786, 925, 949, 973, 980, 1082, 1105, *1119, 1139, *1191
Wilms' Tumour	540

Correlation

All Sites	49, 61, 69, 84, 132, *202, 246, *387, 522, 562, 787, 1057, 1131, 1197, 1238, 1240
Bile Duct	587
Bladder	84, 513, 787
Brain	84, *203
Breast (F)	63, 127, *203, 996, 1197
Childhood Neoplasms	905, 1165
Colon	63, 84, 127, *203, 526, *527, 737, 787, 996, 1057, 1197
Gallbladder	737
Inapplicable	48
Kidney	84, 513
Larynx	513
Leukaemia	84, 127, *385, 829, *886, 905, 1165
Liver	127, 168, 175, 182, 191, 587, 737, 1165
Lung	63, 127, *197, 513, 787, 841, 944, 996, 1131, 1197
Lymphoma	84, 829, *886
Melanoma	20, *269
Mesothelioma	530
Nasopharynx	127
Nervous System	*203
Non–Hodgkin's Lymphoma	1165
Oesophagus	63, 127, 513, 737
Oral Cavity	513
Ovary	996
Pancreas	513, 737, 1165
Peritoneum	530
Pharynx	513
Pleura	530
Prostate	84, *203, 996
Rectum	84, 127, *203, 737, 787
Retinoblastoma	561
Skin	20, *269
Small Intestine	737
Soft Tissue	1165
Stomach	63, 84, 127, 182, *523, *736, 737, 787, 914
Uterus (Cervix)	*615, 1185, 1197
Uterus (Corpus)	996

Cross–Sectional

All Sites	88, 458, *520, 524, 533, 682, 1197
Bladder	965, 1016
Brain	965
Breast (F)	276, 374, *515, 533, *919, 962, 1197
Childhood Neoplasms	646
Colon	374, 533, *594, *1010, *1017, 1197, 1215
Gallbladder	124
Haemopoietic	965
Inapplicable	77, *233, 249, 681, *1020
Larynx	664
Liver	163, 234, *293, 664, 729, 956
Lung	*671, 965, 1197, 1232
Lymphoma	965
Melanoma	39
Mesothelioma	965
Nasal Cavity	278
Oesophagus	130, 163, 664
Oral Cavity	419, *420, 664
Ovary	741, *865
Pharynx	664
Prostate	374, 469, 1033
Rectum	374, *594, *1010
Sarcoma	965
Skin	39, 374, 528, 833, 961
Stomach	163, 533, 571, 912, 913, 965

Testis 212
Thyroid 662
Uterus (Cervix) 55, 274, 369, 374, 645, 1197

Genetic Epidemiology

All Sites 805, *1052
Bone 805
Brain *1056
Breast (F) 25, *78, 146, 170, 401, *703, 805, 962, 1053, 1054, *1063, 1078, 1111, *1112
Breast (M) 1053
Childhood Neoplasms 805, 904
Colon 51, 146, 216, 366, *778, *842, *933, 954
Gastrointestinal 805, 954
Head and Neck *553
Leukaemia 805, 937
Liver 146
Lung 146, 1079
Lymphoma 805
Melanoma 47, *703
Oesophagus 133, *134, 361
Oral Cavity *553
Ovary *78, *703, 741, 805, 868, *1011, 1054
Rectum 51, 146, 216, 366, *842, *933, 954
Retinoblastoma *706, 904
Stomach 146
Testis 805
Thyroid *313, 365
Uterus (Corpus) *703
Wilms' Tumour *15

Incidence

All Sites 1, 14, 42, *54, 67, 69, 82, 90, 100, *109, 122, 176, 196, 220, 245, 246, 289, 296, 312, 344, *387, *411, *418, 426, 440, 448, 449, 450, *453, 503, 525, 562, 582, 643, 648, 663, 674, 713, 718, 719, 720, *780, *814, 820, 839, 869, 875, 876, 889, 981, 1041, 1166, 1170, 1197, 1238, 1239, 1240
Appendix 398
Bile Duct 398
Bladder 104, 1035
Bone *292, 674, *845
Brain 22, 85, 979
Breast (F) 23, *195, 276, *291, 399, 417, *605, 630, 674, 679, 889, 973, 1197
Breast (M) *54, 211
Childhood Neoplasms *54, 307, 375, 376, 590, 598, 646, 732, 823, 824, *852, 904, 905, 922, 1113, 1165
Choriocarcinoma 6
Colon 5, 24, 258, *291, 398, 973, 1035, 1197
Eye 1035
Gallbladder 258
Gastrointestinal 72, 192, 193
Haemopoietic 716
Hodgkin's Disease *292, 479, *605
Hydatidiform Mole 6
Hypopharynx 425
Kaposi's Sarcoma 479, 529, 1041
Larynx 59, 425, 869
Leukaemia 192, 287, *292, 307, *350, 376, 633, 730, 823, *850, 869, 905, 922, 1165
Leukaemia (ALL) 633, 847
Leukaemia (AML) 847
Leukaemia (CML) 847
Lip 98
Liver 258, 345, 869, 1035, 1165

Lung	72, 104, 192, 193, 217, 425, 447, 456, 496, *605, *812, 869, 973, 1035, 1197
Lymphoma	287, *350, 376, *850, 869
Melanoma	20, 100, *291, *292, 720, *1030, 1035
Mesothelioma	42, 59, 72, 496, *536, 1035
Myelodysplastic Syndrome	922
Nasopharynx	*2, 59, 314, 447
Nervous System	*292, 398
Neuroblastoma	*292, 376, 591
Non–Hodgkin's Lymphoma	*292, 479, *605, 823, 1165
Oesophagus	187, 258, 398, *605, 694, 1035
Oral Cavity	66, *195, 398, 425
Oropharynx	*54, 66, 425
Ovary	534, *865, 973
Pancreas	104, 258, 398, 1165
Peritoneum	42
Pleura	42, 496
Prostate	*291, 398
Rectum	5, 8, 24, 258, *605, 1035
Retinoblastoma	561, 904
Salivary Gland	1035
Sarcoma	*292, 889
Skin	12, 20, 720, *1030
Small Intestine	398
Soft Tissue	*292, 398, 889, 1165
Spinal Cord	22
Stomach	258, *291, 576, *605, 1035
Thyroid	318, *453
Tongue	425
Uterus (Cervix)	*195, 447, *605, 674, 679, 1197
Uterus (Corpus)	679, 973
Wilms' Tumour	376

Intervention

All Sites	*256, 344, 986, *987, 995, *1117, 1197
Bladder	642
Breast (F)	174, 309, 518, 630, 717, 723, 791, 942, *987, 1031, 1197, *1206, *1223
Colon	28, 147, 261, 373, 892, 986, *987, 1045, 1060, 1197, 1212, *1223
Female Genital	252
Gastrointestinal	880
Inapplicable	*817
Larynx	*1117
Liver	164, 174, 345
Lung	35, 174, 642, *671, 705, 986, *987, 1197, 1200, 1224, 1232
Male Genital	252
Melanoma	666
Mesothelioma	35, 1200
Oesophagus	*189, 190
Oral Cavity	*416, *421, 429, *430, 444, *1117
Ovary	*1223
Pharynx	*1117
Rectum	28, 147, 174, 261, 373, 892, 1212, *1223
Skin	1055, 1216, 1217, 1220
Stomach	174, 285
Uterus (Cervix)	*37, 174, *294, 882, 994, 1197
Uterus (Corpus)	*1223

Methodology

All Sites	69, 117, 145, *418, 477, 500, 519, *520, 759, 809, 899, 970, 978, *1052, 1123, 1131, 1166, *1172, 1178, 1197
Appendix	398
Benign Tumours	1115
Bile Duct	398
Bladder	*548, 899, 965, 1035

Brain	965, 1091
Breast (F)	11, 60, 288, 970, *1063, 1115, 1197, 1242
Colon	398, 1035, 1197
Eye	1035
Gastrointestinal	1091
Haemopoietic	965, 1091
Hodgkin's Disease	1192
Inapplicable	201, *233, 249, 354, *355, 626, *853, 966, 1001, 1009, 1049
Leukaemia	899, 970, 1064
Liver	1035, 1091
Lung	*43, 497, 613, 960, 965, 970, 1035, 1091, 1131, 1162, 1197
Lymphoma	965
Melanoma	16, 1035
Mesothelioma	965, 1035
Nervous System	398
Oesophagus	398, 1035
Oral Cavity	398
Pancreas	398
Prostate	398
Rectum	1035
Salivary Gland	1035
Sarcoma	965
Small Intestine	398
Soft Tissue	398
Stomach	965, 1035
Thyroid	970
Urinary Tract	828
Uterus (Cervix)	10, *294, 1197

Molecular Epidemiology

All Sites	62, 551, 572, 578, 586, 684, 753, 759, 1123, 1131
Angiosarcoma	738
Bile Duct	*514, *734
Bladder	*120, *543, 642, 738
Brain	*514
Breast (F)	170, *514, 558, 1116
Colon	558
Gallbladder	*734
Haemopoietic	677
Hodgkin's Disease	*676
Inapplicable	*4, 77, 198, 199, 201, 249, 354, *355, 390, 626, 681, *817, 830, 1001, 1003, 1009, *1020
Leukaemia	62, 539, 977
Leukaemia (ALL)	*676
Liver	163, *514, 738, 956
Lung	135, 239, 343, *459, 481, 642, 683, 738, 960, *1051, 1131
Lymphoma	62, 977
Myelodysplastic Syndrome	539
Non–Hodgkin's Lymphoma	*514, *676
Oesophagus	163
Oral Cavity	402
Pancreas	*734
Rectum	558
Respiratory	*514
Retinoblastoma	1148
Sarcoma	*514
Skin	*514, 1184
Soft Tissue	*514
Stomach	163, 558
Testis	391, *514
Uterus (Cervix)	608

Mortality

All Sites 49, 61, 69, 90, 132, *142, 176, 194, 196, 266, 289, 312, *341, *387, *418, *535, 550, 563, 568, 648, 688, 689, 713, 719, 809, 820, 869, 876, 899, 1146, 1170, *1172, 1178, 1188, 1237, 1238, 1240
Bladder 104, 513, 899, 985
Breast (F) 65, 127, 194, *195, 630, 636, 941
Childhood Neoplasms 823
Colon 65, 127, 1106
Gastrointestinal 64, 72, 192, 193, 1034
Kidney 513
Larynx 513, 869
Leukaemia 127, 192, *342, 633, 823, 869, 899, 985
Leukaemia (ALL) 633
Liver 65, 127, 869
Lung 65, 72, 104, 127, 192, 193, 194, *197, 513, 841, 869, 985, 1034
Lymphoma 869
Mesothelioma 72
Nasopharynx 127
Non–Hodgkin's Lymphoma 823, 1034
Oesophagus 65, 127, 187, 513
Oral Cavity *195, 513
Pancreas 104, 513
Pharynx 513
Prostate 65
Rectum 65, 127, 1106
Stomach 65, 127, *523
Uterus (Cervix) 65, 156, *195

Registry

All Sites 42, *54, 312, 344, *411, 440, 449, 525, 643, 805, *814, *838, 839, 981, 1126, 1188
Bone 805, *845
Brain 22, 432
Breast (F) 235, 276, *703, 805
Breast (M) *54
Childhood Neoplasms *54, 307, 321, 375, 598, 599, 732, 805, 824, *852, 922, 1150
Choriocarcinoma 6, *549
Colon 5, 24, 28
Gastrointestinal 805
Hodgkin's Disease 431
Hydatidiform Mole 6, *549
Inapplicable 58
Leukaemia 307, 431, 805, *850, 922
Leukaemia (ALL) 847
Leukaemia (AML) 847
Leukaemia (CML) 847
Lung 496, *812
Lymphoma 431, 805, *850
Melanoma *703
Mesothelioma 42, 496, *536
Multiple Myeloma 431
Myelodysplastic Syndrome 922
Oropharynx *54
Ovary 534, *703, 805
Peritoneum 42
Pleura 42, 496
Rectum 5, 24, 28
Spinal Cord 22
Testis 805
Thyroid 318
Uterus (Corpus) *703

TYPE

Relative Frequency

INDEX OF CHEMICALS

The CHEMICALS index includes many individual substances which are members of a class; eight such classes have been identified:

Drugs	DR	Pesticides	PE
Dusts	DU	Plastics, resins and monomers	PL
Metals and metal compounds	ME	Solvents	SO
Mycotoxins	MY	Sweeteners	SW

The code for each class is shown next to each chemical: projects indexed to a chemical in any class are cross–indexed to the class name in the TERMS index. Thus a study of exposure to nickel compounds will be indexed to "Nickel" in the CHEMICALS index and "Metals" in the TERMS index.

Chemicals for which an abbreviation is given in the "List of Abbreviations" are now listed in abbreviated form in the index, e.g. PAH ("Polycyclic Aromatic Hydrocarbons").

New chemicals in this issue are:

Adriamycin	DR
Endrin	
Ferritin	
Isocyanate	
Stibnite	
Tungsten Carbide	

The list below identifies all entries in current use in this Directory.

Acrylamides PL
Acrylic Acid PL
Acrylonitrile PL
Adriamycin DR
Aetiocholanolone
Aflatoxin MY
Alachlor PE
Aldrin PE
Alkylating Agents DR
Aluminium ME
Amines, Aromatic
Androgens
Androsterone
Antimony ME
Arsenic
Asbestos DU
Asbestos, Chrysotile DU
Asbestos, Crocidolite DU
Ascorbic Acid DR
Aspirin DR
Atrazine PE
Attapulgite DU
Azathioprine DR

Barbiturates DR
BCME
Benzene SO
Benzidine
Benzo(a)pyrene
Benzoyl Chloride
Beryllium ME
Beta Carotene
Butadiene PL

Cadmium ME
Caffeine
Calcium
Caprylyl Chloride
Carbon Monoxide
Chloramphenicol DR
Chlordimeform
Chlorine
Chlorobenzene
Chloroform
Chlorophenols PE
Chloroprene
Chloropyrenes
Chromium ME
Cimetidine DR
Cisplatin DR
CMME
Copper ME
Cyclophosphamide DR

2,4–D PE
Dacarbazine DR
DBCP PE
DDE PE
DDT PE
Dehydroepiandrosterone Sulphate
DES DR
Dieldrin PE
Diesel Exhaust

Diisobutyl Ketone	SO
Diisocyanates	PL
Dimethylsulphate	
Dioxins	
Endrin	PE
Epichlorohydrin	PL
Epoxy Resins	PL
Ethylene Oxide	
Ferro–Alloys	ME
Ferric Oxide	
Ferritin	
Fluorides	
Folic Acid	
Formaldehyde	PL
Furans	
Gasoline	
Glass Fibres	DU
Glycol Ethers	SO
Gold	ME
Hexane	SO
Hydrocarbons	
Hydrocarbons, Chlorinated	
Hydrocarbons, Halogenated	
Iodine	
Iron	ME
Isocyanate	
Kerosene	
Lead	ME
Lindane	PE
Magnesium	ME
Malathion	PE
Melphalan	DR
Mercury	ME
Methanol	
Methoxsalen	DR
Methyl Bromide	
Methyl Ethyl Ketone	SO
Methyl Isobutyl Ketone	SO
Methyl Methacrylate	
4,4'–Methylene–bis(2–chloroaniline)	
Methylxanthines	
Mineral Fibres	
Mineral Oil	
Mustard Gas	
Naphthalenes, Chlorinated	
1–Naphthylamine	
2–Naphthylamine	DR
Naphthylthiourea, Alpha	
Nickel	ME
Nicotine	DR
Nitrates	
Nitrites	
Nitrogen Oxides	
N–Nitroso Compounds	
Oestradiol	
Oestriol	
Oestrogens	
Oestrone	
Oil Mist	
Organophosphates	PE
Oxygen	
PAH	
Paracetamol	DR
PCB	
Perchloroethylene	SO
Pethidine	DR
Phenacetin	DR
Phenol	PL
Phenothiazines	DR
Phenoxy Acids	PE
Phenylbutazone	DR
Phenytoin	DR
Phosphates, Inorganic	
Piperazine	DR
Plutonium	ME
Polyethylenes	
Polystyrene	PL
Potassium	ME
Procarbazine	DR
Progesterone	
Progestogens	
Prolactin	
PVC	PL
Radium	ME
Radon	
Reserpine	DR
Retinoids	
Selenium	ME
Silica	DU
Silvex	PE
Sodium Chloride	
Steel	ME
Steroids	
Styrene	PL
Sulphur Dioxide	
Sulphuric Acid	
2,4,5–T	PE
Talc	DU
Tamoxifen	DR
Tars	
Technetium	ME

Terpenes	
Testosterone	
Tetrachloroethylene	SO
Thorium	ME
Thoron	ME
Thorotrast	DR
Tocopherol	
Toluene	SO
Trichloroethylene	SO
Trihalomethanes	
Tungsten Carbide	
Uranium	ME
Urethane	PL
Vehicle Exhaust	
Vinyl Acetate	PL
Vinyl Chloride	PL
Vinylidene Chloride	PL
Warfarin	DR
Xylene	SO
Zinc	ME
Zircon	ME

CHEM

EXPERIMENTAL AND EPIDEMIOLOGICAL STUDIES OF CHEMICALS

A number of chemicals being investigated in epidemiological studies are also being tested for carcinogenicity in animals. 47 such agents listed in the latest Directory of Agents Being Tested for Carcinogenicity, also published by IARC (see list of IARC publications), were the subject of 241 current epidemiological studies (see below).

For further information on the Directory of Agents Being Tested For Carcinogenicity, please contact Ms M.–J. Ghess, Unit of Carcinogens Identification and Evaluation, IARC.

	Agent	Studies
1.	Aflatoxin	10
2.	Aluminium	1
3.	Antimony	2
4.	Asbestos, amosite	2
5.	Asbestos, chrysotile	6
6.	Atrazine	1
7.	Attpulgite	1
8.	Benzene	11
9.	Benzidine	3
10.	Benzo[a]pyrene	5
11.	Beryllium	1
12.	Butadiene	4
13.	Cadmium	10
14.	Chloramphenicol	1
15.	Chlorine	1
16.	Chloroform	1
17.	Chromium	17
18.	Diesel exhaust	4
19.	Dioxins	9
20.	Formaldehyde	16
21.	Furans	2
22.	Glass fibres	3
23.	Hexane	1
24.	Isopropyl alcohol	1
25.	Lead	10
26.	Lindane	1
27.	Mercury	1
28.	Mineral fibres	19
29.	N–Nitroso compounds	17
30.	2–Naphthylamine	2
31.	Nickel	16
32.	Nicotine	3
33.	Paracetamol	2
34.	Phenytoin	1
35.	Plutonium	1
36.	Potassium	1
37.	Reserpine	1
38.	Styrene	12
39.	2,4,5–T	3
40.	Talc	3
41.	Tetrachloroethylene	1
42.	Tocopherol	6
43.	Trichloroethylene	4
44.	Urethane	1
45.	Vinyl acetate	1
46.	Vinyl chloride	13
47.	Zinc	8

Acrylamide

All Sites	1230
Brain	1091
Colon	1092
Gastrointestinal	1091
Haemopoietic	1091
Liver	1091, 1092
Lung	1091, 1092

Acrylic Acid

Colon	1092
Liver	1092
Lung	1092

Acrylonitrile

All Sites	623, 1171, 1173, 1230
Bladder	965
Brain	38, 965, 1091
Colon	1092
Gastrointestinal	1091
Haemopoietic	965, 1091
Liver	1091, 1092
Lung	623, 965, 1091, 1092, 1173
Lymphoma	965
Mesothelioma	965
Prostate	623
Sarcoma	965
Stomach	965

Adriamycin

All Sites	*1175
Ovary	*1175

Aetiocholanolone

Ovary	635

Aflatoxin

Liver	164, 191, *293, 345, 726, 811, 956
Stomach	912

Alachlor

Inapplicable	966

Aldrin

Liver	*983

Alkylating Agents

Inapplicable	1001

Aluminium

All Sites	484
Bladder	484
Lung	484
Lymphoma	484
Pancreas	484

Amines, Aromatic

All Sites 682, 695
Bladder 178, 541, 695, 1025
Brain *362
Colon *976
Larynx *362
Lung *362, 683, 695
Lymphoma *362
Oesophagus *362
Oropharynx *362
Stomach *362, 695
Tongue *362
Tonsil *362

Androgens

Breast (F) 644, 1130

Androsterone

Ovary 635

Antimony

All Sites *485
Lung *485, 945

Arsenic

All Sites 792, 872, 1171
Bladder 213, 487, 652
Brain 213
Gastrointestinal 213
Haemopoietic 213
Kidney 487, 506
Larynx 59
Leukaemia *106, 487
Liver 487
Lung *106, 487, 792, 872, 945, 968
Lymphoma 506
Male Genital 213
Mesothelioma 59
Nasopharynx 59
Prostate 872
Respiratory 213
Sarcoma 213
Skin 213
Stomach *106
Thyroid 662

Asbestos

All Sites 42, 90, 489, 500, 501, 522, 545, *669, 684, *686, 754, 856, 1171, 1197, *1235
Bladder 73, *490, 965
Brain 965
Breast (F) 1197
Colon 73, *323, 748, *758, 1197
Gastrointestinal 691, *692
Haemopoietic *692, 965
Hypopharynx 315
Kidney 73, 506
Larynx 59, 315
Lung 73, 239, *320, *322, *323, 343, 353, 489, 496, 604, *671, *686, 691, *692, 742, 748, 965, 1197, 1200, 1224, *1226

Lymphoma 73, 506, 965
Melanoma 73
Mesothelioma 42, 59, 353, 496, 530, *536, *686, 748, 965, 1200
Nasopharynx 59
Oesophagus 73
Pancreas 73
Peritoneum 42, 530
Pleura 42, 496, 530
Prostate 73
Rectum 73, 748, *758
Respiratory 488, 856
Sarcoma 965
Stomach 73, 748, 965
Testis 907
Uterus (Cervix) *692, 1197

Asbestos, Chrysotile
All Sites 52
Gastrointestinal 72, 76, 101
Kidney 76
Larynx 76
Lung 52, 72, 76
Mesothelioma 52, 72, 101
Peritoneum 76
Pleura 76
Respiratory 101

Asbestos, Crocidolite
All Sites 52
Colon 34
Gastrointestinal 101
Lung 34, 35, 52
Mesothelioma 34, 35, 52, 101
Peritoneum 34
Respiratory 101
Stomach 34

Ascorbic Acid
All Sites 625
Uterus (Cervix) 994, *1048

Aspirin
All Sites *987, 992, 995
Breast (F) *987
Colon *987, *1127
Lung *987
Oesophagus *1074
Rectum *1127
Stomach *1074

Atrazine
Inapplicable 966

Attapulgite
All Sites 1019

Azathioprine
All Sites 227

Barbiturates

All Sites 221
Brain 44, 303
Childhood Neoplasms 221
Leukaemia 221
Liver 221

BCME

Inapplicable 198
Lung 1153

Benzene

All Sites 305, 501, *535, *696, *697, *698, 869, 1019, 1157, 1171
Bladder 305, 965, 1016, 1157
Brain *362, 965, 1104
Childhood Neoplasms 1165
Haemopoietic 677, 965, *984
Hodgkin's Disease *311
Inapplicable 354, *355
Kidney 1104
Larynx *362, 869
Leukaemia 136, 334, *864, 869, *886, 950, *984, 1165
Liver *698, 869, 1104, 1165
Lung *362, *698, 869, 965, 1104
Lymphoma *362, 869, *886, 965, *984
Mesothelioma 965
Nasal Cavity 305, 306
Non–Hodgkin's Lymphoma *311, 1165
Oesophagus *362
Oropharynx *362
Pancreas 1165
Respiratory *864
Sarcoma 965
Soft Tissue *984, 1165
Stomach *362, 965
Tongue *362
Tonsil *362

Benzidine

All Sites 509, 1157, 1171
Bladder 509, 1157
Nasal Cavity 509
Pharynx 509
Respiratory 509

Benzo(a)pyrene

Gastrointestinal 192
Leukaemia 192
Lung 135, *183, 192
Stomach *183

Benzoyl Chloride

Kidney 506
Lymphoma 506

Beryllium

Kidney 506
Lymphoma 506

Beta Carotene

All Sites	241, 533, 593, 625, 884, 986, *987, 992, 995, 1044, *1117, 1197
Bladder	642
Breast (F)	127, 241, 347, *436, 533, 612, 639, 640, 711, 797, *919, *987, 1197, 1222
Colon	127, 241, 533, 640, 894, 986, *987, 1197
Inapplicable	*817
Larynx	*1117
Leukaemia	127
Liver	127
Lung	35, 97, 127, 241, 640, 642, 986, *987, *999, 1197, 1200, *1209, 1224
Melanoma	1221
Mesothelioma	35, 1200
Nasopharynx	127
Oesophagus	127, 187
Oral Cavity	444, *1117
Pharynx	*1117
Prostate	92, 241
Rectum	8, 127, 241, 640, 894
Skin	1221
Stomach	127, 241, 285, 533, 640
Uterus (Cervix)	*37, 241, 284, 611, 994, *1048, 1197
Uterus (Corpus)	1047

Butadiene

All Sites	1019
Brain	1091
Colon	1092
Gastrointestinal	1091
Haemopoietic	*984, 1091
Inapplicable	249
Leukaemia	*984
Liver	1091, 1092
Lung	1091, 1092
Lymphoma	*984
Soft Tissue	*984

Cadmium

All Sites	872, 1019
Bladder	487
Kidney	487, 506, 873
Leukaemia	487
Liver	175, 234, 487
Lung	487, *768, 872, 873, 968
Lymphoma	506
Prostate	872, 873, 918, 943

Caffeine

Breast (F)	1072

Calcium

All Sites	*669
Breast (F)	63, *1223
Colon	63, 373, 641, 1045, 1058, 1060, *1223, 1228
Lung	63
Oesophagus	63, 188, 190
Ovary	*1223
Rectum	373, 1058, *1223
Stomach	63
Uterus (Corpus)	*1223

Caprylyl Chloride

Colon 1092
Liver 1092
Lung 1092

Carbon Monoxide

All Sites 884
Bladder 513, 1134
Kidney 513, 1134
Larynx 513, 1134
Lung 513, 1134
Oesophagus 513, 1134
Oral Cavity 513, 1134
Pancreas 513, 1134
Pharynx 513

Chloramphenicol

Childhood Neoplasms 647
Leukaemia (ALL) 647

Chlordimeform

Bladder 162

Chlorine

All Sites 282, 787
Bladder 787
Colon 787
Lung 787
Rectum 787
Stomach 787

Chlorobenzene

Bladder 1016

Chloroform

All Sites 519

Chlorophenols

All Sites 282, 296, 371, 755
Kidney 506
Lymphoma 506

Chloroprene

All Sites 308, *698, *700
Liver *698, *700
Lung *698, *700

Chloropyrenes

Kidney 506
Lymphoma 506

Chromium

All Sites 126, 501, 509, *555, 573, 755, 1171, 1227
Bladder 73, 213, 392, 509

Brain 213
Colon 73, *323
Gastrointestinal 213, 392
Haemopoietic 213
Hypopharynx 315
Inapplicable 354, *355
Kidney 73, 506
Larynx 59, 315
Leukaemia *106
Liver 556
Lung 73, *106, 126, *320, *322, *323, 343, *555, 556, 1227
Lymphoma 73, 506
Male Genital 213
Melanoma 73
Mesothelioma 59
Nasal Cavity 509
Nasopharynx 59
Oesophagus 73
Pancreas 73
Pharynx 509
Prostate 73
Rectum 73
Respiratory 213, 509
Sarcoma 213
Skin 213
Stomach 73, *106

Cimetidine
All Sites 625
Gastrointestinal 927

Cisplatin
All Sites *1175
Ovary *1175

CMME
Lung 1153

Cobalt
Lung *265, *325, *785

Copper
Liver 175, 234
Lung *158
Stomach 725

Cyclophosphamide
Childhood Neoplasms 1225
Inapplicable 626

2,4–D
All Sites 247, 763, 1180
Brain 1180
Inapplicable 966
Leukaemia 1180
Lung 1180
Lymphoma 247
Multiple Myeloma 1180

Non-Hodgkin's Lymphoma 1180
Sarcoma 247
Soft Tissue 538, 1180

Dacarbazine
All Sites 270
Hodgkin's Disease 270

DBCP
Liver *983

DDE
Breast (F) *1043, *1144

DDT
Breast (F) *1043, *1144

Dehydroepiandrosterone Sulphate
All Sites 684
Ovary 635

DES
Breast (F) 926
Female Genital 926
Vagina 1102

Dieldrin
Liver *983

Diesel Exhaust
All Sites *535, *935
Inapplicable 48, 626
Leukaemia *106
Lung *106, *774, 788
Stomach *106

Diisobutyl Ketone
All Sites 1015

Diisocyanates
All Sites 751, 825

Dimethylsulphate
Kidney 506
Lymphoma 506

Dioxins
All Sites 282, 296, *357, 371, 508, 622, 749, 1094
Bile Duct *514
Brain *514
Breast (F) *357, *514
Childhood Neoplasms 1165

Leukaemia	508, 1165
Liver	*514, 1165
Lymphoma	508
Non–Hodgkin's Lymphoma	*290, *514, 622, 1165
Pancreas	1165
Respiratory	*514
Sarcoma	*290, *514, 622
Skin	*514
Soft Tissue	*290, 508, *514, 622, 1165
Testis	*514

Endrin

Liver	*983

Epichlorohydrin

Kidney	506
Lymphoma	506

Epoxy Resins

Brain	1104
Kidney	1104
Liver	1104
Lung	1104

Ethylene Oxide

All Sites	752, 757, *781, 1231
Inapplicable	354, *355
Leukaemia	752, *781, 1026, 1231
Lymphoma	752, 757, *781
Multiple Myeloma	757

Ferric Oxide

Lung	*320

Ferritin

Colon	*1017
Stomach	137

Ferro–Alloys

All Sites	*669

Fluorides

All Sites	229
Gastrointestinal	193
Lung	193

Folic Acid

Colon	*1017
Uterus (Cervix)	611

Formaldehyde

All Sites	90, 282, 305, 509, 755, 1158, 1159, 1171
Bladder	73, 305, 509, 965
Brain	38, 965
Colon	73, 1032

Gastrointestinal 691
Haemopoietic 965
Kidney 73, 506
Lung 73, 691, 965, *1209, *1226
Lymphoma 73, 506, 965
Melanoma 73, 874
Mesothelioma 965
Nasal Cavity 305, 306, *316, 509, 874
Nasopharynx 1159, 1208
Oesophagus 73
Pancreas 73
Pharynx 509
Prostate 73
Rectum 73, 1032
Respiratory 509
Sarcoma 965
Stomach 73, 965

Furans
All Sites 296, 749, 1094
Brain *362
Larynx *362
Lung *362
Lymphoma *362
Oesophagus *362
Oropharynx *362
Stomach *362
Tongue *362
Tonsil *362

Gasoline
All Sites *535

Glass Fibres
All Sites 90
Lung *1226
Respiratory 1156

Glycol Ethers
Leukaemia (AML) 280

Gold
All Sites 90

Hexane
Brain 1091
Gastrointestinal 1091
Haemopoietic 1091
Liver 1091
Lung 1091

Hydrocarbons
All Sites 19, 501, 1019, 1158
Brain 1091
Childhood Neoplasms 222
Colon 1092
Gastrointestinal 1091
Haemopoietic 19, 1091

Kidney *219, 506, 1037
Liver 1091, 1092
Lung 1091, 1092
Lymphoma 19, 506

Hydrocarbons, Chlorinated

All Sites 1171

Hydrocarbons, Halogenated

All Sites 266, 333, 1019, 1158
Brain 1091
Colon *53, 1092
Gastrointestinal 1091
Haemopoietic 1091
Kidney 506
Liver 1091, 1092
Lung 1091, 1092
Lymphoma 506
Rectum *53

Iodine

All Sites *338, *775
Leukaemia *775
Stomach *523
Thyroid 662, *775, 1194

Iron

All Sites 1168
Brain *362
Colon *1017
Larynx *362
Liver 175
Lung *158, *183, *322, 326, 343, *362, 721
Lymphoma *362
Oesophagus *362
Oropharynx *362
Stomach *183, 326, *362, 725
Tongue *362
Tonsil *362

Isocyanate

Brain *362
Larynx *362
Lung *362
Lymphoma *362
Oesophagus *362
Oropharynx *362
Stomach *362
Tongue *362
Tonsil *362

Kerosene

Lung *428

Lead

All Sites 117, 271, 750, 872
Bladder 487
Kidney 487, 506, 790

Leukaemia 487
Liver 175, 234, 487
Lung 271, 487, 790, 872
Lymphoma 506
Prostate 872
Stomach 271

Lindane
All Sites 928

Magnesium
Brain 1104
Kidney 1104
Liver 1104
Lung 1104

Malathion
Inapplicable 966

Melphalan
All Sites 746, *1175
Inapplicable 354, *355
Ovary *1175

Mercury
All Sites *272, *715
Brain *272, *715, 1091
Gastrointestinal 1091
Haemopoietic 1091
Kidney *272, *715
Liver 1091
Lung *272, *715, 1091

Methanol
Colon 1092
Liver 1092
Lung 1092

Methoxsalen
All Sites 993
Eye 993
Skin 993

Methyl Bromide
Inapplicable *1020

Methyl Ethyl Ketone
All Sites 1015

Methyl Isobutyl Ketone
All Sites 1015

Methyl Methacrylate

All Sites 1230

4,4'-Methylene-bis(2-chloroaniline)

All Sites 1027
Bladder 1027

Methylxanthines

All Sites 512
Choriocarcinoma 516
Hydatidiform Mole 516

Mineral Fibres

All Sites 42, 52, 298, *686, 1019
Colon *758
Gastrointestinal 76, 691
Hypopharynx *317
Kidney 76
Larynx 76, *317
Lung 52, 76, 239, 298, *317, 496, 579, *686, 691
Mesothelioma 42, 52, *324, 496, 579, *686, 851
Nasal Cavity *317
Peritoneum 42, 76
Pleura 42, 76, *317, 496
Rectum *758
Respiratory *324, 1156

Mineral Oil

All Sites 266, *697, 1095
Bladder 73
Brain 1095
Colon 73, 860, 1032
Gastrointestinal *649
Inapplicable 198
Kidney 73
Larynx 59, 1095
Lung 73, *649, 860
Lymphoma 73
Melanoma 73
Mesothelioma 59
Nasopharynx 59
Oesophagus 73
Pancreas 73
Prostate 73
Rectum 73, 860, 1032
Stomach 73, 860
Testis 907

Mustard Gas

Stomach 552

N-Nitroso Compounds

All Sites *687
Bladder 392, 541, *687
Brain 44, 303, 304, *362, *473, 765, 1087, 1151
Breast (F) 127
Childhood Neoplasms 1151
Colon 127
Gastrointestinal 392

Larynx *362
Leukaemia 127, *687
Liver 127, *293, 811
Lung 127, *362, *687
Lymphoma *362
Nasopharynx 127
Neuroblastoma 1008
Oesophagus 127, 130, 361, *362, *1103
Oropharynx *362
Rectum 127
Stomach 127, 137, *362, 914, *1103
Tongue *362
Tonsil *362

Naphthalenes, Chlorinated

Liver 1028
Lymphoma 1028
Soft Tissue 1028

Naphthylthiourea, Alpha

All Sites 928

1–Naphthylamine

Bladder 178

2–Naphthylamine

All Sites 1157
Bladder 178, 1157

Nickel

All Sites 90, 501, 1161, 1171
Bladder 73, 213, 392
Brain 213
Colon 73, *323, 1032, 1161
Gastrointestinal 213, 392
Haemopoietic 213
Hypopharynx 315
Inapplicable 354, *355
Kidney 73, 506
Larynx 59, 315
Liver 1161
Lung 73, *320, *322, *323, 343, *768, 968
Lymphoma 73, 506
Male Genital 213
Melanoma 73
Mesothelioma 59
Nasal Cavity 1161
Nasopharynx 59
Oesophagus 73
Pancreas 73
Prostate 73, 918
Rectum 73, 1032, 1161
Respiratory 213, 1161
Sarcoma 213
Skin 213
Stomach 73

Nicotine

Bladder	513, 1134
Childhood Neoplasms	831
Kidney	513, 1134
Larynx	513, 1134
Lung	513, 1134
Oesophagus	513, 1134
Oral Cavity	513, 1134
Pancreas	513, 1134
Pharynx	513

Nitrates

All Sites	*701, 787, *1021, 1158
Bladder	392, 787, 910
Brain	*1021
Colon	787
Gastrointestinal	193, 392
Liver	*293
Lung	193, 756, 787, 910
Oesophagus	910, *1103
Rectum	787
Stomach	725, 756, 787, 910, *1103, 1177

Nitrites

All Sites	218, 787
Bladder	392, 787
Colon	787
Gastrointestinal	392
Hodgkin's Disease	218
Lung	787
Oesophagus	*1103
Rectum	787
Stomach	725, 787, *1103, 1177

Nitrogen Mustard

All Sites	270
Hodgkin's Disease	270

Nitrogen Oxide

All Sites	*696, *701

Oestradiol

Breast (F)	299, *436
Colon	*407
Ovary	635
Rectum	*407

Oestriol

Breast (F)	*436
Ovary	635

Oestrogens

All Sites	991, 1082
Benign Tumours	1140
Breast (F)	*46, *436, 618, 620, 650, 803, 991, 1082, 1083, *1119, 1130, 1140, *1195, *1223
Colon	*1119, *1223
Female Genital	803

Kidney 45, 795
Ovary 534, *1223, 1229
Rectum *1119, *1223
Thyroid *1119
Uterus (Corpus) 1082, *1119, 1190, *1223

Oestrone
Breast (F) *436
Ovary 635

Oil Mist
Lung 861
Stomach 861

Organophosphates
Liver *983

Oxygen
Inapplicable 1001

PAH
All Sites 483, 484, *669, *687, *696, *697, 772, 872, 884, 982, 1171
Bladder 73, *204, 213, 484, 487, *687, 1134
Brain 38, 213
Colon 73, *323, 860, *976
Gastrointestinal 192, 213
Haemopoietic 213
Inapplicable 48, *233, 626, 1001
Kidney 73, 487, 506, 1134, 1189
Larynx 59, *204, 1134
Leukaemia 192, 349, 487, *687, *864
Liver 487
Lung 73, *183, 192, *204, *320, *322, *323, 343, 481, 483, 484, 487, 613, *687, 772, 860, 872, 960, 982, 1134, *1209
Lymphoma 73, 349, 484, 506
Male Genital 213
Melanoma 73
Mesothelioma 59
Nasopharynx 59
Oesophagus 73, 1134
Oral Cavity 1134
Pancreas 73, 484, 1134
Prostate 73, 872
Rectum 73, 860
Respiratory 213, *864
Sarcoma 213
Skin 213
Stomach 73, *183, 860, 982

Paracetamol
Colon *1127
Kidney 45, 795
Rectum *1127

PCB
All Sites 749, 1019, 1094
Breast (F) *1043, *1144
Hodgkin's Disease *311

Liver 1028
Lymphoma 1028
Non–Hodgkin's Lymphoma *311
Soft Tissue 1028

Perchloroethylene
All Sites *520

Pethidine
Childhood Neoplasms 832

Phenacetin
Bladder 310

Phenol
All Sites 1171
Brain 38, *362, 1091
Colon 1092
Gastrointestinal 691, 1091
Haemopoietic 1091
Larynx *362
Liver 1091, 1092
Lung *362, 691, 1091, 1092
Lymphoma *362
Oesophagus *362
Oropharynx *362
Stomach *362
Tongue *362
Tonsil *362

Phenothiazines
Bladder 232
Breast (F) 232
Lung 232
Prostate 232
Uterus (Cervix) 232

Phenoxy Acids
All Sites 86, 247, 296, 371, 622, 654, 763, 1180
Brain 86, 1180
Hodgkin's Disease *311, 542
Kidney 506
Leukaemia 86, 542, 1180
Lung 1180
Lymphoma 13, 247, 506, 654
Multiple Myeloma 86, 542, 1180
Non–Hodgkin's Lymphoma 86, *311, 542, 622, 1180
Prostate 86
Sarcoma 13, 247, 622
Soft Tissue 538, 622, 654, 1068, 1180

Phenylbutazone
Leukaemia (AML) 1086

Phenytoin
Childhood Neoplasms 647
Leukaemia (ALL) 647

Phosphates, Inorganic

Gastrointestinal 193
Lung 193

Piperazine

All Sites 757
Lymphoma 757
Multiple Myeloma 757

Plutonium

All Sites 1071

Polyethylenes

Brain 1104
Kidney 1104
Liver 1104
Lung 1104

Polystyrene

All Sites 990

Potassium

All Sites 1057
Colon 1057

Procarbazine

All Sites 270
Hodgkin's Disease 270

Progesterone

Breast (F) 299, 309, *436, 618, 650, 1204
Colon *407
Rectum *407

Progestogens

Breast (F) 650, 803, 1083, *1119
Colon *1119
Female Genital 803
Rectum *1119
Thyroid *1119
Uterus (Corpus) *1119, 1190

Prolactin

Breast (F) 299, 1130
Colon *407
Rectum *407

PVC

All Sites 754
Brain 1091
Gastrointestinal 1091
Haemopoietic 1091
Liver 1091
Lung 1091

Radium

Bone	102

Radon

All Sites	87, 185, 263, *272, *387, 483, 685
Brain	*272
Gastrointestinal	192
Haemopoietic	393
Inapplicable	*4, *355, 830
Kidney	*272
Leukaemia	*106, 192, 393, 849
Lung	87, 94, 95, *106, 139, 159, 192, 263, *264, *272, *379, 393, 483, *670, 685, 710, 721, *773, 782, 798, 844, 903, 944, 947, 963, 968, 1022, 1024, 1121, 1182
Lymphoma	1024
Stomach	*106

Reserpine

Bladder	232
Breast (F)	232
Lung	232
Prostate	232
Uterus (Cervix)	232

Retinoids

All Sites	241, 884, 992, 995, 1044
Breast (F)	127, 241, 518, 612, 797, 1222
Colon	127, 241, 894, 1228
Inapplicable	*817
Leukaemia	127
Liver	127
Lung	35, 127, 241, 705, 1200, 1224
Melanoma	1221
Mesothelioma	35, 1200
Nasopharynx	127
Oesophagus	127
Prostate	241
Rectum	127, 241, 894
Skin	1220, 1221
Stomach	127, 241
Uterus (Cervix)	241, 284, 547, *1048

Selenium

All Sites	241, 586, 991, 992
Bladder	392
Breast (F)	74, 241, 347, 467, 468, 624, 639, 640, 797, 991, 1222
Colon	74, 147, 241, 468, 526, 624, 638, 640, *1017, 1215
Gastrointestinal	392
Liver	175
Lung	241, 624, 640
Melanoma	1221
Prostate	74, 241
Rectum	147, 241, 468, 624, 638, 640
Skin	1216, 1217, 1221
Stomach	241, 624, 640, 725
Thyroid	662
Uterus (Cervix)	241, 611

Silica

All Sites	*272, 483, *669, 685, *715, 939, 1171
Bladder	73, 487, 965
Brain	*272, *715, 965
Colon	73, *323
Haemopoietic	965
Hypopharynx	315
Kidney	73, *272, 487, *715
Larynx	315
Leukaemia	*106, 487, *864
Liver	487
Lung	36, *43, 73, *106, *158, *272, *320, *323, 483, 487, *580, 685, *715, 857, 939, 965, 968, 1024, *1226
Lymphoma	73, 965, 1024
Melanoma	73
Mesothelioma	965
Oesophagus	73
Pancreas	73
Prostate	73
Rectum	73
Respiratory	*864
Sarcoma	965
Stomach	73, *106, 965

Silvex

All Sites	763

Sodium Chloride

All Sites	1057
Colon	1057
Stomach	1233

Steel

All Sites	863, 1160
Genitourinary	1160
Haemopoietic	1160
Lung	*183
Lymphoma	1160
Respiratory	1160
Stomach	*183

Steroids

Benign Tumours	*1202
Breast (F)	*515, *1202
Liver	729, 811
Ovary	1229

Styrene

All Sites	281, 779, 990, 1019, 1158, 1171
Brain	1091, 1104
Gastrointestinal	1091
Haemopoietic	*984, 1091
Inapplicable	201, 1009
Kidney	506, 1104
Leukaemia	779, *984
Liver	1091, 1104
Lung	1091, 1104
Lymphoma	506, 779, *984
Soft Tissue	*984

Sulphur Dioxide

All Sites	*701
Gastrointestinal	192
Haemopoietic	744
Leukaemia	192
Lung	172, 192, 744
Lymphoma	744
Stomach	744

Sulphuric Acid

All Sites	117
Bladder	73
Colon	73
Gastrointestinal	193
Hypopharynx	315
Kidney	73
Larynx	315, 931
Lung	73, 193
Lymphoma	73
Melanoma	73
Oesophagus	73
Pancreas	73
Prostate	73
Rectum	73
Stomach	73

2,4,5–T

All Sites	247, 763
Lymphoma	247
Sarcoma	247
Soft Tissue	538, 1068

Talc

Bladder	965
Brain	965
Haemopoietic	965
Lung	353, 965, *1226
Lymphoma	965
Mesothelioma	353, 965
Ovary	534
Sarcoma	965
Stomach	965

Tamoxifen

Breast (F)	*302, 309, *1191
Ovary	*1191
Uterus (Corpus)	*302, *1191

Tars

All Sites	484, 573, 884
Bladder	484, 513, 1134
Kidney	513, 1134
Larynx	59, 513, 1134
Lung	484, 513, 1134
Lymphoma	484
Mesothelioma	59
Nasopharynx	59
Oesophagus	513, 1134
Oral Cavity	513, 1134
Pancreas	484, 513, 1134
Pharynx	513

Technetium
Inapplicable 77

Terpenes
Haemopoietic 744
Lung 744, *1209
Lymphoma 744
Stomach 744

Testosterone
Breast (F) 299, *436
Colon *407
Rectum *407

Tetrachloroethylene
Kidney 506
Lymphoma 506

Thorium
All Sites *776
Gastrointestinal 193
Liver 138
Lung 193, *670
Stomach 138

Thoron
Haemopoietic 393
Leukaemia 393
Lung 393, 844

Thorotrast
All Sites 210, 221, *776
Bone 370
Childhood Neoplasms 221
Leukaemia 221, 370
Liver 221, 370
Lung 370
Lymphoma 370
Spleen 370

Tocopherol
All Sites 241
Breast (F) 74, 127, 241, 347, 639
Colon 74, 127, 241, 1228
Leukaemia 127
Liver 127
Lung 127, 241, 705
Nasopharynx 127
Oesophagus 127
Prostate 74, 241
Rectum 127, 241
Stomach 127, 241
Uterus (Cervix) 241, 994, *1048

Toluene
Brain 1091
Colon 1092

Gastrointestinal 1091
Haemopoietic 677, 1091
Inapplicable 354
Liver 1091, 1092
Lung 1091, 1092

Trichloroethylene
All Sites 266, 1015
Bladder 1016
Kidney 506
Lymphoma 506

Trihalomethanes
All Sites 519
Bile Duct 587
Bladder *110
Colon *110
Liver 587
Rectum *110

Tungsten Carbide
Lung *265

Uranium
All Sites 90, 185, 263
Gastrointestinal 192, 193
Leukaemia 192
Lung 3, 159, 192, 193, 263, 947, 1024
Lymphoma 1024

Urethane
All Sites 757
Lymphoma 757
Multiple Myeloma 757

Vehicle Exhaust
All Sites 511, *535
Brain *205
Hypopharynx 315
Larynx 315, 511
Leukaemia *205, *886
Lung 511, *774
Lymphoma *205, *886
Prostate 629

Vinyl Acetate
Brain 1091
Gastrointestinal 1091
Haemopoietic 1091
Liver 1091
Lung 1091

Vinyl Chloride
All Sites 90, 128, 333, 754, 869, 1019, 1158
Angiosarcoma 738
Bladder 738
Brain 38, 1091, 1104

Colon 1092
Gastrointestinal 1091
Haemopoietic 1091
Kidney 506, 1104
Larynx 869
Leukaemia 869
Liver 128, 738, 869, 1091, 1092, 1104
Lung 738, 869, 1091, 1092, 1104
Lymphoma 506, 869

Vinylidene Chloride
Brain 1091
Gastrointestinal 1091
Haemopoietic 1091
Liver 1091
Lung 1091

Warfarin
All Sites 928

Xylene
Haemopoietic 677

Zinc
All Sites 117
Bladder 487
Colon 1228
Kidney 487
Leukaemia 487
Liver 175, 487
Lung 487
Oesophagus 188
Prostate 943

Zircon
Lung 945

INDEX OF OCCUPATIONS

The use of occupational titles is very variable between countries, but every effort has been made to classify occupational cancer studies to the specific group identified by the principal investigator.

All studies indexed to a specific occupation in this index can also be found under the general heading "Occupation" in the TERMS index.

There is inevitably a degree of overlap with the CHEMICALS index; thus for example, a study of vinyl chloride workers can be found in both indexes.

New entries for this issue are "Fumigation Workers", "Furnacemen" and "Millers".

The list below identifies all entries in use in this Directory.

Acrylonitrile Workers
Administrative Workers
Agricultural Workers
Aircraft Workers
Aluminium Workers
Antimony Process Workers
Asbestos Textile Workers
Asbestos Workers
Automobile Workers

Battery Plant Workers
Bus Drivers

Cabinet Makers
Cable Manufacturers
Carpenters, Joiners
Cement Workers
Ceramic Workers
Chemical Industry Workers
Chemists
Chromate Producing Workers
Chromium Plating Workers
Coke–Oven Workers
Construction and Maintenance Workers
Cooks
Cryolite Workers

Dockers
Drivers
Dry Cleaners
Dyestuff Workers

Electrical Workers
Electrode Manufacturers
Electronics Workers
Engineering Workers

Farmers
Ferro–Alloy Workers
Fertilizer Workers
Firemen
Fishermen
Forest Workers
Foundry Workers
Fumigation Workers
Furnacemen

Gas Workers
Glass Workers
Grain Millers

Health Care Workers
Heavy Equipment Operators
Herbicide Manufacturers
Herbicide Sprayers
Horticulturists

Jewellery Manufacturers

Laboratory Workers
Lacquerers
Laminators
Leather Workers

Machinists
Meat Workers
Mechanics
Metal Workers
Military Servicemen
Millers
Mineral Fibre Workers
Miners
Miners, Asbestos
Miners, Coal
Miners, Copper
Miners, Fluorspar
Miners, Gold
Miners, Iron
Miners, Uranium
Miners, Zinc–Lead

OCCU

Model and Pattern Makers
Morticians

Nickel Workers

Office Workers

Painters
Paper and Pulp Workers
Pesticide Workers
Petrochemical Workers
Petrol Stations Attendants
Petroleum Workers
Physicists
Plastics Workers
Plutonium Workers
Potters
Power Plant Workers
Printers

Quarry Workers

Radiation Workers
Railroad Workers
Rubber Workers

Sailors
Sawmill Workers
Screw Cutters
Sewage Workers
Shipyard Workers
Shoemakers–Repairers
Smelters
Smelters, Aluminium
Smelters, Copper
Smelters, Lead
Steel (Stainless) Workers
Steel Workers

Taxi Drivers
Textile Dyers
Textile Workers
Tool Makers
Transport Workers

Varnishers
Vinyl Chloride Workers

Waiters
Waste Incineration Workers
Welders
Wood Workers

Acrylonitrile Workers
All Sites 623, 1173
Bladder 965
Brain 965
Colon 1092
Haemopoietic 965
Liver 1092
Lung 623, 965, 1092, 1173
Lymphoma 965
Mesothelioma 965
Prostate 623
Sarcoma 965
Stomach 965

Administrative Workers
All Sites 451

Agricultural Workers
All Sites 296, 532, 537
Bladder 310, 965
Brain 532, *665, 965
Gastrointestinal 193
Haemopoietic *727, 965
Hodgkin's Disease *311
Leukaemia *665
Leukaemia (AML) *727
Leukaemia (CLL) *727
Lung 94, 193, 965
Lymphoma 965
Mesothelioma 965
Nasal Cavity *316
Non–Hodgkin's Lymphoma *311, 532, *727
Sarcoma 965
Soft Tissue 532
Stomach 965

Aircraft Workers
All Sites *967
Multiple Myeloma *967
Non–Hodgkin's Lymphoma *967

Aluminium Workers
All Sites 484
Bladder 484
Lung 484, 613
Lymphoma 484
Pancreas 484

Antimony Process Workers
Lung 945

Asbestos Textile Workers
Gastrointestinal 691
Lung 691

Asbestos Workers
All Sites 52, 90, 500, 856, 1197
Breast (F) 1197

Colon	748, *758, 1197
Gastrointestinal	691, *692
Haemopoietic	*692
Lung	35, 52, 691, *692, 748, 1197
Mesothelioma	35, 52, 530, 748
Peritoneum	530
Pleura	530
Rectum	748, *758
Respiratory	856
Stomach	748
Uterus (Cervix)	*692, 1197

Automobile Workers

All Sites	982, 1095
Bladder	965
Brain	965, 1095
Colon	860
Gastrointestinal	1034
Haemopoietic	965
Larynx	1095
Lung	860, 861, 965, 982, 1034
Lymphoma	965
Mesothelioma	965
Non–Hodgkin's Lymphoma	1034
Rectum	860
Sarcoma	965
Stomach	860, 861, 965, 982

Battery Plant Workers

All Sites	271
Larynx	931
Lung	271, *768
Stomach	271

Bus Drivers

All Sites	709
Bladder	*204
Larynx	*204
Lung	*204, *774

Cabinet Makers

All Sites	500
Bladder	965
Brain	965
Haemopoietic	965
Leukaemia	208
Lung	965
Lymphoma	965
Mesothelioma	965
Nasal Cavity	208
Respiratory	208
Sarcoma	965
Stomach	965

Cable Manufacturers

Liver	1028
Lymphoma	1028
Soft Tissue	1028

Carpenters, Joiners

Leukaemia 208
Lung *1209
Nasal Cavity 208
Respiratory 208

Cement Workers

All Sites *686
Colon *758
Gastrointestinal 101
Lung *686
Mesothelioma 101, *686
Rectum *758
Respiratory 101, 488

Ceramic Workers

Mesothelioma *324
Respiratory *324

Chemical Industry Workers

All Sites 90, 117, 296, 371, 622, 695, *700, 757, *781, 990, 1158, 1230
Bladder 695, 900
Brain 1091, 1104
Gastrointestinal 1091
Haemopoietic 1091
Inapplicable 198, 626
Kidney 1104
Leukaemia *781
Liver *700, *983, 1091, 1104
Lung 177, 695, *700, 1091, 1104
Lymphoma 757, *781
Multiple Myeloma 757
Non–Hodgkin's Lymphoma 622
Sarcoma 622
Soft Tissue 622
Stomach 695
Wilms' Tumour 71

Chemists

Bladder 965
Brain 965
Haemopoietic 965
Leukaemia 1179
Lung 965
Lymphoma 965, 1179
Melanoma 1069
Mesothelioma 965
Pancreas 1179
Sarcoma 965
Stomach 965

Chromate Producing Workers

All Sites 573, 1227
Lung 1227

Chromium Plating Workers

All Sites *555
Liver 556
Lung *555, 556

Coke–Oven Workers

All Sites 573, *696, 1160
Genitourinary 1160
Haemopoietic 1160
Inapplicable *355
Lung 481
Lymphoma 1160
Respiratory 1160

Construction and Maintenance Workers

All Sites 1171

Cooks

Larynx 664
Liver 664
Oesophagus 664
Oral Cavity 664
Pharynx 664

Cryolite Workers

All Sites 229

Dockers

Bladder *204
Larynx *204
Lung *204, 788
Skin 959

Drivers

All Sites 511
Inapplicable 626
Larynx 511
Lung 511, *774

Dry Cleaners

All Sites *520
Bladder 965
Brain 965
Haemopoietic 965
Lung 965
Lymphoma 965
Mesothelioma 965
Sarcoma 965
Stomach 965

Dyestuff Workers

Angiosarcoma 738
Bladder 310, 738
Liver 738
Lung 738
Wilms' Tumour 71

Electrical Workers

All Sites 312
Breast (M) *1004
Hodgkin's Disease 542
Leukaemia 542, 653, 655, 849

Melanoma *1004
Multiple Myeloma 542
Non–Hodgkin's Lymphoma 542
Prostate *1004

Electrode Manufacturers

All Sites 573, 772
Lung 772

Electronics Workers

Brain 1012
Childhood Neoplasms 1012

Engineering Workers

All Sites 1171
Gastrointestinal *649
Lung *649

Farmers

All Sites 33, 86, 90, 452, 504, 532, *1021
Bladder 965
Brain 86, 96, 251, 504, 532, *665, 965, *1021
Childhood Neoplasms 1147
Haemopoietic 504, *727, 965
Hodgkin's Disease 93, 251, *311, 542
Inapplicable 966
Leukaemia 33, 86, 251, 542, 653, *665, 977
Leukaemia (AML) *727
Leukaemia (CLL) *727
Lung 94, 251, 965
Lymphoma 965, 977
Mesothelioma 965
Multiple Myeloma 86, 93, 96, 542
Non–Hodgkin's Lymphoma 86, 93, 96, *311, 532, 542, *727
Prostate 86, 251, 629
Sarcoma 93, 96, 965
Skin 504
Soft Tissue 33, 93, 96, 532
Stomach 965

Ferro–Alloy Workers

All Sites *669

Fertilizer Workers

All Sites 117, *701
Bladder 910
Lung 756, 910
Oesophagus 910
Stomach 756, 910

Firemen

All Sites 90, 1181
Bladder 965
Brain 965, 1181
Colon 1181
Haemopoietic 965
Leukaemia 1181
Lung 960, 965

Lymphoma 965
Mesothelioma 965
Rectum 1181
Sarcoma 965
Stomach 965

Fishermen

All Sites 749
Colon *758
Rectum *758
Skin 959

Forest Workers

All Sites 296, 763
Hodgkin's Disease 93
Leukaemia 653
Multiple Myeloma 93
Non–Hodgkin's Lymphoma 93
Sarcoma 93
Soft Tissue 93

Foundry Workers

All Sites 982, 1123
Brain *362
Gastrointestinal *649
Inapplicable *233
Larynx *362
Lung *183, *362, *649, 982
Lymphoma *362
Oesophagus *362
Oropharynx *362
Stomach *183, *362, 982
Tongue *362
Tonsil *362

Fumigation Workers

Inapplicable *1020

Furnacemen

Lung *183
Stomach *183

Gas Workers

All Sites 312

Glass Workers

All Sites 500
Inapplicable 198

Grain Millers

Leukaemia 977
Lymphoma 977

Health Care Workers

All Sites *396, 583, 714, 986, 991, 1076, 1107
Brain 96, 583

Breast (F)	*396, 991, 1107
Colon	986
Inapplicable	199, 354, *355, 626, 681
Leukaemia	1026, 1107
Lung	986, 1107
Multiple Myeloma	96
Non–Hodgkin's Lymphoma	96
Sarcoma	96
Soft Tissue	96
Thyroid	1107

Heavy Equipment Operators

Bladder	965
Brain	965
Haemopoietic	965
Lung	965
Lymphoma	965
Mesothelioma	965
Sarcoma	965
Stomach	965

Herbicide Manufacturers

All Sites	247, 296, *357, 654, 1180
Brain	1180
Breast (F)	*357
Leukaemia	1180
Lung	1180
Lymphoma	13, 247, 654
Multiple Myeloma	1180
Non–Hodgkin's Lymphoma	1180
Sarcoma	13, 247
Soft Tissue	654, 1180

Herbicide Sprayers

All Sites	247, 296, 654, 1180
Bladder	965
Brain	965, 1180
Haemopoietic	965
Leukaemia	653, 1180
Lung	965, 1180
Lymphoma	13, 247, 654, 965
Mesothelioma	965
Multiple Myeloma	1180
Non–Hodgkin's Lymphoma	1180
Sarcoma	13, 247, 965
Soft Tissue	654, 1180
Stomach	965

Horticulturists

All Sites	532, *699
Brain	96, 532
Hodgkin's Disease	93
Multiple Myeloma	93, 96
Non–Hodgkin's Lymphoma	93, 96, 532
Sarcoma	93, 96
Soft Tissue	93, 96, 532

Jewellery Manufacturers

Bladder	965
Brain	965
Haemopoietic	965

Lung 965
Lymphoma 965
Mesothelioma 965
Sarcoma 965
Stomach 965

Laboratory Workers
All Sites 300, *332, 335, 451, *621, 764, 767
Bone *621
Brain *621
Leukaemia *621
Lymphoma *621
Melanoma 1069
Pancreas *621

Lacquerers
Haemopoietic 677
Leukaemia 208
Nasal Cavity 208
Respiratory 208

Laminators
All Sites 281

Leather Workers
All Sites 305, 500, 755
Bladder 305
Hodgkin's Disease 542
Leukaemia 542
Multiple Myeloma 542
Nasal Cavity 305, 306
Non–Hodgkin's Lymphoma 542
Wilms' Tumour 71

Machinists
Colon 860
Gastrointestinal 1034
Lung 858, 860, 1034
Non–Hodgkin's Lymphoma 1034
Rectum 860
Stomach 860

Meat Workers
Leukaemia 653
Lung 932

Mechanics
Hodgkin's Disease 542
Leukaemia 542
Multiple Myeloma 542
Non–Hodgkin's Lymphoma 542
Prostate 629

Metal Workers
All Sites *693
Childhood Neoplasms 1147
Colon 1093

Gastrointestinal	1034
Kidney	1093
Liver	1093
Lung	*183, *265, *325, *785, 1034, 1093
Non–Hodgkin's Lymphoma	1034
Prostate	629
Retinoblastoma	1148
Skin	1093
Stomach	*183

Military Servicemen

All Sites	*935, 1163

Millers

All Sites	*272, *715
Brain	*272, *715
Kidney	*272, *715
Lung	*272, *715

Mineral Fibre Workers

All Sites	90, 298
Inapplicable	201
Lung	298, *1226
Respiratory	1156

Miners

All Sites	90, 185, *272, *485, *715, 1171
Bladder	965
Brain	*272, *715, 965
Haemopoietic	965
Kidney	*272, *715
Leukaemia	*106
Lung	*106, 129, *272, *485, 579, *670, *715, 798, 965
Lymphoma	965
Mesothelioma	579, 965
Sarcoma	965
Stomach	*106, 965

Miners, Asbestos

Colon	34
Gastrointestinal	76, *692
Haemopoietic	*692
Kidney	76
Larynx	76
Lung	34, 35, 76, *692
Mesothelioma	34, 35
Peritoneum	34, 76
Pleura	76
Stomach	34
Uterus (Cervix)	*692

Miners, Coal

All Sites	*29
Gastrointestinal	192
Hodgkin's Disease	40
Inapplicable	48, 198
Leukaemia	192
Leukaemia (CLL)	40
Lung	192, 328, 844

Non–Hodgkin's Lymphoma 40

Miners, Copper
Lung *158
Skin 961

Miners, Fluorspar
All Sites 87
Lung 87

Miners, Gold
Lung 36

Miners, Iron
Lung 326, 721
Stomach 326

Miners, Uranium
All Sites 90, 263
Inapplicable *4
Lung 3, 159, 263, 1022, 1024
Lymphoma 1024

Miners, Zinc–Lead
All Sites 117, 483
Lung 483

Model and Pattern Makers
Colon 1032, 1212
Rectum 1032, 1212

Morticians
All Sites 90
Bladder 965
Brain 965
Haemopoietic 965
Lung 965
Lymphoma 965
Mesothelioma 965
Sarcoma 965
Stomach 965

Nickel Workers
All Sites 90, 1161
Colon 1161
Liver 1161
Nasal Cavity 1161
Rectum 1161
Respiratory 1161

Office Workers
Respiratory 488

Painters

Leukaemia	653

Paper and Pulp Workers

All Sites	*103, *250, 282, *327
Haemopoietic	744
Lung	744, *1209
Lymphoma	744
Respiratory	743
Stomach	743, 744

Pesticide Workers

All Sites	*357, 928
Bladder	162
Breast (F)	*357

Petrochemical Workers

Inapplicable	198, 249
Kidney	1037
Skin	*1038

Petrol Station Attendants

All Sites	511, *535
Larynx	511
Lung	511

Petroleum Workers

All Sites	19, 501, 1018, 1171, 1188
Haemopoietic	19
Kidney	45, 795, 1189
Lymphoma	19
Wilms' Tumour	71

Physicists

Melanoma	1069

Plastics Workers

All Sites	281, 751, 754, 779, 825, 1158
Angiosarcoma	738
Bladder	738
Inapplicable	249, 1009
Leukaemia	779
Liver	738
Lung	738, *1226
Lymphoma	779

Plutonium Workers

All Sites	1071

Potters

All Sites	500, 939, 1171
Lung	939

Power Plant Workers

All Sites 56, 185
Childhood Neoplasms *934
Leukaemia 56, 859, *934
Lymphoma 56, *934
Non–Hodgkin's Lymphoma 859

Printers

All Sites *697
Lung 858

Quarry Workers

Lung 857

Radiation Workers

All Sites 81, 90, *262, *277, 360, *412, *422, *453, 819, 837, *838, 839, 896, 897, 1013, 1167
Childhood Neoplasms 890
Inapplicable 77, *355
Leukaemia *262, *277, *936
Lymphoma *936
Thyroid *453

Railroad Workers

All Sites 296, 545
Bladder *204
Larynx *204
Lung *204

Rubber Workers

All Sites *687, 1171
Bladder 310, *687
Haemopoietic *984
Leukaemia *687, *984
Lung 353, *687
Lymphoma *984
Melanoma 874
Mesothelioma 353
Nasal Cavity 874
Soft Tissue *984

Sailors

All Sites *661

Sawmill Workers

Leukaemia 653
Lung *1209

Screw Cutters

All Sites 266

Sewage Workers

All Sites 799

Shipyard Workers

All Sites 502, 1171
Lung *668

Shoemakers – Repairers

All Sites *698
Inapplicable 354
Liver *698
Lung *698

Smelters

All Sites *485
Brain *362
Larynx *362
Lung *183, *362, *485
Lymphoma *362
Oesophagus *362
Oropharynx *362
Stomach *183, *362
Tongue *362
Tonsil *362

Smelters, Aluminium

All Sites 484
Bladder 484
Lung 484
Lymphoma 484
Pancreas 484

Smelters, Copper

All Sites 792
Lung 792

Smelters, Lead

All Sites 117, 271, 750
Bladder 487
Kidney 487, 790
Leukaemia 487
Liver 487
Lung 271, 487, 790
Stomach 271

Steel (Stainless) Workers

Colon *323
Lung *320, *322, *323

Steel Workers

All Sites 863, 1015, 1160
Genitourinary 1160
Haemopoietic 1160
Leukaemia *864
Lung *183, *320
Lymphoma 1160
Respiratory *864, 1160
Stomach *183

Taxi Drivers
Lung *774

Textile Dyers
All Sites 509
Bladder 509
Nasal Cavity 509
Pharynx 509
Respiratory 509

Textile Workers
All Sites 500, 1171
Angiosarcoma 738
Bladder 738
Liver 738
Lung 738

Tool Makers
Gastrointestinal 1034
Lung 1034
Non–Hodgkin's Lymphoma 1034

Transport Workers
Bladder *204
Larynx *204
Lung 35, *204
Mesothelioma 35

Varnishers
Haemopoietic 677

Vinyl Chloride Workers
All Sites 128
Angiosarcoma 738
Bladder 738
Liver 128, 738
Lung 738

Waiters
Larynx 664
Liver 664
Oesophagus 664
Oral Cavity 664
Pharynx 664

Waste Incineration Workers
All Sites 1123

Welders
Childhood Neoplasms 1147
Gastrointestinal 1034
Inapplicable 354
Liver 234
Lung *667, *668, 1034
Non–Hodgkin's Lymphoma 1034
Retinoblastoma 1148

INDEX OF COUNTRIES

All studies are indexed, by site, under the country (or countries), territory or region where the data are being collected, so that, say, studies of cancer of the pancreas in Poland or nasopharyngeal cancer in Greenland can be readily identified.

Country names used in the Directory conform as far as possible to the UN list of member states.

When data are being collected in a different country from that where the principal investigator is based, or in more than one country, the countries concerned are listed beneath the study abstract. The study will be indexed under each of the countries where data are being collected.

Large international studies are coded to the country of the principal investigator co–ordinating the project.

New entries this year are:

Armenia
Belarus
Bosnia–Herzegovina
Croatia
Czech Republic
Kyrgyzstan
Lithuania
Russian Federation
Slovak Republic
Slovenia

"Czechoslovakia" and "USSR" have been deleted.

The list below identifies all entries in use in this Directory.

Algeria

All Sites	1
Nasopharynx	*2
Stomach	914

Argentina

All Sites	289
Oral Cavity	363
Oropharynx	363
Respiratory	363
Uterus (Cervix)	363

Armenia

All Sites	*700
Liver	*700
Lung	*700

Australia

All Sites	14, 19, *29, 33, 42, *277, 289, 296
Benign Tumours	31
Brain	22, 38, 44, 303, 304, 1104
Breast (F)	11, 21, 23, 25, *26, *27, *46, *291
Choriocarcinoma	6
Colon	5, 21, 24, *27, 28, 34, *291
Haemopoietic	19
Hodgkin's Disease	40
Hydatidiform Mole	6
Inapplicable	*4, 48, 1001
Kaposi's Sarcoma	41
Kidney	45, 795, 1104
Leukaemia	*18, 33, *277
Leukaemia (CLL)	40
Liver	1104
Lung	3, 34, 35, 36, *43, 1104
Lymphoma	13, *18, 19, 41
Melanoma	7, 9, 16, 20, 31, 39, 47, *269, *291
Mesothelioma	30, 34, 35, 42
Multiple Myeloma	*18
Myelodysplastic Syndrome	17
Non-Hodgkin's Lymphoma	40
Peritoneum	34, 42
Pleura	42
Prostate	21, *291
Rectum	5, 8, 21, 24, *27, 28
Sarcoma	13
Skin	12, 20, 32, 39, *269
Soft Tissue	33
Spinal Cord	22
Stomach	34, *291
Uterus (Cervix)	10, *37
Wilms' Tumour	*15

Austria

All Sites	49, 52, 296, 378, 688
Colon	51, *53
Leukaemia	287
Lung	*50, 52
Lymphoma	287
Melanoma	346
Mesothelioma	52
Ovary	534
Rectum	51, *53

COUNTRY

Bangladesh

All Sites *54
Breast (M) *54
Childhood Neoplasms *54
Oropharynx *54

Belarus

Leukaemia 287
Lymphoma 287

Belgium

All Sites 56, 61, 62, 270, *277, 282
Brain 57
Breast (F) 60
Colon 260
Gastrointestinal 340
Head and Neck 632
Hodgkin's Disease 270
Inapplicable 58
Larynx 59
Leukaemia 56, 57, 62, *277, 287, 340
Lung 340
Lymphoma 56, 57, 62, 287
Mesothelioma 59
Nasopharynx 59
Neuroblastoma 57
Non–Hodgkin's Lymphoma 340
Ovary 534
Stomach 914
Uterus (Cervix) 55

Bolivia

Bile Duct 1152
Gallbladder 1152

Bosnia–Herzegovina

All Sites 688

Brazil

All Sites 282, 863, *1174
Breast (F) 63, 65
Colon 63, 65
Gastrointestinal 64
Liver 65
Lung 63, 65
Lymphoma *1174
Non–Hodgkin's Lymphoma *1174
Oesophagus 63, 65
Oral Cavity 66
Oropharynx 66
Prostate 65
Rectum 65
Stomach 63, 65, 602
Uterus (Cervix) 65, *615
Wilms' Tumour 71

Bulgaria

All Sites 688
Leukaemia 287

Lymphoma 287
Ovary 534

Canada

All Sites 67, 68, 69, 80, 81, 82, 83, 84, 86, 87, 88, 89, 90, 91, 100, *103, *107, *109, *111, 112, 116, 117, 122, 220, *277, 282, 289, 296, 970, 1080, *1101, 1160, *1174
Bladder 73, 84, 99, 104, *110
Bone 102
Brain 84, 85, 86, 96, 303, 304, *1056, 1104, 1151
Breast (F) 74, *78, *79, 89, 105, 114, 121, 510, 970, 1078, *1119
Childhood Neoplasms *111, 1042, 1147, 1150, 1151
Colon 73, 74, 84, 89, 105, *110, 510, *1119
Gastrointestinal 72, 76, 101
Genitourinary 1160
Haemopoietic *984, 1160
Hodgkin's Disease 93, 1154
Inapplicable 77
Kidney 73, 76, 84, 105, 1104
Larynx 70, 76
Leukaemia 84, 86, *106, 118, 119, *277, 970, *984
Leukaemia (ALL) 75
Lip 70, 98
Liver 1104
Lung 72, 73, 76, 87, 94, 95, 97, 104, *106, 123, 297, 970, 1104
Lymphoma 73, 84, *984, 1160, *1174
Melanoma 73, 100
Mesothelioma 72, 101
Multiple Myeloma 86, 93, 96
Myelodysplastic Syndrome 118
Neuroblastoma 1008, 1114
Non–Hodgkin's Lymphoma 86, 93, 96, *1174
Oesophagus 73
Oral Cavity 70
Ovary *78, 105, 108, *1120
Pancreas 73, 104, 105
Peritoneum 76
Pharynx 70
Pleura 76
Prostate 73, 74, 84, 86, 92, 1214
Rectum 73, 84, 105, *110, 510, *1119
Respiratory 101, 1160
Retinoblastoma 1148
Sarcoma 83, 93, 96
Soft Tissue 93, 96, *984
Stomach 73, 84, *106
Testis 108
Thyroid 970, *1119
Uterus (Cervix) 113, 115
Uterus (Corpus) 105, *1119
Wilms' Tumour 1007

Chile

Bile Duct 125
Gallbladder 124, 125

China

All Sites 126, 128, 132, 140, 141, *142, 145, 176, 185, 378, 970
Benign Tumours *1202
Bladder 162, 178, 965
Bone 171
Brain 171, 965
Breast (F) 127, 131, 140, 146, 157, 170, 174, 299, 970, *1202

Colon 127, 146, 147, 148, 149, 150, 161, 166, 167
Haemopoietic 965
Larynx 181
Leukaemia 127, 136, 171, 180, 970
Liver 127, 128, 138, 146, *160, 163, 164, 165, 168, 171, 174, 175, 182
Lung 126, 127, 129, 135, 139, 146, 152, 154, *155, *158, 159, 172, *173, 174, 177, *183, 965, 968, 970
Lymphoma 171, 965
Mesothelioma 965
Nasal Cavity 181, 306
Nasopharynx 127, 143, 144, 151, 179, *184
Oesophagus 127, 130, 133, *134, 163, 167, 187, 188, *189, 190, 361
Oral Cavity 181
Pancreas 167
Pharynx 181
Rectum 127, 146, 147, 148, 149, 150, 161, 166, 167, 174
Sarcoma 965
Stomach 127, 137, 138, 140, 146, 153, 163, 174, 182, *183, 186, 965
Thyroid 970
Uterus (Cervix) 156, 169, 174

Colombia

Uterus (Cervix) 284

Costa Rica

Liver 191

Croatia

All Sites 688
Gastrointestinal 192, 193
Leukaemia 192
Lung 192, 193

Cuba

All Sites 194
Breast (F) 194, *195
Lung 194
Oral Cavity *195
Uterus (Cervix) *195

Czech Republic

All Sites 196, 270, 378, 688, 970
Breast (F) 970
Childhood Neoplasms 831
Hodgkin's Disease 270
Inapplicable 198, 199, *853
Leukaemia 287, 970
Lung *197, 970
Lymphoma 287
Ovary 534
Thyroid 970
Uterus (Cervix) 200

Denmark

All Sites *202, 210, 218, 221, 223, 225, 227, 229, 270, 281, 282, 296, 298, 970
Bladder *204, 213, 232, 965
Brain *203, *205, 213, 965
Breast (F) *203, *214, 223, 224, 230, 232, 970, 973
Breast (M) 211

Childhood Neoplasms 221, 222
Colon *203, 216, 226, 973
Gastrointestinal 213, 230
Haemopoietic 213, 965
Hodgkin's Disease 218, 270
Inapplicable 58, 201, *233, *853
Kidney 45, *219, 795
Larynx *204, 231
Leukaemia *205, 208, 221, 224, 230, 287, 672, 970
Leukaemia (CLL) 224
Liver 221
Lung *204, 217, 223, 231, 232, 298, 965, 970, 973
Lymphoma *205, 287, 965
Male Genital 213
Melanoma 230
Mesothelioma 965
Nasal Cavity 208
Nervous System *203
Oral Cavity 207
Ovary 973
Prostate *203, 228, 232
Rectum *203, 216
Respiratory 208, 213
Sarcoma 213, 965
Skin 213
Small Intestine 226
Stomach 914, 965
Testis *206, 209, 212, 230
Thyroid 970
Uterus (Cervix) 215, 223, 230, 232, *615
Uterus (Corpus) 230, 973

Egypt
Bladder *120
Liver 234

Estonia
All Sites 970
Breast (F) 235, 236, 510, 970
Colon 510
Leukaemia 287, 970
Lung 970
Lymphoma 287
Rectum 510
Thyroid 970

Finland
All Sites 227, 240, 241, 245, 246, 247, 248, *250, *256, 270, *277, 281, 282, 296, 298, 300, 378, 753, 970
Benign Tumours 253
Brain 251
Breast (F) 241, 242, 253, *255, 510, 970
Breast (M) 211
Colon 241, 242, 510
Female Genital 252
Hodgkin's Disease 251, 270
Inapplicable 249, 1001
Leukaemia 251, *277, 287, 672, 970
Lung 239, 241, 242, 243, 251, 298, 970
Lymphoma 247, 287
Male Genital 252
Ovary 534
Pancreas 244

Prostate 241, 242, 251
Rectum 241, 242, 510
Sarcoma 247
Stomach 237, 238, 241, 242
Thyroid 970
Uterus (Cervix) 241, 254

France

All Sites 80, *262, 263, 266, 270, 271, *277, 282, 289, 295, 300, 305, 308, 312, *319, *327, *332, 333, 335, *336, *338, 339, *341
Bladder 279, 305, 310
Brain 303, 304
Breast (F) *78, 276, *302, 309, 510
Childhood Neoplasms 307, 321, 1150
Colon 258, 260, 261, *323, 510
Gallbladder 258
Gastrointestinal 340
Hodgkin's Disease 270, *311
Hypopharynx 268, 315
Inapplicable 58
Kidney 330
Larynx 315, 807
Leukaemia *262, *277, 287, 307, 334, 339, 340, *342
Leukaemia (AML) 280
Liver 258
Lung 263, *264, *265, 271, 279, 297, *320, *322, *323, *325, 326, 328, 331, 339, 340, 343, 807
Lymphoma 287
Melanoma *269
Mesothelioma *324
Myelodysplastic Syndrome 267
Nasal Cavity 278, 305, 306, *316
Nasopharynx 314
Non–Hodgkin's Lymphoma *311, 339, 340
Oesophagus 258, *259
Oral Cavity 268, 807
Oropharynx 268
Ovary *78, 534
Pancreas 258
Pharynx 807
Prostate 257, 329
Rectum 258, 261, 510
Respiratory *324
Skin *269, *337
Stomach 258, 271, 326
Testis 283
Thyroid *313, 318, *319, *337
Uterus (Cervix) 274
Uterus (Corpus) *302

Gambia

All Sites 344, 524
Liver 345, 956

Germany

All Sites 270, *277, 282, 295, 296, 298, *357, 359, 360, 371, 372, 378, 688, 970
Bladder *352
Bone 370, 377
Brain 304, 348, *362
Breast (F) 347, *357, 374, 639, 970
Childhood Neoplasms 375, 376
Colon 260, 366, 373, 374
Gastrointestinal 340

Haemopoietic 358
Head and Neck 632
Hodgkin's Disease 270
Inapplicable 58, 354, *355
Kidney 45, 368, 377, 795
Larynx *362, 364
Leukaemia *277, 287, 340, 349, *350, 367, 370, 376, 970
Liver 370, 377
Lung 297, 298, 340, 351, 353, *362, 370, *379, 970
Lymphoma 287, 349, *350, *362, 370, 376
Melanoma 346
Mesothelioma *324, 353
Nasal Cavity 306, *316
Neuroblastoma 376
Non–Hodgkin's Lymphoma 340
Oesophagus *362
Oropharynx *362
Ovary 534
Prostate 374
Rectum 366, 373, 374
Respiratory *324
Skin 374
Spleen 370
Stomach *362, 914
Thyroid 365, 970
Tongue *362
Tonsil *362
Ureter *352
Urethra *352
Uterus (Cervix) 374
Wilms' Tumour 376

Ghana

Retinoblastoma 877

Greece

All Sites 270, 295, 378
Breast (F) 510
Childhood Neoplasms 831
Colon 510
Hodgkin's Disease 270
Inapplicable *853
Kaposi's Sarcoma 382
Leukaemia *380, *384, *385
Liver 383
Ovary 534
Rectum 510
Stomach 914
Thyroid 381

Greenland

All Sites 220
Uterus (Cervix) *615

Honduras

Uterus (Cervix) *627

Hong Kong

All Sites *387
Lung 386

Hungary

All Sites	378, 688
Bladder	392
Breast (F)	395
Gastrointestinal	392
Haemopoietic	393
Head and Neck	389
Inapplicable	390
Leukaemia	287, 393
Lung	388, 393
Lymphoma	287
Oral Cavity	389
Ovary	534
Stomach	394
Testis	391

Iceland

All Sites	227, *396, *397
Appendix	398
Bile Duct	398
Breast (F)	*396, 399, 400, 401
Colon	398
Leukaemia	672
Lung	*397
Nervous System	398
Oesophagus	398
Oral Cavity	398
Pancreas	398
Prostate	398
Small Intestine	398
Soft Tissue	398
Stomach	*397, 914
Uterus (Cervix)	*397

India

All Sites	301, *411, *412, *415, *418, *422, 426, 440
Brain	432
Breast (F)	408, 417, 423, 433, *436, 442, 446
Childhood Neoplasms	441
Colon	*407, 424
Head and Neck	*553
Hodgkin's Disease	431
Hypopharynx	425, 439
Larynx	425, 439
Leukaemia	431, 435
Lung	297, 425, *428
Lymphoma	431, 435
Multiple Myeloma	431
Nasopharynx	409
Oesophagus	424, 439
Oral Cavity	402, 413, 414, *416, 419, *420, *421, 425, 429, *430, 439, 444, 445, *553
Oropharynx	405, 425, 439
Ovary	404, 406, *427
Prostate	*427
Rectum	*407, 424
Stomach	424
Tongue	425
Uterus (Cervix)	*294, 403, 410, *434, 437, 438, 443

Indonesia

All Sites	448, 449, 450
Breast (F)	565

Lung 447
Nasopharynx 447
Uterus (Cervix) 447

Ireland

All Sites 300, 451
Breast (F) 510
Colon 510
Inapplicable 58
Rectum 510

Israel

All Sites 289, 452, *453, 458, 465, 466, 471, 472, *475, 970
Benign Tumours 472
Bone *292
Brain 303, 304, *462, *473
Breast (F) *291, *461, *462, 463, 464, 467, 468, *474, 970
Childhood Neoplasms 457
Colon *291, 455, 463, 465, 468, 470
Haemopoietic *476
Head and Neck 463
Hodgkin's Disease *292
Leukaemia *292, *462, 463, 970
Liver *460
Lung 456, *459, *462, 465, 970
Lymphoma 463
Melanoma *291, *292
Nervous System *292
Neuroblastoma *292
Non–Hodgkin's Lymphoma *292
Oesophagus 454
Ovary 534
Prostate *291, 469
Rectum 455, 468
Sarcoma *292
Soft Tissue *292
Stomach *291
Thyroid *453, *462, 463, 970

Italy

All Sites 270, *272, 281, 282, 295, 296, 298, 300, 305, *338, 378, 477, 483, 484, *485, 486, 489, 498, 500, 501, 502, 503, 504, 508, 509, 511, 512, 519, *520, 522, 525, 532, 533, *535, 537, 545, *1174
Benign Tumours *494, 517
Bile Duct *514
Bladder 305, 484, 487, *490, 509, 513, 541, *543
Brain *272, 303, 504, *514, 532
Breast (F) *478, 480, 499, 507, 510, *514, *515, 517, 518, 533, 544
Childhood Neoplasms 1150
Choriocarcinoma 516
Colon 260, *478, 480, 510, 521, 526, *527, 533
Endocrine Glands *494
Gastrointestinal 340
Haemopoietic 504
Hodgkin's Disease 270, 479, 540, 542
Hydatidiform Mole 516
Inapplicable 58
Kaposi's Sarcoma 479, 529
Kidney *272, 487, 506, 513
Larynx *478, 511, 513
Leukaemia 287, 340, 487, 508, 539, 542
Leukaemia (ALL) 540
Leukaemia (ANLL) 540

Liver 487, 491, *514
Lung *197, *272, 297, 298, 340, 480, 481, 483, 484, *485, 487, 489, 496, 497, 511, 513
Lymphoma 287, 484, 506, 508, *1174
Melanoma 495
Mesothelioma 496, 530, *536
Multiple Myeloma 542
Myelodysplastic Syndrome 539
Nasal Cavity 305, 306, *316, 509
Neuroblastoma 540
Non–Hodgkin's Lymphoma 340, 479, *514, 532, 540, 542, *1174
Oesophagus *478, 513
Oral Cavity *478, 513
Ovary *478, 517, 534
Pancreas 484, 513
Peritoneum 530
Pharynx 509, 513
Pleura 496, 530
Rectum *478, 510, 521
Respiratory 488, 509, *514
Sarcoma *514
Skin 504, *514, 528
Soft Tissue 508, *514, 532, 538
Stomach *493, 505, *523, 531, 533, 913, 914
Testis *514
Thyroid 492
Uterus (Cervix) *482, 517
Uterus (Corpus) 517
Vulva 517
Wilms' Tumour 540

Japan

All Sites *277, 282, 550, 551, *555, 562, 563, 564, 568, 572, 573, 575, 578, 582, 583, 586, 588, 589, 593, 601, 603, 970
Bile Duct 587
Bladder 546, *548
Brain 583
Breast (F) *436, 546, 558, 565, *567, 970
Childhood Neoplasms 590, 598, 599, 600
Choriocarcinoma *549
Colon 546, 558, *567, *594, 595, 601
Head and Neck *553
Hydatidiform Mole *549
Kidney 560
Leukaemia *277, 570, 581, 600, 970
Liver 556, 559, 569, 574, *577, 587, 595, 597, 601
Lung 546, *555, 556, 566, *567, 574, 579, *580, 595, 596, 601, 604, 970
Lymphoma 570
Mesothelioma 579
Nasal Cavity 306
Neuroblastoma 591
Oesophagus 546
Oral Cavity *553
Rectum 546, 558, *567, *594, 601
Retinoblastoma 561
Stomach 546, 552, 557, 558, *567, 571, 576, 584, *585, 592, 595, 601, 914
Thyroid 970
Uterus (Cervix) 547, *567
Uterus (Corpus) 595

Kenya

Stomach 912
Uterus (Cervix) 55

Kyrgyzstan

Breast (F)	*605
Hodgkin's Disease	*605
Lung	*605
Non–Hodgkin's Lymphoma	*605
Oesophagus	*605
Rectum	*605
Stomach	*605
Uterus (Cervix)	*605

Lithuania

Brain	303, 304
Leukaemia	287
Lymphoma	287

Luxembourg

All Sites	270
Breast (F)	510
Colon	510
Hodgkin's Disease	270
Inapplicable	58
Rectum	510

Malaysia

Nasopharynx	606, 607
Uterus (Cervix)	608

Malta

Lung	683

Mexico

Bile Duct	1152
Breast (F)	*609, 612
Gallbladder	1152
Leukaemia	1026
Stomach	*610
Uterus (Cervix)	611

Morocco

Nasopharynx	314

Namibia

Childhood Neoplasms	732

Netherlands

All Sites	270, 282, 295, 296, 300, 619, *621, 622, 623, 625, 1230
Benign Tumours	616, 617
Bile Duct	958
Bladder	642
Bone	*621
Brain	*621, 1104
Breast (F)	616, 617, 618, 619, 620, 624, 630, 636, 637, 639, 640
Childhood Neoplasms	1150
Colon	624, 638, 640, 641, 958
Gastrointestinal	340
Head and Neck	632
Hodgkin's Disease	270

Inapplicable 58, 626
Kidney 1104
Leukaemia 287, 340, 619, *621, 633
Leukaemia (ALL) 633
Liver 1104
Lung 340, 613, 619, 623, 624, 640, 642, 1104
Lymphoma 287, *621
Melanoma 628
Myelodysplastic Syndrome 619
Nasal Cavity *316
Non–Hodgkin's Lymphoma 340, 619, 622
Ovary 534, 634, 635
Pancreas *621, 958
Prostate 623, 629
Rectum 624, 638, 640, 958
Sarcoma 622
Soft Tissue 622
Stomach 624, 640, 958
Uterus (Cervix) 614, *615, 631

New Caledonia

All Sites 643
Hypopharynx *317
Larynx *317
Lung *317
Nasal Cavity *317
Pleura *317

New Zealand

All Sites 282, 296, 648, 654
Bladder 652
Breast (F) 644, 650, 651
Childhood Neoplasms 646, 647
Gastrointestinal *649
Leukaemia 653, 655
Leukaemia (ALL) 647
Lung *649
Lymphoma 654
Soft Tissue 654
Uterus (Cervix) 645

Nigeria

All Sites 378
Breast (F) 659
Burkitt's Lymphoma *658
Hodgkin's Disease *658
Kaposi's Sarcoma 657, *658
Liver 656
Non–Hodgkin's Lymphoma *658

Norway

All Sites 227, 270, 281, 282, 298, 300, 660, *661, 663, *669, 753, 970
Brain *665
Breast (F) 660, 970
Breast (M) 211
Gastrointestinal 660
Hodgkin's Disease 270
Larynx 664
Leukaemia 287, *665, 672, 970
Liver 664
Lung 298, *667, *668, *670, *671, 970
Lymphoma 287

Melanoma 666
Oesophagus 664
Oral Cavity 664
Ovary 660
Pharynx 664
Thyroid 662, 970
Uterus (Cervix) 660
Uterus (Corpus) 660

Pakistan
All Sites 674
Bone 674
Breast (F) 674
Colon *675
Gallbladder 673
Hodgkin's Disease *676
Leukaemia (ALL) *676
Non–Hodgkin's Lymphoma *676
Rectum *675
Uterus (Cervix) 674

Philippines
Breast (F) 288
Uterus (Cervix) *294

Poland
All Sites 270, 282, 680, 682, 684, 685, *686, *687, 688, 689
Bladder *687
Breast (F) 510, 679, 680
Colon 510
Haemopoietic 677
Hodgkin's Disease 270
Inapplicable 681
Leukaemia 287, *687
Lung 678, 680, 683, 685, *686, *687
Lymphoma 287
Mesothelioma *686
Ovary 534
Rectum 510
Stomach 680, 914
Uterus (Cervix) 679, 680
Uterus (Corpus) 679

Portugal
All Sites 282
Colon 260
Inapplicable 249
Lung 297
Stomach 914

Puerto Rico
All Sites 981

Romania
All Sites 688
Liver *690

Russian Federation

All Sites	220, 270, *693, 695, *696, *697, *698, *699, *701, 702, 708, 709, 713, 714, 718
Bladder	695
Breast (F)	510, *703, 707, 711, 717
Childhood Neoplasms	831
Colon	510, 712
Gastrointestinal	691, *692
Haemopoietic	*692, 716
Hodgkin's Disease	270
Inapplicable	*355
Leukaemia	287, *704
Liver	*698
Lung	691, *692, 695, *698, 705, 710
Lymphoma	287
Melanoma	*703
Oesophagus	694
Ovary	534, *703, 707
Rectum	510, 712
Retinoblastoma	*706
Stomach	695
Thyroid	*704
Uterus (Cervix)	*692, 707
Uterus (Corpus)	*703, 707

Rwanda

Kaposi's Sarcoma	*898
Non–Hodgkin's Lymphoma	*898
Uterus (Cervix)	*898

Slovak Republic

All Sites	270, 378, 688, 719, 720
Breast (F)	510
Colon	510
Hodgkin's Disease	270
Inapplicable	249
Leukaemia	287
Lung	*197, 721
Lymphoma	287
Melanoma	720
Rectum	510
Skin	720

Slovenia

All Sites	*272, 970
Brain	*272
Breast (F)	510, 722, 723, 724, 970
Colon	510
Kidney	*272
Leukaemia	287, 970
Lung	*272, 970
Lymphoma	287
Rectum	510
Stomach	914
Thyroid	970

South Africa

All Sites	282
Childhood Neoplasms	732
Haemopoietic	*727
Leukaemia	730

Leukaemia (AML) *727
Leukaemia (CLL) *727
Liver 726, 729
Mesothelioma 731
Non–Hodgkin's Lymphoma *727
Stomach 725
Uterus (Cervix) 728

Spain

All Sites 270, *272, *277, 282, 295, *1174
Angiosarcoma 738
Bile Duct *734
Bladder 738
Brain *272, 303
Breast (F) 510, 639
Childhood Neoplasms 831
Colon 275, 510, 737
Gallbladder *734, 737
Hodgkin's Disease 270
Kidney *272
Leukaemia *277
Liver 273, 737, 738
Lung *272, 297, 738
Lymphoma *1174
Non–Hodgkin's Lymphoma *1174
Oesophagus 733, 737
Pancreas 733, *734, 737
Rectum 275, 510, 737
Small Intestine 737
Stomach 733, 735, *736, 737
Uterus (Cervix) 284

Sri Lanka

Oral Cavity 739

Sweden

All Sites 227, 270, *277, 281, 282, 295, 296, 298, 300, 740, 746, 749, 750, 751, 752, 753, 754, 755, 757, 759, 763, 764, 767, 771, 772, *775, *776, 777, 779, *780, *781, 787, 792, 799, 821, 970, *1174
Bladder 787
Brain 304, 765, 766
Breast (F) 770, 786, 791, 797, 802, 803, 804, 970, 973
Breast (M) 211
Childhood Neoplasms 766, 769, 783
Colon 747, 748, *758, 762, *778, 787, 800, 821, 973
Female Genital 803
Haemopoietic 744
Hodgkin's Disease 270
Inapplicable 1001
Kidney 45, 790, 795
Larynx 784
Leukaemia *277, 287, 672, 752, 766, *775, 777, 779, *781, 970, 1064
Lung 297, 298, 742, 744, 748, 756, *768, 772, *773, *774, 782, *785, 787, 788, 790, 792, 798, 970, 973
Lymphoma 287, 744, 752, 757, 779, *781, *1174
Mesothelioma 748
Multiple Myeloma 757
Nasal Cavity 306
Non–Hodgkin's Lymphoma *761, *1174
Oesophagus 784
Oral Cavity 784
Ovary 534, 741, 786, 804, 973
Pharynx 784

LOCA

Prostate *760, 793
Rectum 747, 748, *758, 762, 787, 800, 821
Respiratory 743
Stomach 743, 744, 748, 756, 787, 794
Thyroid *775, 796, *801, 970
Uterus (Cervix) 745, 786, *789
Uterus (Corpus) 786, 973

Switzerland

All Sites 270, *277, 805, 806, 809
Bone 805
Breast (F) 510, 639, 805, *808
Childhood Neoplasms 805
Colon 510, 806
Gastrointestinal 805
Hodgkin's Disease 270
Inapplicable 58
Leukaemia *277, 287, 805
Lung 297, 806
Lymphoma 287, 805
Melanoma 346
Ovary 534, 805
Prostate 329
Rectum 510
Stomach 806
Testis 805
Uterus (Corpus) 810

Tanzania

Uterus (Cervix) 369

Thailand

Liver *293, 811
Lung *812
Uterus (Cervix) 813, *1207

Turkey

All Sites *814, 1171
Bladder 965
Brain 965
Haemopoietic 965
Lung 965
Lymphoma 965
Mesothelioma 965
Prostate 816
Sarcoma 965
Stomach 965
Testis 815

Uganda

Penis 356
Uterus (Cervix) 356
Vulva 356

Ukraine

All Sites *272, *715
Brain *272, *715
Kidney *272, *715
Lung *272, *715
Retinoblastoma *706

United Kingdom

Site	References
All Sites	270, *277, 281, 282, 289, 295, 296, 298, 300, 305, *336, 471, 819, 820, 821, 825, 837, *838, 839, 840, 854, 856, 869, 870, 872, 875, 876, 881, 884, 887, 888, 889, 896, 897, 899, 901, 902, 921, 924, 928, *929, *935, 938, 939, 940, 970, *1174
Bladder	305, 866, 899, 900, 910
Bone	*845
Breast (F)	*46, *78, 510, 639, 836, 848, 854, 855, 879, 888, 889, 908, *919, 923, 924, 925, 926, 941, 942, 970
Childhood Neoplasms	*822, 823, 824, 827, 831, 832, *852, 890, 891, 904, 905, *906, 916, 917, 922, *934, 1150
Colon	260, 510, *818, 821, *842, *843, 860, *862, 885, 892, 893, 894, 908, *933
Female Genital	888, 926
Gastrointestinal	880, 923, 927
Hodgkin's Disease	270, *929
Inapplicable	58, *817, 830, *853
Kidney	873
Larynx	869, 931
Leukaemia	*277, 287, *822, 823, 827, 829, 849, *850, 859, *864, 869, 871, *886, 899, 901, 902, 905, 915, 922, *929, *934, *936, 937, 970
Leukaemia (ALL)	847
Leukaemia (AML)	847
Leukaemia (CML)	847
Liver	869
Lung	298, 841, 844, 854, 857, 858, 860, 861, 869, 872, 873, 903, 910, 932, 939, 944, 945, 970
Lymphoma	287, 829, *850, 869, *886, *934, *936, *1174
Melanoma	874, 911
Mesothelioma	*324, 851
Myelodysplastic Syndrome	835, 922
Nasal Cavity	305, 306, 874
Non–Hodgkin's Lymphoma	823, 859, *929, *1174
Oesophagus	910
Ovary	*78, 534, *865, 868
Prostate	872, 873, 895, 908, 918, 943
Rectum	510, *818, 821, *842, *843, 860, *862, 885, 892, 893, 894, *933
Respiratory	*324, 856, *864
Retinoblastoma	877, 904
Sarcoma	889
Skin	833, *878
Soft Tissue	889
Stomach	846, 854, 860, 861, 908, 910, 913, 914, 930
Testis	834, 907, 909, 940
Thyroid	970
Urinary Tract	828
Uterus (Cervix)	867, 882, 883, 920, 924
Uterus (Corpus)	925

United States of America

Site	References
All Sites	90, 220, 223, 282, 289, 296, 305, 471, 826, 953, *967, 970, 978, 981, 982, 986, *987, *989, 990, 991, 992, 993, 995, 1013, 1015, 1018, 1019, *1021, 1023, 1027, 1039, 1041, 1044, 1046, 1050, *1052, 1057, 1065, 1070, 1071, 1075, 1076, *1077, 1080, 1082, 1094, 1095, 1099, *1101, 1105, 1107, *1110, *1117, 1123, 1126, 1131, 1137, 1138, 1146, 1157, 1158, 1159, 1160, 1161, 1163, 1166, 1167, 1168, 1170, *1172, 1173, *1174, *1175, 1178, 1180, 1181, 1188, 1197, 1201, 1227, 1230, 1231
Benign Tumours	1059, 1115, 1140
Bile Duct	1152
Bladder	305, 965, 985, 1016, 1025, 1027, 1035, 1096, 1134, 1157
Brain	303, 304, 955, 965, 979, 1012, *1021, *1056, 1087, 1088, 1091, 1095, 1104, 1151, 1180, 1181

Breast (F)	*78, 223, 463, *948, 949, 962, 970, 974, 980, *987, 991, 996, 997, 1000, *1005, *1006, 1031, 1040, *1043, 1046, 1053, 1054, 1059, *1063, 1065, *1067, 1072, 1078, 1082, 1083, 1097, 1105, 1107, 1111, *1112, 1115, 1116, *1124, 1130, 1132, 1139, 1140, *1142, *1143, *1144, *1191, *1195, *1196, 1197, 1204, *1206, 1213, 1218, 1222, *1223
Breast (M)	*1004, 1053
Childhood Neoplasms	1012, 1042, *1110, 1113, 1147, 1150, 1151, 1165, 1225
Colon	463, 949, 954, *976, 980, 986, *987, 996, 997, *1010, *1014, *1017, 1032, 1035, 1045, 1057, 1058, 1060, 1065, 1066, 1092, 1093, 1098, 1106, *1127, 1128, 1145, 1161, 1181, 1183, 1197, *1199, 1210, 1212, 1213, 1215, *1223, 1228
Eye	993, 1035
Female Genital	949
Gallbladder	1081, 1152
Gastrointestinal	954, 1034, 1091
Genitourinary	1160
Haemopoietic	965, *984, 1091, 1160
Head and Neck	463
Hodgkin's Disease	*1125, 1154, 1192
Inapplicable	554, 966, 1001, 1003, 1009, *1020, 1049
Kaposi's Sarcoma	*952, 957, 964, 1041
Kidney	45, 795, 1037, 1093, 1096, 1104, 1134, 1141, 1164, 1189
Larynx	1095, *1117, 1134
Leukaemia	463, 950, 970, 977, *984, 985, 1026, 1046, 1064, 1107, 1165, 1179, 1180, 1181, 1231
Leukaemia (ALL)	1108
Leukaemia (AML)	1086
Leukaemia (ANLL)	1109
Leukaemia (CLL)	*1125
Lip	1085
Liver	656, *983, 1028, 1035, 1091, 1092, 1093, 1104, 1149, 1161, 1165
Lung	223, 947, 960, 963, 965, 970, 980, 982, 985, 986, *987, 996, *999, 1022, 1024, 1034, 1035, *1051, 1061, 1065, 1079, 1091, 1092, 1093, 1104, 1107, 1121, 1131, 1134, 1135, 1153, 1162, 1173, 1180, 1182, 1197, 1200, *1209, 1213, 1224, *1226, 1227, 1232
Lymphoma	463, 964, 965, 977, *984, 1024, 1028, *1062, 1073, 1100, 1160, *1174, 1179
Melanoma	16, *269, *988, *1004, *1030, 1035, 1069, 1122, 1176, 1221
Mesothelioma	965, 1035, 1200
Multiple Myeloma	*967, 1036, 1180
Nasal Cavity	305, *316, 1084, 1161
Nasopharynx	1084, 1159, 1208
Neuroblastoma	1008, 1114
Non-Hodgkin's Lymphoma	*967, 1034, *1125, 1165, *1174, 1180, 1186, 1193
Oesophagus	*969, 1035, 1036, *1074, *1103, *1118, 1134
Oral Cavity	1089, 1090, *1117, 1134, 1136, 1203
Oropharynx	1136, 1203
Ovary	*78, 949, 980, 996, 997, 1000, *1011, 1054, 1133, 1139, *1175, *1191, 1198, *1223, 1229
Pancreas	1000, 1036, 1134, 1165, 1179, 1213
Pharynx	1089, 1090, *1117
Prostate	996, 997, 1000, *1002, *1004, 1033, 1036, 1065, 1129, *1205, 1213, 1214
Rectum	954, *1010, *1014, 1032, 1035, 1058, 1066, 1098, 1106, *1127, 1128, 1145, 1161, 1181, *1199, 1212, 1213, *1223
Respiratory	1156, 1160, 1161, 1219
Retinoblastoma	1148
Rhabdomyosarcoma	*1155
Salivary Gland	1035
Sarcoma	965
Skin	*269, 959, 961, *988, 993, *1030, *1038, 1055, 1093, 1184, 1216, 1217, 1220, 1221
Soft Tissue	*984, 1028, 1068, 1165, 1180
Stomach	576, 914, 965, *969, *975, 982, *1014, 1035, *1074, *1103, *1118, 1169, 1177
Thyroid	463, 970, 1107, 1194

Uterus (Cervix) 223, 946, 949, 951, 994, *998, 1029, *1048, 1139, 1185, 1187, 1197, 1211
Uterus (Corpus) 949, 972, 980, 996, 1000, 1047, 1082, 1105, 1139, 1190, *1191, *1223
Vagina 971, 1102
Vulva 971
Wilms' Tumour 1007

Uruguay

All Sites *1235, 1237, 1238, 1239, 1240
Larynx 1236
Lung 1234
Oral Cavity 1236
Pharynx 1236
Stomach 1233

Venezuela

Stomach 285, 286

Viet Nam

Non–Hodgkin's Lymphoma *290
Sarcoma *290
Soft Tissue *290

Yugoslavia

All Sites 378, 688
Breast (F) 1242
Gastrointestinal 1241
Genitourinary 1241
Hodgkin's Disease *1244
Inapplicable 58
Larynx *1246
Ovary 534, 1243
Thyroid *1245

INDEX OF CANCER REGISTRIES

The index of Cancer Registries identifies each project in which a registry is involved. Each registry has been given a short title for indexing. The index is in alphabetical order by country and shortname.

The index is followed by an address list, giving the name of the director or contact person and the full address of the registry. Registries appear in alphabetical order by country and short name. Registries which are members of the International Association of Cancer Registries are marked with an asterisk.

In the body of this Directory, studies with cancer registry involvementare identified beneath each abstract under the heading "REGI", followed by the short title and the country name in abbreviated form.

New registries are described as project abstracts in the body of the Directory for two years. After that, they are indexed only to projects in which they are involved.

We would be grateful if readers could notify us of any errors or omissions in the list or the index. Registry assignation to studies may be incomplete or incorrect, or the short title we have chosen may be inappropriate. We need your help to create a complete and correct index which reflects the full range of research in which cancer registries are involved.

Algeria

Sétif (Alg) *2

Australia

Brisbane II (Aus) 5, 28
NSW (Aus) 9, *29, 39, 40, 44, 45, *46
Queensland (Aus) 9, 12
S. Australia (Aus) 3
Sydney (Aus) 30, 42
Tasmania (Aus) 17
Victoria (Aus) 13, 20, 21, 22, 23, 24, 25, *26, *27
W. Australia (Aus) 30, 36

Belgium

Belgium (Bel) 55, 56, 57, 60

Canada

Alberta (Can) 67, 68, *277
Br. Columbia (Can) 116, 121, *277, 1214
Canada (Can) 81, 89, 90, 104, 220, 282, 1114
Manitoba (Can) *109, 122, 123, *277
N.W. Territory (Can) 82, *277
New Brunswick (Can) *277
Newfoundland (Can) 82, *277
Nova Scotia (Can) *277
Ontario (Can) 69, 108, *110, *111, *277, *984, *1120, 1214
Quebec (Can) 82, *277
Saskatchewan (Can) 94, 95, *277, *1119

China

Shanghai (Chi) 166, 167, 170, 171, 176, 179, 180, 181

Cuba

Cuba (Cub) *195

Czechoslovakia

Czech (Cze) 196

Denmark

Denmark (Den) *202, *203, *204, *205, *206, 207, 208, 210, 211, 213, *214, 215, 216, 217, *219, 220, 221, 222, 223, 224, 225, 227, 231, 232, 270, 281, 282, 283, 287, 296, 298, 973

Estonia

Estonia (Est) 235, 236, 287

Finland

Finland (Fin) 211, 227, 238, 240, 241, 242, 244, 245, 246, 247, 248, *250, 253, 254, *255, 270, *277, 281, 282, 287, 296, 298

France

Côte d'Or (Fra) *259
Caen (Fra) 258
Calvados (Fra) *259

Dijon I (Fra) 261
Doubs (Fra) *311
Haute–Garonne (Fra) *259
Isère (Fra) 287, 308
Lorraine (Fra) 321
Marseille (Fra) 307
Nouméa (Fra) *317
Pacific Islands (Fra) 643
Rhône (Fra) 276
Tarn (Fra) 257, 329

Gambia

Gambia (Gam) 344, 345

Germany

Berlin (Ger) 287, 348
Mainz (Ger) 287, 367, 375, 376
Saarland (Ger) *362

Hong Kong

Hong Kong (HK) *387

Hungary

Szabolcs–Szatmár (Hun) 394, 395

Iceland

Iceland (Ice) 227, *396, *397, 399, 400, 401

India

Bombay (Ind) *415, 417, *418, 425, 426, *427
Madras (Ind) 433

Israel

Israel (Isr) *291, *292, 452, *453, 454, 456, 463, 464, 465, 466, 472, *475

Italy

Genoa (Ita) *478, 503
Latina (Ita) *478, 532, 533
Modena (Ita) 521, *527
Padova (Ita) 270
Piedmont (Ita) 287
Ragusa (Ita) *527
Varese (Ita) 506, 507, 509

Japan

Aichi (Jap) 564
Chiba (Jap) 546
Kanagawa (Jap) 604
Miyagi (Jap) 593
Nagasaki (Jap) 562
Osaka (Jap) 576, *580

Lithuania

Lithuania (Lit) 287

Martinique

Martinique (Mar) 257, 329

Netherlands

Amsterdam (Net) 620, 624, 625, 626, 640
Eindhoven (Net) 620, 624, 625, 626
Groningen (Net) 624, 625, 626, 640
Leeuwarden (Net) 624, 625, 626
Leiden (Net) 620, 624, 625, 626, 640
Leiderdorp (Net) 623, 624, 625, 626
Maastricht (Net) 624, 625, 626, 640
Nijmegen (Net) 620, 624, 625, 626, 628, 629, 640
Rotterdam (Net) 620, 624, 625, 626, 640
The Hague (Net) 287, 633
Tilburg (Net) 624, 625, 626, 640
Utrecht (Net) 624, 625, 626, 637, 640, 641

New Zealand

New Zealand (NZ) 282, 296, 646, 647, 648, *649, 651, 654, 655

Norway

Norway (Nor) 211, 227, 270, 281, 282, 287, 298, 662, 663, 664, *665, 666, *667, *668, *669, *670, *671

Poland

Cracow (Pol) 680
Poland (Pol) 270, 287

Russian Federation

Moscow (Rus) 695, 714
St Petersburg (Rus) 287

Slovak Republic

Slovakia (Slvk) 287, 720

Slovenia

Slovenia (Slvn) *272, 287, 722, 723, 724

Spain

Basque (Spa) 733
Navarra (Spa) 733
Zaragoza (Spa) 733

Sweden

Lund (Swe) 748, 750, 784
Stockholm (Swe) *773, 784
Sweden (Swe) 211, 227, 270, 271, *277, 281, 282, 287, 296, 298, 743, 744, 745, 749, 751, 752, 754, 756, 757, *758, *761, 762, 764, 766, 767, *768, 769, 770, 771, *774, *775, *776, 777, *780, *781, 782, 783, *785, 786, 787, 790, 791, 800, 802, 803, 973
Uppsala (Swe) 795, 796, 804

Switzerland

Basel (Swi) 805
Geneva (Swi) 257, 287, 329, 807
Vaud (Swi) 270, 810

CANCER REGISTRY

Thailand

Lampang (Tha) *812

United Kingdom

Birmingham (UK) 824
E. Anglia (UK) *277, 834, 871
Mersey (UK) *277, 871
N.W. Region (UK) *277
Newcastle (UK) 889, 890, 891
Northern UK (UK) *277
OPCS (UK) 281, 282, 869, 875, 884, 897, 899, 923, 925, 927, 938
Oxford I (UK) *277, 287, 871
Oxford II (UK) *852, 904, 905, *906, 915, 916, 917, 922
S. Western (UK) *277
Scotland (UK) *277, 287, *845
Scotland N. (UK) 879
Scotland N.E. (UK) *818
Thames (UK) *277, *862, 871, 940
Trent (UK) *277
W. Midlands (UK) *277, 828, 871
Wales (UK) *277
Wessex (UK) *277
Yorkshire (UK) *277, 851, *852, 871

United States of America

Arizona (USA) 1218, *1223
Atlanta (USA) 1036
Bay Area (USA) 1041, 1069, 1070, 1087, 1141, 1183, 1214
California (USA) 1141
Connecticut (USA) 98, 463, *1117, *1118, 1170
Delaware (USA) 1114
Florida (USA) 1114
Hawaii (USA) *1051, *1052, 1214
Idaho (USA) 1182
Illinois (USA) 1015, 1016
Iowa (USA) 1016, 1105
Los Angeles (USA) 1070, 1072, 1073, *1074, 1083, 1087, 1088, 1089, 1090, 1214
Maine (USA) 1097
Maryland (USA) 1044
Massachusetts (USA) 1097
Michigan (USA) 1035, 1036
Minnesota (USA) 1114, 1183
Missouri (USA) 963
N. Carolina (USA) *1006, 1165, 1166
Nebraska (USA) 1177
New Hampshire (USA) 1097
New Jersey (USA) 1036
New Mexico (USA) *948
New York (USA) *999
Orange County (USA) 1061, *1062, *1063
San Diego (USA) *1067
Seattle (USA) 1087, 1190, *1191, *1195, *1196, 1198, *1199, 1201, 1203, *1205, *1209, 1210
SEER (USA) 576, 955, 979, 981, 1170, 1200
Tennessee (USA) *1103
Utah (USA) *967, 1182, 1183
Wisconsin (USA) 1016, 1097, 1098

Uruguay

Uruguay (Uru) *1235, 1237, 1238, 1239, 1240

POPULATION-BASED CANCER REGISTRIES

The list below contains the address and the name of the director or contact person for 324 population-based cancer registries in 84 countries around the world. Most registries are general, collecting data on all malignant tumours in both sexes and at all ages for persons resident in their territory, but some are specialised registries, restricted to an age-group (e.g. childhood registries) or a group of sites (e.g. haematological or gynaecological tumours).

Each registry has been assigned a short title for indexing and this will be found below the abstract of each project in which the registry is collaborating.

For each country, the registries are listed in alphabetical order of their short title.

Please contact the registries directly for further information about possible research collaboration.

Incidence data from many of these registries have been published in Cancer Incidence in Five Continents, Vol. VI, eds: D.M. Parkin, C.S. Muir, S.L. Whelan, Y.T. Gao, J. Ferlay and J. Powell, IARC Scientific Publications No. 120, 1992 and in Patterns of Cancer in Five Continents, eds: S.L. Whelan, D.M. Parkin and E. Masuyer, IARC Scientific Publications No. 102, 1990.

Algeria(Alg)

***Algiers**
Dr L. Abid
Registre des Cancers Digestifs d'Alger
Service de Chirurgie
Hôpital de Birtraria
16030 - El Biar
Alger
Algeria

***Sétif**
Dr M. Hamdi-Chérif
Service de Médecine Préventive
& d'Epidémiologie
CHU de Sétif
Hôpital Mère-Enfant
Sétif
Algeria

Argentina(Arg)

***Bahia**
Dr E. Laura
Registro Regional de Tumores del
Sur de la Pcia de Buenos Aires
Roca 571
8000 Bahia Blanca
Argentina

La Plata
Dr E. Perez Arias
Direccion de Planeamiento y Desarollo
Ministerio de Salud de la Placia
de Buenos Aires
1900 La Plata
Argentina

***Parana**
Dr J.E. Cura
Registro de Tumores
Servicio de Oncologia
Pte Peron 450
Hospital 'San Martin'
31000 Parana E.R.
Argentina

Australia(Aus)

***Brisbane I**
Dr W.R. McWhirter
Queensland Childhood Malignancy Registry
Department of Child Health
Royal Childrens Hospital
Herston, QLD 4029
Australia

***Brisbane II**
Ms D. Battistutta
Queensland Familial Adenomatous Polyposis
Register
Bancroft Centre
300 Herston Rd
Herston, QLD 4029
Australia

Brisbane III
Dr S.K. Khoo
Registry of Gestational Trophoblastic Disease
University of Queensland
Department of Obstetrics & Gynaecology
Clinical Sciences Building
Royal Brisbane Hospital
Brisbane, QLD 4029
Australia

Hobart
Director
Registry of Lymphoproliferative
& Myeloproliferative Disorders
University of Tasmania
Department of Community Health
43 Collins Street
Hobart, TAS 7000
Australia

***NSW**
Prof. R. Taylor
NSW Central Cancer Registry
& Cancer Epidemiology
Research Unit
NSW Cancer Council
153 Dowling Street
Woolloomoolloo, NSW 2011
Australia

***N. Territory**
Dr G. Durling
Northern Territory Cancer Registry
Department of Health & Community Services
P.O. Box 1701
Darwin, NT 0801
Australia

Perth I
Dr L.E. Dougan
Cancer Council of Western Australia
Leukaemia & Allied Disorders Registry
184 St. George's Terrace
Perth, WA 6005
Australia

***S. Australia**
Dr A.Z. Bonnet
South Australian Central Cancer Registry &
Cancer Control Unit
South Australian Health Commission
P.O. Box 6, Rundle Mall
Adelaide, SA 5000
South Australia

***Sydney**
Dr J. Leigh
Australian Mesothelioma Register
National Institute for Occupational Health &
Safety
G.P.O. Box 58
Sydney, NSW 2001
Australia

***Queensland**
Dr I. Ring
Epidemiology & Prevention Unit
(Non-communicable Diseases)
Queensland Department of Health
G.P.O. Box 48
147-163 Charlotte Street
Brisbane, QLD 4001
Australia

***Tasmania**
Mrs D. Shugg
Tasmanian Cancer Registry
Menzies Centre for Population Health Research
43 Collins Street
Hobart, Tasmania 7000
Australia

***Victoria**
Dr G.G. Giles
Victorian Cancer Registry
Cancer Epidemiology Center
1 Rathdowne Street
Carlton South, VICT 3053
Australia

***W. Australia**
Dr C.D.J. Holman
Western Australia Cancer Registry
Epidemiology Branch
Health Dept. of Western Australia
189 Royal Street
East Perth, WA 6004
Australia

Austria

***Vienna**
Dr H.P. Friedl
Austrian Cancer Registry
Austrian Central Statistics Office
Hintere Zollamtstrasse 2b
1033 Vienna
Austria

Belarus

Dr A.E. Okeanov
Inst. Oncology & Radiology
Dept. of Epidemiology
223052 Minsk
Belarus

Belgium(Bel)

***Belgium**
Dr M. Haelterman
Belgian Cancer Registry
Belgian Cancer Foundation
217, rue Royale
1210 Brussels
Belgium

Bermuda

***Paget**
Ms K. Scott
Oncology Services Co-ordinator
The Bermuda Hospital Board
King Edward VII Memorial Hospital
Paget
Bermuda

Bolivia(Bol)

***La Paz**
Dr J.L. Rios Dalenz
Registro de Cancer de La Paz
Calle Bollivian 1266
Casilla Postale 2801
La Paz
Bolivia

Brazil(Bra)

***Campinas**
Dr A. Valéria de Britto
Registro de Cáncer do Campinas
Rua Vital Brasil No. 100
13081 Campinas, SP
Brazil

***Fortaleza**
Dr M. Gurgel Carlos da Silva
Registro de Cáncer do Ceará
Rua Papi Junior 1222
Bairro Rodolfo Teófilo
60430 Fortaleza-Ceará
Brazil

***Goiania**
Dr M.P. Curado
Goiania Population-Based Cancer Registry
Estado de Goias
Fundaçao Leide da Neves Ferreira
Rua 16-A, 792 - Setor Aeroporto
74000 Goiania, GO
Brazil

***Porto Alegre**
Dr P.R. Grassi
Cancer Registry of Porto Alegre
Av. Washington Luiz, 868
90310 Porto Alegre, RS
Brazil

***Recife**
Dr M.R. da Costa Carvalho
Registro de Cancer de Pernambuco
Departamento de Patologia
Faculdade de Medicina
de Universidade Federal de Pernambuco
Cidade Universitária
Rua Prof. Moraes Rego, s/n
50730 Recife, Pernambuco
Brazil

***Sao Paulo**
Dr A.P. Mirra
Registro de Cancer de Sao Paulo
Avenida Dr Arnaldo, 715-1°A
01255 Sao Paulo, SP
Brazil

Bulgaria(Bul)

***Bulgaria**
Prof. C.G. Tzvetansky
Bulgarian National Cancer Registry
National Oncological Centre
Dept. of Epidemiology and Cancer Control
Plovdivsko pole 6
1756 Sofia
Bulgaria

Canada(Can)

***Alberta**
Dr H.J. Berkel
Alberta Cancer Registry
Department of Epidemiology
& Preventive Oncology
Alberta Cancer Board
6th Floor, Capital Place
9707 - 110 Street
Edmonton, Alberta
Canada T5K 2L9

***Br. Columbia**
Mr P. Hayles
British Columbia Cancer Registry
Cancer Control Agency of British Columbia
600 West 10th Avenue
Vancouver, British Columbia
Canada V5Z 4E6

CANCER REGISTRY

***Canada**
Mrs L. Gaudette
National Cancer Incidence
Reporting System of Canada
Health Division
Statistics Canada
R.H. Coats Building
Tunney's Pasture
Ottawa, Ontario
Canada K1A 0T6

***Manitoba**
Ms H. Whittaker
Manitoba Cancer Treatment
& Research Foundation
100 Olivia Street
Winnipeg, Manitoba
Canada R3E 0V9

***New Brunswick**
Dr C. Balram
New Brunswick Provincial Tumor Registry
Saint John Regional Hospital
P.O. Box 2100
Saint John, New Brunswick
Canada E2L 4L2

***Newfoundland**
Mrs D.L. Ball
Provincial Tumor Registry
25 Kenmount Rd
Saint John's, Newfoundland
Canada A1B 1W1

N.W. Territories
Dr L. Barreto
Northwest Territories Region
Health & Welfare Canada
900 Liberty Building
10506 Jasper Ave.
Edmonton, Alberta
Canada T5J 2W9

***Nova Scotia**
Mrs T. Croucher
Provincial Cancer Registry of Nova Scotia
Cancer Treatment & Research
Foundation of Nova Scotia
5820 University Avenue
Halifax, Nova Scotia
Canada B3H 1V7

***Ontario**
Dr E.A. Clarke
The Ontario Cancer Registry
7 Overlea Boulevard
Toronto, Ontario
Canada M4H 1A8

***Pr. Edward Is.**
Dr D.E. Dryer
Division of Cancer Control
Dept. of Health & Social Services
P.O. Box 2000
Charlottetown, PEI
Canada C1A 7N8

***Québec**
Mr G.-P. Sanscartier
Fichier des Tumeurs du Québec
Ministère de la Santé
& des Services Sociaux
1075, chemin Ste Foy, 3ème étage
Québec, Québec
Canada G1S 2M1

***Saskatchewan**
Mrs D. Robson
Saskatchewan Cancer Foundation
400-2631 28th Avenue
Regina, Saskatchewan
Canada S4S 6X3

***Toronto**
Dr G. Howe
National Cancer Institute of Canada
Epidemiology Unit
Faculty of Medicine
McMurrich Bldg.
12 Queen's Park Cres. W.
Toronto, Ontario
Canada M5S 1A8

China(Chi)

***Beijing**
Dr Wang Qi-jun
Department of Cancer Epidemiology
Beijing Institute for Cancer Research
Da-Hong-Lu-Chang Street
Western District
Beijing
China

***Guangdong**
Dr Hu Meng Xuan
Cancer Research Institute of Zhongshan City
East-Sun-Wen Road
Zhongshan City
Guangdong 528403
China

***Qidong**
Dr Li Wen Guang
Qidong Cancer Registry
Qidong Cancer Institute
57 Jian Hai Bei Rd
226200 Qidong County
Jiangsu Province
China

***Shanghai**
Dr Gao Yu-Tang
Shanghai Cancer Registry
Shanghai Cancer Institute
2200 Xie Tu Road
Shanghai 200032
China

Taiwan
Dr K. Ting-Yuan Kuan
Central Cancer Registry
Department of Health
Executive Yuan
Taipei, Taiwan
China

***Tianjin**
Dr Zhang Tian-ze
Tianjin Cancer Institute
Huan Hu Xi Road
Ti-Yuan-Bei
Tianjin 300060
China

Colombia(Col)

***Bogota**
Dra M. Ronderos
Registro Nacional de Cancer
Calle la., No. 9-85
Bogota
Colombia

Cali
Dr E. Carrascal Cortes
Cali Cancer Registry
Department of Pathology
University of Valle
Sede San Fernando
P.O. Box 25360
Cali
Colombia

Costa Rica

***Costa Rica**
Dra C. Bratti
National Cancer Registry
Ministerio de Salud
San José
Costa Rica

Croatia

***Croatia**
Dr M. Strnad
National Cancer Registry
Institute of Inst. of Public Health
Rockefellerova 7
41000 Zagreb
Croatia

Cuba(Cub)

***Cuba**
Dra L. Fernandez
Registro Nacional de Cancer
Instituto Nacional de Oncologia
29 y F. Vedado
Habana
Cuba

Cyprus

***Cyprus**
Dr M. Boyiadzis
Cyprus Cancer Registry
c/o Ministry of Health
Nicosia
Cyprus

Czech Republic(Cze)

***Brno**
Dr F. Beska
Cancer Epidemiology Dept.
Mazaryk Memorial Cancer Epidemiology
Zluty Kopec 7
656-53 Brno
Czech Republic

***Prague**
Mgr V. Mazànkovà
Czech Cancer Registry
Institute of Health Information & Statistics
120 55 Prague 2-
Palackého nàm. 4
Box 42
Czech Republic

Denmark(Den)

Copenhagen
Dr K.W. Anderson
Danish Breast Cancer
Cooperative Group
Finsen Institutet
Strandblvd 49
2100 Copenhagen
Denmark

***Denmark**
Dr H. Storm
Danish Cancer Society
Div. of Cancer Epidemiology
(Danish Cancer Registry)
Strandblvd 49
P.O. Box 839
2100 Copenhagen
Denmark

Ecuador

***Ecuador**
Dr F. Corral
National Tumor Registry
Avda Los Shyris 3307 y Tomàs
de Berlanga
P.O. Box 17114965
Quito
Ecuador

Egypt

***Alexandria**
Dr R.N. Bedwani
Medical Research Institute
165 Al Honia Rd
Alexandria
Egypt

Estonia(Est)

***Estonia**
Dr M. Rahu
Estonian Cancer Registry
Inst. of Experimental & Clinical Medicine
42 Hiiu Street
Tallinn 200107
Estonia

Fiji

***Suva**
Dr S. Govind
Cancer Registry
Ministry of Health
Private Bag
Suva
Fiji

Finland(Fin)

***Finland**
Dr L. Teppo
Finnish Cancer Registry
Liisankatu 21 B
00170 Helsinki 17
Finland

France(Fra)

***Bas-Rhin**
Prof. P. Schaffer
Registre Bas-Rhinois des Tumeurs
Institut d'Hygiène
Faculté de Médecine
4, rue Kirschleger
67085 Strasbourg Cédex
France

***Caen**
Dr D. Pottier
Registre des Tumeurs Digestives
du Calvados
CHU Côte de Nacre
Pièce 703
14040 Caen Cédex
France

***Calvados**
Dr J.M. Robillard
Registre Général des Tumeurs du Calvados
Route de Lion sur Mer
14021 Caen Cédex
France

***Côte d'Or**
Prof. P.M. Carli
Registre des Hémopathies Malignes
en Côte d'Or
Laboratoire d'Hématologie
Hôpital du Bocage
2, blvd de Lattre de Tassigny
21034 Dijon Cédex
France

***Dijon I**
Prof. J. Faivre
Registre Bourguignon des Cancers Digestifs
Faculté de Médecine
7, blvd Jeanne d'Arc
21033 Dijon Cédex
France

***Dijon II**
Dr D.B. Chaplain
Registre Bourguignon
de Pathologie Gynécologique
Centre Georges-François Leclerc
1, rue du Professeur Marion
21034 Dijon Cédex
France

***Doubs**
Prof. S. Schraub
Registre des Tumeurs du Doubs
Centre Hospitalier Régional
1, blvd Fleming
25030 Besançon Cédex
France

***Essonne**
Dr H. Gauthier
Registre des Cancers de l'Essonne
2 bis, rue du Ronneau
91150 Etampes
France

Haute-Garonne
Dr P. Pienkowski
Registre des Cancers Digestifs
de la Haute Garonne
CHU Rangueil
Chemin du Vallon
31054 Toulouse Cédex
France

***Haut-Rhin**
Dr A. Buemi
Registre Haut-Rhinois des Tumeurs
Association pour la Recherche Epidémiologique
par les Registres dans le Haut-Rhin
9, rue du Dr Mangeney
B.P. 1370
68070 Mulhouse Cedex
France

***Hérault**
Dr J.-P. Daurès
Registre des Tumeurs de l'Hérault
Centre Epidaure
rue des Apothicaires
Parc Euromédecine
B.P. 4111
34091 Montpellier Cedex
France

***Isère**
Dr F. Ménégoz
Registre du Cancer du
Département de l'Isère
21, chemin des Sources
38240 Meylan
France

***Lille**
Prof. L. Adenis
Registre des Cancers
des Voies Aéro-Digestives Supérieures
des Départements du Nord
et du Pas-de-Calais
Centre Oscar Lambret
B.P. 307
59020 Lille
France

***Limousin**
Dr J. Grasser
Registre Général des Cancers du Limousin
COPAS
6, rue de l'Amphithéâtre
87000 Limoges
France

***Lorraine**
Prof. D. Sommelet-Olive
Registre Pédiâtrique de Lorraine
Hôpital d'Enfants
Service de Pédiatrie II
Allée du Morvan
54511 Vandoeuvre Cédex
France

***Marseille**
Prof. J.-L. Bernard
Registre des Cancers de l'Enfant
Faculté de Médecine
27, blvd Jean Moulin
13385 Marseille Cédex 05
France

***Martinique**
Dr P. Escarmant
Registre du Cancer de la Martinique
A.M.R.E.C.
Service de Radiothérapie
Centre Hospitalier de Fort-de-France
B.P. 632
97261 Fort-de-France Cedex
Martinique, France

Nice
Dr M. Héry
Registre des Tumeurs Cérébrales Primaires de l'Adulte
Centre A. Lacassagne
36 Voie Romaine
06054 Nice Cédex
France

***Réunion**
Dr H. Chamouillet
Registre des Cancers de l'Ile de la Réunion
12 rue Jean Chatel
97400 Sainte Denis
La Réunion
France

***Somme**
Prof. A. Lorriaux
Association pour la Recherche Epidémiologique par les Registres en Picardie
Registre des Cancers de la Somme
CHR Médecine 4ème est
B.P. 3006
80054 Amiens Cedex
France

***Tarn**
Dr M. Roumagnac
Registre des Cancers du Tarn
Chemin des Trois Tarns
81000 Albi
France

French Polynesia

***Tahiti**
Dr F. Laudon
Cancer Registry
Direction de la Santé Publique
Bureau des Statistiques
B.P. 611
Papeete
Tahiti

Gabon(Gab)

Libreville
Université Omar Bongo
Centre Universitaire des Sciences de la Santé
B.P. 4009
Libreville
Gabon

Gambia(Gam)

***Gambia**
Dr A. Jack
The Gambia Cancer Registry
c/o MRC Laboratories
Fajara, P.O. Box 273
Banjul
The Gambia

Germany(Ger)

***Berlin**
Dr B. Eisinger
Gemeinsames Krebsregister der Länder Berlin, Brandenburg, Mecklenburg-Vorpommern, Sachsen-Anhalt, Thüringen und des Freistaates Sachsen beim Bundesgesundheitsamt
Rheinsteinstrasse 17-23
10318 Berlin
Germany

***Hamburg**
Dr C. Baumgardt-Elms
Hamburgisches Krebsregister
Gesundheitsbehörde
Postfach 2524
Tesdorpfstrasse 8
2000 Hamburg 13
Germany

***Mainz**
Prof. P. Schneider
Tumorzentrum Mainz
Am Pulverturm 13
6500 Mainz 2
Germany

***Child**
Prof. J.H. Michaelis
Cooperative Register of Childhood Malignancies
Institut für Medizin, Statistik
& Dokumentation (IMSD)
Universität Mainz
Langenbeckstrasse 1
55101 Mainz
Germany

***Saarland**
Mr H. Ziegler
Statistisches Amt des Saarlandes
Krebsregister
Postfach 409
Hardenbergstrasse 3
6600 Saarbrücken
Germany

Tübingen
Prof R. Schrage
Krebsregister Baden-Württemberg
Universitätsfrauenklinik Tübingen
Postfach
7400 Tübingen
Germany

Greece

***Athens**
Dr E. Stavrakakis
Central Health Council
Greek Cancer Registry
Aristotelous 19
10187 Athens
Greece

***Crete**
Dr I.G. Vlachonikolis
Cancer Registry of Crete
University of Crete
School of Health Sciences
Division of Medicine
P.O.B. 1393
Heraklion
Crete

Guinea

***Guinea**
Dr M. Koulibaly
Guinean Tumour Registry
Centre National d'Anatomie Pathologique
Faculté de Médecine/Pharmacie
Université de Conakry
B.P. 4152
Conakry
Guinea

Honduras

***San Padro Sula**
Dr O. Raudales
Liga Contra el Cancer
11 Ave y 8a Calle S.O.
San Padro Sula
Honduras

Hong Kong (HK)

***Hong Kong**
Dr Y.F. Poon
Hong Kong Cancer Registry
Medical & Health Department
Institute of Radiology & Oncology
Queen Elizabeth Hospital
Wylie Road
Kowloon
Hong Kong

Hungary(Hun)

***Budapest**
Dr D. Schuler
Paediatric Tumour Registry
Hungarian Society for Paediatric Oncology
Semmelweis University Medical School
Budapest IX, Tüzolto u. 7-9
Hungary 1094

***County Vas**
Dr E. Kocsis
County Vas Registry
Hospital "Maarkusovszky"
Semmelweis u. 2
9700 Szombathely
Hungary

***Szabolcs-Szatmar-Bereg**
Dr L. Juhasz
Cancer Registry of the County
Szabolcs-Szatmar-Bereg
Dept. of Oncology
County Hospital
Szent István u. 68
4401 Nyiregyhaza
Hungary

Iceland(Ice)

Iceland
Dr H. Tulinius
Icelandic Cancer Registry
Skogarhlid 8
P.O. Box 5420
125 Reykjavik
Iceland

India(Ind)

***Ahmedadabad**
Dr N.L. Patel
The Gujarat Cancer & Research Inst.
M.P. Shah Cancer Hospital
New Civil Hospital Compound
Asarwa, Ahmedadabad
India

***Bangalore**
Dr N. Anantha
Population-Based Cancer Registry
Kidwai Memorial Institute of Oncology
Hosur Road, Karnataka
Bangalore 560 029
India

***Barshi**
Dr K. Jayant
Rural Cancer Registry
Cancer Hospital
Agalaon Road
Barshi - 413 401 (Dist-Solapur)
Maharashtra
India

***Bhopal**
Dr S. Kanhere
Population-Based Cancer Registry
Department of Pathology
Gandhi Medical College
Bhopal 462 001
India

***Bombay**
Dr D.J. Jussawalla
Bombay Cancer Registry
Indian Cancer Society
74 Jerbai Wadia Rd
P.O. Box 6033
Parel, Bombay 400 012
India

***Madras**
Dr V. Shanta
Madras Metropolitan Tumor Registry
Cancer Institute (W.I.A.)
Adyar, Tamil Nadu
Madras 600 020
India

***New Delhi**
Dr Ramachandran
Indian Council for Medical Research
ansari Nagar
P.O. Box 4508
New Delhi 110 029
India

Ireland(Ire)

Dublin I
Dr F. Breatnach
Irish Paediatric Tumour Registry
Our Lady's Hospital for Sick Children
Crumlin
Dublin 12
Ireland

Dublin II
Dr J.A. Thornhill
Irish Testicular Tumour Registry
5 Northumberland Road
Dublin 4
Ireland

***Ireland**
Dr H. Comber
National Cancer Registry
Elm Court
Boreenmanna Road
Cork
Ireland

Israel(Isr)

***Israel**
Dr J. Iscovich
Israel Cancer Registry
107 Hebron Road
Jerusalem 93480
Israel

Italy(Ita)

***Forli**
Dr D. Amadori
Hospital G.B. Morgagni Pierantoni
Department of Oncology
Romagna Cancer Registry
Via Forlaninì, 11
Forli
Italy

***Genoa**
Prof. L. Santi
Registro Ligure Tumori
Istituto de Oncologia
Viale Benedetto XV, 10
16132 Genoa
Italy

***Latina**
Dr E.M.S. Conti
Centro Oncologico "G. Porfiri"
Registro Tumori di Popolazione
della Provinzia di Latina
Ospedale S.M. Goretti
04100 Latina
Italy

***Modena**
Dr M. Ponz de Leon
Registry of Tumours of the Digestive Organs
Istituto di Patologia Medica
Policlinico
Via del Pozzo, 71
41100 Modena
Italy

***Padova**
Dr L. Simonato
Registro Tumori Veneto
Via Guistiniani 7
c/o Direzzione Amministrativa
35100 Padova
Italy

***Parma**
Prof. G. Cocconi
Registro Tumori della Provincia di Parma
Via Gramsci, 14
43100 Parma
Italy

***Piemonte**
Dr R. Zanetti
Registro dei Tumori per il
Piemonte e la Valle d'Aosta
Via S. Francesco da Paola, 31
10123 Torino
Italy

***Ragusa**
Prof. L. Gafà
Registro Tumori di Ragusa
Via Virgilio 10
97100 Ragusa
Italy

***Torino**
Dr C. Magnani
Childhood Cancer Registry
Cattedra di Epidemiologia dei Tumori
Via Santena 7
10126 Torino
Italy

***Trieste**
Dr G. Stanta
Registro Tumori della Provincia di Trieste
Via Pietà 2-4
34126 Trieste
Italy

***Tuscany**
Dr M. Geddes
Registro Tumori Toscana (R.T.T.)
Unita di Epidemiologia
C.S.P.O. Viale A. Volta, 171
50131 Firenze 1
Italy

***Varese**
Dr F. Berrino
Registro Tumori Lombardia
(Provincia di Varese)
Servizio di Epidemiologia
Istituto Nazionale per lo Studio
e la Cura dei Tumori
Via Venezian, 1
20133 Milan
Italy

Jamaica

***Jamaica**
Dr S.E.H. Brooks
Jamaica Cancer Registry
(Kingston & St. Andrew)
Department of Pathology
University Hospital
University of the West Indies
Mona, Kingston 7
Jamaica

Japan(Jap)

***Aichi**
Dr A. Ikari
Aichi Cancer Registry
Division of Health & Prevention
Department of Health
Aichi Prefectural Government
3-1-2 Sannomaru, Naka-ku
Nagoya 460, Aichi
Japan

***Fukui**
Dr S. Yamazaki
Fukui Cancer Registry
Fukui Medical Association
3-4-10 Doigangi
Fukui City 910
Japan

***Fukuoka**
Dr T. Nagata
Fukuoka Cancer Registry
Fukuoka Medical Association
2-9-30 Hakata-eki-Minami
Hakata-ku
Fukuoka 812
Japan

***Hiroshima**
Dr K. Mabuchi
Hiroshima Tumor Registry
Radiation Effects Research Foundation (RERF)
5-2 Hijiyama Park, Minami ku
Hiroshima 732
Japan

***Hyogo**
Dr T. Ishida
Hyogo Medical Center for Adults
13-70, Kita-Oji-cho
Akashi 673, Hyogo-ken
Japan

***Kanagawa**
Dr N. Okamoto
Kanagawa Cancer Registry
Clinical Research Institute
Kanagawa Cancer Center
54-2 Nakao
Asahi-ku
Yokohama 241, Kanagawa
Japan

***Miyagi**
Dr A. Takano
Miyagi Prefectural Cancer Registry
c/o Miyagi Cancer Society
Kamisugi 6-chome, Aoba-ku
Sendai 980
Japan

***Nagasaki**
Prof. T. Ikeda
Nagasaki Tumor Registry
Radiation Effects Research Foundation (RERF)
1-8-6 Nakagawa
Nagasaki 850, Nagasaki-ken
Japan

Nagoya
Dr Y. Tomoda
Trophoblastic Tumour Registry
Nagoya University
School of Medicine
Dept. of Obstetrics & Gynaecology
Tsurumaicho 65
Showa-ku
Nagoya
Japan

***Osaka**
Dr A. Hanai
Osaka Cancer Registry
Center for Adult Diseases of Osaka
Department of Field Research
3 Nakamichi 1-chome
Higashinari-ku
Osaka 537
Japan

***Saga**
Dr M. Mori
Saga Prefectural Cancer Registry
Saga Medical School
Dept. of Community Health Science
5-1-1 Nabeshima
Saga 849
Japan

Sapporo
Dr Y. Kurimura
Hokkaido Cancer Registry
Adult Health Division
Dept. of Health & Environment
Hokkaido Government
North 3, West 6, Chuo-ku
Sapporo, Hokkaido 060
Japan

Tokyo
Dr K. Minoda
National Registry of Children
with Retinoblastoma
Teikyo University, Ichihara Hospital
Department of Ophthalmology
3426-3 Anesaki, Ichihara-City
Chiba 299-01
Japan

***Tottori**
Dr H. Nakayama
Tottori Cancer Registry
Dept. of Hygiene
Tottori University School of Medicine
86 Nishi-machi
Yonago 683
Japan

***Yamagata**
Dr Y. Satô
Yamagata Cancer Registry
Yamagata Medical Center for Adults
7-17 Sakuracho
Yamagata 990
Japan

Korea

***Seoul**
Dr Il Suh
Dept. of Preventive Med. & Public Health
Yonsei Univ. College of Med.
C.P.O. Box 8044
Seoul
Korea

Kuwait

***Kuwait**
Dr M. Al-Jarallah
Kuwait Cancer Registry
Kuwait Cancer Control Centre
P.O. Box No. 42262
Postal Code No. 70653
Shuwaikh
Kuwait

Latvia(Lat)

***Latvia**
Dr A. Stengrevics
Latvian Cancer Registry
Hipocrate Str.
LV-1079 Riga
Latvia

Lithuania(Lit)

***Lithuania**
Dr J. Kurtinaitis
Lithuanian Cancer Centre
Lithuanian Cancer Registry
Polocko 2, Vilnius 2007
Lithuania

Luxembourg

***Luxembourg**
Dr C. Capesius
Registre Morphologique des Tumeurs
1 A rue Auguste Lumière
1011 Luxembourg
Luxembourg

Malawi

***Malawi**
Dr L.T. Banda
National Cancer Registry
P.O. Box 30067
Chichii Blantyre 3
Malawi

Malaysia

***Malaysia**
Dr G.B. Nainani
Cancer Registry of Malaysia
Ministry of Helath, Malaysia
Tingkat 2 Blok E
Kompleks Pejabat - Pejabat
Jalan Dungun, Bukit Damansara
50490 - Kuala Lumpur
Malaysia

Mali(Mal)

***Bamako**
Dr S. Bayo
Institut National de Recherche en Santé Publique
Section des Maladies Néoplasiques
Route de Koulikoro
B.P. 1771
Bamako
Mali

Malta

***Malta**
Dr H. Agius-Muscat
Malta National Cancer Registry
c/o Health Information Systems Unit
St Luke's Hospital
Guardamangia
Malta

Myanmar

***Yangon**
Dr D. San San Aye
Yangon Cancer Registry
c/o Radiotherapy Dept.
Yangon General Hospital
Yangon
Myanmar

Netherlands(Net)

***Amsterdam**
Dr O. Visser
Comprehensive Cancer Centre
Amsterdam (IKA)
Plesmanlaan 125
1066 CX Amsterdam
The Netherlands

***Eindhoven**
Dr J.W.W. Coebergh
Comprehensive Cancer Centre South (IKZ)
P.O. Box 231
5600 AE Eindhoven
The Netherlands

***Enschede**
Mr R. Terhuerne
Comprehensive Cancer Centre
Stedendriekoek Twente
Lasondesingel 133
7514 BP Enschede
The Netherlands

***Groningen**
Dr R. Otter
Comprehensive Cancer Centre North (IKN)
Waterlolaan 1-13
9725 BE Groningen
The Netherlands

***Leiden**
Dr H. Kruijff
Cancer Registry
Comprehensive Cancer Center West (IKW)
Schipholweg 5A
2316 XB Leiden
The Netherlands

***Maastricht**
Dr L.J. Schouten
Comprehensive Cancer Centre Limburg (IKL)
Postbus 2208
6201 HA Maastricht
The Netherlands

Middelburg
Dr H.T. Planteydt
Mesothelioma Register
Pathologisch Anatomisch Lab.
Streeklaboratorium "Seeland"
Molenwater 47
Middelburg
The Netherlands

***Nijmegen**
Mrs H. de Kok
Comprehensive Cancer Centre East (IKO)
Regional Cancer Registry
P.O. Box 1281
6501 GB Nijmegen
The Netherlands

***Rotterdam**
Dr R.A.M. Damhuis
Comprehensive Cancer Centre Rotterdam (IKR)
P.O. Box 289
3000 AG Rotterdam
The Netherlands

***The Hague**
Dr A. van der Does-van den Berg
Dutch Childhood Leukaemia Study Group
c/o Juliana Kinderziekenhuis
Postbus 60604
2506 The Hague
The Netherlands

Utrecht
Dr C. Gimbrère
Comprehensive Cancer Centre Mid-Nederland (IKMN)
Regional Cancer Registry
Servaasbolwerk 14
3512 NK Utrecht
The Netherlands

New Caledonia(Cal)

***Nouméa**
Dr D. Dubordieu
Registre du Cancer de Nouvelle-Calédonie
Institut Pasteur de Nouméa
B.P. 61
Nouméa, New Caledonia
France

***Pacific Islands**
Dr Y. Souarès
Pacific Island Cancer Registry
South Pacific Commission
Box D5
Nouméa Cédex, New Caledonia
France

New Zealand(NZ)

***Auckland**
Mr H. Gaudin
Herald Centenary Cancer Registry
Auckland Hospital
P.O. Box 5546
Auckland
New Zealand

***New Zealand**
Mr F.J. Findlay
The New Zealand Cancer Registry
National Health Statistics Centre
Ballantrae House
192 Willis Street
Private Bag 2
Cumberland House Post Office
Wellington
New Zealand

***Waikato**
Dr E.H.M. Kerr
Waikato Tumour Registry
Waikato Hospital
Private Bag
Hamilton
New Zealand

Nigeria(Nig)

***Ibadan**
Prof. U. Aghadiuno
Ibadan Cancer Registry
Department of Pathology
University College Hospital
Ibadan
Nigeria

***Ile-Ife**
Dr O.S. Ojo
The IFE/IJESHA Cancer Registry
c/o Dept. of Morbid Anatomy
Obafemo Awolowo University
Ile-Ife
Nigeria

***Zaria**
Dr E.A.O. **Afolayan**
Zaria Cancer Registry
Department of Pathology
Ahmadu Bello University
Teaching Hospital
Zaria
Nigeria

Norway(Nor)

***Norway**
Dr F. Langmark
Cancer Registry of Norway
Inst. for Epidemiological Research
Montebello
0310 Oslo
Norway

Panama

***Panama**
Dr G. Campos
Registro Nacional del Cancer
Instituto Oncologico Nacional
Apartado 6-108
El Dorado
Panama

Paraguay(Par)

***Asunciòn**
Dr P.A. Rolòn
Registro Nacional de Patologia
Tumoral - Anatomia - Patologia
Igatimì 883
Asunciòn
Paraguay

Papua New Guinea(PNG)

Papua New Guinea
Dr J.S. Niblett
Papua New Guinea Tumour Registry
Angau Memorial Hospital
Dept. of Health
Box 457
Lae
Papua New Guinea

Peru(Per)

***Lima**
Dr L.E. Salem
Ministerio de Salud
Instituto Nacional de Enfermedade Neoplasicas
Avenida Angamos Este 2520
Lima
Peru

***Trujillo**
Dr P.F. Albujar
Registro de Cáncer de Base Poblacional
Calle Guillermo Charún 279
Urb. San Andrés
Trujillo
Peru

Philippines(Phi)

***Cebu City**
Dr N. Alsay
Metro Cebu Population-based Cancer Registry
Eduardo J. Aboitiz Cancer Center
Southern Islands Hospital Compound
B. Rodriguez St.
6000 Cebu City
Philippines

***Manila**
Dr A.V. Laudico
Philippine Cancer Society
P.O. Box 3066
310 Sab Rafael
Manila 1005
Philippines

***Rizal**
Dr A.V. Laudico
Rizal Medical Center
Cancer Registry
Department of Health
Pasig Boulevard
Pasig
Metro-Manila 1600
Philippines D. 2801

Poland(Pol)

Biakystok
Dr M. Grygoruk
Specjalistyczny Onkologiczny
Zespok Opieki Zdrowotnej
ul. Ogrodowa 12
15062 Biakystok
Poland

Bydgoszcz
Dr J. Sniegocki
Wojewodzka Przychodnia Onkologiczna
Ul. Ujejskiego 75
85168 Bydgoszcz
Poland

Cracow I
Dr J. Pawlega
Cracow City & District Cancer Registry
Institute of Oncology
Garncarska 11
31115 Cracow
Poland

Cracow II
Prof. J. Skolyszewski
Regional Cancer Registry
Marie Curie Research Institute
of Oncology
Oddziak w Krakowie
35115 Cracow
Poland

Gdynia
Dr E. Jordan
Regional Cancer Registry
Szpital Morski im. PCK
Ul. Powstania Styczniowego 1
81519 Gdynia
Poland

Gliwice
Dr S. Majewski
Centrum Onkologii - Instytut
im. M. Sklodowskiej-Curie
Oddziak w Gliwicach
Ul. Wybrzeze Armii Czerwonej 15
44101 Gliwice
Poland

***Katowice**
Dr B. Zemla
Katowice District Cancer Registry
ul. Armii Czerwonej 15
44101 Gliwice
Poland

Kielce
Dr S. Gozdz
Wojewodzka Przychodnia Onkologiczna
Ul. Kosciuszki 3
25310 Kielce
Poland

Lodz
Dr T. Naganski
Regionalny Osdarek Onkologiczny
Ul. Gagarina 4
93509 Lodz
Poland

Lublin
Dr R. Patyra
Onkologiczny Specjalistyczny
Zespol Opieki Zdrowotnej
ul. Jaczewskiego 7
20090 Lublin
Poland

***Opole**
Dr E. Piasecka
Regional Cancer Registry
Ul. Katowicka 66a
45060 Opole
Poland

***Poland**
Dr W. Zatonski
Polish Cancer Registry
Curie-Sklodowska Institute of Oncology
Wawelska Street 15
00-973 Warsaw
Poland

Poznan
Dr K. Gorny
Specjalistyczny Onkologiczny
Zespol Opieki Zdrowotnej
Ul. Garbary 13/15
61866 Poznan
Poland

Szczecin
Dr A. Niezgoda
Specjalistyczny Onkologiczny
Zespol Opieki Zdrowotnej
Ul. Strzalowska 22
71780 Szczecin
Poland

***Warsaw**
Dr Z. Wronkowski
Regional Cancer Registry
Centrum Onkologii
Ul. Wawelska 13
00973 Warsaw
Poland

***Wroclaw**
Prof. M. Wawrzkiewicz
Lower Silesian Cancer Registry
Ul. Hirszfelda 12
53413 Wroclaw
Poland

Portugal(Por)

***Coimbra**
Dr M.A.L. Silva
Registro Oncologico Regional - Zona Centro
Centro de Oncologia de Coimbra do IPOFG
Av. Bissaia Barreto 3000
Coimbra

***Lisboa**
Dr A. da Costa Miranda
IPORG - Centro de Lisboa
Registro Oncológico Regional Sul
1093 Lisboa Codex
Portugal

Porto
Dr M.J. Bento
Serviço de Epidemiologia
Inst. Português de Oncologia do Porto
Av. Antonio Bernardino de Almeida
4200 Porto
Portugal

Vila Nova
Dr J. Teixeira Gomes
Registro Oncologico de V.N. Gaia
Centro Hospitalar de Vila Nova de Gaia
Vilar de Andorinho
4400 Vila Nova de Gaia
Portugal

Romania(Rom)

***Cluj**
Dr A. Porutiu
Cluj County Cancer Registry
Oncological Institute of Cluj-Napoca
Dept. of Epidemiology & Statistics
Str. Republicii 34-36
3400 Cluj-Napoca
Romania

***Timis**
Dr R. Panaitescu
Service Départemental d'Oncologie
Spitalul Clinic No. 2
Str. Marasesti No. 5
Timisoara
Romania

Russian Federation(Rus)

Moscow
Dr D. Zaridze
Institute of Carcinogenesis
All-Union Cancer Research Centre
Academy of Medical Sciences
of the USSR
Kashirskoye Shosse 24
Moscow 115478
Russia

***St Petersburg**
Dr V.M. Merabishvili
Cancer Registry of St Petersburg
Pesochny 2
Leningradskaya Street 68
189646 - St Petersburg
Russia

Rwanda

***Butare**
Dr B. Sindikubwabo
Registre du Cancer de Butare
Faculté de Médecine
B.P. 221
Butare
Rwanda

Singapore(Sin)

***Singapore**
Prof. K. Shanmugaratnam
Singapore Cancer Registry
Department of Pathology
National University Hospital
Lower Kent Ridge Road
Singapore 0511
Singapore

Slovak Republic

***Bratislava**
Dr I. Plesko
National Cancer Registry of Slovakia
Oncological Center
Spitalskà 21
812 32 Bratislava
Slovak Republic

Slovenia

***Ljubljana**
Dr V. Pompe-Kirn
Cancer Registry of Slovenia
The Institute of Oncology
Zaloska 2
61000 Ljubljana
Slovenia

South Africa

***Johannesburg**
Dr F. Sitas
South African Cancer Registry
Department of Epidemiology
c/o South African Institute for
Medical Research
University of the Witwatersrand
P.O. Box 1038
Johannesburg 2000
South Africa

Spain(Spa)

***Basque**
Dr I. Izarzugaza
Registro de Cancer de Euskadi
Depto. de Sanidad y Consumo
Gobierno Vasco
Dirección de Información Sanitaria y Evaluación
c/Duque de Wellington, 2
01011 Vitoria-Gasteiz
Spain

***Granada**
Dr C.L. Martinez Garcia
Escuela Andaluza
de Salud Publica
Registro de Cáncer de Granada
Avenida del Sur 11
Ap. de Correos, 668
18080 Granada
Spain

***Guipuzcoa**
Dr M.J. Michelena
Registro de Cáncer de Guipuzcoa
Instituto de Oncologia
Aldako-enea
20012 San Sebastian
Spain

***Mallorca**
Dr I. Garau
Registre de Cancer de Mallorca
Paseig Mallorca 42
07012 Palma
Spain

***Murcia**
Dr C. Navarro Sanchez
Registro de Cáncer de Murcia
Servicio de Epidemiologia
Ronda de Levante, 11
30008 Murcia
Spain

***Navarra**
Dr A. Barricarte Gurrea
Registro de Cáncer de Navarra
Departmento de Salud
Instituto de Salud Publica vigilancia
Epidemiologica
Ciudadela, 5-1°
31003 Pamplona
Spain

***Oviedo**
Dr M. Echeverria Rodriguez
Registro de Tumores del Principado de Asturias
Consejeria de Sanidad y Servicios Sociales
General Elorza 32
33001 Oviedo
Spain

***Tarragona**
Dr J. Borras
Registro de Cáncer de Tarragona
Associacion Espanola Contra el Cancer
de Tarragona
Avenida Maria Cristina 54
43002 Tarragona, Catalonia
Spain

***Tenerife**
Dr D. Nunez
Registro de Tumores Poblacional
Rambla de General Franco, 53
Santa Cruz de Tenerife
(Canary Islands)
Spain

***Zaragoza**
Dr P. Moreo
Registro de Cáncer de Zaragoza
C/ Ramon y Cajal, 68
E-50004 Zaragoza
Spain

Saudi Arabia

***Riyadh**
Ms D.K. Michels
Cancer Registry/Oncology Department
King Faisal Specialist Hospital
P.O. Box 3354
Riyadh 11211
Saudi Arabia

Sudan(Sud)

***Khartoum**
Dr A. Hidayatalla
RICK - Cancer Registry
Radiation & Isotope Center
P.O. Box 677
Khartoum
Sudan

Sweden(Swe)

***Göteborg**
Dr I. Branehög
Oncology Centre
Sahlgrenska Hospital
413 45 Göteborg
Sweden

***Linköping**
Dr T. Hatschek
South East Sweden Cancer Registry
Oncological Centre
University Hospital
581 85 Linköping
Sweden

***Lund**
Dr T.R. Möller
Southern Swedish Regional
Tumor Registry
University Hospital
221 85 Lund
Sweden

***Stockholm**
Dr I. Rosendahl
Stockholm-Gotland Regional
Tumor Registry
Karolinska Hospital
104 01 Stockholm
Sweden

***Sweden**
Dr L. Barlow
The Swedish Cancer Registry
The National Board of Health & Welfare
F-enheten
Socialstyrelsen
106 30 Stockholm
Sweden

***Umeå**
Dr L. Damber
Regional Cancer Registry
for Northern Sweden
Oncology Centre
Regional Hospital
901 85 Umeå
Sweden

***Uppsala**
Dr H.O. Isaksson
Regional Tumor Registry
Oncology Centre
University Hospital
750 14 Uppsala
Sweden

Switzerland(Swi)

Basel
Prof. J. Torhorst
Basel Cancer Registry
Institut für Pathologie
Schonbeinstr. 40
4003 Basel
Switzerland

***Bern**
Dr A.B. Gasser
Klinisches Tumorregister
Inselspital
3010 Bern
Switzerland

***Geneva**
Dr C. Bouchardy
Registre Genevois des Tumeurs
55, blvd de la Cluse
1205 Geneva 4
Switzerland

***Neuchâtel**
Dr F.G. Levi
Registre Neuchâtelois des Tumeurs
31, rue Sophie Mairet
2300 La Chaux-de-Fonds
Switzerland

***St. Gallen**
Dr T. Fisch
Krebsregister St-Gallen-Appenzell
Institut für Pathologie
Kantonsspital
Rorschacherstr. 95
9001 St. Gallen
Switzerland

***Sion**
Dr F. Joris
Registre Valaisan des Tumeurs
Institut Central des Hôpitaux
Valaisans
1950 Sion
Switzerland

***Vaud**
Dr F.G. Levi
Registre Vaudois des Tumeurs
CHUV
Falaises 1
1011 Lausanne
Switzerland

***Zürich**
Dr G. Schüler
Kantonalzürcherisches Krebsregister
Institut für Pathologie der
Universität
Universitätsspital
8091 Zürich
Switzerland

Thailand(Tha)

***Chiang Mai**
Dr N.C. Martin
Tumor Registry Cancer Unit
Maharaj Nakorn Chiang Mai Hospital
Faculty of Medicine
Chiang Mai University
Chiang Mai
Thailand 50002

***Khon Kaen**
Dr V. Vatanasapt
Khon Kaen Provincial Registry
Cancer Unit
Srinagarind Hospital
Khon Kaen University
Khon Kaen
Thailand

***Songkla**
Dr H. SripLung
Songkla Cancer Registry
Faculty of Medicine
Prince of Songkla University
Hat Yai
Thailand 90112

Trinidad & Tobago

***Trinidad**
Dr G. Raj-Kumar
Trinidad & Tobago National Cancer Registry
Trinidad & Tobago Cancer Society
157A Western Main Road, St James
Port of Spain
Trinidad, West Indies

Uganda(Uga)

***Kampala**
Dr H. Wabinga
Cancer Registry
Department of Pathology
Makerere University Medical School
P.O. Box 7072
Kampala
Uganda

United Kingdom(UK)

***Birmingham**
Dr J.R. Mann
West Midlands Regional Children's Tumour Registry
Children's Hospital
Ladywood Middleway
Birmingham B16 8ET
United Kingdom

***E. Anglia**
Dr T. Davies
Cancer Registration Bureau of East Anglian Region
Addenbrookes Hospital
Hills Road
Cambridge CB2 2QQ
United Kingdom

Liverpool
Dr J. Osman
Epidemiology & Medical Statistics Unit
Health & Safety Executive
Room 134
Magdalen House
Stanley Precinct
Bootle, Merseyside L20 3QZ
United Kingdom

Manchester I
Dr J.M. Birch
Children's Tumour Registry
Christie Hospital
Kinnaird Road
Manchester M20 9BX
United Kingdom

***Manchester II**
Prof. S.A. Langley
Manchester Ovarian Tumour Register
Department of Obstetrics & Gynaecology
Manchester University
St. Mary's Hospital
Whitworth Park
Manchester M13 0JH
United Kingdom

***Mersey**
Dr E.M.I. Williams
The Mersey Regional Cancer Registry
2nd Floor, Muspratt Building
University of Liverpool
P.O. Box 147
Liverpool L69 3BX
United Kingdom

***N. Ireland**
North of Ireland Cancer Registry
The Lodge
59 Derryhaw Road, Tunan
Armagh
Northern Ireland
United Kingdom

***N.W. Region**
Prof. C. Woodman
North Western Regional
Cancer Epidemiology Unit
Christie Hospital & Holt Radium Inst.
Kinnaird Road, Withington
Manchester M20 9QL
United Kingdom

***Newcastle**
Dr A.W. Craft
Northern Region
Young Persons Cancer Registry
Dept. of Child Health, Medical School
Sir James Spence Bldg
Queen Victoria Road
Newcastle-upon-Tyne NE1 4LP
United Kingdom

Northern UK
Mr R.A. McNay
Northern RHA Cancer Registry
Benfield Road
Walkergate
Newcastle-upon-Tyne NE6 4PY
United Kingdom

***OPCS**
Dr M.J. Quinn
Office of Population Censuses & Surveys
St. Catherine's House
10 Kingsway
London WC2B 6JP
United Kingdom

***Oxford I**
Ms J. Redburn
Oxford Cancer Registry
Oxford Cancer Intelligence Unit
Old Road
Headington
Oxford OX3 7LF
United Kingdom

***Oxford II**
Dr C.A. Stiller
National Registry of Childhood Tumours
Childhood Cancer Research Group
57 Woodstock Road
Oxford OX2 6HJ
United Kingdom

***Scotland**
Dr C.S. Muir
The Scottish Cancer Registry
Scottish Health Service
Information Services Division
Trinity Park House
South Trinity Road
Edinburgh EH5 3SQ
Scotland
United Kingdom

***Scotland E.**
Dr S.N. Das
Scotland-East Cancer Registry
Radiotherapy Department
Ninewells Hospital
Dundee DD2 1UB
Scotland
United Kingdom

***Scotland N.**
Mrs L. Smart
Northern Cancer Registry
(Highlands & Western Isles)
Radiotherapy Department
Raigmore Hospital
Inverness IV2 3UJ
Scotland
United Kingdom

***Scotland N.E.**
Dr N.R. Waugh
Scotland-North East Cancer Registry
Dept. of Clinical Oncology
Aberdeen Royal Infirmary
Foresterhill
Aberdeen AB9 2ZB
Scotland
United Kingdom

***Scotland S.E.**
Dr S. Parker
Scotland-South-Eastern Cancer Registry
Liberton Hospital
Lasswade Road
Edinburgh EH16 6UB
Scotland
United Kingdom

***Scotland W.**
Dr C.R. Gillis
West of Scotland Intelligence Unit
Ruchill Hospital
Glasgow G20 9NB
Scotland
United Kingdom

***S. Western**
Dr D. Pheby
South Western Regional
Dept. of Epidemiology
and Public Health
Canynge Hall
Whiteladies Road
Bristol BS8 2PR
United Kingdom

***Thames**
Dr M.P. Coleman
Thames Cancer Registry
15 Cotswold Road
Sutton, Surrey SM2 5NL
United Kingdom

Trent
Mr D.L. Holmes
Trent Regional Cancer
Registration Bureau
Fulwood House
Old Fulwood Road
Sheffield S10 3TH
United Kingdom

***W. Midlands**
Dr R. Griffiths
Birmingham & West Midlands
Regional Cancer Registry
Queen Elizabeth Medical Centre
Edgbaston
Birmingham B15 2TH
United Kingdom

***Wales**
Dr M. Cotter
Wales Cancer Registry
Health Intelligence Unit
6th Floor, Heron House
35/43 Newport Road
Cardiff CF2 1SB
United Kingdom

***Wessex**
Dr J. Smith
Wessex Cancer Registry
Wessex Cancer Intelligence Unit
Wessex Regional Health Authority
Highcroft - Romsey Road
Winchester
Hants S022 5DH
United Kingdom

***Yorkshire**
Prof. C.A.F. Joslin
Yorkshire Regional Health
Authority Cancer Registry
Cookridge Hospital
Leeds LS16 6QB
United Kingdom

United States of America(USA)

***Alaska**
Dr A. Lanier
Alaskan Native Cancer Surveillance Program
Alaska Native Medical Center
250 Gambell Steet
Anchorage, AK 99501
USA

***Atlanta**
Dr R.S. Greenberg
Georgia Center for Cancer Statistics
Dept. of Epidemiology and Biostatistics
School of Medicine
1599 Clifton Rd NE
Atlanta, GA 30329
USA

***Bay Area**
Dr D.W. West
Northern California Cancer Center
32960 Alvarado Niles Road 600
Union City, CA 94587-3106
USA

***Brockton**
Dr V. O'Sullivan
Southeastern Mass. Cancer Registry
c/o Healthstat 680 Centre St.
Brockton, MA 03402
USA

***California**
Dr P. Reynolds
California Statewide Tumor Registry
Cancer Prevention Section
State Dept. of Health Services
5850 Shellmound Street, Suite 200
Emeyville, CA 94608
USA

***Colorado**
Mrs R. Bott
Colorado Central Cancer Registry
State Department of Health
4210 E. 11th Avenue
Denver, CO 80220
USA

***Connecticut**
Mr J.T. Flannery
Connecticut Tumor Registry
Connecticut State Department of Health Services
150 Washington Street
Hartford, CT 06106
USA

***Delaware**
Mr A. Topham
Delaware Cancer Reporting Service
Delaware State Tumor Registry
3000 Newport Gap Pike B
Wilmington, DE 19808-2378
USA

***Detroit**
Dr G.M. Swanson
Metropolitan Detroit
Cancer Surveillance Section
and Comprehensive Cancer Center
Michigan State University
A-211 East Fee Hall
East Lansing, MI 482824
USA

***Harrisburg**
Dr G. Weinberg
Pennsylvania Cancer Program
Health & Welfare Building
P.O. Box 90
Harrisburg, PA 17120
USA

***Hawaii**
Dr L.N. Kolonel
Hawaii Tumor Registry
Cancer Center of Hawaii
1236 Lauhala Street, Room 402
Honolulu, HI 96813
USA

***Houston**
Dr V.F. Guinee
University of Texas
M.D. Anderson Hospital & Tumor Inst.
Texas Medical Center
6723 Bertner Avenue C6001
Houston, TX 77030
USA

Idaho
Mrs M.L. England
Idaho Central Tumor Registry
Idaho Hospital Association
P.O. Box 7482
Boise, ID 83707
USA

***Illinois**
Dr H. Howe
Illinois State Cancer Registry
Div. of Epidemiological Studies
Illinois State Dept. of Public Health
605 W. Jefferson Street
Springfield, IL 62761
USA

***Indiana**
Dr L. Hathcock
Indiana State Cancer Registry
Indiana State Dept. of Health
P.O. Box 1964
1330 West Michigan Street
Indianapolis, IN 46206-1964
USA

***Iowa**
Dr C.F. Lynch
State Health Registry of Iowa
Dept. of Prevent. Med. & Environ. Health
University of Iowa
2826 SB
Iowa City, IA 52242
USA

***Kansas**
Dr F.F. Holmes
Cancer Data Service
University of Kansas
Medical Center
Kansas City, KS 66103
USA

***Los Angeles**
Dr R.K. Ross
USC School of Medicine
USC Cancer Surveillance Program
of Los Angeles
1721 Griffin Avenue, Room 200
Los Angeles, CA 90031
USA

***Louisiana**
Dr V. Chen
Louisiana State University
Medical Center
1901 Perdido St.
New Orleans, LA 70112-1393
USA

***Lowell**
Dr J. Evjy
Geater Lowell Cancer Program
295 Varnum Avenue
Lowell, MA 01854
USA

***Maine**
Dr G.F. Bogdan
State of Maine Cancer Registry
Maine Bureau of Health
State House Station II
157 Capitol Street
Augusta, ME 04333
USA

***Maryland**
Maryland Cancer Registry
Univ. of Maryland School of Medicine
DepT. of Epidemiol. & Prev. Med.
660 W. Redwood Street
Baltimore, MD 21201
USA

***Massachusetts**
Dr S.T. Gershman
Massachusetts Cancer Registry
Department of Public Health
150 Tremont Street - 5th Floor
Boston, MA 02111-1126
USA

***Michigan**
Mr G. Van Amburg
Michigan Cancer Surveillance Program
P.O. Box 30195
3423 North Logan
Lansing, MI 48909
USA

Minnesota
Dr A.P. Bender
Minnesota Cancer Registry
Minnesota Dept. of Health
Room 408
717 S.E. Delaware Street
P.O. Box 9441
Minneapolis, MN 55440
USA

Missouri
Dr J.C. Chang
Missouri Cancer Registry
Missouri Dept. of Health
Cancer Research Center
201 Business Loop 70 West
Columbia, MO 65205
USA

***Monroe**
Dr L.S. Baum
North Louisiana Tumor Registry
Northeast Louisiana University
NLU Cancer Research Center
Monroe, LA 71209
USA

Montana
Ms D. Hellhake
Montana Central Tumor Registry
Department of Health & Environmental Sciences
Cogswell Building
Helena, MT 59620
USA

***N. Carolina**
Dr T.E. Aldrich
North Carolina Central Cancer Registry
North Carolina Div. of Statistics & Information Services
P.O. Box 27687
Raleigh, NC 27611-7687
USA

Nebraska
Dr F.W. Karrer
Nebraska Cancer Registry
8303 Dodge Street
Omaha, NE 68114
USA

***New Hampshire**
Ms S.K. Foret
New Hampshire State Cancer Registry
Dartmouth Medical School
Hinman Box 7828
Hanover, NH 03756
USA

***New Jersey**
Ms D.M. Harlan
New Jersey State Cancer Registry
New Jersey Department of Health
3635 Quaker Bridge Road, CN 369
Trenton, NJ 08625-0379
USA

***New Mexico**
Dr C.R. Key
New Mexico Tumor Registry
Cancer Research & Treatment Center
900 Camino de Salud, N.E.
Albuquerque, NM 87131
USA

***New York**
Dr M.S. Baptiste
New York State Cancer Registry
Cancer Control Bureau
State of New York Dept. of Health
Corning Tower, Room 565
Empire State Plaza
Albany, NY 12237
USA

***Orange**
Dr H. Anton-Culver
Cancer Surveillance Program
of Orange County
Department of Community & Environmental Medicine
University of California at Irvine
Irvine, CA 92717
USA

***Pennsylvania**
Ms J.L. Phillips
Geisinger Medical Center
Tumor Registry
Medical Records Dept.
13-11 Geisinger Medical Center
Danville, PA 17822
USA

Philadelphia
Dr A.T. Meadows
Greater Delaware Valley
Pediatric Tumor Registry
Children's Hospital
Children's Cancer Research Center
Division of Oncology
34th & Civic Center Blvd
Philadelphia, PA 19104
USA

***Puerto Rico**
Dr I. Martinez
Central Cancer Registry of Puerto Rico
Dept. of Health
P.O. Box 9342
Santurce, PR 00908
USA

***Rhode Island**
Dr J.P. Fulton
Rhode Island Department of Health
Rhode Island Cancer Registry
75 Davis Street, Room 409
Providence, RI 02908
USA

RIMS-CSP
Dr A.R. King
RIMS Cancer Surveillance Program
Loma Linda Univ.
25455 Barton Road Suite B-208
Loma Linda, CA 92354
USA

Rochester
Ms M. Richardson
Rochester Regional Tumor Registry
Univ. of Rochester Cancer Center
P.O. Box 704
601 Elmwood Avenue
Rochester, NY 14620
USA

San Diego
Dr S.L. Saltzstein
SANDIOCC
2251 San Diego Avenue, Suite A-141
San Diego, CA 92110
USA

San Jose
Dr T.E. Davis
Region 1 Cancer Incidence Registry
1762 Technology Drive, Suite 204
San Jose, CA 95110
USA

***Seattle**
Dr D.B. Thomas
Cancer Surveillance System of Western Washington
The Fred Hutchinson Cancer Research Center
1124 Columbia Street
Seattle, WA 98104
USA

***SEER**
Dr B.F. Hankey
National Cancer Institute
Surveillance, Epicemiology & End Results Program (SEER)
Executive Plaza North
Room 343J
Bethesda, MD 20892
USA

Tennessee
Ms D. Weber
Tennessee Cancer Registry
Tennessee Dept. of Health & Environment
C2-233 Cordell Hull Building
Nashville, TN 37219
USA

Tri-Counties
Dr L. Hart
Tri-Counties Regional
Cancer Registry
Santa Barbara County
Health Care Services
300 San Antonio Rd, Room M-340
Santa Barbara, CA 93110
USA

***Utah**
Mrs R. Dibble
Utah Cancer Registry
Research Park
420 Chipeta Way, Suite 190
Salt Lake City, UT 84108
USA

***Virginia**
Dr W.J. Munsie
Virginia Tumor Registry
James Madison Bldg, Room 715
109 Governor Street
Richmond, VA 23219
USA

***Wisconsin**
Ms J.L. Phillips
Wisconsin Cancer Reporting System
Department of Health Statistics
P.O. Box 309
Madison, WI 53701-0309
USA

Uruguay(Uru)

***Uruguay**
Dr E. de Stefani
Registro Nacionale de Cancer
Oncology Institute
Juanico 3265
Montevideo
Uruguay

Vanuatu

***Vanuatu**
Ms Y. Taga
Cancer Registry
Ministry of Health
Private Mail Bag
Port vila
Vanuatu

Viet Nam

***Hanoi**
Dr Phan Thi Hoang Anh
Registre de Cancer
Hopital Benh Vien K.
43 Quan Su
Hanoï
Viet Nam

***Ho Chi Minh**
Dr N.C. Hung
Centre Anticancéreux
3 No Trang Long-Binh-Thanh
Ho Chi Minh Ville
Viet Nam

Yugoslavia

***Voyvodina**
Dr M.M. Mikov
Cancer Registry of Voyvodina
Institute of Oncology
Medical Faculty Novi Sad
Institutski put 4
21204 Sremska Kamenica
Novi Sad
Yugoslavia

Zimbabwe

***Zimbabwe**
Dr M. Bassett
Zimbabwe Cancer Registry
Box A449
Avondale
Harare
Zimbabwe

PUBLICATIONS OF THE INTERNATIONAL AGENCY FOR RESEARCH ON CANCER

Scientific Publications Series

(Available from Oxford University Press through local bookshops)

No. 1 **Liver Cancer**
1971; 176 pages (*out of print*)

No. 2 **Oncogenesis and Herpesviruses**
Edited by P.M. Biggs, G. de-Thé and L.N. Payne
1972; 515 pages (*out of print*)

No. 3 ***N*-Nitroso Compounds: Analysis and Formation**
Edited by P. Bogovski, R. Preussman and E.A. Walker
1972; 140 pages (*out of print*)

No. 4 **Transplacental Carcinogenesis**
Edited by L. Tomatis and U. Mohr
1973; 181 pages (*out of print*)

No. 5/6 **Pathology of Tumours in Laboratory Animals, Volume 1, Tumours of the Rat**
Edited by V.S. Turusov
1973/1976; 533 pages (*out of print*)

No. 7 **Host Environment Interactions in the Etiology of Cancer in Man**
Edited by R. Doll and I. Vodopija
1973; 464 pages (*out of print*)

No. 8 **Biological Effects of Asbestos**
Edited by P. Bogovski, J.C. Gilson, V. Timbrell and J.C. Wagner
1973; 346 pages (*out of print*)

No. 9 ***N*-Nitroso Compounds in the Environment**
Edited by P. Bogovski and E.A. Walker
1974; 243 pages (*out of print*)

No. 10 **Chemical Carcinogenesis Essays**
Edited by R. Montesano and L. Tomatis
1974; 230 pages (*out of print*)

No. 11 **Oncogenesis and Herpesviruses II**
Edited by G. de-Thé, M.A. Epstein and H. zur Hausen
1975; Part I: 511 pages
Part II: 403 pages (*out of print*)

No. 12 **Screening Tests in Chemical Carcinogenesis**
Edited by R. Montesano, H. Bartsch and L. Tomatis
1976; 666 pages (*out of print*)

No. 13 **Environmental Pollution and Carcinogenic Risks**
Edited by C. Rosenfeld and W. Davis
1975; 441 pages (*out of print*)

No. 14 **Environmental *N*-Nitroso Compounds. Analysis and Formation**
Edited by E.A. Walker, P. Bogovski and L. Griciute
1976; 512 pages (*out of print*)

No. 15 **Cancer Incidence in Five Continents, Volume III**
Edited by J.A.H. Waterhouse, C. Muir, P. Correa and J. Powell
1976; 584 pages (*out of print*)

No. 16 **Air Pollution and Cancer in Man**
Edited by U. Mohr, D. Schmähl and L. Tomatis
1977; 328 pages (*out of print*)

No. 17 **Directory of On-going Research in Cancer Epidemiology 1977**
Edited by C.S. Muir and G. Wagner
1977; 599 pages (*out of print*)

No. 18 **Environmental Carcinogens. Selected Methods of Analysis. Volume 1: Analysis of Volatile Nitrosamines in Food**
Editor-in-Chief: H. Egan
1978; 212 pages (*out of print*)

No. 19 **Environmental Aspects of *N*-Nitroso Compounds**
Edited by E.A. Walker, M. Castegnaro, L. Griciute and R.E. Lyle
1978; 561 pages (*out of print*)

No. 20 **Nasopharyngeal Carcinoma: Etiology and Control**
Edited by G. de-Thé and Y. Ito
1978; 606 pages (*out of print*)

No. 21 **Cancer Registration and its Techniques**
Edited by R. MacLennan, C. Muir, R. Steinitz and A. Winkler
1978; 235 pages (*out of print*)

No. 22 **Environmental Carcinogens. Selected Methods of Analysis. Volume 2: Methods for the Measurement of Vinyl Chloride in Poly(vinyl chloride), Air, Water and Foodstuffs**
Editor-in-Chief: H. Egan
1978; 142 pages (*out of print*)

No. 23 **Pathology of Tumours in Laboratory Animals. Volume II: Tumours of the Mouse**
Editor-in-Chief: V.S. Turusov
1979; 669 pages (*out of print*)

No. 24 **Oncogenesis and Herpesviruses III**
Edited by G. de-Thé, W. Henle and F. Rapp
1978; Part I: 580 pages, Part II: 512 pages (*out of print*)

No. 25 **Carcinogenic Risk. Strategies for Intervention**
Edited by W. Davis and C. Rosenfeld
1979; 280 pages (*out of print*)

No. 26 **Directory of On-going Research in Cancer Epidemiology 1978**
Edited by C.S. Muir and G. Wagner
1978; 550 pages (*out of print*)

No. 27 **Molecular and Cellular Aspects of Carcinogen Screening Tests**
Edited by R. Montesano, H. Bartsch and L. Tomatis
1980; 372 pages £30.00

No. 28 **Directory of On-going Research in Cancer Epidemiology 1979**
Edited by C.S. Muir and G. Wagner
1979; 672 pages (*out of print*)

No. 29 **Environmental Carcinogens. Selected Methods of Analysis. Volume 3: Analysis of Polycyclic Aromatic Hydrocarbons in Environmental Samples**
Editor-in-Chief: H. Egan
1979; 240 pages (*out of print*)

No. 30 **Biological Effects of Mineral Fibres**
Editor-in-Chief: J.C. Wagner
1980; **Volume 1:** 494 pages **Volume 2:** 513 pages (*out of print*)

No. 31 ***N*-Nitroso Compounds: Analysis, Formation and Occurrence**
Edited by E.A. Walker, L. Griciute, M. Castegnaro and M. Börzsönyi
1980; 835 pages (*out of print*)

No. 32 **Statistical Methods in Cancer Research. Volume 1. The Analysis of Case-control Studies**
By N.E. Breslow and N.E. Day
1980; 338 pages £18.00

No. 33 **Handling Chemical Carcinogens in the Laboratory**
Edited by R. Montesano *et al.*
1979; 32 pages (*out of print*)

No. 34 **Pathology of Tumours in Laboratory Animals. Volume III. Tumours of the Hamster**
Editor-in-Chief: V.S. Turusov
1982; 461 pages (*out of print*)

No. 35 **Directory of On-going Research in Cancer Epidemiology 1980**
Edited by C.S. Muir and G. Wagner
1980; 660 pages (*out of print*)

No. 36 **Cancer Mortality by Occupation and Social Class 1851-1971**
Edited by W.P.D. Logan
1982; 253 pages (*out of print*)

No. 37 **Laboratory Decontamination and Destruction of Aflatoxins B_1, B_2, G_1, G_2 in Laboratory Wastes**
Edited by M. Castegnaro *et al.*
1980; 56 pages (*out of print*)

No. 38 **Directory of On-going Research in Cancer Epidemiology 1981**
Edited by C.S. Muir and G. Wagner
1981; 696 pages (*out of print*)

No. 39 **Host Factors in Human Carcinogenesis**
Edited by H. Bartsch and B. Armstrong
1982; 583 pages (*out of print*)

No. 40 **Environmental Carcinogens. Selected Methods of Analysis. Volume 4: Some Aromatic Amines and Azo Dyes in the General and Industrial Environment**
Edited by L. Fishbein, M. Castegnaro, I.K. O'Neill and H. Bartsch
1981; 347 pages (*out of print*)

No. 41 ***N*-Nitroso Compounds: Occurrence and Biological Effects**
Edited by H. Bartsch, I.K. O'Neill, M. Castegnaro and M. Okada
1982; 755 pages £50.00

No. 42 **Cancer Incidence in Five Continents, Volume IV**
Edited by J. Waterhouse, C. Muir, K. Shanmugaratnam and J. Powell
1982; 811 pages (*out of print*)

No. 43 **Laboratory Decontamination and Destruction of Carcinogens in Laboratory Wastes: Some *N*-Nitrosamines**
Edited by M. Castegnaro *et al.*
1982; 73 pages £7.50

No. 44 **Environmental Carcinogens. Selected Methods of Analysis. Volume 5: Some Mycotoxins**
Edited by L. Stoloff, M. Castegnaro, P. Scott, I.K. O'Neill and H. Bartsch
1983; 455 pages £32.50

No. 45 **Environmental Carcinogens. Selected Methods of Analysis. Volume 6: *N*-Nitroso Compounds**
Edited by R. Preussmann, I.K. O'Neill, G. Eisenbrand, B. Spiegelhalder and H. Bartsch
1983; 508 pages £32.50

No. 46 **Directory of On-going Research in Cancer Epidemiology 1982**
Edited by C.S. Muir and G. Wagner
1982; 722 pages (*out of print*)

No. 47 **Cancer Incidence in Singapore 1968–1977**
Edited by K. Shanmugaratnam, H.P. Lee and N.E. Day
1983; 171 pages (*out of print*)

No. 48 **Cancer Incidence in the USSR (2nd Revised Edition)**
Edited by N.P. Napalkov, G.F. Tserkovny, V.M. Merabishvili, D.M. Parkin, M. Smans and C.S. Muir
1983; 75 pages (*out of print*)

No. 49 **Laboratory Decontamination and Destruction of Carcinogens in Laboratory Wastes: Some Polycyclic Aromatic Hydrocarbons**
Edited by M. Castegnaro *et al.*
1983; 87 pages (*out of print*)

No. 50 **Directory of On-going Research in Cancer Epidemiology 1983**
Edited by C.S. Muir and G. Wagner
1983; 731 pages (*out of print*)

No. 51 **Modulators of Experimental Carcinogenesis**
Edited by V. Turusov and R. Montesano
1983; 307 pages (*out of print*)

* Available from booksellers through the network of WHO Sales agents.

† Available directly from IARC

No. 52 **Second Cancers in Relation to Radiation Treatment for Cervical Cancer: Results of a Cancer Registry Collaboration**
Edited by N.E. Day and J.C. Boice, Jr
1984; 207 pages (*out of print*)

No. 53 **Nickel in the Human Environment**
Editor-in-Chief: F.W. Sunderman, Jr
1984; 529 pages (*out of print*)

No. 54 **Laboratory Decontamination and Destruction of Carcinogens in Laboratory Wastes: Some Hydrazines**
Edited by M. Castegnaro *et al.*
1983; 87 pages (*out of print*)

No. 55 **Laboratory Decontamination and Destruction of Carcinogens in Laboratory Wastes: Some *N*-Nitrosamides**
Edited by M. Castegnaro *et al.*
1984; 66 pages (*out of print*)

No. 56 **Models, Mechanisms and Etiology of Tumour Promotion**
Edited by M. Börzsönyi, N.E. Day, K. Lapis and H. Yamasaki
1984; 532 pages (*out of print*)

No. 57 ***N*-Nitroso Compounds: Occurrence, Biological Effects and Relevance to Human Cancer**
Edited by I.K. O'Neill, R.C. von Borstel, C.T. Miller, J. Long and H. Bartsch
1984; 1013 pages (*out of print*)

No. 58 **Age-related Factors in Carcinogenesis**
Edited by A. Likhachev, V. Anisimov and R. Montesano
1985; 288 pages (*out of print*)

No. 59 **Monitoring Human Exposure to Carcinogenic and Mutagenic Agents**
Edited by A. Berlin, M. Draper, K. Hemminki and H. Vainio
1984; 457 pages (*out of print*)

No. 60 **Burkitt's Lymphoma: A Human Cancer Model**
Edited by G. Lenoir, G. O'Conor and C.L.M. Olweny
1985; 484 pages (*out of print*)

No. 61 **Laboratory Decontamination and Destruction of Carcinogens in Laboratory Wastes: Some Haloethers**
Edited by M. Castegnaro *et al.*
1985; 55 pages (*out of print*)

No. 62 **Directory of On-going Research in Cancer Epidemiology 1984**
Edited by C.S. Muir and G. Wagner
1984; 717 pages (*out of print*)

No. 63 **Virus-associated Cancers in Africa**
Edited by A.O. Williams, G.T. O'Conor, G.B. de-Thé and C.A. Johnson
1984; 773 pages (*out of print*)

No. 64 **Laboratory Decontamination and Destruction of Carcinogens in Laboratory Wastes: Some Aromatic Amines and 4-Nitrobiphenyl**
Edited by M. Castegnaro *et al.*
1985; 84 pages (*out of print*)

No. 65 **Interpretation of Negative Epidemiological Evidence for Carcinogenicity**
Edited by N.J. Wald and R. Doll
1985; 232 pages (*out of print*)

No. 66 **The Role of the Registry in Cancer Control**
Edited by D.M. Parkin, G. Wagner and C.S. Muir
1985; 152 pages £10.00

No. 67 **Transformation Assay of Established Cell Lines: Mechanisms and Application**
Edited by T. Kakunaga and H. Yamasaki
1985; 225 pages (*out of print*)

No. 68 **Environmental Carcinogens. Selected Methods of Analysis. Volume 7. Some Volatile Halogenated Hydrocarbons**
Edited by L. Fishbein and I.K. O'Neill
1985; 479 pages (*out of print*)

No. 69 **Directory of On-going Research in Cancer Epidemiology 1985**
Edited by C.S. Muir and G. Wagner
1985; 745 pages (*out of print*)

No. 70 **The Role of Cyclic Nucleic Acid Adducts in Carcinogenesis and Mutagenesis**
Edited by B. Singer and H. Bartsch
1986; 467 pages (*out of print*)

No. 71 **Environmental Carcinogens. Selected Methods of Analysis. Volume 8: Some Metals: As, Be, Cd, Cr, Ni, Pb, Se, Zn**
Edited by I.K. O'Neill, P. Schuller and L. Fishbein
1986; 485 pages (*out of print*)

No. 72 **Atlas of Cancer in Scotland, 1975–1980. Incidence and Epidemiological Perspective**
Edited by I. Kemp, P. Boyle, M. Smans and C.S. Muir
1985; 285 pages (*out of print*)

No. 73 **Laboratory Decontamination and Destruction of Carcinogens in Laboratory Wastes: Some Antineoplastic Agents**
Edited by M. Castegnaro *et al.*
1985; 163 pages £12.50

No. 74 **Tobacco: A Major International Health Hazard**
Edited by D. Zaridze and R. Peto
1986; 324 pages £22.50

No. 75 **Cancer Occurrence in Developing Countries**
Edited by D.M. Parkin
1986; 339 pages £22.50

No. 76 **Screening for Cancer of the Uterine Cervix**
Edited by M. Hakama, A.B. Miller and N.E. Day
1986; 315 pages £30.00

No. 77 **Hexachlorobenzene: Proceedings of an International Symposium**
Edited by C.R. Morris and J.R.P. Cabral
1986; 668 pages (*out of print*)

No. 78 **Carcinogenicity of Alkylating Cytostatic Drugs**
Edited by D. Schmähl and J.M. Kaldor
1986; 337 pages (*out of print*)

No. 79 **Statistical Methods in Cancer Research. Volume III: The Design and Analysis of Long-term Animal Experiments**
By J.J. Gart, D. Krewski, P.N. Lee, R.E. Tarone and J. Wahrendorf
1986; 213 pages £22.00

No. 80 **Directory of On-going Research in Cancer Epidemiology 1986**
Edited by C.S. Muir and G. Wagner
1986; 805 pages (*out of print*)

No. 81 **Environmental Carcinogens: Methods of Analysis and Exposure Measurement. Volume 9: Passive Smoking**
Edited by I.K. O'Neill, K.D. Brunnemann, B. Dodet and D. Hoffmann
1987; 383 pages £35.00

No. 82 **Statistical Methods in Cancer Research. Volume II: The Design and Analysis of Cohort Studies**
By N.E. Breslow and N.E. Day
1987; 404 pages £25.00

No. 83 **Long-term and Short-term Assays for Carcinogens: A Critical Appraisal**
Edited by R. Montesano, H. Bartsch, H. Vainio, J. Wilbourn and H. Yamasaki
1986; 575 pages £35.00

No. 84 **The Relevance of *N*-Nitroso Compounds to Human Cancer: Exposure and Mechanisms**
Edited by H. Bartsch, I.K. O'Neill and R. Schulte-Hermann
1987; 671 pages (*out of print*)

No. 85 **Environmental Carcinogens: Methods of Analysis and Exposure Measurement. Volume 10: Benzene and Alkylated Benzenes**
Edited by L. Fishbein and I.K. O'Neill
1988; 327 pages £40.00

No. 86 **Directory of On-going Research in Cancer Epidemiology 1987**
Edited by D.M. Parkin and J. Wahrendorf
1987; 676 pages (*out of print*)

No. 87 **International Incidence of Childhood Cancer**
Edited by D.M. Parkin, C.A. Stiller, C.A. Bieber, G.J. Draper, B. Terracini and J.L. Young
1988; 401 pages £35.00

No. 88 **Cancer Incidence in Five Continents Volume V**
Edited by C. Muir, J. Waterhouse, T. Mack, J. Powell and S. Whelan
1987; 1004 pages £55.00

No. 89 **Method for Detecting DNA Damaging Agents in Humans: Applications in Cancer Epidemiology and Prevention**
Edited by H. Bartsch, K. Hemminki and I.K. O'Neill
1988; 518 pages £50.00

No. 90 **Non-occupational Exposure to Mineral Fibres**
Edited by J. Bignon, J. Peto and R. Saracci
1989; 500 pages £50.00

No. 91 **Trends in Cancer Incidence in Singapore 1968–1982**
Edited by H.P. Lee , N.E. Day and K. Shanmugaratnam
1988; 160 pages (*out of print*)

No. 92 **Cell Differentiation, Genes and Cancer**
Edited by T. Kakunaga, T. Sugimura, L. Tomatis and H. Yamasaki
1988; 204 pages £27.50

No. 93 **Directory of On-going Research in Cancer Epidemiology 1988**
Edited by M. Coleman and J. Wahrendorf
1988; 662 pages (*out of print*)

No. 94 **Human Papillomavirus and Cervical Cancer**
Edited by N. Muñoz, F.X. Bosch and O.M. Jensen
1989; 154 pages £22.50

No. 95 **Cancer Registration: Principles and Methods**
Edited by O.M. Jensen, D.M. Parkin, R. MacLennan, C.S. Muir and R. Skeet
1991; 288 pages £28.00

No. 96 **Perinatal and Multigeneration Carcinogenesis**
Edited by N.P. Napalkov, J.M. Rice, L. Tomatis and H. Yamasaki
1989; 436 pages £50.00

No. 97 **Occupational Exposure to Silica and Cancer Risk**
Edited by L. Simonato, A.C. Fletcher, R. Saracci and T. Thomas
1990; 124 pages £22.50

No. 98 **Cancer Incidence in Jewish Migrants to Israel, 1961–1981**
Edited by R. Steinitz, D.M. Parkin, J.L. Young, C.A. Bieber and L. Katz
1989; 320 pages £35.00

No. 99 **Pathology of Tumours in Laboratory Animals, Second Edition, Volume 1, Tumours of the Rat**
Edited by V.S. Turusov and U. Mohr
740 pages £85.00

No. 100 **Cancer: Causes, Occurrence and Control**
Editor-in-Chief L. Tomatis
1990; 352 pages £24.00

No. 101 **Directory of On-going Research in Cancer Epidemiology 1989/90**
Edited by M. Coleman and J. Wahrendorf
1989; 818 pages £36.00

No. 102 **Patterns of Cancer in Five Continents**
Edited by S.L. Whelan, D.M. Parkin & E. Masuyer
1990; 162 pages £25.00

No. 103 **Evaluating Effectiveness of Primary Prevention of Cancer**
Edited by M. Hakama, V. Beral, J.W. Cullen and D.M. Parkin
1990; 250 pages £32.00

No. 104 **Complex Mixtures and Cancer Risk**
Edited by H. Vainio, M. Sorsa and A.J. McMichael
1990; 442 pages £38.00

No. 105 **Relevance to Human Cancer of *N*-Nitroso Compounds, Tobacco Smoke and Mycotoxins**
Edited by I.K. O'Neill, J. Chen and H. Bartsch
1991; 614 pages £70.00

No. 106 **Atlas of Cancer Incidence in the Former German Democratic Republic**
Edited by W.H. Mehnert, M. Smans, C.S. Muir, M. Möhner & D. Schön
1992; 384 pages £55.00

* Available from booksellers through the network of WHO Sales agents.

† Available directly from IARC

No. 107 **Atlas of Cancer Mortality in the European Economic Community**
Edited by M. Smans, C.S. Muir and P. Boyle
1992; 280 pages £35.00

No. 108 **Environmental Carcinogens: Methods of Analysis and Exposure Measurement. Volume 11: Polychlorinated Dioxins and Dibenzofurans**
Edited by C. Rappe, H.R. Buser, B. Dodet and I.K. O'Neill
1991; 426 pages £45.00

No. 109 **Environmental Carcinogens: Methods of Analysis and Exposure Measurement. Volume 12: Indoor Air Contaminants**
Edited by B. Seifert, H. van de Wiel, B. Dodet and I.K. O'Neill
1993; 384 pages £45.00

No. 110 **Directory of On-going Research in Cancer Epidemiology 1991**
Edited by M. Coleman and J. Wahrendorf
1991; 753 pages £38.00

No. 111 **Pathology of Tumours in Laboratory Animals, Second Edition, Volume 2, Tumours of the Mouse**
Edited by V.S. Turusov and U. Mohr
1993; 776 pages; £90.00

No. 112 **Autopsy in Epidemiology and Medical Research**
Edited by E. Riboli and M. Delendi
1991; 288 pages £25.00

No. 113 **Laboratory Decontamination and Destruction of Carcinogens in Laboratory Wastes: Some Mycotoxins**
Edited by M. Castegnaro, J. Barek, J.-M. Frémy, M. Lafontaine, M. Miraglia, E.B. Sansone and G.M. Telling
1991; 64 pages £11.00

No. 114 **Laboratory Decontamination and Destruction of Carcinogens in Laboratory Wastes: Some Polycyclic Heterocyclic Hydrocarbons**
Edited by M. Castegnaro, J. Barek J. Jacob, U. Kirso, M. Lafontaine, E.B. Sansone, G.M. Telling and T. Vu Duc
1991; 50 pages £8.00

No. 115 **Mycotoxins, Endemic Nephropathy and Urinary Tract Tumours**
Edited by M. Castegnaro, R. Plestina, G. Dirheimer, I.N. Chernozemsky and H Bartsch
1991; 340 pages £45.00

No. 116 **Mechanisms of Carcinogenesis in Risk Identification**
Edited by H. Vainio, P.N. Magee, D.B. McGregor & A.J. McMichael
1992; 616 pages £65.00

No. 117 **Directory of On-going Research in Cancer Epidemiology 1992**
Edited by M. Coleman, J. Wahrendorf & E. Démaret
1992; 773 pages £42.00

No. 118 **Cadmium in the Human Environment: Toxicity and Carcinogenicity**
Edited by G.F. Nordberg, R.F.M. Herber & L. Alessio
1992; 470 pages £60.00

No. 119 **The Epidemiology of Cervical Cancer and Human Papillomavirus**
Edited by N. Muñoz, F.X. Bosch, K.V. Shah & A. Meheus
1992; 288 pages £28.00

No. 120 **Cancer Incidence in Five Continents, Volume VI**
Edited by D.M. Parkin, C.S. Muir, S.L. Whelan, Y.T. Gao, J. Ferlay & J.Powell
1992; 1080 pages £120.00

No. 121 **Trends in Cancer Incidence and Mortality**
M.P. Coleman, J. Estève, P. Damiecki, A. Arslan and H. Renard
1993; 806 pages, £120.00

No. 122 **International Classification of Rodent Tumours.** Part 1. **The Rat**
Editor-in-Chief: U. Mohr
1992/93; 10 fascicles of 60–100 pages, £120.00

No. 123 **Cancer in Italian Migrant Populations**
Edited by M. Geddes, D.M. Parkin, M. Khlat, D. Balzi and E. Buiatti
1993; 292 pages, £40.00

No. 124 **Postlabelling Methods for Detection of DNA Adducts**
Edited by D.H. Phillips, M. Castegnaro and H. Bartsch
1993; 392 pages; £46.00

No. 125 **DNA Adducts: Identification and Biological Significance.**
Edited by K. Hemminki, A. Dipple, D. Shuker, F.F. Kadlubar, D. Segerbäck and H. Bartsch
1993; 480 pages; £52.00

No. 127 **Butadiene and Styrene: Assessment of Health Hazards.**
Edited by M. Sorsa, K. Peltonen, H. Vainio and K. Hemminki
1993; 412 pages; £46.00

No. 130 **Directory of On-going Research in Cancer Epidemiology 1994**
Edited by R. Sankaranarayanan, J. Wahrendorf and E. Démaret
1994; approx. 800 pages, £46.00

IARC MONOGRAPHS ON THE EVALUATION OF CARCINOGENIC RISKS TO HUMANS

(Available from booksellers through the network of WHO Sales Agents)

Volume 1 **Some Inorganic Substances, Chlorinated Hydrocarbons, Aromatic Amines, *N*-Nitroso Compounds, and Natural Products**
1972; 184 pages (*out of print*)

Volume 2 **Some Inorganic and Organometallic Compounds**
1973; 181 pages (*out of print*)

Volume 3 **Certain Polycyclic Aromatic Hydrocarbons and Heterocyclic Compounds**
1973; 271 pages (*out of print*)

Volume 4 **Some Aromatic Amines, Hydrazine and Related Substances, *N*-Nitroso Compounds and Miscellaneous Alkylating Agents**
1974; 286 pages Sw. fr. 18.-

Volume 5 **Some Organochlorine Pesticides**
1974; 241 pages (*out of print*)

Volume 6 **Sex Hormones**
1974; 243 pages (*out of print*)

Volume 7 **Some Anti-Thyroid and Related Substances, Nitrofurans and Industrial Chemicals**
1974; 326 pages (*out of print*)

Volume 8 **Some Aromatic Azo Compounds**
1975; 357 pages Sw. fr. 36.-

Volume 9 **Some Aziridines, *N*-, *S*- and *O*-Mustards and Selenium**
1975; 268 pages Sw.fr. 27.-

Volume 10 **Some Naturally Occurring Substances**
1976; 353 pages (*out of print*)

Volume 11 **Cadmium, Nickel, Some Epoxides, Miscellaneous Industrial Chemicals and General Considerations on Volatile Anaesthetics**
1976; 306 pages (*out of print*)

Volume 12 **Some Carbamates, Thiocarbamates and Carbazides**
1976; 282 pages Sw. fr. 34.-

Volume 13 **Some Miscellaneous Pharmaceutical Substances**
1977; 255 pages Sw. fr. 30.-

Volume 14 **Asbestos**
1977; 106 pages (*out of print*)

Volume 15 **Some Fumigants, The Herbicides 2,4-D and 2,4,5-T, Chlorinated Dibenzodioxins and Miscellaneous Industrial Chemicals**
1977; 354 pages Sw. fr. 50.-

Volume 16 **Some Aromatic Amines and Related Nitro Compounds - Hair Dyes, Colouring Agents and Miscellaneous Industrial Chemicals**
1978; 400 pages Sw. fr. 50.-

Volume 17 **Some *N*-Nitroso Compounds**
1978; 365 pages Sw. fr. 50.-

Volume 18 **Polychlorinated Biphenyls and Polybrominated Biphenyls**
1978; 140 pages Sw. fr. 20.-

Volume 19 **Some Monomers, Plastics and Synthetic Elastomers, and Acrolein**
1979; 513 pages (*out of print*)

Volume 20 **Some Halogenated Hydrocarbons**
1979; 609 pages (*out of print*)

Volume 21 **Sex Hormones (II)**
1979; 583 pages Sw. fr. 60.-

Volume 22 **Some Non-Nutritive Sweetening Agents**
1980; 208 pages Sw. fr. 25.-

Volume 23 **Some Metals and Metallic Compounds**
1980; 438 pages (*out of print*)

Volume 24 **Some Pharmaceutical Drugs**
1980; 337 pages Sw. fr. 40.-

Volume 25 **Wood, Leather and Some Associated Industries**
1981; 412 pages Sw. fr. 60.-

Volume 26 **Some Antineoplastic and Immunosuppressive Agents**
1981; 411 pages Sw. fr. 62.-

Volume 27 **Some Aromatic Amines, Anthraquinones and Nitroso Compounds, and Inorganic Fluorides Used in Drinking Water and Dental Preparations**
1982; 341 pages Sw. fr. 40.-

Volume 28 **The Rubber Industry**
1982; 486 pages Sw. fr. 70.-

Volume 29 **Some Industrial Chemicals and Dyestuffs**
1982; 416 pages Sw. fr. 60.-

Volume 30 **Miscellaneous Pesticides**
1983; 424 pages Sw. fr. 60.-

Volume 31 **Some Food Additives, Feed Additives and Naturally Occurring Substances**
1983; 314 pages Sw. fr. 60.-

Volume 32 **Polynuclear Aromatic Compounds, Part 1: Chemical, Environmental and Experimental Data**
1983; 477 pages Sw. fr. 60.-

Volume 33 **Polynuclear Aromatic Compounds, Part 2: Carbon Blacks, Mineral Oils and Some Nitroarenes**
1984; 245 pages Sw. fr. 50.-

Volume 34 **Polynuclear Aromatic Compounds, Part 3: Industrial Exposures in Aluminium Production, Coal Gasification, Coke Production, and Iron and Steel Founding**
1984; 219 pages Sw. fr. 48.-

Volume 35 **Polynuclear Aromatic Compounds, Part 4: Bitumens, Coal-tars and Derived Products, Shale-oils and Soots**
1985; 271 pages Sw. fr. 70.-

* Available from booksellers through the network of WHO Sales agents.

† Available directly from IARC

Volume 36 **Allyl Compounds, Aldehydes, Epoxides and Peroxides**
1985; 369 pages Sw. fr. 70.–

Volume 37 **Tobacco Habits Other than Smoking: Betel-quid and Areca-nut Chewing; and some Related Nitrosamines**
1985; 291 pages Sw. fr. 70.–

Volume 38 **Tobacco Smoking**
1986; 421 pages Sw. fr. 75.–

Volume 39 **Some Chemicals Used in Plastics and Elastomers**
1986; 403 pages Sw. fr. 60.–

Volume 40 **Some Naturally Occurring and Synthetic Food Components, Furocoumarins and Ultraviolet Radiation**
1986; 444 pages Sw. fr. 65.–

Volume 41 **Some Halogenated Hydrocarbons and Pesticide Exposures**
1986; 434 pages Sw. fr. 65.–

Volume 42 **Silica and Some Silicates**
1987; 289 pages Sw. fr. 65.

Volume 43 **Man-Made Mineral Fibres and Radon**
1988; 300 pages Sw. fr. 65.–

Volume 44 **Alcohol Drinking**
1988; 416 pages Sw. fr. 65.

Volume 45 **Occupational Exposures in Petroleum Refining; Crude Oil and Major Petroleum Fuels**
1989; 322 pages Sw. fr. 65.–

Volume 46 **Diesel and Gasoline Engine Exhausts and Some Nitroarenes**
1989; 458 pages Sw. fr. 65.–

Volume 47 **Some Organic Solvents, Resin Monomers and Related Compounds, Pigments and Occupational Exposures in Paint Manufacture and Painting**
1989; 536 pages Sw. fr. 85.–

Volume 48 **Some Flame Retardants and Textile Chemicals, and Exposures in the Textile Manufacturing Industry**
1990; 345 pages Sw. fr. 65.–

Volume 49 **Chromium, Nickel and Welding**
1990; 677 pages Sw. fr. 95.-

Volume 50 **Pharmaceutical Drugs**
1990; 415 pages Sw. fr. 65.-

Volume 51 **Coffee, Tea, Mate, Methylxanthines and Methylglyoxal**
1991; 513 pages Sw. fr. 80.-

Volume 52 **Chlorinated Drinking-water; Chlorination By-products; Some Other Halogenated Compounds; Cobalt and Cobalt Compounds**
1991; 544 pages Sw. fr. 80.-

Volume 53 **Occupational Exposures in Insecticide Application and some Pesticides**
1991; 612 pages Sw. fr. 95.-

Volume 54 **Occupational Exposures to Mists and Vapours from Strong Inorganic Acids; and Other Industrial Chemicals**
1992; 336 pages Sw. fr. 65.-

Volume 55 **Solar and Ultraviolet Radiation**
1992; 316 pages Sw. fr. 65.-

Volume 56 **Some Naturally Occurring Substances: Food Items and Constituents, Heterocyclic Aromatic Amines and Mycotoxins**
1993; 600 pages Sw. fr. 95.-

Volume 57 **Occupational Exposures of Hairdressers and Barbers and Personal Use of Hair Colourants; Some Hair Dyes, Cosmetic Colourants, Industrial Dyestuffs and Aromatic Amines**
1993; 428 pages Sw. fr. 75.-

Volume 58 **Beryllium, Cadmium, Mercury and Exposures in the Glass Manufacturing Industry**
1993; 426 pages Sw. fr. 75.-

Supplement No. 1
Chemicals and Industrial Processes Associated with Cancer in Humans (IARC Monographs, Volumes 1 to 20)
1979; 71 pages (*out of print*)

Supplement No. 2
Long-term and Short-term Screening Assays for Carcinogens: A Critical Appraisal
1980; 426 pages Sw. fr. 40.-

Supplement No. 3
Cross Index of Synonyms and Trade Names in Volumes 1 to 26
1982; 199 pages (*out of print*)

Supplement No. 4
Chemicals, Industrial Processes and Industries Associated with Cancer in Humans (IARC Monographs, Volumes 1 to 29)
1982; 292 pages (*out of print*)

Supplement No. 5
Cross Index of Synonyms and Trade Names in Volumes 1 to 36
1985; 259 pages (*out of print*)

Supplement No. 6
Genetic and Related Effects: An Updating of Selected IARC Monographs from Volumes 1 to 42
1987; 729 pages Sw. fr. 80.-

Supplement No. 7
Overall Evaluations of Carcinogenicity: An Updating of IARC Monographs Volumes 1-42
1987; 440 pages Sw. fr. 65.-

Supplement No. 8
Cross Index of Synonyms and Trade Names in Volumes 1 to 46
1990; 346 pages Sw. fr. 60.-

IARC TECHNICAL REPORTS*

No. 1 **Cancer in Costa Rica**
Edited by R. Sierra, R. Barrantes, G. Muñoz Leiva, D.M. Parkin, C.A. Bieber and N. Muñoz Calero
1988; 124 pages Sw. fr. 30.-

No. 2 **SEARCH: A Computer Package to Assist the Statistical Analysis of Case-control Studies**
Edited by G.J. Macfarlane, P. Boyle and P. Maisonneuve
1991; 80 pages (*out of print*)

No. 3 **Cancer Registration in the European Economic Community**
Edited by M.P. Coleman and E. Démaret
1988; 188 pages Sw. fr. 30.-

No. 4 **Diet, Hormones and Cancer: Methodological Issues for Prospective Studies**
Edited by E. Riboli and R. Saracci
1988; 156 pages Sw. fr. 30.-

No. 5 **Cancer in the Philippines**
Edited by A.V. Laudico, D. Esteban and D.M. Parkin
1989; 186 pages Sw. fr. 30.-

No. 6 **La genèse du Centre International de Recherche sur le Cancer**
Par R. Sohier et A.G.B. Sutherland
1990; 104 pages Sw. fr. 30.-

No. 7 **Epidémiologie du cancer dans les pays de langue latine**
1990; 310 pages Sw. fr. 30.-

No. 8 **Comparative Study of Anti-smoking Legislation in Countries of the European Economic Community**
Edited by A. Sasco, P. Dalla Vorgia and P. Van der Elst
1992; 82 pages Sw. fr. 30.-

No. 9 **Epidemiologie du cancer dans les pays de langue latine**
1991 346 pages Sw. fr. 30.-

No. 11 **Nitroso Compounds: Biological Mechanisms, Exposures and Cancer Etiology**
1991; 346 pages Sw. fr. 30.-
Edited by I.K. O'Neill & H. Bartsch
1992; 149 pages Sw. fr. 30.-

No. 12 **Epidémiologie du cancer dans les pays de langue latine**
1992; 375 pages Sw. fr. 30.-

No. 13 **Health, Solar UV Radiation and Environmental Change**
Edited by A. Kricker, B.K. Armstrong, M.E. Jones and R.C. Burton
1993; 216 pages Sw.fr. 30.–

No. 14 **Epidémiologie du cancer dans les pays de langue latine**
1993; 385 pages Sw. fr. 30.-

No. 15 **Cancer in the African Population of Bulawayo, Zimbabwe, 1963–1977: Incidence, Time Trends and Risk Factors**
By M.E.G. Skinner, D.M. Parkin, A.P. Vizcaino and A. Ndhlovu
1993; 123 pages Sw. fr. 30.-

No. 16 **Cancer in Thailand, 1988–1991**
By V. Vatanasapt, N. Martin, H. Sriplung, K. Vindavijak, S. Sontipong, S. Sriamporn, D.M. Parkin and J. Ferlay
1993; 164 pages Sw. fr. 30.-

DIRECTORY OF AGENTS BEING TESTED FOR CARCINOGENICITY (Until Vol. 13 Information Bulletin on the Survey of Chemicals Being Tested for Carcinogenicity)*

No. 8 Edited by M.-J. Ghess, H. Bartsch and L. Tomatis
1979; 604 pages Sw. fr. 40.-

No. 9 Edited by M.-J. Ghess, J.D. Wilbourn, H. Bartsch and L. Tomatis
1981; 294 pages Sw. fr. 41.-

No. 10 Edited by M.-J. Ghess, J.D. Wilbourn and H. Bartsch
1982; 362 pages Sw. fr. 42.-

No. 11 Edited by M.-J. Ghess, J.D. Wilbourn, H. Vainio and H. Bartsch
1984; 362 pages Sw. fr. 50.-

No. 12 Edited by M.-J. Ghess, J.D. Wilbourn, A. Tossavainen and H. Vainio
1986; 385 pages Sw. fr. 50.-

No. 13 Edited by M.-J. Ghess, J.D. Wilbourn and A. Aitio 1988; 404 pages Sw. fr. 43.-

No. 14 Edited by M.-J. Ghess, J.D. Wilbourn and H. Vainio
1990; 370 pages Sw. fr. 45.-

No. 15 Edited by M.-J. Ghess, J.D. Wilbourn and H. Vainio
1992; 318 pages Sw. fr. 45.-

NON-SERIAL PUBLICATIONS

Alcool et Cancer†
By A. Tuyns (in French only)
1978; 42 pages Fr. fr. 35.-

Cancer Morbidity and Causes of Death Among Danish Brewery Workers†
By O.M. Jensen
1980; 143 pages Fr. fr. 75.-

Directory of Computer Systems Used in Cancer Registries†
By H.R. Menck and D.M. Parkin
1986; 236 pages Fr. fr. 50.-

Facts and Figures of Cancer in the European Community*
Edited by J. Estève, A. Kricker, J. Ferlay and D.M. Parkin
1993; 52 pages Sw. fr. 10.-

* Available from booksellers through the network of WHO Sales agents.

† Available directly from IARC